McGraw-Hill

Dictionary of

Bioscience

McGraw-Hill Dictionary of Bioscience

Sybil P. Parker
Editor in Chief

McGraw-Hill

New York San Francisco Washington, D.C. Auckland
Bogotá Caracas Lisbon London Madrid Mexico City
Milan Montreal New Delhi San Juan Singapore
Sydney Tokyo Toronto

Library of Congress Cataloging in Publication Data

McGraw-Hill dictionary of bioscience / Sybil P. Parker, editor in chief.
p. cm.
ISBN 0-07-052430-0
1. Biology—Dictionaries. 2. Life sciences—Dictionaries.
I. Parker, Sybil P. II. McGraw-Hill Book Company.
QH302.5.M382 1997
574′.03—dc20 96-46183

McGraw-Hill 

A Division of The ***McGraw-Hill*** *Companies*

1 2 3 4 5 6 7 8 9 0 DOC/DOC 9 0 1 0 9 8 7 6

ISBN 0-07-052430-0

INTERNATIONAL EDITION

When ordering this title, use ISBN 0-07-114919-8.

This book is printed on recycled, acid-free paper containing a minimum of 50% recycled, de-inked fiber.

This book was set in Helvetica Bold and Novarese Book by Progressive Information Technologies, Emigsville, Pennsylvania. It was printed and bound by R. R. Donnelley & Sons Company, The Lakeside Press.

Preface

The *McGraw-Hill Dictionary of Bioscience* concentrates on the vocabulary of those disciplines that constitute bioscience and related fields. With more than 16,000 terms, it serves as a major compendium of the specialized language that is essential to understanding bioscience. The language of bioscience embraces many unique disciplines which are usually represented in specialized dictionaries and glossaries. Biologists, researchers, students, teachers, librarians, writers, and the general public will appreciate the convenience of a single comprehensive reference.

Terms and definitions in the Dictionary represent 23 fields: anatomy, biochemistry, biology, biophysics, botany, cytology, ecology, embryology, evolution, genetics, histology, immunology, invertebrate zoology, microbiology, molecular biology, mycology, paleobotany, paleontology, physiology, systematics, vertebrate zoology, virology, and zoology. Each definition is identified by the field in which it is primarily used. When a definition applies to more than one field, it is given a more general field label. For example, a definition that applies to both botany and zoology is assigned to BIOLOGY.

The terms selected for this Dictionary are fundamental to understanding bioscience. All definitions were drawn from the *McGraw-Hill Dictionary of Scientific and Technical Terms* (5th ed., 1994). Along with definitions and pronunciations, terms also include synonyms, acronyms, and abbreviations where appropriate. Such synonyms, acronyms, and abbreviations also appear in the alphabetical sequence as cross references to the defining terms.

The *McGraw-Hill Dictionary of Bioscience* is a reference that the editors hope will facilitate the communication of ideas and information, and thus serve the needs of readers with either professional or pedagogical interests in bioscience.

Sybil P. Parker
EDITOR IN CHIEF

Editorial Staff

Sybil P. Parker, Editor in Chief

Arthur Biderman, Senior editor
Jonathan Weil, Editor
Betty Richman, Editor
Patricia W. Albers, Editorial administrator

Dr. Henry F. Beechhold, Pronunciation Editor
Professor of English
Chairman, Linguistics Program
College of New Jersey
Trenton, New Jersey

Joe Faulk, Editing manager
Frank Kotowski, Jr., Senior editing supervisor
Ruth W. Mannino, Senior editing supervisor

Suzanne W. B. Rapcavage, Senior production supervisor

How to Use the Dictionary

ALPHABETIZATION. The terms in the *McGraw-Hill Dictionary of Bioscience* are alphabetized on a letter-by-letter basis; word spacing, hyphen, comma, solidus, and apostrophe in a term are ignored in the sequencing. For example, an ordering of terms would be:

allelotropism	**all-or-none law**
Allen-Doisy unit	**allosteric transition**
Allen's rule	**allostery**

FORMAT. The basic format for a defining entry provides the term in boldface, the field in small capitals, and the single definition in lightface:

term [FIELD] Definition.

A field may be followed by multiple definitions, each introduced by a boldface number:

term [FIELD] **1.** Definition. **2.** Definition. **3.** Definition.

A term may have definitions in two or more fields:

term [ECOLOGY] Definition. [GENETICS] Definition.

A simple cross-reference entry appears as:

term [FIELD] *See* another term.

A cross reference may also appear in combination with definitions:

term [ECOLOGY] Definition. [GENETICS] *See* another term.

CROSS REFERENCING. A cross-reference entry directs the user to the defining entry. For example, the user looking up "aiophyllous" finds:

aiophyllous [BOTANY] *See* evergreen.

The user then turns to the "E" terms for the definition. Cross references are also made from variant spellings, acronyms, abbreviations, and symbols.

aesthacyte [INVERTEBRATE ZOOLOGY] *See* esthacyte.
AMP [BIOCHEMISTRY] *See* adenylic acid.
D-loop [MOLECULAR BIOLOGY] *See* displacement loop.

ALSO KNOWN AS . . ., etc. A definition may conclude with a mention of a synonym of the term, a variant spelling, an abbreviation for the term, or other such information, introduced by "Also known as . . .," "Also spelled . . .," "Abbreviated . . .," "Symbolized . . .," "Derived from . . .." When a term has more than one definition, the positioning of any of these phrases conveys the extent of applicability. For example:

term [ECOLOGY] **1.** Definition. Also known as synonym. **2.** Definition. Symbolized T.

In the above arrangement, "Also known as . . ." applies only to the first definition. "Symbolized . . ." applies only to the second definition.

term [ECOLOGY] **1.** Definition. **2.** Definition. [GENETICS] Definition. Also known as synonym.

In the above arrangement, "Also known as . . ." applies only to the second field.

term [ECOLOGY] Also known as synonym. **1.** Definition. **2.** Definition. [GENETICS] Definition.

In the above arrangement, "Also known as . . ." applies to both definitions in the first field.

term Also known as synonym. [ECOLOGY] **1.** Definition. **2.** Definition. [GENETICS] Definition.

In the above arrangement, "Also known as . . ." applies to all definitions in both fields.

CHEMICAL FORMULAS Chemistry definitions may include either an empirical formula (say, for abscisic acid, $C_{16}H_{20}O_4$) or a line formula (for succinamide, $H_2NCOCH_2CONH_2$), whichever is appropriate.

Scope of Fields

Anatomy The branch of morphology concerned with the gross and microscopic structure of animals, especially humans.

Biochemistry The study of the chemical substances that occur in living organisms, the processes by which these substances enter into or are formed in the organisms and react with each other and the environment, and the methods by which the substances and processes are identified, characterized, and measured.

Biology The science of living organisms, concerned with the study of embryology, anatomy, physiology, cytology, morphology, taxonomy, genetics, evolution, and ecology.

Biophysics The hybrid science involving the methods and ideas of physics and chemistry to study and explain the structures of living organisms and the mechanics of life processes.

Botany That branch of biological science which embraces the study of plants and plant life, including algae; deals with taxonomy, morphology, physiology, and other aspects.

Cytology The branch of biological science which deals with the structure, behavior, growth, and reproduction of cells and the function and chemistry of cells and cell components.

Ecology The study of the interrelationships between organisms and their environment.

Embryology The study of the development of the organism from the zygote, or fertilized egg.

Evolution The processes of biological and organic change in organisms by which descendants come to differ from their ancestors, and a history of the sequence of such change.

Genetics The science concerned with biological inheritance, that is, with the causes of the resemblances and differences among related individuals.

Histology The study of the structure and chemical composition of animal tissues as related to their function.

Immunology The division of biological science concerned with the native or acquired resistance of higher animal forms and humans to infection with microorganisms.

Invertebrate zoology A branch of zoology concerned with the taxonomy, behavior, and morphology of invertebrate animals.

Microbiology The science and study of microorganisms, especially bacteria and rickettsiae, and of antibiotic substances.

Molecular biology That branch of biology which attempts to interpret biological events in terms of the molecules in the cell.

Mycology A branch of biological science concerned with the study of fungi.

Paleobotany The study of fossil plants and vegetation of the geologic past.

Paleontology The study of life in the geologic past as recorded by fossil remains.

Physiology The branch of biological science concerned with the basic activities that occur in cells and tissues of living organisms and involving physical and chemical studies of these organisms.

Systematics The science of animal and plant classification.

Vertebrate zoology A branch of zoology concerned with the taxonomy, behavior, and morphology of vertebrate animals.

Virology The science that deals with the study of viruses.

Zoology The science that deals with the taxonomy, behavior, and morphology of animal life.

Pronunciation Key

Vowels

a as in b**a**t, th**a**t
ā as in b**ai**t, cr**a**te
ä as in b**o**ther, f**a**ther
e as in b**e**t, n**e**t
ē as in b**ee**t, tr**ea**t
i as in b**i**t, sk**i**t
ī as in b**i**te, l**igh**t
ō as in b**oa**t, n**o**te
ȯ as in b**ough**t, t**au**t
u̇ as in b**oo**k, p**u**ll
ü as in b**oo**t, p**oo**l
ə as in b**u**t, sof**a**
au̇ as in cr**ow**d, p**ow**er
ȯi as in b**oi**l, sp**oi**l
yə as in form**u**la, spectac**u**lar
yü as in f**ue**l, m**u**le

Semivowels/Semiconsonants

w as in **w**ind, t**w**in
y as in **y**et, on**ion**

Stress (Accent)

ˈ precedes syllable with primary stress

ˌ precedes syllable with secondary stress

¦ precedes syllable with variable or indeterminate primary/secondary stress

Consonants

b as in **b**i**b**, dri**bb**le
ch as in **ch**arge, stre**tch**
d as in **d**og, ba**d**
f as in **f**ix, sa**f**e
g as in **g**ood, si**g**nal
h as in **h**and, be**h**ind
j as in **j**oint, di**g**it
k as in **c**ast, bri**ck**
k̲ as in Ba**ch** (used rarely)
l as in **l**oud, be**ll**
m as in **m**ild, su**mm**er
n as in **n**ew, de**n**t
n̲ indicates nasalization of preceding vowel
ŋ as in ri**ng**, si**ng**le
p as in **p**ier, sli**p**
r as in **r**ed, sca**r**
s as in **s**ign, po**s**t
sh as in **su**gar, **sh**oe
t as in **t**imid, ca**t**
th as in **th**in, brea**th**
t̲h̲ as in **th**en, brea**th**e
v as in **v**eil, wea**v**e
z as in **z**oo, crui**s**e
zh as in bei**g**e, trea**s**ure

Syllabication

· Indicates syllable boundary when following syllable is unstressed

Contents

A

aapamoor [ECOLOGY] A moor with elevated areas or mounds supporting dwarf shrubs and sphagnum, interspersed with low areas containing sedges and sphagnum, thus forming a mosaic. { 'äp·ə,mür }

aardvark [VERTEBRATE ZOOLOGY] A nocturnal, burrowing, insectivorous mammal of the genus *Orycteropus* in the order Tubulidentata. Also known as earth pig. { 'ärd,värk }

aardwolf [VERTEBRATE ZOOLOGY] *Proteles cristatus*. A hyenalike African mammal of the family Hyaenidae. { 'ärd,wu̇lf }

ABA [BIOCHEMISTRY] *See* abscisic acid.

abaca [BOTANY] *Musa textilis*. A plant of the banana family native to Borneo and the Philippines, valuable for its hard fiber. Also known as Manila hemp. { 'ä·bä,kä *or* 'ä·bə,kä }

abactinal [INVERTEBRATE ZOOLOGY] In radially symmetrical animals, pertaining to the surface opposite the side where the mouth is located. { a'bak·tin·əl }

abalone [INVERTEBRATE ZOOLOGY] A gastropod mollusk composing the single genus *Haliotis* of the family Haliotidae. Also known as ear shell; ormer; paua. { ,ab·ə'lō·nē }

abambulacral [INVERTEBRATE ZOOLOGY] Pertaining to that part of the surface of an echinoderm that lacks tube feet. { ab,am·byə'lak-rəl }

A band [HISTOLOGY] The region between two adjacent I bands in a sarcomere; characterized by partial overlapping of actin and myosin filaments. { 'ā band }

abapertural [INVERTEBRATE ZOOLOGY] Away from the shell aperture, referring to mollusks. { ab 'ap·ər,chu̇r·əl }

abapical [BIOLOGY] On the opposite side to, or directed away from, the apex. { ab'ap·i·kəl }

abaxial [BIOLOGY] On the opposite side to, or facing away from, the axis of an organ or organism. { ab'ak·sē·əl }

abb [VERTEBRATE ZOOLOGY] A coarse wool from the fleece areas of lesser quality. { ab }

abdomen [ANATOMY] **1.** The portion of the vertebrate body between the thorax and the pelvis. **2.** The cavity of this part of the body. [INVERTEBRATE ZOOLOGY] The elongate region posterior to the thorax in arthropods. { ab'dōm·ən *or* 'ab·də,mən }

abdominal gills [INVERTEBRATE ZOOLOGY] Paired, segmental, leaflike, filamentous expansions of the abdominal cuticle for respiration in the aquatic larvae of many insects. { ab'däm·ə·nəl 'gilz }

abdominal limb [INVERTEBRATE ZOOLOGY] In most crustaceans, any of the segmented abdominal appendages. { ab'däm·ə·nəl 'lim }

abdominal regions [ANATOMY] Nine theoretical areas delineated on the abdomen by two horizontal and two parasagittal lines: above, the right hypochondriac, epigastric, and left hypochondriac; in the middle, the right lateral, umbilical, and left lateral; and below, the right inguinal, hypogastric, and left inguinal. { ab'däm·ə·nəl 'rē·jənz }

abducens [ANATOMY] The sixth cranial nerve in vertebrates; a paired, somatic motor nerve arising from the floor of the fourth ventricle of the brain and supplying the lateral rectus eye muscles. { ab'dyü·sənz }

abduction [PHYSIOLOGY] Movement of an extremity or other body part away from the axis of the body. { ab'dək·shən }

abductor [PHYSIOLOGY] Any muscle that draws a part of the body or an extremity away from the body axis. { ab'dək·tər }

aberrant [BIOLOGY] An atypical group, individual, or structure, especially one with an aberrant chromosome number. { ə'ber·ənt }

Abies [BOTANY] The firs, a genus of trees in the pine family characterized by erect cones, absence of resin canals in the wood, and flattened needlelike leaves. { 'ā·bē,ēz }

abiocoen [ECOLOGY] A nonbiotic habitat. { 'ā,bī·ō,sēn }

abiogenesis [BIOLOGY] The obsolete concept that plant and animal life arise from nonliving organic matter. Also known as autogenesis; spontaneous generation. { ¦ā,bī·ō'jen·ə·sis }

abiotic [BIOLOGY] Referring to the absence of living organisms. { ¦a,bī'äd·ik }

abiotic environment [ECOLOGY] All physical and nonliving chemical factors, such as soil, water, and atmosphere, which influence living organisms. { ¦a,bī'äd·ik in'vī·rən,mənt }

abiotic substance [ECOLOGY] Any fundamental chemical element or compound in the environment. { ¦a,bī'äd·ik 'səb·stəns }

abjection [MYCOLOGY] The discharge or casting off of spores by the spore-bearing structure of a fungus. { ab'jek·shən }

ablastin [IMMUNOLOGY] An antibodylike substance elicited by *Trypanosoma lewisi* in the blood serum of infected rats that inhibits reproduction of the parasite. { ə'blas·tən }

ABO blood group [IMMUNOLOGY] An immunologically distinct, genetically determined group of human erythrocyte antigens represented by two blood factors (A and B) and four blood types (A, B, AB, and O). { ā·bē'ō 'bləd ,grüp }

ABO blood group system [IMMUNOLOGY] A set of multiple alleles found on a single locus on human chromosome 9 that specifies the presence or absence of certain red cell antigens, which determines the ABO blood group. { ,ā,bē 'ō 'bləd ,grüp ,sis·təm }

abomasum [VERTEBRATE ZOOLOGY] The final chamber of the complex stomach of ruminants; has a glandular wall and corresponds to a true stomach. { ,ab·ō'mā·səm }

aboral [INVERTEBRATE ZOOLOGY] Opposite to the mouth. { a'bór·əl }

abortive [BIOLOGY] Imperfectly formed or developed. { ə'bórd·iv }

abortive infection [VIROLOGY] The viral infection of a cell in which viral components may be synthesized without the production of infective viruses. Also known as nonproductive infection. { ə'bórd·iv in'fek·shən }

abortive transduction [MICROBIOLOGY] Failure of exogenous fragments that were introduced into a bacterial cell by viruses to become inserted into the bacterial chromosome. { ə'bórd·iv tranz'dək·shən }

abranchiate [ZOOLOGY] Without gills. { ā'braŋk·ē·ət }

abrupt [BOTANY] Ending suddenly, as though broken off. { ə'brəpt }

abscisic acid [BIOCHEMISTRY] $C_{16}H_{20}O_4$ A plant hormone produced by fruits and leaves that promotes abscission and dormancy and retards vegetative growth. Abbreviated ABA. Formerly known as abscisin. { ab'sis·ik 'as·əd }

abscisin [BIOCHEMISTRY] *See* abscisic acid. { ab 'sis·ən }

abscission [BOTANY] A physiological process promoted by abscisic acid whereby plants shed a part, such as a leaf, flower, seed, or fruit. { ab 'sizh·ən }

abscission layer [BOTANY] A zone of cells whose breakdown causes separation of a leaf or other structure from the stem. { ab'sizh·ən ,lā·ər }

absolute plating efficiency [CELL BIOLOGY] The percentage of individual cells that give rise to colonies when the cells are inoculated into culture media. { 'ab·sə,lüt 'plād·iŋ i,fish·ən·sē }

absolute refractory period [PHYSIOLOGY] A period ranging from 0.5 to 2 milliseconds during which neural tissue is totally unresponsive. { 'ab·sə ,lüt ri'frak·trē,pir·ē·əd }

absolute threshold [PHYSIOLOGY] The minimum stimulus energy that an organism can detect. { 'ab·sə,lüt 'thresh·hóld }

absorption [BIOLOGY] The net movement (transport) of water and solutes from outside a cell or an organism to the interior. [IMMUNOLOGY] **1.** Removal of antibodies from an antiserum by addition of antigen. **2.** Removal of antigens from a mixture by addition of antibodies. [PHYSIOLOGY] Passage of a chemical substance, a pathogen, or radiant energy through a body membrane. { əb'sórp·shən }

absorption test [IMMUNOLOGY] Analysis of the antigenic components of bacterial cells and large macromolecules by a series of precipitation or agglutination reactions with specific antibodies. { əb'sórp·shən ,test }

abstriction [MYCOLOGY] In fungi, the cutting off of spores in hyphae by formation of septa followed by abscission of the spores, especially by constriction. { ab'strik·shən }

abterminal [BIOLOGY] Referring to movement from the end toward the middle; specifically, describing the mode of electric current flow in a muscle. { ab'tərm·ən·əl }

abzyme [IMMUNOLOGY] Any of a class of monoclonal antibodies that bind to and stabilize molecules as the molecules pass through transition states to form products. { 'ab,zīm }

Acala [BOTANY] A type of cotton indigenous to Mexico and cultivated in Texas, Oklahoma, and Arkansas. { ə'kal·ə }

acalyculate [BOTANY] Lacking a calyx. { ¦ā·kə ¦lik·yü,lāt }

Acalyptratae [INVERTEBRATE ZOOLOGY] *See* Acalyptreatae. { ¦ā·kə'lip·trəd,ē }

Acalyptreatae [INVERTEBRATE ZOOLOGY] A large group of small, two-winged flies in the suborder Cyclorrhapha characterized by small or rudimentary calypters. Also spelled Acalyptratae. { ¦ā·kə·lip'trē·ə,dē }

acantha [BIOLOGY] A sharp spine; a spiny process, as on vertebrae. { ə'kan·thə }

Acanthaceae [BOTANY] A family of dicotyledonous plants in the order Scrophulariales distinguished by their usually herbaceous habit, irregular flowers, axile placentation, and dry, dehiscent fruits. { ə,kan'thās·ē,ē }

acanthaceous [BOTANY] Having sharp points or prickles; prickly. { ə,kan'thā·shəs }

Acantharia [INVERTEBRATE ZOOLOGY] A subclass of essentially pelagic protozoans in the class Actinopodea characterized by skeletal rods constructed of strontium sulfate (celestite). { ə,kan 'tha·rē·ə }

Acanthaster [INVERTEBRATE ZOOLOGY] A genus of Indo-Pacific starfishes, including the crown-of-thorns, of the family Asteriidae; economically important as a destroyer of oysters in fisheries. { ə,kan'thas·tər }

acanthella [INVERTEBRATE ZOOLOGY] A transitional larva of the phylum Acanthocephala in which rudiments of reproductive organs, lemnisci, a proboscis, and a proboscis receptacle are formed. { ə,kan'thel·ə }

acanthocarpous [BOTANY] Having spiny fruit. { ə ,kan·thə'kär·pəs }

Acanthocephala [INVERTEBRATE ZOOLOGY] The spiny-headed worms, a phylum of helminths; adults are parasitic in the alimentary canal of vertebrates. { ə,kan·thō'sef·ə·lə }

Acanthocheilonema perstans [INVERTEBRATE ZOOLOGY] A tropical filarial worm, parasitic in humans. { ə,kan·thə,kī·lə'nē·mə 'pərs·təns }

acanthocladous [BOTANY] Having spiny branches. { ə‚kan·thə'klad·əs }

Acanthodes [PALEONTOLOGY] A genus of Carboniferous and Lower Permian eellike acanthodian fishes of the family Acanthodidae. { ə‚kan 'thō·dēz }

Acanthodidae [PALEONTOLOGY] A family of extinct acanthodian fishes in the order Acanthodiformes. { ə‚kan'thō·də‚dē }

Acanthodiformes [PALEONTOLOGY] An order of extinct fishes in the class Acanthodii having scales of acellular bone and dentine, one dorsal fin, and no teeth. { ə‚kan·thō·də'fòr‚mēz }

Acanthodii [PALEONTOLOGY] A class of extinct fusiform fishes, the first jaw-bearing vertebrates in the fossil record. { ə‚kan'thō·dē‚ī }

acanthoid [BIOLOGY] Shaped like a spine. { ə 'kan·thòid }

Acanthometrida [INVERTEBRATE ZOOLOGY] An order of marine protozoans in the subclass Acantharia with 20 or less skeletal rods. { ə‚kan·thə 'met·rə·də }

Acanthophis antarcticus [VERTEBRATE ZOOLOGY] The death adder, a venomous snake found in Australia and New Guinea; venom is neurotoxic. { ə'kan·thə·fəs ant'ärk·tə·kəs }

Acanthophractida [INVERTEBRATE ZOOLOGY] An order of marine protozoans in the subclass Acantharia; skeleton includes a latticework shell and skeletal rods. { ə‚kan·thə'frak·tə·də }

acanthopodia [INVERTEBRATE ZOOLOGY] The long subpseudopodia of amoebas of the suborder Acanthopodina, order Amoebida. { ə‚kan·thə 'pōd·ē·ə }

acanthopodous [BOTANY] Having a spiny or prickly petiole or peduncle. { ¦ā‚kan¦thä·pə·dəs }

acanthopore [PALEONTOLOGY] A tubular spine in some fossil bryozoans. { ə'kan·thə‚pòr }

Acanthopteri [VERTEBRATE ZOOLOGY] An equivalent name for the Perciformes. { a‚kan'thäp·tə·rī }

Acanthopterygii [VERTEBRATE ZOOLOGY] An equivalent name for the Perciformes. { a‚kan ‚thäp·tə'rē·jē·ī }

acanthosoma [INVERTEBRATE ZOOLOGY] The last primitive larval stage, the mysis, in the family Sergestidae. { ə‚kan·thə'sō·mə }

Acanthosomatidae [INVERTEBRATE ZOOLOGY] A small family of insects in the order Hemiptera. { ə‚kan·thə·sə'mad·ə‚dē }

acanthosphere [BOTANY] A specialized ciliated body in *Nitella* cells. { ə'kan·thə‚sfir }

acanthostegous [INVERTEBRATE ZOOLOGY] Being overlaid with two series of spines, as the ovicell or ooecium of certain bryozoans. { ə‚kan·thə 'steg·əs }

acanthozooid [INVERTEBRATE ZOOLOGY] A specialized individual in a bryozoan colony that secretes tubules which project as spines above the colony's outer surface. { ə‚kan·thə'zō·òid }

Acanthuridae [VERTEBRATE ZOOLOGY] The surgeonfishes, a family of perciform fishes in the suborder Acanthuroidei. { ə‚kan'thù·rə·dē }

Acanthuroidei [VERTEBRATE ZOOLOGY] A suborder of chiefly herbivorous fishes in the order Perciformes. { ə‚kan·thə'ròid·ē‚ī }

Acari [INVERTEBRATE ZOOLOGY] The equivalent name for Acarina. { 'a·kə·rē }

Acaridiae [INVERTEBRATE ZOOLOGY] A group of pale, weakly sclerotized mites in the suborder Sarcoptiformes, including serious pests of stored food products and skin parasites of warm-blooded vertebrates. { ‚a·kə'rid·ē‚ē }

Acarina [INVERTEBRATE ZOOLOGY] The ticks and mites, a large order of the class Arachnida, characterized by lack of body demarcation into cephalothorax and abdomen. { ‚a·kə'rēn·ə }

acarology [INVERTEBRATE ZOOLOGY] A branch of zoology dealing with the mites and ticks. { ‚a·kə 'rä·lə·je }

acarophily [ECOLOGY] A symbiotic relationship between plants and mites. { ¦a·kə·rō¦fil·ē }

acarpellous [BOTANY] Lacking carpels. { ā'kär·pə·ləs }

acarpous [BOTANY] Not producing fruit. { ā'kär·pəs }

acaulous [BOTANY] **1.** Lacking a stem. **2.** Being apparently stemless but having a short underground stem. { ¦ā'kòl·əs }

acceleration globulin [BIOCHEMISTRY] A globulin that acts to accelerate the conversion of prothrombin to thrombin in blood clotting; found in blood plasma in an inactive form. { ak‚sel·ə'rā·shən 'gläb·yə·lən }

acceleration tolerance [PHYSIOLOGY] The maximum *g* forces an individual can withstand without losing control or consciousness. { ak‚sel·ə 'rā·shən 'täl·ər·əns }

acceleratory reflex [PHYSIOLOGY] Any reflex originating in the labyrinth of the inner ear in response to a change in the rate of movement of the head. { ak'sel·ə·rə‚tòr·ē 'rē·fleks }

accessorius [ANATOMY] Any muscle that reinforces the action of another. { ak·sə'sòr·ē·əs }

accessory bud [BOTANY] An embryonic shoot occurring above or to the side of an axillary bud. Also known as supernumerary bud. { ak'ses·ə·rē ‚bəd }

accessory cell [BOTANY] A morphologically distinct epidermal cell adjacent to, and apparently functionally associated with, guard cells on the leaves of many plants. [IMMUNOLOGY] Any nonlymphocytic cell that helps in the induction of the immune response by presenting antigen to a helper T lymphocyte. { ak'ses·ə·rē ‚sel }

accessory chromosome [CELL BIOLOGY] *See* supernumerary chromosome. { ak'ses·ə·rē 'krōm·ə‚sōm }

accessory fruit [BOTANY] A fruit in which the conspicuous portion consists of tissue other than that of the ripened ovary. Also known as pseudocarp. { ak'ses·ə·rē ‚früt }

accessory gland [ANATOMY] A mass of glandular tissue separate from the main body of a gland. [INVERTEBRATE ZOOLOGY] A gland associated with the male reproductive organs in insects. { ak 'ses·ə·rē ‚gland }

accessory movement [PHYSIOLOGY] *See* synkinesia. { ak'ses·ə·rē ‚müv·mənt }

accessory nerve [ANATOMY] The eleventh cranial

nerve in tetrapods, a paired visceral motor nerve; the bulbar part innervates the larynx and pharynx, and the spinal part innervates the trapezius and sternocleidomastoid muscles. { ak'ses·ə·rē ˌnərv }

accessory sexual characters [ANATOMY] Those structures and organs (excluding the gonads) composing the genital tract and including accessory glands and external genitalia. { ak'ses·ə·rē 'seksh·ə·wəl 'kar·ik·tərz }

accessory species [ECOLOGY] A species comprising 25-50% of a community. { ak'ses·ə·rē 'spē·shēz }

accidental species [ECOLOGY] A species which constitutes less than one-fourth of the population of a stand. { ¦ak·sə¦den·təl 'spē·shēz }

accidental whorl [ANATOMY] A type of whorl fingerprint pattern which is a combination of two different types of pattern, with the exception of the plain arch, with two or more deltas; or a pattern which possesses some of the requirements for two or more different types; or a pattern which conforms to none of the definitions; in accidental whorl tracing three types appear: an outer (O), inner (I), or meeting (M). { ¦ak·sə ¦den·təl 'wərl }

Accipitridae [VERTEBRATE ZOOLOGY] The diurnal birds of prey, the largest and most diverse family of the order Falconiformes, including hawks, eagles, and kites. { ˌak·sə'pi·trəˌdē }

acclimated microorganism [ECOLOGY] Any microorganism that is able to adapt to environmental changes such as a change in temperature, or a change in the quantity of oxygen or other gases. { ə'klīm·əd·əd ˌmī·krō'ór·gə·niz əm }

acclimation [BIOLOGY] *See* acclimatization. { ˌak·lə'mā·shən }

acclimatization [BIOLOGY] Physiological, emotional, and behavioral adjustment by an individual to changes in the environment. [EVOLUTION] Adaptation of a species or population to a changed environment over several generations. Also known as acclimation. { əˌklī·mə·tə'zā·shən }

accommodation [ECOLOGY] A population's location within a habitat. [PHYSIOLOGY] A process in most vertebrates whereby the focal length of the eye is changed by automatic adjustment of the lens to bring images of objects from various distances into focus on the retina. { əˌkäm·ə'dā·shən }

accommodation reflex [PHYSIOLOGY] Changes occurring in the eyes when vision is focused from a distant to a near object; involves pupil contraction, increased lens convexity, and convergence of the eyes. { əˌkäm·ə'dā·shən 'rē·fleks }

accrescent [BOTANY] Growing continuously with age, especially after flowering. { ə'krēs·ənt }

accretion line [HISTOLOGY] A microscopic line on a tooth, marking the addition of a layer of enamel or dentin. { ə'krē·shən ˌlīn }

accumbent [BOTANY] Describing an organ that leans against another; specifically referring to cotyledons having their edges folded against the hypocotyl. { ə'kəm·bənt }

accumulated dose [PHYSIOLOGY] The total amount of radiation absorbed by an organism as a result of exposure to radiation. { ə'kyü·myə ˌlād·əd 'dōs }

accumulator plant [BOTANY] A plant or tree that grows in a metal-bearing soil and accumulates an abnormal content of the metal. { ə'kyü·myə ˌlād·ər ˌplant }

acellular [BIOLOGY] Not composed of cells. { 'ā·sel·yə·lər }

acellular gland [PHYSIOLOGY] A gland, such as intestinal glands, the pancreas, and the parotid gland, that secretes a noncellular product. { 'ā·sel·yə·lər ˌgland }

acellular slime mold [MYCOLOGY] The common name for members of the Myxomycetes. { 'ā·sel·yə·lər 'slīm ˌmōld }

acentric [BIOLOGY] Not oriented around a middle point. [GENETICS] A chromosome or chromosome fragment lacking a centromere. { ˌā'sen·trik }

acentrous [VERTEBRATE ZOOLOGY] Lacking vertebral centra and having the notochord persistent throughout life, as in certain primitive fishes. { ˌā'sen·trəs }

Acephalina [INVERTEBRATE ZOOLOGY] A suborder of invertebrate parasites in the protozoan order Eugregarinida characterized by nonseptate trophozoites. { āˌsef·ə'līn·ə }

acephalocyst [INVERTEBRATE ZOOLOGY] An abnormal cyst of the *Echinococcus granulosus* larva, lacking a head and brood capsules, found in human organs. { ā'sef·ə·ləˌsist }

acephalous [BOTANY] Having the style originate at the base instead of at the apex of the ovary. [ZOOLOGY] Lacking a head. { ā'sef·ə·ləs }

Acer [BOTANY] A genus of broad-leaved, deciduous trees of the order Sapindales, commonly known as the maples; the sugar or rock maple (*A. saccharum*) is the most important commercial species. { 'ā·sər *or* 'äˌkər }

acerate [BOTANY] Needle-shaped, specifically referring to leaves. { 'as·əˌrāt }

Acerentomidae [INVERTEBRATE ZOOLOGY] A family of wingless insects belonging to the order Protura; the body lacks tracheae and spiracles. { ˌa·sə·rən'tōm·ə·dē }

acervate [BIOLOGY] Growing in heaps or dense clusters. { 'a·sərˌvāt }

acervulus [MYCOLOGY] A cushion- or disk-shaped mass of hyphae, peculiar to the Melanconiales, on which there are dense aggregates of conidiophores. { ə'sər·vyə·ləs }

acetabulum [ANATOMY] A cup-shaped socket on the hipbone that receives the head of the femur. [INVERTEBRATE ZOOLOGY] **1.** A cavity on an insect body into which a leg inserts for articulation. **2.** The sucker of certain invertebrates such as trematodes and tapeworms. { ˌa·sə'tab·yə·ləm }

acetaldehydase [BIOCHEMISTRY] An enzyme that catalyzes the oxidation of acetaldehyde to acetic acid. { ¦as·ədˌal·də'hīˌdās }

acetic fermentation [MICROBIOLOGY] Oxidation of alcohol to produce acetic acid by the action of bacteria of the genus *Acetobacter*. { ə'sēd·ik fər·mən'tā·shən }

acetic thiokinase [BIOCHEMISTRY] An enzyme that catalyzes the formation of acetyl coenzyme A from acetate and adenosinetriphosphate. { ə 'sēd·ik ,thī·ə'kīn,ās }

acetoacetyl coenzyme A [BIOCHEMISTRY] $C_{25}H_{41}O_{18}N_7P_3S$ An intermediate product in the oxidation of fatty acids. { ¦as·ə,tō·ə'sēd·əl ,kō 'en·zīm 'ā }

Acetobacter [MICROBIOLOGY] A genus of gram-negative, aerobic bacteria of uncertain affiliation comprising ellipsoidal to rod-shaped cells as singles, pairs, or chains; they oxidize ethanol to acetic acid. Also known as acetic acid bacteria; vinegar bacteria. { ə'sēd·ō,bak·tər }

Acetobacter aceti [MICROBIOLOGY] An aerobic, rod-shaped bacterium capable of efficient oxidation of glucose, ethyl alcohol, and acetic acid; found in vinegar, beer, and souring fruits and vegetables. { ə'sēd·ō,bak·tər ə'sēd·ē }

Acetobacter suboxydans [MICROBIOLOGY] A short, nonmotile vinegar bacterium that can oxidize ethanol to acetic acid; useful for industrial production of ascorbic and tartaric acids. { ə 'sēd·ō,bak·tər səb'äks·ə·dəns }

acetolactic acid [BIOCHEMISTRY] $C_5H_8O_4$ A monocarboxylic acid formed as an intermediate in the synthesis of the amino acid valine. { ,as·ə·tə'lak·tik 'as·əd }

Acetomonas [MICROBIOLOGY] A genus of aerobic, polarly flagellated vinegar bacteria in the family Pseudomonadaceae; used industrially to produce vinegar, gluconic acid, and L-sorbose. { ə,sed·ə'mōn·əs }

acetone body [BIOCHEMISTRY] *See* ketone body. { 'as·ə,tōn ,bäd·ē }

acetone fermentation [MICROBIOLOGY] Formation of acetone by the metabolic action of certain anaerobic bacteria on carbohydrates. { 'as·ə,tōn fər·mən'tā·shən }

acetylase [BIOCHEMISTRY] Any enzyme that catalyzes the formation of acetyl esters. { ə'sed·əl ,ās }

acetylcholine [BIOCHEMISTRY] $C_7H_{17}O_3N$ A compound released from certain autonomic nerve endings which acts in the transmission of nerve impulses to excitable membranes. { ə,sed·əl'kō ,lēn }

acetylcholinesterase [BIOCHEMISTRY] An enzyme found in excitable membranes that inactivates acetylcholine. { ə,sed·əl·kō·lən'es·tər,ās }

acetyl coenzyme A [BIOCHEMISTRY] $C_{23}H_{39}O_{17}N_7P_3S$ A coenzyme, derived principally from the metabolism of glucose and fatty acids, that takes part in many biological acetylation reactions; oxidized in the Krebs cycle. { ə,sed·əl ,kō'en,zīm 'ā }

acetyl phosphate [BIOCHEMISTRY] $C_2H_5O_5P$ The anhydride of acetic and phosphoric acids occurring in the metabolism of pyruvic acid by some bacteria; phosphate is used by some microorganisms, in place of adenosinetriphosphate, for the phosphorylation of hexose sugars. { ə'sed·əl 'fäs,fāt }

Achaenodontidae [PALEONTOLOGY] A family of Eocene dichobunoids, piglike mammals belonging to the suborder Palaeodonta. { ə,kēn·ə 'dän·tə·dē }

achaetous [INVERTEBRATE ZOOLOGY] Without setae. Also known as asetigerous. { ə'kēd·əs }

A chain [IMMUNOLOGY] *See* heavy chain. { 'ā ,chān }

acheb [ECOLOGY] Short-lived vegetation regions of the Sahara composed principally of mustards (Cruciferae) and grasses (Gramineae). { ə 'cheb }

achene [BOTANY] A small, dry, indehiscent fruit formed from a simple ovary bearing a single seed. { ə'kēn }

achiasmate [CELL BIOLOGY] Pertaining to meiosis that lacks chiasma. { ,ā·kī'az,māt }

achilary [BOTANY] In flowers, having the lip (labellum) undeveloped or lacking. { ə'kil·ə·rē }

Achilles jerk [PHYSIOLOGY] A reflex action seen as plantar flection in response to a blow to the Achilles tendon. Also known as Achilles tendon reflex. { ə'kil·ēz ,jərk }

Achilles tendon [ANATOMY] The tendon formed by union of the tendons of the calf muscles, the soleus and gastrocnemius, and inserted into the heel bone. { ə'kil·ēz 'ten·dən }

Achilles tendon reflex [PHYSIOLOGY] *See* Achilles jerk. { ə'kil·ēz 'ten·dən 'rē·fleks }

achlamydeous [BOTANY] Lacking a perianth. { ¦ā·klə'mid·ē·əs }

Acholeplasma [MICROBIOLOGY] The single genus of the family Acholeplasmataceae, comprising spherical and filamentous cells. { ə,kōl·ə'plaz·mə }

Acholeplasmataceae [MICROBIOLOGY] A family of the order Mycoplasmatales; characters same as for the order and class (Mollicutes); members do not require sterol for growth. { ə,kōl·ə,plaz·mə'tās·ē,ē }

achordate [VERTEBRATE ZOOLOGY] Lacking a notochord. { ¦ā'kȯr,dāt }

achroglobin [BIOCHEMISTRY] A colorless respiratory pigment present in some mollusks and urochordates. { ¦ak·rə'glōb·ən }

Achromatiaceae [MICROBIOLOGY] A family of gliding bacteria of uncertain affiliation; cells are spherical to ovoid or cylindrical, movements are slow and jerky, and microcysts are not known. { a,krō·mə·dē'ās·ē,ē }

achromatic interval [PHYSIOLOGY] The difference between the achromatic threshold and the smallest light stimulus at which the hue is detectable. { ¦a·krə¦mad·ik 'int·ər·vəl }

achromatic threshold [PHYSIOLOGY] The smallest light stimulus that can be detected by a dark-adapted eye, so called because all colors lose their hue at this illumination. { ¦a·krə¦mad·ik 'thresh,hōld }

achromatin [CELL BIOLOGY] The portion of the cell nucleus which does not stain easily with basic dyes. { ,ā'krō·mə·tən }

Achromatium [MICROBIOLOGY] The type genus of the family Achromatiaceae. { ,a·krə'māsh·ē·əm }

achromatophilia [BIOLOGY] The property of not staining readily. { ¦ā·krō,mad·ə'fil·ē·ə }

achromic [BIOLOGY] Colorless; lacking normal pigmentation. { ¦ā'krō·mik }
Achromobacter [MICROBIOLOGY] A genus of motile and nonmotile, gram-negative, rod-shaped bacteria in the family Achromobacteraceae. { ,ā 'krō·mə,bak·tər }
Achromobacteraceae [MICROBIOLOGY] Formerly a family of true bacteria, order Eubacteriales, characterized by aerobic metabolism. { ,ā,krō·mə,bak·tər'ās·ē,ē }
Achroonema [MICROBIOLOGY] A genus of bacteria in the family Pelonemataceae; cells have smooth, delicate, porous walls. { ,ak·rō'ōn·ə·mə }
aciculignosa [ECOLOGY] Narrow sclerophyll or coniferous vegetation that is mostly subalpine, subarctic, or continental. { ə,sik·yə·lig'nōs·ə }
Acidaminococcus [MICROBIOLOGY] A genus of bacteria in the family Veillonellaceae; cells are often oval or kidney-shaped and occur in pairs; amino acids can supply the single energy source. { ,as·əd,a·mə·nō'käk·əs }
acid-base balance [PHYSIOLOGY] Physiologically maintained equilibrium of acids and bases in the body. { 'as·əd 'bās 'bal·əns }
acid cell [HISTOLOGY] A parietal cell of the stomach. { 'as·əd ,sel }
acid-fast bacteria [MICROBIOLOGY] Bacteria, especially mycobacteria, that stain with basic dyes and fluorochromes and resist decoloration by acid solutions. { 'as·əd ¦fast bak'tir·ē·ə }
acid-fast stain [MICROBIOLOGY] A differential stain used in identifying species of *Mycobacterium* and one species of *Nocardia*. { 'as·əd ,fast 'stān }
acidophil [BIOLOGY] **1.** Any substance, tissue, or organism having an affinity for acid stains. **2.** An organism having a preference for an acid environment. [HISTOLOGY] **1.** An alpha cell of the adenohypophysis. **2.** *See* eosinophil. { ə'sid·ə,fil }
acidophilic erythroblast [HISTOLOGY] *See* normoblast. { ə'sid·ə,fil·ik ə'rith·rə,blast }
acidotrophic [BIOLOGY] Having an acid nutrient requirement. { ə,sid·ə'trōf·ik }
acid phosphatase [BIOCHEMISTRY] An enzyme in blood which catalyzes the release of phosphate from phosphate esters; optimum activity at pH 5. { 'as·əd 'fäs·fə,tās }
acinar [ANATOMY] Pertaining to an acinus. { 'as·ə·nər }
acinar cell [ANATOMY] Any of the cells lining an acinous gland. { 'as·ə·nər 'sel }
Acinetobacter [MICROBIOLOGY] A genus of nonmotile, short, plump, almost spherical rods in the family Neisseriaceae; strictly aerobic; resistant to penicillin. { ,as·ə,nēd·ə'bak·tər }
aciniform [ZOOLOGY] Shaped like a berry or a bunch of grapes. { ə'sin·ə,fórm }
acinotubular gland [ANATOMY] *See* tubuloalveolar gland. { ,as·ə,nō'tü·byə·lər 'gland }
acinous [BIOLOGY] Of or pertaining to acini. { 'as·ə,nəs }
acinous gland [ANATOMY] A multicellular gland with sac-shaped secreting units. Also known as alveolar gland. { 'as·ə,nəs ,gland }
acinus [ANATOMY] The small terminal sac of an acinous gland, lined with secreting cells. [BOTANY] An individual drupelet of a multiple fruit. { 'as·ə·nəs }
Acipenser [VERTEBRATE ZOOLOGY] A genus of actinopterygian fishes in the sturgeon family, Acipenseridae. { 'as·ə,pen·sər }
Acipenseridae [VERTEBRATE ZOOLOGY] The sturgeons, a family of actinopterygian fishes in the order Acipenseriformes. { ,as·ə,pen'ser·ə·dē }
Acipenseriformes [VERTEBRATE ZOOLOGY] An order of the subclass Actinopterygii represented by the sturgeons and paddlefishes. { ,as·ə·pen ,ser·ə'fór,mēz }
Acmaeidae [INVERTEBRATE ZOOLOGY] A family of gastropod mollusks in the order Archaeogastropoda; includes many limpets. { ak'mē·ə,dē }
acme [PALEONTOLOGY] The time of largest abundance or variety of a fossil taxon; the taxon may be either general or local. { 'ak·mē }
acmic [ECOLOGY] A phase or period in which an aquatic population undergoes seasonal changes. { 'ak·mik }
Acnidosporidia [INVERTEBRATE ZOOLOGY] An equivalent name for the Haplosporea. { ak,nī·də ,spō'rid·ē·ə }
Acoela [INVERTEBRATE ZOOLOGY] An order of marine flatworms in the class Turbellaria characterized by the lack of a digestive tract and coelomic cavity. { ā'sēl·ə }
Acoelea [INVERTEBRATE ZOOLOGY] An order of gastropod mollusks in the subclass Opistobranchia; includes many sea slugs. { ,ā·sə'lē·ə }
Acoelomata [INVERTEBRATE ZOOLOGY] A subdivision of the animal kingdom; individuals are characterized by lack of a true body cavity. { ā ,sēl·ə'mäd·ə }
acoelomate [ZOOLOGY] Pertaining to an animal that lacks a coelom. { ā'sēl·ə,māt }
acoelous [ZOOLOGY] **1.** Lacking a true body cavity or coelom. **2.** Lacking a true stomach or digestive tract. { ,ā'sēl·əs }
acolpate [BOTANY] Of pollen grains, lacking furrows or grooves. { ,ā'kōl,pāt }
Aconchulinida [INVERTEBRATE ZOOLOGY] An order of protozoans in the subclass Filosia comprising a small group of naked amebas having filopodia. { ə,käŋ·kə'lin·ə·də }
aconitase [BIOCHEMISTRY] An enzyme involved in the Krebs citric acid cycle that catalyzes the breakdown of citric acid to *cis*-aconitic and isocitric acids. { ə'kän·ə,tās }
aconite [BOTANY] Any plant of the genus *Aconitum*. Also known as friar's cowl; monkshood; mousebane; wolfsbane. { 'ak·ə,nīt }
acorn [BOTANY] The nut of the oak tree, usually surrounded at the base by a woody involucre. { 'ā,kórn }
acorn barnacle [INVERTEBRATE ZOOLOGY] Any of the sessile barnacles that are enclosed in conical, flat-bottomed shells and attach to ships and near-shore rocks and piles. { 'ā,kórn ,bär·nə·kəl }
acorn worm [INVERTEBRATE ZOOLOGY] Any member of the class Enteropneusta, free-living ani-

mals that usually burrow in sand or mud. Also known as tongue worm. { 'ā,kȯrn ,wərm }

acotyledon [BOTANY] A plant without cotyledons. { ā,käd·əl'ēd·ən }

acouchi [VERTEBRATE ZOOLOGY] A hystricomorph rodent represented by two species in the family Dasyproctidae; believed to be a dwarf variety of the agouti. { ə'kü·shē }

acoustic nerve [ANATOMY] *See* auditory nerve. { ə 'küs·tik ,nərv }

acoustic reflex [PHYSIOLOGY] Brief, involuntary closure of the eyes due to stimulation of the acoustic nerve by a sudden sound. { ə'küs·tik 'rē ,fleks }

ACP [BIOCHEMISTRY] *See* acyl carrier protein.

acquired [BIOLOGY] Not present at birth, but developed by an individual in response to the environment and not subject to hereditary transmission. { ə'kwīrd }

acquired immunity [IMMUNOLOGY] Resistance to a microbial or other antigenic substance taken on by a naturally susceptible individual; may be either active or passive. { ə'kwīrd ə'myün·ə·dē }

acquired immunological tolerance [IMMUNOLOGY] Failure of immunological responsiveness, that is, inability of antigen-sensitive cells to synthesize antibodies; induced by exposure to large amounts of an antigen. Also known as immunological paralysis. { ə'kwīrd ,im·yü·nə'laj·ə·kəl 'täl·ə·rəns }

Acrania [ZOOLOGY] A group of lower chordates with no cranium, jaws, vertebrae, or paired appendages; includes the Tunicata and Cephalochordata. { ā'krān·ē·ə }

Acrasiales [BIOLOGY] A group of microorganisms that have plant and animal characteristics; included in the phylum Myxomycophyta by botanists and Mycetozoia by zoologists. { ə'krāzh·ē 'ā·lēz }

Acrasida [MYCOLOGY] An order of Mycetozoia containing cellular slime molds. { ə'kras·ə·də }

Acrasieae [BIOLOGY] An equivalent name for the Acrasiales. { ə,krāz·ē'ē,ē }

acrasin [BIOCHEMISTRY] The chemotactic substance thought to be secreted by, and to effect aggregation of, myxamebas during their fruiting phase. { ə'krāz·ən }

Acrasiomycota [MYCOLOGY] The phylum containing the cellular slime molds. { ə¦krā·zē·ō·mī 'käd·ə }

acraspedote [INVERTEBRATE ZOOLOGY] Describing tapeworm segments which are not overlapping. { ə'kras·pə,dōt }

acritarch [PALEONTOLOGY] A unicellular microfossil of unknown or uncertain biological origin that occurs abundantly in strata from the Precambrian and Paleozoic. { 'ak·rə,tark }

acroblast [CELL BIOLOGY] A vesicular structure in the spermatid formed from Golgi material. { 'ak·rə,blast }

acrocarpous [BOTANY] In some mosses of the subclass Eubrya, having the sporophyte at the end of a stem and therefore exhibiting the erect habit. { ,ak·rə'kär·pəs }

acrocentric chromosome [CELL BIOLOGY] A chromosome having the centromere close to one end. { ¦ak·rə¦sen·trik 'krō·mə·sōm }

Acroceridae [INVERTEBRATE ZOOLOGY] The humpbacked flies, a family of orthorrhaphous dipteran insects in the series Brachycera. { a,krä 'ser·ə·de }

acrodomatia [ECOLOGY] Specialized structures on certain plants adapted to shelter mites; relationship is presumably symbiotic. { ,ak·rə·də 'māsh·ē·ə }

acrodont [ANATOMY] Having teeth fused to the edge of the supporting bone. { 'ak·rə,dänt }

acromere [HISTOLOGY] The distal portion of a rod or cone in the retina. { 'ak·rō,mēr }

acromion [ANATOMY] The flat process on the outer end of the scapular spine that articulates with the clavicle and forms the outer angle of the shoulder. { ə'krō·me,än }

acron [EVOLUTION] Unsegmented head of the ancestral arthropod. [INVERTEBRATE ZOOLOGY] **1.** The preoral, nonsegmented portion of an arthropod embryo. **2.** The prostomial region of the trochophore larva of some mollusks. { 'ak,rän }

acronematic [BIOLOGY] Referring to a flagellum without hairs. { ,ak·rō·nə'mad·ik }

acropetal [BOTANY] From the base toward the apex, as seen in the formation of certain organs or the spread of a pathogen. { ə'krä·pəd·əl }

Acrosaleniidae [PALEONTOLOGY] A family of Jurassic and Cretaceous echinoderms in the order Salenoida. { ¦ak·rō,sal·ə'nī·ə·dē }

acroscopic [BOTANY] Facing, or on the side toward, the apex. { ,ak·rə'skäp·ik }

acrosin [BIOCHEMISTRY] A proteolytic enzyme located in the acrosome of a spermatozoon; thought to be involved in penetration of the egg. { 'ak·rə·sin }

acrosome [CELL BIOLOGY] The anterior, crescent-shaped body of spermatozoon, formed from Golgi material of the spermatid. Also known as perforatorium. { 'ak·rə,sōm }

acrospore [MYCOLOGY] In fungi, a spore formed at the outer tip of a hypha. { 'ak·rə,spȯr }

acrotarsium [ANATOMY] Instep of the foot. { ,ak·rō'tär·sē·əm }

Acrothoracica [INVERTEBRATE ZOOLOGY] A small order of burrowing barnacles in the subclass Cirripedia that inhabit corals and the shells of mollusks and barnacles. { ,ak·rə·thə'ras·ik·ə }

Acrotretacea [PALEONTOLOGY] A family of Cambrian and Ordovician inarticulate brachipods of the suborder Acrotretidina. { ,ak·rō·tre'tās·ē·ə }

Acrotretida [INVERTEBRATE ZOOLOGY] An order of brachiopods in the class Inarticulata; representatives are known from Lower Cambrian to the present. { ,ak·rō'tred·ə·də }

Acrotretidina [INVERTEBRATE ZOOLOGY] A suborder of inarticulate brachiopods of the order Acrotretida; includes only species with shells composed of calcium phosphate. { ,ak·rō·tre'tī·də·nə }

Actaeonidae [INVERTEBRATE ZOOLOGY] A family of gastropod mollusks in the order Tectibranchia. { ,ak·tē'än·ə·dē }

Actaletidae [INVERTEBRATE ZOOLOGY] A family of

insects belonging to the order Collembola characterized by simple tracheal systems. { ˌak·tə 'led·ə·dē }

ACTH [BIOCHEMISTRY] *See* adrenocorticotropic hormone.

Actidione [MICROBIOLOGY] Trade name for the antibiotic cyclohexamide. { ˌak·tə'dīˌōn }

actin [BIOCHEMISTRY] A muscle protein that is the chief constituent of the Z-band myofilaments of each sarcomere. { 'ak·tən }

actinal [INVERTEBRATE ZOOLOGY] In radially symmetrical animals, referring to the part from which the tentacles or arms radiate or to the side where the mouth is located. { 'ak·tə·nəl }

Actiniaria [INVERTEBRATE ZOOLOGY] The sea anemones, an order of cnidarians in the subclass Zoantharia. { akˌtin·ē'a·rē·ə }

Actinobacillus [MICROBIOLOGY] A species of gram-negative, oval, spherical, or rod-shaped bacteria that are of uncertain affiliation; coccal and bacillary cells are often interspersed, giving a "Morse code" form; species are pathogens of animals, occasionally of humans. { ˌak·tə·nō·bə 'sil·əs }

Actinobifida [MICROBIOLOGY] A genus of bacteria in the family Micromonosporaceae with a dichotomously branched substrate; an aerial mycelium is formed which produces single spores. { ˌak·tə·nō'bī·fə·də }

actinocarpous [BOTANY] Having flowers and fruit radiating from one point. { ˌak·tə·nō'kär·pəs }

actinochitin [BIOCHEMISTRY] A form of birefringent or anisotropic chitin found in the seta of certain mites. { ˌak·tə·nō'kī·tən }

Actinochitinosi [INVERTEBRATE ZOOLOGY] A group name for two closely related suborders of mites,the Trombidiformes and the Sarcoptiformes. { ˌak·tə·nōˌkī·tə'nō·sē }

Actinolaimoidea [INVERTEBRATE ZOOLOGY] A superfamily of nematodes in the order Dorylaimida, containing some species with remarkable elaborations of the stoma and the characteristic axial spear. { ¦ak·tə·nō·lə'mȯid·ē·ə }

actinomere [INVERTEBRATE ZOOLOGY] One of the segments composing the body of a radially symmetrical animal. { ˌak'tin·əˌmir }

actinomorphic [BIOLOGY] Descriptive of an organism, organ, or part that is radially symmetrical. { ˌak·tə·nō'mȯr·fik }

Actinomyces [MICROBIOLOGY] The type genus of the family Actinomycetaceae; anaerobic to facultatively anaerobic; includes human and animal pathogens. { ˌak·tə·nō'mī·sēs }

Actinomycetaceae [MICROBIOLOGY] A family of bacteria in the order Actinomycetales; gram-positive, diphtheroid cells which form filaments but not mycelia; chemoorganotrophs that ferment carbohydrates. { ˌak·tə·nōˌmī·sə'tās·ēˌē }

Actinomycetales [MICROBIOLOGY] An order of bacteria; cells form branching filaments which develop into mycelia in some families. { ˌak·tə·nōˌmī·sə'tā·lēz }

actinomycete [MICROBIOLOGY] Any member of the bacterial family Actinomycetaceae. { ˌak·tə·nō'mīˌsēt }

actinomycin [MICROBIOLOGY] The collective name for a large number of red chromoprotein antibiotics elaborated by various strains of *Streptomyces*. { ˌak·tə·nō'mī·sən }

actinomyosin [BIOCHEMISTRY] A protein complex formed by the combination of actin and myosin during muscle contraction. { ˌak·tə·nō'mī·əs·ən }

Actinomyxida [INVERTEBRATE ZOOLOGY] An order of protozoan invertebrate parasites of the class Myxosporidea characterized by trivalved spores with three polar capsules. { ˌak·tə·nō'mik·sə·də }

actinophage [MICROBIOLOGY] A bacteriophage that infects and lyses members of the order Actinomycetales. { ak'tin·əˌfāj }

Actinophryida [INVERTEBRATE ZOOLOGY] An order of protozoans in the subclass Heliozoia; individuals lack an organized test, a centroplast, and a capsule. { ˌak·tə·nō'frī·ə·də }

Actinoplanaceae [MICROBIOLOGY] A family of bacteria in the order Actinomycetales with well-developed mycelia and spores formed on sporangia. { ˌak·tə·nō·plə'nās·ēˌē }

Actinoplanes [MICROBIOLOGY] A genus of bacteria in the family Actinoplanaceae having aerial mycelia and spherical to subspherical sporangia; spores are spherical and motile by means of a tuft of polar flagella. { ˌak·tə·nō'plā·nēz }

Actinopodea [INVERTEBRATE ZOOLOGY] A class of protozoans belonging to the superclass Sarcodina; most are free-floating, with highly specialized pseudopodia. { ˌak·tə·nō'pōd·ē·ə }

Actinopteri [VERTEBRATE ZOOLOGY] An equivalent name for the Actinopterygii. { ˌak·tə'näp·tə ˌrī }

Actinopterygii [VERTEBRATE ZOOLOGY] The ray-fin fishes, a subclass of the Osteichthyes distinguished by the structure of the paired fins, which are supported by dermal rays. { ˌak·təˌnäp·tə 'rij·ēˌī }

actinostele [BOTANY] A protostele characterized by xylem that is either star-shaped in cross section or has ribs radiating from the center. { ak 'tin·əˌstēl }

actinostome [BIOLOGY] **1.** The mouth of a radiate animal. **2.** The peristome of an echinoderm. { ak 'tin·əˌsōm }

Actinostromariidae [PALEONTOLOGY] A sphaeractinoid family of extinct marine hydrozoans. { ˌak·tə·nōˌstrō·mə'rī·əˌdē }

actinotrocha [INVERTEBRATE ZOOLOGY] The free-swimming larva of *Phoronis*, a genus of small, marine, tubicolous worms. { ˌak·tə·nō'trō·kə }

actinula [INVERTEBRATE ZOOLOGY] A larval stage of some hydrozoans that has tentacles and a mouth; attaches and develops into a hydroid in some species, or metamorphoses into a medusa. { ak'tin·yə·lə }

action current [PHYSIOLOGY] The electric current accompanying membrane depolarization and repolarization in an excitable cell. { 'ak·shən ˌkə·rənt }

action potential [PHYSIOLOGY] A transient change in electric potential at the surface of a nerve or muscle cell occurring at the moment of excitation. { 'ak·shən pəˌten·chəl }

action spectrum [PHYSIOLOGY] Graphic representation of the comparative effects of different wavelengths of light on living systems or their components. { 'ak·shən ˌspek·trəm }

actium [ECOLOGY] A rocky seashore community. { 'ak·tē·əm }

activated macrophage [IMMUNOLOGY] A macrophage whose ability to destroy microbes or other cells has been enhanced because of stimulation by a lymphokine. { 'ak·təˌvād·əd 'mak·rəˌfāj }

activating enzyme [BIOCHEMISTRY] An enzyme that catalyzes a reaction involving adenosinetriphosphate and a specific amino acid to give a product that subsequently reacts with a specific transfer ribonucleic acid. { 'ak·təˌvād·iŋ 'en ˌzīm }

activating receptor [PHYSIOLOGY] A sense organ at the end of a nerve that triggers a specific response when it is stimulated. { 'ak·təˌvād·iŋ rə 'sep·tər }

activation [MOLECULAR BIOLOGY] A change that is induced in an amino acid before it is utilized for protein synthesis. [PHYSIOLOGY] The designation for all changes in the ovum during fertilization, from sperm contact to the dissolution of nuclear membranes. { ˌak·tə'vā·shən }

activator [GENETICS] A molecule that modifies a repressor in a way that enables it to stimulate operon transcription. { 'ak·təˌvād·ər }

activator ribonucleic acid [GENETICS] Ribonucleic acid molecules which form a sequence-specific complex with receptor genes linked to producer genes. { 'ak·təˌvā·tər ¦rībō¦nü¦klē·ik 'as·əd }

active anaphylaxis [IMMUNOLOGY] The allergic response following reintroduction of an antigen into a hypersensitive individual. { 'ak·tiv 'an·ə·fə'lak·səs }

active center [BIOCHEMISTRY] **1.** A flexible portion of an enzyme that binds to the substrate and converts it into the reaction product. **2.** In carrier and receptor proteins, the portion of the molecule that interacts with the specific target compounds. { 'ak·tiv 'sen·tər }

active immunity [IMMUNOLOGY] Disease resistance in an individual due to antibody production after exposure to a microbial antigen following disease, inapparent infection, or inoculation. { 'ak·tiv im'yü·nət·ē }

active site [MOLECULAR BIOLOGY] The region of an enzyme molecule at which binding with the substrate occurs. Also known as binding site; catalytic site. { 'ak·tiv 'sīt }

active transport [PHYSIOLOGY] The pumping of ions or other substances across a cell membrane against an osmotic gradient, that is, from a lower to a higher concentration. { 'ak·tiv 'tranzˌpȯrt }

actomyosin [BIOCHEMISTRY] A protein complex consisting of myosin and actin; the major constituent of a contracting muscle fibril. { ˌak·tə 'mī·ə·sən }

actophilous [ECOLOGY] Having a seashore growing habit. { ˌak'tä·fə·ləs }

acuate [BIOLOGY] **1.** Having a sharp point. **2.** Needle- shaped. { 'ak·yəˌwāt }

acuity [BIOLOGY] Sharpness of sense perception, as of vision or hearing. { ə'kyü·ə·dē }

Aculeata [INVERTEBRATE ZOOLOGY] A group of seven superfamilies that constitute the stinging forms of hymenopteran insects in the suborder Apocrita. { əˌkyü·lē'ä·də }

aculeus [INVERTEBRATE ZOOLOGY] **1.** A sharp, hairlike spine, as on the wings of certain lepidopterans. **2.** An insect stinger modified from an ovipositor. { ə'kyü·lē·əs }

Aculognathidae [INVERTEBRATE ZOOLOGY] The ant-sucking beetles, a family of coleopteran insects in the superfamily Cucujoidea. { əˌkyü·ləg 'nath·əˌdē }

acuminate [BOTANY] Tapered to a slender point, especially referring to leaves. { ə'kyüm·ə·nət }

acute [BIOLOGY] Ending in a sharp point. { ə 'kyüt }

acute-phase protein [IMMUNOLOGY] Any of a group of proteins that are produced by the liver and appear in the blood in increased amounts shortly after the onset of infection or tissue damage; they include C-reactive protein, fibrinogen, proteolytic enzyme inhibitors, and transferrin. { ə ¦kyüt ¦fāz 'prōˌtēn }

acute-phase reaction [IMMUNOLOGY] During inflammation, change in the rates of synthesis of certain serum proteins that are important in nonspecific defense reactions. { ə¦kyüt ¦fāz rē'ak·shən }

acute transfection [GENETICS] Short-term deoxyribonucleic acid infection of cells. { ə'kyüt tranz 'fek·shən }

acutifoliate [BOTANY] Having sharply pointed leaves. { əˌkyüd·ə'fō·lē·āt }

acutilobate [BOTANY] Having sharply pointed lobes. { əˌkyüd·ə'lōˌbāt }

acyclic [BOTANY] Having flowers arranged in a spiral instead of a whorl. { ā'sik·lik }

acyl carnitine [BIOCHEMISTRY] *See* fatty acyl carnitine. { 'a·səl 'kär·nəˌtēn }

acyl carrier protein [BIOCHEMISTRY] A protein in fatty acid synthesis that picks up aceytl and malonyl groups from acetyl coenzyme A and malonyl coenzyme A and links them by condensation to form β-keto acid acyl carrier protein, releasing carbon dioxide and the sulfhydryl form of acyl carrier protein. Abbreviated ACP. { 'a·səl 'kar·ē·ər 'prōˌtēn }

acyl-coenzyme A [BIOCHEMISTRY] *See* fatty acyl-coenzyme A. { 'a·səl kō'enˌzim 'ā }

adambulacral [INVERTEBRATE ZOOLOGY] Lying adjacent to the ambulacrum. { ¦ad·am·byə'lāk·rəl }

adapertural [INVERTEBRATE ZOOLOGY] Near the aperture, specifically of a conch. { ˌad'ap·ə ˌchər·əl }

adapical [BOTANY] Near or toward the apex or tip. { ˌad'a·pi·kəl }

adaptation [GENETICS] The occurrence of genetic changes in a population or species as the result of natural selection so that it adjusts to new or altered environmental conditions. [PHYSIOLOGY] The occurrence of physiological changes in an individual exposed to changed conditions; for example, tanning of the skin in sunshine, or increased red blood cell counts at high altitudes. { ˌaˌdap'tā·shən }

adaptive disease [PHYSIOLOGY] The physiologic changes impairing an organism's health as the result of exposure to an unfamiliar environment. { ə'dap·tiv diˌzēz }

adaptive divergence [EVOLUTION] Divergence of new forms from a common ancestral form due to adaptation to different environmental conditions. { ə'dap·tiv də'vər·jəns }

adaptive enzyme [MICROBIOLOGY] Any bacterial enzyme formed in response to the presence of a substrate specific for that enzyme. { ə'dap·tiv 'enˌzīm }

adaptive norm [GENETICS] The various genotypes of a species evident in a given population. { ə 'dap·tiv 'nȯrm }

adaptive radiation [EVOLUTION] Diversification of a dominant evolutionary group into a large number of subsidiary types adapted to more restrictive modes of life (different adaptive zones) within the range of the larger group. { ə'dap·tiv ˌrād·ē'ā·shən }

adaptive value [GENETICS] The property of a given genotype that confers fitness to an organism in a given environment. { ə'dap·tiv 'val·yü }

adaxial [BIOLOGY] On the same side as or facing toward the axis of an organ or organism. { ˌad 'ak·sē·əl }

adder [VERTEBRATE ZOOLOGY] Any of the venomous viperine snakes included in the family Viperidae. { 'ad·ər }

additive factor [GENETICS] Any of a group of nonallelic genes that affect the same phenotypic characteristics. { 'ad·ə·div 'fak·tər }

additive gene action [GENETICS] **1.** A form of allelic interaction in which dominance is absent, resulting in a heterozygote that is intermediate in phenotype between homozygotes for the alternative alleles. **2.** The cumulative contribution made by all loci in a group of nonallelic genes to a polygenic trait. { ¦ad·ə·div ¦jēn ˌak·shən }

additive genetic variance [GENETICS] That part of the genetic variance of a quantitative character attributed to the average effects of substituting one allele for another at a given locus or at the multiple loci governing a polygenic trait. { ¦ad·ə·div jə¦ned·ik 'ver·ē·əns }

adduction [PHYSIOLOGY] Movement of one part of the body toward another or toward the median axis of the body. { ə'dək·shən }

adductor [ANATOMY] Any muscle that draws a part of the body toward the median axis. { ə 'dək·tər }

Adeleina [INVERTEBRATE ZOOLOGY] A suborder of protozoan invertebrate parasites in the order Eucoccida in which the sexual and asexual stages are in different hosts. { ˌad·ə'līn·ə }

adelphous [BOTANY] Having stamens fused together by their filaments. { ə'del·fəs }

adenase [BIOCHEMISTRY] An enzyme that catalyzes the hydrolysis of adenine to hypoxanthine and ammonia. { 'ad·ənˌās }

adenine [BIOCHEMISTRY] $C_5H_5N_5$ A purine base, 6-aminopurine, occurring in ribonucleic acid and deoxyribonucleic acid and as a component of adenosinetriphosphate. { 'ad·ənˌēn }

adeno-associated satellite virus [VIROLOGY] A defective virus that is unable to reproduce without the help of an adenovirus. { ¦ad·ənˌō·ə'sō·shēˌād·əd 'sad·əˌlīt ˌvī·rəs }

adenohypophysis [ANATOMY] The glandular part of the pituitary gland, composing the anterior and intermediate lobes. { ¦ad·ənˌōˌhī'pä·fə·səs }

adenoid [ANATOMY] **1.** A mass of lymphoid tissue. **2.** Lymphoid tissue of the nasopharynx. Also known as pharyngeal tonsil. { 'adˌnȯid }

adenomere [EMBRYOLOGY] The embryonic structure which will become the functional portion of a gland. { ˌad·ən'ō·mir }

Adenophorea [INVERTEBRATE ZOOLOGY] A class of unsegmented worms in the phylum Nematoda. { ˌad·ən·ə'fȯr·ē·ə }

adenophyllous [BOTANY] Having leaves with glands. { ˌad·ən'ä·fə·ləs }

adenosine [BIOCHEMISTRY] $C_{10}H_{13}N_5O_4$ A nucleoside composed of adenine and D-ribose. { ə 'den·əˌsēn }

adenosine 3′,5′-cyclic monophosphate [BIOCHEMISTRY] *See* cyclic adenylic acid. { ə¦den·ə·sēn ¦thrē¦prīm ¦fīv¦prīm 'sīk·lik 'mä·nō'fäs·fāt }

adenosine 3′,5′-cyclic phosphate [BIOCHEMISTRY] *See* cyclic adenylic acid. { ə¦den·ə·sēn ¦thrē ¦prīm ¦fīv¦prīm 'sīk·lik 'fäs·fāt }

adenosinediphosphatase [BIOCHEMISTRY] An enzyme that catalyzes the hydrolysis of adenosinediphosphate. Abbreviated ADPase. { ə¦den·əˌsēnˌdī'fäs·fəˌtās }

adenosinediphosphate [BIOCHEMISTRY] $C_{10}H_{15}N_5O_{10}P_2$ A coenzyme composed of adenosine and two molecules of phosphoric acid that is important in intermediate cellular metabolism. Abbreviated ADP. { ə¦den·əˌsēnˌdī'fäs·fāt }

adenosinemonophosphate [BIOCHEMISTRY] *See* adenylic acid. { ə¦den·əˌsēn'mä·nō'fäs·fāt }

adenosine 3′,5′-monophosphate [BIOCHEMISTRY] *See* cyclic adenylic acid. { ə¦den·əˌsēn ¦thrē¦prīm ¦fīv¦prīm 'mä·nō'fäs·fāt }

adenosinetriphosphatase [BIOCHEMISTRY] An enzyme that catalyzes the hydrolysis of adenosinetriphosphate. Abbreviated ATPase. { ə¦den·əˌsēnˌtrī'fäs·fəˌtās }

adenosinetriphosphate [BIOCHEMISTRY] $C_{10}H_{16}N_5O_{12}P_3$ A coenzyme composed of adenosinediphosphate with an additional phosphate group; an important energy compound in metabolism. Abbreviated ATP. { ə¦dēn·əˌsēnˌtri·'fäs ˌfāt }

adeno-SV40 hybrid virus [VIROLOGY] A defective virus particle in which part of the genetic material of papovavirus SV40 is encased in an adenovirus protein coat. { ¦ad·ənˌō ¦es¦vē¦fȯr·tē 'hī·brəd 'vī·rəs }

Adenoviridae [VIROLOGY] A family of double-stranded DNA viruses with icosahedral symmetry; usually found in the respiratory tract of the host species and often associated with respiratory diseases. Also known as adenovirus. { ˌad·ən·ō'vīr·əˌdē }

adenovirus [VIROLOGY] *See* Adenoviridae. { ¦ad·ənˌo'vī·rəs }

adenylcyclase [BIOCHEMISTRY] The catalyzing enzyme in the conversion of adenosinetriphosphate to cyclic adenosinemonophosphate during metabolism. { ¦ad·ənˌil'sīˌklās }

adenylic acid [BIOCHEMISTRY] **1.** A generic term for a group of isomeric nucleotides. **2.** The phosphoric acid ester of adenosine. Also known as adenosinemonophosphate (AMP). { ¦ad·ən¦il·ik 'as·əd }

adeoniform [INVERTEBRATE ZOOLOGY] **1.** A lobate, bilamellar zooarium. **2.** Resembling the fossil bryozoan *Adeona*. { ˌad·ē'ä·nəˌfȯrm }

Adephaga [INVERTEBRATE ZOOLOGY] A suborder of insects in the order Coleoptera characterized by fused hind coxae that are immovable. { ə 'def·ə·gə }

adequate stimulus [PHYSIOLOGY] The energy of any specific mode that is sufficient to elicit a response in an excitable tissue. { 'ad·ə·kwət 'stim·yə·ləs }

ADH [BIOCHEMISTRY] *See* vasopressin.

adhering junction [CELL BIOLOGY] An intercellular junction that promotes adhesion between cells. Also known as desmosome. { ad¦hir·iŋ 'jəŋk·shən }

adhesion [BOTANY] Growing together of members of different and distinct whorls. { ad'hē·zhən }

adhesive cell [INVERTEBRATE ZOOLOGY] Any of various glandular cells in ctenophores, turbellarians, and hydras used for adhesion to a substrate and for capture of prey. Also known as colloblast; glue cell; lasso cell. { ad'hēz·iv 'sel }

Adimeridae [INVERTEBRATE ZOOLOGY] An equivalent name for the Colydiidae. { ˌad·ə'mer·ə·dē }

adipocellulose [BIOCHEMISTRY] A type of cellulose found in the cell walls of cork tissue. { ˌad·ə·pō'sel·yəˌlōs }

adipogenesis [PHYSIOLOGY] The formation of fat or fatty tissue. { ˌad·ə·pō'jen·ə·səs }

adipose [BIOLOGY] Fatty; of or relating to fat. { 'ad·əˌpōs }

adipose fin [VERTEBRATE ZOOLOGY] A modified posterior dorsal fin that is fleshy and lacks rays; found in salmon and typical catfishes. { 'ad·ə ˌpōs ˌfin }

adipose tissue [HISTOLOGY] A type of connective tissue specialized for lipid storage. { 'ad·əˌpōs 'tish·ü }

adjustor neuron [ANATOMY] Any of the interconnecting nerve cells between sensory and motor neurons of the central nervous system. { ə'jəs·tər 'nüˌrän }

ad lib [BIOLOGY] Shortened form for ad libitum; without limit or restraint. { ˌad 'lib }

adnate [BIOLOGY] United through growth; used especially for unlike parts. [BOTANY] Pertaining to growth with one side adherent to a stem. { 'adˌnāt }

adnexa [BIOLOGY] Subordinate or accessory parts, such as eyelids, Fallopian tubes, and extraembryonic membranes. { ad'neks·ə }

adonite [BIOCHEMISTRY] *See* adonitol. { 'ad·ə ˌnīt }

adonitol [BIOCHEMISTRY] $C_5H_{12}O_5$ A pentitol from the dicotyledenous plant *Adonis vernalis*; large crystals that are optically inactive and melt at 102°C; it does not reduce Fehling's solution, and is freely soluble in water and hot alcohol. Also known as adonite; ribitol. { ə'dän·əˌtōl }

adont hinge [INVERTEBRATE ZOOLOGY] A type of ostracod hinge articulation which either lacks teeth and has overlapping valves or has a ridge and groove. { 'āˌdänt ˌhinj }

adoptive immunity [IMMUNOLOGY] Immunity resulting from the transfer of an immune function from one organism to another through the transfer of immunologically competent cells. Also known as transfer immunity. { ə¦däp·təv ə 'myü·nəd·ē }

adoral [ZOOLOGY] Near the mouth. { ˌa'dȯr·əl }

ADP [BIOCHEMISTRY] *See* adenosinediphosphate.

ADPase [BIOCHEMISTRY] *See* adenosinediphosphatase.

adrenal cortex [ANATOMY] The cortical moiety of the suprarenal glands which secretes glucocorticoids, mineralocorticoids, androgens, estrogens, and progestagens. { ə'drēn·əl 'kȯr·teks }

adrenal cortex hormone [BIOCHEMISTRY] Any of the steroids produced by the adrenal cortex. Also known as adrenocortical hormone; corticoid. { ə 'drēn·əl 'kȯr·teks 'hȯrˌmōn }

adrenal gland [ANATOMY] An endocrine organ located close to the kidneys of vertebrates and consisting of two morphologically distinct components, the cortex and medulla. Also known as suprarenal gland. { ə'drēn·əl ˌgland }

adrenaline [BIOCHEMISTRY] *See* epinephrine. { ə 'dren·əl·ən }

adrenal medulla [ANATOMY] The hormone-secreting chromaffin cells of the adrenal gland that produce epinephrine and norepinephrine. { ə'drēn·əl mə'dəl·ə }

adrenergic [PHYSIOLOGY] Describing the chemical activity of epinephrine or epinephrine-like substances. { ˌad·rə'nər·jik }

adrenergic blocking agent [BIOCHEMISTRY] Any substance that blocks the action of epinephrine or an epinephrine-like substance. { ˌad·rə'nər·jik 'bläk·iŋ ˌā·jənt }

adrenochrome [BIOCHEMISTRY] $C_9H_9O_3N$ A brick red oxidation product of epinephrine which can convert hemoglobin into methemoglobin. { ə'dren·əˌkrōm }

adrenocortical hormone [BIOCHEMISTRY] *See* adrenal cortex hormone. { ə¦drēn·ō'kȯrd·ə·kəl 'hȯrˌmōn }

adrenocorticosteroid [BIOCHEMISTRY] **1.** A steroid that is obtained from the adrenal cortex. **2.** A steroid that resembles adrenal cortex steroids or has physiological effects like them. { ə ¦drē·nōˌkȯrd·ə·kō'stirˌȯid }

adrenocorticotropic hormone [BIOCHEMISTRY] The chemical secretion of the adenohypophysis that stimulates the adrenal cortex. Abbreviated ACTH. Also known as adrenotropic hormone. { ə ¦drēn·ō'kȯrd·ə·kō'träp·ik 'hȯr‚mōn }

adrenomedullary [PHYSIOLOGY] Pertaining to the adrenal gland medulla. { ə¦drē·nō·mə'dəl·ə·rē }

adrenotropic [PHYSIOLOGY] Of or pertaining to an effect on the adrenal cortex. { ə¦drēn·ə'träp·ik }

adrenotropic hormone [BIOCHEMISTRY] *See* adrenocorticotropic hormone. { ə¦drēn·ə'träp·ik 'hȯr‚mōn }

adret [ECOLOGY] The sunny (usually south) face of a mountain featuring high timber and snow lines. { 'ad·rət }

advanced [EVOLUTION] Denoting a later stage within a lineage that demonstrates evolutionary progression. { əd'vanst }

adventitia [ANATOMY] The external, connective-tissue covering of an organ or blood vessel. Also known as tunica adventitia. { ‚ad·ven'tish·ə }

adventitious [BIOLOGY] Also known as adventive. **1.** Acquired spontaneously or accidentally, not by heredity. **2.** Arising, as a tissue or organ, in an unusual or abnormal place. { ‚ad·ven'tish·əs }

adventitious bud [BOTANY] A bud that arises at points on the plant other than at the stem apex or a leaf axil. { ‚ad·ven'tish·əs 'bəd }

adventitious root [BOTANY] A root that arises from any plant part other than the primary root (radicle) or its branches. { ‚ad·ven'tish·əs 'rüt }

adventitious vein [INVERTEBRATE ZOOLOGY] The vessel between the intercalary and accessory veins on certain insect wings. { ‚ad·ven'tish·əs 'vān }

adventitious virus [VIROLOGY] A contaminant virus present by chance in a virus preparation. { ‚ad·ven'tish·əs 'vī·rəs }

adventive [BIOLOGY] **1.** An organism that is introduced accidentally and is imperfectly naturalized; not native. **2.** *See* adventitious. { ad 'ven·tiv }

advolution [BIOLOGY] Development or growth with increasing similarities; growth toward; the opposite of evolution. { ‚ad·və'lü·shən }

aebi [BIOLOGY] A unit for the standardization of a phosphatase. { ‚ā'ē·bē }

Aechminidae [PALEONTOLOGY] A family of extinct ostracods in the order Paleocopa in which the hollow central spine is larger than the valve. { ēk'min·ə‚dē }

aeciospore [MYCOLOGY] A spore produced by an aecium. { 'ēsh·ē·ə‚spȯr }

aecium [MYCOLOGY] The fruiting body or sporocarp of rust fungi. { 'ēsh·ē·əm }

aedeagus [INVERTEBRATE ZOOLOGY] The copulatory organ of a male insect. { ¦ē·dē'ā·gəs }

Aedes [INVERTEBRATE ZOOLOGY] A genus of the dipterous subfamily Culicinae in the family Culicidae, with species that are vectors for many diseases of humans. { ā'ē·dēz }

Aeduellidae [PALEONTOLOGY] A family of Lower Permian palaeoniscoid fishes in the order Palaeonisciformes. { ‚ē·dü'el·ə‚dī }

Aegeriidae [INVERTEBRATE ZOOLOGY] The clear-wing moths, a family of lepidopteran insects in the suborder Heteroneura characterized by the lack of wing scales. { ‚ē·jə'rē·ə‚dē }

Aegialitidae [INVERTEBRATE ZOOLOGY] An equivalent name for the Salpingidae. { ‚ē·jyə'lid·ə ‚dē }

Aegidae [INVERTEBRATE ZOOLOGY] A family of isopod crustaceans in the suborder Flabellifera whose members are economically important as fish parasites. { 'ē·jə·dē }

aegithognathous [VERTEBRATE ZOOLOGY] Referring to a bird palate in which the vomers are completely fused and truncate in appearance. { ‚ē·gə‚thäg'nā·thəs }

Aegothelidae [VERTEBRATE ZOOLOGY] A family of small Australo-Papuan owlet-nightjars in the avian order Caprimulgiformes. { ‚ē·gə'thel·ə ‚dē }

Aegypiinae [VERTEBRATE ZOOLOGY] The Old World vultures, a subfamily of diurnal carrion feeders of the family Accipitridae. { ‚ē·jə'pī·ə·nē }

Aegyptianella [MICROBIOLOGY] A genus of the family Anaplasmataceae; organisms from inclusions in red blood cells of birds. { ə‚jip·shə 'nel·ə }

Aegyptopithecus [PALEONTOLOGY] A primitive primate that is thought to represent the common ancestor of both the human and ape families. { ə‚jip·tō'pith·e‚kəs }

aelophilous [BOTANY] Describing a plant whose disseminules are dispersed by wind. { ‚ē'lä·fə·ləs }

Aelosomatidae [INVERTEBRATE ZOOLOGY] A family of microscopic fresh-water annelid worms in the class Oligochaeta characterized by a ventrally ciliated prostomium. { ‚e‚lä·sə'mad·ə‚dē }

Aepophilidae [INVERTEBRATE ZOOLOGY] A family of bugs in the hemipteran superfamily Saldoidea. { ‚ē·pō'fil·ə‚dē }

Aepyornis [PALEONTOLOGY] A genus of extinct ratite birds representing the family Aepyornithidae. { ‚ē·pē'ȯrn·əs }

Aepyornithidae [PALEONTOLOGY] The single family of the extinct avian order Aepyornithiformes. { ‚ē·pē‚ȯr'nith·ə‚dē }

Aepyornithiformes [PALEONTOLOGY] The elephant birds, an extinct order of ratite birds in the superorder Neognathae. { ‚ē·pē‚ȯr‚nith·ə'fȯr ‚mēz }

aequorin [BIOCHEMISTRY] A bioluminescent protein that is produced by jellyfish of the genus *Aequorea* and emits light in the presence of calcium or strontium. { 'ē·kwə‚rin }

aerenchyma [BOTANY] A specialized tissue in some water plants characterized by thin-walled cells and large intercellular air spaces. { ‚a 'reŋk·ə·mə }

aerial [BIOLOGY] Of, in, or belonging to the air or atmosphere. { 'e·rē·əl }

aerial mycelium [MYCOLOGY] A mass of hyphae that occurs above the surface of a substrate. { 'e·rē·əl mī'sē·lē·əm }

aerial root [BOTANY] A root exposed to the air, usually anchoring the plant to a tree, and often functioning in photosynthesis. { 'e·rē·əl 'rüt }

aerial stem [BOTANY] A stem with an erect or vertical growth habit above the ground. { 'e·rē·əl 'stem }

aerobe [BIOLOGY] An organism that requires air or free oxygen to maintain its life processes. { 'e,rōb }

aerobic bacteria [MICROBIOLOGY] Any bacteria requiring free oxygen for the metabolic breakdown of materials. { e'rōb·ik ,bak'tir·ē·ə }

aerobic process [BIOLOGY] A process requiring the presence of oxygen. { e'rōb·ik 'präs·əs }

aerobiology [BIOLOGY] The study of the atmospheric dispersal of airborne fungus spores, pollen grains, and microorganisms; and, more broadly, of airborne propagules of algae and protozoans, minute insects such as aphids, and pollution gases and particles which exert specific biologic effects. { ,e·rō,bī'äl·ə·jē }

aerobioscope [MICROBIOLOGY] An apparatus for collecting and determining the bacterial content of a sample of air. { ,e·rō'bi·ə,skōp }

aerobiosis [BIOLOGY] Life existing in air or oxygen. { ,e·rō,bi'ō·səs }

Aerococcus [MICROBIOLOGY] A genus of bacteria in the family Streptococcaceae; spherical cells have the tendency to form tetrads; they ferment glucose with production of dextrorotatory lactic acid (homofermentative). { 'e·rō,käk·əs }

aerocyst [BOTANY] An air vesicle in certain species of algae. { 'e·rō,sist }

Aeromonas [MICROBIOLOGY] A genus of bacteria in the family Vibrionaceae; straight, motile rods with rounded ends; most species are pathogenic to marine and fresh-water animals. { e·rō'mōn·əs }

aerophyte [ECOLOGY] *See* epiphyte. { 'e·rō,fīt }

aeroplankton [ECOLOGY] Small airborne organisms such as insects. { ¦e·rō'plaŋk·tən }

Aerosporin [MICROBIOLOGY] Trade name for the antibiotic polymyxin B. { ¦e·rō¦spȯr·ən }

aerotaxis [BIOLOGY] The movement of an organism, especially aerobic and anaerobic bacteria, with reference to the direction of oxygen or air. { ,e·rō'tak·səs }

aerotropism [BOTANY] A response in which the growth direction of a plant component changes due to modifications in oxygen tension. { ,e·rō 'trō,piz·əm }

aeschynomenous [BOTANY] Having sensitive leaves that droop when touched, such as members of the Leguminosae. { ,es·kə'näm·ə·nəs }

Aesculus [BOTANY] A genus of deciduous trees or shrubs belonging to the order Sapindales. Commonly known as buckeye. { ,es·kyə·ləs }

Aeshnidae [INVERTEBRATE ZOOLOGY] A family of odonatan insects in the suborder Anisoptera distinguished by partially fused eyes. { 'esh·nə·dē }

aesthacyte [INVERTEBRATE ZOOLOGY] *See* esthacyte. { 'es·thə,sīt }

aesthesia [PHYSIOLOGY] *See* esthesia. { es'thē·zhə }

aesthete [BOTANY] A plant organ with the capacity to respond to definite physical stimuli. { 'es ,thēt }

aestidurilignosa [ECOLOGY] A mixed woodland of evergreen and deciduous hardwoods. { ,es·tə·də,ril·əg'nōs·ə }

aestilignosa [ECOLOGY] A woodland of tropophytic vegetation in temperate regions. { es·tə·lig'nōs·ə }

aestivation [BOTANY] The arrangement of floral parts in a bud. [PHYSIOLOGY] The condition of dormancy or torpidity. { ,es·tə'vā·shən }

Aetosauria [PALEONTOLOGY] A suborder of Triassic archosaurian quadrupedal reptiles in the order Thecodontia armored by rings of thick, bony plates. { ā¦et·ə'sȯr·ē·ə }

afferent [PHYSIOLOGY] Conducting or conveying inward or toward the center, specifically in reference to nerves and blood vessels. { 'af·ə·rənt }

afferent neuron [ANATOMY] A nerve cell that conducts impulses toward a nerve center, such as the central nervous system. { 'af·ə·rənt 'nü ,rän }

affinity [IMMUNOLOGY] The strength of the attractive forces between an antigen and an antibody. { ə'fin·əd·ē }

affinity labeling [BIOCHEMISTRY] A method for introducing a label into the active site of an enzyme by relying on the tight binding between the enzyme and its substrate (or cofactors). { ə'fin·əd·ē 'lā·bə·liŋ }

aflatoxin [BIOCHEMISTRY] The toxin produced by some strains of the fungus *Aspergillus flavus*, the most potent carcinogen yet discovered. { ,af·lə 'täk·sin }

African violet [BOTANY] *Saintpaulia ionantha*. A flowering plant typical of the family Gesneriaceae. { 'af·ri·kən 'vī·ə·lət }

afterbirth [EMBRYOLOGY] The placenta and fetal membranes expelled from the uterus following birth of offspring in viviparous mammals. { 'af·tər,bərth }

afterimage [PHYSIOLOGY] A visual sensation occurring after the stimulus to which it is a response has been removed. { 'af·tər,im·əj }

afterpotential [PHYSIOLOGY] A small positive or negative wave that follows and is dependent on the main spike potential, seen in the oscillograph tracing of an action potential passing along a nerve. { 'af·tərpə¦ten·chəl }

afterripening [BOTANY] A period of dormancy after a seed is shed during which the synthetic machinery of the seed is prepared for germination and growth. { 'af·tər,rī·pən·iŋ }

aftershaft [VERTEBRATE ZOOLOGY] An accessory, plumelike feather near the upper umbilicus on the feathers of some birds. { 'af·tər,shaft }

agameon [BIOLOGY] An organism which reproduces only by asexual means. Also known as agamospecies. { ā'gam·ē·ən }

agamete [BIOLOGY] An asexual reproductive cell that develops into an adult individual. { ā'ga ,mēt }

agamic [BIOLOGY] Referring to a species or generation which does not reproduce sexually. { ā 'gam·ik }

Agamidae [VERTEBRATE ZOOLOGY] A family of

Old World lizards in the suborder Sauria that have acrodont dentition. { ə'gam·ə,dē }

agamogony [BIOLOGY] Asexual reproduction, specifically schizogony. { ,ā·gə'mäg·ə·nē }

agamospecies [BIOLOGY] *See* agameon. { ,a·gə·mō'spē·shēz }

agamospermy [BOTANY] Apogamy in which sexual union is incomplete because of abnormal development of the pollen and the embryo sac. { 'ā·gam·ə,spərm·ē }

Agaontidae [INVERTEBRATE ZOOLOGY] A family of small hymenopteran insects in the superfamily Chalcidoidea; commonly called fig insects for their role in cross-pollination of figs. { ,a·gā'än·tə,dē }

agar-gel reaction [IMMUNOLOGY] A precipitin type of antigen-antibody reaction in which the reactants are introduced into different regions of an agar gel and allowed to diffuse toward each other. { 'äg·ər ,jel ri'ak·shən }

Agaricales [MYCOLOGY] An order of fungi in the class Basidiomycetes containing all forms of fleshy, gilled mushrooms. { ə,gar·ə'kā·lēz }

agarophyte [BOTANY] Any seaweed that yields agar. { ə'gar·ə,fīt }

agarose [BIOCHEMISTRY] The gelling component of agar; possesses a double-helical structure which forms a three-dimensional framework capable of holding water molecules in the interstices. { 'ag·ə,rōs }

Agavaceae [BOTANY] A family of flowering plants in the order Liliales characterized by parallel, narrow-veined leaves, a more or less corolloid perianth, and an agavaceous habit. { 'ag·ə'vās·ē,ē }

age [BIOLOGY] Period of time from origin or birth to a later time designated or understood; length of existence. { āj }

age distribution [ECOLOGY] The distribution of different age groups in a population. { 'āj dis·trə'byü·shən }

agenesis [BIOLOGY] Absence of a tissue or organ due to lack of development. { 'ā·jen·ə·səs }

agglutination reaction [IMMUNOLOGY] Clumping of a particulate suspension of antigen by a reagent, usually an antibody. { ə,glüt·ən'ā·shən rē'ak·shən }

agglutinin [IMMUNOLOGY] An antibody from normal or immune serum that causes clumping of its complementary particulate antigen, such as bacteria or erythrocytes. { ə'glüt·ən·ən }

agglutinogen [IMMUNOLOGY] An antigen that stimulates production of a specific antibody (agglutinin) when introduced into an animal body. { ə,glü'tin·ə·jən }

agglutinoid [IMMUNOLOGY] An agglutin that lacks the power to agglutinate but has the ability to unite with its agglutinogen. { ə'glüt·ən,òid }

aggregate [BOTANY] Referring to fruit formed in a cluster, from a single flower, such as raspberry, or from several flowers, such as pineapple. { 'ag·rə·gət }

aggregate fruit [BOTANY] A type of fruit composed of a number of small fruitlets all derived from the ovaries of a single flower. { 'ag·rə·gət 'früt }

aggregation [BIOLOGY] A grouping or clustering of separate organisms. { ,ag·rə'gā·shən }

aggressive mimicry [ZOOLOGY] Mimicry used to attract or deceive a species in order to prey upon it. { ə'gres·iv 'mim·ə·krē }

aging [BIOLOGY] Growing older. { 'āj·iŋ }

Aglaspida [PALEONTOLOGY] An order of Cambrian and Ordovician merostome arthropods in the subclass Xiphosurida characterized by a phosphatic exoskeleton and vaguely trilobed body form. { ə'glas·pə·də }

aglomerular [HISTOLOGY] Lacking glomeruli. { ¦ā·glə'mər·yə·lər }

Aglossa [VERTEBRATE ZOOLOGY] A suborder of anuran amphibians represented by the single family Pipidea and characterized by the absence of a tongue. { ā'gläs·ə }

aglycon [BIOCHEMISTRY] The nonsugar compound resulting from the hydrolysis of glycosides; an example is 3,5,7,3′,4′-pentahydroxyflavylium, or cyanidin. { ə'glī,kän }

aglyphous [VERTEBRATE ZOOLOGY] Having solid teeth. { 'a·glə'fəs }

agmatine [BIOCHEMISTRY] $C_5H_{14}N_4$ Needlelike crystals with a melting point of 231°C; soluble in water; a product of the enzymatic decarboxylation of argenine. { 'ag·mə,tēn }

agnate [BIOLOGY] Related exclusively through male descent. { 'ag,nāt }

Agnatha [VERTEBRATE ZOOLOGY] The most primitive class of vertebrates, characterized by the lack of true jaws. { 'ag·nə'thə }

Agonidae [VERTEBRATE ZOOLOGY] The poachers, a small family of marine perciform fishes in the suborder Cottoidei. { ə'gän·ə·dē }

agonist [BIOCHEMISTRY] A chemical substance that can combine with a cell receptor and cause a reaction or create an active site. [PHYSIOLOGY] A contracting muscle that is resisted or counteracted by another muscle, called an antagonist, with which it is paired. { 'ag·ə,nist }

agouti [VERTEBRATE ZOOLOGY] A hystricomorph rodent, *Dasyprocta*, in the family Dasyproctidae, with 13 species. { ə'güd·ē }

agranular leukocyte [HISTOLOGY] A type of white blood cell, including lymphocytes and monocytes, characterized by the absence of cytoplasmic granules and by a relatively large spherical or indented nucleus. { ,ā'gran·yə·lər 'lü·kə,sīt }

agranular reticulum [CELL BIOLOGY] Endoplasmic reticulum lacking ribosomes. { ā'gran·yə·lər ri'tik·yə·ləm }

agrestal [ECOLOGY] Growing wild in the fields. { ə'grest·əl }

agretope [IMMUNOLOGY] In antigen presentation, the part of an antigen that interacts with a class II histocompatibility molecule. { 'ag·rə ,tōp }

agriculture [BIOLOGY] The production of plants and animals useful to man, involving soil cultivation and the breeding and management of crops and livestock. { 'ag·rə,kəl·chər }

Agriochoeridae [PALEONTOLOGY] A family of extinct tylopod ruminants in the superfamily Merycoidodontoidea. { ,ag·rē·ō'kir·ə,dē }

agrioecology [ECOLOGY] The ecology of cultivated plants. { ¦ag·rē·ō,ē'käl·ə·jē }

Agrionidae [INVERTEBRATE ZOOLOGY] A family of odonatan insects in the suborder Zygoptera characterized by black or red markings on the wings. { ,ag·rē'än·ə,dē }

Agrobacterium [MICROBIOLOGY] A genus of bacteria in the family Rhizobiaceae; cells do not fix free nitrogen, and three of the four species are plant pathogens, producing galls and hairy root. { ¦ag·rō,bak'tir·e·əm }

agroecosystem [ECOLOGY] A model for the functionings of an agricultural system with all its inputs and outputs. { ¦ag·rō'ek·ō,sis·təm }

Agromyzidae [INVERTEBRATE ZOOLOGY] A family of myodarian cyclorrhaphous dipteran insects of the subsection Acalypteratae; commonly called leaf-miner flies because the larvae cut channels in leaves. { ¦ag·rō'mīz·ə,dē }

agrophilous [ECOLOGY] Having a natural habitat in grain fields. { ə'gräf·ə·ləs }

agrostology [BOTANY] A division of systematic botany concerned with the study of grasses. { ¦ag·rə¦stä·lə·jē }

ahermatypic [INVERTEBRATE ZOOLOGY] Non-reef-building, as applied to corals. { ¦ā,hər·mə¦tip·ik }

AIA [IMMUNOLOGY] *See* anti-immunoglobulin antibody.

aiophyllous [BOTANY] *See* evergreen. { ,ī·ō'fil·əs }

air cell [ZOOLOGY] A cavity or receptacle for air such as an alveolus, an air sac in birds, or a dilation of the trachea in insects. { 'er ,sel }

air layering [BOTANY] A method of vegetative propagation, usually of a wounded part, in which the branch or shoot is enclosed in a moist medium until roots develop, and then it is severed and cultivated as an independent plant. { 'er ,lā·ər·iŋ }

air pollution [ECOLOGY] The presence in the outdoor atmosphere of one or more contaminants such as dust, fumes, gas, mist, odor, smoke, or vapor in quantities and of characteristics and duration such as to be injurious to human, plant, or animal life or to property, or to interfere unreasonably with the comfortable enjoyment of life and property. { ¦er pə'lü·shən }

air sac [INVERTEBRATE ZOOLOGY] One of large, thin-walled structures associated with the tracheal system of some insects. [VERTEBRATE ZOOLOGY] In birds, any of the small vesicles that are connected with the respiratory system and located in bones and muscles to increase buoyancy. { 'er ,sak }

air spora [BIOLOGY] Airborne fungus spores, pollen grains, and microorganisms. { 'er ,spȯr·ə }

Aistopoda [PALEONTOLOGY] An order of Upper Carboniferous amphibians in the subclass Lepospondyli characterized by reduced or absent limbs and an elongate, snakelike body. { ,ā·ə'stäp·ə·də }

Aizoaceae [BOTANY] A family of flowering plants in the order Caryophyllales; members are unarmed leaf-succulents, chiefly of Africa. { ā,īz·ə'wā·sē,ē }

akaryote [CELL BIOLOGY] A cell that lacks a nucleus. { ,ā'ka·rē,ōt }

akinete [BOTANY] A thick-walled resting cell of unicellular and filamentous green algae. { ,a'kī,nēt }

ala [BIOLOGY] A wing or winglike structure. { 'ā·lə }

alang-alang [BOTANY] *See* cogon. { 'ä,läŋ'a,läŋ }

alanine [BIOCHEMISTRY] $C_3H_7NO_2$ A white, crystalline, nonessential amino acid of the pyruvic acid family. { 'al·ə,nēn }

alar [BIOLOGY] Winglike or pertaining to a wing. { 'ā·lər }

alarm reaction [BIOLOGY] The sum of all nonspecific phenomena which are elicited by sudden exposure to stimuli, which affect large portions of the body, and to which the organism is quantitatively or qualitatively not adapted. { ə'lärm rē'ak·shən }

alarm song [INVERTEBRATE ZOOLOGY] A stress signal occurring in many families of beetles. { ə'lärm ,sȯŋ }

alate [BIOLOGY] Possessing wings or winglike structures. { 'ā,lāt }

Alaudidae [VERTEBRATE ZOOLOGY] The larks, a family of Oscine birds in the order Passeriformes. { ə'lau̇·də,dē }

albatross [VERTEBRATE ZOOLOGY] Any of the large, long-winged oceanic birds composing the family Diomedeidae of the order Procellariformes. { 'al·bə,trȯs }

albinism [BIOLOGY] The state of having colorless chromatophores, which results in the absence of pigmentation in animals that are normally pigmented. { 'al·bə,niz·əm }

albino [BIOLOGY] A human or animal with a congenital deficiency of pigment in the skin, hair, and eyes. [BOTANY] An abnormal plant with colorless chromatophores. { al'bī·nō }

albomaculatus [BOTANY] A variegation consisting of irregularly distributed white and green regions on plants due to the mitotic segregation of genes or plastids. { ¦al·bō,ma·kyə'läd·əs }

albomycin [MICROBIOLOGY] An antibiotic produced by *Actinomyces subtropicus*; effective against penicillin-resistant pneumococci and staphylococci. { ,al·bō'mīs·ən }

albuginea [HISTOLOGY] A layer of white, fibrous connective tissue investing an organ or other body part. { ,al·byü'jin·ē·ə }

albumen [CELL BIOLOGY] The white of an egg, composed principally of albumin. { ,al'byü·mən }

albumin [BIOCHEMISTRY] Any of a group of plant and animal proteins which are soluble in water, dilute salt solutions, and 50% saturated ammonium sulfate. { ,al'byü·mən }

albumin-globulin ratio [BIOCHEMISTRY] The ratio of the concentrations of albumin to globulin in blood serum. { ,al'byü·mən 'gläb·yə·lən ,rā·shō }

albuminoid [BIOCHEMISTRY] *See* scleroprotein. [BIOLOGY] Having the characteristics of albumin. { ,al'byü·mə,nȯid }

albumose [BIOCHEMISTRY] A protein derivative

formed by the action of a hydrolytic enzyme, such as pepsin. { 'al·byə,mōs }

alburnum [BOTANY] *See* sapwood. { al'bər·nəm }

Alcaligenes [MICROBIOLOGY] A genus of gram-negative, aerobic rods and cocci of uncertain affiliation; cells are motile, and species are commonly found in the intestinal tract of vertebrates. { ,al·kə'lij·ə,nēz }

Alcedinidae [VERTEBRATE ZOOLOGY] The kingfishers, a worldwide family of colorful birds in the order Coraciiformes; characterized by large heads, short necks, and heavy, pointed bills. { ,al·sə'din·ə,dē }

Alcidae [VERTEBRATE ZOOLOGY] A family of shorebirds, predominantly of northern coasts, in the order Charadriiformes, including auks, puffins, murres, and guillemots. { 'al·sə,dē }

Alciopidae [INVERTEBRATE ZOOLOGY] A pelagic family of errantian annelid worms in the class Polychaeta. { ,al·sē'äp·ə,dē }

alcohol dehydrogenase [BIOCHEMISTRY] The enzyme that catalyzes the oxidation of ethanol to acetaldehyde. { 'al·kə,hȯl ,dē·hī'drä·jə,nās }

alcoholic fermentation [MICROBIOLOGY] The process by which certain yeasts decompose sugars in the absence of oxygen to form alcohol and carbon dioxide; method for production of ethanol, wine, and beer. { ,al·kə'hȯl·ik ,fər·mən'tā·shən }

Alcyonacea [INVERTEBRATE ZOOLOGY] The soft corals, an order of littoral anthozoans of the subclass Alcyonaria. { ,al·sī·ə'nās·ē·ə }

Alcyonaria [INVERTEBRATE ZOOLOGY] A subclass of the Anthozoa; members are colonial cnidarians, most of which are sedentary and littoral. { ,al·sī·ə'ner·ē·ə }

aldehyde dehydrogenase [BIOCHEMISTRY] An enzyme that catalyzes the conversion of an aldehyde to its corresponding acid. { 'al·də,hīd ,dē'hī·drə·jə,nās }

aldehyde lyase [BIOCHEMISTRY] Any enzyme that catalyzes the nonhydrolytic cleavage of an aldehyde. { 'al·də,hīd 'lī,ās }

alder [BOTANY] The common name for several trees of the genus *Alnus*. { 'ȯl·dər }

aldolase [BIOCHEMISTRY] An enzyme in anaerobic glycolysis that catalyzes the cleavage of fructose 1,6-diphosphate to glyceraldehyde 3-phosphate; used also in the reverse reaction. { 'al·də ,lās }

aldosterone [BIOCHEMISTRY] $C_{21}H_{28}O_5$ A steroid hormone extracted from the adrenal cortex that functions chiefly in regulating sodium and potassium metabolism. { al'däs·tə,rōn }

alecithal [CELL BIOLOGY] Referring to an egg without yolk, such as the eggs of placental mammals. { ā'les·ə·thəl }

Alepocephaloidei [VERTEBRATE ZOOLOGY] The slickheads, a suborder of deap-sea teleostean fishes of the order Salmoniformes. { ə,lep·ō·ə·fə'lȯi·de,ī }

aletophyte [ECOLOGY] A weedy plant growing on the roadside or in fields where natural vegetation has been disrupted by humans. { ə'lēd·ə,fīt }

aleuron [BOTANY] Protein in the form of grains stored in the embryo, endosperm, or perisperm of many seeds. { 'al·yə,rän }

aleurospore [MYCOLOGY] A simple terminal or lateral, thick-walled, nondeciduous spore produced by some fungi of the order Moniliales. { ə 'lyür·ə,spȯr }

alewife [VERTEBRATE ZOOLOGY] *Pomolobus pseudoharengus*. A food fish of the herring family that is very abundant on the Atlantic coast. { 'āl,wīf }

Alexinic unit [BIOLOGY] A unit for the standardization of blood serum. { ¦a·lek¦sin·ik 'yü·nət }

Aleyrodidae [INVERTEBRATE ZOOLOGY] The whiteflies, a family of homopteran insects included in the series Sternorrhyncha; economically important as plant pests. { ,al·ə'räd·ə ,dē }

alfalfa [BOTANY] *Medicago sativa*. A herbaceous perennial legume in the order Rosales, characterized by a deep taproot. Also known as lucerne. { al'fal·fə }

algae [BOTANY] General name for the chlorophyll-bearing organisms in the plant subkingdom Thallobionta. { 'al·jē }

algae bloom [ECOLOGY] A heavy growth of algae in and on a body of water as a result of high phosphate concentration from farm fertilizers and detergents. { 'al·jē ,blüm }

algae wash [ECOLOGY] A shoreline drift consisting almost entirely of filamentous algae. { 'al·jē ,wash }

algal [BOTANY] Of or pertaining to algae. { 'al·gəl }

algesia [PHYSIOLOGY] Sensitivity to pain. { al 'jēz·ē·ə }

algesimeter [PHYSIOLOGY] A device used to determine pain thresholds. { ,al·jə'sim·əd·ər }

algesiroreceptor [PHYSIOLOGY] A pain-sensitive cutaneous sense organ. { ,al·jə¦si·rō·ri¦sep·tər }

alginate [BOTANY] An algal polysaccharide that is a major constituent of the cell walls of brown algae. { 'al·jə,nāt }

algology [BOTANY] The study of algae. Also known as phycology. { al'gäl·ə·jē }

algophage [VIROLOGY] *See* cyanophage. { 'al·gə ,fāj }

alien substitution [GENETICS] The replacement of one or more chromosomes by those from a different species. { ¦āl·ē·ən ,səb·stə'tü·shən }

aliesterase [BIOCHEMISTRY] Any one of the lipases or nonspecific esterases. { al·ē'es·tə,rās }

alimentary [BIOLOGY] Of or relating to food, nutrition, or diet. { ¦al·ə¦men·trē }

alimentary canal [ANATOMY] The tube through which food passes; in humans, includes the mouth, pharynx, esophagus, stomach, and intestine. { ¦al·ə¦men·trē kə'nal }

alimentation [BIOLOGY] Providing nourishment by feeding. { ,al·ə·mən'tā·shən }

Alismataceae [BOTANY] A family of flowering plants belonging to the order Alismatales characterized by schizogenous secretory cells, a horseshoe-shaped embryo, and one or two ovules. { ə,liz·mə'tās·ē,ē }

Alismatales [BOTANY] A small order of flowering

plants in the subclass Alismatidae, including aquatic and semiaquatic herbs. { ə,liz·mə'tā·lēz }

Alismatidae [BOTANY] A relatively primitive subclass of aquatic or semiaquatic herbaceous flowering plants in the class Liliopsida, generally having apocarpous flowers, and trinucleate pollen and lacking endosperm. { ə,liz'mad·ə,dē }

alisphenoid [ANATOMY] **1.** The bone forming the greater wing of the sphenoid in adults. **2.** Of or pertaining to the sphenoid wing. { ¦al·ə¦sfē,nȯid }

alivincular [INVERTEBRATE ZOOLOGY] In some bivalves, having the long axis of the short ligament transverse to the hinge line. { ¦al·ə¦viŋ·kyə·lər }

alkaline phosphatase [BIOCHEMISTRY] A phosphatase active in alkaline media. { 'al·kə,līn 'fäs·fə,tās }

alkaline tide [PHYSIOLOGY] The temporary decrease in acidity of urine and body fluids after eating, attributed by some to the withdrawal of acid from the body due to gastric digestion. { 'al·kə,līn 'tīd }

allantoic acid [BIOCHEMISTRY] $C_4H_8N_4O_4$ A crystalline acid obtained by hydrolysis of allantoin; intermediate product in nucleic acid metabolism. { ¦al·ən¦tō·ik 'as·əd }

allantoin [BIOCHEMISTRY] $C_4H_6N_4O_3$ A crystallizable oxidation product of uric acid found in allantoic and amniotic fluids and in fetal urine. { ə'lan·tə'wən }

allantoinase [BIOCHEMISTRY] An enzyme, occurring in nonmammalian vertebrates, that catalyzes the hydrolysis of allantoin. { ə'lan·tə·wə,nās }

allantois [EMBRYOLOGY] A fluid-filled, saclike, extraembryonic membrane lying between the chorion and amnion of reptilian, bird, and mammalian embryos. { ə'lan·tə'wəs }

allantoxanic acid [BIOCHEMISTRY] $C_4H_3N_3O_4$ An acid formed by oxidation of uric acid or allantoin. { ,a,lan,täk'san·ik 'as·əd }

allanturic acid [BIOCHEMISTRY] $C_3H_4N_2O_3$ An acid formed principally by the oxidation of allantoin. { ¦al·ən¦tür·ik 'as·əd }

Alleculidae [INVERTEBRATE ZOOLOGY] The comb claw beetles, a family of mostly tropical coleopteran insects in the superfamily Tenebrionoidea. { ,ȯl·ə'kyü·lə,dē }

Allee's principle [GENETICS] The concept of an intermediate optimal population density by which groups of organisms often flourish best if neither too few nor too many individuals are present. { a'lēz ,prin·sə·bəl }

Alleghenian life zone [ECOLOGY] A biome that includes the eastern mixed coniferous and deciduous forests of New England. { ¦al·ə¦gān·ē·ən 'līf ,zōn }

allele [GENETICS] One of a pair of genes, or of multiple forms of a gene, located at the same locus of homologous chromosomes. Also known as allelomorph. { ə'lēl }

allele frequency [GENETICS] The proportion of a particular allele in a population. Also known as gene frequency. { ə'lēl ¦frē·kwən·sē }

allelic mutant [GENETICS] A cell or organism with characters different from those of the parent due to alterations of one or more alleles. { ə¦lēl·ik ¦myüt·ənt }

allelochemic [PHYSIOLOGY] Pertaining to a semiochemical that acts as an interspecific agent of communication. { ə¦lē·lō¦kem·ik }

allelomorph [GENETICS] *See* allele. { ə'lē·lə,mȯrf }

allelotropism [BIOLOGY] A mutual attraction between two cells or organisms. { ə¦lē·lō¦trä,piz·əm }

Allen-Doisy unit [BIOLOGY] A unit for the standardization of estrogens. { ¦al·ən ¦dȯiz·ē ,yü·nət }

Allen's rule [VERTEBRATE ZOOLOGY] The generalization that the protruding parts of a warm-blooded animal's body, such as the tail, ears, and limbs, are shorter in animals from cold parts of the species range than from warm parts. { 'al·ənz ,rül }

allergen [IMMUNOLOGY] Any antigen, such as pollen, a drug, or food, that induces an allergic state in humans or animals. { 'al·ər,jen }

alliance [SYSTEMATICS] A group of related families ranking between an order and a class. { ə'lī·əns }

alligator [VERTEBRATE ZOOLOGY] Either of two species of archosaurian reptiles in the genus *Alligator* of the family Alligatoridae. { 'al·ə,gād·ər }

Alligatorinae [VERTEBRATE ZOOLOGY] A subgroup of the crocodilian family Crocodylidae that includes alligators, caimans, *Melanosuchus*, and *Paleosuchus*. { ,al·ə·gə'tȯr·ə,nē }

Allium [BOTANY] A genus of bulbous herbs in the family Liliaceae including leeks, onions, and chives. { 'al·ē·əm }

alloantibody [IMMUNOLOGY] Antibody that reacts with an antigen occurring in a genetically different member of the same species. { ¦a·lō¦ant·i,bäd·ē }

alloantigen [IMMUNOLOGY] *See* isoantigen. { ¦a·lō¦ant·i·jən }

allochoric [BOTANY] Describing a species that inhabits two or more closely related communities, such as forest and grassland, in the same region. { ,a·lə'kȯr·ik }

allochthonous [ECOLOGY] Pertaining to organisms or organic sediments in a given ecosystem that originated in another system. { ə'läk·thə·nəs }

Alloeocoela [INVERTEBRATE ZOOLOGY] An order of platyhelminthic worms of the class Turbellaria distinguished by a simple pharynx and a diverticulated intestine. { ə,lē·ə'sēl·ə }

allogeneic [IMMUNOLOGY] Referring to a transplant made to a different genotype within the same species. Also spelled allogenic. { ¦al·ə·jə¦nē·ik }

allogeneic graft [BIOLOGY] *See* allograft. { ¦al·ə·jə¦nē·ik 'graft }

allogenic [ECOLOGY] Caused by external factors, as in reference to the change in habitat of a natural community resulting from drought. [IMMUNOLOGY] *See* allogeneic. { ¦a·lə¦jen·ik }

allograft [BIOLOGY] Graft from a donor transplanted to a genetically dissimilar recipient of the same species. Also known as allogeneic graft. { 'a·lō,graft }

Allogromiidae [INVERTEBRATE ZOOLOGY] A little-known family of protozoans in the order Foraminiferida; adults are characterized by a chitinous test. { ,a·lə·grə'mī·ə,dē }

Allogromiina [INVERTEBRATE ZOOLOGY] A suborder of marine and fresh-water protozoans in the order Foraminiferida characterized by an organic test of protein and acid mucopolysaccharide. { ,a·lə·grə'mī·ə·nə }

Alloionematoidea [INVERTEBRATE ZOOLOGY] A superfamily of parasitic nematodes belonging to the order Rhabditida, having either no lips or six small amalgamated lips, and a rhabditiform esophagus with a weakly developed valve in the posterior bulb. { ə,lȯi·ō,nem·ə'tȯid·ē·ə }

allometry [BIOLOGY] **1.** The quantitative relation between a part and the whole or another part as the organism increases in size. Also known as heterauxesis; heterogony. **2.** The quantitative relation between the size of a part and the whole or another part, in a series of related organisms that differ in size. { ə'läm·ə·trē }

allomone [PHYSIOLOGY] A chemical produced by an organism which induces in a member of another species a behavioral or physiological reaction favorable to the emitter; may be mutualistic or antagonistic. { 'a·lə,mōn }

allomorphosis [EVOLUTION] Allometry in phylogenetic development. { ¦al·ō·mȯr'fō·səs }

Allomyces [MYCOLOGY] A genus of aquatic phycomycetous fungi in the order Blastocladiales characterized by basal rhizoids, terminal hyphae, and zoospores with a single posterior flagellum. { a·lō'mī,sēz }

alloparapatric speciation [EVOLUTION] A mode of gradual speciation in which new species originate through populations that are initially allopatric but eventually become parapatric before effective reproductive isolation has evolved. { ¦al·ə,pa·rə¦pa·trik ,spē·sē'ā·shən }

allopatric [ECOLOGY] Referring to populations or species that occupy naturally exclusive, but usually adjacent, geographical areas. { ¦a·lō¦pa·trik }

allopatric speciation [ECOLOGY] Differentiation of populations in geographical isolation to the point where they are recognized as separate species. { ¦al·ō¦pa·trik ,spē·sē'ā·shən }

allopelagic [ECOLOGY] Relating to organisms living at various depths in the sea in response to influences other than temperature. { ¦a·lō·pə'laj·ik }

allophene [GENETICS] A mutant phenotype that can revert to a normal phenotype if it is transplanted to a wild-type host. { 'al·ə,fēn }

allophore [HISTOLOGY] A chromatophore which contains a red pigment that is soluble in alcohol; found in the skin of fishes, amphibians, and reptiles. { 'a·lō,fȯr }

alloploid [GENETICS] *See* allopolyploid. { 'al·ə,plȯid }

allopolyploid [GENETICS] An organism or strain arising from a combination of genetically distinct chromosome sets from two diploid organisms. Also known as alloploid. { ,a·lə'päl·ə,plȯid }

allopregnancy [IMMUNOLOGY] A pregnancy in which the male partner is allogeneic with respect to the female. { ¦al·ō'preg·nən·sē }

all-or-none law [PHYSIOLOGY] The principle that transmission of a nerve impulse is either at full strength or not at all. { ¦ȯl ər ¦nən ,lȯ }

allosome [GENETICS] **1.** Sex chromosome. **2.** Any atypical chromosome. { 'a·lō,sōm }

allosteric effector [BIOCHEMISTRY] A small molecule that reacts either with a nonbinding site of an enzyme molecule, or with a protein molecule, and causes a change in the function of the molecule. Also known as allosteric modulator. { ¦a·lə¦stir·ik ə'fek·tər }

allosteric enzyme [BIOCHEMISTRY] Any of the regulatory bacterial enzymes, such as those involved in end-product inhibition. { ¦a·lə¦stir·ik 'en,zīm }

allosteric modulator [BIOCHEMISTRY] *See* allosteric effector. { ¦a·lə¦stir·ik 'mäd·yə,lād·ər }

allosteric site [BIOCHEMISTRY] The inactive (or less active) region of an enzyme molecule. { ¦a·lə¦ster·ik 'sīt }

allosteric transition [BIOCHEMISTRY] A reversible exchange of one base pair for another on a protein molecule that alters the properties of the active site and changes the biological activity of the protein. { ¦a·lə¦stir·ik tranz'ish·ən }

allostery [BIOCHEMISTRY] The property of an enzyme able to shift reversibly between an active and an inactive configuration. { 'a·lō,stir·ē }

allotetraploid [GENETICS] *See* amphidiploid. { ,a·lō'te·trə,plȯid }

Allotheria [PALEONTOLOGY] A subclass of Mammalia that appeared in the Upper Jurassic and became extinct in the Cenozoic. { ,a·lō'thir·ē·ə }

Allotriognathi [VERTEBRATE ZOOLOGY] An equivalent name for the Lampridiformes. { ə'lä·trē'äg·nə,thī }

allotype [SYSTEMATICS] A paratype of the opposite sex to the holotype. { 'a·lə,tīp }

alloxan [BIOCHEMISTRY] $C_4H_2N_2O_4$ Crystalline oxidation product of uric acid; induces diabetes experimentally by selective destruction of pancreatic beta cells. Also known as mesoxyalyurea. { ə'läk·sən }

allozygote [GENETICS] An individual that is homozygous at a given locus with the two homologous genes being of independent origin. { ,al·ə'zī,gōt }

allozyme [GENETICS] One of two or more forms of an enzyme that are specified by allelic genes. { 'al·ə,zīm }

allspice [BOTANY] The dried, unripe berries of a small, tropical evergreen tree, *Pimenta officinalis*, of the myrtle family; yields a pungent, aromatic spice. { 'ȯl,spīs }

alm [ECOLOGY] A meadow in alpine or subalpine mountain regions. { älm }

almond [BOTANY] *Prunus amygdalus*. A small deciduous tree of the order Rosales; it produces a drupaceous edible fruit with an ellipsoidal, slightly compressed nutlike seed. { 'ä·mənd }

Almquist unit [BIOLOGY] A unit for the standardization of vitamin K. { 'äm,kwist ,yü·nət }

Alopiidae [VERTEBRATE ZOOLOGY] A family of pelagic isurid elasmobranchs commonly known as thresher sharks because of their long, whiplike tail. { ,al·ə'pī·ə,dē }

alpaca [VERTEBRATE ZOOLOGY] *Lama pacos*. An artiodactyl of the camel family (Camelidae); economically important for its long, fine wool. { al 'pak·ə }

alpage [ECOLOGY] A summer grazing area composed of natural plant pasturage in upland or mountainous regions. { 'al·pəj }

alpestrine [ECOLOGY] Referring to organisms that live at high elevation but below the timberline. Also known as subalpine. { ,al'pes·trən }

alpha-adrenergic receptor [CELL BIOLOGY] *See* alpha receptor. { ¦al·fə ,ad·rə¦nər·jik ri'sep·tər }

alpha cell [HISTOLOGY] Any of the acidophilic chromophiles in the anterior lobe of the adenohypophysis. { 'al·fə ,sel }

alpha fetoprotein [IMMUNOLOGY] A serum protein that has become associated with the detection of certain types of cancer and fetal abnormalities. Abbreviated AFP. { 'al·fə ¦fēd·ō'prō ,tēn }

alpha globulin [BIOCHEMISTRY] A heterogeneous fraction of serum globulins containing the proteins of greatest electrophoretic mobility. { 'al·fə 'gläb·yə·lən }

alpha helix [MOLECULAR BIOLOGY] A spatial configuration of the polypeptide chains of proteins in which the chain assumes a helical form, 0.54 nanometer in pitch, 3.6 amino acids per turn, presenting the appearance of a hollow cylinder with radiating side groups. { 'al·fə 'hē·liks }

alpha hemolysis [MICROBIOLOGY] Partial hemolysis of red blood cells with green discoloration in a blood agar medium by certain hemolytic streptococci. { 'al·fə hi'mäl·ə·səs }

alpha receptor [CELL BIOLOGY] Any of a group of receptors on cell membranes that are thought to be associated with vasoconstriction, relaxation of intestinal muscle, and contraction of the nictitating membrane, iris dilator muscle, smooth muscle of the spleen, and muscular layer of the uterine wall. Also called alpha-adrenergic receptor. { ¦al·fə ri¦sep·tər }

alpha rhythm [PHYSIOLOGY] An electric current from the occipital region of the brain cortex having a pulse frequency of 8 to 13 per second; associated with a relaxed state in normal human adults. { 'al·fə ,rith·əm }

alpha taxonomy [SYSTEMATICS] The level of taxonomic study concerned with characterizing and naming species. { 'al·fə tak·'sän·ə·mē }

Alpheidae [INVERTEBRATE ZOOLOGY] The snapping shrimp, a family of decapod crustaceans that is included in the section Caridea. { ,al'fē·ə ,dē }

alpine [ECOLOGY] Any plant native to mountain peaks or boreal regions. { 'al,pīn }

alpine tundra [ECOLOGY] Large, flat or gently sloping, treeless tracts of land above the timberline. { 'al,pīn 'tən,drə }

alteration enzyme [IMMUNOLOGY] A protein of bacteriophage T4 that is injected into a host bacterium along with the deoxyribonucleic acid of the bacteriophage and that modifies the ribonucleic acid polymerase of the host so that the ribonucleic acid is unable to initiate transcription at host promoters. { ȯl·tə'rā·shən ,en ,zīm }

alternate [BOTANY] **1.** Of the arrangement of leaves on opposite sides of the stem at different levels. **2.** Of the arrangement of the parts of one whorl between members of another whorl. { 'ȯl·tər·nət }

alternate phyllotaxy [BOTANY] **1.** An arrangement of leaves that occur individually at nodes and on opposite sides of the stem. **2.** A spiral arrangement of leaves on a stem, with one leaf at a node. { 'ȯl·tər·nət 'fil·ə,tak·sē }

alternation of generations [BIOLOGY] *See* metagenesis. { ,ȯl·tər'nā·shən əv ,jen·ə'rā·shənz }

alterne [ECOLOGY] A community exhibiting alternating dominance with other communities in the same area. { 'ȯl,tərn }

altherbosa [ECOLOGY] Communities of tall herbs, usually succeeding where forests have been destroyed. { ¦al·thər¦bōs·ə }

altitude acclimatization [PHYSIOLOGY] A physiological adaptation to reduced atmospheric and oxygen pressure. { 'al·tə,tüd ə,klī·məd·ə'zā·shən }

altitudinal vegetation zone [ECOLOGY] A geographical band of physiognomically similar vegetation correlated with vertical and horizontal gradients of environmental conditions. { ¦al·tə ¦tüd·ən·əl ,vej·ə'tā·shən ,zōn }

altricial [VERTEBRATE ZOOLOGY] Pertaining to young that are born or hatched immature and helpless, thus requiring extended development and parental care. { ,al'trish·əl }

alula [ZOOLOGY] **1.** Digit of a bird wing homologous to the thumb. **2.** *See* calypter. { 'al·yə·lə }

alvar [ECOLOGY] Dwarfed vegetation characteristic of certain Scandinavian steppelike communities with a limestone base. { 'al,vär }

alveator [INVERTEBRATE ZOOLOGY] A type of pedicellaria in echinoderms. { 'al·vē,ād·ər }

alveolar [BIOLOGY] Of or relating to an alveolus. { al'vē·ə·lər }

alveolar gland [ANATOMY] *See* acinous gland. { al 'vē·ə·lər ,gland }

alveolar oxygen pressure [PHYSIOLOGY] The oxygen pressure in the alveoli; the value is about 105 mmHg. { al'vē·ə·lər 'äk·sə·jən ,presh·ər }

alveolar process [ANATOMY] The ridge of bone surrounding the alveoli of the teeth. { al¦vē·ə·lər ¦prä,ses }

alveolar ridge [ANATOMY] The bony remains of the alveolar process of the maxilla or mandible. { al¦vē·ə·lər 'rij }

alveolated cell [HISTOLOGY] *See* epithelioid cell. { al'vē·ə,lād·əd ,sel }

alveolitoid [INVERTEBRATE ZOOLOGY] A type of tabulate coral having a vaulted upper wall and a lower wall parallel to the surface of attachment. { al'vē·ə·lə,tȯid }

alveolus [ANATOMY] **1.** A tiny air sac of the lung. **2.** A tooth socket. **3.** A sac of a compound gland. { al'vē·ə·ləs }

Alydidae [INVERTEBRATE ZOOLOGY] A family of hemipteran insects in the superfamily Coreoidea. { ə'lid·ə,dē }

Alysiella [MICROBIOLOGY] A genus of bacteria in the family Simonsiellaceae; cells are arranged in pairs within the filaments. { ə,lis·ē'el·ə }

amanthophilous [BOTANY] Of plants having a habitat in sandy plains or hills. { ,a·mən'thä·fə·ləs }

Amaranthaceae [BOTANY] The characteristic family of flowering plants in the order Caryophyllales; they have a syncarpous gynoecium, a monochlamydeous perianth that is more or less scarious, and mostly perfect flowers. { ,a·mə·rə 'thā·sē,ē }

Amaryllidaceae [BOTANY] The former designation for a family of plants now included in the Liliaceae. { ,a·mə,ri·lə'dā·sē,ē }

amatoxin [BIOCHEMISTRY] Any of a group of toxic peptides that selectively inhibit ribonucleic acid polymerase in mammalian cells; produced by the mushroom *Amanita phalloides*. { ¦am·ə¦täk·sən }

amber codon [VIROLOGY] The polypeptide chain-termination messenger-RNA codon UAG, which brings about the termination of protein translation. { 'am·bər 'kō,dän }

ambergris [PHYSIOLOGY] A fatty substance formed in the intestinal tract of the sperm whale; used in the manufacture of perfume. { 'am·bə ,gris }

amber mutation [GENETICS] Alteration of a codon to UAG, a codon that results in premature polypeptide chain termination in bacteria. { 'am·bər myü'tā·shən }

ambidextrous [PHYSIOLOGY] Capable of using both hands with equal skill. { ¦am·bə¦dek·strəs }

ambigenous [BOTANY] Of a perianth whose outer leaves resemble the calyx while the inner leaves resemble the corolla. { ,am'bi·jən·əs }

ambiguous codon [GENETICS] A codon capable of coding for more than one amino acid sequence. { am'big·yə·wəs 'kō,dän }

ambitus [BIOLOGY] The periphery or external edge, as of a mollusk shell or leaf. { 'am·bə·təs }

Amblyopsidae [VERTEBRATE ZOOLOGY] The cave fishes, a family of actinopterygian fishes in the order Percopsiformes. { am·blē'äp·sə,dē }

Amblyopsiformes [VERTEBRATE ZOOLOGY] An equivalent name for the Percopsiformes. { am·blē¦äp·sə'fȯr·mēz }

Amblypygi [INVERTEBRATE ZOOLOGY] An order of chelicerate arthropods in the class Arachnida, commonly known as the tailless whip scorpions. { am'blip·i,jī }

amboceptor [IMMUNOLOGY] According to P. Ehrlich, an antibody present in the blood of immunized animals which contains two specialized elements: a cytophil group that unites with a cellular antigen, and a complementophil group that joins with the complement. { 'am·bō,sep·tər }

amboceptor unit [BIOLOGY] A unit for the standardization of blood serum. { 'am·bō,sep·tər ,yü·nət }

ambulacrum [INVERTEBRATE ZOOLOGY] In echinoderms, any of the radial series of plates along which the tube feet are arranged. { ,am·byə 'lak·rəm }

ambulatorial [ZOOLOGY] **1.** Capable of walking. **2.** In reference to a forest animal, having adapted to walking, as opposed to running, crawling, or leaping. { ,am·byə·lə'tȯr·ē·əl }

Ambystoma [VERTEBRATE ZOOLOGY] A genus of common salamanders; the type genus of the family Ambystomatidae. { am'bis·tə·mə }

Ambystomatidae [VERTEBRATE ZOOLOGY] A family of urodele amphibians in the suborder Salamandroidea; neoteny occurs frequently in this group. { am,bis·tə'mad·ə,dē }

Ambystomoidea [VERTEBRATE ZOOLOGY] A suborder to which the family Ambystomatidae is sometimes elevated. { am,bis·tə'mȯid·ē·ə }

ameba [INVERTEBRATE ZOOLOGY] The common name for a number of species of naked unicellular protozoans of the order Amoebida; an example is a member of the genus *Amoeba*. { ə'mē·bə }

Amebelodontinae [PALEONTOLOGY] A subfamily of extinct elephantoid proboscideans in the family Gomphotheriidae. { ,a·mə,bel·ə'dän·tə,nē }

amebocyte [INVERTEBRATE ZOOLOGY] One of the wandering ameboid cells in the tissues and fluids of many invertebrates that function in assimilation and excretion. { ə'mēb·ə,sīt }

ameboid movement [CELL BIOLOGY] A type of cellular locomotion involving the formation of pseudopodia. { ə'mēb·ȯid 'müv·mənt }

ameiosis [GENETICS] Nonreduction of chromosome number due to suppression of one of the miotic divisions, as in parthenogenesis. { ,ā·mī 'ō·səs }

Ameiuridae [VERTEBRATE ZOOLOGY] A family of North American catfishes belonging to the suborder Siluroidei. { ,a·mī'yür·ə,dē }

ameloblast [EMBRYOLOGY] One of the columnar cells of the enamel organ that form dental enamel in developing teeth. { 'am·ə·lō,blast }

amensalism [ECOLOGY] An interaction between species in which one species is harmed and the other species is unaffected. { ā'men·sə,liz·əm }

ament [BOTANY] A catkin. { 'ā,ment }

Amera [INVERTEBRATE ZOOLOGY] One of the three divisions of the phylum Vermes proposed by O. Bütschli in 1910 and given the rank of a subphylum. { 'am·ə·rə }

American boreal faunal region [ECOLOGY] A zoogeographic region comprising marine littoral animal communities of the coastal waters off east-central North America. { ə'mer·ə·kən 'bȯr·ē·əl 'fȯn·əl ,rē·jən }

American-Egyptian cotton [BOTANY] A type of cotton developed by hybridization of Egyptian and American plants. { ə'mer·ə·kən i¦jip·shən 'kät·ən }

American lion [VERTEBRATE ZOOLOGY] *See* puma. { ə'mer·ə·kən 'lī·ən }

Amerosporae [MYCOLOGY] A spore group of the Fungi Imperfecti characterized by one-celled or threadlike spores. { ˌam·ə'räs·pəˌrē }

ametabolous metamorphosis [INVERTEBRATE ZOOLOGY] A growth stage of certain insects characterized by an increase in size without distinct external changes. { ¦ā·mə¦tab·ə·ləs ˌmed·ə 'mȯr·fə·səs }

ametoecious [ECOLOGY] Of a parasite that remains with the same host. { ˌam·ə'tēsh·əs }

amictic [INVERTEBRATE ZOOLOGY] **1.** In rotifers, producing diploid eggs that are incapable of being fertilized. **2.** Pertaining to the egg produced by the amictic female. { ə'mik·tik }

amidase [BIOCHEMISTRY] Any enzyme that catalyzes the hydrolysis of nonpeptide C=N linkages. { 'am·əˌdās }

amidohydrolase [BIOCHEMISTRY] An enzyme that catalyzes deamination. { ə¦mē·dō¦nī·drəˌlās }

Amiidae [VERTEBRATE ZOOLOGY] A family of actinopterygian fishes in the order Amiiformes represented by a single living species, the bowfin (*Amia calva*). { ə'mī·əˌdē }

Amiiformes [VERTEBRATE ZOOLOGY] An order of actinopterygian fishes characterized by an abbreviate heterocercal tail, fusiform body, and median fin rays. { əˌmī·ə'fȯrˌmēz }

amine oxidase [BIOCHEMISTRY] An enzyme that catalyzes the oxidation of tyramine and tryptamine to aldehyde. { ə'mēn 'äk·səˌdās }

aminoacetic acid [BIOCHEMISTRY] *See* glycine.

amino acid [BIOCHEMISTRY] Any of the organic compounds that contain one or more basic amino groups and one or more acidic carboxyl groups and that are polymerized to form peptides and proteins; only 20 of the more than 80 amino acids found in nature serve as building blocks for proteins; examples are tyrosine and lysine. { ə'mēˌnō 'as·əd }

para-aminobenzoic acid [BIOCHEMISTRY] $C_7H_7O_2N$ A yellow-red, crystalline compound that is part of the folic acid molecule; essential in metabolism of certain bacteria. Abbreviated PABA. { ¦par·ə ə¦mē·nōˌben¦zō·ik 'as·əd }

α-aminohydrocinnamic acid [BIOCHEMISTRY] *See* phenylalanine. { ¦al·fə ə¦mē·nō¦hī·drō·sə¦nam·ik 'as·əd }

aminopeptidase [BIOCHEMISTRY] An enzyme which catalyzes the liberation of an amino acid from the end of a peptide having a free amino group. { əˌmē·nō'pep·təˌdās }

α-amino-β-phenylpropionic acid [BIOCHEMISTRY] *See* phenylalanine. { ¦al·fe ə¦mē·nō ¦bād·ə ¦fen·əl¦prō·pē¦an·ik 'as·əd }

aminopolypeptidase [BIOCHEMISTRY] A proteolytic enzyme that cleaves polypeptides containing either a free amino group or a basic nitrogen atom having at least one hydrogen atom. { ə¦mē·nōˌpä·lə'pep·təˌdās }

aminoprotease [BIOCHEMISTRY] An enzyme that hydrolyzes a protein and unites with its free amino group. { ə¦mē·nō'prō·tēˌās }

amino sugar [BIOCHEMISTRY] A monosaccharide in which a nonglycosidic hydroxyl group is replaced by an amino or substituted amino group; an example is D-glucosamine. { ə¦mē·nō ¦shu̇g·ər }

amino terminal [BIOCHEMISTRY] The end part of a polypeptide chain which contains a free alpha-amino group. { ə¦mē·nō ¦tərm·ən·əl }

aminotransferase [BIOCHEMISTRY] *See* transaminase. { əˌmē·nō'tranz·fəˌrās }

amitosis [CELL BIOLOGY] Cell division by simple fission of the nucleus and cytoplasm without chromosome differentiation. { ¦āˌmī'tō·səs }

ammocoete [ZOOLOGY] A protracted larval stage of lampreys. { 'a·məˌsēt }

ammocolous [ECOLOGY] Describing plants having a habitat in dry sand. { ə'mä·kə·ləs }

Ammodiscacea [INVERTEBRATE ZOOLOGY] A superfamily of foraminiferal protozoans in the suborder Textulariina, characterized by a simple to labyrinthic test wall. { ˌa·məˌdis'kāsh·ə }

Ammodytoidei [VERTEBRATE ZOOLOGY] The sand lances, a suborder of marine actinopterygian fishes in the order Perciformes, characterized by slender, eel-shaped bodies. { əˌmäd·i'tȯi·dēˌī }

ammonifiers [ECOLOGY] Fungi, or actinomycetous bacteria, that participate in the ammonification part of the nitrogen cycle and release ammonia (NH_3) by decomposition of organic matter. { ə'män·əˌfī·ərz }

ammonite [PALEONTOLOGY] A fossil shell of the cephalopod order Ammonoidea. { 'a·məˌnīt }

ammonoid [PALEONTOLOGY] A cephalopod of the order Ammonoidea. { 'a·məˌnȯid }

Ammonoidea [PALEONTOLOGY] An order of extinct cephalopod mollusks in the subclass Tetrabranchia; important as index fossils. { ˌa·mə 'nȯid·ē·ə }

ammonotelic [BIOLOGY] Pertaining to the excretion of nitrogen primarily as ammonium ion, [NH_4^+]. { ə¦mä·nō'tēl·ik }

Ammotheidae [INVERTEBRATE ZOOLOGY] A family of marine arthropods in the subphylum Pycnogonida. { ˌa·mə'thē·əˌdē }

amnicolous [ECOLOGY] Describing plants having a habitat on sandy riverbanks. { ˌam·nə'kä·ləs }

amniochorionic [EMBRYOLOGY] Relating to both amnion and chorion. { ¦am·nē·ōˌkȯr·ē'än·ik }

amniogenesis [EMBRYOLOGY] The development or formation of the amnion. { ¦am·nē·ō'jen·ə·səs }

amnion [EMBRYOLOGY] A thin extraembryonic membrane forming a closed sac around the embryo in birds, reptiles, and mammals. { 'am·nē ˌän }

Amniota [VERTEBRATE ZOOLOGY] A collective term for the Reptilia, Aves, and Mammalia, all of which have an amnion during development. { ˌam·nē'äd·ə }

amniote [ZOOLOGY] An animal that develops an amnion during its embryonic stage; includes birds, reptiles, and mammals. { 'am·nēˌōt }

amniotic fluid [PHYSIOLOGY] A substance that fills the amnion to protect the embryo from desiccation and shock. { ¦am·nē¦äd·ik 'flü·əd }

Amoeba [INVERTEBRATE ZOOLOGY] A genus of naked, rhizopod protozoans in the order Amoebida characterized by a thin pellicle and thick, irregular pseudopodia. { ə'mē·bə }

Amoebida [INVERTEBRATE ZOOLOGY] An order of rhizopod protozoans in the subclass Lobosia characterized by the absence of a protective covering (test). { ˌam·ə'bī·də }

Amoebobacter [MICROBIOLOGY] A genus of bacteria in the family Chromatiaceae; cells are spherical and nonmotile, have gas vacuoles, and contain bacteriochlorophyll *a* on vesicular photosynthetic membranes. { əˌmē·bə'bak·tər }

amorphic allele [GENETICS] An allele that lacks gene activity. { ə'mȯr·fik ə'lēl }

Amorphosporangium [MICROBIOLOGY] A genus of bacteria in the family Actinoplanaceae with irregular sporangia and rod-shaped spores; they are motile by means of polar flagella. { ə¦mȯr·fə·spə'ran·jəm }

AMP [BIOCHEMISTRY] *See* adenylic acid.

3′,5′-AMP [BIOCHEMISTRY] *See* cyclic adenylic acid.

Ampeliscidae [INVERTEBRATE ZOOLOGY] A family of tube-dwelling amphipod crustaceans in the suborder Gammaridea. { ˌam·pə'lis·əˌdē }

Ampharetidae [INVERTEBRATE ZOOLOGY] A large, deep-water family of polychaete annelids belonging to the Sedentaria. { ˌam·fə'red·əˌdē }

Ampharetinae [INVERTEBRATE ZOOLOGY] A subfamily of annelids belonging to the family Ampharetidae. { ˌam·fə'ret·əˌnē }

amphiarthrosis [ANATOMY] An articulation of limited movement in which bones are connected by fibrocartilage, such as that between vertebrae or that at the tibiofibular junction. { ˌam·fī·är'thrō·səs }

amphiaster [INVERTEBRATE ZOOLOGY] Type of spicule found in some sponges. { 'am·fēˌas·tər }

Amphibia [VERTEBRATE ZOOLOGY] A class of vertebrate animals in the superclass Tetrapoda characterized by a moist, glandular skin, gills at some stage of development, and no amnion during the embryonic stage. { am'fib·ē·ə }

Amphibicorisae [INVERTEBRATE ZOOLOGY] A subdivision of the insect order Hemiptera containing surface water bugs with exposed antennae. { amˌfib·ə'kȯr·əˌsē }

Amphibioidei [INVERTEBRATE ZOOLOGY] A family of tapeworms in the order Cyclophyllidea. { amˌfəˌbī'ȯid·ēˌī }

amphibiotic [ZOOLOGY] Being aquatic during the larval stage and terrestrial in the adult stage. { ¦am·fəˌbī¦äd·ik }

amphibious [BIOLOGY] Capable of living both on dry or moist land and in water. { ˌam'fib·ē·əs }

amphiblastic cleavage [EMBRYOLOGY] The unequal but complete cleavage of telolecithal eggs. { ¦am·fə¦blas·tik 'klēv·ij }

amphiblastula [EMBRYOLOGY] A blastula resulting from amphiblastic cleavage. [INVERTEBRATE ZOOLOGY] The free-swimming flagellated larva of many sponges. { ¦am·fə¦blas·chə·lə }

amphibolic [ZOOLOGY] Possessing the ability to turn either backward or forward, as the outer toe of certain birds. { ¦am·fə¦bäl·ik }

amphibolic pathway [BIOCHEMISTRY] A microbial biosynthetic and energy-producing pathway, such as the glycolytic pathway. { ¦am·fə¦bäl·ik 'pathˌwā }

Amphibolidae [INVERTEBRATE ZOOLOGY] A family of gastropod mollusks in the order Basommatophora. { ˌam·fə'bäl·əˌdē }

amphicarpic [BOTANY] Having two types of fruit, differing either in form or ripening time. { ¦am·fə¦kär·pik }

Amphichelydia [PALEONTOLOGY] A suborder of Triassic to Eocene anapsid reptiles in the order Chelonia; these turtles did not have a retractable neck. { ˌam·fə·kə'lid·ē·ə }

amphichrome [BOTANY] A plant that produces flowers of different colors on the same stalk. { 'am·fəˌkrōm }

Amphicoela [VERTEBRATE ZOOLOGY] A small suborder of amphibians in the order Anura characterized by amphicoelous vertebrae. { ¦am·fə¦sēl·ə }

amphicoelous [VERTEBRATE ZOOLOGY] Describing vertebrae that have biconcave centra. { ¦am·fə¦sēl·əs }

amphicribral [BOTANY] Having the phloem surrounded by the xylem, as seen in certain vascular bundles. { ¦am·fə¦krib·rəl }

amphicryptophyte [BOTANY] A marsh plant with amphibious vegetative organs. { ¦am·fə¦krip·təˌfīt }

Amphicyonidae [PALEONTOLOGY] A family of extinct giant predatory carnivores placed in the infraorder Miacoidea by some authorities. { ¦am·fə·sī¦än·əˌdē }

amphicytula [EMBRYOLOGY] A zygote that is capable of holoblastic unequal cleavage. { ˌam·fə'sich·ə·lə }

amphid [INVERTEBRATE ZOOLOGY] Either of a pair of sensory receptors in nematodes, believed to be chemoreceptors and situated laterally on the anterior end of the body. { 'am·fəd }

amphidetic [INVERTEBRATE ZOOLOGY] Of a bivalve ligament, extending both before and behind the beak. { ¦am·fə¦ded·ik }

amphidiploid [GENETICS] An organism having a diploid set of chromosomes from each parent. Also known as allotetraploid. { ¦am·fə¦diˌplȯid }

Amphidiscophora [INVERTEBRATE ZOOLOGY] A subclass of sponges in the class Hexactinellida characterized by an anchoring tuft of spicules and no hexasters. { ¦am·fə·di'skäf·ə·rə }

Amphidiscosa [INVERTEBRATE ZOOLOGY] An order of hexactinellid sponges in the subclass Amphidiscophora characterized by amphidisc spicules, that is, spicules having a stellate disk at each end. { ¦am·fə·di'skō·sə }

amphigastrium [BOTANY] Any of the small appendages located ventrally on the stem of some liverworts. { ¦am·fə'gas·trē·əm }

amphigean [ECOLOGY] An organism that is native to both Old and New Worlds. { ¦am·fə¦jē·ən }

Amphilestidae [PALEONTOLOGY] A family of Jurassic triconodont mammals whose subclass is uncertain. { ˌam·fə'les·təˌdē }

Amphilinidea [INVERTEBRATE ZOOLOGY] An order

of tapeworms in the subclass Cestodaria characterized by a protrusible proboscis, anterior frontal glands, and no holdfast organ; they inhabit the coelom of sturgeon and other fishes. { ˌam·fə·lə′nid·ē·ə }

Amphimerycidae [PALEONTOLOGY] A family of late Eocene to early Oligocene tylopod ruminants in the superfamily Amphimerycoidea. { ˌam·fə·mə′ris·əˌdē }

Amphimerycoidea [PALEONTOLOGY] A superfamily of extinct ruminant artiodactyls in the infraorder Tylopoda. { ˌam·fəˌmir·ə′kȯid·ē·ə }

amphimixis [PHYSIOLOGY] The union of egg and sperm in sexual reproduction. { ˌam·fə′mik·səs }

Amphimonadidae [INVERTEBRATE ZOOLOGY] A family of zoomastigophorean protozoans in the order Kinetoplastida. { ˌam·fə·mə′näd·əˌdē }

Amphineura [INVERTEBRATE ZOOLOGY] A class of the phylum Mollusca; members are bilaterally symmetrical, elongate marine animals, such as the chitons. { ˌam·fə′nu̇r·ə }

Amphinomidae [INVERTEBRATE ZOOLOGY] The stinging or fire worms, a family of amphinomorphan polychaetes belonging to the Errantia. { ˌam·fə′näm·əˌdē }

Amphinomorpha [INVERTEBRATE ZOOLOGY] Group name for three families of errantian polychaetes: Amphenomidae, Euphrosinidae, and Spintheridae. { ¦am·fə·nə¦mȯr·fə }

amphioxus [ZOOLOGY] Former designation for the lancelet, *Branchiostoma*. { ˌam·fē′äk·səs }

amphiphloic [BOTANY] Pertaining to the central vascular cylinder of stems having phloem on both sides of the xylem. { ˌam·fə′flō·ik }

amphiphyte [ECOLOGY] A plant growing on the boundary zone of wet land. { ′am·fəˌfīt }

amphiplatyan [ANATOMY] Describing vertebrae having centra that are flat both anteriorly and posteriorly. { ¦am·fə¦plad·ē·ən }

amphiploid [GENETICS] A polyploid produced when chromosomes of a species hybrid double. { ′am·fəˌplȯid }

amphipneustic [VERTEBRATE ZOOLOGY] Having both gills and lungs through all life stages, as in some amphibians. { ¦am·fə¦nüs·tik }

Amphipoda [INVERTEBRATE ZOOLOGY] An order of crustaceans in the subclass Malacostraca; individuals lack a carapace, bear unstalked eyes, and respire through thoracic branchiae or gills. { am′fip·ə·də }

amphipodous [INVERTEBRATE ZOOLOGY] Having both walking and swimming legs. { am′fip·ə·dəs }

amphisarca [BOTANY] An indehiscent fruit characterized by many cells and seeds, pulpy flesh, and a hard rind; melon is an example. { ˌam·fə′sär·kə }

Amphisbaenidae [VERTEBRATE ZOOLOGY] A family of tropical snakelike lizards in the suborder Sauria. { ¦am·fəs¦bēn·əˌdē }

Amphisopidae [INVERTEBRATE ZOOLOGY] A family of isopod crustaceans in the suborder Phreactoicoidea. { ¦am·fə¦säp·əˌdē }

amphispore [MYCOLOGY] A specialized urediospore with a thick, colorful wall; a resting spore. { ′am·fəˌspȯr }

Amphissitidae [PALEONTOLOGY] A family of extinct ostracods in the suborder Beyrichicopina. { ¦am·fə¦sid·əˌdē }

Amphistaenidae [VERTEBRATE ZOOLOGY] The worm lizards, a family of reptiles in the suborder Sauria; structural features are greatly reduced, particularly the limbs. { ¦am·fə¦stēn·əˌdē }

amphistome [INVERTEBRATE ZOOLOGY] An adult type of digenetic trematode having a well-developed ventral sucker (acetabulum) on the posterior end. { ′am·fəˌstōm }

amphistylic [VERTEBRATE ZOOLOGY] Having the jaw suspended from the brain case and the hyomandibular cartilage, as in some sharks. { ¦am·fə¦stīl·ik }

amphitene [CELL BIOLOGY] *See* zygotene. { ′am·fəˌtēn }

amphithecium [BOTANY] The external cell layer during development of the sporangium in mosses. { ˌam·fə′thē·shē·əm }

Amphitheriidae [PALEONTOLOGY] A family of Jurassic therian mammals in the infraclass Pantotheria. { ˌam·fə·thə′rī·əˌdē }

amphitriaene [INVERTEBRATE ZOOLOGY] A poriferan spicule having three divergent rays at each end. { ˌam·fə′trīˌēn }

amphitrichous [BIOLOGY] Having flagella at both ends, as in certain bacteria. { ˌam′fī·trə·kəs }

Amphitritinae [INVERTEBRATE ZOOLOGY] A subfamily of sedentary polychaete worms in the family Terebellidae. { ˌam·fə′trī·təˌnē }

amphitropical distribution [ECOLOGY] Distribution of mostly temperate organisms which are discontinuous between the Northern and Southern hemispheres. { ¦am·fə¦träp·ə·kəl ˌdis·trə′byü·shən }

amphitropous [BOTANY] Having a half-inverted ovule with the funiculus attached near the middle. { ˌam′fi·trə·pəs }

Amphiumidae [VERTEBRATE ZOOLOGY] A small family of urodele amphibians in the suborder Salamandroidea composed of three species of large, eellike salamanders with tiny limbs. { ˌam·fē′yü·məˌdē }

amphivasal [BOTANY] Having the xylem surrounding the phloem, as seen in certain vascular bundles. { ¦am·fə¦vā·zəl }

Amphizoidae [INVERTEBRATE ZOOLOGY] The trout stream beetles, a small family of coleopteran insects in the suborder Adephaga. { ˌam·fə′zō·əˌdē }

Amphoriscidae [INVERTEBRATE ZOOLOGY] A family of calcareous sponges in the order Sycettida. { ¦am·fə¦ris·əˌdē }

amphotericin [MICROBIOLOGY] An amphoteric antifungal antibiotic produced by *Streptomyces nodosus* and having of two components, A and B. { ˌam·fə′ter·ə·sən }

amphotericin A [MICROBIOLOGY] The relatively inactive component of amphotericin. { ˌam·fə′ter·ə·sən ′ā }

amphotericin B [MICROBIOLOGY] $C_{46}H_{73}O_{20}N$ The active component of amphotericin, suitable for systemic therapy of deep or superficial mycotic infections. { ˌam·fə'ter·ə·sən 'bē }

amplexicaul [BOTANY] Pertaining to a sessile leaf with the base or stipules embracing the stem. { am'plek·səˌkȯl }

amplexus [BOTANY] Having the edges of a leaf overlap the edges of a leaf above it in vernation. [VERTEBRATE ZOOLOGY] The copulatory embrace of frogs and toads. { am'plek·səs }

ampliate [BIOLOGY] Widened or enlarged. { 'am·plēˌāt }

amplification [GENETICS] **1.** Treatment with an antibiotic or other agent to increase the relative proportion of plasmid to bacterial deoxyribonucleic acid. **2.** Bulk replication of a gene library. { ˌam·plə·fə'kā·shən }

amplitude of accommodation [PHYSIOLOGY] The range of focal powers of which the eye is capable, expressed in diopters. { 'am·pləˌtüd əv əˌkäm·ə 'dā·shən }

Ampulicidae [INVERTEBRATE ZOOLOGY] A small family of hymenopteran insects in the superfamily Sphecoidea. { ˌam·pyə'lis·əˌdē }

ampulla [ANATOMY] A dilated segment of a gland or tubule. [BOTANY] A small air bladder in some aquatic plants. [INVERTEBRATE ZOOLOGY] The sac at the base of a tube foot in certain echinoderms. { am'pu̇l·ə }

ampulla of Lorenzini [VERTEBRATE ZOOLOGY] Any of the cutaneous receptors in the head region of elasmobranchs; thought to have a thermosensory function. { am'pu̇l·ə əv ˌlȯ·rent'zēˌnē }

ampulla of Vater [ANATOMY] Dilation at the junction of the bile and pancreatic ducts and the duodenum in humans. Also known as papilla of Vater. { am'pu̇l·ə əv 'fät·ər }

Ampullariella [MICROBIOLOGY] A genus of bacteria in the family Actinoplanaceae having cylindrical or bottle-shaped sporangia and rod-shaped spores arranged in parallel chains; motile by means of a polar tuft of flagella. { ˌam·pyəˌlar·ē'el·ə }

ampullary organ [PHYSIOLOGY] An electroreceptor most sensitive to direct-current and low-frequency electric stimuli; found over the body surface of electric fish and also in certain other fish such as sharks, rays, and catfish. { ¦am·pyəˌler·ē 'ȯr·gən }

amygdalin [BIOCHEMISTRY] $C_6H_5CH(CN)OC_{12}H_{21}O_{10}$ A glucoside occurring in the kernels of certain plants of the genus *Prunus*. { ə'mig·də ˌlən }

amylase [BIOCHEMISTRY] An enzyme that hydrolyzes reserve carbohydrates, starch in plants and glycogen in animals. { 'am·əˌlās }

amylolysis [BIOCHEMISTRY] The enzyme-catalyzed hydrolysis of starch to soluble products. { ˌam·ə'läl·ə·səs }

amylolytic enzyme [BIOCHEMISTRY] A type of enzyme capable of denaturing starch molecules; used in textile manufacture to remove starch added to slash sizing agents. { ˌam·ə'läd·ik 'en ˌzīm }

amylopectin [BIOCHEMISTRY] A highly branched, high-molecular-weight carbohydrate polymer composed of about 80% corn starch. { ˌam·ə·lō 'pek·tən }

amyloplast [BOTANY] A colorless cell plastid packed with starch grains and occurring in cells of plant storage tissue. { 'am·ə·lōˌplast }

amylopsin [BIOCHEMISTRY] An enzyme in pancreatic juice that acts to hydrolyze starch into maltose. { ˌam·ə'läp·sən }

amylose [BIOCHEMISTRY] A linear starch polymer. { 'am·əˌlōs }

Amynodontidae [PALEONTOLOGY] A family of extinct hippopotamuslike perissodactyl mammals in the superfamily Rhinocerotoidea. { ˌa·mə·nə 'dän·təˌdē }

ANA [IMMUNOLOGY] *See* antinuclear antibody.

Anabaena [BOTANY] A genus of blue-green algae in the class Cyanophycaea; members fix atmospheric nitrogen. { ˌan·ə'bēn·ə }

Anabantidae [VERTEBRATE ZOOLOGY] A fresh-water family of actinopterygian fishes in the order Perciformes, including climbing perches and gourami. { ˌan·ə'ban·təˌdē }

Anabantoidei [VERTEBRATE ZOOLOGY] A suborder of fresh-water labyrinth fishes in the order Perciformes. { ˌan·əˌban'tȯi·dēˌī }

anabiosis [BIOLOGY] State of suspended animation induced by desiccation and reversed by addition of moisture; can be achieved in rotifers. { ˌan·əˌbī'ō·səs }

anabolic [BIOCHEMISTRY] Pertaining to anabolism. [EVOLUTION] Pertaining to anaboly. { ˌan·ə'bäl·ik }

anabolic steroid [BIOCHEMISTRY] Any of a group of steroid hormones that increase anabolism. { ˌan·ə¦bäl·ik 'stiˌrȯid }

anabolism [BIOCHEMISTRY] A part of metabolism involving the union of smaller molecules into larger molecules; the method of synthesis of tissue structure. { ˌan'ab·əˌliz·əm }

anaboly [EVOLUTION] The addition, through evolutionary differentiation, of a new terminal stage to the morphogenetic pattern. { ə'nab·ə· lī }

Anacanthini [VERTEBRATE ZOOLOGY] An equivalent name for the Gadiformes. { ˌan·ə'kan·thə ˌnī }

Anacardiaceae [BOTANY] A family of flowering plants, the sumacs, in the order Sapindales; many species are allergenic to humans. { ˌan·ə ˌkärd·ē'ās·ēˌē }

anaconda [VERTEBRATE ZOOLOGY] *Eunectes murinus*. The largest living snake, an arboreal-aquatic member of the boa family (Boidae). { ˌan·ə'kän·də }

Anactinochitinosi [INVERTEBRATE ZOOLOGY] A group name for three closely related suborders of mites and ticks: Onychopalpida, Mesostigmata, and Ixodides. { ə¦nak·tə·nə¦kīt·ən'ōˌsī }

Anacystis [BOTANY] A genus of blue-green algae in the class Cyanophycea. { ˌan·ə'sis·təs }

anadromous [VERTEBRATE ZOOLOGY] Said of a fish, such as the salmon and shad, that ascends fresh-water streams from the sea to spawn. { ə 'na·drə·məs }

Anadyomenaceae [BOTANY] A family of green marine algae in the order Siphonocladales characterized by the expanded blades of the thallus. { ə,na·dyə,men'ās·ē,ē }

anaerobe [BIOLOGY] An organism that does not require air or free oxygen to maintain its life processes. { 'an·ə,rōb }

anaerobic bacteria [MICROBIOLOGY] Any bacteria that can survive in the partial or complete absence of air; two types are facultative and obligate. { ¦an·ə¦rōb·ik ,bak'tir·ē·ə }

anaerobic condition [BIOLOGY] The absence of oxygen, preventing normal life for organisms that depend on oxygen. { ¦an·ə¦rōb·ik kən'dish·ən }

anaerobic glycolysis [BIOCHEMISTRY] A metabolic pathway in plants by which, in the absence of oxygen, hexose is broken down to lactic acid and ethanol with some adenosinetriphosphate synthesis. { ¦an·ə¦rōb·ik glī'käl·ə·səs }

anaerobic petri dish [MICROBIOLOGY] A glass laboratory dish for plate cultures of anaerobic bacteria; a thioglycollate agar medium and restricted air space give proper conditions. { ¦an·ə¦rōb·ik 'pē·trē ,dish }

anaerobiosis [BIOLOGY] A mode of life carried on in the absence of molecular oxygen. { ,an·ə,rō 'bī·ə·səs }

anaerophyte [ECOLOGY] A plant that does not need free oxygen for respiration. { ə'ner·ə,fīt }

anakinesis [BIOCHEMISTRY] A process in living organisms by which energy-rich molecules, such as adenosinetriphosphate, are formed. { ,an·ə·kə'nē·səs }

anal [ANATOMY] Relating to or located near the anus. { 'ān·əl }

anal fin [VERTEBRATE ZOOLOGY] An unpaired fin located medially on the posterior ventral part of the fish body. { 'ān·əl ,fin }

analgesia [PHYSIOLOGY] Insensibility to pain with no loss of consciousness. { ,an·əl'jēzh·ə }

anal gland [INVERTEBRATE ZOOLOGY] A gland in certain mollusks that secretes a purple substance. [VERTEBRATE ZOOLOGY] A gland located near the anus or opening into the rectum in many vertebrates. { 'ān·əl ,gland }

analogous [BIOLOGY] Referring to structures that are similar in function and general appearance but not in origin, such as the wing of an insect and the wing of a bird. { ə'nal·ə·gəs }

anal plate [EMBRYOLOGY] An embryonic plate formed of endoderm and ectoderm through which the anus later ruptures. [VERTEBRATE ZOOLOGY] **1.** One of the plates on the posterior portion of the plastron in turtles. **2.** A large scale anterior to the anus of most snakes. { 'ān·əl ,plāt }

anal sphincter [ANATOMY] Either of two muscles, one voluntary and the other involuntary, controlling closing of the anus in vertebrates. { 'ān·əl 'sfiŋk·tər }

anamnestic response [IMMUNOLOGY] A rapidly increased antibody level following renewed contact with a specific antigen, even after several years. Also known as booster response. { ,an·əm 'nes·tik ri'späns }

Anamnia [VERTEBRATE ZOOLOGY] Vertebrate animals which lack an amnion in development, including Agnatha, Chondrichthyes, Osteichthyes, and Amphibia. { a'nam·nē·ə }

Anamniota [VERTEBRATE ZOOLOGY] The equivalent name for Anamnia. { ¦a,nam·nē'ōd·ə }

anamniote [ZOOLOGY] An animal that does not develop an amnion during its embryonic stage. { an'am·nē,ōt }

anamorphism [EVOLUTION] *See* anamorphosis. { ,an·ə'mȯr·fiz·əm }

anamorphosis [EVOLUTION] Gradual increase in complexity of form and function during evolution of a group of animals or plants. Also known as anamorphism. { ,an·ə'mȯr·fə·səs }

Anancinae [PALEONTOLOGY] A subfamily of extinct proboscidean placental mammals in the family Gomphotheriidae. { ə'nan·sə,nē }

anaphase [CELL BIOLOGY] **1.** The stage in mitosis and in the second meiotic division when the centromere splits and the chromatids separate and move to opposite poles. **2.** The stage of the first meiotic division when the two halves of a bivalent chromosome separate and move to opposite poles. { 'an·ə,fāz }

anaphoresis [PHYSIOLOGY] Movement of positively charged ions into tissues under the influence of an electric current. { ¦an·ə·fə'rē·səs }

anaphylatoxin [IMMUNOLOGY] The vasodilator principle, a toxic substance released by tissues of sensitized animals when antigen and antibody react. { ¦an·ə,fil·ə'täk·sən }

Anaplasma [MICROBIOLOGY] A genus of the family Anaplasmataceae; organisms form inclusions in red blood cells of ruminants. { ,an·ə'plaz·mə }

Anaplasmataceae [MICROBIOLOGY] A family of the order Rickettsiales; obligate parasites, either in or on red blood cells or in the plasma of various vertebrates. { ,an·ə,plaz·mə'tās·ē,ē }

Anaplotheriidae [PALEONTOLOGY] A family of extinct tylopod ruminants in the superfamily Anaplotherioidea. { ,an·ə,pläth·ə'rī·ə,dē }

Anaplotherioidea [PALEONTOLOGY] A superfamily of extinct ruminant artiodactyls in the infraorder Tylopoda. { ,an·ə,pläth·ə,rē'ȯid·ē·ə }

anapolysis [INVERTEBRATE ZOOLOGY] Lifetime retention of ripe proglottids in some tapeworms. { ,an·ə'päl·ə·səs }

anapophysis [ANATOMY] An accessory process on the dorsal side of the transverse process of the lumbar vertebrae in humans and other mammals. { ,an·ə'päf·ə·səs }

Anapsida [VERTEBRATE ZOOLOGY] A subclass of reptiles characterized by a roofed temporal region in which there are no temporal openings. { ə'nap·sə·də }

Anasca [PALEONTOLOGY] A suborder of extinct bryozoans in the order Cheilostomata. { ə'nas·kə }

Anaspida [PALEONTOLOGY] An order of extinct fresh- or brackish-water vertebrates in the class Agnatha. { ə'nas·pə·də }

Anaspidacea [INVERTEBRATE ZOOLOGY] An order of the crustacean superorder Syncarida. { ə ¦nas·pə'dās·ē·ə }

Anaspididae [INVERTEBRATE ZOOLOGY] A family

of crustaceans in the order Anaspidacea. { ə ¦nas·pə'dī,dē }

anastral [CELL BIOLOGY] Lacking asters. { a'nas·trəl }

anastral mitosis [CELL BIOLOGY] Mitosis in which a spindle forms but no centrioles or asters are observed; typically occurs in plants. { ,ā¦nas·trəl mī'tō·səs }

Anatidae [VERTEBRATE ZOOLOGY] A family of waterfowl, including ducks, geese, mergansers, pochards, and swans, in the order Anseriformes. { ə 'nad·ə,dē }

anatomical dead space [ANATOMY] *See* dead space. { ,an·ə'täm·ə·kəl 'ded ,spās }

anatomical landmark [ANATOMY] *See* anatomical reference point. { ,an·ə¦täm·ə·kəl 'lan,märk }

anatomical position [ANATOMY] A reference posture used in anatomical description in which the subject stands erect against a wall with feet parallel and touching, and arms adducted and supinated, with palms facing forward. { ,an·ə ¦täm·ə·kəl pə'zi·shən }

anatomical reference point [ANATOMY] A prominent structure or feature of the human body that can be located and described by visual inspection or palpation at the body's surface; used to define movements and postures. Also known as anatomical landmark. { ,an·ə¦täm·ə·kəl 'ref·rəns ,pȯint }

anatomy [BIOLOGY] A branch of morphology dealing with the structure of animals and plants. { ə'nad·ə·mē }

anatomy of function [PHYSIOLOGY] A description of the changes in the configuration of limbs during the performance of a task, considered a subdiscipline of kinesiology. { ə¦nad·ə·mē əv 'fəŋk·shən }

anatropous [BOTANY] Having the ovule fully inverted so that the micropyle adjoins the funiculus. { ə'na·trə·pəs }

anautogenous insect [INVERTEBRATE ZOOLOGY] Any insect in which the adult female must feed before producing eggs. { ¦ā,nȯ¦täj·ə·nəs 'in ,sekt }

anaxial [BIOLOGY] Lacking an axis, therefore being irregular in form. { a'nak·sē·əl }

Ancalomicrobium [MICROBIOLOGY] A genus of prosthecate bacteria; nonmotile, unicellar forms with two to eight prosthecae per cell; reproduction is by budding of cells. { ,an¦kal·ō,mī¦krob·ē·əm }

ancestroecium [INVERTEBRATE ZOOLOGY] The tube that encloses an ancestrula. { ,an·səs 'trēsh·əm }

ancestrula [INVERTEBRATE ZOOLOGY] The first polyp of a bryozoan colony. { ,an'ses·trə·lə }

anchor [INVERTEBRATE ZOOLOGY] **1.** An anchor-shaped spicule in the integument of sea cucumbers. **2.** An anchor-shaped ossicle in echinoderms. { 'aŋ·kər }

anchorage-dependent cell [CELL BIOLOGY] A cell that grows, survives, or maintains function only when attached to an inert surface, such as glass or plastic. { ¦aŋ·kə,rij di¦pen·dənt 'sel }

anchovy [VERTEBRATE ZOOLOGY] Any member of the Engraulidae, a family of herringlike fishes harvested commercially for human consumption. { 'an,chō·vē }

ancipital [BOTANY] Having two edges, specifically referring to flattened stems, as of certain grasses. { an'sip·əd·əl }

ancora [INVERTEBRATE ZOOLOGY] The initial, anchor-shaped growth stage of graptolithinids. { 'aŋ·kə·rə }

ancyloid [INVERTEBRATE ZOOLOGY] A limpet-shaped or patelliform shell with the apex directed anteriorly. { 'an·sə,lȯid }

ancylopoda [PALEONTOLOGY] A suborder of extinct herbivorous mammals in the order Perissodactyla. { ,an·sə'lä·pə·də }

Ancylostoma [INVERTEBRATE ZOOLOGY] A genus of roundworms, commonly known as hookworms, in the order Ancylostomidae; parasites of humans, dogs, and cats. { ,aŋ·kə'läs·tə·mə }

Ancylostoma duodenale [INVERTEBRATE ZOOLOGY] The Old World hookworm, a human intestinal parasite that causes microcytic hypochromic anemia. { ,aŋ·kə'läs·tə·mə ¦dü·ə·di ¦näl }

Ancylostomidae [INVERTEBRATE ZOOLOGY] A family of nematodes belonging to the group Strongyloidea. { ¦aŋ·kə·lō·stä'mad·ə,dē }

Andreaeales [BOTANY] The single order of mosses of the subclass Andreaeobrya. { ,an·drē·ē'ā·lēz }

Andreaeceae [BOTANY] The single family of the Andreaeales, an order of mosses. { ,an·drē'ē·sē,ē }

Andreaeobrya [BOTANY] The granite mosses, a subclass of the class Bryopsida. { ,an·drē·ē'äb·rē·ə }

Andreaeopsida [BOTANY] A class of the plant division Bryophyta distinguished by longitudinal splitting of the mature capsule into four valves; commonly known as granite mosses. { ,an·drē·ē'äp·səd·ə }

Andrenidae [INVERTEBRATE ZOOLOGY] The mining or burrower bees, a family of hymenopteran insects in the superfamily Apoidea. { ,an'dren·ə ,dē }

androecious [BOTANY] Pertaining to plants that have only male flowers. { an'drē·shəs }

androecium [BOTANY] The aggregate of stamens in a flower. { ,an'drēsh·ē·əm }

androgen [BIOCHEMISTRY] A class of steroid hormones produced in the testis and adrenal cortex which act to regulate masculine secondary sexual characteristics. { 'an·drə·jən }

androgenesis [EMBRYOLOGY] Development of an embryo from a fertilized irradiated egg, involving only the male nucleus. { ,an·drə'jen·ə·səs }

androgenetic merogony [EMBRYOLOGY] The fertilization of egg fragments that lack a nucleus. { ¦an·drə·jə¦ned·ik mə'rä·gə·nē }

androgenic gland [INVERTEBRATE ZOOLOGY] A gland found in most malacostracan crustaceans and producing hormones that control the development of the testes and male sexual characteristics. { ¦an·drə¦jen·ik 'gland }

androgen unit [BIOLOGY] A unit for the standardization of male sex hormones. { 'an·drə·jən ,yü·nət }

android pelvis [ANATOMY] *See* masculine pelvis. { 'an,drȯid 'pel·vəs }

andromerogony [EMBRYOLOGY] Development of an egg fragment following cutting, shaking, or centrifugation of a fertilized or unfertilized egg. { ¦an,drō·mə'räg·ə·nē }

androphile [ECOLOGY] An organism, such as a mosquito, showing a preference for humans as opposed to animals. { 'an·drō,fīl }

androphore [BOTANY] A stalk that supports stamens or antheridia. [INVERTEBRATE ZOOLOGY] A gonophore in cnidarians in which only male elements develop. { 'an·drō,fȯr }

androsin [BIOCHEMISTRY] $C_{15}H_{20}O_8$ A glucoside found in the herb *Apocynum androsaemifolium*; yields glucose and acetovanillone on hydrolysis. { 'an·drə·sən }

androsperm [BIOLOGY] A sperm cell carrying a Y chromosome. { 'an·drə,spərm }

androstane [BIOCHEMISTRY] $C_{19}H_{32}$ The parent steroid hydrocarbon for all androgen hormones. Also known as etioallocholane. { 'an·drə,stān }

androstenedione [BIOCHEMISTRY] $C_{19}H_{26}O_2$ Any one of three isomeric androgens produced by the adrenal cortex. { ¦an·drō·stə¦ned·ē,ōn }

androsterone [BIOCHEMISTRY] $C_{19}H_{30}O_2$ An androgenic hormone occurring as a hydroxy ketone in the urine of men and women. { ,an'dräs·tə ,rōn }

Anelytropsidae [VERTEBRATE ZOOLOGY] A family of lizards represented by a single Mexican species. { ə,nel·ə'träp·sə,dē }

anemochory [ECOLOGY] Wind dispersal of plant and animal disseminules. { ə'nēm·ə,kȯr·ē }

anemophilous [BOTANY] Pollinated by wind-carried pollen. { ¦an·ə¦mäf·ə·ləs }

anemotaxis [BIOLOGY] Orientation movement of a free-living organism in response to wind. { ,an·ə·mō'tak·səs }

anemotropism [BIOLOGY] Orientation response of a sessile organism to air currents and wind. { ,an·ə'mä·trə,piz·əm }

anenterous [ZOOLOGY] Having no intestine, as a tapeworm. { a'nen·tə·rəs }

Anepitheliocystidia [INVERTEBRATE ZOOLOGY] A superorder of digenetic trematodes proposed by G. LaRue. { ¦an·ə·pə¦thē·lī·ō,sis'tid·ē·ə }

anergy [IMMUNOLOGY] The condition of exhibiting no response to an antigen or antibody. { 'a ,nər·jē }

anesthesia [PHYSIOLOGY] **1.** Insensibility, general or local, induced by anesthetic agents. **2.** Loss of sensation, of neurogenic or psychogenic origin. { ,an·əs'thēzh·ə }

anestrus [VERTEBRATE ZOOLOGY] A prolonged period of inactivity between two periods of heat in cyclically breeding female mammals. { a'nes·trəs }

aneucentric aberration [GENETICS] An aberration that results in a chromosome with more than one centromere. { ¦an·yə¦sen·trik ab·ə'rā·shən }

aneuploidy [GENETICS] Deviation from a normal haploid, diploid, or polyploid chromosome complement by the presence in excess of, or in defect of, one or more individual chromosomes. { 'a·nyü,plȯid·ē }

aneurine [BIOCHEMISTRY] *See* thiamine. { 'an·yə ,rēn }

aneusomatic organism [GENETICS] An organism whose cells contain variable numbers of individual chromosomes. { ¦an·yə·sō¦mad·ik 'ȯr·gə ,niz·əm }

angioblast [EMBRYOLOGY] A mesenchyme cell derived from extraembryonic endoderm that differentiates into embryonic blood cells and endothelium. { 'an·jē·ə,blast }

angiogenesis [EMBRYOLOGY] The origin and development of blood vessels. { ¦an·jē·ō¦jen·ə·səs }

angiokinesis [PHYSIOLOGY] *See* vasomotion. { ¦an·jē·ō·kə'nē·səs }

angiosperm [BOTANY] The common name for members of the plant division Magnoliophyta. { 'an·jē·ō,spərm }

Angiospermae [BOTANY] An equivalent name for the Magnoliophyta. { ¦an·jē·ə¦spər·mē }

angiotensin [BIOCHEMISTRY] A decapeptide hormone that influences blood vessel constriction and aldosterone secretion by the adrenal cortex. Also known as hypertensin. { ,an·jē·ə'ten·sən }

anglerfish [VERTEBRATE ZOOLOGY] Any of several species of the order Lophiiformes characterized by remnants of a dorsal fin seen as a few rays on top of the head that are modified to bear a terminal bulb. { 'aŋ·glər,fish }

Anguidae [VERTEBRATE ZOOLOGY] A family of limbless, snakelike lizards in the suborder Sauria, commonly known as slowworms or glass snakes. { 'aŋ·gwə,dē }

Anguilliformes [VERTEBRATE ZOOLOGY] A large order of actinopterygian fishes containing the true eels. { aŋ,gwil·ə'fȯr,mēz }

Anguilloidei [VERTEBRATE ZOOLOGY] The typical eels, a suborder of actinopterygian fishes in the order Anguilliformes. { aŋ·gwə'lȯid·ē,ī }

Anhimidae [VERTEBRATE ZOOLOGY] The screamers, a family of birds in the order Anseriformes characterized by stout bills, webbed feet, and spurred wings. { an'him·ə,dē }

Anhingidae [VERTEBRATE ZOOLOGY] The anhingas or snakebirds, a family of swimming birds in the order Pelecaniformes. { an,hin·jə,dē }

anhydrase [BIOCHEMISTRY] Any enzyme that catalyzes the removal of water from a substrate. { an 'hī,drās }

anhydrobiosis [PHYSIOLOGY] A type of cryptobiosis induced by dehydration. { ¦an,hī·drō ,bī 'ō·səs }

Aniliidae [VERTEBRATE ZOOLOGY] A small family of nonvenomous, burrowing snakes in the order Squamata. { ,an·əl'ī·ə,dē }

animal [ZOOLOGY] Any living organism distinguished from plants by the lack of chlorophyll, the requirement for complex organic nutrients, the lack of a cell wall, limited growth, mobility, and greater irritability. { 'an·ə·məl }

animal community [ECOLOGY] An aggregation of animal species held together in a continuous or discontinuous geographic area by ties to the same physical environment, mainly vegetation. { 'an·ə·məl kə'myü·nəd·ē }

animal ecology [ECOLOGY] A study of the relationships of animals to their environment. { 'an·ə·məl i'käl·ə·jē }

Animalia [SYSTEMATICS] The animal kingdom. { ˌan·ə'māl·yə }

animal kingdom [SYSTEMATICS] One of the two generally accepted major divisions of living organisms which live or have lived on earth (the other division being the plant kingdom). { 'an·ə·məl ˌkiŋ·dəm }

animal locomotion [ZOOLOGY] Progressive movement of an animal body from one point to another. { 'an·ə·məl ˌlō·kə'mō·shən }

animal pole [CELL BIOLOGY] The region of an ovum which contains the least yolk and where the nucleus gives off polar bodies during meiosis. { 'an·ə·məl ˌpōl }

animal virus [VIROLOGY] A small infectious agent able to propagate only within living animal cells. { 'an·ə·məl 'vī·rəs }

Anisakidae [INVERTEBRATE ZOOLOGY] A family of parasitic roundworms in the superfamily Ascaridoidea. { ˌan·ə'säk·əˌdē }

anise [BOTANY] The small fruit of the annual herb *Pimpinella anisum* in the family Umbelliferae; fruit is used for food flavoring, and oil is used in medicines, soaps, and cosmetics. { 'an·əs }

anisocarpous [BOTANY] Referring to a flower whose number of carpels is different from the number of stamens, petals, and sepals. { ¦a ˌnis·ə'kär·pəs }

anisochela [INVERTEBRATE ZOOLOGY] A chelate sponge spicule with dissimilar ends. { ¦aˌnis·ə 'kēl·ə }

anisodactylous [VERTEBRATE ZOOLOGY] Having unequal digits, especially referring to birds with three toes forward and one backward. { ¦aˌnis·ə 'dak·tə·ləs }

anisogamete [BIOLOGY] *See* heterogamete. { ¦a ˌnis·ə'gaˌmēt }

anisogamy [BIOLOGY] *See* heterogamy. { ¦aˌnis 'äg·ə·mē }

anisomerous [BOTANY] Referring to flowers that do not have the same number of parts in each whorl. { ˌaˌnī'säm·ə·rəs }

anisometric particle [VIROLOGY] Any unsymmetrical, rod-shaped plant virus. { ¦aˌnī·sə¦me·trik 'pärd·ə·kəl }

Anisomyaria [INVERTEBRATE ZOOLOGY] An order of mollusks in the class Bivalvia containing the oysters, scallops, and mussels. { ˌaˌnī·səˌmī'a·rē·ə }

anisophyllous [BOTANY] Having leaves of two or more shapes and sizes. { ¦aˌnī·sə¦fil·əs }

Anisoptera [INVERTEBRATE ZOOLOGY] The true dragonflies, a suborder of insects in the order Odonata. { ˌaˌnī'säp·tə·rə }

anisostemonous [BOTANY] Referring to a flower whose number of stamens is different from the number of carpels, petals, and sepals. { ¦aˌnī·sä 'stem·ə·nəs }

Anisotomidae [INVERTEBRATE ZOOLOGY] An equivalent name for Leiodidae. { ¦aˌnī·sə'täm·ə ˌdē }

anisotropy [BOTANY] The property of a plant that assumes a certain position in response to an external stimulus. [ZOOLOGY] The property of an egg that has a definite axis or axes. { ¦aˌnī'sä·trə·pē }

ankle [ANATOMY] The joint formed by the articulation of the leg bones with the talus, one of the tarsal bones. { 'aŋ·kəl }

Ankylosauria [PALEONTOLOGY] A suborder of Cretaceous dinosaurs in the reptilian order Ornithischia characterized by short legs and flattened, heavily armored bodies. { ¦aŋ·kə·lə'sȯr·ē·ə }

anlage [EMBRYOLOGY] Any group of embryonic cells when first identifiable as a future organ or body part. Also known as blastema; primordium. { 'änˌläg·ə }

annatto [BOTANY] *Bixa orellana*. A tree found in tropical America, characterized by cordate leaves and spinose, seed-filled capsules; a yellowish-red dye obtained from the pulp around the seeds is used as a food coloring. { ə'näd·ō }

anneal [GENETICS] To recombine strands of denatured bacterial deoxyribonucleic acid that were separated. { ə'nēl }

Annedidae [VERTEBRATE ZOOLOGY] A small family of limbless, snakelike, burrowing lizards of the suborder Sauria. { ə'ned·əˌdē }

Annelida [INVERTEBRATE ZOOLOGY] A diverse phylum comprising the multisegmented wormlike animals. { ə'nel·ə·də }

annidation [ECOLOGY] The phenomenon whereby a mutant is maintained in a population because it can flourish in an available ecological niche that the parent organisms cannot utilize. { ˌan·ə'dā·shən }

Anniellidae [VERTEBRATE ZOOLOGY] A family of limbless, snakelike lizards in the order Squamata. { ˌan·ē'el·əˌdē }

Annonaceae [BOTANY] A large family of woody flowering plants in the order Magnoliales, characterized by hypogynous flowers, exstipulate leaves, a trimerous perianth, and distinct stamens with a short, thick filament. { ˌa·nə'nās·ēˌē }

annual growth ring [BOTANY] *See* annual ring. { 'an·yə·wəl 'grōth ˌriŋ }

annual plant [BOTANY] A plant that completes its growth in one growing season and therefore must be planted annually. { 'an·yə·wəl 'plant }

annual ring [BOTANY] A line appearing on tree cross sections marking the end of a growing season and showing the volume of wood added during the year. Also known as annual growth ring. { 'an·yə·wəl 'riŋ }

annuation [ECOLOGY] The annual variation in the presence, absence, or abundance of a member of a plant community. { ˌan·yə'wā·shən }

annular budding [BOTANY] Budding by replacement of a ring of bark on a stock with a ring bearing a bud from a selected species or variety. { 'an·yə·lər 'bəd·iŋ }

annular vessel [BOTANY] A xylem tube or duct with internal lignified rings. { 'an·yə·lər 'ves·əl }

annulus [ANATOMY] Any ringlike anatomical part. [BOTANY] **1.** An elastic ring of cells between the operculum and the mouth of the capsule in mosses. **2.** A line of cells, partly or entirely surrounding the sporangium in ferns, which constricts, thus causing rupture of the sporangium to release spores. **3.** A whorl resembling a calyx at the base of the strobilus in certain horsetails. [MYCOLOGY] A ring of tissue representing the remnant of the veil around the stipe of some agarics. { 'an·yə·ləs }

Anobiidae [INVERTEBRATE ZOOLOGY] The deathwatch beetles, a family of coleopteran insects of the superfamily Bostrichoidea. { ˌan·ə'bī·əˌdē }

anole [VERTEBRATE ZOOLOGY] Any arboreal lizard of the genus *Anolis*, characterized by flattened adhesive digits and a prehensile outer toe. { ə'nō·lē }

Anomalinacea [INVERTEBRATE ZOOLOGY] A superfamily of marine and benthic sarcodinian protozoans in the order Foraminiferida. { əˌnäm·ə·lə'nās·ē·ə }

anomalous trichromatism [PHYSIOLOGY] A mild defect in red-green color vision in which the subject, when asked to mix red and green light to match a certain shade of yellow, produces a different shade than does someone with normal color vision. { ə'näm·ə·ləs trī'krōm·əˌtiz·əm }

Anomaluridae [VERTEBRATE ZOOLOGY] The African flying squirrels, a small family in the order Rodentia characterized by the climbing organ, a series of scales at the root of the tail. { ə¦näm·ə¦lür·əˌdē }

anomaly [BIOLOGY] An abnormal deviation from the characteristic form of a group. { ə'näm·ə·lē }

Anomocoela [VERTEBRATE ZOOLOGY] A suborder of toadlike amphibians in the order Anura characterized by a lack of free ribs. { ¦an·ə·mō¦sē·lə }

anomocoelous [ANATOMY] Describing a vertebra with a centrum that is concave anteriorly and flat or convex posteriorly. { ¦an·ə·mō¦sē·ləs }

Anomphalacea [PALEONTOLOGY] A superfamily of extinct gastropod mollusks in the order Aspidobranchia. { əˌnäm·fə' lāsh·ə }

Anomura [VERTEBRATE ZOOLOGY] A section of the crustacean order Decapoda that includes lobsterlike and crablike forms. { ˌan·ə'mür·ə }

Anopheles [INVERTEBRATE ZOOLOGY] A genus of mosquitoes in the family Culicidae; members are vectors of malaria, dengue, and filariasis. { ə 'näf·əˌlēz }

anopheline [INVERTEBRATE ZOOLOGY] Pertaining to mosquitoes of the genus *Anopheles* or a closely related genus. { ə'näf·ə·lən }

Anopla [INVERTEBRATE ZOOLOGY] A class or subclass of the phylum Rhynchocoela characterized by a simple tubular proboscis and by having the mouth opening posterior to the brain. { 'an·ə·plə }

Anoplocephalidae [INVERTEBRATE ZOOLOGY] A family of tapeworms in the order Cyclophyllidea. { ¦an·əˌplä·sə'fal·əˌdē }

Anoplura [INVERTEBRATE ZOOLOGY] The sucking lice, a small group of mammalian ectoparasites usually considered to constitute an order in the class Insecta. { ˌan·ə'plür·ə }

Anostraca [INVERTEBRATE ZOOLOGY] An order of shrimplike crustaceans generally referred to the subclass Branchiopoda. { ə'näs·trə·kə }

anoxybiosis [PHYSIOLOGY] A type of cryptobiosis induced by lack of oxygen. { aˌnäk·səˌbī'ō·səs }

Ansbacher unit [BIOLOGY] A unit for the standardization of vitamin K. { 'änzˌbäk·ər ˌyü·nət }

Anser [VERTEBRATE ZOOLOGY] A genus of birds in the family Anatidae comprising the typical geese. { 'an·sər }

Anseranatini [VERTEBRATE ZOOLOGY] A subfamily of aquatic birds in the family Anatidae represented by a single species, the magpie goose. { ˌan·sər·ə'nat·əˌnī }

Anseriformes [VERTEBRATE ZOOLOGY] An order of birds, including ducks, geese, swans, and screamers, characterized by a broad, flat bill and webbed feet. { ˌan·sər·ə'förˌmēz }

Anson unit [BIOLOGY] A unit for the standardization of trypsin and proteinases. { 'an·sən ˌyü·nət }

ant [INVERTEBRATE ZOOLOGY] The common name for insects in the hymenopteran family Formicidae; all are social, and colonies exhibit a highly complex organization. { ant }

antagonism [BIOLOGY] **1.** Mutual opposition as seen between organisms, muscles, physiologic actions, and drugs. **2.** Opposing action between drugs and disease or drugs and functions. { an 'tag·əˌniz·əm }

antagonist [BIOCHEMISTRY] A molecule that bears sufficient structural similarity to a second molecule to compete with that molecule for binding sites on a third molecule. [PHYSIOLOGY] A muscle that contracts with, and limits the action of, another muscle, called an agonist, with which it is paired. { an'tag·əˌnist }

antarctic faunal region [ECOLOGY] A zoogeographic region describing both the marine littoral and terrestrial animal communities on and around Antarctica. { ˌant'ärd·ik 'fön·əl ˌrē·jən }

anteater [VERTEBRATE ZOOLOGY] Any of several mammals, in five orders, which live on a diet of ants and termites. { 'antˌēd·ər }

antebrachium [ANATOMY] *See* forearm. { ˌan·tə 'brāk·ē·əm }

antecosta [INVERTEBRATE ZOOLOGY] The internal, anterior ridge of the tergum or sternum of many insects that provides a surface for attachment of the longitudinal muscles. { 'an·təˌkäs·tə }

antelope [VERTEBRATE ZOOLOGY] Any of the hollow-horned, hoofed ruminants assigned to the artiodactyl subfamily Antilopinae; confined to Africa and Asia. { 'an·təlˌōp }

antenna [INVERTEBRATE ZOOLOGY] Any one of the paired, segmented, and movable sensory appendages occurring on the heads of many arthropods. { an'ten·ə }

antenna chlorophyll [BIOCHEMISTRY] Chlorophyll molecules which collect light quanta. { an'ten·ə 'klȯr·ə,fil }

antennal gland [INVERTEBRATE ZOOLOGY] An excretory organ in the cephalon of adult crustaceans and best developed in the Malacostraca. Also known as green gland. { an'ten·ə ,gland }

Antennata [INVERTEBRATE ZOOLOGY] An equivalent name for the Mandibulata. { ,an·tə'näd·ə }

antenodal [INVERTEBRATE ZOOLOGY] Before or in front of the nodus, a cross vein near the middle of the costal border of the wing of dragonflies. { ,an·tē'nōd·əl }

anter [INVERTEBRATE ZOOLOGY] Part of a bryozoan operculum which serves to close off a portion of the operculum. { 'an·tər }

anteriad [ZOOLOGY] Toward the anterior portion of the body. { an'tir·ē,ad }

anterior [ZOOLOGY] Situated near or toward the front or head of an animal body. { an'tir·ē·ər }

anterior commissure [ANATOMY] A bundle of nerve fibers that cross the midline of the brain in front of the third ventricle and serve to connect parts of the cerebral hemispheres. { an'tir·ē·ər 'käm·ə·shu̇r }

anteromedial [ZOOLOGY] Anterior and toward the middle of the body. { ¦an·tə·rə¦mēd·ē·əl }

antetheca [PALEONTOLOGY] The last or exposed septum at any stage of fusulinid growth. { ¦an·tē¦thek·ə }

anthelate [BOTANY] An open, paniculate cyme. { 'an·thə,lāt }

anther [BOTANY] The pollen-producing structure of a flower. { 'an·thər }

antheraxanthin [BIOCHEMISTRY] A neutral yellow plant pigment unique to the Euglenophyta. { ¦an·thər,aks¦an·thən }

anther culture [BOTANY] A haploid tissue culture derived from anthers or pollen cells. { 'an·thər ,kəl·chər }

antheridiophore [BOTANY] A specialized stemlike structure that supports an antheridium in some mosses and liverworts. { ,an·thə'rid·ē·ə·fȯr }

antheridium [BOTANY] **1.** The sex organ that produces male gametes in cryptogams. **2.** A minute structure within the pollen grain of seed plants. { ,an·thə'rid·ē·əm }

antheriferous [BOTANY] Anther-bearing. { ,an·thə'rif·ə·rəs }

antherozoid [BOTANY] A motile male gamete produced by plants. { ,an·thə·rə'zō·əd }

anther smut [MYCOLOGY] *Ustilago violacea*. A smut fungus that attacks certain plants and forms spores in the anthers. { 'an·thər ,smət }

anthesis [BOTANY] The flowering period in plants. { an'thē·səs }

Anthicidae [INVERTEBRATE ZOOLOGY] The antlike flower beetles, a family of coleopteran insects in the superfamily Tenebrionoidea. { an'this·ə ,dē }

anthill [INVERTEBRATE ZOOLOGY] A mound of earth deposited around the entrance to ant and termite nests in the ground. { 'ant,hil }

anthoblast [INVERTEBRATE ZOOLOGY] A developmental stage of some corals; produced by budding. { 'an·thə,blast }

anthocarpous [BOTANY] Describing fruits having accessory parts. { ¦an·thə¦kär·pəs }

anthocaulis [INVERTEBRATE ZOOLOGY] The stemlike basal portion of some solitary corals; the oral portion becomes a new zooid. { ¦an·thə'kȯl·əs }

Anthocerotae [BOTANY] A small class of the plant division Bryophyta, commonly known as hornworts or horned liverworts. { ,an·thə'ser·ə ,tē }

anthocodium [INVERTEBRATE ZOOLOGY] The free, oral end of an anthozoan polyp. { ,an·thə'kōd·ē·əm }

Anthocoridae [INVERTEBRATE ZOOLOGY] The flower bugs, a family of hemipteran insects in the superfamily Cimicimorpha. { ,an·thə'kȯr·ə,dē }

anthocyanidin [BIOCHEMISTRY] Any of the colored aglycone plant pigments obtained by hydrolysis of anthocyanins. { ,an·thə,sī'an·ə·dən }

anthocyanin [BIOCHEMISTRY] Any of the intensely colored, sap-soluble glycoside plant pigments responsible for most scarlet, purple, mauve, and blue coloring in higher plants. { ,an·thə'sī·ə·nən }

Anthocyathea [PALEONTOLOGY] A class of extinct marine organisms in the phylum Archaeocyatha characterized by skeletal tissue in the central cavity. { ,an·thə,sī'ā·thē·ə }

anthocyathus [INVERTEBRATE ZOOLOGY] The oral portion of the anthocaulus of some solitary corals that becomes a new zooid. { an·thə,sī'ā·thəs }

Anthomedusae [INVERTEBRATE ZOOLOGY] A suborder of hydrozoan cnidarians in the order Hydroida characterized by athecate polyps. { ,an·thō·mi'dü·sē }

Anthomyzidae [INVERTEBRATE ZOOLOGY] A family of cyclorrhaphous myodarian dipteran insects belonging to the subsection Acalypteratae. { ,an·thō'mīz·ə,dē }

anthophagous [ZOOLOGY] Feeding on flowers. { an'thä·fə·gəs }

anthophore [BOTANY] A stalklike extension of the receptacle bearing the pistil and corolla in certain plants. { 'an·thə,fȯr }

Anthosomidae [INVERTEBRATE ZOOLOGY] A family of fish ectoparasites in the crustacean suborder Caligoida. { ,an·thə'säm·ə,dē }

anthostele [INVERTEBRATE ZOOLOGY] A thick-walled, nonretractile aboral region of certain cnidarians. { 'an·thə,stēl }

Anthozoa [INVERTEBRATE ZOOLOGY] A class of marine organisms in the phylum Cnidaria including the soft, horny, stony, and black corals, the sea pens, and the sea anemones. { ,an·thō 'zō·ə }

anthozooid [INVERTEBRATE ZOOLOGY] Any of the individual zooids of a compound anthozoan. { ,an·thō'zō,ȯid }

Anthracosauria [PALEONTOLOGY] An order of Carboniferous and Permian labyrinthodont amphibians that includes the ancestors of living reptiles. { ,an·thrə·kə'sȯr·ē·ə }

Anthracotheriidae [PALEONTOLOGY] A family of middle Eocene and early Pleistocene artiodactyl mammals in the superfamily Anthracotherioidea. { ˌan·thrə·kə·thəˈrī·əˌdē }

Anthracotherioidea [PALEONTOLOGY] A superfamily of extinct artiodactyl mammals in the suborder Paleodonta. { ˈan·thrə·kə·thəˌrīˈȯid·ē·ə }

anthraquinone pigments [BIOCHEMISTRY] Coloring materials which occur in plants, fungi, lichens, and insects; consists of about 50 derivatives of the parent compound, anthraquinone. { ˌan·thrə·kwiˈnōn ˈpig·məns }

Anthribidae [INVERTEBRATE ZOOLOGY] The fungus weevils, a family of coleopteran insects in the superfamily Curculionoidea. { anˈthrib·əˌdē }

anthropochory [ECOLOGY] Dispersal of plant and animal disseminules by humans. { ¦an·thrə·pə¦kȯr·ē }

anthropodesoxycholic acid [BIOCHEMISTRY] *See* chenodeoxycholic acid. { ¦an·thrəˌpō·de¦zäk·səˈkäl·ik ˈas·əd }

anthropogenic [ECOLOGY] Referring to environmental alterations resulting from the presence or activities of humans. { ¦an·thrə·pə¦jen·ik }

anthropoid [VERTEBRATE ZOOLOGY] Pertaining to or resembling the Anthropoidea. { ˈan·thrəˌpȯid }

Anthropoidea [VERTEBRATE ZOOLOGY] A suborder of mammals in the order Primates including New and Old World monkeys. { ˌan·thrəˈpȯid·ē·ə }

anthropology [BIOLOGY] The study of the interrelations of biological, cultural, geographical, and historical aspects of humankind. { ˌan·thrəˈpäl·ə·jē }

anthroposphere [ECOLOGY] The biosphere of the great geological activities of humankind. Also known as noosphere. { anˈthrä·pəˌsfir }

Anthuridea [INVERTEBRATE ZOOLOGY] A suborder of crustaceans in the order Isopoda characterized by slender, elongate, subcylindrical bodies, and by the fact that the outer branch of the paired tail appendage (uropod) arches over the base of the terminal abdominal segment, the telson. { ˌan·thəˈrīd·ē·ə }

antiagglutinin [IMMUNOLOGY] A substance that neutralizes a corresponding agglutinin. { ˌan·tē·əˈglüt·ən·ən }

antiaggressin [IMMUNOLOGY] An antibody that neutralizes aggressin, a substance produced by pathogenic microorganisms to enhance virulence. { ˌan·tē·əˈgres·ən }

antianaphylaxis [IMMUNOLOGY] A condition in which a sensitized animal resists anaphylaxis. { ¦an·tēˌan·ə·fəˈlak·səs }

Antiarchi [PALEONTOLOGY] A division of highly specialized placoderms restricted to fresh-water sediments of the Middle and Upper Devonian. { ˌan·tēˈärˌkī }

antiauxin [BIOCHEMISTRY] A molecule that competes with an auxin for receptor sites. { ¦an·tēˈȯk·sən }

antibacterial agent [MICROBIOLOGY] A synthetic or natural compound which inhibits the growth and division of bacteria. { ¦an·tēˌbakˈtir·ē·əl ˈā·jənt }

antibiosis [ECOLOGY] Antagonistic association between two organisms in which one is adversely affected. { ¦an·tēˌbī¦ō·səs }

antibiotic [MICROBIOLOGY] A chemical substance, produced by microorganisms and synthetically, that has the capacity in dilute solutions to inhibit the growth of, and even to destroy, bacteria and other microorganisms. { ¦an·tēˌbī¦äd·ik }

antibiotic assay [MICROBIOLOGY] A method for quantitatively determining the concentration of an antibiotic by its effect in inhibiting the growth of a susceptible microorganism. { ¦an·tēˌbī¦äd·ik ˈaˌsā }

antibody [IMMUNOLOGY] A protein, found principally in blood serum, originating either normally or in response to an antigen and characterized by a specific reactivity with its complementary antigen. Also known as immune body. { ˈan·təˌbäd·ē }

antibody binding site [MOLECULAR BIOLOGY] *See* antibody combining site. { ¦an·təˌbäd·ē ˈbīnd·iŋ ˌsīt }

antibody combining site [MOLECULAR BIOLOGY] **1.** The portion of an antibody molecule that makes physical contact with the corresponding antigenic determinant. **2.** The portion of an antigen that makes physical contact with the corresponding antibody. Also known as antibody binding site. { ¦an·təˌbäd·ē kəmˈbīn·iŋ ˌsīt }

antibody-mediated immunity [IMMUNOLOGY] *See* humoral immunity. { ¦an·təˌbäd·ē ¦mēd·ēˌād·əd iˈmyün·əd·ē }

antiboreal faunal region [ECOLOGY] A zoogeographic region including marine littoral faunal communities at the southern end of South America. { ˈan·tēˌbȯr·ē·əl ˈfȯn·əl ˌrē·jən }

anticholinesterase [BIOCHEMISTRY] Any agent, such as a nerve gas, that inhibits the action of cholinesterase and thereby destroys or interferes with nerve conduction. { ˌan·tiˌkō·ləˈnes·təˌrās }

anticlinal [BOTANY] Pertaining to a cell layer that runs at right angles across the circumference of a plant part. { ¦an·tē¦klīn·əl }

anticoding strand [MOLECULAR BIOLOGY] *See* antisense strand. { ˌan·tēˈkōd·iŋ ˌstrand }

anticodon [GENETICS] A three-nucleotide sequence of transfer RNA that complements the codon in messenger RNA. { ¦an·tē¦kōˌdän }

anticryptic [ECOLOGY] Pertaining to protective coloration that makes an animal resemble its surroundings so that it is inconspicuous to its prey. { ¦an·tē¦krip·tik }

anticusp [INVERTEBRATE ZOOLOGY] An anterior, downward projection in conodonts. { ˈan·tēˌkəsp }

antidiuretic hormone [BIOCHEMISTRY] *See* vasopressin. { ¦an·tēˌdī·yə¦red·ik ˈhȯrˌmōn }

antienzyme [BIOCHEMISTRY] An agent that selectively inhibits the action of an enzyme. { ˌan·tēˈenˌzīm }

antifertilizin [BIOCHEMISTRY] An immunologically specific substance produced by animal sperm to implement attraction by the egg before fertilization. { ¦an·tē·fərˈtil·ə·zən }

antifibrinolysin [BIOCHEMISTRY] Any substance

that inhibits the proteolytic action of fibrinolysin. { ¦an·tē‚fī·brə'näl·ə·sən }

antigen [IMMUNOLOGY] A substance which reacts with the products of specific humoral or cellular immunity, even those induced by related heterologous immunogens. { 'an·tə·jən }

antigenaemia [IMMUNOLOGY] A condition in which viral antigen is present in the blood; occurs in viral hepatitis and may occur in smallpox, myxomatosis, and yellow fever. { ‚an·tə·jə'nē·mē·ə }

antigen-antibody reaction [IMMUNOLOGY] The combination of an antigen with its antibody. { 'an·tə·jən ¦an·tə¦bäd·ē rē'ak·shən }

antigenic competition [IMMUNOLOGY] A decrease in immune response to one antigenic peptide due to a concurrent immune response to a different antigenic peptide. { ‚an·tə¦jen·ik ‚käm·pə'tish·ən }

antigenic determinant [IMMUNOLOGY] The portion of an antigen molecule that determines the specificity of the antigen-antibody reaction. { ‚an·tə¦jen·ik di'tər·mə·nənt }

antigenic drift [IMMUNOLOGY] Minor change of an antigen on the surface of a pathogenic microorganism. { ‚an·tə¦jen·ik 'drift }

antigenicity [IMMUNOLOGY] Ability of an antigen to induce an immune response and combine with specific antibodies or T-cell receptors. { ‚an·tə·jə'nis·əd·ē }

antigenic mimicry [IMMUNOLOGY] Acquisition or production of host antigens by a parasite, enabling it to avoid detection by the host's immune system. { 'an·tə¦jen·ik 'mim·ə·krē }

antigenic modulation [IMMUNOLOGY] Loss of detectable antigen from the surface of a cell after incubation with antibodies. { ‚an·tə¦jen·ik ‚mäj·ə'lā·shən }

antigenic shift [VIROLOGY] An abrupt major change in the antigenicity of a virus; believed to result from recombination of genes. { ‚an·tə ¦jen·ik 'shift }

antigenic variation [IMMUNOLOGY] Alteration of an antigen on the surface of a microorganism; may enable a pathogenic mocroorganism to evade destruction by the host's immune system. { ‚an·tə¦jen·ik ‚ver·ē'ā·shən }

antigen presentation [IMMUNOLOGY] The process whereby a cell expresses antigen on its surface in a form that can be recognized by a T lymphocyte. { ¦an·tə·jən ‚prē·zən'tā·shən }

antihemophilic factor [BIOCHEMISTRY] A soluble protein clotting factor in mammalian blood. Also known as factor VIII; thromboplastinogen. { ¦an·tē‚hē·mə'fil·ik ‚fak·tər }

antihemorrhagic vitamin [BIOCHEMISTRY] *See* vitamin K. { ¦an·tē‚hem·ə'raj·ik 'vīd·ə·mən }

anti-immunoglobulin antibody [IMMUNOLOGY] An antibody produced in response to a foreign antibody introduced into an experimental animal. Abbreviated AIA. { ¦an·tē‚im·yə·nō¦gläb·yə·lən 'an·tē‚bäd·ē }

anti-infective vitamin [BIOCHEMISTRY] *See* vitamin A. { ‚an·tē·in'fek·div ‚vīd·ə·mən }

Antilocapridae [VERTEBRATE ZOOLOGY] A family of artiodactyl mammals in the superfamily Bovoidea; the pronghorn is the single living species. { ‚an·tə‚lō'kap·rə‚dē }

Antilopinae [VERTEBRATE ZOOLOGY] The antelopes, a subfamily of artiodactyl mammals in the family Bovidae. { ‚an·tə'lōp·ə‚nē }

antilymphocyte serum [IMMUNOLOGY] An immunosuppressive agent effective in prolonging the lives of homografts in experimental animals by reducing the circulating lymphocytes. { ¦an·tē'lim·fə‚sīt ‚sir·əm }

antilysin [IMMUNOLOGY] A substance antagonistic to the action of a lysin. { 'an·tē¦lī·sən }

antimere [INVERTEBRATE ZOOLOGY] Any one of the equivalent parts into which a radially symmetrical animal may be divided. { 'an·tē‚mir }

antimicrobial agent [MICROBIOLOGY] A chemical compound that either destroys or inhibits the growth of microscopic and submicroscopic organisms. { ‚an·tē‚mī'krōb·ē·əl ‚ā·jənt }

antimutagen [GENETICS] A compound that is antagonistic to the action of mutagenic agents on bacteria. { ¦an·tē'myüd·ə·jən }

antinuclear antibody [IMMUNOLOGY] Antibody to deoxyribonucleic acid, ribonucleic acid, histone, or nonhistone proteins found in the serum of individuals with certain autoimmune diseases. Abbreviated ANA. { ¦an·tē¦nü·klē·ər 'an·tə‚bäd·ē }

antioncogene [GENETICS] Any of a class of genes that are involved in the negative regulation of normal growth; the loss of these genes leads to malignant growth. { ¦an·tē'aŋ·kə‚jēn }

antiparallel [GENETICS] Pertaining to parallel molecules that point in opposite directions, as the strands of deoxyribonucleic acid. { ¦an·tē 'par·ə‚lel }

Antipatharia [INVERTEBRATE ZOOLOGY] The black or horny corals, an order of tropical and subtropical cnidarians in the subclass Zoantharia. { ‚an·tē·pə'thar·ē·ə }

antipetalous [BOTANY] Having stamens positioned opposite to, rather than alternating with, the petals. { ¦an·ti¦ped·ə·ləs }

antipodal [BOTANY] Any of three cells grouped at the base of the embryo sac, that is, at the end farthest from the micropyle, in most angiosperms. { an'tip·əd·əl }

anti-Rh agglutinin [IMMUNOLOGY] An antibody against any Rh antigen; it must be acquired and is never natural. { 'an·tē‚är¦āch ə'glüt·ən·ən }

anti-Rh immunoglobulin [IMMUNOLOGY] A serum protein that destroys Rh-positive fetal erythrocytes in an Rh-negative mother when administered after delivery. { ¦an·tē‚är¦āch ‚im·yə·nə 'gläb·yə·lən }

anti-Rh serum [IMMUNOLOGY] A blood serum containing anti-Rh antibodies. { ¦an·tē‚är¦āch 'sir·əm }

antirostrum [INVERTEBRATE ZOOLOGY] The terminal segment of the appendages of certain mites. { ‚an·tē'räs·trəm }

antisense [GENETICS] A strand of deoxyribonucleic acid having a sequence identical to messenger ribonucleic acid. { ¦an·tē'sens }

antisense strand [MOLECULAR BIOLOGY] The strand of a double-stranded deoxyribonucleic acid molecule from which ribonucleic acid is transcribed. Also known as anticoding strand. { ¦an·tē'sens ˌstrand }

antiserum [IMMUNOLOGY] Any immune serum that contains antibodies active chiefly in destroying a specific infecting virus or bacterium. { 'an·tēˌsir·əm }

antismallpox vaccine [IMMUNOLOGY] *See* smallpox vaccine. { ¦an·tē¦smȯlˌpäks ˌvak'sēn }

antistreptolysin [IMMUNOLOGY] The antibody that neutralizes the streptolysin of group A hemolytic streptococci. { ˌan·tēˌstrep·tə'līs·ən }

antitermination factor [BIOCHEMISTRY] Protein that interferes with normal termination of ribonucleic acid synthesis. { ˌan·tēˌtər·mə'nā·shən 'fak·tər }

antithrombin [BIOCHEMISTRY] A substance in blood plasma that inactivates thrombin. { ˌan·tē 'thräm·bən }

antitoxin [IMMUNOLOGY] An antibody elaborated by the body in response to a bacterial toxin that will combine with and generally neutralize the toxin. { ˌan·tē'täk·sən }

antitumor antibiotic [MICROBIOLOGY] A substance, such as actinomycin, luteomycin, or mitomycin C, which is produced by microorganisms and is effective against some forms of cancer. { ˌan·tē'tüm·ər ˌan·tēˌbī'äd·ik }

antivenin [IMMUNOLOGY] An immune serum that neutralizes the venoms of certain poisonous snakes and black widow spiders. { ¦an·tē¦ven·ən }

antivernalization [BOTANY] Delayed flowering in plants due to treatment with heat. { ˌan·tē ˌvərn·əl·ə'zā·shən }

antivitamin [BIOCHEMISTRY] Any substance that prevents a vitamin from normal metabolic functioning. { ˌan·tē'vīd·ə·mən }

antixerophthalmic vitamin [BIOCHEMISTRY] *See* vitamin A. { ¦an·tēˌzirˌäf'thal·mik 'vīd·ə·mən }

antler [VERTEBRATE ZOOLOGY] One of a pair of solid bony, usually branched outgrowths on the head of members of the deer family (Cervidae); shed annually. { 'ant·lər }

ant lion [INVERTEBRATE ZOOLOGY] The common name for insects of the family Myrmeleontidae in the order Neuroptera; larvae are commonly called doodlebugs. { 'ant ˌlī·ən }

antrorse [BIOLOGY] Turned or directed forward or upward. { 'anˌtrȯrs }

antrum [ANATOMY] A cavity of a hollow organ or a sinus. { 'an·trəm }

Anura [VERTEBRATE ZOOLOGY] An order of the class Amphibia comprising the frogs and toads. { ə'nu̇r·ə }

anus [ANATOMY] The posterior orifice of the alimentary canal. { 'ā·nəs }

anvil [ANATOMY] *See* incus. { 'an·vəl }

aorta [ANATOMY] The main vessel of systemic arterial circulation arising from the heart in vertebrates. [INVERTEBRATE ZOOLOGY] The large dorsal or anterior vessel in many invertebrates. { ā 'ȯrd·ə }

aortic arch [ANATOMY] The portion of the aorta extending from the heart to the third thoracic vertebra; single in warm-blooded vertebrates and paired in fishes, amphibians, and reptiles. { ā'ȯrd·ik 'ärch }

aortic paraganglion [ANATOMY] A structure in vertebrates belonging to the chromaffin system and found on the front of the abdominal aorta near the mesenteric arteries. Also known as aortic body; organs of Zuckerkandl. { ā'ȯrd·ik ˌpa·rə'gaŋ·glēˌän }

aortic valve [ANATOMY] A heart valve comprising three flaps which guards the passage from the left ventricle to the aorta and prevents the backward flow of blood. { ā'ȯrd·ik 'valv }

apandrous [BOTANY] Lacking male organs or having nonfunctional male organs. { ˌa'pan·drəs }

Apatemyidae [PALEONTOLOGY] A family of extinct rodentlike insectivorous mammals belonging to the Proteutheria. { əˌpad·ə'mī·əˌdē }

apatetic [ECOLOGY] Pertaining to the imitative protective coloration of an animal subject to being preyed upon. { ¦a·pə¦ted·ik }

Apathornithidae [PALEONTOLOGY] A family of Cretaceous birds, with two species, belonging to the order Ichthyornithiformes. { ˌa·pəˌthȯr'nith·əˌdē }

ape [VERTEBRATE ZOOLOGY] Any of the tailless primates of the families Hylobatidae and Pongidae in the same superfamily as humans. { āp }

apetalous [BOTANY] Lacking petals. { ˌā'ped·əl·əs }

apex [ANATOMY] **1.** The upper portion of a lung extending into the root. **2.** The pointed end of the heart. **3.** The tip of the root of a tooth. [BOTANY] The pointed tip of a leaf. { 'āˌpeks }

apex impulse [PHYSIOLOGY] The point of maximum outward movement of the left ventricle of the heart during systole, normally localized in the fifth left intercostal space in the midclavicular line. Also known as left ventricular thrust. { 'āˌpeks ¦imˌpəls }

Aphanomyces [MYCOLOGY] A genus of fungi in the phycomycetous order Saprolegniales; species cause root rot in plants. { 'af·ə·nə'mīˌsēz }

Aphasmidea [INVERTEBRATE ZOOLOGY] An equivalent name for the Adenophorea. { ¦aˌfaz'mid·ē·ə }

Aphelenchoidea [INVERTEBRATE ZOOLOGY] A superfamily of plant and insect-associated nematodes in the order Tylenchida. { ˌaf·əˌleŋ'kȯid·ē·ə }

Aphelenchoidoidea [INVERTEBRATE ZOOLOGY] A superfamily of parasitic nematodes containing only one family, characterized by the lack of an isthmus in the esophagus and, in males, thorn-shaped spicules. { ˌaf·əˌleŋˌkȯi'dȯid·ē·ə }

Aphelocheiridae [INVERTEBRATE ZOOLOGY] A family of hemipteran insects belonging to the superfamily Naucoroidea. { ¦af·əˌläk·ə'rīˌdē }

aphid [INVERTEBRATE ZOOLOGY] The common name applied to the soft-bodied insects of the family Aphididae; they are phytophagous plant pests and vectors for plant viruses and fungal parasites. { ā·fəd }

Aphididae [INVERTEBRATE ZOOLOGY] The true

aphids, a family of homopteran insects in the superfamily Aphidoidea. { ə'fid·ə,dē }

Aphidoidea [INVERTEBRATE ZOOLOGY] A superfamily of sternorrhynchan insects in the order Homoptera. { ,a·fə'dȯid·ē·ə }

Aphis [INVERTEBRATE ZOOLOGY] A genus of aphid, the type genus of the family Aphididae. { 'ā·fəs }

Aphredoderidae [VERTEBRATE ZOOLOGY] A family of actinopterygian fishes in the order Percopsiformes containing one species, the pirate perch. { ,a·frə·də'der·ə,dē }

aphrodisiac [PHYSIOLOGY] Any chemical agent or odor that stimulates sexual desires. { ,af·rə'dē·zē,ak }

Aphroditidae [INVERTEBRATE ZOOLOGY] A family of scale-bearing polychaete worms belonging to the Errantia. { ,af·rə'did·ə,dē }

Aphrosalpingoidea [PALEONTOLOGY] A group of middle Paleozoic invertebrates classified with the calcareous sponges. { ¦af·rō,sal,piŋ'gȯid·ē·ə }

Aphylidae [INVERTEBRATE ZOOLOGY] An Australian family of hemipteran insects composed of two species; not placed in any higher taxonomic group. { ə'fil·ə,dē }

aphyllous [BOTANY] Lacking foliage leaves. { ā'fil·əs }

aphytic zone [ECOLOGY] The part of a lake floor that lacks plants because it is too deep for adequate light penetration. { ā'fid·ik ,zōn }

apical [BOTANY] Relating to the apex or tip. { 'ap·i·kəl }

apical bud [BOTANY] *See* terminal bud. { 'ap·i·kəl ,bəd }

apical dominance [BOTANY] Inhibition of lateral bud growth by the apical bud of a shoot, believed to be a response to auxins produced by the apical bud. { 'ap·i·kəl 'däm·ə·nəns }

apicalia [INVERTEBRATE ZOOLOGY] Paired sensory cilia on the head of gnathostomulids. { ¦ap·ə¦kal·yə }

apical meristem [BOTANY] A region of embryonic tissue occurring at the tips of roots and stems. Also known as promeristem. { 'ap·i·kəl 'mer·ə,stem }

apical plate [INVERTEBRATE ZOOLOGY] A group of cells at the anterior end of certain trochophore larvae; believed to have nervous and sensory functions. { 'ap·i·kəl 'plāt }

apiculate [BOTANY] Ending abruptly in a short, sharp point. { ə'pik·yə·lət }

Apidae [INVERTEBRATE ZOOLOGY] A family of hymenopteran insects in the superfamily Apoidea including the honeybees, bumblebees, and carpenter bees. { 'a·pə,dē }

Apioceridae [INVERTEBRATE ZOOLOGY] A family of orthorrhaphous dipteran insects in the series Brachycera. { ,ap·ē·ō'ser·ə,dē }

apiology [INVERTEBRATE ZOOLOGY] The scientific study of bees, particularly honeybees. { ,ā·pē'äl·ə·jē }

Apis [INVERTEBRATE ZOOLOGY] A genus of bees, the type genus of the Apidae. { 'ā·pəs }

Apistobranchidae [INVERTEBRATE ZOOLOGY] A family of spioniform annelid worms belonging to the Sedentaria. { ə¦pis·tə¦braŋk·ə,dē }

Aplacophora [INVERTEBRATE ZOOLOGY] A subclass of vermiform mollusks in the class Amphineura characterized by no shell and calcareous integumentary spicules. { ¦ā,pla'käf·ə·rə }

aplanogamete [BIOLOGY] A gamete that lacks motility. { ā'plan·ə·gə,mēt }

aplanospore [MYCOLOGY] A nonmotile, asexual spore, usually a sporangiospore, common in the Phycomycetes. { ā'plan·ə,spȯr }

aplysiatoxin [BIOCHEMISTRY] A bislactone toxin produced by the blue-green alga *Lyngbya majuscula*. { ə¦plīzh·ə¦tak·sən }

Apneumonomorphae [INVERTEBRATE ZOOLOGY] A suborder of arachnid arthropods in the order Araneida characterized by the lack of book lungs. { ā,nü·mə,nō'mȯr,fē }

apneusis [PHYSIOLOGY] In certain lower vertebrates, sustained tonic contraction of the respiratory muscles to allow prolonged inspiration. { ap'nü·səs }

apocarpous [BOTANY] Having carpels separate from each other. { ¦ap·ə¦kär·pəs }

apocrine gland [PHYSIOLOGY] A multicellular gland, such as a mammary gland or an axillary sweat gland, that extrudes part of the cytoplasm with the secretory product. { 'ap·ə·krən ,gland }

Apocynaceae [BOTANY] A family of tropical and subtropical flowering trees, shrubs, and vines in the order Gentianales, characterized by a well-developed latex system, granular pollen, a poorly developed corona, and the carpels often united by the style and stigma; well-known members are oleander and periwinkle. { ə,päs·ə'nās·ē,ē }

Apoda [VERTEBRATE ZOOLOGY] The caecilians, a small order of wormlike, legless animals in the class Amphibia. { 'a·pəd·ə }

Apodacea [INVERTEBRATE ZOOLOGY] A subclass of echinoderms in the class Holothuroidea characterized by simple or pinnate tentacles and reduced or absent tube feet. { ,a·pə'dās·ē·ə }

apodeme [INVERTEBRATE ZOOLOGY] An internal ridge or process on an arthropod exoskeleton to which organs and muscles attach. { 'ap·ə,dēm }

Apodes [VERTEBRATE ZOOLOGY] An equivalent name for the Anguilliformes. { 'ap·ə,dēz }

Apodi [VERTEBRATE ZOOLOGY] The swifts, a suborder of birds in the order Apodiformes. { 'ap·ə,dī }

Apodida [INVERTEBRATE ZOOLOGY] An order of worm-shaped holothurian echinoderms in the subclass Apodacea. { ə'päd·ə·də }

Apodidae [VERTEBRATE ZOOLOGY] The true swifts, a family of apodiform birds belonging to the suborder Apodi. { ə'päd·ə,dē }

Apodiformes [VERTEBRATE ZOOLOGY] An order of birds containing the hummingbirds and swifts. { ə,päd·ə'fȯr,mēz }

apoenzyme [BIOCHEMISTRY] The protein moiety of an enzyme; determines the specificity of the enzyme reaction. { ¦a·pō¦en,zīm }

apoferritin [BIOCHEMISTRY] A protein found in intestinal mucosa cells that has the ability to combine with ferric ion. { ¦ap·ə¦fer·əl·ən }

apogamy [BIOLOGY] Asexual, parthenogenetic development of diploid cells, such as the development of a sporophyte from a gametophyte without fertilization. { ə'päg·ə·mē }

apogeny [BOTANY] Loss of the function of reproduction. { ə'päj·ə·nē }

apogeotropism [BOTANY] Negative geotropism; growth up or away from the soil. { ¦a·pō,jē·ō'trä ,piz·əm }

Apogonidae [VERTEBRATE ZOOLOGY] The cardinal fishes, a family of tropical marine fishes in the order Perciformes; males incubate eggs in the mouth. { ,ap·ə'gän·ə,dē }

Apoidea [INVERTEBRATE ZOOLOGY] The bees, a superfamily of hymenopteran insects in the suborder Apocrita. { ə'pȯid·ē·ə }

apoinducer [BIOCHEMISTRY] A protein that, when bound to deoxyribonucleic acid, activates transcription by ribonucleic acid polymerase. { ¦a·pō·in¦dü·sər }

apolipoprotein [BIOCHEMISTRY] A protein that combines with a lipid to form a lipoprotein. { ¦a·pō¦li·pō'prō,tēn }

apolysis [INVERTEBRATE ZOOLOGY] In most tapeworms, the shedding of ripe proglottids. { ə 'päl·ə·səs }

apomeiosis [CELL BIOLOGY] Meiosis that is either suppressed or imperfect. { ¦ap·ə,mī¦ō·səs }

apomixis [EMBRYOLOGY] Parthenogenetic development of sex cells without fertilization. { ,ap·ə 'mik·səs }

apomorph [SYSTEMATICS] Any derived character occurring at a branching point and carried through one descending group in a phyletic lineage. { 'ap·ə,mȯrf }

apomyoglobin [BIOCHEMISTRY] Myoglobin that lacks its heme group. { ¦ap·ə¦mī·e,glōb·ən }

aponeurosis [ANATOMY] A broad sheet of regularly arranged connective tissue that covers a muscle or serves to connect a flat muscle to a bone. { ¦ap·ə,nu̇'rō·səs }

apophyllous [BOTANY] Having the parts of the perianth distinct. { ə'päf·ə·ləs }

apophysis [ANATOMY] An outgrowth or process on an organ or bone. [MYCOLOGY] A swollen filament in fungi. { ə'päf·ə·səs }

apoprotein [BIOCHEMISTRY] The protein portion of a conjugated protein exclusive of the prosthetic group. { ¦ap·ə¦prō,tēn }

apopyle [INVERTEBRATE ZOOLOGY] Any one of the large pores in a sponge by which water leaves a flagellated chamber to enter the exhalant system. { 'ap·ə,pīl }

Aporidea [INVERTEBRATE ZOOLOGY] An order of tapeworms of uncertain composition and affinities; parasites of anseriform birds. { ,ap·ə'rīd·ē·ə }

aporogamy [BOTANY] Entry of the pollen tube into the embryo sac through an opening other than the micropyle. { ,ap·ə'räg·ə·mē }

aposematic [ECOLOGY] Pertaining to colors or structures on an organism that provide a special means of defense against enemies. Also known as sematic. { ¦ap·ə·sə¦mad·ik }

apospory [MYCOLOGY] Suppression of spore formation with development of the haploid (sexual) generation directly from the diploid (asexual) generation. { 'ap·ə,spȯr·ē }

apostatic selection [ECOLOGY] Predation on the most abundant forms in a population, leading to balanced distribution of a variety of forms. { ¦ap·ə¦stad·ik sə'lek·shən }

Apostomatida [INVERTEBRATE ZOOLOGY] An order of ciliated protozoans in the subclass Holotrichia; majority are commensals on marine crustaceans. { ə,päs·tə'mad·ə·də }

apotele [ANATOMY] A scalloped ridge around the edge of an otolith. { 'ap·ə,tēl }

apothecium [MYCOLOGY] A spore-bearing structure in some Ascomycetes and lichens in which the fruiting surface or hymenium is exposed during spore maturation. { ¦ap·ə¦thēsh·əm }

apozymase [BIOCHEMISTRY] The protein component of a zymase. { ,ap·ə'zī,mās }

appendage [BIOLOGY] Any subordinate or nonessential structure associated with a major body part. [ZOOLOGY] Any jointed, peripheral extension, especially limbs, of arthropod and vertebrate bodies. { ə'pen·dij }

appendicular skeleton [ANATOMY] The bones of the pectoral and pelvic girdles and the paired appendages in vertebrates. { ,ap·ən'dik·yə·lər 'skel·ə·tən }

appendiculate [BIOLOGY] Having or forming appendages. { ,ap·ən'dik·yə,lāt }

appendix [ANATOMY] **1.** Any appendage. **2.** *See* vermiform appendix. { ə'pen·diks }

appestat [PHYSIOLOGY] The center for appetite regulation in the hypothalamus. { 'ap·ə ,stat }

appetitive behavior [ZOOLOGY] Any behavior that increases the probability that an animal will be able to satisfy a need; for example, a hungry animal will move around to find food. { ə'ped·ə·tiv bi'hāv·yər }

apple [BOTANY] *Malus domestica*. A deciduous tree in the order Rosales which produces an edible, simple, fleshy, pome-type fruit. { 'ap·əl }

apple of Peru [BOTANY] *See* jimsonweed. { 'ap·əl əv pə'rü }

applied anatomy [ANATOMY] **1.** A discipline that considers problems involving the biomechanical functions of a body. **2.** The application of anatomical principles to specific fields of human endeavor, for example, surgical anatomy. { ə'plīd ə 'nad·ə·mē }

applied ecology [ECOLOGY] Activities involved in the management of natural resources. { ə'plīd i 'käl·ə·jē }

apposition eye [INVERTEBRATE ZOOLOGY] A compound eye found in diurnal insects and crustaceans in which each ommatidium focuses on a small part of the whole field of light, producing a mosaic image. { ,ap·ə'zish·ən ,ī }

appressed [BIOLOGY] Pressed close to or lying flat against something. { ə'prest }

apricot [BOTANY] *Prunus armeniaca*. A deciduous

tree in the order Rosales which produces a simple fleshy stone fruit. { 'ap·rə‚kät }

aproctous [ZOOLOGY] Lacking an anus. { ā 'präk·təs }

Apsidospondyli [VERTEBRATE ZOOLOGY] A term used to include, as a subclass, amphibians in which the vertebral centra are formed from cartilaginous arches. { ¦ap·sə·də'spän·də‚lī }

apterium [VERTEBRATE ZOOLOGY] A bare space between feathers on a bird's skin. { ap'tir·ē·əm }

apterous [BIOLOGY] Lacking wings, as in certain insects, or winglike expansions, as in certain seeds. { 'ap·tə·rəs }

Apterygidae [VERTEBRATE ZOOLOGY] The kiwis, a family of nocturnal ratite birds in the order Apterygiformes. { ¦ap·tə¦rij·ə‚dē }

Apterygiformes [VERTEBRATE ZOOLOGY] An order of ratite birds containing three living species, the kiwis, characterized by small eyes, limited eyesight, and nostrils at the tip of the bill. { ‚ap·tə‚rij·ə'fȯr‚mēz }

Apterygota [INVERTEBRATE ZOOLOGY] A subclass of the Insecta characterized by being primitively wingless. { ‚ap·tə·rə'gōd·ə }

Apus [VERTEBRATE ZOOLOGY] A genus of birds comprising the Old World swifts. { 'ā·pəs }

apyrase [BIOCHEMISTRY] Any enzyme that hydrolyzes adenosinetriphosphate, with liberation of phosphate and energy, and that is believed to be associated with actomyosin activity. { 'ap·ə‚rās }

aquaculture [BIOLOGY] *See* aquiculture. { 'ak·wə‚kəl·chər }

aquatic [BIOLOGY] Living or growing in, on, or near water; having a water habitat. { ə'kwäd·ik }

aqueous desert [ECOLOGY] A marine bottom environment with little or no macroscopic invertebrate shelled life. { 'āk·wē·əs 'dez·ərt }

aqueous humor [PHYSIOLOGY] The transparent fluid filling the anterior chamber of the eye. { 'āk·wē·əs 'yü·mər }

aquiculture [BIOLOGY] Cultivation of natural faunal resources of water. Also spelled aquaculture. { 'ak·wə‚kəl·chər }

Aquifoliaceae [BOTANY] A family of woody flowering plants in the order Celastrales characterized by pendulous ovules, alternate leaves, imbricate petals, and drupaceous fruit; common members include various species of holly (*Ilex*). { ‚ak·wə‚fōl·ē'ās·ē‚ē }

aquiherbosa [ECOLOGY] Herbaceous plant communities in wet areas, such as swamps and ponds. { ‚ak·wē‚hər'bōs·ə }

aquiprata [ECOLOGY] Communities of plants which are found in areas such as wet meadows where groundwater is a factor. { ə'kwip·rəd·ə }

Arabellidae [INVERTEBRATE ZOOLOGY] A family of polychaete worms belonging to the Errantia. { ‚ar·ə'bel·ə‚dē }

arabinose [BIOCHEMISTRY] $C_5IH_{10}O_5$ A pentose sugar obtained in crystalline form from plant polysaccharides such as gums, hemicelluloses, and some glycosides. { ə'rab·ə‚nōs }

Araceae [BOTANY] A family of herbaceous flowering plants in the order Arales; plants have stems, roots, and leaves, the inflorescence is a spadix, and the growth habit is terrestrial or sometimes more or less aquatic; well-known members include dumb cane (*Dieffenbachia*), jack-in-the-pulpit (*Arisaema*), and *Philodendron*. { ə'rās·ē‚ē }

arachidonate [BIOCHEMISTRY] A salt or ester of arachidonic acid. { ‚a¦rak·ə¦dän‚āt }

arachidonic acid [BIOCHEMISTRY] $C_{20}H_{32}O_2$ An essential unsaturated fatty acid that is a precursor in the biosynthesis of prostaglandins, thromboxanes, and leukotrienes. { ə¦rak·ə¦dan·ik 'as·əd }

Arachnia [MICROBIOLOGY] A genus of bacteria in the family Actinomycetaceae; branched diphtheroid rods and branched filaments form filamentous microcolonies; facultatively anaerobic; the single species is a human pathogen. { ə'rak·nē·ə }

Arachnida [INVERTEBRATE ZOOLOGY] A class of arthropods in the subphylum Chelicerata characterized by four pairs of thoracic appendages. { ə'rak·nəd·ə }

arachnoid [ANATOMY] A membrane that covers the brain and spinal cord and lies between the pia mater and dura mater. [BOTANY] Of cobweblike appearance, caused by fine white hairs. Also known as araneose. [INVERTEBRATE ZOOLOGY] Any invertebrate related to or resembling the Arachnida. { ə'rak‚nȯid }

arachnoidal granulations [ANATOMY] Projections of the arachnoid layer of the cerebral meninges through the dura mater. Also known as arachnoid villi; Pacchionian bodies. { ¦a‚rak¦nȯid·əl ‚gran·yə'lā·shənz }

Arachnoidea [INVERTEBRATE ZOOLOGY] The name used in some classification schemes to describe a class of primitive arthropods. { ¦a‚rak'nȯid·ē·ə }

arachnoid villi [ANATOMY] *See* arachnoidal granulations. { ə'rak‚nȯid'vil·ē }

arachnology [INVERTEBRATE ZOOLOGY] The study of arachnids. { ‚a‚rak'näl·ə·jē }

Aradidae [INVERTEBRATE ZOOLOGY] The flat bugs, a family of hemipteran insects in the superfamily Aradoidea. { ə'rad·ə‚dē }

Aradoidea [INVERTEBRATE ZOOLOGY] A small superfamily of hemipteran insects belonging to the subdivision Geocorisae. { ‚a·rə'dȯid·ē·ə }

Araeoscelidia [PALEONTOLOGY] A provisional order of extinct reptiles in the subclass Euryapsida. { ə¦rē·ə·sə'lid·ē·ə }

Arales [BOTANY] An order of monocotyledonous plants in the subclass Arecidae. { ə'rā‚lēz }

Araliaceae [BOTANY] A family of dicotyledonous trees and shrubs in the order Umbellales; there are typically five carpels and the fruit, usually a berry, is fleshy or dry; well-known members are ginseng (*Panax*) and English ivy (*Hedera helix*). { ə‚rāl·ē'ās·ē‚ē }

Aramidae [VERTEBRATE ZOOLOGY] The limpkins, a family of birds in the order Gruiformes. { ə'ram·ə‚dē }

Araneae [INVERTEBRATE ZOOLOGY] An equivalent name for Araneida. { ə'rān·ē‚ē }

Araneida [INVERTEBRATE ZOOLOGY] The spiders, an order of arthropods in the class Arachnida. { ˌa·rəˈnē·ə·də }

araneology [INVERTEBRATE ZOOLOGY] The study of spiders. { əˌrān·ēˈäl·ə·jē }

araneose [BOTANY] *See* arachnoid. { əˈrān·ēˌōs }

Arbacioida [INVERTEBRATE ZOOLOGY] An order of echinoderms in the superorder Echinacea. { är ˌbās·ēˈȯid·ə }

arboreal Also known as arboreous. [BOTANY] Relating to or resembling a tree. [ZOOLOGY] Living in trees. { ärˈbȯr·ē·əl }

arboreous [BOTANY] **1.** Wooded. **2.** *See* arboreal. { ärˈbȯr·ē·əs }

arborescence [BIOLOGY] The state of being treelike in form and appearance. { ¦är·bə¦res·əns }

arboretum [BOTANY] An area where trees and shrubs are cultivated for educational and scientific purposes. { ˌär·bəˈrēd·əm }

arboriculture [BOTANY] The cultivation of ornamental trees and shrubs. { ¦är·bə·rə¦kəl·chər }

arborization [BIOLOGY] A treelike arrangement, such as a branched dendrite or axon. { ˌär·bə·rəˈzā·shən }

arborvitae [BOTANY] Any of the ornamental trees, sometimes called the tree of life, in the genus *Thuja* of the order Pinales. { ¦ar·bər¦vīd·ē }

arbor vitae [ANATOMY] The treelike arrangement of white nerve tissue seen in a median section of the cerebellum. { ˈär·bər ˈvīd·ē }

arbovirus [VIROLOGY] Small, arthropod-borne animal viruses that are unstable at room temperature and inactivated by sodium deoxycholate; cause several types of encephalitis. Also known as arthropod-borne virus. { ˈär·bəˌvī·rəs }

arbuscule [MYCOLOGY] A treelike haustorial organ in certain mycorrhizal fungi. { ärˈbə·skyül }

arcade [INVERTEBRATE ZOOLOGY] A type of cell associated with the pharyngeal region of nematodes and united with like cells by an arch. { är ˈkād }

Arcellinida [INVERTEBRATE ZOOLOGY] An order of rhizopodous protozoans in the subclass Lobosia characterized by lobopodia and a well-defined aperture in the test. { ˌär·səˈlin·ə·də }

archaebacteria [MICROBIOLOGY] A group of unusual prokaryotic organisms that microscopically resemble true bacteria but differ biochemically and genetically, and form a distinct evolutionary group; some occur widely in oxygen-free environments and produce methane, while others are found in extreme salty or acidic conditions or grow at high temperatures. { ˌar·kē·bakˈtir·ē·ə }

Archaeoceti [PALEONTOLOGY] The zeuglodonts, a suborder of aquatic Eocene mammals in the order Cetacea; the oldest known cetaceans. { ˌärk·ē·əˈsēˌtī }

Archaeocidaridae [PALEONTOLOGY] A family of Carboniferous echinoderms in the order Cidaroida characterized by a flexible test and more than two columns of interambulacral plates. { ˌärk·ē·əˌsəˈdar·əˌdē }

Archaeocopida [PALEONTOLOGY] An order of Cambrian crustaceans in the subclass Ostracoda characterized by only slight calcification of the carapace. { ˌärk·e·əˈkäp·ə·də }

Archaeogastropoda [INVERTEBRATE ZOOLOGY] An order of gastropod mollusks that includes the most primitive snails. { ˌärk·ē·əˌgasˈträp·ə·də }

Archaeopteridales [PALEOBOTANY] An order of Upper Devonian sporebearing plants in the class Polypodiopsida characterized by woody trunks and simple leaves. { ˌärk·ēˌäp·təˈrīd·ə·lēz }

Archaeopteris [PALEOBOTANY] A genus of fossil plants in the order Archaeopteridales; used sometimes as an index fossil of the Upper Devonian. { ˌärk·ēˈäp·tə·rəs.}

Archaeopterygiformes [PALEONTOLOGY] The single order of the extinct avian subclass Archaeornithes. { ˌärk·ēˌäp·təˌrij·əˈfȯrˌmēz }

Archaeopteryx [PALEONTOLOGY] The earliest known bird; a genus of fossil birds in the order Archaeopterygiformes characterized by flight feathers like those of modern birds. { ˌärk·ēˈäp·tə·riks }

Archaeornithes [PALEONTOLOGY] A subclass of Upper Jurassic birds comprising the oldest fossil birds. { ˌärk·ēˈȯr·nəˌthēz }

archallaxis [BIOLOGY] Deviation from an ancestral pattern early in development, eliminating duplication of the phylogenetic history. { ˌärk·ə ˈlak·səs }

Archangiaceae [MICROBIOLOGY] A family of bacteria in the order Myxobacterales; microcysts are rod-shaped, ovoid, or spherical and are not enclosed in sporangia, and fruiting bodies are irregular masses. { ¦ärkˌan·jēˈās·ēˌē }

Archangium [MICROBIOLOGY] The single genus of the family Archangiaceae; sporangia are lacking, and there is no definite slime wall. { ärk ˈan·jē·əm }

Archanthropinae [PALEONTOLOGY] A subfamily of the Hominidae, set up by F. Weidenreich, which is no longer used. { ˌärk·ənˈthräp·əˌnē }

archegoniophore [BOTANY] The stalk supporting the archegonium in liverworts and ferns. { ¦ärk·əˈgōn·ē·əˌfȯr }

archegonium [BOTANY] The multicellular female sex organ in all plants of the Embryobionta except the Pinophyta and Magnoliophyta. { ˌark·ə ˈgōn·ē·əm }

archencephalon [EMBRYOLOGY] The primitive embryonic forebrain from which the forebrain and midbrain develop. { ¦ärkˌinˈsef·ə·län }

archenteron [EMBRYOLOGY] The cavity of the gastrula formed by ingrowth of cells in vertebrate embryos. Also known as gastrocoele; primordial gut. { ¦ärkˈen·təˌrän }

archeocyte [INVERTEBRATE ZOOLOGY] A type of ovoid amebocyte in sponges, characterized by large nucleolate nuclei and blunt pseudopodia; gives rise to germ cells. { ˈär·kē·əˌsīt }

archerfish [VERTEBRATE ZOOLOGY] The common name for any member of the fresh-water family Toxotidae in the order Perciformes; individuals eject a stream of water from the mouth to capture insects. { ˈär·chərˌfish }

Archeria [PALEONTOLOGY] Genus of amphibians, order Embolomeri, in early Permian in Texas; fish eaters. { ˌärˈkir·ē·ə }

archespore [BOTANY] A cell from which the spore mother cell develops in either the pollen sac or the ovule of an angiosperm. { 'är·kə,spór }

archetype [EVOLUTION] A hypothetical ancestral type conceptualized by eliminating all specialized character traits. { 'är·ki,tīp }

Archiacanthocephala [INVERTEBRATE ZOOLOGY] An order of worms in the phylum Acanthocephala; adults are endoparasites of terrestrial vertebrates. { ¦är·kē·ə,kan·thə'sef·ə·lə }

Archiannelida [INVERTEBRATE ZOOLOGY] A group name applied to three families of unrelated annelid worms: Nerillidae, Protodrilidae, and Dinophilidae. { ¦är·kē·ə'nel·ə·də }

Archichlamydeae [BOTANY] An artificial group of flowering plants, in the Englerian system of classification, consisting of those families of dicotyledons that lack petals or have petals separate from each other. { ¦är·kē·klə'mid·ē,ē }

archicoel [ZOOLOGY] The segmentation cavity persisting between the ectoderm and endoderm as a body cavity in certain lower forms. { 'är·kē ,sēl }

Archidiidae [BOTANY] A subclass of the plant class Bryopsida; consists of a single genus, *Archidium*, unique in having spores scattered in a single layer of the endothecium and having no quadrant stage in the early ontogeny of the capsule. { ,är·kə'dī·ə,dē }

archigastrula [EMBRYOLOGY] A gastrula formed by invagination, as opposed to ingrowth of cells. { ¦är·kē'gas·trə·lə }

Archigregarinida [INVERTEBRATE ZOOLOGY] An order of telosporean protozoans in the subclass Gregarinia; endoparasites of invertebrates and lower chordates. { ¦är·kē,greg·ə'rin·ə·də }

archinephridium [INVERTEBRATE ZOOLOGY] One of a pair of primitive nephridia found in each segment of some annelid larvae. { ¦är·kē·nə¦frid·ē·əm }

archinephros [VERTEBRATE ZOOLOGY] The paired excretory organ of primitive vertebrates and the larvae of hagfishes and caecilians. { ¦är·kē'ne ,frōs }

archipallium [PHYSIOLOGY] The olfactory pallium or the olfactory cerebral cortex; phylogenetically, the oldest part of the cerebral cortex. { ¦är·ki ¦pal·ē·əm }

Archosauria [VERTEBRATE ZOOLOGY] A subclass of reptiles composed of five orders: Thecodontia, Saurischia, Ornithschia, Pterosauria, and Crocodylia. { ,är·kə'sór·ē·ə }

Archostemata [INVERTEBRATE ZOOLOGY] A suborder of insects in the order Coleoptera. { ,är·kə·stə'mäd·ə }

arch pattern [ANATOMY] A fingerprint pattern in which ridges enter on one side of the impression, form a wave or angular upthrust, and flow out the other side. { 'ärch ,pad·ərn }

arcocentrum [ANATOMY] A centrum formed of modified, fused mesial parts of the neural or hemal arches. { ¦ar·kō¦sen·trəm }

arctic-alpine [ECOLOGY] Of or pertaining to areas above the timberline in mountainous regions. { ¦ärd·ik ¦al,pīn }

arctic tree line [ECOLOGY] The northern limit of tree growth; the sinuous boundary between tundra and boreal forest. { 'ärd·ik ¦trē ,līn }

Arctiidae [INVERTEBRATE ZOOLOGY] The tiger moths, a family of lepidopteran insects in the suborder Heteroneura. { ärk'tī·ə,dē }

Arctocyonidae [PALEONTOLOGY] A family of extinct carnivorelike mammals in the order Condylarthra. { 'ärk·tō,sī'än·ə,dē }

Arctolepiformes [PALEONTOLOGY] A group of the extinct joint-necked fishes belonging to the Arthrodira. { ,ärk·tō,lep·ə'fór,mēz }

Arcturidae [INVERTEBRATE ZOOLOGY] A family of isopod crustaceans in the suborder Valvifera characterized by an almost cylindrical body and extremely long antennae. { ,ärk'tür·ə,dē }

arcuale [EMBRYOLOGY] Any of the four pairs of primitive cartilages from which a vertebra is formed. { ,ärk·yə'wā·lē }

arcuate [ANATOMY] Arched or curved; bow-shaped. { 'ärk·yə·wət }

Arcyzonidae [PALEONTOLOGY] A family of Devonian paleocopan ostracods in the superfamily Kirkbyacea characterized by valves with a large central pit. { ¦är,sī'zän·ə,dē }

Ardeidae [VERTEBRATE ZOOLOGY] The herons, a family of wading birds in the order Ciconiiformes. { är'dē·ə,dē }

area amniotica [EMBRYOLOGY] The transparent part of the blastodisc in mammals. { 'er·ē·ə ,am·nē'äd·ə·kə }

area opaca [EMBRYOLOGY] The opaque peripheral area of the blastoderm of birds and reptiles, continuous with the yolk. { 'er·ē·ə ō'päk·ə }

area pellucida [EMBRYOLOGY] The central transparent area of the blastoderm of birds and reptiles, overlying the subgerminal cavity. { 'er·ē·ə pə'lü·səd·ə }

area placentalis [EMBRYOLOGY] The part of the trophoblast in immediate contact with the uterine mucosa in the embryos of early placental vertebrates. { 'er·ē·ə pla·sən'tāl·əs }

area vitellina [EMBRYOLOGY] The outer nonvascular zone of the area opaca; consists of ectoderm and endoderm. { 'er·ē·ə ,vid·ə'līn·ə }

Arecaceae [BOTANY] The palms, the single family of the order Arecales. { ,ar·ə'ka·sē,ē }

Arecales [BOTANY] An order of flowering plants in the subclass Arecidae composed of the palms. { ,ar·ə'kā,lēz }

Arecidae [BOTANY] A subclass of flowering plants in the class Liliopsida characterized by numerous, small flowers subtended by a prominent spathe and often aggregated into a spadix, and broad, petiolate leaves without typical parallel venation. { ə'res·ə,dē }

areg [ECOLOGY] A sand desert. { 'a,reg }

arena [ZOOLOGY] *See* lek. { ə'rēn·ə }

Arenaviridae [VIROLOGY] A family of ribonucleic acid animal viruses consisting of a single genus, Arenavirus, having an enveloped, spherical pleomorphic form. { ə,rēn·ə'vī·rə,dē }

Arenicolidae [INVERTEBRATE ZOOLOGY] The lugworms, a family of mud-swallowing worms belonging to the Sedentaria. { ə,ren·ə'käl·ə,dē }

arenicolous [ZOOLOGY] Living or burrowing in sand. { ,a·rə'nik·ə·ləs }

areography [ECOLOGY] Descriptive biogeography. { ˌar·ē′äg·rə·fē }

areola [ANATOMY] **1.** The portion of the iris bordering the pupil of the eye. **2.** A pigmented ring surrounding a nipple, vesicle, or pustule. **3.** A small space, interval, or pore in a tissue. { ə′rē·ə·lə }

areola mammae [ANATOMY] The circular pigmented area surrounding the nipple of the breast. Also known as areola papillaris; mammary areola. { ə′rē·ə·lə ′mam·ē }

areola papillaris [ANATOMY] *See* areola mammae. { ə′rē·ə·lə pap·ə′lär·əs }

areolar tissue [HISTOLOGY] A loose network of fibrous tissue and elastic fiber that connects the skin to the underlying structures. { ə′rē·ə·lər ′tish·ü }

Argasidae [INVERTEBRATE ZOOLOGY] The soft ticks, a family of arachnids in the suborder Ixodides; several species are important as ectoparasites and disease vectors for humans and domestic animals. { är′gas·əˌdē }

argentaffin cell [HISTOLOGY] Any of the cells of the gastrointestinal tract that are thought to secrete serotonin. { är′jen·tə·fən ˌsel }

argentaffin fiber [HISTOLOGY] *See* reticular fiber. { är′jen·tə·fən ˌfī·bər }

Argentinoidei [VERTEBRATE ZOOLOGY] A family of marine deepwater teleostean fishes, including deep-sea smelts, in the order Salmoniformes. { ¦ärˌjen·tə′nȯid·ēˌī }

argentophil [BIOLOGY] Of cells, tissues, or other structures, having an affinity for silver. { är′jen·təˌfil }

Argidae [INVERTEBRATE ZOOLOGY] A small family of hymenopteran insects in the superfamily Tenthredinoidea. { ′är·jəˌdē }

arginase [BIOCHEMISTRY] An enzyme that catalyzes the splitting of urea from the amino acid arginine. { ′ar·jəˌnās }

arginine [BIOCHEMISTRY] $C_6H_{14}N_4O_2$ A colorless, crystalline, water-soluble, essential amino acid of the α-ketoglutaric acid family. { ′ar·jəˌnēn }

Arguloida [INVERTEBRATE ZOOLOGY] A group of crustaceans known as the fish lice; taxonomic status is uncertain. { ˌär·gə′lȯid·ə }

argyrophil lattice fiber [HISTOLOGY] *See* reticular fiber. { ′är·jə·rōˌfil ′lad·əs ′fī·bər }

Arhynchobdellae [INVERTEBRATE ZOOLOGY] An order of annelids in the class Hirudinea characterized by the lack of an eversible proboscis; includes most of the important leech parasites of human and warm-blooded animals. { ¦āˌriŋ ′käb dəˌlē }

Arhynchodina [INVERTEBRATE ZOOLOGY] A suborder of ciliophoran protozoans in the order Thigmotrichida. { ¦āˌriŋ′kä·də·nə }

arhythmicity [BIOLOGY] A condition that is characterized by the absence of an expected behavioral or physiologic rhythm. { ¦āˌrith′mis·əd·ē }

arid biogeographic zone [ECOLOGY] Any region of the world that supports relatively little vegetation due to lack of water. { ′ar·əd ¦bī·ōˌgē·ō ′graf·ik ˌzōn }

Arid Transition life zone [ECOLOGY] The zone of climate and biotic communities occurring in the chaparrals and steppes from the Rocky Mountain forest margin to California. { ′ar·əd trans′ish·ən ′līf ˌzōn }

arietiform [VERTEBRATE ZOOLOGY] Shaped like a ram's horns; specifically, describing the dark facial marking that extends across the nose of kangaroo rats. { ˌar·ē′ed·əˌfȯrm }

Ariidae [VERTEBRATE ZOOLOGY] A family of tropical salt-water catfishes in the order Siluriformes. { ə′rī·əˌdē }

aril [BOTANY] An outgrowth of the funiculus in certain seeds that either remains as an appendage or envelops the seed. { ′ar·əl }

arilode [BOTANY] An aril originating from tissues in the micropyle region; a false aril. { ′ar·ə ˌlōd }

Arionidae [INVERTEBRATE ZOOLOGY] A family of mollusks in the order Stylommatophora, including some of the pulmonate slugs. { ˌar·ē′än·ə ˌdē }

arista [INVERTEBRATE ZOOLOGY] The bristlelike or hairlike structure in many organisms, especially at or near the tip of the antenna of many Diptera. { ə′ris·tə }

Aristolochiaceae [BOTANY] The single family of the plant order Aristolochiales. { əˌris·təˌlō·kē ′ās·ēˌē }

Aristolochiales [BOTANY] An order of dicotyledonous plants in the subclass Magnoliidae; species are herbaceous to woody, often climbing, with perigynous to epigynous, apetalous flowers, uniaperturate or nonaperturate pollen, and seeds with a small embryo and copious endosperm. { əˌris·təˌlō·kē′āˌlēz }

aristopedia [INVERTEBRATE ZOOLOGY] Replacement of the arista by a nearly perfect leg. { əˌris·tə′pēd·ē·ə }

Aristotle's lantern [INVERTEBRATE ZOOLOGY] A five-sided feeding and locomotor apparatus surrounding the esophagus of most sea urchins. { ′ar·əˌstäd·əlz ′lant·ərn }

arm [ANATOMY] The upper or superior limb in humans which comprises the upper arm with one bone and the forearm with two bones. { ärm }

armadillo [VERTEBRATE ZOOLOGY] Any of 21 species of edentate mammals in the family Dasypodidae. { ˌär·mə′dil·ō }

Armilliferidae [INVERTEBRATE ZOOLOGY] A family of pentastomid arthropods belonging to the suborder Porocephaloidea. { ˌär·mə·lə′fer·əˌdē }

Armour unit [BIOLOGY] A unit for the standardization of adrenal cortical hormones and trypsin. { ′är·mər ˌyü·nət }

armyworm [INVERTEBRATE ZOOLOGY] Any of the larvae of certain species of noctuid moths composing the family Phalaenidae; economically important pests of corn and other grasses. { ′är·mē ˌwərm }

Arneth's classification [HISTOLOGY] *See* Arneth's index. { ′ärˌnets ˌklas·ə·fə′kā·shən }

Arneth's count [HISTOLOGY] *See* Arneth's index. { ′ärˌnets ˌkau̇nt }

Arneth's formula [HISTOLOGY] *See* Arneth's index. { ′ärˌnets ˌfȯr·myə·lə }

Arneth's index [HISTOLOGY] A system for dividing peripheral blood granulocytes into five classes according to the number of nuclear lobes, the least mature cells being tabulated on the left, giving rise to the terms "shift to left" and "shift to right" as an indication of granulocytic immaturity or hypermaturity respectively. Also known as Arneth's classification; Arneth's count; Arneth's formula. { 'är,nets ,in,deks }

Arnold sterilizer [MICROBIOLOGY] An apparatus that employs steam under pressure at 212°F (100°C) for fractional sterilization of specialized bacteriological culture media. { ärn·əld 'ster·ə ,liz·ər }

Arodoidea [INVERTEBRATE ZOOLOGY] A superfamily of hemipteran insects belonging to the subdivision Geocorisae. { ,a·rə'dȯid·ē·ə }

arolium [INVERTEBRATE ZOOLOGY] A pad projecting between the tarsal claws of some insects and arachnids. { ə'rōl·ē·əm }

aromatic amino acid [BIOCHEMISTRY] An organic acid containing at least one amino group and one or more aromatic groups; for example, phenylalanine, one of the essential amino acids. { ¦ar·ə¦mad·ik ə'mēn·ō 'as·əd }

arrested evolution [EVOLUTION] Evolution that was extremely slow in comparison with that characteristic of most organic lineages. { ə'res·təd ,ev·ə'lü·shən }

arrhenotoky [BIOLOGY] Production of only male offspring by a parthenogenetic female. { ,a·rə 'näd·ə·kē }

Arridae [VERTEBRATE ZOOLOGY] A family of catfishes in the suborder Siluroidei found from Cape Cod to Panama. { 'a·rə,dē }

arrowroot [BOTANY] Any of the tropical American plants belonging to the genus *Maranta* in the family Marantaceae. { 'ar·ō,rüt }

arrowworm [INVERTEBRATE ZOOLOGY] Any member of the phylum Chaetognatha; useful indicator organism for identifying displaced masses of water. { 'ar·ō,wərm }

Artacaminae [INVERTEBRATE ZOOLOGY] A subfamily of polychaete annelids in the family Terebellidae of the Sedentaria. { ,är·tə'kam·ə,nē }

arteriole [ANATOMY] An artery of small diameter that terminates in capillaries. { är'tir·ē,ōl }

arteriovenous anastomosis [ANATOMY] A blood vessel that connects an arteriole directly to a venule without capillary intervention. { är,tir·ē·ō'vē·nəs ə,nas·tə'mō·səs }

artery [ANATOMY] A vascular tube that carries blood away from the heart. { 'ärd·ə·rē }

arthochromatic erythroblast [HISTOLOGY] *See* normoblast. { ¦är·thrō,krō ¦mad·ik ə'rith·rə ,blast }

Arthoniaceae [BOTANY] A family of lichens in the order Hysteriales. { är,thän·ē'ās·ē,ē }

Arthrobacter [MICROBIOLOGY] A genus of gram-positive, aerobic rods in the coryneform group of bacteria; metabolism is respiratory, and cellulose is not attached. { 'är·thrō,bak·tər }

arthrobranch [INVERTEBRATE ZOOLOGY] In Malacostraca, the gill attached to the joint between the body and the first leg segment. { 'är·thrō ,braŋk }

arthrodia [ANATOMY] A diarthrosis permitting only restricted motion between a concave and a convex surface, as in some wrist and ankle articulations. Also known as gliding joint. { är 'thrōd·ē·ə }

Arthrodira [PALEONTOLOGY] The joint-necked fishes, an Upper Silurian and Devonian order of the Placodermi. { ,är·thrō'dī·rə }

Arthrodonteae [BOTANY] A family of mosses in the subclass Eubrya characterized by thin, membranous peristome teeth composed of cell walls. { ,är·thrō'dänt·ē,ē }

Arthromitaceae [MICROBIOLOGY] Formerly a family of nonmotile bacteria in the order Caryophanales found in the intestine of millipedes, cockroaches, and toads. { ,är·thräm·ə 'tās·ē,ē }

Arthropoda [INVERTEBRATE ZOOLOGY] The largest phylum in the animal kingdom; adults typically have a segmented body, a sclerotized integument, and many-jointed segmental limbs. { är 'thräp·ə·də }

arthropod-borne virus [VIROLOGY] *See* arbovirus. { ¦är·thrə,päd ¦bȯrn 'vī·rəs }

arthropodin [BIOCHEMISTRY] A water-soluble protein which forms part of the endocuticle of insects. { är'thräp·ə·dən }

arthrosis [ANATOMY] An articulation or suture uniting two bones. { är'thrō·səs }

arthrospore [BOTANY] A jointed, vegetative resting spore resulting from filament segmentation in some blue-green algae and hypha segmentation in many Basidiomycetes. { 'är·thrō'spȯr }

Arthrotardigrada [INVERTEBRATE ZOOLOGY] A suborder of microscopic invertebrates in the order Heterotardigrada characterized by toelike terminations on the legs. { ¦är·thrō,tard·ə 'gräd·ə }

Arthur unit [BIOLOGY] A unit for the standardization of splenin A. { 'är·thər ,yü·nət }

Arthus reaction [IMMUNOLOGY] An allergic reaction of the immediate hypersensitive type that results from the union of antigen and antibody, with complement present, in blood vessel walls. { 'är·thəs rē'ak·shən }

artichoke [BOTANY] *Cynara scolymus.* A herbaceous perennial plant belonging to the order Asterales; the flower head is edible. { 'ärd·ə ,chōk }

article [INVERTEBRATE ZOOLOGY] A segment of an arthropod leg between two articulations. { 'ärd·ə·kəl }

articulamentum [INVERTEBRATE ZOOLOGY] The innermost layer of a calcareous plate in a chiton. { är,tik·yə·lə'men·təm }

articular cartilage [ANATOMY] Cartilage that covers the articular surfaces of bones. { är'tik·yə·lər 'kärt·lij }

articular disk [ANATOMY] A disk of fibrocartilage, dividing the cavity of certain joints. { är'tik·yə·lər 'disk }

articular membrane [INVERTEBRATE ZOOLOGY] A flexible region of the cuticle between sclerotized areas of the exoskeleton of an arthropod; functions as a joint. { är'tik·yə·lər 'mem,brān }

Articulata [INVERTEBRATE ZOOLOGY] **1.** A class of the Brachiopoda having hinged valves that usually bear teeth. **2.** The only surviving subclass of the echinoderm class Crinoidea. { är,tik·yə'läd·ə }

articulation [ANATOMY] See joint. [BOTANY] A joint between two parts of a plant that can separate spontaneously. [INVERTEBRATE ZOOLOGY] A joint between rigid parts of an animal body, such as the segments of an appendage in insects. [PHYSIOLOGY] The act of enunciating speech. { är,tik·yə'lā·shən }

artifact [HISTOLOGY] A structure in a fixed cell or tissue formed by manipulation or by the reagent. { 'ärd·ə,fakt }

artificial parthenogenesis [PHYSIOLOGY] Activation of an egg by chemical and physical stimuli in the absence of sperm. { ¦ärd·ə¦fish·əl ¦pär·thə·nō¦gen·ə·səs }

artificial selection [GENETICS] A breeding method whereby particular genetic traits are selected by human manipulation. { ¦ärd·ə¦fish·əl si'lek·shən }

Artiodactyla [VERTEBRATE ZOOLOGY] An order of terrestrial, herbivorous mammals characterized by having an even number of toes and by having the main limb axes pass between the third and fourth toes. { ,ärd·ē·ō'dak·tə·lə }

arytenoid [ANATOMY] Relating to either of the paired, pyramid-shaped, pivoting cartilages on the dorsal aspect of the larynx, in man and most other mammals, to which the vocal cords and arytenoid muscles are attached. { ,ar·ə'tē,nȯid }

Ascaphidae [VERTEBRATE ZOOLOGY] A family of amphicoelous frogs in the order Anura, represented by four living species. { ə'skaf·ə,dē }

ascarid [INVERTEBRATE ZOOLOGY] The common name for any roundworm belonging to the superfamily Ascaridoidea. { 'as·kə·rəd }

Ascaridata [INVERTEBRATE ZOOLOGY] An equivalent name for the Ascaridina. { ə,skar·ə'däd·ə }

Ascaridida [INVERTEBRATE ZOOLOGY] An order of parasitic nematodes in the subclass Phasmidia. { ə,skar·ə'dī·də }

Ascarididae [INVERTEBRATE ZOOLOGY] A family of parasitic nematodes in the superfamily Ascaridoidea. { ,as·kə'rid·ə,dē }

Ascaridina [INVERTEBRATE ZOOLOGY] A suborder of parasitic nematodes in the order Ascaridida. { ə,skar·ə'dī·nə }

Ascaridoidea [INVERTEBRATE ZOOLOGY] A large superfamily of parasitic nematodes of the suborder Ascaridina. { ə,skar·ə'dȯid·ē·ə }

Ascaris [INVERTEBRATE ZOOLOGY] A genus of roundworms that are intestinal parasites in mammals, including humans. { 'as·kə·rəs }

Ascaroidea [INVERTEBRATE ZOOLOGY] An equivalent name for Ascaridoidea. { ,as·kə'rȯid·ē·ə }

ascending aorta [ANATOMY] The first part of the aorta, extending from its origin in the heart to the aortic arch. { ə'send·iŋ,ā'ȯrd·ə }

ascending colon [ANATOMY] The portion of the colon that extends from the cecum to the bend on the right side below the liver. { ə'send·iŋ 'kōl·ən }

Aschelmintha [INVERTEBRATE ZOOLOGY] A theoretical grouping erected by B. G. Chitwood as a series that includes the phylum Nematoda. { ,ask,hel'min thə }

Aschelminthes [INVERTEBRATE ZOOLOGY] A heterogeneous phylum of small to microscopic wormlike animals; individuals are pseudocoelomate and mostly unsegmented and are covered with a cuticle. { ,ask,hel'min,thēz }

Ascidiacea [INVERTEBRATE ZOOLOGY] A large class of the phylum Tunicata; adults are sessile and may be solitary or colonial. { ə,sid·ē'āsh·ē·ə }

ascidiform [BOTANY] Pitcher-shaped, as certain leaves. { ə'sid·ə,fȯrm }

ascidium [BOTANY] A pitcher-shaped plant organ or part. { ə'sid·ē·əm }

Asclepiadaceae [BOTANY] A family of tropical and subtropical flowering plants in the order Gentianales characterized by a well-developed latex system; milkweed (*Asclepias*) is a well-known member. { ə,sklēp·ē·ə'dās·ē,ē }

ascocarp [MYCOLOGY] The mature fruiting body bearing asci with ascospores in higher Ascomycetes. { 'as·kə,kärp }

ascogenous [MYCOLOGY] Pertaining to or producing asci. { a'skäj·ə·nəs }

ascogonium [MYCOLOGY] The specialized female sexual organ in higher Ascomycetes. { ,as·kə'gōn·ē·əm }

Ascolichenes [BOTANY] A class of the lichens characterized by the production of asci similar to those produced by Ascomycetes. { ¦as·kə,lī'kē·nēz }

Ascomycetes [MYCOLOGY] A class of fungi in the subdivision Eumycetes, distinguished by the ascus. { ,as·kō,mī'sēd·ēz }

ascon [INVERTEBRATE ZOOLOGY] A sponge or sponge larva having incurrent canals leading directly to the paragaster. { 'a,skän }

ascorbic acid [BIOCHEMISTRY] $C_6H_8O_6$ A white, crystalline, water-soluble vitamin found in many plant materials, especially citrus fruit. Also known as vitamin C. { ə'skȯr·bik 'as·əd }

ascospore [MYCOLOGY] An asexual spore representing the final product of the sexual process, borne on an ascus in Ascomycetes. { 'as·kə,spȯr }

Ascothoracica [INVERTEBRATE ZOOLOGY] An order of marine crustaceans in the subclass Cirripedia occurring as endo- and ectoparasites of echinoderms and cnidarians. { ¦as·kə·thə'ras·ə·kə }

ascus [MYCOLOGY] An oval or tubular spore sac bearing ascospores in members of the Ascomycetes. { 'as·kəs }

A selection [ECOLOGY] Selection that favors species adapted to consistently adverse environments. { 'ā si,lek·shən }

Aselloidea [INVERTEBRATE ZOOLOGY] A group of free-living, fresh-water isopod crustaceans in the suborder Asellota. { ,a·sə'lȯid·ē·ə }

Asellota [INVERTEBRATE ZOOLOGY] A suborder of morphologically and ecologically diverse aquatic crustaceans in the order Isopoda. { ə'sel·ə·də }

asetigerous [INVERTEBRATE ZOOLOGY] See achaetous. { ā·sə'tij·ə·rəs }

asexual [BIOLOGY] **1.** Not involving sex.

2. Exhibiting absence of sex or of functional sex organs. { ā'seksh·ə·wəl }

asexual reproduction [BIOLOGY] Formation of new individuals from a single individual without the involvement of gametes. { ā'seksh·ə·wəl ˌrē·prə'dək·shən }

ash [BOTANY] **1.** A tree of the genus *Fraxinus*, deciduous trees of the olive family (Oleaceae) characterized by opposite, pinnate leaflets. **2.** Any of various Australian trees having wood of great toughness and strength; used for tool handles and in work requiring flexibility. { ash }

Asilidae [INVERTEBRATE ZOOLOGY] The robber flies, a family of predatory, orthorrhaphous, dipteran insects in the series Brachycera. { ə'sil·ə ˌdē }

Asopinae [INVERTEBRATE ZOOLOGY] A family of hemipteran insects in the superfamily Pentatomoidea including some predators of caterpillars. { ə'sōp·əˌnē }

asparaginase [BIOCHEMISTRY] An enzyme that catalyzes the hydrolysis of asparagine to asparaginic acid and ammonia. { ˌas·pə'raj·əˌnās }

asparagine [BIOCHEMISTRY] $C_4H_8N_2O_3$ A white, crystalline amino acid found in many plant seeds. { ə'spar·əˌjēn }

asparagus [BOTANY] *Asparagus officinalis*. A dioecious, perennial monocot belonging to the order Liliales; the shoot of the plant is edible. { ə 'spar·ə·gəs }

aspartase [BIOCHEMISTRY] A bacterial enzyme that catalyzes the deamination of aspartic acid to fumaric acid and ammonia. { ə'spärˌtās }

aspartate [BIOCHEMISTRY] A compound that is an ester or salt of aspartic acid. { ə'spärˌtāt }

aspartic acid [BIOCHEMISTRY] $C_4H_7NO_4$ A nonessential, crystalline dicarboxylic amino acid found in plants and animals, especially in molasses from young sugarcane and sugarbeet. { ə 'spärd·ik 'as·əd }

aspartokinase [BIOCHEMISTRY] An enzyme that catalyzes the reaction of aspartic acid with adenosinetriphosphate to give aspartyl phosphate. { əˌspärd·ō'kīˌnās }

aspartoyl [BIOCHEMISTRY] $—COCH_2CH(NH_2)-CO—$ A bivalent radical derived from aspartic acid. { ə'spärd·əˌwil }

aspartyl phosphate [BIOCHEMISTRY] $H_2O_3PO-OCH_2CHNH_2COOH$ An intermediate in the biosynthesis of pyrimidines. { ə'spärd·əl 'fäsˌfāt }

aspect [ECOLOGY] Seasonal appearance. { 'a ˌspekt }

aspection [ECOLOGY] Seasonal change in appearance or constitution of a plant community. { a'spek·shən }

aspen [BOTANY] Any of several species of poplars (*Populus*) characterized by weak, flattened leaf stalks which cause the leaves to flutter in the slightest breeze. { 'as·pən }

Aspergillaceae [MYCOLOGY] Former name for the Eurotiaceae. { ˌas·pər·jə'lās·ēˌē }

Aspergillales [MYCOLOGY] Former name for the Eurotiales. { ˌas·pər·jə'lā·lēz }

aspergillic acid [BIOCHEMISTRY] $C_{12}H_{20}O_2N_2$ A diketopiperazine-like antifungal antibiotic produced by certain strains of *Aspergillus flavus*. { ¦as·pər¦jil·ik 'as·əd }

aspergillin [MYCOLOGY] **1.** A black pigment found in spores of some molds of the genus *Aspergillus*. **2.** A broad-spectrum antibacterial antibiotic produced by the molds *Aspergillus flavus* and *A. fumigatus*. { ˌas·pər'jil·ən }

Aspergillus [MYCOLOGY] A genus of fungi including several species of common molds and some human and plant pathogens. { ˌas·pər'jil·əs }

asperifoliate [BOTANY] Rough-leaved. { ¦as·pər·ə¦fōl·ēˌāt }

asperulate [BOTANY] Delicately roughened. { a 'sper·əˌlāt }

asphradium [INVERTEBRATE ZOOLOGY] An organ, believed to be a chemoreceptor, in mollusks. Also spelled osphradium. { a'sfrād·ē·əm }

aspiculate [INVERTEBRATE ZOOLOGY] Lacking spicules, referring to Porifera. { a'spik·yə·lət }

Aspidiotinae [INVERTEBRATE ZOOLOGY] A subfamily of homopteran insects in the superfamily Coccoidea. { aˌspid·ē'ä·təˌnē }

Aspidiphoridae [INVERTEBRATE ZOOLOGY] An equivalent name for the Sphindidae. { aˌspid·ə 'fȯr·əˌdē }

Aspidobothria [INVERTEBRATE ZOOLOGY] An equivalent name for the Aspidogastrea. { ¦as·pə ˌdō'bäth·rē·ə }

Aspidobothroidea [INVERTEBRATE ZOOLOGY] A group of trematodes accorded class rank by W. J. Hargis. { ¦as·pəˌdō·bə'thrȯid·ē·ə }

Aspidobranchia [INVERTEBRATE ZOOLOGY] An equivalent name for the Archaeogastropoda. { ¦as·pəˌdō'braŋk·ē·ə }

Aspidochirotacea [INVERTEBRATE ZOOLOGY] A subclass of bilaterally symmetrical echinoderms in the class Holothuroidea characterized by tube feet and 10-30 shield-shaped tentacles. { ¦as·pə ˌdōˌkī·rə'tās·ē·ə }

Aspidochirotida [INVERTEBRATE ZOOLOGY] An order of holothurioid echinoderms in the subclass Aspidochirotacea characterized by respiratory trees and dorsal tube feet converted into tactile warts. { ¦as·pəˌdōˌkī'räd·ə·də }

Aspidocotylea [INVERTEBRATE ZOOLOGY] An equivalent name for the Aspidogastrea. { ¦as·pə ˌdōˌkäd·ə'lē·ə }

Aspidodiadematidae [INVERTEBRATE ZOOLOGY] A small family of deep-sea echinoderms in the order Diadematoida. { ¦as·pəˌdōˌdī·ə·də'mad·ə ˌdē }

Aspidogastrea [INVERTEBRATE ZOOLOGY] An order of endoparasitic worms in the class Trematoda having strongly developed ventral holdfasts. { ¦as·pəˌdō'gas·trē·ə }

Aspidogastridae [INVERTEBRATE ZOOLOGY] A family of trematode worms in the order Aspidogastrea occurring as endoparasites of mollusks. { ¦as·pəˌdō'gas·trəˌde }

Aspidorhynchidae [PALEONTOLOGY] The single family of the Aspidorhynchiformes, an extinct order of holostean fishes. { ¦as·pəˌdō'riŋ·kəˌdē }

Aspidorhynchiformes [PALEONTOLOGY] A small, extinct order of specialized holostean fishes. { ¦as·pəˌdōˌriŋk·ə'fȯrˌmēz }

Aspinothoracida [PALEONTOLOGY] The equivalent name for Brachythoraci. { a‚spīn·ō·thə'ras·əd·ə }

aspiration [MICROBIOLOGY] The use of suction to draw up a sample in a pipette. { ‚as·pə'rā·shən }

asporogenic mutant [MICROBIOLOGY] A bacillus that is unable to form spores due to alterations at any of several gene loci. { ¦ā‚spȯr·ə¦jen·ik 'myüt·ənt }

asporogenous [BOTANY] Not producing spores, especially of certain yeasts. { ¦ā·spə'räj·ə·nəs }

Aspredinidae [VERTEBRATE ZOOLOGY] A family of salt-water catfishes in the order Siluriformes found off the coast of South America. { ‚a·sprə 'din·ə·dē }

ass [VERTEBRATE ZOOLOGY] Any of several perissodactyl mammals in the family Equidae belonging to the genus *Equus*, especially *E. hemionus* and *E. asinus*. { as }

assemblage [ECOLOGY] A group of organisms sharing a common habitat by chance. [PALEONTOLOGY] A group of fossils occurring together at one stratigraphic level. { ə'sem·blij }

assemblage zone [PALEONTOLOGY] A biostratigraphic unit defined and identified by a group of associated fossils rather than by a single index fossil. { ə'sem·blij ‚zōn }

assimilation [PHYSIOLOGY] The conversion of nutritive materials into protoplasm. { ə‚sim·ə'lā·shən }

assimilative nitrate reduction [MICROBIOLOGY] The reduction of nitrates by some aerobic bacteria for purposes of assimilation. { ə‚sim·ə 'lād·iv 'nī‚trāt ri‚dək·shən }

assimilative sulfate reduction [MICROBIOLOGY] The reduction of sulfates by certain obligate anaerobic bacteria for purposes of assimilation. { ə ‚sim·ə'lād·iv 'səl‚fāt ri‚dək·shən }

associated automatic movement [PHYSIOLOGY] *See* synkinesia. { ə'sō·sē‚ād·əd ¦ȯd·ə'mad·ik 'müv·mənt }

association [ECOLOGY] Major segment of a biome formed by a climax community, such as an oak-hickory forest of the deciduous forest biome. { ə‚sō·sē'ā·shən }

association area [PHYSIOLOGY] An area of the cerebral cortex that is thought to link and coordinate activities of the projection areas. { ə‚sō·sē 'ā·shən ‚er·ē·ə }

association center [INVERTEBRATE ZOOLOGY] In invertebrates, a nervous center coordinating and distributing stimuli from sensory receptors. { ə ‚sō·sē'ā·shən ‚sen·tər }

association fiber [ANATOMY] One of the white nerve fibers situated just beneath the cortical substance and connecting the adjacent cerebral gyri. { ə‚sō·sē'ā·shən ‚fi·bər }

association neuron [ANATOMY] A neuron, usually within the central nervous system, between sensory and motor neurons. { ə‚sō·sē'ā·shən 'nü‚rän }

assortative mating [GENETICS] Nonrandom mating with respect to phenotypes. { ə'sȯrd·əd·iv ¦mād·iŋ }

Astacidae [INVERTEBRATE ZOOLOGY] A family of fresh-water crayfishes belonging to the section Macrura in the order Decapoda, occurring in the temperate regions of the Northern Hemisphere. { ‚as·tə'sī‚dē }

astacin [BIOCHEMISTRY] $C_{40}H_{48}O_4$ A red carotenoid ketone pigment found in crustaceans, as in the shell of a boiled lobster. { 'as·tə·sən }

Astacinae [INVERTEBRATE ZOOLOGY] A subfamily of crayfishes in the family Astacidae including all North American species west of the Rocky Mountains. { ‚as·tə'sī‚nē }

astaxanthin [BIOCHEMISTRY] $C_{40}H_{52}O_4$ A violet carotenoid pigment found in combined form in certain crustacean shells and bird feathers. { ¦as·tə'zan·thən }

Asteidae [INVERTEBRATE ZOOLOGY] A small, obscure family of cyclorrhaphous myodarian dipteran insects in the subsection Acalypteratae. { ‚as·tē'ī‚dē }

astelic [BOTANY] Lacking a stele or having a discontinuous arrangement of vascular bundles. { ā 'stēl·ik }

aster [BOTANY] Any of the herbaceous ornamental plants of the genus *Aster* belonging to the family Compositae. [CELL BIOLOGY] The star-shaped structure that encloses the centrosome at the end of the spindle during mitosis. { 'as·tər }

Asteraceae [BOTANY] An equivalent name for the Compositae. { ‚as·tə'rās·ē‚ē }

Asterales [BOTANY] An order of dicotyledonous plants in the subclass Asteridae, including aster, sunflower, zinnia, lettuce, artichoke, and dandelion; the ovary is inferior, flowers are borne in involucrate, centripetally flowering heads, and the calyx, when present, is modified into a set of scale-, hair-, or bristlelike structures called the pappus. { ‚as·tə'rāl·ēz }

Asteridae [BOTANY] A large subclass of dicotyledonous plants in the class Magnoliopsida; plants are sympetalous, with unitegmic, tenuinucellate ovules and with the stamens usually as many as, or fewer than, the corolla lobes and alternate with them. { ‚as·tə'rī‚dē }

Asteriidae [INVERTEBRATE ZOOLOGY] A large family of echinoderms in the order Forcipulatida, including many predatory sea stars. { ‚as·tə'ri·ə ‚dē }

Asterinidae [INVERTEBRATE ZOOLOGY] The starlets, a family of echinoderms in the order Spinulosida. { ‚as·tə'rin·ə‚dē }

asternal [ANATOMY] **1.** Not attached to the sternum. **2.** Without a sternum. { ā'stərn·əl }

Asteroidea [INVERTEBRATE ZOOLOGY] The starfishes, a subclass of echinoderms in the subphylum Asterozoa characterized by five radial arms. { ‚as·tə'rȯid·ē·ə }

Asteroschematidae [INVERTEBRATE ZOOLOGY] A family of ophiuroid echinoderms in the order Phrynophiurida with individuals having a small disk and stout arms. { ‚as·tə·rō‚skē'mad·ə‚dē }

Asterozoa [INVERTEBRATE ZOOLOGY] A subphylum of echinoderms characterized by a star-shaped body and radially divergent axes of symmetry. { ‚as·tə·rə'zō·ə }

Asticcacaulis [MICROBIOLOGY] A genus of prosthecate bacteria; cells are rod-shaped with an appendage (pseudostalk), and reproduction is by binary fission of cells. { ə‚stik·ə'kōl·əs }

astichous [BOTANY] Not arranged in rows. { 'as·tə·kəs }

astogeny [INVERTEBRATE ZOOLOGY] Morphological and size changes associated with zooids of aging colonial animals. { a'stäj·ə·nē }

astomatal [BOTANY] Lacking stomata. Also known as astomous. { ā'stōm·əd·əl }

Astomatida [INVERTEBRATE ZOOLOGY] An order of mouthless protozoans in the subclass Holotrichia; all species are invertebrate parasites, typically in oligochaete annelids. { as·tō'mad·ə·də }

astomatous [INVERTEBRATE ZOOLOGY] Lacking a mouth, especially a cytostome, as in certain ciliates. { a'stam·əd·əs }

astomocnidae nematocyst [INVERTEBRATE ZOOLOGY] A stinging cell whose thread has a closed end and either is adhesive or acts as a lasso to entangle prey. { ‚as·tə'mäk·nə‚dē ni 'mad·ə‚sist }

astomous [BOTANY] **1.** Having a capsule that bursts irregularly and is not dehiscent by an operculum. **2.** *See* astomatal. { 'as·tə·məs }

astraeid [INVERTEBRATE ZOOLOGY] Of a group of corals that are imperforate. { a'strē·əd }

astragalus [ANATOMY] The bone of the ankle which articulates with the bones of the leg. Also known as talus. { ə'strag·ə·ləs }

Astrapotheria [PALEONTOLOGY] A relatively small order of large, extinct South American mammals in the infraclass Eutheria. { ‚as·trə·pə 'thir·ē·ə }

Astrapotheroidea [PALEONTOLOGY] A suborder of extinct mammals in the order Astrapotheria, ranging from early Eocene to late Miocene. { ‚as·trə·pə·thə'rȯid·ē·ə }

Astrea [INVERTEBRATE ZOOLOGY] A genus of mollusks in the class Gastropoda. { 'as·trē·ə }

astrobiology [BIOLOGY] The study of living organisms on celestial bodies other than the earth. { ¦as·trō·bī'äl·ə·jē }

astrocyte [HISTOLOGY] A star-shaped cell; specifically, a neuroglial cell. { 'as·trə‚sīt }

astroglia [HISTOLOGY] Neuroglia composed of astrocytes. { ə'sträg·lē·ə }

Astropectinidae [INVERTEBRATE ZOOLOGY] A family of echinoderms in the suborder Paxillosina occurring in all seas from tidal level downward. { ‚as·trō‚pek'tin·ə‚dē }

astropyle [INVERTEBRATE ZOOLOGY] A small, rounded projection from the central capsule of some radiolarians. { 'as·trō‚pīl }

astrosphere [CELL BIOLOGY] The center of the aster exclusive of the rays. { 'as·trō‚sfir }

asty [INVERTEBRATE ZOOLOGY] A bryozoan colony. { 'a‚stī }

asulcal [BIOLOGY] Without a sulcus. { ā'səl·kəl }

asynapsis [CELL BIOLOGY] Absence of pairing of homologous chromosomes during meiosis. { ‚ā·si'nap·səs }

atactostele [BOTANY] A type of monocotyledonous siphonostele in which the vascular bundles are dispersed irregularly throughout the center of the stem. { ə'tak·tə‚stēl }

atavism [EVOLUTION] Appearance of a distant ancestral form of an organism or one of its parts due to reactivation of ancestral genes. { 'ad·ə ‚viz·əm }

Ateleopoidei [VERTEBRATE ZOOLOGY] A family of oceanic fishes in the order Cetomimiformes characterized by an elongate body, lack of a dorsal fin, and an anal fin continuous with the caudal fin. { ə¦tel·ē·ə'pȯid·ē‚ī }

Atelopodidae [VERTEBRATE ZOOLOGY] A family of small, brilliantly colored South and Central American frogs in the suborder Procoela. { ə ‚tel·ə'päd·ə‚dē }

Atelostomata [INVERTEBRATE ZOOLOGY] A superorder of echinoderms in the subclass Euechinoidea characterized by a rigid, exocylic test and lacking a lantern, or jaw, apparatus. { ə‚tel·ə 'stōm·əd·ə }

Athalamida [INVERTEBRATE ZOOLOGY] An order of naked amebas of the subclass Granuloreticulosia in which pseudopodia are branched and threadlike (reticulopodia). { ‚ath·ə'läm·əd·ə }

Athecanephria [INVERTEBRATE ZOOLOGY] An order of tube-dwelling, tentaculate animals in the class Pogonophora characterized by a saclike anterior coelom. { ‚ath·ə·kə'nef·rē·ə }

athecate [INVERTEBRATE ZOOLOGY] Lacking a theca. { 'ath·ə‚kāt }

Atherinidae [VERTEBRATE ZOOLOGY] The silversides, a family of actinopterygian fishes of the order Atheriniformes. { ‚ath·ə'rin·ə‚dē }

Atheriniformes [VERTEBRATE ZOOLOGY] An order of actinopterygian fishes in the infraclass Teleostei, including flyingfishes, needlefishes, killifishes, silversides, and allied species. { ‚ath·ə ‚rin·ə'fȯr‚mēz }

Athiorhodaceae [MICROBIOLOGY] Formerly the nonsulfur photosynthetic bacteria, a family of small, gram-negative, nonsporeforming, motile bacteria in the suborder Rhodobacteriineae. { ā ‚thī·ə‚rō'dās·ē‚ē }

athrocyte [HISTOLOGY] A cell that engulfs extraneous material and stores it as granules in the cytoplasm. { 'ath·rə‚sīt }

Athyrididina [PALEONTOLOGY] A suborder of fossil articulate brachiopods in the order Spiriferida characterized by laterally or, more rarely, ventrally directed spires. { ‚ath·ə·rə'də'dī·nə }

Atlantacea [INVERTEBRATE ZOOLOGY] A superfamily of mollusks in the subclass Prosobranchia. { ¦at‚lan¦tās·ē·ə }

atlas [ANATOMY] The first cervical vertebra. { 'at·ləs }

ATP [BIOCHEMISTRY] *See* adenosinetriphosphate.

ATPase [BIOCHEMISTRY] *See* adenosinetriphosphatase.

Atractidae [INVERTEBRATE ZOOLOGY] A family of parasitic nematodes in the superfamily Oxyuroidea. { a'trak·tə‚dē }

atrial septum [ANATOMY] The muscular wall between the atria of the heart. Also known as interatrial septum. { 'ā·trē·əl 'sep·təm }

Atrichornithidae [VERTEBRATE ZOOLOGY] The scrubbirds, a family of suboscine perching birds in the suborder Menurae. { a‚trī·kȯr'nith·ə‚dē }

atrichous [CELL BIOLOGY] Lacking flagella. { 'a·trə·kəs }

atriopore [ZOOLOGY] The opening of an atrium as seen in lancelets and tunicates. { 'ā·trē·ə‚pȯr }

atrioventricular bundle [ANATOMY] *See* bundle of His. { ¦ā·trē·ō‚ven'trik·yə·lər 'bən·dəl }

atrioventricular canal [EMBRYOLOGY] The common passage between the atria and ventricles in the heart of the mammalian embryo before division of the organ into right and left sides. { ¦ā·trē·ō‚ven'trik·yə·lər kə'nal }

atrioventricular node [ANATOMY] A group of slow-conducting fibers in the atrium of the vertebrate heart that are stimulated by impulses originating in the sinoatrial node and conduct impulses to the bundle of His. { ¦ā·trē·o‚ven¦trik·yə·lər 'nōd }

atrioventricular valve [ANATOMY] A structure located at the orifice between the atrium and ventricle which maintains a unidirectional blood flow through the heart. { ¦a·trē·ō‚ven'trik·yə·lər 'valv }

atrium [ANATOMY] **1.** The heart chamber that receives blood from the veins. **2.** The main part of the tympanic cavity, below the malleus. **3.** The external chamber to receive water from the gills in lancelets and tunicates. { 'ā·trē·əm }

atropinization [PHYSIOLOGY] The physiological condition of being under the influence of atropine. { ə¦trō·pə·nə'zā·shən }

Atrypidina [PALEONTOLOGY] A suborder of fossil articulate brachiopods in the order Spiriferida. { a·trī'pid·ə·nə }

attachment [VIROLOGY] The initial stage in the infection of a cell by a virus that follows a chance collision by the virus with a suitable receptor area on the cell. { ə'tach·mənt }

attenuated vaccine [IMMUNOLOGY] A suspension of weakened bacteria, viruses, or fractions thereof used to produce active immunity. { ə'ten·yə‚wād·əd ‚vak'sēn }

attenuation [BOTANY] Tapering, sometimes to a long point. [MICROBIOLOGY] Weakening or reduction of the virulence of a microorganism. { ə‚ten·yə'wā·shən }

Atyidae [INVERTEBRATE ZOOLOGY] A family of decapod crustaceans belonging to the section Caridea. { a'tī·ə‚dē }

A-type virus particles [VIROLOGY] A morphologically defined group of double-shelled spherical ribonucleic acid virus particles, often found in tumor cells. { 'ā ¦tīp 'vī·rəs ‚pard·ə·kəlz }

Auberger blood group system [IMMUNOLOGY] An immunologically distinct, genetically determined human erythrocyte antigen, demonstrated by reaction with anti-Au[2] (anti-Auberger) antibody. { ¦ō·bər‚zhā 'bləd ‚grüp ‚sis·təm }

Auchenorrhyncha [INVERTEBRATE ZOOLOGY] A group of homopteran families and one superfamily, in which the beak arises at the anteroventral extremity of the face and is not sheathed by the propleura. { ‚ȯk·ə·nə'riŋ·kə }

audition [PHYSIOLOGY] Ability to hear. { ȯ'dish·ən }

auditory [PHYSIOLOGY] Pertaining to the act or the organs of hearing. { 'ȯd·ə‚tȯr·ē }

auditory association area [PHYSIOLOGY] The cortical association area in the brain just inferior to the auditory projection area, related to it anatomically and functionally by association fibers. { 'ȯd·ə‚tȯr·ē ə‚sō·sē'ā·shən ‚er·ē·ə }

auditory impedance [PHYSIOLOGY] The acoustic impedance of the ear. { 'ȯd·ə‚tȯr·ē im'pēd·əns }

auditory nerve [ANATOMY] The eighth cranial nerve in vertebrates; either of a pair of sensory nerves composed of two sets of nerve fibers, the cochlear nerve and the vestibular nerve. Also known as acoustic nerve; vestibulocochlear nerve. { 'ȯd·ə‚tȯr·ē 'nərv }

auditory placode [EMBRYOLOGY] An ectodermal thickening from which the inner ear develops in vertebrates. { 'ȯd·ə‚tȯr·ē 'pla‚kōd }

Auerbach's plexus [ANATOMY] *See* myenteric plexus. { 'au̇r‚baks ¦plek·səs }

aufwuch [ECOLOGY] A plant or animal organism which is attached or clings to surfaces of leaves or stems of rooted plants above the bottom stratum. { 'ȯf‚wək }

auk [VERTEBRATE ZOOLOGY] Any of several large, short-necked diving birds (*Alca*) of the family Alcidae found along North Atlantic coasts. { ȯk }

Aulodonta [INVERTEBRATE ZOOLOGY] An order of echinoderms proposed by R. Jackson in 1921. { ‚ȯl·ə'dän·tə }

aulodont dentition [INVERTEBRATE ZOOLOGY] In echinoderms, grooved teeth with epiphyses that do not meet, so the foramen magnum of the jaw is open. { 'ȯl·ə‚dänt ‚den'tish·ən }

Aulolepidae [PALEONTOLOGY] A family of marine fossil teleostean fishes in the order Ctenothrissiformes. { ‚ȯl·ə'lep·ə‚dē }

aulophyte [ECOLOGY] A nonparasitic plant that lives in the cavity of another plant for shelter. { 'ȯl·ə‚fīt }

Auloporidae [PALEONTOLOGY] A family of Paleozoic corals in the order Tabulata. { ‚ȯl·ə'pȯr·ə‚dē }

aural [BIOLOGY] Pertaining to the ear or the sense of hearing. { 'ȯr·əl }

aurelia [INVERTEBRATE ZOOLOGY] A morphological grouping of paramecia, including the elongate, cigar-shaped species which appear to be nearly circular in cross section. { ȯ'rēl·yə }

Aurelia [INVERTEBRATE ZOOLOGY] A genus of scyphozoans. { ȯ'rēl·yə }

aureofacin [MICROBIOLOGY] An antifungal antibiotic produced by a strain of *Streptomyces aureofaciens*. { ‚ȯr·ē·ō'fās·ən }

aureothricin [MICROBIOLOGY] $C_9H_{10}O_2N_2S_2$ An antibacterial antibiotic produced by a strain of *Actinomyces*. { ‚ȯr·ē·ō'thrīs·ən }

aureusidin [BIOCHEMISTRY] $C_{15}H_{11}O_5$ A yellow flavonoid pigment found typically in the yellow snapdragon. Also known as 4,6,3′,4′-tetrahydroxyaurone. { ˌȯr·e·ə′sīd·ən }

auricle [ANATOMY] **1.** An ear-shaped appendage to an atrium of the heart. **2.** Any ear-shaped structure. **3.** *See* pinna. { ′ȯr·ə·kəl }

auricularia larva [INVERTEBRATE ZOOLOGY] A barrel-shaped, food-gathering larval form with a winding ciliated band, common to holothurians and asteroids. { ȯˌrik·yə′lar·ē·ə ′lär·və }

auricularis [ANATOMY] Any of the three muscles attached to the cartilage of the external ear. { ȯ ¦rik·yə′lär·əs }

aurophore [INVERTEBRATE ZOOLOGY] A bell-shaped structure which is part of the float of certain cnidarians. { ′ȯr·əˌfȯr }

Australia antigen [IMMUNOLOGY] An infectious agent that causes hepatitis in some people; similar to an inherited serum protein in being polymorphic. { ȯ′strāl·yə ′ant·i·jən }

Australian faunal region [ECOLOGY] A zoogeographic region that includes the terrestrial animal communities of Australia and all surrounding islands except those of Asia. { ȯ′strāl·yən ′fȯn·əl ˌrē·jən }

Australopithecinae [PALEONTOLOGY] The near-men, a subfamily of the family Hominidae composed of the single genus *Australopithecus.* { ȯ ˌstrā·lōˌpith·ə′sī·nē }

Australopithecus [PALEONTOLOGY] A genus of near-men in the subfamily Australopithecinae representing a side branch of human evolution. { ȯˌstrā·lō′pith·ə·kəs }

austral region [ECOLOGY] A North American biogeographic region including the region between transitional and tropical zones. { ′ȯs·trəl ˌrē·jən }

Austroastacidae [INVERTEBRATE ZOOLOGY] A family of crayfish in the order Decapoda found in temperate regions of the Southern Hemisphere. { ¦ȯs·trō·ə′stās·əˌdē }

Austrodecidae [INVERTEBRATE ZOOLOGY] A monogeneric family of marine arthropods in the subphylum Pycnogonida. { ˌȯs·trə′des·əˌdē }

Austroriparian life zone [ECOLOGY] The zone in which occurs the climate and biotic communities of the southeastern coniferous forests of North America. { ¦ȯs·trōˌrī′per·ē·ən ′līf ˌzōn }

autecology [ECOLOGY] The ecological relations between a plant species and its environment. { ˌaud·i′käl·ə·jē }

autoagglutination [IMMUNOLOGY] Agglutination of an individual's erythrocytes by his own serum. Also known as autohemagglutination. { ¦ȯd·ō·ə ˌglüt·ən′ā·sh ppn }

autoagglutinin [IMMUNOLOGY] An antibody in an individual's blood serum that causes agglutination of his own erythrocytes. { ¦ȯd·ō·ə′glüt·ən·ən }

autoantibody [IMMUNOLOGY] An antibody formed by an individual against his own tissues; common in hemolytic anemias. { ¦ȯd·ō′ant·i ˌbäd·ē }

autoantigen [IMMUNOLOGY] A tissue within the body which acquires the ability to incite the formation of complementary antibodies. { ¦ȯd·ō ′ant·i·jən }

autoasphyxiation [PHYSIOLOGY] Asphyxiation by the products of metabolic activity. { ¦ȯd·ō·as ˌfik·sē′ā·shən }

autobasidium [MYCOLOGY] An undivided basidium typically found in higher Basidiomycetes. { ¦ȯd·ō·bə′sid·ē·əm }

autocarp [BOTANY] **1.** A fruit formed as the result of self-fertilization. **2.** A fruit consisting of the ripened pericarp without adnate parts. { ′ȯd·ō ˌkärp }

autocarpy [BOTANY] Production of fruit by self-fertilization. { ′ȯd·ōˌkärp·ē }

autochory [ECOLOGY] Active self-dispersal of individuals or their disseminules. { ′ȯd·ōˌkȯr·ē }

autochthon [PALEONTOLOGY] A fossil occurring where the organism once lived. { ȯ′täk·thən }

autochthonous [ECOLOGY] Pertaining to organisms or organic sediments that are indigenous to a given ecosystem. { ȯ′täk·thə·nas }

autochthonous microorganism [MICROBIOLOGY] An indigenous form of soil microorganisms, responsible for chemical processes that occur in the soil under normal conditions. { ȯ′täk·thə·nas ˌmī·krō′ȯr·gəˌniz·əm }

autocoenobium [INVERTEBRATE ZOOLOGY] An asexually produced coenobium that is a miniature of the parent. { ¦ȯd·ō·sē′nō·bē·əm }

autocopulation [INVERTEBRATE ZOOLOGY] Self-copulation; sometimes occurs in certain hermaphroditic worms. { ¦ȯd·ōˌkäp·yə′lā·shan }

autocrine signaling [PHYSIOLOGY] Signaling in which cells respond to substances that they themselves release. { ¦ȯd·əˌkrīn ′sig·nəl·iŋ }

autodeme [ECOLOGY] A plant population in which most individuals are self-fertilized. { ′ȯd·ōˌdēm }

autoecious [BOTANY] *See* autoicous. [MYCOLOGY] Referring to a parasitic fungus that completes its entire life cycle on a single host. { ȯ ′tēsh·əs }

autogamy [BIOLOGY] A process of self-fertilization that results in homozygosis; occurs in some flowering plants, fungi, and protozoans. { ′ȯd·ō ˌgam·ē }

autogenesis [BIOLOGY] *See* abiogenesis. { ¦ȯd·ō′jen·ə·səs }

autogenous [BIOLOGY] Originating or derived from sources within the same individual. { ȯ ′täj·ə·nəs }

autogenous control [MOLECULAR BIOLOGY] Regulation of gene expression by a product of the gene itself that either inhibits or enhances the gene's activity. { ȯ′täj·ə·nəs kən′trōl }

autogenous insect [INVERTEBRATE ZOOLOGY] Any insect in which adult females can produce eggs without first feeding. { ȯ′täj·ə·nəs ′in ˌsekt }

autogenous vaccine [IMMUNOLOGY] A vaccine prepared from a culture of microorganisms taken directly from the infected person. { ȯ′täj·ə·nəs ˌvak′sēn }

autograft [BIOLOGY] *See* autotransplant. { ′ȯd·ō ˌgraft }

autohemagglutination [IMMUNOLOGY] *See* autoagglutination. { ¦ȯd·ōˌhēm·əˌglüt·ən′ā·shən }

autohemolysis [PHYSIOLOGY] The spontaneous lysis of blood which occurs during an incubation in a hematological procedure. { ¦ȯd·ō·hə'mäl·ə·səs }

autohemorrhage [INVERTEBRATE ZOOLOGY] Voluntary exudation or ejection of nauseous or poisonous blood by certain insects as a defense against predators. { ¦ȯd·ō'hem·rij }

autoicous [BOTANY] Having male and female organs on the same plant but on different branches. Also spelled autoecious. { ȯ'tō·ə·kəs }

autoimmune disease [IMMUNOLOGY] An illness involving the formation of autoantibodies which appear to cause pathological damage to the host. { ¦ȯd·ō·ə'myün di,zēz }

autoimmunity [IMMUNOLOGY] An immune state in which antibodies are formed against the person's own body tissues. { ¦ȯd·ō·ə'myün·əd·ē }

autoimmunization [IMMUNOLOGY] Immunization obtained by natural processes within the body. { ¦ȯd·ō,i·myə·nə'zā·shən }

autologous [BIOLOGY] Derived from or produced by the individual in question, such as an autologous protein or an autologous graft. { ȯ'täl·ə·gəs }

autolysosome [CELL BIOLOGY] *See* autophagic vacuole. { ¦ȯd·ō'lī·sə,sōm }

autolytic enzyme [BIOCHEMISTRY] A bacterial enzyme, located in the cell wall, that causes disintegration of the cell following injury or death. { ¦ȯd·əl¦id·ik 'en,zīm }

Autolytinae [INVERTEBRATE ZOOLOGY] A subfamily of errantian polychaetes in the family Syllidae. { ,ȯd·ə'lid·ə·nē }

automatism [BIOLOGY] Spontaneous activity of tissues or cells. { ȯ'täm·ə,tiz·əm }

automutagen [GENETICS] Any mutagenic chemical formed as a product of metabolism. { ¦ȯd·ō 'myü·də·jən }

autonomic movement [BOTANY] A plant movement that results from internal growth changes and is independent of changes in the external environment. { ¦ȯd·ə¦näm·ik 'müv·mənt }

autonomic nervous system [ANATOMY] The visceral or involuntary division of the nervous system in vertebrates, which enervates glands, viscera, and smooth, cardiac, and some striated muscles. { ¦ȯd·ə¦näm·ik 'nər·vəs ,sis·təm }

autonomic reflex system [PHYSIOLOGY] An involuntary biological control system characterized by the uncontrolled functioning of smooth muscles and glands to maintain an acceptable internal environment. { ¦ȯd·ə¦näm·ik 'rē,fleks ,sis·təm }

autophagic vacuole [CELL BIOLOGY] A membrane-bound cellular organelle that engulfs pieces of the substance of the cell itself. Also known as autolysosome. { ¦ȯd·ō¦fā·jik 'vak·yə·wōl }

autophagocytosis [CELL BIOLOGY] The cellular process of phagocytizing a portion of protoplasm by a vacuole within the cell. { ¦ȯd·ō,fag·ə·sī'tō·səs }

autophagy [CELL BIOLOGY] The cellular process of self-digestion. { 'ȯd·ə,fā·jē }

autopolyploid [GENETICS] A cell or organism having three or more sets of chromosomes all derived from the same species. { ¦ȯd·ō¦päl·i ,plȯid }

autoserum [IMMUNOLOGY] A serum obtained from a patient used for treatment of that patient. { 'ȯd·ō,sir·əm }

autosexing [BIOLOGY] Displaying differential sex characters at birth, noted particularly in fowl bred for sex-specific colors and patterns. { 'ȯd·ō,seks·iŋ }

autoskeleton [INVERTEBRATE ZOOLOGY] The endoskeleton of a sponge. { 'ȯd·ō,skel·ə·tən }

autosomal trait [GENETICS] Any characteristic determined by autosomal genes. { ¦ȯd·ə¦sō·məl 'trāt }

autosome [GENETICS] Any chromosome other than a sex chromosome. { 'ȯd·ō,sōm }

autospore [BOTANY] In algae, a nonmotile asexual reproductive cell or a nonmotile spore that is a miniature of the cell that produces it. { 'ȯd·ō,spȯr }

autostylic [VERTEBRATE ZOOLOGY] Having the jaws attached directly to the cranium, as in chimeras, amphibians, and higher vertebrates. { ¦ȯd·ō¦stīl·ik }

autosyndesis [CELL BIOLOGY] The act of pairing of homologous chromosomes from the same parent during meiosis in polyploids. { ¦ȯd·ō ¦sin·də·səs }

autotomy [ZOOLOGY] The process of self-amputation of appendages in crabs and other crustaceans and tails in some salamanders and lizards under stress. { ȯ'täd·ə·mē }

autotransplant [BIOLOGY] Tissue removed from an organism and grafted on another site of the same organism. Also known as autograft. { ¦ȯd·ō'tranz,plant }

autotroph [BIOLOGY] An organism capable of synthesizing organic nutrients directly from simple inorganic substances, such as carbon dioxide and inorganic nitrogen. { 'ȯd·ō,träf }

autozooecium [INVERTEBRATE ZOOLOGY] The tube enclosing an autozooid. { ¦ȯd·ō,zō'ēsh·ē·əm }

autozooid [INVERTEBRATE ZOOLOGY] An unspecialized feeding individual in a bryozoan colony, possessing fully developed organs and exoskeleton. { ¦ȯd·ō¦zō,ȯid }

auxanogram [MICROBIOLOGY] A plate culture provided with variable growth conditions to determine the effects of specific environmental factors. { ȯg'zan·ə,gram }

auxanography [MICROBIOLOGY] The study of growth-inhibiting or growth-promoting agents by means of auxanograms. { ,ȯg·zə'näg·rə·fē }

auxesis [PHYSIOLOGY] Growth resulting from increase in cell size. { ,ȯg'zē·səs }

auximone [BIOCHEMISTRY] Any of certain growth-promoting substances occurring principally in sphagnum peat decomposed by nitrogen bacteria. { 'ȯk·sə,mōn }

auxin [BIOCHEMISTRY] Any organic compound which promotes plant growth along the longitudinal axis when applied to shoots free from indigenous growth-promoting substances. { 'ȯk·sən }

auxoautotrophic [BIOLOGY] Requiring no exogenous growth factors. { ¦ȯk·sō¦ȯd·ə¦trä·fik }

auxocyte [BIOLOGY] A gamete-forming cell, such as an oocyte or spermatocyte, or a sporocyte during its growth period. Also known as gonotocont. { 'ȯk·sə,sīt }

auxoheterotrophic [BIOLOGY] Requiring exogenous growth factors. { ¦ȯk·sō¦hed·ə·rə¦trä·fik }

auxospore [INVERTEBRATE ZOOLOGY] A reproductive cell in diatoms formed in association with rejuvenescence by the union of two cells that have diminished in size through repeated divisions. { 'ȯk·sə,spȯr }

auxotonic [BOTANY] Induced by growth rather than by exogenous stimuli. { ,ȯk·sə'tän·ik }

auxotrophic mutant [GENETICS] An organism that requires a specific growth factor, such as an amino acid, for its growth. { ¦ȯk·sə¦trä·fik 'myüt·ənt }

available-chlorine method [MICROBIOLOGY] A technique for the standardization of chlorine disinfectants intended for use as germicidal rinses on cleaned surfaces; increments of bacterial inoculum are added to different disinfectant concentrations, and after incubation the results indicate the capacity of the disinfectant to handle an increasing bacterial load before exhaustion of available chlorine, the germicidal principle. { ə 'vāl·ə·bəl 'klȯr,ēn ,meth·əd }

avalanche conduction [PHYSIOLOGY] Conduction of a nerve impulse through several neurons which converge, increasing the discharge intensity by summation. { 'av·ə,lanch kən'dək·shən }

Avena [BOTANY] A genus of grasses (family Gramineae), including oats, characterized by an inflorescence that is loosely paniculate, two-toothed lemmas, and deeply furrowed grains. { ə 'vēn·ə }

avenin [BIOCHEMISTRY] The glutelin of oats. { ə 'vēn·ən }

Aves [VERTEBRATE ZOOLOGY] A class of animals composed of the birds, which are warm-blooded, egg-laying vertebrates primarily adapted for flying. { 'ā,vēz }

avianize [VIROLOGY] To attenuate a virus by repeated culture on chick embryos. { 'av·ē·ə,nīz }

avicolous [ECOLOGY] Living on birds, as of certain insects. { ā'vik·ə·ləs }

avicularium [INVERTEBRATE ZOOLOGY] A specialized individual in a bryozoan colony with a beak that keeps other animals from settling on the colony. { ə,vik·yə'lar·ē·ən }

aviculture [VERTEBRATE ZOOLOGY] Care and breeding of birds, especially wild birds, in captivity. { 'ā·və,kəl·chər }

avidin [BIOCHEMISTRY] A protein constituting 0.2% of the total protein in egg white; molecular weight is 70,000; combines firmly with biotin but loses this ability when subjected to heat. { 'av·əd·ən }

avifauna [VERTEBRATE ZOOLOGY] **1.** Birds, collectively. **2.** Birds characterizing a period, region, or environment. { ¦ā·və¦fȯn·ə }

avocado [BOTANY] *Persea americana.* A subtropical evergreen tree of the order Magnoliales that bears a pulpy pear-shaped edible fruit. { ,av·ə 'käd·ō }

awn [BOTANY] Any of the bristles at the ends of glumes or bracts on the spikelets of oats, barley, and some wheat and grasses. Also known as beard. { ȯn }

axenic culture [BIOLOGY] The growth of organisms of a single species in the absence of cells or living organisms of any other species. { ā'zen·ik 'kəl·chər }

axial filament [CELL BIOLOGY] The central microtubule elements of a cilium or flagellum. [INVERTEBRATE ZOOLOGY] An organic fiber which serves as the core for deposition of mineral substance to form a ray of a sponge spicule. { 'ak·sē·əl 'fil·ə·mənt }

axial gland [INVERTEBRATE ZOOLOGY] A structure enclosing the stone canal in certain echinoderms; its function is uncertain. { 'ak·sē·əl 'gland }

axial gradient [EMBRYOLOGY] In some invertebrates, a graded difference in metobolic activity along the anterior-posterior, dorsal-ventral, and medial-lateral embryonic axes. { 'ak·sē·əl 'grād·ē·ənt }

axial musculature [ANATOMY] The muscles that lie along the longitudinal axis of the vertebrate body. { 'ak·sē·əl 'məs·kyə·lə·chər }

axial skeleton [ANATOMY] The bones composing the skull, vertebral column, and associated structures of the vertebrate body. { 'ak·sē·əl 'skel·i·tən }

axiation [EMBRYOLOGY] The formation or development of axial structures, such as the neural tube. { ,ak·sē'ā·shən }

Axiidae [INVERTEBRATE ZOOLOGY] A family of decapod crustaceans, including the hermit crabs, in the suborder Reptantia. { ,ak'sī·ə,dē }

axil [BIOLOGY] The angle between a structure and the axis from which it arises, especially for branches and leaves. { 'ak·səl }

axilla [ANATOMY] The depression between the arm and the thoracic wall; the armpit. [BOTANY] An axil. { ak'sil·ə }

axillary [ANATOMY] Of, pertaining to, or near the axilla or armpit. [BOTANY] Placed or growing in the axis of a branch or leaf. { 'ak·sə,ler·ē }

axillary bud [BOTANY] A lateral bud borne in the axil of a leaf. { 'ak·sə,ler·ē 'bəd }

axillary sweat gland [ANATOMY] An apocrine gland located in the axilla. { 'ak·sə,ler·ē 'swet ,gland }

Axinellina [INVERTEBRATE ZOOLOGY] A suborder of sponges in the order Clavaxinellida. { ,ak·sə 'nə'lī·nə }

axis [ANATOMY] **1.** The second cervical vertebra in higher vertebrates; the first vertebra of amphibians. **2.** The center line of an organism, organ, or other body part. { 'ak·səs }

axis cylinder [CELL BIOLOGY] **1.** The central mass of a nerve fiber. **2.** The core of protoplasm in a medullated nerve fiber. { 'ak·səs ¦sil·ən·dər }

axis of pelvis [ANATOMY] A curved line which

forms right angles to the pelvic-cavity planes. { 'ak·səs əv 'pel·vəs }

axoblast [INVERTEBRATE ZOOLOGY] **1.** The germ cell in mesozoans; cells are linearly arranged in the longitudinal axis and produce the primary nematogens. **2.** The individual scleroblast of the axis epithelium which produces spicules in octocorals. { 'ak·sə,blast }

axocoel [INVERTEBRATE ZOOLOGY] The anterior pair of coelomic sacs in the dipleurula larval ancestral stage of echinoderms. { 'ak·sə,sēl }

axogamy [BOTANY] Having sex organs on a leafy stem. { ak'säg·ə·mē }

axolemma [CELL BIOLOGY] The plasma membrane of an axon. { ,ak·sə'lem·ə }

axolotl [VERTEBRATE ZOOLOGY] The neotenous larva of some salamanders in the family Ambystomidae. { ¦ak·sə¦läd·əl }

axolotl unit [BIOLOGY] A unit for the standardization of thyroid extracts. { ¦ak·sə¦läd·əl 'yü·nət }

axon [ANATOMY] The process or nerve fiber of a neuron that carries the unidirectional nerve impulse away from the cell body. Also known as neuraxon; neurite. { 'ak,sän }

axoneme [CELL BIOLOGY] A bundle of fibrils enclosed by a membrane that is continuous with the plasma membrane. { 'ak·sə,nēm }

Axonolaimoidea [INVERTEBRATE ZOOLOGY] A superfamily of free-living nematodes with species inhabiting marine and brackish-water environments. { ¦ak·sə·nō·lə'mȯid·ē·ə }

axoplasm [CELL BIOLOGY] The protoplasm of an axon. { 'ak·sə,plaz·əm }

axopodium [INVERTEBRATE ZOOLOGY] A semipermanent pseudopodium composed of axial filaments surrounded by a cytoplasmic envelope. { ,ak·sə'pōd·ē·əm }

aye-aye [VERTEBRATE ZOOLOGY] *Daubentonia madagascariensis*. A rare prosimian primate indigenous to eastern Madagascar; the single species of the family Daubentoniidae. { 'ī,ī }

2-azetidinecarboxylic acid [BIOCHEMISTRY] $C_4H_7NO_2$ Crystals which discolor at 200°C and darken until 310°C; soluble in water; a specific antagonist of L-proline; used in the production of abnormally high-molecular-weight polypeptides. { ¦tü ə,zed·ə,dēn·ə,kär,bäk'sil·ik 'as·əd }

Azomonas [MICROBIOLOGY] A genus of large, motile, oval to spherical bacteria in the family Axotobacteraceae; members produce no cysts and secrete large quantities of capsular slime. { ,az·ə'mō·nəs }

azomycin [MICROBIOLOGY] $C_3H_3O_2N_3$ An antimicrobial antibiotic produced by a strain of *Nocardia mesenterica*. { ā·zō'mis·ən }

azoospermia [PHYSIOLOGY] **1.** Absence of motile sperm in the semen. **2.** Failure of formation and development of sperm. { ¦ā,zō·ō'spər·mē·ə }

Azotobacter [MICROBIOLOGY] A genus of large, usually motile, rod-shaped, oval, or spherical bacteria in the family Azotobacteraceae; form thick-walled cysts, and may produce large quantities of capsular slime. { ə'zōd·ə,bak·tər }

Azotobacteraceae [MICROBIOLOGY] A family of large, bluntly rod-shaped, gram-negative, aerobic bacteria capable of fixing molecular nitrogen. { ə¦zōd·ə,bak·tə'rās·ē,ē }

azure B [CELL BIOLOGY] $C_{15}H_{16}ClN_3S$ A metachromatic basic dye that imparts a green color to chromosomes, a blue color to nucleoli and cytoplasmic ribosomes, and a red color to deposits containing mucopolysaccharides. { ¦azh·ər 'bē }

azygospore [MYCOLOGY] A spore which is morphologically similar to a zygospore but is formed parthenogenetically. Also known as parthenospore. { ā'zī·gə,spȯr }

azygos vein [ANATOMY] A branch of the right precava which drains the intercostal muscles and empties into the superior vena cava. { ā'zī·gəs ,vān }

azygote [BIOLOGY] An individual produced by haploid parthenogenesis. { ā'zī,gōt }

B

Babesia [INVERTEBRATE ZOOLOGY] The type genus of the Babesiidae, a protozoan family containing red blood cell parasites. { bə'bezh·ə }

Babesiidae [INVERTEBRATE ZOOLOGY] A family of protozoans in the suborder Haemosporina containing parasites of vertebrate red blood cells. { ˌbab·ə'zī·əˌdē }

baboon [VERTEBRATE ZOOLOGY] Any of five species of large African and Asian terrestrial primates of the genus *Papio*, distinguished by a doglike muzzle, a short tail, and naked callosities on the buttocks. { ba'bün }

babulna [VERTEBRATE ZOOLOGY] A female baboon. { ˌbab·ə'wēn·ə }

baccate [BOTANY] **1.** Bearing berries. **2.** Having pulp like a berry. { 'bakˌāt }

bacciferous [BOTANY] Bearing berries. { bak'sif·ə·rəs }

Bacillaceae [MICROBIOLOGY] The single family of endospore-forming rods and cocci. { ˌbas·ə'lās·ēˌē }

Bacillariophyceae [BOTANY] The diatoms, a class of algae in the division Chrysophyta. { ˌbas·əˌler·ē·ə'fīs·ēˌē }

Bacillariophyta [BOTANY] An equivalent name for the Bacillariophyceae. { ˌbas·əˌler·ē'ä·fəd·ə }

bacillary [MICROBIOLOGY] **1.** Rod-shaped. **2.** Produced by, pertaining to, or resembling bacilli. { 'bas·əˌler·ē }

bacillus [MICROBIOLOGY] Any rod-shaped bacterium. { bə'sil·əs }

Bacillus [MICROBIOLOGY] A genus of bacteria in the family Bacillaceae; rod-shaped cells are aerobes or facultative anaerobes and usually produce catalase. { bə'sil·əs }

Bacillus Calmette-Guérin vaccine [IMMUNOLOGY] A vaccine prepared from attenuated human tubercle bacilli and used to immunize humans against tuberculosis. Abbreviated BCG vaccine. { bə'sil·əs ˌkal'met ˌgā'ran ˌvak'sēn }

bacitracin [MICROBIOLOGY] A group of polypeptide antibiotics produced by *Bacillus licheniformis*. { ˌbas·ə'trās·ən }

back [ANATOMY] The part of the human body extending from the neck to the base of the spine. { bak }

backbone [ANATOMY] *See* spine. { 'bakˌbōn }

back bulb [BOTANY] A pseudobulb on certain orchid plants that remains on the plant after removal of the terminal growth, and that is used for propagation. { 'bak ˌbəlb }

backcross [GENETICS] A cross between an F_1 heterozygote and an individual of P_1 genotype. { 'bakˌkrȯs }

backcross parent [GENETICS] *See* recurrent parent. { 'bakˌkrȯs ˌper·ənt }

background genotype [GENETICS] The genotype of the organism in addition to the genetic loci responsible for the phenotype. { ¦baˌkraund 'jē·nəˌtīp }

backmarsh [ECOLOGY] Marshland formed in poorly drained areas of an alluvial floodplain. { 'bakˌmärsh }

backswamp depression [ECOLOGY] A low swamp found adjacent to river levees. { 'bakˌswamp di'presh·ən }

bacteria [MICROBIOLOGY] Extremely small, relatively simple prokaryotic microorganisms traditionally classified with the fungi as Schizomycetes. { bak'tir·ē·ə }

Bacteriaceae [MICROBIOLOGY] A former designation for Brevibacteriaceae. { bakˌtir·ē'ās·ēˌē }

bacterial capsule [MICROBIOLOGY] A thick, mucous envelope, composed of polypeptides or carbohydrates, surrounding some bacteria. { bak'tir·ē·əl 'kap·səl }

bacterial coenzyme [MICROBIOLOGY] Organic molecules that participate directly in a bacterial enzymatic reaction and may be chemically altered during the reaction. { bak'tir·ē·əl kō'enˌzīm }

bacterial competence [MICROBIOLOGY] The ability of cells in a bacterial culture to accept and be transformed by a molecule of transforming deoxyribonucleic acid. { bak'tir·ē·əl 'käm·pə·təns }

bacterial endoenzyme [MICROBIOLOGY] An enzyme produced and active within the bacterial cell. { bak'tir·ē·əl ˌen·dō'enˌzīm }

bacterial endospore [MICROBIOLOGY] A body, resistant to extremes of temperature and to dehydration, produced within the cells of gram-positive, sporeforming rods of *Bacillus* and *Clostridium* and by the coccus *Sporosarcina*. { bak'tir·ē·əl 'en·dōˌspȯr }

bacterial enzyme [MICROBIOLOGY] Any of the metabolic catalysts produced by bacteria. { bak'tir·ē·əl 'enˌzīm }

bacterial genetics [GENETICS] The study of inheritance and variation patterns in bacteria. { bak'tir·ē·əl jə'ned·iks }

bacterial luminescence [MICROBIOLOGY] A light-producing phenomenon exhibited by certain bacteria. { bak'tir·ē·əl lü·mə'nes·əns }

bacterial metabolism [MICROBIOLOGY] Total chemical changes carried out by living bacteria. { bak'tir·ē·əl mə'tab·ə,liz·əm }

bacterial motility [MICROBIOLOGY] Self-propulsion in bacteria, either by gliding on a solid surface or by moving the flagella. { bak'tir·ē·əl mō'til·əd·ē }

bacterial photosynthesis [MICROBIOLOGY] Use of light energy to synthesize organic compounds in green and purple bacteria. { bak'tir·ē·əl ,fōd·ō'sin·thə·səs }

bacterial pigmentation [MICROBIOLOGY] The organic compounds produced by certain bacteria which give color to both liquid cultures and colonies. { bak'tir·ē·əl ,pig·mən'tā·shən }

bacterial vaccine [IMMUNOLOGY] A preparation of living, attenuated, or killed bacteria used to enhance the immune reaction in an individual already infected with the same bacteria. { bak'tir·ē·əl vak'sēn }

bactericidin [IMMUNOLOGY] An antibody that kills bacteria in the presence of complement. { bak¦tir·ə¦sīd·ən }

bacterin [IMMUNOLOGY] A suspension of killed or weakened bacteria used in artificial immunization. { 'bak·tə·rən }

bacteriochlorophyll [BIOCHEMISTRY] $C_{52}H_{70}O_6N_4Mg$ A tetrahydroporphyrin chlorophyll compound occurring in the forms *a* and *b* in photosynthetic bacteria; there is no evidence that *b* has the empirical formula given. { bak¦tir·ē·ə'klȯr·ə,fil }

bacteriocin [MICROBIOLOGY] Any of a group of proteins produced by various strains of gram-negative bacteria that may inhibit the growth of other strains of the same or related species. { bak'tir·ē·ə,sīn }

bacteriocinogen [GENETICS] A plasmid deoxyribonucleic acid found in some strains of bacteria which specifies production of a bacteriocin. { bak,tir·ē·ə'sin·ə·jən }

bacteriocyte [INVERTEBRATE ZOOLOGY] A modified fat cell found in certain insects that contains bacterium-shaped rods believed to be symbiotic bacteria. { bak'tir·ē·ə,sīt }

bacteriogenic [MICROBIOLOGY] Caused by bacteria. { bak¦tir·ē·ə'jen·ik }

bacteriologist [MICROBIOLOGY] A specialist in the study of bacteria. { bak,tir·ē'äl·ə·jəst }

bacteriology [MICROBIOLOGY] The science and study of bacteria; a specialized branch of microbiology. { bak,tir·ē'äl·ə·jē }

bacteriolysin [MICROBIOLOGY] An antibody that is active against and causes lysis of specific bacterial cells. { bak,tir·ē·ə'līs·ən }

bacteriolysis [MICROBIOLOGY] Dissolution of bacterial cells. { bak,tir·ē'äl·ə·səs }

Bacterionema [MICROBIOLOGY] A genus of bacteria in the family Actinomycetaceae; characteristically cells are rods with filaments attached and produce filamentous microcolonies; facultative anaerobes; carbohydrates are fermented. { bak,tir·ē'än·ə·mə }

bacteriophage [VIROLOGY] Any of the viruses that infect bacterial cells; each has a narrow host range. Also known as phage. { bak'tir·ē·ə,fāj }

bacteriostasis [MICROBIOLOGY] Inhibition of bacterial growth and metabolism. { bak,tir·ē·ō'stā·səs }

bacteriostatic agent [MICROBIOLOGY] A substance that inhibits the growth of bacteria. { ,bak¦tir·ē·ō¦stad·ik 'ā·jənt }

bacteriotoxin [MICROBIOLOGY] **1.** Any toxin that destroys or inhibits growth of bacteria. **2.** A toxin produced by bacteria. { bak,tir·ē·ō'täk·sən }

bacteriotropin [IMMUNOLOGY] An antibody that is increased in amount during specific immunization and that renders the corresponding bacterium more susceptible to phagocytosis. { bak,tir·ē'ä·trə·pən }

bacterioviridin [BIOCHEMISTRY] *See* chlorobium chlorophyll. { bak,tir·ē·ə'vir·ə·dən }

bacteroid [MICROBIOLOGY] A bacterial form of irregular shape, frequently associated with special conditions. { 'bak·tə,rȯid }

Bacteroidaceae [MICROBIOLOGY] The single family of gram-negative anaerobic bacteria; cells are nonsporeforming rods; some species are pathogenic. { ,bak·tə,rȯi'dās·ē,ē }

Bactrian camel [VERTEBRATE ZOOLOGY] *Camelus bactrianus*. The two-humped camel. { 'bak·trē·ən 'kam·əl }

baculum [VERTEBRATE ZOOLOGY] The penis bone in lower mammals. Also known as os priapi. { 'bak·yə·ləm }

badger [VERTEBRATE ZOOLOGY] Any of eight species of carnivorous mammals in six genera comprising the subfamily Melinae of the weasel family (Mustelidae). { 'baj·ər }

baeocyte [INVERTEBRATE ZOOLOGY] A motile or nonmotile blue-green bacterial endospore. { 'bē·ə,sīt }

Bagridae [VERTEBRATE ZOOLOGY] A family of semitropical catfishes in the suborder Siluroidei. { 'bag·rə,dē }

Bainbridge reflex [PHYSIOLOGY] A poorly understood reflex acceleration of the heart rate due to rise of pressure in the right atrium and vena cavae, possibly mediated through afferent vagal fibers. { 'bān,brij 'rē,fləks }

Bairdiacea [INVERTEBRATE ZOOLOGY] A superfamily of ostracod crustaceans in the suborder Podocopa. { ,ber·dē'ās·ē·ə }

Balaenicipitidae [VERTEBRATE ZOOLOGY] A family of wading birds composed of a single species, the shoebill stork (*Balaeniceps rex*), in the order Ciconiiformes. { bə,lēn·ə·sə'pid·ə,dē }

Balaenidae [VERTEBRATE ZOOLOGY] The right whales, a family of cetacean mammals composed of five species in the suborder Mysticeti. { bə'lēn·ə,dē }

balanced polymorphism [GENETICS] Maintenance in a population of two or more alleles in equilibrium at frequencies too high to be explained, particularly for the rarer of them, by mutation balanced by selection; for example, the selective advantage of heterozygotes over both homozygotes. { 'bal·ənst ¦päl·i'mȯr‚fiz·əm }

balanced translocation [GENETICS] Positional change of one or more chromosome segments in cells or gametes without alteration of the normal diploid or haploid complement of genetic material. { 'bal·ənst tranz·lō'kā·shən }

balancer [INVERTEBRATE ZOOLOGY] *See* haltere. [VERTEBRATE ZOOLOGY] Either of a pair of rodlike lateral appendages on the heads of some larval salamanders. { 'bal·ən·sər }

Balanidae [INVERTEBRATE ZOOLOGY] A family of littoral, sessile barnacles in the suborder Balanomorpha. { bə'lan·ə‚dē }

Balanoglossus [INVERTEBRATE ZOOLOGY] A cosmopolitan genus of tongue worms belonging to the class Enteropneusta. { ‚bal·ə·nō'glä·səs }

Balanomorpha [INVERTEBRATE ZOOLOGY] The symmetrical barnacles, a suborder of sessile crustaceans in the order Thoracica. { ‚bal·ə·nō 'mȯr·fə }

Balanopaceae [BOTANY] A small family of dioecious dicotyledonous plants in the order Fagales characterized by exstipulate leaves, seeds with endosperm, and the pistillate flower solitary in a multibracteate involucre. { ‚bal·ə·nō'pās·ē‚ē }

Balanopales [BOTANY] An ordinal name suggested for the Balanopaceae in some classifications. { ‚bal·ə'näp·ə‚lēz }

Balanophoraceae [BOTANY] A family of dicotyledonous terrestrial plants in the order Santalales characterized by dry nutlike fruit, one to five ovules, unisexual flowers, attachment to the stem of the host, and the lack of chlorophyll. { ‚bal·ə‚näf·ə'rās·ē‚ē }

Balanopsidales [BOTANY] An order in some systems of classification which includes only the Balanopaceae of the Fagales. { ‚bal·ən‚äp·sə 'dā·lēz }

Balantidium [INVERTEBRATE ZOOLOGY] A genus of protozoans in the order Trichostomatida containing the only ciliated protozoan species parasitic in humans, *Balantidium coli*. { ‚bal·ən'tid·ē·əm }

Balanus [INVERTEBRATE ZOOLOGY] A genus of barnacles composed of sessile acorn barnacles; the type genus of the family Balanidae. { 'bal·ə·nəs }

Balbiani chromosome [GENETICS] *See* polytene chromosome. { ‚bäl·bē'än·ē 'krō·mə‚sōm }

Balbiani rings [CELL BIOLOGY] Localized swellings of a polytene chromosome. { ‚bäl·bē'än·ē ‚riŋz }

baleen [VERTEBRATE ZOOLOGY] A horny substance, growing as fringed filter plates suspended from the upper jaws of whalebone whales. Also known as whalebone. { be'lēn }

Balfour's law [EMBRYOLOGY] The law that the speed with which any part of the ovum segments is roughly proportional to the protoplasm's concentration in that area; the segment's size is inversely proportional to the protoplasm's concentration. { 'bal·fərz ‚lȯ }

ball-and-socket joint [ANATOMY] *See* enarthrosis. { ¦bȯl ən 'säk·ət ‚jȯint }

ballistospore [MYCOLOGY] A type of fungal spore that is forcibly discharged at maturity. { bə'lis·tə ‚spȯr }

balsa [BOTANY] *Ochroma lagopus*. A tropical American tree in the order Malvales; its wood is strong and lighter than cork. { 'bȯl·sə }

Balsaminaceae [BOTANY] A family of flowering plants in the order Geraniales, including touch-me-not (*Impatiens*); flowers are irregular with five stamens and five carpels, leaves are simple with pinnate venation, and the fruit is an elastically dehiscent capsule. { ‚bȯl·sə·mə'nās·ē‚ē }

bamboo [BOTANY] The common name of various tropical and subtropical, perennial, ornamental grasses in five genera of the family Gramineae characterized by hollow woody stems up to 6 inches (15 centimeters) in diameter. { bam'bü }

Bambusoideae [BOTANY] A subfamily of grasses, composed of bamboo species, in the family Gramineae. { ‚bam·bə'sȯid·ē‚ē }

banana [BOTANY] Any of the treelike, perennial plants of the genus *Musa* in the family Musaceae; fruit is a berry characterized by soft, pulpy flesh and a thin rind. { bə'nan·ə }

band [CELL BIOLOGY] Any of the characteristic transverse stripes exhibited by polytene or metaphase chromosomes that are stained. { band }

banded anteater [VERTEBRATE ZOOLOGY] *See* marsupial anteater. { 'ban·dəd 'ant‚ēd·ər }

bandicoot [VERTEBRATE ZOOLOGY] **1.** Any of several large Indian rats of the genus *Nesokia* and related genera. **2.** Any of several small insectivorous and herbivorous marsupials comprising the family Peramelidae and found in Tasmania, Australia, and New Guinea. { 'ban·di‚küt }

Bangiophyceae [BOTANY] A class of red algae in the plant division Rhodophyta. { ‚baŋ·ē·ə'fīs·ē‚ē }

banner [BOTANY] The fifth or posterior petal of a butterfly-shaped (papilionaceous) flower. { 'ban·ər }

barb [VERTEBRATE ZOOLOGY] A side branch on the shaft of a bird's feather. { bärb }

barbel [VERTEBRATE ZOOLOGY] **1.** A slender, tactile process near the mouth in certain fishes, such as catfishes. **2.** Any European fresh-water fish in the genus *Barbus*. { 'bär·bəl }

barbellate [BIOLOGY] Having short, stiff, hooked bristles. { 'bär·bə‚lāt }

barbicel [VERTEBRATE ZOOLOGY] One of the small, hook-bearing processes on a barbule of the distal side of a barb or a feather. { 'bär·bə ‚sel }

bark [BOTANY] The tissues external to the cambium in a stem or root. { bärk }

bark graft [BOTANY] A graft made by slipping the scion beneath a slit in the bark of the stock. { 'bärk ‚graft }

bark pocket [BOTANY] An opening between tree annual rings which contains bark. { 'bärk ‚päk·ət }

barley [BOTANY] A plant of the genus *Hordeum* in the order Cyperales that is cultivated as a grain crop; the seed is used to manufacture malt beverages and as a cereal. { 'bär·lē }

barnacle [INVERTEBRATE ZOOLOGY] The common

name for a number of species of crustaceans which compose the subclass Cirripedia. { 'bär·nə·kəl }

baroduric bacteria [MICROBIOLOGY] Bacteria that can tolerate conditions of high hydrostatic pressure. { ¦bar·ə¦dùr·ik bak'tir·ē·ə }

barophile [MICROBIOLOGY] An organism that thrives under conditions of high hydrostatic pressure. { 'bar·ə,fīl }

barotaxis [BIOLOGY] Orientation movement of an organism in response to pressure changes. { ¦bar·ə¦tak·səs }

barracuda [VERTEBRATE ZOOLOGY] The common name for about 20 species of fishes belonging to the genus *Sphyraena* in the order Perciformes. { ,bar·ə'küd·ə }

Barr body [CELL BIOLOGY] A condensed, inactivated X chromosome inside the nuclear membrane in interphase somatic cells of women and most female mammals. Also known as sex chromatin. { 'bär ,bäd·ē }

barrier [ECOLOGY] Any physical or biological factor that restricts the migration or free movement of individuals or populations. { 'bar·ē·ər }

barrier marsh [ECOLOGY] A type of marsh that restricts or prevents invasion of the area beyond it by new species of animals. { 'bar·ē·ər ,märsh }

barrier zone [BOTANY] In a tree, new tissue formed by the cambium after it has been wounded; serves as both an anatomical and a chemical wall. { 'bar·ē·ər ,zōn }

Bartonella [MICROBIOLOGY] A genus of the family Bartonellaceae; parasites in or on red blood cells and within fixed tissue cells; found in humans and in the arthropod genus *Phlebotomus*. { ,bärt·ən'el·ə }

Bartonellaceae [MICROBIOLOGY] A family of the order Rickettsiales; rod-shaped, coccoid, ring- or disk-shaped cells; parasites of human and other vertebrate red blood cells. { ,bärt·ən,e'lās·ē,ē }

Barychilinidae [PALEONTOLOGY] A family of Paleozoic crustaceans in the suborder Platycopa. { ,bar·ə·kə'lin·ə,dē }

Barylambdidae [PALEONTOLOGY] A family of late Paleocene and early Eocene aquatic mammals in the order Pantodonta. { ,bar·ə'lam·də,dē }

Barytheriidae [PALEONTOLOGY] A family of extinct proboscidean mammals in the suborder Barytherioidea. { ,bar·ə·thə'rī·ə,dē }

Barytherioidea [PALEONTOLOGY] A suborder of extinct mammals of the order Proboscidea, in some systems of classification. { ,bar·ə,thir·ē'ȯid·ē·ə }

basal [BIOLOGY] Of, pertaining to, or located at the base. [PHYSIOLOGY] Being the minimal level for, or essential for maintenance of, vital activities of an organism, such as basal metabolism. { 'bā·səl }

basal body [CELL BIOLOGY] A cellular organelle that induces the formation of cilia and flagella and is similar to and sometimes derived from a centriole. Also known as kinetosome. { 'bā·səl ,bäd·ē }

basal disc [BIOLOGY] The expanded basal portion of the stalk of certain sessile organisms, used for attachment to the substrate. { 'bā·səl 'disk }

basal ganglia [ANATOMY] The corpus striatum, or the corpus striatum and the thalamus considered together as the important subcortical centers. { 'bā·səl 'gaŋ·glē·ə }

basalia [VERTEBRATE ZOOLOGY] The cartilaginous rods that support the base of the pectoral and pelvic fins in elasmobranchs. { bə'sal·ē·ə }

basalis [HISTOLOGY] The basal portion of the endometrium; it is not shed during menstruation. { bə'sal·əs }

basal lamina [EMBRYOLOGY] The portion of the gray matter of the embryonic neural tube from which motor nerve roots develop. { 'bā·səl 'lam·ə·nə }

basal membrane [ANATOMY] The tissue beneath the pigment layer of the retina that forms the outer layer of the choroid. { 'bā·səl 'mem ,brān }

basal metabolic rate [PHYSIOLOGY] The amount of energy utilized per unit time under conditions of basal metabolism; expressed as calories per square meter of body surface or per kilogram of body weight per hour. Abbreviated BMR. { 'bā·səl med·ə'bäl·ik 'rāt }

basal metabolism [PHYSIOLOGY] The sum total of anabolic and catabolic activities of an organism in the resting state providing just enough energy to maintain vital functions. { 'bā·səl mə'tab·ə ,liz·əm }

basal wall [BOTANY] The wall dividing the oospore into an anterior and a posterior half in plants bearing archegonia. { 'bā·səl 'wȯl }

base [GENETICS] *See* nitrogenous base. { bās }

base analog [MOLECULAR BIOLOGY] A molecule similar enough to a purine or pyrimidine base to substitute for the normal bases, resulting in abnormal base pairing. { 'bās 'an·ə,läg }

basement membrane [HISTOLOGY] A delicate connective-tissue layer underlying the epithelium of many organs. { 'bās·mənt 'mem,brān }

basendite [INVERTEBRATE ZOOLOGY] In crustaceans, either of a pair of lobes at the end of each specialized paired appendage. { 'bā'sen,dīt }

base pair [MOLECULAR BIOLOGY] Two nitrogenous bases, one purine and one pyrimidine, that pair in double-stranded deoxyribonucleic acid. { 'bās ,per }

base pairing [MOLECULAR BIOLOGY] The hydrogen bonding of complementary purine and pyrimidine bases—adenine with thymine, guanine with cytosine—in double-stranded deoxyribonucleic acids or ribonucleic acids or in DNA/RNA hybrid molecules. { 'bās 'per·iŋ }

base sequence [GENETICS] The specific order of purine and pyrimidine bases in a polynucleotide. { 'bās 'sē·kwəns }

base stacking [MOLECULAR BIOLOGY] The orientation of adjacent base pairs such that their planes are parallel and their surfaces almost touch, as occurs in double-stranded deoxyribonucleic acid molecules. { 'bās ,stak·iŋ }

basichromatic [BIOLOGY] Staining readily with basic dyes. { ¦bā·si¦krō¦mad·ik }

basidiocarp [MYCOLOGY] The fruiting body of a fungus in the class Basidiomycetes. { bə'sid·ē·ə ˌkärp }

Basidiolichenes [BOTANY] A class of the Lichenes characterized by the production of basidia. { bəˌsid·ē·ōˌlī'kēˌnēz }

Basidiomycetes [MYCOLOGY] A class of fungi in the subdivision Eumycetes; important as food and as causal agents of plant diseases. { bə ˌsid·ē·ōˌmī'sēdˌēz }

basidiophore [MYCOLOGY] A basidia-bearing sporophore. { bə'sid·ē·əˌfȯr }

basidiospore [MYCOLOGY] A spore produced by a basidium. { bə'sid·ē·əˌspȯr }

basidium [MYCOLOGY] A cell, usually terminal, occurring in Basidiomycetes and producing spores (basidiospores) by nuclear fusion followed by meiosis. { bə'sid·ē·əm }

basifixed [BOTANY] Attached at or near the base. { ¦bās·ə¦fikst }

basil [BOTANY] The common name for any of the aromatic plants in the genus *Ocimum* of the mint family; leaves of the plant are used for food flavoring. { 'bāz·əl *or* 'baz·əl }

basilar [BIOLOGY] Of, pertaining to, or situated at the base. { 'bas·ə·lər }

basilar groove [ANATOMY] The cavity which is located on the upper surface of the basilar process of the brain and upon which the medulla rests. { 'bas·ə·lər ¦grüv }

basilar membrane [ANATOMY] A membrane of the mammalian inner ear supporting the organ of Corti and separating two cochlear channels, the scala media and scala tympani. { 'bas·ə·lər 'memˌbrān }

basilar papilla [ANATOMY] **1.** A sensory structure in the lagenar portion of an amphibian's membranous labyrinth between the oval and round windows. **2.** The organ of Corti in mammals. { 'bas·ə·lər pə'pil·ə }

basilar plate [EMBRYOLOGY] An embryonic cartilaginous plate in vertebrates that is formed from the parachordals and anterior notochord and gives rise to the ethmoid and other bones of the skull. { 'bas·ə·lər ¦plāt }

basilar process [ANATOMY] A strong, quadrilateral plate of bone forming the anterior portion of the occipital bone, in front of the foramen magnum. { 'bas·ə·lər 'präs·əs }

basilic vein [ANATOMY] The large superficial vein of the arm on the medial side of the biceps brachii muscle. { bə'sil·ik 'vān }

basin swamp [ECOLOGY] A fresh-water swamp at the margin of a small calm lake, or near a large lake protected by shallow water or a barrier. { 'bās·ən ˌswämp }

basion [ANATOMY] In craniometry, the point on the anterior margin of the foramen magnum where the midsagittal plane of the skull intersects the plane of the foramen magnum. { 'bās·ēˌän }

basipetal [BIOLOGY] Movement or growth from the apex toward the base. { bā'sip·əd·əl }

basipodite [INVERTEBRATE ZOOLOGY] The distal segment of the protopodite of a biramous appendage in arthropods. { bā'sip·əˌdīt }

basipterygium [VERTEBRATE ZOOLOGY] A basal bone or cartilage supporting one of the paired fins in fishes. { bəˌsip·tə'rij·ē·əm }

basisternum [INVERTEBRATE ZOOLOGY] In insects, the anterior one of the two sternal skeletal plates. { ¦bās·ə¦stər·nəm }

basistyle [INVERTEBRATE ZOOLOGY] Either of a pair of flexible processes on the hypopygium of certain male dipterans. { ¦bās·ə¦stīl }

basitarsus [INVERTEBRATE ZOOLOGY] The basal segment of the tarsus in arthropods. { ¦bās·ə ¦tär·səs }

basket cell [HISTOLOGY] A type of cell in the cerebellum whose axis-cylinder processes terminate in a basketlike network around the cells of Purkinje. { 'bas·kət ˌsel }

basket star [INVERTEBRATE ZOOLOGY] The common name for ophiuroid echinoderms belonging to the family Gorgonocephalidae. { 'bas·kət ˌstär }

Basommatophora [INVERTEBRATE ZOOLOGY] An order of mollusks in the subclass Pulmonata containing many aquatic snails. { bəˌsäm·ə'täf·rə }

basonym [SYSTEMATICS] The original, validly published name of a taxon. { 'bās·əˌnim }

basophil [HISTOLOGY] A white blood cell with granules that stain with basic dyes and are water-soluble. { 'bās·əˌfil }

basophilia [BIOLOGY] An affinity for basic dyes. { ˌbās·ə'fil·ē·ə }

basophilous [BIOLOGY] Staining readily with the basic dyes. [ECOLOGY] Of plants, growing best in alkaline soils. { bə'säf·ə·ləs }

bass [VERTEBRATE ZOOLOGY] The common name for a number of fishes assigned to two families, Centrarchidae and Serranidae, in the order Perciformes. { bas }

basset [VERTEBRATE ZOOLOGY] A French breed of short-legged hunting dogs with long ears and a typical hound coat. { 'bas·ət }

basswood [BOTANY] A common name for trees of the genus *Tilia* in the linden family of the order Malvales. Also known as linden. { 'basˌwu̇d }

bast [BOTONY] *See* phloem. { bast }

bast fiber [BOTANY] Any fiber stripped from the inner bark of plants, such as flax, hemp, jute, and ramie; used in textile and paper manufacturing. { 'bast ˌfī·bər }

bat [VERTEBRATE ZOOLOGY] The common name for all members of the mammalian order Chiroptera. { bat }

Batales [BOTANY] A small order of dicotyledonous plants in the subclass Caryophillidae of the class Magnoliopsida containing a single family with only one genus, *Batis*. { bə'tāˌlēz }

Batesian mimicry [ECOLOGY] Resemblance of an innocuous species to one that is distasteful to predators. { 'bāt·sē·ən 'mim·ə·krē }

Bathornithidae [PALEONTOLOGY] A family of Oligocene birds in the order Gruiformes. { ˌba·thȯr 'nith·əˌdē }

Bathyctenidae [INVERTEBRATE ZOOLOGY] A family of bathypelagic coelenterates in the phylum Ctenophora. { ˌba·thik'ten·əˌdē }

Bathyergidae [VERTEBRATE ZOOLOGY] A family of

mammals, including the South African mole rats, in the order Rodentia. { ˌbath·ē'ər·jəˌdē }

Bathylaconoidei [VERTEBRATE ZOOLOGY] A suborder of deep-sea fishes in the order Salmoniformes. { ¦bath·əˌla·kə'nòi·dēˌī }

Bathynellacea [INVERTEBRATE ZOOLOGY] An order of crustaceans in the superorder Syncarida found in subterranean waters in England and central Europe. { ¦bath·əˌnel'ās·ē·ə }

Bathynellidae [INVERTEBRATE ZOOLOGY] The single family of the crustacean order Bathynellacea. { ˌbath·ə'nel·əˌdē }

Bathyodontoidea [INVERTEBRATE ZOOLOGY] A superfamily of nematodes of the order Mononchida containing the single family Bathyodontidae, characterized by a high, usually well-developed lip region, a hexaradiate oral opening with cuticularized walls, and a twofold stoma; eight longitudinal rows of pores occur along the length of the body. { ¦bath·ē·ō·dän'tòid·ē·ə }

Bathypteroidae [VERTEBRATE ZOOLOGY] A family of benthic, deep-sea fishes in the order Salmoniformes. { bəˌthip·tə'rói·dē }

Bathysquillidae [INVERTEBRATE ZOOLOGY] A family of mantis shrimps, with one genus (*Bathysquilla*) and two species, in the order Stomatopoda. { ¦bath·ə'skwil·əˌdē }

Batoidea [VERTEBRATE ZOOLOGY] The skates and rays, an order of the subclass Elasmobranchii. { bə'tòid·ē·ə }

Batrachoididae [VERTEBRATE ZOOLOGY] The single family of the order Batrachoidiformes. { ˌba·trə'kòi·dəˌdē }

Batrachoidiformes [VERTEBRATE ZOOLOGY] The toadfishes, an order of teleostean fishes in the subclass Actinopterygii. { ˌba·trəˌkòi·də 'fòrˌmēz }

battery of genes [GENETICS] The set of producer genes which is activated when a particular sensor gene activates its set of integrator genes. { 'bad·ə·rē əv 'jēnz }

bay [BOTANY] *Laurus nobilis*. An evergreen tree of the laurel family. { bā }

bayberry [BOTANY] **1.** *Pimenta acris*. A West Indian tree related to the allspice; a source of bay oil. Also known as bay-rum tree; Jamaica bayberry; wild cinnamon. **2.** Any tree of the genus *Myrica*. { 'bāˌber·ē }

bay-rum tree [BOTONY] *See* bayberry. { ¦bā 'rəm ˌtrē }

B cell [IMMUNOLOGY] One of a heterogeneous population of bone-marrow-derived lymphocytes which participates in the immune responses. Also known as B lymphocyte. { 'bē ˌsel }

BCG vaccine [IMMUNOLOGY] *See* Bacillus Calmette-Guérin vaccine. { ¦bē¦sē¦jē vak'sēn }

B chain [IMMUNOLOGY] *See* light chain. { 'bē ˌchān }

Bdelloidea [INVERTEBRATE ZOOLOGY] An order of the class Rotifera comprising animals which resemble leeches in body shape and manner of locomotion. { də'lòid·ē·ə }

Bdellomorpha [INVERTEBRATE ZOOLOGY] An order of ribbonlike worms in the class Enopla containing the single genus *Malacobdella*. { ˌdel·ə 'mòr·fə }

Bdellonemertini [INVERTEBRATE ZOOLOGY] An equivalent name for the Bdellomorpha. { ˌdel·ə·nə'mer·təˌnī }

Bdellovibrio [MICROBIOLOGY] A genus of bacteria of uncertain affiliation; parasites of other bacteria; curved, motile rods with a polar flagellum in parasitic state; in nonparasitic state, cells are helical. { ˌdel·ə'vī·brēˌō }

beak [BOTANY] Any pointed projection, as on some fruits, that resembles a bird bill. [INVERTEBRATE ZOOLOGY] The tip of the umbo in bivalves. [VERTEBRATE ZOOLOGY] **1.** The bill of a bird or some other animal, such as the turtle. **2.** A projecting jawbone element of certain fishes, such as the sawfish and pike. { bēk }

bean [BOTANY] The common name for various leguminous plants used as food for humans and livestock; important commercial beans are true beans (*Phaseolus*) and California blackeye (*Vigna sinensis*). { bēn }

bear [VERTEBRATE ZOOLOGY] The common name for a few species of mammals in the family Ursidae. { ber }

beard [BOTANY] *See* awn. { bird }

beaver [VERTEBRATE ZOOLOGY] The common name for two different and unrelated species of rodents, the mountain beaver (*Aplodontia rufa*) and the true or common beaver (*Castor canadensis*). { 'bē·vər }

bedbug [INVERTEBRATE ZOOLOGY] The common name for a number of species of household pests in the insect family Cimicidae that infest bedding, and by biting humans obtain blood for nutrition. { 'bedˌbəg }

Bedsonia [MICROBIOLOGY] The psittacosis-lymphogranuloma-trachoma (PLT) group of bacteria belonging to the Chlamydozoaceae; all are obligatory intracellular parasites. { bed'sō·nē·ə }

bee [INVERTEBRATE ZOOLOGY] Any of the membranous-winged insects which compose the superfamily Apoidea in the order Hymenoptera characterized by a hairy body and by sucking and chewing mouthparts. { bē }

beech [BOTANY] Any of various deciduous trees of the genus *Fagus* in the beech family (Fagaceae) characterized by smooth gray bark, triangular nuts enclosed in burs, and hard wood with a fine grain. { bēch }

bee dance [INVERTEBRATE ZOOLOGY] Circling and wagging movements exhibited by worker bees to give other bees in the hive information about the location of a new source of food. { 'bē ˌdans }

beehive Also known as hive. [INVERTEBRATE ZOOLOGY] A colony of bees. { 'bēˌhīv }

beet [BOTANY] *Beta vulgaris*.The red or garden beet, a cool-season biennial of the order Caryophyllales grown for its edible, enlarged fleshy root. { bēt }

beetle [INVERTEBRATE ZOOLOGY] The common name given to members of the insect order Coleoptera. { 'bēd·əl }

Beggiatoa [MICROBIOLOGY] A genus of bacteria in the family Beggiatoaceae; filaments are individual, and cells contain sulfur granules when grown on media containing hydrogen sulfide. { bə'jad·ə·wə }

Beggiatoaceae [MICROBIOLOGY] A family of bacteria in the order Cytophagales; cells are in chains in colorless, flexible, motile filaments; microcysts are not known. { bə¦jad·ə'wās·ē,ē }

Beggiatoales [MICROBIOLOGY] Formerly an order of motile, filamentous bacteria in the class Schizomycetes. { bə,jad·ə'wā,lēz }

Begoniaceae [BOTANY] A family of dicotyledonous plants in the order Violales characterized by an inferior ovary, unisexual flowers, stipulate leaves, and two to five carpels. { bə,gō·nē'ās·ē,ē }

behavioral ecology [ECOLOGY] The branch of ecology that focuses on the evolutionary causes of variation in behavior among populations and species. { bi'hāv·yə·rəl ē'käl·ə·jē }

behavioral isolation [BIOLOGY] An isolating mechanism in which two allopatric species do not mate because of differences in courtship behavior. Also known as ethological isolation. { bi'hāv·yə·rəl ī·sə'lā·shən }

behavior genetics [GENETICS] A branch of genetics that is concerned with the inheritance of forms of behavior, intelligence, and personality traits. { bi'hāv·yər jə,ned·iks }

Beijerinckia [MICROBIOLOGY] A genus of bacteria in the family Azotobacteraceae; slightly curved or pear-shaped rods with large, refractile lipoid bodies at the ends of the cell. { bī·zhə'riŋ·kyə }

Belemnoidea [PALEONTOLOGY] An order of extinct dibranchiate mollusks in the class Cephalopoda. { ,bə·ləm'nȯid·ē·ə }

Belinuracea [PALEONTOLOGY] An extinct group of horseshoe crabs; arthropods belonging to the Limulida. { ,bel·ə·nu̇'rās·ē·ə }

belladonna [BOTANY] *Atropa belladonna*. A perennial poisonous herb that belongs to the family Solanaceae; atropine is produced from the roots and leaves; used as an antispasmodic, as a cardiac and respiratory stimulant, and to check secretions. Also known as deadly nightshade. { ,bel·ə'dän·ə }

Bellerophontacea [PALEONTOLOGY] A superfamily of extinct gastropod mollusks in the order Aspidobranchia. { bə,ler·ə,fän'tās·ē·ə }

Bell's law [PHYSIOLOGY] **1.** The law that in the spinal cord the ventral roots are motor and the dorsal roots sensory in function. **2.** The law that in a reflex arc the nerve impulse can be conducted in one direction only. { 'belz ,lȯ }

belly [ANATOMY] **1.** The abdominal cavity or the abdomen. **2.** The most prominent, fleshy, central portion of a muscle. { 'bel·ē }

Belondiroidea [INVERTEBRATE ZOOLOGY] A diverse superfamily of nematodes belonging to the order Dorylaimida, consisting of a diverse group whose principal common characteristic is a thick sheath of spiral muscles enclosing the basal swollen portion of the esophagus. { bə,län·də'rȯid·ē·ə }

Beloniformes [VERTEBRATE ZOOLOGY] The former ordinal name for a group of fishes now included in the order Atheriniformes. { ,be·lə·nə'fȯr,mēz }

Belostomatidae [INVERTEBRATE ZOOLOGY] The giant water bugs, a family of hemipteran insects in the subdivision Hydrocorisae. { ,bel·ə·stō'mad·ə,dē }

belt [ECOLOGY] **1.** Any altitudinal vegetation zone or band from the base to the summit of a mountain. **2.** Any benthic vegetation zone or band from sea level to the ocean depths. **3.** Any of the concentric vegetation zones around bodies of fresh water. { belt }

bending [MOLECULAR BIOLOGY] A conformational change characterized by a localized bend or kink in deoxyribonucleic acid due to heterogeneities in local structural composition. { 'ben·diŋ }

Bennettitales [PALEOBOTANY] An equivalent name for the Cycadeoidales. { bə,ned·ə'tā,lēz }

Bennettitatae [PALEOBOTANY] A class of fossil gymnosperms in the order Cycadeoidales. { be,ned·ə'tā,dē }

benthos [ECOLOGY] Bottom-dwelling forms of marine life. Also known as bottom fauna. { 'ben,thäs }

benzyl penicillin potassium [MICROBIOLOGY] $C_{16}H_{17}KN_2O_4S$ Moderately hygroscopic crystals; soluble in water; inactivated by acids and alkalies; obtained from fermentation of *Penicillium chrysogenum*; used as an antimicrobial drug in human and animal disease. { 'ben·zəl ,pen·ə'sil·ən pə'tas·ē·əm }

benzyl penicillin sodium [MICROBIOLOGY] $C_{16}H_{17}N_2NaO_4S$ Crystals obtained from a methanol-ethyl acetate acidified extract of fermentation broth of *Penicillium chrysogenum*; used as an antimicrobial in human and animal disease. { 'ben·zəl ,pen·ə'sil·ən 'sōd·ē·əm }

Berberidaceae [BOTANY] A family of dicotyledonous herbs and shrubs in the order Ranunculales characterized by alternate leaves, perfect, well-developed flowers, and a seemingly solitary carpel. { ,bər·bə·rə'dās·ē,ē }

Bergmann's rule [ECOLOGY] The principle that in a polytypic wide-ranging species of warm-blooded animals the average body size of members of each geographic race varies with the mean environmental temperature. { 'bərg·mənz ,rül }

Bergman-Turner unit [BIOLOGY] A unit for the standardization of thyroid extract. { ¦bərg·mən ¦tər·nər ,yü·nət }

Berkefeld filter [MICROBIOLOGY] A diatomaceous-earth filter that is used for the sterilization of heat-labile liquids, such as blood serum, enzyme solutions, and antibiotics. { 'berk·ə,feld ,fil·tər }

Bermuda grass [BOTANY] *Cynodon dactylon*. A long-lived perennial in the order Cyperales. { bər'myüd·ə ,gras }

Beroida [INVERTEBRATE ZOOLOGY] The single order of the class Nuda in the phylum Ctenophora. { bə'rō·ə·də }

berry [BOTANY] A usually small, simple, fleshy or pulpy fruit, such as a strawberry, grape, tomato, or banana. { 'ber·ē }

Bertrand's rule [MICROBIOLOGY] The rule stating that in those compounds, and only in those compounds, having cis secondary alcoholic groups containing at least one carbon atom of D configuration which is subtended by a primary alcohol group, or having a methyl-substituted primary alcohol group of D configuration, the D-carbon atom will be dehydrogenated by the vinegar bacteria *Acetobacter suboxydans*, yielding a ketone. { 'ber,tränz ,rül }

Beryciformes [VERTEBRATE ZOOLOGY] An order of actinopterygian fishes in the infraclass Teleostei. { bə,ris·ə'fór,mēz }

Berycomorphi [VERTEBRATE ZOOLOGY] An equivalent name for the Beryciformes. { bə,rik·ō 'mór·fī }

Berytidae [INVERTEBRATE ZOOLOGY] The stilt bugs, a small family of hemipteran insects in the superfamily Pentatomorpha. { bə'rid·ə,dē }

Bessey unit [BIOLOGY] A unit for the standardization of phosphatase. { 'bes·ē ,yü·nət }

beta blocker [PHYSIOLOGY] An adrenergic blocking agent capable of blocking nerve impulses to special sites (beta receptors) in the cerebellum; reduces the rate of heartbeats and the force of heart contractions. { 'bād·ə ¦bläk·ər }

beta carotene [BIOCHEMISTRY] $C_{40}H_{56}$ A carotenoid hydrocarbon pigment found widely in nature, always associated with chlorophylls; converted to vitamin A in the liver of many animals. { 'bād·ə 'kar·ə,tēn }

beta cell [HISTOLOGY] **1.** Any of the basophilic chromophiles in the anterior lobe of the adenohypophysis. **2.** One of the cells of the islets of Langerhans which produce insulin. { 'bād·ə ,sel }

betacyanin [BIOCHEMISTRY] A group of purple plant pigments found in leaves, flowers, and roots of members of the order Caryophyllales. { ¦bād·ə¦sī·ə·nən }

beta globulin [BIOCHEMISTRY] A heterogeneous fraction of serum globulins containing transferrin and various complement components. { 'bād·ə 'gläb·yə·lən }

beta hemolysis [MICROBIOLOGY] A sharply defined, clear, colorless zone of hemolysis surrounding certain streptococci colonies growing on blood agar. { 'bād·ə hə'mäl·ə·səs }

Betaherpesvirinae [VIROLOGY] A subfamily of animal, double-stranded linear deoxyribonucleic acid viruses of the family Herpesviridae, which are enveloped by a lipid bilayer and several glycoproteins. Also known as cytomegalovirus group. { ¦bād·ə¦hər·pēz'vī·rə,nē }

betalain [BIOCHEMISTRY] The name for a group of 35 red or yellow compounds found only in plants of the family Caryophyllales, including red beets, red chard, and cactus fruits. { 'bed·ə,lān }

betanin [BIOCHEMISTRY] An anthocyanin that contains nitrogen and constitutes the principal pigment of garden beets. { 'bēd·ə·nən }

beta oxidation [BIOCHEMISTRY] Catabolism of fatty acids in which the fatty acid chain is shortened by successive removal of two carbon fragments from the carboxyl end of the chain. { 'bād·ə äks·ə'dā·shən }

beta rhythm [PHYSIOLOGY] An electric current of low voltage from the brain, with a pulse frequency of 13-30 per second, encountered in a person who is aroused and anxious. { 'bād·ə ,rith·əm }

beta taxonomy [SYSTEMATICS] The level of taxonomic study dealing with the arrangement of species into lower and higher taxa. { 'bād·ə tak 'sän·ə·mē }

betaxanthin [BIOCHEMISTRY] The name given to any of the yellow pigments found only in plants of the family Caryophyllales; they always occur with betacyanins. { ¦bād·ə'zan·thən }

betel nut [BOTANY] A dried, ripe seed of the palm tree *Areca catechu* in the family Palmae; contains a narcotic. { 'bēd·əl ,nət }

Bethylidae [INVERTEBRATE ZOOLOGY] A small family of hymenopteran insects in the superfamily Bethyloidea. { bə'thil·ə,dē }

Bethyloidea [INVERTEBRATE ZOOLOGY] A superfamily of hymenopteran insects in the suborder Apocrita. { ,beth·ə'lȯi·dē·ə }

Betula [BOTANY] The birches, a genus of deciduous trees composing the family Betulaceae. { 'bech·ə·lə }

Betulaceae [BOTANY] A small family of dicotyledonous plants in the order Fagales characterized by stipulate leaves, seeds without endosperm, and by being monoecious with female flowers mostly in catkins. { ,bech·ə'lās·ē,ē }

Betz cell [HISTOLOGY] Any of the large conical cells composing the major histological feature of the precentral motor cortex in humans. { 'bets ,sel }

Beyrichacea [PALEONTOLOGY] A superfamily of extinct ostracods in the suborder Beyrichicopina. { ,bī·rə'kās·ē·ə }

Beyrichicopina [PALEONTOLOGY] A suborder of extinct ostracods in the order Paleocopa. { ,bī·rə·kə,kō'pī·nə }

Beyrichiidae [PALEONTOLOGY] A family of extinct ostracods in the superfamily Beyrichacea. { ,bī·rə'kī·ə,dē }

Bezold-Abney phenomenon [PHYSIOLOGY] The perception of light at very high intensities as colorless. { 'bāt,zōlt 'ab·nē fə,näm·ə,nän }

Bibionidae [INVERTEBRATE ZOOLOGY] The March flies, a family of orthorrhaphous dipteran insects in the series Nematocera. { ,bib·ē'än·ə,dē }

bicameral [BIOLOGY] Having two chambers, as the heart of a fish. { bī'kam·ə·rəl }

bicapsular [BIOLOGY] **1.** Having two capsules. **2.** Having a capsule with two locules. { bī'kap·sə·lər }

bicarinate [BIOLOGY] Having two keellike projections. { bī'kar·ə,nāt }

bicaudal [ZOOLOGY] Having two tails. { bī'kȯd·əl }

bicellular [BIOLOGY] Having two cells. { bī'sel·yə·lər }

bicephalous [ZOOLOGY] Having two heads. { bī 'sef·ə·ləs }

biceps [ANATOMY] **1.** A bicipital muscle. **2.** The

large muscle of the front of the upper arm that flexes the forearm; biceps brachii. **3.** The thigh muscle that flexes the knee joint and extends the hip joint; biceps femoris. { 'bī,seps }

biciliate [BIOLOGY] Having two cilia. { bī'sil·ē ,āt }

bicipital [ANATOMY] **1.** Pertaining to muscles having two origins. **2.** Pertaining to ribs having double articulation with the vertebrae. [BOTANY] Having two heads or two supports. { bī'sip·əd·əl }

bicipital groove [ANATOMY] *See* intertubercular sulcus. { bī'sip·əd·əl ,grüv }

bicipital tuberosity [ANATOMY] An eminence on the anterior inner aspect of the neck of the radius; the tendon of the biceps muscle is inserted here. { bī'sip·əd·əl ,tüb·ə'räs·əd·ē }

bicollateral bundle [BOTANY] A vascular bundle in which phloem is located both externally and internally with respect to the xylem, with all tissues lying on the same radius and with the internal phloem lying next to the pith. { ,bī·kə 'lad·ə·rəl 'bənd·əl }

bicornuate uterus [ANATOMY] A uterus with two horn-shaped processes on the superior aspect. { bī'kȯr·yə,nāt 'yüd·ə·rəs }

Bicosoecida [INVERTEBRATE ZOOLOGY] An order of colorless, free-living protozoans, each member having two flagella, in the class Zoomastigophorea. { bī·kō'se·shə·də }

bicostate [BOTANY] Of a leaf, having two principal longitudinal ribs. { 'bī·kō,stāt }

bicuspid [ANATOMY] Any of the four double-pointed premolar teeth in humans. [BIOLOGY] Having two points or prominences. { bī'kəs·pəd }

Bidder's organ [VERTEBRATE ZOOLOGY] A structure in the males of some toad species that may develop into an ovary in older individuals. { 'bid·ərz ,ȯr·gən }

bidentate [BIOLOGY] Having two teeth or toothlike processes. { bī'den,tāt }

bidirectional replication [MOLECULAR BIOLOGY] A mechanism of replication of deoxyribonucleic acid that involves two replicating forks moving in opposite directions away from the same origin. { ,bī·də,rek·shən·əl ,rep·lə'kā·shən }

biennial plant [BOTANY] A plant that requires two growing seasons to complete its life cycle. { bī 'en·ē·əl ¦plant }

bifacial [BOTANY] Of a leaf, having dissimilar tissues on the upper and lower surfaces. { bī'fā·shəl }

bifanged [ANATOMY] Of a tooth, having two roots. { bī'faŋgd }

bifid [BIOLOGY] Divided into two equal parts by a median cleft. { 'bī,fid }

Bifidobacterium [MICROBIOLOGY] A genus of bacteria in the family Actinomycetaceae; branched, bifurcated, club-shaped or spatulate rods forming smooth microcolonies; metabolism is saccharoclastic. { ,bī·fə·dō·bak'tir·ē·əm }

biflabellate [INVERTEBRATE ZOOLOGY] The shape of certain insect antennae, characterized by short joints with long, flattened processes on opposite sides. { ¦bī·flə'bel·ət }

biflagellate [BIOLOGY] Having two flagella. { bī 'flaj·ə,lāt }

bifoliate [BOTANY] Two-leaved. { bī'fōl·ē·ət }

biforate [BIOLOGY] Having two perforations. { 'bī·fə,rāt }

bifunctional vector [MOLECULAR BIOLOGY] *See* shuttle vector. { ¦bī¦fəŋk·shən·əl 'vek·tər }

bigeminal pulse [PHYSIOLOGY] A pulse that is characterized by a double impulse produced by coupled heartbeats, that is, an extra heartbeat occurs just after the normal beat. { bī'jem·ə·nəl 'pəls }

Bignoniaceae [BOTANY] A family of dicotyledonous trees or shrubs in the order Scrophulariales characterized by a corolla with mostly five lobes, mature seeds with little or no endosperm and with wings, and opposite or whorled leaves. { ,big·nō·nē'ās·ē,ē }

bijugate [BOTANY] Of a pinnate leaf, having two pairs of leaflets. { 'bī·jə,gāt }

bilabiate [BOTANY] Having two lips, such as certain corollas. { bī'lāb·ē·ət }

bilateral [BIOLOGY] Of or relating to both right and left sides of an area, organ, or organism. { bī 'lad·ə·rəl }

bilateral cleavage [EMBRYOLOGY] The division pattern of a zygote that results in a bilaterally symmetrical embryo. { bī'lad·ə·rəl 'klēv·ij }

bilateral hermaphroditism [ZOOLOGY] The presence of an ovary and a testis on each side of the animal body. { bī'lad·ə·rəl hər'maf·rə,dīd,iz·əm }

bilateral symmetry [BIOLOGY] Symmetry such that the body can be divided by one median, or sagittal, dorsoventral plane into equivalent right and left halves, each a mirror image of the other. { bī'lad·ə·rəl 'sim·ə·trē }

Bilateria [ZOOLOGY] A major division of the animal kingdom embracing all forms with bilateral symmetry. { ,bī·lə'tir·ē·ə }

bile [PHYSIOLOGY] An alkaline fluid secreted by the liver and delivered to the duodenum to aid in the emulsification, digestion, and absorption of fats. Also known as gall. { bīl }

bile acid [BIOCHEMISTRY] Any of the liver-produced steroid acids, such as taurocholic acid and glycocholic acid, that appear in the bile as sodium salts. { 'bīl 'as·əd }

bile duct [ANATOMY] Any of the major channels in the liver through which bile flows toward the hepatic duct. { 'bīl ,dəkt }

bile pigment [BIOCHEMISTRY] Either of two colored organic compounds found in bile: bilirubin and biliverdin. { 'bīl 'pig·mənt }

bile salt [BIOCHEMISTRY] The sodium salt of glycocholic and taurocholic acids found in bile. { 'bīl ,sȯlt }

biliary system [ANATOMY] The complex of canaliculi, or microscopic bile ducts, that empty into the larger intrahepatic bile ducts. { 'bil·ē,er·ē ,sis·təm }

bilicyanin [BIOCHEMISTRY] A blue pigment found

in gallstones; an oxidation product of biliverdin or bilirubin. { ¦bil·ə¦sī·ə·nən }

bilification [PHYSIOLOGY] Formation and excretion of bile. { ˌbil·ə·fə'kā·shən }

biliprotein [BIOCHEMISTRY] The generic name for the organic compounds in certain algae that are composed of phycobilin and a conjugated protein. { ˌbil·ə'prōˌtēn }

bilirubin [BIOCHEMISTRY] $C_{33}H_{36}N_4O_6$ An orange, crystalline pigment occurring in bile; the major metabolic breakdown product of heme. { ˌbil·ə'rü·bən }

biliverdin [BIOCHEMISTRY] $C_{33}H_{34}N_4O_6$ A green, crystalline pigment occurring in the bile of amphibians, birds, and humans; oxidation product of bilirubin in humans. { ˌbil·ə'vərd·ən }

bill [INVERTEBRATE ZOOLOGY] A flattened portion of the shell margin of the broad end of an oyster. [VERTEBRATE ZOOLOGY] The jaws, together with the horny covering, of a bird. [ZOOLOGY] Any jawlike mouthpart. { bil }

Billingsellacea [PALEONTOLOGY] A group of extinct articulate brachiopods in the order Orthida. { ˌbil·iŋ·sə'lās·ē·ə }

bilobate [BIOLOGY] Divided into two lobes. { 'bī'lōˌbāt }

bilobular [BIOLOGY] Having two lobules. { ˌbī'läb·yə·lər }

bilocular [BIOLOGY] Having two cells or compartments. { ˌbī'läk·yə·lər }

bilophodont [ZOOLOGY] Having two transverse ridges, as the molar teeth of certain animals. { bī'läf·əˌdänt }

bimanous [ANATOMY] Having the distal part of the two forelimbs modified as hands, as in primates. { bī'man·əs }

bimaxillary [ZOOLOGY] Pertaining to the two halves of the maxilla. { ¦bī'max·əˌler·ē }

binary fission [BIOLOGY] A method of asexual reproduction accomplished by the splitting of a parent cell into two equal, or nearly equal, parts, each of which grows to parental size and form. { 'bīn·ə·rē 'fish·ən }

binate [BOTANY] Growing in pairs. { 'bīˌnāt }

binaural hearing [PHYSIOLOGY] The perception of sound by stimulation of two ears. { bī'nȯr·əl 'hir·iŋ }

binding site [MOLECULAR BIOLOGY] *See* active site. { 'bīn·diŋ ˌsīt }

binocular [BIOLOGY] **1.** Of, pertaining to, or used by both eyes. **2.** Of a type of visual perception which provides depth-of-field focus due to angular difference between the two retinal images. { bī'näk·yə·lər }

binocular accommodation [PHYSIOLOGY] Automatic lens adjustment by both eyes simultaneously for focusing on distant objects. { bī'näk·yə·lər əˌkäm·ə'dā·shən }

binomen [SYSTEMATICS] A binomial name assigned to species, as *Canis familiaris* for the dog. { bī'nō·mən }

binomial nomenclature [SYSTEMATICS] The Linnean system of classification requiring the designation of a binomen, the genus and species name, for every species of plant and animal. { bī'nō·mē·əl ˌnō·mən'klā·chər }

binuclear [CELL BIOLOGY] Having two nuclei. { bī'nü·klē·ər }

binucleolate [CELL BIOLOGY] Having two nucleoli. { ¦bī·nü'klē·əˌlāt }

bioacoustics [BIOLOGY] The study of the relation between living organisms and sound. { ¦bī·ō·ə'kü·stiks }

bioastronautics [BIOLOGY] The study of biological, behavioral, and medical problems pertaining to astronautics. { ¦bī·ōˌas·trə'nȯd·iks }

bioavailability [PHYSIOLOGY] The extent and rate at which a substance, such as a drug, is absorbed into a living system or is made available at the site of physiological activity. { ˌbī·ō·əˌvāl·ə'bil·əd·ē }

biobubble [ECOLOGY] A model concept of the ecosphere in which all living things are considered as particles held together by nonliving forces. { 'bī·ōˌbəb·əl }

biocatalyst [BIOCHEMISTRY] A biochemical catalyst, especially an enzyme. { ˌbī·ō'kad·əl·ist }

biocenology [ECOLOGY] The study of natural communities and of interactions among the members of these communities. { ˌbī·ō·sə'näl·ə·jē }

biocenose [ECOLOGY] *See* biotic community. { ˌbī·ō'sēˌnōs }

biochemical oxygen demand [MICROBIOLOGY] The amount of dissolved oxygen required to meet the metabolic needs of aerobic microorganisms in water rich in organic matter, such as sewage. Abbreviated BOD. Also known as biological oxygen demand. { ¦bī·ō'kem·ə·kəl 'äk·sə·jən di'mand }

biochemical oxygen demand test [MICROBIOLOGY] A standard laboratory procedure for measuring biochemical oxygen demand; standard measurement is made for 5 days at 20°C. Abbreviated BOD test. { ¦bī·ō'kem·ə·kəl 'äk·sə·jən di'mand ˌtest }

biochemorphology [BIOCHEMISTRY] The science dealing with the chemical structure of foods and drugs and their reactions on living organisms. { ¦bī·ō¦ke·mȯr¦fäl·ə·jē }

biochore [ECOLOGY] A group of similar biotopes. { 'bī·ōˌkȯr }

biochrome [BIOCHEMISTRY] Any naturally occurring plant or animal pigment. { 'bī·ōˌkrōm }

biochron [PALEONTOLOGY] A fossil of relatively short range of time. { 'bī·ōˌkrän }

biociation [ECOLOGY] A subdivision of a biome distinguished by the predominant animal species. { bīˌäs·ē'ā·shən }

bioclimatic law [ECOLOGY] The law which states that phenological events are altered by about 4 days for each 5° change of latitude northward or longitude eastward; events are accelerated in spring and retreat in autumn. { ¦bī·ōˌklī'mad·ik 'lȯ }

bioclimatograph [ECOLOGY] A climatograph

showing the relation between climatic conditions and some living organisms. { ¦bī·ō‚klī 'mad·ə‚graf }

bioclimatology [ECOLOGY] The study of the effects of the natural environment on living organisms. { ¦bī·ō‚klī·mə'täl·ə·jē }

biocoenosis [ECOLOGY] A group of organisms that live closely together and form a natural ecologic unit. { ‚bī·ō·sə'nō·səs }

biocompatibility [PHYSIOLOGY] The condition of being compatible with living tissue by virtue of a lack of toxicity or ability to cause immunological rejection. { ‚bī·ō·kəm‚pad·ə'bil·əd·ē }

biocycle [ECOLOGY] A group of similar biotopes composing a major division of the biosphere; there are three biocycles: terrestrial, marine, and fresh-water. { 'bī·ō‚sī·kəl }

biocytin [BIOCHEMISTRY] $C_{16}H_{28}N_4O_4S$ Crystals with a melting point of 241-243°C; obtained from dilute methanol or acetone solutions; characterized by its utilization by *Lactobacillus casei* and *L. delbrückii* LD5 as a biotin source, and by its unavailability as a biotin source to *L. arabinosus*. Also known as biotin complex of yeast. { ‚bī·ō 'sīt·ən }

biocytinase [BIOCHEMISTRY] An enzyme present in the blood and liver which hydrolyzes biocytin into biotin and lysine. { ‚bī·ō'sīt·ən‚ās }

biodynamic [BIOPHYSICS] Of or pertaining to the dynamic relation between an organism and its environment. { ‚bī·ō·dī'nam·ik }

biodynamics [BIOPHYSICS] The study of the effects of dynamic processes (motion, acceleration, weightlessness, and so on) on living organisms. { ‚bī·ō·dī'nam·iks }

bioelectric current [PHYSIOLOGY] A self-propagating electric current generated on the surface of nerve and muscle cells by potential differences across excitable cell membranes. { ¦bī·ō·i'lek·trik 'kər·ənt }

bioelectricity [PHYSIOLOGY] The generation by and flow of an electric current in living tissue. { ¦bī·ō‚i‚lek'tris·əd·ē }

bioelectric model [PHYSIOLOGY] A conceptual model for the study of animal electricity in terms of physical principles. { ¦bī·ō·i'lek·trik 'mäd·əl }

bioelectrochemistry [PHYSIOLOGY] The study of the control of biological growth and repair processes by electrical stimulation. { ¦bī·ō·i¦lek·trō 'kem·ə·strē }

bioelectronics [BIOPHYSICS] The application of electronic theories and techniques to the problems of biology. { ¦bī·ō‚i‚lek'trän·iks }

bioenergetics [BIOCHEMISTRY] The branch of biology dealing with energy transformations in living organisms. { ¦bī·ō‚en·ər'jed·iks }

bioethics [BIOLOGY] A discipline concerned with the application of ethics to biological problems, especially in the field of medicine. { ‚bī·ō'eth·iks }

bioflavonoid [BIOCHEMISTRY] A group of compounds obtained from the rinds of citrus fruits and involved with the homeostasis of the walls of small blood vessels; in guinea pigs a marked reduction of bioflavonoids results in increased fragility and permeability of the capillaries; used to decrease permeability and fragility in capillaries in certain conditions. Also known as citrus flavonoid compound; vitamin P complex. { ¦bī·ō 'flav·ə‚nȯid }

biogenesis [BIOLOGY] Development of a living organism from a similar living organism. { ¦bī·ō 'jen·ə·səs }

biogenetic law [BIOLOGY] *See* recapitulation theory. { ¦bī·ō·jə'ned·ik 'lȯ }

biogenic [BIOLOGY] **1.** Essential to the maintenance of life. **2.** Produced by actions of living organisms. { ¦bī·ō¦jen·ik }

biogenic amine [BIOCHEMISTRY] Any of a group of organic compounds that contain one or more amine groups ($—NH_2$) and have a possible role in brain functioning, including catecholamines and indoles. { ¦bī·ō¦jen·ik 'a·mēn }

biogeographic realm [ECOLOGY] Any of the divisions of the landmasses of the world according to their distinctive floras and faunas. { ‚bī·ō‚jē·ə‚graf·ik 'relm }

biogeography [ECOLOGY] The science concerned with the geographical distribution of animal and plant life. { ¦bī·ō·jē'äg·rə·fē }

biogeosphere [ECOLOGY] The region of the earth extending from the surface of the upper crust to the maximum depth at which organic life exists. { ‚bī·ō'jē·ə‚sfir }

biohazard [BIOLOGY] Any biological agent or condition that presents a hazard to life. { 'bī·ō ‚haz·ərd }

biohydrology [ECOLOGY] Study of the interactions between water, plants, and animals, including the effects of water on biota as well as the physical and chemical changes in water or its environment produced by biota. { ¦bī·ō‚hī'dräl·ə·jē }

biological [BIOLOGY] Of or pertaining to life or living organisms. [IMMUNOLOGY] A biological product used to induce immunity to various infectious diseases or noxious substances of biological origin. { ¦bī·ə¦läj·ə·kəl }

biological balance [ECOLOGY] Dynamic equilibrium that exists among members of a stable natural community. { ¦bī·ə¦läj·ə·kəl 'bal·əns }

biological clock [PHYSIOLOGY] Any physiologic factor that functions in regulating body rhythms. { ¦bī·ə¦läj·ə·kəl 'kläk }

biological containment [GENETICS] A technique by which the genetic constitution of an organism is altered in order to minimize its ability to grow outside the laboratory. { ¦bī·ə¦läj·ə·kəl kən 'tān·mənt }

biological control [ECOLOGY] Natural or applied regulation of populations of pest organisms, especially insects, through the role or use of natural enemies. { ¦bī·ə¦läj·ə·kəl kən'trōl }

biological equilibrium [BIOPHYSICS] A state of body balance for an actively moving animal, when internal and external forces are in equilibrium. { ¦bī·ə¦läj·ə·kəl ‚ē·kwə'lib·rē·əm }

biological half-life [PHYSIOLOGY] The time required by the body to eliminate half of the amount of an administered substance through normal channels of elimination. { ¦bī·ə¦läj·ə·kəl 'haf ˌlīf }

biological indicator [BIOLOGY] An organism that can be used to determine the concentration of a chemical in the environment. { ˌbī·ə¦läj·ə·kəl 'in·dəˌkād·ər }

biological magnification [ECOLOGY] The increasing concentration of toxins from pesticides, herbicides, and various types of waste in living organisms that accompanies cycling of nutrients through the trophic levels of food webs. { ˌbī·ə ¦läj·ə·kəl ˌmag·nə·fə'kā·shən }

biological oil-spill control [ECOLOGY] The use of cultures of microorganisms capable of living on oil as a means of degrading an oil slick biologically. { ¦bī·ə¦läj·ə·kəl 'ȯil ˌspil kən'trōl }

biological oxidation [BIOCHEMISTRY] Energy-producing reactions in living cells involving the transfer of hydrogen atoms or electrons from one molecule to another. { ¦bī·ə¦läj·ə·kəl ˌäk·sə 'dā·shən }

biological oxygen demand [MICROBIOLOGY] *See* biochemical oxygen demand. { ¦bī·ə¦läj·ə·kəl 'äk·sə·jən di'mand }

biological productivity [ECOLOGY] The quantity of organic matter or its equivalent in dry matter, carbon, or energy content which is accumulated during a given period of time. { ¦bī·ə¦läj·ə·kəl prəˌdək'tiv·əd·ē }

biological shield [MICROBIOLOGY] A structure designed to prevent the migration of living organisms from one part of a system to another; used on sterilized space probes. { ¦bī·ə¦läj·ə·kəl 'shēld }

biological specificity [BIOLOGY] The principle that defines the orderly patterns of metabolic and developmental reactions giving rise to the unique characteristics of the individual and of its species. { ¦bī·ə¦läj·ə·kəl spes·ə'fis·əd·ē }

biological value [BIOCHEMISTRY] A measurement of the efficiency of the protein content in a food for the maintenance and growth of the body tissues of an individual. { ¦bī·ə¦läj·ə·kəl 'val·yü }

bioluminescence [BIOLOGY] The emission of visible light by living organisms. { ˌbī·ōˌlü·mə 'nes·əns }

biolysis [BIOLOGY] **1.** Death and the following tissue disintegration. **2.** Decomposition of organic materials, such as sewage, by living organisms. { bī'äl·ə·səs }

biomagnetism [BIOPHYSICS] The production of a magnetic field by a living organism. { ¦bī·ō 'mag·nəˌtiz·əm }

biomass [ECOLOGY] The dry weight of living matter, including stored food, present in a species population and expressed in terms of a given area or volume of the habitat. { 'bī·ōˌmas }

biome [ECOLOGY] A complex biotic community covering a large geographic area and characterized by the distinctive life-forms of important climax species. { 'bīˌōm }

biomechanics [BIOPHYSICS] The study of the mechanics of living things. { ¦bī·ō·mə'kan·iks }

biomere [ECOLOGY] A biostratigraphic unit bounded by abrupt nonevolutionary changes in the dominant elements of a single phylum. { 'bī·ōˌmir }

biometeorology [BIOLOGY] The study of the relationship between living organisms and atmospheric phenomena. { ¦bī·ōˌmēd·ē·ə'räl·ə·jē }

biomimetics [BIOCHEMISTRY] A branch of science in which synthetic systems are developed by using information obtained from biological systems. { ¦bī·ō·mə¦med·iks }

biomineralization [PHYSIOLOGY] A mineralization process carried out within a living organism, such as formation of bone in vertebrates. { ˌbī·ō ˌmin·rəl·ə'zā·shən }

biomolecular [BIOCHEMISTRY] Pertaining to organic molecules occurring in living organisms, especially macromolecules. { ˌbī·ō·mə'le·kyə·lər }

bion [ECOLOGY] An independent, individual organism. { 'bīˌän }

bionavigation [VERTEBRATE ZOOLOGY] The ability of animals such as birds to find their way back to their roost, even if the landmarks on the outward-bound trip were effectively concealed from them. { ˌbī·ōˌnav·ə'gā·shən }

biophage [ECOLOGY] *See* macroconsumer. { 'bī·ō ˌfāj }

biophagous [ZOOLOGY] Feeding on living organisms. { bī'äf·ə·gəs }

biophile [BIOCHEMISTRY] Any element concentrated or found in the bodies of living organisms and organic matter; examples are carbon, nitrogen, and oxygen. { 'bī·əˌfīl }

biopolymer [BIOCHEMISTRY] A biological macromolecule such as a protein or nucleic acid. { ¦bī·ō'päl·ə·mər }

biopotency [BIOCHEMISTRY] Capacity of a chemical substance, as a hormone, to function in a biological system. { ¦bī·ō'pōt·ən·sē }

biopotential [PHYSIOLOGY] Voltage difference measured between points in living cells, tissues, and organisms. { ¦bī·ō·pə'ten·chəl }

biorheology [BIOPHYSICS] The study of the deformation and flow of biological fluids, such as blood, mucus, and synovial fluid. { ¦bī·ō·rē'äl·ə·jē }

biorhythm [PHYSIOLOGY] A biologically inherent cyclic variation or recurrence of an event or state, such as a sleep cycle or circadian rhythm. { 'bī·ōˌrith·əm }

bioseries [EVOLUTION] A historical sequence produced by the changes in a single hereditary character. { ¦bī·ōˌsir·ēz }

biosocial [ZOOLOGY] Pertaining to the interplay of biological and social influences. { ˌbī·ō'sō·shəl }

biosonar [PHYSIOLOGY] A guidance system in certain animals, such as bats, utilizing the reflection of sounds that they produce as they move about. { ¦bī·ō¦sōˌnär }

biospeleology [BIOLOGY] The study of cave-dwelling organisms. { ˌbī·ōˌspē·lē'äl·ə·jē }

biosphere [ECOLOGY] The life zone of the earth, including the lower part of the atmosphere, the hydrosphere, soil, and the lithosphere to a depth of about 1.2 miles (2 kilometers). { 'bī·əˌsfir }

biostasy [ECOLOGY] Maximum development of

organisms when, during tectonic repose, residual soils form extensively on the land and calcium carbonate deposition is widespread in the sea. { bī'äs·tə·sē }

biostratigraphy [PALEONTOLOGY] A part of paleontology concerned with the study of the conditions and deposition order of sedimentary rocks. { ¦bī·ō·strə'tig·rə·fē }

biosynthesis [BIOCHEMISTRY] Production, by synthesis or degradation, of a chemical compound by a living organism. { ˌbī·ō'sin·thə·səs }

biota [BIOLOGY] **1.** Animal and plant life characterizing a given region. **2.** Flora and fauna, collectively. { bī'ōd·ə }

biotechnology [GENETICS] The use of advanced genetic techniques to construct novel microbial and plant strains and obtain site-directed mutants to improve the quantity or quality of products. { ¦bī·ō·tek'näl·ə·jē }

biotic [BIOLOGY] **1.** Of or pertaining to life and living organisms. **2.** Induced by the actions of living organisms. { bī'äd·ik }

biotic community [ECOLOGY] An aggregation of organisms characterized by a distinctive combination of both animal and plant species in a particular habitat. Also known as biocenose. { bī'äd·ik kə'myün·əd·ē }

biotic district [ECOLOGY] A subdivision of a biotic province. { bī'äd·ik 'dis·trikt }

biotic environment [ECOLOGY] That environment comprising living organisms, which interact with each other and their abiotic environment. { bī'äd·ik in'vī·ərn·mənt }

biotic isolation [ECOLOGY] The occurrence of organisms in isolation from others of their species. { bī'äd·ik ˌi·sə'lā·shən }

biotic potential [ECOLOGY] The maximum possible growth rate of living things under ideal conditions. { bī'äd·ik pə'ten·chəl }

biotic province [ECOLOGY] A community, according to some systems of classification, occupying an area where similarity of climate, physiography, and soils leads to the recurrence of similar combinations of organisms. { bī'äd·ik 'präv·əns }

biotin [BIOCHEMISTRY] $C_{10}H_{16}N_2O_3S$ A colorless, crystalline vitamin of the vitamin B complex occurring widely in nature, mainly in bound form. { 'bī·ə·tən }

biotin carboxylase [BIOCHEMISTRY] An enzyme which condenses bicarbonate with biotin to form carboxybiotin. { 'bī·ə·tən kär'bäk·səˌlās }

biotin complex of yeast [BIOCHEMISTRY] *See* biocytin. { 'bī·ə·tən 'kämˌpleks əv 'yēst }

biotope [ECOLOGY] An area of uniform environmental conditions and biota. { 'bī·əˌtōp }

biotype [GENETICS] A group of organisms having the same genotype. { 'bī·əˌtīp }

biozone [PALEONTOLOGY] The range of a single taxonomic entity in geologic time as reflected by its occurrence in fossiliferous rocks. { 'bī·ō ˌzōn }

biparasitic [ECOLOGY] Parasitic upon or in a parasite. { ¦bīˌpar·ə'sid·ik }

biparous [BOTANY] Bearing branches on dichotomous axes. [VERTEBRATE ZOOLOGY] Bringing forth two young at a birth. { 'bī·pə·rəs }

bipartite uterus [ANATOMY] A uterus divided into two parts almost to the base. { bī'pärˌtīt 'yüd·ə·rəs }

bipectinate [INVERTEBRATE ZOOLOGY] Of the antennae of certain moths, having two margins with comblike teeth. [ZOOLOGY] Branching like a feather on both sides of a main shaft. { bī'pek·təˌnāt }

biped [VERTEBRATE ZOOLOGY] **1.** A two-footed animal. **2.** Any two legs of a quadruped. { 'bīˌped }

bipedal [BIOLOGY] Having two feet. { bī'ped·əl }

bipedal dinosaur [PALEONTOLOGY] A dinosaur having two long, stout hindlimbs for walking and two relatively short forelimbs. { bī'ped·əl 'dīn·ə ˌsȯr }

bipeltate [BOTANY] Having two shield-shaped parts. [ZOOLOGY] Having a shell or other covering resembling a double shield. { bī'pelˌtāt }

bipenniform [ANATOMY] Of the arrangement of muscle fibers, resembling a feather barbed on both sides. { bī'pen·əˌfȯrm }

biphasic [BOTANY] Possessing both a sporophyte and a gametophyte generation in the life cycle. { bī'fāz·ik }

biphyletic [EVOLUTION] Descended in two branches from a common ancestry. { ¦bī·fī'led·ik }

Biphyllidae [INVERTEBRATE ZOOLOGY] The false skin beetles, a family of coleopteran insects in the superfamily Cucujoidea. { ˌbī'fil·əˌdē }

bipinnaria [INVERTEBRATE ZOOLOGY] The complex, bilaterally symmetrical, free-swimming larval stage of most asteroid echinoderms. { ¦bī·pi'ner·ē·ə }

bipinnate [BOTANY] Pertaining to a leaf that is pinnate for both its primary and secondary divisions. { bī'pinˌāt }

bipolar flagellation [MICROBIOLOGY] The presence of flagella at both poles in certain bacteria. { bī'pō·lər ˌflaj·ə'lā·shən }

Bipolarina [INVERTEBRATE ZOOLOGY] A suborder of protozoan parasites in the order Myxosporida. { ˌbī·pō·lə'rī·nə }

bipotential [BIOLOGY] Having the potential to develop in either of two mutually exclusive directions. { ¦bī·pə'ten·chəl }

bipotentiality [BIOLOGY] **1.** Capacity to function either as male or female. **2.** Hermaphroditism. { ¦bī·pəˌten·chē'al·əd·ē }

biradial symmetry [BIOLOGY] Symmetry both radial and bilateral. Also known as disymmetry. { bī'rād·ē·əl 'sim·ə·trē }

biramous [BIOLOGY] Having two branches, such as an arthropod appendage. { bī'rā·məs }

birch [BOTANY] The common name for all deciduous trees of the genus *Betula* that compose the family Betulaceae in the order Fagales. { bərch }

bird [VERTEBRATE ZOOLOGY] Any of the warm-blooded vertebrates which make up the class Aves. { bərd }

bird-hipped dinosaur [PALEONTOLOGY] Any

member of the order Ornithischia, distinguished by the birdlike arrangement of their hipbones. { 'bərd ˌhipt 'dīn·əˌsȯr }

bird louse [INVERTEBRATE ZOOLOGY] The common name for any insect of the order Mallophaga. Also known as biting louse. { 'bərd ˌlau̇s }

bird of prey [VERTEBRATE ZOOLOGY] Any of various carnivorous birds of the orders Falconiformes and Strigiformes which feed on meat taken by hunting. { 'bərd əv 'prā }

birimose [BOTANY] Opening by two slits, as an anther. { bī'rīˌmōs }

birotulate [INVERTEBRATE ZOOLOGY] A sponge spicule characterized by two wheel-shaped ends. { bī'räch·əˌlāt }

birth [BIOLOGY] The emergence of a new individual from the body of its parent. { bərth }

birth canal [ANATOMY] The channel in mammals through which the fetus is expelled during parturition; consists of the cervix, vagina, and vulva. { 'bərth kə'nal }

birth rate [BIOLOGY] The ratio between the number of live births and a specified number of people in a population over a given period of time. { 'bərth ˌrāt }

biserial [BIOLOGY] Arranged in two rows or series. { ˌbī'sir·ē·əl }

biserrate [BIOLOGY] **1.** Having serrated serrations. **2.** Serrate on both sides. { ˌbī'serˌāt }

bisexual [BIOLOGY] Of or relating to two sexes. { ˌbī'sek·shə·wəl }

bison [VERTEBRATE ZOOLOGY] The common name for two species of the family Bovidae in the order Artiodactyla; the wisent or European bison (*Bison bonasus*), and the American species (*Bison bison*). { 'bīs·ən }

bisporangiate [BOTANY] Having two different types of sporangia. { ¦bī·spə'ran·jēˌāt }

bispore [BOTANY] In certain red algae, an asexual spore that is produced in pairs. { 'bīˌspȯr }

bite [BIOLOGY] **1.** To seize with the teeth. **2.** Closure of the lower teeth against the upper teeth. { bīt }

bitegmic [BOTANY] Having two integuments, especially in reference to ovules. { bī'teg·mik }

biternate [BOTANY] Of a ternate leaf, having each division ternate. { bī'tərˌnāt }

biting louse [INVERTEBRATE ZOOLOGY] *See* bird louse. { 'bīd·iŋ ˌlau̇s }

bittern [VERTEBRATE ZOOLOGY] Any of various herons of the genus *Botaurus* characterized by streaked and speckled plumage. { 'bid·ərn }

Bittner milk factor [BIOCHEMISTRY] *See* milk factor. { 'bit·nər 'milk ˌfak·tər }

bivalent antibody [IMMUNOLOGY] An antibody possessing two antibody combining sites. { bī 'vāl·ənt 'ant·iˌbäd·ē }

bivalent chromosome [CELL BIOLOGY] The structure formed following synapsis of a pair of homologous chromosomes from the zygotene stage of meiosis up to the beginning of anaphase. { bī'vā·lənt 'krō·məˌsōm }

bivalve [INVERTEBRATE ZOOLOGY] The common name for a number of diverse, bilaterally symmetrical animals, including mollusks, ostracod crustaceans, and brachiopods, having a soft body enclosed in a calcareous two-part shell. { 'bīˌvalv }

Bivalvia [INVERTEBRATE ZOOLOGY] A large class of the phylum Mollusca containing the clams, oysters, and other bivalves. { bī'val·vē·ə }

biventer [ANATOMY] A muscle having two bellies. { bī'ven·tər }

bivittate [ZOOLOGY] Having a pair of longitudinal stripes. { bī'viˌtāt }

bivium [INVERTEBRATE ZOOLOGY] The pair of starfish rays that extend on either side of the madreporite. { 'bī·vē·əm }

bivoltine [INVERTEBRATE ZOOLOGY] **1.** Having two broods in a season, used especially of silkworms. **2.** Of insects, producing two generations a year. { bī'vōlˌtēn }

blackberry [BOTANY] Any of the upright or trailing shrubs of the genus *Rubus* in the order Rosales; an edible berry is produced by the plant. { 'blakˌber·ē }

blackbird [VERTEBRATE ZOOLOGY] Any bird species in the family Icteridae, of which the males are predominantly or totally black. { 'blakˌbərd }

black coral [INVERTEBRATE ZOOLOGY] The common name for antipatharian cnidarians having black, horny axial skeletons. { ¦blak 'kär·əl }

blackeye bean [BOTONY] *See* cowpea. { 'blakˌī ¦bēn }

black mold [MYCOLOGY] Any dark fungus belonging to the order Mucorales. { 'blak ˌmōld }

bladder [ANATOMY] Any saclike structure in humans and animals, such as a swimbladder or urinary bladder, that contains a gas or functions as a receptacle for fluid. { 'blad·ər }

bladder cell [INVERTEBRATE ZOOLOGY] Any of the large vacuolated cells in the outer layers of the tunic in some tunicates. { 'blad·ər ˌsel }

blade [BOTANY] The broad, flat portion of a leaf. Also known as lamina. [VERTEBRATE ZOOLOGY] A single plate of baleen. { blād }

blast cell [HISTOLOGY] An undifferentiated precursor of a human blood cell in the reticuloendothelial tissue. { 'blast ˌsel }

blastema [EMBRYOLOGY] **1.** A mass of undifferentiated protoplasm capable of growth and differentiation. **2.** *See* anlage. { bla'stēma }

blasto- [EMBRYOLOGY] A germ or bud, with reference to early embryonic stages of development. { 'blas·tō }

Blastobacter [MICROBIOLOGY] A genus of budding bacteria; cells are rod-, wedge-, or club-shaped, and several attach at the narrow end to form a rosette; reproduce by budding at the rounded end. { ¦blas·tō¦bak·tər }

Blastobasidae [INVERTEBRATE ZOOLOGY] A family of lepidopteran insects in the superfamily Tineoidea. { ¦blas·tō'bas·əˌdē }

blastocarpous [BOTANY] Germinating in the pericarp. { ¦blas·tō¦kär·pəs }

blastochyle [EMBRYOLOGY] The fluid filling the blastocoele. { 'blas·təˌkīl }

Blastocladiales [MYCOLOGY] An order of aquatic fungi in the class Phycomycetes. { ¦blas·tō¦klā·dē'ā·lēz }

blastocoele [EMBRYOLOGY] The cavity of a blastula. Also known as segmentation cavity. { 'blas·tə,sēl }

blastocone [EMBRYOLOGY] An incomplete blastomere. { 'blas·tō,kōn }

blastocyst [EMBRYOLOGY] A modified blastula characteristic of placental mammals. { 'blas·tə ,sist }

blastocyte [EMBRYOLOGY] An embryonic cell that is undifferentiated. [INVERTEBRATE ZOOLOGY] An undifferentiated cell capable of replacing damaged tissue in certain lower animals. { 'blas·tə,sīt }

blastoderm [EMBRYOLOGY] The blastodisk of a fully formed vertebrate blastula. { 'blas·tə ,dərm }

blastodisk [EMBRYOLOGY] The embryo-forming, protoplasmic disk on the surface of a yolk-filled egg, such as in reptiles, birds, and some fish. { 'blas·tə,disk }

Blastoidea [PALEONTOLOGY] A class of extinct pelmatozoan echinoderms in the subphylum Crinozoa. { bla'stȯid·ē·ə }

blastokinesis [INVERTEBRATE ZOOLOGY] Movement of the embryo into the yolk in some insect eggs. { ,blas·tō,kə'nē·səs }

blastomere [EMBRYOLOGY] A cell of a blastula. { 'blas·tə,mir }

blastophore [CELL BIOLOGY] The cytoplasm that is detached from a spermatid during its transformation to a spermatozoon. [INVERTEBRATE ZOOLOGY] An amorphose core of cytoplasm connecting cells of the male morula of developing germ cells in oligochaetes. { 'blas·tə,fȯr }

blastopore [EMBRYOLOGY] The opening of the archenteron. { 'blas·tə,pȯr }

blastospore [MYCOLOGY] A fungal resting spore that arises by budding. { 'blas·tə,spȯr }

blastostyle [INVERTEBRATE ZOOLOGY] A zooid on certain hydroids that lacks a mouth and tentacles and functions to produce medusoid buds. { 'blas·tə,stīl }

blastotomy [EMBRYOLOGY] Separation of cleavage cells during early embryogenesis. { ¦blas·'täd·ə·mē }

blastozooid [INVERTEBRATE ZOOLOGY] A zooid produced by budding. { ,blas·tə'zō,ȯid }

blastula [EMBRYOLOGY] A hollow sphere of cells characteristic of the early metazoan embryo. { 'blas·chə·lə }

blastulation [EMBRYOLOGY] Formation of a blastula from a solid ball of cleaving cells. { ,blas·chə'lā·shən }

Blattabacterium [MICROBIOLOGY] A genus of the tribe Wolbachiae; straight or slightly curved rod-shaped cells; symbiotic in cockroaches. { ,blad·ə·bak'tir·ē·əm }

Blattidae [INVERTEBRATE ZOOLOGY] The cockroaches, a family of insects in the order Orthoptera. { 'blad·ə·dē }

bleeding time [PHYSIOLOGY] The time required for bleeding to stop after a small puncture wound. { 'blēd·iŋ ,tīm }

blending inheritance [GENETICS] Inheritance in which the character of the offspring is a blend of those in the parents; a common feature for quantitative characters, such as stature, determined by large numbers of genes and affected by environmental variation. { 'blen·diŋ in'her·ə·təns }

Blenniidae [VERTEBRATE ZOOLOGY] The blennies, a family of carnivorous marine fishes in the suborder Blennioidei. { ble'nī·ə,dē }

Blennioidei [VERTEBRATE ZOOLOGY] A large suborder of small marine fishes in the order Perciformes that live principally in coral and rock reefs. { ,ble·nē'ȯi·dē,ī }

Blephariceridae [INVERTEBRATE ZOOLOGY] A family of dipteran insects in the suborder Orthorrhapha. { ,blef·ə·ri'ser·ə,dē }

blepharoplast [MICROBIOLOGY] A cytoplasmic granule bearing a bacterial flagellum. { 'blef·ə·rō,plast }

blind passage [MICROBIOLOGY] Transfer of some material from an inoculated animal or cell culture that does not exhibit evidence of infection, to a fresh animal or cell culture. { 'blīnd 'pas·ij }

blind spot [PHYSIOLOGY] A place on the retina of the eye that is insensitive to light, where the optic nerve passes through the eyeball's inner surface. { 'blīnd ,spät }

blocked reading frame [MOLECULAR BIOLOGY] *See* closed reading frame. { ¦bläkt 'rēd·iŋ ,frām }

blocking [HISTOLOGY] **1.** The process of embedding tissue in a solid medium, such as paraffin. **2.** A histochemical process in which a portion of a molecule is treated to prevent it from reacting with some other agent. { 'bläk·iŋ }

blocking antibody [IMMUNOLOGY] Antibody that inhibits the reaction between antigen and other antibodies or sensitized T lymphocytes. { ¦bläk·iŋ ¦ant·i,bäd·e }

blood [HISTOLOGY] A fluid connective tissue consisting of the plasma and cells that circulate in the blood vessels. { bləd }

blood agar [MICROBIOLOGY] A nutrient microbiologic culture medium enriched with whole blood and used to detect hemolytic strains of bacteria. { 'bləd ,äg·ər }

blood cell [HISTOLOGY] An erythrocyte or a leukocyte. { 'bləd ,sel }

blood chimerism [GENETICS] Having red blood cells of two genetic types. { 'bləd ka'mir,iz·əm }

blood group [IMMUNOLOGY] An immunologically distinct, genetically determined class of human erythrocyte antigens, identified as A, B, AB, and O. Also known as blood type. { 'bləd ,grüp }

blood island [EMBRYOLOGY] One of the areas in the yolk sac of vertebrate embryos allocated to the production of the first blood cells. { 'bləd ,ī·lənd }

bloodline [BIOLOGY] A line of direct ancestors, especially in a pedigree. { 'bləd,līne }

blood-plate hemolysis [MICROBIOLOGY] Destruction of red blood cells in a blood agar medium by a bacterial toxin. { 'bləd ,plāt hə'mäl·ə·səs }

blood platelet [HISTOLOGY] *See* thrombocyte. { 'bləd ,plāt·lət }

blood pressure [PHYSIOLOGY] Pressure exerted

by blood on the walls of the blood vessels. { 'bləd ˌpresh·ər }

bloodstream [PHYSIOLOGY] The flow of blood in its circulation through the body. { 'bləd ˌstrēm }

blood sugar [BIOCHEMISTRY] The carbohydrate, principally glucose, of the blood. { 'bləd ˌshu̇g·ər }

blood type [IMMUNOLOGY] *See* blood group. { 'bləd ˌtīp }

blood typing [IMMUNOLOGY] Determination of an individual's blood group. { 'bləd ˌtīp·iŋ }

blood vessel [ANATOMY] A tubular channel for blood transport. { 'bləd ˌves·əl }

bloom [BOTANY] **1.** An individual flower. Also known as blossom. **2.** To yield blossoms. **3.** The waxy coating that appears as a powder on certain fruits, such as plums, and leaves, such as cabbage. [ECOLOGY] A colored area on the surface of bodies of water caused by heavy planktonic growth. { blüm }

blossom [BOTANY] *See* bloom. { 'bläs·əm }

blowball [BOTANY] A fluffy seed ball, as of the dandelion. { 'blōˌbȯl }

blowhole [VERTEBRATE ZOOLOGY] The nostril on top of the head of cetacean mammals. { 'blō ˌhōl }

blowpipe [BIOLOGY] A small tube, tapering to a straight or slightly curved tip, used in anatomy and zoology to reveal or clean a cavity. { 'blō ˌpīp }

blubber [INVERTEBRATE ZOOLOGY] A large sea nettle or medusa. [VERTEBRATE ZOOLOGY] A thick insulating layer of fat beneath the skin of whales and other marine mammals. { 'bləb·ər }

blueberry [BOTANY] Any of several species of plants in the genus *Vaccinium* of the order Ericales; the fruit, a berry, occurs in clusters on the plant. { 'blüˌber·ē }

bluefish [VERTEBRATE ZOOLOGY] *Pomatomus saltatrix.* A predatory fish in the order Perciformes. Also known as skipjack. { 'blüˌfish }

bluegrass [BOTANY] The common name for several species of perennial pasture and lawn grasses in the genus *Poa* of the order Cyperales. { 'blüˌgras }

blue-green algae [MICROBIOLOGY] *See* cyanobacteria. { ¦blü¦grēn 'al·jē }

blue-green algal virus [VIROLOGY] *See* cyanophage. { ¦blü¦grēn ¦al·gəl 'vī·rəs }

blue mold [MYCOLOGY] Any fungus of the genus *Penicillium.* { 'blü ˌmōld }

bluestem grass [BOTANY] The common name for several species of tall, perennial grasses in the genus *Andropogon* of the order Cyperales. { 'blü ˌstem ¦gras }

B lymphocyte [IMMUNOLOGY] *See* B cell. { ¦bē 'lim·fəˌsīt }

BMR [PHYSIOLOGY] *See* basal metabolic rate.

boa [VERTEBRATE ZOOLOGY] Any large, nonvenomous snake of the family Boidae in the order Squamata. { 'bō·ə }

boar [VERTEBRATE ZOOLOGY] *See* wild boar. { bȯr }

Bobasatranidae [PALEONTOLOGY] A family of extinct palaeonisciform fishes in the suborder Platysomoidei. { bəˌbas·ə'tran·əˌdē }

BOD [MICROBIOLOGY] *See* biochemical oxygen demand.

Bodonidae [INVERTEBRATE ZOOLOGY] A family of protozoans in the order Kinetoplastida characterized by two unequally long flagella, one of them trailing. { bə'dän·əˌdē }

BOD test [MICROBIOLOGY] *See* biochemical oxygen demand test. { ¦bē¦ō'dē ˌtest }

body cavity [ANATOMY] The peritoneal, pleural, or pericardial cavities, or the cavity of the tunica vaginalis testis. { 'bäd·ē ˌkav·əd·ē }

body rhythm [PHYSIOLOGY] Any bodily process having some degree of regular periodicity. { 'bäd·ē ˌrith·əm }

body-righting reflex [PHYSIOLOGY] A postural reflex, initiated by the asymmetric stimulation of the body surface by the weight of the body, so that the head tends to assume a horizontal position. { 'bäd·ē ˌrīd·iŋ 'rēˌfleks }

bog [ECOLOGY] A plant community that develops and grows in areas with permanently waterlogged peat substrates. Also known as moor; quagmire. { bäg }

bog moss [ECOLOGY] Moss of the genus *Sphagnum* occurring as the characteristic vegetation of bogs. { 'bäg ˌmȯs }

Bohr effect [BIOCHEMISTRY] The effect of carbon dioxide and pH on the oxygen equilibrium of hemoglobin; increase in carbon dioxide prevents an increase in the release of oxygen from oxyhemoglobin. { 'bȯr i'fekt }

Boidae [VERTEBRATE ZOOLOGY] The boas, a family of nonvenomous reptiles of the order Squamata, having teeth on both jaws and hindlimb rudiments. { 'bō·əˌdē }

boll [BOTANY] A pod or capsule (pericarp), as of cotton and flax. { bōl }

boll weevil [INVERTEBRATE ZOOLOGY] A beetle, *Anthonomus grandis,* of the order Coleoptera; larvae destroy cotton plants and are the most important pests in agriculture. { 'bōl ˌwē·vəl }

bolus [PHYSIOLOGY] The mass of food prepared by the mouth for swallowing. { 'bō·ləs }

Bombacaceae [BOTANY] A family of dicotyledonous tropical trees in the order Malvales with dry or fleshy fruit usually having woolly seeds. { ˌbäm·bə'kās·ēˌē }

Bombay blood group system [IMMUNOLOGY] A system comprising an immunologically distinct, genetically determined group of human erythrocytes characterized by the lack of A, B, or H antigens. { bäm'bā 'bləd ˌgrüp ˌsis·təm }

Bombidae [INVERTEBRATE ZOOLOGY] A family of relatively large, hairy, black and yellow bumblebees in the hymenopteran superfamily Apoidea. { 'bäm·bəˌdē }

Bombycidae [INVERTEBRATE ZOOLOGY] A family of lepidopteran insects of the superorder Heteroneura that includes only the silkworms. { bäm 'bis·əˌdē }

Bombyliidae [INVERTEBRATE ZOOLOGY] The bee flies, a family of dipteran insects in the suborder Orthorrhapha. { ˌbäm·bə'lī·əˌdē }

Bombyx [INVERTEBRATE ZOOLOGY] The type genus of Bombycidae. { 'bäm,biks }

Bombyx mori [INVERTEBRATE ZOOLOGY] The commercial silkworm. { 'bäm,biks 'mȯr·ē }

bone [ANATOMY] One of the parts constituting a vertebrate skeleton. [HISTOLOGY] A hard connective tissue that forms the major portion of the vertebrate skeleton. { bōn }

bone conduction [BIOPHYSICS] Transmission of sound vibrations to the internal ear via the bones of the skull. { 'bōn kən'dək·shən }

Bonellidae [INVERTEBRATE ZOOLOGY] A family of wormlike animals belonging to the order Echiuroinea. { bō'nel·ə,dē }

bone marrow [HISTOLOGY] A vascular modified connective tissue occurring in the long bones and certain flat bones of vertebrates. { 'bōn ,mar·ō }

bonsai [BOTANY] The production of a mature, very dwarfed tree in a relatively small container. { bōn'sī }

bony fish [VERTEBRATE ZOOLOGY] The name applied to all members of the class Osteichthyes. { 'bō·nē ,fish }

bony labyrinth [ANATOMY] The system of canals within the otic bones of vertebrates that houses the membranous labyrinth of the inner ear. { 'bō·nē 'lab·ə,rinth }

Boodleaceae [BOTANY] A family of green marine algae in the order Siphonocladales. { bō,äd·lē 'ās·ē,ē }

book gill [INVERTEBRATE ZOOLOGY] A type of gill in king crabs consisting of folds of membranous tissue arranged like the leaves of a book. { 'bu̇k ,gil }

book louse [INVERTEBRATE ZOOLOGY] A common name for a number of insects belonging to the order Psocoptera; important pests in herbaria, museums, and libraries. { 'bu̇k ,lau̇s }

book lung [INVERTEBRATE ZOOLOGY] A saccular respiratory organ in many arachnids consisting of numerous membranous folds arranged like the pages of a book. { 'bu̇k ,ləŋ }

Boöpidae [INVERTEBRATE ZOOLOGY] A family of lice in the order Mallophaga, parasitic on Australian marsupials. { bō'äp·ə,dē }

booster [IMMUNOLOGY] The dose of an immunizing agent given to stimulate the effects of a previous dose of the same agent. { 'büs·tər }

booster response [IMMUNOLOGY] *See* anamnestic response. { 'büs·tər ri'späns }

Bopyridae [INVERTEBRATE ZOOLOGY] A family of epicaridean isopods in the tribe Bopyrina known to parasitize decapod crustaceans. { bō'pī·rə ,dē }

Bopyrina [INVERTEBRATE ZOOLOGY] A tribe of dioecious isopods in the suborder Epicaridea. { bō·pī'rə·nə }

Bopyroidea [INVERTEBRATE ZOOLOGY] An equivalent name for Epicaridea. { ,bō·pi'rȯid·ē·ə }

Boraginaceae [BOTANY] A family of flowering plants in the order Lamiales comprising mainly herbs and some tropical trees. { bə¦ra·jə'nās· ē,ē }

bordered [BOTANY] Having a margin with a distinctive color or texture; used especially of a leaf. { 'bȯrd·ərd }

bordered pit [BOTANY] A wood-cell pit having the secondary cell wall arched over the cavity of the pit. { 'bȯrd·ərd 'pit }

Bordetella [MICROBIOLOGY] A genus of gram-negative, aerobic bacteria of uncertain affiliation; minute coccobacilli, parasitic and pathogenic in the respiratory tract of mammals. { ,bȯr·də 'tel·ə }

boreal [ECOLOGY] Of or relating to northern geographic regions. { 'bȯr·ē·əl }

boreal forest [ECOLOGY] *See* taiga. { 'bȯr·ē·əl 'fär·əst }

Boreal life zone [ECOLOGY] The zone comprising the climate and biotic communities between the Arctic and Transitional zones. { 'bȯr·ē·əl 'līf ,zōn }

borer [INVERTEBRATE ZOOLOGY] Any insect or other invertebrate that burrows into wood, rock, or other substances. { 'bȯr·ər }

Borhyaenidae [VERTEBRATE ZOOLOGY] A family of carnivorous mammals in the superfamily Borhyaenoidea. { ,bȯr·ē'ēn·ə,dē }

Borhyaenoidea [VERTEBRATE ZOOLOGY] A superfamily of carnivorous mammals in the order Marsupialia. { ,bȯr·ē·ə'nȯid·ē·ə }

boring sponge [INVERTEBRATE ZOOLOGY] Marine sponge of the family Clionidae represented by species which excavate galleries in mollusks, shells, corals, limestone, and other calcareous matter. { 'bȯr·iŋ ,spənj }

Borrelia [MICROBIOLOGY] A genus of bacteria in the family Spirochaetaceae; helical cells with uneven coils and parallel fibrils coiled around the cell body for locomotion; many species cause relapsing fever in humans. { bə'rel·ē·ə }

bosque [ECOLOGY] *See* temperate and cold scrub. { 'bäsk, 'bä·skā }

Bostrichidae [INVERTEBRATE ZOOLOGY] The powder-post beetles, a family of coleopteran insects in the superfamily Bostrichoidea. { bä'strik·ə ,dē }

Bostrichoidea [INVERTEBRATE ZOOLOGY] A superfamily of beetles in the coleopteran suborder Polyphaga. { ,bä·strə'kȯid·ē·ə }

botanical garden [BOTANY] An institution for the culture of plants collected chiefly for scientific and educational purposes. { bə'tan·ə·kəl 'gär· dən }

botany [BIOLOGY] A branch of the biological sciences which embraces the study of plants and plant life. { 'bät·ən·ē }

bothridium [INVERTEBRATE ZOOLOGY] A muscular holdfast organ, often with hooks, on the scolex of tetraphyllidean tapeworms. { bä'thrid·ē·əm }

Bothriocephaloidea [INVERTEBRATE ZOOLOGY] The equivalent name for the Pseudophyllidea. { ,bä·thrē·ō,sef·ə'lȯid·ē·ə }

Bothriocidaroida [PALEONTOLOGY] An order of extinct echinoderms in the subclass Perischoechinoidea in which the ambulacra consist of two columns of plates, the interambulacra of one column, and the madreporite is placed radially. { ,bä·thrē·ō,sik·ə'rȯid·ē·ə }

bothrium [INVERTEBRATE ZOOLOGY] A suction

groove on the scolex of pseudophyllidean tapeworms. { 'bäth·rē·əm }

bottle graft [BOTANY] A plant graft in which the scion is a detached branch and is protected from wilting by keeping the base of the branch in a bottle of water until union with the stock. { 'bäd·əl ˌgraft }

bottom break [BOTANY] A branch that arises from the base of a plant stem. { 'bäd·əm ˌbrāk }

bottom fauna [ECOLOGY] *See* benthos. { 'bäd·əm ˌfȯn·ə }

botulin [MICROBIOLOGY] The neurogenic toxin which is produced by *Clostridium botulinum* and *C. parabotulinum* and causes botulism. Also known as botulinus toxin. { 'bäch·ə·lən }

botulinus [MICROBIOLOGY] A bacterium that causes botulism. { 'bäch·ə'lī·nəs }

botulinus toxin [MICROBIOLOGY] *See* botulin. { 'bäch·ə'lī·nəs 'täk·sən }

bough [BOTANY] A main branch on a tree. { bau̇ }

bouton [ANATOMY] A club-shaped enlargement at the end of a nerve fiber. Also known as end bulb. { bü'tōn }

Bovidae [VERTEBRATE ZOOLOGY] A family of pecoran ruminants in the superfamily Bovoidea containing the true antelopes, sheep, and goats. { 'bō·vəˌdē }

bovine [VERTEBRATE ZOOLOGY] **1.** Any member of the genus *Bos*. **2.** Resembling or pertaining to a cow or ox. { 'bōˌvīn }

Bovoidea [VERTEBRATE ZOOLOGY] A superfamily of pecoran ruminants in the order Artiodactyla comprising the pronghorns and bovids. { bō 'vȯid·ē·ə }

bowel [ANATOMY] The intestine. { bau̇l }

bowfin [VERTEBRATE ZOOLOGY] *Amia calva*. A fish recognized as the only living species of the family Amiidae. Also known as dogfish; grindle; mudfish. { 'bōˌfin }

Bowman's capsule [ANATOMY] A two-layered membranous sac surrounding the glomerulus and constituting the closed end of a nephron in the kidneys of all higher vertebrates. { ¦bō·mənz 'kap·səl }

brace root [BOTANY] *See* prop root. { 'brās ˌrüt }

brachial [ZOOLOGY] Of or relating to an arm or armlike process. { 'brā·kē·əl }

brachial artery [ANATOMY] An artery which originates at the axillary artery and branches into the radial and ulnar arteries; it distributes blood to the various muscles of the arm, the shaft of the humerus, the elbow joint, the forearm, and the hand. { 'brā·kē·əl 'ärd·ə·rē }

brachial cavity [INVERTEBRATE ZOOLOGY] The anterior cavity which is located inside the valves of brachiopods and into which the brachia are withdrawn. { 'brā·kē·əl 'kav·əd·ē }

brachial plexus [ANATOMY] A plexus of nerves located in the neck and axilla and composed of the anterior rami of the lower four cervical and first thoracic nerves. { 'brā·kē·əl 'plek·səs }

Brachiata [INVERTEBRATE ZOOLOGY] A phylum of deuterostomous, sedentary bottom-dwelling marine animals that live encased in tubes. { ˌbra·kē'ad·ə }

brachiate [BOTANY] Possessing widely divergent branches. [ZOOLOGY] Having arms. { 'bra·kē ˌāt }

brachiolaria [INVERTEBRATE ZOOLOGY] A transitional larva in the development of certain starfishes that is distinguished by three anterior processes homologous with those of the adult. { ˌbra·kē·ō'la·re·ə }

Brachiopoda [INVERTEBRATE ZOOLOGY] A phylum of solitary, marine, bivalved coelomate animals. { ˌbrā·kē'ä·pə·də }

brachioradialis [ANATOMY] The muscle of the arm that flexes the elbow joint; origin is the lateral supracondylar ridge of the humerus, and insertion is the lower end of the radius. { ¦brā·kē·ōˌrā·dē'äl·əs }

brachium [ANATOMY] The upper arm or forelimb, from the shoulder to the elbow. [INVERTEBRATE ZOOLOGY] **1.** A ray of a crinoid. **2.** A tentacle of a cephalopod. **3.** Either of the paired appendages constituting the lophophore of a brachiopod. { 'brā·kē·əm }

Brachyarchus [MICROBIOLOGY] A genus of bacteria of uncertain affiliation; rod-shaped cells bent in a bowlike configuration and usually arranged in groups of two, four, or more cells. { ¦bra·kē'är·kəs }

brachyblast [BOTANY] A short shoot often bearing clusters of leaves. { 'bra·kəˌblast }

brachydont [ANATOMY] Of teeth, having short crowns, well-developed roots, and narrow root canals; characteristic of humans. { 'brak·ə ˌdänt }

Brachygnatha [INVERTEBRATE ZOOLOGY] A subsection of brachyuran crustaceans to which most of the crabs are assigned. { bra'kig·nə·thə }

Brachypsectridae [INVERTEBRATE ZOOLOGY] A family of coleopteran insects in the superfamily Cantharoidea represented by a single species. { ˌbra·kip'sek·trəˌdē }

Brachypteraciidae [VERTEBRATE ZOOLOGY] The ground rollers, a family of colorful Madagascan birds in the order Coraciiformes. { brəˌkip·ter·ə 'sī·əˌdē }

brachypterous [INVERTEBRATE ZOOLOGY] Having rudimentary or abnormally small wings, referring to certain insects. { brə'kip·tə·rəs }

brachysclereid [BOTANY] A sclereid that is more or less isodiametric and is found in certain fruits and in the pith, cortex, and bark of many stems. Also known as stone cell. { ˌbrak·i'sklir·ē·əd }

brachysm [BOTANY] Plant dwarfing in which there is shortening of the internodes only. { 'braˌkiz·əm }

Brachythoraci [PALEONTOLOGY] An order of the joint-neckfishes, now extinct. { ˌbrak·i'thȯr·ə·sī }

Brachyura [INVERTEBRATE ZOOLOGY] The section of the crustacean order Decapoda containing the true crabs. { ˌbra·kē'yu̇r·ə }

bracket fungus [MYCOLOGY] A basidiomycete characterized by shelflike sporophores, sometimes seen on tree trunks. { 'brak·ət ˌfəŋ·gəs }

Braconidae [INVERTEBRATE ZOOLOGY] The braconid wasps, a family of hymenopteran insects in the superfamily Ichneumonoidea. { brə'kän·ə ˌdē }

bract [BOTANY] A modified leaf associated with plant reproductive structures. { brakt }

bracteolate [BOTANY] Having bracteoles. { brak 'tē·ə·lət }

bracteole [BOTANY] A small bract, especially if on the floral axis. Also known as bractlet. { 'brak·tē ˌōl }

bractlet [BOTANY] *See* bracteole. { 'brak·lət }

bradyauxesis [BIOLOGY] Allometric growth in which a part lags behind the body as a whole in development. { ¦brā·dē·ȯg'zē·səs }

bradykinin [BIOCHEMISTRY] $C_{50}H_{73}N_{15}O_{11}$ A polypeptide kinin; forms an amorphous precipitate in glacial acetic acid; released from plasma precursors by plasmin; acts as a vasodilator. Also known as callideic I; kallidin I. { ¦brād·i'kī·nən }

Bradyodonti [PALEONTOLOGY] An order of Paleozoic cartilaginous fishes (Chondrichthyes), presumably derived from primitive sharks. { ˌbrā·dē·ō'dänˌtī }

Bradypodidae [VERTEBRATE ZOOLOGY] A family of mammals in the order Edentata comprising the true sloths. { ˌbrād·i'pä·dəˌdē }

bradytely [EVOLUTION] Evolutionary change that is either arrested or occurring at a very slow rate over long geologic periods. { 'brād·əˌte·lē }

brain [ANATOMY] The portion of the vertebrate central nervous system enclosed in the skull. [ZOOLOGY] The enlarged anterior portion of the central nervous system in most bilaterally symmetrical animals. { brān }

braincase [ANATOMY] *See* cranium. { 'brānˌkās }

brain coral [INVERTEBRATE ZOOLOGY] A reef-building coral resembling the human cerebrum in appearance. { 'brān ˌkär·əl }

brain hormone [INVERTEBRATE ZOOLOGY] A neurohormone secreted by the insect brain that regulates the release of ecdysone from the prothoracic glands. { 'brān ¦hȯrˌmōn }

brainstem [ANATOMY] The portion of the brain remaining after the cerebral hemispheres and cerebellum have been removed. { 'brānˌstem }

brain wave [PHYSIOLOGY] A rhythmic fluctuation of voltage between parts of the brain, ranging from about 1 to 60 hertz and 10 to 100 microvolts. { 'brān ˌwāv }

bramble [BOTANY] **1.** A plant of the genus *Rubus*. **2.** A rough, prickly vine or shrub. { 'bram·bəl }

branch [BOTANY] A shoot or secondary stem on the trunk or a limb of a tree. { branch }

branched acinous gland [ANATOMY] A multicellular structure with saclike glandular portions connected to the surface of the containing organ or structure by a common duct. { ¦brancht 'as·ə·nəs ˌgland }

branched tubular gland [ANATOMY] A multicellular structure with tube-shaped glandular portions connected to the surface of the containing organ or structure by a common secreting duct. { ¦brancht ¦tüb·yə·lər 'gland }

branchia [VERTEBRATE ZOOLOGY] *See* gill. { 'braŋ·kē·ə }

branchial [ZOOLOGY] Of or pertaining to gills. { 'braŋ·kē·əl }

branchial arch [VERTEBRATE ZOOLOGY] One of the series of paired arches on the sides of the pharynx which support the gills in fishes and amphibians. { 'braŋ·kē·əl 'ärch }

branchial basket [ZOOLOGY] A cartilaginous structure that supports the gills in protochordates and certain lower vertebrates such as cyclostomes. { 'braŋ·kē·əl 'bas·kət }

branchial cleft [EMBRYOLOGY] A rudimentary groove in the neck region of air-breathing vertebrate embryos. [VERTEBRATE ZOOLOGY] One of the openings between the branchial arches in fishes and amphibians. { 'braŋ·kē·əl 'kleft }

branchial heart [INVERTEBRATE ZOOLOGY] A muscular enlarged portion of a vein of a cephalopod that contracts and forces the blood into the gills. { 'braŋ·kē·əl 'härt }

branchial plume [INVERTEBRATE ZOOLOGY] An accessory respiratory organ that extends out under the mantle in certain Gastropoda. { 'braŋ·kē·əl 'plüm }

branchial pouch [ZOOLOGY] In cyclostomes and some sharks, one of the respiratory cavities occurring in the branchial clefts. { 'braŋ·kē·əl 'pau̇ch }

branchial sac [INVERTEBRATE ZOOLOGY] In tunicates, the dilated pharyngeal portion of the alimentary canal that has vascular walls pierced with clefts and serves as a gill. { 'braŋ·kē·əl 'sak }

branchial segment [EMBRYOLOGY] Any of the paired pharyngeal segments indicating the visceral arches and clefts posterior to and including the third pair in air-breathing vertebrate embryos. { 'braŋ·kē·əl 'seg·mənt }

branchiate [VERTEBRATE ZOOLOGY] Having gills. { 'braŋ·kēˌāt }

branching adaptation [EVOLUTION] *See* divergent adaptation. { 'branch·iŋ ˌadˌap'tā·shən }

branchiocranium [VERTEBRATE ZOOLOGY] The division of the fish skull constituting the mandibular and hyal regions and the branchial arches. { ¦braŋ·kē·ō'krā·nē·əm }

branchiomere [EMBRYOLOGY] An embryonic metamere that will differentiate into a visceral arch and cleft; a branchial segment. { 'braŋ·kē·əˌmir }

branchiomeric musculature [VERTEBRATE ZOOLOGY] Those muscles derived from branchial segments in vertebrates. { ¦braŋ·kē·ə¦mer·ik 'məs·kyə·ləˌchər }

Branchiopoda [INVERTEBRATE ZOOLOGY] A subclass of crustaceans containing small or moderate-sized animals commonly called fairy shrimps, clam shrimps, and water fleas. { ˌbraŋ·kē'äp·ə·də }

branchiostegite [INVERTEBRATE ZOOLOGY] A gill cover and chamber in certain malacostracan crustaceans, formed by lateral expansion of the carapace. { ˌbraŋ·kē'äs·təˌjīt }

Branchiostoma [ZOOLOGY] A genus of lancelets formerly designated as amphioxus. { ˌbraŋ·kē 'äs·tə·mə }

Branchiotremata [INVERTEBRATE ZOOLOGY] The

hemichordates, a branch of the subphylum Oligomera. { ¦braŋ·kē·ə'trem·əd·ə }

Branchiura [INVERTEBRATE ZOOLOGY] The fish lice, a subclass of fish ectoparasites in the class Crustacea. { ,braŋ·kē'yür·ə }

Branhamella [MICROBIOLOGY] A genus of bacteria in the family Neisseriaceae; cocci occur in pairs with flattened adjacentsides; parasites of mammalian mucous membranes. { ,bran·ə 'mel·ə }

Brassica [BOTANY] A large genus of herbs in the family Cruciferae of the order Capparales, including cabbage, watercress, and sweet alyssum. { 'bras·ə·kə }

Brassicaceae [BOTANY] An equivalent name for the Cruciferae. { ,bras·ə'kās·ē,ē }

brassin [BIOCHEMISTRY] Any of a class of plant hormones characterized as long-chain fatty-acid esters; brassins act to induce both cell elongation and cell division in leaves and stems. { 'bra·sən }

Brathinidae [INVERTEBRATE ZOOLOGY] The grass root beetles, a small family of coleopteran insects in the superfamily Staphylinoidea. { brə 'thī·nə,dē }

Braulidae [INVERTEBRATE ZOOLOGY] The bee lice, a family of cyclorrhaphous dipteran insects in the section Pupipara. { 'braủl·ə,dē }

Brazil nut [BOTANY] *Bertholletia excelsa.* A large broad-leafed evergreen tree of the order Lecythedales; an edible seed is produced by the tree fruit. { brə'zil ,nət }

breadfruit [BOTANY] *Artocarpus altilis.* An Indo-Malaysian tree, a species of the mulberry family (Moraceae).The tree produces a multiple fruit which is edible. { 'bred,früt }

bread mold [MYCOLOGY] Any fungus belonging to the family Mucoraceae in the order Mucorales. { 'bred ,mōld }

breakage and reunion [CELL BIOLOGY] The classical model of crossing over by means of physical breakage and crossways reunion of completed chromatids during the process of meiosis. { 'brāk·ij ən rē'yün·yən }

breast [ANATOMY] The human mammary gland. { brest }

breathing [PHYSIOLOGY] Inhaling and exhaling. { 'brēth·iŋ }

breeding [GENETICS] Controlled mating and selection, or hybridization of plants and animals in order to improve the species. { 'brēd·iŋ }

bregma [ANATOMY] The point at which the coronal and sagittal sutures of the skull meet. { 'breg·mə }

Brentidae [INVERTEBRATE ZOOLOGY] The straight-snouted weevils, a family of coleopteran insects in the superfamily Curculionoidea. { 'bren·tə ,dē }

Brevibacteriaceae [MICROBIOLOGY] Formerly a family of gram-positive, rod-shaped, schizomycetous bacteria in the order Eubacteriales. { ,brev·ə,bak,tir·ē'ās·ē,ē }

Brevibacterium [MICROBIOLOGY] A genus of short, unbranched, rod-shaped bacteria in the coryneform group. { ,brev·ə,bak'tir·ē·əm }

brevitoxin [BIOCHEMISTRY] One of several ichthyotoxins produced by the dinoflagellate *Ptychodiscus brevis.* { ,brev·ə'täk·sən }

Brewer anaerobic jar [MICROBIOLOGY] A glass container in which petri dish cultures are stacked and maintained under anaerobic conditions. { 'brü·ər an·ə'rō·bik 'jär }

bridge graft [BOTANY] A plant graft in which each of several scions is grafted in two positions on the stock, one above and the other below an injury. { 'brij ,graft }

Brisingidae [INVERTEBRATE ZOOLOGY] A family of deep-water echinoderms with as many as 44 arms, belonging to the order Forcipulatida. { brə 'sin·jə,dē }

bristle [BIOLOGY] A short stiff hair or hairlike structure on an animal or plant. { 'bris·əl }

brittle star [INVERTEBRATE ZOOLOGY] The common name for all members of the echinoderm class Ophiuroidea. { ¦brid·əl 'stär }

broadleaf tree [BOTANY] Any deciduous or evergreen tree having broad, flat leaves. { 'bròd,lēf ,trē }

broad-spectrum antibiotic [MICROBIOLOGY] An antibiotic that is effective against both gram-negative and gram-positive bacterial species. { ¦bròd ¦spek·trəm ,ant·i·bī'äd·ik }

broccoli [BOTANY] *Brassica oleracea* var. *italica.* A biennial crucifer of the order Capparales which is grown for its edible stalks and buds. { 'brak·ə·lē }

Brodmann's area 4 [PHYSIOLOGY] *See* motor area. { 'bräd·mənz ,er·ē·ə 'fòr }

Brodmann's area 17 [PHYSIOLOGY] *See* visual projection area. { 'bräd·mənz ,er·ē·ə sev·ən 'tēn }

Brodmann's areas [PHYSIOLOGY] Numbered regions of the cerebral cortex used to identify cortical functions. { 'bräd·mənz ,er·ē·əz }

bromatium [ECOLOGY] A swollen hyphal tip on fungi growing in ant nests that is eaten by the ants. { brō'māsh·əm }

bromegrass [BOTANY] The common name for a number of forage grasses of the genus *Bromus* in the order Cyperales. { 'brōm,gras }

bromelain [BIOCHEMISTRY] An enzyme that digests protein and clots milk; prepared by precipitation by acetone from pineapple juice; used to tenderize meat, to chill-proof beer, and to make protein hydrolysates. Also spelled bromelin. { 'brō·mə,lān }

Bromeliaceae [BOTANY] The single family of the flowering plant order Bromeliales. { brō,mel·ē 'ās·ē,ē }

Bromeliales [BOTANY] An order of monocotyledonous plants in the subclass Commelinidae, including terrestrial xerophytes and some epiphytes. { brō,mel·ē'ā·lēz }

bromelin [BIOCHEMISTRY] *See* bromelain. { 'brō·mə·lən }

5-bromodeoxyruridine [BIOCHEMISTRY] $C_9H_{11}O_5NBr$ A thymidine analog that can be incorporated into deoxyribonucleic acid during its replication; induces chromosomal breakage in regions rich in heterochromatin. Abbreviated BUDR. { ¦fīv ¦brō·mō·dē,äk·sē'rur·ə,dēn }

bromouracil [BIOCHEMISTRY] $C_4H_3N_2O_2Br$

5-Bromouracil, an analog of thymine that can react with deoxyribonucleic acid to produce a polymer with increased susceptibility to mutation. { ˌbrō·mō'yu̇r·ə·səl }

bromuration [HISTOLOGY] A process in which a tissue section is treated with a solution of bromine or a bromine compound. { ˌbräm·yə'rā·shən }

bronchial tree [ANATOMY] The arborization of the bronchi of the lung, considered as a structural and functional unit. { 'bräŋ·kē·əl 'trē }

bronchiole [ANATOMY] A small, thin-walled branch of a bronchus, usually terminating in alveoli. { 'bräŋ·kēˌōl }

bronchoconstriction [PHYSIOLOGY] Narrowing of the air passages in bronchi and bronchioles. { ¦bräŋ·kō·kən¦strik·shən }

bronchodilation [PHYSIOLOGY] Widening of the air passages in bronchi and bronchioles. { ¦bräŋ·kōˌdī'lā·shən }

bronchus [ANATOMY] Either of the two primary branches of the trachea or any of the bronchi's pulmonary branches having cartilage in their walls. { 'bräŋ·kəs }

Brontotheriidae [PALEONTOLOGY] The single family of the extinct mammalian superfamily Brontotherioidea. { ¦brän·tō·thə'rī·əˌdē }

Brontotherioidea [PALEONTOLOGY] The titanotheres, a superfamily of large, extinct perissodactyl mammals in the suborder Hippomorpha. { ¦brän·tōˌthe·rē'ȯid·ē·ə }

brood [BOTANY] Heavily infested by insects. [ZOOLOGY] **1.** The young of animals. **2.** To incubate eggs or cover the young for warmth. **3.** An animal kept for breeding. { brüd }

brood capsule [INVERTEBRATE ZOOLOGY] A secondary scolex-containing cyst constituting the infective agent of a tapeworm. { 'brüd ˌkap·səl }

brood parasitism [ECOLOGY] A type of social parasitism among birds characterized by a bird of one species laying and abandoning its eggs in the nest of a bird of another species. { ¦brüd ˌpar·ə·səˌtiz·əm }

brood pouch [VERTEBRATE ZOOLOGY] A pouch of an animal body where eggs or embryos undergo certain stages of development. { 'brüd ˌpau̇ch }

Brotulidae [VERTEBRATE ZOOLOGY] A family of benthic teleosts in the order Perciformes. { brō 'tü·ləˌdē }

brown algae [BOTANY] The common name for members of the Phaeophyta. { ¦brau̇n ¦al·jē }

brown fat cell [HISTOLOGY] A moderately large, generally spherical cell in adipose tissue that has small fat droplets scattered in the cytoplasm. { ¦brau̇n 'fat ˌsel }

brown seaweed [BOTANY] A common name for the larger algae of the division Phaeophyta. { ¦brau̇n 'sēˌwēd }

browse [BIOLOGY] **1.** Twigs, shoots, and leaves eaten by livestock and other grazing animals. **2.** To feed on this vegetation. { brau̇z }

Brucella [MICROBIOLOGY] A genus of gram-negative, aerobic bacteria of uncertain affiliation; single, nonmotile coccobacilli or short rods, all of which are parasites and pathogens of mammals. { brü'sel·ə }

Brucellaceae [MICROBIOLOGY] Formerly a family of small, coccoid to rod-shaped, gram-negative bacteria in the order Eubacteriales. { ˌbrü·sə 'lās·ēˌē }

brucellergen [BIOCHEMISTRY] A nucleoprotein fraction of brucellae used in skin tests to detect the presence of *Brucella* infections. { ˌbrü'sel·ər·jen }

brucellergen test [IMMUNOLOGY] A diagnostic skin test for detection of *Brucella* infections. { ˌbrü'sel·ər·jen ˌtest }

Bruchidae [INVERTEBRATE ZOOLOGY] The pea and bean weevils, a family of coleopteran insects in the superfamily Chrysomeloidea. { 'brü·kəˌdē }

Bruch's membrane [ANATOMY] The membrane of the retina that separates the pigmented layer of the retina from the choroid coat of the eye. { 'brüks ˌmemˌbrān }

Brücke-Abney phenomenon [PHYSIOLOGY] The inability to distinguish colors other than blue-violet, green, and red at very low light levels. { 'bru̇k·ə 'ab·nē fəˌnäm·əˌnän }

Brunner's glands [ANATOMY] Simple, branched, tubular mucus-secreting glands in the submucosa of the duodenum in mammals. Also known as duodenal glands; glands of Brunner. { 'brən·ərz ˌglanz }

brush [ECOLOGY] *See* tropical scrub. { brəsh }

brush border [CELL BIOLOGY] A superficial protoplasic modification in the form of filiform processes or microvilli; present on certain absorptive cells in the intestinal epithelium and the proximal convolutions of nephrons. { 'brəsh ˌbȯr·dər }

brussels sprouts [BOTANY] *Brassica oleracea* var. *gemmifera*. A biennial crucifer of the order Capparales cultivated for its small, edible, headlike buds. { ¦brəs·əlz 'sprau̇ts }

Bryales [BOTANY] An order of the subclass Bryidae; consists of mosses which often grow in disturbed places. { brī'ā·lēz }

Bryidae [BOTANY] A subclass of the class Bryopsida; includes most genera of the true mosses. { 'brī·əˌdē }

bryology [BOTANY] The study of bryophytes. { brī 'äl·ə·jē }

Bryophyta [BOTANY] A small phylum of the plant kingdom, including mosses, liverworts, and hornworts, characterized by the lack of true roots, stems, and leaves. { brī'ä·fə·də }

Bryopsida [BOTANY] The mosses, a class of small green plants in the phylum Bryophyta. Also known as Musci. { brī'äp·sə·də }

Bryopsidaceae [BOTANY] A family of green algae in the order Siphonales. { brīˌäp·sə'dās·ēˌē }

Bryoxiphiales [BOTANY] An order of the class Bryopsida in the subclass Bryidae; consists of a single genus and species, *Bryoxiphium norvegicum*, the sword moss. { brīˌäk·sə·fē'ā·lēz }

Bryozoa [INVERTEBRATE ZOOLOGY] The moss animals, a major phylum of sessile aquatic invertebrates occurring in colonies with hardened exoskeleton. { ˌbrī·ə'zō·ə }

buccal cavity [ANATOMY] The space anterior to

the teeth and gums in the mouths of all vertebrates having lips and cheeks. Also known as vestibule. { 'bək·əl ¦kav·əd·ē }

buccal gland [ANATOMY] Any of the mucous glands in the membrane lining the cheeks of mammals, except aquatic forms. { 'bək·əl ˌgland }

Buccinacea [INVERTEBRATE ZOOLOGY] A superfamily of gastropod mollusks in the order Prosobranchia. { ˌbək·sin'ās·ē·ə }

Buccinidae [INVERTEBRATE ZOOLOGY] A family of marine gastropod mollusks in the order Neogastropoda containing the whelks in the genus *Buccinum*. { bək'sin·əˌdē }

Bucconidae [VERTEBRATE ZOOLOGY] The puffbirds, a family of neotropical birds in the order Piciformes. { bə'kän·əˌdē }

Bucerotidae [VERTEBRATE ZOOLOGY] The hornbills, a family of Old World tropical birds in the order Coraciiformes. { ˌbyü·sə'räd·əˌdē }

buck [VERTEBRATE ZOOLOGY] A male deer. { bək }

bucket [BOTANY] *See* calyx. { 'bək·ət }

buckeye [BOTANY] The common name for deciduous trees composing the genus *Aesculus* in the order Sapindales; leaves are opposite and palmately compound, and the seed is large with a firm outer coat. { 'bəkˌī }

buckwheat [BOTANY] The common name for several species of annual herbs in the genus *Fagopyrum* of the order Polygonales; used for the starchy seed. { 'bəkˌwēt }

bud [BOTANY] An embryonic shoot containing the growing stem tip surrounded by young leaves or flowers or both and frequently enclosed by bud scales. { bəd }

budbreak [BOTANY] Initiation of growth from a bud. { ¦bədˌbrāk }

budding [BIOLOGY] A form of asexual reproduction in which a new individual arises as an outgrowth of an older individual. Also known as gemmation. [BOTANY] A method of vegetative propagation in which a single bud is grafted laterally onto a stock. [VIROLOGY] A form of virus release from the cell in which replication has occurred, common to all enveloped animal viruses; the cell membrane closes around the virus and the particle exits from the cell. { 'bəd·iŋ }

budding bacteria [MICROBIOLOGY] Bacteria that reproduce by budding. { 'bəd·iŋ bak'tir·ē·ə }

bud grafting [BOTANY] Grafting a plant by budding. { 'bəd ˌgraf·tiŋ }

budling [BOTANY] The shoot that develops from the bud which was the scion of a bud graft. { 'bəd·liŋ }

BUDR [BIOCHEMISTRY] *See* 5-bromodeoxyuridine.

bud scale [BOTANY] One of the modified leaves enclosing and protecting buds in perennial plants. { 'bəd ˌskāl }

buffalo [VERTEBRATE ZOOLOGY] The common name for several species of artiodactyl mammals in the family Bovidae, including the water buffalo and bison. { 'bəf·əˌlō }

buffer [ECOLOGY] An animal that is introduced to serve as food for other animals to reduce the losses of more desirable animals. { 'bəf·ər }

Bufonidae [VERTEBRATE ZOOLOGY] A family of toothless frogs in the suborder Procoela including the true toads (*Bufo*). { byü'fän·əˌdē }

bug [INVERTEBRATE ZOOLOGY] Any insect in the order Hemiptera. { bəg }

buildup [ECOLOGY] A significant increase in a natural population, usually as a result of progressive changes in ecological relations. { 'bil ˌdəp }

bulb [BOTANY] A short, subterranean stem with many overlapping fleshy leaf bases or scales, such as in the onion and tulip. { bəlb }

bulbil [BOTANY] A secondary bulb usually produced on the aerial part of a plant. { 'bəl·bəl }

bulbocavernosus [ANATOMY] *See* bulbospongiosus. { ¦bəl·bōˌka·vər'nō·səs }

bulb of the penis [ANATOMY] The expanded proximal portion of the corpus spongiosum of the penis. { ¦bəlb əv thə 'pē·nəs }

bulbospongiosus [ANATOMY] A muscle encircling the bulb and adjacent proximal parts of the penis in the male, and encircling the orifice of the vagina and covering the lateral parts of the vestibular bulbs in the female. Also known as bulbocavernosus. { ¦bəl·bōˌspən·jē'ō·səs }

bulbourethral gland [ANATOMY] Either of a pair of compound tubular glands, anterior to the prostate gland in males, which discharge into the urethra. Also known as Cowper's gland. { ¦bəl·bō·yu̇'rēth·rəl ˌgland }

Buliminacea [INVERTEBRATE ZOOLOGY] A superfamily of benthic, marine foraminiferans in the suborder Rotaliina. { ˌbyüˌlim·ə'nās·ē·ə }

bullate [BIOLOGY] Appearing blistered or puckered, especially of certain leaves. { 'bəˌlāt }

bulliform [BOTANY] Type of plant cell involved in tissue contraction or water storage, or of uncertain function. { 'bu̇l·əˌfȯrm }

bulliform cell [BOTANY] One of the large, highly vacuolated cells occurring in the epidermis of grass leaves. Also known as motor cell. { 'bu̇l·ə ˌfȯrm ˌsel }

bumblebee [INVERTEBRATE ZOOLOGY] The common name for several large, hairy social bees of the genus *Bombus* in the family Apidae. { 'bəm·bəlˌbē }

bundle branch [ANATOMY] Either of the components of the atrioventricular bundle passing to the right and left ventricles of the heart. { 'bənd·əl ˌbranch }

bundle of His [ANATOMY] A small region of heart muscle located in the right auricle and specialized to relay contraction impulses from the right auricle to the ventricles. Also known as atrioventricular bundle. { 'bənd·əl əv ˌächˌī'es }

bundle scar [BOTANY] A mark within a leaf scar that shows the point of an abscised vascular bundle. { 'bənd·əl ˌskär }

bundle sheath [BOTANY] A sheath around a vascular bundle that consists of a layer of parenchyma. { 'bənd·əl ˌshēth }

bunodont [ANATOMY] Having tubercles or rounded cusps on the molar teeth, as in humans. { ¦byü·nəˌdänt }

bunolophodont [VERTEBRATE ZOOLOGY] **1.** Of teeth, having the outer cusps in the form of blunt cones and the inner cusps as transverse ridges. **2.** Having such teeth, as in tapirs. { ¦byü·nə ¦läf·ə,dänt }

Bunonematoidea [INVERTEBRATE ZOOLOGY] A superfamily of nematodes in the order Rhabditida, characterized by asymmetrical bodies in both the labia and the distribution of sensory organs. { ,bü·nō,nem·ə'tȯid·ē·ə }

bunoselenodont [VERTEBRATE ZOOLOGY] **1.** Of teeth, having the inner cusps in the form of blunt cones and the outer ones as longitudinal crescents. **2.** Having such teeth, as in the extinct titanotheres. { ¦byü·nō·sə¦lē·nə,dänt }

Bunyaviridae [VIROLOGY] A family of enveloped spherical viruses whose lipid envelopes contain at least one virus-specific glycopeptide; members develop in the cytoplasm and mature by budding. { ¦bən·yə'vī·rə,dē }

Bunyavirus [VIROLOGY] A genus of viruses in the family Bunyaviridae; contains a minimum of 87 species which exhibit some degree of antigenic relationship. { 'bən·yə,vī·rəs }

Buprestidae [INVERTEBRATE ZOOLOGY] The metallic wood-boring beetles, the large, single family of the coleopteran superfamily Buprestoidea. { byü'pres·tə,dē }

Buprestoidea [INVERTEBRATE ZOOLOGY] A superfamily of coleopteran insects in the suborder Polyphaga including many serious pests of fruit trees. { ,byü·pres'tȯid·ē·ə }

Burhinidae [VERTEBRATE ZOOLOGY] The thick-knees or stone curlews, a family of the avian order Charadriiformes. { byü'rin·ə,dē }

burl [BOTANY] A hard, woody outgrowth on a tree, usually resulting from the entwined growth of a cluster of adventitious buds. { bərl }

Burmanniaceae [BOTANY] A family of monocotyledonous plants in the order Orchidales characterized by regular flowers, three or six stamens opposite the petals, and ovarian nectaries. { bər ,man·ē'ās·ē,ē }

burr [BOTANY] **1.** A rough or prickly envelope on a fruit. **2.** A fruit so characterized. { bər }

burr ball [ECOLOGY] *See* lake ball. { 'bər ,bȯl }

burro [VERTEBRATE ZOOLOGY] A small donkey used as a pack animal. { 'bu̇r·ō }

bursa [ANATOMY] A simple sac or cavity with smooth walls containing a clear, slightly sticky fluid and interposed between two moving surfaces of the body to reduce friction. { 'bər·sə }

bursa of Fabricius [VERTEBRATE ZOOLOGY] A thymuslike organ in the form of a diverticulum at the lower end of the alimentary canal in birds. { 'bər·sə əv fə'brēsh·əs }

Burseraceae [BOTANY] A family of dicotyledonous plants in the order Sapindales characterized by an ovary of two to five cells, prominent resin ducts in the bark and wood, and an intrastaminal disk. { ,bər·sə'rās·ē,ē }

bursicle [BOTANY] A purse or pouchlike receptacle. { ,bər·sə·kəl }

burster [BOTANY] An abnormally double flower having the calyx split or fragmented. { 'bər·stər }

bushbaby [VERTEBRATE ZOOLOGY] Any of six species of African arboreal primates in two genera (*Galago* and *Euoticus*) of the family Lorisidae. Also known as galago; night ape. { 'bu̇sh,bā·bē }

butanol fermentation [MICROBIOLOGY] Butanol production as a result of the fermentation of corn and molasses by the anaerobic bacterium *Clostridium acetobutylicum*. { 'byüt·ən,ȯl ,fər·mən'tā·shən }

Butomaceae [BOTANY] A family of monocotyledonous plants in the order Alismatales characterized by secretory canals, linear leaves, and a straight embryo. { ,byüd·ə'mās·ē,ē }

butt [BOTANY] The portion of a plant from which the roots extend, for example, the base of a tree trunk. { bət }

butterfly [INVERTEBRATE ZOOLOGY] Any insect belonging to the lepidopteran suborder Rhopalocera, characterized by a slender body, broad colorful wings, and club-shaped antennae. { 'bəd·ər,flī }

buttocks [ANATOMY] The two fleshy parts of the body posterior to the hip joints. { 'bəd·əks }

buttress [BOTANY] A ridge of wood developed in the angle between a lateral root and the butt of a tree. [PALEONTOLOGY] A ridge on the inner surface of a pelecypod valve which acts as a support for part of the hinge. { 'bə·trəs }

butyric fermentation [BIOCHEMISTRY] Fermentation in which butyric acid is produced by certain anaerobic bacteria acting on organic substances, such as butter; occurs in putrefaction and in digestion in herbivorous mammals. { byü'tir·ik fər·mən'tā·shən }

butyrinase [BIOCHEMISTRY] An enzyme that hydrolyzes butyrin, found in the blood serum. { 'byüd·ə·rə,nās }

Butyrivibrio [MICROBIOLOGY] A genus of gram-negative, strictly anaerobic bacteria of uncertain affiliation; motile curved rods occur singly, in chains, or in filaments; ferment glucose to produce butyrate. { ¦byüd·ə·rə'vī·brē,ō }

Buxbaumiales [BOTANY] An order of very small, atypical mosses (Bryopsida) composed of three genera and found on soil, rock, and rotten wood. { ,bəks,bōm·ē'ā·lēz }

B virus [VIROLOGY] An animal virus belonging to subgroup A of the herpesvirus group. { 'bē ,vī·rəs }

Byrrhidae [INVERTEBRATE ZOOLOGY] The pill beetles, the single family of the coleopteran insect superfamily Byrrhoidea. { 'bir·ə,dē }

Byrrhoidea [INVERTEBRATE ZOOLOGY] A superfamily of coleopteran insects in the superorder Polyphaga. { bə'rȯid·ē·ə }

Byturidae [INVERTEBRATE ZOOLOGY] The raspberry fruitworms, a small family of coleopteran insects in the superfamily Cucujoidea. { bī'tu̇r·ə ,dē }

C

caatinga [ECOLOGY] A sparse, stunted forest in areas of little rainfall in northeastern Brazil; trees are leafless in the dry season. { kä'tiŋ·gə }

cabbage [BOTANY] *Brassica oleracea* var. *capitata*. A biennial crucifer of the order Capparales grown for its head of edible leaves. { 'kab·ij }

cabezon [VERTEBRATE ZOOLOGY] *Scorpaenichthys marmoratus*. A fish that is the largest of the sculpin species, weighing as much as 25 pounds (11.3 kilograms) and reaching a length of 30 inches (76 centimeters). { 'kab·ə‚zōn, *or* ‚ka·bə 'zōn }

Cabot's ring [CELL BIOLOGY] A ringlike body in immature erythrocytes that may represent the remains of the nuclear membrane. { 'kab·əts 'riŋ }

cacao [BOTANY] *Theobroma cacao*. A small tropical tree of the order Theales that bears capsular fruits which are a source of cocoa powder and chocolate. Also known as chocolate tree. { kə 'kaů }

cacomistle [VERTEBRATE ZOOLOGY] *Bassariscus astutus*. A raccoonlike mammal that inhabits the southern and southwestern United States; distinguished by a bushy black-and-white ringed tail. Also known as civet cat; ringtail. { 'kak·ə ‚mis·əl }

caconym [SYSTEMATICS] A taxonomic name that is linguistically unacceptable. { 'kak·ə‚nim }

Cactaceae [BOTANY] The cactus family of the order Caryophyllales; represented by the American stem succulents, which are mostly spiny with reduced leaves. { kak'tās·ē‚ē }

cactus [BOTANY] The common name for any member of the family Cactaceae, a group characterized by a fleshy habit, spines and bristles, and large, brightly colored, solitary flowers. { 'kak·təs }

cadaverine [BIOCHEMISTRY] $C_5H_{14}N_2$ A nontoxic, organic base produced as a result of the decarboxylation of lysine by the action of putrefactive bacteria on flesh. { kə'dav·ə‚rēn }

caddis fly [INVERTEBRATE ZOOLOGY] The common name for all members of the insect order Trichoptera; adults are mothlike and the immature stages are aquatic. { 'kad·əs ‚flī }

caducicorn [VERTEBRATE ZOOLOGY] Having deciduous horns, as certain deer. { kə'dü·sə ‚kȯrn }

caducous [BOTANY] Lasting on a plant only a short time before falling off. { 'kad·ə·kəs }

caecilian [VERTEBRATE ZOOLOGY] The common name for members of the amphibian order Apoda. { sē'sil·yən }

caecum [ANATOMY] *See* cecum. { 'sē·kəm }

Caenolestidae [PALEONTOLOGY] A family of extinct insectivorous mammals in the order Marsupialia. { ‚sē·nə'les·tə‚de }

Caenolestoidea [VERTEBRATE ZOOLOGY] A superfamily of marsupial mammals represented by the single living family Caenolestidae. { ‚sē·nə·le 'stȯid·ē·ə }

caenostylic [VERTEBRATE ZOOLOGY] Having the first two visceral arches attached to the cranium and functioning in food intake; a condition found in sharks, amphibians, and chimaeras. { ¦sē·nə ¦stīl·ik }

Caesalpinoidea [BOTANY] A subfamily of dicotyledonous plants in the legume family, Leguminosae. { ‚sē‚zal·pə'nȯid·ē·ə }

caiman [VERTEBRATE ZOOLOGY] Any of five species of reptiles of the genus *Caiman* in the family Alligatoridae, differing from alligators principally in having ventral armor and a sharper snout. { 'kā·mən }

Cainotheriidae [PALEONTOLOGY] The single family of the extinct artiodactyl superfamily Cainotherioidea. { ‚kān·ə·thə'rī·ə‚dē }

Cainotherioidea [PALEONTOLOGY] A superfamily of extinct, rabbit-sized tylopod ruminants in the mammalian order Artiodactyla. { ‚kān·ə·ther·ē 'ȯid·ē·ə }

Calamitales [PALEOBOTANY] An extinct group of reedlike plants of the subphylum Sphenopsida characterized by horizontal rhizomes and tall, upright, grooved, articulated stems. { kə‚lam·ə 'tā·lēz }

Calanoida [INVERTEBRATE ZOOLOGY] A suborder of the crustacean order Copepoda, including the larger and more abundant of the pelagic species. { ‚kal·ə'nȯid·ə }

Calappidae [INVERTEBRATE ZOOLOGY] The box crabs, a family of reptantian decapods in the subsection Oxystomata of the section Brachyura. { kə'lap·ə‚dē }

calathiform [BIOLOGY] Cup-shaped, being almost hemispherical. { kə'lath·ə‚fȯrm }

calcaneocuboid ligament [ANATOMY] The ligament that joins the calcaneus and the cuboid bones. { kal¦kan·ē·ō¦kyü‚bȯid ¦lig·ə·mənt }

calcaneum [ANATOMY] **1.** A bony projection of

the metatarsus in birds. **2.** *See* calcaneus. { kal 'kan·ē·əm }

calcaneus [ANATOMY] A bone of the tarsus, forming the heel bone in humans. Also known as calcaneum. { kal'kan·ē·əs }

calcar [ZOOLOGY] A spur or spurlike process, especially on an appendage or digit. { 'kal,kär }

calcarate [BOTANY] Having spurs. { 'kal·kə,rāt }

Calcarea [INVERTEBRATE ZOOLOGY] A class of the phylum Porifera, including sponges with a skeleton composed of calcium carbonate spicules. { kal'kar·ē·ə }

calcareous algae [BOTANY] Algae that grow on limestone or in soil impregnated with lime. { kal 'ker·ē·əs 'al·jē }

calcarine fissure [ANATOMY] *See* calcarine sulcus. { 'kal·kə,rēn 'fish·ər }

calcarine sulcus [ANATOMY] A sulcus on the medial aspect of the occipital lobe of the cerebrum, between the lingual gyrus and the cuneus. Also known as calcarine fissure. { 'kal·kə,rēn 'səl·kəs }

Calcaronea [INVERTEBRATE ZOOLOGY] A subclass of sponges in the class Calcarea in which the larva are amphiblastulae. { kal·kə'rō·nē·ə }

calcicole [BOTANY] Requiring soil rich in calcium carbonate for optimum growth. { 'kal·sə,kōl }

calciferous [BIOLOGY] Containing or producing calcium or calcium carbonate. { kal'sif·ə·rəs }

calciferous gland [INVERTEBRATE ZOOLOGY] One of a series of glands that secrete calcium carbonate into the esophagus of certain oligochaetes. { kal'sif·ə·rəs ,gland }

calcification [PHYSIOLOGY] The deposit of calcareous matter within the tissues of the body. { ,kal·sə·fə'kā·shən }

calcifuge [ECOLOGY] A plant that grows in an acid medium that is poor in calcareous matter. { 'kal·sə,fyüj }

Calcinea [INVERTEBRATE ZOOLOGY] A subclass of sponges in the class Calcarea in which the larvae are parenchymulae. { kal'sin·ē·ə }

calciphylaxis [IMMUNOLOGY] A sudden local calcification in tissues in response to induced hypersensitivity following systemic sensitization by a calcifying factor. { ,kal·sə·fə'lak·səs }

calcitonin [BIOCHEMISTRY] A polypeptide, calcium-regulating hormone produced by the ultimobranchial bodies in vertebrates. Also known as thyrocalcitonin. { kal·sə'tō·nən }

calcium metabolism [BIOCHEMISTRY] Biochemical and physiological processes involved in maintaining the concentration of calcium in plasma at a constant level and providing a sufficient supply of calcium for bone mineralization. { 'kal·sē·əm mə'tab·ə,liz·əm }

Calclamnidae [PALEONTOLOGY] A family of Paleozoic echinoderms of the order Dendrochirotida. { kal'klam·nə,dē }

calculus [ANATOMY] A small and cuplike structure. { 'kal·kyə·ləs }

calf [VERTEBRATE ZOOLOGY] The young of the domestic cow, elephant, rhinoceros, hippopotamus, moose, whale, and others. { kaf }

Caliciaceae [BOTANY] A family of lichens in the order Caliciales in which the disk of the apothecium is borne on a short stalk. { kə,lē·sē'ās·ē,ē }

Caliciales [BOTANY] An order of lichens in the class Ascolichenes characterized by an unusual apothecium. { kə,lē·se'ā,lēz }

Caliciviridae [VIROLOGY] A family of nonenveloped ribonucleic acid viruses with characteristic hollow surface structures that resemble cups. { kə,lē·sē'vī·rə,dē }

Caligidae [INVERTEBRATE ZOOLOGY] A family of fish ectoparasites belonging to the crustacean suborder Caligoida. { kə'lij·ə,dē }

Caligoida [INVERTEBRATE ZOOLOGY] A suborder of the crustacean order Copepoda, including only fish ectoparasites and characterized by a sucking mouth with styliform mandibles. { kal·ə 'gȯid·ə }

Callichthyidae [VERTEBRATE ZOOLOGY] A family of tropical catfishes in the suborder Siluroidei. { ,ka,lik'thī·ə,dē }

callideic I [BIOCHEMISTRY] *See* bradykinin. { ¦kal·i¦dē·ik ¦wən }

calling song [INVERTEBRATE ZOOLOGY] A high-intensity insect sound which may play a role in habitat selection among certain species. { 'kȯl·iŋ ,sȯŋ }

Callionymoidei [VERTEBRATE ZOOLOGY] A suborder of fishes in the order Perciformes, including two families of colorful marine bottom fishes known as dragonets. { kə¦län·ē¦mȯid·ē,ī }

Callipallenidae [INVERTEBRATE ZOOLOGY] A family of marine arthropods in the subphylum Pycnogonida lacking palpi and having chelifores and 10-jointed ovigers. { ,kal·ə·pə'len·ə,dē }

Calliphoridae [INVERTEBRATE ZOOLOGY] The blow flies, a family of myodarian cyclorrhaphous dipteran insects in the subsection Calypteratae. { ,kal·ə'fȯr·ə,dē }

Callithricidae [VERTEBRATE ZOOLOGY] The marmosets, a family of South American mammals in the order Primates. { kal·ə'thris·ə,dē }

Callitrichales [BOTANY] An order of flowering plants, division Magnoliophyta (Angiospermae), in the subclass Asteridae of the class Magnoliopsida (dicotyledons); consists of three small families with about 50 species, most of which are aquatics or small herbs of wet places. { kə,lit·rə 'kā·lēz }

Callorhinchidae [VERTEBRATE ZOOLOGY] A family of ratfishes of the chondrichthyan order Chimaeriformes. { kal·ə'riŋk·ə,dē }

callose [BIOCHEMISTRY] A carbohydrate component of plant cell walls; associated with sieve plates where calluses are formed. [BIOLOGY] Having hardened protuberances, as on the skin or on leaves and stems. { 'ka,lōs }

callus [BOTANY] **1.** A thickened callose deposit on sieve plates. **2.** A hard tissue that forms over a damaged plant surface. { 'kal·əs }

calmodulin [BIOCHEMISTRY] A calcium-modulated protein consisting of a single polypeptide with 148 amino acids and a molecular weight of 16,700; found in all eukaryotes. { kal'mäj·ə·lən }

Calobryales [BOTANY] An order of liverworts;

characterized by prostrate, simple or branched, leafless stems and erect, leafy branches of a radial organization. { ˌkal·ō·brī'ā·lēz }

caloreceptor [PHYSIOLOGY] A cutaneous sense organ that is stimulated by heat. { ¦kal·ō·ri 'sep·tər }

calotte [BIOLOGY] A cap or caplike structure. [INVERTEBRATE ZOOLOGY] **1.** The four-celled polar cap in Dicyemidae. **2.** A ciliated, retractile disc in certain bryozoan larva. **3.** A dark-colored anterior area in certain nematomorphs. { kə'lät }

calsequestrin [BIOCHEMISTRY] An acidic protein, with a molecular weight of 44,000, which binds calcium inside the sarcoplasmic reticulum. { ¦kal·sə¦kwes·trən }

calthrop [INVERTEBRATE ZOOLOGY] A sponge spicule having four axes in which the rays are equal or almost equal in length. { 'kal·thrəp }

calvarium [ANATOMY] A skull lacking facial parts and the lower jaw. { kal'ver·ē·əm }

calving [VERTEBRATE ZOOLOGY] Giving birth to a calf. { 'kav·iŋ }

Calycerales [BOTANY] An order of flowering plants, division Magnoliophyta (Angiospermae), in the subclass Asteridae of the class Magnoliopsida (dicotyledons); consists of a single family with about 60 species native to tropical America. { 'kal·ə·sə'rā·lēz }

calyculate [BOTANY] Having bracts that imitate a second, external calyx. { kə'lik·yə·lət }

calymma [INVERTEBRATE ZOOLOGY] The outer, vacuolated protoplasmic layer of certain radiolarians. { kə'lim·ə }

Calymmatobacterium [MICROBIOLOGY] A genus of gram-negative, usually encapsulated, pleomorphic rods of uncertain affiliation; the single species causes granuloma inguinale in humans. { kə¦lim·əd·ōˌbak'tir·ē·əm }

Calymnidae [INVERTEBRATE ZOOLOGY] A family of echinoderms in the order Holasteroida characterized by an ovoid test with a marginal fasciole. { kə'lim·nəˌdē }

calypter [INVERTEBRATE ZOOLOGY] A scalelike or lobelike structure above the haltere of certain two-winged flies. Also known as alula; squama. { kə'lip·tər }

Calypteratae [INVERTEBRATE ZOOLOGY] A subsection of dipteran insects in the suborder Cyclorrhapha characterized by calypters associated with the wings. { kə'lip·trə·dē }

Calyptoblastea [INVERTEBRATE ZOOLOGY] A suborder of cnidarians in the order Hydroida, including the hydroids with protective cups around the hydranths and gonozooids. { kə¦lip·tō'blas·tē·ə }

calyptra [BOTANY] **1.** A membranous cap or hoodlike covering, especially the remains of the archegonium over the capsule of a moss. **2.** Tissue surrounding the archegonium of a liverwort. **3.** Root cap. { kə'lip·trə }

calyptrate [BOTANY] Having a calyptra. { kə'lip ˌtrāt }

calyptrogen [BOTANY] The specialized cell layer from which a root cap originates. { kə'lip·trə·jən }

Calyssozoa [INVERTEBRATE ZOOLOGY] The single class of the bryozoan subphylum Entoprocta. { kəˌlis·ə'zō·ə }

calyx [BOTANY] The outermost whorl of a flower; composed of sepals. [INVERTEBRATE ZOOLOGY] A cup-shaped structure to which the arms are attached in crinoids. { 'kāˌliks }

calyx tube [BOTANY] A tube formation resulting from fusion of the lateral edges of a group of sepals. { 'kā·liks ˌtüb }

Camacolaimoidea [INVERTEBRATE ZOOLOGY] A superfamily of nematodes consisting of a single family, the Camacolaimidae; they occur in marine or brackish-water environments. { ˌkam·ə·kō·lə'mȯid·ē·ə }

Camallanida [INVERTEBRATE ZOOLOGY] An order of phasmid nematodes in the subclass Spiruria, including parasites of domestic animals. { kam·ə'lan·ə·də }

Camallanoidea [INVERTEBRATE ZOOLOGY] A superfamily of parasitic nematodes in the subclass Spiruria. { kəˌmal·ə'nȯid·ē·ə }

Camarodonta [INVERTEBRATE ZOOLOGY] An order of Euechinoidea proposed by R. Jackson and abandoned in 1957. { kam·ə·rə'dän·tə }

camarodont dentition [INVERTEBRATE ZOOLOGY] In echinoderms, keeled teeth meeting the epiphyses so that the foramen magnum of the jaw is closed. { 'kam·ə·rəˌdänt den'tish·ən }

Cambaridae [INVERTEBRATE ZOOLOGY] A family of crayfishes belonging to the section Macrura in the crustacean order Decapoda. { kam'bär·ə ˌdē }

Cambarinae [INVERTEBRATE ZOOLOGY] A subfamily of crayfishes in the family Astacidae, including all North American species east of the Rocky Mountains. { kam'bär·əˌnē }

cambium [BOTANY] A layer of cells between the phloem and xylem of most vascular plants that is responsible for secondary growth and for generating new cells. { 'kam·be·əm }

camel [VERTEBRATE ZOOLOGY] The common name for two species of artiodactyl mammals, the bactrian camel (*Camelus bactrianus*) and the dromedary camel (*C. dromedarius*), in the family Camelidae. { 'kam·əl }

Camelidae [VERTEBRATE ZOOLOGY] A family of tylopod ruminants in the superfamily Cameloidea of the order Artiodactyla, including four species of camels and llamas. { ka'mel·əˌdē }

Cameloidea [VERTEBRATE ZOOLOGY] A superfamily of tylopod ruminants in the order Artiodiodactyla. { kam·ə'lȯid·ē·ə }

Camerata [PALEONTOLOGY] A subclass of extinct stalked echinoderms of the class Crinoidea. { ˌkam·ə'räd·ə }

cAMP [BIOCHEMISTRY] *See* cyclic adenylic acid.

Campanulaceae [BOTANY] A family of dicotyledonous plants in the order Campanulales characterized by a style without an indusium but with well-developed collecting hairs below the stigmas, and by a well-developed latex system. { kamˌpan·yə'lās·ēˌē }

Campanulales [BOTANY] An order of dicotyledonous plants in the subclass Asteridae distinguished by a chiefly herbaceous habit, alternate leaves, and inferior ovary. { kam,pan·yə'lā,lēz }

campanulate [BOTANY] Bell-shaped; applied particularly to the corolla. { kam'pan·yə·lət }

campestrian [ECOLOGY] Of or pertaining to the northern Great Plains area. { kam'pes·trē·ən }

camphor tree [BOTANY] *Cinnamomum camphora*. A plant of the laurel family (Lauraceae) in the order Magnoliales from which camphor is extracted. { 'kam·fər ,trē }

Campodeidae [INVERTEBRATE ZOOLOGY] A family of primarily wingless insects in the order Diplura which are most numerous in the Temperate Zone of the Northern Hemisphere. { kam·pə'dē·ə ,dē }

campodeiform [INVERTEBRATE ZOOLOGY] Elongate, flattened, and narrowed posteriorly. { kam 'pō·dē·ə,fȯrm }

campos [ECOLOGY] The savanna of South America. { 'käm,pōs }

Campylobacter [MICROBIOLOGY] A genus of bacteria in the family Spirillaceae; spirally curved rods that are motile by means of a polar flagellum at one or both poles. { ,kam·pə·lə'bak·tər }

campylotropous [BOTANY] Having the ovule symmetrical but half inverted, with the micropyle and funiculus at right angles to each other. { ¦kam·pə¦lä·trə·pəs }

Canaceidae [INVERTEBRATE ZOOLOGY] The seashore flies, a family of myodarian cyclorrhaphous dipteran insects in the subsection Acalypteratae. { ,kan·ə'sē·ə,dē }

Canadian life zone [ECOLOGY] The zone comprising the climate and biotic communities of the portion of the Boreal life zone exclusive of the Hudsonian and Arctic-Alpine zones. { kə 'nād·ē·ən 'līf ,zōn }

canal [BIOLOGY] A tubular duct or passage in bone or soft tissues. { kə'nal }

canal cell [BOTANY] One of the row of cells that make up the axial row within the neck of an archegonium. { kə'nal ,sel }

canaliculate [BIOLOGY] Having small channels, canals, or grooves. { ,kan·əl'ik·yə,lāt }

canaliculus [HISTOLOGY] **1.** One of the minute channels in bone radiating from a Haversian canal and connecting lacunae with each other and with the canal. **2.** A passage between the cells of the cell cords in the liver. { ,kan·əl'ik·yə·ləs }

canalization [EVOLUTION] The effect of natural selection on development to produce pathways that are insensitive to minor genetic or environmental variation; results in the phenotypic norm of the species. [PHYSIOLOGY] The formation of new channels in tissues, such as the formation of new blood vessels through a thrombus. { ,kan·əl·ə'zā·shən }

canalized character [GENETICS] A trait whose variability is restricted within narrow boundaries even when the species is subjected to environmental pressures or mutations. { ¦kan·ə,līzd 'kar·ik·tər }

canal of Schlemm [ANATOMY] An irregular channel at the junction of the sclera and cornea in the eye that drains aqueous humor from the anterior chamber. { kə'nal əv 'shlem }

canal valve [ANATOMY] The semilunar valve in the right atrium of the heart between the orifice of the inferior vena cava and the right atrioventricular orifice. Also known as eustachian valve. { kə'nal ,valv }

canary-pox virus [VIROLOGY] An avian poxvirus that causes canary pox, a disease closely related to fowl pox. { kə'ner·ē ,päks ,vī·rəs }

canavanine [BIOCHEMISTRY] $C_5H_{12}O_3N_4$ An amino acid found in the jack bean. { kə'nav·ə ,nēn }

cancellate [BIOLOGY] Lattice-shaped. Also known as clathrate. { 'kan·sə,lāt }

cancellous [BIOLOGY] Having a reticular or spongy structure. { kan'sel·əs }

cancellous bone [HISTOLOGY] A form of bone near the ends of long bones having a cancellous matrix composed of rods, plates, or tubes; spaces are filled with marrow. { kan'sel·əs ,bōn }

Candida [MYCOLOGY] A genus of yeastlike, pathogenic imperfect fungi that produce very small mycelia. { 'kan·də·də }

cane [BOTANY] **1.** A hollow, usually slender, jointed stem, such as in sugarcane or the bamboo grasses. **2.** A stem growing directly from the base of the plant, as in most Rosaceae, such as blackberry and roses. { kān }

canescent [BOTANY] Having a grayish epidermal covering of short hairs. { kə'nes·ənt }

Canidae [VERTEBRATE ZOOLOGY] A family of carnivorous mammals in the superfamily Canoidea, including dogs and their allies. { 'kan·ə,dē }

canine [ANATOMY] A conical tooth, such as one located between the lateral incisor and first premolar in humans and many other mammals. Also known as cuspid. [VERTEBRATE ZOOLOGY] Pertaining or related to dogs or to the family Canidae. { 'kā,nīn }

Canis [VERTEBRATE ZOOLOGY] The type genus of the dog family (Canidae), including dogs, wolves, and jackals. { 'kā·nəs }

cankerworm [INVERTEBRATE ZOOLOGY] Any of several lepidopteran insect larvae in the family Geometridae which cause severe plant damage by feeding on buds and foliage. { 'kaŋ·kər ,wərm }

Cannabaceae [BOTANY] A family of dicotyledonous herbs in the order Urticales, including Indian hemp (*Cannabis sativa*) and characterized by erect anthers, two styles or style branches, and the lack of milky juice. { kan·ə'bās·ē,ē }

Cannabis [BOTANY] A genus of tall annual herbs in the family Cannabaceae having erect stems, leaves with three to seven elongate leaflets, and pistillate flowers in spikes along the stem. { 'kan·ə·bəs }

Cannaceae [BOTANY] A family of monocotyledonous plants in the order Zingiberales characterized by one functional stamen, a single functional pollen sac in the stamen, mucilage canals in the stem, and numerous ovules in each of the one to three locules. { kə'nās·ē,ē }

Canoidea [VERTEBRATE ZOOLOGY] A superfamily

belonging to the mammalian order Carnivora, including all dogs and doglike species such as seals, bears, and weasels. { kə'nȯid·ē·ə }

canonical sequence [MOLECULAR BIOLOGY] An archetypical nucleotide or amino acid sequence to which all variants are compared. { kə'nän·ə·kəl 'sē·kwəns }

cantala [BOTANY] A fiber produced from agave (*Agave cantala*) leaves; used to make twine. Also known as Cebu maguey; maguey; Manila maguey. { kan'täl·ə }

cantaloupe [BOTANY] The fruit (pepo) of *Cucumis malo*, a small, distinctly netted, round to oval muskmelon of the family Cucurbitaceae in the order Violales. { 'kant·əl,ōp }

Cantharidae [INVERTEBRATE ZOOLOGY] The soldier beetles, a family of coleopteran insects in the superfamily Cantharoidea. { kan'thar·ə,dē }

Cantharoidea [INVERTEBRATE ZOOLOGY] A superfamily of coleopteran insects in the suborder Polyphaga. { kan·thə'rȯid·ē·ə }

canthus [ANATOMY] Either of the two angles formed by the junction of the eyelids, designated outer or lateral, and inner or medial. { 'kan·thəs }

cap [GENETICS] In many eukaryotic messenger ribonucleic acids, the structure at the 5' end consisting of 7'-methyl-guanosine-pppX, where X is the first nucleotide encoded in the deoxyribonucleic acid; it is added posttranscriptionally. { kap }

cap-binding protein [MOLECULAR BIOLOGY] A protein that specifically recognizes the methylated cap of eukaryotic messenger ribonucleic acid (mRNA) and is essential in the regulation of mRNA translation. { ¦kap ,bīnd·iŋ 'prō,tēn }

capillary [ANATOMY] The smallest vessel of both the circulatory and lymphatic systems; the walls are composed of a single cell layer. { 'kap·ə,ler·ē }

capillary bed [ANATOMY] The capillaries, collectively, of a given area or organ. { 'kap·ə,ler·ē ,bed }

capillary pressure [PHYSIOLOGY] Pressure exerted by blood against capillary walls. { 'kap·ə,ler·ē ,presh·ər }

capillitium [MYCOLOGY] A network of threadlike tubes or filaments in which spores are embedded within sporangia of certain fungi, such as the slime molds. { ,kap·ə'lish·ē·əm }

capitate [BIOLOGY] Enlarged and swollen at the tip. [BOTANY] Forming a head, as certain flowers of the Compositae. { 'kap·ə,tāt }

capitellate [BOTANY] **1.** Having a small knoblike termination. **2.** Grouped to form a capitulum. { ¦kap·ə¦te,lāt }

Capitellidae [INVERTEBRATE ZOOLOGY] A family of mud-swallowing annelid worms, sometimes called bloodworms, belonging to the Sedentaria. { ,kap·ə'te·lə,dē }

capitellum [ANATOMY] A small head or rounded process of a bone. { ,kap·ə'te·ləm }

Capitonidae [VERTEBRATE ZOOLOGY] The barbets, a family of pantropical birds in the order Piciformes. { ,kap·ə'tä·nə,dē }

capitulum [BIOLOGY] A rounded, knoblike, usually terminal protuberance on a structure. [BOTANY] One of the rounded cells on the manubrium in the antheridia of lichens belonging to the Caliciales. { kə'pich·ə·ləm }

Capnocytophaga [MICROBIOLOGY] A genus of bacteria comprising fusiform, fermentative, gram-negative rods which require carbon dioxide for growth and show gliding motility. { ,kap·nō ,sī'täf·ə·gə }

Caponidae [INVERTEBRATE ZOOLOGY] A family of arachnid arthropods in the order Araneida characterized by having tracheae instead of book lungs. { kə'pä·nə,dē }

Capparaceae [BOTANY] A family of dicotyledonous herbs, shrubs, and trees in the order Capparales characterized by parietal placentation; hypogynous, mostly regular flowers; four to many stamens; and simple to trifoliate or palmately compound leaves. { ,kap·ə'rās·ē,ē }

Capparales [BOTANY] An order of dicotyledonous plants in the subclass Dilleniidae. { ,kap·ə 'rā,lēz }

capping [MOLECULAR BIOLOGY] Addition of a methylated cap to eukaryotic messenger ribonucleic acid molecules. { 'kap·iŋ }

Caprellidae [INVERTEBRATE ZOOLOGY] The skeleton shrimps, a family of slender, cylindrical amphipod crustaceans in the suborder Caprellidea. { kə'prel·ə,dē }

Caprellidea [INVERTEBRATE ZOOLOGY] A suborder of marine and brackish-water animals of the crustacean order Amphipoda. { ,kap·rə'lid·ē·ə }

Caprifoliaceae [BOTANY] A family of dicotyledonous, mostly woody plants in the order Dipsacales, including elderberry and honeysuckle; characterized by distinct filaments and anthers, typically five stamens and five corolla lobes, more than one ovule per locule, and well-developed endosperm. { ,kap·rə,fōl·ē'ās·e,ē }

Caprimulgidae [VERTEBRATE ZOOLOGY] A family of birds in the order Caprimulgiformes, including the nightjars, or goatsuckers. { ,kap·rə'məl·jə ,dē }

Caprimulgiformes [VERTEBRATE ZOOLOGY] An order of nocturnal and crepuscular birds, including nightjars, potoos, and frog-mouths. { ,kap·rə ,məl·jə'fȯr,mēz }

Capripoxvirus [VIROLOGY] A genus of viruses belonging to the subfamily Chordopoxvirinae; natural hosts are ungulates; these viruses can be mechanically transmitted by arthropods. { 'kap·ri,päks,vī·rəs }

Capsaloidea [INVERTEBRATE ZOOLOGY] A superfamily of ectoparasitic trematodes in the subclass Monogenea characterized by a sucker-shaped holdfast with anchors and hooks. { ,kap·sə'lȯid·ē·ə }

capsanthin [BIOCHEMISTRY] $C_{40}H_{58}IO_3$ Carmine-red carotenoid pigment occurring in paprika. { kap'san·thən }

capsicum [BOTANY] The fruit of a plant of the genus *Capsicum*, especially *C. frutescens*, cultivated in southern India and the tropics; a strong irritant to mucous membranes and eyes. { 'kap·sə·kəm }

capsid [INVERTEBRATE ZOOLOGY] The name applied to all members of the family Miridae. [VIROLOGY] In a virus, the protein shell surrounding the nucleic acid and its associated protein core. Also known as protein coat. { 'kap·səd }

capsomere [VIROLOGY] An individual protein subunit of a capsid. { 'kap·sə,mir }

capsular ligament [ANATOMY] A saclike ligament surrounding the articular cavity of a freely movable joint and attached to the bones. { 'kap·sə·lər 'lig·ə·mənt }

capsulate [BIOLOGY] Enclosed in a capsule. { 'kap·sə,lāt }

capsule [ANATOMY] A membranous structure enclosing a body part or organ. [BOTANY] A closed structure bearing seeds or spores; it is dehiscent at maturity. [MICROBIOLOGY] A thick, mucous envelope, composed of polypeptide or carbohydrate, surrounding certain microorganisms. { 'kap·səl }

Captorhinomorpha [PALEONTOLOGY] An extinct subclass of primitive lizardlike reptiles in the order Cotylosauria. { ,kap·tə¦rī·nə¦mȯr·fə }

capybara [VERTEBRATE ZOOLOGY] *Hydrochoerus capybara*. An aquatic rodent (largest rodent in existence) found in South America and characterized by partly webbed feet, no tail, and coarse hair. { ,kap·ə'bar·ə }

Carabidae [INVERTEBRATE ZOOLOGY] The ground beetles, a family of predatory coleopteran insects in the suborder Adephaga. { kə'rab·ə,dē }

caraboid larva [INVERTEBRATE ZOOLOGY] The morphologically distinct larva of certain beetles, characterized by a narrow, elongate body with long legs, a head that occasionally bears a single ocellus on each side, and three-segmented antennae with a well-developed sensorium. { 'kar·ə ,bȯid 'lar·və }

Caracarinae [VERTEBRATE ZOOLOGY] The caracaras, a subfamily of carrion-feeding birds in the order Falconiformes. { ,karə'kar·ə,nē }

Carangidae [VERTEBRATE ZOOLOGY] A family of perciform fishes in the suborder Percoidei, including jacks, scads, and pompanos. { kə'ran·jə ,dē }

carapace [INVERTEBRATE ZOOLOGY] A dorsolateral, chitinous case covering the cephalothorax of many arthropods. [VERTEBRATE ZOOLOGY] The bony, dorsal part of a turtle shell. { 'kar·ə ,pās }

Carapidae [VERTEBRATE ZOOLOGY] The pearlfishes, a family of sinuous, marine shore fishes in the order Gadiformes that live as commensals in the body cavity of holothurians. { kə'rap·ə ,dē }

caraway [BOTANY] *Carum carvi*. A white-flowered perennial herb of the family Umbelliferae; the fruit is used as a spice and flavoring agent. { 'kar·ə,wā }

carbamic acid [BIOCHEMISTRY] NH_2COOH An amino acid known for its salts, such as urea and carbamide. { kar'bam·ik 'as·əd }

carbamino [BIOCHEMISTRY] A compound formed by the combination of carbon dioxide with a free amino group in an amino acid or a protein. { kär 'bam·ə,nō }

carbamyl phosphate [BIOCHEMISTRY] $NH_2CO\text{-}PO_4H_2$ The ester formed from reaction of phosphoric acid and carbamyl acid. { 'kär·bə,mil 'fäs ,fāt }

carbohydrase [BIOCHEMISTRY] Any enzyme that catalyzes the hydrolysis of disaccharides and more complex carbohydrates. { ,kär·bō'hī ,drās }

carbohydrate [BIOCHEMISTRY] Any of the group of organic compounds composed of carbon, hydrogen, and oxygen, including sugars, starches, and celluloses. { ,kär·bō'hī,drāt }

carbohydrate metabolism [BIOCHEMISTRY] The sum of the biochemical and physiological processes involved in the breakdown and synthesis of simple sugars, oligosaccharides, and polysaccharides and in the transport of sugar across cell membranes. { ,kär·bō'hī,drāt me'tab·ə,liz·əm }

carbomycin [MICROBIOLOGY] $C_{42}H_{67}O_{16}N$ Colorless, crystalline antibiotic produced by *Streptomyces halstedii*; principally active against gram-positive bacteria. { kär·bō'mīs·ən }

carbomycin B [MICROBIOLOGY] $C_{42}H_{67}O_{15}N$ A colorless, crystalline antibiotic differing from carbomycin only in having one less oxygen atom in its molecule. { kär·bō'mīs·ən 'bē }

carbonic anhydrase [BIOCHEMISTRY] An enzyme which aids carbon dioxide transport and release by catalyzing the synthesis, and the dehydration, of carbonic acid from, and to, carbon dioxide and water. { kär'bän·ik an'hī,drās }

carbonmonoxyhemoglobin [BIOCHEMISTRY] A stable combination of carbon monoxide and hemoglobin formed in the blood when carbon monoxide is inhaled. Also known as carbonylhemoglobin; carboxyhemoglobin. { ¦kär·bən·mə ¦näk·sē'hē·mə,glō·bən }

carbonylhemoglobin [BIOCHEMISTRY] *See* carbonmonoxyhemoglobin. { 'kär·bə,nil¦hēm·ə 'glō·bən }

carboxyhemoglobin [BIOCHEMISTRY] *See* carbonmonoxyhemoglobin. { kär¦bäk·sē¦hē·mə,glō·bən }

carboxylase [BIOCHEMISTRY] Any enzyme that catalyzes a carboxylation or decarboxylation reaction. { kär'bäk·sə,lās }

carboxyl terminal [BIOCHEMISTRY] The end of a polypeptide chain with a free carboxyl group. { kär'bäk·səl 'tər·mən·əl }

carboxypeptidase [BIOCHEMISTRY] Any enzyme that catalyzes the hydrolysis of a peptide at the end containing the free carboxyl group. { kär ,bäk·sē'pep·tə,dās }

Carcharhinidae [VERTEBRATE ZOOLOGY] A large family of sharks belonging to the charcharinid group of galeoids, including the tiger sharks and blue sharks. { ,kär·kə'rīn·ə,dē }

Carchariidae [VERTEBRATE ZOOLOGY] A family of shallow-water predatory sharks belonging to the isurid group of galeoids. { ,kär·kə'rī·ə,dē }

carcharodont [VERTEBRATE ZOOLOGY] Possessing sharp, flat, triangular teeth with serrated margins, like those of the human-eating sharks. { kär'kar·ə,dänt }

carcinoembryonic antigen [IMMUNOLOGY] A glycoprotein found in tissues of the fetal gut during the first two trimesters of pregnancy and in the peripheral blood of individuals with some forms of cancer, such as digestive-system or breast cancer. { ¦kärs·ən·ō,em·brē¦än·ik 'ant·i·jən }

cardamon [BOTANY] *Elettaria cardamomum.* A perennial herbaceous plant in the family Zingiberaceae; the seed of the plant is used as a spice. Also spelled cardamom. { 'kärd·ə·mən }

cardia [ANATOMY] **1.** The orifice where the esophagus enters the stomach. **2.** The large, blind diverticulum of the stomach adjoining the orifice. [INVERTEBRATE ZOOLOGY] Anterior enlargement of the ventriculus in some insects. { 'kärd·ē·ə }

cardiac [ANATOMY] **1.** Of, pertaining to, or situated near the heart. **2.** Of or pertaining to the cardia of the stomach. { 'kärd·ē,ak }

cardiac cycle [PHYSIOLOGY] The sequence of events in the heart between the start of one contraction and the start of the next. { 'kärd·ē,ak ,sī·kəl }

cardiac electrophysiology [PHYSIOLOGY] The science that is concerned with the mechanism, spread, and interpretation of the electric currents which arise within heart muscle tissue and initiate each heart muscle contraction. { 'kärd·ē,ak i,lek·trō,fiz·ē'äl·ə·jē }

cardiac gland [ANATOMY] Any of the mucus-secreting, compound tubular structures near the esophagus or in the cardia of the stomach of vertebrates; capable of secreting digestive enzymes. { 'kärd·ē,ak ,gland }

cardiac input [PHYSIOLOGY] The amount of venous blood returned to the heart during a specified period of time. { 'kärd·ē,ak 'in,pu̇t }

cardiac loop [EMBRYOLOGY] The embryonic heart formed by bending and twisting of the cardiac tube. { 'kärd·ē,ak 'lüp }

cardiac muscle [HISTOLOGY] The principal tissue of the vertebrate heart; composed of a syncytium of striated muscle fibers. { 'kärd·ē,ak ¦məs·əl }

cardiac output [PHYSIOLOGY] The total blood flow from the heart during a specified period of time. { 'kärd·ē,ak 'au̇t,pu̇t }

cardiac plexus [ANATOMY] A network of visceral nerves situated at the base of the heart; contains both sympathetic and vagal nerve fibers. { 'kärd·ē,ak 'plek·səs }

cardiac sphincter [ANATOMY] The muscular ring at the orifice between the esophagus and stomach. { 'kärd·ē,ak 'sfiŋk·tər }

cardiac valve [ANATOMY] Any of the structures located within the orifices of the heart that maintain unidirectional blood flow. { 'kärd·ē,ak 'valv }

cardinal teeth [INVERTEBRATE ZOOLOGY] Ridges and grooves on the inner surfaces of both valves of a bivalve mollusk near the anterior end of the hinge. { 'kärd·nəl ,tēth }

cardinal vein [EMBRYOLOGY] Any of four veins in the vertebrate embryo which run along each side of the vertebral column; the paired veins on each side discharge blood to the heart through the duct of Cuvier. { 'kärd·nəl ,vān }

Cardiobacterium [MICROBIOLOGY] A genus of gram-negative, rod-shaped bacteria of uncertain affiliation; facultative anaerobes that ferment fructose, glucose, mannose, sorbitol, and sucrose; the single species causes endocarditis in humans. { ¦kärd·ē·ō,bak'tir·ē·əm }

cardioblast [INVERTEBRATE ZOOLOGY] Any of certain early embryonic cells in insects from which the heart develops. { 'kärd·ē·ə,blast }

cardiocirculatory [PHYSIOLOGY] *See* cardiovascular. { ¦kärd·ē·ō'sər·kyə·lə,tȯr·ē }

cardiodynamics [PHYSIOLOGY] The dynamics of the heart's action in pumping blood. { ¦kärd·ē·ō·dī'nam·iks }

cardiogenic plate [EMBRYOLOGY] An area of splanchnic mesoderm in the early mammalian embryo from which the heart develops. { ¦kärd·ē·ō¦jen·ik 'plāt }

cardiolipin [BIOCHEMISTRY] A complex phospholipid found in the ether alcohol extract of powdered beef heart; mixed with leutin and cholesterol, it functions as the antigen in the Wassermann complement-fixation test for syphilis. Also known as diphosphatidyl glycerol. { ,kärd·ē·ō'lip·ən }

cardiopulmonary [PHYSIOLOGY] Pertaining to the heart and lungs. { ¦kärd·ē·ō'pu̇l·mə,ner·ē }

cardiovascular [PHYSIOLOGY] Pertaining to the heart and circulatory system. Also known as cardiocirculatory. { ¦kärd·ē·ō'vas·kyə·lər }

cardiovascular system [ANATOMY] Those structures, including the heart and blood vessels, which provide channels for the flow of blood. { ¦kärd·ē·ō'vas·kyə·lər ,sis·təm }

Cardiovirus [VIROLOGY] A genus of viruses of the family Picornaviridae; consists of strains of encephalomyocarditis virus and mouse encephalomyelitis. { 'kär·dē·ō,vī·rəs }

Carettochelyidae [VERTEBRATE ZOOLOGY] A family of reptiles in the order Chelonia containing only one species, the New Guinea plateless turtle (*Carettochelys insculpta*). { kə¦red·ō·kə'lī·ə,dē }

Cariamidae [VERTEBRATE ZOOLOGY] The long-legged cariamas, a family of birds in the order Gruiformes. { ,kär·ē'am·ə,dē }

Caribosireninae [VERTEBRATE ZOOLOGY] A subfamily of trichechiform sirenean mammals in the family Dugongidae. { ,kar·ə,bäs·ə'ren·ə,nē }

Caridea [INVERTEBRATE ZOOLOGY] A large section of decapod crustaceans in the suborder Natantia including many diverse forms of shrimps and prawns. { kə'rid·ē·ə }

carina [BIOLOGY] A ridge or a keel-shaped anatomical structure. [VERTEBRATE ZOOLOGY] *See* keel. { kə'rī·nə }

carinate [BIOLOGY] Having a ridge or keel, as the breastbone of certain birds. { 'kar·ə,nāt }

Carinomidae [INVERTEBRATE ZOOLOGY] A monogeneric family of littoral ribbonlike worms in the order Palaeonemertini. { ,kär·ə'näm·ə,dē }

cariostatic [PHYSIOLOGY] Acting to halt bone or tooth decay. { ¦kar·ē·ə¦stad·ik }

carnassial [ANATOMY] Of or pertaining to molar or premolar teeth specialized for cutting and shearing. { kär'nas·ē·əl }

carnitine [BIOCHEMISTRY] $C_7H_{15}NO_3$ α-Amino-β-hydroxybutyric acid trimethylbetaine; a constituent of striated muscle and liver, identical with vitamin B_T. { 'kär·nə,tēn }

Carnivora [VERTEBRATE ZOOLOGY] A large order of placental mammals, including dogs, bears, and cats, that is primarily adapted for predation as evidenced by dentition and jaw articulation. { kär'niv·ə·rə }

carnivorous [BIOLOGY] Eating flesh or, as in plants, subsisting on nutrients obtained from animal protoplasm. { kär'niv·ə·rəs }

carnivorous plant [BOTANY] *See* insectivorous plant. { kär'niv·ə·rəs 'plant }

Carnosauria [PALEONTOLOGY] A group of large, predacious saurischian dinosaurs in the suborder Theropoda having short necks and large heads. { ,kär·nə'sȯr·ē·ə }

carnosine [BIOCHEMISTRY] $C_9H_{14}N_4O_3$ A colorless, crystalline dipeptide occurring in the muscle tissue of vertebrates. { 'kär·nə,sēn }

Carolinian life zone [ECOLOGY] A zone comprising the climate and biotic communities of the oak savannas of eastern North America. { ¦kar·ə ¦lin·ē·ən 'līf ,zōn }

carotenase [BIOCHEMISTRY] An enzyme that effects the hydrolysis of carotenoid compounds, used in bleaching of flour. { kə'rät·ən,ās }

carotene [BIOCHEMISTRY] $C_{40}H_{56}$ Any of several red, crystalline, carotenoid hydrocarbon pigments occurring widely in nature, convertible in the animal body to vitamin A, and characterized by preferential solubility in petroleum ether. Also known as carotin. { 'kar·ə,tēn }

carotenoid [BIOCHEMISTRY] A class of labile, easily oxidizable, yellow, orange, red, or purple pigments that are widely distributed in plants and animals and are preferentially soluble in fats and fat solvents. { kə'rät·ən,ȯid }

carotenol [BIOCHEMISTRY] *See* xanthophyll. { kə 'rät·ən,ȯl }

carotid artery [ANATOMY] Either of the two principal arteries on both sides of the neck that supply blood to the head and neck. Also known as common carotid artery. { kə'räd·əd 'ärd·ə·rē }

carotid body [ANATOMY] Either of two chemoreceptors sensitive to changes in blood chemistry which lie near the bifurcations of the carotid arteries. Also known as glomus caroticum. { kə 'räd·əd ,bäd·ē }

carotid ganglion [ANATOMY] A group of nerve cell bodies associated with each carotid artery. { kə 'räd·əd 'gaŋ·glē,än }

carotid sinus [ANATOMY] An enlargement at the bifurcation of each carotid artery that is supplied with sensory nerve endings and plays a role in reflex control of blood pressure. { kə'räd·əd 'sī·nəs }

carotin [BIOCHEMISTRY] *See* carotene. { 'kar·ə ,tin }

carotol [BIOCHEMISTRY] $C_{15}H_{25}OH$ A sesquiterpenoid alcohol in carrots. { 'kar·ə,tȯl }

carp [VERTEBRATE ZOOLOGY] The common name for a number of fresh-water, cypriniform fishes in the family Cyprinidae, characterized by soft fins, pharyngeal teeth, and a suckerlike mouth. { kärp }

carpel [BOTANY] The basic specialized leaf of the female reproductive structure in angiosperms; a megasporophyll. { 'kär·pəl }

carpogonium [BOTANY] The basal, egg-bearing portion of the female reproductive organ in some thallophytes, especially red algae. { ,kär·pə'gō·nē·əm }

Carpoidea [PALEONTOLOGY] Former designation for a class of extinct homalozoan echinoderms. { kär'pȯid·ē·ə }

carpoids [PALEONTOLOGY] An assemblage of three classes of enigmatic, rare Paleozoic echinoderms formerly grouped together as the class Carpoidea. { 'kär,pȯidz }

carpology [BOTANY] The study of the morphology of fruit and seeds. { kär'päl·ə·jē }

carpophagous [ZOOLOGY] Feeding on fruits. { kär'pa·fə·gəs }

carpophore [BOTANY] The portion of a flower receptacle that extends between and attaches to the carpels. [MYCOLOGY] The stalk of a fruiting body in fungi. { 'kär·pə,fȯr }

carpophyte [BOTANY] A thallophyte that forms a sporocarp following fertilization. { 'kär·pə,fīt }

carposporangium [BOTANY] In red algae, a sporangium that forms the cystocarp and contains carpospores. { ,kär·pō·spə'ran·jē·əm }

carpospore [BOTANY] In red algae, a diploid spore produced terminally by a gonimoblast, giving rise to the diploid tetrasporic plant. { 'kär·pə ,spȯr }

carpus [ANATOMY] **1.** The wrist in humans or the corresponding part in other vertebrates. **2.** The eight bones of the human wrist. [INVERTEBRATE ZOOLOGY] The fifth segment from the base of a generalized crustacean appendage. { 'kär·pəs }

carrageen [BOTANY] *Chondrus crispus.* A cartilaginous red algae harvested in the northern Atlantic as a source of carrageenan. Also known as Irish moss; pearl moss. { 'kar·ə,gēn }

carrier [GENETICS] An individual who is heterozygous for a recessive gene. [IMMUNOLOGY] A protein to which a hapten becomes attached, thereby rendering the hapten immunogenic. { 'kar·ē·ər }

carrier culture [VIROLOGY] A cell culture exhibiting a persistent infection; only a small fraction of the cell population is infected, but the viruses released when the infected cells die infect a small number of other cells. { 'kar·ē·ər ,kəl·chər }

carrier molecule [IMMUNOLOGY] An immunogenic molecule such as a foreign protein to which a hapten is coupled, thus enabling the hapten to induce an immune response. { 'kar·ē·ər 'mäl·ə,kyül }

carrot [BOTANY] *Daucus carota.* A biennial umbellifer of the order Umbellales with a yellow or orange-red edible root. { 'kar·ət }

carrying capacity [ECOLOGY] The maximum population size that the environment can support without deterioration. { 'kar·ē·iŋ kə'pas·əd·ē }

Carterinacea [INVERTEBRATE ZOOLOGY] A monogeneric superfamily of marine, benthic foraminiferans in the suborder Rotaliina characterized by a test with monocrystal calcite spicules in a granular groundmass. { ,kärd·ər·ə'nās·ē·ə }

cartilage [HISTOLOGY] A specialized connective tissue which is bluish, translucent, and hard but yielding. { 'kärd·əl·ij }

cartilage bone [HISTOLOGY] Bone formed by ossification of cartilage. { 'kärd·əl·ij 'bōn }

cartilaginous fish [VERTEBRATE ZOOLOGY] The common name for all members of the class Chondrichthyes. { ¦kärd·əl¦aj·ə·nas 'fish }

caruncle [ANATOMY] Any normal or abnormal fleshy outgrowth, such as the comb and wattles of fowl or the mass in the inner canthus of the eye. [BOTANY] A fleshy outgrowth developed from the seed coat near the hilum in some seeds, such as the castor bean. { 'ka,rəŋ·kəl }

Caryophanaceae [MICROBIOLOGY] Formerly a family of large, gram-negative bacteria belonging to the order Caryophanales and having disklike cells arranged in chains. { ¦kar·ē·ō·fə'nās·ē,ē }

Caryophanales [MICROBIOLOGY] Formerly an order of bacteria in the class Schizomycetes occurring as trichomes which produce short structures that function as reproductive units. { ,kar·ē,äf·ə'nā·lēz }

Caryophanon [MICROBIOLOGY] A genus of gram-positive, large, rod-shaped or filamentous bacteria of uncertain affiliation. { ,kar·ē'äf·ə,nän }

Caryophyllaceae [BOTANY] A family of dicotyledonous plants in the order Caryophyllales differing from the other families in lacking betalains. { ¦kar·ē·ō·fə'lās·ē,ē }

Caryophyllales [BOTANY] An order of dicotyledonous plants in the subclass Caryophyllidae characterized by free-central or basal placentation. { ¦kar·ē·ō·fə'lā·lēz }

Caryophyllidae [BOTANY] A relatively small subclass of plants in the class Magnoliopsida characterized by trinucleate pollen, ovules with two integuments, and a multilayered nucellus. { ,kar·ē·ō'fil·ə,dē }

caryopsis [BOTANY] A small, dry, indehiscent fruit having a single seed with such a thin, closely adherent pericarp that a single body, a grain, is formed. { ,kar·ē'äp·səs }

casaba melon [BOTANY] *Cucumis melo.* A winter muskmelon with a yellow rind and sweet flesh belonging to the family Cucurbitaceae of the order Violales. { kə,säb·ə ,mel·ən }

cascade regulation [MOLECULAR BIOLOGY] **1.** In prokaryotes, a form of genetic regulation in which one operon codes for the production of an internal inducer that turns on one or more operons. **2.** In eukaryotes, a multistep model of genetic regulation involving mechanisms that interface with messenger ribonucleic acid formation, transport, and translation. { ka'skād ,reg·yə'lā·shən }

cashew [BOTANY] *Anacardium occidentale.* An evergreen tree of the order Sapindales grown for its kidney-shaped edible nuts and resinous oil. { 'kash·ü }

Casparian strip [BOTANY] A thin band of suberin- or lignin-like deposition in the radial and transverse walls of certain plant cells during the primary development phase of the endodermis. { ka 'spar·ē·ən ,strip }

cassava [BOTANY] *Manihot esculenta.* A shrubby perennial plant grown for its starchy, edible tuberous roots. Also known as manihot; manioc. { kə'sav·ə }

cassette [MYCOLOGY] In yeast, any of the sites lying in tandem that contain nucleotide sequences that can be substituted for one another. { kə'set }

Cassidulinacea [INVERTEBRATE ZOOLOGY] A superfamily of marine, benthic foraminiferans in the suborder Rotaliina, characterized by a test of granular calcite with monolamellar septa. { ,kas·ə,dü·lə'nās·ē·ə }

Cassiduloida [INVERTEBRATE ZOOLOGY] An order of exocyclic Euechinoidea possessing five similar ambulacra which form petal-shaped areas (phyllodes) around the mouth. { ,kas·ə·də'lȯid·ē·ə }

cassowary [VERTEBRATE ZOOLOGY] Any of three species of large, heavy, flightless birds composing the family Casuariidae in the order Casuariiformes. { 'kas·ə,wer·ē }

cast [PALEONTOLOGY] A fossil reproduction of a natural object formed by infiltration of a mold of the object by waterborne minerals. [PHYSIOLOGY] A mass of fibrous material or exudate having the form of the body cavity in which it has been molded; classified from its source, such as bronchial, renal, or tracheal. { kast }

castaneous [BIOLOGY] Chestnut-colored. { ka 'stān·ē·əs }

caste [INVERTEBRATE ZOOLOGY] One of the levels of mature social insects in a colony that carry out a specific function; examples are workers and soldiers. { kast }

Castle's intrinsic factor [BIOCHEMISTRY] *See* intrinsic factor. { 'kas·əlz in¦trin·zik 'fak·tər }

Castniidae [INVERTEBRATE ZOOLOGY] The castniids; large diurnal, butterflylike moths composing the single, small family of the lepidopteran superfamily Castnioidea. { ,kast'nī·ə,dē }

Castnioidea [INVERTEBRATE ZOOLOGY] A superfamily of neotropical and Indo-Australian lepidopteran insects in the suborder Heteroneura. { ,kast·nī'ȯid·ē·ə }

castor bean [BOTANY] The seed of the castor oil plant (*Ricinus communis*), a coarse, erect annual herb in the spurge family (Euphorbiaceae) of the order Geraniales. { 'kas·tər ,bēn }

castoreum gland [VERTEBRATE ZOOLOGY] A preputial scent gland in the beaver. { ka'stȯr·ē·əm ,gland }

Casuariidae [VERTEBRATE ZOOLOGY] The cassowaries, a family of flightless birds in the order Casuariiformes lacking head and neck feathers and having bony casques on the head. { ,kazh ə ,wa'rē·ə,dē }

Casuariiformes [VERTEBRATE ZOOLOGY] An order of large, flightless, ostrichlike birds of Australia and New Guinea. { ,kazh·ə,wa·rē·ə'fȯr,mēz }

Casuarinaceae [BOTANY] The single, monogeneric family of the plant order Casuarinales characterized by reduced flowers and green twigs bearing whorls of scalelike, reduced leaves. { ,kazh·ə,wa·rə'nās·ē,ē }

Casuarinales [BOTANY] A monofamilial order of dicotyledonous plants in the subclass Hamamelidae. { ,kazh·ə,wa·rə'nā·lēz }

cat [VERTEBRATE ZOOLOGY] The common name for all members of the carnivoran mammalian family Felidae, especially breeds of the domestic species, *Felis domestica*. { kat }

catabiosis [PHYSIOLOGY] Degenerative changes accompanying cellular senescence. { ˌkad·əˌbī'ō·səs }

catabolism [BIOCHEMISTRY] That part of metabolism concerned with the breakdown of large protoplasmic molecules and tissues, often with the liberation of energy. { kə'tab·əˌliz·əm }

catabolite [BIOCHEMISTRY] Any product of catabolism. { kə'tab·əˌlīt }

catabolite repression [BIOCHEMISTRY] An intracellular regulatory mechanism in bacteria whereby glucose, or any other carbon source that is an intermediate in catabolism, prevents formation of inducible enzymes. { kə'tab əˌlīt ri'presh·ən }

catadromous [VERTEBRATE ZOOLOGY] Pertaining to fishes which live in fresh water and migrate to spawn in salt water. { kə'ta·drə·məs }

catalase [BIOCHEMISTRY] An enzyme that catalyzes the decomposition of hydrogen peroxide into molecular oxygen and water. { 'kad·əlˌās }

catalectrotonus [PHYSIOLOGY] The negative electric potential during the passage of a current on the surface of a nerve or muscle in the region of the cathode. { ˌkad·əlˌek·'träd·ən·əs }

catalytic site [MOLECULAR BIOLOGY] *See* active site. { ¦kad·əl¦id·ik ˌsīt }

catamount [VERTEBRATE ZOOLOGY] *See* puma. { 'kad·ə·maünt }

Catapochrotidae [INVERTEBRATE ZOOLOGY] A monospecific family of coleopteran insects in the superfamily Cucujoidea. { ˌkad·ə·pə'kräd·əˌdē }

catarobic [ECOLOGY] Pertaining to a body of water characterized by the slow decomposition of organic matter, and oxygen utilization which is insufficient to prevent the activity of aerobic organisms. { ¦kad·ə¦rō·bik }

catastrophism [PALEONTOLOGY] The theory that the differences between fossils in successive stratigraphic horizons resulted from a general catastrophe followed by creation of the different organisms found in the next-younger beds. { kə'tas·trəˌfiz·əm }

catecholamine [BIOCHEMISTRY] Any one of a group of sympathomimetic amines containing a catechol moiety, including especially epinephrine, norepinephrine (levarterenol), and dopamine. { ¦kad·ə'käl·əˌmīn }

category [SYSTEMATICS] In a hierarchical classification system, the level at which a particular group is ranked. { 'kad·əˌgȯr·ē }

catenulate [BIOLOGY] Having a chainlike form. { kə'ten·yəˌlāt }

Catenulida [INVERTEBRATE ZOOLOGY] An order of threadlike, colorless fresh-water rhabdocoeles with a simple pharynx and a single, median protonephridium. { kəˌten·yə'lī·də }

caterpillar [INVERTEBRATE ZOOLOGY] **1.** The wormlike larval stage of a butterfly or moth. **2.** The larva of certain insects, such as scorpion flies and sawflies. { 'kad·ərˌpil·ər }

catfish [VERTEBRATE ZOOLOGY] The common name for a number of fishes which constitute the suborder Siluroidei in the order Cypriniformes, all of which have barbels around the mouth. { 'katˌfish }

Cathartidae [VERTEBRATE ZOOLOGY] The New World vultures, a family of large, diurnal predatory birds in the order Falconiformes that lack a voice and have slightly webbed feet. { kə'thär·dəˌdē }

cathepsin [BIOCHEMISTRY] Any of several proteolytic enzymes occurring in animal tissue that hydrolyze high-molecular-weight proteins to proteoses and peptones. { kə'thep·sən }

catkin [BOTANY] An indeterminate type of inflorescence that resembles a scaly spike and sometimes is pendant. { 'kat·kən }

Catostomidae [VERTEBRATE ZOOLOGY] The suckers, a family of cypriniform fishes in the suborder Cyprinoidei. { ˌkad·ə'stäm·əˌdē }

cauda equina [ANATOMY] The roots of the sacral and coccygeal nerves, collectively; so called because of their resemblance to a horse's tail. { 'kaüd·ə i'kwīn·ə }

caudal [ZOOLOGY] Toward, belonging to, or pertaining to the tail or posterior end. { 'kȯd·əl }

caudal artery [VERTEBRATE ZOOLOGY] The extension of the dorsal aorta in the tail of a vertebrate. { 'kȯd·əl 'ard·ə·rē }

caudal vertebra [ANATOMY] Any of the small bones of the vertebral column that support the tail in vertebrates; in humans, three to five are fused to form the coccyx. { 'kȯd·əl 'vər·tə·brə }

Caudata [VERTEBRATE ZOOLOGY] An equivalent name for Urodela. { kaü·dad·ə }

caudate [ZOOLOGY] **1.** Having a tail or taillike appendage. **2.** Any member of the Caudata. { 'kȯˌdāt }

caudate lobe [ANATOMY] The tailed lobe of the liver that separates the right extremity of the transverse fissure from the commencement of the fissure for the inferior vena cava. { 'kȯˌdāt 'lōb }

caudate nucleus [ANATOMY] An elongated arched gray mass which projects into and forms part of the lateral wall of the lateral ventricle. { 'kȯˌdāt 'nü·klē·əs }

caudex [BOTANY] The main axis of a plant, including stem and roots. { 'kȯˌdeks }

caudicle [BOTANY] A slender appendage attaching pollen masses to the stigma in orchids. { 'kȯd·ə·kəl }

Caulerpaceae [BOTANY] A family of green algae in the order Siphonales. { ˌkȯ·lər'pās·ēˌē }

caulescent [BOTANY] Having an aboveground stem. { kȯ'les·ənt }

cauliflorous [BOTANY] Producing flowers on the older branches or main stem. { ¦kȯl·ə¦flȯr·əs }

cauliflory [BOTANY] Of flowers, growth on the main stem of limbs of a tree. { 'kol·əˌflȯr·ē }

cauliflower [BOTANY] *Brassica oleracea* var. *botrytis*. A biennial crucifer of the order Capparales grown for its edible white head or curd, which is a tight mass of flower stalks. { 'kȯl·əˌflaü·ər }

cauline [BOTANY] Belonging to or arising from the stem, particularly if on the upper portion. { 'kȯˌlīn }

Caulobacter [MICROBIOLOGY] A genus of prosthecate bacteria; cells are rod-shaped, fusiform, or vibrioid and stalked, and reproduction is by binary fission of cells. { ˌkȯl·ō'bak·tər }

Caulobacteraceae [MICROBIOLOGY] Formerly a family of aquatic, stalked, gram-negative bacteria in the order Pseudomonadales. { ¦kȯl·əˌbak·tə'rās·ēˌē }

caulocarpic [BOTANY] Having stems that bear flowers and fruit every year. { ¦kȯl·ō¦kär·pik }

Caulococcus [MICROBIOLOGY] A genus of bacteria of uncertain affiliation; coccoid cells may be connected by threads; reproduces by budding. { ¦kȯl·ō¦käk·əs }

caulome [BOTANY] The stem structure or stem axis of a plant as a whole. { 'kȯˌlōm }

Cavellinidae [PALEONTOLOGY] A family of Paleozoic ostracods in the suborder Platycopa. { ˌkav·ə'lin·əˌdē }

caveolae [CELL BIOLOGY] Tiny indentations in the cell surface membrane which trap fluids during the process of micropinocytosis. { kə'vē·ə·lē }

cavernicolous [BIOLOGY] Inhabiting caverns. { ¦kav·ər¦nik·ə·ləs }

cavernous sinus [ANATOMY] Either of a pair of venous sinuses of the dura mater located on the side of the body of the sphenoid bone. { 'kav·ər·nəs 'sī·nəs }

Caviidae [VERTEBRATE ZOOLOGY] A family of large, hystricomorph rodents distinguished by a reduced number of toes and a rudimentary tail. { kə'vī·əˌdē }

CA virus [VIROLOGY] *See* croup-associated virus. { sē'ā ˌvī·rəs }

cavity [BIOLOGY] A hole or hollow space in an organ, tissue, or other body part. { 'kav·əd·ē }

cavy [VERTEBRATE ZOOLOGY] Any of the rodents composing the family Caviidae, which includes the guinea pig, rock cavies, mountain cavies, capybara, salt desert cavy, and mara. { 'kā·vē }

Caytoniales [PALEOBOTANY] An order of Mesozoic plants. { ˌkā·tän·ē'āˌlēz }

cDNA [MOLECULAR BIOLOGY] *See* complementary deoxyribonucleic acid.

Cebidae [VERTEBRATE ZOOLOGY] The New World monkeys, a family of primates in the suborder Anthropoidea including the capuchins and howler monkeys. { 'seb·əˌdē }

Cebochoeridae [PALEONTOLOGY] A family of extinct palaeodont artiodactyls in the superfamily Entelodontoidae. { ¦seb·əˌkō'er·əˌdē }

Cebrionidae [INVERTEBRATE ZOOLOGY] The robust click beetles, a family of cosmopolitan coleopteran insects in the superfamily Elateroidea. { ˌseb·rē'än·əˌdē }

Cecropidae [INVERTEBRATE ZOOLOGY] A family of crustaceans in the suborder Caligoida which are external parasites on fish. { sə'kräp·əˌdē }

cecum [ANATOMY] The blind end of a cavity, duct, or tube, especially the sac at the beginning of the large intestine. Also spelled caecum. { 'sē·kəm }

cedar [BOTANY] The common name for a large number of evergreen trees in the order Pinales having fragrant, durable wood. { 'sē·dər }

ceiba [BOTANY] *See* kapok. { 'sā·bə }

Celastraceae [BOTANY] A family of dicotyledonous plants in the order Celastrales characterized by erect and basal ovules, a flower disk that surrounds the ovary at the base, and opposite or sometimes alternate leaves. { ˌsel·ə'strās·ēˌē }

Celastrales [BOTANY] An order of dicotyledonous plants in the subclass Rosidae marked by simple leaves and regular flowers. { ˌsel·ə'strā·lēz }

celery [BOTANY] *Apium graveolens* var. *dulce.* A biennial umbellifer of the order Umbellales with edible petioles or leaf stalks. { 'sel·rē }

celiac [ANATOMY] Of, in, or pertaining to the abdominal cavity. { 'sēl·ēˌak }

cell [BIOLOGY] The microscopic functional and structural unit of all living organisms, consisting of a nucleus, cytoplasm, and a limiting membrane. { sel }

cell-associated virus [VIROLOGY] Virus particles that remain attached to or within the host cell after replication. { 'sel ə¦sō·shēˌād·əd 'vī·rəs }

cell constancy [BIOLOGY] The condition in which the entire body, or a part thereof, consists of a fixed number of cells that is the same for all adults of the species. { 'sel 'kän·stən·sē }

cell cycle [CELL BIOLOGY] In eukaryotic cells, the cycle of events consisting of cell division, including mitosis and cytokinesis, and interphase. { 'sel ˌsī·kəl }

cell determination [PHYSIOLOGY] The process by which multipotential cells become committed to a particular development pathway. { ¦sel ˌdiˌtər·mə¦nā·shən }

cell differentiation [CELL BIOLOGY] The series of events involved in the development of a specialized cell having specific structural, functional, and biochemical properties. { 'sel dif·əˌren·chē'ā·shən }

cell division [CELL BIOLOGY] The process by which living cells multiply; may be mitotic or amitotic. { 'sel di'vizh·ən }

cell-free extract [CELL BIOLOGY] A fluid obtained by breaking open cells; contains most of the soluble molecules of a cell. { 'sel ˌfrē 'ekˌstrakt }

cell inclusion [CELL BIOLOGY] A small, nonliving intracellular particle, usually representing a form of stored food, not immediately vital to life processes. { 'sel in'klü·zhan }

cell lineage [EMBRYOLOGY] The developmental history of individual blastomeres from their first cleavage division to their ultimate differentiation into cells of tissues and organs. { 'sel 'lin·yəj }

cell-lineage mutant [CELL BIOLOGY] Any mutation that affects the division of cells or the fates of their progeny. { ¦sel ˌlin·ē·əj 'myüt·ənt }

cell-mediated immunity [IMMUNOLOGY] Immune responses produced by the activities of T cells rather than by immunoglobulins. { ¦sel ¦mē·dē ˌād·əd i'myü·nəd·ē }

cell membrane [CELL BIOLOGY] A thin layer of protoplasm, consisting mainly of lipids and proteins, which is present on the surface of all cells. Also known as plasmalemma; plasma membrane. { 'sel ¦memˌbrān }

cell movement [CELL BIOLOGY] **1.** Intracellular

movement of cellular components. **2.** The movement of a cell relative to its environment. { 'sel ˌmüv·mənt }

cellobiase [BIOCHEMISTRY] An enzyme that participates in the hydrolysis of cellobiose into glucose. { ˌsel·ō'bīˌās }

cell permeability [CELL BIOLOGY] The permitting or activating of the passage of substances into, out of, or through cells. { 'sel pər·mē·ə'bil·əd·ē }

cell plate [CELL BIOLOGY] A membrane-bound disk formed during cytokinesis in plant cells which eventually becomes the middle lamella of the wall formed between daughter cells. { 'sel ˌplāt }

cell recognition [CELL BIOLOGY] The mutual recognition of cells, as expressed by specific cellular adhesion, due to a specific complementary interaction between molecules on adjacent cell surfaces. { 'sel ˌrek·əgˌnish·ən }

cell sap [CELL BIOLOGY] The liquid content of a plant cell vacuole. { 'sel ˌsap }

cells of Paneth [HISTOLOGY] Coarsely granular secretory cells found in the crypts of Lieberkühn in the small intestine. Also known as Paneth cells. { ¦selz əv 'pän·ət }

cell-surface differentiation [CELL BIOLOGY] The specialization or modification of the cell surface. { 'sel ˌsər·fəs dif·əˌren·chē'ā·shən }

cell-surface ionization [PHYSIOLOGY] The presence of a negative charge on the surface of all living cells suspended in aqueous salt solutions at neutral pH. { 'sel ˌsər·fəs ˌī·ən·ə'zā·shən }

cell theory [BIOLOGY] **1.** A principle that describes the cell as the fundamental unit of all living organisms. **2.** A principle that describes the properties of an organism as the sum of the properties of its component cells. { 'sel ˌthē·ə·rē }

cellular [BIOLOGY] Characterized by, consisting of, or pertaining to cells. { 'sel·yə·lər }

cellular affinity [BIOLOGY] The phenomenon of selective adhesiveness observed among the cells of certain sponges, slime molds, and vertebrates. { 'sel·yə·lər ə'fin·əd·ē }

cellular immunity [IMMUNOLOGY] Immune responses carried out by active cells rather than by antibodies. { ¦sel·yə·lər i'myü·nəd·ē }

cellular immunology [IMMUNOLOGY] The study of the cells of the lymphoid organs, which are the main agents of immune reactions in all vertebrates. { ¦sel·yə·lər ˌim·yə'näl·ə·jē }

cellular slime molds [BIOLOGY] A group of funguslike protozoa that form slimy aggregations on decaying organic matter; they differ from true slime molds in that the pseudoplasmodium is made of a group of separate cells rather than an amebalike mass. { 'sel·yə·lər 'slīm ˌmōlz }

cellulase [BIOCHEMISTRY] Any of a group of extracellular enzymes, produced by various fungi, bacteria, insects, and other lower animals, that hydrolyze cellulose. { 'sel·yəˌlās }

cellulolytic [BIOLOGY] Having the ability to hydrolyze cellulose; applied to certain bacteria and protozoans. { ¦sel·yə¦lid·ik }

Cellulomonas [MICROBIOLOGY] A genus of gram-positive, irregular rods in the coryneform group of bacteria; metabolism is respiratory; most strains produce acid from glucose, and cellulose is attacked by all strains. { ˌsel·yə'läm·ə·nəs }

cellulose [BIOCHEMISTRY] $(C_6H_{10}O_5)_n$ The main polysaccharide in living plants, forming the skeletal structure of the plant cell wall; a polymer of β-D-glucose units linked together, with the elimination of water, to form chains comprising 2000-4000 units. { 'sel·yəˌlōs }

cell wall [CELL BIOLOGY] A semirigid, permeable structure that is composed of cellulose, lignin, or other substances and that envelops most plant cells. { ¦sel ¦wȯl }

Celyphidae [INVERTEBRATE ZOOLOGY] A family of myodarian cyclorrhaphous dipteran insects in the subsection Acalypteratae. { se'lif əˌdē }

cement [HISTOLOGY] Calcified tissue which fastens the roots of teeth to the alveolus. Also known as cementum. [INVERTEBRATE ZOOLOGY] Any of the various adhesive secretions, produced by certain invertebrates, that harden on exposure to air or water and are used to bind objects. { si'ment }

cement gland [INVERTEBRATE ZOOLOGY] A structure in many invertebrates that produces cement. { si'ment ˌgland }

cementum [HISTOLOGY] *See* cement. { si'men·təm }

centimorgan [GENETICS] A unit of genetic map distance, equal to the distance along a chromosome that gives a recombination frequency of 1. { 'sent·əˌmȯrg·ən }

centipede [INVERTEBRATE ZOOLOGY] The common name for an arthropod of the class Chilopoda. { 'sent·əˌpēd }

central apparatus [CELL BIOLOGY] The centrosome or centrosomes together with the surrounding cytoplasm. Also known as cytocentrum. { 'sen·trəl ˌap·ə'rad·əs }

central canal [ANATOMY] The small canal running through the center of the spinal cord from the conus medullaris to the lower part of the fourth ventricle; represents the embryonic neural tube. { 'sen·trəl kə'nal }

central dogma [GENETICS] The concept, subject to several exceptions, that genetic information is coded in self-replicating deoxyribonucleic acid and undergoes unidirectional transfer to messenger ribonucleic acids in transcription that act as templates for protein synthesis in translation. { 'sen·trəl 'dȯg·mə }

Centrales [BOTANY] An order of diatoms (Bacillariophyceae) in which the form is often circular and the markings on the valves are radial. { sen 'trā·lēz }

central nervous system [ANATOMY] The division of the vertebrate nervous system comprising the brain and spinal cord. { 'sen·trəl 'nər·vəs ˌsis·təm }

central placentation [BOTANY] Having the ovules located in the center of the ovary. { 'sen·trəl ˌpla·sən'tā·shən }

central pocket loop [ANATOMY] A whorl type of

fingerprint pattern having two deltas and at least one ridge that make a complete circuit. { ¦sen·trəl 'päk·ət ˌlüp }

central sulcus [ANATOMY] A groove situated about the middle of the lateral surface of the cerebral hemisphere, separating the frontal from the parietal lobe. { 'sen·trəl 'səl·kəs }

Centrarchidae [VERTEBRATE ZOOLOGY] A family of fishes in the order Perciformes, including the fresh-water or black basses and several sunfishes. { ˌsen'trär·kəˌdē }

centric [ANATOMY] Having all teeth of both jaws meet normally with perfect distribution of forces in the dental arch. { 'sen·trik }

centriole [CELL BIOLOGY] A complex cellular organelle forming the center of the centrosome in most cells; usually found near the nucleus in interphase cells and at the spindle poles during mitosis. { 'sen·trēˌōl }

Centrohelida [INVERTEBRATE ZOOLOGY] An order of protozoans in the subclass Heliozoia lacking a central capsule and having axopodia or filopodia, and siliceous scales and spines. { ¦sen·trō'hel·ə·də }

centrolecithal ovum [CELL BIOLOGY] An egg cell having the yolk centrally located; occurs in arthropods. { ¦sen·trō'les·ə·thəl 'ō·vəm }

Centrolenidae [VERTEBRATE ZOOLOGY] A family of arboreal frogs in the suborder Procoela characterized by green bones. { ˌsen·trə'len·əˌdē }

centromere [CELL BIOLOGY] A specialized chromomere to which the spindle fibers are attached during mitosis. Also known as kinetochore; kinomere; primary constriction. { 'sen·trəˌmir }

centromere distance [GENETICS] The distance of a gene from a centromere, measured in terms of recombination frequency. { 'sen·trəˌmir ˌdis·təns }

centromere effect [GENETICS] The reduced level of genetic recombination shown by genetic loci close to the centromere. { 'sen·trəˌmir iˌfekt }

Centronellidina [PALEONTOLOGY] A suborder of extinct articulate brachiopods in the order Terebratulida. { ¦sen·trō·nə'lid·ən·ə }

centrosome [CELL BIOLOGY] A spherical hyaline region of the cytoplasm surrounding the centriole in many cells; plays a dynamic part in mitosis as the focus of the spindle pole. { 'sen·trə ˌsōm }

Centrospermae [BOTANY] An equivalent name for the Caryophyllales. { ˌsen·trō'spərˌmē }

Centrospermales [BOTANY] An equivalent name for the Caryophyllales. { ˌsen·trō·spər'mā·lēz }

centrosphere [CELL BIOLOGY] The differentiated layer of cytoplasm immediately surrounding the centriole. { 'sen·trəˌsfir }

centrum [ANATOMY] The main body of a vertebra. [BOTANY] The central space in hollow-stemmed plants. { 'sen·trəm }

Cephalaspida [PALEONTOLOGY] An equivalent name for the Osteostraci. { ˌsef·ə'las·pə·də }

Cephalaspidomorphi [VERTEBRATE ZOOLOGY] An equivalent name for Monorhina. { ˌsef·ə¦las·pə·də'mór·fī }

cephalic [ZOOLOGY] Of or pertaining to the head or anterior end. { sə'fal·ik }

cephalic vein [ANATOMY] A superficial vein located on the lateral side of the arm which drains blood from the radial side of the hand and forearm into the axillary vein. { sə'fal·ik 'vān }

cephalin [BIOCHEMISTRY] Any of several acidic phosphatides whose composition is similar to that of lecithin but having ethanolamine, serine, and inositol instead of choline; found in many living tissues, especially nervous tissue of the brain. { 'sef·ə·lən }

Cephalina [INVERTEBRATE ZOOLOGY] A suborder of protozoans in the order Eugregarinida that are parasites of certain invertebrates. { sef·ə'lī·nə }

cephalization [ZOOLOGY] Anterior specialization resulting in the concentration of sensory and neural organs in the head. { ˌsef·ə·lə'zā·shən }

Cephalobaenida [INVERTEBRATE ZOOLOGY] An order of the arthropod class Pentastomida composed of primitive forms with six-legged larvae. { ˌsef·ə·lō'bēn·ə·də }

Cephaloboidea [INVERTEBRATE ZOOLOGY] A superfamily of free-living nematodes in the order Rhabditida distinguished by cephalic elaborations or ornamentations. { 'sef·ə·lə'bóid·ē·ə }

Cephalocarida [INVERTEBRATE ZOOLOGY] A subclass of Crustacea erected to include the primitive crustacean *Hutchinsoniella macracantha*. { ˌsef·ə·lō'kar·ə·də }

Cephalochordata [VERTEBRATE ZOOLOGY] A subphylum of the Chordata comprising the lancelets, including *Branchiostoma*. { ¦sef·ə·lōˌkór 'däd·ə }

Cephaloidae [INVERTEBRATE ZOOLOGY] The false longhorn beetles, a small family of coleopteran insects in the superfamily Tenebrionoidea. { ˌsef·ə'lóid·ē }

cephalomere [INVERTEBRATE ZOOLOGY] One of the somites that make up the head of an arthropod. { sə'fal·əˌmir }

cephalont [INVERTEBRATE ZOOLOGY] A sporozoan just prior to spore formation. { 'sef·əˌlänt }

Cephalopoda [INVERTEBRATE ZOOLOGY] Exclusively marine animals constituting the most advanced class of the Mollusca, including squids, octopuses, and *Nautilus*. { ˌsef·ə'läp·ə·də }

cephalosporin [MICROBIOLOGY] Any of a group of antibiotics produced by strains of the imperfect fungus *Cephalosporium*. { ˌsef·ə·lə'spór·ən }

cephalothin [MICROBIOLOGY] An antibiotic derived from the fungus *Cephalosporium*, resembling penicillin units in structure and activity, and effective against many gram-positive cocci that are resistant to penicillin. { 'sef·ə·lə·thən }

cephalothorax [INVERTEBRATE ZOOLOGY] The body division comprising the united head and thorax of arachnids and higher crustaceans. { ¦sef·ə·lə'thórˌaks }

Cephalothrididae [INVERTEBRATE ZOOLOGY] A family of ribbonlike worms in the order Palaeonemertini. { ˌsef·ə·lō'thrī·dəˌdē }

cephalotrichous flagellation [CELL BIOLOGY] Insertion of flagella in polar tufts. { ˌsef·ə·lō'trī·kəs ˌflaj·ə'lā·shən }

Cephidae [INVERTEBRATE ZOOLOGY] The stem

sawflies, composing the single family of the hymenopteran superfamily Cephoidea. { 'sef·ə ˌdē }

Cephoidea [INVERTEBRATE ZOOLOGY] A superfamily of hymenopteran insects in the suborder Symphyta. { sa'foid·ē·ə }

Ceractinomorpha [INVERTEBRATE ZOOLOGY] A subclass of sponges belonging to the class Demospongiae. { sə¦rak·tə·nə'mȯr·fə }

Cerambycidae [INVERTEBRATE ZOOLOGY] The longhorn beetles, a family of coleopteran insects in the superfamily Chrysomeloidea. { se·rəm 'bī·səˌdē }

cerambycoid larva [INVERTEBRATE ZOOLOGY] A beetle larva that is morphologically similar to a caraboid larva except for the former's absence of legs. { sə'ram·bəˌkȯid 'lar·və }

Ceramonematoidea [INVERTEBRATE ZOOLOGY] A superfamily of marine nematodes in the order Desmodorida characterized by distinctive cuticular ornamentation, giving the appearance of a body covered with rings of crested tilelike plates. { ˌser·ə·mōˌnem·ə'tȯid·ē·ə }

Ceramoporidae [PALEONTOLOGY] A family of extinct, marine bryozoans in the order Cystoporata. { səˌram·ə'pȯr·əˌdē }

Cerapachyinae [INVERTEBRATE ZOOLOGY] A subfamily of predacious ants in the family Formicidae, including the army ant. { ˌser·ə·pə'kī·ə ˌnē }

Ceraphronidae [INVERTEBRATE ZOOLOGY] A superfamily of hymenopteran insects in the superfamily Proctotrupoidea. { ˌser·ə'frän·əˌdē }

cerata [INVERTEBRATE ZOOLOGY] Respiratory papillae of the mantle in certain nudibranchs. { sa 'räd·ə }

ceratine [INVERTEBRATE ZOOLOGY] A hornlike material secreted by some anthozoans. { 'ser·ə ˌtēn }

Ceratiomyxaceae [MYCOLOGY] The single family of the fungal order Ceratiomyxales. { səˌrāsh·ē·ōˌmik'sās·ēˌē }

Ceratiomyxales [MYCOLOGY] An order of myxomycetous fungi in the subclass Ceratiomyxomycetidae. { səˌrāsh·ē·ōˌmik 'sā·lēz }

Ceratiomyxomycetidae [MYCOLOGY] A subclass of fungi belonging to the Myxomycetes. { sə ˌrāsh·ē·ōˌmik·sōˌmī'sed·əˌ dē }

ceratite [PALEONTOLOGY] A fossil ammonoid of the genus *Ceratites* distinguished by a type of suture in which the lobes are further divided into subordinate crenulations while the saddles are not divided and are smoothly rounded. { 'ser·ə ˌtīt }

ceratitic [PALEONTOLOGY] Pertaining to a ceratite. { ˌser·ə'tid·ik }

Ceratodontidae [PALEONTOLOGY] A family of Mesozoic lungfishes in the order Dipteriformes. { ˌser·ə·tō'dän·təˌdē }

Ceratomorpha [VERTEBRATE ZOOLOGY] A suborder of the mammalian order Perissodactyla including the tapiroids and the rhinoceratoids. { ˌser·ə·tō'mȯr·fə }

Ceratophyllaceae [BOTANY] A family of rootless, free-floating dicotyledons in the order Nymphaeales characterized by unisexual flowers and whorled, cleft leaves with slender segments. { ˌser·ə·tō·fə'lās·ēˌē }

Ceratopsia [PALEONTOLOGY] The horned dinosaurs, a suborder of Upper Cretaceous reptiles in the order Ornithischia. { ˌser·ə'täp·sē·ə }

cercaria [INVERTEBRATE ZOOLOGY] The larval generation which terminates development of a digenetic trematode in the intermediate host. { sər'kar·ē·ə }

Cercopidae [INVERTEBRATE ZOOLOGY] A family of homopteran insects belonging to the series Auchenorrhyncha. { sar'käp·əˌdē }

Cercopithecidae [VERTEBRATE ZOOLOGY] The Old World monkeys, a family of primates in the suborder Anthropoidea. { ¦sər·kō·pə'thē·səˌdē }

cercopod [INVERTEBRATE ZOOLOGY] **1.** Either of two filamentous projections on the posterior end of notostracan crustaceans. **2.** *See* cercus. { 'sər·kəˌpäd }

Cercospora [MYCOLOGY] A genus of imperfect fungi having dark, elongate, multiseptate spores. { sər'käs·pə·rə }

cercus [INVERTEBRATE ZOOLOGY] Either of a pair of segmented sensory appendages on the last abdominal segment of many insects and certain other anthropods. Also known as cercopod. { 'sər·kəs }

cere [VERTEBRATE ZOOLOGY] A soft, swollen mass of tissue at the base of the upper mandible through which the nostrils open in certain birds, such as parrots and birds of prey. { sir }

cereal [BOTANY] Any member of the grass family (Graminae) which produces edible, starchy grains usable as food by humans and livestock. Also known as grain. { 'sir·ē·əl }

cerebellum [ANATOMY] The part of the vertebrate brain lying below the cerebrum and above the pons, consisting of three lobes and concerned with muscular coordination and the maintenance of equilibrium. { ˌser·ə'bel·əm }

cerebral cortex [ANATOMY] The superficial layer of the cerebral hemispheres, composed of gray matter and concerned with coordination of higher nervous activity. { sə'rē·brəl 'kȯrˌteks }

cerebral hemisphere [ANATOMY] Either of the two lateral halves of the cerebrum. { sə'rē·brəl 'hem·əˌsfir }

cerebral localization [ANATOMY] Designation of a specific region of the brain as the area controlling a specific physiologic function or as the site of a lesion. { sə'rē·brəl lō·kə·lə'zā·shən }

cerebral peduncle [ANATOMY] One of two large bands of white matter (containing descending axons of upper motor neurons) which emerge from the underside of the cerebral hemispheres and approach each other as they enter the rostral border of the pons. { sə'rē·brəl pi'dəŋ·kəl }

cerebrose [BIOCHEMISTRY] *See* galactose. { 'ser·ə ˌbrōs }

cerebroside [BIOCHEMISTRY] Any of a complex group of glycosides found in nerve tissue, consisting of a hexose, a nitrogenous base, and a fatty acid. Also known as galactolipid. { 'ser·ə·brōˌsīd }

cerebrospinal axis [ANATOMY] The axis of the body composed of the brain and spinal cord. { sə ¦rē·brō'spīn·əl 'ak·səs }

cerebrospinal fluid [PHYSIOLOGY] A clear liquid that fills the ventricles of the brain and the spaces between the arachnoid mater and pia mater. { sə¦rē·brō'spīn·əl 'flü·əd }

cerebrum [ANATOMY] The enlarged anterior or upper part of the vertebrate brain consisting of two lateral hemispheres. { sə'rē·brəm }

Cerelasmidae [INVERTEBRATE ZOOLOGY] A family of Psamminida with a soft test composed principally of organic cement; xenophyae, when present, are not systematically arranged. { ‚ser·ə'las·mə‚dē }

cerelose [BIOCHEMISTRY] *See* glucose. { 'sir·ə ‚lōs }

Ceriantharia [INVERTEBRATE ZOOLOGY] An order of the Zoantharia distinguished by the elongate form of the anemone-like body. { ¦ser·ē·ən 'thar·ē·ə }

Ceriantipatharia [INVERTEBRATE ZOOLOGY] A subclass proposed by some authorities to include the anthozoan orders Antipatharia and Ceriantharia. { ‚sir·ē‚an·tə·pə'thar·ē·ə }

Cerithiacea [INVERTEBRATE ZOOLOGY] A superfamily of gastropod mollusks in the order Prosobranchia. { ‚ser·ə‚thī'ās·ē·ə }

cernuous [BOTANY] Drooping or inclining. { 'sərn·yə·wəs }

Cerophytidae [INVERTEBRATE ZOOLOGY] A small family of coleopteran insects in the superfamily Elateroidea. { ‚ser·ə'fīd·ə‚dē }

certation [BOTANY] Competition in growth rate between pollen tubes of different genotypes resulting in unequal chances of accomplishing fertilization. { sər'tā·shən }

ceruloplasmin [BIOCHEMISTRY] The copper-binding serum protein in human blood. { sə¦rül·ō ¦plaz·mən }

cerumen [PHYSIOLOGY] The waxy secretion of the ceruminous glands of the external ear. Also known as earwax. { sə'rü·mən }

ceruminous gland [ANATOMY] A modified sweat gland in the external ear that produces earwax. { sə'rü·mə·nəs ‚gland }

cervical [ANATOMY] Of or relating to the neck, a necklike part, or the cervix of an organ. { 'sər·və·kəl }

cervical canal [ANATOMY] Canal of the cervix of the uterus. { 'sər·və·kəl kə'nal }

cervical flexure [EMBRYOLOGY] A ventrally concave flexure of the embryonic brain occurring at the junction of hindbrain and spinal cord. { 'sər·və·kəl 'flek·shər }

cervical ganglion [ANATOMY] Any ganglion of the sympathetic nervous system located in the neck. { 'sər·və·kəl 'gaŋ·glē‚än }

cervical plexus [ANATOMY] A plexus in the neck formed by the anterior branches of the upper four cervical nerves. { 'sər·və·kəl 'plek·səs }

cervical sinus [EMBRYOLOGY] A triangular depression caudal to the hyoid arch containing the posterior visceral arches and grooves. { 'sər·və·kəl 'sī·nəs }

cervical vertebra [ANATOMY] Any of the bones in the neck region of the vertebral column; the transverse process has a characteristic perforation by a transverse foramen. { 'sər·və·kəl 'vərd·ə·brə }

Cervidae [VERTEBRATE ZOOLOGY] A family of pecoran ruminants in the superfamily Cervoidea, characterized by solid, deciduous antlers; includes deer and elk. { 'sər·və‚dē }

cervix [ANATOMY] A constricted or necklike portion of a structure. { 'sər·viks }

Cervoidea [VERTEBRATE ZOOLOGY] A superfamily of tylopod ruminants in infraorder Pecora, including deer, giraffes, and related species. { sər 'vȯid·ē·ə }

cespitose [BOTANY] **1.** Tufted; growing in tufts, as grass. **2.** Having short stems forming a dense turf. { 'ses·pə‚tōs }

Cestida [INVERTEBRATE ZOOLOGY] An order of ribbon-shaped ctenophores having a very short tentacular axis and an elongated pharyngeal axis. { 'ses·tə·də }

Cestoda [INVERTEBRATE ZOOLOGY] A subclass of tapeworms including most members of the class Cestoidea; all are endoparasites of vertebrates. { se'stō·də }

Cestodaria [INVERTEBRATE ZOOLOGY] A small subclass of worms belonging to the class Cestoidea; all are endoparasites of primitive fishes. { ‚ses·tə'dar·ē·ə }

Cestoidea [INVERTEBRATE ZOOLOGY] The tapeworms, endoparasites composing a class of the phylum Platyhelminthes. { se'stȯid·ē·ə }

Cetacea [VERTEBRATE ZOOLOGY] An order of aquatic mammals, including the whales, dolphins, and porpoises. { sē'tā·shə }

cetology [VERTEBRATE ZOOLOGY] The study of whales. { sē'täl·ə·jē }

Cetomimiformes [VERTEBRATE ZOOLOGY] An order of rare oceanic, deepwater, soft-rayed fishes that are structurally diverse. { ‚sēd·ə‚mim·ə'fȯr ‚mēz }

Cetomimoidei [VERTEBRATE ZOOLOGY] The whale-fishes, a suborder of the Cetomimiformes, including bioluminescent, deep-sea species. { ‚sēd·ə·mə'mȯid·ē‚ī }

Cetorhinidae [VERTEBRATE ZOOLOGY] The basking sharks, a family of large, galeoid elasmobranchs of the isurid line. { ‚sēd·ə'rīn·ə‚dē }

chaeta [BIOLOGY] *See* seta. { 'kēd·ə }

Chaetetidae [PALEONTOLOGY] A family of Paleozoic corals of the order Tabulata. { ‚kē'tē·də ‚dē }

Chaetodontidae [VERTEBRATE ZOOLOGY] The butterflyfishes, a family of perciform fishes in the suborder Percoidei. { ‚kēd·ō'dän·tə‚dē }

Chaetognatha [INVERTEBRATE ZOOLOGY] A phylum of abundant planktonic arrowworms. { ‚kē 'täg·nə·thə }

Chaetonotoidea [INVERTEBRATE ZOOLOGY] An order of the class Gastrotricha characterized by two adhesive tubes connected with the distinctive paired, posterior tail forks. { ‚kē‚tän·ō'tȯid· ē·ə }

Chaetophoraceae [BOTANY] A family of algae in the order Ulotrichales characterized as branched filaments which taper toward the apices, sometimes bearing terminal setae. { ˌkēd·ō·fə′rās·ēˌē }

Chaetopteridea [INVERTEBRATE ZOOLOGY] A family of spioniform polychaete annelids belonging to the Sedentaria. { ˌkēˌtäp·tə′rīd·ē·ə }

chainette [INVERTEBRATE ZOOLOGY] In some cestodes, a longitudinal row of similar spines, usually with lateral winglike expansions, found near the base of the tentacles. { chā′net }

chalaza [BOTANY] The region at the base of the nucellus of an ovule; gives rise to the integuments. [CELL BIOLOGY] One of the paired, spiral, albuminous bands in a bird's egg that attach the yolk to the shell lining membrane at the ends of the egg. { kə′lāz·ə }

chalazogamy [BOTANY] A process of fertilization in which the pollen tube passes through the chalaza to reach the embryo sac. { ˌkal·ə′zäg·ə·mē }

Chalcididae [INVERTEBRATE ZOOLOGY] The chalcids, a family of hymenopteran insects in the superfamily Chalcidoidea. { kal′sid·əˌdē }

Chalcidoidea [INVERTEBRATE ZOOLOGY] A superfamily of hymenopteran insects in the suborder Apocrita, including primarily insect parasites. { ˌkal·sə′dȯid·ē·ə }

chalice cell [HISTOLOGY] *See* goblet cell. { ′chal·əs ˌsel }

Chalicotheriidae [PALEONTOLOGY] A family of extinct perissodactyl mammals in the superfamily Chalicotherioidea. { ˌkal·əˌkō·thə′rī·əˌdē }

Chalicotherioidea [PALEONTOLOGY] A superfamily of extinct, specialized perissodactyls having claws rather than hooves. { ¦kal·əˌkōˌthi·rē′ȯid·ē·ə }

challenge [IMMUNOLOGY] Administration of an antigen to ascertain state of immunity. { ′chal·ənj }

chalones [BIOCHEMISTRY] Substances thought to be molecules of the protein-polypeptide class that are produced as part of the growth-control systems of tissues; known to inhibit cell division by acting on several phases in the mitotic cycle. { ′kaˌlōnz }

Chamaeleontidae [VERTEBRATE ZOOLOGY] The chameleons, a family of reptiles in the suborder Sauria. { kəˌmēl·ē′än·təˌdē }

Chamaemyidae [INVERTEBRATE ZOOLOGY] The aphid flies, a family of myodarian cyclorrhaphous dipteran insects of the subsection Acalypteratae. { ˌkam·ə′mī·əˌdē }

chamaephyte [ECOLOGY] Any perennial plant whose winter buds are within 10 inches (25 centimeters) of the soil surface. { ′kam·əˌfīt }

Chamaesiphonales [BOTANY] An order of blue-green algae of the class Cyanophyceae; reproduce by cell division, colony fragmentation, and endospores. { ˌkam·ēˌsī·fə′nā·lēz }

Chambersielloidea [INVERTEBRATE ZOOLOGY] A superfamily of nematodes in the order Rhabdita characterized by highly unusual elaborations of the oral opening in the form of six mandibles. { chāmˌbər·sē·ə′lȯid·ē·ə }

chameleon [VERTEBRATE ZOOLOGY] The common name for about 80 species of small to medium-size lizards composing the family Chamaeleontidae. { kə′mēl·yən }

chamois [VERTEBRATE ZOOLOGY] *Rupicapra rupicapra*. A goatlike mammal included in the tribe Rupicaprini of the family Bovidae. { ′sham·ē }

Chanidae [VERTEBRATE ZOOLOGY] A monospecific family of teleost fishes in the order Gonorynchiformes which contain the milkfish (*Chanos chanos*), distinguished by the lack of teeth. { ′kan·əˌdē }

Channidae [VERTEBRATE ZOOLOGY] The snakeheads, a family of fresh-water perciform fishes in the suborder Anabantoidei. { ′kan·əˌdē }

Chaoboridae [INVERTEBRATE ZOOLOGY] The phantom midges, a family of dipteran insects in the suborder Orthorrhapha. { ˌkā·ə′bȯr·əˌdē }

chaparral [ECOLOGY] A vegetation formation characterized by woody plants of low stature, impenetrable because of tough, rigid, interlacing branches, which have simple, waxy, evergreen, thick leaves. { ¦shap·ə¦ral }

Characeae [BOTANY] The single family of the order Charales. { kə′rās·ēˌē }

Characidae [VERTEBRATE ZOOLOGY] The characins, the single family of the suborder Characoidei. { kə′ras·əˌdē }

Characoidei [VERTEBRATE ZOOLOGY] A suborder of the order Cypriniformes including fresh-water fishes with toothed jaws and an adipose fin. { ˌkar·ə′kȯid·ēˌī }

character convergence [ECOLOGY] A process whereby two relatively evolved species interact so that one converges toward the other with respect to one or more traits. { ′kar·ik·tər kən ˌvər·jəns }

character displacement [ECOLOGY] An outcome of competition in which two species living in the same area have evolved differences in morphology or other characteristics that lessen competition for food resources. { ′kar·ik·tər dis′plās·mənt }

character progression [ECOLOGY] The geographic gradation of expression of specific characters over the range of distribution of a race or species. { ′kar·ik·tər prəˌgresh·ən }

character stasis [GENETICS] Long-term constancy in a phenotypic character within a lineage. { ′kar·ik·tər ˌstā·səs }

Charadrii [VERTEBRATE ZOOLOGY] The shore birds, a suborder of the order Charadriiformes. { kə′rad·rēˌī }

Charadriidae [VERTEBRATE ZOOLOGY] The plovers, a family of birds in the superfamily Charadrioidea. { kə·rə′drī·əˌdē }

Charadriiformes [VERTEBRATE ZOOLOGY] An order of cosmopolitan birds, most of which live near water. { kəˌrad·rē·ə′fȯrˌmēz }

Charadrioidea [VERTEBRATE ZOOLOGY] A superfamily of the suborder Charadrii, including plovers, sandpipers, and phalaropes. { kəˌrad·rē′ȯid·ē·ə }

Charales [BOTANY] Green algae composing the single order of the class Charophyceae. { kə′rā·lēz }

Chareae [BOTANY] A tribe of green algae belonging to the family Characeae. { ′kar·ēˌē }

Charophyceae [BOTANY] A class of green algae in the division Chlorophyta. { ˌkar·əˈfīs·ēˌē }

Charophyta [BOTANY] A group of aquatic plants, ranging in size from a few inches to several feet in height, that live entirely submerged in water. { kəˈräf·əd·ə }

chartaceous [BOTANY] Resembling paper. { chär ˈtā·shəs }

chartreusin [MICROBIOLOGY] $C_{18}H_{18}O_{18}$ Crystalline, greenish-yellow antibiotic produced by a strain of *Streptomyces chartreusis*; active against gram-positive microorganisms, acid-fast bacilli, and phage of *Staphylococcus pyogenes*. { shärˈtrü·zən }

chasmogamy [BOTANY] The production of a hermaphroditic floral type that opens at anthesis and may be visited by an insect vector, providing a means of cross-pollination. { kazˈmäg·ə·mē }

chasmophyte [ECOLOGY] A plant that grows in rock crevices. { ˈkaz·məˌfīt }

chE [BIOCHEMISTRY] *See* cholinesterase.

check cross [GENETICS] The crossing of an unknown genotype with a phenotypically similar individual of known genotype. { ˈchek ˌkrȯs }

check ligament [ANATOMY] A thickening of the orbital fascia running from the insertion of the lateral rectus muscle to the medial orbital wall (medial check ligament) or from the insertion of the lateral rectus muscle to the lateral orbital wall (lateral check ligament). { ˈchek ˌlig·ə·mənt }

cheek [ANATOMY] The wall of the mouth in humans and other mammals. [ZOOLOGY] The lateral side of the head in submammalian vertebrates and in invertebrates. { chēk }

cheek pouch [VERTEBRATE ZOOLOGY] A saclike dilation of the cheeks in certain animals, such as rodents, in which food is held. { ˈchēk ˌpau̇ch }

cheetah [VERTEBRATE ZOOLOGY] *Acinonyx jubatus*. A doglike carnivoran mammal belonging to the cat family, having nonretractile claws and long legs. { ˈchēd·ə }

Cheilostomata [INVERTEBRATE ZOOLOGY] An order of ectoproct bryozoans in the class Gymnolaemata possessing delicate erect or encrusting colonies composed of loosely grouped zooecia. { ˌkī·ləˈstō·məd·ə }

Cheiracanthidae [PALEONTOLOGY] A family of extinct acanthodian fishes in the order Acanthodiformes. { ˌkī·rəˈkan·thəˌdē }

chela [INVERTEBRATE ZOOLOGY] **1.** A claw or pincer on the limbs of certain crustaceans and arachnids. **2.** A sponge spicule with talonlike terminal processes. { ˈkē·lə }

chelate [INVERTEBRATE ZOOLOGY] Pertaining to an appendage with a pincerlike organ or claw. { ˈkēˌlāt }

chelicera [INVERTEBRATE ZOOLOGY] Either appendage of the first pair in arachnids, usually modified for seizing, crushing, or piercing. { kə ˈlis·ə·rə }

Chelicerata [INVERTEBRATE ZOOLOGY] A subphylum of the phylum Arthropoda; chelicerae are characteristically modified as pincers. { kəˌlis·ə ˈräd·ə }

Chelidae [VERTEBRATE ZOOLOGY] The side-necked turtles, a family of reptiles in the suborder Pleurodira. { ˈkel·əˌdē }

chelifore [INVERTEBRATE ZOOLOGY] Either of the first pair of appendages on the cephalic segment of pycnogonids. { ˈkel·əˌfȯr }

cheliform [INVERTEBRATE ZOOLOGY] Having a forcepslike organ formed by a movable joint closing against an adjacent segment; referring especially to a crab's claw. { ˈkel·əˌfȯrm }

cheliped [INVERTEBRATE ZOOLOGY] Either of the paired appendages bearing chelae in decapod crustaceans. { ˈkel·əˌped }

Chelonariidae [INVERTEBRATE ZOOLOGY] A family of coleopteran insects in the superfamily Dryopoidea. { keˌlän·əˈrī·əˌdē }

Chelonethida [INVERTEBRATE ZOOLOGY] An equivalent name for the Pseudoscorpionida. { ˌkel·əˈneth·ə·də }

Chelonia [VERTEBRATE ZOOLOGY] An order of the Reptilia, subclass Anapsida, including the turtles, terrapins, and tortoises. { keˈlōn·ē·ə }

Cheloniidae [VERTEBRATE ZOOLOGY] A family of reptiles in the order Chelonia including the hawksbill, loggerhead, and green sea turtles. { ˌkel·əˈnī·əˌdē }

Cheluridae [INVERTEBRATE ZOOLOGY] A family of amphipod crustaceans in the suborder Gammaridea. { kəˈlür·əˌdē }

Chelydridae [VERTEBRATE ZOOLOGY] The snapping turtles, a small family of reptiles in the order Chelonia. { kəˈlid·rəˌdē }

chemical sense [PHYSIOLOGY] A process of the nervous system for reception of and response to chemical stimulation by excitation of specialized receptors. { ˈkem·i·kəl ˈsens }

chemiosmotic coupling [BIOCHEMISTRY] The mechanism by which adenosinediphosphate is phosphorylated to adenosinetriphosphate in mitochondria and chloroplasts. { ¦kem·ēˌäs ¦mäd·ik ˈkəp·liŋ }

chemoautotroph [MICROBIOLOGY] Any of a number of autotrophic bacteria and protozoans which do not carry out photosynthesis. { ˌkē·mō ˌȯd·əˈträf·ik }

chemodifferentiation [EMBRYOLOGY] The process of cellular differentiation at the molecular level by which embryonic cells become specialized as tissues and organs. { ¦kē·mōˌdif·əˌren·chēˈā·shən }

chemoheterotroph [BIOLOGY] An organism that derives energy and carbon from the oxidation of preformed organic compounds. { ¦kē·mōˈhed·ə·rəˌträf }

chemoorganotroph [BIOLOGY] An organism that requires an organic source of carbon and metabolic energy. { ¦kē·mōˌȯrˈgan·əˌträf }

chemoreception [PHYSIOLOGY] Reception of a chemical stimulus by an organism. { ˌkē·mō·ri ˈsep·shən }

chemoreceptor [PHYSIOLOGY] Any sense organ that responds to chemical stimuli. { ˌkē·mō·ri ˈsep·tər }

chemostat [MICROBIOLOGY] An apparatus, and a principle, for the continuous culture of bacterial populations in a steady state. { ˈkē·məˌstat }

chemosynthesis [BIOCHEMISTRY] The synthesis

of organic compounds from carbon dioxide by microorganisms using energy derived from chemical reactions. { ˌkē·mō'sin·thə·səs }

chemotaxin [BIOCHEMISTRY] A chemical that promotes movement of a cell or microorganism in the process of chemotaxis. { 'kē·mōˌtak·sən }

chemotaxis [BIOLOGY] The orientation or movement of a motile organism with reference to a chemical agent. { ˌkē·mō'tak·səs }

chemotaxonomy [BOTANY] The classification of plants based on natural products. { ˌkē·mōˌtak'sän·ə·mē }

chemotropism [BIOLOGY] Orientation response of a sessile organism with reference to chemical stimuli. { ˌkē·mō'trōˌpiz·əm }

chenic acid [BIOCHEMISTRY] *See* chenodeoxycholic acid. { 'kēn·ik 'as·əd }

chenodeoxycholic acid [BIOCHEMISTRY] $C_{24}H_{40}O_4$ A constituent of bile; needlelike crystals with a melting point of 119°C; soluble in alcohol, methanol, and acetic acid; used on an experimental basis to prevent and dissolve gallstones. Also known as anthropodesoxycholic acid; chenic acid; gallodesoxycholic acid. { ¦kē·nōˌdēˌäk·sē¦kō·lik 'as·əd }

Chenopodiaceae [BOTANY] A family of dicotyledonous plants in the order Caryophyllales having reduced, mostly greenish flowers. { ˌkē·nəˌpō·dē'ās·ēˌē }

Chermidae [INVERTEBRATE ZOOLOGY] A small family of minute homopteran insects in the superfamily Aphidoidea. { 'kər·mə·dē }

Cherminae [INVERTEBRATE ZOOLOGY] A subfamily of homopteran insects in the family Chermidae; all forms have a beak and an open digestive tract. { 'kər·mə·nē }

cherry [BOTANY] **1.** Any trees or shrub of the genus *Prunus* in the order Rosales. **2.** The simple, fleshy, edible drupe or stone fruit of the plant. { 'cher·ē }

chersophyte [ECOLOGY] A plant that grows in dry waste lands. { 'kərz·əˌfīt }

chestnut [BOTANY] The common name for several species of large, deciduous trees of the genus *Castanea* in the order Fagales, which bear sweet, edible nuts. { 'chesˌnət }

chevron [VERTEBRATE ZOOLOGY] The bone forming the hemal arch of a caudal vertebra. { 'shev·rən }

chevrotain [VERTEBRATE ZOOLOGY] The common name for four species of mammals constituting the family Tragulidae in the order Artiodactyla. Also known as mouse deer. { 'shev·rəˌtān }

chiasma [ANATOMY] A cross-shaped point of intersection of two parts, especially of the optic nerves. [CELL BIOLOGY] The point of junction and fusion between paired chromatids or chromosomes, first seen during diplotene of meiosis. { kī'az·mə }

Chiasmodontidae [VERTEBRATE ZOOLOGY] A family of deep-sea fishes in the order Perciformes. { kīˌaz·mə'dän·təˌdē }

chicken [VERTEBRATE ZOOLOGY] *Galus galus.* The common domestic fowl belonging to the order Galliformes. { 'chik·ən }

chick unit [BIOLOGY] A unit for the standardization of pantothenic acid. { 'chick ˌyü·nət }

chicory [BOTANY] *Cichorium intybus.* A perennial herb of the order Campanulales grown for its edible green leaves. { 'chik·ə·rē }

chief cell [HISTOLOGY] **1.** A parenchymal, secretory cell of the parathyroid gland. **2.** A cell in the lumen of the gastric fundic glands. { ¦chēf 'sel }

chi element [GENETICS] Any of the special sites in bacterial deoxyribonucleic acid near which enhancement of genetic recombination occurs. { 'kī ˌel·ə·mənt }

chigger [INVERTEBRATE ZOOLOGY] The common name for bloodsucking larval mites of the Trombiculidae which parasitize vertebrates. { 'chig·ər }

chilarium [INVERTEBRATE ZOOLOGY] One of a pair of processes between the bases of the fourth pair of walking legs in the king crab. { kī'lar·ē·əm }

Chilobolbinidae [PALEONTOLOGY] A family of extinct ostracods in the superfamily Hollinacea showing dimorphism of the velar structure. { ˌkī·ləˌbäl'bīn·əˌdē }

Chilopoda [INVERTEBRATE ZOOLOGY] The centipedes, a class of the Myriapoda that is exclusively carnivorous and predatory. { ˌkī'läp·ə·də }

Chimaeridae [VERTEBRATE ZOOLOGY] A family of the order Chimaeriformes. { kī'mir·əˌdē }

Chimaeriformes [VERTEBRATE ZOOLOGY] The single order of the chondrichthyan subclass Holocephali comprising the ratfishes, marine bottom-feeders of the Atlantic and Pacific oceans. { kī·mir·ə'fórˌmēz }

chimera [BIOLOGY] An organism or a part made up of tissues or cells exhibiting chimerism. { kī'mir·ə }

chimeric deoxyribonucleic acid [GENETICS] A recombinant deoxyribonucleic acid (DNA) molecule that contains sequences from more than one organism. [MOLECULAR BIOLOGY] A DNA molecule produced by recombinant DNA technology that consists of a unique combination of unrelated genetic material. { kī'mir·ik ˌdē¦äk·se¦rī·bō¦nü¦klē·ik 'as·əd }

chimerism [BIOLOGY] The admixture of cell populations from more than one zygote. { 'kī·məˌriz·əm }

chimopelagic [ECOLOGY] Pertaining to, belonging to, or being marine organisms living at great depths throughout most of the year; during the winter they move to the surface. { ¦kī·mō·pə'laj·ik }

chimpanzee [VERTEBRATE ZOOLOGY] Either of two species of Primates of the genus *Pan* indigenous to central-west Africa. { ˌchim ˌpan'zē }

chin [ANATOMY] The lower part of the face, at or near the symphysis of the lower jaw. { chin }

China grass [BOTANY] *See* ramie. { 'chī·nə ˌgras }

chinchilla [VERTEBRATE ZOOLOGY] The common name for two species of rodents in the genus *Chinchilla* belonging to the family Chinchillidae. { chin'chil·ə }

Chinchillidae [VERTEBRATE ZOOLOGY] A family of

rodents comprising the chinchillas and viscachas. { chin'chil·ə‚dē }

Chionididae [VERTEBRATE ZOOLOGY] The white sheathbills, a family of birds in the order Charadriiformes. { ‚kī·ə'nid·ə‚dē }

chionophile [ECOLOGY] Having a preference for snow. { ‚kī'än·ə‚fīl }

chipmunk [VERTEBRATE ZOOLOGY] The common name for 18 species of rodents belonging to the tribe Marmotini in the family Sciuridae. { 'chip ‚məŋk }

Chiridotidae [INVERTEBRATE ZOOLOGY] A family of holothurians in the order Apodida having six-spoked, wheel-shaped spicules. { ‚kī·rə'däd·ə ‚dē }

Chirodidae [PALEONTOLOGY] A family of extinct chondrostean fishes in the suborder Platysomoidei. { ‚kī'räd·ə‚dē }

Chirognathidae [PALEONTOLOGY] A family of conodonts in the suborder Neurodontiformes. { ‚kī·rəg'näth·ə‚dē }

Chiroptera [VERTEBRATE ZOOLOGY] The bats, an order of mammals having the front limbs modified as wings. { kī'räp·tə·rə }

chiropterophilous [BIOLOGY] Pollinated by bats. { kī¦räp·tə¦räf·ə·ləs }

chiropterygium [VERTEBRATE ZOOLOGY] A typical vertebrate limb, thought to have evolved from a finlike appendage. { ¦kī·räp·tə¦rij·ē·əm }

Chirotheuthidae [INVERTEBRATE ZOOLOGY] A family of mollusks comprising several deep-sea species of squids. { ‚kī·rō'thyüd·ə‚dē }

chitin [BIOCHEMISTRY] A white or colorless amorphous polysaccharide that forms a base for the hard outer integuments of crustaceans, insects, and other invertebrates. { 'kīt·ən }

chitinase [BIOCHEMISTRY] An externally secreted digestive enzyme produced by certain microorganisms and invertebrates that hydrolyzes chitin. { 'kīt·ən‚ās }

chitinivorous bacterium [MICROBIOLOGY] Any bacterium which secretes chitinase and can digest chitin; organisms extract chitin from lobster exoskeletons, causing an infection called soft-shell disease. { ‚kīt·ən'iv·ə·rəs bak'tir·ē·əm }

Chitinozoa [PALEONTOLOGY] An extinct group of unicellular microfossils of the kingdom Protista. { ¦kīt·ən·ə¦zō·ə }

chiton [INVERTEBRATE ZOOLOGY] The common name for over 600 extant species of mollusks which are members of the class Polyplacophora. { 'kīt·ən }

chlamydeous [BOTANY] **1.** Pertaining to the floral envelope. **2.** Having a perianth. { klə'mid·ē·əs }

Chlamydia [MICROBIOLOGY] The single genus of the family Chlamydiaceae. { klə'mid·ē·ə }

Chlamydiaceae [MICROBIOLOGY] The single family of the order Chlamydiales; characterized by a developmental cycle from a small elementary body to a larger initial body which divides, with daughter cells becoming elementary bodies. { klə‚mid·ē'ās·ē‚ē }

Chlamydiales [MICROBIOLOGY] An order of coccoid rickettsias; gram-negative, obligate, intracellular parasites of vertebrates. { klə‚mid·ē'ā·lēz }

Chlamydobacteriaceae [MICROBIOLOGY] Formerly a family of gram-negative bacteria in the order Chlamydobacteriales possessing trichomes in which false branching may occur. { ¦klam·ə‚dō‚bak·tir·ē'ās·ē‚ē }

Chlamydobacteriales [MICROBIOLOGY] Formerly an order comprising colorless, gram-negative, algae-like bacteria of the class Schizomycetes. { ¦klam·ə‚dō‚bak·tir·ē'ā·lēz }

Chlamydomonadidae [INVERTEBRATE ZOOLOGY] A family of colorless, flagellated protozoans in the order Volvocida considered to be close relatives of protozoans that possess chloroplasts. { ¦klam·ə‚dō·mə'näd·ə‚dē }

Chlamydoselachidae [VERTEBRATE ZOOLOGY] The frilled sharks, a family of rare deep-water forms having a combination of hybodont and modern shark characteristics. { ¦klam·ə‚dō·se 'lak·ə‚dē }

chlamydospore [MYCOLOGY] A thick-walled, unicellar resting spore developed from vegetative hyphae in almost all parasitic fungi. { klə 'mid·ə‚spȯr }

Chlamydozoaceae [MICROBIOLOGY] A family of small, gram-negative, coccoid bacteria in the order Rickettsiales; members are obligate intracytoplasmic parasites, or saprophytes. { ¦klam·ə ‚dō·zō'ās·ē‚ē }

Chloracea [MICROBIOLOGY] The green sulfur bacteria, a family of photosynthetic bacteria in the suborder Rhodobacteriineae. { klȯr'ās·ē·ə }

chloragogen [INVERTEBRATE ZOOLOGY] Of or pertaining to certain specialized cells forming the outer layer of the alimentary tract in earthworms and other annelids. { ¦klȯr·ə¦gō·jən }

chloramphenicol [MICROBIOLOGY] $C_{11}H_{12}O_2N_2Cl_2$ A colorless, crystalline, broad-spectrum antibiotic produced by *Streptomyces venezuelae*; industrial production is by chemical synthesis. Also known as chloromycetin. { ‚klȯr‚am'fen·ə‚kȯl }

Chlorangiaceae [BOTANY] A primitive family of colonial green algae belonging to the Tetrasporales in which the cells are directly attached to each other. { ‚klȯr‚an·jē'ās·ē‚ē }

chloranthy [BOTANY] A reverting of normally colored floral leaves or bracts to green foliage leaves. { 'klȯr‚an·thē }

chlorenchyma [BOTANY] Chlorophyll-containing tissue in parts of higher plants, as in leaves. { klȯr'eŋ·kə·mə }

chloride shift [PHYSIOLOGY] The reversible exchange of chloride and bicarbonate ions between erythrocytes and plasma to effect transport of carbon dioxide and maintain ionic equilibrium during respiration. { 'klȯr‚īd ‚shift }

chlorin [BIOCHEMISTRY] A saturated porphyrin for which one double bond at a single pyrrole ring has been reduced. { 'klȯr·ən }

Chlorobacteriaceae [MICROBIOLOGY] The equivalent name for Chlorobiaceae. { ¦klȯr·ō‚bak·tir·ē'ās·ē‚ē }

Chlorobiaceae [MICROBIOLOGY] A family of bacteria in the suborder Chlorobiineae; cells are nonmotile and contain bacteriochlorophylls *c*, *d*, or *e* in chlorobium vesicles attached to the cytoplasmic membrane. { ‚klȯr·ō‚bī'ās·ē‚ē }

Chlorobiineae [MICROBIOLOGY] The green sulfur bacteria, a suborder of the order Rhodospirillales; contains the families Chlorobiaceae and Chloroflexaceae. { ˌklȯr·ōˌbī'i·əˌnē }

Chlorobium [MICROBIOLOGY] A genus of bacteria in the family Chlorobiaceae; cells are ovoid, rod- or vibrio-shaped, and nonmotile, do not have gas vacuoles, contain bacteriochlorophyll *c* or *d*, and are free-living. { klȯr'ō·bē·əm }

chlorobium chlorophyll [BIOCHEMISTRY] $C_{51}H_{67}O_4N_4Mg$ Either of two spectral forms of chlorophyll occurring as esters of farnesol in certain (*Chlorobium*) photosynthetic bacteria. Also known as bacterioviridin. { klȯr'ō·bē·əm 'klȯr·əˌfil }

Chlorococcales [BOTANY] A large, highly diverse order of unicellular or colonial, mostly freshwater green algae in the class Chlorophyceae. { ˌklȯr·ō·kä'kā·lēz }

chlorocruorin [BIOCHEMISTRY] A green metalloprotein respiratory pigment found in body fluids or tissues of certain sessile marine annelids. { ¦klȯr·ə'krȯr·ən }

Chlorodendrineae [BOTANY] A suborder of colonial green algae in the order Volvocales comprising some genera with individuals capable of detachment and motility. { ˌklȯr·ōˌden'drin·ēˌē }

Chloroflexaceae [MICROBIOLOGY] A family of phototrophic bacteria in the suborder Chlorobiineae; cells possess chlorobium vesicles and bacteriochlorophyll *a* and *c*, are filamentous, and show gliding motility; capable of anaerobic, phototrophic growth or aerobic chemotrophic growth. { ¦klȯr·ō·fleks'as·ēˌē }

Chloroflexus [MICROBIOLOGY] The single genus of the family Chloroflexaceae. { ˌklȯr·ō'flek ˌsēz }

chlorogenic acid [BIOCHEMISTRY] $C_{16}H_{18}O_9$ An important factor in plant metabolism; isolated from green coffee beans; the hemihydrate crystallizes in needlelike crystals from water. { ¦klȯr·ə¦jen·ik 'as·əd }

Chloromonadida [INVERTEBRATE ZOOLOGY] An order of flattened, grass-green or colorless, flagellated protozoans of the class Phytamastigophorea. { ¦klȯr·ō·mə'näd·ə·də }

Chloromonadina [INVERTEBRATE ZOOLOGY] The equivalent name for Chloromonadida. { ¦klȯr·ō·mə'näd·ə·nə }

Chloromonadophyceae [BOTANY] A group of algae considered by some to be a class of the division Chrysophyta. { ¦klȯr·ō·məˌnäd·ə'fīs·ēˌē }

Chloromonadophyta [BOTANY] A division of algae in the plant kingdom considered by some to be a class, Chloromondophyceae. { ¦klȯr·ōˌmä·nə'dä·fə·də }

chloromycetin [MICROBIOLOGY] *See* chloramphenicol. { ˌklȯr·ōˌmī'sēt·ən }

Chlorophyceae [BOTANY] A class of microscopic or macroscopic green algae, division Chlorophyta, composed of fresh- or salt-water, unicellular or multicellular, colonial, filamentous or sheetlike forms. { ˌklȯr·ō'fīs·ēˌē }

chlorophyll [BIOCHEMISTRY] The generic name for any of several oil-soluble green tetrapyrrole plant pigments which function as photoreceptors of light energy for photosynthesis. { 'klȯr·ə ˌfil }

chlorophyll a [BIOCHEMISTRY] $C_{55}H_{72}O_5N_4Mg$ A magnesium chelate of dihydroporphyrin that is esterified with phytol and has a cyclopentanone ring; occurs in all higher plants and algae. { 'klȯr·əˌfil 'ā }

chlorophyllase [BIOCHEMISTRY] An enzyme that splits or hydrolyzes chlorophyll. { 'klȯr·ə·fə ˌlās }

chlorophyll b [BIOCHEMISTRY] $C_{55}H_{70}O_6N_4Mg$ An ester similar to cholorphyll *a* but with a —CHO substituted for a $—CH_3$; occurs in small amounts in all green plants and algae. { 'klȯr·ə ˌfil 'bē }

Chlorophyta [BOTANY] The green algae, a highly diversified plant division characterized by chloroplasts, having chlorophyll *a* and *b* as the predominating pigments. { klō'räf·ə·də }

Chloropidae [INVERTEBRATE ZOOLOGY] The chloropid flies, a family of myodarian cyclorrhaphous dipteran insects in the subsection Acalypteratae. { klō'räp·əˌdē }

chloroplast [BOTANY] A type of cell plastid occurring in the green parts of plants, containing chlorophyll pigments, and functioning in photosynthesis and protein synthesis. { 'klȯr·ə ˌplast }

chloroplast deoxyribonucleic acid [BIOCHEMISTRY] The circular deoxyribonucleic acid duplex, generally 40 to 80 copies, contained within a chloroplast. Abbreviated ctDNA. Also known as chloroplast genome. { ¦klȯr·əˌplast dēˌäk·sēˌrī·bō·nü¦klē·ik 'as·əd }

chloroplast genome [BIOCHEMISTRY] *See* chloroplast deoxyribonucleic acid. { 'klȯr·əˌplast 'jē ˌnōm }

Chloropseudomonas [MICROBIOLOGY] An invalid genus of bacteria; originally described as a member of the family Chlorobiaceae, with motile rod-shaped cells, without gas vacuoles, containing bacteriochlorophyll *c*, and capable of photoheterotrophic growth; cultures now known to be symbiotic of one of a number of typical Chlorobiaceae and a chemoorganotrophic bacterium capable of reducing elemental sulfur to sulfide. { ˌklȯr·ōˌsüd·ə'mō·nəs }

chlortetracycline [MICROBIOLOGY] $C_{22}H_{23}O_8N_2Cl$ Yellow, crystalline, broad-spectrum antibiotic produced by a strain of *Streptomyces aureofaciens*. { ¦klȯrˌte·trə'sīˌklēn }

choana [ANATOMY] A funnel-shaped opening, especially the posterior nares. [INVERTEBRATE ZOOLOGY] A protoplasmic collar surrounding the basal ends of the flagella in certain flagellates and in the choanocytes of sponges. { 'kō·ə·nə }

choanate fish [PALEONTOLOGY] Any of the lobefins composing the subclass Crossopterygii. { 'kō·əˌnāt ˌfish }

Choanichthyes [VERTEBRATE ZOOLOGY] An equivalent name for the Sarcopterygii. { ˌkō·ə 'nik·thēˌēz }

choanocyte [INVERTEBRATE ZOOLOGY] Any of the

choanate, flagellate cells lining the cavities of a sponge. Also known as collar cell. { kō'an·ə ˌsīt }

Choanoflagellida [INVERTEBRATE ZOOLOGY] An order of single-celled or colonial, colorless flagellates in the class Zoomastigophorea; distinguished by a thin protoplasmic collar at the anterior end. { ¦kō·ə·nō·flə'jel·ə·də }

Choanolaimoidea [INVERTEBRATE ZOOLOGY] A superfamily of marine nematodes in the order Chromadoria distinguished by a complex stoma in two parts. { ¦kō·ə·nō·lə'mȯid·ē·ə }

choanosome [INVERTEBRATE ZOOLOGY] The inner layer of a sponge; composed of choanocytes. { kō 'an·əˌsōm }

chocolate tree [BOTANY] *See* cacao. { 'chäk·lət ˌtrē }

Choeropotamidae [PALEONTOLOGY] A family of extinct palaeodont artiodactyls in the superfamily Entelodontoidae. { ˌkir·ə·pə'täm·əˌdē }

cholagogic [PHYSIOLOGY] Inducing the flow of bile. { ¦käl·ə¦gäj·ik }

cholaic acid [BIOCHEMISTRY] *See* taurocholic acid. { kō'lā·ik 'as·əd }

cholane [BIOCHEMISTRY] $C_{24}H_{42}$ A tetracyclic hydrocarbon which may be considered as the parent substance of sterols, hormones, bile acids, and digitalis aglycons. { 'kōˌlān }

cholate [BIOCHEMISTRY] Any salt of cholic acid. { 'kōˌlāt }

cholecystokinin [BIOCHEMISTRY] A hormone produced by the mucosa of the upper intestine which stimulates contraction of the gallbladder. { ˌkō·ləˌsis·tə'kī·nən }

choledochoduodenal junction [ANATOMY] The point where the common bile duct enters the duodenum. { ¦kō·lə¦däk·əˌdü'wäd·ən·əl 'jəŋk·shən }

choleglobin [BIOCHEMISTRY] Combined native protein (globin) and open-ring iron-porphyrin, which is bile pigment hemoglobin; a precursor of biliverdin. { ¦kō·lə¦glō·bən }

cholera vibrio [MICROBIOLOGY] *Vibrio comma*, the bacterium that causes cholera. { 'käl·ə·rə 'vib·rēˌō }

cholesterol [BIOCHEMISTRY] $C_{27}H_{46}O$ A sterol produced by all vertebrate cells, particularly in the liver, skin, and intestine, and found most abundantly in nerve tissue. { kə'les·təˌrȯl }

cholic acid [BIOCHEMISTRY] $C_{24}H_{40}O_5$ An unconjugated, crystalline bile acid. { 'kō·lik 'as·əd }

choline [BIOCHEMISTRY] $C_5H_{15}O_2N$ A basic hygroscopic substance constituting a vitamin of the B complex; used by most animals as a precursor of acetylcholine and a source of methyl groups. { 'kōˌlēn }

cholinergic [PHYSIOLOGY] Liberating, activated by, or resembling the physiologic action of acetylcholine. { ¦kō·lə¦nər·jik }

cholinergic nerve [PHYSIOLOGY] Any nerve, such as autonomic preganglionic nerves and somatic motor nerves, that releases a cholinergic substance at its terminal points. { ¦kō·lə¦nər·jik 'nərv }

cholinesterase [BIOCHEMISTRY] An enzyme found in blood and in various other tissues that catalyzes hydrolysis of choline esters, including acetylcholine. Abbreviated chE. { 'ko·lə'nes·tə ˌrās }

cholytaurine [BIOCHEMISTRY] *See* taurocholic acid. { ˌkäl·ə'tȯˌrēn }

Chondrichthyes [VERTEBRATE ZOOLOGY] A class of vertebrates comprising the cartilaginous, jawed fishes characterized by the absence of true bone. { kän'drik·thēˌēz }

chondrification [PHYSIOLOGY] Formation of or conversion into cartilage. { ˌkän·drə·fə'kā·shən }

chondrin [BIOCHEMISTRY] A horny gelatinous protein substance obtainable from the collagen component of cartilage. { 'kän·drən }

chondrioid [MICROBIOLOGY] A cell organelle in bacteria that is functionally equivalent to the mitochondrion of eukaryotes. { 'kän·drēˌȯid }

chondriome [CELL BIOLOGY] Referring collectively to the chondriosomes (mitochondria) of a cell as a functional unit. { 'kän·drēˌōm }

chondriosome [CELL BIOLOGY] Any of a class of self-perpetuating lipoprotein complexes in the form of grains, rods, or threads in the cytoplasm of most cells; thought to function in cellular metabolism and secretion. { 'kän·drē·əˌsōm }

chondroblast [HISTOLOGY] A cell that produces cartilage. { 'kän·drōˌblast }

Chondrobrachii [VERTEBRATE ZOOLOGY] The equivalent name for Ateleopoidei. { ¦kän·drō 'brā·kēˌī }

chondroclast [HISTOLOGY] A cell that absorbs cartilage. { 'kän·drōˌklast }

chondrocranium [ANATOMY] The part of the adult cranium derived from the cartilaginous cranium. [EMBRYOLOGY] The cartilaginous, embryonic cranium of vertebrates. { ¦kän·drō'krā·nē·əm }

chondrocyte [HISTOLOGY] A cartilage cell. { 'kän·drōˌsīt }

chondrogenesis [EMBRYOLOGY] The development of cartilage. { ¦kän·drō·jə'nē·səs }

chondroitin [BIOCHEMISTRY] A nitrogenous polysaccharide occurring in cartilage in the form of condroitinsulfuric acid. { kän'drō·ə·tən }

chondrology [ANATOMY] The anatomical study of cartilage. { kän'dräl·ə·jē }

chondromucoid [BIOCHEMISTRY] A mucoid found in cartilage; a glycoprotein in which chondroitinsulfuric acid is the prosthetic group. { ¦kän·drō 'myüˌkȯid }

Chondromyces [MICROBIOLOGY] A genus of bacteria in the family Polyangiaceae; sporangia are stalked, and vegetative cells are short rods or spheres. { ˌkän·drō'mīˌsēz }

chondrophone [INVERTEBRATE ZOOLOGY] In bivalve mollusks, a structure or cavity supporting the internal hinge cartilage. { 'kän·drōˌfōn }

Chondrophora [INVERTEBRATE ZOOLOGY] A suborder of polymorphic, colonial, free-floating cnidarians of the class Hydrozoa. { kän'drä·fə·rə }

chondroprotein [BIOCHEMISTRY] A protein (glycoprotein) occurring normally in cartilage. { ¦kän·drō'prōˌtēn }

chondroskeleton [ANATOMY] **1.** The parts of the

bony skeleton which are formed from cartilage. **2.** Cartilaginous parts of a skeleton. [VERTEBRATE ZOOLOGY] A cartilaginous skeleton, as in Chondrostei. { ¦kän·drō'skel·ə·tən }

Chondrostei [PALEONTOLOGY] The most archaic infraclass of the subclass Actinopterygii, or rayfin fishes. { kän'dräs·tē,ī }

Chondrosteidae [PALEONTOLOGY] A family of extinct actinopterygian fishes in the order Acipenseriformes. { ,kän·drə'stē·ə,dē }

Chonetidina [PALEONTOLOGY] A suborder of extinct articulate brachiopods in the order Strophomenida. { ,kän·ə·tə·dī·nə }

Chonotrichida [INVERTEBRATE ZOOLOGY] A small order of vase-shaped ciliates in the subclass Holotrichia; commonly found as ectocommensals on marine crustaceans. { ,kän·ə'trik·ə·də }

chordamesoderm [EMBRYOLOGY] The portion of the mesoderm in the chordate embryo from which the notochord and related structures arise, and which induces formation of ectodermal neural structures. { ,kȯrd·ə'mes·ə,dərm }

Chordata [ZOOLOGY] The highest phylum in the animal kingdom, characterized by a notochord, nerve cord, and gill slits; includes the urochordates, lancelets, and vertebrates. { kȯr'däd·ə }

Chordodidae [INVERTEBRATE ZOOLOGY] A family of worms in the order Gordioidea distinguished by a rough cuticle containing thickenings called areoles. { kȯr'däd·ə,dē }

Chordopoxivirinae [VIROLOGY] A subfamily of vertebrate deoxyribonucleic acid viruses of the Poxviridae family whose members replicate within the cytoplasm. { ¦kȯr·dō,päk·sə'vī·rə ,nē }

chorioallantois [EMBRYOLOGY] A vascular fetal membrane that is formed by the close association or fusion of the chorion and allantois. { ¦kȯr·ē·ō·ə'lant·ə·wəs }

chorion [EMBRYOLOGY] The outermost of the extraembryonic membranes of amniotes, enclosing the embryo and all of its other membranes. { 'kȯr·ē·än }

chorionic gonadotropin [BIOCHEMISTRY] *See* human chorionic gonadotropin. { ,kȯr·ē'än·ik gō ,nad·ə'trō·pən }

chorioretinal [ANATOMY] Pertaining to the choroid and retina of the eye. { ,kȯr·ē·ə'ret·ən·əl }

choripetalous [BOTANY] *See* polypetalous. { ¦kȯr·ə'ped·əl·əs }

chorisepalous [BOTANY] *See* polysepalous. { ¦kȯr·ə'sep·əl·əs }

chorisis [BOTANY] Separation of a leaf or floral part into two or more parts during development. { 'kȯr·ə·səs }

Choristida [INVERTEBRATE ZOOLOGY] An order of sponges in the class Demospongiae in which at least some of the megascleres are tetraxons. { kə 'ris·tə·də }

Choristodera [PALEONTOLOGY] A suborder of extinct reptiles of the order Eosuchia composed of a single genus, *Champsosaurus*. { ,kȯr·ə'städ·ə·rə }

choroid [ANATOMY] The highly vascular layer of the vertebrate eye, lying between the sclera and the retina. { 'kȯr,ȯid }

choroid plexus [ANATOMY] Any of the highly vascular, folded processes that project into the third, fourth, and lateral ventricles of the brain. { ¦kȯr,ȯid 'plek·səs }

chorology [ECOLOGY] The study of how organisms are distributed geographically. { kə'räl·ə·jē }

Christmas factor [BIOCHEMISTRY] A soluble protein blood factor involved in blood coagulation. Also known as factor IX; plasma thromboplastin component (PTC). { 'kris·məs ,fak·tər }

Chromadoria [INVERTEBRATE ZOOLOGY] A subclass of nematode worms in the class Adenophorea. { ,krō·mə'dȯr·ē·ə }

Chromadorida [INVERTEBRATE ZOOLOGY] An order of principally aquatic nematode worms in the subclass Chromadoria. { ,krō·mə'dȯr·ə·də }

Chromadoridae [INVERTEBRATE ZOOLOGY] A family of soil and fresh-water, free-living nematodes in the superfamily Chromadoroidea; generally associated with algal substances. { ,krō·mə 'dȯr·ə,dē }

Chromadoroidea [INVERTEBRATE ZOOLOGY] A superfamily of small to moderate-sized, free-living nematodes with spiral, transversely ellipsoidal amphids and a striated cuticle. { ,krō·mə·də 'rȯid·ē·ə }

chromaffin [BIOLOGY] Staining with chromium salts. { krō'ma·fən }

chromaffin body [ANATOMY] *See* paraganglion. { krō'ma·fən ,bäd·ē }

chromaffin cell [HISTOLOGY] Any cell of the suprarenal organs in lower vertebrates, of the adrenal medulla in mammals, of the paraganglia, or of the carotid bodies that stains with chromium salts. { krō'ma·fən ,sel }

chromaffin system [PHYSIOLOGY] The endocrine organs and tissues of the body that secrete epinephrine; characterized by an affinity for chromium salts. { krō'ma·fən ,sis·təm }

Chromatiaceae [MICROBIOLOGY] A family of bacteria in the suborder Rhodospirillineae; motile cells have polar flagella, photosynthetic membranes are continuous with the cytoplasmic membrane, all except one species are anaerobic, and bacteriochlorophyll *a* or *b* is present. { ,krō· mad·ē'as·ē,ī }

chromatic adaptation [PHYSIOLOGY] A decrease in sensitivity to a color stimulus with prolonged exposure. { krō'mad·ik ,ad,ap'tā·shən }

chromatic valence [PHYSIOLOGY] A relative measure of the hue-producing effectiveness of a chromatic stimulus. { krō'mad·ik 'vāl·əns }

chromatic vision [PHYSIOLOGY] Vision pertaining to the color sense, that is, the perception and evaluation of the colors of the spectrum. { krō 'mad·ik 'vizh·ən }

chromatid [CELL BIOLOGY] **1.** One of the pair of strands formed by longitudinal splitting of a chromosome which are joined by a single centromere in somatic cells during mitosis. **2.** One of a tetrad of strands formed by longitudinal splitting of paired chromosomes during diplotene of meiosis. { 'krō·mə·təd }

chromatin [BIOCHEMISTRY] The deoxyribonucleoprotein complex forming the major portion of the nuclear material and of the chromosomes. { 'krō·mɔ·tən }

Chromatium [MICROBIOLOGY] A genus of bacteria in the family Chromatiaceae; cells are ovoid to rod-shaped, are motile, do not have gas vacuoles, and contain bacteriochlorophyll *a* on vesicular photosynthetic membranes. { krō'māsh·ē·əm }

chromatophore [HISTOLOGY] A type of pigment cell found in the integument and certain deeper tissues of lower animals that contains color granules capable of being dispersed and concentrated. { krō'mad·ə‚fȯr }

chromatophorotrophin [BIOCHEMISTRY] Any crustacean neurohormone which controls the movement of pigment granules within chromatophores. { krō¦mad·ə¦fȯr·ə'trō·fən }

chromatoplasm [BOTANY] The peripheral protoplasm in blue-green algae containing chlorophyll, accessory pigments, and stored materials. { krō'mad·ə‚plaz·əm }

Chromobacterium [MICROBIOLOGY] A genus of gram-negative, aerobic or facultatively anaerobic, motile, rod-shaped bacteria of uncertain affiliation; they produce violet colonies and violacein, a violet pigment with antibiotic properties. { ¦krō·mō‚bak'tir·ē·əm }

chromocenter [CELL BIOLOGY] An irregular, densely staining mass of heterochromatin in the chromosomes, with six armlike extensions of euchromatin, in the salivary glands of *Drosophila*. { 'krō·mō‚sen·tər }

chromocyte [HISTOLOGY] A pigmented cell. { 'krō·mə‚sīt }

chromogen [BIOCHEMISTRY] A pigment precursor. [MICROBIOLOGY] A microorganism capable of producing color under suitable conditions. { 'krō·mə‚jen }

chromogenesis [BIOCHEMISTRY] Production of colored substances as a result of metabolic activity; characteristic of certain bacteria and fungi. { ¦krō·mō'jen·ə·səs }

chromolipid [BIOCHEMISTRY] *See* lipochrome. { ¦krō·mō'lip·id }

chromomere [CELL BIOLOGY] Any of the linearly arranged chromatin granules in leptotene and pachytene chromosomes and in polytene nuclei. { 'krō·mō‚mir }

chromomycin [MICROBIOLOGY] Any of five components of an antibiotic complex produced by a strain of *Streptomyces griseus*; components are designated A_1 to A_5, of which A_3 ($C_{51}H_{72}O_{32}$) is biologically active. { ¦krō·mō'mī·sən }

chromonema [CELL BIOLOGY] The coiled core of a chromatid; it is thought to contain the genes. { ‚krō·mō'nē·mə }

chromoneme [GENETICS] The genetic material of a bacterium or virus, as distinguished from true chromosomes in plant or animal cells. { 'krō·mə‚nēm }

chromophile [BIOLOGY] Staining readily. { 'krō·mō‚fīl }

chromophobe [BIOLOGY] Not readily absorbing a stain. { 'krō·mə‚fōb }

Chromophycota [BOTANY] A division of the plant kingdom comprising nine classes of algae ranging in size and complexity from unicellular flagellates to gigantic kelps, distinguished by the presence (in almost all) of chlorophyll *c* to complement chlorophyll *a*, and usually having brownish or yellowish chloroplasts. Also known as Chromophyta. { ¦krō·mō'fī·kəd·ə }

chromophyll [BIOCHEMISTRY] Any plant pigment. { 'krō·mə‚fil }

Chromophyta [BOTANY] *See* Chromophycota. { krō'mäf·əd·ə }

chromoplasm [BOTANY] The pigmented, peripheral protoplasm of blue-green algae cells; contains chlorophyll, carotenoids, and phycobilins. { 'krō·mō‚plaz·əm }

chromoplast [CELL BIOLOGY] Any colored cell plastid, excluding chloroplasts. { 'krō·mō‚plast }

chromoprotein [BIOCHEMISTRY] Any protein, such as hemoglobin, with a metal-containing pigment. { ¦krō·mō'prō‚tēn }

chromosomal hybrid sterility [GENETICS] Sterility caused by inability of homologous chromosomes to pair during meiosis due to a chromosome aberration. { ‚krō·mə'sō·məl 'hī·brəd stə'ril·əd·ē }

chromosomal mosaic [CELL BIOLOGY] An individual showing at least two cell lines with different karyotypes. { ¦krō·mə‚sō·məl mō'zā·ik }

chromosome [CELL BIOLOGY] Any of the complex, threadlike structures seen in animal and plant nuclei during karyokinesis which carry the linearly arranged genetic units. { 'krō·mə‚sōm }

chromosome aberration [GENETICS] Modification of the normal chromosome complement due to deletion, duplication, or rearrangement of genetic material. { 'krō·mə‚sōm ab·ə'rā·shən }

chromosome arm [CELL BIOLOGY] One of the two main segments of the chromosome that are separated by the centromere. { 'krō·mə‚sōm 'ärm }

chromosome complement [GENETICS] The species-specific, normal diploid set of chromosomes in somatic cells. { 'krō·mə‚sōm 'käm·plə‚mənt }

chromosome condensation [CELL BIOLOGY] The process whereby chromosomes become shorter and thicker during prophase as a consequence of coiling and supercoiling of chromatic strands. { ¦krō·mə‚sōm ‚kän·dən'sā·shən }

chromosome congression [CELL BIOLOGY] *See* congression. { ¦krō·mə‚sōm kəŋ'gresh·ən }

chromosome diminution [EMBRYOLOGY] During embryogenesis, the elimination of certain chromosomes from cells that form somatic tissues. Also known as chromosome elimination. { ¦krō·mə‚sōm ‚dim·ə'nyü·shən }

chromosome elimination [EMBRYOLOGY] *See* chromosome diminution. { ¦krō·mə‚sōm i‚lim·ə'nā·shən }

chromosome loss [CELL BIOLOGY] Failure of a chromosome to become incorporated into a daughter nucleus at cell division. { ¦krō·mə‚sōm 'lȯs }

chromosome map [GENETICS] *See* genetic map. { 'krō·mə‚sōm ‚map }

chromosome puff [CELL BIOLOGY] Chromatic material accumulating at a restricted site on a chromosome; thought to reflect functional activity of the gene at that site during differentiation. { 'krō·mə,sōm ,pəf }

chromosome transfer [GENETICS] The transfer of isolated metaphase chromosomes into cultured mammalian cells. { ¦krō·mə,sōm ,tranz·fər }

chromosome walking [GENETICS] Sequential isolation of overlapping molecular clones in order to span large intervals on the chromosome. { 'krō·mə,sōm ,wȯk·iŋ }

chronaxie [PHYSIOLOGY] The time interval required to excite a tissue by an electric current of twice the galvanic threshold. { 'krä,nak·sē }

chronocline [PALEONTOLOGY] A cline shown by successive morphological changes in the members of a related group, such as a species, in successive fossiliferous strata. { 'krän·ō,klīn }

Chroococcales [BOTANY] An order of blue-green algae (Cyanophyceae) that reproduce by cell division and colony fragmentation only. { ,krō·ō·kə'kā·lēz }

Chryomyidae [INVERTEBRATE ZOOLOGY] A family of myodarian cyclorrhaphous dipteran insects in the subsection Acalypteratae. { ,krī·ō'mī·ə,dē }

Chrysididae [INVERTEBRATE ZOOLOGY] The cuckoo wasps, a family of hymenopteran insects in the superfamily Bethyloidea having brilliant metallic blue and green bodies. { krə'sid·ə,dē }

chrysocarpous [BOTANY] Bearing yellow fruits. { ¦kris·ə¦kär·pəs }

Chrysochloridae [PALEONTOLOGY] The golden moles, a family of extinct lipotyphlan mammals in the order Insectivora. { ¦kris·ə'klȯr·ə,dē }

Chrysomelidae [INVERTEBRATE ZOOLOGY] The leaf beetles, a family of coleopteran insects in the superfamily Chrysomeloidea. { ,kris·ə'mel·ə ,dē }

Chrysomeloidea [INVERTEBRATE ZOOLOGY] A superfamily of coleopteran insects in the suborder Polyphaga. { ¦kris·ə·mə'lȯid·ē·ə }

Chrysomonadida [INVERTEBRATE ZOOLOGY] An order of yellow to brown, flagellated colonial protozoans of the class Phytamastigophorea. { ¦kris·ə·mə'näd·ə·də }

Chrysomonadina [INVERTEBRATE ZOOLOGY] The equivalent name for the Chrysomonadida. { ¦kris·ə,män·ə'dī·nə }

Chrysopetalidae [INVERTEBRATE ZOOLOGY] A small family of scale-bearing polychaete worms belonging to the Errantia. { ,kris·ō·pə'tal·ə ,dē }

Chrysophyceae [BOTANY] Golden-brown algae making up a class of fresh- and salt-water unicellular forms in the division Chrysophyta. { ,kris·ō'fīs·ē,ē }

Chrysophyta [BOTANY] The golden-brown algae, a division of plants with a predominance of carotene and xanthophyll pigments in addition to chlorophyll. { krə'säf·ə·də }

Chthamalidae [INVERTEBRATE ZOOLOGY] A small family of barnacles in the suborder Thoracica. { thə'mal·ə,dē }

chyle [PHYSIOLOGY] Lymph containing emulsified fat, present in the lacteals of the intestine during digestion of ingested fats. { kīl }

chylomicron [BIOCHEMISTRY] One of the extremely small lipid droplets, consisting chiefly of triglycerides, found in blood after ingestion of fat. { ,kīl·ə'mī,krän }

chylophyllous [BOTANY] Having succulent or fleshy leaves. { ¦kīl·ō¦fil·əs }

chyme [PHYSIOLOGY] The semifluid, partially digested food mass that is expelled into the duodenum by the stomach. { kīm }

chymopapain [BIOCHEMISTRY] Any one of several proteolytic enzymes obtained from papaya. { ,kī·mō·pə'pī·ən }

chymosin [BIOCHEMISTRY] *See* rennin. { 'kī·mə·sən }

chymotrypsin [BIOCHEMISTRY] A proteinase in the pancreatic juice that clots milk and hydrolyzes casein and gelatin. { ,kī·mə'trip·sən }

chymotrypsinogen [BIOCHEMISTRY] An inactive proteolytic enzyme of pancreatic juice; converted to the active form, chymotrypsin, by trypsin. { ¦kī·mō,trip'sin·ə·jən }

Chytridiales [MYCOLOGY] An order of mainly aquatic fungi of the class Phycomycetes having a saclike to rhizoidal thallus and zoospores with a single posterior flagellum. { kī,trid·ē'ā·lēz }

Chytridiomycetes [MYCOLOGY] A class of true fungi. { kī,trid·ē·ō,mī'sēd·ēz }

cibarium [INVERTEBRATE ZOOLOGY] In insects, the space anterior to the mouth cavity in which food is chewed. { sə'bar·ē·əm }

Cicadellidae [INVERTEBRATE ZOOLOGY] Large family of homopteran insects belonging to the series Auchenorrhyncha; includes leaf hoppers. { ,sik·ə'del·ə,dē }

Cicadidae [INVERTEBRATE ZOOLOGY] A family of large homopteran insects belonging to the series Auchenorrhyncha; includes the cicadas. { sə 'kad·ə,dē }

cicatrix [BIOLOGY] A scarlike mark, usually caused by previous attachment of a part or organ. { 'sik·ə,triks }

Cichlidae [VERTEBRATE ZOOLOGY] The cichlids, a family of perciform fishes in the suborder Percoidei. { 'sik·lə,dē }

Cicindelidae [INVERTEBRATE ZOOLOGY] The tiger beetles, a family of coleopteran insects in the suborder Adephaga. { ,si·sən'del·ə,dē }

Ciconiidae [VERTEBRATE ZOOLOGY] The tree storks, a family of wading birds in the order Ciconiiformes. { ,si·kə'nī·ə,dē }

Ciconiiformes [VERTEBRATE ZOOLOGY] An order of predominantly long-legged, long-necked birds, including herons, storks, ibises, spoonbills, and their relatives. { sə,kōn·ē·ə'fȯr,mēz }

Cidaroida [INVERTEBRATE ZOOLOGY] An order of echinoderms in the subclass Perischoechinoidea in which the ambulacra comprise two columns of simple plates. { ,sid·ə'rȯi·də }

ciguatoxin [BIOCHEMISTRY] A toxin produced by the benthic dinoflagellate *Gambierdiscus toxicus*. { ¦sēg·wə¦täk·sən }

Ciidae [INVERTEBRATE ZOOLOGY] The minute, tree-fungus beetles, a family of coleopteran insects in the superfamily Cucujoidea. { 'sī·ə,dē }

cilia [ANATOMY] Eyelashes. [CELL BIOLOGY] Relatively short, centriole-based, hairlike processes on certain anatomical cells and motile organisms. { 'sil·ē·ə }

ciliary body [ANATOMY] A ring of tissue lying just anterior to the retinal margin of the eye. { ¦sil·ē ,er·ē ¦bäd·ē }

ciliary movement [BIOLOGY] A type of cellular locomotion accomplished by the rhythmical beat of cilia. { ¦sil·ē,er·ē 'müv·mənt }

ciliary muscle [ANATOMY] The smooth muscle of the ciliary body. { ¦sil·ē,er·ē 'məs·əl }

ciliary process [ANATOMY] Circularly arranged choroid folds continuous with the iris in front. { ¦sil·ē,er·ē 'präs·əs }

Ciliatea [INVERTEBRATE ZOOLOGY] The single class of the protozoan subphylum Ciliophora. { ,sil·ē'ad·ē·ə }

ciliated epithelium [HISTOLOGY] Epithelium composed of cells bearing cilia on their free surfaces. { 'sil·ē,ād·əd ep·ə'thēl·ē·əm }

ciliolate [BIOLOGY] Ciliated to a very minute degree. { 'sil·ē·ə,lāt }

Ciliophora [INVERTEBRATE ZOOLOGY] The ciliated protozoans, a homogeneous subphylum of the Protozoa distinguished principally by a mouth, ciliation, and infraciliature. { ,sil·ē'äf·ə·rə }

Cimbicidae [INVERTEBRATE ZOOLOGY] The cimbicid sawflies, a family of hymenopteran insects in the superfamily Tenthredinoidea. { ,sim'bis·ə ,dē }

Cimex [INVERTEBRATE ZOOLOGY] The type genus of Cimicidae, including bedbugs and related forms. { 'sī,meks }

Cimicidae [INVERTEBRATE ZOOLOGY] The bat, bed, and bird bugs, a family of flattened, wingless, parasitic hemipteran insects in the superfamily Cimicimorpha. { sī'mis·ə,dē }

Cimicimorpha [INVERTEBRATE ZOOLOGY] A superfamily, or group according to some authorities, of hemipteran insects in the subdivision Geocorisae. { ,sī·mə·sə'mȯr·fə }

Cimicoidea [INVERTEBRATE ZOOLOGY] A superfamily of the Cimicimorpha in some systems of classification. { ,sī·mə'kȯid·ē·ə }

cinchona [BOTANY] The dried, alkaloid-containing bark of trees of the genus *Cinchona*. { siŋ'kō·nə }

Cinclidae [VERTEBRATE ZOOLOGY] The dippers, a family of insect-eating songbirds in the order Passeriformes. { 'siŋ·klə,dē }

cinclides [INVERTEBRATE ZOOLOGY] Pores in the body wall of some sea anemones for the release of water and stinging cells. { siŋ'klī,dēz }

cinclis [INVERTEBRATE ZOOLOGY] Singular of cinclides. { 'siŋ·kləs }

cinereous [BIOLOGY] **1.** Ashen in color. **2.** Having the inert and powdery quality of ashes. { sə'nir·ē·əs }

Cingulata [VERTEBRATE ZOOLOGY] A group of xenarthran mammalsin the order Edentata, including the armadillos. { ,siŋ·gyə'läd·ə }

cingulate [BIOLOGY] Having a girdle of bands or markings. { 'siŋ·gyə·lət }

cingulum [ANATOMY] **1.** The ridge around the base of the crown of a tooth. **2.** The tract of association nerve fibers in the brain, connecting the callosal and the hippocampal convolutions. [BOTANY] The part of a plant between stem and root. [INVERTEBRATE ZOOLOGY] **1.** Any girdlelike structure. **2.** A band of color or a raised line on certain bivalve shells. **3.** The outer zone of cilia on discs of certain rotifers. **4.** The clitellum in annelids. { 'sin·gyə·ləm }

cinnamon [BOTANY] *Cinnamomum zeylanicum*. An evergreen shrub of the laurel family (Lauraceae) in the order Magnoliales; a spice is made from the bark. { 'sin·ə·mən }

circadian rhythm [PHYSIOLOGY] A rhythmic process within an organism occurring independently of external synchronizing signals. { sər'kād·ē·ən 'rith·əm }

circinate [BIOLOGY] Having the form of a flat coil with the apex at the center. { 'sərs·ən,āt }

circinate vernation [BOTANY] Uncoiling of new leaves from the base toward the apex, as in ferns. { 'sərs·ən,āt vər'nā·shən }

circle of Willis [ANATOMY] A ring of arteries at the base of the cerebrum. { 'sər·kəl əv 'wil·əs }

circular deoxyribonucleic acid [BIOCHEMISTRY] A single- or double-stranded ring of deoxyribonucleic acid found in certain bacteriophages and in human wart virus. Also known as ring deoxyribonucleic acid. { 'sər·kyə·lər dē¦äk·sē¦rī,bō ¦nü¦klē·ik 'as·əd }

circulation [PHYSIOLOGY] The movement of blood through defined channels and tissue spaces; movement is through a closed circuit in vertebrates and certain invertebrates. { ,sər·kyə'lā·shən }

circulatory system [ANATOMY] The vessels and organs composing the lymphatic and cardiovascular systems. { 'sər·kyə·lə,tȯr·ē ,sis·təm }

circulin [MICROBIOLOGY] Any of a group of peptide antibiotics produced by *Bacillus circulans* which are related to polymixin and are active against both gram-negative and gram-positive bacteria. { 'sər·kyə·lən }

circulus [BIOLOGY] Any of various ringlike structures, such as the vascular circle of Willis or the concentric ridges on fish scales. { 'sər·kyə·ləs }

circumboreal distribution [ECOLOGY] The distribution of a Northern Hemisphere organism whose habitat includes North American, European, and Asian stations. { ¦sər·kəm'bȯr·ē·əl ,dis·trə'byü·shən }

circumduction [ANATOMY] Movement of the distal end of a body part in the form of an arc; performed at ball-and-socket and saddle joints. { ,sər·kəm'dək·shən }

circumflex artery [ANATOMY] Any artery that follows a curving or winding course. { 'sər·kəm ,fleks 'ärd·ə·rē }

circumnutation [BOTANY] The bending or turning of a growing stem tip that occurs as a result of unequal rates of growth along the stem. { ¦sər·kəm·nü'tā·shən }

circumpharyngeal connective [INVERTEBRATE ZOOLOGY] One of a pair of nerve strands passing around the esophagus in annelids and anthropods, connecting the brain and subesophageal ganglia. { ¦sər·kəm·fə'rin·jē·əl kə'nek·tiv }

circumscissile [BOTANY] Dehiscing along the

line of a circumference, as exhibited by a pyxidium. { ¦sər·kəm'sis·əl }

circumvallate papilla [ANATOMY] *See* vallate papilla. { ¦sər·kəm¦va,lāt pa'pil·ə }

Cirolanidae [INVERTEBRATE ZOOLOGY] A family of isopod crustaceans in the suborder Flabellifera composed of actively swimming predators and scavengers with biting mouthparts. { ,sir·ə'lan·ə,dē }

Cirratulidae [INVERTEBRATE ZOOLOGY] A family of fringe worms belonging to the Sedentaria which are important detritus feeders in coastal waters. { ,sir·ə'tül·ə,dē }

cirriform [ZOOLOGY] Having the form of a cirrus; generally applied to a prolonged, slender process. { 'sir·ə,fórm }

Cirripedia [INVERTEBRATE ZOOLOGY] A subclass of the Crustacea, including the barnacles and goose barnacles; individuals are free-swimming in the larval stages but permanently fixed in the adult stage. { ,sir·ə'pēd·ē·ə }

Cirromorpha [INVERTEBRATE ZOOLOGY] A suborder of cephalopod mollusks in the order Octopoda. { ¦sir·ō¦mór·fə }

cirrus [INVERTEBRATE ZOOLOGY] **1.** The conical locomotor structure composed of fused cilia in hypotrich protozoans. **2.** Any of the jointed thoracic appendages of barnacles. **3.** Any hairlike tuft on insect appendages. **4.** The male copulatory organ in some mollusks and trematodes. [VERTEBRATE ZOOLOGY] Any of the tactile barbels of certain fishes. [ZOOLOGY] A tendrillike animal appendage. { 'sir·əs }

cirrus sac [INVERTEBRATE ZOOLOGY] A pouch or channel containing the copulatory organ (cirrus) in certain invertebrates. { 'sir·əs ,sak }

cistern [ANATOMY] A closed, fluid-filled sac or vesicle, such as the subarachnoid spaces or the vesicles comprising the dictyosomes of a Golgi apparatus. { 'sis·tərn }

cistron [MOLECULAR BIOLOGY] The genetic unit (deoxyribonucleic acid fragment) that codes for a particular polypeptide; mutants do not complement each other within a cistron. Also known as structural gene. { 'sis,trän }

Citheroniinae [INVERTEBRATE ZOOLOGY] Subfamily of lepidopteran insects in the family Saturniidae, including the regal moth and the imperial moth. { ,sith·ə·rō'nī·ə,nē }

citramalase [BIOCHEMISTRY] An enzyme that is involved in the fermentation of glutamate by *Clostridium tetanomorphum*; catalyzes the breakdown of citramalic acid to acetate and pyruvate. { ,si·trə'ma,lās }

citrate [BIOCHEMISTRY] A salt or ester of citric acid. { 'si,trāt }

citrate test [MICROBIOLOGY] A differential cultural test to identify genera within the bacterial family Enterobacteriaceae that are able to utilize sodium citrate as a sole source of carbon. { 'si ,trāt ,test }

citric acid [BIOCHEMISTRY] $C_6H_8O_7 \cdot H_2O$ A colorless crystalline or white powdery organic, tricarboxylic acid occurring in plants, especially citrus fruits, and used as a flavoring agent, as an antioxidant in foods, and as a sequestering agent; the commercially produced form melts at 153°C. { 'si,trik 'as·əd }

citric acid cycle [BIOCHEMISTRY] *See* Krebs cycle. { 'si,trik 'as·əd 'sī·kəl }

citriculture [BOTANY] The cultivation of citrus fruits. { 'si·trə,kəl·chər }

Citrobacter [MICROBIOLOGY] A genus of bacteria in the family Enterobacteriaceae; motile rods that utilize citrate as the only carbon source. { ,si·trō'bak·tər }

citron [BOTANY] *Citrus medica.* A shrubby, evergreen citrus tree in the order Sapindales cultivated for its edible, large, lemonlike fruit. { 'si·trən }

citronella [BOTANY] *Cymbopogon nardus.* A tropical grass; the source of citronella oil. { ,si·trə'nel·ə }

citrulline [BIOCHEMISTRY] $C_6H_{13}O_3N_3$ An amino acid formed in the synthesis of arginine from ornithine. { 'si·trə,lēn }

citrus flavonoid compound [BIOCHEMISTRY] *See* bioflavonoid. { 'si·trəs 'flav·ə,nóid ,käm ,paünd }

citrus fruit [BOTANY] Any of the edible fruits having a pulpy endocarp and a firm exocarp that are produced by plants of the genus *Citrus* and related genera. { 'si·trəs ,früt }

civet [PHYSIOLOGY] A fatty substance secreted by the civet gland; used as a fixative in perfumes. [VERTEBRATE ZOOLOGY] Any of 18 species of catlike, nocturnal carnivores assigned to the family Viverridae, having a long head, pointed muzzle, and short limbs with nonretractile claws. { 'siv·ət }

civet cat [VERTEBRATE ZOOLOGY] *See* cacomistle. { 'siv·ət ,kat }

civet gland [VERTEBRATE ZOOLOGY] A large anal scent gland in civet cats that secretes civet. { 'siv·ət ,gland }

civetone [BIOCHEMISTRY] $C_{17}H_{30}O$ 9-Cycloheptadecen-1-one, a macrocyclic ketone component of civet used in perfumes because of its pleasant odor and lasting quality; believed to function as a sex attractant among civet cats. { 'siv·ə ,tōn }

clade [SYSTEMATICS] A taxonomic group containing a common ancestor and its descendants. { klād }

cladism [SYSTEMATICS] A theory of taxonomy by which organisms are grouped and ranked on the basis of the most recent phylogenetic branching point. { 'kla,diz·əm }

Cladistia [VERTEBRATE ZOOLOGY] The equivalent name for Polypteriformes. { kla'dis·tē·ə }

Cladocera [INVERTEBRATE ZOOLOGY] An order of small, fresh-water branchiopod crustaceans, commonly known as water fleas, characterized by a transparent bivalve shell. { kla'däs·ə·rə }

Cladocopa [INVERTEBRATE ZOOLOGY] A suborder of the order Myodocopida including marine animals having a carapace that lacks a permanent aperture when the two valves are closed. { kla 'däk·ə·pə }

Cladocopina [INVERTEBRATE ZOOLOGY] The equivalent name for Cladocopa. { ¦klad·ə¦käp·ə·nə }

cladode [BOTANY] *See* cladophyll. { 'kla,dōd }

cladodont [PALEONTOLOGY] Pertaining to sharks of the most primitive evolutionary level. { 'klad·ə,dänt }

cladogenesis [EVOLUTION] Evolution associated with altered habit and habitat, usually in isolated species populations. { ,klad·ə'jen·ə·səs }

cladogenic adaptation [EVOLUTION] *See* divergent adaptation. { ¦klad·ə¦jen·ik ,ad,ap'tā·shən }

cladogram [EVOLUTION] A dendritic diagram which shows the evolution and descent of a group of organisms. { 'klad·ə,gram }

Cladoniaceae [BOTANY] A family of lichens in the order Lecanorales, including the reindeer mosses and cup lichens, in which the main thallus is hollow. { kla¦dō·nē'as·ē,ē }

Cladophorales [BOTANY] An order of coarse, wiry, filamentous, branched and unbranched algae in the class Chlorophyceae. { kla,däf·ə'rā·lēz }

cladophyll [BOTANY] A branch arising from the axil of a true leaf and resembling a foliage leaf. Also known as cladode. { 'klad·ə,fil }

cladoptosis [BOTANY] The annual abscission of twigs or branches instead of leaves. { ,kla,däp'tō·səs }

Cladoselachii [PALEONTOLOGY] An order of extinct elasmobranch fishes including the oldest and most primitive of sharks. { ,klad·ō·sə'lāk·ē,ī }

cladus [BOTANY] A branch of a ramose spicule. { 'klā·dəs }

clam [INVERTEBRATE ZOOLOGY] The common name for a number of species of bivalve mollusks, many of which are important as food. { klam }

Clambidae [INVERTEBRATE ZOOLOGY] The minute beetles, a family of coleopteran insects in the superfamily Dascilloidea. { 'klam·bə,dē }

clammy [BIOLOGY] Moist and sticky, as the skin or a stem. { 'klam·ē }

clamp connection [MYCOLOGY] In the Basidiomycetes, a lateral connection formed between two adjoining cells of a filament and arching over the septum between them and permitting a type of pseudosexual activity. { 'klamp kə,nek·shən }

clam worm [INVERTEBRATE ZOOLOGY] The common name for a number of species of dorsoventrally flattened annelid worms composing the large family Nereidae in the class Polychaeta; all have a distinct head, with numerous appendages. { 'klam ,wərm }

clan [ECOLOGY] A very small community, perhaps a few square yards in area, in climax formation, and dominated by one species. { klan }

Clariidae [VERTEBRATE ZOOLOGY] A family of Asian and African catfishes in the suborder Siluroidei. { kla'rī·ə,dē }

Clarkecarididae [PALEONTOLOGY] A family of extinct crustaceans in the order Anaspidacea. { ,klär·kə'rid·ə,dē }

clasper [VERTEBRATE ZOOLOGY] A modified pelvic fin of male elasmobranchs and holocephalians used for the transmission of sperm. { 'klasp·ər }

class [SYSTEMATICS] A taxonomic category ranking above the order and below the phylum or division. { klas }

classification [SYSTEMATICS] A systematic arrangement of plants and animals into categories based on a definite plan, considering evolutionary, physiologic, cytogenetic, and other relationships. { ,klas·ə·fə'kā·shən }

class switch [IMMUNOLOGY] A switch of B-lymphocyte expression from one antibody class to another. { 'klas ,swich }

clathrate [BIOLOGY] *See* cancellate. { 'kla,thrāt }

Clathrinida [INVERTEBRATE ZOOLOGY] A monofamilial order of sponges in the subclass Calcinea having an asconoid structure and lacking a true dermal membrane or cortex. { kla'thrin·ə·də }

Clathrinidae [INVERTEBRATE ZOOLOGY] The single family of the order Clathrinida. { kla'thrin·ə,dē }

Clathrochloris [MICROBIOLOGY] A genus of bacteria in the family Chlorobiaceae; cells are spherical to ovoid and arranged in chains united in trellis-like aggregates, are nonmotile, contain gas vacuoles, and are free-living. { ¦klath·rō¦klȯr·əs }

claustrum [ANATOMY] A thin layer of gray matter in each cerebral hemisphere between the lenticular nucleus and the island of Reil. { 'klȯ,strəm }

clava [BIOLOGY] A club-shaped structure, as the tip on the antennae of certain insects or the fruiting body of certain fungi. { 'klā·va }

clavate [BIOLOGY] Club-shaped. Also known as claviform. { 'klā,vāt }

Clavatoraceae [PALEOBOTANY] A group of middle Mesozoic algae belonging to the Charophyta. { ,klav·əd·ə'rās·ē,ē }

Clavaxinellida [INVERTEBRATE ZOOLOGY] An order of sponges in the class Demospongiae; members have monaxonid megascleres arranged in radial or plumose tracts. { kla¦vak·sə'nel·ə·də }

clavicle [ANATOMY] A bone in the pectoral girdle of vertebrates with articulation occurring at the sternum and scapula. { 'klav·ə·kal }

claviculate [ANATOMY] Having a clavicle. { kla'vik·yə·lət }

claviform [BIOLOGY] *See* clavate. { 'klav·ə,fȯrm }

clavus [INVERTEBRATE ZOOLOGY] Any of several rounded or fingerlike processes, such as the club of an insect antenna or the pointed anal portion of the hemelytron in hemipteran insects. { 'klāv·əs }

claw [ANATOMY] A sharp, slender, curved nail on the toe of an animal, such as a bird. [INVERTEBRATE ZOOLOGY] A sharp-curved process on the tip of the limb of an insect. { klȯ }

cleavage [EMBRYOLOGY] The subdivision of activated eggs into blastomeres. { 'klēv·ij }

cleavage nucleus [EMBRYOLOGY] The nucleus of

a zygote formed by fusion of male and female pronuclei. { 'klēv·ij ˌnü·klē·əs }

cleft grafting [BOTANY] A top-grafting method in which the scion is inserted into a cleft cut into the top of the stock. { 'kleft ˌgraft·iŋ }

cleistocarp [MYCOLOGY] *See* cleistothecium. { 'klī·stəˌkärp }

cleistocarpous [BOTANY] Of mosses, having the capsule opening irregularly without an operculum. [MYCOLOGY] Forming or having cleistothecia. { ¦klī·stə¦kär·pəs }

cleistogamy [BOTANY] The production of small closed flowers that are self-pollinating and contain numerous seeds. { ˌklī'stäg·ə·mē }

cleistothecium [MYCOLOGY] A closed sporebearing structure in Ascomycetes; asci and spores are freed of the fruiting body by decay or desiccation. Also known as cleistocarp. { ¦klī·stə'thē·sē·əm }

cleithrum [VERTEBRATE ZOOLOGY] A bone external and adjacent to the clavicle in certain fishes, stegocephalians, and primitive reptiles. { 'klī·thrəm }

Cleridae [INVERTEBRATE ZOOLOGY] The checkered beetles, a family of coleopteran insects in the superfamily Cleroidea. { 'kler·əˌdē }

Cleroidea [INVERTEBRATE ZOOLOGY] A superfamily of coleopteran insects in the suborder Polyphaga. { klə'ròid·ē·ə }

climacteric [PHYSIOLOGY] *See* menopause. { klī 'mak·tə·rik }

climatic climax [ECOLOGY] A climax community viewed, by some authorities, as controlled by climate. { klī'mad·ik 'klīˌmaks }

Climatiidae [PALEONTOLOGY] A family of archaic tooth-bearing fishes in the suborder Climatioidei. { ˌklī·mə'tī·əˌdē }

Climatiiformes [PALEONTOLOGY] An order of extinct fishes in the class Acanthodii having two dorsal fins and large plates on the head and ventral shoulder. { ˌklī·məˌtī·ə'fòrˌmēz }

Climatioidei [PALEONTOLOGY] A suborder of extinct fishes in the order Climatiiformes. { ˌklī·mə ˌtī'òid·ēˌī }

climatophysiology [PHYSIOLOGY] The study of the interaction of the natural environment with physiologic factors. { ¦klī·mə·tōˌfiz·ē'äl·ə·jē }

climax [ECOLOGY] A mature, relatively stable community in an area, which community will undergo no further change under the prevailing climate; represents the culmination of ecological succession. { 'klīˌmaks }

climax plant formation [ECOLOGY] A mature, stable plant population in a climax community. { 'klīˌmaks 'plant fòr'mā·shən }

climbing bog [ECOLOGY] An elevated boggy area on a swamp margin, usually occurring where there is a short summer and considerable rainfall. { 'klīm·iŋ 'bäg }

climbing stem [BOTANY] A long, slender stem that climbs up a support or along the tops of other plants by using spines, adventitious roots, or tendrils for attachment. { 'klīm·iŋ 'stem }

cline [BIOLOGY] A graded series of morphological or physiological characters exhibited by a natural group (as a species) of related organisms, generally along a line of environmental or geographic transition. { klīn }

clinical genetics [GENETICS] The study of biological inheritance by direct observation of the living patient. { 'klin·ə·kəl jə'ned·iks }

Clionidae [INVERTEBRATE ZOOLOGY] The boring sponges, a family of marine sponges in the class Demospongiae. { ˌklī'än·əˌdē }

clisere [ECOLOGY] The succession of ecological communities, especially climax formations, as a consequence of intense climatic changes. { klī ˌsir }

clitellum [INVERTEBRATE ZOOLOGY] The thickened, glandular, saddlelike portion of the body wall of some annelid worms. { klə'tel·əm }

clitoris [ANATOMY] The homolog of the penis in females, located in the anterior portion of the vulva. { 'klid·ə·rəs }

cloaca [INVERTEBRATE ZOOLOGY] The chamber which functions as a respiratory, excretory, and reproductive duct in certain invertebrates. [VERTEBRATE ZOOLOGY] The chamber which receives the discharges of the intestine, urinary tract, and reproductive canals in monotremes, amphibians, birds, reptiles, and many fish. { klō'ā·kə }

cloacal bladder [VERTEBRATE ZOOLOGY] A diverticulum of the cloacal wall in monotremes, amphibians, and some fish, into which urine is forced from the cloaca. { klō'ā·kəl 'blad·ər }

cloacal gland [VERTEBRATE ZOOLOGY] Any of the sweat glands in the cloaca of lower vertebrates, as snakes or amphibians. { klō'ā·kəl 'gland }

clone [BIOLOGY] All individuals, considered collectively, produced asexually or by parthenogenesis from a single individual. [GENETICS] A copy of genetically engineered DNA sequences. { klōn }

Clonothrix [MICROBIOLOGY] A genus of sheathed bacteria; cells are attached and encrusted with iron and manganese oxides, and filaments are tapered. { 'klän·əˌthriks }

clonus [PHYSIOLOGY] Irregular, alternating muscular contractions and relaxations. { 'klō·nəs }

closed ecological system [ECOLOGY] A community into which a new species cannot enter due to crowding and competition. { ¦klōzd ek·ə 'läj·ə·kəl ˌsis·təm }

closed reading frame [MOLECULAR BIOLOGY] A reading frame containing terminator codons that prevent the translation of subsequent nucleotides into protein. Also known as blocked reading frame. { ¦klōzd 'rēd·iŋ ˌfrām }

Clostridium [MICROBIOLOGY] A genus of bacteria in the family Bacillaceae; usually motile rods which form large spores that distend the cell; anaerobic and do not reduce sulfate. { klä'strid·ē·əm }

clot [PHYSIOLOGY] A semisolid coagulum of blood or lymph. { klät }

clotting factor [PHYSIOLOGY] Any of several plasma components that are involved in the clotting of blood, such as fibrinogen, prothrombin, and thromboplastin. { 'kläd·iŋ ˌfak·tər }

clotting time [PHYSIOLOGY] The length of time required for shed blood to coagulate under standard conditions. Also known as coagulation time. { 'kläd·iŋ ,tīm }

cloud forest [ECOLOGY] *See* temperate rainforest. { 'klau̇d ,fär·əst }

clove [BOTANY] **1.** The unopened flower bud of a small, conical, symmetrical evergreen tree, *Eugenia caryophyllata*, of the myrtle family (Myrtaceae); the dried buds are used as a pungent, strongly aromatic spice. **2.** A small bulb developed within a larger bulb, as in garlic. { klōv }

clover [BOTANY] **1.** A common name designating the true clovers, sweet clovers, and other members of the Leguminosa. **2.** A herb of the genus *Trifolium*. { 'klō·vər }

club fungi [MYCOLOGY] The common name for members of the class Basidiomycetes. { 'kləb ,fən,jī }

club moss [BOTANY] The common name for members of the class Lycopodiatae. { 'kləb ,mȯs }

Clupeidae [VERTEBRATE ZOOLOGY] The herrings, a family of fishes in the suborder Clupoidea composing the most primitive group of higher bony fishes. { klü'pē·ə,dē }

Clupeiformes [VERTEBRATE ZOOLOGY] An order of teleost fishes in the subclass Actinopterygii, generally having a silvery, compressed body. { ,klu·pe·ə'tȯr,mēz }

clupeine [BIOCHEMISTRY] A protamine found in salmon sperm, mainly composed of arginine (74.1%) and small percentages of threonine, serine, proline, alanine, valine, and isoleucine. { 'klü·pē,ēn }

Clupoidea [VERTEBRATE ZOOLOGY] A suborder of fishes in the order Clupeiformes comprising the herrings and anchovies. { ,klü'pȯid·ē·ə }

Clusiidae [INVERTEBRATE ZOOLOGY] A family of myodarian cyclorrhaphous dipteran insects in the subsection Acalypteratae. { ,klü'sī·ə,dē }

cluster gene [GENETICS] A gene that codes for a multifunctional enzyme. { 'kləs·tər ,jēn }

clutch [VERTEBRATE ZOOLOGY] A nest of eggs or a brood of chicks. { kləch }

Clypeasteroida [INVERTEBRATE ZOOLOGY] An order of exocyclic Euechinoidea having a monobasal apical system in which all the genital plates fuse together. { ,klip·ē,as·tə'rȯid·ə }

clypeus [INVERTEBRATE ZOOLOGY] An anterior medial plate on the head of an insect, commonly bearing the labrum on its anterior margin. [MYCOLOGY] A disk of black tissue about the mouth of the perithecia in certain ascomycetes. { 'klip·ē·əs }

Clythiidae [INVERTEBRATE ZOOLOGY] The flat-footed flies, a family of cyclorrhaphous dipteran insects in the series Aschiza characterized by a flattened distal end on the hind tarsus. { klə 'thī·ə,dē }

Cnidaria [INVERTEBRATE ZOOLOGY] A phylum of the Radiata whose members typically bear tentacles and possess intrinsic nematocysts. Also known as Coelenterata. { nī'dar·ē·ə }

cnidoblast [INVERTEBRATE ZOOLOGY] A cell that produces nematocysts. Also known as cnidocyte; nettle cell; stinging cell. { 'nīd·ə,blast }

cnidocil [INVERTEBRATE ZOOLOGY] The trigger on a cnidoblast that activates discharge of the nematocyst when touched. { 'nīd·ə,sil }

cnidocyte [INVERTEBRATE ZOOLOGY] *See* cnidoblast. { 'knīd·ə,sīt }

cnidophore [INVERTEBRATE ZOOLOGY] A modified structure bearing nematocysts in certain cnidarians. { 'nīd·ə,fȯr }

Cnidospora [INVERTEBRATE ZOOLOGY] A subphylum of spore-producing protozoans that are parasites in cells and tissues of invertebrates, fishes, a few amphibians, and turtles. { nī'däs·pə·rə }

CoA [BIOCHEMISTRY] *See* coenzyme A.

coadaptation [EVOLUTION] The selection process that tends to accumulate favorably interacting genes in the gene pool of a population. { ,kō ,ad·əp'tā·shən }

coagulase [BIOCHEMISTRY] Any enzyme that causes coagulation of blood plasma. { kō'ag·yə ,lās }

coagulation time [PHYSIOLOGY] *See* clotting time. { kō,ag·yə'lā·shən ,tīm }

coalescence [BOTANY] The union of plant parts of the same kind such as the united sepals of flowering plants. { ,kō·ə'les·əns }

coal paleobotany [PALEOBOTANY] A branch of the paleobotanical sciences concerned with the origin, composition, mode of occurrence, and significance of fossil plant materials that occur in or are associated with coal seams. { 'kōl ,pā·lē·ō'bät·ən·ē }

coancestry [GENETICS] The degree of relationship between two parents of a diploid individual. { ,kō'an,səs·trē }

coated pit [CELL BIOLOGY] A cell surface depression that is coated with clathrin on its cytoplasmic surface and functions in receptor-mediated endocytosis. { 'kōd·əd 'pit }

coati [VERTEBRATE ZOOLOGY] The common name for three species of carnivorous mammals assigned to the raccoon family (Procyonidae) characterized by their elongated snout, body, and tail. { kə'wäd·ē }

cobalamin [BIOCHEMISTRY] *See* vitamin B. { kə 'bȯl·ə·mən }

Cobitidae [VERTEBRATE ZOOLOGY] The loaches, a family of small fishes, many eel-shaped, in the suborder Cyprinoidei, characterized by barbels around the mouth. { kə'bid·ə,dē }

cobra [VERTEBRATE ZOOLOGY] Any of several species of venomous snakes in the reptilian family Elaphidae characterized by a hoodlike expansion of skin on the anterior neck that is supported by a series of ribs. { 'kō·brə }

coca [BOTANY] *Erythroxylon coca*. A shrub in the family Erythroxylaceae; its leaves are the source of cocaine. { 'kō·kə }

cocarboxylase [BIOCHEMISTRY] *See* thiamine pyrophosphate. { ¦kō·kär'bäk·sə,lās }

Coccidia [INVERTEBRATE ZOOLOGY] A subclass of protozoans in the class Telosporea; typically intracellular parasites of epithelial tissue in vertebrates and invertebrates. { käk'sid·ē·ə }

coccine [INVERTEBRATE ZOOLOGY] For protozoa, denoting the sessile state during which reproduction does not occur. { 'käk,sēn }

Coccinellidae [INVERTEBRATE ZOOLOGY] The ladybird beetles, a family of coleopteran insects in the superfamily Cucujoidea. { käk·sə'nel·ə ˌdē }

coccobacillus [MICROBIOLOGY] A short, thick, oval bacillus, midway between the coccus and the bacillus in appearance. { ¦kä·kō·ba'sil·əs }

coccoid [MICROBIOLOGY] A spherical bacterial cell. { 'käˌkȯid }

Coccoidea [INVERTEBRATE ZOOLOGY] A superfamily of homopteran insects belonging to the Sternorrhyncha; includes scale insects and mealy bugs. { kä'kȯid·ē·ə }

coccolith [BOTANY] One of the small, interlocking calcite plates covering members of the Coccolithophorida. { 'käk·əˌlith }

Coccolithophora [INVERTEBRATE ZOOLOGY] An order of phytoflagellates in the protozoan class Phytamastigophorea. { ˌkäk·ō·li'thäf·ə·rə }

Coccolithophorida [BOTANY] A group of unicellular, biflagellate, golden-brown algae characterized by a covering of coccoliths. { ˌkäk·ōˌlith·ə 'fȯr·ə·də }

Coccomyxaceae [BOTANY] A family of algae belonging to the Tetrasporales composed of elongate cells which reproduce only by vegetative means. { ˌkäk·ōˌmik'sās·ēˌē }

coccosphere [PALEOBOTANY] The fossilized remains of a member of Coccolithophorida. { 'käk ·əˌsfir }

Coccosteomorphi [PALEONTOLOGY] An aberrant lineage of the joint-necked fishes. { kä¦kä·stē·ə ¦mȯr·fē }

cocculin [BIOCHEMISTRY] *See* picrotoxin. { 'käk·yə·lən }

coccus [MICROBIOLOGY] A form of eubacteria which are more or less spherical in shape. { 'käk·əs }

coccygeal body [ANATOMY] A small mass of vascular tissue near the tip of the coccyx. { käk'sij·ē·əl 'bäˌdē }

coccyx [ANATOMY] The fused vestige of caudal vertebrae forming the last bone of the vertebral column in man and certain other primates. { 'käkˌsiks }

cochlea [ANATOMY] The snail-shaped canal of the mammalian inner ear; it is divided into three channels and contains the essential organs of hearing. { 'kōk·lē·ə }

cochlear duct [ANATOMY] *See* scala media. { 'kōk·lē·ər 'dəkt }

Cochleariidae [VERTEBRATE ZOOLOGY] A family of birds in the order Ciconiiformes composed of a single species, the boatbill. { ˌkōk·lē·ə'rī·əˌdē }

cochlear nerve [ANATOMY] A sensory branch of the auditory nerve which receives impulses from the organ of Corti. { 'kok·lē·ər 'nərv }

cochlear nucleus [ANATOMY] One of the two nuclear masses in which the fibers of the cochlear nerve terminate; located ventrad and dorsad to the inferior cerebellar peduncle. { 'kok·lē·ər 'nük·lē·əs }

cochleate [BIOLOGY] Spiral; shaped like a snail shell. { 'kōk·lēˌāt }

Cochliodontidae [PALEONTOLOGY] A family of extinct chondrichthian fishes in the order Bradyodonti. { ˌkōk·lē·ō'dän·təˌdē }

cock [VERTEBRATE ZOOLOGY] The adult male of the domestic fowl and of gallinaceous birds. { käk }

cockle [INVERTEBRATE ZOOLOGY] The common name for a number of species of marine mollusks in the class Bivalvia characterized by a shell having convex radial ribs. { 'käk·əl }

cockroach [INVERTEBRATE ZOOLOGY] *See* roach. { 'käkˌrōch }

coconut [BOTANY] *Cocos nucifera.* A large palm in the order Arecales grown for its fiber and fruit, a large, ovoid, edible drupe with a fibrous exocarp and a hard, bony endocarp containing fleshy meat (endosperm). { 'kō·kəˌnət }

cocoon [INVERTEBRATE ZOOLOGY] **1.** A protective case formed by the larvae of many insects, in which they pass the pupa stage. **2.** Any of the various protective egg cases formed by invertebrates. { kə'kün }

cod [VERTEBRATE ZOOLOGY] The common name for fishes of the subfamily Gadidae, especially the Atlantic cod (*Gadus morrhua*). { käd }

codecarboxylase [BIOCHEMISTRY] The prosthetic component of the enzyme carboxylase which catalyzes decarboxylation of D-amino acids. Also known as pyridoxal phosphate. { ¦kō·də·kär 'bäk·səˌlās }

codehydrogenase I [BIOCHEMISTRY] *See* diphosphopyridine nucleotide. { ¦kō·dē'hī·drə·jəˌnās ¦wən }

codehydrogenase II [BIOCHEMISTRY] *See* triphosphopyridine nucleotide. { ¦kō·dē'hī·drə·jəˌnās ¦tü }

Codiaceae [BOTANY] A family of green algae in the order Siphonales having macroscopic thalli composed of aggregates of tubes. { ˌkō·dē'as·ēˌē }

coding ratio [BIOCHEMISTRY] The number of bases in nucleic acids divided by the number of amino acids whose sequence the bases determine in a particular polypeptide. { 'kōd·iŋ ˌrā·shō }

coding strand [MOLECULAR BIOLOGY] *See* sense strand. { 'kōd·iŋ ˌstrand }

codominance [GENETICS] A condition in which each allele of a heterozygous pair expresses itself fully, as in the human blood group AB. { kō 'däm·ə·nəns }

codon [GENETICS] The basic unit of the genetic code, comprising three-nucleotide sequences of messenger ribonucleic acid, each of which is translated into one amino acid in protein synthesis. { 'kōˌdän }

codon family [GENETICS] A group of four codons that code for the same amino acid and differ only in the nucleotide that occupies the third codon position. { 'kōˌdän ˌfam·lē }

codon fidelity [MOLECULAR BIOLOGY] The constancy of the genetic coding process as maintained during deoxyribonucleic acid replication and the synthesis of proteins by a series of proofreading reactions that remove errors. { 'kōˌdän fi ˌdel·əd·ē }

codon misreading [MOLECULAR BIOLOGY] The

mistranslation of a codon in messenger ribonucleic acid that increases errors in protein synthesis by generation of amino acid substitutions. { ′kō‚dän mis‚rēd·iŋ }

coelacanth [VERTEBRATE ZOOLOGY] Any member of the Coelacanthiformes, an order of lobefin fishes represented by a single living genus, *Latimeria*. { ′sē·lə‚kanth }

Coelacanthidae [PALEONTOLOGY] A family of extinct lobefin fishes in the order Coelacanthiformes. { ‚sē·lə′kan·thə‚dē }

Coelacanthiformes [VERTEBRATE ZOOLOGY] An order of lobefin fishes in the subclass Crossopterygii which were common fresh-water animals of the Carboniferous and Permian; one genus, *Latimeria*, exists today. { ¦sē·lə¦kan·thə′fór‚mēz }

Coelacanthini [VERTEBRATE ZOOLOGY] The equivalent name for Coelacanthiformes. { ‚sē·lə′kan·thə‚nī }

Coelenterata [INVERTEBRATE ZOOLOGY] *See* Cnidaria. { sə‚len·tə′räd·ə }

coelenteron [INVERTEBRATE ZOOLOGY] The internal cavity of cnidarians. { sə′len·tə‚rän }

coeloblastula [EMBRYOLOGY] A simple, hollow blastula with a single-layered wall. { ‚sē·lō′blas·chə·lə }

Coelolepida [PALEONTOLOGY] An order of extinct jawless vertebrates (Agnatha) distinguished by skin set with minute, close-fitting scales of dentine, similar to placoid scales of sharks. { ‚sē·lō′lep·ə·da }

coelom [ZOOLOGY] The mesodermally lined body cavity of most animals higher on the evolutionary scale than flatworms and nonsegmented roundworms. { ′sē·ləm }

Coelomata [ZOOLOGY] The equivalent name for Eucoelomata. { ‚sē·lə′mad·ə }

coelomocyte [INVERTEBRATE ZOOLOGY] A corpuscle, including amebocytes and eleocytes, in the coelom of certain animals, especially annelids. { ′sē·lō·mə‚sīt }

coelomoduct [INVERTEBRATE ZOOLOGY] Either of a pair of ciliated excretory and reproductive channels passing from the coelom to the exterior in certain invertebrates, including annelids and mollusks. { ′sē·lō·mə‚dəkt }

Coelomomycetaceae [MYCOLOGY] A family of entomophilic fungi in the order Blastocladiales which parasitize primarily mosquito larvae. { ¦sē·lō·mə‚mī·sə′tās·ē‚ē }

coelomostome [INVERTEBRATE ZOOLOGY] The opening of a coelomoduct into the coelom. { ′sē·lō·mə‚stōm }

Coelomycetes [MYCOLOGY] A group set up by some authorities to include the Sphaerioidaceae and the Melanconiales. { ¦sē·lō·mī′sēd‚ēz }

Coelopidae [INVERTEBRATE ZOOLOGY] The seaweed flies, a family of myodarian cyclorrhaphous dipteran insects in the subsection Acalypteratae whose larvae breed on decomposing seaweed. { sə′lō·pə‚dē }

coeloplanula [INVERTEBRATE ZOOLOGY] A hollow planula having a wall of two layers of cells. { ‚sē·lə′plan·yə·lə }

Coelurosauria [PALEONTOLOGY] A group of small, lightly built saurischian dinosaurs in the suborder Theropoda having long necks and narrow, pointed skulls. { sə‚lúr·ə′sór·ē·ə }

Coenagrionidae [INVERTEBRATE ZOOLOGY] A family of zygopteran insects in the order Odonata. { ‚sē‚nag·rē′än·ə‚dē }

coenenchyme [INVERTEBRATE ZOOLOGY] The mesagloea surrounding and uniting the polyps in compound anthozoans. Also known as coenosarc. { sə′neŋ‚kīm }

Coenobitidae [INVERTEBRATE ZOOLOGY] A family of terrestrial decapod crustaceans belonging to the Anomura. { ‚sē·nə′bid·ə‚dē }

coenobium [INVERTEBRATE ZOOLOGY] A colony of protozoans having a constant size, shape, and cell number, but with undifferentiated cells. { sə′nō·bē·əm }

coenocyte [BIOLOGY] A multinucleate mass of protoplasm formed by repeated nucleus divisions without cell fission. { ′sē·nə‚sīt }

Coenomyidae [INVERTEBRATE ZOOLOGY] A family of orthorrhaphous dipteran insects in the series Brachycera. { ‚sē·nə′mī·ə‚dē }

Coenopteridales [PALEOBOTANY] A heterogeneous group of fernlike fossil plants belonging to the Polypodiophyta. { ‚sē·näp‚ter·ə′dā·lēz }

coenosarc [INVERTEBRATE ZOOLOGY] **1.** The living axial part of a hydroid colony. **2.** *See* coenenchyme. { ′sē·nə‚särk }

coenosteum [INVERTEBRATE ZOOLOGY] The calcareous skeleton of a compound coral or bryozoan colony. { sə′näs·tē·əm }

Coenothecalia [INVERTEBRATE ZOOLOGY] An order of the class Alcyonaria that forms colonies; lacks spicules but has a skeleton composed of fibrocrystalline argonite. { ‚sē·nō·thə′kāl·ē·ə }

coenotype [BIOLOGY] An organism having the characteristic structure of the group to which it belongs. { ′sē·nə‚tīp }

coenzyme [BIOCHEMISTRY] The nonprotein portion of an enzyme; a prosthetic group which functions as an acceptor of electrons or functional groups. { kō′en‚zīm }

coenzyme A [BIOCHEMISTRY] $C_{21}H_{36}O_{16}N_7P_3S$ A coenzyme in all living cells; required by certain condensing enzymes to act in acetyl or other acyl-group transfer and in fatty-acid metabolism. Abbreviated CoA. { kō′en‚zīm ′ā }

coenzyme I [BIOCHEMISTRY] *See* diphosphopyridine nucleotide. { kō′en‚zīm ′wən }

coenzyme II [BIOCHEMISTRY] *See* triphosphopyridine nucleotide. { kō′en‚zīm ′tü }

coevolution [EVOLUTION] An evolutionary pattern based on the interaction among major groups or organisms with an obvious ecological relationship; for example, plant and plant-eater, flower and pollinator. { ¦kō‚ev·ə′lü·shən }

cofactor [BIOCHEMISTRY] A specific substance required for the activity of an enzyme, such as a coenzyme or metal ion. { ′kō‚fak·tər }

coffee [BOTANY] Any of various shrubs or small trees of the genus *Coffea* (family Rubiaceae) cultivated for the seeds (coffee beans) of its fruit; most coffee beans are obtained from the Arabian species, *C. arabica*. { kóf·ē }

cogon [BOTANY] *Imperate cylindrica.* A grass found in rainforests. Also known as alang-alang. { kō 'gōn }

cog region [BIOCHEMISTRY] Any group of similar sequences of nucleotides that occurs in deoxyribonucleic acid molecules and may specifically be recognized by endonucleases or other enzymes. { 'käg ˌrē·jən }

cohesion [BOTANY] The union of similar plant parts or organs, as of the petals to form a corolla. { kō'hē·zhən }

cohesive end [BIOCHEMISTRY] *See* sticky end. { kō 'hē·siv ˌend }

cohesive terminus [MOLECULAR BIOLOGY] Either of the ends of single-stranded deoxyribonucleic acid that are complementary in the nucleotide sequences and can join, by base pairing, to form circular molecules. { kō'hē·siv 'tər·mə·nəs }

cohort selection [EVOLUTION] A type of natural selection due to interactions among groups of similar ages in a population. { 'kōˌhȯrt siˌlek·shən }

coiled tubular gland [ANATOMY] A structure having a duct interposed between the surface opening and the coiled glandular portion; an example is a sweat gland. { ¦kȯild ¦tü·byə·lər 'gland }

coincidence [GENETICS] A numerical value equal to the number of double crossovers observed, divided by the number expected. { kō'in·sə·dəns }

coincidental evolution [EVOLUTION] The maintenance of sequence homology among nonallelic members of a multigene family within a species. Also known as concerted evolution; horizontal evolution. { ˌkōˌin·sə¦dent·əl ˌev·ə'lü·shən }

co-inducer [BIOCHEMISTRY] A molecule that interacts with a repressor to free the operon from restraints on its transcription into messenger ribonucleic acid. { ˌkō·in'dü·sər }

cointegrate structure [MOLECULAR BIOLOGY] The circular molecule formed by fusing two replicons, one possessing a transposon, the other lacking it. { kō¦in·tə·grət 'strək·chər }

coisogenic strain [BIOLOGY] A strain that differs from another one by only a single gene as the result of a mutation. { ¦kō;am'i·səˌjen·ik 'strān }

coitus [ZOOLOGY] The act of copulation. Also known as intercourse. { 'kō·əd·əs }

cola [BOTANY] *Cola acuminata.* A tree of the sterculia family (Sterculiaceae) cultivated for cola nuts, the seeds of the fruit; extract of cola nuts is used in the manufacture of soft drinks. { 'kō·lə }

cold agglutination phenomenon [IMMUNOLOGY] Clumping of human blood group O erythrocytes at 0-4°C, but not at body temperature; occurs in primary atypical pneumonia, trypanosomiasis, and other unidentifiable states. { 'kōld əˌglüt·ən 'ā·shən fə'näm·əˌnän }

cold agglutinin [IMMUNOLOGY] A nonspecific panagglutinin found in many normal human serums which produce maximum clumping of erythrocytes at 4°C and none at 37°C. { 'kōld ə 'glüt·ən·ən }

cold-blooded [PHYSIOLOGY] Having body temperature approximating that of the environment and not internally regulated. { 'kōld ¦bləd·əd }

cold desert [ECOLOGY] *See* tundra. { ¦kōld 'dez·ərt }

cold hemagglutination [IMMUNOLOGY] A phenomenon caused by the presence of cold agglutinin. { 'kōld ˌhēm·əˌglüt·ən'ā·shən }

cold-sensitive mutation [GENETICS] An alteration that causes a gene to be inactive at low temperature. { 'kōld ¦sen·səd·iv myü'tā·shən }

cold torpor [PHYSIOLOGY] Condition of reduced body temperature in poikilotherms. { 'kōld ˌtȯr·pər }

Coleochaetaceae [BOTANY] A family of green algae in the suborder Ulotrichineae; all occur as attached, disklike, or parenchymatous thalli. { ˌkō·lēˌō·kē'tās·ēˌē }

Coleodontidae [PALEONTOLOGY] A family of conodonts in the suborder Neurodontiformes. { ˌkō·lē·ō'dän·təˌdē }

Coleoidea [INVERTEBRATE ZOOLOGY] A subclass of cephalopod mollusks including all cephalopods except *Nautilus*, according to certain systems of classification. { ˌkō·lē'ȯid·ē·ə }

Coleophoridae [INVERTEBRATE ZOOLOGY] The case bearers, moths with narrow wings composing a family of lepidopteran insects in the suborder Heteroneura; named for the silk-and-leaf shell carried by larvae. { ˌkō·lē·ō'fȯr·əˌdē }

Coleoptera [INVERTEBRATE ZOOLOGY] The beetles, holometabolous insects making up the largest order of the animal kingdom; general features of the Insecta are found in this group. { ˌkō·lē 'äp·tə·rə }

coleoptile [BOTANY] The first leaf of a monocotyledon seedling. { ˌkō·lē'äp·təl }

coleorhiza [BOTANY] The sheath surrounding the radicle in monocotyledons. { ˌkō·lē·ə'rīz·ə }

Coleorrhyncha [INVERTEBRATE ZOOLOGY] A monofamilial group of homopteran insects in which the beak is formed at the anteroventral extremity of the face and the propleura form a shield for the base of the beak. { ˌkō·lē·ə'riŋ·kə }

Coleosporaceae [MYCOLOGY] A family of parasitic fungi in the order Uredinales. { ˌkō·lē·ō·spə 'rās·ēˌē }

colic artery [ANATOMY] Any of the three arteries that supply the colon. { 'käl·ik 'ärd·ə·rē }

colicin [MICROBIOLOGY] A bacteriocin produced by coliform bacteria, such as *Escherichia coli.* { 'kä·lə·sən }

coliform bacteria [MICROBIOLOGY] Colon bacilli, or forms which resemble or are related to them. { 'kä·ləˌfȯrm bak'tir·ē·ə }

Coliidae [VERTEBRATE ZOOLOGY] The colies or mousebirds, composing the single family of the avian order Coliiformes. { kə'lī·əˌdē }

Coliiformes [VERTEBRATE ZOOLOGY] A monofamilial order of birds distinguished by long tails, short legs, and long toes, all four of which are directed forward. { kəˌlī·ə'fȯrˌmēz }

colinearity [MOLECULAR BIOLOGY] The relationship between the linear sequence of codons in deoxyribonucleic acid and the order of amino acids in the polypeptide product that it specifies. Also spelled collinearity. { kō,lin·ē'ar·əd·e }

coliphage [VIROLOGY] Any bacteriophage able to infect *Escherichia coli*. { 'kä·lə,fāj }

colistin [MICROBIOLOGY] $C_{45}H_{85}O_{10}N_{13}$ A basic polypeptide antibiotic produced by *Bacillus colistinus*; consists of an A and B component, active against a broad spectrum of gram-positive microorganisms and some gram-negative microorganisms. { kə'lis·tən }

collagen [BIOCHEMISTRY] A fibrous protein found in all multicellular animals, especially in connective tissue. { 'kä·lə·jən }

collagenase [BIOCHEMISTRY] Any proteinase that decomposes collagen and gelatin. { 'kä·lə·jə ,nās }

collar cell [INVERTEBRATE ZOOLOGY] *See* choanocyte. { 'käl·ər ,sel }

collard [BOTANY] *Brassica oleracea* var. *acephala*. A biennial crucifer of the order Capparales grown for its rosette of edible leaves. { 'käl·ərd }

collateral bud [BOTANY] An accessory bud produced beside an axillary bud. { kə¦lad·ə·rəl 'bəd }

collateral bundle [BOTANY] A vascular bundle in which the phloem and xylem lie on the same radius, with the phloem located toward the periphery of the stem and the xylem toward the center. { kə¦lad·ə·rəl 'bənd·əl }

collateral circulation [PHYSIOLOGY] The circulation established for an organ or a part of an organ through the intercommunication of blood vessels when the original direct blood supply is obstructed or abolished. { kə¦lad·ə·rəl ,sər·kyə 'lā·shən }

collateral fiber [ANATOMY] A lateral branch of an axon. { kə¦lad·ə·rəl 'fīb·ər }

collateral ligament [ANATOMY] Any of various stabilizing ligaments on either side of a hinge joint such as the knee or elbow. { kə¦lad·ə·rəl 'lig·ə·mənt }

collateral respiration [PHYSIOLOGY] The passage of air between lobules within the same lobe of a lung, enabling ventilation of a lobule whose branchiole is obstructed. { ke¦lad·ə·rəl ,res·pə 'rā·shən }

collecting tubule [ANATOMY] One of the ducts conveying urine from the renal tubules (nephrons) to the minor calyces of the renal pelvis. { kə'lek·tiŋ ,tü,byül }

Collembola [INVERTEBRATE ZOOLOGY] The springtails, an order of primitive insects in the subclass Apterygota having six abdominal segments. { kə'lem·bə·lə }

collenchyma [BOTANY] A primary, or early-differentiated, subepidermal supporting tissue in leaf petioles and vein ribs formed before vascular differentiation. { kə'leŋ·kə·mə }

collenchyme [INVERTEBRATE ZOOLOGY] A loose mesenchyme that fills the space between ectoderm and endoderm in the body wall of many lower invertebrates, such as sponges. { 'kä·lən ,kīm }

collenia [PALEOBOTANY] A convex, slightly arched, or turbinate stromatolite produced by late Precambrian blue-green algae of the genus *Collenia*. { kə'len·ē·ə }

Colletidae [INVERTEBRATE ZOOLOGY] The colletid bees, a family of hymenopteran insects in the superfamily Apoidea. { kə'led·ə,dē }

colliculus [ANATOMY] **1.** Any of the four prominences of the corpora quadrigemina. **2.** The elevation of the optic nerve where it enters the retina. **3.** The anterolateral, apical elevation of the arytenoid cartilages. { kə'lik·yə·ləs }

collinearity [MOLECULAR BIOLOGY] *See* colinearity. { kō,lin·ē'ar·əd·ē }

colloblast [INVERTEBRATE ZOOLOGY] An adhesive cell on the tentacles of ctenophores. { 'käl·ə ,blast }

colloidal osmotic pressure [PHYSIOLOGY] *See* oncotic pressure. { kə'lòid·əl äz'mäd·ik ,presh·ər }

Collothecacea [INVERTEBRATE ZOOLOGY] A monofamilial suborder of mostly sessile rotifers in the order Monogonata; many species are encased in gelatinous tubes. { ,käl·ə·thə'kās·ē·ə }

Collothecidae [INVERTEBRATE ZOOLOGY] The single family of the Collothecacea. { ,käl·ə'thes·ə ,dē }

Collyritidae [PALEONTOLOGY] A family of extinct, small, ovoid, exocyclic Euechinoidea with fascioles or a plastron. { ,käl·ə'rid·ə,dē }

colon [ANATOMY] The portion of the human intestine extending from the cecum to the rectum; it is divided into four sections: ascending, transverse, descending, and sigmoid. Also known as large intestine. { 'kō·lən }

colon bacillus [MICROBIOLOGY] *See* Escherichia coli. { 'kō·lən bə'sil·əs }

colonization [ECOLOGY] The establishment of an immigrant species in a peripherally unsuitable ecological area; occasional gene exchange with the parental population occurs, but generally the colony evolves in relative isolation and in time may form a distinct unit. { ,käl·ə·nə'zā·shən }

colony [BIOLOGY] A localized population of individuals of the same species which are living either attached or separately. [MICROBIOLOGY] A cluster of microorganisms growing on the surface of or within a solid medium; usually cultured from a single cell. { 'käl·ə·nē }

colony count [MICROBIOLOGY] The number of colonies of bacteria growing on the surface of a solid medium. { 'käl·ə·nē ,kaůnt }

color vision [PHYSIOLOGY] The ability to discriminate light on the basis of wavelength composition. { 'kəl·ər ,vizh·ən }

Colossendeidae [INVERTEBRATE ZOOLOGY] A family of deep-water marine arthropods in the subphylum Pycnogonida, having long palpi and lacking chelifores, except in polymerous forms. { ,käl·ə·sen'dā·ə,dē }

colostrum [PHYSIOLOGY] The first milk secreted by the mammary gland during the first days following parturition. { kə'las·trəm }

Colubridae [VERTEBRATE ZOOLOGY] A family of cosmopolitan snakes in the order Squamata. { kə 'lü·brə,dē }

Columbidae [VERTEBRATE ZOOLOGY] A family of

birds in the order Columbiformes composed of the pigeons and doves. { kə'ləm·bə,dē }

Columbiformes [VERTEBRATE ZOOLOGY] An order of birds distinguished by a short, pointed bill, imperforate nostrils, and short legs. { kə,ləm·bə 'fór,mēz }

columella [ANATOMY] *See* stapes. [BIOLOGY] Any part shaped like a column. [BOTANY] A sterile axial body within the capsules of certain mosses, liverworts, and many fungi. { ,käl·yə 'mel·ə }

columnar epithelium [HISTOLOGY] Epithelium distinguished by elongated, columnar, or prismatic cells. { kə'ləm·nər ep·ə'thēl·ē·əm }

columnar stem [BOTANY] An unbranched, cylindrical stem bearing a set of large leaves at its summit, as in palms, or no leaves, as in cacti. { kə'ləm·nər ,stem }

Colydiidae [INVERTEBRATE ZOOLOGY] The cylindrical bark beetles, a large family of coleopteran insects in the superfamily Cucujoidea. { käl·ə 'dī·ə,dē }

Comasteridae [INVERTEBRATE ZOOLOGY] A family of radially symmetrical Crinozoa in the order Comatulida. { ,kō·mə'ster·ə,dē }

Comatulida [INVERTEBRATE ZOOLOGY] The feather stars, an order of free-living echinoderms in the subclass Articulata. { ,kō·mə'tül·ə·də }

comb [INVERTEBRATE ZOOLOGY] **1.** A system of hexagonal cells constructed of beeswax by a colony of bees. **2.** A comblike swimming plate in ctenophores. [VERTEBRATE ZOOLOGY] A crest of naked tissue on the head of many male fowl. { kōm }

comb growth unit [BIOLOGY] A unit for the standardization of male sex hormones. { 'kōm ,grōth ,yü·nət }

Comesomatidae [INVERTEBRATE ZOOLOGY] A family of free-living nematodes in the superfamily Comesomatoidea found as deposit feeders on soft bottom sediments. { ,kō·mə·sō'mad·ə ,dē }

Comesomatoidea [INVERTEBRATE ZOOLOGY] A superfamily of marine nematodes in the order Chromadorida distinguished by their wide multispiral amphids that make at least two complete turns. { ,kō·mə·sō·mə'tóid·ē·ə }

Commelinaceae [BOTANY] A family of monocotyledonous plants in the order Commelinales characterized by differentiation of the leaves into a closed sheath and a well-defined, commonly somewhat succulent blade. { ,kä·mə·lə'nās· ē,ē }

Commelinales [BOTANY] An order of monocotyledonous plants in the subclass Commelinidae marked by having differentiated sepals and petals but lacking nectaries and nectar. { ,kä·mə·lə 'nā·lēz }

Commelinidae [BOTANY] A subclass of flowering plants in the class Liliopsida. { ,kä·mə'lī·nə ,dē }

commensal [ECOLOGY] An organism living in a state of commensalism. { kə'men·səl }

commensalism [ECOLOGY] An interspecific, symbiotic relationship in which two different species are associated, wherein one is benefited and the other neither benefited nor harmed. { kə'men·sə ,liz·əm }

commissure [BIOLOGY] A joint, seam, or closure line where two structures unite. { 'käm·ə,shür }

commitment [CELL BIOLOGY] The establishment of a unique developmental sequence in a cell that differs from the prior state. { kə'mit·mənt }

common bile duct [ANATOMY] The duct formed by the union of the hepatic and cystic ducts. { ¦käm·ən 'bīl ,dəkt }

common carotid artery [ANATOMY] *See* carotid artery. { ¦käm·ən kə'räd·əd 'ärd·ə·rē }

common hepatic duct [ANATOMY] *See* hepatic duct. { ¦käm·ən he'pad·ik 'dəkt }

common iliac artery [ANATOMY] *See* iliac artery. { ¦käm·ən ,il·ē,ak 'ärd·ə·rē }

communicating junction [CELL BIOLOGY] *See* gap junction. { kə'myü·nə,kād·iŋ ,jəŋk·shən }

community [ECOLOGY] Aggregation of organisms characterized by a distinctive combination of two or more ecologically related species; an example is a deciduous forest. Also known as ecological community. { kə'myü·nə·dē }

community classification [ECOLOGY] Arrangement of communities into classes with respect to their complexity and extent, their stage of ecological succession, or their primary production. { kə'myü·nə·dē ,klas·ə·fə'kā·shən }

comose [BOTANY] Having a tuft of soft hairs. { 'kō,mōs }

companion cell [BOTANY] A specialized parenchyma cell occurring in close developmental and physiologic association with a sieve-tube member. { kəm'pan·yən ,sel }

comparative embryology [EMBRYOLOGY] A branch of embryology that deals with the similarities and differences in the development of animals or plants of different orders. { kəm'par· əd·iv ,em·brē'äl·ə·jē }

compartment [CELL BIOLOGY] Any of the membrane-bound organelles within cells. { kəm 'pärt·mənt }

compatibility [IMMUNOLOGY] Ability of two bloods or other tissues to unite and function together. { kəm,pad·ə'bil·ə·dē }

compensation point [BOTANY] The light intensity at which the amount of carbon dioxide released in respiration equals the amount used in photosynthesis, and the amount of oxygen released in photosynthesis equals the amount used in respiration. { ,käm·pən'sā·shən ,póint }

competence [EMBRYOLOGY] The ability of a reacting system to respond to the inductive stimulus during early developmental stages. { 'käm·pəd·əns }

competition [ECOLOGY] The inter- or intraspecific interaction resulting when several individuals share an environmental necessity. { ,käm· pə'tish·ən }

competitive enzyme inhibition [BIOCHEMISTRY] Prevention of an enzymatic process resulting from the reversible interaction of an inhibitor with a free enzyme. { kəm'ped·əd·iv 'en,zīm ,in·ə'bish·ən }

competitive exclusion [ECOLOGY] The result of a

competition in which one species is forced out of part of the available habitat by a more efficient species. { kəm'ped·əd·iv iks'klüzh·ən }

competitive-exclusion principle [ECOLOGY] *See* Gause's principle. { kəm'ped·əd·iv iks'klüzh·ən ˌprin·sə·pəl }

competitive inhibition [BIOCHEMISTRY] Enzyme inhibition in which the inhibitor competes with the natural substrate for the active site of the enzyme; may be overcome by increasing substrate concentration. { kəm'ped·əd·iv ˌin·ə 'bish·ən }

compital [BOTANY] **1.** Of the vein of a leaf, intersecting at a wide angle. **2.** Of a fern, bearing sori at the intersection of two veins. { 'käm·pəd·əl }

complement [IMMUNOLOGY] A heat-sensitive, complex system in fresh human and other sera which, in combination with antibodies, is important in the host defense mechanism against invading microorganisms. { 'käm·plə·mənt }

complemental air [PHYSIOLOGY] The amount of air that can still be inhaled after a normal inspiration. { ˌkäm·plə'ment·əl 'er }

complementary deoxyribonucleic acid [MOLECULAR BIOLOGY] A deoxyribonucleic acid molecule that is synthesized by reverse transcriptase from a ribonucleic acid template.Abbreviated cDNA. Also known as copy DNA. { ˌkäm·plə ¦men·trē ˌdē¦äk·sē¦rī·bōˌnü¦klē·ik 'as·əd }

complementary deoxyribonucleic acid library [MOLECULAR BIOLOGY] A collection of complementary deoxyribonucleic acid molecules, representative of all the various messenger ribonucleic acid molecules produced by a specific type of cell of a given species, spliced into a corresponding collection of DNA vectors. { ˌkäm·plə ¦men·trē dē¦äk·sē¦rīb·ōˌnü¦klē·ik 'as·əd }

complementary genes [GENETICS] Nonallelic genes that complement one another's expression in a trait. { ˌkäm·plə'men·trē 'jēnz }

complementation [GENETICS] The action of complementary genes. { ˌkäm·plə·mən'tā·shən }

complementation group [GENETICS] A group having mutations in the same cistron. { ˌkäm·plə·mən'tā·shən ˌgrüp }

complementation test [GENETICS] An analytic procedure for determining whether two mutants are defective in the same cistron. { ˌkäm·plə·mən'tā·shən ˌtest }

complement cascade [IMMUNOLOGY] The sequential activation of complement proteins resulting in lysis of a target cell. { 'käm·plə·mənt kas'kād }

complement fixation [IMMUNOLOGY] The binding of complement to an antigen-antibody complex so that the complement is unavailable for subsequent reaction. { 'käm·plə·mənt ˌfik'sā·shən }

complement-fixation test [IMMUNOLOGY] A diagnostic test to determine the presence of antigen or antibody in the blood by adding complement to the test system; used especially in diagnosing syphilis. { 'käm·plə·mənt ˌfik'sā·shən ˌtest }

complete flower [BOTANY] A flower having all four floral parts, that is, having sepals, petals, stamens, and carpels. { kəm'plēt 'flau̇·ər }

complete leaf [BOTANY] A dicotyledon leaf consisting of three parts: blade, petiole, and a pair of stipules. { kəm'plēt 'lēf }

complicate [INVERTEBRATE ZOOLOGY] Folded lengthwise several times, as applied to insect wings. { 'käm·pləˌkāt }

Compositae [BOTANY] The single family of the order Asterales; perhaps the largest family of flowering plants, it contains about 19,000 species. { kəm'päz·əˌtē }

composite gene [GENETICS] Any gene arising by recombination between two nonallelic genes, located on two nonhomologous chromosomes, and containing portions of both genes. { kəm 'päz·ət 'jēn }

composite nerve [PHYSIOLOGY] A nerve containing both sensory and motor fibers. { kəm'päz·ət 'nərv }

compound acinous gland [ANATOMY] A structure with spherical secreting units connected to many ducts that empty into a common duct. { 'kämˌpau̇nd 'as·ə·nəs ˌgland }

compound eye [INVERTEBRATE ZOOLOGY] An eye typical of crustaceans, insects, centipedes, and horseshoe crabs, constructed of many functionally independent photoreceptor units (ommatidia) separated by pigment cells. { 'kämˌpau̇nd 'ī }

compound gland [ANATOMY] A secretory structure with many ducts. { 'kämˌpau̇nd 'gland }

compound leaf [BOTANY] A type of leaf with the blade divided into two or more separate parts called leaflets. { 'kämˌpau̇nd 'lēf }

compound pistil [BOTANY] A pistil composed of two or more united carpels. { 'kämˌpau̇nd 'pis·təl }

compound sugar [BIOCHEMISTRY] *See* oligosaccharide. { 'kämˌpau̇nd 'shu̇g·ər }

compound tubular-acinous gland [ANATOMY] A structure in which the secreting units are simple tubes with acinous side chambers and all are connected to a common duct. { 'kämˌpau̇nd ¦tüb·yə·lər ¦as·ə·nəs ˌgland }

compound tubular gland [ANATOMY] A structure having branched ducts between the surface opening and the secreting portion. { 'kämˌpau̇nd 'tüb·yə·lər ˌgland }

compression wood [BOTANY] Dense wood found at the base of some tree trunks and on the undersides of branches. { kəm'presh·ən ˌwu̇d }

conarium [ANATOMY] *See* pineal body. { kō'nar·ē·əm }

concentric bundle [BOTANY] A vascular bundle in which xylem surrounds phloem, or phloem surrounds xylem. { kən'sen·trik 'bən·dəl }

conceptacle [BOTANY] A cavity which is shaped like a flask with a pore opening to the outside, contains reproductive structures, and is bound in a thallus such as in the brown algae. { kən 'sep·tə·kəl }

conception [BIOLOGY] Fertilization of an ovum by the sperm resulting in the formation of a viable zygote. { kən'sep·shən }

conceptus [BIOLOGY] The product of a conception, including the embryo or fetus and extraembryonic membranes, at any stage of development from fertilization to birth. { kən'sep·təs }

concerted evolution [EVOLUTION] *See* coincidental evolution. { kən'sərd·əd ev·ə'lü·shən }

conch [INVERTEBRATE ZOOLOGY] The common name for several species of large, colorful gastropod mollusks of the family Strombidae; the shell is used to make cameos and porcelain. { käŋk }

conchiolin [BIOCHEMISTRY] A nitrogenous substance that is the organic basis of many molluscan shells. { käŋ'kī·ə·lən }

Conchorhagae [INVERTEBRATE ZOOLOGY] A suborder of benthonic wormlike animals in the class Kinorhyncha. { käŋ'kór·ə,gē }

Conchostraca [INVERTEBRATE ZOOLOGY] An order of mussellike crustaceans of moderate size belonging to the subclass Branchiopoda. { käŋ'käs·trə·kə }

concordance [GENETICS] Similarity in appearance of members of a twin pair with respect to one or more specific traits. { kən'kórd·əns }

concordant segregation [GENETICS] The simultaneous appearance or disappearance of gene markers in hybrid cells undergoing chromosome diminution. { kən'kórd·ənt ,seg·rə'gā·shən }

concrescence [BIOLOGY] Convergence and fusion of parts originally separate, as the lips of the blastopore in embryogenesis. { kən'krēs·əns }

conditional lethal mutant [GENETICS] A lethal mutant that expresses characteristics of the wild type when grown under certain conditions, as at a particular temperature, and mutant characteristics under other conditions. { kən'dish·ən·əl 'lē·thəl 'myüt·ənt }

condor [VERTEBRATE ZOOLOGY] *Vultur gryphus*. A large American vulture having a bare head and neck, dull black plumage, and a white neck ruff. { 'kän,dór }

conduplicate [BOTANY] Folded lengthwise and in half with the upper faces together, applied to leaves and petals in the bud. { kən'düp·lə·kət }

Condylarthra [PALEONTOLOGY] A mammalian order of extinct, primitive, hoofed herbivores with five-toed plantigrade to semidigitigrade feet. { ,kän·də'lär·thrə }

condyle [ANATOMY] A rounded bone prominence that functions in articulation. [BOTANY] The antheridium of certain stoneworts. [INVERTEBRATE ZOOLOGY] A rounded, articular process on arthropod appendages. { 'kän,dīl }

condyloid articulation [ANATOMY] A joint, such as the wrist, formed by an ovoid surface that fits into an elliptical cavity, permitting all movement except rotation. { 'kän·də,lóid är,tik·yə'lā·shən }

cone [BOTANY] The ovulate or staminate strobilus of a gymnosperm. [HISTOLOGY] A photoceptor of the vertebrate retina that responds differentially to light across the visible spectrum, providing both color vision and visual acuity in bright light. { kōn }

congeneric [SYSTEMATICS] Referring to the species of a given genus. { ¦kän·jə'ner·ik }

congestin [BIOCHEMISTRY] A toxin produced by certain sea anemones. { kən'jes·tən }

conglutination [IMMUNOLOGY] The completion of an agglutinating system, or the enhancement of an incomplete one, by the addition of certain substances. { kən,glüt·ən'ā·shən }

conglutination phenomenon [IMMUNOLOGY] Clumping of cells or particles, such as red cells or bacteria, when treated with conglutinin in the presence of antibody and nonhemolytic complement. { kən,glüt·ən'ā·shən fə'näm·ə·nən }

conglutinin [IMMUNOLOGY] A heat-stable substance in bovine and other serums that aids or causes agglomeration or lysis of certain sensitized cells or particles. { kən'glüt·ən·ən }

congression [CELL BIOLOGY] The movement of chromosomes to the spindle equator during mitosis. Also known as chromosome congression. { kəŋ'gresh·ən }

Coniconchia [PALEONTOLOGY] A class name proposed for certain extinct organisms thought to have been mollusks; distinguished by a calcareous univalve shell that is open at one end and by lack of a siphon. { ,kän·ə'käŋ·kē·ə }

Conidae [INVERTEBRATE ZOOLOGY] A family of marine gastropod mollusks in the order Neogastropoda containing the poisonous cone shells. { 'kän·ə,dē }

conidiophore [MYCOLOGY] A specialized aerial hypha that produces conidia in certain ascomycetes and imperfect fungi. { kə'nid·ē·ə,fór }

conidiospore [MYCOLOGY] *See* conidium. { kə'nid·ē·ə,spór }

conidium [MYCOLOGY] Unicellular, asexual reproductive spore produced externally upon a conidiophore. Also known as conidiospore. { kə'nid·ē·əm }

conifer [BOTANY] The common name for plants of the order Pinales. { 'kän·ə·fər }

Coniferales [BOTANY] The equivalent name for Pinales. { kə,nif·ə'rā·lēz }

Coniferophyta [BOTANY] The equivalent name for Pinicae. { kə,nif·ə'räf·əd·ə }

coniferous forest [ECOLOGY] An area of wooded land predominated by conifers. { kə'nif·ə·rəs 'fär·əst }

conjoint tendon [ANATOMY] The common tendon of the transverse and internal oblique muscles of the abdomen. { kən'jóint 'ten·dən }

Conjugales [BOTANY] An order of fresh-water green algae in the class Chlorophyceae distinguished by the lack of flagellated cells, and conjugation being the method of sexual reproduction. { ,kän·jə'gā·lēz }

conjugase [BIOCHEMISTRY] Any of a group of enzymes which catalyze the breakdown of pteroylglutamic acid. { 'kän·jə,gās }

conjugate division [MYCOLOGY] Division of dikaryotic cells in certain fungi in which the two haploid nuclei divide independently, each daughter cell receiving one product of each nuclear division. { 'kän·jə·gət də'vizh·ən }

conjugated protein [BIOCHEMISTRY] A protein combined with a nonprotein group, other than a salt or a simple protein. { 'kän·jə,gād·əd 'prō,tēn }

conjugation [BOTANY] Sexual reproduction by fusion of two protoplasts in certain thallophytes to form a zygote. [INVERTEBRATE ZOOLOGY] Sexual reproduction by temporary union of cells with exchange of nuclear material between two individuals, principally ciliate protozoans. [MICROBIOLOGY] A process involving contact between two bacterial cells during which genetic material is passed from one cell to the other. { ˌkän·jə ˈgā·shən }

conjugon [GENETICS] Any of a number of different genetic elements in bacterial deoxyribonucleic acid that promote bacterial conjugation and gene transfer. { ˈkän·jəˌgän }

conjunctiva [ANATOMY] The mucous membrane covering the eyeball and lining the eyelids. { kən ˈjəŋk·tə·və }

connate leaf [BOTANY] A leaf shaped as though the bases of two opposite leaves had fused around the stem. { kəˈnāt ˈlēf }

connective tissue [HISTOLOGY] A primary tissue, distinguished by an abundance of fibrillar and nonfibrillar extracellular components. { kəˈnek·tiv ˈtish·ü }

connexon [CELL BIOLOGY] Any of the cylindrical channels associated with gap junctions. { kəˈnek ˌsän }

connivent [BIOLOGY] Converging so as to meet, but not fused into a single part. { kəˈnīv·ənt }

Conoclypidae [PALEONTOLOGY] A family of Cretaceous and Eocene exocyclic Euechinoidea in the order Holectypoida having developed aboral petals, internal partitions, and a high test. { ˌkän·ō·kləˈpid·ēˌē }

Conocyeminae [PALEONTOLOGY] A subfamily of Mesozoan parasites in the family Dicyemidae. { ˌkän·əˌsīˈem·əˌnē }

conodont [PALEONTOLOGY] A minute, toothlike microfossil, composed of translucent amber-brown, fibrous or lamellar calcium phosphate; taxonomic identity is controversial. { ˈkän·ə ˌdänt }

Conodontiformes [PALEONTOLOGY] A suborder of conodonts from the Ordovician to the Triassic having a lamellar internal structure. { ¦kän·ə ˌdän·təˈfórˌmēz }

Conodontophoridia [PALEONTOLOGY] The ordinal name for the conodonts. { ¦kän·əˌdän·tə·fə ˈrid·ē·ə }

Conopidae [INVERTEBRATE ZOOLOGY] The wasp flies, a family of dipteran insects in the suborder Cyclorrhapha. { kəˈnäp·əˌdē }

conotheca [INVERTEBRATE ZOOLOGY] The thin integument of the phragmocone in certain mollusks. { ¦kō·nəˈthē·kə }

consanguineous [GENETICS] Pertaining to two or more individuals that have a common recent ancestor. { ¦kän·saŋ¦gwin·ē·əs }

consanguinity [GENETICS] Blood relationship arising from common parentage. { ¦kän·saŋ ¦gwin·əd·ē }

consensual eye reflex [PHYSIOLOGY] *See* consensual light reflex. { kən¦sench·yə·wəl ¦ī ˈrē ˌfleks }

consensual light reflex [PHYSIOLOGY] The reaction of both pupils when only one eye is exposed to a change in light intensity. Also known as consensual eye reflex. { kən¦sench·yə·wəl ¦līt ˈrē ˌfleks }

consensus sequence [GENETICS] An average nucleotide sequence; each nucleotide is the most frequent at its position in the sequence. { kən ˈsen·səs ˌsē·kwəns }

conservation [ECOLOGY] Those measures concerned with the preservation, restoration, beneficiation, maximization, reutilization, substitution, allocation, and integration of natural resources. { ˌkän·sərˈvā·shən }

conservative substitution [MOLECULAR BIOLOGY] Replacement of an amino acid in a polypeptide by one with similar characteristics. { kənˈsər·vəd·iv ˌsəb·stəˈtü·shən }

conserved sequence [EVOLUTION] A sequence of nucleotides in genetic material or of amino acids in a polypeptide chain that has changed only slightly or not at all during an evolutionary period of time. { kənˈsərvd ˈsē·kwəns }

consociation [ECOLOGY] A climax community of plants which is dominated by a single species. { kənˌsō·sēˈā·shən }

consortism [ECOLOGY] *See* symbiosis. { ˈkänˌsórd ˌiz·əm }

conspecific [SYSTEMATICS] Referring to individuals or populations of a single species. { ¦kän·spəˈsif·ik }

constant guidance [CELL BIOLOGY] The oriented response of isolated tissue cells in culture according to the topography of their substratum. { ˈkänˌtakt ˌgīd·əns }

Constellariidae [PALEONTOLOGY] A family of extinct, marine bryozoans in the order Cystoporata. { ˌkän·stə·ləˈrī·əˌdē }

constitutive enzyme [BIOCHEMISTRY] An enzyme whose concentration in a cell is constant and is not influenced by substrate concentration. { ˈkän·stəˌtüd·iv ˈenˌzīm }

constitutive gene [GENETICS] A gene that encodes a product required in the maintenance of basic cellular processes or cell architecture. Also known as housekeeping gene. { ˈkän·stəˌtüd·iv ˌjēn }

constitutive heterochromatin [GENETICS] A type of heterochromatin that is always condensed and is often centered on either side of the centromere, and that stains to give a C band. { ˈkän·stəˌtüd·iv ¦hed·ə·rōˈkrō·məd·ən }

constitutive mutation [GENETICS] A mutation that modifies an operator gene or a regulator gene, resulting in unregulated expression of structural genes that are normally regulated. { ˈkän·stəˌtüd·iv myüˈtā·shən }

consumer [ECOLOGY] A nutritional grouping in the food chain of an ecosystem, composed of heterotrophic organisms, chiefly animals, which ingest other organisms or particulate organic matter. { kənˈsüm·ər }

contact inhibition [CELL BIOLOGY] Cessation of cell division when cultured cells are in physical contact with each other. { ˈkänˌtakt in·əˈbish·ən }

contact paralysis [CELL BIOLOGY] The cessation of forward extension of the pseudopods of a cell as a result of its collision with another cell. { 'kän,takt pə,ral·ə·səs }

containment [MOLECULAR BIOLOGY] Prevention of the replication of the products of recombinant deoxyribonucleic acid technology outside the laboratory. { kən'tān·mənt }

contamination [MICROBIOLOGY] The process or act of soiling with bacteria. { kən,tam·ə'nā·shən }

contig [MOLECULAR BIOLOGY] A group of cloned nucleotide sequences that are contiguous. { kən 'tig }

continuous cell line [CELL BIOLOGY] A group of morphologically uniform cells that can be propagated in vitro for an indefinite time. { kən¦tin·yə·wəs 'sel ,līn }

contorted [BOTANY] Twisted; applied to proximate leaves whose margins overlap. { kən'tȯrd·əd }

contour feather [VERTEBRATE ZOOLOGY] Any of the large flight feathers or long tail feathers of a bird. Also known as penna; vane feather. { 'kän ,tu̇r ,feth·ər }

contractile [BIOLOGY] Displaying contraction; having the property of contracting. { kən'trak·təl }

contractile vacuole [CELL BIOLOGY] A tiny, intracellular, membranous bladder that functions in maintaining intra- and extracellular osmotic pressures in equilibrium, as well as excretion of water, such as occurs in protozoans. { kən'trak·təl 'vak·yə·wōl }

contraction [PHYSIOLOGY] Shortening of the fibers of muscle tissue. { kən'trak·shən }

contralateral [PHYSIOLOGY] Opposite; acting in unison with a similar part on the opposite side of the body. { ¦kän·trə'lad·ə·rəl }

control element [MOLECULAR BIOLOGY] A site within a gene or operon that acts to control gene expression. { kən'trōl ,el·ə·mənt }

control gene [GENETICS] A gene that regulates the time and rate at which neighboring structural genes are transcribed in messenger ribonucleic acid. { kən'trōl ,jēn }

controller node [GENETICS] A genetic unit of regulation consisting of a set of regulators, effectors, and receptors located near the gene that the unit controls. { kən'trō·lər ,nōd }

controlling element [GENETICS] Any of a class of transposable genetic elements that have the capacity to control gene expression at several loci, as well as to render target genes extremely likely to mutate. { kən'trōl·iŋ 'el·ə·mənt }

Conularida [PALEONTOLOGY] A small group of extinct invertebrates showing a narrow, four-sided, pyramidal-shaped test. { ,kän·əl'ar·ə·də }

Conulata [INVERTEBRATE ZOOLOGY] A subclass of free-living cnidarians in the class Scyphozoa; individuals are described as tetraramous cones to elongate pyramids having tentacles on the oral margin. { ,kän·əl'äd·ə }

Conulidae [PALEONTOLOGY] A family of Cretaceous exocyclic Euechinoidea characterized by a flattened oral surface. { kə'nü·lə,dē }

conus arteriosus [EMBRYOLOGY] The cone-shaped projection from which the pulmonary artery arises on the right ventricle of the heart in man and mammals. { 'kō·nəs är,tir·ē'ō·səs }

convalescent serum [IMMUNOLOGY] The serum of the blood of one or more patients recovering from an infectious disease; used for prophylaxis of the particular infection. { ¦kän·və¦les·ənt 'sir·əm }

convergence [ANATOMY] The coming together of a group of afferent nerves upon a motoneuron of the ventral horn of the spinal cord. [EVOLUTION] Development of similarities between animals or plants of different groups resulting from adaptation to similar habitats. { kən'vər·jəns }

conversion polarity [GENETICS] A gradient in the frequency of gene conversion from one end of a gene to the other. { kən'vər·zhən pə'lar·əd·ē }

convivium [ECOLOGY] A population exhibiting differentiation within the species and isolated geographically, generally a subspecies or ecotype. { kən'viv·ē·əm }

convolute [BIOLOGY] Twisted or rolled together, specifically referring to leaves, mollusk shells, and renal tubules. { 'kän·və,lüt }

convolution [ANATOMY] A fold, twist, or coil of any organ, especially any one of the prominent convex parts of the brain, separated from each other by depressions or sulci. { ,kän·və'lü·shən }

Convolvulaceae [BOTANY] A large family of dicotyledonous plants in the order Polemoniales characterized by internal phloem, the presence of chlorophyll, two ovules per carpel, and plicate cotyledons. { kən,väl·və'lās·ē,ē }

Cooke unit [BIOLOGY] A unit for the standardization of pollen antigenicity. { 'ku̇k ,yü·nət }

Coombs serum [IMMUNOLOGY] An immune serum containing antiglobulin that is used in testing for Rh and other sensitizations. { 'kümz ,sir·əm }

Copeognatha [INVERTEBRATE ZOOLOGY] An equivalent name for Psocoptera. { ,kō·pē'äg·nə·thə }

Copepoda [INVERTEBRATE ZOOLOGY] An order of Crustacea commonly included in the Entomostraca; contains free-living, parasitic, and symbiotic forms. { kō'pep·ə·də }

Copodontidae [PALEONTOLOGY] An obscure family of Paleozoic fishes in the order Bradyodonti. { ,kō·pə'dän·tə,dē }

copperhead [VERTEBRATE ZOOLOGY] *Agkistrodon contortrix.* A pit viper of the eastern United States; grows to about 3 feet (90 centimeters) in length and is distinguished by its coppery-brown skin with dark transverse blotches. { 'käp·ər,hed }

coppice [ECOLOGY] A growth of small trees that are repeatedly cut down at short intervals; the new shoots are produced by the old stumps. { 'käp·əs }

coproantibody [IMMUNOLOGY] An antibody whose presence in the intestinal tract can be demonstrated by its presence in an extract of the feces. { ¦käp·rō'ant·i,bäd·ē }

coprophagy [ZOOLOGY] Feeding on dung or excrement. { kə'präf·ə·jē }

coprophilous [ECOLOGY] Living in dung. { kə 'präf·ə·ləs }

copulation [ZOOLOGY] The sexual union of two individuals, resulting in insemination or deposition of the male gametes in proximity to the female gametes. { ˌkäp·yəˈlā·shən }

copulatory bursa [INVERTEBRATE ZOOLOGY] **1.** A sac that receives the sperm during copulation in certain insects. **2.** The caudal expansion of certain male nematodes that functions as a clasper during copulation. { ˈkäp·yə·ləˌtȯr·ē ˈbər·sə }

copulatory organ [ANATOMY] An organ employed by certain male animals for insemination. { ˈkäp·yə·ləˌtȯr·ē ˌȯr·gən }

copulatory spicule [INVERTEBRATE ZOOLOGY] *See* spiculum. { ˈkäp·yə·ləˌtȯr·ē ˈspik·yəl }

copy error [MOLECULAR BIOLOGY] A mutation that occurs during deoxyribonucleic acid replication as a result of an error in base pairing. { ˈkäp·ē ˌer·ər }

coquina [INVERTEBRATE ZOOLOGY] A small marine clam of the genus *Donax*. { kōˈkē·nə }

Coraciidae [VERTEBRATE ZOOLOGY] The rollers, a family of Old World birds in the order Coraciiformes. { ˌkȯr·əˈsī·əˌdē }

Coraciiformes [VERTEBRATE ZOOLOGY] An order of predominantly tropical and frequently brightly colored birds. { ˌkȯr·əˌsī·əˈfȯrˌmēz }

coracoid [ANATOMY] One of the paired bones on the posterior-ventral aspect of the pectoral girdle in vertebrates. { ˈkȯr·əˌkȯid }

coracoid ligament [ANATOMY] The transverse ligament of the scapula which crosses over the suprascapular notch. { ˈkȯr·əˌkȯid ˈlig·ə·mənt }

coracoid process [ANATOMY] The beak-shaped process of the scapula. { ˈkȯr·əˌkȯid ˌpräs·əs }

coral [INVERTEBRATE ZOOLOGY] The skeleton of certain solitary and colonial anthozoan cnidarians; composed chiefly of calcium carbonate. { ˈkä·rəl }

Corallanidae [INVERTEBRATE ZOOLOGY] A family of sometimes parasitic, but often free-living, isopod crustaceans in the suborder Flabellifera. { ˌkä·rəˈlan·əˌdē }

Corallidae [INVERTEBRATE ZOOLOGY] A family of dimorphic cnidarians in the order Gorgonacea. { kəˈral·əˌdē }

Corallimorpharia [INVERTEBRATE ZOOLOGY] An order of solitary sea anemones in the subclass Zoantharia resembling coral in many aspects. { kə¦ral·əˌmȯrˈfar·ē·ə }

Corallinaceae [BOTANY] A family of red algae, division Rhodophyta, having compact tissue with lime deposits within and between the cell walls. { kəˌral·əˈnās·ēˌē }

coralline [INVERTEBRATE ZOOLOGY] Any animal that resembles coral, such as a bryozoan or hydroid. { ˈkär·əˌlēn }

coralline algae [BOTANY] Red algae belonging to the family Corallinaceae. { ˈkär·əˌlēn ˈal·jē }

corallite [INVERTEBRATE ZOOLOGY] Skeleton of an individual polyp in a compound coral. { ˈkär·əˌlīt }

coralloid [BIOLOGY] Resembling coral, or branching like certain coral. { ˈkär·əˌlȯid }

corallum [INVERTEBRATE ZOOLOGY] Skeleton of a compound coral. { kəˈral·əm }

Corbiculidae [INVERTEBRATE ZOOLOGY] A family of fresh-water bivalve mollusks in the subclass Eulamellibranchia; an important food in the Orient. { kȯr·bəˈkyül·əˌdē }

Cordaitaceae [PALEOBOTANY] A family of fossil plants belonging to the Cordaitales. { ˌkȯr·dāˌīˈtās·ēˌē }

Cordaitales [PALEOBOTANY] An extensive natural grouping of forest trees of the late Paleozoic. { ˌkȯr·dāˌīˈtā·lēz }

cordate [BOTANY] Heart-shaped; generally refers to a leaf base. { ˈkȯrˌdāt }

cord factor [MICROBIOLOGY] A toxic glycolipid found as a surface component of tubercle bacilli that is responsible for virulence and serpentine growth. { ˈkȯrd ˌfak·tər }

cordon [BOTANY] A plant trained to grow flat against a vertical structure, in a single horizontal shoot or two opposed horizontal shoots. { ˈkȯrd·ən }

Cordulegasteridae [INVERTEBRATE ZOOLOGY] A family of anisopteran insects in the order Odonata. { ¦kȯrd·yə·ləˌgaˈster·əˌdē }

Coreidae [INVERTEBRATE ZOOLOGY] The squash bugs and leaf-footed bugs, a family of hemipteran insects belonging to the superfamily Coreoidea. { kəˈrē·əˌdē }

coremium [MYCOLOGY] A small bundle of conidiophores in certain imperfect fungi. { kəˈrē·mē·əm }

corepressor [MOLECULAR BIOLOGY] A certain metabolite which, through combination with a repressor apoprotein produced by a regulator gene, can cause the binding of the protein to the operator gene region of a deoxyribonucleic acid chain. { ¦kō·riˌpres·ər }

coriaceous [BIOLOGY] Leathery, applied to leaves and certain insects. { ¦kȯr·ē¦ā·shəs }

coriander [BOTANY] *Coriandrum sativum*. A strong-scented perennial herb in the order Umbellales; the dried fruit is used as a flavoring. { ˌkȯr·ēˈan·dər }

Cori ester [BIOCHEMISTRY] *See* glucose-1-phosphate. { ˈkȯr·ē ˌes·tər }

Coriolis effect [PHYSIOLOGY] The physiological effects (nausea, vertigo, dizziness, and so on) felt by a person moving radially in a rotating system, as a rotating space station. { kȯr·ēˈō·ləs iˈfekt }

corium [ANATOMY] *See* dermis. [INVERTEBRATE ZOOLOGY] Middle portion of the forewing of hemipteran insects. { ˈkȯr·ē·əm }

Corixidae [INVERTEBRATE ZOOLOGY] The water boatmen, the single family of the hemipteran superfamily Corixoidea. { kəˈrik·səˌdē }

Corixoidea [INVERTEBRATE ZOOLOGY] A superfamily of hemipteran insects belonging to the subdivision Hydrocorisae that lack ocelli. { kəˌrik ˈsȯid·ē·ə }

cork [BOTANY] A protective layer of cells that replaces the epidermis in older plant stems. { kȯrk }

corm [BOTANY] A short, erect, fleshy underground stem, usually broader than high and covered with membranous scales. { kȯrm }

cormatose [BOTANY] Having or producing a corm. { ˈkȯr·məˌtōs }

cormidium [INVERTEBRATE ZOOLOGY] The assemblage of individuals dangling in clusters from the main stem of pelagic siphonophores. { kȯr'mid·ē·əm }

corn [BOTANY] *Zea mays.* A grain crop of the grass order Cyperales grown for its edible seeds (technically fruits). { kȯrn }

Cornaceae [BOTANY] A family of dicotyledonous plants in the order Cornales characterized by perfect or unisexual flowers, a single ovule in each locule, as many stamens as petals, and opposite leaves. { kȯr'nās·ē,ē }

Cornales [BOTANY] An order of dicotyledonous plants in the subclass Rosidae marked by a woody habit, simple leaves, well-developed endosperm, and fleshy fruits. { kȯr'nā·lēz }

cornea [ANATOMY] The transparent anterior portion of the outer coat of the vertebrate eye covering the iris and the pupil. [INVERTEBRATE ZOOLOGY] The outer transparent portion of each ommatidium of a compound eye. { 'kȯr·nē·ə }

cornicle [INVERTEBRATE ZOOLOGY] Either of two protruding horn-shaped dorsal tubes in aphids which secrete a waxy fluid. { 'kȯr·nə·kəl }

corniculate [BIOLOGY] Possessing small horns or hornlike processes. { kȯr'nik·yə·lət }

corniculate cartilage [ANATOMY] The cartilaginous nodule on the tip of the arytenoid cartilage. { kȯr'nik·yə·lət ¦kärt·lij }

cornification [PHYSIOLOGY] Conversion of stratified squamous epithelial cells into a horny layer and into derivatives such as nails, hair, and feathers. { ,kȯr·nə·fə'kā·shən }

corn sugar [BIOCHEMISTRY] *See* dextrose. { 'kȯrn ,shu̇g·ər }

cornu [ANATOMY] A horn or hornlike structure. { 'kȯr·nü }

corolla [BOTANY] Collectively, the petals of a flower. { kə'räl·ə }

corollate [BOTANY] Having a corolla. { kə'rä ,lāt }

corolline [BOTANY] Relating to, resembling, or being borne on a corolla. { kȯr·ə,līn }

corona [BOTANY] **1.** An appendage or series of fused appendages between the corolla and stamens of some flowers. **2.** The region where stem and root of a seed plant merge. Also known as crown. [INVERTEBRATE ZOOLOGY] **1.** The anterior ring of cilia in rotifers. **2.** A sea urchin test. **3.** The calyx and arms of a crinoid. { kə'rō·nə }

coronal suture [ANATOMY] The union of the frontal with the parietal bones transversely across the vertex of the skull. { kə'rō·nəl 'sü·chər }

corona radiata [HISTOLOGY] The layer of cells immediately surrounding a mammalian ovum. { kə 'rō·nə ,rā·dē'äd·ə }

coronary artery [ANATOMY] Either of two arteries arising in the aortic sinuses that supply the heart tissue with blood. { 'kär·ə,ner·ē 'ärd·ə·rē }

coronary sinus [ANATOMY] A venous sinus opening into the heart's right atrium which drains the cardiac veins. { 'kär·ə,ner·ē 'sī·nəs }

coronary sulcus [ANATOMY] A groove in the external surface of the heart separating the atria from the ventricles, containing the trunks of the nutrient vessels of the heart. { 'kär·ə,ner·ē 'səl·kəs }

coronary valve [ANATOMY] A semicircular fold of the endocardium of the right atrium at the orifice of the coronary sinus. { 'kär·ə,ner·ē 'valv }

coronary vein [ANATOMY] **1.** Any of the blood vessels that bring blood from the heart and empty into the coronary sinus. **2.** A vein along the lesser curvature of the stomach. { 'kär·ə ,ner·ē 'vān }

Coronatae [INVERTEBRATE ZOOLOGY] An order of the class Scyphozoa which includes mainly abyssal species having the exumbrella divided into two parts by a coronal furrow. { ,kȯr·ə'näd·ē }

Coronaviridae [VIROLOGY] A family of vertebrate viruses consisting of the single genus Coronavirus; the prototype, avian infectious virus, has an enveloped spherical form with large spikes and a helical nucleocapsid with single-stranded ribonucleic acid. { kə,rō·nə'vī·rə,dē }

coronavirus [VIROLOGY] A major group of animal viruses including avian infectious bronchitis virus and mouse hepatitis virus. { kə¦rō·nə¦vī·rəs }

coronoid [BIOLOGY] Shaped like a beak. { 'kȯr·ə ,nȯid }

coronoid fossa [ANATOMY] A depression in the humerus into which the apex of the coronoid process of the ulna fits in extreme flexion of the forearm. { 'kȯr·ə,nȯid ¦fäs·ə }

coronoid process [ANATOMY] **1.** A thin, flattened process projecting from the anterior portion of the upper border of the ramus of the mandible, and serving for the insertion of the temporal muscle. **2.** A triangular projection from the upper end of the ulna, forming the lower part of the radial notch. { 'kȯr·ə,nȯid ¦präs·əs }

coronule [INVERTEBRATE ZOOLOGY] A peripheral ring of spines on some diatom shells. { 'kȯr·ə ,nyül }

Corophiidae [INVERTEBRATE ZOOLOGY] A family of amphipod crustaceans in the suborder Gammaridea. { kȯr·ə'fī·ə,dē }

corpora quadrigemina [ANATOMY] The inferior and superior colliculi collectively. Also known as quadrigeminal body. { 'kȯr·pə·rə ,kwäd·rə'jem·ə·nə }

corpus albicans [HISTOLOGY] The white fibrous scar in an ovary; produced by the involution of the corpus luteum. { 'kȯr·pəs 'al·bə,kanz }

corpus allatum [INVERTEBRATE ZOOLOGY] An endocrine structure near the brain of immature arthropods that secretes a juvenile hormone, neotenin. { 'kȯr·pəs ə'lād·əm }

corpus callosum [ANATOMY] A band of nerve tissue connecting the cerebral hemispheres in humans and higher mammals. { 'kȯr·pəs kə'lō·səm }

corpus cardiacum [INVERTEBRATE ZOOLOGY] One of a pair of separate or fused bodies of nervous tissue in many insects that lie posterior to the brain and dorsal to the esophagus and that function in the storage and secretion of brain hormone. { 'kȯr·pəs ,kärd·ē'ak·əm }

corpus cavernosum [ANATOMY] The cylinder of

erectile tissue forming the clitoris in the female and the penis in the male. { 'kȯr·pəs ˌka·vər 'nō·səm }

corpus cerebelli [ANATOMY] The central lobe or zone of the cerebellum; regulates reflex tonus of postural muscles in mammals. { 'kȯr·pəs ˌser·ə 'bel·ē }

corpuscle [ANATOMY] **1.** A small, rounded body. **2.** An encapsulated sensory-nerve end organ. { 'kȯr·pəs·əl }

corpus luteum [HISTOLOGY] The yellow endocrine body formed in the ovary at the site of a ruptured Graafian follicle. { 'kȯr·pəs 'lüd·ē·əm }

corpus striatum [ANATOMY] The caudate and lenticular nuclei, together with the internal capsule which separates them. { 'kȯr·pəs ˌstrī'ād·əm }

corresponding points [PHYSIOLOGY] Any two retinal areas in the respective eyes so that the area in one eye has an identical direction in the opposite retina. { ˌkär·ə'spänd·iŋ 'pȯins }

corridor [ECOLOGY] A land bridge that allows free migration of fauna in both directions. { 'kär·ə·dər }

Corrodentia [INVERTEBRATE ZOOLOGY] The equivalent name for Psocoptera. { ˌkȯr·ə'dench·ə }

cortex [ANATOMY] The outer portion of an organ or structure, such as of the brain and adrenal glands. [BOTANY] A primary tissue in roots and stems of vascular plants that extends inward from the epidermis to the phloem. [CELL BIOLOGY] A peripheral layer in many cells that includes the plasma membrane and associated cytoskeletal and extracellular components. [INVERTEBRATE ZOOLOGY] The peripheral layer of certain protozoans. { 'kȯrˌteks }

cortical granule [CELL BIOLOGY] Any of the round to elliptical membrane-bound bodies that occur in the cortex of animal oocytes, contain mucopolysaccharides, and participate in formation of the fertilization membrane. { 'kȯrd·ə·kəl 'gran·yəl }

corticoid [BIOCHEMISTRY] *See* adrenal cortex hormone. { 'kȯrd·əˌkȯid }

corticosteroid [BIOCHEMISTRY] **1.** Any steroid hormone secreted by the adrenal cortex of vertebrates. **2.** Any steroid with properties of an adrenal cortex steroid. { ¦kȯrd·əˌkō'stirˌȯid }

corticosterone [BIOCHEMISTRY] $C_{21}H_{30}O_4$ A steroid hormone produced by the adrenal cortex of vertebrates that stimulates carbohydrate synthesis and protein breakdown and is antagonistic to the action of insulin. { ˌkȯrd·ə'käs·təˌrōn }

corticotrophic [PHYSIOLOGY] Having an effect on the adrenal cortex. { ¦kȯrd·əˌkō¦trä·fik }

corticotropin [BIOCHEMISTRY] A hormonal preparation having adrenocorticotropic activity, derived from the adenohypophysis of certain domesticated animals. { ˌkȯrd·əˌkō'trō·pən }

corticotropin-releasing hormone [BIOCHEMISTRY] A substance produced by the hypothalamus that stimulates the pituitary gland to produce adrenocorticotropic hormone (ACTH). Abbreviated CRH. { ˌkȯrd·ə·kō'trō·pən ri¦lēs·iŋ ˌhȯrˌmōn }

cortin unit [BIOLOGY] A unit for the standardization of adrenal cortical hormones. { 'kȯrt·ən ˌyü·nət }

cortisol [BIOCHEMISTRY] *See* hydrocortisone. { 'kȯrd·əˌsȯl }

cortisone [BIOCHEMISTRY] $C_{21}H_{28}O_5$ A steroid hormone produced by the adrenal cortex of vertebrates that acts principally in carbohydrate metabolism. { 'kȯrd·əˌsōn }

Corvidae [VERTEBRATE ZOOLOGY] A family of large birds in the order Passeriformes having stout, long beaks; includes the crows, jays, and magpies. { 'kȯr·vəˌdē }

Corylophidae [INVERTEBRATE ZOOLOGY] The equivalent name for Orthoperidae. { ˌkȯr·ə'läf·əˌdē }

corymb [BOTANY] An inflorescence in which the flower stalks arise at different levels but reach the same height, resulting in a flat-topped cluster. { 'kōˌrim }

corymbose [BOTANY] Resembling or pertaining to a corymb. { kə'rimˌbōs }

Corynebacteriaceae [MICROBIOLOGY] Formerly a family of nonsporeforming, usually nonmotile rod-shaped bacteria in the order Eubacteriales including animal and plant parasites and pathogens. { ¦kȯr·əˌnēˌbakˌtir·ē'ās·ēˌē }

corynebacteriophage [VIROLOGY] Any bacteriophage able to infect *Corynebacterium* species. { ¦kȯr·əˌnē·bak'tir·ē·əˌfāzh }

Corynebacterium [MICROBIOLOGY] A genus of gram-positive, straight or slightly curved rods in the coryneform group of bacteria; club-shaped swellings are common; includes human and animal parasites and pathogens, and plant pathogens. { ¦kȯr·əˌnē·bak'tir·ē·əm }

Corynebacterium diphtheriae [MICROBIOLOGY] A facultatively aerobic, nonmotile species of bacteria that causes diphtheria in humans. Also known as Klebs-Loeffler bacillus. { ¦kȯr·əˌnē·bak'tir·ē·əm dif'thir·ēˌī }

Coryphaenidae [VERTEBRATE ZOOLOGY] A family of pelagic fishes in the order Perciformes characterized by a blunt nose and deeply forked tail. { ˌkȯr·ə'fēn·əˌdē }

Coryphodontidae [PALEONTOLOGY] The single family of the Coryphodontoidea, an extinct superfamily of mammals. { ˌkȯr·ə·fə'dän·təˌdē }

Coryphodontoidea [PALEONTOLOGY] A superfamily of extinct mammals in the order Pantodonta. { ˌkȯr·ə·fəˌdän'tȯid·ē·ə }

Cosmocercidae [INVERTEBRATE ZOOLOGY] A group of nematodes assigned to the suborder Oxyurina by some authorities and to the suborder Ascaridina by others. { ˌkäz·mə'sər·səˌdē }

Cosmocercoidea [INVERTEBRATE ZOOLOGY] A superfamily of parasitic nematodes having either three or six lips surrounding a weakly developed stoma. { ˌkäz·mō·sər'kȯid·ē·ə }

cosmoid scale [VERTEBRATE ZOOLOGY] A structure in the skin of primitive rhipidistians and dipnoans that is composed of enamel, a dentine layer (cosmine), and laminated bone. { 'käz ˌmȯid 'skāl }

cosmopolitan [ECOLOGY] Having a worldwide distribution wherever the habitat is suitable, with reference to the geographical distribution of a taxon. { ¦käz·mə¦päl·ət·ən }

Cossidae [INVERTEBRATE ZOOLOGY] The goat or carpenter moths, a family of heavy-bodied lepidopteran insects in the superfamily Cossoidea having the abdomen extending well beyond the hindwings. { 'käs·ə,dē }

Cossoidea [INVERTEBRATE ZOOLOGY] A monofamilial superfamily of lepidopteran insects belonging to suborder Heteroneura. { kə 'sȯid·ē·ə }

Cossuridae [INVERTEBRATE ZOOLOGY] A family of fringe worms belonging to the Sedentaria. { kə 'syu̇r·ə,dē }

Cossyphodidae [INVERTEBRATE ZOOLOGY] The lively ant guest beetles, a small family of coleopteran insects in the superfamily Tenebrionoidea. { ,käs·ə'fä·də,dē }

costa [BIOLOGY] A rib or riblike structure. [BOTANY] The midrib of a leaf. [INVERTEBRATE ZOOLOGY] The anterior vein of an insect's wing. { 'käs·tə }

Costaceae [BOTANY] A family of monocotyledonous plants in the order Zingiberales distinguished by having one functional stamen with two pollen sacs and spirally arranged leaves and bracts. { kȯs'tās·ē,ē }

costal cartilage [ANATOMY] The cartilage occupying the interval between the ribs and the sternum or adjacent cartilages. { 'käst·əl ¦kärd·əl·ij }

costal process [ANATOMY] An anterior or ventral projection on the lateral part of a cervical vertebra. [EMBRYOLOGY] An embryonic rib primordium, the ventrolateral outgrowth of the caudal, denser half of a sclerotome. { 'käst·əl ¦präs·əs }

costate [BIOLOGY] Having ribs or ridges. { 'kä ,stāt }

Cotingidae [VERTEBRATE ZOOLOGY] The cotingas, a family of neotropical suboscine birds in the order Passeriformes. { kō'tin·jə,dē }

Cottidae [VERTEBRATE ZOOLOGY] The sculpins, a family of perciform fishes in the suborder Cottoidei. { 'käd·ə,dē }

Cottiformes [VERTEBRATE ZOOLOGY] An order set up in some classification schemes to include the Cottoidei. { ,käd·ə'fȯr,mēz }

Cottoidei [VERTEBRATE ZOOLOGY] The mail-cheeked fishes, a suborder of the order Perciformes characterized by the expanded third infraorbital bone. { kə'tȯid·ē,ī }

cotton [BOTANY] Any plant of the genus *Gossypium* in the order Malvales; cultivated for the fibers obtained from its encapsulated fruits or bolls. { 'kät·ən }

cottonmouth [VERTEBRATE ZOOLOGY] *See* water moccasin. { 'kät·ən,mau̇th }

cottonwood [BOTANY] Any of several poplar trees (*Populus*) having hairy, encapsulated fruit. { 'kät·ən,wu̇d }

cotyledon [BOTANY] The first leaf of the embryo of seed plants. { ,käd·əl'ēd·ən }

cotylocercous cercaria [INVERTEBRATE ZOOLOGY] A digenetic trematode larva characterized by a sucker or adhesive gland on the tail. { käd·əl'äs·ə·rəs ,sər'kar·ē·ə }

Cotylosauria [PALEONTOLOGY] An order of primitive reptiles in the subclass Anapsida, including the stem reptiles, ancestors of all of the more advanced Reptilia. { ¦käd·əl·ə¦sȯr·ē·ə }

cotype [SYSTEMATICS] *See* syntype. { 'kō,tīp }

cougar [VERTEBRATE ZOOLOGY] *See* puma. { 'kü·gər }

Couinae [VERTEBRATE ZOOLOGY] The couas, a subfamily of Madagascan birds in the family Cuculidae. { 'kü·ə,nē }

Coulter counter [MICROBIOLOGY] An electronic device for counting the number of cells in a liquid culture. { 'kōl·tər ¦kau̇nt·ər }

coumarin glycoside [BIOCHEMISTRY] Any of several glycosidic aromatic principles in many plants; contains coumaric acid as the aglycon group. { 'kü·mə·rən 'glī·kə,sīd }

counterimmunoelectrophoresis [IMMUNOLOGY] Immunoelectrophoresis which uses two wells of application, one above the other, along the electrical axis—the anodal well filled with antibody and the cathodal with a negatively charge antigen; electrophoresis results in the antigen and antibody migrating cathodally and anodally, respectively, and a line of precipitation appears where the two meet. { ¦kau̇nt·ər¦im·yə·nō·i ,lek·trō·fə'r ē·səs }

counterstain [BIOLOGY] A second stain applied to a biological specimen to color elements other than those demonstrated by the principal stain. { 'kau̇nt·ər,stān }

counting chamber [MICROBIOLOGY] An accurately dimensioned chamber in a microslide which can hold a specific volume of fluid and which is usually ruled into units to facilitate the counting under the microscope of cells, bacteria, or other structures in the fluid. { 'kau̇nt·iŋ ,chām·bər }

covert [ECOLOGY] A refuge or shelter, such as a coppice, for game animals. { 'kō·vərt }

covey [VERTEBRATE ZOOLOGY] **1.** A brood of birds. **2.** A small flock of birds of one kind, used typically of partridge and quail. { 'kəv·ē }

cow [VERTEBRATE ZOOLOGY] A mature female cattle of the genus *Bos*. { kau̇ }

Cowdria [MICROBIOLOGY] A genus of the tribe Ehrlichieae; coccoid to ellipsoidal, pleomorphic, or rod-shaped cells; intracellular parasites in cytoplasm and vacuoles of vascular endothelium of ruminants. { 'kau̇·drē·ə }

cowpea [BOTANY] *Vigna sinensis*. An annual legume in the order Rosales cultivated for its edible seeds. Also known as blackeye bean. { 'kau̇,pē }

Cowper's gland [ANATOMY] *See* bulbourethral gland. { 'küp·ərz ,gland }

cowpox virus [VIROLOGY] The causative agent of cowpox in cattle. { 'kau̇,päks ,vī·rəs }

coxa [INVERTEBRATE ZOOLOGY] The proximal or basal segment of the leg of insects and certain other arthropods which articulates with the body. { 'käk·sə }

coxal cavity [INVERTEBRATE ZOOLOGY] A cavity in

which the coxa of an arthropod limb articulates. { 'käk·səl 'kav·əd·ē }

coxal gland [INVERTEBRATE ZOOLOGY] One of certain paired glands with ducts opening in the coxal region of arthropods. { 'käk·səl 'gland }

Coxiella [MICROBIOLOGY] A genus of the tribe Rickettsieae; short rods which grow preferentially in host cell vacuoles. { ,käk·sē'el·ə }

coxopodite [INVERTEBRATE ZOOLOGY] The basal joint of a crustacean limb. { käk'säp·ə,dīt }

coxsackievirus [VIROLOGY] A large subgroup of the enteroviruses in the picornavirus group including various human pathogens. { ku̇k'säk·ē ,vī·rəs }

coyote [VERTEBRATE ZOOLOGY] *Canis latrans*. A small wolf native to western North America but found as far eastward as New York State. Also known as prairie wolf. { 'kī,ōd·ē }

C_3 plant [BOTANY] A plant that produces the 3-carbon compound phosphoglyceric acid as the first stage of photosynthesis. { ¦sē'thrē ,plant }

C_4 plant [BOTANY] A plant that produces the 4-carbon compound oxalocethanoic (oxaloacetic) acid as the first stage of photosynthesis. { ¦sē'fȯr ,plant }

crab [INVERTEBRATE ZOOLOGY] **1.** The common name for a number of crustaceans in the order Decapoda having five pairs of walking legs, with the first pair modified as chelipeds. **2.** The common name for members of the Merostoma. { krab }

crabapple [BOTANY] Any of several trees of the genus *Malus*, order Rosales, cultivated for their small, edible pomes. { 'krab,ap·əl }

Cracidae [VERTEBRATE ZOOLOGY] A family of New World tropical upland game birds in the order Galliformes; includes the chachalacas, guans, and curassows. { 'kra·sə,dē }

Crambiinae [INVERTEBRATE ZOOLOGY] The snout moths, a subfamily of lepidopteran insects in the family Pyralididae containing small marshland and grassland forms. { kram'bī·ə,nē }

cranberry [BOTANY] Any of several plants of the genus *Vaccinium*, especially *V. macrocarpon*, in the order Ericales, cultivated for its small, edible berries. { 'kran,ber·ē }

Cranchiidae [INVERTEBRATE ZOOLOGY] A family of cephalopod mollusks in the subclass Dibranchia. { ,kraŋ'kī·ə,dē }

crane [VERTEBRATE ZOOLOGY] The common name for the long-legged wading birds composing the family Gruidae of the order Gruiformes. { krān }

Craniacea [INVERTEBRATE ZOOLOGY] A family of inarticulate branchiopods in the suborder Craniidina. { ,krā·nē'ās·ē·ə }

cranial capacity [ANATOMY] The volume of the cranial cavity. { 'krān·ē·əl kə'pas·əd·ē }

cranial flexure [EMBRYOLOGY] A flexure of the embryonic brain. { 'krān·ē·əl 'flek·shər }

cranial fossa [ANATOMY] Any of the three depressions in the floor of the interior of the skull. { 'krān·ē·əl 'fäs·ə }

cranial nerve [ANATOMY] Any of the paired nerves which arise in the brainstem of vertebrates and pass to peripheral structures through openings in the skull. { 'krān·ē·əl 'nərv }

Craniata [VERTEBRATE ZOOLOGY] A major subdivision of the phylum Chordata comprising the vertebrates, from cyclostomes to mammals, distinguished by a cranium. { ,krā·nē'ād·ə }

Craniidina [INVERTEBRATE ZOOLOGY] A subdivision of inarticulate branchiopods in the order Acrotretida known to possess a pedicle; all forms are attached by cementation. { krā·ne'īd·ən·ə }

craniobuccal pouch [EMBRYOLOGY] A diverticulum from the buccal cavity in the embryo from which the anterior lobe of the hypophysis is developed. Also known as Rathke's pouch. { ¦krā·nē·ō'bək·əl 'pau̇ch }

cranium [ANATOMY] That portion of the skull enclosing the brain. Also known as braincase. { 'krā·nē·əm }

craspedon [INVERTEBRATE ZOOLOGY] A cnidarian medusa stage possessing a velum. { 'kras·pə ,dän }

craspedote [INVERTEBRATE ZOOLOGY] Having a velum, used specifically for velate hydroid medusae. { 'kras·pə,dōt }

Crassulaceae [BOTANY] A family of dicotyledonous plants in the order Rosales notable for their succulent leaves and resistance to desiccation. { ,kras·ə'las·e,e }

craw [ZOOLOGY] **1.** The crop of a bird or insect. **2.** The stomach of a lower animal. { krȯ }

crayfish [INVERTEBRATE ZOOLOGY] The common name for a number of lobsterlike fresh-water decapod crustaceans in the section Astacura. { 'krā,fish }

C-reactive protein [IMMUNOLOGY] A plasma protein that is present normally in low concentration, and after trauma or infection in much higher concentration; the biological function is unknown. { ¦sē rē,ak·tiv 'prō,tēn }

creatine [BIOCHEMISTRY] $C_4H_9O_2N_3$ α-Methylguanidine-acetic acid; a compound present in vertebrate muscle tissue, principally as phosphocreatine. { 'krē·ə,tēn }

creatine kinase [BIOCHEMISTRY] An enzyme of vertebrate skeletal and myocardial muscle that catalyzes the transfer of a high-energy phosphate group from phosphocreatine to adenosinediphosphate with the formation of adenosinetriphosphate and creatine. { ¦krē·ə,tēn 'kī,nās }

creatinine [BIOCHEMISTRY] $C_4H_7ON_3$ A compound present in urine, blood, and muscle that is formed from the dehydration of creatine. { ,krē'at·ən,ēn }

creeping disk [INVERTEBRATE ZOOLOGY] The smooth and adhesive undersurface of the foot or body of a mollusk or of certain other invertebrates, on which the animal creeps. { 'krē·piŋ ,disk }

cremocarp [BOTANY] A dry dehiscent fruit consisting of two indehiscent one-seeded mericarps which separate at maturity and remain pendant from the carpophore. { 'krem·ə,kärp }

crenate [BIOLOGY] Having a scalloped margin; used specifically for foliar structures, shrunken erythrocytes, and shells of certain mollusks. { 'krē,nāt }

crenation [PHYSIOLOGY] A notched appearance of shrunken erythrocytes; seen when they are exposed to the air or to strong saline solutions. { krə'nā·shən }

Crenothrix [MICROBIOLOGY] A genus of sheathed bacteria; cells are nonmotile, sheaths are attached and encrusted with iron and manganese oxides, and filaments may have swollen tips. { 'kren·ə,thriks }

Crenotrichaceae [MICROBIOLOGY] Formerly a family of bacteria in the order Chlamydobacteriales having trichomes that are differentiated at the base and tip and attached to a firm substrate. { ¦kren·ə·trə'kās·ē,ē }

crenulate [BIOLOGY] Having a minutely crenate margin. { 'kren·əl,āt }

Creodonta [PALEONTOLOGY] A group formerly recognized as a suborder of the order Carnivora. { ,krē·ə'dän·tə }

creosote bush [BOTANY] *Larrea divaricata*. A bronze-green, xerophytic shrub characteristic of all the American warm deserts. { 'krē·ə,sōt ,bu̇sh }

crepuscular [ZOOLOGY] Active during the hours of twilight or preceding dawn. { krə'pəs·kyə·lər }

cress [BOTANY] Any of several prostrate crucifers belonging to the order Capparales and grown for their flavorful leaves; includes watercress (*Nasturtium officinale*), garden cress (*Lepidium sativum*), and upland or spring cress (*Barbarea verna*). { kres }

CRH [BIOCHEMISTRY] *See* corticotropin-releasing hormone.

cribellum [INVERTEBRATE ZOOLOGY] **1.** A small accessory spinning organ located in front of the ordinary spinning organ in certain spiders. **2.** A chitinous plate perforated with the openings of certain gland ducts in insects. { krə'bel·əm }

Cribrariaceae [BOTANY] A family of true slime molds in the order Liceales. { krə,brer·ē'ās·ē,ē }

cribriform [BIOLOGY] Perforated, like a sieve. { 'krib·rə,fȯrm }

cribriform fascia [ANATOMY] The sievelike covering of the fossa ovalis of the thigh. { 'krib·rə ,fȯrm ¦fā·shə }

cribriform plate [ANATOMY] **1.** The horizontal plate of the ethmoid bone, part of the floor of the anterior cranial fossa. **2.** The bone lining a dental alveolus. { 'krib·rə,fȯrm ¦plāt }

Cricetidae [VERTEBRATE ZOOLOGY] A family of the order Rodentia including hamsters, voles, and some mice. { krə'sed·ə,dē }

Cricetinae [VERTEBRATE ZOOLOGY] A subfamily of mice in the family Cricetidae. { krə'set·ən,ē }

cricket [INVERTEBRATE ZOOLOGY] **1.** The common name for members of the insect family Gryllidae. **2.** The common name for any of several related species of orthopteran insects in the families Tettigoniidae, Gryllotalpidae, and Tridactylidae. { 'krik·ət }

cricoid [ANATOMY] The signet-ring-shaped cartilage forming the base of the larynx in humans and most other mammals. { 'krī,kȯid }

Criconematoidea [INVERTEBRATE ZOOLOGY] A superfamily of plant parasitic nematodes of the order Diplogasterida distinguished by their ectoparasitic habit and males that have atrophied mouthparts and do not feed. { ,krī·kō,nem·ə 'tȯid·ē·ə }

Crinoidea [INVERTEBRATE ZOOLOGY] A class of radially symmetrical crinozoans in which the adult body is flower-shaped and is either carried on an anchored stem or is free-living. { krə'nȯid·ē·ə }

Crinozoa [INVERTEBRATE ZOOLOGY] A subphylum of the Echinodermata comprising radially symmetrical forms that show a partly meridional pattern of growth. { 'krī·nə,zō·ə }

crissum [VERTEBRATE ZOOLOGY] **1.** The region surrounding the cloacal opening in birds. **2.** The vent feathers covering the circumcloacal region. { 'kris·əm }

crista [BIOLOGY] A ridge or crest. [CELL BIOLOGY] A fold on the inner membrane of a mitochondrion. { 'kris·tə }

cristate [BIOLOGY] Having a crista. { 'kri,stāt }

Cristispira [MICROBIOLOGY] A genus of bacteria in the family Spirochaetaceae; helical cells with 3-10 complete turns; they have ovoid inclusion bodies and bundles of axial fibrils; commensals in mollusks. { ,kris·tə'spī·rə }

crocodile [VERTEBRATE ZOOLOGY] The common name for about 12 species of aquatic reptiles included in the family Crocodylidae. { 'kräk·ə ,dīl }

Crocodylia [VERTEBRATE ZOOLOGY] An order of the class Reptilia which is composed of large, voracious, aquatic species, including the alligators, caimans, crocodiles, and gavials. { 'kräk·ə ¦dil·ē·ə }

Crocodylidae [VERTEBRATE ZOOLOGY] A family of reptiles in the order Crocodylia including the true crocodiles, false gavial, alligators, and caimans. { ,kräk·ə'dil·ə,dē }

Crocodylinae [VERTEBRATE ZOOLOGY] A subfamily of reptiles in the family Crocodylidae containing the crocodiles, *Osteolaemus*, and the false gavial. { ,kräk·ə'dil·ə,nē }

crocus [BOTANY] A plant of the genus *Crocus*, comprising perennial herbs cultivated for their flowers. { 'krō·kəs }

Cro-Magnon man [PALEONTOLOGY] **1.** A race of tall, erect Caucasoid men having large skulls; identified from skeletons found in southern France. **2.** A general term to describe all fossils resembling this race that belong to the upper Paleolithic (35,000-8000 B.C.) in Europe. { krō 'mag·nən 'man }

cron [EVOLUTION] A time unit equal to 10^6 years; used in reference to evolutionary processes. { krän }

crop [VERTEBRATE ZOOLOGY] A distensible saccular diverticulum near the lower end of the esophagus of birds which serves to hold and soften food before passage into the stomach. { kräp }

crossbreed [BIOLOGY] To propagate new individuals by breeding two distinctive varieties of a species. Also known as outbreed. { 'krȯs,brēd }

crossed electrophoresis [IMMUNOLOGY] Immunoelectrophoresis that uses an initial separation along one axis of the plate, after which only a strip of medium along the axis is preserved; new medium containing antisera is poured beside the strip, the plate is turned 90°, and electrophoresis is resumed. { 'kröst i,lek·trō·fə'rē·səs }

cross-fertilization [BOTANY] Fertilization between two separate plants. [ZOOLOGY] Fertilization between different kinds of individuals. { 'krös ,fərd·əl·ə'zā·shən }

crossing barrier [GENETICS] Any of the genetically controlled mechanisms that either prevent or significantly reduce the ability of individuals in a population to hybridize with individuals of other populations. { 'krös·iŋ ,bar·ē·ər }

crossing-over [GENETICS] The exchange of genetic material between paired homologous chromosomes during meiosis. Also known as crossover. { ¦krös·iŋ 'ō·vər }

crossing-over map [GENETICS] A genetic map made by utilizing the frequency of crossing-over as a measure of the relative distances between genes in one linkage group. { ¦krös·iŋ 'ō·vər ,map }

crossing-over unit [GENETICS] A frequency of genetic exchange of 1% between two pairs of linked genes. { ¦krös·iŋ 'ō·vər ,yü·nət }

crossing-over value [GENETICS] The frequency of crossing-over between two linked genes. { ¦krös·iŋ 'ō·vər ,val·yü }

cross-link [MOLECULAR BIOLOGY] A covalent linkage between the complementary strands of deoxyribonucleic acid (DNA) duplex or between bases of a single strand of DNA. { 'krös ,liŋk }

cross matching [IMMUNOLOGY] Determination of blood compatibility for transfusion by mixing donor cells with recipient serum, and recipient cells with donor serum, and examining for an agglutination reaction. { 'krös ,mach·iŋ }

Crossopterygii [PALEONTOLOGY] A subclass of the class Osteichthyes comprising the extinct lobefins or choanate fishes and represented by one extant species; distinguished by two separate dorsal fins. { krä,säp·tə'rij·ē,ī }

Crossosomataceae [BOTANY] A monogeneric family of xerophytic shrubs in the order Dilleniales characterized by perigynous flowers, seeds with thin endosperm, and small, entire leaves. { ¦krä·sə,sō·mə¦tās·ē,ē }

crossover [GENETICS] *See* crossing over. { 'krös ,ō·vər }

cross-pollination [BOTANY] Transfer of pollen from the anthers of one plant to the stigmata of another plant. { ¦krös ,pä·lə,nā·shən }

cross-reacting antibody [IMMUNOLOGY] Antibody that reacts with an antigen that did not stimulate the production of that antibody. { 'krös rē,ak·tiŋ 'ant·i,bäd·ē }

cross-reacting antigen [IMMUNOLOGY] Antigen that reacts with an antibody whose production was induced by a different antigen. { 'krös rē ,ak·tiŋ 'ant·i·jən }

cross-reacting material [BIOCHEMISTRY] A protein produced by a mutant gene that is enzymatically inactive but shows serological properties similar to the protein of the wild-type gene. { 'krös rē,ak·tiŋ mə'tir·ē·əl }

cross-reaction [IMMUNOLOGY] Reaction between an antibody and a closely related, but not complementary, antigen. { 'krös rē'ak·shən }

Crotalidae [VERTEBRATE ZOOLOGY] A family of proglyphodont venomous snakes in the reptilian suborder Serpentes. { krō'tal·ə,dē }

croup-associated virus [VIROLOGY] A virus belonging to subgroup 2 of the parainfluenza viruses and found in children with croup. Also known as CA virus; laryngotracheobronchitis virus. { ¦krüp ə¦sō·sē,ād·əd 'vī·rəs }

crow [VERTEBRATE ZOOLOGY] The common name for a number of predominantly black birds in the genus *Corvus* comprising the most advanced members of the family Corvidae. { krō }

crown [ANATOMY] **1.** The top of the skull. **2.** The portion of a tooth above the gum. [BOTANY] **1.** The topmost part of a plant or plant part. **2.** *See* corona. { kraün }

crown grafting [BOTANY] A method of vegetative propagation whereby a scion 3-6 inches (8-15 centimeters) long is grafted at the root crown, just below ground level. { 'kraün ,graf·tiŋ }

cruciate [ANATOMY] Resembling a cross. { 'krü·shē,āt }

Cruciferae [BOTANY] A large family of dicotyledonous herbs in the order Capparales characterized by parietal placentation; hypogynous, mostly regular flowers; and a two-celled ovary with the ovules attached to the margins of the partition. { krü'sif·ə,rē }

cruciform structure [MOLECULAR BIOLOGY] A cross-shaped configuration of deoxyribonucleic acid produced by intrastrand base pairing of complementary inverted repeats. { 'krü·sə,förm ,strək·chər }

crura [ANATOMY] Plural of crus. { 'krür·ə }

crus [ANATOMY] **1.** The shank of the hindleg, that portion between the femur and the ankle. **2.** Any of various parts of the body resembling a leg or root. { krüs }

Crustacea [INVERTEBRATE ZOOLOGY] A class of arthropod animals in the subphylum Mandibulata having jointed feet and mandibles, two pairs of antennae, and segmented, chitin-encased bodies. { krə'stā·shə }

crustecdysone [BIOCHEMISTRY] $C_{27}H_{44}O_7$ 20-Hydroxyecdysone, the molting hormone produced by Y organs in crustaceans. { krəs'tek·də ,sōn }

crustose [BOTANY] Of a lichen, forming a thin crustlike thallus which adheres closely to the substratum of rock, bark, or soil. { 'krəs,tōs }

crust vegetation [ECOLOGY] Zonal growths of algae, mosses, lichens, or liverworts having variable coverage and a thickness of only a few centimeters. { 'krəst ,vej·ə'tā·shən }

cryesthesia [PHYSIOLOGY] **1.** The sensation of coldness. **2.** Exceptional sensitivity to low temperatures. { ,krī·əs'thē·zhə }

cryobiology [BIOLOGY] The use of low-temperature environments in the study of living plants and animals. { ¦krī·ō·bī'äl·ə·jē }

cryobiosis [PHYSIOLOGY] A type of cryptobiosis induced by low temperatures. { ¦krī·ō·bī'ō·səs }

cryophilic [ECOLOGY] *See* cryophilous. { ˌkrī·ə'fil·ik }

cryophilous [ECOLOGY] Having a preference for low temperatures. Also known as cryophilic. { krī'äf·ə·ləs }

cryophyte [ECOLOGY] A plant that forms winter buds below the soil surface. { 'krī·əˌfīt }

cryoprecipitate [BIOCHEMISTRY] The precipitate of a cryoglobulin. { ¦krī·ō·prə'sip·əˌtāt }

Cryphaeaceae [BOTANY] A family of mosses in the order Isobryales distinguished by a rough calyptra. { ˌkrī·fē'ās·ēˌē }

crypt [ANATOMY] **1.** A follicle or pitlike depression. **2.** A simple glandular cavity. { kript }

cryptic behavior [ZOOLOGY] A behavior pattern that maximizes an organism's ability to conceal itself. { 'krip·tik bə'hāv·yər }

cryptic coloration [ZOOLOGY] A phenomenon of protective coloration by which an animal blends with the background through color matching or countershading. { 'krip·tik kəl·ə'rā·shən }

Cryptobiidae [INVERTEBRATE ZOOLOGY] A family of flagellate protozoans in the order Kinetoplastida including organisms with two flagella, one free and one with an undulating membrane. { ˌkrip·tō'bī·əˌdē }

cryptobiosis [PHYSIOLOGY] A state in which metabolic rate of the organism is reduced to an imperceptible level. { ¦krip·tō·bī'ō·səs }

cryptobiotic [ECOLOGY] Living in concealed or secluded situations. { ¦krip·tō·bī'äd·ik }

Cryptobranchidae [VERTEBRATE ZOOLOGY] The giant salamanders and hellbenders, a family of tailed amphibians in the suborder Cryptobranchoidea. { ˌkrip·tə'braŋ·kəˌdē }

Cryptobranchoidea [VERTEBRATE ZOOLOGY] A primitive suborder of amphibians in the order Urodela distinguished by external fertilization and aquatic larvae. { ˌkrip·təˌbraŋ'kȯid·ē·ə }

Cryptocerata [INVERTEBRATE ZOOLOGY] A division of hemipteran insects in some systems of classification that includes the water bugs (Hydrocorisae). { ˌkrip·tō·sə'räd·ə }

Cryptochaetidae [INVERTEBRATE ZOOLOGY] A family of myodarian cyclorrhaphous dipteran insects in the subsection Acalypteratae. { ˌkrip·tə'kēd·əˌdē }

Cryptococcaceae [MYCOLOGY] A family of imperfect fungi in the order Moniliales in some systems of classification; equivalent to the Cryptococcales in other systems. { ˌkrip·tə·käk'sā·sēˌē }

Cryptococcales [MYCOLOGY] An order of imperfect fungi, in some systems of classification, set up to include the yeasts or yeastlike organisms whose perfect or sexual stage is not known. { ˌkrip·tə·kä'kāˌlēz }

Cryptococcus [MYCOLOGY] A genus of encapsulated pathogenic yeasts in the order Moniliales. { ˌkrip·tə'käk·əs }

Cryptodira [VERTEBRATE ZOOLOGY] A suborder of the reptilian order Chelonia including all turtles in which the cervical spines are uniformly reduced and the head folds directly back into the shell. { ˌkrip·tə'dī·rə }

cryptogam [BOTANY] An old term for nonflowering plants. { 'krip·təˌgam }

cryptomedusa [INVERTEBRATE ZOOLOGY] The final stage in the reduction of a hydroid medusa to a rudiment having sex cells within the gonophore. { ¦krip·tō·mə'dü·sə }

cryptomitosis [INVERTEBRATE ZOOLOGY] Cell division in certain protozoans in which a modified spindle forms, and chromatin assembles with no apparent chromosome differentiation. { ¦krip·tō·mī'tō·səs }

Cryptomonadida [BIOLOGY] An order of the class Phytamastigophorea considered to be protozoans by biologists and algae by botanists. { ˌkrip·tō·mə'näd·ə·də }

Cryptomonadina [BIOLOGY] The equivalent name for Cryptomonadida. { ˌkrip·tō·män·ə'dī·nə }

cryptonephridic [INVERTEBRATE ZOOLOGY] In certain insects, referring to Malpighian tubules independently attached to the hindgut (in contrast to being free). { ¦krip·tō·ne¦frid·ik }

Cryptophagidae [INVERTEBRATE ZOOLOGY] The silken fungus beetles, a family of coleopteran insects in the superfamily Cucujoidea. { ˌkrip·tə'fāj·əˌdē }

Cryptophyceae [BOTANY] A class of algae of the Pyrrhophyta in some systems of classification; equivalent to the division Cryptophyta. { ˌkrip·tə'fīs·ēˌē }

Cryptophyta [BOTANY] A division of the algae in some classification schemes; equivalent to the Cryptophyceae. { ˌkrip·tə'fīd·ə }

cryptophyte [BOTANY] A plant that produces buds either underwater or underground on corms, bulbs, or rhizomes. { 'krip·təˌfīt }

Cryptopidae [INVERTEBRATE ZOOLOGY] A family of epimorphic centipedes in the order Scolopendromorpha. { krip'täp·əˌdē }

Cryptostomata [PALEONTOLOGY] An order of extinct bryozoans in the class Gymnolaemata. { ˌkrip·tə'stō·məd·ə }

cryptotope [IMMUNOLOGY] A determinant (or epitope) of an immunological antigen or immunogen which is initially hidden and becomes functional only when the molecule is broken or degraded. { 'krip·tōˌtōp }

cryptovirogenic [VIROLOGY] Possessing the ability to produce infective virus particles after derepression of the viral genome within the cell. { ¦krip·tōˌvī·rə'jen·ik }

cryptoxanthin [BIOCHEMISTRY] $C_{40}H_{57}O$ A xanthophyll carotenoid pigment found in plants; convertible to vitamin A by many animal livers. { ¦krip·tō'zan·thən }

cryptozoon [PALEOBOTANY] A hemispherical or cabbagelike reef-forming fossil algae, probably from the Cambrian and Ordovician. { ¦krip·tō'zō·ən }

crypts of Lieberkühn [ANATOMY] Simple, tubular glands which arise as evaginations into the mucosa of the small intestine. { 'krips əv 'lē·bər ˌkyün }

crystalliferous bacteria [MICROBIOLOGY] *Bacillus thuringiensis* and related species characterized by the formation of a protein crystal in the sporangium at the time of spore formation. { ¦kris·tə ¦lif·ə·rəs bak'tir·ē·ə }

crystalline lens [ANATOMY] *See* lens. { 'kris·tə·lən 'lenz }

crystallin protein [BIOCHEMISTRY] Any of a group of stable structural components distributed nonuniformly in the lens of the eye of vertebrates. { 'krist·əl·ən 'prōˌtēn }

ctenidium [INVERTEBRATE ZOOLOGY] **1.** The comb- or featherlike respiratory apparatus of certain mollusks. **2.** A row of spines on the head or thorax of some fleas. { tə'nid·ē·əm }

Ctenodrilidae [INVERTEBRATE ZOOLOGY] A family of fringe worms belonging to the Sedentaria. { ˌten·ə'drī·ləˌdē }

ctenoid scale [VERTEBRATE ZOOLOGY] A thin, acellular structure composed of bonelike material and characterized by a serrated margin; found in the skin of advanced teleosts. { 'tenˌȯid ˌskāl }

Ctenophora [INVERTEBRATE ZOOLOGY] The comb jellies, a phylum of marine organisms having eight rows of comblike plates as the main locomotory structure. { tə'näf·ə·rə }

Ctenostomata [INVERTEBRATE ZOOLOGY] An order of bryozoans in the class Gymnolaemata recognized as inconspicuous, delicate colonies made up of relatively isolated, short, tubular zooecia with chitinous walls. { ˌten·ə'stäm·ə·də }

Ctenostomatida [INVERTEBRATE ZOOLOGY] The equivalent name for Odontostomatida. { ˌten·ə·stə'mad·ə·də }

Ctenothrissidae [PALEONTOLOGY] A family of extinct teleostean fishes in the order Ctenothrissiformes. { ten·ə'thris·əˌdē }

Ctenothrissiformes [PALEONTOLOGY] A small order of extinct teleostean fishes; important as a group on the evolutionary line leading from the soft-rayed to the spiny-rayed fishes. { ˌten·ə ˌthris·ə'fȯrˌmēz }

C-type virus particle [VIROLOGY] One of a morphologically similar group of enveloped virus particles having a central, spherical ribonucleic acid-containing nucleoid; associated with certain cancers, as sarcomas and leukemias. { 'sē ¦tīp 'vī·rəs ˌpärd·ə·kəl }

cubeb [BOTANY] The dried, nearly ripe fruit (berries) of a climbing vine, *Piper cubeba*, of the pepper family (Piperaceae). { 'kyüˌbeb }

cuboid [ANATOMY] The outermost distal tarsal bone in vertebrates. [INVERTEBRATE ZOOLOGY] Main vein of the wing in many insects, particularly the flies (Diptera). { 'kyüˌbȯid }

cuboidal epithelium [HISTOLOGY] A single-layered epithelium made up of cubelike cells. { kyü 'bȯid·əl ˌep·ə'thēl·ē·əm }

Cubomedusae [INVERTEBRATE ZOOLOGY] An order of cnidarians in the class Scyphozoa distinguished by a cubic umbrella. { ¦kyü·bo·mə'dü·sē }

cuckoo [VERTEBRATE ZOOLOGY] The common name for about 130 species of primarily arboreal birds in the family Cuculidae; some are social parasites. { 'küˌkü }

Cucujidae [INVERTEBRATE ZOOLOGY] The flat-back beetles, a family of predatory coleopteran insects in the superfamily Cucujoidea. { kə'kü·yə ˌdē }

Cucujoidea [INVERTEBRATE ZOOLOGY] A large superfamily of coleopteran insects in the suborder Polyphaga. { ˌkü·kə'yȯid·ē·ə }

Cuculidae [VERTEBRATE ZOOLOGY] A family of perching birds in the order Cuculiformes, including the cuckoos and the roadrunner, characterized by long tails, heavy beaks and conspicuous lashes. { kə'kyü·ləˌdē }

Cuculiformes [VERTEBRATE ZOOLOGY] An order of birds containing the cuckoos and allies, characterized by the zygodactyl arrangement of the toes. { kəˌkyü·lə'fȯrˌmēz }

cucullus [INVERTEBRATE ZOOLOGY] A transverse flap at the anterior edge of the carapace that hangs over the mouthparts of certain arachnids. { kyü'kəl·əs }

Cucumariidae [INVERTEBRATE ZOOLOGY] A family of dendrochirotacean holothurian echinoderms in the order Dendrochirotida. { ˌkü·kə·mə'rī·ə ˌdē }

cucumber [BOTANY] *Cucumis sativus*. An annual cucurbit, in the family Cucurbitaceae grown for its edible, immature fleshy fruit. { 'kyü·kəm·bər }

cul-de-sac [ANATOMY] Blind pouch or diverticulum. { 'kəl·dəˌsak }

Culex [INVERTEBRATE ZOOLOGY] A genus of mosquitoes important as vectors for malaria and several filarial parasites. { 'kyüˌleks }

Culicidae [INVERTEBRATE ZOOLOGY] The mosquitoes, a family of slender, orthorrhaphous dipteran insects in the series Nematocera having long legs and piercing mouthparts. { kyü'lis·ə ˌdē }

Culicinae [INVERTEBRATE ZOOLOGY] A subfamily of the dipteran family Culicidae. { kyü'lis·əˌnē }

culm [BOTANY] **1.** A jointed and usually hollow grass stem. **2.** The solid stem of certain monocotyledons, such as the sedges. { kəlm }

culmen [VERTEBRATE ZOOLOGY] The edge of the upper bill in birds. { 'kəl·mən }

cultigen [BIOLOGY] A cultivated variety or species of organism for which there is no known wild ancestor. Also known as cultivar. { 'kəl·tə·jən }

cultivar [BIOLOGY] *See* cultigen. { 'kəl·təˌvär }

cultural ecology [ECOLOGY] The branch of ecology that involves the study of the interaction of human societies with one another and with the natural environment. { ¦kəl·chər·əl ē'käl·ə·jē }

culture [BIOLOGY] A growth of living cells or microorganisms in a controlled artificial environment. { 'kəl·chər }

culture alteration [CELL BIOLOGY] A persistent change in the properties of cultured cells, such as altered morphology, virus susceptibility, nutritional requirements, or proliferative capacity. { 'kəl·chər ˌȯl·təˌrā·shən }

culture community [ECOLOGY] A plant commu-

nity which is established or modified through human intervention; for example, a fencerow, hedgerow, or windbreak. { 'kəl·chər kə'myü·nəd·ē }
cultured pearl [INVERTEBRATE ZOOLOGY] A natural pearl grown by means of controlled stimulation of the oyster. { ¦kəl·chərd 'pərl }
culture medium [MICROBIOLOGY] The nutrients and other organic and inorganic materials used for the growth of microorganisms and plant and animal tissue in culture. { ¦kəl·chər ˌmēd·ē·əm }
Cumacea [INVERTEBRATE ZOOLOGY] An order of the class Crustacea characterized by a well-developed carapace which is fused dorsally with at least the first three thoracic somites and overhangs the sides. { kyü'mās·ē·ə }
cumatophyte [ECOLOGY] A plant that grows under surf conditions. { kyü'mad·əˌfīt }
cumin [BOTANY] *Cuminum cyminum* An annual herb in the family Umbelliferae; the fruit is valuable for its edible, aromatic seeds. { 'kyü·mən }
cumulus oophorus [HISTOLOGY] The layer of gelatinous, follicle cells surrounding the ovum in a Graafian follicle. { 'kyü·myə·ləs ¦ō·ə'fór·əs }
cuneate [BIOLOGY] Wedge-shaped with the acute angle near the base, as in certain insect wings and the leaves of various plants. { 'kyü·nē·āt }
cuneiform [ANATOMY] **1.** Any of three wedge-shaped tarsal bones. **2.** Either of a pair of cartilages lying dorsal to the thyroid cartilage of the larynx. **3.** Wedge-shaped, chiefly referring to skeletal elements. { 'kyü·nē·əˌfórm }
cunnus [ANATOMY] The vulva. { 'kən·əs }
Cupedidae [INVERTEBRATE ZOOLOGY] The reticulated beetles, the single family of the coleopteran suborder Archostemata. { kyü'ped·əˌdē }
cupule [BOTANY] **1.** The cup-shaped involucre characteristic of oaks. **2.** A cup-shaped corolla. **3.** The gemmae cup of the Marchantiales. [INVERTEBRATE ZOOLOGY] A small sucker on the feet of certain male flies. { 'kyüˌpyül }
Curculionidae [INVERTEBRATE ZOOLOGY] The true weevils or snout beetles, a family of coleopteran insects in the superfamily Curculionoidea. { kər ˌkyü·lē'än·əˌdē }
Curculionoidea [INVERTEBRATE ZOOLOGY] A superfamily of coleopteran insects in the suborder Polyphaga. { kərˌkyü·lē·ə'nóid·ē·ə }
curculionoid larva [INVERTEBRATE ZOOLOGY] A kind of beetle larva having a highly reduced and grublike body. { kər'kyü·lē·əˌnóid 'lär·və }
Curcurbitaceae [BOTANY] A family of dicotyledonous herbs or herbaceous vines in the order Violales characterized by an inferior ovary, unisexual flowers, one to five stamens but typically three, and a sympetalous corolla. { kərˌkər·bə 'tās·ēˌē }
Curcurbitales [BOTANY] The ordinal name assigned to the Curcurbitaceae in some systems of classification. { kərˌkər·bə'tā·lēz }
curd [BOTANY] The edible flower heads of members of the mustard family such as broccoli. { kərd }
curling factor [MICROBIOLOGY] *See* griseofulvin. { 'kərl·iŋ ˌfak·tər }
currant [BOTANY] A shrubby, deciduous plant of the genus *Ribes* in the order Rosales; the edible fruit, a berry, is borne in clusters on the plant. { 'kər·ənt }
cursorial [VERTEBRATE ZOOLOGY] Adapted for running. { kər'sór·ē·əl }
Cuscutaceae [BOTANY] A family of parasitic dicotyledonous plants in the order Polemoniales which lack internal phloem and chlorophyll, have capsular fruit, and are not rooted to the ground at maturity. { kəˌskyü'tās·ēˌē }
cusp [ANATOMY] **1.** A pointed or rounded projection on the masticating surface of a tooth. **2.** One of the flaps of a heart valve. { kəsp }
cuspid [ANATOMY] *See* canine. { 'kəs·pəd }
cuspidate [BIOLOGY] Having a cusp; terminating in a point. { 'kəs·pəˌdāt }
cut [BIOCHEMISTRY] A double-strand incision in a duplex deoxyribonucleic acid molecule. [MOLECULAR BIOLOGY] A double-strand incision in a duplex deoxyribonucleic acid molecule. { kət }
cutaneous anaphylaxis [IMMUNOLOGY] Hypersensitivity that is marked by an intense skin reaction following parenteral contact with a sensitizing agent. { kyü'tā·nē·əs an·ə·fə'lak·səs }
cutaneous appendage [ANATOMY] Any of the epidermal derivatives, including the nails, hair, sebaceous glands, mammary glands, and sweat glands. { kyü'tā·nē·əs ə'pen·dij }
cutaneous pain [PHYSIOLOGY] A sensation of pain arising from the skin. { kyü'tā·nē·əs 'pān }
cutaneous sensation [PHYSIOLOGY] Any feeling originating in sensory nerve endings of the skin, including pressure, warmth, cold, and pain. { kyü 'tā·nē·əs sen'sā·shən }
Cuterebridae [INVERTEBRATE ZOOLOGY] The robust botflies, a family of myodarian cyclorrhaphous dipteran insects in the subsection Calypteratae. { kyü·də'reb·rəˌdē }
cuticle [ANATOMY] The horny layer of the nail fold attached to the nail plate at its margin. [BIOLOGY] A noncellular, hardened or membranous secretion of an epithelial sheet, such as the integument of nematodes and annelids, the exoskeleton of arthropods, and the continuous film of cutin on certain plant parts. { 'kyüd·ə·kəl }
cutin [BIOCHEMISTRY] A mixture of fatty substances characteristically found in epidermal cell walls and in the cuticle of plant leaves and stems. { 'kyüt·ən }
cutis [ANATOMY] *See* dermis. { 'kyüd·əs }
cutting [BOTANY] A piece of plant stem with one or more nodes, which, when placed under suitable conditions, will produce roots and shoots resulting in a complete plant. { 'kəd·iŋ }
cuttlefish [INVERTEBRATE ZOOLOGY] An Old World decapod mollusk of the genus *Sepia*; shells are used to manufacture dentifrices and cosmetics. { 'kəd·əlˌfish }
Cuvieroninae [PALEONTOLOGY] A subfamily of extinct proboscidean mammals in the family Gomphotheriidae. { küv·yə'rän·əˌnē }
Cyamidae [INVERTEBRATE ZOOLOGY] The whale lice, a family of amphipod crustaceans in the suborder Caprellidea that bear a resemblance to insect lice. { sī'am·əˌdē }

cyanobacteria [MICROBIOLOGY] A group of one-celled to many-celled aquatic organisms. Also known as blue-green algae. { ¦sī·ə·nō,bak'tir·ē·ə }

cyanocobalamin [BIOCHEMISTRY] *See* vitamin B. { ¦sī·ə·nō·kō'bal·ə·mən }

cyanophage [VIROLOGY] A virus that replicates in blue-green algae. Also known as algophage; blue-green algal virus. { sī'an·ə,fāj }

cyanophilous [BIOLOGY] Having an affinity for blue or green dyes. { ¦sī·ə¦näf·ə·ləs }

Cyanophyceae [BOTANY] A class of photosynthetic monerans distinguished by their algalike biology and bacteriumlike cell organization. { ,sī·ə·nō'fīs·ē,ē }

cyanophycin [BIOCHEMISTRY] A granular protein food reserve in the cells of blue-green algae, especially in the peripheral cytoplasm. { ,sī·ə·nō'fīs·ən }

Cyanophyta [BOTANY] An equivalent name for the Cyanophyceae. { ,sī·ə'näf·ə·də }

Cyatheaceae [BOTANY] A family of tropical and pantropical tree ferns distinguished by the location of sori along the veins. { sī·,ath·ə'ās·ē,ē }

cyathium [BOTANY] An inflorescence in which the flowers arise from the base of a cuplike involucre. { sī'ath·ē·əm }

Cyathoceridae [INVERTEBRATE ZOOLOGY] The equivalent name for the Lepiceridae. { sī,ath·ə'räs·ə,dē }

Cyatholaimoidea [INVERTEBRATE ZOOLOGY] A superfamily of nematodes of the order Chromadorida, distinguished by tightly coiled multispiral amphids located a short distance posterior to the cephalic sensilla. { ,sī·ə·thō·lə'mȯid·ē·ə }

cybrid [GENETICS] An individual produced following fusion of protoplasts from different species with complete elimination of the chromosomes of one of the species. { 'sī·brəd }

Cycadales [BOTANY] An ancient order of plants in the class Cycadopsida characterized by tuberous or columnar stems that bear a crown of large, usually pinnate leaves. { ,sī·kə'dā·lēz }

Cycadeoidaceae [PALEOBOTANY] A family of extinct plants in the order Cycadeoidales characterized by sparsely branched trunks and a terminal crown of leaves. { sī,kad·ē·ȯid'ās·ē,ē }

Cycadeoidales [PALEOBOTANY] An order of extinct plants that were abundant during the Triassic, Jurassic, and Cretaceous periods. { sī,kad·ē·ȯid'ā·lēz }

Cycadicae [BOTANY] A subdivision of large-leaved gymnosperms with stout stems in the plant division Pinophyta; only a few species are extant. { sī'kad·ə,sē }

Cycadofilicales [PALEOBOTANY] The equivalent name for the extinct Pteridospermae. { ¦sī·kə·dō,fil·ə'kā·lēz }

Cycadophyta [BOTANY] An equivalent name for Cycadecae elevated to the level of a division. { sī·kə'däf·əd·ə }

Cycadophytae [BOTANY] An equivalent name for Cycadicae. { sī·kə'däf·ə,tē }

Cycadopsida [BOTANY] A class of gymnosperms in the plant subdivision Cycadicae. { sī·kə'däp·sə·də }

Cyclanthaceae [BOTANY] The single family of the order Cyclanthales. { ,sī,klan'thās·ē,ē }

Cyclanthales [BOTANY] An order of monocotyledonous plants composed of herbs; or, seldom, composed of more or less woody plants with leaves that usually have a bifid, expanded blade. { ,sī,klan'thā·lēz }

cyclase [BIOCHEMISTRY] An enzyme that catalyzes cyclization of a compound. { 'sī,klās }

cyclic adenylic acid [BIOCHEMISTRY] $C_{10}H_{12}N_5O_6P$ An isomer of adenylic acid; crystal platelets with a melting point of 219-220°C; a key regulator which acts to control the rate of a number of cellular processes in bacteria, most animals, and some higher plants. Abbreviated cAMP. Also known as adenosine 3′,5′-cyclic monophosphate; adenosine 3′,5′-cyclic phosphate; adenosine 3′,5′-monophosphate; 3′,5′-AMP; cyclic AMP. { 'sīk·lik ¦ad·ən¦il·ik 'as·əd }

cyclic AMP [BIOCHEMISTRY] *See* cyclic adenylic acid. { 'sīk·lik ¦ā¦em¦pē }

cycloamylose [BIOCHEMISTRY] A member of a group of cyclic oligomers of glucose in which the individual glucose units are connected by 1,4 bonds. Also known as cyclodextrin; Schardinger dextrin. { ¦sī·klō'am·ə,lōs }

Cyclocystoidea [PALEONTOLOGY] A class of small, disk-shaped, extinct echinozoans in which the lower surface of the body probably consisted of a suction cup. { ¦sī·klō·si'stȯid·ē·ə }

cyclodextrin [BIOCHEMISTRY] *See* cycloamylose. { ¦sī·klō'dek·strən }

cycloheximide [MICROBIOLOGY] $C_{15}H_{23}NO_4$ Colorless crystals with a melting point of 119.5-121°C; soluble in water, in amyl acetate, and in common organic solvents such as ether, acetone, and chloroform; used as an agricultural fungicide. { ¦sī·klō'hek·sə,mīd }

cycloid scale [VERTEBRATE ZOOLOGY] A thin, acellular structure which is composed of a bonelike substance and shows annual growth rings; found in the skin of soft-rayed fishes. { 'sī,klȯid ,skāl }

cyclomorphosis [ECOLOGY] Cyclic recurrent polymorphism in certain planktonic fauna in response to seasonal temperature or salinity changes. { ,sī·klō'mȯr·fə·səs }

cyclooxygenase [BIOCHEMISTRY] An enzyme that catalyzes the conversion of arachidonic acid into prostaglandins. { ,sī·klō'äks·ə·jə,nās }

Cyclophoracea [INVERTEBRATE ZOOLOGY] A superfamily of gastropod mollusks in the order Prosobranchia. { ,sī·klō·fə'rās·ē·ə }

Cyclophoridae [INVERTEBRATE ZOOLOGY] A family of land snails in the order Pectinibranchia. { ,sī·klō'fȯr·ə,dē }

Cyclophyllidea [INVERTEBRATE ZOOLOGY] An order of platyhelminthic worms comprising most tapeworms of warm-blooded vertebrates. { ,sī·klō·fə'lid·ē·ə }

Cyclopinidae [INVERTEBRATE ZOOLOGY] A family of copepod crustaceans in the suborder Cyclopoida, section Gnathostoma. { ˌsī·klō'pin·ə ˌdē }

Cyclopoida [INVERTEBRATE ZOOLOGY] A suborder of small copepod crustaceans. { ˌsī·klō'pȯid·ē·ə }

Cyclopteridae [VERTEBRATE ZOOLOGY] The lumpfishes and snailfishes, a family of deep-sea forms in the suborder Cottoidei of the order Perciformes. { ˌsīˌkläp'ter·əˌdē }

Cyclorhagae [INVERTEBRATE ZOOLOGY] A suborder of benthonic, microscopic marine animals in the class Kinorhyncha of the phylum Aschelminthes. { sī'klȯr·əˌgē }

Cyclorrhapha [INVERTEBRATE ZOOLOGY] A suborder of true flies, order Diptera, in which developing adults are always formed in a puparium from which they emerge through a circular opening. { sī'klȯr·ə·fə }

cycloserine [MICROBIOLOGY] $C_3H_6O_2N_2$ Broad-spectrum, crystalline antibiotic produced by several species of *Streptomyces*; useful in the treatment of tuberculosis and urinary-tract infections caused by resistant gram-negative bacteria. { ¦sī·klō'seˌrēn }

cyclosis [CELL BIOLOGY] Massive rotational streaming of cytoplasm in certain vacuolated cells, such as the stonewort *Nitella* and *Paramecium*. { sī'klō·səs }

Cyclosporeae [BOTANY] A class of brown algae, division Phaeophyta, in which there is only a free-living diploid generation. { ¦sī·klō'spȯr·ēˌē }

Cyclosteroidea [PALEONTOLOGY] A class of Middle Ordovician to Middle Devonian echinoderms in the subphylum Echinozoa. { ¦sī·klō·stə'rȯid·ē·ə }

Cyclostomata [INVERTEBRATE ZOOLOGY] An order of bryozoans in the class Stenolaemata. [VERTEBRATE ZOOLOGY] A subclass comprising the simplest and most primitive of living vertebrates characterized by the absence of jaws and the presence of a single median nostril and an uncalcified cartilaginous skeleton. { ¦sī·klō'stō·mə·də }

Cydippida [INVERTEBRATE ZOOLOGY] An order of the pelagic ctenophores; members retain the cydippid state (resemble the cydippid larva) until the adult stage is reached in development. { sī 'dip·ə·də }

Cydippidea [INVERTEBRATE ZOOLOGY] An order of the Ctenophora having well-developed tentacles. { ˌsī·də'pid·ē·ə }

Cydnidae [INVERTEBRATE ZOOLOGY] The ground or burrower bugs, a family of hemipteran insects in the superfamily Pentatomorpha. { 'sid·nə ˌdē }

cylindrarthrosis [ANATOMY] A joint characterized by rounded articular surfaces. { ¦sil·ənˌdrär 'thrō·səs }

Cylindrocapsaceae [BOTANY] A family of green algae in the suborder Ulotrichineae comprising thick-walled, sheathed cells having massive chloroplasts. { səˌlin·drōˌkap'sās·ēˌē }

Cylindrocorporoidea [INVERTEBRATE ZOOLOGY] A superfamily of both free-living and parasitic nematodes of the order Diplogasterida having well-developed lips surrounding the oral opening and lateral lips bearing small amphids. { sə ˌlin·drəˌkȯr·pə'rȯid·ē·ə }

cyme [BOTANY] An inflorescence in which each main axis terminates in a single flower; secondary and tertiary axes may also have flowers, but with shorter flower stalks. { sīm }

cymose [BOTANY] Of, pertaining to, or resembling a cyme. { 'sīˌmōs }

Cymothoidae [INVERTEBRATE ZOOLOGY] A family of isopod crustaceans in the suborder Flabellifera; members are fish parasites with reduced maxillipeds ending in hooks. { ˌsī·mə'thȯiˌdē }

Cynipidae [INVERTEBRATE ZOOLOGY] A family of hymenopteran insects in the superfamily Cynipoidea { sə'nip·əˌdē }.

Cynipoidea [INVERTEBRATE ZOOLOGY] A superfamily of hymenopteran insects in the suborder Apocrita. { ˌsin·ə'pȯid·ē·ə }

Cynoglossidae [VERTEBRATE ZOOLOGY] The tonguefishes, a family of Asiatic flatfishes in the order Pleuronectiformes. { ˌsin·ə'gläs·əˌdē }

Cyperaceae [BOTANY] The sedges, a family of monocotyledonous plants in the order Cyperales characterized by spirally arranged flowers on a spike or spikelet; a usually solid, often triangular stem; and three carpels. { ˌsip·ə'rās·ēˌē }

Cyperales [BOTANY] An order of monocotyledonous plants in the subclass Commelinidae with reduced, mostly wind-pollinated or self-pollinated flowers that have a unilocular, two- or three-carpellate ovary bearing a single ovule. { sip·ə'rā·lēz }

Cypheliaceae [BOTANY] A family of typically crustose lichens with sessile apothecia in the order Caliciales. { səˌfel·ē'ās·ēˌē }

cyphonautes [INVERTEBRATE ZOOLOGY] The free-swimming bivalve larva of certain bryozoans. { ˌsī·fə'nōd·ēz }

Cyphophthalmi [INVERTEBRATE ZOOLOGY] A family of small, mitelike arachnids in the order Phalangida. { ˌsī·fə'thalˌmī }

Cypraecea [INVERTEBRATE ZOOLOGY] A superfamily of gastropod mollusks in the order Prosobranchia. { sī'prēsh·ē·ə }

Cypraeidae [INVERTEBRATE ZOOLOGY] A family of colorful marine snails in the order Pectinibranchia. { sī'prē·əˌdē }

cypress [BOTANY] The common name for members of the genus *Cupressus* and several related species in the order Pinales. { 'sī·prəs }

Cypridacea [INVERTEBRATE ZOOLOGY] A superfamily of mostly fresh-water ostracods in the suborder Podocopa. { ˌsī·prə'dās·ē·ə }

Cypridinacea [INVERTEBRATE ZOOLOGY] A superfamily of ostracods in the suborder Myodocopa characterized by a calcified carapace and having a round back with a downward-curving rostrum. { ˌsīˌprid·ə'nās·ē·ə }

Cyprinidae [VERTEBRATE ZOOLOGY] The largest family of fishes, including minnows and carps in the order Cypriniformes. { sī'prin·əˌdē }

Cypriniformes [VERTEBRATE ZOOLOGY] An order of actinopterygian fishes in the suborder Ostariophysi. { sīˌprin·ə'fōrˌmēz }

Cyprinodontidae [VERTEBRATE ZOOLOGY] The killifishes, a family of actinopterygian fishes in the order Atheriniformes that inhabit ephemeral tropical ponds. { sī͵prin·ə'dän·tə͵dē }

Cyprinoidei [VERTEBRATE ZOOLOGY] A suborder of primarily fresh-water actinopterygian fishes in the order Cypriniformes having toothless jaws, no adipose fin, and faliciform lower pharyngeal bones. { ͵sī·prə'nȯi·dē͵ī }

cypris [INVERTEBRATE ZOOLOGY] An ostracodlike, free-swimming larval stage in the development of Cirripedia. { 'sī·prəs }

Cyrtophorina [INVERTEBRATE ZOOLOGY] The equivalent name for Gymnostomatida. { ͵sərd·ə 'fə'rī·nə }

cyrtopia [INVERTEBRATE ZOOLOGY] A type of crustacean larva (Ostracoda) characterized by an elongation of the first pair of antennae and loss of swimming action in the second pair. { sər'tō·pē·ə }

cystacanth [INVERTEBRATE ZOOLOGY] The infective larva of the Acanthocephala; lies in the hemocele of the intermediate host. { 'sis·tə͵kanth }

L-cystathionine [BIOCHEMISTRY] $C_7H_{14}N_2O_4S$ An amino acid formed by condensation of homocysteine with serine, catalyzed by an enzyme transsulfurase; found in high concentration in the brain of primates. { ¦el ͵sis·tə'thī·ə͵nēn }

cysteine [BIOCHEMISTRY] $C_3H_7O_2NS$ A crystalline amino acid occurring as a constituent of glutathione and cystine. { 'si͵stēn }

cystic duct [ANATOMY] The duct of the gallbladder. { 'sis·tik 'dəkt }

cysticercus [INVERTEBRATE ZOOLOGY] A larva of tapeworms in the order Cyclophyllidea that has a bladder with a single invaginated scolex. { ¦sis·tə'sər·kəs }

cystine [BIOCHEMISTRY] $C_6H_{12}N_2S_2$ A white, crystalline amino acid formed biosynthetically from cysteine. { 'si͵stēn }

Cystobacter [MICROBIOLOGY] A genus of bacteria in the family Cystobacteraceae; vegetative cells are tapered, sporangia are sessile, and microcysts are rigid rods. { ¦sis·tə'bak·tər }

Cystobacteraceae [MICROBIOLOGY] A family of bacteria in the order Myxobacterales; vegetative cells are tapered, and microcysts are rod-shaped and enclosed in sporangia. { ¦sis·tə͵bak·tə'rās·ē͵ē }

cystocarp [BOTANY] A fruiting structure with a special protective envelope, produced after fertilization in red algae. { 'sis·tə͵kärp }

cystocercous cercaria [INVERTEBRATE ZOOLOGY] A digenetic trematode larva that can withdraw the body into the tail. { ¦sis·tə¦sər·kəs sər'kar·ē͵ə }

Cystoidea [PALEONTOLOGY] A class of extinct crinozoans characterized by an ovoid body that was either sessile or attached by a short aboral stem. { si'stȯid·ē·ə }

cystolith [BOTANY] A concretion of calcium carbonate arising from the cell walls of modified epidermal cells in some flowering plants. { 'sis·tə͵lith }

Cystoporata [PALEONTOLOGY] An order of extinct, marine bryozoans characterized by cystopores and minutopores. { ͵sis·tə'pȯr·əd·ə }

cytase [BIOCHEMISTRY] Any of several enzymes in the seeds of cereals and other plants, which hydrolyze the cell-wall material. { 'sī͵tās }

Cytheracea [INVERTEBRATE ZOOLOGY] A superfamily of ostracods in the suborder Podocopa comprising principally crawling and digging marine forms. { ͵sith·ə'rās·ē·ə }

Cytherellidae [INVERTEBRATE ZOOLOGY] The family comprising all living members of the ostracod suborder Platycopa. { ͵sith·ə'rel·ə͵dē }

cytidine [BIOCHEMISTRY] $C_9H_{13}N_3O_5$ Cytosine riboside, a nucleoside composed of one molecule each of cytosine and D-ribose. { 'sid·ə͵dēn }

cytidylic acid [BIOCHEMISTRY] $C_9H_{14}O_8N_3P$ A nucleotide synthesized from the base cytosine and obtained by hydrolysis of nucleic acid. { ¦sid·ə¦dil·ik 'as·əd }

cytocentrum [CELL BIOLOGY] *See* central apparatus. { ¦sīd·ō'sen·trəm }

cytochalasin [BIOCHEMISTRY] One of a series of structurally related fungal metabolic products which, among other effects on biological systems, selectively and reversibly block cytokinesis while not affecting karyokinesis; the molecule with minor variations consists of a benzyl-substituted hydroaromatic isoindolone system, which in turn is fused to a small macrolide-like cyclic ring. { ¦sīd·ō·kə'lā·sən }

cytochemistry [CELL BIOLOGY] The science concerned with the chemistry of cells and cell components, primarily with the location of chemical constituents and enzymes. { ¦sīd·ō'kem·ə·strē }

cytochrome [BIOCHEMISTRY] Any of the complex protein respiratory pigments occurring within plant and animal cells, usually in mitochondria, that function as electron carriers in biological oxidation. { 'sīd·ə͵krōm }

cytochrome a_3 [BIOCHEMISTRY] *See* cytochrome oxidase. { 'sīd·ə͵krōm ¦ā səb'thrē }

cytochrome oxidase [BIOCHEMISTRY] Any of a family of respiratory pigments that react directly with oxygen in the reduced state. Also known as cytochrome a_3. { 'sīd·ə͵krōm 'äk·sə͵dās }

cytocidal [CELL BIOLOGY] Causing cell death. { sī 'täs·əd·əl }

cytocidal unit [BIOLOGY] A unit for the standardization of adrenal cortical hormones. { sī'täs·əd·əl ͵yü·nət }

cytocrine gland [CELL BIOLOGY] A cell, especially a melanocyte, that passes its secretion directly to another cell. { 'sīd·ə·krən ͵gland }

cytoduction [MYCOLOGY] In yeast, the production of cells with mixed cytoplasm but with the nucleus of one or the other parent. { 'sīd·ə ͵dək·shən }

cytogamy [CELL BIOLOGY] Fusion or conjugation of cells. { sī'täg·ə·mē }

cytogenetics [CELL BIOLOGY] The comparative study of the mechanisms and behavior of chromosomes in populations and taxa, and the effect of chromosomes on inheritance and evolution. { ¦sīd·ō·jə'ned·iks }

cytogenous gland [PHYSIOLOGY] A structure that secretes living cells; an example is the testis. { sī 'tä·jə·nəs ͵gland }

cytohet [CELL BIOLOGY] A cell containing two genetically distinct types of a specific organelle. { 'sīd·ə,het }

cytokine [IMMUNOLOGY] Any of a group of peptides involved in sending signals that affect interactions between cells in the immune response. { 'sīd·ə,kīn }

cytokinesis [CELL BIOLOGY] Division of the cytoplasm following nuclear division. { ¦sīd·ō·kə'nē·səs }

cytokinin [BIOCHEMISTRY] Any of a group of plant hormones which elicit certain plant growth and development responses, especially by promoting cell division. { ¦sīd·ō'kī·nən }

cytology [BIOLOGY] A branch of the biological sciences which deals with the structure, behavior, growth, and reproduction of cells and the function and chemistry of cell components. { sī'täl·ə·jē }

cytolysosome [CELL BIOLOGY] An enlarged lysosome that contains organelles such as mitochondria. { ¦sīd·ō'lī·sə,sōm }

cytomegalovirus [VIROLOGY] An animal virus belonging to subgroup B of the herpesvirus group; causes cytomegalic inclusion disease and pneumonia. { ¦sīd·ō¦meg·ə·lō'vī·rəs }

cytomegalovirus group [VIROLOGY] *See* Betaherpesvirinae. { ¦sīd·ō¦meg·ə·lō'vī·rəs ,grüp }

cytomixis [CELL BIOLOGY] Extrusion of chromatin from one cell into the cytoplasm of an adjoining cell. { ,sīd·ə'mik·səs }

cytomorphosis [CELL BIOLOGY] All the structural alterations which cells or successive generations of cells undergo from the earliest undifferentiated stage to their final destruction. { ,sīd·ə'mȯr·fə·səs }

cyton [CELL BIOLOGY] The central body of a neuron containing the nucleus and excluding its processes. { 'sī,tän }

cytopathic effect [CELL BIOLOGY] A change in the microscopic appearance of cells in a culture after being infected with a virus. { ¦sīd·ə¦path·ik i,fekt }

Cytophaga [MICROBIOLOGY] A genus of bacteria in the family Cytophagaceae; cells are unsheathed, unbranched rods or filaments and are motile; microcysts are not known; decompose agar, cellulose, and chitin. { sī'täf·ə·gə }

Cytophagaceae [MICROBIOLOGY] A family of bacteria in the order Cytophagales; cells are rods or filaments, unsheathed cells are motile, filaments are not attached, and carotenoids are present. { ,sīd·ō·fə'gās·ē,ē }

Cytophagales [MICROBIOLOGY] An order of gliding bacteria; cells are rods or filaments and motile by gliding, and fruiting bodies are not produced. { ,sīd·ō·fə'gā·lēz }

cytopharynx [INVERTEBRATE ZOOLOGY] A channel connecting the surface with the protoplasm in certain protozoans; functions as a gullet in ciliates. { ¦sīd·ō'far·iŋks }

cytophilic antibody [IMMUNOLOGY] A substance capable of combining directly with the receptors of a corresponding antigenic cell. Also known as cytotropic antibody. { ¦sid·ə¦fil·ik 'ant·i,bäd·ē }

cytoplasm [CELL BIOLOGY] The protoplasm of an animal or plant cell external to the nucleus. { 'sīd·ə,plaz·əm }

cytoplasmic inheritance [GENETICS] The control of genetic difference by hereditary units carried in cytoplasmic organelles. Also known as extrachromosomal inheritance. { ,sīd·ə'plaz·mik in'her·ə·təns }

cytoplasmic male sterility [BOTANY] The maternally inherited inability of a higher plant to produce viable pollen. { ¦sīd·ə,plaz·mik ¦māl stə'ril·əd·ē }

cytoplasmic streaming [CELL BIOLOGY] Intracellular movement involving irreversible deformation of the cytoplasm produced by endogenous forces. { ,sīd·ə'plaz·mik 'strem·iŋ }

cytoplast [CELL BIOLOGY] The cytoplasmic substance of eukaryotic cells, including a network of proteins forming an internal skeleton and the attached nucleus and organelles. { 'sīd·ə,plast }

cytopyge [INVERTEBRATE ZOOLOGY] A fixed point for waste discharge in the body of a protozoan, especially a ciliate. { 'sīd·ə,pīj }

cytosine [BIOCHEMISTRY] $C_4H_5ON_3$ A pyrimidine occurring as a fundamental unit or base of nucleic acids. { 'sīd·ə,sēn }

cytoskeleton [CELL BIOLOGY] Protein fibers composing the structural framework of a cell. { ¦sīd·ō'skel·ə·tən }

cytosol [CELL BIOLOGY] The fluid portion of the cytoplasm, that is, the cytoplasm exclusive of organelles and membranes. { 'sīd·ə,säl *or* 'sīd·ə,sōl }

cytosome [CELL BIOLOGY] The cytoplasm of the cell, as distinct from the nucleus. { 'sīd·ə,sōm }

cytostome [INVERTEBRATE ZOOLOGY] The mouthlike opening in many unicellular organisms, particularly Ciliophora. { 'sīd·ə,stōm }

cytotaxis [PHYSIOLOGY] Attraction of motile cells by specific diffusible stimuli emitted by other cells. { ¦sīd·ō'tak·səs }

cytotoxic [CELL BIOLOGY] Pertaining to an agent, such as a drug or virus, that exerts a toxic effect on cells. { ¦sīd·ə'täk·sik }

cytotrophoblast [EMBRYOLOGY] The inner, cellular layer of a trophoblast, covering the chorion and the chorionic villi during the first half of pregnancy. { ¦sīd·ō'trō·fə,blast }

cytotropic antibody [IMMUNOLOGY] *See* cytophilic antibody. { ¦sīd·ō'trä·pik 'an·tə,bäd·ē }

cytotropism [BIOLOGY] The tendency of individual cells and groups of cells to move toward or away from each other. { sī'tä·trə,piz·əm }

Czapek's agar [MICROBIOLOGY] A nutrient culture medium consisting of salt, sugar, water, and agar; used for certain mold cultures. { 'chä·peks ,äg·ər }

D

Dacromycetales [MYCOLOGY] An order of jelly fungi in the subclass Heterobasidiomycetidae having branched basidia with the appearance of a tuning fork. { ˌdak·rəˌmī·sə'tā·lēz }

dacryocyst [ANATOMY] *See* lacrimal sac. { 'dak·rə ˌsist }

dacryon [ANATOMY] The point of the face where the frontomaxillary, the maxillolacrimal, and frontolacrimal sutures meet. { 'dak·rēˌän }

Dactylochirotida [INVERTEBRATE ZOOLOGY] An order of dendrochirotacean holothurians in which there are 8-30 digitate or digitiform tentacles, which sometimes bifurcate. { ˌdak·tə·lō·kə'räd·ə·də }

dactylognathite [INVERTEBRATE ZOOLOGY] The distal segment of a maxilliped in crustaceans. { ˌdak·tə·lō'naˌthīt }

Dactylogyroidea [INVERTEBRATE ZOOLOGY] A superfamily of trematodes in the subclass Monogenea; all are fish ectoparasites. { ¦dak·tə·loˌji 'ròid·ē·ə }

dactylopodite [INVERTEBRATE ZOOLOGY] The distal segment of ambulatory limbs in decapods and of certain limbs in other arthropods. { ˌdak·tə'läp·əˌdīt }

dactylopore [INVERTEBRATE ZOOLOGY] Any of the small openings on the surface of Milleporina through which the bodies of the polyps are extended. { dak'til·əˌpòr }

Dactylopteridae [VERTEBRATE ZOOLOGY] The flying gurnards, the single family of the perciform suborder Dactylopteroidei. { ˌdak·tə·lō'ter·ə ˌdē }

Dactylopteroidei [VERTEBRATE ZOOLOGY] A suborder of marine shore fishes in the order Perciformes, characterized by tremendously expansive pectoral fins. { ˌdak·tə·lō·tə'ròid·ēˌī }

Dactyloscopidae [VERTEBRATE ZOOLOGY] The sand stargazers, a family of small tropical and subtropical perciform fishes in the suborder Blennioidei. { ˌdak·tə·'skäp·əˌdē }

Dactylosporangium [MICROBIOLOGY] A genus of bacteria in the family Actunoplanaceae; fingerlike sporangia are formed in clusters, each containing a single row of three or four motile spores. { ˌdak·tə·lō·spə'ran·jē·əm }

dactylosternal [VERTEBRATE ZOOLOGY] Of turtles, having marginal fingerlike processes in joining the plastron to the carapace. { ¦dak·tə·lō'stərn·əl }

dactylozooid [INVERTEBRATE ZOOLOGY] One of the long defensive polyps of the Milleporina, armed with stinging cells. { ¦dak·tə·lə'zōˌòid }

dactylus [INVERTEBRATE ZOOLOGY] The structure of the tarsus of certain insects which follows the first joint; usually consists of one or more joints. { 'dak·tə·ləs }

Da Fano bodies [VIROLOGY] Minute basophilic areas of abnormal staining found within cells infected with human herpesvirus 1 or 2. { dä'fän·ō ˌbäd·ēz }

Dalatiidae [VERTEBRATE ZOOLOGY] The spineless dogfishes, a family of modern sharks belonging to the squaloid group. { ˌdal·ə'tī·əˌdē }

Dallis grass [BOTANY] The common name for the tall perennial forage grasses composing the genus *Paspalum* in the order Cyperales. { 'da·ləs ˌgras }

Danaidae [INVERTEBRATE ZOOLOGY] A family of large tropical butterflies, order Lepidoptera, having the first pair of legs degenerate. { də'nā·ə ˌdē }

Dane particle [VIROLOGY] The causative virus of type B viral hepatitis visualized ultrastructurally in its complete form. { 'dān ˌpard·ə·kəl }

Danysz reaction [IMMUNOLOGY] A toxin-antitoxin reaction that occurs when an exact equivalence of toxin is added to antitoxin, not in one portion but in successive increments. { 'dä·nish rē'ak·shən }

DAP [BIOCHEMISTRY] *See* diallyl phthalate; diaminopimelate.

Daphniphyllales [BOTANY] An order of dicotyledonous plants in the subclass Hamamelidae, consisting of a single family with one genus, *Daphniphyllum*, containing about 35 species; dioecious trees or shrubs native to eastern Asia and the Malay region, they produce a unique type of alkaloid and often accumulate aluminum and sometimes produce iridoid compounds. { ˌdaf·ni·fə'lā·lēz }

Daphoenidae [PALEONTOLOGY] A family of extinct carnivoran mammals in the superfamily Miacoidea. { də'fēn·əˌdē }

dart [INVERTEBRATE ZOOLOGY] A small sclerotized structure ejected from the dart sac of certain snails into the body of another individual as a stimulant before copulation. { därt }

dart sac [INVERTEBRATE ZOOLOGY] A dart-forming pouch associated with the reproductive system of certain snails. { 'därt ˌsak }

darwin [EVOLUTION] A unit of evolutionary rate

of change; if some dimension of a part of an animal or plant, or of the whole animal or plant, changes from l_o to l_t over a time of t years according to the formula $l_t = l_o \exp(Et/10^6)$, its evolutionary rate of change is equal to E darwins. { 'där·wən }

Darwinism [BIOLOGY] The theory of the origin and perpetuation of new species based on natural selection of those offspring best adapted to their environment because of genetic variation and consequent vigor. Also known as Darwin's theory. { 'där·wə,niz·əm }

Darwin's finch [VERTEBRATE ZOOLOGY] A bird of the subfamily Fringillidae; Darwin studied the variation of these birds and used his data as evidence for his theory of evolution by natural selection. { ¦där·winz 'finch }

Darwin's theory [BIOLOGY] *See* Darwinism. { ¦där·winz 'thē·ə·rē }

Darwinulacea [INVERTEBRATE ZOOLOGY] A small superfamily of nonmarine, parthenogenetic ostracods in the suborder Podocopa. { där,win·ə 'lās·ē·ə }

Dasayatidae [VERTEBRATE ZOOLOGY] The stingrays, a family of modern sharks in the batoid group having a narrow tail with a single poisonous spine. { ,da·sā'ad·ə,dē }

Dascillidae [INVERTEBRATE ZOOLOGY] The soft-bodied plant beetles, a family of coleopteran insects in the superfamily Dascilloidea. { ,də'sil·ə ,dē }

Dascilloidea [INVERTEBRATE ZOOLOGY] Superfamily of coleopteran insects in the suborder Polyphaga. { ,das·ə'lòid·ē·ə }

dasheen [BOTANY] *Colocasia esculenta.* A plant in the order Arales, grown for its edible corm. { da 'shēn }

Dasycladaceae [BOTANY] A family of green algae in the order Dasycladales comprising plants formed of a central stem from which whorls of branches develop. { ,das·ə·klə'dās·ē,ē }

Dasycladales [BOTANY] An order of lime-encrusted marine algae in the division Chlorophyta, characterized by a thallus composed of nonseptate, highly branched tubes. { ,das·ə·klə 'dā·lēz }

Dasyonygidae [INVERTEBRATE ZOOLOGY] A family of biting lice, order Mallophaga, that are confined to rodents of the family Procaviidae. { ,das·ē·ə'nij·ə,dē }

Dasypodidae [VERTEBRATE ZOOLOGY] The armadillos, a family of edentate mammals in the infraorder Cingulata. { ,das·ə'päd·ə,dē }

Dasytidae [INVERTEBRATE ZOOLOGY] An equivalent name for Melyridae. { də'sid·ə,dē }

Dasyuridae [VERTEBRATE ZOOLOGY] A family of mammals in the order Marsupialia characterized by five toes on each hindfoot. { das·ē'yůr·ə ,dē }

Dasyuroidea [VERTEBRATE ZOOLOGY] A superfamily of marsupial mammals. { ,das·ē·yə'ròid·ē·ə }

Daubentoniidae [VERTEBRATE ZOOLOGY] A family of Madagascan prosimian primates containing a single species, the aye-aye. { ,dō·bən·tō'nī·ə ,dē }

daughter nucleus [CELL BIOLOGY] One of the two cell nuclei resulting from a nuclear division. { 'dȯd·ər ,nü·klē·əs }

Dawsoniales [BOTANY] An order of mosses comprising rigid plants with erect stems rising from a rhizomelike base. { dȯ,sō·nē'ā·lēz }

day neutral [BOTANY] Reaching maturity regardless of relative length of light and dark periods. { 'dā ,nü·trəl }

deadly nightshade [BOTANY] *See* belladonna. { ¦ded·lē 'nīt,shād }

dead space [ANATOMY] The space in the trachea, bronchi, and other air passages which contains air that does not reach the alveoli during respiration, the amount of air being about 140 milliliters. Also known as anatomical dead space. [PHYSIOLOGY] A calculated expression of the anatomical dead space plus whatever degree of overventilation or underperfusion is present; it is alleged to reflect the relationship of ventilation to pulmonary capillary perfusion. Also known as physiological dead space. { 'ded ,spās }

deamidase [BIOCHEMISTRY] An enzyme that catalyzes the removal of an amido group from a compound. { dē'am·ə,dās }

deaminase [BIOCHEMISTRY] An enzyme that catalyzes the hydrolysis of amino compounds, removing the amino group. { de'am·ə,nās }

death assemblage [PALEONTOLOGY] *See* thanatocoenosis. { 'deth ə,sem·blij }

death point [PHYSIOLOGY] The limit (as of extremes of temperature) beyond which an organism cannot survive. { 'deth ,pȯint }

debromoaplysiatoxin [BIOCHEMISTRY] A bislactone toxin related to aplysiatoxin and produced by the blue-green alga *Lyngbya majuscula.* { dē ,brō·mō·ə'plizh·ə,täk·sən }

decanth larva [INVERTEBRATE ZOOLOGY] *See* lycophore larva. { ,dē'kanth ,lär·və }

Decapoda [INVERTEBRATE ZOOLOGY] **1.** A diverse order of the class Crustacea including the shrimps, lobsters, hermit crabs, and true crabs; all members have a carapace, well-developed gills, and the first three pairs of thoracic appendages specialized as maxillipeds. **2.** An order of dibranchiate cephalopod mollusks containing the squids and cuttle fishes, characterized by eight arms and two long tentacles. { də'kap· əd·ə }

decarboxylase [BIOCHEMISTRY] An enzyme that hydrolyzes the carboxyl radical, COOH. { ,dē·kär 'bäk·sə,lās }

deciduous [BIOLOGY] Falling off or being shed at the end of the growing period or season. { di 'sij·ə·wəs }

deciduous teeth [ANATOMY] Teeth of a young mammal which are shed and replaced by permanent teeth. Also known as milk teeth. { di 'sij·ə·wəs 'tēth }

deckzelle [INVERTEBRATE ZOOLOGY] In certain hydroids, one of the supporting or epithelial cells which are usually columnar or cuboidal. { ¦dek ¦zel }

declinate [BIOLOGY] Curved toward one side or downward. { 'dek·lə,nāt }

declining population [ECOLOGY] A population in

which old individuals outnumber young individuals. { də'klīn·iŋ ,päp·yə'lā·shən }

decomposer [ECOLOGY] A heterotrophic organism (including bacteria and fungi) which breaks down the complex compounds of dead protoplasm, absorbs some decomposition products, and releases substances usable by consumers. Also known as microcomposer; microconsumer; reducer. { de·kəm'pō·zər }

decompound [BOTANY] Divided or compounded several times, with each division being compound. { dē'käm,paủnd }

decorticate [BIOLOGY] Lacking a cortical layer. { dē'kȯrd·ə,kāt }

decumbent [BOTANY] Lying down on the ground but with an ascending tip, specifically referring to a stem. { di'kəm·bənt }

decurrent [BOTANY] Running downward, especially of a leaf base extended past its insertion in the form of a winged expansion. { di'kər·ənt }

decussate [BOTANY] Of the arrangement of leaves, occurring in alternating pairs at right angles. { 'dek·ə,sāt }

dedifferentiation [BIOLOGY] Disintegration of a specialized habit or adaptation. [CELL BIOLOGY] Loss of recognizable specializations that define a differentiated cell. [PHYSIOLOGY] Return of a specialized cell or structure to a more general or primitive condition. { dē,dif·ə,ren·chē'ā·shən }

deep fascia [ANATOMY] The fibrous tissue between muscles and forming the sheaths of muscles, or investing other deep, definitive structures, as nerves and blood vessels. { 'dēp 'fā·shə }

deep hibernation [PHYSIOLOGY] Profound decrease in metabolic rate and physiological function during winter, with a body temperature near 0°C, in certain warm-blooded vertebrates. Also known as hibernation. { 'dēp ,hī·bər'nā·shən }

deep pain [PHYSIOLOGY] Pattern of somesthetic sensation of pain, usually indefinitely localized, originating in the viscera, muscles, and other deep tissues. { 'dēp 'pān }

deep palmar arch [ANATOMY] The anastomosis between the terminal part of the radial artery and the deep palmar branch of the ulnar artery. Also known as deep volar arch; palmar arch. { 'dēp ¦pä·mər ,ärch }

deep volar arch [ANATOMY] *See* deep palmar arch. { ¦dēp ¦vō·lər ,ärch }

deer [VERTEBRATE ZOOLOGY] The common name for 41 species of even-toed ungulates that compose the family Cervidae in the order Artiodactyla; males have antlers. { dir }

defecation [PHYSIOLOGY] The process by which fecal wastes that reach the lower colon and rectum are evacuated from the body. { ,def·ə'kā·shən }

defective interfering virus [VIROLOGY] A virus generated at the peak of an infection that can interfere with replication of the normal virus and may modify the outcome of the disease. { di'fek·tiv ,in·tər,fir·iŋ 'vī·rəs }

defective virus [VIROLOGY] A virus, such as adeno-associated satellite virus, that can grow and reproduce only in the presence of another virus. { di'fek·tiv 'vī·rəs }

defeminization [PHYSIOLOGY] Loss or reduction of feminine attributes, usually caused by ovarian dysfunction or removal. { dē,fem·ə·nə'zā·shən }

definitive host [BIOLOGY] The host in which a parasite reproduces sexually. { də'fin·əd·iv 'hōst }

deflexed [BIOLOGY] Turned sharply downward. { dē'flekst }

defoliate [BOTANY] To remove leaves or cause leaves to fall, especially prematurely. { dē'fō·lē ,āt }

degenerate code [GENETICS] A genetic code in which more than one triplet sequence of nucleotides (codon) can specify the insertion of the same amino acid into a polypeptide chain. { di 'jen·ə·rət 'kōd }

Degeneriaceae [BOTANY] A family of dicotyledonous plants in the order Magnoliales characterized by laminar stamens; a solitary, pluriovulate, unsealed carpel; and ruminate endosperm. { ,dē·jen·ə,rī'ās·ē,ē }

deglutition [PHYSIOLOGY] Act of swallowing. { ,dē,glü'tish·ən }

degradative plasmid [GENETICS] A type of plasmid that specifies a set of genes involved in biodegradation of an organic compound. { ¦deg·rə,dād·iv 'plaz·mid }

dehiscence [BOTANY] Spontaneous bursting open of a mature plant structure, such as fruit, anther, or sporangium, to discharge its contents. { də'his·əns }

dehydrase [BIOCHEMISTRY] An enzyme which catalyzes the removal of water from a substrate. { dē'hī,drās }

dehydrochlorinase [BIOCHEMISTRY] An enzyme that dechlorinates a chlorinated hydrocarbon such as the insecticide DDT; found in some insects that are resistant to DDT. { dē,hī·drō 'klȯr·ə,nās }

dehydrochlorination [BIOCHEMISTRY] Removal of hydrogen and chlorine or hydrogen chloride from a compound. { dē,hī·drō,klȯr·ə'nā·shən }

dehydrocholesterol [BIOCHEMISTRY] $C_{27}H_{43}OH$ A provitamin of animal origin found in the skin of humans, in milk, and elsewhere, which upon irradiation with ultraviolet rays becomes vitamin D. { dē¦hī·drō·kə'les·tə,rȯl }

dehydrogenase [BIOCHEMISTRY] An enzyme which removes hydrogen atoms from a substrate and transfers it to an acceptor other than oxygen. { dē'hī·drə·jə,nās }

Deinotheriidae [PALEONTOLOGY] A family of extinct proboscidean mammals in the suborder Deinotherioidea; known only by the genus *Deinotherium*. { ,dī·nō·thə'rī·ə,dē }

Deinotherioidea [PALEONTOLOGY] A monofamilial suborder of extinct mammals in the order Proboscidea. { ,dī·nō,ther·ē'oid·ē·ə }

delamination [BIOLOGY] The separation of cells into layers. [EMBRYOLOGY] Gastrulation in which the endodermal layer splits off from the inner surface of the blastoderm and the space between this layer and the yolk represents the archenteron. { dē,lam·ə'nā·shən }

delayed hypersensitivity [IMMUNOLOGY] Abnormal reactivity in a sensitized individual beginning several hours after contact with the allergen. { di'lād ,hī·pər,sen·sə'tiv·əd·ē }

deletion [GENETICS] Loss of a chromosome segment of any size, down to a part of a single gene. { di'lē·shən }

deliquescence [BOTANY] The condition of repeated divisions ending in fine divisions; seen especially in venation and stem branching. { del·ə'kwes·əns }

delomorphous cell [HISTOLOGY] *See* parietal cell. { ¦dē·lō¦mȯr·fəs 'sel }

Delphinidae [VERTEBRATE ZOOLOGY] A family of aquatic mammals in the order Cetacea; includes the dolphins. { del'fin·ə,dē }

delphinidin [BIOCHEMISTRY] $C_{15}H_{11}O_7Cl$ A purple or brownish-red anthocyanin compound occurring widely in plants. { del'fin·ə·dən }

Delphinus [VERTEBRATE ZOOLOGY] A genus of cetacean mammals, including the dolphin. { del 'fē·nəs }

delta rhythm [PHYSIOLOGY] An electric current generated in slow waves with frequencies of 0.5-3 per second from the forward portion of the brain of normal subjects when asleep. { 'del·tə ,rith·əm }

Deltatheridia [PALEONTOLOGY] An order of mammals that includes the dominant carnivores of the early Cenozoic. { ,del·tə·thə'rid·ē·ə }

deltoid [ANATOMY] The large triangular shoulder muscle; originates on the pectoral girdle and inserts on the humerus. [BIOLOGY] Triangular in shape. { 'del,tȯid }

deltoid ligament [ANATOMY] The ligament on the medial side of the ankle joint; the fibers radiate from the medial malleolus to the talus, calcaneus, and navicular bones. { 'del,tȯid 'lig·ə·mənt }

demarcation potential [PHYSIOLOGY] *See* injury potential. { dē,mär'kā·shən pə,ten·chəl }

Dematiaceae [MYCOLOGY] A family of fungi in the order Moniliales; sporophores are not grouped, hyphae are always dark, and the spores are hyaline or dark. { də,mad·ē'ās·ē,ē }

deme [ECOLOGY] A local population in which the individuals freely interbreed among themselves but not with those of other demes. { dēm }

demersal [BIOLOGY] Living at or near the bottom of the sea. { də'mər·səl }

demethylchlortetracycline [MICROBIOLOGY] $C_{21}H_{21}O_8N_2Cl$ A broad-spectrum tetracycline antibiotic produced by a mutant strain of *Streptomyces aureofaciens.* { dē¦meth·əl,klȯr,te·trə'sī ,klēn }

demilune [BIOLOGY] Crescent-shaped. { 'dem·i ,lün }

Demodicidae [INVERTEBRATE ZOOLOGY] The pore mites, a family of arachnids in the suborder Trombidiformes. { ,dem·ə'dis·ə,dē }

demographic genetics [BIOLOGY] A branch of population genetics and ecology concerned with genetic differences related to age, population size, genetic alteration in competitive ability, and viability. { ¦dem·ə,graf·ik jə¦ned·iks }

demography [ECOLOGY] The statistical study of populations with reference to natality, mortality, migratory movements, age, and sex, among other social, ethnic, and economic factors. { də 'mäg·rə·fē }

Demospongiae [INVERTEBRATE ZOOLOGY] A class of the phylum Porifera, including sponges with a skeleton of one- to four-rayed siliceous spicules, or of spongin fibers, or both. { dem·ə'spən·jē,ē }

denaturation map [MOLECULAR BIOLOGY] A map that shows the positions of denaturation loops of deoxyribonucleic acid and that provides a unique way to distinguish different molecules of deoxyribonucleic acid. { di'nā·chə,rā·shən ,map }

dendrite [ANATOMY] The part of a neuron that carries the unidirectional nerve impulse toward the cell body. Also known as dendron. { 'den ,drīt }

Dendrobatinae [VERTEBRATE ZOOLOGY] A subfamily of anuran amphibians in the family Ranidae, including the colorful poisonous frogs of Central and South America. { ,den·drō'bat·ən·ē }

dendrobranchiate gill [INVERTEBRATE ZOOLOGY] A respiratory structure of certain decapod crustaceans, characterized by extensive branching of the two primary series. { ¦den·drō'braŋ·kē,āt 'gil }

Dendroceratida [INVERTEBRATE ZOOLOGY] A small order of sponges of the class Demospongiae; members have a skeleton of spongin fibers or lack a skeleton. { ,den·drō·sə'räd·əd·ə }

Dendrochirotacea [INVERTEBRATE ZOOLOGY] A subclass of echinoderms in the class Holothuroidea. { ,den·drō,kī·rō'tās·ē·ə }

Dendrochirotida [INVERTEBRATE ZOOLOGY] An order of dendrochirotacean holothurian echinoderms with 10-30 richly branched tentacles. { ,den·drō,kī'räd·əd·ə }

Dendrocolaptidae [VERTEBRATE ZOOLOGY] The woodcreepers, a family of passeriform birds belonging to the suboscine group. { ,den·drō·kə 'lap·tə,dē }

dendrogram [BIOLOGY] A genealogical tree; the trunk represents the oldest common ancestor, and the branches indicate successively more recent divisions of a lineage for a group. { 'den·drə,gram }

dendroid [BIOLOGY] Branched or treelike in form. { 'den,drȯid }

Dendroidea [PALEONTOLOGY] An order of extinct sessile, branched colonial animals in the class Graptolithina occurring among typical benthonic fauna. { den'drȯid·ē·ə }

Dendromurinae [VERTEBRATE ZOOLOGY] The African tree mice and related species, a subfamily of rodents in the family Muridae. { ¦den·drō 'myu̇r·ə·nē }

dendron [ANATOMY] *See* dendrite. { 'den,drän }

dendrophagous [ZOOLOGY] Feeding on trees, referring to insects. { den'dräf·ə·gəs }

dendrophysis [MYCOLOGY] A hyphal thread with arboreal branching in certain fungi. { den'dräf·ə·səs }

denervation hypersensitivity [PHYSIOLOGY] Extreme sensitivity of an organ that has recovered from the removal or interruption of its nerve supply. { ˌdē·nər'vā·shən ˌhīp·ərˌsen·sə'tiv·ədē }

denitrification [MICROBIOLOGY] The reduction of nitrate or nitrite to gaseous products such as nitrogen, nitrous oxide, and nitric oxide; brought about by denitrifying bacteria. { dēˌnī·trə·fə'kā·shən }

denitrifying bacteria [MICROBIOLOGY] Bacteria that reduce nitrates to nitrites or nitrogen gas; most are found in soil. { dē'nī·trəˌfī·iŋ bak'tir·ē·ə }

denitrogenate [PHYSIOLOGY] To remove nitrogen from the body by breathing nitrogen-free gas. { dē'nī·trə·jəˌnāt }

dense connective tissue [HISTOLOGY] A fibrous connective tissue with an abundance of enlarged collagenous fibers which tend to crowd out the cells and ground substance. { ¦dens kə¦nek·tiv 'tish·yü }

dense fibrillar component [CELL BIOLOGY] A component of the nucleolus that lacks granules and stains more intensely than other nucleolar components. { ¦dens ¦fi·brə·lər kəm¦pōn·ənt }

density-dependent factor [ECOLOGY] A factor that affects the birth rate or mortality rate of a population in ways varying with the population density. { ¦den·səd·ē ˌdi¦pen·dənt ˌfak·tər }

density-independent factor [ECOLOGY] A factor that affects the birth rate or mortality rate of a population in ways that are independent of the population density. { ¦den·səd·ē ˌin·də¦pen·dənt ˌfak·tər }

Densovirus [VIROLOGY] A genus of the animal virus family Parvoviridae whose virion is nonenveloped, with deoxyribonucleic acid single-stranded; replicates autonomously. { ˌden·sō 'vī·rəs }

dental [ANATOMY] Pertaining to the teeth. { 'dent·əl }

dental arch [ANATOMY] The parabolic curve formed by the cutting edges and masticating surfaces of the teeth. { ¦dent·əl 'ärch }

dental epithelium [HISTOLOGY] The cells forming the boundary of the enamel organ. { 'dent·əl ep·ə'thē·lē·əm }

dental follicle [EMBRYOLOGY] *See* dental sac. { ¦dent·əl 'fäl·ə·kəl }

dental formula [VERTEBRATE ZOOLOGY] An expression of the number and kind of teeth in each half jaw, both upper and lower, of mammals. { ¦dent·əl 'fór·myə·lə }

Dentaliidae [INVERTEBRATE ZOOLOGY] A family of mollusks in the class Scaphopoda; members have pointed feet. { ˌdent·əl'ī·əˌdē }

dental pad [VERTEBRATE ZOOLOGY] A firm ridge that replaces incisors in the maxilla of cud-chewing herbivores. { 'dent·əl ˌpad }

dental papilla [EMBRYOLOGY] The mass of connective tissue located inside the enamel organ of a developing tooth, and forming the dentin and dental pulp of the tooth. { 'dent·əl pə 'pil·ə }

dental plate [INVERTEBRATE ZOOLOGY] A flat plate that replaces teeth in certain invertebrates, such as some worms. [VERTEBRATE ZOOLOGY] A flattened plate that represents fused teeth in parrot fishes and related forms. { 'dent·əl ˌplāt }

dental pulp [HISTOLOGY] The vascular connective tissue of the roots and pulp cavity of a tooth. { 'dent·əl ˌpəlp }

dental ridge [EMBRYOLOGY] An elevation of the embryonic jaw that forms a cusp or margin of a tooth. { 'dent·əl ˌrij }

dental sac [EMBRYOLOGY] The connective tissue that encloses the developing tooth. Also known as dental follicle. { 'dent·əl ˌsak }

dentate [BIOLOGY] **1.** Having teeth. **2.** Having toothlike or conical marginal projections. { 'den ˌtāt }

dentate fissure [ANATOMY] *See* hippocampal sulcus. { 'denˌtāt 'fish·ər }

dentate nucleus [ANATOMY] An ovoid mass of nerve cells located in the center of each cerebellar hemisphere, which give rise to fibers found in the superior cerebellar peduncle. { 'denˌtāt 'nü·klē·əs }

denticle [ZOOLOGY] A small tooth or toothlike projection, as the type of scale of certain elasmobranchs. { 'dent·ə·kəl }

denticulate [ZOOLOGY] Having denticles; serrate. { den'tik·yə·lət }

dentigerous [BIOLOGY] Having teeth or toothlike structures. { den'tij·ə·rəs }

dentin [HISTOLOGY] A bonelike tissue composing the bulk of a vertebrate tooth; consists of 70% inorganic materials and 30% water and organic matter. { 'dent·ən }

dentinoblast [HISTOLOGY] A mesenchymal cell that forms dentin. { den'tēn·əˌblast }

dentinogenesis [PHYSIOLOGY] The formation of dentin. { den¦tēn·ə¦jen·ə·səs }

dentition [VERTEBRATE ZOOLOGY] The arrangement, type, and number of teeth which are variously located in the oral or in the pharyngeal cavities, or in both, in vertebrates. { den'tish·ən }

deoperculate [BOTANY] Of mosses and liverworts, to shed the operculum. { dē·ō'pər·kyə ˌlāt }

deoxycholate [BIOCHEMISTRY] A salt or ester of deoxycholic acid. { dē¦äk·sə'kōˌlāt }

deoxycholic acid [BIOCHEMISTRY] $C_{24}H_{40}O_4$ One of the unconjugated bile acids; in bile it is largely conjugated with glycine or taurine. { dē ¦äk·sə'käl·ik 'as·əd }

deoxycorticosterone [BIOCHEMISTRY] $C_{21}H_{30}O_3$ A steroid hormone secreted in small amounts by the adrenal cortex. { dē¦äk·sēˌkórd·ə'kä·stə ˌrōn }

deoxyribonuclease [BIOCHEMISTRY] An enzyme that catalyzes the hydrolysis of deoxyribonucleic acid to nucleotides. Abbreviated DNase. { dē ¦äk·sēˌrī·bō'nü·klēˌās }

deoxyribonucleic acid [BIOCHEMISTRY] A linear polymer made up of deoxyribonucleotide repeating units (composed of the sugar 2-deoxyribose, phosphate, and a purine or pyrimidine base) linked by the phosphate group joining the 3′ position of one sugar to the 5′ position of the next; most molecules are double-stranded and antiparallel, resulting in a right-handed helix structure kept together by hydrogen bonds between a purine on one chain and a pyrimidine on another; carrier of genetic information, which is encoded in the sequence of bases; present in chromosomes and chromosomal material of cell organelles such as mitochondria and chloroplasts, and also present in some viruses. Abbreviated DNA. { dē¦äk·sē‚rī·bō·nü¦klē·ik ′as·əd }

deoxyribonucleic acid clone [MOLECULAR BIOLOGY] A deoxyribonucleic acid segment inserted via a vector into a host cell and replicated along with the vector to form many copies per cell. { dē¦äk·sē‚rī·bō·nü¦klē·ik ¦as·əd ′klōn }

deoxyribonucleic acid complexity [MOLECULAR BIOLOGY] A measure of the fraction of nonrepetitive deoxyribonucleic acid that is characteristic of a given sample. { dē¦äk·sē‚rī·bō·nü¦klē·ik ¦as·əd kəm′plek·səd·ē }

deoxyribonucleic acid-directed ribonucleic acid polymerase [BIOCHEMISTRY] An enzyme which transcribes a ribonucleic acid (RNA) molecule complementary to deoxyribonucleic acid (DNA); required for initiation of DNA replication as well as transcription of RNA. { dē¦äk·sē‚rī·bō·nü¦klē·ik ¦as·əd də′rek·təd ¦rī·bō·nü¦kle·ik ¦as·əd pə′lim·ə‚rās }

deoxyribonucleic acid footprinting [MOLECULAR BIOLOGY] A method for determining the sequence of deoxyribonucleic acid-binding proteins. { dē¦äk·sē‚rī·bō·nü¦klē·ik ¦as·əd ′fu̇t‚print·iŋ }

deoxyribonucleic acid hybridization [MOLECULAR BIOLOGY] A technique for selectively binding specific segments of single-stranded deoxyribonucleic acid (DNA) or ribonucleic acid by base pairing to complementary sequences on single-stranded DNA molecules that are trapped on a nitrocellulose filter. { dē¦äk·sē‚rī·bō·nü¦klē·ik ¦as·əd ‚hī·brə·də′zā·shən }

deoxyribonucleic acid lesion [MOLECULAR BIOLOGY] Deoxyribonucleic acid deformations that may result in gene mutation or changes in chromosome structure. { de¦ak·sē‚rī·bō·nü¦klē·ik ¦as·əd ′lē·zhən }

deoxyribonucleic acid ligase [BIOCHEMISTRY] An enzyme which joins the ends of two deoxyribonucleic acid chains by catalyzing the synthesis of a phosphodiester bond between a 3′-hydroxyl group at the end of one chain and a 5′-phosphate at the end of the other. { dē¦äk·sē‚rī·bō·nü¦klē·ik ¦as·əd ′lī‚gās }

deoxyribonucleic acid polymerase I [BIOCHEMISTRY] An enzyme which catalyzes the addition of deoxyribonucleotide residues to the end of a deoxyribonucleic acid (DNA) strand; generally considered to function in the repair of damaged DNA. { dē¦äk·sē‚rī·bō·nü¦klē·ik ¦as·əd pə′lim·ə‚rās ¦wən }

deoxyribonucleic acid polymerase II [BIOCHEMISTRY] An enzyme similar in action to DNA polymerase I but with lower activity. { dē¦äk·sē‚rī·bō·nü¦klē·ik ¦as·əd pə′lim·ə‚rās ¦tü }

deoxyribonucleic acid polymerase III [BIOCHEMISTRY] An enzyme thought to be the primary enzyme involved in deoxyribonucleic acid replication. { dē¦äk·sē‚rī·bō·nü¦klē·ik ¦as·əd pə′lim·ə‚rās ¦thrē }

deoxyribonucleoprotein [BIOCHEMISTRY] A protein containing molecules of deoxyribonucleic acid in close association with protein molecules. { dē¦äk·sē‚rī·bō‚nü·klē·ō ′prō‚tēn }

deoxyribonucleotide [BIOCHEMISTRY] A nucleotide that contains deoxyribose and is a constituent of deoxyribonucleic acid. { dē¦äk·sē‚rī·bō ′nü·klē·ə‚tīd }

deoxyribose [BIOCHEMISTRY] $C_5H_{10}O_4$ A pentose sugar in which the hydrogen replaces the hydroxyl groups of ribose; a major constituent of deoxyribonucleic acid. { dē¦äk·sē′rī‚bōs }

deoxyribovirus [VIROLOGY] Any virus that contains deoxyribonucleic acid. { dē¦äk·sē‚rī·bō′vī·rəs }

deoxy sugar [BIOCHEMISTRY] A substance which has the characteristics of a sugar, but which shows a deviation from the required hydrogen-to-oxygen ratio. { dē¦äk·sē ′shu̇g·ər }

depauperate [BIOLOGY] Inferiority of natural development or size. { dē′pȯ·pə·rət }

Depertellidae [PALEONTOLOGY] A family of extinct perissodactyl mammals in the superfamily Tapiroidea. { de·pər′tel·ə‚dē }

depletion [ECOLOGY] Using a resource, such as water or timber, faster than it is replenished. { də ′plē·shən }

deposit feeder [INVERTEBRATE ZOOLOGY] Any animal that feeds on the detritus that collects on the substratum at the bottom of water. Also known as detritus feeder. { də′päz·ət ‚fēd·ər }

depressor [ANATOMY] A muscle that draws a part down. { di′pres·ər }

depressor nerve [PHYSIOLOGY] A nerve which, upon stimulation, lowers the blood pressure either in a local part or throughout the body. { di ′pres·ər ‚nərv }

depth perception [PHYSIOLOGY] Ability to judge spatial relationships. { ′depth pər′sep·shən }

depurination [BIOCHEMISTRY] Detachment of guanine from sugar in a deoxyribonucleic acid molecule. { dē‚pyu̇r·ə′nā·shən }

derelict land [ECOLOGY] Land that, because of mining, drilling, or other industrial processes, or by serious neglect, is unsightly and cannot be beneficially utilized without treatment. { ′der·ə‚likt ‚land }

derepression [MICROBIOLOGY] Transfer of microbial cells from an enzyme-repressing medium to a nonrepressing medium. [MOLECULAR BIOLOGY] Increased production of a gene product due to interference with the action of a repressor on the operator portion of the operon. { dē·ri ′presh·ən }

Dermacentor [INVERTEBRATE ZOOLOGY] A genus of ticks, important as vectors of disease. { ′dər·mə‚sen·tər }

Dermacentor andersoni [INVERTEBRATE ZOOL-

OGY] The wood tick, which is the vector of Rocky Mountain spotted fever and tularemia. { 'dər·mə ˌsen·tər an·dər'sō·nē }

Dermacentor variabilis [INVERTEBRATE ZOOLOGY] A North American tick which is parasitic primarily on dogs but may attack humans and other mammals. { 'dər·məˌsen·tər ver·ē'ab·ə·ləs }

dermal [ANATOMY] Pertaining to the dermis. { 'dər·məl }

dermal bone [ANATOMY] A type of bone that ossifies directly from membrane without a cartilaginous predecessor; occurs only in the skull and shoulder region. Also known as investing bone; membrane bone. { ¦dər·məl 'bōn }

dermal denticle [VERTEBRATE ZOOLOGY] A toothlike scale composed mostly of dentine with a large central pulp cavity, found in the skin of sharks. { ¦dər·məl 'dent·i·kəl }

dermalia [INVERTEBRATE ZOOLOGY] Dermal microscleres in sponges. { dər'mal·yə }

dermal pore [INVERTEBRATE ZOOLOGY] One of the minute openings on the surface of poriferans leading to the incurrent canals. { ¦dər·məl 'pōr }

Dermaptera [INVERTEBRATE ZOOLOGY] An order of small or medium-sized, slender insects having incomplete metamorphosis, chewing mouthparts, short forewings, andcerci. { dər'map·tə·rə }

Dermatemydinae [VERTEBRATE ZOOLOGY] A family of reptiles in the order Chelonia; includes the river turtles. { ˌdər·mə·tə'mī·dəˌnē }

Dermatocarpaceae [BOTANY] A family of lichens in the order Pyrenulales having an umbilicate or squamulose growth form; most members grow on limestone or calcareous soils. { dər¦mad·ō ˌkär'pās·ēˌē }

dermatocranium [ANATOMY] Bony parts of the skull derived from ossifications in the dermis of the skin. { dər¦mad·ə'krā·nē·əm }

dermatogen [BOTANY] The outer layer of primary meristem or the primordial epidermis in embryonic plants. Also known as protoderm. { dər 'mad·əˌjen }

dermatoglyphics [ANATOMY] **1.** The integumentary patterns on the surface of the fingertips, palms, and soles. **2.** The study of these patterns. { dər¦mad·ə¦glif·iks }

dermatome [ANATOMY] An area of skin delimited by the supply of sensory fibers from a single spinal nerve. [EMBRYOLOGY] Lateral portion of an embryonic somite from which the dermis will develop. { 'dər·məˌtōm }

Dermatophilaceae [MICROBIOLOGY] A family of bacteria in the order Actinomycetales; cells produce mycelial filaments or muriform thalli; includes human and mammalian pathogens. { dər ¦mad·ō·fə'lās·ēˌē }

Dermatophilus [MICROBIOLOGY] A genus of bacteria in the family Dermatophilaceae; mycelial filaments are long, tapering, and branched, and are divided transversely and longitudinally (in two planes) by septa; spherical spores are motile. { dər¦mad·ə'fil·əs }

dermatophyte [MYCOLOGY] A fungus parasitic on skin or its derivatives. { dər'mad·əˌfīt }

dermatoplast [BOTANY] In angiosperms, a cell with a cell wall. { dər'mad·əˌplast }

Dermestidae [INVERTEBRATE ZOOLOGY] The skin beetles, a family of coleopteran insects in the superfamily Dermestoidea, including serious pests of stored agricultural grain products. { dər 'mes·tə·dē }

Dermestoidea [INVERTEBRATE ZOOLOGY] A superfamily of coleopteran insects in the suborder Polyphaga. { ˌdər·mə'stȯid·ē·ə }

dermis [ANATOMY] The deep layer of the skin, a dense connective tissue richly supplied with blood vessels, nerves, and sensory organs. Also known as corium; cutis. { 'dər·məs }

Dermochelidae [VERTEBRATE ZOOLOGY] A family of reptiles in the order Chelonia composed of a single species, the leatherback turtle. { ˌdər·mə 'kel·əˌdē }

dermoepidermal junction [HISTOLOGY] The area of separation between the stratum basale of the epidermis and the papillary layer of the dermis. { ¦dər·mōˌep·ə¦dər·məl 'jəŋk·shən }

dermometer [PHYSIOLOGY] An instrument used to measure the electrical resistance of the skin. { dər'mäm·əd·ər }

Dermoptera [VERTEBRATE ZOOLOGY] The flying lemurs, an ancient order of primatelike herbivorous and frugivorous gliding mammals confined to southeastern Asia and eastern India. { dər 'mäp·tə·rə }

Derodontidae [INVERTEBRATE ZOOLOGY] The tooth-necked fungus beetles, a small family of coleopteran insects in the superfamily Dermestoidea. { ˌder·ə'dän·təˌdē }

derris [BOTANY] Any of certain tropical shrubs in the genus *Derris* in the legume family (Leguminosae), having long climbing branches. { 'der·əs }

Derxia [MICROBIOLOGY] A genus of bacteria in the family Azotobacteraceae; rod-shaped, pleomorphic, motile cells, older cells contain large refractive bodies. { 'dərk·sē·ə }

DES [BIOCHEMISTRY] *See* diethylstilbesterol.

Descemet's membrane [HISTOLOGY] A layer of the cornea between the posterior surface of the stroma and the anterior surface of the endothelium which contains collagen arranged on a crystalline lattice. { des'māz ˌmemˌbrān }

descending [ANATOMY] Extending or directed downward or caudally, as the descending aorta. [PHYSIOLOGY] In the nervous system, efferent; conducting impulses or progressing down the spinal cord or from central to peripheral. { di 'sen·diŋ }

descending colon [ANATOMY] The portion of the colon on the left side, extending from the bend below the spleen to the sigmoid flexure. { di 'send·iŋ 'kōl·ən }

descriptive anatomy [ANATOMY] Study of the separate and individual portions of the body, with regard to form, size, character, and position. { di'skrip·tiv ə'nad·ə·mē }

descriptive botany [BOTANY] The branch of botany that deals with diagnostic characters or systematic description of plants. { di'skrip·tiv 'bät·ən·ē }

desensitization [IMMUNOLOGY] Loss or reduction of sensitivity to infection or an allergen accomplished by means of frequent, small doses of the antigen. Also known as hyposensitization. { dē‚sen·sə·tə'zā·shən }

deserticolous [ECOLOGY] Living in a desert. { ¦dez·ər¦tik·ə·ləs }

desertification [ECOLOGY] The creation of desiccated, barren, desertlike conditions due to natural changes in climate or possibly through mismanagement of the semiarid zone. { də‚zərd·ə·fə'kā·shən }

desexualization [PHYSIOLOGY] Depriving an organism of sexual characters or power, as by spaying or castration. { dē‚seksh·ə·lə'zā·shən }

desma [INVERTEBRATE ZOOLOGY] A branched, knobby spicule in some Demospongiae. { 'dez·mə }

desmacyte [INVERTEBRATE ZOOLOGY] A bipolar collencyte found in the cortex of certain sponges. { 'dez·mə‚sīt }

Desmanthos [MICROBIOLOGY] A genus of bacteria in the family Pelonemataceae; unbranched, relatively straight filaments with a thickened base which are arranged in bundles partially enclosed in a sheath. { dez'man·thȯs }

desmid [BOTANY] Any member of a group of microscopic, unicellular green algae of the family Desmidiaceae, having cells of varying shapes but always composed of mirror-image semicells, often demarcated by a median constriction or incision, and a cell wall that has pores, which are frequently ornamented. { 'dez·məd }

Desmidiaceae [BOTANY] A family of desmids, mostly unicellular algae in the order Conjugales. { dez‚mid·ē'ās·ē‚ē }

desmin [BIOCHEMISTRY] The muscle protein forming the Z lines in striated muscle. { 'dez·mən }

desmochore [ECOLOGY] A plant having sticky or barbed disseminules. { 'dez·mə‚kȯr }

Desmodonta [PALEONTOLOGY] An order of extinct bivalve, burrowing mollusks. { ‚dez·mə'dän·tə }

Desmodontidae [VERTEBRATE ZOOLOGY] A small family of chiropteran mammals comprising the true vampire bats. { ‚dez·mə'dän·tə‚dē }

Desmodoroidea [INVERTEBRATE ZOOLOGY] A superfamily of marine- and brackish-water-inhabiting nematodes with an annulated, usually smooth cuticle. { 'dez·mə·də'rȯid·ē·ə }

Desmokontae [BOTANY] The equivalent name for Desmophyceae. { ‚dez·mə'kän·tē }

desmolase [BIOCHEMISTRY] Any of a group of enzymes which catalyze rupture of atomic linkages that are not cleaved through hydrolysis, such as the bonds in the carbon chain of D-glucose. { 'dez·mə‚lās }

desmoneme [INVERTEBRATE ZOOLOGY] A nematocyst having a long coiled tube which is extruded and wrapped around the prey. { ‚dez·mə'nēm }

desmopelmous [VERTEBRATE ZOOLOGY] A type of bird foot in which the hindtoe cannot be bent independently because planter tendons are united. { ‚dez·mə'pel·məs }

Desmophyceae [BOTANY] A class of rare, mostly marine algae in the division Pyrrhophyta. { ‚dez·mə'fīs·ē‚ē }

Desmoscolecida [INVERTEBRATE ZOOLOGY] An order of the class Nematoda. { ‚dez·mə·skə'les·ə·də }

Desmoscolecidae [INVERTEBRATE ZOOLOGY] A family of nematodes in the superfamily Desmoscolecoidea; individuals resemble annelids in having coarseannulation. { ‚dez·mə·skə'les·ə‚dē }

Desmoscolecoidea [INVERTEBRATE ZOOLOGY] A small superfamily of free-living nematodes characterized by a ringed body, an armored head set, and hemispherical amphids. { ‚dez·mə‚skō·lə'kȯid·ē·ə }

desmose [INVERTEBRATE ZOOLOGY] A fibril connecting the centrioles during mitosis in certain protozoans. { 'dez‚mōs }

Desmostylia [PALEONTOLOGY] An extinct order of large hippopotamuslike, amphibious, gravigrade, shellfish-eating mammals. { ‚dez·mə'stīl·ē·ə }

Desmostylidae [PALEONTOLOGY] A family of extinct mammals in the order Desmostylia. { ‚dez·mə'stīl·ə‚dē }

Desmothoracida [INVERTEBRATE ZOOLOGY] An order of sessile and free-living protozoans in the subclass Heliozoia having a spherical body with a perforate, chitinous test. { ‚dez·mə·thə'ras·ə·də }

Desor's larva [INVERTEBRATE ZOOLOGY] An oval, ciliated larva of certain nemertineans in which the gastrula remains inside the egg membrane. { də'zȯrz ‚lär·va }

desquamation [PHYSIOLOGY] Shedding; a peeling and casting off, as of the superficial epithelium, mucous membranes, renal tubules, and the skin. { dē·skwə'mā·shən }

Desulfotomaculum [MICROBIOLOGY] A genus of bacteria in the family Bacillaceae; motile, straight or curved rods with terminal to subterminal spores; anaerobic and reduce sulfate. { dē ‚səl·fəd·ə'mak·yə·ləm }

Desulfovibrio [MICROBIOLOGY] A genus of gram-negative, strictly anaerobic bacteria of uncertain affiliation; motile, curved rods reduce sulfates and other sulfur compounds to hydrogen sulfide. { dē‚səl·fə'vib·rē‚ō }

detached meristem [BOTANY] A meristematic region originating from apical meristem but becoming discontinuous with it because of differentiation of intervening tissue. { di'tacht 'mer·ə‚stem }

determinate cleavage [EMBRYOLOGY] A type of cleavage which separates portions of the zygote with specific and distinct potencies for development as specific parts of the body. { də'tər·mə·nət 'klē·vij }

determinate growth [BOTANY] Growth in which the axis, or central stem, being limited by the development of the floral reproductive structure, does not grow or lengthen indefinitely. { də'tər·mə·nət 'grōth }

detorsion [INVERTEBRATE ZOOLOGY] Untwisting

of the 180° visceral twist imposed by embryonic torsion on many gastropod mollusks. { dē'tȯr·shən }

detoxification [BIOCHEMISTRY] The act or process of removing a poison or the toxic properties of a substance in the body. { dē,täk·sə·fə'kā·shən }

detritus [ECOLOGY] Dead plants and corpses or cast-off parts of various organisms. { də'trīd·əs }

detritus feeder [INVERTEBRATE ZOOLOGY] *See* deposit feeder. { də'trīd·əs 'fēd·ər }

detrivorous [BIOLOGY] Referring to an organism that feeds on dead animals or partially decomposed organic matter. { də'triv·ə·rəs }

deutencephalon [EMBRYOLOGY] *See* epichordal brain. { 'düt'en'sef·ə,län }

deuterogamy [BOTANY] Secondary pairing of sexual cells or nuclei replacing direct copulation in many fungi, algae, and higher plants. { ,düd·ə'räg·ə·mē }

Deuteromycetes [MYCOLOGY] The equivalent name for Fungi Imperfecti. { ¦düd·ə·rō,mī'sēd·ēz }

Deuterophlebiidae [INVERTEBRATE ZOOLOGY] A family of dipteran insects in the suborder Cyclorrhapha. { ,düd·ə·rō·flə'bī·ə,dē }

Deuterostomia [ZOOLOGY] A division of the animal kingdom which includes the phyla Echinodermata, Chaetognatha, Hemichordata, and Chordata. { ,düd·ə·rō'stō·mē·ə }

deutocerebrum [INVERTEBRATE ZOOLOGY] The median lobes of the insect brain. { ¦düd·ō'ser·ə·brəm }

deutoplasm [EMBRYOLOGY] The nutritive yolk granules in egg cells. { 'düd·ə,plaz·əm }

development [BIOLOGY] A process of regulated growth and differentiation that results from interaction of the genome with the cytoplasm, the internal cellular environment, and the external environment. { də'vel·əp·mənt }

developmental control gene [GENETICS] A gene whose primary function is the regulation of cell fates during development. { di,vel·əp¦men·təl kən'trōl ,jēn }

developmental genetics [GENETICS] A branch of genetics primarily concerned with the manner in which genes control or regulate development. { də¦vel·əp,ment·əl jə'ned·iks }

developmental instability [GENETICS] Variation of development within a genotype due to local fluctuations in internal or external environmental conditions. { di,vel·əp¦men·təl ,in·stə'bil·əd·ē }

developmental noise [GENETICS] Any uncontrollable variation in phenotype due to random events during development. { di,vel·əp¦men·təl 'nȯiz }

devernalization [BOTANY] Annulment of the vernalization effect. { dē,vərn·əl·ə'zā·shən }

deviation [EVOLUTION] Evolutionary differentiation involving interpolation of new stages in the ancestral pattern of morphogenesis. { ,dēv·ē'ā·shən }

dewclaw [VERTEBRATE ZOOLOGY] **1.** A vestigial digit on the foot of a mammal which does not reach the ground. **2.** A claw or hoof terminating such a digit. { 'dü,klȯ }

dewlap [ANATOMY] A fleshy or fatty fold of skin on the throat of some humans. [BOTANY] One of a pair of hinges at the joint of a sugarcane leaf blade. [VERTEBRATE ZOOLOGY] A fold of skin hanging from the neck of some reptiles and bovines. { 'dü,lap }

Dexaminidae [INVERTEBRATE ZOOLOGY] A family of amphipod crustaceans in the suborder Gammeridea. { dek'sam·in·ə,dē }

dexterotropic [BIOLOGY] Turning toward the right; applied to cleavage, shell formation, and whorl patterns. { ¦dek·stə·rō ¦träp·ik }

dextran [BIOCHEMISTRY] Any of the several polysaccharides, $(C_5H_{10}O_5)_n$, that yield glucose units on hydrolysis. { 'dek ,stran }

dextranase [BIOCHEMISTRY] An enzyme that hydrolyzes 1,6-α-glucosidic linkages in dextran. { 'dek·strə,nās }

dextrin [BIOCHEMISTRY] A polymer of D-glucose which is intermediate in complexity between starch and maltose. { 'dek·strən }

dextrorse [BOTANY] Twining toward the right. { 'dek,strȯrs }

dextrose [BIOCHEMISTRY] $C_6H_{12}O_6 \cdot H_2O$ A dextrorotatory monosaccharide obtained as a white, crystalline, odorless, sweet powder, which is soluble in about one part of water; an important intermediate in carbohydrate metabolism; used for nutritional purposes, for the temporary increase of blood volume, and as a diuretic. Also known as corn sugar; grape sugar. { 'dek,strōs }

dextrotopic cleavage [EMBRYOLOGY] A clockwise spiral cleavage pattern. { ¦dek·strə¦täp·ik 'klē·vij }

Diacodectidae [PALEONTOLOGY] A family of extinct artiodactyl mammals in the suborder Palaeodonta. { ,dī·ə·kə'dek·tə,dē }

diactine [INVERTEBRATE ZOOLOGY] A type of sponge spicule which develops in two directions from a central point. { dī'ak,tēn }

diadelphous stamen [BOTANY] A stamen that has its filaments united into two sets. { ¦dī·ə¦del·fəs 'stā·mən }

Diadematacea [INVERTEBRATE ZOOLOGY] A superorder of Euchinoidea having a rigid or flexible test, perforate tubercles, and branchial slits. { ,dī·ə,dē·mə'tās·ē·ə }

Diadematidae [INVERTEBRATE ZOOLOGY] A family of large euechinoid echinoderms in the order Diadematoida having crenulate tubercles and long spines. { ,dī·ə·də'mad·ə,dē }

Diadematoida [INVERTEBRATE ZOOLOGY] An order of echinoderms in the superorder Diadematacea with hollow primary radioles and crenulate tubercles. { ,dī·ə·dē·mə'tȯid·ə }

diadromous [BOTANY] Having venation in the form of fanlike radiations. [VERTEBRATE ZOOLOGY] Of fish, migrating between salt and fresh waters. { dī'ad·rə·məs }

Diadumenidae [INVERTEBRATE ZOOLOGY] A family of anthozoans in the order Actiniaria. { dī·ə·dü'men·ə,dē }

diageotropism [BIOLOGY] Growth orientation of

a sessile organism or structure perpendicular to the line of gravity. { ¦dī·ə·jē′ä·trə,piz·əm }

diagnosis [SYSTEMATICS] In taxonomic study, a statement of the characters that distinguish a taxon from coordinate taxa. { ,dī·əg′nō·səs }

diaheliotropism [BOTANY] Movement of plant leaves which follow the sun such that they remain perpendicular to the sun's rays throughout the day. { ,dī·ə,hē·lē·ə′trä,piz·əm }

diakinesis [CELL BIOLOGY] The last stage of meiotic prophase, when the chromatids attain maximum contraction and the bivalents move apart and position themselves against the nuclear membrane. { dī·ə·kə′nē·səs }

diamine oxidase [BIOCHEMISTRY] A flavoprotein which catalyzes the aerobic oxidation of amines to the corresponding aldehyde and ammonia. { ′dī·ə,mēn ′äk·sə,dās }

diaminopimelate [BIOCHEMISTRY] $C_7H_{14}O_4N_2$ A compound that serves as a component of cell wall mucopeptide in some bacteria and as a source of lysine in all bacteria. Abbreviated DAP. { dī¦am·ə,nō′pim·ə,lāt }

diandrous [BOTANY] Having two stamens. { dī ′an·drəs }

Dianemaceae [MICROBIOLOGY] A family of slime molds in the order Trichales. { ,dī·ə·nə′mās·ē,ē }

Dianulitidae [PALEONTOLOGY] A family of extinct, marine bryozoans in the order Cystoporata. { dī ,an·yə′lid·ə,dē }

diapause [PHYSIOLOGY] A period of spontaneously suspended growth or development in certain insects, mites, crustaceans, and snails. { ′dī·ə,pȯz }

Diapensiaceae [BOTANY] The single family of the Diapensiales, an order of flowering plants. { ,dī·ə,pen·sē′ās·ē,ē }

Diapensiales [BOTANY] A monofamilial order of dicotyledonous plants in the subclass Dilleniidae comprising certain herbs and dwarf shrubs in temperate and arctic regions of the Northern Hemisphere. { ,dī·ə,pen·sē′ā·lēz }

Diaphanocephalidae [INVERTEBRATE ZOOLOGY] A family of parasitic roundworms belonging to the Strongyloidea; snakes are the principal host. { di ¦af·ə·nō·sə′fal·ə,dē }

Diaphanocephaloidea [INVERTEBRATE ZOOLOGY] A superfamily of nematodes represented by a single family, Diaphanocephalidae, distinguished by the modification of the stoma into two massive lateral jaws and the absence of a corona radiata or lips. { dī¦äf·ə·nō,sef·ə′lȯid·ē·ə }

diaphorase [BIOCHEMISTRY] Mitochondrial flavoprotein enzymes which catalyze the reduction of dyes, such as methylene blue, by reduced pyridine nucleotides such as reduced diphosphopyridine nucleotide. { dī′af·ə,rās }

diaphragm [ANATOMY] The dome-shaped partition composed of muscle and connective tissue that separates the abdominal and thoracic cavities in mammals. { ′dī·ə,fram }

diaphragmatic respiration [PHYSIOLOGY] Respiration effected primarily by movement of the diaphragm, changing the intrathoracic pressure. { ¦di·ə,frag¦mad·ik ,res·pə′rā·shən }

diaphysis [ANATOMY] The shaft of a longbone. { dī′af·ə·səs }

diapophysis [ANATOMY] The articular portion of a transverse process of a vertebra. { ,dī·ə′päf·ə·səs }

Diapriidae [INVERTEBRATE ZOOLOGY] A family of hymenopteran insects in the superfamily Proctotrupoidea. { ,dī·ə′prī·ə,dē }

diarch [BOTANY] Of a plant, having two protoxylem points or groups. { ′dī,ärch }

diarthrosis [ANATOMY] A freely moving articulation, characterized by a synovial cavity between the bones. { ¦dī·är′thrō·səs }

diastase [BIOCHEMISTRY] An enzyme that catalyzes the hydrolysis of starch to maltose. Also known as vegetable diastase. { ′di·ə,stās }

diastasis [PHYSIOLOGY] The final phase of diastole, the phase of slow ventricular filling. { dī ′as·tə·səs }

diastema [ANATOMY] A space between two types of teeth, as between an incisor and premolar. [CELL BIOLOGY] Modified cytoplasm of the equatorial plane prior to cell division. { ,dī·ə′stē·mə }

diastole [PHYSIOLOGY] The rhythmic relaxation and dilation of a heart chamber, especially a ventricle. { dī′as·tə·lē }

diastolic pressure [PHYSIOLOGY] The lowest arterial blood pressure during the cardiac cycle; reflects relaxation and dilation of a heart chamber. { ¦dī·ə¦stäl·ik ′presh·ər }

diatom [INVERTEBRATE ZOOLOGY] The common name for algae composing the class Bacillariophyceae; noted for the symmetry and sculpturing of the siliceous cell walls. { ′dī·ə,täm }

diatropism [BOTANY] Growth orientation of certain plant organs that is transverse to the line of action of a stimulus. { dī′a·trə,piz·əm }

Diatrymiformes [PALEONTOLOGY] An order of extinct large, flightless birds having massive legs, tiny wings, and large heads and beaks. { dī,a·trə·mə′fōr,mēz }

diauxic growth [MICROBIOLOGY] The diphasic response of a culture of microorganisms based on a phenotypic adaptation to the addition of a second substrate; characterized by a growth phase followed by a lag after which growth is resumed. { dī′ȯk,sik ′grōth }

diazotroph [MICROBIOLOGY] An organism that carries out nitrogen fixation; examples are *Clostridium* and *Azotobacter*. { dī′az·ə,träf }

Dibamidae [VERTEBRATE ZOOLOGY] The flap-legged skinks, a small family of lizards in the suborder Sauria comprising three species confined to southeastern Asia. { dī′bäm·ə,dē }

Dibranchia [INVERTEBRATE ZOOLOGY] A subclass of the Cephalopoda containing all living cephalopods except *Nautilus*; members possess two gills and, when present, an internal shell. { dī ′braŋ·kē·ə }

dicarpellate [BOTANY] Having two carpels. { dī ′kär·pə,lāt }

dicaryon [MYCOLOGY] *See* dikaryon. { dī′kar·ē ,än }

dicentric [CELL BIOLOGY] Having two centromeres. { dī'sen·trik }

dicerous [INVERTEBRATE ZOOLOGY] Having two tentacles or two antennae. { 'dī·sə·rəs }

Dice's life zones [ECOLOGY] Biomes proposed by L.R. Dice based on the concept of the biotic province. { 'dīs·əz 'līf ˌzōnz }

dichasium [BOTANY] A cyme producing two main axes from the primary axis or shoot. { dī'kā·zhē·əm }

Dichelesthiidae [INVERTEBRATE ZOOLOGY] A family of parasitic copepods in the suborder Caligoida; individuals attach to the gills of various fishes. { dī¦ke·ləs'thī·əˌdē }

dichlamydeous [BOTANY] Having both calyx and corolla. { dī·klə'mid·ē·əs }

Dichobunidae [PALEONTOLOGY] A family of extinct artiodactyl mammals in the superfamily Dichobunoidea. { ˌdī·kə'byün·əˌdē }

Dichobunoidea [PALEONTOLOGY] A superfamily of extinct artiodactyl mammals in the suborder Paleodonta composed of small to medium-size forms with tri- to quadritubercular bunodont upper teeth. { ˌdī·kə·byə'nȯid·ē·ə }

dichogamous [BOTANY] Referring to a type of flower in which the pistils and stamens reach maturity at different times. { dī'käg·ə·məs }

dichogamy [BIOLOGY] Producing mature male and female reproductive structures at different times. { dī'käg·ə·mē }

dichoptous [INVERTEBRATE ZOOLOGY] Having the margins of the compound eyes separate. { dī'käp·təs }

dichotomy [BIOLOGY] **1.** Divided in two parts. **2.** Repeated branching or forking. { dī'käd·ə·mē }

dichotriaene [INVERTEBRATE ZOOLOGY] A type of sponge spicule with three rays. { ¦di·kō'trīˌēn }

dichromatic [BIOLOGY] Having or exhibiting two color phases independently of age or sex. { dī·krə'mād·ik }

Dickinsoniidae [PALEONTOLOGY] A family that comprises extinct flat-bodied, multisegmented coelomates; identified as ediacaran fauna. { ˌdik·ən·sə'nī·əˌdē }

Dicksoniaceae [BOTANY] A family of tree ferns characterized by marginal sori which are terminal on the veins and protected by a bivalved indusium. { ˌdik·sə·nē'ās·ēˌē }

Dick test [IMMUNOLOGY] A skin test to determine immunity to scarlet fever; *Streptococcus pyogenes* toxin is injected intracutaneously and produces a reaction if there is no circulating antitoxin. { 'dik ˌtest }

diclinous [BOTANY] Having stamens and pistils on different flowers. { dī'klī·nəs }

dicoccous [BOTANY] Composed of two adherent one-seeded carpels. { dī'käk·əs }

dicotyledon [BOTANY] Any plant of the class Magnoliopsida, all having two cotyledons. { ˌdī ˌkäd·əl'ēd·ən }

Dicotyledoneae [BOTANY] The equivalent name for Magnoliopsida. { dīˌkäd·əl·ə'dän·ēˌē }

Dicranales [BOTANY] An order of mosses having erect stems, dichotomous branching, and dense foliation. { ˌdī·krə'nā·lēz }

dictyoblastospore [MYCOLOGY] A blastospore with both cross and longitudinal septa. { ¦dik·tē·ō¦blas·təˌspȯr }

Dictyoceratida [INVERTEBRATE ZOOLOGY] An order of sponges of the class Demospongiae; includes the bath sponges of commerce. { ¦dik·tē·ō·sə'rad·əd·ə }

Dictyonellidina [PALEONTOLOGY] A suborder of extinct articulate brachiopods. { ¦dik·tē·ō·ne'lid·ən·ə }

dictyosome [CELL BIOLOGY] A stack of two or more cisternae; a component of the Golgi apparatus. { 'dik·tē·əˌsōm }

Dictyospongiidae [PALEONTOLOGY] A family of extinct sponges in the subclass Amphidiscophora having spicules resembling a one-ended amphidisc (paraclavule). { ¦dik·tē·ōˌspən'jī·əˌdē }

Dictyosporae [MYCOLOGY] A spore group of the imperfect fungi characterized by multicelled spores with cross and longitudinal septae. { ˌdik·tē·ə'spȯr·ē }

dictyospore [MYCOLOGY] A multicellular spore in certain fungi characterized by longitudinal walls and cross septa. { 'dik·tē·əˌspȯr }

dictyostele [BOTANY] A modified siphonostele in which the vascular tissue is dissected into a network of distinct strands; found in certain fern stems. { 'dik·tē·əˌstēl }

Dictyosteliaceae [MICROBIOLOGY] A family of microorganisms belonging to the Acrasiales and characterized by strongly differentiated fructifications. { ¦dik·tē·ōˌstel·ē'ās·ēˌē }

Dicyemida [INVERTEBRATE ZOOLOGY] An order of mesozoans comprising minute, wormlike parasites of the renal organs of cephalopod mollusks. { ˌdīˌsī'em·ə·də }

didelphic [ANATOMY] Having a double uterus or genital tract. { dī'del·fik }

Didelphidae [VERTEBRATE ZOOLOGY] The opossums, a family of arboreal mammals in the order Marsupialia. { dī'del·fəˌdē }

Didolodontidae [PALEONTOLOGY] A family consisting of extinct medium-sized herbivores in the order Condylarthra. { dīd·əl·ō'dänt·əˌdē }

Didymelales [BOTANY] An order of dicotyledonous plants in the subclass Hamamelidae, characterized by the primitive nature of the wood, which has vessels with scalariform perforations, and a pistil, which has one carpel; dioecious, evergreen trees restricted to Madagascar. { ˌdī·də·mə'lā·lēz }

Didymiaceae [MICROBIOLOGY] A family of slime molds in the order Physarales. { ˌdī·də·mī'as·ēˌē }

Didymosporae [MYCOLOGY] A spore group of the imperfect fungi characterized by two-celled spores. { dī·də·mə'spȯr·ē }

didymous [BIOLOGY] Occurring in pairs. { 'did·ə·məs }

didynamous [BOTANY] Having four stamens occurring in two pairs, one pair long and the other short. { dī'din·ə·məs }

dieback [ECOLOGY] A large area of exposed, unprotected swamp or marsh deposits resulting from the salinity of a coastal lagoon. { 'dīˌbak }

die down [BOTANY] Normal seasonal death of aboveground parts of herbaceous perennials. { ′dī ‚dau̇n }

Diego blood group [IMMUNOLOGY] A genetically determined, immunologically distinct group of human erythrocyte antigens recognized by reaction with a specific antibody. { dē′ā·gō ′bləd ‚grüp }

diencephalon [EMBRYOLOGY] The posterior division of the embryonic forebrain in vertebrates. { ¦dī·en′sef·ə‚län }

diesterase [BIOCHEMISTRY] An enzyme such as a nuclease which splits the linkages binding individual nucleotides of a nucleic acid. { dī′es·tə ‚rās }

diestrus [PHYSIOLOGY] The long, quiescent period following ovulation in the estrous cycle in mammals; the stage in which the uterus prepares for the reception of a fertilized ovum. { dī ′es·trəs }

diet [BIOLOGY] The food or drink regularly consumed. { ′dī·ət }

diethylstilbesterol [BIOCHEMISTRY] $C_{18}H_{20}O_2$ A white, crystalline, nonsteroid estrogen that is used therapeutically as a substitute for natural estrogenic hormones. Also known as stilbestrol. Abbreviated DES. { ‚dī¦eth·əl·stil′bes·tə‚rȯl }

differential centrifugation [CELL BIOLOGY] The separation of mixtures such as cellular particles in a medium at various centrifugal forces to separate particles of different density, size, and shape from each other. { ‚dif·ə′ren·chəl ‚sen·trə·fə′gā·shən }

differential diagnosis [SYSTEMATICS] In taxonomic study, a statement of the characters that distinguish a given taxon from other, specifically mentioned equivalent taxa. { ‚dif·ə′ren·chəl ‚dī·əg′nō·səs }

differentiation antigen [IMMUNOLOGY] A cell surface antigen that is expressed only during a specific period of embryological differentiation. { dif·ə‚ren·chē¦ā·shən ′ant·i·jən }

diffuse placenta [EMBRYOLOGY] A placenta having villi diffusely scattered over most of the surface of the chorion; found in whales, horses, and other mammals. { də′fyüs plə′sent·ə }

diffusion respiration [PHYSIOLOGY] Exchange of gases through the cell membrane, between the cells of unicellular or other simple organisms and the environment. { də′fyü·zhən res·pə′rā·shən }

digastric [ANATOMY] Of a muscle, having a fleshy part at each end and a tendinous part in the middle. { dī′gas·trik }

Digenea [INVERTEBRATE ZOOLOGY] A group of parasitic flatworms or flukes constituting a subclass or order of the class Trematoda and having two types of generations in the life cycle. { dī ′jē·nē·ə }

digenesis [BIOLOGY] Sexual and asexual reproduction in succession. { dī′jen·ə·səs }

digestion [PHYSIOLOGY] The process of converting food to an absorbable form by breaking it down to simpler chemical compounds. { də′jes·chən }

digestive enzyme [BIOCHEMISTRY] Any enzyme that causes or aids in digestion. { də′jes·tiv ′en ‚zīm }

digestive gland [PHYSIOLOGY] Any structure that secretes digestive enzymes. { də′jes·tiv ‚gland }

digestive system [ANATOMY] A system of structures in which food substances are digested. { də ′jes·tiv ‚sis·təm }

digestive tract [ANATOMY] The alimentary canal. { də′jes·tiv ‚trakt }

digicitrin [BIOCHEMISTRY] $C_{21}H_{21}O_{10}$ A flavone compound that is found in foxglove leaves. { ‚dīj·ə′si·trən }

Digitalis [BOTANY] A genus of herbs in the figwort family, Scrophulariaceae. { dij·ə′tal·əs }

digitate [ANATOMY] Having digits or digitlike processes. { ′dij·ə‚tāt }

digitellum [INVERTEBRATE ZOOLOGY] A tentacle-like gastric filament in scyphozoans. { ‚dij·ə′tel·əm }

digitigrade [VERTEBRATE ZOOLOGY] Pertaining to animals, such as dogs and cats, which walk on the digits with the posterior part of the foot raised from the ground. { ′dij·ə·də‚grād }

digitinervate [BOTANY] Having straight veins extending from the petiole like fingers. { ¦dij·ə·də ′nər‚vāt }

digitipinnate [BOTANY] Having digitate leaves with pinnate leaflets. { ¦dij·ə·də′pin‚āt }

digitus [INVERTEBRATE ZOOLOGY] In insects, the claw-bearing terminal segment of the tarsus. { ′dij·əd·əs }

diglucoside [BIOCHEMISTRY] A compound containing two glucose molecules. { dī′glü·kə‚sīd }

dihydrostreptomycin [MICROBIOLOGY] $C_{21}H_{41}O_{12}N_7$ A hydrogenated derivative of streptomycin having the same action as streptomycin. { dī ¦hī·drō‚strep·tə′mī·sən }

dihydroxyacetonephosphoric acid [BIOCHEMISTRY] $C_3H_7O_6P$ A phosphoric acid ester of dehydroxyacetone, produced as an intermediate substance in the conversion of glycogen to lactic acid during muscular contraction. { ¦dī‚hī¦dräk·sē¦as·ə‚tōn·fäs′fȯr·ik ′as·əd }

dihydroxyphenylalanine [BIOCHEMISTRY] $C_9H_{11}NO_4$ An amino acid that can be formed by oxidation of tyrosine; it is converted by a series of biochemical transformations, utilizing the enzyme dopa oxidase, to melanins. Also known as dopa. { ¦dī‚hī¦dräk·sē‚fen·əl′al·ə‚nēn }

dikaryon [MYCOLOGY] Also spelled dicaryon. **1.** A pair of distinct, unfused nuclei in the same cell brought together by union of plus and minus hyphae in certain mycelia. **2.** Mycelium containing dikaryotic cells. { dī′kar·ē‚än }

dilator [PHYSIOLOGY] Any muscle, instrument, or drug causing dilation of an organ or part. { dī ′lād·ər }

dill [BOTANY] *Anethum graveolens.* A small annual or biennial herb in the family Umbelliferae; the aromatic leaves and seeds are used for food flavoring. { dil }

Dilleniaceae [BOTANY] A family of dicotyledonous trees, woody vines, and shrubs in the order Dilleniales having hypogynous flowers and mostly entire leaves. { di‚len·ē′ās·ē‚ē }

Dilleniales [BOTANY] An order of dicotyledonous plants in the subclass Dilleniidae characterized by separate carpels and numerous stamens. { di ˌlen·ē'ā·lēz }

Dilleniidae [BOTANY] A subclass of plants in the class Magnoliopsida distinguished by being syncarpous, having centrifugal stamens, and usually having bitegmic ovules and binucleate pollen. { dil·ə'nī·əˌdē }

dilution gene [GENETICS] Any modifier gene that acts to reduce the effect of another gene. { də 'lü·shən ˌjēn }

dilution method [MICROBIOLOGY] A technique in which a series of cultures is tested with various concentrations of an antibiotic to determine the minimum inhibiting concentration of antibiotic. { də'lü·shən ˌmeth·əd }

dimerous [BIOLOGY] Composed of two parts. { 'di·mər·əs }

diminution [BOTANY] Increasing simplification of inflorescences on successive branches. { ˌdim·ə 'nyü·shən }

Dimylidae [PALEONTOLOGY] A family of extinct lipotyphlan mammals in the order Insectivora; a side branch in the ancestry of the hedgehogs. { dī 'mil·əˌdē }

Dinidoridae [INVERTEBRATE ZOOLOGY] A family of hemipteran insects in the superfamily Pentatomoidea. { ˌdī·nə'dȯr·əˌdē }

dinitrogen fixation [MICROBIOLOGY] *See* nitrogen fixation. { dī'nī·trə·jən fik'sā·shən }

Dinocerata [PALEONTOLOGY] An extinct order of large, herbivorous mammals having semigraviportal limbs and hoofed, five-toed feet; often called uintatheres. { ˌdī·nō'ser·ə·də }

Dinoflagellata [INVERTEBRATE ZOOLOGY] The equivalent name for Dinoflagellida. { ¦dī·nō ˌflaj·ə'läd·ə }

Dinoflagellida [INVERTEBRATE ZOOLOGY] An order of flagellate protozoans in the class Phytamastigophorea; most members have fixed shapes determined by thick covering plates. { ¦dī·nōˌflə 'jel·ə·də }

dinokaryon [CELL BIOLOGY] Nuclear organization peculiar to dinoflagellates and characterized by the absence of a chromosome coiling cycle. { ˌdī·nə'kar·ēˌän }

Dinophilidae [INVERTEBRATE ZOOLOGY] A family of annelid worms belonging to the Archiannelida. { ˌdī·nō'fil·əˌdē }

Dinophyceae [BOTANY] The dinoflagellates, a class of thallophytes in the division Pyrrhophyta. { ˌdī·nō'fīs·ēˌē }

Dinornithiformes [PALEONTOLOGY] The moas, an order of extinct birds of New Zealand; all had strong legs with four-toed feet. { ˌdīn·ȯrˌnith·ə 'fȯrˌmēz }

dinosaur [PALEONTOLOGY] The name, meaning terrible lizard, applied to the fossil bones of certain large, ancient bipedal and quadripedal reptiles placed in the orders Saurischia and Ornithischia. { 'dī·nəˌsȯr }

diocoel [EMBRYOLOGY] The cavity of the diencephalon, which becomes the third brain ventricle. { 'dī·əˌsēl }

Dioctophymatida [INVERTEBRATE ZOOLOGY] An order of parasitic nematode worms in the subclass Enoplia. { dīˌäk·tə·fə'mad·ə·də }

Dioctophymoidea [INVERTEBRATE ZOOLOGY] An order or superfamily of parasitic nematodes characterized by the peculiar structure of the copulatory bursa of the male. { dīˌäk·tə·fə 'mȯid·ē·ə }

dioecious [BIOLOGY] Having the male and female reproductive organs on different individuals. Also known as dioic. { dī'ē·shəs }

Diomedeidae [VERTEBRATE ZOOLOGY] The albatrosses, a family of birds in the order Procellariiformes. { ˌdī·ə·mə'dī·əˌdē }

Diopsidae [INVERTEBRATE ZOOLOGY] The stalk-eyed flies, a family of myodarian cyclorrhaphous dipteran insects in the subsection Acalypteratae. { dī'äp·səˌdē }

diorchism [ANATOMY] Having two testes. { dī'ȯr ˌkiz·əm }

Dioscoreaceae [BOTANY] A family of monocotyledonous, leafy-stemmed, mostly twining plants in the order Liliales, having an inferior ovary and septal nectaries and lacking tendrils. { ˌdī·ə ˌskȯr·ē'ās·ēˌē }

dioxygenase [BIOCHEMISTRY] Any of a group of enzymes which catalyze the insertion of both atoms of an oxygen molecule into an organic substrate according to the generalized formula $AH_2 + O_2 \rightarrow A(OH)_2$. { dī'äk·sə·jəˌnās }

dipeptidase [BIOCHEMISTRY] An enzyme that hydrolyzes a dipeptide. { dī'pep·təˌdās }

diphosphatidyl glycerol [BIOCHEMISTRY] *See* cardiolipin. { dīˌfäs'fad·əd·əl 'glis·əˌrȯl }

diphosphopyridine nucleotide [BIOCHEMISTRY] $C_{21}H_{27}O_{14}N_7P_2$ An organic coenzyme that functions in enzymatic systems concerned with oxidation-reduction reactions. Abbreviated DPN. Also known as codehydrogenase I; coenzyme I; nicotinamide adenine dinucleotide (NAD). { dī ˌfäs·fə'pir·əˌdēn 'nü·klē·əˌtīd }

diphycercal [VERTEBRATE ZOOLOGY] Pertaining to a tail fin, having symmetrical upper and lower parts, and with the vertebral column extending to the tip without upturning. { ¦dī·fə'ser·kəl }

diphyletic [EVOLUTION] Originating from two lines of descent. { dī·fə'led·ik }

Diphyllidea [INVERTEBRATE ZOOLOGY] A monogeneric order of tapeworms in the subclass Cestoda; all species live in the intestine of elasmobranch fishes. { dī·fə'lid·ē·ə }

Diphyllobothrium [INVERTEBRATE ZOOLOGY] A genus of tapeworms; including parasites of humans, dogs, and cats. Formerly known as *Dibothriocephalus*. { dīˌfil·ō'bäth·rē·əm }

Diphyllobothrium latum [INVERTEBRATE ZOOLOGY] A large tapeworm that infects humans, dogs, and cats; causes anemia and disorders of the nervous and digestive systems in humans. { dīˌfil·ō'bäth·rē·əm 'lād·əm }

diphyllous [BOTANY] Having two leaves. { dī'fil·əs }

diphyodont [ANATOMY] Having two successive sets of teeth, deciduous followed by permanent, as in humans. { dī'fī·əˌdänt }

dipicolinic acid [BIOCHEMISTRY] $C_7H_5O_4N$ ·

$1^1/_2$ H_2O A chelating agent composing 5-15% of the dry weight of bacterial spores. { dī¦pik·ə¦lin·ik 'as·əd }

Diplacanthidae [PALEONTOLOGY] A family of extinct acanthodian fishes in the suborder Diplacanthoidei. { ˌdip·lə'kan·thəˌdē }

Diplacanthoidei [PALEONTOLOGY] A suborder of extinct acanthodian fishes in the order Climatiiformes. { ˌdip·ləˌkan'thȯid·ēˌī }

Diplasiocoela [VERTEBRATE ZOOLOGY] A suborder of amphibians in the order Anura typically having the eighth vertebra biconcave. { diˌplā·zē·ō'sē·lə }

dipleurula [INVERTEBRATE ZOOLOGY] **1.** A hypothetical bilaterally symmetrical larva postulated to be an ancestral form of echinoderms and chordates. **2.** Any bilaterally symmetrical, ciliated echinoderm larva. { dī'plu̇r·ə·lə }

Diplobathrida [PALEONTOLOGY] An order of extinct, camerate crinoids having two circles of plates beneath the radials. { ˌdip·lō'bath·rə·də }

diplobiont [BIOLOGY] An organism characterized by alternating, morphologically dissimilar haploid and diploid generations. { ¦dip·lō¦bīˌänt }

diploblastic [ZOOLOGY] Having two germ layers, referring to embryos and certain lower invertebrates. { ¦dip·lō¦blas·tik }

diploblastula [INVERTEBRATE ZOOLOGY] A two-layered, flagellated larva of certain ceractinomorph sponges. Also known as parenchymella. { ¦dip·lō¦blas·chə·lə }

diplococci [MICROBIOLOGY] A pair of micrococci. { ˌdīp·lō'kä·kē *or* 'käk·sī }

Diplogasteroidea [INVERTEBRATE ZOOLOGY] A superfamily of nematodes in the subclass Diplogasteria, having a stoma of variable and often complex shape, a very distinctive esophagus, and a muscular corpus with well-developed valve. { ˌdip·lōˌgas·tə'rȯid·ē·ə }

diploglossate [VERTEBRATE ZOOLOGY] Pertaining to certain lizards, having the ability to retract the end of the tongue into the basal portion. { ¦dip·lō'gläˌsāt }

diplohaplont [BIOLOGY] An organism characterized by alternating, morphologically similar haploid and diploid generations. { ¦dip·lō¦hap ˌlänt }

diploidization [GENETICS] The process of attaining the diploid state. { ˌdiˌplȯid·ə'zā·shən }

diploid merogony [EMBRYOLOGY] Development of a part of an egg in which the nucleus is the normal diploid fusion product of egg and sperm nuclei. { 'diˌplȯid mə'räg·ə·nē }

diploid state [GENETICS] A condition in which a chromosome set is present in duplicate in a nucleus (2N). { 'diˌplȯid ˌstāt }

diplolepidious [BOTANY] Double-scaled, specifically referring to the peristome of mosses with two rows of scales on the outside and one row on the inner. { ¦dip·lō·lə'pid·ē·əs }

Diplomonadida [INVERTEBRATE ZOOLOGY] An order of small, colorless protozoans in the class Zoomastigophorea, having a bilaterally symmetrical body with four flagella on each side. { ˌdip·lō·mə'nad·ə·də }

Diplomystidae [VERTEBRATE ZOOLOGY] A family of catfishes in the suborder Siluroidei confined to the waters of Chile and Argentina. { ˌdip·lō 'mis·təˌdē }

diplont [BIOLOGY] An organism with diploid somatic cells and haploid gametes. { 'dipˌlänt }

Diplopoda [INVERTEBRATE ZOOLOGY] The millipeds, a class of terrestrial tracheate, oviparous arthropods; each body segment except the first few bears two pairs of walking legs. { də'plä·pə·də }

Diploporita [PALEONTOLOGY] An extinct order of echinoderms in the class Cystoidea in which the thecal canals were associated in pairs. { ˌdip·lə 'pȯr·əd·ə }

Diplorhina [VERTEBRATE ZOOLOGY] The subclass of the class Agnatha that includes the jawless vertebrates with paired nostrils. { ˌdip·lə'rī·nə }

diplosome [CELL BIOLOGY] A double centriole. { 'dip·ləˌsōm }

diplospondyly [ANATOMY] Having two centra in one vertebra. { ˌdip·lə'spän·də·lē }

diplotene [CELL BIOLOGY] The stage of meiotic prophase during which pairs of nonsister chromatids of each bivalent repel each other and are kept from falling apart by the chiasmata. { 'dip·lōˌtēn }

Diplura [INVERTEBRATE ZOOLOGY] An order of small, primarily wingless insects of worldwide distribution. { də'plu̇r·ə }

Dipneumonomorphae [INVERTEBRATE ZOOLOGY] A suborder of the order Araneida comprising the spiders common in the United States, including grass spiders, hunting spiders, and black widows. { dīˌnü·mən·ō'mȯr·fē }

Dipneusti [VERTEBRATE ZOOLOGY] The equivalent name for Dipnoi. { dip'nüˌstī }

Dipnoi [VERTEBRATE ZOOLOGY] The lungfishes, a subclass of the Osteichthyes having lungs that arise from a ventral connection in the gut. { 'dipˌnȯi }

Dipodidae [VERTEBRATE ZOOLOGY] The Old World jerboas, a family of mammals in the order Rodentia. { də'päd·əˌdē }

diporpa larva [INVERTEBRATE ZOOLOGY] A developmental stage of a monogenean trematode. { 'diˌpȯr·pə 'lär·və }

Diprionidae [INVERTEBRATE ZOOLOGY] The conifer sawflies, a family of hymenopteran insects in the superfamily Tenthredinoidea. { ˌdip·rē'än·ə ˌdē }

Diprotodonta [VERTEBRATE ZOOLOGY] A proposed order of marsupial mammals to include the phalangers, wombats, koalas, and kangaroos. { dīˌprōd·ə'dän·tə }

Diprotodontidae [PALEONTOLOGY] A family of extinct marsupial mammals. { dīˌprōd·ə'dän·tə ˌdē }

Dipsacales [BOTANY] An order of dicotyledonous herbs and shrubs in the subclass Asteridae characterized by an inferior ovary and usually opposite leaves. { ˌdip·sə'kā·lēz }

Dipsocoridae [INVERTEBRATE ZOOLOGY] A family of hemipteran insects in the superfamily Dipsocoroidea; members are predators on small insects under bark or in rotten wood. { ˌdip·sə ˈkȯr·əˌdē }

Dipsocoroidea [INVERTEBRATE ZOOLOGY] A superfamily of minute, ground-inhabiting hemipteran insects belonging to the subdivision Geocorisae. { ˌdip·sə·kəˈrȯid·ē·ə }

Diptera [INVERTEBRATE ZOOLOGY] The true flies, an order of the class Insecta characterized by possessing only two wings and a pair of balancers. { ˈdip·tə·rə }

Dipteriformes [VERTEBRATE ZOOLOGY] The single order of the subclass Dipnoi, the lungfishes. { ˌdip·tə·rəˈfȯr·mēz }

Dipterocarpaceae [BOTANY] A family of dicotyledonous plants in the order Theales having mostly stipulate, alternate leaves, a prominently exserted connective, and a calyx that is mostly winged in fruit. { ¦dip·tə·rōˌkärˈpās·ēˌē }

dipterous [BIOLOGY] **1.** Of, related to, or characteristic of Diptera. **2.** Having two wings or winglike structures. { ˈdip·tə·rəs }

direct immunofluorescence [IMMUNOLOGY] The use of labeled reactant to reveal the presence of an unlabeled one. { dəˈrekt ¦im·yəˌnō·flu̇rˈes·əns }

directive [INVERTEBRATE ZOOLOGY] Any of the dorsal and ventral paired mesenteries of certain anthozoan cnidarians. { dəˈrek·tiv }

direct repeat [MOLECULAR BIOLOGY] Identical or closely related nucleotide sequences present in two or more copies in the same orientation within the same molecule. { di¦rekt riˈpēt }

disaccharide [BIOCHEMISTRY] Any of the class of compound sugars which yield two monosaccharide units upon hydrolysis. { dīˈsak·əˌrīd }

Disasteridae [PALEONTOLOGY] A family of extinct burrowing, exocyclic Euechinoidea in the order Holasteroida comprising mainly small, ovoid forms without fascioles or a plastron. { ˌdis·ə ˈster·əˌdē }

Discellaceae [MYCOLOGY] A family of fungi of the order Sphaeropsidales, including saprophytes and some plant pathogens. { ˌdis·əˈlās·ēˌē }

discifloral [BOTANY] Having flowers with enlarged, disklike receptacles. { ¦dis·kə¦flȯr·əl }

disciform [BIOLOGY] Disk-shaped. { ˈdis·kə ˌfȯrm }

Discinacea [INVERTEBRATE ZOOLOGY] A family of inarticulate brachiopods in the suborder Acrotretidina. { ˌdis·kəˈnās·ē·ə }

disclimax [ECOLOGY] A climax community that includes foreign species following a disturbance of the natural climax by humans or domestic animals. Also known as disturbance climax. { dis ˈklī·maks }

discoaster [BOTANY] A star-shaped coccolith. { disˈkō·ə·stər }

discoblastula [EMBRYOLOGY] A blastula formed by cleavage of a meroblastic egg; the blastoderm is disk-shaped. { ¦dis·kō¦blas·chə·lə }

discocephalous [INVERTEBRATE ZOOLOGY] Having a sucker on the head. { ¦dis·kō¦sef·ə·ləs }

discoctaster [INVERTEBRATE ZOOLOGY] A type of spicule with eight rays terminating in discs in hexactinellid sponges. { disˈkäk·tə·stər }

discodactylous [VERTEBRATE ZOOLOGY] Having sucking disks on the toes. { ¦dis·kō¦dak·tə·ləs }

discogastrula [EMBRYOLOGY] A gastrula formed from a blastoderm. { ¦dis·kō¦gas·trə·lə }

Discoglossidae [VERTEBRATE ZOOLOGY] A family of anuran amphibians in and typical of the suborder Opisthocoela. { ¦dis·kō¦gläs·əˌdē }

discoid [BIOLOGY] **1.** Being flat and circular in form. **2.** Any structure shaped like a disc. { ˈdis ˌkȯid }

discoidal cleavage [EMBRYOLOGY] A type of cleavage producing a disc of cells at the animal pole. { disˈkȯid·əl ˈklē·vij }

Discoidiidae [PALEONTOLOGY] A family of extinct conical or globular, exocyclic Euechinoidea in the order Holectypoida distinguished by the rudiments of internal skeletal partitions. { disˌkȯi ˈdī·əˌdē }

Discolichenes [BOTANY] The equivalent name for Lecanorales. { ¦dis·kō·lī¦kē·nēz }

Discolomidae [INVERTEBRATE ZOOLOGY] The tropical log beetles, a family of coleopteran insects in the superfamily Cucujoidea. { ˌdis·kō ˈläm·əˌdē }

Discomycetes [MYCOLOGY] A group of fungi in the class Ascomycetes in which the surface of the fruiting body is exposed during maturation of the spores. { ˌdis·kōˌmīˈsēd·ēz }

discontinuous coding sequence [MOLECULAR BIOLOGY] The coding sequence in deoxyribonucleic acid of eukaryotic split genes consisting of exons and introns. { ˌdis·kənˈtin·yə·wəs ˈkōd·iŋ ˌsē·kwəns }

discopodous [INVERTEBRATE ZOOLOGY] Having a disk-shaped foot. { diˈskäp·ə·dəs }

Discorbacea [INVERTEBRATE ZOOLOGY] A superfamily of foraminiferan protozoans in the suborder Rotaliina characterized by a radial, perforate, calcite test and a monolamellar septa. { ˌdis·kərˈbās·ē·ə }

disjunct endemism [PALEONTOLOGY] A type of regionally restricted distribution of a fossil taxon in which two or more component parts are separated by a major physical barrier and hence not readily explicable in terms of present-day geography. { ˈdisˌjəŋkt ˈen·dəˌmiz·əm }

disjunction [CELL BIOLOGY] Separation of chromatids or homologous chromosomes during anaphase. { disˈjəŋk·shən }

disjunctor [MYCOLOGY] A small cellulose body between the conidia of certain fungi, which eventually breaks down and thus frees the conidia. { disˈjəŋk·tər }

disk Also spelled disc. [BIOLOGY] Any of various rounded and flattened animal and plant structures. { disk }

disk flower [BOTANY] One of the flowers on the disk of a composite plant. { ˈdisk ˌflau̇·ər }

disomaty [CELL BIOLOGY] Duplication of chromosomes unaccompanied by nuclear division. { dīˈsō·məd·ē }

Disomidae [INVERTEBRATE ZOOLOGY] A family of spioniform annelid worms belonging to the Sedentaria. { dəˈsäm·əˌdē }

dispermy [PHYSIOLOGY] Entrance of two spermatozoa into an ovum. { ˈdīˌspər·mē }

displacement loop |MOLECULAR BIOLOGY| In circular deoxyribonucleic acid (DNA), a small region in which ribonucleic acid is paired with one strand of DNA, effectively displacing the other DNA strand. Also known as D-loop. { dis'plās·mənt ,lüp }

dissect |BIOLOGY| To divide, cut, and separate into different parts. { də'sekt }

disseminule |BIOLOGY| An individual organism or part of an individual adapted for the dispersal of a population of the same organisms. { də'sem·ə,nyül }

dissepiment |BOTANY| A partition which divides a fruit or an ovary into chambers. |PALEONTOLOGY| One of the vertically positioned thin plates situated between the septa in extinct corals of the order Rugosa. { də'sep·ə·mənt }

dissociation |MICROBIOLOGY| The appearance of a novel colony type on solid media after one or more subcultures of the microorganism in liquid media. { də,sō·sē'ā·shən }

dissogeny |ZOOLOGY| Having two sexually mature stages, larva and adult, in the life of an individual. { də'sä·jə·nē }

Distacodidae |PALEONTOLOGY| A family of conodonts in the suborder Conodontiformes characterized as simple curved cones with deeply excavated attachment scars. { dis·tə'käd·ə,dē }

distal |BIOLOGY| Located away from the point of origin or attachment. { 'dist·əl }

distal convoluted tubule |ANATOMY| The portion of the nephron in the vertebrate kidney lying between the loop of Henle and the collecting tubules. { 'dist·əl ,kän·və'lüd·əd 'tü·byül }

distichous |BIOLOGY| Occurring in two vertical rows. { 'dis·tə·kəs }

distome |INVERTEBRATE ZOOLOGY| A digenetic trematode characterized by possession of an oral and a ventral sucker. { 'dī,stōm }

disturbance climax |ECOLOGY| *See* disclimax. { də'stər·bəns 'klī,maks }

ditokous |VERTEBRATE ZOOLOGY| Producing two eggs or giving birth to two young at one time. { 'did·ə·kəs }

diuretic hormone |BIOCHEMISTRY| A neurohormone that promotes water loss in insects by increasing the volume of fluid secreted into the Malpighian tubules. { ,dī·yü'red·ik 'hör,mōn }

diurnal |BIOLOGY| Active during daylight hours. { dī'ərn·əl }

diurnal migration |BIOLOGY| The daily rhythmic movements of organisms in the sea from deeper water to the surface at the approach of darkness and their return to deeper water before dawn. { dī'ərn·əl mī'grā·shən }

divaricate |BIOLOGY| Broadly divergent and spread apart. { dī'var·ə,kāt }

divaricator |ZOOLOGY| A muscle that causes separation of parts, as of brachipod shells. { dī'var·ə,kād·ər }

divergent adaptation |EVOLUTION| Adaptation to different kinds of environment that results in divergence from a common ancestral form. Also known as branching adaptation; cladogenic adaptation. { də'vər·jənt ,ad,ap'tā·shən }

divergent transcription |MOLECULAR BIOLOGY| The initiation of genetic transcription at two promoters that are facing in opposite directions. { də'vər·jənt ,tran'skrip·shən }

diving bird |VERTEBRATE ZOOLOGY| Any bird adapted for diving and swimming, including loons, grebes, and divers. { 'dīv·iŋ ,bərd }

Dixidae |INVERTEBRATE ZOOLOGY| A family of orthorrhaphous dipteran insects in the series Nematocera. { 'dik·sə,dē }

dizygotic twins |BIOLOGY| Twins derived from two eggs. Also known as fraternal twins. { ¦dī·zī'gäd·ik 'twinz }

D-loop |MOLECULAR BIOLOGY| *See* displacement loop. { 'de ,lüp }

DNA |BIOCHEMISTRY| *See* deoxyribonucleic acid.

DNA fingerprinting |GENETICS| *See* genetic fingerprinting. { ,dē,en'ā 'fiŋ·gər,print·iŋ }

DNase |BIOCHEMISTRY| *See* deoxyribonuclease.

Docodonta |PALEONTOLOGY| A primitive order of Jurassic mammals of North America and England. { ,däk·ə'dän·tə }

dodecamerous |BOTANY| Having the whorls of floral parts in multiples of 12. { ¦dō·də¦kam·ə·rəs }

dodo |VERTEBRATE ZOOLOGY| *Raphus calcullatus.* A large, flightless, extinct bird of the family Raphidae. { 'dō,dō }

doe |VERTEBRATE ZOOLOGY| The adult female deer, antelope, goat, rabbit, or any other mammal of which the male is referred to as buck. { dō }

dog |VERTEBRATE ZOOLOGY| Any of various wild and domestic animals identified as *Canis familiaris* in the family Canidae; all are carnivorous and digitigrade, are adapted to running, and have four toes with nonretractable claws on each foot. { dȯg }

dogfish |VERTEBRATE ZOOLOGY| *See* bowfin. { 'dȯg,fish }

Doisy unit |BIOLOGY| A unit for standardization of vitamin K. { 'dȯi·zē ,yü·nət }

dolabriform |BIOLOGY| Shaped like an ax head. { dō'lab·rə,fȯrm }

dolichol |BIOCHEMISTRY| Any of a group of long-chain unsaturated isoprenoid alcohols containing up to 84 carbon atoms; found free or phosphorylated in membranes of the endoplasmic reticulum and Golgi apparatus. { 'däl·ə,kȯl }

Dolichopodidae |INVERTEBRATE ZOOLOGY| The long-legged flies, a family of orthorrhaphous dipteran insects in the series Brachycera. { ,däl·ə·kō'päd·ə,dē }

Dolichothoraci |PALEONTOLOGY| A group of joint-necked fishes assigned to the Arctolepiformes in which the pectoral appendages are represented solely by large fixed spines. { ¦däl·ə·kō 'thȯr·ə,sī }

dolioform |BIOLOGY| Barrel-shaped. { 'dō·lē·ə ,fȯrm }

doliolaria larva |INVERTEBRATE ZOOLOGY| A free-swimming larval stage of crinoids and holothurians having an apical tuft and four or five bands of cilia. { ,dō·lē·ə'lar·ē·ə 'lär·və }

Doliolida |INVERTEBRATE ZOOLOGY| An order of pelagic tunicates in the class Thaliacea; transparent forms, partly or wholly ringed by muscular bands. { ˌdō·lēʹä·lə·də }

dolphin [VERTEBRATE ZOOLOGY] The common name for about 33 species of cetacean mammals included in the family Delphinidae and characterized by the pronounced beak-shaped mouth. { ʹdäl·fən }

domestication [BIOLOGY] The adaptation of an animal or plant through breeding in captivity to a life intimately associated with and advantageous to humans. { dəˌmes·təʹkā·shən }

dominance [ECOLOGY] The influence that a controlling organism has on numerical composition or internal energy dynamics in a community. [GENETICS] The expression of a heritable trait in the heterozygote such as to make it phenotypically indistinguishable from the homozygote. { ʹdäm·ə·nəns }

dominant allele [GENETICS] The member of a pair of alleles which is phenotypically indistinguishable in both the homozygous and heterozygous condition. { ʹdäm·ə·nənt əʹlēl }

dominant hemisphere [PHYSIOLOGY] The cerebral hemisphere which controls certain motor activities; usually the left hemisphere in right-handed individuals. { ʹdäm·ə·nənt ʹhem·əˌsfir }

dominant species [ECOLOGY] A species of plant or animal that is particularly abundant or controls a major portion of the energy flow in a community. { ʹdäm·ə·nənt ʹspēˌshēz }

donation [MYCOLOGY] In conjugation, a process involving a nonconjugative plasmid and a conjugative plasmid in which the latter provides the missing conjugative function to the former so that the former may be transferred. { dōʹnā·shən }

donkey [VERTEBRATE ZOOLOGY] A domestic ass (*Equus asinus*); a perissodactyl mammal in the family Equidae. { ʹdəŋ·kē }

donor splicing site [MOLECULAR BIOLOGY] The boundary between the left (5′) end of an intron and the right (3′) end of an exon in messenger ribonucleic acid. Also known as left splicing junction. { ʹdō·nər ˌsplīs·iŋ ˌsīt }

doodlebug [INVERTEBRATE ZOOLOGY] The larva of an ant lion. { ʹdüd·əlˌbəg }

dopa [BIOCHEMISTRY] *See* dihydroxyphenylalanine. { ʹdō·pə }

dopamine [BIOCHEMISTRY] $C_8H_{11}O_2N$ An intermediate in epinephrine and norepinephrine biosynthesis; the decarboxylation product of dopa. { ʹdō·pəˌmēn }

dopa oxidase [BIOCHEMISTRY] An enzyme that catalyzes the oxidation of dihydroxyphenylalanine to melanin; occurs in the skin. { ʹdō·pə ʹäk·səˌdās }

Doradidae [VERTEBRATE ZOOLOGY] A family of South American catfishes in the suborder Siluroidei. { dəʹra·dəˌdē }

Dorilaidae [INVERTEBRATE ZOOLOGY] The big-headed flies, a family of cyclorrhaphous dipteran insects in the series Aschiza. { dȯr·əʹlā·əˌdē }

Dorippidae [INVERTEBRATE ZOOLOGY] The mask crabs, a family of brachyuran decapods in the subsection Oxystomata. { dəʹrip·əˌdē }

dormancy [BOTANY] A state of quiescence during the development of many plants characterized by their inability to grow, though continuing their morphological and physiological activities. { ʹdȯr·mən·sē }

dormant bud [BOTANY] *See* latent bud. { ʹdȯr·mənt ʹbəd }

dormouse [VERTEBRATE ZOOLOGY] The common name applied to members of the family Gliridae; they are Old World arboreal rodents intermediate between squirrels and rats. { ʹdȯrˌmaủs }

Dorngeholz [ECOLOGY] *See* thornbush. { ʹdȯrn·gəˌhōlts }

Dorngestrauch [ECOLOGY] *See* thornbush. { ʹdorn·gəˌstraủk }

dornveld [ECOLOGY] *See* thornbush. { ʹdȯrnˌfelt }

dorsal [ANATOMY] Located near or on the back of an animal or one of its parts. { ʹdȯr·səl }

dorsal aorta [ANATOMY] The portion of the aorta extending from the left ventricle to the first branch. [INVERTEBRATE ZOOLOGY] The large, dorsal blood vessel in many invertebrates. { ʹdȯr·səl āʹȯrd·ə }

dorsal column [ANATOMY] A column situated dorsally in each lateral half of the spinal cord which receives the terminals of some afferent fibers from the dorsal roots of the spinal nerves. { ʹdȯr·səl ʹkäl·əm }

dorsal fin [VERTEBRATE ZOOLOGY] A median longitudinal vertical fin on the dorsal aspect of a fish or other aquatic vertebrate. { ʹdȯr·səl ʹfin }

dorsalia [INVERTEBRATE ZOOLOGY] Paired sensory bristles on the dorsal aspect of the head of gnathostomalids. { dȯrʹsal·yə }

dorsal lip [EMBRYOLOGY] In an amphibian embryo, the margin or lip of the fold of blastula wall marking the dorsal limit of the blastopore during gastrulation and constituting the primary organizer, is necessary to the development of neural tissue, and forms the originating point of chordamesoderm. { ʹdȯr·səl ʹlip }

dorsiferous [BOTANY] Of ferns, bearing sori on the back of the frond. [ZOOLOGY] Bearing the eggs or young on the back. { dȯrʹsif·ə·rəs }

dorsiflex [ZOOLOGY] To flex or cause to flex in a dorsal direction. { ʹdȯr·səˌfleks }

dorsigrade [VERTEBRATE ZOOLOGY] Walking on the back of the toes. { ʹdȯr·səˌgrād }

dorsocaudad [ANATOMY] To or toward the dorsal surface and caudal end of the body. { ¦dȯr·sō¦kȯˌdad }

dorsomedial [ANATOMY] Located on the back, toward the midline. { ¦dȯr·sō¦mēd·ē·əl }

dorsoposteriad [ANATOMY] To or toward the dorsal surface and posterior end of the body. { ¦dȯr·sō·pōʹstir·ēˌad }

dorsum [ANATOMY] **1.** The entire dorsal surface of the animal body. **2.** The upper part of the tongue, opposite the velum. { ʹdȯr·səm }

Dorvilleidae [INVERTEBRATE ZOOLOGY] A family of minute errantian annelids in the superfamily Eunicea. { ˌdȯr·vəʹlē·əˌdē }

Dorylaimoidea [INVERTEBRATE ZOOLOGY] An order or superfamily of nematodes inhabiting soil and fresh water. { ˌdor·ə·ləʹmȯid·ē·ə }

Dorylinae [INVERTEBRATE ZOOLOGY] A subfamily of predacious ants in the family Formicidae, including the army ant (*Eciton hamatum*). { dȯ'rī·lə·nē }

Dorypteridae [PALEONTOLOGY] A family of Permian palaeonisciform fishes sometimes included in the suborder Platysomoidei. { dȯ,rip 'ter·ə,dē }

dosage [GENETICS] The number of genes with a similar action that control a given character. { 'dō·sij }

dosage compensation [GENETICS] A mechanism that equalizes the phenotypic effects of genes located on the X chromosome, which genes therefore exist in two doses in the homogametic sex and one dose in the heterogametic sex. { 'dō·sij ,käm·pən,sā·shən }

dose fractionation [BIOPHYSICS] The application of a radiation dose in two or more fractions separated by a certain minimal time interval. { 'dōs ,frak·shə,nā·shən }

double circulation [PHYSIOLOGY] A circulatory system in which blood flows through two separate circuits, as pulmonary and systemic. { ¦dəb·əl sər·kyə'lā·shən }

double fertilization [BOTANY] In most seed plants, fertilization involving fusion between the egg nucleus and one sperm nucleus, and fusion between the other sperm nucleus and the polar nuclei. { ¦dəb·əl ,fərd·əl·ə'zā·shən }

double-loop pattern [ANATOMY] A whorl type of fingerprint pattern consisting of two separate loop formations and two deltas. { ¦dəb·əl ,lüp 'pad·ərn }

double-work [BOTANY] In plant propagation, to graft or bud a scion to an intermediate variety that is itself grafted on a stock of still another variety. { ¦dəb·əl ¦wərk }

doubling dose [GENETICS] The radiation dose that would double the rate of spontaneous mutation. { ¦dəb·liŋ ¦dōs }

Douglas-fir [BOTANY] *Pseudotsuga menziesii.* A large coniferous tree in the order Pinales; cones are characterized by bracts extending beyond the scales. Also known as red fir. { ¦dəg·ləs 'fər }

Dounce homogenizer [BIOLOGY] An apparatus consisting of a glass tube with a tight-fitting glass pestle used manually to disrupt tissue suspensions to obtain single cells or subcellular fractions. { 'daủns hə'mäj·ə,nīz·ər }

dove [VERTEBRATE ZOOLOGY] The common name for a number of small birds of the family Columbidae. { dəv }

DPN [BIOCHEMISTRY] *See* diphosphopyridine nucleotide.

Draconematoidea [INVERTEBRATE ZOOLOGY] A superfamily of marine nematodes in the order Desmodorida distinguished by a body that, when relaxed, is dorsally and then ventrally arched into a shallow sigmoid shape. { ,drā·kō,nem·ə'tȯid·ē·ə }

Dracunculoidea [INVERTEBRATE ZOOLOGY] An order or superfamily of parasitic nematodes characterized by their habitat in host tissues and by the way larvae leave the host through a skin lesion. { drə,kəŋ·kyə'lōid·ē·ə }

dragonfly [INVERTEBRATE ZOOLOGY] Any of the insects composing six families of the suborder Anisoptera and having four large, membranous wings and compound eyes that provide keen vision. { 'drag·ən,flī }

Drepanellacea [PALEONTOLOGY] A monomorphic superfamily of extinct paleocopan ostracods in the suborder Beyrichicopina having a subquadrate carapace, many with a marginal rim. { drə ,pan·əl'ās·ē·ə }

Drepanellidae [PALEONTOLOGY] A monomorphic family of extinct ostracods in the superfamily Drepanellacea. { ,dre·pə'nel·ə,dē }

Drepanidae [INVERTEBRATE ZOOLOGY] The hooktips, a small family of lepidopteran insects in the suborder Heteroneura. { dre'pan·ə,dē }

Drilidae [INVERTEBRATE ZOOLOGY] The false firefly beetles, a family of coleopteran insects in the superfamily Cantharoidea. { 'dril·ə,dē }

Drilonematoidea [INVERTEBRATE ZOOLOGY] A superfamily of parasitic nematodes in the subclass Spiruria. { ,drī·lō,nem·ə'tȯid·ē·ə }

Dromadidae [VERTEBRATE ZOOLOGY] A family of the avian order Charadriiformes containing a single species, the crab plover (*Dromas ardeola*). { drō 'mad·ə,dē }

dromedary [VERTEBRATE ZOOLOGY] *Camelus dromedarius.* The Arabian camel, distinguished by a single hump. { 'dräm·ə,der·ē }

Dromiacea [INVERTEBRATE ZOOLOGY] The dromiid crabs, a subsection of the Brachyura in the crustacean order Decapoda. { ,drō·mē'ā·shē·ə }

Dromiceidae [VERTEBRATE ZOOLOGY] The emus, a monospecific family of flightless birds in the order Casuariiformes. { ,drō·mə'sē·ə,dē }

drone [INVERTEBRATE ZOOLOGY] A haploid male bee or ant; one of the three castes in a colony. { drōn }

Droseraceae [BOTANY] A family of dicotyledonous plants in the order Sarraceniales, distinguished by leaves that do not form pitchers, parietal placentation, and several styles. { ,drä·sə 'rās·ē,ē }

Drosophilidae [INVERTEBRATE ZOOLOGY] The vinegar flies, a family of myodarian cyclorrhaphous dipteran insects in the subsection Acalypteratae, including the fruit fly (*Drosophila melanogaster*). { ,drä·sə'fil·ə,dē }

drug resistance [MICROBIOLOGY] A decreased reactivity of living organisms to the injurious actions of certain drugs and chemicals. { ¦drəg ri 'zis·təns }

drupaceous [BOTANY] Of, pertaining to, or characteristic of a drupe. { drü'pā·shəs }

drupe [BOTANY] A fruit, such as a cherry, having a thin or leathery exocarp, a fleshy mesocarp, and a single seed with a stony endocarp. Also known as stone fruit. { drüp }

drupelet [BOTANY] An individual drupe of an aggregate fruit. Also known as grain. { 'drüp·lət }

Dryinidae [INVERTEBRATE ZOOLOGY] A family of hymenopteran insects in the superfamily Bethyloidea. { drī'in·ə,dē }

Dryomyzidae [INVERTEBRATE ZOOLOGY] A family

of myodarian cyclorrhaphous dipteran insects in the subsection Acalypteratae. { ˌdrī·ō'mīz·ə ˌdē }

Dryopidae [INVERTEBRATE ZOOLOGY] The long-toed water beetles, a family of coleopteran insects in the superfamily Dryopoidea. { drī'äp·ə ˌdē }

Dryopoidea [INVERTEBRATE ZOOLOGY] A superfamily of coleopteran insects in the suborder Polyphaga, including the nonpredatory aquatic beetles. { drī·ə'pȯid·ē·ə }

dry rot [MICROBIOLOGY] A rapid decay of seasoned timber caused by certain fungi which cause the wood to be reduced to a dry, friable texture. { 'drī ˌrät }

duck [VERTEBRATE ZOOLOGY] The common name for a number of small waterfowl in the family Anatidae, having short legs, a broad, flat bill, and a dorsoventrally flattened body. { dək }

duck-billed dinosaur [PALEONTOLOGY] Any of several herbivorous, bipedal ornithopods having the front of the mouth widened to form a duck-like beak. { ¦dək ˌbild 'dīn·əˌsȯr }

duckbill platypus [VERTEBRATE ZOOLOGY] *See* platypus. { ¦dəkˌbil 'plad·ə·pu̇s }

duck wheat [BOTANY] *See* tartary buckwheat. { 'dək ˌwēt }

Ducrey test [IMMUNOLOGY] A skin test to determine past or present infection with *Hemophilus ducreyi*. { dü'krā ˌtest }

duct [ANATOMY] An enclosed tubular channel for conducting a glandular secretion or other body fluid. { dəkt }

ductless gland [PHYSIOLOGY] *See* endocrine gland. { ¦dək·ləs 'gland }

duct of Cuvier [EMBRYOLOGY] Either of the paired common cardinal veins in a vertebrate embryo. { ¦dəkt əv küv'yā }

duct of Santorini [ANATOMY] The dorsal pancreatic duct in a vertebrate embryo; persists in adult life in some species and serves as the pancreatic duct in the adult elasmobranch, pig, and ox. { ¦dəkt əv san·tə'rē·nē }

ductus arteriosus [EMBRYOLOGY] Blood shunt between the pulmonary artery and the aorta of the mammalian embryo. { 'dək·təs ˌär·tir·ē'ō·səs }

ductus deferens [ANATOMY] *See* vas deferens. { 'dək·təs 'def·əˌrenz }

ductus venosus [EMBRYOLOGY] Blood shunt between the left umbilical vein and the right sinus venosus of the heart in the mammalian embryo. { 'dək·təs ve'nō·səs }

Duffy blood group [IMMUNOLOGY] A genetically determined, immunologically distinct group of human erythrocyte antigens defined by their reaction with anti-Fy[a] serum. { 'dəf·ē 'bləd ˌgrüp }

Dugongidae [VERTEBRATE ZOOLOGY] A family of aquatic mammals in the order Sirenia comprising two species, the dugong and the sea cow. { dü'gän·jəˌdē }

Dugonginae [VERTEBRATE ZOOLOGY] The dugongs, a subfamily of sirenian mammals in the family Dugongidae characterized by enlarged, sharply deflected premaxillae and the absence of nasal bones. { dü'gän·jəˌnē }

dulse [BOTANY] Any of several species of red algae of the genus *Rhodymenia* found below the intertidal zone in northern latitudes; an important food plant. { dəls }

duodenal glands [ANATOMY] *See* Brunner's glands. { dü¦äd·ən·əl 'glanz }

duodenum [ANATOMY] The first section of the small intestine of mammals, extending from the pylorus to the jejunum. { dü'äd·ən·əm *or* dü·ə 'dē·nəm }

duplex deoxyribonucleic acid [MOLECULAR BIOLOGY] The deoxyribonucleic acid double helix. { ¦düˌpleks dē¦äk·seˌrī·bō·nü¦klē·ik 'as·əd }

duplex uterus [ANATOMY] A condition in certain primitive mammals, such as rodents and bats, that have two distinct uteri opening separately into the vagina. { ¦düˌpleks 'yüd·ə·rəs }

duplicate genes [GENETICS] Two identical genes that occur on different chromosomes and display the same phenotypic action. { 'düp·lə·kət ˌjēnz }

dura mater [ANATOMY] The fibrous membrane forming the outermost covering of the brain and spinal cord. Also known as endocranium. { 'du̇r·ə ˌmā·dər }

Durham fermentation tube [MICROBIOLOGY] A test tube containing lactose or lauryl tryptose and an inverted vial for gas collection; used to test for the presence of coliform bacteria. { 'du̇r·əm ˌfər·mən'tā·shən ˌtüb }

dwarf [BIOLOGY] Being an atypically small form or variety of something. { dwȯrf }

dwarf mouse unit [BIOLOGY] A unit for the standardization of somatotropin. { ¦dwȯrf 'mau̇s ˌyü·nət }

dyad [CELL BIOLOGY] Either of the two pair of chromatids produced by separation of a tetrad during the first meiotic division. { 'dīˌad }

dynamic work [BIOPHYSICS] Performance of work by a muscle in which one end of the muscle moves with respect to the other, resulting in external movement. { dī'nam·ik ˌwərk }

Dysideidae [INVERTEBRATE ZOOLOGY] A family of sponges in the order Dictyoceratida. { ˌdis·ə 'dē·əˌdē }

Dysodonta [PALEONTOLOGY] An order of extinct bivalve mollusks with a nearly toothless hinge and a ligament in grooves or pits. { ˌdis·ə'dän·tə }

dystonia [PHYSIOLOGY] Disorder or lack of muscle tonicity. { di'stōn·ē·ə }

dystophic [BIOLOGY] Pertaining to an environment that does not supply adequate nutrition. { di'stäf·ik }

Dytiscidae [INVERTEBRATE ZOOLOGY] The predacious diving beetles, a family of coleopteran insects in the suborder Adephaga. { dī'tis·əˌdē }

E

eagle [VERTEBRATE ZOOLOGY] Any of several large, strong diurnal birds of prey in the family Accipitridae. { 'ē·gəl }

Eagle's medium [MICROBIOLOGY] A tissue-culture medium, developed by H. Eagle, containing vitamins, amino acids, inorganic salts and serous enrichments, and dextrose. { 'ē·gəlz ˌmēd·ē·əm }

ear [ANATOMY] The receptor organ that sends both auditory information and space orientation information to the brain in vertebrates. { ir }

eardrum [ANATOMY] *See* tympanic membrane. { 'ir,drəm }

earlobe [ANATOMY] The pendulous, fleshy lower portion of the auricle or external ear. { 'ir,lōb }

early enzyme [BIOCHEMISTRY] Any of the enzymes that are synthesized in a bacterial cell under the direction of an invading bacteriophage. { ¦ər·lē 'en,zīm }

early gene [GENETICS] Any gene expressed early in development. { 'ər·lē ˌjēn }

earlywood [BOTANY] The portion of the annual ring that is formed during the early part of a tree's growing season. { 'ər·lē,wu̇d }

ear shell [INVERTEBRATE ZOOLOGY] *See* abalone. { 'ir ˌshel }

earth pig [INVERTEBRATE ZOOLOGY] *See* aardvark. { 'ərth ˌpig }

earthstar [MYCOLOGY] A fungus of the genus *Geastrum* that resembles a puffball with a double peridium, the outer layer of which splits into the shape of a star. { 'ərth,stär }

earthworm [INVERTEBRATE ZOOLOGY] The common name for certain terrestrial members of the class Oligochaeta, especially forms belonging to the family Lumbricidae. { 'ərth,wərm }

earwax [PHYSIOLOGY] *See* cerumen. { 'ir,waks }

earwig [INVERTEBRATE ZOOLOGY] The common name for members of the insect order Dermaptera. { 'ir,wig }

Eaton agent [MICROBIOLOGY] The name applied to *Mycoplasma pneumoniae* when it was regarded as a virus. { 'ēt·ən ˌā·jənt }

Ebenaceae [BOTANY] A family of dicotyledonous plants in the order Ebenales, in which a latex system is absent and flowers are mostly unisexual with the styles separate, at least distally. { ˌeb·ə'nās·ē,ē }

Ebenales [BOTANY] An order of woody, sympetalous dicotyledonous plants in the subclass Dilleniidae, having axile placentation and usually twice as many stamens as corolla lobes. { ˌeb·ə'nā·lēz }

ebony [BOTANY] Any of several African and Asian trees of the genus *Diospyros*, providing a hard, durable wood. { 'eb·ə·nē }

ebracteate [BOTANY] Without bracts, or much reduced leaves. { ē'brak·tē,āt }

ebracteolate [BOTANY] Without bracteoles. { ē'brak·tē·ə,lāt }

Ebriida [INVERTEBRATE ZOOLOGY] An order of flagellate protozoans in the class Phytamastigophorea characterized by a solid siliceous skeleton. { ē'brī·ə·də }

ebullism [PHYSIOLOGY] The formation of bubbles, especially of water vapor bubbles in biological fluids, owing to reduced ambient pressure. { 'eb·yə,liz·əm }

ecad [ECOLOGY] A type of plant which is altered by its habitat and possesses nonheritable characteristics. { 'ē,kad }

eccentric contraction [BIOPHYSICS] The increase in tension that occurs in a muscle as it lengthens. { ek¦sen·trik kən'trak·shən }

eccrine gland [PHYSIOLOGY] One of the small sweat glands distributed all over the human body surface; they are tubular coiled merocrine glands that secrete clear aqueous sweat. { 'ek·rən ˌgland }

ecdysis [INVERTEBRATE ZOOLOGY] Molting of the outer cuticular layer of the body, as in insects and crustaceans. { 'ek·də·səs }

ecdysone [BIOCHEMISTRY] The molting hormone of insects. { 'ek·də,sōn }

ecesis [ECOLOGY] Successful naturalization of a plant or animal population in a new environment. { ə'sē·səs }

Echeneidae [VERTEBRATE ZOOLOGY] The remoras, a family of perciform fishes in the suborder Percoidei. { ˌek·ə'nā·ə,dē }

echidna [VERTEBRATE ZOOLOGY] A spiny anteater; any member of the family Tachyglossidae. { ə'kid·nə }

Echinacea [INVERTEBRATE ZOOLOGY] A suborder of echinoderms in the order Euechinoidea; individuals have a rigid test, keeled teeth, and branchial slits. { ˌek·ə'nā·shə }

echinate [ZOOLOGY] Having a dense covering of spines or bristles. { ə'kī,nāt }

Echinidae [INVERTEBRATE ZOOLOGY] A family of echinacean echinoderms in the order Echinoida possessing trigeminate or polyporous plates with the pores in a narrow vertical zone. { ə'kī·nə,dē }

Echiniscoidea [INVERTEBRATE ZOOLOGY] A suborder of tardigrades in the order Heterotardigrada characterized by terminal claws on the legs. { ə,kī·nə'skoid·ē·ə }

Echinococcus [INVERTEBRATE ZOOLOGY] A genus of tapeworms. { ə,kī·nə'kä·kəs }

echinococcus cyst [INVERTEBRATE ZOOLOGY] A cyst formed in host tissues by the larva of *Echinococcus granulosus*. { ə,kī·nə'kä·kəs 'sist }

Echinocystitoida [PALEONTOLOGY] An order of extinct echinoderms in the subclass Perischoechinoidea. { ,ek·ə·nō,sis·tə'toid·ə }

Echinodera [INVERTEBRATE ZOOLOGY] The equivalent name for Kinorhyncha. { ek·ə'nä·də·rə }

Echinodermata [INVERTEBRATE ZOOLOGY] A phylum of exclusively marine coelomate animals distinguished from all others by an internal skeleton composed of calcite plates, and a water-vascular system to serve the needs of locomotion, respiration, nutrition, or perception. { ,ek·ə·nə'dər·məd·ə }

Echinoida [INVERTEBRATE ZOOLOGY] An order of Echinacea with a camarodont lantern, smooth test, and imperforate noncrenulate tubercles. { ,ek·ə'noid·ə }

Echinoidea [INVERTEBRATE ZOOLOGY] The sea urchins, a class of Echinozoa having a compact body enclosed in a hard shell, or test, formed by regularly arranged plates which bear movable spines. { ,ek·ə'noid·e·ə }

Echinometridae [INVERTEBRATE ZOOLOGY] A family of echinoderms in the order Echinoida, including polyporous types with either an oblong or a spherical test. { ,ek·ə·nō'me·trə,dē }

echinomycin [MICROBIOLOGY] $C_{50}H_{60}O_{12}N_{12}S_2$ A toxic polypeptide antibiotic produced by species of *Streptomyces*. { ,ek·ə·nō'mīs·ən }

echinopluteus [INVERTEBRATE ZOOLOGY] The bilaterally symmetrical larva of sea urchins. { ,ek·ə·nō'plüd·ē·əs }

Echinosteliaceae [MYCOLOGY] A family of slime molds in the order Echinosteliales. { ,ek·ə·nō ,ste·lē'ās·ē,ē }

Echinosteliales [MYCOLOGY] An order of slime molds in the subclass Myxogastromycetidae. { ,ek·ə·nō,ste·lē'ā·lēz }

echinostome cercaria [INVERTEBRATE ZOOLOGY] A digenetic trematode larva characterized by the large anterior acetabulum and a collar with spines. { ə'kī·nə,stōm sər'kar·ē·ə }

Echinothuriidae [INVERTEBRATE ZOOLOGY] A family of deep-water echinoderms in the order Echinothurioida in which the large, flexible test collapses into a disk at atmospheric pressure. { ,ek·ə·nō·thə'rī·ə,dē }

Echinothurioida [INVERTEBRATE ZOOLOGY] An order of echinoderms in the superorder Diadematacea with solid or hollow primary radioles, diademoid ambulacral plates, noncrenulate tubercles, and the anus within the apical system. { ,ek·ə·nō·thu̇·rē'ȯid·ə }

Echinozoa [INVERTEBRATE ZOOLOGY] A subphylum of free-living echinoderms having the body essentially globoid with meridional symmetry and lacking appendages. { ,ek·ə·nō'zō·ə }

Echiurida [INVERTEBRATE ZOOLOGY] A small group of wormlike organisms regarded as a separate phylum of the animal kingdom; members have a saclike or sausage-shaped body with an anterior, detachable prostomium. { ,ek·ē'yu̇r·ə·də }

Echiuridae [INVERTEBRATE ZOOLOGY] A small family of the order Echiuroinea characterized by a flaplike prostomium. { ,ek·ē'yu̇r·ə,dē }

Echiuroidea [INVERTEBRATE ZOOLOGY] A phylum of schizocoelous animals. { ,ek·ē·yə'rȯid·ē·ə }

Echiuroinea [INVERTEBRATE ZOOLOGY] An order of the Echiurida. { ,ek·ē·yə'rȯi·nē·ə }

echo ranging [VERTEBRATE ZOOLOGY] An auditory feedback mechanism in bats, porpoises, seals, and certain other animals whereby reflected ultrasonic sounds are utilized in orientation. { 'ek·ō ,rānj·iŋ }

echovirus [VIROLOGY] Any member of the Picornaviridae family, genus *Enterovirus*; the name is derived from the group designation enteric cytopathogenic human orphan virus. { 'ek·ō,vī·rəs }

echylosis [CELL BIOLOGY] The release of nonparticulate material from the cell through an apparently intact cell membrane. { ē'kī·lə·səs }

eclipsed antigen [IMMUNOLOGY] An antigenic determinant of parasitic origin resembling an antigenic determinant of the parasite's host to such a degree that it does not elicit the formation of antibody by the host. { i'klipst 'ant·i·jen }

eclipse period [VIROLOGY] A phase in the proliferation of viral particles during which the virus cannot be detected in the host cell. { i'klips ,pir·ē·əd }

eclosion [INVERTEBRATE ZOOLOGY] The process of an insect hatching from its egg. { ē'klō·zhən }

ECM [CELL BIOLOGY] *See* extracellular matrix.

ecocline [ECOLOGY] A genetic gradient of adaptability to an environmental gradient; formed by the merger of ecotypes. { 'ek·ō,klīn }

ecological association [ECOLOGY] A complex of communities, such as an elm-hackberry association, which develops in accord with variations in physiography, soil, and successional history within the major subdivision of a biotic realm. { ek·ə'läj·ə·kəl ə,sō·shē'ā·shən }

ecological climatology [BIOLOGY] A branch of bioclimatology, including the physiological adaptation of plants and animals to their climate, and the geographical distribution of plants and animals in relation to climate. { ek·ə'läj·ə·kəl klī·mə'täl·ə·jē }

ecological community [ECOLOGY] *See* community. { ek·ə'läj·ə·kəl kə'myün·əd·ē }

ecological interaction [ECOLOGY] The relation between species that live together in a community; specifically, the effect an individual of one species may exert on an individual of another species. { ek·ə'läj·ə·kəl in·tər'ak·shən }

ecological physiology [BIOLOGY] The science of

the interrelationships between the physiology of organisms and their environment. { ˌē·kə¦läj·ə·kəl fiz·ē'äl·ə·jē }

ecological pyramid [ECOLOGY] A pyramid-shaped diagram representing quantitatively the numbers of organisms, energy relationships, and biomass of an ecosystem; numbers are high for the lowest trophic levels (plants) and low for the highest trophic level (carnivores). { ek·ə'läj·ə·kəl 'pir·ə·mid }

ecological succession [ECOLOGY] A gradual process incurred by the change in the number of individuals of each species of a community and by establishment of new species populations that may gradually replace the original inhabitants. { ek·ə'läj·ə·kəl sək'sesh·ən }

ecology [BIOLOGY] A study of the interrelationships which exist between organisms and their environment. Also known as environmental biology. { ē'käl·ə·jē }

economic entomology [BIOLOGY] A branch of entomology concerned with the study of economic losses of commercially important animals and plants due to insect predation. { ˌek·ə'näm·ik ˌen·tə'mäl·ə·jē }

ecophene [GENETICS] The range of phenotypic modifications produced by one genotype within the limits of the habitat under which the genotype is found in nature. { 'ē·kəˌfēn }

ecophenotype [ECOLOGY] A nongenetic phenotypic modification in response to environmental conditions. { ˌē·kō'phēn·əˌtīp }

ecospecies [ECOLOGY] A group of ecotypes capable of interbreeding without loss of fertility or vigor in the offspring. { 'ē·kōˌspē·shēz }

ecosystem [ECOLOGY] A functional system which includes the organisms of a natural community together with their environment. Derived from ecological system. { 'ek·ōˌsis·təm *or* 'ē·kōˌsis·təm }

ecosystem mapping [ECOLOGY] The drawing of maps that locate different ecosystems in a geographic area. { 'ek·ōˌsis·təm ˌmap·iŋ }

ecotone [ECOLOGY] A zone of intergradation between ecological communities. { 'ek·əˌtōn }

ecotrine [ECOLOGY] A metabolite produced by one kind of organism and utilized by another. { 'ek·əˌtrēn }

ecotype [ECOLOGY] A subunit, race, or variety of a plant ecospecies that is restricted to one habitat; equivalent to a taxonomic subspecies. { 'ek·əˌtīp }

Ecterocoelia [INVERTEBRATE ZOOLOGY] The equivalent name for Protostomia. { ˌek·tə·rō 'sēl·yə }

ectethmoid [ANATOMY] Either one of the lateral cellular masses of the ethmoid bone. { ek'teth ˌmȯid }

ectocommensal [ECOLOGY] An organism living on the outer surface of the body of another organism, without affecting its host. { ¦ek·tō·kə 'men·səl }

ectocornea [ANATOMY] The outer layer of the cornea. { ˌek·tō'kȯr·nē·ə }

ectocyst [INVERTEBRATE ZOOLOGY] **1.** The outer layer of the wall of a zooecium. **2.** *See* epicyst. { 'ek·təˌsist }

ectoderm [EMBRYOLOGY] The outer germ layer of an animal embryo. Also known as epiblast. [INVERTEBRATE ZOOLOGY] The outer layer of a diploblastic animal. { 'ek·təˌdərm }

ectoenzyme [BIOCHEMISTRY] An enzyme which is located on the external surface of a cell. { ¦ek·tō 'enˌzīm }

ectogenesis [EMBRYOLOGY] Development of an embryo or of embryonic tissue outside the body in an artificial environment. { ˌek·tō'jen·ə·səs }

ectogony [BOTANY] The influence of pollination and fertilization on structures outside the embryo and endosperm; effect may be on color, chemical composition, ripening, or abscission. { ek'täg·ə·nē }

ectomere [EMBRYOLOGY] A blastomere that will differentiate into ectoderm. { 'ek·təˌmir }

ectomesoblast [EMBRYOLOGY] An undifferentiated layer of embryonic cells from which arises the epiblast and mesoblast. { ¦ek·tō'me·zō ˌblast }

ectomesoderm [EMBRYOLOGY] Mesoderm which is derived from ectoderm and is always mesenchymal; a type of primitive connective tissue. { ¦ek·tō'me·zōˌdərm }

ectoparasite [ECOLOGY] A parasite that lives on the exterior of its host. { ¦ek·tō'par·əˌsīt }

ectophagous [INVERTEBRATE ZOOLOGY] The larval stage of a parasitic insect which is in the process of development externally on a host. { ek 'täf·ə·gəs }

ectophloic siphonostele [BOTANY] A type of stele with pith that has the phloem only on the outside of the xylem. { ˌek·tə'flō·ək sī'fän·ə ˌstēl }

ectophyte [ECOLOGY] A plant which lives externally on another organism. { 'ek·təˌfīt }

ectopic pairing [CELL BIOLOGY] Pairing between nonhomologous segments of the salivary gland chromosomes in *Drosophila*, presumably involving mainly heterochromatic regions. { ek'täp·ik 'per·iŋ }

ectoplasm [CELL BIOLOGY] The outer, gelled zone of the cytoplasmic ground substance in many cells. Also known as ectosarc. { 'ek·təˌplaz·əm }

Ectoprocta [INVERTEBRATE ZOOLOGY] A subphylum of colonial bryozoans having eucoelomate visceral cavities and the anus opening outside the circlet of tentacles. { ek·tō'präk·tə }

ectopterygoid [VERTEBRATE ZOOLOGY] A membrane bone located ventrally on the skull, situated behind the palate and extending to the quadrate; found in some fishes and reptiles. { ¦ek·tō'ter·əˌgȯid }

ectosarc [CELL BIOLOGY] *See* ectoplasm. { 'ek·tə ˌsärk }

ectosome [INVERTEBRATE ZOOLOGY] The outer, cortical layer of a sponge. { 'ek·təˌsōm }

ectostosis [PHYSIOLOGY] Formation of bone immediately beneath the perichondrium and surrounding and replacing underlying cartilage. { ¦ek·tə'stō·səs }

ectosymbiont [ECOLOGY] A symbiont that lives

on the surface of or is physically separated from its host. { ¦ek·tō′sim·bē‚änt }

ectotherm [PHYSIOLOGY] An animal that obtains most of its heat from the environment and therefore has a body temperature very close to that of its environment. { ′ek·tə‚thərm }

Ectothiorhodospira [MICROBIOLOGY] A genus of bacteria in the family Chromatiaceae; cells are spiral to slightly bent rods, are motile, contain bacteriochlorophyll *a* on lamellar stock membranes, and produce and deposit sulfur as globules outside the cells. { ¦ek·tō‚thī·ə‚rō′däs·pə·rə }

ectotrophic [BIOLOGY] Obtaining nourishment from outside; applied to certain parasitic fungi that live on and surround the roots of the host plant. { ¦ek·tə′träf·ik }

ectozoa [ECOLOGY] Animals which live externally on other organisms. { ‚ek·tə′zō·ə }

Ectrephidae [INVERTEBRATE ZOOLOGY] An equivalent name for Ptinidae. { ek′tref·ə‚dē }

ectromelia virus [VIROLOGY] A member of subgroup I of the poxvirus group; causes mousepox. { ‚ek·tra′mē·lē·ə ′vī·rəs }

edaphic community [ECOLOGY] A plant community that results from or is influenced by soil factors such as salinity and drainage. { ē′daf·ik kə′myün·əd·ē }

edaphon [BIOLOGY] Flora and fauna in soils. { ′ed·ə‚fän }

Edaphosuria [PALEONTOLOGY] A suborder of extinct, lowland, terrestrial, herbivorous reptiles in the order Pelycosauria. { ‚ed·ə·fō′sór·ē·ə }

Edentata [VERTEBRATE ZOOLOGY] An order of mammals characterized by the absence of teeth or the presence of simple prismatic, unspecialized teeth with no enamel. { ‚ē‚den′tä·də }

edentate [VERTEBRATE ZOOLOGY] **1.** Lacking teeth. **2.** Any member of the Edentata. { ē′den‚tāt }

edentulous [VERTEBRATE ZOOLOGY] Having no teeth; especially, having lost teeth that were present. { ē′den·chə·ləs }

edge effect [ECOLOGY] The influence of adjacent plant communities on the number of animal species present in the direct vicinity. { ′ej i‚fekt }

Ediacaran fauna [PALEONTOLOGY] The oldest known assemblage of fossil remains of soft-bodied marine animals; first discovered in the Ediacara Hills, Australia. { ‚ēd·ē·ə′kar·ən ′fón·ə }

editing [MOLECULAR BIOLOGY] *See* proofreading. { ′ed·əd·iŋ }

Edman degradation technique [BIOCHEMISTRY] In protein analysis, an approach to amino-end-group determination involving the use of a reagent, phenylisothiocyanate, that can be applied to the liberation of a derivative of the amino-terminal residue without hydrolysis of the remainder of the peptide chain. { ′ed·mən ‚deg·rə′dā·shən tek‚nēk }

Edrioasteroidea [PALEONTOLOGY] A class of extinct Echinozoa having ambulacral radial areas bordered by tube feet, and the mouth and anus located on the upper side of the theca. { ‚ed·rē·ō‚as·tə′ròid·ē·ə }

eduction [MOLECULAR BIOLOGY] Loss of host genetic material when the plasmid that had been integrated into the host chromosome exits. { ē′dək·shən }

Edwardsiella [MICROBIOLOGY] A genus of bacteria in the family Enterobacteriaceae; motile rods that produce hydrogen sulfide from TSI agar. { e¦dwärd·zē′el·ə }

eel [VERTEBRATE ZOOLOGY] The common name for a number of unrelated fishes included in the orders Anguilliformes and Cypriniformes; all have an elongate, serpentine body. { ēl }

eel grass [BOTANY] *See* tape grass. { ēl ‚gras }

effective lethal phase [GENETICS] The developmental stage at which a lethal gene generally causes death of the organism carrying it. { ə¦fek·tiv ¦lē·thəl ′fāz }

effector [BIOCHEMISTRY] An activator of an allosteric enzyme. [PHYSIOLOGY] A structure that is sensitive to a stimulus and causes an organism or part of an organism to react to the stimulus, either positively or negatively. { ə′fek·tər }

effector organ [PHYSIOLOGY] Any muscle or gland that mediates overt behavior, that is, movement or secretion. { ə′fek·tər ‚ór·gən }

effector system [PHYSIOLOGY] A system of effector organs in the animal body. { ə‚fek·tər ‚sis·təm }

efferent [PHYSIOLOGY] Carrying or conducting away, as the duct of an exocrine gland or a nerve. { ′ef·ə·rənt }

efflorescence [BOTANY] The period or process of flowering. { ‚ef·lə′res·əns }

effuse [BOTANY] Expanded; spread out in a definite form. { e′fyüz }

egest [PHYSIOLOGY] **1.** To discharge indigestible matter from the digestive tract. **2.** To rid the body of waste. { ē′jest }

egg [CELL BIOLOGY] **1.** A large, female sex cell enclosed in a porous, calcareous or leathery shell, produced by birds and reptiles. **2.** *See* ovum. { eg }

egg apparatus [BOTANY] A group of three cells, consisting of the egg and two synergid cells, in the micropylar end of the embryo sac in seed plants. { ′eg ‚ap·ə‚rad·əs }

egg capsule [INVERTEBRATE ZOOLOGY] *See* egg case. { ′eg ‚kap·səl }

egg case [INVERTEBRATE ZOOLOGY] **1.** A protective capsule containing the eggs of certain insects and mollusks. Also known as egg capsule. **2.** A silk pouch in which certain spiders carry their eggs. Also known as egg sac. [VERTEBRATE ZOOLOGY] A soft, gelatinous (amphibians) or strong, horny (skates) envelope containing the egg of certain vertebrates. { ′eg ‚kās }

eggplant [BOTANY] *Solanum melongena.* A plant of the order Polemoniales grown for its edible egg-shaped, fleshy fruit. { ′eg‚plant }

egg raft [ZOOLOGY] A floating mass of eggs; produced by a variety of aquatic organisms. { ′eg ‚raft }

egg sac [INVERTEBRATE ZOOLOGY] **1.** The structure containing the eggs of certain microcrustaceans. **2.** *See* egg case. { ′eg ‚sak }

egg tooth [VERTEBRATE ZOOLOGY] A toothlike prominence on the tip of the beak of a bird embryo and the tip of the nose of an oviparous reptile, which is used to break the eggshell. { 'eg ˌtüth }

eglandular [BIOLOGY] Without glands. { ē'glan·dyə·lər }

Egyptian cotton [BOTANY] Long-staple, high-quality cotton grown in Egypt. { i'jip·shən 'kät·ən }

Egyptian henna [BOTANY] *See* henna. { i'jip·shən 'hen·ə }

Ehrlichia [MICROBIOLOGY] A genus of the tribe Ehrlichieae; coccoid to ellipsoidal or pleomorphic cells; intracellular parasites in cytoplasm of host leukocytes. { er'lik·ē·ə }

Ehrlichieae [MICROBIOLOGY] A tribe of the family Rickettsiaceae; spherical and occasionally pleomorphic cells; pathogenic for some mammals, not including humans. { ˌer·lə'kī·ēˌē }

eicosanoid [BIOCHEMISTRY] Any member of a group of naturally occurring substances composed of prostaglandins, thromboxanes, and leukotrienes that are derived from polyunsaturated fatty acids, particularly arachidonic acid, and exhibit various types of biological activity. { ī'käs·əˌnȯid }

Eimeriina [INVERTEBRATE ZOOLOGY] A suborder of coccidian protozoans in the order Eucoccida in which there is no syzygy and the microgametocytes produce a large number of microgametes. { ˌī·mə'rī·ə·nə }

ejaculation [PHYSIOLOGY] The act or process of suddenly discharging a fluid from the body; specifically, the ejection of semen during orgasm. { iˌjak·yə'lā·shən }

ejaculatory duct [ANATOMY] The terminal part of the ductus deferens after junction with the duct of a seminal vesicle, embedded in the prostate gland and opening into the urethra. { i'jak·yə·lə ˌtȯr·ē 'dəkt }

ejecta [PHYSIOLOGY] Excrement. { ē'jek·tə }

Elaeagnaceae [BOTANY] A family of dicotyledonous plants in the order Proteales, noted for peltate leaf scales which often give the leaves a silvery-gray appearance. { ˌel·ē·ag'nās·ēˌē }

elaioplast [HISTOLOGY] An oil-secreting leucoplast. { ə'lī·əˌplast }

Elaphomycetaceae [MYCOLOGY] A family of underground, saprophytic or mycorrhiza-forming fungi in the order Eurotiales characterized by ascocarps with thick, usually woody walls. { ˌel·ə·fōˌmī·sə'tās·ēˌē }

Elapidae [VERTEBRATE ZOOLOGY] A family of poisonous reptiles, including cobras, kraits, mambas, and coral snakes; all have a pteroglyph fang arrangement. { ə'lap·əˌdē }

Elasipodida [INVERTEBRATE ZOOLOGY] An order of deep-sea aspidochirotacean holothurians in which there are no respiratory trees and bilateral symmetry is often quite conspicuous. { əˌlaz·ə 'päd·ə·də }

Elasmidae [INVERTEBRATE ZOOLOGY] A family of hymenopteran insects in the superfamily Chalcidoidea. { ə'laz·məˌdē }

Elasmobranchii [VERTEBRATE ZOOLOGY] The sharks and rays, a subclass of the class Chondrichthyes distinguished by separate gill openings, amphistylic or hyostylic jaw suspension, and ampullae of Lorenzini in the head region. { əˌlaz·mə'braŋ·kēˌī }

Elassomatidae [VERTEBRATE ZOOLOGY] The pygmy sunfishes, a family of the order Perciformes. { əˌlas·ō'mad·əˌdē }

elastase [BIOCHEMISTRY] An enzyme which acts on elastin to change it chemically and render it soluble. { i'laˌstās }

elastic cartilage [HISTOLOGY] A type of cartilage containing elastic fibers in the matrix. { i'las·tik 'kärt·lij }

elastic fiber [HISTOLOGY] A homogeneous, fibrillar connective tissue component that is highly refractile and appears yellowish when arranged in bundles. { i'las·tik 'fī·bər }

elastic tissue [HISTOLOGY] A type of connective tissue having a predominance of elastic fibers, bands, or lamellae. { i'las·tik 'tish·ü }

elastin [BIOCHEMISTRY] An elastic protein composing the principal component of elastic fibers. { i'las·tən }

elater [BOTANY] A spiral, filamentous structure that functions in the dispersion of spores in certain plants, such as liverworts and slime molds. { 'el·ə·tər }

Elateridae [INVERTEBRATE ZOOLOGY] The click beetles, a large family of coleopteran insects in the superfamily Elateroidea; many have light-producing organs. { ˌel·ə'ter·əˌdē }

Elateroidea [INVERTEBRATE ZOOLOGY] A superfamily of coleopteran insects in the suborder Polyphaga. { iˌlad·ə'rȯid·ē·ə }

elaterophore [BOTANY] A tissue bearing elaters, found in some liverworts. { i'lad·ə·rəˌfȯr }

elbow [ANATOMY] The arm joint formed at the junction of the humerus, radius, and ulna. { 'el ˌbō }

elective culture [MICROBIOLOGY] A type of microorganism grown selectively from a mixed culture by culturing in a medium and under conditions selective for only one type of organism. { i¦lek·tiv 'kəl·chər }

electric eel [VERTEBRATE ZOOLOGY] *Electrophorus electricus.* An eellike cypriniform electric fish of the family Gymnotidae. { i¦lek·trik 'ēl }

electric fish [VERTEBRATE ZOOLOGY] Any of several fishes capable of producing electric discharges from an electric organ. { i¦lek·trik 'fish }

electric organ [VERTEBRATE ZOOLOGY] An organ consisting of rows of electroplaques which produce an electric discharge. { i¦lek·trik 'ȯr·gən }

electric shock [PHYSIOLOGY] The sudden pain, convulsion, unconsciousness, or death produced by the passage of electric current through the body. { i¦lek·trik 'shäk }

electroculogram [PHYSIOLOGY] A record of the standing voltage between the front and back of the eye that is correlated with eyeball movement and obtained by electrodes placed on the skin near the eye. { iˌlek·trə·'kyül·əˌgram }

electroculography [PHYSIOLOGY] The production and study of electroculograms. { i¦lek·trə·kyü 'läg·rə·fē }

electrogenesis [PHYSIOLOGY] The generation of electric current by living tissue. { i¦lek·trə'jen·ə·səs }

electrogram [PHYSIOLOGY] The graphic representation of electric events in living tissues; commonly, an electrocardiogram or electroencephalogram. { i'lek·trə‚gram }

electroinjection [BIOLOGY] The use of electric-field impulses to introduce foreign deoxyribonucleic acid directly into intact cells. { i‚lek·trō·in'jek·shən }

electron transport system [BIOCHEMISTRY] The components of the final sequence of reactions in biological oxidations; composed of a series of oxidizing agents arranged in order of increasing strength and terminating in oxygen. { i'lek‚trän 'trans‚pȯrt ‚sis·təm }

electrophonic effect [BIOPHYSICS] The sensation of hearing produced when an alternating current of suitable frequency and magnitude is passed through a person. { i‚lek·trə'fän·ik i'fekt }

electrophoretic mobility [BIOCHEMISTRY] A characteristic of living cells in suspension and biological compounds (proteins) in solution to travel in an electric field to the positive or negative electrode, because of the charge on these substances. { i¦lek·trō·fə'red·ik mō'bil·əd·ē }

electrophoretic variants [BIOCHEMISTRY] Phenotypically different proteins that are separable into distinct electrophoretic components due to differences in mobilities; an example is erythrocyte acid phosphatase. { i¦lek·trō·fə'red·ik 'ver·ē·əns }

electrophysiology [PHYSIOLOGY] The branch of physiology concerned with determining the basic mechanisms by which electric currents are generated within living organisms. { i‚lek·trō‚fiz·ē'ä·lə·jē }

electroplax [VERTEBRATE ZOOLOGY] One of the structural units of an electric organ of some fishes, composed of thin, flattened plates of modified muscle that appear as two large, waferlike, roughly circular or rectangular surfaces. { i'lek·trō‚plaks }

electroporation [BIOLOGY] The application of electric pulses to increase the permeability of cell membranes. [CELL BIOLOGY] The application of electric pulses to animal cells or plant protoplasts to increase membrane permeability. { i‚lek·trō·pə'rā·shən }

electrotaxis [BIOLOGY] Movement of an organism in response to stimulation by electric charges. { i¦lek·trō'tak·səs }

electrotonus [PHYSIOLOGY] The change of condition in a nerve or a muscle during the passage of a current of electricity. { i‚lek'trät·ən·əs }

electrotropism [BIOLOGY] Orientation response of a sessile organism to stimulation by electric charges. { i‚lek'trä·trə‚piz·əm }

elephant [VERTEBRATE ZOOLOGY] The common name for two living species of proboscidean mammals in the family Elephantidae; distinguished by the elongation of the nostrils and upper lip into a sensitive, prehensile proboscis. { 'el·ə·fənt }

Elephantidae [VERTEBRATE ZOOLOGY] A family of mammals in the order Proboscidea containing the modern elephants and extinct mammoths. { el·ə'fan·tə‚dē }

elfinwood [ECOLOGY] *See* krummholz. { 'el·fən ‚wu̇d }

elimination coefficient [GENETICS] The frequency with which certain genotypes die prematurely or are hindered during reproduction and are genetically eliminated as a consequence. { ə¦lim·ə'nā·shən ‚kō·ə‚fish·ənt }

elk [VERTEBRATE ZOOLOGY] *Alces alces.* A mammal (family Cervidae) in Europe and Asia that resembles the North American moose but is smaller; it is the largest living deer. { elk }

elm [BOTANY] The common name for hardwood trees composing the genus *Ulmus*, characterized by simple, serrate, deciduous leaves. { elm }

Elmidae [INVERTEBRATE ZOOLOGY] The drive beetles, a small family of coleopteran insects in the superfamily Dryopoidea. { 'el·mə‚dē }

elongation factor [BIOCHEMISTRY] Any of several proteins required for elongation of growing polypeptide chains during protein synthesis. { ē‚lȯŋ'gā·shən ‚fak·tər }

Elopidae [VERTEBRATE ZOOLOGY] A family of fishes in the order Elopiformes, including the tarpon, ladyfish, and machete. { e'läp·ə‚dē }

Elopiformes [VERTEBRATE ZOOLOGY] A primitive order of actinopterygian fishes characterized by a single dorsal fin composed of soft rays only, cycloid scales, and toothed maxillae. { e‚läp·ə'fȯr‚mēz }

El Tor vibrio [MICROBIOLOGY] Any of the rod-shaped paracholera vibrios; many strains can be agglutinated with anticholera serum. { el 'tȯr 'vib·rē·ō }

elytron [INVERTEBRATE ZOOLOGY] **1.** One of the two sclerotized or leathery anterior wings of beetles which serve to cover and protect the membranous hindwings. **2.** A dorsal scale of certain Polychaeta. { 'el·ə‚trän }

emarginate [BIOLOGY] Having a margin that is notched or slightly forked. { ē'mär·jə‚nāt }

EMB agar [MICROBIOLOGY] A culture medium containing sugar, eosin, and methylene blue, used in the confirming test for coliform bacteria. { ¦ē¦em¦bē 'äg·ər }

Emballonuridae [VERTEBRATE ZOOLOGY] The sheath-tailed bats, a family of mammals in the order Chiroptera. { em‚bal·ə'nu̇r·ə‚dē }

embed [BIOLOGY] To prepare a specimen for sectioning for microscopic examination by infiltrating with or enclosing in paraffin or other supporting material. Also spelled imbed { em'bed }

Embden-Meyerhof pathway [BIOCHEMISTRY] *See* glycolytic pathway. { ¦em·dən ¦mī·ər‚hȯf 'path ‚wā }

Embiidina [INVERTEBRATE ZOOLOGY] An equivalent name for Embioptera. { em‚bē·ə'dī·nə }

Embioptera [INVERTEBRATE ZOOLOGY] An order of silk-spinning, orthopteroid insects resembling the grasshoppers; commonly called the embiids or webspinners. { ‚em·bē'äp·tə·rə }

Embiotocidae [VERTEBRATE ZOOLOGY] The surfperches, a family of perciform fishes in the suborder Percoidei. { ‚em·bē·ə'täs·ə‚dē }

Embolomeri [PALEONTOLOGY] An extinct side branch of slender-bodied, fish-eating aquatic anthracosaurs in which intercentra as well as centra form complete rings. { ˌem·bə'lä·məˌrī }

emboly [EMBRYOLOGY] Formation of a gastrula by the process of invagination. { 'em·bə·lē }

Embrithopoda [PALEONTOLOGY] An order established for the unique Oligocene mammal *Arsinoitherium*, a herbivorous animal that resembled the modern rhinoceros. { ˌem·brə'thä·pə·də }

embryo [BOTANY] The young sporophyte of a seed plant. [EMBRYOLOGY] **1.** An early stage of development in multicellular organisms. **2.** The product of conception up to the third month of human pregnancy. { 'em·brē·ō }

Embryobionta [BOTANY] The land plants, a subkingdom of the Plantae characterized by having specialized conducting tissue in the sporophyte (except bryophytes), having multicellular sex organs, and producing an embryo. { ¦em·brē·ō·bī'än·tə }

embryogenesis [EMBRYOLOGY] The formation and development of an embryo from an egg. { ˌem·brē·ō'jen·ə·səs }

embryoid [BOTANY] An embryolike structure originating from somatic cells, such as immature plant embryos, inflorescences, or leaves cultivated in culture. { 'em·brēˌȯid }

embryology [BIOLOGY] The study of the development of the organism from the zygote, or fertilized egg. { em·brē'äl·ə·jē }

embryonate [EMBRYOLOGY] **1.** To differentiate into a zygote. **2.** Containing an embryo. { 'em·brē·ə'nāt }

embryonated egg culture [VIROLOGY] Embryonated hen's eggs inoculated with animal viruses for the purpose of identification, isolation, titration, or for quantity cultivation in the production of viral vaccines. { 'em·brē·əˌnād·əd 'eg ˌkəl·chər }

embryonic differentiation [EMBRYOLOGY] The process by which specialized and diversified structures arise during embryogenesis. { ˌem·brē'än·ik ˌdif·əˌren·chē'ā·shən }

embryonic inducer [EMBRYOLOGY] The acting system in embryos, which contributes to the formation of specialized tissues by controlling the mode of development of the reacting system. { ˌem·brē'än·ik in'dü·sər }

embryonic induction [EMBRYOLOGY] The influence of one cell group (inducer) over a neighboring cell group (induced) during embryogenesis. Also known as induction. { ˌem·brē'än·ik in'dək·shən }

Embryophyta [BOTANY] The equivalent name for Embryobionta. { ˌem·brē'äf·əd·ə }

embryo rescue [GENETICS] A technique for crossing wild and domestic species of plants in which the wild species is used as the male parent, and the embryos are excised approximately one month after pollination and placed on an artificial medium, where a small fraction survive. { 'em·brē·ō 'res·kyü }

embryo sac [BOTANY] The female gametophyte of a seed plant, containing the egg, synergids, and polar and antipodal nuclei; fusion of the antipodals and a pollen generative nucleus forms the endosperm. { 'em·brē·ō ˌsak }

emerged bog [ECOLOGY] A bog which grows vertically above the water table by drawing water up through the mass of plants. { ə¦mərjd 'bäg }

emigration [ECOLOGY] The movement of individuals or their disseminules out of a population or population area. { ˌem·ə'grā·shən }

emiocytosis [CELL BIOLOGY] Fusion of intracellular granules with the cell membrane, followed by discharge of the granules outside of the cell; applied chiefly to the mechanism of insulin secretion. Also known as reverse pinocytosis. { ˌem·ēˌsī'tō·səs }

emission inventory [ECOLOGY] A quantitative detailed compilation of pollutants emitted into the atmosphere of a given community. { i'mish·ən 'in·vənˌtȯr·ē }

Empididae [INVERTEBRATE ZOOLOGY] The dance flies, a family of orthorrhaphous dipteran insects in the series Nematocera. { em'pid·əˌdē }

empodium [INVERTEBRATE ZOOLOGY] A small peripheral part located between the claws of the tarsi of many insects and arachnids. { em'pōd·ē·əm }

emu [VERTEBRATE ZOOLOGY] *Dromiceius novaehollandiae.* An Australian ratite bird, the second largest living bird, characterized by rudimentary wings and a feathered head and neck without wattles. { 'ēˌmyü }

Emydidae [VERTEBRATE ZOOLOGY] A family of aquatic and semiaquatic turtles in the suborder Cryptodira. { e'mid·əˌdē }

Enaliornithidae [PALEONTOLOGY] A family of extinct birds assigned to the order Hesperornithiformes, having well-developed teeth found in grooves in the dentary and maxillary bones of the jaws. { e¦nal·ē·ȯr¦nith·əˌdē }

enamel organ [EMBRYOLOGY] The epithelial ingrowth from the dental lamina which covers the dental papilla, furnishes a mold for the shape of a developing tooth, and forms the dental enamel. { i'nam·əl ˌȯr·gən }

Enantiozoa [INVERTEBRATE ZOOLOGY] The equivalent name for Parazoa. { ə¦nan·te·ə'zō·ə }

enarthrosis [ANATOMY] A freely movable joint that allows a wide range of motion on all planes. Also known as ball-and-socket joint. { ˌeˌnär 'thrō·səs }

Encalyptales [BOTANY] An order of true mosses (subclass Bryidae) characterized by broad papillose leaves and erect capsules covered by very long calyptrae. { enˌka·lip'tā·lēz }

Encholaimoidea [INVERTEBRATE ZOOLOGY] A superfamily of nematodes of the order Dorylaimida, characterized by two circlets of cephalic sense organs on the lips, pouchlike amphids with slitlike openings, a stoma armed with an axial spear, and a body cuticle marked by widely spaced annulations giving a platelike appearance. { ˌen·kō·lə'mȯid·ē·ə }

enchymatous [PHYSIOLOGY] Of gland cells, distended with secreted material. { en'kim·əd·əs }

encrinite [PALEONTOLOGY] One of certain fossil crinoids, especially of the genus *Encrinus.* { 'eŋ·krəˌnīt }

Encyrtidae [INVERTEBRATE ZOOLOGY] A family of hymenopteran insects in the superfamily Chalcidoidea. { en'sərd·ə,dē }

encystment [BIOLOGY] The process of forming or becoming enclosed in a cyst or capsule. { en 'sist·mənt }

Endamoeba [INVERTEBRATE ZOOLOGY] The type genus of the Endamoebidae comprising insect parasites and, in some systems of classification, certain vertebrate parasites. { ¦end·ə'mē·bə }

endarch [BOTANY] Formed outward from the center, referring to xylem or its development. { 'en ,därk }

end bulb [ANATOMY] *See* bouton. { 'end ,bəlb }

end bulb of Krause [ANATOMY] *See* Krause's corpuscle. { 'end ,bəlb əv 'kraüs }

Endeidae [INVERTEBRATE ZOOLOGY] A family of marine arthropods in the subphylum Pycnogonida. { en'dē·ə,dē }

endergonic [BIOCHEMISTRY] Of or pertaining to a biochemical reaction in which the final products possess more free energy than the starting materials; usually associated with anabolism. { ¦en·dər¦gän·ik }

endexine [BOTANY] An inner membranous layer of the exosporium. { en'dek,sēn }

endite [INVERTEBRATE ZOOLOGY] **1.** One of the appendages on the inner aspect of an arthropod limb. **2.** A ridgelike chewing surface on the inner part of the pedipalpus or maxilla of many arachnids. { 'en,dīt }

end labeling [BIOCHEMISTRY] The addition of a radioactively labeled group to one end of a deoxyribonucleic acid strand. { 'end ,lab·əl·iŋ }

endobasion [ANATOMY] The anteriormost point of the margin of the foramen magnum at the level of its smallest diameter. { ¦en·dō'bā·sē ,än }

endobiotic [ECOLOGY] Referring to an organism living in the cells or tissues of a host. { ¦en·dō·bī'äd·ik }

endobranchiate [ZOOLOGY] Animal form with endodermal gills. { ¦en·dō'braŋ·kē,āt }

endocardium [ANATOMY] The membrane lining the heart. { ,en·dō'kärd·ē·əm }

endocarp [BOTANY] The inner layer of the wall of a fruit or pericarp. { 'en·dō,kärp }

endocervix [ANATOMY] The glandular mucous membrane of the cervix uteri. { ¦en·dō'sər·viks }

endochondral ossification [PHYSIOLOGY] The conversion of cartilage into bone. Also known as intracartilaginous ossification. { ¦en·dō'kän·drəl ,äs·ə·fə'kā·shən }

endocommensal [ECOLOGY] A commensal that lives within the body of its host. { ¦en·dō·kə 'men·səl }

endocorpuscular [CELL BIOLOGY] Located within an erythrocyte. { ¦en·dō·kór'pəs·kyə·lər }

endocranium [ANATOMY] **1.** The inner surface of the cranium. **2.** *See* dura mater. [INVERTEBRATE ZOOLOGY] The processes on the inner surface of the head capsule of certain insects. { ¦en·dō 'krā·nē·əm }

endocrine gland [PHYSIOLOGY] A ductless structure whose secretion (hormone) is passed into adjacent tissue and then to the bloodstream either directly or by way of the lymphatics. Also known as ductless gland. { 'en·də·krən ,gland }

endocrine signaling [PHYSIOLOGY] Signaling in which endocrine cells release hormones that act on distant target cells. { 'en·də·krən ,sig·nəl·iŋ }

endocrine system [PHYSIOLOGY] The chemical coordinating system in animals, that is, the endocrine glands that produce hormones. { 'en·də·krən ,sis·təm }

endocrinology [PHYSIOLOGY] The study of the endocrine glands and the hormones that they synthesize and secrete. { ,en·də·krə'näl·ə·jē }

endocuticle [INVERTEBRATE ZOOLOGY] The inner, elastic layer of an insect cuticle. { ¦en·dō'kyüd·i·kəl }

endocyst [INVERTEBRATE ZOOLOGY] The soft layer consisting of ectoderm and mesoderm, lining the ectocyst of bryozoans. { 'en·də,sist }

endocytic vacuole [CELL BIOLOGY] A membrane-bound cellular organelle containing extracellular particles engulfed by the mechanisms of endocytosis. { ¦en·də¦sīd·ik 'vak·yə,wōl }

endocytobiosis [ECOLOGY] Symbiosis in which the symbionts live within host cells. { ,en·dō ,sī·tō·bī'ō·səs }

endocytosis [CELL BIOLOGY] **1.** An active process in which extracellular materials are introduced into the cytoplasm of cells by either phagocytosis or pinocytosis. **2.** The process by which animal cells internalize large molecules and large collections of fluid. { ¦en·do·sī'tō·səs }

endoderm [EMBRYOLOGY] The inner, primary germ layer of an animal embryo; sometimes referred to as the hypoblast. Also known as entoderm; hypoblast. { 'en·dō,dərm }

endodermis [BOTANY] The innermost tissue of the cortex of most plant roots and certain stems consisting of a single layer of at least partly suberized or cutinized cells; functions to control the movement of water and other substances into and out of the stele. { ¦en·dō¦dər·məs }

endodeoxyribonuclease [BIOCHEMISTRY] *See* restriction endonuclease. { ¦en·dō·dē¦äk·sē,rī·bō 'nü·klē,ās }

endoenzyme [BIOCHEMISTRY] An intracellular enzyme, retained and utilized by the secreting cell. { ¦en·dō¦en,zīm }

endogamy [BIOLOGY] Sexual reproduction between organisms which are closely related. [BOTANY] Pollination of a flower by another flower of the same plant. { en'däg·ə·mē }

endogenote [MICROBIOLOGY] The genetic complement of the partial zygote formed as a result of gene transfer during the process of recombination in bacteria. { en'däj·ə,nōt }

endogenous [BIOCHEMISTRY] Relating to the metabolism of nitrogenous tissue elements. { en 'däj·ə·nəs }

endogenous pyrogen [BIOCHEMISTRY] A fever-inducing substance (protein) produced by cells of the host body, such as leukocytes and macrophages. { en'däj·ə·nəs 'pī·rə·jən }

endogenous virus [GENETICS] An inactive virus that is integrated into the chromosome of its host cell and can, therefore, exhibit vertical transmission. { en¦däj·ən·əs 'vī,rəs }

endognath [INVERTEBRATE ZOOLOGY] The inner and main branch of a crustacean's oral appendage. { 'en·dəg,nath }

endolecithal [INVERTEBRATE ZOOLOGY] A type of egg found in turbellarians with yolk granules in the cytoplasm of the egg. Also spelled entolecithal. { ¦en·dō'les·ə·thəl }

endolithic [ECOLOGY] Living within rocks, as certain algae and coral. { ¦en·də¦lith·ik }

endolymph [PHYSIOLOGY] The lymph fluid found in the membranous labyrinth of the ear. { 'en·də ,limf }

endomembrane system [CELL BIOLOGY] In eukaryotes, the functional continuum of membraneous cell components consisting of the nuclear envelope, endoplastic reticulum, and Golgi apparatus as well as vesicles and other structures derived from these major components. { ,en·dō 'mem,brān ,sis·təm }

endomeninx [EMBRYOLOGY] The internal part of the meninx primitiva that differentiates into the pia mater and arachnoid membrane. { ¦en·dō ¦mē·niŋks }

endomere [EMBRYOLOGY] A blastomere that forms endoderm. { 'en·də,mir }

endometrium [ANATOMY] The mucous membrane lining the uterus. { ¦en·dō¦mē·trē·əm }

endomitosis [CELL BIOLOGY] Division of the chromosomes without dissolution of the nuclear membrane; results in polyploidy or polyteny. { ¦en·dō,mī'tō·səs }

endomixis [INVERTEBRATE ZOOLOGY] Periodic division and reorganization of the nucleus in certain ciliated protozoans. { ¦en·dō¦mik·səs }

Endomycetales [MICROBIOLOGY] Former designation for Saccharomycetales. { ,en·də,mī·sə 'tā·lēz }

Endomycetoideae [MICROBIOLOGY] A subfamily of ascosporogenous yeasts in the family Saccharomycetaceae. { ,en·də,mī·sə'tȯid·ē,ē }

Endomychidae [INVERTEBRATE ZOOLOGY] The handsome fungus beetles, a family of coleopteran insects in the superfamily Cucujoidea. { ,en·də'mīk·ə,dē }

endomysium [HISTOLOGY] The connective tissue layer surrounding an individual skeletal muscle fiber. { ,en·də'miz·ē·əm }

endoneurium [HISTOLOGY] Connective tissue fibers surrounding and joining the individual fibers of a nerve trunk. { ,en·dō'nu̇r·ē·əm }

endonuclease [BIOCHEMISTRY] Any of a group of enzymes which degrade deoxyribonucleic acid or ribonucleic acid molecules by attaching nucleotide linkages within the polynucleotide chain. { ¦en·dō'nü·klē,ās }

endoparasite [ECOLOGY] A parasite that lives inside its host. { ¦en·dō'par·ə,sīt }

endopeptidase [BIOCHEMISTRY] An enzyme that acts upon the centrally located peptide bonds of a protein molecule. { ¦en·dō'pep·tə,dās }

endoperoxide [BIOCHEMISTRY] Any of various intermediates in the biosynthesis of prostaglandins. { ,en·dō·pə'räk,sīd }

endophagous [INVERTEBRATE ZOOLOGY] Of an insect larva, living within and feeding upon the host tissues. { en'däf·ə·gəs }

endophallus [INVERTEBRATE ZOOLOGY] Inner wall of the phallus of insects. { ¦en·dō'fal·əs }

endophyte [ECOLOGY] A plant that lives within, but is not necessarily parasitic on, another plant. { 'en·də,fīt }

endoplasm [CELL BIOLOGY] The inner, semifluid portion of the cytoplasm. { 'en·də,plaz·əm }

endoplasmic reticulum [CELL BIOLOGY] A vacuolar system of the cytoplasm in differentiated cells that functions in protein synthesis and sequestration. Abbreviated ER. { ¦en·də¦plaz·mik rə'tik·yə·ləm }

endopleurite [INVERTEBRATE ZOOLOGY] **1.** The portion of a crustacean apodeme which develops from the interepimeral membrane. **2.** One of the laterally located parts on the thorax of an insect which fold inward, extending into the body cavity. { ,en·də'plu̇r,īt }

endopodite [INVERTEBRATE ZOOLOGY] The inner branch of a biramous crustacean appendage. { en'däp·ə,dīt }

endopolyploid cell [CELL BIOLOGY] Any cell whose chromosome number has been increased by endomitosis and for which the degree of ploidy is proportional to the number of times that endomitosis has taken place. { ,en·dō'päl·ə,plȯid ,sel }

Endoprocta [INVERTEBRATE ZOOLOGY] The equivalent name for Entoprocta. { ,en·də'präk·tə }

endopterygoid [VERTEBRATE ZOOLOGY] A paired dermal bone of the roof of the mouth in fishes. { ¦en·dō'ter·ə,gȯid }

Endopterygota [INVERTEBRATE ZOOLOGY] A division of the insects in the subclass Pterygota, including those orders which undergo a holometabolous metamorphosis. { ¦en·dō,ter·ə 'gäd·ə }

endoreduplication [CELL BIOLOGY] Appearance in mitotic cells of certain chromosomes or chromosome sets in the form of multiples. { ¦en·dō·rē,dü·plə'kā·shən }

end organ [ANATOMY] The expanded termination of a nerve fiber in muscle, skin, mucous membrane, or other structure. { 'end ,ȯr·gən }

***β*-endorphin** [BIOCHEMISTRY] A 31-amino acid peptide fragment of pituitary *β*-lipotropic hormone having morphinelike activity. { ¦bād·ə en 'dȯr·fən }

endosalpinx [ANATOMY] The mucous membrane that lines the fallopian tube. { ¦en·dō¦sal ,piŋks }

endosarc [INVERTEBRATE ZOOLOGY] The inner, relatively fluid part of the protoplasm of certain unicellular organisms. { 'en·də,särk }

endoskeleton [ZOOLOGY] An internal skeleton or supporting framework in an animal. { ¦en·dō 'skel·ə·tən }

endosmosis [PHYSIOLOGY] The passage of a liquid inward through a cell membrane. { ¦en·dō·äs'mō·səs }

endosome [CELL BIOLOGY] A mass of chromatin near the center of a vesicular nucleus. [INVERTEBRATE ZOOLOGY] The inner layer of certain sponges. { 'en·də,sōm }

endosperm [BOTANY] **1.** The nutritive protein material within the embryo sac of seed plants. **2.** Storage tissue in the seeds of gymnosperms. { 'en·də,spərm }

endosperm nucleus [BOTANY] The triploid nucleus formed within the embryo sac of most seed plants by fusion of the polar nuclei with one sperm nucleus. { 'en·də,spərm 'nü·klē·əs }

endospore [BIOLOGY] An asexual spore formed within a cell. { 'en·də,spȯr }

endostome [BOTANY] The opening in the inner integument of a bitegmic ovule. { 'en·də ,stōm }

endostyle [INVERTEBRATE ZOOLOGY] A ciliated groove or pair of grooves in the pharynx of lower chordates. { 'en·də,stīl }

endosymbiont [ECOLOGY] A symbiont that lives within the body of the host without deleterious effect on the host. { ¦en·dō'sim·bē,änt }

endosymbiont theory [CELL BIOLOGY] A theory that the mitochondria of eukaryotes and the chloroplasts of green plants and flagellates originated as free-living prokaryotes that invaded primitive eukaryotic cells and became established as permanent symbionts in the cytoplasm. { ,en·dō'sim·bē,änt ,thē·ə·rē }

endosymbiotic infection [VIROLOGY] A virus infection in which virus replication occurs in cells without a cytopathic effect. { ¦en·dō,sim·bē 'äd·ik in'fek·shən }

endotergite [INVERTEBRATE ZOOLOGY] A dorsal plate to which muscles are attached in the insect skeleton. { ¦en·dō'tər,jīt }

endotesta [BOTANY] An inner layer of the testa in various seeds. { 'en·dō,tes·tə }

endothecium [BOTANY] The middle of three layers that make up an immature anther; becomes the inner layer of a mature anther. { ,en·də 'thē·shē·əm }

endothelial cell [HISTOLOGY] A type of squamous epithelial cell composing the endothelium. { ,en·də¦thē·lē·əl 'sel }

endotheliochorial placenta [EMBRYOLOGY] A type of placenta in which the maternal blood is separated from the chorion by the maternal capillary endothelium; occurs in dogs. { ¦en·də ,thē·lē·ə'kȯr·ē·əl plə'sen·tə }

endothelium [HISTOLOGY] The epithelial layer of cells lining the heart and vessels of the circulatory system. { ,en·də'thē·lē·əm }

Endotheriidae [PALEONTOLOGY] A family of Cretaceous insectivores from China belonging to the Proteutheria. { ,en·dō·thə'rī·ə,dē }

endotherm [PHYSIOLOGY] An animal that produces enough heat from its own metabolism and employs devices to retard heat loss so that it is able to keep its body temperature higher than that of its environment. { 'en·də,thərm }

Endothyracea [PALEONTOLOGY] A superfamily of extinct benthic marine foraminiferans in the suborder Fusulinina, having a granular or fibrous wall. { ,en·dō·thə'rās·ē·ə }

endotoxin [MICROBIOLOGY] A toxin that is produced within a microorganism and can be isolated only after the cell is disintegrated. { ¦en·dō'täk·sən }

endotracheal [ANATOMY] Within the trachea. { ¦en·dō'trā·kē·əl }

endotrophic [BIOLOGY] Obtaining nourishment from within; applied to certain parasitic fungi that live in the root cortex of the host plant. { ¦en·də¦trä·fik }

end-product inhibition [BIOCHEMISTRY] In sequential enzyme systems, a control mechanism in which accumulation of final product from a metabolic reaction causes inhibition of product formation. { ¦end ,präd·əkt ,in·ə'bish·ən }

energy balance [PHYSIOLOGY] The relation of the amount of utilizable energy taken into the body to that which is employed for internal work, external work, and the growth and repair of tissues. { 'en·ər·jē ,bal·əns }

energy metabolism [BIOCHEMISTRY] The chemical reactions involved in energy transformations within cells. { 'en·ər·jē mə'tab·ə,liz·əm }

energy pyramid [ECOLOGY] An ecological pyramid illustrating the energy flow within an ecosystem. { 'en·ər·jē ,pir·ə·mid }

engram [PHYSIOLOGY] A memory imprint; the alteration that has occurred in nervous tissue as a result of an excitation from a stimulus, which hypothetically accounts for retention of that experience. Also known as memory trace. { 'en ,gram }

Engraulidae [VERTEBRATE ZOOLOGY] The anchovies, a family of herringlike fishes in the suborder Clupoidea. { ,en'grȯl·ə,dē }

enhancer gene [GENETICS] Any modifier gene that acts to enhance the action of a nonallelic gene. { en'han·sər ,jēn }

Enicocephalidae [INVERTEBRATE ZOOLOGY] The gnat bugs, a family of hemipteran insects in the superfamily Enicocephaloidea. { ,en·ə·kō· sə'fal·ə,dē }

Enicocephaloidea [INVERTEBRATE ZOOLOGY] A superfamily of the Hemiptera in the subdivision Geocorisae containing a single family. { ,en·ə· kō·sef·ə'lȯid·ē·ə }

enkephalin [BIOCHEMISTRY] A mixture of two polypeptides isolated from the brain; central mode of action is an inhibition of neurotransmitter release. { en'kef·ə·lən }

enolase [BIOCHEMISTRY] An enzyme that catalyzes the reversible dehydration of phosphoglyceric acid to phosphopyruvic acid. { 'ē·nə,lās }

Enopla [INVERTEBRATE ZOOLOGY] A class or subclass of ribbonlike worms of the phylum Rhynchocoela. { e'näp·ē·ə }

Enoplia [INVERTEBRATE ZOOLOGY] A subclass of nematodes in the class Adenophorea. { e'näp· lē·ə }

Enoplida [INVERTEBRATE ZOOLOGY] An order of nematodes in the subclass Enoplia. { e'näp· lə·də }

Enoplidae [INVERTEBRATE ZOOLOGY] A family of free-living marine nematodes in the superfamily Enoploidea, characterized by a complex arrangement of teeth and mandibles. { e'näp·lə,dē }

Enoploidea [INVERTEBRATE ZOOLOGY] A superfamily of small to very large free-living marine nematodes having pocketlike amphids opening to the exterior via slitlike apertures. { e·nə 'plȯid·ē·ə }

Enoploteuthidae [INVERTEBRATE ZOOLOGY] A molluscan family of deep-sea squids in the class Cephalopoda. { e,näp·lə'tü·thə,dē }

enrichment culture [MICROBIOLOGY] A medium of known composition and specific conditions of incubation which favors the growth of a particular type or species of bacterium. { in'rich·mənt ,kəl·chər }

ensiform [BIOLOGY] Sword-shaped. { 'en·sə ,fȯrm }

Entamoeba [INVERTEBRATE ZOOLOGY] A genus of parasite amebas in the family Endamoebidae, including some species of the genus *Endamoeba* which are parasites of humans and other vertebrates. { ¦ent·ə'mē·bə }

Enteletacea [PALEONTOLOGY] A group of extinct articulate brachiopods in the order Orthida. { ,en·tə·lə'tās·ē·ə }

Entelodontidae [PALEONTOLOGY] A family of extinct palaeodont artiodactyls in the superfamily Entelodontoidea. { ,en·tə·lə'dän·tə,dē }

Entelodontoidea [PALEONTOLOGY] A superfamily of extinct piglike mammals in the suborder Palaeodonta having huge skulls and enlarged incisors. { ,en·tə·lə,dän'tȯid·ē·ə }

enteric bacilli [MICROBIOLOGY] Microorganisms, especially the gram-negative rods, found in the intestinal tract of humans and animals. { en 'ter·ik bə'sil·ī }

Enterobacter [MICROBIOLOGY] A genus of bacteria in the family Enterobacteriaceae; motile rods found in the intestine of humans and other animals; some strains are encapsulated. { ,ent·ə·rō 'bak·tər }

Enterobacteriaceae [MICROBIOLOGY] A family of gram-negative, facultatively anaerobic rods; cells are nonsporeforming and may be nonmotile or motile with peritrichous flagella; includes important human and plant pathogens. { ,ent·ə·rō ,bak·tir·ē'ās·ē,ē }

enterocoel [ZOOLOGY] A coelom that arises by mesodermal outpocketing of the archenteron. { 'ent·ə·rō,sēl }

Enterocoela [SYSTEMATICS] A section of the animal kingdom that includes the Echinodermata, Chaetognatha, Hemichordata, and Chordata. { ,ent·ə·rō'sēl·ə }

enterohydrocoel [INVERTEBRATE ZOOLOGY] In crinoids, an anterior cavity derived from the archenteron. { ,ent·ə·rō'hī·drə,sēl }

enterokinase [BIOCHEMISTRY] An enzyme which catalyzes the conversion of trypsinogen to trypsin. { ,ent·ə·rō'kī,nās }

enteron [ANATOMY] The alimentary canal. { 'ent·ə,rän }

Enteropneusta [INVERTEBRATE ZOOLOGY] The acorn worms or tongue worms, a class of the Hemichordata; free-living solitary animals with no exoskeleton and with numerous gill slits and a straight gut. { ,ent·ə·rə'nüs·tə }

enterotoxin [MICROBIOLOGY] A toxin produced by *Micrococcus pyogenes* var. *aureus* (*Staphylococcus aureus*) which gives rise to symptoms of food poisoning in humans and monkeys. { ,ent·ə·rō 'täk·sən }

enterovirus [VIROLOGY] One of the two subgroups of human picornaviruses; includes the polioviruses, the coxsackieviruses, and the echoviruses. { ,ent·ə·rō'vī·rəs }

Enterozoa [ZOOLOGY] Animals with a digestive tract or cavity; includes all animals except Protozoa, Mesozoa, and Parazoa. { ,ent·ə·rə'zō·ə }

entire [BIOLOGY] Having a continuous, unimpaired margin. { en'tīr }

Entner-Doudoroff pathway [BIOCHEMISTRY] A sequence of reactions for glucose degradation, with the liberation of energy; the distinguishing feature is the formation of 2-keto-3-deoxy-6-phosphogluconate from 6-phosphogluconate and the cleaving of this compound to yield pyruvate and glyceraldehyde-3-phosphate. { ¦ent·nər ¦dō·də,rȯf ,path,wā }

entoblast [EMBRYOLOGY] A blastomere that differentiates into endoderm. { 'ent·ə,blast }

Entodiniomorphida [INVERTEBRATE ZOOLOGY] An order of highly evolved ciliated protozoans in the subclass Spirotrichia, characterized by a smooth, firm pellicle and the lack of external ciliature. { ,ent·ə,dī·nē·ə'mȯr·fə·də }

entolecithal [INVERTEBRATE ZOOLOGY] *See* endolecithal. { ,ent·ə'les·ə·thəl }

Entomoconchacea [PALEONTOLOGY] A superfamily of extinct marine ostracods in the suborder Myodocopa that are without a rostrum above the permanent aperture. { ,ent·ə·mō,käŋ'kās·ē·ə }

entomogenous [BIOLOGY] Growing on or in an insect body, as certain fungi. { ,ent·ə'mäj·ə·nəs }

entomology [INVERTEBRATE ZOOLOGY] A branch of the biological sciences that deals with the study of insects. { ,ent·ə'mäl·ə·jē }

entomophagous [ZOOLOGY] Feeding on insects. { ,ent·ə'mäf·ə·gəs }

entomophilic fungi [MYCOLOGY] Species of fungi that are insect pathogens. { ,ent·ə·mō'fil·ik 'fən,jī }

entomophilous [ECOLOGY] Pollinated by insects. { ,ent·ə'mäf·ə·ləs }

Entomophthoraceae [MYCOLOGY] The single family of the order Entomophthorales. { ,ent·ə ,mäf·thə'rās·ē,ē }

Entomophthorales [MYCOLOGY] An order of mainly terrestrial fungi in the class Phycomycetes having a hyphal thallus and nonmotile sporangiospores, or conidia. { ,ent·ə,mäf·thə 'rā lēz }

Entomostraca [INVERTEBRATE ZOOLOGY] A group of Crustacea comprising the orders Cephalocarida, Branchiopoda, Ostracoda, Copepoda, Branchiura, and Cirripedia. { ,ent·ə'mä·strə·kə }

Entoniscidae [INVERTEBRATE ZOOLOGY] A family of isopod crustaceans in the tribe Bopyrina that are parasitic in the visceral cavity of crabs and porcellanids. { ,ent·ə'nis·ə,dē }

entoplastron [VERTEBRATE ZOOLOGY] The anterior median bony plate of the plastron of chelonians. { ¦en·tō¦plas,trän }

Entoprocta [INVERTEBRATE ZOOLOGY] A group of

bryozoans, sometimes considered to be a subphylum, having a pseudocoelomate visceral cavity and the anus opening inside the circlet of tentacles. { ˌent·ə'präk·tə }

entoderm [EMBRYOLOGY] *See* endoderm. { 'ent·ə ˌdərm }

entry site [MOLECULAR BIOLOGY] The ribosome site available for initial binding of transfer ribonucleic acid during genetic translation. { 'en·trē ˌsīt }

entypy [EMBRYOLOGY] The formation of the amnion in certain mammals by the invagination of the embryonic knob into the yolk sac, without the formation of any amniotic folds. { 'ent·ə·pē }

enucleate [CELL BIOLOGY] To remove the nucleus from a cell. { ē'nü·klēˌāt }

envelope [CELL BIOLOGY] The sum of all cell-surface elements that are located outside the plasma membrane. [VIROLOGY] The outer membranous lipoprotein coat of certain viruses. Also known as bulb. { 'en·vəˌlōp }

environment [ECOLOGY] The sum of all external conditions and influences affecting the development and life of organisms. { in'vī·ərn·mənt *or* in'vī·rən·ment }

environmental biology [BIOLOGY] *See* ecology. { in¦vī·ərn¦ment·əl bī'äl·ə·jē }

environmental resistance [ECOLOGY] The effect of physical and biological factors in preventing a species from reproducing at its maximum rate. { inˌvī·ərn¦men·təl ri'zis·təns }

environmental variance [GENETICS] That portion of the phenotypic variance caused by differences in the environments to which the individuals in a population have been exposed. { in¦vī·ərn ˌment·əl 'ver·ē·əns }

enzyme [BIOCHEMISTRY] Any of a group of catalytic proteins that are produced by living cells and that mediate and promote the chemical processes of life without themselves being altered or destroyed. { 'enˌzīm }

enzyme induction [MICROBIOLOGY] The process by which a microbial cell synthesizes an enzyme in response to the presence of a substrate or of a substance closely related to a substrate in the medium. { 'enˌzīm in'dək·shən }

enzyme inhibition [BIOCHEMISTRY] Prevention of an enzymic process as a result of the interaction of some substance with the enzyme so as to decrease the rate of reaction. { 'enˌzīm ˌin·ə·'bish·ən }

enzyme repression [BIOCHEMISTRY] The process by which the rate of synthesis of an enzyme is reduced in the presence of a metabolite, often the end product of a chain of reactions in which the enzyme in question operates near the beginning. { 'enˌzīm ri'presh·ən }

enzyme unit [BIOCHEMISTRY] The amount of an enzyme that will catalyze the transformation of 10^{-6} mole of substrate per minute or, when more than one bond of each substrate is attacked, 10^{-6} of 1 gram equivalent of the group concerned, under specified conditions of temperature, substrate concentration, and pH number. { 'enˌzīm ˌyü·nət }

enzymology [BIOCHEMISTRY] A branch of science dealing with the chemical nature, biological activity, and biological significance of enzymes. { ˌen·zə'mäl·ə·jē }

Eocanthocephala [INVERTEBRATE ZOOLOGY] An order of the Acanthocephala characterized by the presence of a small number of giant subcuticular nuclei. { ¦ē·ō¦kan·thō'sef·ə·lə }

Eocrinoidea [PALEONTOLOGY] A class of extinct echinoderms in the subphylum Crinozoa that had biserial brachioles like those of cystoids combined with a theca like that of crinoids. { ˌē·ō·krə'nȯid·ē·ə }

Eohippus [PALEONTOLOGY] The earliest, primitive horse, included in the genus *Hyracotherium*; described as a small, four-toed species. { ˌē·ō 'hip·əs }

Eomoropidae [PALEONTOLOGY] A family of extinct perissodactyl mammals in the superfamily Chalicotherioidea. { ˌē·ō·mə'räp·əˌdē }

Eosentomidae [INVERTEBRATE ZOOLOGY] A family of primitive wingless insects in the order Protura that possess spiracles and tracheae. { ¦ē·ō ˌsen¦täm·əˌdē }

eosinophil [HISTOLOGY] A granular leukocyte having cytoplasmic granules that stain with acid dyes and a nucleus with two lobes connected by a thin thread of chromatin. { ˌē·ə'sin·əˌfil }

eosinophil chemotactic factor [IMMUNOLOGY] A peptide released from mast cell granules that stimulates chemotaxis of eosinophils; may be responsible for accumulation of eosinophils at sites of inflammation and allergic reactions. Abbreviated ECF. { ˌē·ə¦sin·ə·fil ¦kē·mō¦tak·tik 'fak·tər }

eosinophilic erythroblast [HISTOLOGY] *See* normoblast. { ¦ē·ə¦sin·ə¦fil·ik ə'rith·rəˌblast }

Eosuchia [PALEONTOLOGY] The oldest, most primitive, and only extinct order of lepidosaurian reptiles. { ˌē·ō'sü·kē·ə }

Epacridaceae [BOTANY] A family of dicotyledonous plants in the order Ericales, distinguished by palmately veined leaves, and stamens equal in number with the corolla lobes. { ˌep·ə·krə'dās·ēˌē }

epaulette [INVERTEBRATE ZOOLOGY] **1.** Any of the branched or knobbed processes on the oral arms of many Scyphozoa. **2.** The first haired scale at the base of the costal vein in Diptera. { ¦ep·ə ¦let }

epaxial [BIOLOGY] Above or dorsal to an axis. { e 'pak·sē·əl }

epaxial muscle [ANATOMY] Any of the dorsal trunk muscles of vertebrates. { e¦pak·sē·əl ¦məs·əl }

ependyma [HISTOLOGY] The layer of epithelial cells lining the cavities of the brain and spinal cord. Also known as ependymal layer. { e'pen·də·mə }

ependymal layer [HISTOLOGY] *See* ependyma. { e'pen·də·məl ˌlā·ər }

Eperythrozoon [MICROBIOLOGY] A genus of the family Anaplasmataceae; rings and coccoids occur on erythrocytes and in the plasma of various vertebrates. { ˌep·əˌrith·rə'zō·ən }

ephapse [ANATOMY] A contact point between neurons. { e'faps }
ephaptic transmission [PHYSIOLOGY] Electrical transfer of activity to a postephaptic unit by the action current of a preephaptic cell. { e'fap·tik tranz'mish·ən }
Ephedra [BOTANY] A genus of low, leafless, green-stemmed shrubs belonging to the order Ephedrales; source of the drug ephedrine. { ə 'fed·rə }
Ephedrales [BOTANY] A monogeneric order of gymnosperms in the subdivision Gneticae. { ˌe·fə'drā·lēz }
ephemeral plant [BOTANY] An annual plant that completes its life cycle in one short moist season; desert plants are examples. { ə'fem·ə·rəl 'plant }
Ephemerida [INVERTEBRATE ZOOLOGY] An equivalent name for the Ephemeroptera. { ˌe·fə'mer·ə·də }
Ephemeroptera [INVERTEBRATE ZOOLOGY] The mayflies, an order of exopterygote insects in the subclass Pterygota. { əˌfem·ə'räp·tə·rə }
Ephydridae [INVERTEBRATE ZOOLOGY] The shore flies, a family of myodarian cyclorrhaphous dipteran insects in the subsection Acalypteratae. { ə'fid·rəˌdē }
ephyra [INVERTEBRATE ZOOLOGY] A larval, free-swimming medusoid stage of scyphozoans; arises from the scyphistoma by transverse fission. Also known as ephyrula. { 'e·fə·rə }
ephyrula [INVERTEBRATE ZOOLOGY] *See* ephyra. { e'fir·ə·lə }
epiandrum [INVERTEBRATE ZOOLOGY] The genital orifice of a male arachnid. { ˌep·ē'an·drəm }
epibasidium [MYCOLOGY] A lengthening of the upper part of each cell of the basidium of various heterobasidiomycetes. { ˌep·ə·bə'sid·ē·əm }
epibiotic [ECOLOGY] Living, usually parasitically, on the surface of plants or animals; used especially of fungi. { ˌep·ə·bī'äd·ik }
epiblem [BOTANY] A tissue that replaces the epidermis in most roots and in stems of submerged aquatic plants. { 'ep·əˌblem }
epiboly [EMBRYOLOGY] The growing or extending of one part, such as the upper hemisphere of a blastula, over and around another part, such as the lower hemisphere, in embryogenesis. { ə 'pib·ə·lē }
epibranchial [ANATOMY] Of or pertaining to the segment below the pharyngobranchial region in a branchial arch. { ¦ep·ə'bräŋ·kē·əl }
epicalyx [BOTANY] A ring of fused bracts below the calyx forming a structure that resembles the calyx. { ¦ep·ə'kā·liks }
epicardium [ANATOMY] The inner, serous portion of the pericardium that is in contact with the heart. [INVERTEBRATE ZOOLOGY] A tubular prolongation of the branchial sac in certain ascidians which takes part in the process of budding. { ˌep·ə'kärd·ē·əm }
Epicaridea [INVERTEBRATE ZOOLOGY] A suborder of the Isopoda whose members are parasitic on various marine crustaceans. { ˌep·ə·kə'rid·ē·ə }
epicarp [BOTANY] The outer layer of the pericarp. Also known as exocarp. { 'ep·əˌkärp }
epichordal [VERTEBRATE ZOOLOGY] Located upon or above the notochord. { ¦ep·ə'kȯrd·əl }
epichordal brain [EMBRYOLOGY] The area of origin of the hindbrain or rhombencephalon, located on the dorsal side of the notochord. Also known as deutencephalon. { ¦ep·ə'kȯrd·əl 'brān }
epicnemial [ANATOMY] Of or pertaining to the anterior portion of the tibia. { ¦ep·ək¦nē·mē·əl }
epicondyle [ANATOMY] An eminence on the condyle of a bone. { ¦ep·ə'känˌdīl }
epicone [INVERTEBRATE ZOOLOGY] The part anterior to the equatorial groove in a dinoflagellate. { 'ep·əˌkōn }
epicotyl [BOTANY] The embryonic plant stem above the cotyledons. { ˌep·ə'käd·əl }
epicranium [INVERTEBRATE ZOOLOGY] The dorsal wall of an insect head. [VERTEBRATE ZOOLOGY] The structures covering the cranium in vertebrates. { ¦ep·ə'krā·nē·əm }
epicuticle [INVERTEBRATE ZOOLOGY] The outer, waxy layer of an insect cuticle or exoskeleton. { ¦ep·ə'kyüd·i·kəl }
epicyst [INVERTEBRATE ZOOLOGY] The outer layer of a cyst wall in encysted protozoans. Also known as ectocyst. { 'ep·əˌsist }
epidermal ridge [ANATOMY] Any of the minute corrugations of the skin on the palmar and plantar surfaces of humans and other primates. { ˌep·ə'dər·məl 'rij }
epidermis [BOTANY] The outermost layer (sometimes several layers) of cells on the primary plant body. [HISTOLOGY] The outer nonsensitive, nonvascular portion of the skin comprising two strata of cells, the stratum corneum and the stratum germinativum. { ˌep·ə'dər·məs }
epididymis [ANATOMY] The convoluted efferent duct lying posterior to the testis and connected to it by the efferent ductules of the testis. { ˌep·ə'did·ə·məs }
epidural [ANATOMY] Located on or over the dura mater. { ¦ep·ə¦dür·əl }
epifauna [ZOOLOGY] Benthic fauna that live on a surface, such as the sea floor, other organisms, or objects. { 'ep·əˌfȯn·ə }
epigaster [EMBRYOLOGY] The portion of the intestine in vertebrate embryos which gives rise to the colon. { 'ep·əˌgas·tər }
epigastric region [ANATOMY] The upper and middle part of the abdominal surface between the two hypochondriac regions. Also known as epigastrium. { ¦ep·ə¦gas·trik ˌrē·jən }
epigastrium [ANATOMY] *See* epigastric region. [INVERTEBRATE ZOOLOGY] The ventral side of mesothorax and metathorax in insects. { ˌep·ə 'gas·trē·əm }
epigean [BOTANY] Pertaining to a plant or plant part that grows above the ground surface. [ZOOLOGY] Living near or on the ground surface, applied especially to insects. { ¦ep·ə¦jē·ən }
epigenesis [EMBRYOLOGY] Development in gradual stages of differentiation. { ˌep·ə'jen·ə·səs }
epigenetics [GENETICS] The study of those processes by which genetic information ultimately results in distinctive physical and behavioral characteristics. { ¦ep·ə·jə¦ned·iks }

epigenotype [GENETICS] The total developmental system through which the adult form of an organism is realized, comprising the interactions among genes and between genes and the nongenetic environment. { ˌep·ə'jēn·əˌtīp }

epigenous [BOTANY] Developing or growing on a surface, especially of a plant or plant part. { ə'pij·ə·nəs }

epiglottis [ANATOMY] A flap of elastic cartilage covered by mucous membrane that protects the glottis during swallowing. { ˌep·ə'gläd·əs }

epigynous [BOTANY] Having the perianth and stamens attached near the top of the ovary; that is, the ovary is inferior. { ə'pij·ə·nəs }

epigynum [INVERTEBRATE ZOOLOGY] **1.** The genital pore of female arachnids. **2.** The plate covering this opening. { ə'pij·ə·nəm }

epilemma [HISTOLOGY] The perineurium of very small nerves. { ˌep·ə'lem·ə }

epimerase [BIOCHEMISTRY] A type of enzyme that catalyzes the rearrangement of hydroxyl groups on a substrate. { ə'pim·əˌrās }

epimere [ANATOMY] The dorsal muscle plate of the lining of a coelomic cavity. [EMBRYOLOGY] The dorsal part of a mesodermal segment in the embryo of chordates. { 'ep·əˌmir }

epimeron [INVERTEBRATE ZOOLOGY] **1.** The posterior plate of the pleuron in insects. **2.** The portion of a somite between the tergum and the insertion of a limb in arthropods. { ˌep·ə'mirˌän }

epimorphosis [PHYSIOLOGY] Regeneration in which cell proliferation precedes differentiation. { ˌep·ə'mȯr·fə·səs }

epimyocardium [EMBRYOLOGY] The layer of the embryonic heart from which both the myocardium and epicardium develop. { ˌep·əˌmī·ə'kärd·ē·əm }

epimysium [ANATOMY] The connective-tissue sheath surrounding a skeletal muscle. { ˌep·ə'mī·sē·əm }

epinasty [BOTANY] Growth changes in which the upper surface of a leaf grows, thus bending the leaf downward. { 'ep·əˌnas·tē }

epinephrine [BIOCHEMISTRY] $C_9H_{13}O_3N$ A hormone secreted by the adrenal medulla that acts to increase blood pressure due to stimulation of heart action and constriction of peripheral blood vessels. Also known as adrenaline. { ˌep·ə'ne·frən }

epineural [ANATOMY] **1.** Arising from the neural arch. **2.** Any process arising from the neural arch. [INVERTEBRATE ZOOLOGY] The nervous tissue dorsal to the ventral nerve cord in arthropods. { ˌep·ə'nu̇r·əl }

epineural canal [INVERTEBRATE ZOOLOGY] A canal that runs between the radial nerve and the epithelium in echinoids and ophiuroids. { ˌep·ə'nu̇r·əl kə'nal }

epineurium [ANATOMY] The connective-tissue sheath of a nerve trunk. { ˌep·ə'nu̇r·ē·əm }

epipetalous [BOTANY] Having stamens located on the corolla. { ¦ep·ə'ped·əl·əs }

epiphallus [INVERTEBRATE ZOOLOGY] A sclerite in some orthopterans in the floor of the genital chamber. { ¦ep·ə'fal·əs }

epipharynx [INVERTEBRATE ZOOLOGY] An organ attached beneath the labrium of many insects. { ¦ep·ə'far·iŋks }

epiphragm [BOTANY] A membrane covering the aperture of the capsule in certain mosses. [INVERTEBRATE ZOOLOGY] A membranous or calcareous partition that covers the aperture of certain hibernating land snails. { 'ep·əˌfram }

epiphyll [ECOLOGY] A plant that grows on the surface of leaves. { 'ep·əˌfil }

epiphyseal arch [EMBRYOLOGY] The arched structure in the third ventricle of the embryonic brain, which marks the site of development of the pineal body. { ə¦pif·ə¦sē·əl 'ärch }

epiphyseal plate [ANATOMY] **1.** The broad, articular surface on each end of a vertebral centrum. **2.** The thin layer of cartilage between the epiphysis and the shaft of a long bone. Also known as metaphysis. { ə¦pif·ə¦sē·əl 'plāt }

epiphysis [ANATOMY] **1.** The end portion of a long bone in vertebrates. **2.** *See* pineal body. { ə'pif·ə·səs }

epiphyte [ECOLOGY] A plant which grows nonparasitically on another plant or on some nonliving structure, such as a building or telephone pole, deriving moisture and nutrients from the air. Also known as aerophyte. { 'ep·əˌfīt }

epiplankton [BIOLOGY] Plankton occurring in the sea from the surface to a depth of about 100 fathoms (180 meters). { ¦ep·ə'plaŋk·tən }

epipleural [ANATOMY] Arising from a rib. [VERTEBRATE ZOOLOGY] An intramuscular bone arising from and extending between some of the ribs in certain fishes. { ¦ep·ə'plu̇r·əl }

epiploic foramen [ANATOMY] An aperture of the peritoneal cavity, formed by folds of the peritoneum and located between the liver and the stomach. Also known as foramen of Winslow. { ¦ep·ə¦plō·ik fə'rā·mən }

epipodite [INVERTEBRATE ZOOLOGY] A branch of the basal joint of the protopodite of thoracic limbs of many arthropods. { ə'pip·əˌdīt }

epipodium [BOTANY] The apical portion of an embryonic phyllopodium. [INVERTEBRATE ZOOLOGY] **1.** A ridge or fold on the lateral edges of each side of the foot of certain gastropod mollusks. **2.** The elevated ring on an ambulacral plate in Echinoidea. { ˌep·ə'pōd·ē·əm }

Epipolasina [INVERTEBRATE ZOOLOGY] A suborder of sponges in the order Clavaxinellida having radially arranged monactinal or diactinal megascleres. { ˌep·ə·pə'laz·ə·nə }

epiproct [INVERTEBRATE ZOOLOGY] A plate above the anus forming the dorsal part of the tenth or eleventh somite of certain insects. { 'ep·əˌpräkt }

epipubis [VERTEBRATE ZOOLOGY] A single cartilage or bone located in front of the pubis in some vertebrates, particularly in some amphibians. { ¦ep·ə'pyü·bəs }

episclera [ANATOMY] The loose connective tissue lying between the conjunctiva and the sclera. { ¦ep·ə'skler·ə }

episepalous [BOTANY] Having stamens growing on or adnate to the sepals. { ¦ep·ə'sep·ə·ləs }

episome [GENETICS] A circular genetic element

in bacteria, presumably a deoxyribonucleic acid fragment, which is not necessary for survival of the organism and which can be integrated in the bacterial chromosome or remain free. { 'ep·ə ˌsōm }

episperm [BOTANY] *See* testa. { 'ep·əˌspərm }

epistasis [GENETICS] The suppression of the effect of one gene by another. { ə'pis·tə·səs }

episternum [VERTEBRATE ZOOLOGY] A dermal bone or pair of bones ventral to the sternum of certain fishes and reptiles. { ¦ep·ə'stər·nəm }

epistome [INVERTEBRATE ZOOLOGY] **1.** The area between the mouth and the second antennae in crustaceans. **2.** The plate covering this region. **3.** The area between the labrum and the epicranium in many insects. **4.** A flap covering the mouth of certain bryozoans. **5.** The area just above the labrum in certain dipterans. { 'ep·ə ˌstōm }

epithalamus [ANATOMY] A division of the vertebrate diencephalon including the habenula, the pineal body, and the posterior commissure. { ¦ep·ə'thal·ə·məs }

epitheca [INVERTEBRATE ZOOLOGY] **1.** An external, calcareous layer around the basal portion of the theca of many corals. **2.** A protective covering of the epicone. **3.** The outer portion of a diatom frustule. { ¦ep·ə'thē·kə }

epitheliochorial placenta [EMBRYOLOGY] A type of placenta in which the maternal epithelium and fetal epithelium are in contact. Also known as villous placenta. { ˌep·ə¦thē·lē·ō'kȯr·ē·əl plə 'sen·tə }

epithelioid cell [HISTOLOGY] A macrophage that resembles an epithelial cell. Also known as alveolated cell. { ˌep·ə'thē·lēˌȯid ˌsel }

epitheliomuscular cell [INVERTEBRATE ZOOLOGY] An epithelial cell with an elongate base that contains contractile fibrils; common among cnidarians. { ˌep·ə¦thē·lē·ō'məs·kyə· lər 'sel }

epithelium [HISTOLOGY] A primary animal tissue, distinguished by cells being close together with little intercellular substance; covers free surfaces and lines body cavities and ducts. { ˌep·ə'thē· lē·əm }

epithema [VERTEBRATE ZOOLOGY] A horny outgrowth on the beak of certain birds. { ˌep·ə 'thē·mə }

epitoke [INVERTEBRATE ZOOLOGY] The posterior portion of marine polychaetes; contains the gonads. { 'ep·əˌtōk }

epitoky [INVERTEBRATE ZOOLOGY] In certain polychaetes, development of the posterior sexual part from the anterior sexless part. { 'ep·əˌtäk·ē }

epitope [IMMUNOLOGY] The portion of the antigen molecule that determines its capacity to combine with the specific combining site of its corresponding antibody in an antigen-antibody interaction. { 'ep·əˌtōp }

epitrichium [EMBRYOLOGY] The outer layer of the fetal epidermis of many mammals. { ¦ep·ə'trik· ē·əm }

epitrochlear [ANATOMY] Of or pertaining to a lymph node that lies above the trochlea of the elbow joint. { ¦ep·ə'trō·klē·ər }

epitympanum [ANATOMY] The attic of the middle ear, or tympanic cavity. { ¦ep·ə'tim·pə·nəm }

epivalve [INVERTEBRATE ZOOLOGY] **1.** The upper or apical shell of certain dinoflagellates. **2.** The upper shell of a diatom. { 'ep·əˌvalv }

epixylous [ECOLOGY] Growing on wood; used especially of fungi. { ¦ep·ə¦zī·ləs }

epizoic [BIOLOGY] Living on the body of an animal. { ¦ep·ə¦zō·ik }

eponychium [ANATOMY] The horny layer of the nail fold attached to the nail plate at its margin; represents the remnant of the embryonic condition. [EMBRYOLOGY] A horny condition of the epidermis from the second to the eighth month of fetal life, indicating the position of the future nail. { ˌep·ə'nik·ē·əm }

epoophoron [ANATOMY] A blind longitudinal duct and 10-15 transverse ductules in the mesosalpinx near the ovary which represent remnants of the reproductive part of the mesonephros in the female; homolog of the head of the epididymis in the male. Also known as parovarium; Rosenmueller's organ. { ¦ep·ō'äf·əˌrän }

Epsilonematoidea [INVERTEBRATE ZOOLOGY] A superfamily of small (0.5-millimeter) marine nematodes in the order Desmodorida; the body is strongly arched in a sigmoid manner when relaxed. { ¦ep·siˌlän·ə·mə'tȯid·ē·ə }

Epstein-Barr virus [VIROLOGY] Herpeslike virus particles first identified in cultures of cells from Burkett's malignant lymphoma. { ¦ep·stīn ¦bär ˌvī·rəs }

equatorial plane [CELL BIOLOGY] The plane in a cell undergoing mitosis that is midway between the centrosomes and perpendicular to the spindle fibers. { ˌe·kwə'tȯr·ē·əl 'plān }

Equidae [VERTEBRATE ZOOLOGY] A family of perissodactyl mammals in the superfamily Equoidea, including the horses, zebras, and donkeys. { 'ēk·wəˌdē }

equilibrium population [EVOLUTION] A population in which the gene frequencies have reached an equilibrium between mutation pressure and selection pressure. { ˌē·kwə¦lib·rē·əm ˌpäp·yə ¦lā·shən }

equine [VERTEBRATE ZOOLOGY] **1.** Resembling a horse. **2.** Of or related to the Equidae. { 'ē ˌkwīn }

Equisetales [BOTANY] The horsetails, a monogeneric order of the class Equisetopsida; the only living genus is *Equisetum*. { ˌe·kwə·sə'tā· lēz }

Equisetineae [BOTANY] The equivalent name for the Equisetophyta. { ˌe·kwə·sə'tin·ēˌē }

Equisetophyta [BOTANY] A division of the subkingdom Embryobionta represented by a single living genus, *Equisetum*. { ˌe·kwə·sə'täf·ə·də }

Equisetopsida [BOTANY] A class of the division Equisetophyta whose members made up a major part of the flora, especially in moist or swampy places, during the Carboniferous Period. { ˌe· kwə·sə'täp·sə·də }

equitant [BOTANY] Of leaves, overlapping transversely at the base. { 'e·kwəd·ənt }

equivalence zone [IMMUNOLOGY] *See* zone of optimal proportion. { i'kwiv·ə·ləns ˌzōn }

Equoidea [VERTEBRATE ZOOLOGY] A superfamily of perissodactyl mammals in the suborder Hippomorpha comprising the living and extinct horses and their relatives. { ē'kwȯid·ē·ə }

Equus [VERTEBRATE ZOOLOGY] The genus comprising the large, one-toed modern horses, including donkeys and zebras. { 'e·kwəs }

ER [CELL BIOLOGY] *See* endoplasmic reticulum.

erection [PHYSIOLOGY] The enlarged state of erectile tissue when engorged with blood, as of the penis or clitoris. { i'rek·shən }

erector [PHYSIOLOGY] Any muscle that produces erection of a part. { i'rek·tər }

erect stem [BOTANY] A stem that stands, having a vertical or upright habit. { i'rekt 'stem }

Eremascoideae [BOTANY] A monogeneric subfamily of ascosporogenous yeasts characterized by mostly septate mycelia, and spherical asci with eight oval to round ascospores. { er·ə·mə 'skȯid·ē,ē }

Erethizontidae [VERTEBRATE ZOOLOGY] The New World porcupines, a family of rodents characterized by sharply pointed, erectile hairs and four functional digits. { ,er·ə·thə'zänt·ə,dē }

Ergasilidae [INVERTEBRATE ZOOLOGY] A family of copepod crustaceans in the suborder Cyclopoida in which the females are parasitic on aquatic animals, while the males are free-swimming. { ər·gə'sil·ə,dē }

ergastic [CELL BIOLOGY] Pertaining to the nonliving components of protoplasm. { ər'gas·tik }

ergastoplasm [CELL BIOLOGY] A cytoplasm component which shows an affinity for basic dyes; a form of the endoplasmic reticulum. { ,ər'gas·tə ,plaz·əm }

ergosterin [BIOCHEMISTRY] *See* ergosterol. { ər 'gäs·tə·rən }

ergosterol [BIOCHEMISTRY] $C_{28}H_{44}O$ A crystalline, water-insoluble, unsaturated sterol found in ergot, yeast, and other fungi, and which may be converted to vitamin D_2 on irradiation with ultraviolet light or activation with electrons. Also known as ergosterin. { ər'gäs·tə,rȯl }

ergot [MYCOLOGY] The dark purple or black sclerotium of the fungus *Claviceps purpurea*. { 'ər·gət }

Ericaceae [BOTANY] A large family of dicotyledonous plants in the order Ericales distinguished by having twice as many stamens as corolla lobes. { ,er·ə'kās·ē,ē }

Ericales [BOTANY] An order of dicotyledonous plants in the subclass Dilleniidae; plants are generally sympetalous with unitegmic ovules and they have twice as many stamens as petals. { ,er·ə'kā·lēz }

ericophyte [ECOLOGY] A plant that grows on a heath or moor. { 'er·ək·ə,fīt }

Erinaceidae [VERTEBRATE ZOOLOGY] The hedgehogs, a family of mammals in the order Insectivora characterized by dorsal and lateral body spines. { ,er·ə·nə'sē·ə,dē }

Erinnidae [INVERTEBRATE ZOOLOGY] A family of orthorrhaphous dipteran insects in the series Brachycera. { ə'rin·ə,dē }

Eriocaulaceae [BOTANY] The single family of the order Eriocaulales. { ,er·ē·ō,kȯ'lās·ē,ē }

Eriocaulales [BOTANY] An order of monocotyledonous plants in the subclass Commelinidae, having a perianth reduced or lacking and having unisexual flowers aggregated on a long peduncle. { ,er·ē·ō,kȯ'lā·lēz }

Eriococcinae [INVERTEBRATE ZOOLOGY] A family of homopteran insects in the superfamily Coccoidea; adult females and late instar nymphs have an anal ring. { ,er·ē·ō'käk·sə,nē }

Eriocraniidae [INVERTEBRATE ZOOLOGY] A small family of lepidopteran insects in the superfamily Eriocranioidea. { ,er·ē·ō·krə'nī·ə,dē }

Eriocranioidea [INVERTEBRATE ZOOLOGY] A superfamily of lepidopteran insects in the suborder Homoneura comprising tiny moths with reduced, untoothed mandibles. { ,er·ē·ō,krā·nē 'ȯid·ē·ə }

Eriophyidae [INVERTEBRATE ZOOLOGY] The bud mites or gall mites, a family of economically important plant-feeding mites in the suborder Trombidiformes. { ,er·ē·ō'fī·ə,dē }

eriophyllous [BOTANY] Having leaves covered by a cottony pubescence. { er·ē'äf·ə·ləs }

erose [BIOLOGY] Having an irregular margin. { ē 'rōs }

Erotylidae [INVERTEBRATE ZOOLOGY] The pleasing fungus beetles, a family of coleopteran insects in the superfamily Cucujoidea. { ,er·ə'tī·lə ,dē }

Errantia [INVERTEBRATE ZOOLOGY] A group of 34 families of polychaete annelids in which the anterior region is exposed and the linear body is often long and is dorsoventrally flattened. { ə 'ran·chə }

eruciform [INVERTEBRATE ZOOLOGY] In certain insect larvae, having a soft cylindrical body with a well-defined head and usually short thoracic legs. { ə'rüs·ə,fȯrm }

Erwinia [MICROBIOLOGY] A genus of motile, rod-shaped bacteria in the family Enterobacteriaceae; these organisms invade living plant tissues and cause dry necroses, galls, wilts, and soft rots. { ər'win·ē·ə }

Erwinieae [MICROBIOLOGY] Formerly a tribe of phytopathogenic bacteria in the family Enterobacteriaceae, including the single genus *Erwinia*. { ər'win·ē,ē }

Erysipelothrix [MICROBIOLOGY] A genus of gram-positive, rod-shaped bacteria of uncertain affiliation; cells have a tendency to form long filaments. { ,er·ə'sip·ə·lō,thriks }

Erysiphaceae [MYCOLOGY] The powdery mildews, a family of ascomycetous fungi in the order Erysiphales with light-colored mycelia and conidia. { ,er·ə·sə'fās·ē,ē }

Erysiphales [MYCOLOGY] An order of ascomycetous fungi which are obligate parasites of seed plants, causing powdery mildew and sooty mold. { ,er·ə·sə'fā·lēz }

erythroblast [HISTOLOGY] A nucleated cell occurring in bone marrow as the earliest recognizable cell of the erythrocytic series. { ə'rith·rə,blast }

erythrocruorin [BIOCHEMISTRY] Any of the iron-porphyrin protein respiratory pigments found in the blood and tissue fluids of certain invertebrates; corresponds to hemoglobin in vertebrates. { ə‚rith·rə'krü·ə·rən }

erythrocyte [HISTOLOGY] A type of blood cell that contains a nucleus in all vertebrates but humans and that has hemoglobin in the cytoplasm. Also known as red blood cell. { ə'rith·rə‚sīt }

erythrocytopoiesis [PHYSIOLOGY] *See* erythropoiesis. { ə‚rith·rə‚sīd·ə‚pöi'ē·səs }

erythromycin [MICROBIOLOGY] A crystalline antibiotic produced by *Streptomyces erythreus* and used in the treatment of gram-positive bacterial infections. { ə‚rith·rə'mīs·ən }

D-erythropentose [BIOCHEMISTRY] *See* ribulose. { ¦dē ə‚rith·rə'pen‚tōs }

erythrophilous [BIOLOGY] Having an affinity for red dyes and other coloring matter. { ¦er·ə¦thräf·ə·ləs }

erythrophore [ZOOLOGY] A chromatophore containing a red pigment, especially a carotenoid. { ə'rith·rə‚fȯr }

erythropoiesis [PHYSIOLOGY] The process by which erythrocytes are formed. Also known as erythrocytopoiesis. { ə‚rith·rə‚pöi'ē·səs }

erythropoietin [BIOCHEMISTRY] A hormone, thought to be produced by the kidneys, that regulates erythropoiesis, at least in higher vertebrates. { ə‚rith·rə'pöi·ət·ən }

Erythroxylaceae [BOTANY] A homogeneous family of dicotyledonous woody plants in the order Linales characterized by petals that are internally appendiculate, three carpels, and flowers without a disk. { ‚er·ə‚thräk·sə'lās·ē‚ē }

erythrulose [BIOCHEMISTRY] $C_4H_8O_4$ A ketose sugar occurring as an oxidation product of erythritol due to the action of certain bacteria. { ə'rith·rə‚lōs }

Escherichia [MICROBIOLOGY] A genus of bacteria in the family Enterobacteriaceae; straight rods occurring singly or in pairs. { ‚esh·ə'rik·ē·ə }

Escherichia coli [MICROBIOLOGY] The type species of the genus, occurring as part of the normal intestinal flora in vertebrates. Also known as colon bacillus. { ‚esh·ə'rik·ē·ə 'kō‚lī }

Escherichieae [MICROBIOLOGY] Formerly a tribe of bacteria in the family Enterobacteriaceae defined by the ability to ferment lactose, with the rapid production of acid and visible gas. { ‚esh·ə·rə‚kī·ē‚ē }

Esocidae [VERTEBRATE ZOOLOGY] The pikes, a family of fishes in the order Clupeiformes characterized by an elongated beaklike snout and sharp teeth. { ə'säs·ə‚dē }

Esocoidei [VERTEBRATE ZOOLOGY] A small suborder of fresh-water fishes in the order Salmoniformes; includes the pikes, mudminnows, and pickerels. { ‚es·ə'kȯid·ē‚ī }

esophageal gland [ANATOMY] Any of the digestive glands within the submucosa of the esophagus; secretions are chiefly mucus and serve to lubricate the esophagus. { ə¦säf·ə¦jē·əl 'gland }

esophageal hiatus [ANATOMY] The opening in the diaphragm for passage of the esophagus. { ə¦säf·ə¦jē·əl hī'ād·əs }

esophageal teeth [VERTEBRATE ZOOLOGY] The enamel-tipped hypapophyses of the posterior cervical vertebrae of certain snakes, which penetrate the esophagus and function to break eggshells. { ə¦säf·ə¦jē·əl 'tēth }

esophagus [ANATOMY] The tubular portion of the alimentary canal interposed between the pharynx and the stomach. Also known as gullet. { ə'säf·ə·gəs }

essential amino acid [BIOCHEMISTRY] Any of eight of the 20 naturally occurring amino acids that are indispensable for optimum animal growth but cannot be formed in the body and must be supplied in the diet. { i'sen·chəl ə'mē·nō ‚as·əd }

essential fatty acid [BIOCHEMISTRY] Any of the polyunsaturated fatty acids which are required in the diet of mammals; they are probably precursors of prostaglandins. { i'sen·chəl 'fad·ē ‚as·əd }

esterase [BIOCHEMISTRY] Any of a group of enzymes that catalyze the synthesis and hydrolysis of esters. { 'es·tə‚rās }

esthacyte [INVERTEBRATE ZOOLOGY] A simple sensory cell occurring in certain lower animals, such as sponges. Also spelled aesthacyte. { 'es·thə‚sīt }

esthesia [PHYSIOLOGY] The capacity for sensation, perception, or feeling. Also spelled aesthesia. { es'thē·zhə }

estivation [PHYSIOLOGY] **1.** The adaptation of certain animals to the conditions of summer, or the taking on of certain modifications, which enables them to survive a hot, dry summer. **2.** The dormant condition of an organism during the summer. { ‚es·tə'vā·shən }

estradiol [BIOCHEMISTRY] $C_{18}H_{24}O_2$ An estrogenic hormone produced by follicle cells of the vertebrate ovary; provokes estrus and proliferation of the human endometrium, and stimulates ICSH (interstitial-cell-stimulating hormone) secretion. { ‚es·trə'dī‚ȯl }

estriol [BIOCHEMISTRY] $C_{18}H_{24}O_3$ A crystalline estrogenic hormone obtained from human pregnancy urine. { 'e‚strī‚ȯl }

estrogen [BIOCHEMISTRY] Any of various natural or synthetic substances possessing the biologic activity of estrus-producing hormones. { 'es·trə·jən }

estrogenic hormone [BIOCHEMISTRY] A hormone, found principally in ovaries and also in the placenta, which stimulates the accessory sex structures and the secondary sex characteristics in the female. { ¦es·trə¦jen·ik 'hȯr‚mōn }

estrone [BIOCHEMISTRY] $C_{18}H_{22}O_2$ An estrogenic hormone produced by follicle cells of the vertebrate ovary; functions the same as estradiol. { 'e‚strōn }

estrous cycle [PHYSIOLOGY] The physiological changes that take place between periods of estrus in the female mammal. { 'es·trəs ‚sī·kəl }

estrus [PHYSIOLOGY] The period in female mammals during which ovulation occurs and the animal is receptive to mating. { 'es·trəs }

ethionine [BIOCHEMISTRY] $C_5H_{13}O_2N$ An amino acid that is the ethyl analog of and the biological antagonist of methionine. { e'thī·ə‚nēn }

Ethiopian zoogeographic region [ECOLOGY] A

geographic unit of faunal homogeneity including all of Africa south of the Sahara. { ˌē·thē′ō·pē·ən ¦zō·ōˌjē·ə·¦graf·ik ′rē·jən }

ethmoid bone [ANATOMY] An irregularly shaped cartilage bone of the skull, forming the medial wall of each orbit and part of the roof and lateral walls of the nasal cavities. { ′ethˌmȯid ˌbōn }

ethmoturbinate [ANATOMY] Of or pertaining to the masses of ethmoid bone which form the lateral and superior portions of the turbinate bones in mammals. { ¦eth·mō′tər·bəˌnāt }

ethogram [ECOLOGY] An extensive list, inventory, or description of the behavior of an organism. { ′ē·thəˌgram }

ethological isolation [BIOLOGY] *See* behavioral isolation. { ˌē·thə′läj·ə·kəl ī·sə′lā·shən }

ethology [VERTEBRATE ZOOLOGY] The study of animal behavior in a natural context. { ē′thäl·ə·jē }

ethyl chlorophyllide [BIOCHEMISTRY] A compound formed by replacing the phytyl ($C_{20}H_{39}$) tail of the chlorophyll molecule with a short ethyl (C_2H_5) tail; crystallizes easily and has an absorption spectrum and electrochemical properties similar to those of chlorophyll. { ¦eth·əl ˌklȯr·ō′filˌīd }

etioallocholane [BIOCHEMISTRY] *See* androstane. { ˌēd·ē·ōˌal·ə′kōˌlān }

etioblast [BOTANY] An immature chloroplast, containing prolamellar bodies. { ′ēd·ē·ōˌblast }

etiolation [BOTANY] The yellowing or whitening of green plant parts grown in darkness. { ˌed·ē·ə′lā·shən }

etioplast [BOTANY] The plastid of a dark-grown plant that contains crystalline prolamellar bodies. { ′ēd·ē·ōˌplast }

Eubacteriales [MICROBIOLOGY] Formerly an order of the class Schizomycetes; considered the true bacteria and characterized by simple, undifferentiated, rigid cells of either spherical or straight, rod-shaped form. { ˌyü·bakˌtir·ē′ā·lēz }

Eubacterium [MICROBIOLOGY] A genus of bacteria in the family Propionibacteriaceae; obligate anaerobes producing a mixture of organic acids (butyric, acetic, formic, and lactic) from carbohydrates and peptone. { ˌyü·bak′tir·ē·əm }

Eubasidiomycetes [MYCOLOGY] An equivalent name for Homobasidiomycetidae. { ˌyü·bəˌsid·ē·ōˌmī′sēd·ēz }

Eubrya [BOTANY] A subclass of the mosses (Bryopsida); the leafy gametophytes arise from buds on the protonemata, which are nearly always filamentous or branched green threads attached to the substratum by rhizoids. { yü′brī·ə }

Eubryales [BOTANY] An order of mosses (Bryatae); plants have the sporophyte at the end of a stem, vary in size from small to robust, and generally grow in tufts. { ¦yüˌbrī′ā·lēz }

Eucalyptus [BOTANY] A large genus of evergreen trees belonging to the myrtle family (Myrtaceae) and occurring in Australia and New Guinea. { ˌyü·kə′lip·təs }

Eucarida [INVERTEBRATE ZOOLOGY] A large superorder of the decapod crustaceans, subclass Malacostraca, including shrimps, lobsters, hermit crabs, and crabs; characterized by having the shell and thoracic segments fused dorsally and the eyes on movable stalks. { ˌyü′kar·ə·də }

Eucaryota [BIOLOGY] Primitive, unicellular organisms having a well-defined nuclear membrane, chromosomes, and mitotic cell division. { yü·kar·ē′ōd·ə }

eucaryote [BIOLOGY] *See* eukaryote. { yü′kar·ēˌōt }

Eucestoda [INVERTEBRATE ZOOLOGY] The true tapeworms, a subclass of the class Cestoda. { yü′ses·tə·də }

Eucharitidae [INVERTEBRATE ZOOLOGY] A family of hymenopteran insects in the superfamily Chalcidoidea. { yü·kə′rid·əˌdē }

euchromatin [CELL BIOLOGY] The portion of the chromosomes that stains with low intensity, uncoils during interphase, and condenses during cell division. { yü′krō·mə·tən }

Eucinetidae [INVERTEBRATE ZOOLOGY] The plate thigh beetles, a family of coleopteran insects in the superfamily Dascilloidea. { yü·sə′ned·əˌdē }

Euclasterida [INVERTEBRATE ZOOLOGY] An order of asteroid echinoderms in which the arms are sharply distinguished from a small, central disk-shaped body. { ˌyü·klə′ster·ə·də }

Eucleidae [INVERTEBRATE ZOOLOGY] The slug moths, a family of lepidopteran insects in the suborder Heteroneura. { yü′klē·əˌdē }

Euclymeninae [INVERTEBRATE ZOOLOGY] A subfamily of annelids in the family Maldonidae of the Sedentaria, having well-developed plaques and an anal pore within the plaque. { ˌyü·klə′men·əˌnē }

Eucnemidae [INVERTEBRATE ZOOLOGY] The false click beetles, a family of coleopteran insects in the superfamily Elateroidea. { yük′nem·əˌdē }

Eucoccida [INVERTEBRATE ZOOLOGY] An order of parasitic protozoans in the subclass Coccidia characterized by alternating sexual and asexual phases; stages of the life cycle occur intracellularly in vertebrates and invertebrates. { yü′käk·sə·də }

Eucoelomata [ZOOLOGY] A large sector of the animal kingdom including all forms in which there is a true coelom or body cavity; includes all phyla above Aschelminthes. { yüˌsē·lə′mäd·ə }

Eucommiales [BOTANY] A monotypic order of dicotyledonous plants in the subclass Hamamelidae; plants have two, unitegmic ovules and lack stipules. { yüˌkäm·ē′ā·lēz }

Eudactylinidae [INVERTEBRATE ZOOLOGY] A family of parasitic copepod crustaceans in the suborder Caligoida; found as ectoparasites on the gills of sharks. { yüˌdak·tə′lin·əˌdē }

eudoxid [INVERTEBRATE ZOOLOGY] *See* eudoxome. { yü′däkˌsīd }

eudoxome [INVERTEBRATE ZOOLOGY] Cormidium of most calycophoran siphonophores which lead a free existence. Also known as eudoxid. { yü′däkˌsōm }

Euechinoidea [INVERTEBRATE ZOOLOGY] A subclass of echinoderms in the class Echinoidea;

distinguished by the relative stability of ambulacra and interambulacra. { yü,ek·ə'nȯid·ē·ə }

eugenics [GENETICS] The use of practices that influence the hereditary qualities of future generations, with the aim of improving the genetic future of humanity. { yü'jen·iks }

Euglena [BIOLOGY] A genus of organisms with one or two flagella, chromatophores in most species, and a generally elongate, spindle-shaped body; classified as algae by botanists (Euglenophyta) and as protozoans by zoologists (Euglenida). { yü'glē·nə }

Euglenida [INVERTEBRATE ZOOLOGY] An order of protozoans in the class Phytamastigophorea, including the largest green, noncolonial flagellates. { yü'glen·ə·də }

Euglenidae [INVERTEBRATE ZOOLOGY] The antlike leaf beetles, a family of coleopteran insects in the superfamily Tenebrionoidea. { yü'glen·ə ,dē }

Euglenoidina [INVERTEBRATE ZOOLOGY] The equivalent name for Euglenida. { ,yü·glə,nȯi 'dī·nə }

Euglenophyceae [BOTANY] The single class of the plant division Euglenophyta. { yü,glē·nə 'fīs·ē,ē }

Euglenophyta [BOTANY] A division of the plant kingdom including one-celled, chiefly aquatic flagellate organisms having a spindle-shaped or flattened body, naked or with a pellicle. { ,yü·glə 'näf·əd·ə }

euglobulin [BIOCHEMISTRY] True globulin; a simple protein that is soluble in distilled water and dilute salt solutions. { yü'gläb·yə·lən }

Eugregarinida [INVERTEBRATE ZOOLOGY] An order of protozoans in the subclass Gregarinia; parasites of certain invertebrates. { ,yü·grə,gar·ə 'nīd·ə }

Eukaryotae [BIOLOGY] A superkingdom that includes living and fossil organisms comprising all taxonomic groups above the primitive unicellular prokaryotic level. { yü,kar·ē'ō,tē }

eukaryote [BIOLOGY] A cell with a definitive nucleus. Also spelled eucaryote. { yü'kar·ē,ōt }

Eulamellibranchia [INVERTEBRATE ZOOLOGY] The largest subclass of the molluscan class Bivalvia, having a heterodont shell hinge, leaflike gills, and well-developed siphons. { ¦yü·lə,mel·ə 'braŋ·kē·ə }

Eulophidae [INVERTEBRATE ZOOLOGY] A family of hymenopteran insects in the superfamily Chalcidoidea including species that are parasitic on the larvae of other insects. { yü'läf·ə,dē }

Eumalacostraca [INVERTEBRATE ZOOLOGY] A series of the class Crustacea comprising shrimplike crustaceans having eight thoracic segments, six abdominal segments, and a telson. { yü,mal·ə 'käs·trə·kə }

Eumetazoa [ZOOLOGY] A section of the animal kingdom that includes the phyla above the Porifera; contains those animals which have tissues or show some tissue formation and organ systems. { yü,med·ə'zō·ə }

eumitosis [CELL BIOLOGY] Typical mitosis. { ¦yü ,mī'tō·səs }

Eumycetes [MYCOLOGY] The true fungi, a large group of microorganisms characterized by cell walls, lack of chlorophyll, and mycelia in most species; includes the unicellular yeasts. { ,yü,mī 'sēd·ēz }

Eumycetozoida [INVERTEBRATE ZOOLOGY] An order of protozoans in the subclass Mycetozoia; includes slime molds which form a plasmodium. { ¦yü,mī,sed·ə'zȯid·ə }

Eumycophyta [MYCOLOGY] An equivalent name for the Eumycetes. { ,yü,mī'käf·əd·ə }

Eumycota [MYCOLOGY] An equivalent name for Eumycetes. { ,yü,mī'kōd·ə }

Eunicea [INVERTEBRATE ZOOLOGY] A superfamily of polychaete annelids belonging to the Errantia. { yü'nis·ē·ə }

Eunicidae [INVERTEBRATE ZOOLOGY] A family of polychaete annelids in the superfamily Eunicea having characteristic pharyngeal armature consisting of maxillae and mandibles. { yü'nis·əd·ē }

Euomphalacea [PALEONTOLOGY] A superfamily of extinct gastropod mollusks in the order Aspidobranchia characterized by shells with low spires, some approaching bivalve symmetry. { yü ,äm·fə'lās·ē·ə }

Eupelmidae [INVERTEBRATE ZOOLOGY] A family of hymenopteran insects in the superfamily Chalcidoidea. { yü'pel·mə,dē }

Euphausiacea [INVERTEBRATE ZOOLOGY] An order of planktonic malacostracans in the class Crustacea possessing photophores which emit a brilliant blue-green light. { yü,fȯ·zē'ās·ē·ə }

euphenics [GENETICS] The production of a satisfactory phenotype by means other than eugenics. { yü'fen·iks }

Eupheterochlorina [INVERTEBRATE ZOOLOGY] A suborder of flagellate protozoans in the order Heterochlorida. { yü¦fed·ə·rō,klə'rī·nə }

Euphorbiaceae [BOTANY] A family of dicotyledonous plants in the order Euphorbiales characterized by dehiscent fruit having more than one seed and by epitropous ovules. { yü,fȯr·bē 'ās·ē,ē }

Euphorbiales [BOTANY] An order of dicotyledonous plants in the subclass Rosidae having simple leaves and unisexual flowers that are aggregated and reduced. { yü,fȯr·bē'ā·lēz }

Euphrosinidae [INVERTEBRATE ZOOLOGY] A family of amphinomorphan polychaete annelids with short, dorsolaterally flattened bodies. { yü·frə 'zin·ə,dē }

Euplexoptera [INVERTEBRATE ZOOLOGY] The equivalent name for Dermaptera. { yü,plek'säp·tə·rə }

euploid [GENETICS] Having a chromosome complement that is an exact multiple of the haploid complement. { 'yü,plȯid }

eupnea [PHYSIOLOGY] Normal or easy respiration rhythm. { 'yüp·nē·ə }

Eupodidae [INVERTEBRATE ZOOLOGY] A family of mites in the suborder Trombidiformes. { yü 'päd·ə,dē }

Euproopacea [PALEONTOLOGY] A group of Paleozoic horseshoe crabs belonging to the Limulida. { yü,prō·ə'pās·ē·ə }

European boreal faunal region [ECOLOGY] A

zoogeographic region describing marine littoral faunal regions of the northern Atlantic Ocean between Greenland and the northwestern coast of Europe. { ¦yu̇r·ə¦pē·ən ¦bōr·ē·əl 'fōn·əl 'rē·jən }

Eurotiaceae [MYCOLOGY] A family of ascomycetous fungi of the order Eurotiales in which the asci are borne in cleistothecia or closed fruiting bodies. { yə,rōd·ē'ās·ē,ē }

Eurotiales [MYCOLOGY] An order of fungi in the class Ascomycetes bearing ascospores in globose or broadly oval, delicate asci which lack a pore. { yə,rōd·ē'ā·lēz }

Euryalae [INVERTEBRATE ZOOLOGY] The basket fishes, a family of echinoderms in the subclass Ophiuroidea. { yə'rī·ə,lē }

Euryalina [INVERTEBRATE ZOOLOGY] A suborder of ophiuroid echinoderms in the order Phrynophiurida characterized by a leathery integument. { yu̇r·ē'a·lə·nə }

Euryapsida [PALEONTOLOGY] A subclass of fossil reptiles distinguished by an upper temporal opening on each side of the skull. { yu̇r·ē'ap·sə·də }

eurybathic [ECOLOGY] Living at the bottom of a body of water. { ¦yu̇r·ə¦bath·ik }

Eurychilinidae [PALEONTOLOGY] A family of extinct dimorphic ostracods in the superfamily Hollinacea. { ,yu̇r·ə·kə'lin·ə,dē }

eurygamous [INVERTEBRATE ZOOLOGY] Mating in flight, as in many insect species. { yu̇'rig·ə·məs }

euryhaline [ECOLOGY] Pertaining to the ability of marine organisms to tolerate a wide range of saline conditions, and therefore a wide variation of osmotic pressure, in the environment. { ¦yu̇r·ə¦ha,līn }

Eurylaimi [VERTEBRATE ZOOLOGY] A monofamilial suborder of suboscine birds in the order Passeriformes. { ,yu̇r·ə'lā,mī }

Eurylaimidae [VERTEBRATE ZOOLOGY] The broadbills, the single family of the avian suborder Eurylaimi. { ,yu̇r·ə'lā·mə,dē }

Eurymylidae [PALEONTOLOGY] A family of extinct mammals presumed to be the ancestral stock of the order Lagomorpha. { ,yu̇r·ə'mil·ə,dē }

euryon [ANATOMY] One of the two lateral points functioning as end points to measure the greatest transverse diameter of the skull. { 'yu̇r·ē ,än }

Euryphoridae [INVERTEBRATE ZOOLOGY] A family of copepod crustaceans in the order Caligoida; members are fish ectoparasites. { ,yu̇r·ə'fōr·ə ,dē }

euryplastic [BIOLOGY] Referring to an organism with a marked ability to change and adapt to a wide spectrum of environmental conditions. { ¦yu̇r·ə¦plas·tik }

Eurypterida [PALEONTOLOGY] A group of extinct aquatic arthropods in the subphylum Chelicerata having elongate-lanceolate bodies encased in a chitinous exoskeleton. { ,yu̇r·əp'ter·ə·də }

Eurypygidae [VERTEBRATE ZOOLOGY] The sun bitterns, a family of tropical and subtropical New World birds belonging to the order Gruiformes. { ,yu̇r·ə'pij·ə,dē }

eurypylous [INVERTEBRATE ZOOLOGY] Having a wide opening; applied to sponges with wide apopyles opening directly into excurrent canals, and wide prosopyles opening directly from incurrent canals. { ¦yu̇r·ə¦pī·ləs }

eurytherm [BIOLOGY] An organism that is tolerant of a wide range of temperatures. { 'yu̇r·ə ,thərm }

Eurytomidae [INVERTEBRATE ZOOLOGY] The seed and stem chalcids, a family of hymenopteran insects in the superfamily Chalcidoidea. { ,yu̇r·ə 'täm·ə,dē }

eurytopic [ECOLOGY] Referring to organisms which are widely distributed. { ,yu̇r·ə'täp·ik }

Eusiridae [INVERTEBRATE ZOOLOGY] A family of pelagic amphipod crustaceans in the suborder Gammaridea. { yü'sir·ə,dē }

eusocial [ZOOLOGY] Pertaining to animal societies, such as those of certain insects, in which sterile individuals work on behalf of reproductive individuals. { ,yü'sō·shəl }

eusporangiate [BOTANY] Having sporogenous tissue derived from a group of epidermal cells. { ¦yü·spə¦ran·jē,āt }

eustachian tube [ANATOMY] A tube composed of bone and cartilage that connects the nasopharynx with the middle ear cavity. { yü'stā·shən ,tüb }

eustachian valve [ANATOMY] *See* canal valve. { yü 'stā·shən ,valv }

eustele [BOTANY] A modified siphonostele containing collateral or bicollateral vascular bundles; found in most gymnosperm and angiosperm stems. { yü'stēl }

eusternum [INVERTEBRATE ZOOLOGY] The anterior sternal plate in insects. { yü'stər·nəm }

Eustigmatophyceae [BOTANY] A small class of mostly nonmotile, photosynthetic, unicellular algae in the division Chromophycota, characterized by the unique organization of motile cells, photosynthetic pigments including chlorophyll *a*, beta-carotene, and violaxanthin, and a single parietal yellow-green chloroplast; live chiefly in fresh water but also in marine and soil habitats. { ,yüs·tig,ma·də'fī·sē·ē }

Eusuchia [VERTEBRATE ZOOLOGY] The modern crocodiles, a suborder of the order Crocodylia characterized by a fully developed secondary palate and procoelous vertebrae. { yü'sü·kē·ə }

Eusyllinae [INVERTEBRATE ZOOLOGY] A subfamily of polychaete annelids in the family Syllidae having a thick body and unsegmented cirri. { yü 'sil·ə,nē }

Eutardigrada [INVERTEBRATE ZOOLOGY] An order of tardigrades which lack both a sensory cephalic appendage and a club-shaped appendage. { yü ,tärd·ə'grād·ə }

eutely [BIOLOGY] Having the body composed of a constant number of cells, as in certain rotifers. { 'yüd·əl·ē }

Euthacanthidae [PALEONTOLOGY] A family of extinct acanthodian fishes in the order Climatiiformes. { ,yü·thə'kan·thə,dē }

euthenics [BIOLOGY] The science that deals with the improvement of the future of humanity by changing the environment. { yü'then·iks }

Eutheria [VERTEBRATE ZOOLOGY] An infraclass of therian mammals including all living forms except the monotremes and marsupials. { yü'thir·ē·ə }

Eutrichosomatidae [INVERTEBRATE ZOOLOGY] Small family of hymenopteran insects in the superfamily Chalcidoidea. { yü¦trik·ə,sō'mad·ə,dē }

eutrophication [ECOLOGY] The process by which a body of water becomes, either by natural means or by pollution, excessively rich in dissolved nutrients, resulting in increased primary productivity that often leads to a seasonal deficiency in dissolved oxygen. { yü·trə·fə'kā·shən }

Evaniidae [INVERTEBRATE ZOOLOGY] The ensign flies, a family of hymenopteran insects in the superfamily Proctotrupoidea. { ,ev·ə'nī·ə,dē }

evaporative heat regulation [PHYSIOLOGY] The composite process by which an animal body is cooled by evaporation of sensible perspiration; this avenue of heat loss serves as a physical means of regulating the body temperature. { i 'vap·ə,rād·iv 'hēt ,reg·yə,lā·shən }

Eventognathi [VERTEBRATE ZOOLOGY] The equivalent name for Cypriniformes. { ,e,ven'täg·nə,thī }

event-related potential [PHYSIOLOGY] Electrical activity produced by the brain in response to a sensory stimulus or associated with the execution of a motor, cognitive, or psychophysiologic task. { i¦vent ri,lād·əd pə'ten·chəl }

everglade [ECOLOGY] A type of wetland in southern Florida usually containing sedges and at least seasonally covered by slowly moving water. { 'ev·ər,glād }

evergreen [BOTANY] Pertaining to a perennially green plant. Also known as aiophyllous. { 'ev·ər ,grēn }

evoked potential [PHYSIOLOGY] Electrical response of any neuron to stimuli. { ē'vōkt pə 'ten·chəl }

evolution [BIOLOGY] The processes of biological and organic change in organisms by which descendants come to differ from their ancestors. { ,ev·ə'lü·shən }

evolutionary distance [EVOLUTION] The number of base substitutions per homologous site that have occurred since the divergence of two deoxyribonucleic acid sequences. { ,ev·ə¦lü·s;a.hə ,ner·ē 'dis·təns }

evolutionary divergence [EVOLUTION] The degree of divergence, at the intra- and interspecific levels, of two or more populations, which presumably have evolved from a common ancestor. { ,ev·ə¦lü·shə,ner·ē də'vər·jəns }

evolutionary force [EVOLUTION] Any factor that brings about changes in gene frequencies or chromosome frequencies in a population and is thus capable of causing evolutionary change. { ,ev·ə¦lü·shə,ner·ē 'fors }

evolutionary plasticity [EVOLUTION] The genetic adaptibility of populations or lines of descent. { ,ev·ə¦lü·shə,ner·ē plas'tis·əd·ē }

evolutionary progress [EVOLUTION] The acquisition of new macromolecular and metabolic processes by which competitive superiority is achieved. { ,ev·ə¦lü·shə,ner ē 'prä,gres }

evolutionary rate [EVOLUTION] The amount of evolutionary change per unit of time. { ,ev·ə¦lü·shə,ner·ē 'rāt }

evolutionary tree [EVOLUTION] **1.** A diagram that portrays the hypothesized genealogical ties and sequence of evolutionary relationships linking individual organisms, populations, or taxa. Also known as phylogenetic tree. **2.** A diagram that depicts the evolutionary relationship of protein or nucleic acid sequences. { ,ev·ə¦lü·shə,ner·ē 'trē }

evolutionary trend [EVOLUTION] Any trend in the evolution of phyletic lines that is a consequence of genotypic cohesion. { ,ev·ə¦lü·shə,ner·ē 'trend }

evolution pressure [EVOLUTION] The result of the combined action of mutation pressure, immigration and hybridization pressure, and selection pressure, giving rise to systematic changes in the gene frequency of a population. { ,ev·ə¦lü·shən ¦presh·ər }

evolvon [EVOLUTION] The operational unit in evolution, assumed to consist of a deoxyribonucleic acid master sequence with a series of redundant sequences that constitute a repository of genetic information. { 'ev·ə,län }

ewe [VERTEBRATE ZOOLOGY] A mature female sheep, goat, or related animal, as the smaller antelopes. { yü }

exalate [BOTANY] Being without winglike appendages. { 'ek·sə,lāt }

exalbuminous [BOTANY] *See* exendospermous. { ¦eks,al'byü·mə·nəs }

exarch [BOTANY] A vascular bundle in which the primary wood is centripetal. { 'ek,särk }

exasperate [BIOLOGY] Having a surface roughened by stiff elevations or bristles. { ig'zas·pə·rət }

exaspidean [VERTEBRATE ZOOLOGY] Of the tarsal envelope of birds, being continuous around the outer edge of the tarsus. { ,eg·zə'spid·ē·ən }

Excipulaceae [MYCOLOGY] The equivalent name for Discellaceae. { ,ek·sə·pə'lās·ē,ē }

excision [GENETICS] Recombination involving removal of a genetic element. { ek'sizh·ən }

excision enzyme [BIOCHEMISTRY] A bacterial enzyme that removes damaged dimers from the deoxyribonucleic acid molecule of a bacterial cell following light or ultraviolet radiation or nitrogen mustard damage. { ek'sizh·ən 'en,zīm }

excitable [BIOLOGY] Referring to a tissue or organism that exhibits irritability. { ek'sīd·ə·bəl }

exclusion principle [ECOLOGY] The principle according to which two species cannot coexist in the same locality if they have identical ecological requirements. { ik'sklü·zhən ,prin·sə·pəl }

exclusive species [ECOLOGY] A species which is completely or nearly limited to one community. { ik'sklü·siv 'spē·shēz }

Excorallanidae [INVERTEBRATE ZOOLOGY] A family of free-living and parasitic isopod crustaceans in the suborder Flabellifera which have mandibles and first maxillae modified as hooklike piercing organs. { ek,skór·ə'lan·ə,dē }

excrement [PHYSIOLOGY] An excreted substance; the feces. { 'ek·skrə·mənt }

excrescence [BIOLOGY] **1.** Abnormal or excessive increase in growth. **2.** An abnormal outgrowth. { ek'skrē·səns }

excretion [PHYSIOLOGY] The removal of unusable or excess material from a cell or a living organism. { ek'skrē·shən }

excretory system [ANATOMY] Those organs concerned with solid, fluid, or gaseous excretion. { 'ek·skrə‚tór·ē ‚sis·təm }

excurrent [BIOLOGY] Flowing out. [BOTANY] **1.** Having an undivided main stem or trunk. **2.** Having the midrib extending beyond the apex. { eks'kə·rənt }

exendospermous [BOTANY] Lacking endosperm. Also known as exalbuminous. { eks¦en·də¦spər·məs }

exergonic [BIOCHEMISTRY] Of or pertaining to a biochemical reaction in which the end products possess less free energy than the starting materials; usually associated with catabolism. { ‚ek·sər'gän·ik }

exhalation [PHYSIOLOGY] The giving off or sending forth in the form of vapor; expiration. { ‚eks·ə'lā·shən }

exine [BOTANY] *See* exosporium. { 'ek‚sēn }

exite [INVERTEBRATE ZOOLOGY] A movable appendage or lobe located on the external side of the limb of a generalized arthropod. { 'ek‚sīt }

exobiology [BIOLOGY] The search for and study of extraterrestrial life. { ¦ek·sō·bī'äl·ə·jē }

exocarp [BOTANY] *See* epicarp. { 'ek·sō‚kärp }

exoccipital [ANATOMY] Lying to the side of the foramen magnum, as the exoccipital bone. { ¦eks·äk'sip·əd·əl }

exochorion [INVERTEBRATE ZOOLOGY] The outer of two layers forming the covering of an insect egg. { ¦ek·sō'kór·ē‚än }

exocoel [INVERTEBRATE ZOOLOGY] The space between pairs of adjacent mesenteries in anthozoan polyps. { 'ek·sə‚sēl }

Exocoetidae [VERTEBRATE ZOOLOGY] The halfbeaks, a family of actinopterygian fishes in the order Atheriniformes. { ‚ek·sə'sēd·ə‚dē }

exocrine gland [PHYSIOLOGY] A structure whose secretion is passed directly or by ducts to its exterior surface, or to another surface which is continuous with the external surface of the gland. { 'ek·sə·krən ‚gland }

exocuticle [INVERTEBRATE ZOOLOGY] The middle layer of the cuticle of insects. { ¦ek·sō'kyüd·ə·kəl }

exocytosis [CELL BIOLOGY] The extrusion of material from a cell. { ¦ek·sō·sī'tō·səs }

exodermis [BOTANY] *See* hypodermis. { ‚ek·sō'dər·məs }

exoenzyme [BIOCHEMISTRY] An enzyme that functions outside the cell in which it was synthesized. { ¦ek·sō¦wen‚zīm }

exogamy [GENETICS] Union of gametes from organisms that are not closely related. Also known as outbreeding. { ek'säg·ə·mē }

exogastrula [EMBRYOLOGY] An abnormal gastrula that is unable to undergo invagination or further development because of a quantitative increase of presumptive endoderm. { ¦ek·sō'gas·trə·lə }

exogenote [GENETICS] The genetic fragment transferred from the donor to the recipient cell during the process of recombination in bacteria. { ‚ek'säj·ə‚nōt }

exogenous [BIOLOGY] **1.** Due to an external cause; not arising within the organism. **2.** Growing by addition to the outer surfaces. [PHYSIOLOGY] Pertaining to those factors in the metabolism of nitrogenous substances obtained from food. { ‚ek'säj·ə·nəs }

exognathite [INVERTEBRATE ZOOLOGY] The external branch of an oral appendage of a crustacean. { ek'säg·nə‚thīt }

Exogoninae [INVERTEBRATE ZOOLOGY] A subfamily of polychaete annelids in the family Syllidae having a short, small body of few segments. { ¦ek·sō'gä·nə‚nē }

exogynous [BOTANY] Having the style longer than and exserted beyond the corolla. { ek'säj·ə·nəs }

exon [GENETICS] That portion of deoxyribonucleic acid which codes for the final messenger ribonucleic acid. { 'ek‚sän }

exonephric [INVERTEBRATE ZOOLOGY] Having the excretory organs discharge through the body wall. { ¦ek·sō'ne‚frik }

exon shuffling [GENETICS] In eukaryotic split genes, the creation of new genes by the recombination of exons through crossing over. { 'ek ‚sän ‚shəf·liŋ }

exonuclease [BIOCHEMISTRY] Any of a group of enzymes which catalyze hydrolysis of single nucleotide residues from the end of a deoxyribonucleic acid chain. { ¦ek·sō'nü·klē‚ās }

exopeptidase [BIOCHEMISTRY] An enzyme that acts on the terminal peptide bonds of a protein chain. { ¦ek·sō'pep·tə‚dās }

exopodite [INVERTEBRATE ZOOLOGY] The outer branch of a biramous crustacean appendage. { ek'säp·ə‚dīt }

Exopterygota [INVERTEBRATE ZOOLOGY] A division of the insect subclass Pterygota including those insects which undergo a hemimetabolous metamorphosis. { ¦ek‚säp‚ter·ə¦gōd·ə }

exoskeleton [INVERTEBRATE ZOOLOGY] The external supportive covering of certain invertebrates, such as arthropods. [VERTEBRATE ZOOLOGY] Bony or horny epidermal derivatives, such as nails, hoofs, and scales. { ¦ek·sō'skel·ə·tən }

exosmosis [PHYSIOLOGY] Passage of a liquid outward through a cell membrane. { ¦ek·sō·äs'mō·səs }

exospore [MYCOLOGY] An asexual spore formed by abstriction, as in certain Phycomycetes. { 'ek·sō‚spór }

exosporium [BOTANY] The outer of two layers forming the wall of spores such as pollen and bacterial spores. Also known as exine. { ‚ek·sə'spór·ē·əm }

exostome [BOTANY] The opening through the outer integument of a bitegmic ovule. { 'ek·sə‚stōm }

exotheca [INVERTEBRATE ZOOLOGY] The tissue external to the theca of corals. { ˌek·sə'thē·kə }

exotic [ECOLOGY] Not endemic to an area. { ig 'zäd·ik }

exotoxin [MICROBIOLOGY] A toxin that is excreted by a microorganism. { ¦ek·sə¦täk·sən }

expanding population [ECOLOGY] A population containing a large proportion of young individuals. { ik'spand·iŋ ˌpäp·yə'lā·shən }

expectorate [PHYSIOLOGY] To eject phlegm or other material from the throat or lungs. { ik 'spek·təˌrāt }

expiratory reserve volume [PHYSIOLOGY] At the end of a normal expiration, the quantity of air that can be expelled by forcible expiration. { ek 'spī·rəˌtȯr·ē ri¦zərv ˌväl·yəm }

expiratory standstill [PHYSIOLOGY] Suspension of action at the end of expiration. { ek'spī·rə ˌtȯr·ē 'standˌstil }

explant [CELL BIOLOGY] An excised fragment of a tissue or an organ used to start a cell culture. { 'eksˌplant }

explosive evolution [EVOLUTION] Rapid diversification of a group of fossil organisms in a short geological time. { ik'splō·siv ev·ə'lü·shən }

exponential growth [MICROBIOLOGY] The period of bacterial growth during which cells divide at a constant rate. Also known as logarithmic growth. { ˌek·spə'nen·chəl 'grōth }

expression vector [MOLECULAR BIOLOGY] A cloning vector that promotes the expression of foreign gene inserts. { ik'spresh·ən ˌvek·tər }

exserted [BIOLOGY] Protruding beyond the enclosing structure, such as stamens extending beyond the margin of the corolla. { ek'sərd·əd }

exsheath [INVERTEBRATE ZOOLOGY] To escape from the residual membrane of a previous developmental stage, as pertaining to the larva of certain nematodes, microfilaria, and so on. { ek 'shēth }

exstipulate [BOTANY] Lacking stipules. { ek'stip·yəˌlāt }

extension [PHYSIOLOGY] A movement which has the effect of straightening a limb. { ik'sten·chən }

external auditory meatus [ANATOMY] The external passage of the ear, leading to the tympanic membrane in reptiles, birds, and mammals. { ek 'stərn·əl 'ȯd·əˌtȯr·ē mē'ād·əs }

external carotid artery [ANATOMY] An artery which originates at the common carotid and distributes blood to the anterior part of the neck, face, scalp, side of the head, ear, and dura mater. { ek'stərn·əl kə'räd·əd'ärd·ə·rē }

external ear [ANATOMY] The portion of the ear that receives sound waves, including the pinna and external auditory meatus. { ek¦stərn·əl 'ēr }

external fertilization [PHYSIOLOGY] Those processes involved in the union of male and female sex cells outside the body of the female. { ek ¦stərn·əl ˌfərd·əl·ə'zā·shən }

external gill [ZOOLOGY] A gill that is external to the body wall, as in certain larval fishes and amphibians, and in many aquatic insects. { ek ¦stərn·əl 'gil }

external respiration [PHYSIOLOGY] The processes by which oxygen is carried into living cells from the outside environment and by which carbon dioxide is carried in the reverse direction. { ek ¦stərn·əl ˌres·pə'rā·shən }

exteroceptor [PHYSIOLOGY] Any sense receptor at the surface of the body that transmits information about the external environment. { ¦ek·stə·rō¦sep·tər }

extinction [EVOLUTION] The worldwide death and disappearance of a specific organism or group of organisms. { ek'stiŋk·shən }

extirpate [BIOLOGY] To uproot, destroy, make extinct, or exterminate. { 'ek·stərˌpāt }

extracellular [BIOLOGY] Outside the cell. { ¦ek·strə'sel·yə·lər }

extracellular matrix [CELL BIOLOGY] A filamentous structure that is attached to the outer cell surface and provides anchorage, traction, and positional recognition to the cell. [HISTOLOGY] A filamentous structure of glycoproteins and proteoglycans that is attached to the cell surface and provides cells with anchorage, traction for movement, and positional recognition. Abbreviated ECM. { ¦ek·strə¦sel·yə·lər 'mā·triks }

extrachromosomal inheritance [GENETICS] *See* cytoplasmic inheritance. { ¦ek·strəˌkrō·mə'sō·məl in'her·ət·əns }

extraembryonic coelom [EMBRYOLOGY] The cavity in the extraembryonic mesoderm; it is continuous with the embryonic coelom in the region of the umbilicus, and is obliterated by growth of the amnion. { ¦ek·strəˌem·brē'än·ik 'sē·ləm }

extraembryonic membrane [EMBRYOLOGY] *See* fetal membrane. { ¦ek·strəˌem·brē'än·ik 'mem ˌbrān }

extrajunctional receptor [PHYSIOLOGY] An acetylcholine receptor which occurs randomly over a muscle fiber surface outside the area of the neuromuscular junction. { ¦ek·strə¦jənk·shən·əl ri'sep·tər }

extrapyramidal system [ANATOMY] Descending tracts of nerve fibers arising in the cortex and subcortical motor areas of the brain. { ¦ek·strə ˌpir·ə'mid·əl 'sis·təm }

extrinsic factor [BIOCHEMISTRY] *See* vitamin B. { ek¦strinz·ik ¦fak·tər }

extrorse [BIOLOGY] Directed outward or away from the axis of growth. { ek'strȯrs }

extroversion [BIOLOGY] A turning outward. { ¦ek·strə¦vər·zhən }

exumbrella [INVERTEBRATE ZOOLOGY] The outer, convex surface of the umbrella of jellyfishes. { ˌek·səm'brel·ə }

eye [ZOOLOGY] A photoreceptive sense organ that is capable of forming an image in vertebrates and in some invertebrates such as the squids and crayfishes. { ī }

eyeball [ANATOMY] The globe of the eye. { 'ī ˌbȯl }

eyeball potential [PHYSIOLOGY] Very small electrical potentials at the eyeball surface resulting from depolarization of muscles controlling eye position. { 'īˌbȯl pəˌten·chəl }

eyelid [ANATOMY] A movable, protective section

of skin that covers and uncovers the eyeball of many terrestrial animals. { 'ī͵lid }

eye socket [ANATOMY] *See* orbit. { 'ī ͵säk·ət }

eyespot [BOTANY] **1.** A small photosensitive pigment body in certain unicellular algae. **2.** A dark area around the hilum of certain seeds, as some beans. [INVERTEBRATE ZOOLOGY] A simple organ of vision in many invertebrates consisting of pigmented cells overlying a sensory termination. { 'ī͵spät }

eyestalk [INVERTEBRATE ZOOLOGY] A movable peduncle bearing a terminal eye in decapod crustaceans. { 'ī͵stȯk }

F

F_1 [GENETICS] *See* first filial generation.

F_2 [GENETICS] Notation for the progeny produced by intercrossing members of the first filial generation. Also known as second generation.

Fabales [BOTANY] An order of dicotyledonous plants whose members typically have stipulate, compound leaves, ten to many stamens which are often united by the filaments, and a single carpel which gives rise to a legume; many harbor symbiotic nitrogen-fixing bacteria in the roots. { fə'bā·lēz }

Fab region [IMMUNOLOGY] Region of an antibody molecule that contains the antigen binding site; Fab is derived from the term antigen binding fragment. { 'fab ‚rē·jən }

Fabriciinae [INVERTEBRATE ZOOLOGY] A subfamily of small to minute, colonial, sedentary polychaete annelids in the family Sabellidae. { ‚fa·brə'sī·ə‚nē }

face [ANATOMY] The anterior portion of the head, including the forehead and jaws. { fās }

facet [ANATOMY] A small plane surface, especially on a bone or a hard body; may be produced by wear, as a worn spot on the surface of a tooth. [INVERTEBRATE ZOOLOGY] The surface of a simple eye in the compound eye of arthropods and certain other invertebrates. { 'fas·ət }

facial artery [ANATOMY] The external branch of the external carotid artery. { ¦fā·shəl 'ärd·ə·rē }

facial bone [ANATOMY] The bone comprising the nose and jaws, formed by the maxilla, zygoma, nasal, lacrimal, palatine, inferior nasal concha, vomer, mandible, and parts of the ethmoid and sphenoid. { 'fā·shəl ‚bōn }

facial nerve [ANATOMY] The seventh cranial nerve in vertebrates; a paired composite nerve, with motor elements supplying muscles of facial expression and with sensory fibers from the taste buds of the anterior two-thirds of the tongue and from other sensory endings in the anterior part of the throat. { 'fā·shəl ‚nərv }

facies [ANATOMY] Characteristic appearance of the face in association with a disease or abnormality. [ECOLOGY] The makeup or appearance of a community or species population. { 'fā·shēz }

facilitated transport [PHYSIOLOGY] The transport of certain materials across a cell membrane, down a concentration gradient, assisted by enzymelike carrier proteins embedded in the membrane and without the explicit provision of energy. { fə'sil·ə‚tād·əd 'trans‚pȯrt }

factor II [BIOCHEMISTRY] *See* prothrombin. { 'fak·tər 'tü }

factor III [BIOCHEMISTRY] *See* thromboplastin. { 'fak·tər 'thrē }

factor IV [BIOCHEMISTRY] Calcium ions involved in the mechanism of blood coagulation. { 'fak·tər 'fōr }

factor V [BIOCHEMISTRY] *See* proaccelerin. { 'fak·tər 'fīv }

factor VII [BIOCHEMISTRY] A procoagulant, related to prothrombin, that is involved in the formation of a prothrombin-converting principle which transforms prothrombin to thrombin. Also known as stable factor. { 'fak·tər 'se·vən }

factor VIII [BIOCHEMISTRY] *See* antihemophilic factor. { 'fak·tər 'āt }

factor IX [BIOCHEMISTRY] *See* Christmas factor. { 'fak·tər 'nīn }

factor X [BIOCHEMISTRY] *See* Stuart factor. { 'fak·tər 'ten }

factor XI [BIOCHEMISTRY] A procoagulant present in normal blood but deficient in hemophiliacs. Also known as plasma thromboplastin antecedent (PTA). { 'fak·tər ə'le·vən }

factor XII [BIOCHEMISTRY] A blood clotting factor effective experimentally only in vitro; deficient in hemophiliacs. Also known as Hageman factor. { 'fak·tər 'twelv }

facultative aerobe [MICROBIOLOGY] An anaerobic microorganism which can grow under aerobic conditions. { 'fa·kəl‚tād·iv 'er‚ōb }

facultative anaerobe [MICROBIOLOGY] A microorganism that grows equally well under aerobic and anaerobic conditions. { 'fak·əl‚tād·iv 'an·ə‚rōb }

facultative heterochromatin [GENETICS] Chromosomal material that may alternate in form between euchromatin and heterochromatin. { 'fak·əl‚tād·iv ¦ hed·ə‚rō'krō·mə·tən }

facultative parasite [ECOLOGY] An organism that can exist independently but may be parasitic on certain occasions, such as the flea. { 'fak·əl‚tād·iv 'par·ə‚sīt }

facultative photoheterotroph [MICROBIOLOGY] Any bacterium that usually grows anaerobically in light but can also grow aerobically in the dark. { 'fak·əl‚tād·iv ¦fōd·ō¦hed·ə·rə‚träf }

FAD [BIOCHEMISTRY] *See* flavin adenine dinucleotide.

Fagaceae [BOTANY] A family of dicotyledonous plants in the order Fagales characterized by stipulate leaves, seeds without endosperm, female flowers generally not in catkins, and mostly three styles and locules. { fə'gās·ē,ē }

Fagales [BOTANY] An order of dicotyledonous woody plants in the subclass Hamamelidae having simple leaves and much reduced, mostly unisexual flowers. { fə'gā·lēz }

Fahrenholz's rule [ECOLOGY] The rule that in groups of permanent parasites the classification of the parasites usually corresponds directly to the natural relationships of the hosts. { 'fär·ən ,hōlt·səz ,rül }

falcate [BIOLOGY] Shaped like a sickle. { 'fal ,kat }

falciform [BIOLOGY] Sickle-shaped. { 'fal·sə ,fȯrm }

falciform ligament [ANATOMY] The ventral mesentery of the liver; its peripheral attachment extends from the diaphragm to the umbilicus and contains the round ligament of the liver. { 'fal·sə,fȯrm 'lig·ə·mənt }

falciger [INVERTEBRATE ZOOLOGY] Seta with a distally blunt and curved tip. { 'fal·sə·gər }

falcon [VERTEBRATE ZOOLOGY] Any of the highly specialized diurnal birds of prey composing the family Falconidae; these birds have been captured and trained for hunting. { 'fal·kən }

Falconidae [VERTEBRATE ZOOLOGY] The falcons, a family consisting of long-winged predacious birds in the order Falconiformes. { fal'kän·ə ,dē }

Falconiformes [VERTEBRATE ZOOLOGY] An order of birds containing the diurnal birds of prey, including falcons, hawks, vultures, and eagles. { fal ,kän·ə'fōr·mēz }

falculate [ZOOLOGY] Curved and with a sharp point. { 'fal· kyə,lāt }

Fallopian tube [ANATOMY] Either of the paired oviducts that extend from the ovary to the uterus for conduction of the ovum in mammals. { fə 'lō·pē·ən 'tüb }

false ligament [ANATOMY] Any peritoneal fold which is not a true supporting ligament. { ¦fȯls 'lig·ə·mənt }

false rib [ANATOMY] A rib that is not attached to the sternum directly; any of the five lower ribs on each side in humans. { ¦fȯls 'rib }

false ring [BOTANY] A layer of wood that is less than a full season's growth and often does not form a complete ring. { ¦fȯls 'riŋ }

falx [ANATOMY] A sickle-shaped structure. { falks }

familial [BIOLOGY] Of, pertaining to, or occurring among the members of a family. { fə'mil·yəl }

family [SYSTEMATICS] A taxonomic category based on the grouping of related genera. { 'fam·lē }

fan [BIOLOGY] Any structure, such as a leaf or the tail of a bird, resembling an open fan. { fan }

fang [ANATOMY] The root of a tooth. [VERTEBRATE ZOOLOGY] A long, pointed tooth, especially one of a venomous serpent. { faŋ }

Fantl unit [BIOLOGY] A unit for the standardization of thrombin. { 'fant·əl ,yü·nət }

faradization [PHYSIOLOGY] Use of a faradic current to stimulate muscles and nerves. { ,far·əd·ə'zā·shən }

farinaceous [BIOLOGY] Having a mealy surface covering. { ¦far·ə¦nā·shəs }

Farinales [BOTANY] An order that includes several groups regarded as orders of the Commelinidae in other systems of classification. { ,far·ə 'nā·lēz }

Farinosae [BOTANY] The equivalent name for Farinales. { ,far·ə'nō·sē }

farinose [BIOLOGY] Covered with a white powdery substance. { 'far·ə,nōs }

farnesol [BIOCHEMISTRY] $C_{15}H_{25}OH$ A colorless liquid extracted from oils of plants such as citronella, neroli, cyclamen, and tuberose; it has a delicate floral odor, and is an intermediate step in the biological synthesis of cholesterol from mevalonic acid in vertebrates; used in perfumery. { 'fär·nə,sȯl }

fascia [HISTOLOGY] Layers of areolar connective tissue under the skin and between muscles, nerves, and blood vessels. { 'fā·shə }

fasciate [BOTANY] Having bands or stripes. { 'fa·she,āt }

fascicle [BOTANY] A small bundle, as of fibers or leaves. { fas·i·kəl }

fasciculate [BOTANY] Arranged in tufts or fascicles. { fə'sik·yə·lət }

fasciculation potential [PHYSIOLOGY] An action potential which is quantitatively comparable to that of a motor unit and which represents spontaneous contraction of a bundle of muscle fibers. { fə,sik·yə'lā·shən pə,ten·chəl }

fasciculus [ANATOMY] A bundle or tract of nerve, muscle, or tendon fibers isolated by a sheath of connective tissues and having common origins, innervation, and functions. { fə'sik·yə·ləs }

Fasciola hepatica [INVERTEBRATE ZOOLOGY] A digenetic trematode which parasitizes sheep, cattle, and occasionally humans. { fə·'sē·ə·lə he 'pad·ə·kə }

fasciole [INVERTEBRATE ZOOLOGY] A band of cilia on the test of certain sea urchins. { 'fas·ē,ōl }

Fasciolopsis buski [INVERTEBRATE ZOOLOGY] A large, fleshy trematode, native to eastern Asia and the southwestern Pacific, which parasitizes humans. { fə,sē·ə'läp·səs 'bəs·kē }

fastigiate [BOTANY] **1.** Having erect branches that are close to the stem. **2.** Becoming narrower at the top. [ZOOLOGY] Arranged in a conical bundle. { fa'stij·e·āt }

fat [ANATOMY] Pertaining to an obese person. [BIOCHEMISTRY] Any of the glyceryl esters of fatty acids which form a class of neutral organic compounds. [PHYSIOLOGY] The chief component of fat cells and other animal and plant tissues. { fat }

fat body [INVERTEBRATE ZOOLOGY] A nutritional reservoir of fatty tissue surrounding the viscera or forming a layer beneath the integument in the immature larval stages of many insects. [VERTEBRATE ZOOLOGY] A mass of adipose tissue attached to each genital gland in amphibians. { 'fat ,bäd·ē }

fat cell [HISTOLOGY] The principal component of

adipose connective tissue; two types are yellow fat cells and brown fat cells. { 'fat ˌsel }

fate map [EMBRYOLOGY] A graphic scheme indicating the definite spatial arrangement of undifferentiated embryonic cells in accordance with their destination to become specific tissues. { 'fāt ˌmap }

fatigue [PHYSIOLOGY] Exhaustion of strength or reduced capacity to respond to stimulation following a period of activity. { fə'tēg }

fat-metabolizing hormone [BIOCHEMISTRY] *See* ketogenic hormone. { 'fat mə¦tab·əˌlīz·iŋ ˌhȯrˌmōn }

fatty acid peroxidase [BIOCHEMISTRY] An enzyme present in germinating plant seeds which catalyzes the oxidation of the carboxyl carbon of fatty acids to carbon dioxide. { ¦fad·ē 'as·əd pə'räk·səˌdās }

fatty acyl carnitine [BIOCHEMISTRY] Transport form of fatty acids which allows them to cross the mitochondrial membrane; formed by reaction of fatty acyl-coenzyme A with carnitine by employing the enzyme carnitine acyltransferase. Also known as acyl carnitine. { 'fad·ē 'as·əl 'kär·nəˌtēn }

fatty acyl-coenzyme A [BIOCHEMISTRY] Activated form of fatty acids formed by the enzyme acyl-coenzyme A synthetase at the expense of adenosinetriphosphate. Also known as acyl-coenzyme A. { 'fad·ē 'as·əl kō¦enˌzīm 'ā }

fatty infiltration [PHYSIOLOGY] Infiltration of an organ or tissue with excessive amounts of fats. { 'fad·ē ˌin·fil'trā·shən }

faucal [BIOLOGY] Of or pertaining to the fauces. [INVERTEBRATE ZOOLOGY] The opening of a spiral shell. { 'fȯ·kəl }

fauces [ANATOMY] The passage in the throat between the soft palate and the base of the tongue. [BOTANY] The throat of a calyx, corolla, or similar part. { 'fȯˌsēz }

faucial tonsil [ANATOMY] *See* palatine tonsil. { 'fȯ·shəl 'tän·səl }

fauna [ZOOLOGY] **1.** Animals. **2.** The animal life characteristic of a particular region or environment. { 'fȯn·ə }

faunal extinction [EVOLUTION] The worldwide death and disappearance of diverse animal groups under circumstances that suggest common and related causes. Also known as mass extinction. { 'fȯn·əl ik'stiŋk·shən }

faunal region [ECOLOGY] A division of the zoosphere, defined by geographic and environmental barriers, to which certain animal communities are bound. { 'fȯn·əl ˌrē·jən }

faunule [PALEONTOLOGY] The localized stratigraphic and geographic distribution of a particular taxon. { 'fȯˌnyül }

Favositidae [PALEONTOLOGY] A family of extinct Paleozoic corals in the order Tabulata. { ˌfav·ə'sid·əˌdē }

Fc region [IMMUNOLOGY] Region of an antibody molecule that binds to antibody receptors on the surface of cells such as macrophages and mast cells, and to complement protein; Fc is derived from the term crystallizable fragment. { ¦ef'sē ˌrē·jən }

feather [VERTEBRATE ZOOLOGY] An ectodermal derivative which is a specialized keratinous outgrowth of the epidermis of birds; functions in flight and in providing insulation and protection. { 'feth·ər }

feathering [VERTEBRATE ZOOLOGY] Plumage. { 'feth·ə·riŋ }

feces [PHYSIOLOGY] The waste material eliminated by the gastrointestinal tract. { 'fē·sēz }

Fechner fraction [PHYSIOLOGY] The smallest difference in the brightness of two sources that can be detected by the human eye divided by the brightness of one of them. { 'fek·ner ˌfrak·shən }

Fechner law [PHYSIOLOGY] The intensity of a sensation produced by a stimulus varies directly as the logarithm of the numerical value of that stimulus. { 'fek·nər ˌlȯ }

fecundity [BIOLOGY] The innate potential reproductive capacity of the individual organism, as denoted by its ability to form and separate from the body the mature germ cells. { fə'kən·dəd·ē }

feedback inhibition [BIOCHEMISTRY] A cellular control mechanism by which the end product of a series of metabolic reactions inhibits the activity of the first enzyme in the sequence. { 'fēdˌbak ˌin·əˌbish·ən }

feeding mechanism [ZOOLOGY] A mechanism by which an animal obtains and utilizes food materials. { 'fēd·iŋ ˌmek·əˌniz·əm }

Felidae [VERTEBRATE ZOOLOGY] The cats and saber-toothed cats, a family of mammals in the superfamily Feloidea. { 'fel·əˌdē }

feline [VERTEBRATE ZOOLOGY] **1.** Of or relating to the genus *Felis*. **2.** Catlike. { 'fēˌlīn }

Felis [VERTEBRATE ZOOLOGY] The type genus of the Felidae, comprising the true or typical cats, both wild and domestic. { 'fē·ləs }

fell-field [ECOLOGY] A culture community of dwarfed, scattered plants or grasses above the timberline. { 'fel ˌfēld }

Feloidea [VERTEBRATE ZOOLOGY] A superfamily of catlike mammals in the order Carnivora. { fə'lȯid·ē·ə }

Felon's unit [BIOLOGY] A unit for the standardization of antipneumococcic serum. { 'fel·ənz ˌyü·nət }

female [BOTANY] A flower lacking stamens. [ZOOLOGY] An individual that bears young or produces eggs. { 'fēˌmāl }

female heterogamety [GENETICS] The occurrence, in females of a species, of an unequal pair of sex chromosomes. { ¦fēˌmāl ¦hed·ə·rō·gə'mēd·ē }

female homogamety [GENETICS] The occurrence, in females of a species, of an equal pair of sex chromosomes. { ¦fēˌmāl ¦hō·mō·gə'mēd·ē }

female pseudohermaphroditism [PHYSIOLOGY] *See* gynandry. { ¦fēˌmāl¦sü·dō·hər¦maf·rə'dīdˌiz·əm }

femoral artery [ANATOMY] The principal artery of the thigh; originates as a continuation of the external iliac artery. { ¦fem·ə·rəl 'ärd·ə·rē }

femoral nerve [ANATOMY] A mixed nerve of the leg; the motor portion innervates muscles of the thigh, and the sensory portion innervates portions of the skin of the thigh, leg, hip, and knee. { ¦fem·ə·rəl 'nərv }

femoral ring [ANATOMY] The abdominal opening of the femoral canal. { ¦fem·ə·rəl 'riŋ }

femoral vein [ANATOMY] A vein accompanying the femoral artery. { ¦fem·ə·rəl 'vān }

femur [ANATOMY] **1.** The proximal bone of the hind or lower limb in vertebrates. **2.** The thigh bone in humans, articulating with the acetabulum and tibia. { 'fē·mər }

Fenestellidae [PALEONTOLOGY] A family of extinct fenestrated, cryptostomatous bryozoans which abounded during the Silurian. { ˌfen·ə 'stel·əˌdē }

fenestra [ANATOMY] An opening in the medial wall of the middle ear. { fə'nes·trə }

fenestrated membrane [HISTOLOGY] One of the layers of elastic tissue in the tunica media and tunica intima of large arteries. { 'fen·əˌstrād·əd 'memˌbrān }

fenestration [BIOLOGY] **1.** A transparent or windowlike break or opening in the surface. **2.** The presence of windowlike openings. { ˌfen·ə'strā·shən }

fennel [BOTANY] *Foeniculum vulgare.* A tall perennial herb of the family Umbelliferae; a spice is derived from the fruit. { 'fen·əl }

ferment [BIOCHEMISTRY] An agent that can initiate fermentation and other metabolic processes. { ¦fər¦ment }

fermentation [MICROBIOLOGY] An enzymatic transformation of organic substrates, especially carbohydrates, generally accompanied by the evolution of gas; a physiological counterpart of oxidation, permitting certain organisms to live and grow in the absence of air; used in various industrial processes for the manufacture of products such as alcohols, acids, and cheese by the action of yeasts, molds, and bacteria; alcoholic fermentation is the best-known example. Also known as zymosis. { ˌfər·mən'tā·shən }

fermentation tube [MICROBIOLOGY] A culture tube with a vertical closed arm to collect gas formed in a broth culture by microorganisms. { ˌfər·mən'tā·shən ˌtüb }

fern [BOTANY] Any of a large number of vascular plants composing the division Polypodiophyta. { fərn }

ferredoxins [BIOCHEMISTRY] Iron-containing proteins that transfer electrons, usually at a low potential, to flavoproteins; the iron is not present as a heme. { ˌfer·ə'däk·sənz }

ferret [VERTEBRATE ZOOLOGY] *Mustela nigripes.* The largest member of the weasel family, Mustelidae, and a relative of the European polecat; has yellowish fur with black feet, tail, and mask. { 'fer·ət }

ferrichrome [MICROBIOLOGY] A cyclic hexapeptide that is a microbial hydroxamic acid and is involved in iron transport and metabolism in microorganisms. { 'fer·əˌkrōm }

ferrihemoglobin [BIOCHEMISTRY] Hemoglobin in the oxidized state. Also known as methemoglobin. { ¦fe·riˌhē·mə'glō·bən }

ferrimycin [MICROBIOLOGY] The representative antibiotic of the sideromycin group; a hydroxamic acid compound. { ˌfe·ri'mīs·ən }

ferriporphyrin [BIOCHEMISTRY] A red-brown to black complex of iron and porphyrin in which the iron is in the 3+ oxidation state. { ¦fe·ri'pȯr·fə·rəŋ }

ferritin [BIOCHEMISTRY] An iron-protein complex occurring in tissues, probably as a storage form of iron. { 'fer·ət·ən }

ferrochelatase [BIOCHEMISTRY] A mitochondrial enzyme which catalyzes the incorporation of iron into the protoporphyrin molecule. { ¦fe·rō'kel·ə ˌtās }

ferroporphyrin [BIOCHEMISTRY] A red complex of porphyrin and iron in which the iron is in the 2+ oxidation state. { ¦fe·rō'pȯr·fə·rən }

fertility [BIOLOGY] The state of or capacity for abundant productivity. { fər'til·əd·ē }

fertility factor [GENETICS] An episomal bacterial sex factor which determines the role of a bacterium as either a male donor or as a female recipient of genetic material. Also known as F factor; sex factor. { fər'til·əd·ē ˌfak·tər }

fertilization [PHYSIOLOGY] The physicochemical processes involved in the union of the male and female gametes to form the zygote. { 'fərd·əl·ə 'zā·shən }

fertilization membrane [CELL BIOLOGY] A membrane that separates from the surface of and surrounds many eggs following activation by the sperm; prevents multiple fertilization. { 'fərd·əl·ə'zā·shən ˌmemˌbrān }

fertilizin [BIOCHEMISTRY] A mucopolysaccharide, derived from the jelly coat of an egg, that plays a role in sperm recognition and the stimulation of sperm motility and metabolic activity. { fər 'til·ə·zən }

fervenulin [MICROBIOLOGY] $C_7H_7N_5O_2$ An antibiotic from culture filtrates of *Streptomyces fervens*; yellow, orthorhombic crystals can be formed; melting point is 178-179°C. Also known as planomycin. { fər'ven·ə·lin }

fescue [BOTANY] A group of grasses of the genus *Festuca*, used for both hay and pasture. { 'fes ˌkyü }

fetal hemoglobin [BIOCHEMISTRY] A normal embryonic hemoglobin having alpha chains identical to those of normal adult human hemoglobin, and gamma chains similar to adult beta chains. { 'fēd·əl 'hē·məˌglō·bən }

fetal membrane [EMBRYOLOGY] Any one of the membranous structures which surround the embryo during its development period. Also known as extraembryonic membrane. { 'fēd·əl 'mem ˌbrān }

fetometamorphism [INVERTEBRATE ZOOLOGY] A life cycle variation in the Cantharidae (Coleoptera); the larvae hatch prematurely as legless, immature prelarvae. { ¦fē·dōˌmed·ə'mȯrˌfiz·əm }

fetus [EMBRYOLOGY] **1.** The unborn offspring of viviparous mammals in the later stages of development. **2.** In human beings, the developing body in utero from the beginning of the ninth week after fertilization through the fortieth week of intrauterine gestation, or until birth. { 'fēd·əs }

Feyliniidae [VERTEBRATE ZOOLOGY] The limbless skinks, a family of reptiles in the suborder Sauria represented by four species in tropical Africa. { ˌfā·lə'nī·əˌdē }

F factor [GENETICS] *See* fertility factor. { 'ef ˌfak·tər }

fiber [BOTANY] **1.** An elongate, thick-walled, tapering plant cell that lacks protoplasm and has a small lumen. **2.** A very slender root. { 'fī·bər }

fiber flax [BOTANY] The flax plant grown in fertile, well-drained, well-prepared soil and cool, humid climate; planted in the early spring and harvested when half the seed pods turn yellow; used in the manufacture of linen. { 'fī·bər ˌflaks }

fibril [BIOLOGY] A small thread or fiber, as a root hair or one of the structural units of a striated muscle. { 'fī·brəl }

fibrillation [PHYSIOLOGY] An independent, spontaneous, local twitching of muscle fibers. { ˌfib·rə'lā·shən }

fibrillose [BIOLOGY] Having fibrils. { 'fīb·rəˌlōs }

fibrin [BIOCHEMISTRY] The fibrous, insoluble protein that forms the structure of a blood clot; formed by the action of thrombin. { 'fī·brən }

fibrinase [BIOCHEMISTRY] An enzyme that catalyzes the formation of covalent bonds between fibrin molecules. Also known as fibrin-stabilizing factor. { 'fī·brəˌnās }

fibrinogen [BIOCHEMISTRY] A plasma protein synthesized by the parenchymal cells of the liver; the precursor of fibrin. Also known as factor I. { fī 'brin·ə·jən }

fibrinoid [BIOCHEMISTRY] A homogeneous, refractile, oxyphilic substance occurring in degenerating connective tissue, as in term placentas, rheumatoid nodules, and Aschoff bodies, and in pulmonary alveoli in some prolonged pneumonitides. { 'fī·brəˌnȯid }

fibrinolysin [BIOCHEMISTRY] *See* plasmin. { ˌfī·brə 'näl·ə·sən }

fibrinolysis [PHYSIOLOGY] Liquefaction of coagulated blood by the action of plasmin on fibrin. { ˌfī·brə'näl·ə·səs }

fibrin-stabilizing factor [BIOCHEMISTRY] *See* fibrinase. { ¦fī·brən 'stā·bəˌlīz·iŋ ˌfak·tər }

fibroblast [HISTOLOGY] A stellate connective tissue cell found in fibrous tissue. Also known as fibrocyte. { 'fī·brəˌblast }

fibrocartilage [HISTOLOGY] A form of cartilage rich in dense, closely opposed bundles of collagen fibers; occurs in intervertebral disks, in the symphysis pubis, and in certain tendons. { ¦fī·brō'kärd·əl·ij }

fibrocyte [HISTOLOGY] *See* fibroblast. { 'fī·brə ˌsīt }

fibroid [HISTOLOGY] Composed of fibrous tissue. { 'fīˌbrȯid }

fibroid deoxyribonucleic acid [MOLECULAR BIOLOGY] Sections of relatively uncoiled double-stranded deoxyribonucleic acid thought to be regions of specific base sequences for coding rather than gene control. { 'fīˌbrȯid dē¦äk·sēˌrī·bō·nü¦klē·ik 'as·əd }

fibroin [BIOCHEMISTRY] A protein secreted by spiders and silkworms which rapidly solidifies into strong, insoluble thread that is used to form webs or cocoons. { 'fī·brə·wən }

fibronectin [BIOCHEMISTRY] A type of large glycoprotein that is found on the surface of cells and mediates cellular adhesion, control of cell shape, and cell migration. { 'fī·brə'nek·tən }

fibrous protein [BIOCHEMISTRY] Any of a class of highly insoluble proteins representing the principal structural elements of many animal tissues. { 'fī·brəs 'prōˌtēn }

fibula [ANATOMY] The outer and usually slender bone of the hind or lower limb below the knee in vertebrates; it articulates with the tibia and astragalus in humans, and is ankylosed with the tibia in birds and some mammals. { 'fib·yə·lə }

Ficus [BOTANY] A genus of tropical trees in the family Moraceae including the rubber tree and the fig tree. [INVERTEBRATE ZOOLOGY] A genus of gastropod mollusks having pear-shaped, spirally ribbed sculptured shells. { 'fī·kəs }

fig [BOTANY] *Ficus carica.* A deciduous tree of the family Moraceae cultivated for its edible fruit, which is a syconium, consisting of a fleshy hollow receptacle lined with pistillate flowers. { fig }

Figitidae [INVERTEBRATE ZOOLOGY] A family of hymenopteran insects in the superfamily Cynipoidea. { fə'jid·əˌdē }

filament [BOTANY] **1.** The stalk of a stamen which supports the anther. **2.** A chain of cells joined end to end, as in certain algae. [INVERTEBRATE ZOOLOGY] A single silk fiber in the cocoon of a silkworm. { 'fil·ə·mənt }

filamentous bacteria [MICROBIOLOGY] Bacteria, especially in the order Actinomycetales, whose cells resemble filaments and are often branched. { ˌfil·ə'men·təs bak'tir·ē·ə }

filaria [INVERTEBRATE ZOOLOGY] A parasitic filamentous nematode belonging to the order Filaroidea. { fə'lar·ē·ə }

Filarioidea [INVERTEBRATE ZOOLOGY] An order of the class Nematoda comprising highly specialized parasites of humans and domestic animals. { ˌfil·ə'rōid·ē·ə }

filbert [BOTANY] Either of two European plants belonging to the genus *Corylus* and producing a thick-shelled, edible nut. Also known as hazelnut. { 'fil·bərt }

filial generation [GENETICS] Any generation following the parental generation. { ¦fil·ē·əl ˌjen·ə ¦rā·shən }

Filicales [BOTANY] The equivalent name for Polypodiales. { ˌfil·ə'kā·lēz }

Filicineae [BOTANY] The equivalent name for Polypodiatae. { ˌfil·ə'sin·ēˌē }

Filicornia [INVERTEBRATE ZOOLOGY] A group of hyperiid amphipod crustaceans in the suborder Genuina having the first antennae inserted anteriorly. { ˌfil·ə'kȯr·nē·ə }

filiform [BIOLOGY] Threadlike or filamentous. { 'fil·əˌfȯrm }

filiform papilla [ANATOMY] Any one of the papillae occurring on the dorsum and margins of the oral part of the tongue, consisting of an elevation of connective tissue covered by a layer of epithelium. { 'fil·ə,fȯrm pə'pil·ə }

film [BIOLOGY] A thin, membranous skin, such as a pellicle. { film }

filoplume [VERTEBRATE ZOOLOGY] A specialized feather that may be decorative, sensory, or both; it is always associated with papillae of contour feathers. { 'fil·ə,plüm }

filopodia [INVERTEBRATE ZOOLOGY] Filamentous pseudopodia. { ,fil·ə'pōd·ē·ə }

filoreticulopodia [INVERTEBRATE ZOOLOGY] Branched, filamentous pseudopodia. { ,fil·ə ,red·ə,kyül·ə'pōd·ē·ə }

Filosia [INVERTEBRATE ZOOLOGY] A subclass of the class Rhizopodea characterized by slender filopodia which rarely anastomose. { fī'lō·shə }

filterable virus [VIROLOGY] Virus particles that remain in a fluid after passing through a diatomite or glazed porcelain filter with pores too minute to allow the passage of bacterial cells. { 'fil·trə·bəl 'vī·rəs }

filter feeder [INVERTEBRATE ZOOLOGY] A microphagous organism that uses complex filtering mechanisms to trap particles suspended in water. { 'fil·tər ,fēd·ər }

fimbria [MICROBIOLOGY] *See* pilus. { 'fim·brē·ə }

fimbriate [BIOLOGY] Having a fringe along the edge. { 'fim·brē,āt }

fin [VERTEBRATE ZOOLOGY] A paddle-shaped appendage on fish and other aquatic animals that is used for propulsion, balance, and guidance. { fin }

final common pathway [ANATOMY] *See* lower motor neuron. { ¦fīn·əl ¦käm·ən 'path,wā }

finch [VERTEBRATE ZOOLOGY] The common name for birds composing the family Fringillidae. { finch }

fin fold [EMBRYOLOGY] A median integumentary fold extending along the body of a fish embryo which gives rise to the dorsal, caudal, and anal fins. { 'fin ,fōld }

finger [ANATOMY] Any of the four digits on the hand other than the thumb. { 'fiŋ·gər }

fingerprint [ANATOMY] A pattern of distinctive epidermal ridges on the bulbs of the inside of the end joints of fingers and thumbs. { 'fiŋ·gər ,print }

fin spine [VERTEBRATE ZOOLOGY] A bony process that supports the fins of certain fishes. { 'fin ,spīn }

fir [BOTANY] The common name for any tree of the genus *Abies* in the pine family; needles are characteristically flat. { fər }

fire disclimax [ECOLOGY] A community that is perpetually maintained at an early stage of succession through recurrent destruction by fire followed by regeneration. { 'fīr dis'klī,maks }

firefly [INVERTEBRATE ZOOLOGY] Any of various flying insects which produce light by bioluminescence. { 'fīr,flī }

first filial generation [GENETICS] The first generation resulting from a cross with all members being heterozygous for characters which differ from those of the parents. Symbolized F_1. { ¦fərst ¦fil·ē·əl jen·ə'rā·shən }

fish [VERTEBRATE ZOOLOGY] The common name for the cold-blooded aquatic vertebrates belonging to the groups Cyclostomata, Chondrichthyes, and Osteichthyes. { fish }

fisher [VERTEBRATE ZOOLOGY] *Martes pennanti*. An arboreal, carnivorous mammal of the family Mustelidae; a relatively large weasellike animal with dark fur, found in northern North America. { 'fish·ər }

fisheries conservation [ECOLOGY] Those measures concerned with the protection and preservation of fish and other aquatic life, particularly in sea waters. { 'fish·ə·rēz ,kän·sər'vā·shən }

fishery [ECOLOGY] A place for harvesting fish or other aquatic life, particularly in sea waters. { 'fish·ə·rē }

fish lice [INVERTEBRATE ZOOLOGY] The common name for all members of the crustacean group Arguloida. { 'fish ,līs }

Fissidentales [BOTANY] An order of the Bryopsida having erect to procumbent, simple or branching stems and two rows of leaves arranged in one plane. { ,fis·ə,den'tā·lēz }

fission [BIOLOGY] A method of asexual reproduction among bacteria, algae, and protozoans by which the organism splits into two or more parts, each part becoming a complete organism. { 'fish·ən }

fission fungi [MICROBIOLOGY] A misnomer once used to describe the Schizomycetes. { ¦fish·ən 'fən,jī }

fissiped [VERTEBRATE ZOOLOGY] **1.** Having the toes separated to the base. **2.** Of or relating to the Fissipeda. { 'fis·ə,ped }

Fissipeda [VERTEBRATE ZOOLOGY] Former designation for a suborder of the Carnivora. { fə'sip·ə·də }

Fissurellidae [INVERTEBRATE ZOOLOGY] The keyhole limpets, a family of gastropod mollusks in the order Archeogastropoda. { ,fis·ə'rel·ə,dē }

Fistuliporidae [PALEONTOLOGY] A diverse family of extinct marine bryozoans in the order Cystoporata. { ,fis·chə·lə'pȯr·ə,dē }

fitness [GENETICS] A measure of reproductive success for a genotype, based on the average number of surviving progeny of this genotype as compared to the average number of other, competing genotypes. { 'fit·nəs }

fix [BIOLOGY] To kill, harden, or preserve a tissue, organ, or organism by immersion in dilute acids, alcohol, or solutions of coagulants. { fiks }

fixator [PHYSIOLOGY] A muscle whose action tends to hold a body part in a certain position or limit its movement. { 'fik,sād·ər }

fixed allele [GENETICS] An allele that is homozygous in all members of a population. { ,fikst ə'lēl }

flabellate [BIOLOGY] Fan-shaped. { flə'bel·ət }

Flabellifera [INVERTEBRATE ZOOLOGY] The largest and morphologically most generalized suborder of isopod crustaceans; the biramous uropods are attached to the sides of the abdomen and may form, with the last abdominal fragment, a caudal fan. { ,flab·ə'lif·ə·rə }

Flabelligeridae [INVERTEBRATE ZOOLOGY] The cage worms, a family of spioniform worms belonging to the Sedentaria; the anterior part of the body is often concealed by a cage of setae arising from the first few segments. { flə‚bel·ə′jer·ə ‚dē }

flabellum [INVERTEBRATE ZOOLOGY] Any structure resembling a fan, as the epipodite of certain crustacean limbs. { flə′bel·əm }

flaccid [BOTANY] Deficient in turgor. [PHYSIOLOGY] Soft, flabby, or relaxed. { ′flas·əd *or* ′flak·səd }

flacherie [INVERTEBRATE ZOOLOGY] A fatal bacterial disease of caterpillars, especially silkworms, marked by loss of appetite, dysentery, and flaccidity of the body; after death the body darkens and liquefies. { ‚flash·ə′rē }

Flacourtiaceae [BOTANY] A family of dicotyledonous plants in the order Violales having the characteristics of the more primitive members of the order. { flə‚kůrd·ē′ās·ē‚ē }

flagella [BIOLOGY] Relatively long, whiplike, centriole-based locomotor organelles on some motile cells. { flə′jel·ə }

Flagellata [INVERTEBRATE ZOOLOGY] The equivalent name for Mastigophora. { ‚flaj·ə′läd·ə }

flagellate [BIOLOGY] **1.** Having flagella. **2.** An organism that propels itself by means of flagella. **3.** Resembling a flagellum. [INVERTEBRATE ZOOLOGY] Any member of the protozoan superclass Mastigophora. { ′flaj·ə‚lāt }

flagellated chamber [INVERTEBRATE ZOOLOGY] An outpouching of the wall of the central cavity in Porifera that is lined with choanocytes; connects with incurrent canals through prosophyles. { ¦flaj·ə‚lād·əd ′chām·bər }

flagellation [BIOLOGY] The arrangement of flagella on an organism. { ‚flaj·ə′lā·shən }

flagellin [MICROBIOLOGY] The protein component of bacterial flagella. { flə′jel·ən }

flagilliflory [BOTANY] Of flowers, hanging down freely from ropelike twigs. { flə′jil·ə‚flȯr·ē }

flame bulb [INVERTEBRATE ZOOLOGY] The enlarged terminal part of the flame cell of a protonephridium, consisting of a tuft of cilia. { ′flām ‚bəlb }

flame cell [INVERTEBRATE ZOOLOGY] A hollow cell that contains the terminal branches of excretory vessels in certain flatworms and rotifers and some other invertebrates. { ′flām ‚sel }

flamingo [VERTEBRATE ZOOLOGY] Any of various long-legged and long-necked aquatic birds of the family Phoenicopteridae characterized by a broad bill resembling that of a duck but abruptly bent downward and rosy-white plumage with scarlet coverts. { flə′miŋ·gō }

flank [VERTEBRATE ZOOLOGY] The part of a quadruped mammal between the ribs and the pelvic girdle. { flaŋk }

flash pasteurization [MICROBIOLOGY] A pasteurization method in which a heat-labile liquid, such as milk, is briefly subjected to temperatures of 230°F (110°C). { ¦flash pas·chə·rə′zā·shən }

flatfish [VERTEBRATE ZOOLOGY] Any of a number of asymmetrical fishes which compose the order Pleuronectiformes; the body is laterally compressed, and both eyes are on the same side of the head. { ′flat‚fish }

flatwood [ECOLOGY] An almost-level zone containing mostly imperfectly drained, acid soils and vegetation consisting of wiregrass and saw palmetto at ground level, shrubs such as gallberry and waxmyrtle, and trees such as longleaf and slash pines. { ′flat‚wůd }

flatworm [INVERTEBRATE ZOOLOGY] The common name for members of the phylum Platyhelminthes; individuals are dorsoventrally flattened. { ′flat‚wərm }

flavan [BIOCHEMISTRY] $C_{15}H_{14}O$ 2-Phenylbenzopyran, an aromatic heterocyclic compound from which all flavonoids are derived. { ′fla·vən }

flavanol [BIOCHEMISTRY] Yellow needles with a melting point of 169°C, derived from flavanone; a flavanoid pigment used as a dye. Also known as 3-hydroxyflavone. { ′fla·və‚nȯl }

flavanone [BIOCHEMISTRY] $C_{15}H_{12}O_2$ A colorless crystalline ketone that often occurs in plants in the form of a glycoside. { ′fla·və‚nōn }

flavin [BIOCHEMISTRY] **1.** A yellow dye obtained from the bark of quercitron trees. **2.** Any of several water-soluble yellow pigments occurring as coenzymes of flavoproteins. { ′fla·vən }

flavin adenine dinucleotide [BIOCHEMISTRY] $C_{27}H_{33}N_9O_{15}P_2$ A coenzyme that functions as a hydrogen acceptor in aerobic dehydrogenases (flavoproteins). Abbreviated FAD. { ′fla·vən ′ad·ən‚ēn dī′nü·klē·ə‚tīd }

flavin mononucleotide [BIOCHEMISTRY] *See* riboflavin 5′-phosphate. { ′fla·vən ¦mä·nō′nü·klē·ə‚tīd }

flavin phosphate [BIOCHEMISTRY] *See* riboflavin 5′-phosphate. { ′fla·vən ′fäs‚fāt }

Flavobacterium [MICROBIOLOGY] A genus of bacterium of uncertain affiliation; gram-negative coccobacilli or slender rods producing pigmented (yellow, red, orange, or brown) growth on solid media. { ¦fla·vō·bak′tir·ē·əm }

flavone [BIOCHEMISTRY] **1.** Any of a number of ketones composing a class of flavonoid compounds. **2.** $C_{15}H_{10}O_2$ A colorless crystalline compound occurring as dust on the surface of many primrose plants. { ′fla‚vōn }

flavonoid [BIOCHEMISTRY] Any of a series of widely distributed plant constituents related to the aromatic heterocyclic skeleton of flavan. { ′fla·və‚nȯid }

flavonol [BIOCHEMISTRY] **1.** Any of a class of flavonoid compounds that are hydroxy derivatives of flavone. **2.** $C_{16}H_{10}O_2$ A colorless, crystalline compound from which many yellow plant pigments are derived. { ′fla·və‚nȯl }

flavoprotein [BIOCHEMISTRY] Any of a number of conjugated protein dehydrogenases containing flavin that play a role in biological oxidations in both plants and animals; a yellow enzyme. { ¦fla·vō¦prō‚tēn }

flax [BOTANY] *Linum usitatissimum*. An erect annual plant with linear leaves and blue flowers; cultivated as a source of flaxseed and fiber. { flaks }

flaxseed [BOTANY] The seed obtained from the seed flax plant; a source of linseed oil. { ′flak ‚sēd }

flea [INVERTEBRATE ZOOLOGY] Any of the wingless

insects composing the order Siphonaptera; most are ectoparasites of mammals and birds. { flē }

fleece [VERTEBRATE ZOOLOGY] Coat of wool shorn from sheep; usually taken off the animal in one piece. { flēs }

flesh [ANATOMY] The soft parts of the body of a vertebrate, especially the skeletal muscle and associated connective tissue and fat. { flesh }

fleshy fruit [BOTANY] A fruit having a fleshy pericarp that is usually soft and juicy, but sometimes hard and tough. { ¦flesh·ē ¦früt }

Flexibacter [MICROBIOLOGY] A genus of bacteria in the family Cytophagaceae; cells are unsheathed rods or filaments and are motile; microcysts are not known. { ¦flek·sə¦bak·tər }

Flexibilia [PALEONTOLOGY] A subclass of extinct stalked or creeping Crinoidea; characteristics include a flexible tegmen with open ambulacral grooves, uniserial arms, a cylindrical stem, and five conspicuous basals and radials. { ˌflek·sə ˈbil·ē·ə }

flexion [BIOLOGY] Act of bending, especially of a joint. { ˈflek·shən }

flexion reflex [PHYSIOLOGY] An unconditioned, segmental reflex elicited by noxious stimulation and consisting of contraction of the flexor muscles of all joints on the same side. Also known as the nocioceptive reflex. { ˈflek·shən ˌrē ˌfleks }

Flexithrix [MICROBIOLOGY] A genus of bacteria in the family Cytophagaceae; cells are usually sheathed filaments, and unsheathed cells are motile; microcysts are not known. { ˈflek·sə ˌthriks }

flexor [PHYSIOLOGY] A muscle that bends or flexes a limb or a part. { ˈflek·sər }

flexuous [BIOLOGY] **1.** Flexible. **2.** Bending in a zigzag manner. **3.** Wavy. { ˈflek·shə·wəs }

flexure [EMBRYOLOGY] A sharp bend of the anterior part of the primary axis of the vertebrate embryo. [VERTEBRATE ZOOLOGY] The last joint of a bird's wing. { ˈflek·shər }

flicker fusion [PHYSIOLOGY] The tendency to perceive an oscillating or flickering sensory input signal as continuous when the frequency is above a specific threshold frequency. { ˈflik·ər ˌfyü·zhən }

flight feather [VERTEBRATE ZOOLOGY] Any of the long contour feathers on the wing of a bird. Also known as remex. { ˈflīt ˌfeth·ər }

flipper [VERTEBRATE ZOOLOGY] A broad, flat appendage used for locomotion by aquatic mammals and sea turtles. { ˈflip·ər }

float [BIOLOGY] An air-filled sac in many pelagic flora and fauna that serves to buoy up the body of the organism. { flōt }

floating rib [ANATOMY] One of the last two ribs in humans which have the anterior end free. { ¦flōd·iŋ ˈrib }

floatoblast [INVERTEBRATE ZOOLOGY] A free-floating statoblast having a float of air cells. { ˈflōd·əˌblast }

floccose [BOTANY] Covered with tufts of woollike hairs. { ˈfläˌkōs }

flocculate [BIOLOGY] Having small tufts of hairs. { ˈfläk·yəˌlət (adjective) *or* ˈfläk·yəˌlāt (verb) }

flocculonodular lobes [ANATOMY] The pair of lateral cerebellar lobes in vertebrates which function to regulate vestibular reflexes underlying posture; referred to functionally as the vestibulocerebellum. { ¦fläk·yə·lō¦näj·ə·lər ˈlōbz }

flocculus [ANATOMY] A prominent lobe of the cerebellum situated behind and below the middle cerebellar peduncle on each side of the median fissure. { ˈfläk·yə·ləs }

floccus [BOTANY] A tuft of woolly hairs. { ˈfläk·əs }

flora [BOTANY] **1.** Plants. **2.** The plant life characterizing a specific geographic region or environment. { ˈflȯr·ə }

floral axis [BOTANY] A flower stalk. { ¦flȯr·əl ˈak·səs }

floral diagram [BOTANY] A diagram of a flower in cross section showing the number and arrangement of floral parts. { ¦flȯr·əl ˈdī·əˌgram }

floret [BOTANY] A small individual flower that is part of a compact group of flowers, such as the head of a composite plant or inflorescence. { ˈflȯr·ət }

Florey unit [BIOLOGY] A unit for the standardization of penicillin. { ˈflȯr·ē ˌyü·nət }

floricome [INVERTEBRATE ZOOLOGY] A type of branched hexaster spicule. { ˈflȯr·əˌkōm }

Florideophyceae [BOTANY] A class of red algae, division Rhodophyta, having prominent pit connections between cells. { flə¦rid·ē·ō¦fīs·ēˌē }

floriferous [BOTANY] Blooming freely, used principally of ornamental plants. { flȯˈrif·ə·rəs }

florigen [BIOCHEMISTRY] A plant hormone that stimulates buds to flower. { ˈflȯr·ə·jən }

florivorous [ZOOLOGY] Feeding on flowers. { flȯ ˈriv·ə·rəs }

florula [ECOLOGY] Plants which grow in a small, confined habitat, for example, a pond. { ˈflȯr·yə·lə }

floscelle [INVERTEBRATE ZOOLOGY] A flowerlike structure around the mouth of some echinoids. { flȯˈsel }

Floscularíacea [INVERTEBRATE ZOOLOGY] A suborder of rotifers in the order Monogononta having a malleoramate mastax. { ¦fläs·kyəˌlar·ēˈās·ē·ə }

Flosculariidae [INVERTEBRATE ZOOLOGY] A family of sessile rotifers in the suborder Flosculariacea. { ˌfläs·kyə·ləˈrī·əˌdē }

flosculous [BOTANY] **1.** Composed of florets. **2.** Of a floret, tubular in form. { ˈfläs·kyə·ləs }

flosculus [BOTANY] A floret. { ˈfläs·kyə·ləs }

flounder [VERTEBRATE ZOOLOGY] Any of a number of flatfishes in the families Pleuronectidae and Bothidae of the order Pleuronectiformes. { ˈflau̇n·dər }

flow bog [ECOLOGY] A peat bog with a surface level that fluctuates in accordance with rain and tides. { ˈflō ˌbäg }

flow cytometry [CELL BIOLOGY] A technique for optical analysis and separation of cells and metaphase chromosomes based on light scattering and fluorescence. { ¦flō sī¦täm·ə·trē }

flower [BOTANY] The characteristic reproductive

structure of a seed plant, particularly if some or all of the parts are brightly colored. { 'flau̇·ər }

flow karyotype [CELL BIOLOGY] A karyotype that is based on flow cytometry measurements. { ¦flō 'kar·ē·ə‚tīp }

flow of variability [GENETICS] The movement of genetic variability within a population as a result of hybridization and segregation. { ¦flō əv ‚ver·ē·ə'bil·əd·ē }

fluctuation test [MICROBIOLOGY] A method of demonstrating that bacterial mutations preexist in a population before they are selected; a large parent population is divided into small parts which are grown independently and the number of mutants in each subculture determined; the number of mutants in the subculture will fluctuate because in some a mutant arises early (giving a large number of progeny), while in others the mutant arises late and gives few progeny. { ‚flək·chə'wā·shən ‚test }

fluke [INVERTEBRATE ZOOLOGY] The common name for more than 40,000 species of parasitic flatworms that form the class Trematoda. [VERTEBRATE ZOOLOGY] A flatfish, especially summer flounder. { flük }

fluorescent antibody [IMMUNOLOGY] An antibody that is labeled by a fluorescent dye, such as fluorescein. { flu̇'res·ənt 'an·tē‚bäd·ē }

fluorescent antibody test [IMMUNOLOGY] A clinical laboratory test based on the antigen used in the diagnosis of syphilis and lupus erythematosus and for identification of certain bacteria and fungi, including the tubercle bacillus. { flu̇'res·ənt 'an·tē‚bäd·ē ‚test }

fluorescent staining [CELL BIOLOGY] The use of fluorescent dyes to mark specific cell structures, such as chromosomes. { flu̇¦res·ənt 'stān·iŋ }

fluorochromasia [CELL BIOLOGY] The immediate appearance of fluorescence inside viable cells on exposure to a fluorogenic substrate. { ¦flu̇r·ō·krə'mā·zhə }

flush [ECOLOGY] An evergreen herbaceous or nonflowering vegetation growing in habitats where seepage water causes the surface to be constantly wet but rarely flooded. { fləsh }

fly [INVERTEBRATE ZOOLOGY] The common name for a number of species of the insect order Diptera characterized by a single pair of wings, antennae, compound eyes, and hindwings modified to form knoblike balancing organs, the halters. { flī }

flying fish [VERTEBRATE ZOOLOGY] Any of about 65 species of marine fishes which form the family Exocoetidae in the order Atheriniformes; characteristic enlarged pectoral fins are used for gliding. { ¦flī·iŋ ¦fish }

flyway [VERTEBRATE ZOOLOGY] A geographic migration route for birds, including the breeding and wintering areas that it connects. { 'flī‚wā }

FMN [BIOCHEMISTRY] *See* riboflavin 5′-phosphate.

foal [VERTEBRATE ZOOLOGY] A young horse, especially one under 1 year of age. { fōl }

focal adhesion plaque [CELL BIOLOGY] Points of attachment that form when cells attach to a substrate. { ¦fō·kəl əd'hē·zhən ‚plak }

fog climax [ECOLOGY] A community that deviates from a climatic climax because of the persistent occurrence of a controlling fog blanket. { ¦fäg 'klī‚maks }

fog forest [ECOLOGY] The dense, rich forest growth which is found at high or medium-high altitudes on tropical mountains; occurs when the tropical rain forest penetrates altitudes of cloud formation, and the climate is excessively moist and not too cold to prevent plant growth. { 'fäg ‚fär·əst }

fold [ANATOMY] A plication or doubling, as of various parts of the body such as membranes and other flat surfaces. { fōld }

foliaceous [BOTANY] Consisting of or having the form or texture of a foliage leaf. [ZOOLOGY] Resembling a leaf in growth form or mode. { ‚fō·lē'ā·shəs }

foliage [BOTANY] The leaves of a plant. { 'fō·lē·ij }

foliar [BOTANY] Of, pertaining to, or consisting of leaves. { 'fō·lē·ər }

foliate papilla [VERTEBRATE ZOOLOGY] One of the papillae found on the posterolateral margin of the tongue of many mammals, but vestigial or absent in humans. { 'fō·lē·ət pə'pil·ə }

foliation [BOTANY] **1.** The process of developing into a leaf. **2.** The state of being in leaf. { ‚fō·lē'ā·shən }

folic acid [BIOCHEMISTRY] $C_{19}H_{19}N_7O_6$ A yellow, crystalline vitamin of the B complex; it is slightly soluble in water, usually occurs in conjugates containing glutamic acid residues, and is found especially in plant leaves and vertebrate livers. Also known as pteroylglutamic acid (PGA). { 'fō·lik 'as·əd }

foliferous [BOTANY] Producing leaves. { fə'lif·ə·rəs }

foliicolous [BIOLOGY] Growing or parasitic upon leaves, as certain fungi. { 'fä·lē·ə'kə·ləs }

foliobranchiate [VERTEBRATE ZOOLOGY] Having leaflike gills. { ¦fō·lē·ō¦braŋ·kē‚āt }

foliolate [BOTANY] Having leaflets. { 'fō·lē·ə‚lāt }

follicle [BIOLOGY] A deep, narrow sheath or a small cavity. [BOTANY] A type of dehiscent fruit composed of one carpel opening along a single suture. { 'fäl·ə·kəl }

follicle-stimulating hormone [BIOCHEMISTRY] A protein hormone released by the anterior pituitary of vertebrates which stimulates growth and secretion of the Graafian follicle and also promotes spermatogenesis. Abbreviated FSH. { ¦fäl·ə·kəl ¦stim·yə‚lād·iŋ 'hȯr‚mōn }

folliculate [BIOLOGY] Having or composed of follicles. { fə'lik·yə‚lāt }

fontanelle [ANATOMY] A membrane-covered space between the bones of a fetal or young skull. [INVERTEBRATE ZOOLOGY] A depression on the head of termites. { fänt·ən'el }

Fontéchevade man [PALEONTOLOGY] A fossil man representing the third interglacial *Homo sapiens* and having browridges and a cranial vault similar to those of modern *Homo sapiens*. { fōn·te·che'väd ‚man }

food [BIOLOGY] A material that can be ingested

and utilized by the organism as a source of nutrition and energy. { füd }

food allergy [IMMUNOLOGY] A hypersensitivity to certain foods. { 'füd ˌal·ər·jē }

food chain [ECOLOGY] The scheme of feeding relationships by trophic levels which unites the member species of a biological community. { 'füd ˌchān }

food pyramid [ECOLOGY] An ecological pyramid representing the food relationship among the animals in a community. { 'füd ˌpir·əˌmid }

food vacuole [CELL BIOLOGY] A membrane-bound organelle in which digestion occurs in cells capable of phagocytosis. Also known as heterophagic vacuole; phagocytic vacuole. { 'füd ˌvak·yəˌwōl }

food web [ECOLOGY] A modified food chain that expresses feeding relationships at various, changing trophic levels. { 'füd ˌweb }

foot [ANATOMY] Terminal portion of a vertebrate leg. [BOTANY] In a fern, moss, or liverwort, the basal part of the young sporophyte that attaches it to the gametophyte. [INVERTEBRATE ZOOLOGY] An organ for locomotion or attachment. { fu̇t }

foot gland [INVERTEBRATE ZOOLOGY] A glandular structure which secretes an adhesive substance in many animals. Also known as pedal gland. { 'fu̇t ˌgland }

foramen [BIOLOGY] A small opening, orifice, pore, or perforation. { fə'rā·mən }

foramen magnum [ANATOMY] A large oval opening in the occipital bone at the base of the cranium that allows passage of the spinal cord, accessory nerves, and vertebral arteries. { fə'rā·mən 'mag·nəm }

foramen of Magendie [ANATOMY] The median aperture of the fourth ventricle of the brain. { fə 'rā·mən əv məˌzhän'dē }

foramen of Monro [ANATOMY] *See* interventricular foramen. { fə'rā·mən əv mən'rō }

foramen of Winslow [ANATOMY] *See* epiploic foramen. { fə'rā·mən əv 'winz·lō }

foramen ovale [ANATOMY] An opening in the sphenoid for the passage of nerves and blood vessels. [EMBRYOLOGY] An opening in the fetal heart partition between the two atria. { fə'rā·mən ō'vä·lē }

foramen primum [EMBRYOLOGY] A temporary embryonic interatrial opening. { fə'rā·mən 'prī·məm }

Foraminiferida [INVERTEBRATE ZOOLOGY] An order of dominantly marine protozoans in the subclass Granuloreticulosia having a secreted or agglutinated shell enclosing the ameboid body. { fə ¦ram·ə·nə'fer·ə·də }

forb [BOTANY] A weed or broadleaf herb. { fȯrb }

forceps [INVERTEBRATE ZOOLOGY] A pair of curved, hard, movable appendages at the end of the abdomen of certain insects, for example, the earwig. { 'fȯr·səps }

forcipate [BIOLOGY] Shaped like forceps; deeply forked. { 'fȯr·səˌpāt }

forcipate trophus [INVERTEBRATE ZOOLOGY] A type of masticatory apparatus in certain predatory rotifers which resembles forceps and is used for grasping. { 'fȯr·səˌpāt 'trō·fəs }

Forcipulatida [INVERTEBRATE ZOOLOGY] An order of echinoderms in the subclass Asteroidea characterized by crossed pedicellariae. { fȯr¦sip·ə 'lad·ə·də }

forearm [ANATOMY] The part of the upper extremity between the wrist and the elbow. Also known as antebrachium. { 'fȯrˌärm }

forebrain [EMBRYOLOGY] The most anterior expansion of the neural tube of a vertebrate embryo. [VERTEBRATE ZOOLOGY] The part of the adult brain derived from the embryonic forebrain; includes the cerebrum, thalamus, and hypothalamus. { 'fȯrˌbrān }

forefinger [ANATOMY] The index finger; the first finger next to the thumb. { 'fȯrˌfiŋ·gər }

forefoot [VERTEBRATE ZOOLOGY] An anterior foot of a quadruped. { 'fȯrˌfu̇t }

foregut [EMBRYOLOGY] The anterior alimentary canal in a vertebrate embryo, including those parts which will develop into the pharynx, esophagus, stomach, and anterior intestine. { 'fȯr ˌgət }

forehead [ANATOMY] The part of the face above the eyes. { 'fär·əd }

forelimb [ANATOMY] An appendage (as a wing, fin, or arm) of a vertebrate that is, or is homologous to, the foreleg of a quadruped. { 'fȯr ˌlim }

forest [ECOLOGY] An ecosystem consisting of plants and animals and their environment, with trees as the dominant form of vegetation. { 'fär·əst }

forest conservation [ECOLOGY] Those measures concerned with the protection and preservation of forest lands and resources. { 'fär·əst ˌkän·sər 'vā·shən }

forest ecology [ECOLOGY] The science that deals with the relationship of forest trees to their environment, to one another, and to other plants and to animals in the forest. { 'fär·əst iˌkäl·ə·jē }

forestry [ECOLOGY] The management of forest lands for wood, forages, water, wildlife, and recreation. { 'fär·ə·strē }

forest-tundra [ECOLOGY] A temperate and cold savanna which occurs at high altitudes and consists of scattered or clumped trees and a shrub layer of varying coverage. { ¦fär·əst ¦tən·drə }

formamidase [BIOCHEMISTRY] An enzyme involved in tryptophane catabolism; catalyzes the conversion of N-formylkynurenine to kynurenine and formate. { fȯr'mam·əˌdās }

Formicariidae [VERTEBRATE ZOOLOGY] The antbirds, a family of suboscine birds in the order Passeriformes. { ˌfȯr·mə·kə'rī·əˌdē }

Formicidae [INVERTEBRATE ZOOLOGY] The ants, social insects composing the single family of the hymenopteran superfamily Formicoidea. { fȯr 'mis·əˌdē }

formicivorous [ZOOLOGY] Feeding on ants. { ˌfȯr·mə'siv·ə·rəs }

Formicoidea [INVERTEBRATE ZOOLOGY] A monofamilial superfamily of hymenopteran insects in the suborder Apocrita, containing the ants. { ˌfȯr·mə'kȯid·ē·ə }

formyl methionine [BIOCHEMISTRY] Formylated methionine; initiates peptide chain synthesis in bacteria. { 'fȯrˌmil mə'thī·əˌnēn }

fornix [ANATOMY] A structure that is folded or arched. [BOTANY] A small scale, especially in the corolla tube of some plants. { 'fȯrˌniks }

Forssman antibody [IMMUNOLOGY] A heterophile antibody that reacts with Forssman antigen. { 'fȯrs·mən 'an·təˌbäd·ē }

Forssman antigen [IMMUNOLOGY] A heterophile antigen, occurring in a variety of unrelated animals, which elicits production of hemolysin (Forssman antibody) for sheep red blood cells. { 'fȯrs·mən 'an·tə·jən }

fossa [ANATOMY] A pit or depression. [VERTEBRATE ZOOLOGY] *Cryptoprocta ferox.* A Madagascan carnivore related to the civets. { 'fäs·ə }

fossil [PALEONTOLOGY] The organic remains, traces, or imprint of an organism preserved in the earth's crust since some time in the geologic past. { 'fäs·əl }

fossil man [PALEONTOLOGY] Ancient human identified from prehistoric skeletal remains which are archeologically earlier than the Neolithic. { ¦fäs·əl 'man }

fossorial [VERTEBRATE ZOOLOGY] Adapted for digging. { fä'sȯr·ē·əl }

foulbrood [INVERTEBRATE ZOOLOGY] The common name for three destructive bacterial diseases of honeybee larvae. { 'fau̇lˌbrüd }

fouling organism [ECOLOGY] Any aquatic organism with a sessile adult stage that attaches to and fouls underwater structures of ships. { 'fau̇l·iŋ ˌȯr·gəˌniz·əm }

founder effect [GENETICS] The principle that when a new population is started by a small number of individuals its gene pool carries only a fraction of the genetic variability of the parental population and hence may be unrepresentative. { 'fau̇n·dər iˌfekt }

fourré [ECOLOGY] *See* temperate and cold scrub; tropical scrub. { fu̇'rā }

fovea [BIOLOGY] A small depression or pit. { 'fō·vē·ə }

fovea centralis [ANATOMY] A small, rodless depression of the retina in line with the visual axis, which affords acute vision. { 'fō·vē·ə sen'tral·əs }

foveal vision [PHYSIOLOGY] Vision achieved by looking directly at objects in the daylight so that the image falls on or near the fovea centralis. Also known as photopic vision. { ¦fō·vē·əl 'vizh·ən }

foveola [BIOLOGY] A small pit, especially one in the embryonic gastric mucosa from which gastric glands develop. { fō'vē·ə·lə }

foveolate [BIOLOGY] Having small depressions; pitted. { 'fō·vē·əˌlāt }

fox [VERTEBRATE ZOOLOGY] The common name for certain members of the dog family (Canidae) having relatively short legs, long bodies, large erect ears, pointed snouts, and long bushy tails. { fäks }

fragmentation [CELL BIOLOGY] Amitotic division; a type of asexual reproduction. { ˌfrag·mən'tā·shən }

frameshift mutation [GENETICS] A mutation caused by addition or deletion of nucleotides in numbers other than three which shifts the translation reading frame so a new set of codons beyond the point of abnormality in the messenger ribonucleic acid is read. Also known as phase-shift mutation. { ¦frāmˌshift myü'tā·shən }

frameshift suppression [GENETICS] Reversion of a frameshift mutation by a second mutation in the same gene. { 'frāmˌshift səˌpresh·ən }

Francisella [MICROBIOLOGY] A genus of gram-negative, aerobic bacteria of uncertain affiliation; cells are small, coccoid to ellipsoidal, pleomorphic rods and can be parasitic on mammals, birds, and arthropods. { ˌfran·si'sel·ə }

Frankia [MICROBIOLOGY] The single genus of the family Frankiaceae. { ˌfraŋ'kē·ə }

Frankiaceae [MICROBIOLOGY] A family of bacteria in the order Actinomycetales; filamentous cells form true mycelia; they are symbiotic and found in active, nitrogen-fixing root nodules. { ˌfraŋ·kē'ās·ēˌē }

fraternal twins [BIOLOGY] *See* dizygotic twins. { frə¦tərn·əl 'twinz }

freemartin [VERTEBRATE ZOOLOGY] An intersexual, usually sterile female calf twin born with a male. { 'frēˌmärt·ən }

free recombination [GENETICS] Genetic recombination having a frequency of 50%, that is, occurring by independent reassortment. { ¦frē ˌrē·käm·bə'nā·shən }

freestone [BOTANY] A fruit stone to which the fruit does not cling, as in certain varieties of peach. { 'frēˌstōn }

Fregatidae [VERTEBRATE ZOOLOGY] Frigate birds or man-o'-war birds, a family of fish-eating birds in the order Pelecaniformes. { fre'gad·əˌdē }

Frenatae [INVERTEBRATE ZOOLOGY] The equivalent name for Heteroneura. { 'frē·nəˌtē }

frenulum [ANATOMY] **1.** A small fold of integument or mucous membrane. **2.** A small ridge on the upper part of the anterior medullary velum. [INVERTEBRATE ZOOLOGY] A spine on most moths that projects from the hindwings and is held to the forewings by a clasp, thus coupling the wings together. { 'fren·yə·ləm }

frenum [ANATOMY] A fold of tissue that restricts the movements of an organ. { 'frē·nəm }

frequency coding [PHYSIOLOGY] A means by which the central nervous system analyzes the content of a receptor message; the frequency of discharged impulses is a function of the rate of rise of the generator current and indirectly of the stimulus strength; the greater the stimulus intensity, the higher the impulse frequency of the message. { 'frē·kwən·sē ˌkōd·iŋ }

frequency-dependent selection [EVOLUTION] A type of natural selection that decreases the frequency of more common phenotypes in a population and increases the frequency of less common phenotypes. { ¦frē·kwən·sē di¦pen·dənt si'lek·shən }

frequency theory [PHYSIOLOGY] A theory of human hearing according to which every specific frequency of sound energy is represented by nerve impulses of the same frequency, and pitch differentiation and analysis are carried out by the brain centers. Also known as telephone theory. { 'frē·kwən·sē ˌthē·ə·rē }

fresh-water ecosystem [ECOLOGY] The living organisms and nonliving materials of an inland aquatic environment. { 'fresh ˌwȯd·ər 'ek·ōˌsis·təm }

Freund's adjuvant [IMMUNOLOGY] A water-oil emulsion containing a killed microorganism (usually *Mycobacterium tuberculosis*) which enhances antigenicity. { 'frȯinz 'a·jə·vənt }

friar's cowl [BOTANY] *See* aconite. { 'frī·ərz ˌkau̇l }

friction ridge [ANATOMY] One of the integumentary ridges on the plantar and palmar surfaces of primates. { 'frik·shən ˌrij }

Friedlander's bacillus [MICROBIOLOGY] *See* Klebsiella pneumoniae. { 'frētˌlan·dərz bəˌsil·əs }

frigidoreceptor [PHYSIOLOGY] A cutaneous sense organ which is sensitive to cold. { ¦frij·ə·dō·ri ¦sep·tər }

Fringillidae [VERTEBRATE ZOOLOGY] The finches, a family of oscine birds in the order Passeriformes. { frin'jil·əˌdē }

fritillary [BOTANY] The common name for plants of the genus Fritillaria. [INVERTEBRATE ZOOLOGY] The common name for butterflies in several genera of the subfamily Nymphalinae. { frə 'til·ə·rē }

frog [VERTEBRATE ZOOLOGY] The common name for a number of tailless amphibians in the order Anura; most have hindlegs adapted for jumping, scaleless skin, and large eyes. { fräg }

frond [BOTANY] **1.** The leaf of a palm or fern. **2.** A foliaceous thallus or thalloid shoot. { fränd }

frontal bone [ANATOMY] Either of a pair of flat membrane bones in vertebrates, and a single bone in humans, forming the upper frontal portion of the cranium; the forehead bone. { 'frənt·əl ˌbōn }

frontal crest [ANATOMY] A median ridge on the internal surface of the frontal bone in humans. { 'frənt·əl ˌkrest }

frontal eminence [ANATOMY] The prominence of the frontal bone above each superciliary ridge in humans. { ¦frənt·əl 'em·ə·nəns }

frontalia [INVERTEBRATE ZOOLOGY] Paired sensory bristles on the frontal aspect of the head of gnathostomulids. { frən'tal·yə }

frontal lobe [ANATOMY] The anterior portion of a cerebral hemisphere, bounded behind by the central sulcus and below by the lateral cerebral sulcus. { ¦frənt·əl ¦lōb }

frontal nerve [ANATOMY] A somatic sensory nerve, attached to the ophthalmic nerve, which innervates the skin of the upper eyelid, the forehead, and the scalp. { ¦frənt·əl ¦nərv }

frontal plane [ANATOMY] Any plane parallel with the long axis of the body and perpendicular to the sagittal plane. { ¦frənt·əl ¦plān }

frontal sinus [ANATOMY] Either of a pair of air spaces within the frontal bone above the nasal bridge. { ¦frənt·əl 'sī·nəs }

frost cracks [BOTANY] Cracks in wood that have split outward from ray shakes. { 'frȯst ˌkraks }

frost ring [BOTANY] A false annual growth ring in the trunk of a tree due to out-of-season defoliation by frost and subsequent regrowth of foliage. { 'frȯst ˌriŋ }

frozen section [BIOLOGY] A thin slice of material cut from a frozen sample of tissue or organ. { ¦frōz·ən 'sek·shən }

fructescence [BOTANY] The period of fruit maturation. { ˌfrək'tes·əns }

fructification [BOTANY] **1.** The process of producing fruit. **2.** A fruit and its appendages. [MYCOLOGY] A sporogenous structure. { 'frək·tə·fə'kā·shən }

D-fructopyranose [BIOCHEMISTRY] *See* fructose. { ¦dē ¦frək·tō'pī·rəˌnōs }

fructivorous [ZOOLOGY] *See* frugivorous. { ¦frək ¦tiv·ə·rəs }

fructose [BIOCHEMISTRY] $C_6H_{12}O_5$ The commonest of ketoses and the sweetest of sugars, found in the free state in fruit juices, honey, and nectar of plant glands. Also known as D-fructopyranose. { 'frůkˌtōs }

frugivorous [ZOOLOGY] Fruit-eating. Also known as fructivorous. { frü'jiv·ə·rəs }

fruit [BOTANY] A fully matured plant ovary with or without other floral or shoot parts united with it at maturity. { früt }

fruit bud [BOTANY] A fertilized flower bud that matures into a fruit. { 'früt ˌbəd }

fruit fly [INVERTEBRATE ZOOLOGY] **1.** The common name for those acalypterate insects composing the family Tephritidae. **2.** Any insect whose larvae feed on fruit or decaying vegetable matter. { 'früt ˌflī }

fruiting body [BOTANY] A specialized, spore-producing organ. { 'früd·iŋ ˌbäd·ē }

frustule [INVERTEBRATE ZOOLOGY] **1.** The shell and protoplast of a diatom. **2.** A nonciliated planulalike bud in some hydrozoans. { 'frəs ˌchül }

frutescent [BIOLOGY] *See* fruticose. [BOTANY] Shrublike in habit. { frü'tes·ənt }

fruticose [BIOLOGY] Resembling a shrub; applied especially to lichens. Also known as frutescent. { 'früd·əˌkōs }

FSH [BIOCHEMISTRY] *See* follicle-stimulating hormone.

Fucales [BOTANY] An order of brown algae composing the class Cyclosporeae. { fyü'kā·lēz }

fuchsinophile [BIOLOGY] Having an affinity for the dye fuchsin. { fyük'sin·əˌfil }

fucoidin [BIOCHEMISTRY] A gum composed of L-fucose and sulfate acid ester groups obtained from *Fucus* species and other brown algae. { fyü 'kȯid·ən }

Fucophyceae [BOTANY] A class of brown algae. { ˌfyü·kə'fīs·ēˌē }

L-fucopyranose [BIOCHEMISTRY] *See* L-fucose. { ¦el ˌfyü·kō'pī·rəˌnōs }

L-fucose [BIOCHEMISTRY] $C_6H_{12}O_5$ A methyl pentose present in some algae and a number of gums and identified in the polysaccharides of blood groups and certain bacteria. Also known as 6-deoxy-L-galactose; L-fucopyranose; L-galactomethylose; L-rhodeose. { ¦el 'fyü,kōs }

fucoxanthin [BIOCHEMISTRY] $C_{40}H_{60}O_6$ A carotenoid pigment; a partial xanthophyll ester found in diatoms and brown algae. { ¦fyü·kō¦zan·thən }

Fucus [BOTANY] A genus of dichotomously branched brown algae; it is harvested in the kelp industry as a source of algin. { 'fyü·kəs }

fugacious [BOTANY] Lasting a short time; used principally to describe plant parts that fall soon after being formed. { fyü'gā·shəs }

fulcrate [BIOLOGY] Having a fulcrum. { 'fu̇l,krāt }

fulcrate trophus [INVERTEBRATE ZOOLOGY] A type of masticatory apparatus in certain rotifers characterized by an elongate fulcrum. { 'fu̇l,krāt 'trō·fəs }

Fuld-Gross unit [BIOLOGY] A unit for the standardization of trypsin. { ¦fu̇ld ¦grōs ,yü·nət }

Fulgoroidea [INVERTEBRATE ZOOLOGY] The lantern flies, a superfamily of homopteran insects in the series Auchenorrhyncha distinguished by the anterior and middle coxae being of equal length and joined to the body at some distance from the median line. { ,fu̇l·gə'rȯid·ē·ə }

fulmar [VERTEBRATE ZOOLOGY] Any of the oceanic birds composing the family Procellariidae; sometimes referred to as foul gulls because of the foul-smelling substance spat at intruders upon their nests. { 'fu̇l·mər }

fumagillin [MICROBIOLOGY] $C_{26}H_{34}O_7$ An insoluble, crystalline antibiotic produced by a strain of the fungus *Aspergillus fumigatus*. { ,fyü·mə'jil·ən }

fumarase [BIOCHEMISTRY] An enzyme that catalyzes the hydration of fumaric acid to malic acid, and the reverse dehydration. { 'fyü·mə,rās }

Fumariaceae [BOTANY] A family of dicotyledonous plants in the order Papaverales having four or six stamens, irregular flowers, and no latex system. { fyü,ma·rē'ās·ē,ē }

Funariales [BOTANY] An order of mosses; plants are usually annual, are terrestrial, and have stems that are erect, short, simple, or sparingly branched. { fyü,nar·ē'ā·lēz }

functional anatomy [ANATOMY] The study of the human body and its parts with emphasis on those features that are directly involved in physiological function. { 'fəŋk·shə·nəl ə'nad·ə·mē }

functional residual capacity [PHYSIOLOGY] The volume of gas which remains within the lungs at expiratory standstill. { ¦fəŋk·shən·əl ri¦zij·ə·wəl kə'pas·əd·ē }

fundamental number [GENETICS] The number of chromosome arms of a karyotype. { ¦fən·də ¦men·təl 'nəm·bər }

fundic gland [ANATOMY] Any of the glands of the corpus and fundus of the stomach. { 'fən·dik ,gland }

fundus [ANATOMY] The bottom of a hollow organ. { 'fən·dəs }

fungi [MYCOLOGY] Nucleated, usually filamentous, sporebearing organisms devoid of chlorophyll. { 'fən,jī }

fungiform [BIOLOGY] Mushroom-shaped. { 'fən·jə,fȯrm }

fungiform papilla [ANATOMY] One of the low, broad papillae scattered over the dorsum and margins of the tongue. { 'fən·jə,fȯrm pə'pil·ə }

Fungi Imperfecti [MYCOLOGY] A class of the subdivision Eumycetes; the name is derived from the lack of a sexual stage. { 'fən,jī ,im·pər'fek ,tī }

Fungivoridae [INVERTEBRATE ZOOLOGY] The fungus gnats, a family of orthorrhaphous dipteran insects in the series Nematocera; the larvae feed on fungi. { ,fən·jə'vȯr·ə,dē }

fungivorous [ZOOLOGY] Feeding on or in fungi. { fən'jiv·ə·rəs }

fungus [MYCOLOGY] Singular of fungi. { 'fəŋ·gəs }

funicle [ANATOMY] *See* funiculus. { 'fyün·ə·kəl }

funiculus [ANATOMY] Also known as funicle. **1.** Any structure in the form of a chord. **2.** A column of white matter in the spinal cord. [BOTANY] The stalk of an ovule. [INVERTEBRATE ZOOLOGY] A band of tissue extending from the adoral end of the coelom to the adoral body wall in bryozoans. { fə'nik·yə·ləs }

fur [VERTEBRATE ZOOLOGY] The coat of a mammal. { fər }

furanose [BIOCHEMISTRY] A sugar whose cyclic or ring structure resembles that of furan. { 'fyu̇r·ə ,nōs }

furca [INVERTEBRATE ZOOLOGY] A forked process as the last abdominal segment of certain crustaceans, and as part of the spring in collembolans. { 'fər·kə }

furcate [BIOLOGY] Forked. { 'fər,kāt }

furcocercous cercaria [INVERTEBRATE ZOOLOGY] A free-swimming, digenetic trematode larva with a forked tail. { ¦fər·kō¦sər·kəs sər'kar·ē·ə }

furcula [ZOOLOGY] A forked structure, especially the wishbone of fowl. { 'fər·kyə·lə }

Furipteridae [VERTEBRATE ZOOLOGY] The smoky bats, a family of mammals in the order Chiroptera having a vestigial thumb and small ears. { ,fu̇·rip'ter·ə,dē }

Furnariidae [VERTEBRATE ZOOLOGY] The oven birds, a family of perching birds in the superfamily Furnarioidea. { ,fər·nə'rī·ə,dē }

Furnarioidea [VERTEBRATE ZOOLOGY] A superfamily of birds in the order Passeriformes characterized by a predominance of gray, brown, and black plumage. { fər,nar·ē'ȯid·ē·ə }

Fusarium [MYCOLOGY] A genus of fungi in the family Tuberculariaceae having sickle-shaped, multicelled conidia; includes many important plant pathogens. { fyü'za·rē·əm }

Fusarium oxysporum [MYCOLOGY] A pathogenic fungus causing a variety of plant diseases, including cabbage yellows and wilt of tomato, flax, cotton, peas, and muskmelon. { fyü'za·rē·əm ,äk·sə'spȯr·əm }

Fusarium solani [MYCOLOGY] A pathogenic fungus implicated in root rot and wilt diseases of several plants, including sisal and squash. { fyü 'za·rē·əm sō'lan·ē }

fusiform [BIOLOGY] Spindle-shaped; tapering toward the ends. { 'fyü·zə,fȯrm }

fusiform bacillus [MICROBIOLOGY] A bacillus

having one blunt and one pointed end, as *Fusobacterium fusiforme*. { 'fyü·zə‚fórm bə'sil·əs }

fusiform initial cell [BOTANY] A cell type of the vascular cambium that gives rise to all cells in the vertical system of secondary xylem and phloem. { 'fyü·zə‚fórm ə¦nish·əl 'sel }

fusimotoneuron [PHYSIOLOGY] One of the small motor fibers, composing about 30% of the fibers in the ventral root of the spinal cord, which innervate intrafusal fibers. { ¦fyü·zē‚mō·dō'nů ‚rän }

fusion frequency [PHYSIOLOGY] The frequency of a series of retinal images above which their differences in luminosity or color (that is, flicker) can no longer be perceived. { 'fyü·zhən ‚frē·kwən·sē }

fusion nucleus [BOTANY] The triploid, or 3*n*, nucleus which results from double fertilization and which produces the endosperm in some seed plants. { ¦fyü·zhən ¦nü·klē·əs }

fusula [INVERTEBRATE ZOOLOGY] A spindle-shaped, terminal projection of the spinneret of a spider. { 'fyü·zə·lə }

Fusulinacea [PALEONTOLOGY] A superfamily of large, marine extinct protozoans in the order Foraminiferida characterized by a chambered calcareous shell. { ‚fyü·zə·lə'nās·ē·ə }

Fusulinidae [PALEONTOLOGY] A family of extinct protozoans in the superfamily Fusulinacea. { ‚fyü·zə'lin·ə‚dē }

Fusulinina [PALEONTOLOGY] A suborder of extinct rhizopod protozoans in the order Foraminiferida having a monolamellar, microgranular calcite wall. { ‚fyü·zə·lə'nī·nə }

G

Gadidae [VERTEBRATE ZOOLOGY] A family of fishes in the order Gadiformes, including cod, haddock, pollock, and hake. { 'ga·də,dē }

Gadiformes [VERTEBRATE ZOOLOGY] An order of actinopterygian fishes that lack fin spines and a swim bladder duct and have cycloid scales and many-rayed pelvic fins. { ,gad·ə'fȯr,mēz }

gait analysis [PHYSIOLOGY] An aspect of kinesiology that involves the study of walking or other types of ambulation. { 'gāt ə,nal·ə·səs }

galactan [BIOCHEMISTRY] Any of a number of polysaccharides composed of galactose units. Also known as galactosan. { gə'lak·tən }

galactogen [BIOCHEMISTRY] A polysaccharide, in snails, that yields galactose on hydrolysis. { gə'lak·tə·jən }

galactoglucomannan [BIOCHEMISTRY] Any of a group of polysaccharides which are prominent components of coniferous woods; they are soluble in alkali and consist of D-glucopyranose and D-mannopyranose units. { gə¦lak·tō,glü·kə'man·ən }

galactokinase [BIOCHEMISTRY] An enzyme which reacts D-galactose with adenosinetriphosphate to give D-galactose-1-phosphate and adenosinediphosphate. { gə¦lak·tə'kī,nās }

galactolipid [BIOCHEMISTRY] *See* cerebroside. { gə¦lak·tə'lip·id }

galactomannan [BIOCHEMISTRY] Any of a group of polysaccharides which are composed of D-galactose and D-mannose units, are soluble in water, and form highly viscous solutions; they are plant mucilages existing as reserve carbohydrates in the endosperm of leguminous seeds. { gə¦lak·tō'man·ən }

L-galactomethylose [BIOCHEMISTRY] *See* L-fucose. { ¦el gə¦lak·tə'meth·ə,lōs }

galactonic acid [BIOCHEMISTRY] $C_6H_{12}O_7$ A monobasic acid derived from galactose, occurring in three optically different forms, and melting at 97°C. Also known as pentahydroxyhexoic acid. { ¦ga,lak¦tän·ik 'as·əd }

galactophore [ANATOMY] A duct that carries milk. { gə'lak·tə,fȯr }

galactopoiesis [PHYSIOLOGY] Formation of the components of milk by the cells composing the lobuloalveolar glandular structure. { gə,lak·tə,pōi'ē·səs }

galactosamine [BIOCHEMISTRY] $C_6H_{14}O_5N$ A crystalline amino acid derivative of galactose; found in bacterial cell walls. { gə,lak'tō·sə,mēn }

galactosan [BIOCHEMISTRY] *See* galactan. { gə'lak·tə,san }

galactose [BIOCHEMISTRY] $C_6H_{12}O_6$ A monosaccharide occurring in both levo and dextro forms as a constituent of plant and animal oligosaccharides (lactose and raffinose) and polysaccharides (agar and pectin). Also known as cerebrose. { gə'lak,tōs }

galactosidase [BIOCHEMISTRY] An enzyme that hydrolyzes galactosides. { gə,lak·tə'sī,dās }

galactoside [BIOCHEMISTRY] A glycoside formed by the reaction of galactose with an alcohol; yields galactose on hydrolysis. { gə'lak·tə,sīd }

galacturonic acid [BIOCHEMISTRY] The monobasic acid resulting from oxidation of the primary alcohol group of D-galactose to carboxyl; it is widely distributed as a constituent of pectins and many plant gums and mucilages. { gə¦lakt·yə¦rän·ik 'as·əd }

galago [VERTEBRATE ZOOLOGY] *See* bushbaby. { gə'lä·gō }

Galatheidea [INVERTEBRATE ZOOLOGY] A group of decapod crustaceans belonging to the Anomura and having a symmetrical abdomen bent upon itself and a well-developed tail fan. { ,gal·ə·thē'ī·dē·ə }

Galaxioidei [VERTEBRATE ZOOLOGY] A suborder of mostly small, fresh-water fishes in the order Salmoniformes. { gə,lak·sē'ȯid·ē,ī }

Galbulidae [VERTEBRATE ZOOLOGY] The jacamars, a family of highly iridescent birds of the order Piciformes that resemble giant hummingbirds. { ,gal'bu̇l·ə,dē }

galea [ANATOMY] The epicranial aponeurosis linking the occipital and frontal muscles. [BIOLOGY] A helmet-shaped structure. [BOTANY] A helmet-shaped petal near the axis. [INVERTEBRATE ZOOLOGY] **1.** The endopodite of the maxilla of certain insects. **2.** A spinning organ on the movable digit of chelicerae of pseudoscorpions. { 'gā·lē·ə }

galeate [BIOLOGY] **1.** Shaped like a helmet. **2.** Having a galea. { 'ga·lē,āt }

Galen's vein [ANATOMY] One of the two veins running along the roof the third ventricle that drain the interior of the brain. { 'gā·lənz 'vān }

Galeritidae [PALEONTOLOGY] A family of extinct exocyclic Euechinoidea in the order Holectypoida, characterized by large ambulacral plates with small, widely separated pore pairs. { ˌga·lə ˈrid·əˌdē }

gall [PHYSIOLOGY] *See* bile. { gȯl }

gallbladder [ANATOMY] A hollow, muscular organ in humans and most vertebrates which receives dilute bile from the liver, concentrates it, and discharges it into the duodenum. { ˈgȯlˌblad·ər }

galleria forest [ECOLOGY] A modified tropical deciduous forest occurring along stream banks. { ˌgal·əˈrē·ə ˌfär·əst }

Galleriinae [INVERTEBRATE ZOOLOGY] A monotypic subfamily of lepidopteran insects in the family Pyralididae; contains the bee moth or wax worm (*Galleria mellonella*), which lives in beehives and whose larvae feed on beeswax. { ˌgal·əˈrī·əˌnē }

gallicolous [BIOLOGY] Producing or inhabiting galls. { gəˈlik·ə·ləs }

Galliformes [VERTEBRATE ZOOLOGY] An order of birds that includes important domestic and game birds, such as turkeys, pheasants, and quails. { ˌgal·əˈfȯrˌmēz }

gallinaceous [VERTEBRATE ZOOLOGY] Of, pertaining to, or resembling birds of the order Galliformes. { ˌgal·əˈnā·shəs }

Gallionella [MICROBIOLOGY] A genus of appendaged bacteria; cells are kidney-shaped or rounded and occur on stalks; reproduce by binary fission. { ˌgal·yəˈnel·ə }

gallivorous [ZOOLOGY] Feeding on the tissues of galls, especially certain insect larvae. { gȯˈliv·ə·rəs }

gallodesoxycholic acid [BIOCHEMISTRY] *See* chenodeoxycholic acid. { ¦ga·lō·deˌzäk·sēˈkäl·ik ˈas·əd }

Galumnidae [INVERTEBRATE ZOOLOGY] A family of oribatid mites in the suborder Sarcoptiformes. { gəˈləm·nəˌdē }

galvanic skin response [PHYSIOLOGY] The electrical reactions of the skin to any stimulus as detected by a sensitive galvanometer; most often used experimentally to measure the resistance of the skin to the passage of a weak electric current. { galˈvan·ik ˈskin riˌspäns }

galvanism [BIOLOGY] The use of a galvanic current for medical or biological purposes. { ˈgal·vəˌniz·əm }

galvanotaxis [BIOLOGY] Movement of a free-living organism in response to an electrical stimulus. { ¦gal·və·nō¦tak·səs }

galvanotropism [BIOLOGY] Response of an organism to electrical stimulation. { ˌgal·vəˈnä·trəˌpiz·əm }

game bird [BIOLOGY] A bird that is legal quarry for hunters. { ˈgām ˌbərd }

Gamella [MICROBIOLOGY] A genus of bacteria in the family Streptococcaceae; spherical cells occurring singly or in pairs with flattened adjacent sides; ferment glucose. { gəˈmel·ə }

gametangial copulation [MYCOLOGY] Direct fusion of certain fungal gametangia without differentiation of the gametes. { ˌga·məˈtan·jē·əl ˌkäp·yəˈla·shən }

gametangium [BIOLOGY] A cell or organ that produces sex cells; occurs in algae, fungi, and plants. { ˌgam·əˈtan·jē·əm }

gamete [BIOLOGY] A cell which participates in fertilization and development of a new organism. Also known as germ cell; sex cell. { ˈgaˌmēt }

gametic copulation [MYCOLOGY] The fusion of pairs of differentiated, uninucleate sexual cells or gametes formed in specialized gametangia. { gəˈmed·ik ˌkäp·yəˈlā·shən }

gametoblast [BOTANY] An archespore that has not yet undergone differentiation. { gəˈmēd·əˌblast }

gametocyte [HISTOLOGY] An undifferentiated cell from which gametes are produced. { gəˈmēd·əˌsīt }

gametogenesis [BIOLOGY] The formation of gametes, or reproductive cells such as ova or sperm. { gəˌmēd·əˈjen·ə·səs }

gametophore [BOTANY] A branch that bears gametangia. { gəˈmēd·əˌfȯr }

gametophyte [BOTANY] **1.** The haploid generation producing gametes in plants exhibiting metagenesis. **2.** An individual plant of this generation. { gəˈmēd·əˌfīt }

gamma globulin [IMMUNOLOGY] Any of the serum proteins with antibody activity. { ˈgam·ə ˈgläb·yə·lən }

Gammaherpesvirinae [VIROLOGY] A subfamily of animal, double-stranded linear DNA viruses of the family Herpesviridae, which are enveloped by a lipid bilayer and several glycoproteins. Also known as lymphoproliferative virus group. { ¦gam·ə¦hərˌpēzˈvī·rəˌnē }

Gammaridea [INVERTEBRATE ZOOLOGY] The scuds or sand hoppers, a suborder of amphipod crustaceans; individuals are usually compressed laterally, are poor walkers, and lack a carapace. { ˌgam·əˈrid·ē·ə }

gamma taxonomy [SYSTEMATICS] The level of taxonomic study concerned with biological aspects of taxa, including intraspecific populations, speciation, and evolutionary rates and trends. { ˈgam·ə takˈsän·ə·mē }

gamodeme [ECOLOGY] An isolated breeding community. { ˈgam·əˌdēm }

gamogony [INVERTEBRATE ZOOLOGY] Spore formation by multiple fission in sporozoans. [ZOOLOGY] Sexual reproduction. { gəˈmäg·ə·nē }

gamone [PHYSIOLOGY] Any substance released by a gamete that facilitates fertilization processes. { gaˈmōn }

gamont [INVERTEBRATE ZOOLOGY] The gametocyte of sporozoans. { ˈgaˌmänt }

gamopetalous [BOTANY] Having petals united at their edges. Also known as sympetalous. { ¦gam·ə¦ped·əl·əs }

gamophyllous [BOTANY] Having the leaves of the perianth united. { ¦gam·ə¦fil·əs }

gamosepalous [BOTANY] Having sepals united at their edges. Also known as synsepalous. { ¦ga·mō¦sep·ə·ləs }

Gampsonychidae [PALEONTOLOGY] A family of extinct crustaceans in the order Palaeocaridacea. { ˌgam·səˈnī·kəˌdē }

gangliated cord [ANATOMY] One of the two main trunks of the sympathetic nervous system, one trunk running along each side of the spinal column. { 'gaŋ·glē‚ād·əd 'kȯrd }

ganglion [ANATOMY] A group of nerve cell bodies, usually located outside the brain and spinal cord. { 'gaŋ·glē·ən }

ganglioside [BIOCHEMISTRY] One of a group of glycosphingolipids found in neuronal surface membranes and spleen; they contain an N-acyl fatty acid derivative of sphingosine linked to a carbohydrate (glucose or galactose); they also contain N-acetylglucosamine or N-acetylgalactosamine, and N-acetylneuraminic acid. { 'gaŋ·glē·ō‚sīd }

ganoid scale [VERTEBRATE ZOOLOGY] A structure having several layers of enamellike material (ganoin) on the upper surface and laminated bone below. { ¦ga‚nȯid 'skāl }

ganoin [VERTEBRATE ZOOLOGY] The enamellike covering of a ganoid scale. { 'gan·ə·wən }

gap [GENETICS] A short region that is missing in one strand of a double-stranded deoxyribonucleic acid. { gap }

gape [ANATOMY] The margin to margin distance between open jaws. [INVERTEBRATE ZOOLOGY] The space between the margins of a closed mollusk valve. { gāp }

gap junction [CELL BIOLOGY] An intercellular junction composed of cylindrical channels connecting adjacent cells; considered to be a low-resistance pathway for intercellular communication. Also known as communicating junction; nexus. { 'gap ‚jəŋk·shən }

gar [VERTEBRATE ZOOLOGY] The common name for about seven species of bony fishes in the order Semionotiformes having a slim form, an elongate snout, and close-set ganoid scales. { gar }

garigue [ECOLOGY] A low, open scrubland restricted to limestone sites in the Mediterranean area; characterized by small evergreen shrubs and low trees. { gə'rēg }

garlic [BOTANY] *Allium sativum*. A perennial plant of the order Liliales grown for its pungent, edible bulbs. { 'gär·lik }

Gartner's duct [ANATOMY] The remnant of the embryonic Wolffian duct in the adult female mammal. { 'gärt·nərz ‚dəkt }

gas gland [VERTEBRATE ZOOLOGY] A structure inside the swim bladder of many teleosts which secretes gas into the bladder. { 'gas ‚gland }

Gasserian ganglion [ANATOMY] A group of nerve cells of the sensory root of the trigeminal nerve. Also known as semilunar ganglion. { ga'ser·ē·ən 'gaŋ·glē·ən }

gas sterilization [MICROBIOLOGY] Sterilization of heat and liquid-labile materials by means of gaseous agents, such as formaldehyde, ethylene oxide, and β-propiolactone. { ¦gas ‚ster·ə·lə'zā·shən }

Gasteromycetes [MYCOLOGY] A group of basidiomycetous fungi in the subclass Homobasidiomycetidae with enclosed basidia and with basidiospores borne symmetrically on long sterigmata and not forcibly discharged. { ¦gas·tə·rō ‚mī'sēd·ēz }

Gasterophilidae [INVERTEBRATE ZOOLOGY] The horse bots, a family of myodarian cyclorrhaphous dipteran insects in the subsection Calypteratae, including individuals that cause myiasis in horses. { ¦gas·tə·rō'fil·ə‚dē }

Gasterophilus [INVERTEBRATE ZOOLOGY] A large genus of botflies in the family Gasterophilidae. { ‚gas·tə'rä·fə·ləs }

Gasterosteidae [VERTEBRATE ZOOLOGY] The sticklebacks, a family of actinopterygian fishes in the order Gasterosteiformes. { ‚gas·tə·rō'stē·ə ‚dē }

Gasterosteiformes [VERTEBRATE ZOOLOGY] An order of actinopterygian fishes characterized by a ductless swim bladder, a pelvic fin that is abdominal to subthoracic in position, and an elongate snout. { ‚gas·tə‚rä·stē·ə'fȯr‚mēz }

Gasteruptiidae [INVERTEBRATE ZOOLOGY] A family of hymenopteran insects in the superfamily Proctotrupoidea. { ‚gas·tə‚rəp'tī·ə‚dē }

gastralium [INVERTEBRATE ZOOLOGY] A microsclere located just beneath the inner cell layer of hexactinellid sponges. [VERTEBRATE ZOOLOGY] One of the riblike structures in the abdomen of certain reptiles. { ga'strā·lē·əm }

gastric acid [BIOCHEMISTRY] Hydrochloric acid secreted by parietal cells in the fundus of the stomach. { 'gas·trik 'as·əd }

gastric cecum [INVERTEBRATE ZOOLOGY] One of the elongated pouchlike projections of the upper end of the stomach in insects. { 'gas·trik 'sē·kəm }

gastric enzyme [BIOCHEMISTRY] Any digestive enzyme secreted by cells lining the stomach. { 'gas·trik 'en‚zīm }

gastric filament [INVERTEBRATE ZOOLOGY] In scyphozoans, a row of filaments on the surface of the gastric cavity which function to kill or paralyze live prey taken into the stomach. Also known as phacella. { 'gas·trik 'fil·ə·mənt }

gastric gland [ANATOMY] Any of the glands in the wall of the stomach that secrete components of the gastric juice. { 'gas·trik ‚gland }

gastric juice [PHYSIOLOGY] The digestive fluid secreted by gastric glands; contains gastric acid and enzymes. { 'gas·trik ‚jüs }

gastric mill [INVERTEBRATE ZOOLOGY] A grinding apparatus consisting of calcareous or chitinous plates in the pharynx or stomach of certain invertebrates. { 'gas·trik ‚mil }

gastric ostium [INVERTEBRATE ZOOLOGY] The opening into the gastric pouch in scyphozoans. { 'gas·trik 'äs·tē·əm }

gastric pouch [INVERTEBRATE ZOOLOGY] One of the pouchlike diversions of a scyphozoan stomach. { 'gas·trik 'paůch }

gastric shield [INVERTEBRATE ZOOLOGY] A thickening of the stomach wall in some mollusks for mixing the contents. { 'gas·trik 'shēld }

gastrin [BIOCHEMISTRY] A polypeptide hormone secreted by the pyloric mucosa which stimulates the pancreas to release pancreatic fluid and the stomach to release gastric acid. { 'gas·trən }

gastroblast [INVERTEBRATE ZOOLOGY] A feeding zooid of a tunicate colony. { 'ga·strə‚blast }

gastrocnemius [ANATOMY] A large muscle of the posterior aspect of the leg, arising by two heads from the posterior surfaces of the lateral and medial condyles of the femur, and inserted with the soleus muscle into the calcaneal tendon, and through this into the back of the calcaneus. { ga ˌsträk′nē·mē·əs }

gastrocoele [EMBRYOLOGY] *See* archenteron. { ′ga·strəˌsēl }

gastrodermis [INVERTEBRATE ZOOLOGY] The cellular lining of the digestive cavity of certain invertebrates. { ¦ga·strō¦dər·məs }

gastroepiploic artery [ANATOMY] Either of two arteries arising from the gastroduodenal and splenic arteries respectively and forming an anastomosis along the greater curvature of the stomach. { ¦ga·strō¦ep·ə¦plȯik ′ärd·ə·rē }

gastrointestinal hormone [BIOCHEMISTRY] Any hormone secreted by the gastrointestinal system. { ¦ga·strōˌin′tes·tən·əl ′hȯrˌmōn }

gastrointestinal system [ANATOMY] The portion of the digestive system including the stomach, intestine, and all accessory organs. { ¦ga·strōˌin ′tes·tən·əl ′sis·təm }

gastrointestinal tract [ANATOMY] The stomach and intestine. { ¦ga·strōˌin′tes·tən·əl ′trakt }

gastrolith [VERTEBRATE ZOOLOGY] A pebble swallowed by certain animals and retained in the gizzard or stomach, where it serves to grind food. { ′ga·strəˌlith }

Gastromyzontidae [VERTEBRATE ZOOLOGY] A small family of actinopterygian fishes of the suborder Cyprinoidei found in southeastern Asia. { ¦ga·strōˌmī′zän·təˌdē }

Gastropoda [INVERTEBRATE ZOOLOGY] A large, morphologically diverse class of the phylum Mollusca, containing the snails, slugs, limpets, and conchs. { ga′sträp·ə·də }

gastropore [INVERTEBRATE ZOOLOGY] A pore containing a gastrozooid in hydrozoan corals. { ′ga·strəˌpȯr }

gastrosplenic ligament [ANATOMY] The fold of peritoneum passing from the stomach to the spleen. Also known as gastrosplenic omentum. { ¦ga·strō¦splen·ik ′lig·ə·mənt }

gastrosplenic omentum [ANATOMY] *See* gastrosplenic ligament. { ¦ga·strō¦splen·ik ō′men·təm }

gastrostome [INVERTEBRATE ZOOLOGY] The opening of a gastropore. { ′ga·strəˌstōm }

gastrostyle [INVERTEBRATE ZOOLOGY] A spiculated projection that extends into the gastrozooid from the base of the gastropore. { ′ga·strə ˌstīl }

Gastrotricha [INVERTEBRATE ZOOLOGY] A group of microscopic, pseudocoelomate animals considered either to be a class of the Aschelminthes or to constitute a separate phylum. { ga′strä·trə·kə }

gastrozooid [INVERTEBRATE ZOOLOGY] A nutritive polyp of colonial coelenterates, characterized by having tentacles and a mouth. { ¦ga·strə′zō ˌȯid }

gastrula [EMBRYOLOGY] The stage of development in animals in which the endoderm is formed and invagination of the blastula has occurred. { ′ga·strə·lə }

gastrulation [EMBRYOLOGY] The process by which the endoderm is formed during development. { ˌga·strə′lā·shən }

gas vacuole [BIOLOGY] A membrane-bound, gas-filled cavity in some algae and protozoans; thought to control buoyancy. { ¦gas ′vak·yə ˌwōl }

Gaultheria [BOTANY] A genus of upright or creeping evergreen shrubs (Ericaceae). { gȯl′thir·ē·ə }

Gause's principle [ECOLOGY] A statement that two species cannot occupy the same niche simultaneously. Also known as competitive-exclusion principle. { ′gau̇z·əz ˌprin·sə·pəl }

gavial [VERTEBRATE ZOOLOGY] The name for two species of reptiles composing the family Gavialidae. { ′gā·vē·əl }

Gavialidae [VERTEBRATE ZOOLOGY] The gavials, a family of reptiles in the order Crocodylia distinguished by an extremely long, slender snout with an enlarged tip. { ˌgā·vē′al·əˌdē }

Gaviidae [VERTEBRATE ZOOLOGY] The single, monogeneric family of the order Gaviiformes. { gə′vī·əˌdē }

Gaviiformes [VERTEBRATE ZOOLOGY] The loons, a monofamilial order of diving birds characterized by webbed feet, compressed, bladelike tarsi, and a heavy, pointed bill. { ˌgā·vē·ə′fȯrˌmēz }

gDNA [MOLECULAR BIOLOGY] *See* genomic deoxyribonucleic acid.

Gecarcinidae [INVERTEBRATE ZOOLOGY] The true land crabs, a family of decapod crustaceans belonging to the Brachygnatha. { ˌjē·kär′sin·ə ˌdē }

gecko [VERTEBRATE ZOOLOGY] The common name for more than 300 species of arboreal and nocturnal reptiles composing the family Gekkonidae. { ′gek·ō }

geitonogamy [BOTANY] Pollination and fertilization of one flower by another on the same plant. { ˌgīt·ən′äg·ə·mē }

Gekkonidae [VERTEBRATE ZOOLOGY] The geckos, a family of small lizards in the order Squamata distinguished by a flattened body, a long sensitive tongue, and adhesive pads on the toes of many species. { ge′kän·əˌdē }

Gelastocoridae [INVERTEBRATE ZOOLOGY] The toad bugs, a family of tropical and subtropical hemipteran insects in the subdivision Hydrocorisae. { jeˌla·stō′kȯr·əˌdē }

gelatinase [BIOCHEMISTRY] An enzyme, found in some yeasts and molds, that liquefies gelatin. { ′jel·ə·təˌnās }

gelatin liquefaction [MICROBIOLOGY] Reduction of a gelatin culture medium to the liquid state by enzymes produced by bacteria in a stab culture; used in identifying bacteria. { ′jel·ət·ən ˌlik·wə′fak·shən }

Gelechiidae [INVERTEBRATE ZOOLOGY] A large family of minute to small moths in the lepidopteran superfamily Tineoidea, generally having forewings and trapezoidal hindwings. { ˌjel·ə′kī·əˌdē }

Gelocidae [PALEONTOLOGY] A family of extinct pecoran ruminants in the superfamily Traguloidea. { jə′läs·əˌdē }

geminate [BIOLOGY] Growing in pairs or couples. { 'jem·ə·nət }

geminiflorous [BOTANY] Having flowers in pairs. { ˌjem·ə'nif·lə·rəs }

gemma [BOTANY] A small, multicellular, asexual reproductive body of some liverworts and mosses. { 'jem·ə }

gemmation [BIOLOGY] *See* budding. { je'mā·shən }

gemmiform [BOTANY] Resembling a gemma or bud. { 'jem·əˌfórm }

gemmiparous [BIOLOGY] Producing a bud or reproducing by a bud. { je'mip·ə·rəs }

gemmule [ANATOMY] A minute dendritic process functioning as a synaptic contact point. [BIOLOGY] Any bud formed by gemmation. [INVERTEBRATE ZOOLOGY] A cystlike, asexual reproductive structure of many Porifera that germinates when proper environmental conditions exist; it is a protective, overwintering structure which germinates the following spring. { 'je·myül }

Gempylidae [VERTEBRATE ZOOLOGY] The snake mackerels, a family of the suborder Scombroidei comprising compressed, elongate, or eel-shaped spiny-rayed fishes with caniniform teeth. { jem 'pil·əˌdē }

Gemuendinoidei [PALEONTOLOGY] A suborder of extinct raylike placoderm fishes in the order Rhenanida. { je¦myü·ən·də¦nòid·ēˌī }

gena [ANATOMY] Cheek, or side of the head. { 'jē·nə }

gene [GENETICS] The basic unit of inheritance. { jēn }

gene action [GENETICS] The functioning of a gene in determining the phenotype of an individual. { 'jēn ˌak·shən }

genealogy [GENETICS] A record of the descent of a family, group, or person from an ancestor or ancestors. { ˌjē·nē'al·ə·jē }

gene amplification [GENETICS] The process by which a cell increases the number of copies of a particular gene to a greater extent than it increases the genes composing the remainder of the genome, resulting from repeated replication of the deoxyribonucleic acid in a limited portion of the genome. [MOLECULAR BIOLOGY] Any process by which a deoxyribonucleic acid sequence is disproportionately duplicated in comparison with the parent genome. { 'jēn ˌam·plə·fəˌkā·shən }

gene assignment [MOLECULAR BIOLOGY] The physical or functional localization of specific genes to individual chromosomes. { 'jēn əˌsīn·mənt }

gene bank [GENETICS] *See* gene library. { 'jēn ˌbaŋk }

gene cluster [GENETICS] Any group of two or more closely linked genes that encode for the same or similar products. { 'jēn ˌkləs·tər }

genecology [BIOLOGY] The study of species and their genetic subdivisions, their place in nature, and the genetic and ecological factors controlling speciation. { ¦jēn·ə'käl·ə·jē }

gene conversion [GENETICS] A situation in which gametocytes of an individual that is heterozygous for a pair of alleles undergo meiosis, and the gametes produced are in a 3:1 ratio rather than the expected 2:2 ratio, implying that one allele was converted to the other. { 'jēn kən ˌvər·zhən }

gene expression [GENETICS] The phenotypic manifestation of genes. { 'jēn ik¦spresh·ən }

gene flow [GENETICS] The passage and establishment of genes characteristic of a breeding population into the gene complex of another population through hybridization and backcrossing. { 'jēn ˌflō }

gene frequency [GENETICS] *See* allele frequency. { 'jēn ˌfrē·kwən·sē }

gene fusion [GENETICS] The combination of information from two separate regions of deoxyribonucleic acid. { 'jēn ˌfyü·zhən }

gene library [GENETICS] A random collection of cloned deoxyribonucleic acid fragments in a vector; includes all the genetic information of the species. Also known as gene bank. { 'jēn ˌlī·brēr·ē }

gene loss [GENETICS] Gene elimination from differentiating cells in some protozoans, insects, and crustaceans. Also known as genome elimination. { 'jēn ˌlòs }

gene overlap [GENETICS] The ability of a sequence of deoxyribonucleic acid to code for more than one protein by use of different reading frames. { ¦jēn 'ō·vərˌlap }

gene penetrance [GENETICS] *See* penetrance. { 'jēn ˌpen·ə·trəns }

gene pool [GENETICS] The totality of the genes of a specific population at a given time. { 'jēn ˌpül }

generation [BIOLOGY] A group of organisms having a common parent or parents and comprising a single level in line of descent. { ˌjen·ə'rā·shən }

generation time [MICROBIOLOGY] The time interval required for a bacterial cell to divide or for the population to double. { ˌjen·ə'rā·shən ˌtīm }

generative nucleus [BOTANY] A haploid nucleus in a pollen grain that produces two sperm nuclei by mitosis. { 'jen·rəd·iv 'nü·klē·əs }

gene redundancy [GENETICS] The presence of many copies of one gene within a cell. { 'jēn ri 'dən·dən·sē }

generic [BIOLOGY] Pertaining to or having the rank of a biological genus. { jə'ner·ik }

gene scanning [MOLECULAR BIOLOGY] A method by which mutations are inserted at specific sites on a deoxyribonucleic acid (DNA) segment to determine those DNA sequences needed for gene activity. { ¦jēn ˌskan·iŋ }

gene sharing [GENETICS] The acquisition and maintenance of a second function for a gene without duplication and without loss of primary function. { ¦jēn ˌsher·iŋ }

gene splicing [GENETICS] *See* recombinant technology. { 'jēn ˌsplīs·iŋ }

gene substitution [GENETICS] The replacement of an allele with a mutant allele. { ¦jēn ˌsəb·stə ˌtüi·shən }

gene suppression [GENETICS] The development of a normal phenotype in a mutant individual or cell due to a second mutation either in the same gene or in a different gene. { 'jēn sə‚presh·ən }

genet [VERTEBRATE ZOOLOGY] The common name for nine species of small, arboreal African carnivores in the family Viverridae. { 'jen·ət }

gene therapy [GENETICS] An experimental technique in which a normal gene is inserted into an organism to correct a genetic defect. { ¦jēn ‚ther·ə·pē }

genetic block [GENETICS] The reduction in enzyme activity due to a gene mutation. { jə¦ned·ik 'bläk }

genetic carrier [GENETICS] An individual that has a recessive gene that predisposes for a hereditary disease. { jə¦ned·ik kar·ē·ər }

genetic code [MOLECULAR BIOLOGY] The genetic information in the nucleotide sequences in deoxyribonucleic acid represented by a four-letter alphabet that makes up a vocabulary of 64 three-nucleotide sequences, or codons; a sequence of such codons (averaging about 100 codons) constructs a message for a polypeptide chain. { jə 'ned·ik 'kōd }

genetic colonization [GENETICS] Natural introduction of genetic material into the deoxyribonucleic acid of a host cell; for example, transmission of a tumor-inducing plasmid into a plant cell by the bacterium *Agrobacterium tumefaciens*. { jə ¦ned·ik ‚käl·ə·nə'zā·shən }

genetic death [GENETICS] **1.** Preferential elimination of genotypes that are carriers of alleles that reduce the adaptive value or fitness of those genotypes. **2.** The death of an individual before reproducing. { jə¦ned·ik 'deth }

genetic differentiation [GENETICS] The accumulation of differences in allelic frequencies between completely or partially isolated populations due to evolutionary forces such as selection or genetic drift. { jə¦ned·ik ‚dif·ə‚ren·chē'ā·shən }

genetic distance [GENETICS] **1.** A measure of the allelic substitutions per locus that have occurred during the separate evolution of two populations or species. **2.** The distance between linked genes in terms of recombination frequency or map units. { jə¦ned·ik 'dis·təns }

genetic drift [GENETICS] The random fluctuation of gene frequencies from generation to generation that occurs in small populations. { jə¦ned·ik 'drift }

genetic engineering [GENETICS] The intentional production of new genes and alteration of genomes by the substitution or addition of new genetic material. { jə¦ned·ik en·jə'nir·iŋ }

genetic equilibrium [GENETICS] In a population, the condition in which the frequencies of allelic genes are maintained at the same values from generation to generation. { jə¦ned·ik ‚ē·kwə 'lib·rē·əm }

genetic fingerprinting [GENETICS] A forensic identification technique that enables virtually 100% discrimination between individuals from small samples of blood or semen, using probes for hypervariable minisatellite deoxyribonucleic acid. Also known as DNA fingerprinting. [MOLECULAR BIOLOGY] Identification of chemical entities in animal tissues as indicative of the presence of specific genes. { jə¦ned·ik 'fiŋ·gər ‚print·iŋ }

genetic homeostasis [GENETICS] The tendency of Mendelian populations to maintain a constant genetic composition. { jə¦ned·ik ‚hō·mē·ō'stā·səs }

genetic identity [GENETICS] A measure of the proportion of genes that are identical in two populations. { jə¦ned·ik ī'den·əd·ē }

genetic induction [GENETICS] Gene activation by a molecule that inactivates a repressor protein and thereby activates transcription of one or more structural genes. [MOLECULAR BIOLOGY] Gene activation by a chemical inducer; results in transcription of structural genes. { jə¦ned·ik in 'dək·shən }

genetic isolation [GENETICS] The absence of genetic exchange between populations or species as a result of geographic separation or of mechanisms that prevent reproduction. { jə¦ned·ik īs·əl'ā·shən }

genetic load [GENETICS] The abnormalities, deformities, and deaths produced in every generation by defective genetic material carried in the gene pool of the human race. { jə¦ned·ik 'lōd }

genetic map [GENETICS] A graphic presentation of the linear arrangement of genes on a chromosome; gene positions are determined by percentages of recombination in linkage experiments. Also known as chromosome map. { jə ¦ned·ik 'map }

genetic marker [GENETICS] A gene whose phenotypic expression is easily discerned and thereby can be used to identify an individual or a cell that carries it, or as a probe to mark a nucleus, chromosome, or locus. { jə¦ned·ik 'märk·ər }

genetic material [GENETICS] The ultramicroscopic particles or genes, first defined by H. J. Muller, the influences of which permeate the cell and play a fundamental role in determining the nature of all cell substances, cell structures, and cell effects; the genes have properties of self-propagation and variation. { jə¦ned·ik mə'tir·ē·əl }

genetics [BIOLOGY] The science that is concerned with the study of biological inheritance. { jə'ned·iks }

genetic system [GENETICS] For a given species, the organization of genetic material and the ways in which the genetic material is transmitted. { jə ¦ned·ik 'sis·təm }

genetic variance [GENETICS] The phenotypic variance in a population that is due to genetic heterogeneity. { jə¦ned·ik ‚ver·ē·əns }

genial [ANATOMY] *See* mental. { 'jēn·yəl }

genic hybrid sterility [GENETICS] Sterility resulting from the interaction of genes in a hybrid to cause disturbances of sex-cell formation or meiosis. { 'jēn·ik 'hī·brəd stə'ril·əd·ē }

geniculate body [ANATOMY] Any of the four oval, flattened prominences on the posterior inferior aspect of the thalamus; functions as the synaptic center for fibers leading to the cerebral cortex. { jə'nik·yə·lət ‚bäd·ē }

geniculate ganglion [ANATOMY] A mass of sensory and sympathetic nerve cells located along the facial nerve. { jə'nik·yə·lət 'gaŋ·glē·ən }

geniculum [ANATOMY] **1.** A small, kneelike, anatomical structure. **2.** A sharp bend in any small organ. { je'nik·yə·ləm }

genioglossus [ANATOMY] An extrinsic muscle of the tongue, arising from the superior mental spine of the mandible. { ¦jē·nē·ō¦glä·səs }

Geniohyidae [PALEONTOLOGY] A family of extinct ungulate mammals in the order Hyracoidea; all members were medium to large-sized animals with long snouts. { ¦jē·nē·ō'hī·ə,dē }

genital atrium [ZOOLOGY] A common chamber receiving openings of male, female, and accessory organs. { 'jen·ət·əl 'ā·trē·əm }

genital coelom [INVERTEBRATE ZOOLOGY] In mollusks, the lumina of the gonads. { 'jen·ət·əl 'sē·ləm }

genital cord [EMBRYOLOGY] A mesenchymal shelf bridging the coeloms in mammalian embryos, produced by fusion of the caudal part of the urogenital folds; fuses with the urinary bladder in the male, and is the primordium of the broad ligament and the uterine walls in the female. [INVERTEBRATE ZOOLOGY] Strands of cells located in the genital canal which are primordial sex cells in crinoids. Also known as genital rachis. { 'jen·ət·əl 'kȯrd }

genitalia [ANATOMY] The organs of reproduction, especially those which are external. { ,jen·ə'tāl·yə }

genital orifice [INVERTEBRATE ZOOLOGY] *See* genital pore. { 'jen·ət·əl 'ȯr·ə·fəs }

genital pore [INVERTEBRATE ZOOLOGY] A small opening on the side of the head in some gastropods through which the penis is protruded. Also known as genital orifice. { 'jen·ət·əl 'pȯr }

genital rachis [INVERTEBRATE ZOOLOGY] *See* genital cord. { 'jen·ət·əl 'rā·kəs }

genital recess [INVERTEBRATE ZOOLOGY] A depression between the calyx surface and anal cone in entoprocts which serves as a brood chamber. { 'jen·ət·əl 'rē,ses }

genital ridge [EMBRYOLOGY] A medial ridge or fold on the ventromedial surface of the mesonephros in the embryo, produced by growth of the peritoneum; the primordium of the gonads and their ligaments. { 'jen·ət·əl 'rij }

genital scale [INVERTEBRATE ZOOLOGY] Any of the small calcareous plates in ophiuroids associated with the buccal shields. { 'jen·ət·əl 'skāl }

genital segment [INVERTEBRATE ZOOLOGY] *See* gonosomite. { 'jen·ət·əl 'seg·mənt }

genital shield [INVERTEBRATE ZOOLOGY] In ophiuroids, a support of a bursal slit in the arms located near the arm base. { 'jen·ət·əl 'shēld }

genital stolon [INVERTEBRATE ZOOLOGY] Part of the axial complex in ophiuroids. { 'jen·ət·əl 'stō·lən }

genital sucker [INVERTEBRATE ZOOLOGY] In some trematodes, a suckerlike structure surrounding the gonopore. { 'jen·ət·əl 'sək·ər }

genital tract [ANATOMY] The ducts of the reproductive system. { 'jen·ət·əl ,trakt }

genital tube [INVERTEBRATE ZOOLOGY] A blood lacuna in crinoids, connected with the subtegminal plexus and suspended in the genital canal. { 'jen·ət·əl 'tüb }

genitourinary system [ANATOMY] *See* urogenital system. { ¦jen·ə·tō'yür·ə,ner·ē ,sis·təm }

genome [GENETICS] **1.** The genetic endowment of a species. **2.** The haploid set of chromosomes. { 'jē,nōm }

genome elimination [GENETICS] *See* gene loss. { ¦jē,nōm ə,lim·ə'nā·shən }

genomic deoxyribonucleic acid [MOLECULAR BIOLOGY] Fragments of deoxyribonucleic acid (DNA) that are produced by the action of restriction enzymes on the DNA of a cell or organism. Abbreviated gDNA. { jə¦nōm·ik dē,äk·sē,rī·bō·nü¦klē·ik 'as·əd }

genomic imprinting [GENETICS] *See* parental imprinting. { jə¦nō·mik im,print·iŋ }

genomic stress [GENETICS] Any influence that may disrupt the stability of the genome, such as environmental factors or altered genetic background. { jə¦nōm·ik 'stres }

genotoxant [BIOCHEMISTRY] An agent that induces toxic, lethal, or heritable effects to nuclear and extranuclear genetic material in cells. { ,jēn·ɔ'täk·sɔnt }

genotype [GENETICS] The genetic constitution of an organism, usually in respect to one gene or a few genes relevant in a particular context. [SYSTEMATICS] The type species of a genus. { 'jē·nə ,tīp }

genotype frequency [GENETICS] The proportion or frequency of any particular genotype among the individuals of a population. { 'jēn·ə,tīp ,frē·kwən·sē }

genotypic cohesion [EVOLUTION] The process whereby balanced and superior gene combinations are held together under the force of genetic recombination, thus reducing the frequency of deleterious recombinants, and with it the genetic load. { ¦jēn·ə,tip·ik kō'hē·zhən }

genotypic distance [EVOLUTION] For two individuals A and B, the probability that the genotype of A is not the same as that of B for a given locus; the distance is zero when the genotype is the same at the particular locus. { ¦jen·ə,tip·ik 'dis·təns }

genotypic structure [GENETICS] The set of the genotype frequencies of a population. { ¦jēn·ə ,tip·ik 'strək·chər }

gentamicin [MICROBIOLOGY] A broad-spectrum antibiotic produced by a species of *Micromonospora*. { ¦jent·ə¦mīs·ən }

Gentianaceae [BOTANY] A family of dicotyledonous herbaceous plants in the order Gentianales distinguished by lacking stipules and having parietal placentation. { ,jen·chə'nās·ē,ē }

Gentianales [BOTANY] A family of dicotyledonous plants in the subclass Asteridae having well-developed internal phloem and opposite, simple, mostly entire leaves. { ,jen·chə'nā·lēz }

genu [ANATOMY] *See* knee. { 'ge·nü }

genus [SYSTEMATICS] A taxonomic category that includes groups of closely related species; the principal subdivision of a family. { 'jē·nəs }

geobotany [BOTANY] The study of plants as related to their geologic environment. { ¦jē·ō 'bät·ən·ē }

Geocorisae [INVERTEBRATE ZOOLOGY] A subdivision of hemipteran insects containing those land bugs with conspicuous antennae and an ejaculatory bulb in the male. { ˌjē·ə'kȯr·əˌsē }

Geodermatophilus [MICROBIOLOGY] A genus of bacteria in the family Dermatophilaceae; the mycelium is rudimentary and a muriform thallus is produced; motile spores are elliptical to lanceolate. { ˌjē·ōˌdər·mə'täf·ə·ləs }

geographical botany [BOTANY] *See* plant geography. { ¦jē·ə¦graf·ə·kəl 'bät·ən·ē }

geographic speciation [EVOLUTION] Evolution of two or more species from a single species following geographic isolation. { ¦jē·ə¦graf·ik ˌspē·shē'ā·shən }

Geometridae [INVERTEBRATE ZOOLOGY] A large family of lepidopteran insects in the superfamily Geometroidea that have slender bodies and relatively broad wings; includes measuring worms, loopers, and cankerworms. { ˌjē·ə'me·trəˌdē }

Geometroidea [INVERTEBRATE ZOOLOGY] A superfamily of lepidopteran insects in the suborder Heteroneura comprising small to large moths with reduced maxillary palpi and tympanal organs at the base of the abdomen. { ˌjē·ə·mə 'trȯid·ē·ə }

Geomyidae [VERTEBRATE ZOOLOGY] The pocket gophers, a family of rodents characterized by fur-lined cheek pouches which open outward, a stout body with short legs, and a broad, blunt head. { ˌjē·ō'mī·əˌdē }

geophagous [ZOOLOGY] Feeding on soil, as certain worms. { jē'äf·ə·gəs }

Geophilomorpha [INVERTEBRATE ZOOLOGY] An order of centipedes in the class Chilopoda including specialized forms that are blind, epimorphic, and dorsoventrally flattened. { ˌjē·ō ˌfil·ə'mȯr·fə }

geophilous [ECOLOGY] Living or growing in or on the ground. { jē'äf·ə·ləs }

geophyte [ECOLOGY] A perennial plant that is deeply embedded in the soil substrata. { 'jē·ə ˌfīt }

Georyssidae [INVERTEBRATE ZOOLOGY] The minute mud-loving beetles, a family of coleopteran insects belonging to the Polyphaga. { ˌjē·ə'ris·ə ˌdē }

geosensing [BOTANY] The sensing or detecting of gravity by a plant relative to its longitudinal axis. { 'jē·ōˌsens·iŋ }

Geosiridaceae [BOTANY] A monotypic family of monocotyledonous plants in the order Orchidales characterized by regular flowers with three stamens that are opposite the sepals. { ˌjē·əˌsir·ə'dās·ēˌē }

Geospizinae [VERTEBRATE ZOOLOGY] Darwin finches, a subfamily of perching birds in the family Fringillidae. { ˌjē·ō'spiz·əˌnē }

geotaxis [PHYSIOLOGY] Movement of a free-living organism in response to the stimulus of gravity. { ¦jē·ō¦tak·səs }

geotropism [BOTANY] Response of a plant to the force of gravity. Also known as gravitropism. { jē 'ä·trəˌpiz·əm }

Gephyrea [INVERTEBRATE ZOOLOGY] A class of burrowing worms in the phylum Annelida. { jə 'fir·ē·ə }

gephyrocercal [VERTEBRATE ZOOLOGY] Having the dorsal and anal fins coming together smoothly at the aborted end of the vertebral column of a fish's tail. { ¦jef·ə·rō¦sər·kəl }

Geraniaceae [BOTANY] A family of dicotyledonous plants in the order Geraniales in which the fruit is beaked, styles are usually united, and the leaves have stipules. { jəˌrā·nē'ās·ēˌē }

Geraniales [BOTANY] An order of dicotyledonous plants in the subclass Rosidae comprising herbs or soft shrubs with a superior ovary and with compound or deeply cleft leaves. { jəˌrā·nē'ā· lēz }

Geranium [BOTANY] A genus of plants in the family Geraniaceae characterized by regular flowers, and glands alternating with the petals. { jə'rā· nē·əm }

Gerardiidae [INVERTEBRATE ZOOLOGY] A family of anthozoans in the order Zoanthidea. { ˌjər·är 'dī·əˌdē }

gerbil [VERTEBRATE ZOOLOGY] The common name for about 100 species of African and Asian rodents composing the subfamily Gerbillinae. { 'jər·bəl }

Gerbillinae [VERTEBRATE ZOOLOGY] The gerbils, a subfamily of rodents in the family Muridae characterized by hindlegs that are longer than the front ones, and a long, slightly haired, usually tufted tail. { jər'bil·əˌnē }

germ [BIOLOGY] A primary source, especially one from which growth and development are expected. [MICROBIOLOGY] General designation for a microorganism. { jərm }

germarium [INVERTEBRATE ZOOLOGY] The egg-producing portion of an ovary and the sperm-producing portion of a testis in Platyhelminthes and Rotifera. { jər'mar·ē·əm }

germ ball [INVERTEBRATE ZOOLOGY] A group of cells in digenetic trematode miracidial larvae which are embryos. { 'jərm ˌbȯl }

germ cell [BIOLOGY] *See* gamete. { 'jərm ˌsel }

germfree animal [MICROBIOLOGY] An animal having no demonstrable, viable microorganisms living in intimate association with it. { 'jərmˌfrē 'an·ə·məl }

germfree isolator [MICROBIOLOGY] An apparatus that provides a mechanical barrier surrounding the area in which germfree vertebrates and accessory equipment are housed. { 'jərmˌfrē 'ī·sə ˌlād·ər }

germinal epithelium [EMBRYOLOGY] The region of the dorsal coelomic epithelium lying between the dorsal mesentery and the mesonephros. { 'jər·mən·əl ˌep·ə'thē·lē·əm }

germinal spot [CELL BIOLOGY] The nucleolus of an egg cell. { 'jər·mən·əl ˌspät }

germinal vesicle [CELL BIOLOGY] The enlarged nucleus of the primary oocyte before reduction divisions are complete. { 'jər·mən·əl 'ves·ə· kəl }

germination [BOTANY] The beginning or the process of development of a spore or seed. { ˌjer·məˈnā·shən }

germ layer [EMBRYOLOGY] One of the primitive cell layers which appear in the early animal embryo and from which the embryo body is constructed. { ˈjərm ˌlā·ər }

germ-layer theory [EMBRYOLOGY] The theory that three primary germ layers, ectoderm, mesoderm, and endoderm, are established in the early embryo and all organs and structures are derived from a specific germ layer. { ˈjərm ˌlā·ər ˌthē·ə·rē }

germ line [BIOLOGY] A lineage of cells from which gametes are derived. Also known as germ track. { ˈjərm ˌlīn }

germovitellarium [INVERTEBRATE ZOOLOGY] A sex gland which differentiates into a yolk-producing or egg-producing region. { ¦jər·mōˌvīd·əlˈa·rē·əm }

germ plasm [BIOLOGY] The genetic material contained within a germ cell. { ˈjərm ˌplaz·əm }

germ track [BIOLOGY] *See* germ line. { ˈjərm ˌtrak }

gerontology [PHYSIOLOGY] The scientific study of aging processes in biological systems, particularly in humans. { ˌjer·ənˈtäl·ə·jē }

Gerrhosauridae [VERTEBRATE ZOOLOGY] A small family of lizards in the suborder Sauria confined to Africa and Madagascar. { ˌjer·ōˈsȯr·əˌdē }

Gerridae [INVERTEBRATE ZOOLOGY] The water striders, a family of hemipteran insects in the subdivision Amphibicorisae having long middle and hind legs and a median scent gland opening on the metasternum. { ˈjer·əˌdē }

Gerroidea [INVERTEBRATE ZOOLOGY] The single superfamily of the hemipteran subdivision Amphibicorisae; all members have conspicuous antennae and hydrofuge hairs covering the body. { jəˈrȯid·ē·ə }

Gesneriaceae [BOTANY] A family of dicotyledonous plants in the order Scrophulariales characterized by parietal placentation, mostly opposite or whorled leaves, and a well-developed embryo. { geˌsnir·ēˈās·ēˌē }

gestalt vision [PHYSIOLOGY] The visual perception and retention in the memory of an object in terms of its geometric shape. { gə¦shtält ˌvizh·ən }

gestate [EMBRYOLOGY] To carry the young in the uterus from conception to delivery. { ˈjeˌstāt }

gestation period [EMBRYOLOGY] The period in mammals from fertilization to birth. { jəˈstā·shən ˌpir·ē·əd }

GH [BIOCHEMISTRY] *See* growth hormone.

ghost layer [CELL BIOLOGY] A single layer of cultivated animal cells that has been treated with a nonionic detergent in order to disrupt the membranes. { ˈgōst ˌlā·ər }

Giardia [INVERTEBRATE ZOOLOGY] A genus of zooflagellates that inhabit the intestine of numerous vertebrates, and may cause diarrhea in humans. { jeˈärd·ē·ə }

gibberellic acid [BIOCHEMISTRY] $C_{18}H_{22}O_6$ A crystalline acid occurring in plants that is similar to the gibberellins in its growth-promoting effects. { ¦jib·ə¦rel·ik ˈas·əd }

gibberellin [BIOCHEMISTRY] Any member of a family of naturally derived compounds which have a gibbane skeleton and a broad spectrum of biological activity but are noted as plant growth regulators. { ˌjib·əˈrel·ən }

gibbon [VERTEBRATE ZOOLOGY] The common name for seven species of large, tailless primates belonging to the genus *Hylobates*; the face and ears are hairless, and the arms are longer than the legs. { ˈgib·ən }

Giganturoidei [VERTEBRATE ZOOLOGY] A suborder of small, mesopelagic actinopterygian fishes in the order Cetomimiformes having large mouths and strong teeth. { jīˌgan·təˈrȯid·ēˌī }

Gila monster [VERTEBRATE ZOOLOGY] The common name for two species of reptiles in the genus *Heloderma* (Helodermatidae) distinguished by a rounded body that is covered with multicolored beaded tubercles, and a bifid protrusible tongue. { ˈhē·lə ˌmän·stər }

gill [MYCOLOGY] A structure consisting of radially arranged rows of tissue that hang from the underside of the mushroom cap of certain basidiomycetes. [VERTEBRATE ZOOLOGY] The respiratory organ of water-breathing animals. Also known as branchia. { gil }

gill cover [VERTEBRATE ZOOLOGY] The fold of skin providing external protection for the gill apparatus of most fishes; it may be stiffened by bony plates and covered with scales. { ˈgil kəv·ər }

gill raker [VERTEBRATE ZOOLOGY] One of the bony processes on the inside of the branchial arches of fishes which prevents the passage of solid substances through the branchial clefts. { ˈgil ˌrāk·ər }

ginger [BOTANY] *Zingiber officinale*. An erect perennial herb of the family Zingiberaceae having thick, scaly branched rhizomes; a spice oleoresin is made by an organic solvent extraction of the ground dried rhizome. { ˈjin·jər }

gingiva [ANATOMY] The mucous membrane surrounding the teeth sockets. { ˈjin·jə·və }

gingival crevice [ANATOMY] The space between the free margin of the gingiva and the surface of a tooth. Also known as gingival sulcus. { ˈjin·jə·vəl ˈkrev·əs }

gingival sulcus [ANATOMY] *See* gingival crevice. { ˈjin·jə·vəl ˈsəl·kəs }

ginglymoarthrodia [ANATOMY] A composite joint consisting of one hinged and one gliding element. { ¦jiŋ·glə·mōˌärˈthrō·dē·ə }

Ginglymodi [VERTEBRATE ZOOLOGY] An equivalent name for Semionotiformes. { ˌjiŋ·gləˈmō ˌdī }

ginglymus [ANATOMY] A type of diarthrosis permitting motion only in one plane. Also known as hinge joint. { ˈjiŋ·glə·məs }

Ginkgoales [BOTANY] An order of gymnosperms composing the class Ginkgoopsida with one living species, the dioecious maidenhair tree (*Ginkgo biloba*). { ˌgiŋ·kōˈā·lēz }

Ginkgoopsida [BOTANY] A class of the subdivision Pinicae containing the single, monotypic order Ginkgoales. { ˌgiŋ·kōˈäp·sə·də }

Ginkgophyta [BOTANY] The equivalent name for Ginkgoopsida. { ˌgiŋˈkäf·əd·ə }

ginseng [BOTANY] The common name for plants of the genus *Panax*, a group of perennial herbs in the family Araliaceae; the aromatic root of the plant has been used medicinally in China. { ˈjin ˌseŋ }

giraffe [VERTEBRATE ZOOLOGY] *Giraffa camelopardalis.* An artiodactyl mammal in the family Giraffidae characterized by extreme elongation of the neck vertebrae, and two prominent horns on the head. { jəˈraf }

Giraffidae [VERTEBRATE ZOOLOGY] A family of pecoran ruminants in the superfamily Bovoidea including giraffe, okapi, and relatives. { jəˈraf·ə ˌdē }

girdle [ANATOMY] Either of the ringlike groups of bones supporting the forelimbs (arms) and hindlimbs (legs) in vertebrates. [INVERTEBRATE ZOOLOGY] **1.** Either of the hooplike bands constituting the sides of the two valves of a diatom. **2.** The peripheral portion of the mantle in chitons. { ˈgərd·əl }

gizzard [VERTEBRATE ZOOLOGY] The muscular portion of the stomach of most birds where food is ground with the aid of ingested pebbles. { ˈgiz·ərd }

glabella [ANATOMY] The bony prominence on the frontal bone joining the supraorbital ridges. { glə ˈbel·ə }

glabrous [BIOLOGY] Having a smooth surface; specifically, having the epidermis devoid of hair or down. { ˈglab·rəs }

gladiate [BOTANY] Sword-shaped. { ˈglad·ēˌāt }

Gladiolus [BOTANY] A genus of chiefly African plants in the family Iridaceae having erect, sword-shaped leaves and spikes of brightly colored irregular flowers. { ˌglad·ēˈō·ləs }

gladiolus [ANATOMY] *See* mesosternum. { ˌglad· ēˈō·ləs }

gladius [INVERTEBRATE ZOOLOGY] *See* pen. { ˈglād·ē·əs }

gland [ANATOMY] A structure which produces a substance essential and vital to the existence of the organism. { gland }

glands of Brunner [ANATOMY] *See* Brunner's glands. { ¦glanz əv ˈbrün·ər }

glands of Leydig [VERTEBRATE ZOOLOGY] Unicellular, epidermal structures of urodele larvae and the adult *Necturus* that secrete a substance which digests the egg capsule and permits hatching. { ¦glanz əv ˈlī·dig }

glandulomuscular [ANATOMY] Of or pertaining to glands and muscles. { ˈglan·jə·ləˈməs·kyə· lər }

glans [ANATOMY] The conical body forming the distal end of the clitoris or penis. { glanz }

Glareolidae [VERTEBRATE ZOOLOGY] A family of birds in the order Charadriiformes including the ploverlike coursers and the swallowlike pratincoles. { ˌgla·rēˈä·ləˌdē }

glareous [ECOLOGY] Growing in gravelly soil; refers specifically to plants. { ˈgla·rē·əs }

glass sponge [INVERTEBRATE ZOOLOGY] A siliceous sponge belonging to the class Hyalospongiae. { ˈglas ˌspənj }

glaucous [BOTANY] Having a white or grayish powdery coating that gives a frosty appearance and rubs off easily. { ˈglȯ·kəs }

gleba [MYCOLOGY] The central, sporogenous tissue of the sporophore in certain basidiomycetous fungi. { ˈglē·bə }

glenoid [ANATOMY] A smooth, shallow, socketlike depression, particularly of the skeleton. { ˈgleˌnȯid }

glenoid cavity [ANATOMY] The articular surface on the scapula for articulation with the head of the humerus. { ¦gleˌnȯid ˈkav·əd·ē }

gliding bacteria [MICROBIOLOGY] The descriptive term for members of the orders Beggiatoales and Myxobacterales; they are motile by means of creeping movements. { ¦glīd·iŋ bakˈtir·ē·ə }

gliding joint [ANATOMY] *See* arthrodia. { ¦glīd·iŋ ˈjȯint }

gliding motility [MICROBIOLOGY] A means of bacterial self-propulsion by slow gliding or creeping movements on the surface of a substrate. { ¦glīd·iŋ mōˈtil·əd·ē }

gliotoxin [MICROBIOLOGY] $C_{13}H_{14}O_4N_2S_2$ A heat-labile, bacteriostatic antibiotic produced by species of *Trichoderma* and *Cliocladium* and by *Aspergillus fumigatus.* { ¦glī·ōˌtäk·sən }

Gliridae [VERTEBRATE ZOOLOGY] The dormice, a family of mammals in the order Rodentia. { ˈglir·əˌdē }

Glisson's capsule [ANATOMY] The membranous sheet of collagenous and elastic fibers covering the liver. { ˈglis·ənz ˌkap·səl }

Globigerinacea [INVERTEBRATE ZOOLOGY] A superfamily of foraminiferan protozoans in the suborder Rotaliina characterized by a radial calcite test with bilamellar septa and a large aperture. { glōˌbij·ə·rəˈnās·ē·ə }

globin [BIOCHEMISTRY] Any of a class of soluble histone proteins obtained from animal hemoglobins. { ˈglō·bən }

globoside [BIOCHEMISTRY] A glycoside of ceramide containing several sugar residues, but not neuraminic acid; obtained from human, sheep, and hog erythrocytes. { ˈglō·bəˌsīd }

globular deoxyribonucleic acid [MOLECULAR BIOLOGY] A compact arrangement of double-stranded deoxyribonucleic acid formed by additional twisting of the fiber into a helical double helix. { ˈgläb·yə·lər dē¦äk·sēˌrī·bōˌnü¦klē·ik ˈas·əd }

globular protein [BIOCHEMISTRY] Any protein that is readily soluble in aqueous solvents. { ˈgläb·yə·lər ˈprōˌtēn }

globulin [BIOCHEMISTRY] A heat-labile serum protein precipitated by 50% saturated ammonium sulfate and soluble in dilute salt solutions. { ˈgläb·yə·lən }

glochid [BOTANY] *See* glochidium. { ˈglōˌkȯid }

glochidium [BOTANY] A barbed hair. Also known as glochid. [INVERTEBRATE ZOOLOGY] The larva of fresh-water mussels in the family Unionidae. { glōˈkid·ē·əm }

gloea [INVERTEBRATE ZOOLOGY] An adhesive mucoid substance secreted by certain protozoans and other lower organisms. { ˈglē·ə }

glomerulus [ANATOMY] A tuft of capillary loops projecting into the lumen of a renal corpuscle. { glə'mər·yə·ləs }

glomus [ANATOMY] **1.** A fold of the mesothelium arising near the base of the mesentery in the pronephros and containing a ball of blood vessels. **2.** A prominent portion of the choroid plexus of the lateral ventricle of the brain. { 'glō·məs }

glomus aorticum [ANATOMY] *See* paraaortic body. { 'glō·məs ā'ȯrd·ə·kəm }

glomus caroticum [ANATOMY] *See* carotid body. { 'glō·məs kə'räd·ə·kəm }

glossa [INVERTEBRATE ZOOLOGY] A tongue or tonguelike structure in insects, especially the median projection of the labium. { 'gläs·ə }

glossate [INVERTEBRATE ZOOLOGY] Having a glossa or tonguelike structure. { 'glä,sāt }

Glossinidae [INVERTEBRATE ZOOLOGY] The tsetse flies, a family of cyclorrhaphous dipteran insects in the section Pupipara. { glä'sin·ə,dē }

Glossiphoniidae [INVERTEBRATE ZOOLOGY] A family of small leeches with flattened bodies in the order Rhynchobdellae. { ,glä·sə·fə'nī·ə,dē }

glossopalatine nerve [ANATOMY] The intermediate branch of the facial nerve. { ¦gläs·ō'pal·ə ,tēn ,nərv }

glossopharyngeal nerve [ANATOMY] The ninth cranial nerve in vertebrates; a paired mixed nerve that supplies autonomic innervation to the parotid gland and contains sensory fibers from the posterior one-third of the tongue and the anterior pharynx. { ¦gläs·ō·fə'rin·jē·əl ,nərv }

glossopterid flora [PALEOBOTANY] Permian and Triassic fossil ferns of the genus *Glossopteris*. { glä 'säp·tə·rəd 'flȯr·ə }

glottis [ANATOMY] The opening between the margins of the vocal folds. { 'gläd·əs }

glucagon [BIOCHEMISTRY] The protein hormone secreted by α-cells of the pancreas which plays a role in carbohydrate metabolism. Also known as hyperglycemic factor; hyperglycemic glycogenolytic factor. { 'glü·kə,gän }

glucamine [BIOCHEMISTRY] $C_6H_{15}O_4N$ An amine formed by reduction of glucosylamine or of glucose oxime. { 'glü·kə,mēn }

glucan [BIOCHEMISTRY] A polysaccharide composed of the hexose sugar D-glucose. { 'glü ,kan }

glucocerebroside [BIOCHEMISTRY] A glycoside of ceramide that contains glucose. { ¦glü·kō·sə 'rēb·rə,sīd }

glucocorticoid [BIOCHEMISTRY] A corticoid that affects glucose metabolism; secreted principally by the adrenal cortex. { ¦glü·kō¦kȯrd·ə,kȯid }

glucogenesis [BIOCHEMISTRY] Formation of glucose within the animal body from products of glycolysis. { ,glü·kō'jen·ə·səs }

glucokinase [BIOCHEMISTRY] An enzyme that catalyzes the phosphorylation of D-glucose to glucose-6-phosphate. { ¦glü·kō'kī,nās }

glucolipid [BIOCHEMISTRY] A glycolipid that yields glucose on hydrolysis. { ¦glü·kō'lip·əd }

glucomannan [BIOCHEMISTRY] A polysaccharide composed of D-glucose and D-mannose; a prominent component of coniferous trees. { ¦glü·kō ¦man·ən }

gluconeogenesis [BIOCHEMISTRY] Formation of glucose within the animal body from substances other than carbohydrates, particularly proteins and fats. { ¦glü·kō¦nē·ō'jen·ə·səs }

Gluconobacter [MICROBIOLOGY] A genus of bacteria in the family Pseudomonadaceae; ellipsoidal to rod-shaped cells having three to eight polar flagella and occurring singly, in pairs, or in chains. { glü¦kän ·ə¦bak·tər }

D-glucopyranose [BIOCHEMISTRY] *See* glucose. { ¦dē ¦glü·kə'pir·ə,nōs }

glucopyranoside [BIOCHEMISTRY] Any glucoside that contains a six-membered ring. { ¦glü·kō·pir 'an·ə,sīd }

glucosamine [BIOCHEMISTRY] $C_6H_{13}O_5$ An amino sugar; the most abundant in nature, occurring in glycoproteins and chitin. { 'glü,kōs·ə·mēn }

glucose [BIOCHEMISTRY] $C_6H_{12}O_6$ A monosaccharide; occurs free or combined and is the most common sugar. Also known as cerelose; D-glucopyranose. { 'glü,kōs }

glucose-6-phosphatase [BIOCHEMISTRY] An enzyme found in liver which catalyzes the hydrolysis of glucose-6-phosphate to free glucose and inorganic phosphate. { 'glü,kōs ¦siks 'fäs·fə ,tās }

glucose phosphate [BIOCHEMISTRY] A phosphoric derivative of glucose, as glucose-1-phosphate. { 'glü,kōs 'fäs,fāt }

glucose-1-phosphate [BIOCHEMISTRY] $C_6H_{12}O_8P$ An ester of glucopyranose in which a phosphate group is attached to carbon atom 1; there are two types: α-D- and β-D-glucose-1-phosphates. Also known as Cori ester. { 'glü,kōs ¦wən 'fäs,fāt }

glucose-6-phosphate [BIOCHEMISTRY] $C_6H_{13}O_9P$ An ester of glucose with phosphate attached to carbon atom 6. Also known as Robisonester. { 'glü,kōs ¦siks 'fäs,fāt }

glucose-6-phosphate dehydrogenase [BIOCHEMISTRY] The mammalian enzyme that catalyzes the oxidation of glucose-6-phosphate by TPN^+ (triphosphopyridine nucleotide). { 'glü,kōs ¦siks 'fäs,fāt dē,hī'drä·jə,nās }

glucosidase [BIOCHEMISTRY] An enzyme that hydrolyzes glucosides. { glü'kō·sə,dās }

glucoside [BIOCHEMISTRY] One of a group of compounds containing the cyclic forms of glucose, in which the hydrogen of the hemiacetal hydroxyl has been replaced by an alkyl or aryl group. { 'glü·kə,sīd }

glucosyltransferase [BIOCHEMISTRY] An enzyme that catalyzes the glucosylation of hydroxymethyl cytosine; a constituent of bacteriophage deoxyribonucleic acid. { ¦glü·kə,sil'tranz·fə,rās }

glucuronic acid [BIOCHEMISTRY] $C_6H_{10}O_7$ An acid resulting from oxidation of the CH_2OH radical of D-glucose to COOH; a component of many polysaccharides and certain vegetable gums. Also known as glycuronic acid. { ¦glü·kyə¦rän·ik 'as·əd }

glucuronidase [BIOCHEMISTRY] An enzyme that catalyzes hydrolysis of glucuronides. Also known as glycuronidase. { ,glü·kyə'rän·ə,dās }

glucuronide [BIOCHEMISTRY] A compound resulting from the interaction of glucuronic acid with a phenol, an alcohol, or an acid containing a carboxyl group. Also known as glycuronide. { glü'kyůr·ə,nīd }

D-glucuronolactone [BIOCHEMISTRY] $C_6H_8O_6$ A water-soluble crystalline compound found in plant gums in polymers with other carbohydrates, and an important structural component of almost all fibrous and connective tissues in animals; used in medicine as an antiarthritic. { ¦dē glü¦kyůr·ə·nō'lak,tōn }

glue cell [INVERTEBRATE ZOOLOGY] *See* adhesive cell. { 'glü ,sel }

glume [BOTANY] One of two bracts at the base of a spikelet of grass. { glüm }

glumiferous [BOTANY] Bearing glumes. { glü 'mif·ə·rəs }

Glumiflorae [BOTANY] An equivalent name for Cyperales. { ,glü·mə'flȯr,ē }

glutamate [BIOCHEMISTRY] A salt or ester of glutamic acid. { 'glüd·ə,māt }

glutamic acid [BIOCHEMISTRY] $C_5H_9O_4N$ A dicarboxylic amino acid of the α-ketoglutaric acid family occurring widely in proteins. { glü'tam·ik 'as·əd }

glutaminase [BIOCHEMISTRY] The enzyme which catalyzes the conversion of glutamine to glutamic acid and ammonia. { glü'tam·ə,nās }

glutamine [BIOCHEMISTRY] $C_5H_{10}O_3N_2$ An amino acid; the monamide of glutamic acid; found in the juice of many plants and essential to the development of certain bacteria. { 'glüd·ə,mēn }

glutamine synthetase [BIOCHEMISTRY] An enzyme which catalyzes the formation of glutamine from glutamic acid and ammonia, using adenosinetriphosphate as a source of energy. { 'glüd·ə,mēn 'sin·thə,tās }

glutarate [BIOCHEMISTRY] The salt or ester of glutaric acid. { 'glüd·ə,rāt }

glutaric acid [BIOCHEMISTRY] $C_5H_5O_4$ A water-soluble, crystalline acid that occurs in green sugarbeets and in water extracts of crude wool. { glü'tar·ik 'as·əd }

glutathione [BIOCHEMISTRY] $C_{10}H_{17}O_6N_3S$ A widely distributed tripeptide that is important in plant and animal tissue oxidation reactions. { ¦glüd·ə'thī,ōn }

glutelin [BIOCHEMISTRY] A class of simple, heat-labile proteins occurring in seeds of cereals; soluble in dilute acids and alkalies. { 'glüd·əl·ən }

gluten [BIOCHEMISTRY] **1.** A mixture of proteins found in the seeds of cereals; gives dough elasticity and cohesiveness. **2.** An albuminous element of animal tissue. { 'glüt·ən }

glutenin [BIOCHEMISTRY] A glutelin of wheat. { 'glüt·ən·ən }

gluteus maximus [ANATOMY] The largest and most superficial muscle of the buttocks. { 'glüd·ē·əs 'mak·sə·məs }

gluteus medius [ANATOMY] The muscle of the buttocks lying between the gluteus maximus and gluteus minimus. { 'glüd·ē·əs 'mēd·ē·əs }

gluteus minimus [ANATOMY] The smallest and deepest muscle of the buttocks. { 'glüd·ē·əs 'min·ə·məs }

glutinant nematocyst [INVERTEBRATE ZOOLOGY] A nematocyst characterized by an open, sticky tube used for anchoring the cnidarian when walking on its tentacles. { 'glüt·ən·ənt nə'mad·ə,sist }

glutinous [BOTANY] Having a sticky surface. { 'glüt·ən·əs }

glycemia [PHYSIOLOGY] The presence of glucose in the blood. { glī'sē·mē·ə }

glyceraldehyde [BIOCHEMISTRY] $CH_2OHCOHCHO$ A colorless solid, isomeric with dehydroxyacetone; soluble in water and insoluble in organic solvents; an important intermediate in carbohydrate metabolism; used as a chemical intermediate in biochemical research and nutrition. { ,glis·ə'ral·də,hīd }

glycerate [BIOCHEMISTRY] A salt or ester of glyceric acid. { 'glis·ə,rāt }

glyceric acid [BIOCHEMISTRY] $C_3H_6O_4$ A hydroxy acid obtained by oxidation of glycerin. { glə 'ser·ik 'as·əd }

Glyceridae [INVERTEBRATE ZOOLOGY] A family of polychaete annelids belonging to the Errantia and characterized by an enormous eversible proboscis. { glə'ser·ə,dē }

glyceride [BIOCHEMISTRY] An ester of glycerin and an organic acid radical; fats are glycerides of certain long-chain fatty acids. { 'glis·ə,rīd }

glycerinated vaccine virus [IMMUNOLOGY] *See* smallpox vaccine. { 'glis·ə·rə,nād·əd 'vak,sēn ,vī·rəs }

glycerokinase [BIOCHEMISTRY] An enzyme that catalyzes the phosphorylation of glycerol to glycerophosphate during microbial fermentation of propionic acid. { ¦glis·ə·rō'kī,nās }

glycerophosphate [BIOCHEMISTRY] Any salt of glycerophosphoric acid. { ,glis·ə·rō'fäs,fāt }

glycerophosphoric acid [BIOCHEMISTRY] $C_3H_5(OH)_2OPO_3H_2$ Either of two pale-yellow, water-soluble, isomeric dibasic acids occurring in nature in combined form as cephalin and lecithin. { ¦glis·ə·rō,fäs'fȯr·ik 'as·əd }

glycine [BIOCHEMISTRY] $C_2H_5O_2N$ A white, crystalline amino acid found as a constituent of many proteins. Also known as aminoacetic acid. { 'glī,sēn }

glycocalyx [CELL BIOLOGY] The outer component of a cell surface, outside the plasmalemma; usually contains strongly acidic sugars, hence it carries a negative electric charge. { 'glī·kō'kā,liks }

glycocholic acid [BIOCHEMISTRY] $C_{26}H_{43}NO_6$ A bile obtained by the conjugation of cholic acid with glycine. { ¦gli·kō¦käl·ik 'as·əd }

glycocyamine [BIOCHEMISTRY] $C_3H_7N_3O_2$ A product of interaction of aminocetic acid and arginine, which on transmethylation with methionine is converted to creatine. Also known as guanidine-acetic acid. { 'glī·kō'sī·ə,mēn }

glycogen [BIOCHEMISTRY] A nonreducing, white, amorphous polysaccharide found as a reserve carbohydrate stored in muscle and liver cells of all higher animals, as well as in cells of lower animals. { 'glī·kə·jən }

glycogenesis [BIOCHEMISTRY] The metabolic formation of glycogen from glucose. { 'glī·kə 'jen·ə·səs }

glycogenolysis [BIOCHEMISTRY] The metabolic breakdown of glycogen. { 'glī·kə·jə'näl·ə·səs }

glycogen synthetase [BIOCHEMISTRY] An enzyme that catalyzes the synthesis of the amylose chain of glycogen. { 'glī·kə·jən 'sin·thə,tās }

glycolipid [BIOCHEMISTRY] Any of a class of complex lipids which contain carbohydrate residues. { ¦glī·kō¦lip·əd }

glycolysis [BIOCHEMISTRY] The enzymatic breakdown of glucose or other carbohydrate, with the formation of lactic acid or pyruvic acid and the release of energy in the form of adenosinetriphosphate. { ¦glī'käl·ə·səs }

glycolytic pathway [BIOCHEMISTRY] The principal series of phosphorylative reactions involved in pyruvic acid production in phosphorylative fermentations. Also known as Embden-Meyerhof pathway; hexose diphosphate pathway. { ¦glī·kə ¦lid·ik 'path,wā }

glyconeogenesis [BIOCHEMISTRY] The metabolic process of glycogen formation from noncarbohydrate precursors. { ¦glī·kō¦nē·ō'jen·ə·səs }

glycopeptide [BIOCHEMISTRY] *See* glycoprotein. { ,glī·kō'pep,tīd }

glycophyte [BOTANY] A plant requiring more than 0.5% sodium chloride solution in the substratum. { 'gli·kə,fīt }

glycoprotein [BIOCHEMISTRY] Any of a class of conjugated proteins containing both carbohydrate and protein units. Also known as glycopeptide. { ¦glī·kō'prō,tēn }

glycose [BIOCHEMISTRY] A simple sugar whose structure is in the form either of an open-chain aldehyde or ketone or of a cyclic hemiacetal. { 'glī,kōs }

glycosidase [BIOCHEMISTRY] An enzyme that hydrolyzes a glycoside. { glī'kō·sə,dās }

glycoside [BIOCHEMISTRY] A compound that yields on hydrolysis a sugar (glucose, galactose) and an aglycon; many of the glycosides are therapeutically valuable. { 'glī·kə,sīd }

glycosyl [BIOCHEMISTRY] A univalent functional group derived from the cyclic form of glycose by removal of the hemiacetal hydroxyl group. { 'glī·kə,sil }

glycosylation [BIOCHEMISTRY] A chemical reaction in which glycosyl groups are added to a protein to produce a glycoprotein. { glī,käs·ə'lā·shən }

glycotropic [BIOCHEMISTRY] Acting to antagonize the action of insulin. { ¦glī·kō¦träp·ik }

glycuresis [PHYSIOLOGY] Excretion of sugar seen normally in urine. { ,glik·yə'rē·səs }

glycuronic acid [BIOCHEMISTRY] *See* glucuronic acid. { ¦glik·yə¦rän·ik 'as·əd }

glycuronidase [BIOCHEMISTRY] *See* glucuronidase. { ,glik·yə'rän·ə,dās }

glycuronide [BIOCHEMISTRY] *See* glucuronide. { gli'kyu̇r·ə,nīd }

glyoxalase [BIOCHEMISTRY] An enzyme present in various body tissues which catalyzes the conversion of methylglyoxal into lactic acid. { glī'äk·sə,lās }

glyoxylate cycle [BIOCHEMISTRY] A sequence of biochemical reactions related to respiration in germinating fatty seeds by which acetyl coenzyme A is converted to succinic acid and then to hexose. { glī'äk·sə,lāt ,sī·kəl }

glyoxylic acid [BIOCHEMISTRY] $CH(OH)_2COOH$ An aldehyde acid found in many plant and animal tissues, especially unripe fruit. { ¦glī,äk¦sil·ik 'as·əd }

glyoxysome [BOTANY] A specialized type of peroxisome found in plant tissues that is bounded by a single membrane and contains a broad spectrum of enzymes, including those of the glyoxylate cycle and the β-oxidation cycle in addition to catalase and oxidase. { glī'äk·sə ,sōm }

Glyphocyphidae [PALEONTOLOGY] A family of extinct echinoderms in the order Temnopleuroida comprising small forms with a sculptured test, perforate crenulate tubercles, and diademoid ambulacral plates. { ,glif·ō'sīf·ə,dē }

Glyptocrinina [PALEONTOLOGY] A suborder of extinct crinoids in the order Monobathrida. { ¦glip·tō·krə'nī·nə }

gnat [INVERTEBRATE ZOOLOGY] The common name for a large variety of biting insects in the order Diptera. { nat }

Gnathiidea [INVERTEBRATE ZOOLOGY] A suborder of isopod crustaceans characterized by a much reduced second thoracomere, short antennules and antennae, and a straight pleon. { nā'thī·ə·də }

gnathion [VERTEBRATE ZOOLOGY] The most anterior point of the premaxillae on or near the middle line in certain lower mammals. { 'nā·thē ,än }

gnathite [INVERTEBRATE ZOOLOGY] A mouth appendage in arthropods. { 'nā,thīt }

Gnathobdellae [INVERTEBRATE ZOOLOGY] A suborder of leeches in the order Arhynchobdellae having jaws and a conspicuous posterior sucker; it contains most of the important blood-sucking leeches of humans and other warm-blooded animals. { ¦nā,thäb'del·ē }

Gnathobelodontinae [PALEONTOLOGY] A subfamily of extinct elephantoid proboscideans containing the shovel-jawed forms of the family Gomphotheriidae. { nā¦thäb·ə·lō'dän·tə,nē }

gnathocephalon [INVERTEBRATE ZOOLOGY] The part of the insect head lying behind the protocephalon; bears the maxillae and mandibles. { ,nā·thō'sef·ə,län }

gnathochilarium [INVERTEBRATE ZOOLOGY] The lower lip of certain arthropods; thought to be fused maxillae. { ,nā·thō,kī'lar·ē·əm }

Gnathodontidae [PALEONTOLOGY] A family of extinct conodonts having platforms with large, cup-shaped attachment scars. { ,nā·thō'dän·tə,dē }

gnathopod [INVERTEBRATE ZOOLOGY] Any of the crustacean paired thoracic appendages modified for manipulation of food but sometimes functioning in copulatory amplexion. { 'nā·thə ,päd }

gnathopodite [INVERTEBRATE ZOOLOGY] A segmental, modified appendage which serves as a jaw in arthropods. { nā'thä·pə,dīt }

gnathos [INVERTEBRATE ZOOLOGY] A mid-ventral plate on the ninth tergum in lepidopterans. { 'nā,thōs }

gnathostegite [INVERTEBRATE ZOOLOGY] One of a pair of broad plates formed from the outer maxillipeds of some crustaceans, which function to cover other mouthparts. { nə'thäs·ə‚jīt }

Gnathostomata [INVERTEBRATE ZOOLOGY] A suborder of echinoderms in the order Echinoidea characterized by a rigid, exocyclic test and a lantern or jaw apparatus. [VERTEBRATE ZOOLOGY] A group of the subphylum Vertebrata which possess jaws and usually have paired appendages. { ‚nā·thə'stō·məd·ə }

Gnathostomidae [INVERTEBRATE ZOOLOGY] A family of parasitic nematodes in the superfamily Spiruroidea; sometimes placed in the superfamily Physalopteroidea. { ‚nā·thə'stō·məd‚ē }

Gnathostomulida [INVERTEBRATE ZOOLOGY] Microscopic marine worms of uncertain systematic relationship, mainly characterized by cuticular structures in the pharynx and a monociliated skin epithelium. { nə¦thäs·tə'myül·ə·də }

gnathothorax [INVERTEBRATE ZOOLOGY] The thorax and part of the head bearing feeding organs in arthropods, regarded as a primary region of the body. { ¦nā·thō'thȯr‚aks }

Gnetales [BOTANY] A monogeneric order of the subdivision Gneticae; most species are lianas with opposite, oval, entire-margined leaves. { nə'tā·lēz }

Gneticae [BOTANY] A subdivision of the division Pinophyta characterized by vessels in the secondary wood, ovules with two integuments, opposite leaves, and an embryo with two cotyledons. { 'ned·ə‚sē }

Gnetophyta [BOTANY] The equivalent name for Gnetopsida. { nə'täf·əd·ə }

Gnetopsida [BOTANY] A class of gymnosperms comprising the subdivision Gneticae. { nə'täp·səd·ə }

Gnostidae [INVERTEBRATE ZOOLOGY] An equivalent name for Ptinidae. { 'näs·tə‚dē }

gnotobiology [BIOLOGY] That branch of biology dealing with known living forms; the study of higher organisms in the absence of all demonstrable, viable microorganisms except those known to be present. { ¦nō·dō·bī'äl·ə·jē }

gnotobiote [MICROBIOLOGY] **1.** An individual (host) living in intimate association with another known species (microorganism). **2.** The known microorganism living on a host. { ¦nō·dō'bī‚ōt }

gnu [VERTEBRATE ZOOLOGY] Any of several large African antelopes of the genera *Connochaetes* and *Gorgon* having a large oxlike head with horns that characteristically curve downward and outward and then up, with the bases forming a frontal shield in older individuals. { nü }

goat [VERTEBRATE ZOOLOGY] The common name for a number of artiodactyl mammals in the genus *Capra*; closely related to sheep but differing in having a lighter build and hollow, swept-back, sometimes spiral or twisted horns. { gōt }

Gobiatheriinae [PALEONTOLOGY] A subfamily of extinct herbivorous mammals in the family Uintatheriidae known from one late Eocene genus; characterized by extreme reduction of anterior dentition and by lack of horns. { gō¦bī·ə·thə'rī·ə‚nē }

Gobiesocidae [VERTEBRATE ZOOLOGY] The single family of the order Gobiesociformes. { gō¦bī·ə¦säs·ə‚dē }

Gobiesociformes [VERTEBRATE ZOOLOGY] The clingfishes, a monofamilial order of scaleless bony fishes equipped with a thoracic sucking disk which serves for attachment. { gō¦bī·ə‚säs·ə'fōr‚mēz }

Gobiidae [VERTEBRATE ZOOLOGY] A family of perciform fishes in the suborder Gobioidei characterized by pelvic fins united to form a sucking disk on the breast. { gō'bī·ə‚dē }

Gobioidei [VERTEBRATE ZOOLOGY] The gobies, a suborder of morphologically diverse actinopterygian fishes in the order Perciformes; all lack a lateral line. { ‚gō·bē'ȯid·ē‚ī }

goblet cell [HISTOLOGY] A unicellular, mucus-secreting intraepithelial gland that is distended on the free surface. Also known as chalice cell. [INVERTEBRATE ZOOLOGY] Any of the unicellular choanocytes of the genus *Monosiga*. { 'gäb·lət ‚sel }

golden algae [BOTANY] The common name for members of the class Chrysophyceae. { 'gol·dən 'al·jē }

golden-brown algae [BOTANY] The common name for members of the division Chrysophyta. { 'gol·dən ‚braün 'al·jē }

goldfish [VERTEBRATE ZOOLOGY] *Crassius auratus*. An orange cypriniform fish of the family Cyprinidae that can grow to over 18 inches (46 centimeters); closely related to the carps. { 'gōl ‚fish }

Golgi apparatus [CELL BIOLOGY] A cellular organelle that is part of the cytoplasmic membrane system; it is composed of regions of stacked cisternae and it functions in secretory processes. { 'gȯl‚jē ‚ap·ə‚rad·əs }

Golgi cell [ANATOMY] **1.** A nerve cell with long axons. **2.** A nerve cell with short axons that branch repeatedly and terminate near the cell body. { 'gȯl·jē ‚sel }

Golgi-Mazzoni's corpuscle [ANATOMY] A small sensory lamellar corpuscle located in the parietal pleura. { 'gȯl·jē mät'sō·nēz ‚kȯr·pə·səl }

Golgi stack [CELL BIOLOGY] The central structure of the Golgi apparatus consisting of flattened membrane-bounded cisternae. Formerly known as dictyosome. { 'gōl·jē ‚stak }

Golgi tendon organ [PHYSIOLOGY] Any of the kinesthetic receptors situated near the junction of muscle fibers and a tendon which act as muscle-tension recorders. { 'gȯl·jē 'ten·dən ‚ȯr·gən }

Gomphidae [INVERTEBRATE ZOOLOGY] A family of dragonflies belonging to the Anisoptera. { 'gäm·fə‚dē }

gomphosis [ANATOMY] An immovable articulation, as that formed by the insertion of teeth into the bony sockets. { gäm'fō·səs }

Gomphotheriidae [PALEONTOLOGY] A family of extinct proboscidean mammals in the suborder Elephantoidea consisting of species with shoveling or digging specializations of the lower tusks. { ˌgäm·fō·thə'rī·əˌdē }

Gomphotheriinae [PALEONTOLOGY] A subfamily of extinct elephantoid proboscideans in the family Gomphotheriidae containing species with long jaws and bunomastodont teeth. { ˌgäm·fō·thə'rī·əˌnē }

gonad [ANATOMY] A primary sex gland; an ovary or a testis. { 'gōˌnad }

gonadotropic hormone [BIOCHEMISTRY] Either of two adenohypophyseal hormones, FSH (follicle-stimulating hormone) or ICSH (interstitial-cell-stimulating hormone), that act to stimulate the gonads. { gō¦nad·ə¦träp·ik 'hȯrˌmōn }

gonadotropin [BIOCHEMISTRY] A substance that acts to stimulate the gonads. { gōˌnad·ə'trō·pən }

gonapophysis [INVERTEBRATE ZOOLOGY] A paired, modified appendage of the anal region in insects that functions in copulation, oviposition, or stinging. { gän·ə'päf·ə·səs }

gonia [HISTOLOGY] Primordial sex cells, such as oogonia and spermatogonia. { 'gō·nē·ə }

Goniadidae [INVERTEBRATE ZOOLOGY] A family of marine polychaete annelids belonging to the Errantia. { ˌgō·nē'ad·əˌdē }

gonidium [BIOLOGY] An asexual reproductive cell or group of cells arising in a special organ on or in a gametophyte. { gō'nid·ē·əm }

gonimoblast [BOTANY] A filament arising from the fertilized carpogonium of most red algae. { 'gän·ə·mōˌblast }

Gonodactylidae [INVERTEBRATE ZOOLOGY] A family of mantis shrimp in the order Stomatopoda. { ˌgän·ō·dak'til·əˌdē }

gonodendrum [INVERTEBRATE ZOOLOGY] A branched structure which bears clusters of gonophores. { ˌgän·ə'den·drəm }

gonomery [EMBRYOLOGY] In some insect embryos, grouping of maternal and paternal chromosomes separately during the first couple of mitotic divisions after fertilization. { gō'näm·ə·rē }

gonopalpon [INVERTEBRATE ZOOLOGY] Tentacle-like, sensitive structures associated with cnidarian gonophores. { ˌgän·ə'pal·pən }

gonophore [BOTANY] An elongation of the receptacle extending between the stamens and corolla. [INVERTEBRATE ZOOLOGY] Reproductive zooid of a hydroid colony. { 'gän·əˌfȯr }

gonopodium [VERTEBRATE ZOOLOGY] Anal fin modified as a copulatory organ in certain fishes. { ˌgän·ə'pōd·ē·əm }

Gonorhynchiformes [VERTEBRATE ZOOLOGY] A small order of soft-rayed teleost fishes having fusiform or moderately compressed bodies, single short dorsal and anal fins, a forked caudal fin, and weak toothless jaws. { ¦gän·əˌriŋ·kə'fȯrˌmēz }

gonosome [INVERTEBRATE ZOOLOGY] Aggregate of gonophores in a hydroid colony. { 'gän·əˌsōm }

gonosomite [INVERTEBRATE ZOOLOGY] The ninth segment of the abdomen of the male insect. Also known as genital segment. { ˌgän·ə'sōˌmīt }

gonostome [INVERTEBRATE ZOOLOGY] The part of the genital duct of a coelomate invertebrate known as the coelomic funnel. Also known as coelomostome. { 'gän·əˌstōm }

gonostyle [INVERTEBRATE ZOOLOGY] Gonapophysis of dipteran insects. { 'gän·əˌstīl }

gonotocont [BIOLOGY] *See* auxocyte. { gə'näd·əˌkänt }

gonotome [EMBRYOLOGY] The part of an embryonic somite involved in gonad formation. { 'gän·əˌtōm }

gonozooid [INVERTEBRATE ZOOLOGY] A zooid of bryozoans and tunicates which produces gametes. { ¦gän·ə'zōˌȯid }

gonyautoxin [BIOCHEMISTRY] One of a group of saxitoxin-related compounds that are produced by the dinoflagellates *Gonyaulax catenella* and *G. tamarensis*. { ¦gō·nē·ō¦täk·sən }

gonys [VERTEBRATE ZOOLOGY] A ridge along the mid-ventral line of the lower mandible of certain birds. { 'gō·nəs }

goose [VERTEBRATE ZOOLOGY] The common name for a number of waterfowl in the subfamily Anatinae; they are intermediate in size and features between ducks and swans. { güs }

gooseberry [BOTANY] The common name for about six species of thorny, spreading bushes of the genus *Ribes* in the order Rosales, producing small, acidic, edible fruit. { 'güsˌber·ē }

gooseneck barnacle [INVERTEBRATE ZOOLOGY] Any stalked barnacle, especially of the genus *Lepas*. { ¦güsˌnek 'bär·ni·kəl }

gopher [VERTEBRATE ZOOLOGY] The common name for North American rodents composing the family Geomyidae. Also known as pocket gopher. { 'gō·fər }

Gordiidae [INVERTEBRATE ZOOLOGY] A monogeneric family of worms in the order Gordioidea distinguished by a smooth cuticle. { gȯr'dī·əˌdē }

Gordioidea [INVERTEBRATE ZOOLOGY] An order of worms belonging to the Nematomorpha in which there is one ventral epidermal cord, a body cavity filled with mesenchymal tissue, and paired gonads. { ˌgȯr·dē'ȯid·ē·ə }

gordioid larva [INVERTEBRATE ZOOLOGY] The developmental stage of nematomorphs, free-living for a short time. { 'gȯr·dēˌȯid 'lär·və }

Gorgonacea [INVERTEBRATE ZOOLOGY] The horny corals, an order of the cnidarian subclass Alcyonaria; colonies are fanlike or featherlike with branches spread radially or oppositely in one plane. { ˌgȯr·gə'nās·ē·ə }

gorgonin [BIOCHEMISTRY] The protein, frequently containing iodine and bromine, composing the horny skeleton of members of the Gorgonacea; contains iodine and bromine. { 'gȯr·gə·nən }

Gorgonocephalidae [INVERTEBRATE ZOOLOGY] A family of ophiuroid echinoderms in the order Phrynophiurida in which the individuals often have branched arms. { ˌgȯr·gəˌnō·sə'fal·əˌdē }

gorilla [VERTEBRATE ZOOLOGY] *Gorilla gorilla*. An anthropoid ape, the largest living primate; the two African subspecies are the lowland gorilla and the mountain gorilla. { gə'ril·ə }

gossypose [BIOCHEMISTRY] *See* raffinose. { 'gäs·ə‚pōs }

Graafian follicle [HISTOLOGY] The mature mammalian ovum with its surrounding epithelial cells. { 'gräf·ē·ən 'fäl·ə·kəl }

Gracilariidae [INVERTEBRATE ZOOLOGY] A family of small moths in the superfamily Tineoidea; both pairs of wings are lanceolate and widely fringed. { ‚gra·sil·ə'rī·ə‚dē }

gracilis [ANATOMY] A long slender muscle on the medial aspect of the thigh. { 'gras·ə·ləs }

grade [EVOLUTION] A stage of evolution in which a similar level of organization is reached by one or more species in the development of a structure, physiological process, or behavioral character. { grād }

graded topocline [ECOLOGY] A topocline having a wide range, or ranging into different kinds of environment, thus subjecting its members to differential selection so that divergence between local races may become sufficient to warrant creation of varietal, or even specific, names. { ¦grād·əd 'täp·ə‚klīn }

gradualism [EVOLUTION] A model of evolution in which change is slow, steady, and on the whole ameliorative, resulting in a gradual and continuous increase in biological diversity. Also known as phyletic gradualism. { 'graj·ə·wə‚liz·əm }

graft [BIOLOGY] **1.** To unite to form a graft. **2.** A piece of tissue transplanted from one individual to another or to a different place on the same individual. **3.** An individual resulting from the grafting of parts. [BOTANY] To unite a scion to an understock in such manner that the two grow together and continue development as a single plant without change in scion or stock. { graft }

Grahamella [MICROBIOLOGY] A genus of the family Bartonellaceae; intracellular parasites in red blood cells of rodents and other mammals. { ‚grā·ə'mel·ə }

grain [BOTANY] **1.** A rounded, granular prominence on the back of a sepal. **2.** *See* cereal. **3.** *See* drupelet. { grān }

grain sorghum [BOTANY] *Sorghum bicolor.* A grass plant cultivated for its grain and to a lesser extent for forage. Also known as nonsaccharine sorghum. { ¦grān 'sȯr·gəm }

gramagrass [BOTANY] Any grass of the genus *Bouteloua*; pasture grass. { 'gram·ə‚gras }

gramicidin [MICROBIOLOGY] A polypeptide antibacterial antibiotic produced by *Bacillus brevis*; active locally against gram-positive bacteria. { ‚gram·ə'sīd·ən }

Graminales [BOTANY] The equivalent name for Cyperales. { ‚gram·ə'nā·lēz }

Gramineae [BOTANY] The grasses, a family of monocotyledonous plants in the order Cyperales characterized by distichously arranged flowers on the axis of the spikelet. { grə'min·ē‚ē }

graminicolous [ECOLOGY] Living upon grass. { ¦gram·ə¦nik·ə·ləs }

graminivorous [ZOOLOGY] Feeding on grasses. { ¦gram·ə¦niv·ə·rəs }

graminoid [BOTANY] Of or resembling the grasses. { 'gram·ə‚nȯid }

gram-negative [MICROBIOLOGY] Of bacteria, decolorizing and staining with the counterstain when treated with Gram's stain. { 'gram ¦neg·əd·iv }

gram-negative diplococci [MICROBIOLOGY] The three bacteriologic genera composing the family Neisseriaceae: *Gemella*, *Veillonella*, and *Neisseria*. { 'gram ¦neg·əd·iv ‚dip·lə'käk·sē }

gram-positive [MICROBIOLOGY] Of bacteria, holding the color of the primary stain when treated with Gram's stain. { 'gram ¦päs·əd·iv }

Gram's stain [MICROBIOLOGY] A differential bacteriological stain; a fixed smear is stained with a slightly alkaline solution of basic dye, treated with a solution of iodine in potassium iodide, and then with a neutral decolorizing agent, and usually counterstained; bacteria stain either blue (gram-positive) or red (gram-negative). { 'gramz ‚stān }

gram-variable [MICROBIOLOGY] Pertaining to staining inconsistently with Gram's stain. { 'gram ¦ver·ē·ə·bəl }

grana [CELL BIOLOGY] A multilayered membrane unit formed by stacks of the lobes or branches of a chloroplast thylakoid. { 'grän·ə }

granellare [INVERTEBRATE ZOOLOGY] In xenophyophores, that portion of the body consisting of the multinucleate plasmodium and its enclosing, branching organic tube. { 'gran·əl‚är }

granite moss [BOTANY] The common name for a group of the class Bryatae represented by two Arctic genera and distinguished by longitudinal splitting of the mature capsule into four valves. { 'gran·ət ¦mȯs }

Grantiidae [INVERTEBRATE ZOOLOGY] A family of calcareous sponges in the order Sycettida. { gran 'tī·ə‚dē }

granular component [CELL BIOLOGY] The component of the nucleolus that contains the cleaved preribosomal particles. { ¦gran·yə·lər kəm¦pō·nənt }

granular gland [PHYSIOLOGY] A gland that produces and secretes a granular material. { 'gran·yə·lər 'gland }

granular leukocyte [HISTOLOGY] *See* granulocyte. { 'gran·yə·lər 'lü·kə‚sīt }

granulocyte [HISTOLOGY] A leukocyte containing granules in the cytoplasm. Also known as granular leukocyte; polymorph; polymorphonuclear leukocyte. { 'gran·yə·lō‚sīt }

Granuloreticulosia [INVERTEBRATE ZOOLOGY] A subclass of the protozoan class Rhizopodea characterized by reticulopodia which often fuse into networks. { ¦gran·yə·lō·re‚tik·yə'lō·shə }

granulosis [INVERTEBRATE ZOOLOGY] A virus disease of lepidopteran larvae characterized by the accumulation of small granular inclusion bodies (capsules) in the infected cells. { ‚gran·yə'lō·səs }

grape [BOTANY] The common name for plants of the genus *Vitis* characterized by climbing stems with cylindrical-tapering tendrils and polygamodioecious flowers; grown for the edible, pulpy berries. { grāp }

grapefruit [BOTANY] *Citrus paradisi.* An evergreen tree with a well-rounded top cultivated for its edible fruit, a large, globose citrus fruit characterized by a yellow rind and white, pink, or red pulp. { 'grāp,früt }

grape sugar [BIOCHEMISTRY] *See* dextrose. { 'grāp ,shu̇g·ər }

Graphidaceae [BOTANY] A family of mosses formerly grouped with lichenized Hysteriales but now included in the order Lecanorales; individuals have true paraphyses. { ,graf·ə'dās·ē,ē }

Graphiolaceae [MYCOLOGY] A family of parasitic fungi in the order Ustilaginales in which teleutospores are produced in a cuplike fruiting body. { ,graf·ē·ō'lās·ē,ē }

Grapsidae [INVERTEBRATE ZOOLOGY] The square-backed crabs, a family of decapod crustaceans in the section Brachyura. { 'grap·sə,dē }

Graptolithina [PALEONTOLOGY] A class of extinct colonial animals believed to be related to the class Pterobranchia of the Hemichordata. { ,grap·tə·lə'thīn·ə }

Graptoloidea [PALEONTOLOGY] An order of extinct animals in the class Graptolithina including branched, planktonic forms described from black shales. { ,grap·tə'lȯid·ē·ə }

Graptozoa [PALEONTOLOGY] The equivalent name for Graptolithina. { ,grap·tə'zō·ə }

grass [BOTANY] The common name for all members of the family Gramineae; moncotyledonous plants having leaves that consist of a sheath which fits around the stem like a split tube, and a long, narrow blade. { gras }

grasserie [INVERTEBRATE ZOOLOGY] A polyhedrosis disease of silkworms characterized by spotty yellowing of the skin and internal liquefaction. Also known as jaundice. { 'gras·ə·rē }

grasshopper [INVERTEBRATE ZOOLOGY] The common name for a number of plant-eating orthopteran insects composing the subfamily Saltatoria; individuals have hindlegs adapted for jumping, and mouthparts adapted for biting and chewing. { 'gras,häp·ər }

grassland [ECOLOGY] Any area of herbaceous terrestrial vegetation dominated by grasses and graminoid species. { 'gras,land }

gravid [ZOOLOGY] **1.** Of the uterus, containing a fetus. **2.** Pertaining to female animals when carrying young or eggs. { 'grav·əd }

Gravigrada [VERTEBRATE ZOOLOGY] The sloths, a group of herbivorous xenarthran mammals in the order Edentata; members are completely hairy and have five upper and four lower prismatic teeth without enamel. { grə'vig·rə·də }

gravitropism [BOTANY] *See* geotropism. { grə'vi·trə,piz·əm }

gray matter [HISTOLOGY] The part of the central nervous system composed of nerve cell bodies, their dendrites, and the proximal and terminal unmyelinated portions of axons. { 'grā ,mad·ər }

graze [VERTEBRATE ZOOLOGY] To feed by browsing on, cropping, and eating grass. { grāz }

greasewood [BOTANY] Any plant of the genus *Sarcobatus*, especially *S. vermiculatus*, which is a low shrub that grows in alkali soils of the western United States. { 'grēs,wu̇d }

greater omentum [ANATOMY] A fold of peritoneum that is attached to the greater curvature of the stomach and hangs down over the intestine and fuses with the mesocolon. { 'grād·ər ō 'men·təm }

great galago [VERTEBRATE ZOOLOGY] *See* thick-tailed bushbaby. { ¦grāt gə'lā·gō }

grebe [VERTEBRATE ZOOLOGY] The common name for members of the family Podicipedidae; these birds have legs set far posteriorly, compressed bladelike tarsi, individually broadened and lobed toes, and a rudimentary tail. { grēb }

Greeffiellidae [INVERTEBRATE ZOOLOGY] A family of free-living nematodes in the superfamily Desmoscolecoidea. { grē·fē'el·ə,dē }

Greeffielloidea [INVERTEBRATE ZOOLOGY] A superfamily of primarily marine, free-living nematodes in the order Desmoscolecida, distinguished by a prominent nonencrusted annulation that bears a ring of elongate spines or short scales, and large subdorsal and subventral tubular setae along the body. { grē·fə'lȯid·ē·ə }

green algae [BOTANY] The common name for members of the plant division Chlorophyta. { 'grēn ¦al·jē }

green gland [INVERTEBRATE ZOOLOGY] *See* antennal gland. { 'grēn ¦gland }

greenhouse [BOTANY] Glass-enclosed, climate-controlled structure in which young or out-of-season plants are cultivated and protected. { 'grēn,hau̇s }

green mold [MYCOLOGY] Any fungus, especially *Penicillium* and *Aspergillus* species, that is green or has green spores. { ¦grēn ¦mōld }

green sulfur bacteria [MICROBIOLOGY] A physiologic group of green photosynthetic bacteria of the Chloraceae that are capable of using hydrogen sulfide (H_2S) and other inorganic electron donors. { 'grēn ¦səl·fər bak'tir·ē·ə }

Gregarinia [INVERTEBRATE ZOOLOGY] A subclass of the protozoan class Telosporea occurring principally as extracellular parasites of invertebrates. { 'greg·ə'rin·ē·ə }

gressorial [VERTEBRATE ZOOLOGY] Adapted for walking, as certain birds' feet. { gre'sȯr·ē·əl }

Grimmiales [BOTANY] An order of mosses commonly growing upon rock in dense tufts or cushions and having hygroscopic, costate, usually lanceolate leaves arranged in many rows on the stem. { ,grim·ē'ā·lēz }

grindle [VERTEBRATE ZOOLOGY] *See* bowfin. { 'grind·əl }

grisein [MICROBIOLOGY] $C_{40}H_{61}O_{20}N_{10}SFe$ A red, crystalline, water-soluble antibiotic produced by strains of *Streptomyces griseus*. { grə'zēn }

griseofulvin [MICROBIOLOGY] $C_{17}H_{17}O_6Cl$ A colorless, crystalline antifungal antibiotic produced by several species of *Penicillium*. Also known as curling factor. { ,griz·ē·ō'fu̇l·vən }

griseolutein [MICROBIOLOGY] Either of two fractions, A or B, of broad-spectrum antibiotics produced by *Streptomyces griseoluteus*; more active against gram-positive than gram-negative microorganisms. { ,griz·ē·ō'lüd·ē·ən }

griseomycin [MICROBIOLOGY] A white, crystalline antibiotic produced by an actinomycete resembling *Streptomyces griseolus*. { ,griz·ē·ō'mīs·ən }

grizzly bear [VERTEBRATE ZOOLOGY] The common name for a number of species of large carnivorous mammals in the genus *Ursus*, family Ursidae. { 'griz·lē ˌber }

Groeberiidae [PALEONTOLOGY] A family of extinct rodentlike marsupials. { ˌgrə·bə'rī·əˌdē }

groin [ANATOMY] Depression between the abdomen and the thigh. { grȯin }

Gromiida [INVERTEBRATE ZOOLOGY] An order of protozoans in the subclass Filosia; the test, which is chitinous in some species and thin and somewhat flexible in others, is reinforced with sand grains or siliceous particles. { grə'mī·ə·də }

groove [BIOCHEMISTRY] Any of a group of depressions in the double helix of deoxyribonucleic acid that are believed to be sites occupied by nuclear proteins. { grüv }

gross anatomy [ANATOMY] Anatomy that deals with the naked-eye appearance of tissues. { 'grōs ə'nad·ə·mē }

gross primary production [ECOLOGY] The incorporation of organic matter or biocontent by a grassland community over a given period of time. { ¦grōs 'prīˌmer·ē prə'dək·shən }

gross production rate [ECOLOGY] The speed of assimilation of organisms belonging to a specific trophic level. { ¦grōs prə'dək·shən ˌrāt }

ground cover [BOTANY] Prostrate or low plants that cover the ground instead of grass. { 'grau̇nd ˌkəv·ər }

ground meristem [BOTANY] Partially differentiated meristematic tissue derived from the apical meristem that gives rise to ground tissue. { ¦grau̇nd 'mer·əˌstem }

groundplasm [CELL BIOLOGY] A polyphasic system in which the resolvable elements of the cytoplasm are suspended, including the larger organelles, enzymes of intermediate cell metabolism, contractile protein molecules, and the main cellular pool of soluble precursors. { 'grau̇ndˌplaz·əm }

ground substance [HISTOLOGY] *See* matrix. { 'grau̇nd ˌsəb·stəns }

ground tissue [BOTANY] In leaves and young roots and stems, any tissue other than the epidermis and vascular tissues. { 'grau̇nd ˌtish·ü }

group selection [EVOLUTION] Selection in which changes in gene frequency are brought about by the differential extinction and proliferation of the local population. { ¦grüp si¦lek·shən }

grouse [VERTEBRATE ZOOLOGY] Any of a number of game birds in the family Tetraonidae having a plump body and strong, feathered legs. { grau̇s }

growth [PHYSIOLOGY] Increase in the quantity of metabolically active protoplasm, accompanied by an increase in cell number or cell size, or both. { grōth }

growth curve [MICROBIOLOGY] A graphic representation of the growth of a bacterial population in which the log of the number of bacteria or the actual number of bacteria is plotted against time. { 'grōth ˌkərv }

growth factor [PHYSIOLOGY] Any factor, genetic or extrinsic, which affects growth. { 'grōth ˌfak·tər }

growth form [ECOLOGY] The habit of a plant determined by its appearance of branching and periodicity. { 'grōth ˌfȯrm }

growth hormone [BIOCHEMISTRY] **1.** A polypeptide hormone secreted by the anterior pituitary which promotes an increase in body size. Abbreviated GH. **2.** Any hormone that regulates growth in plants and animals. { 'grōth ¦hȯrˌmōn }

growth rate [MICROBIOLOGY] Increase in the number of bacteria in a population per unit time. { 'grōth ˌrāt }

growth regulator [BIOCHEMISTRY] A synthetic substance that produces the effect of a naturally occurring hormone in stimulating plant growth; an example is dichlorophenoxyacetic acid. { 'grōth ¦reg·yəˌlād·ər }

Gruidae [VERTEBRATE ZOOLOGY] The cranes, a family of large, tall, cosmopolitan wading birds in the order Gruiformes. { 'grü·əˌdē }

Gruiformes [VERTEBRATE ZOOLOGY] A heterogeneous order of generally cosmopolitan birds including the rails, coots, limpkins, button quails, sun grebes, and cranes. { 'grü·ə'fȯrˌmēz }

Gryllidae [INVERTEBRATE ZOOLOGY] The true crickets, a family of orthopteran insects in which individuals are dark-colored and chunky with long antennae and long, cylindrical ovipositors. { 'gril·əˌdē }

Grylloblattidae [INVERTEBRATE ZOOLOGY] A monogeneric family of crickets in the order Orthoptera; members are small, slender, wingless insects with hindlegs not adapted for jumping. { ˌgril·ō'blad·əˌdē }

Gryllotalpidae [INVERTEBRATE ZOOLOGY] A family of North American insects in the order Orthoptera which live in sand or mud; they eat the roots of seedlings growing in moist, light soils. { 'gril·ō'tal·pəˌdē }

guanidine [BIOCHEMISTRY] CH_5N_3 Aminomethanamidine, a product of protein metabolism found in urine. { 'gwän·əˌdēn }

guanidine-acetic acid [BIOCHEMISTRY] *See* glycocyamine. { 'gwän·əˌdēn ə¦sēd·ik 'as·əd }

guanine [BIOCHEMISTRY] $C_5H_5ON_5$ A purine base; occurs naturally as a fundamental component of nucleic acids. { 'gwänˌēn }

guanophore [HISTOLOGY] *See* iridocyte. { 'gwän·əˌfȯr }

guanosine [BIOCHEMISTRY] $C_{10}H_{13}O_5N_5$ Guanine riboside, a nucleoside composed of guanine and ribose. Also known as vernine. { 'gwän·əˌsēn }

guanosine tetraphosphate [BIOCHEMISTRY] A nucleotide which participates in the regulation of gene transcription in bacteria by turning off the synthesis of ribosomal ribonucleic acid. { 'gwän·əˌsēn te·trə'fäsˌfāt }

guanylic acid [BIOCHEMISTRY] A nucleotide composed of guanine, a pentose sugar, and phosphoric acid and formed during the hydrolysis of nucleic acid. Abbreviated GMP. Also known as guanosine monophosphate; guanosine phosphoric acid. { gwə'nil·ik 'as·əd }

guard cell [BOTANY] Either of two specialized cells surrounding each stoma in the epidermis of plants; functions in regulating stoma size. { 'gärd ,sel }

guava [BOTANY] *Psidium guajava.* A shrub or low tree of tropical America belonging to the family Myrtaceae; produces an edible, aromatic, sweet, juicy berry. { 'gwäv·ə }

guayule [BOTANY] *Parthenium argentatum.* A subshrub of the family Compositae that is native to Mexico and the southwestern United States; it has been cultivated as a source of rubber. { wī 'yü·lē }

gubernaculum [ANATOMY] A guiding structure, as the fibrous cord extending from the fetal testes to the scrotal swellings. [INVERTEBRATE ZOOLOGY] **1.** A posterior flagellum of certain protozoans. **2.** A sclerotized structure associated with the copulatory spicules of certain nematodes. { ,gü·bər'nak·yə·ləm }

Guest unit [BIOLOGY] A unit for the standardization of plasmin. { 'gest ,yü·nət }

guide ribonucleic acid [MOLECULAR BIOLOGY] A ribonucleic acid sequence that provides a template for the alignment of splice junctions. { ¦gīd ,rī·bō·nü¦klē·ik 'as·əd }

guild [ECOLOGY] A group of species that utilize the same kinds of resources, such as food, nesting sites, or places to live, in a similar manner. { gild }

guinea fowl [VERTEBRATE ZOOLOGY] The common name for plump African game birds composing the family Numididae; individuals have few feathers on the head and neck, but may have a crest of feathers and various fleshy appendages. { 'gin·ē ,faül }

guinea pig [VERTEBRATE ZOOLOGY] The common name for several species of wild and domestic hystricomorph rodents in the genus *Cavia*, family Caviidae; individuals are stocky, short-eared, short-legged, and nearly tailless. { 'gin·ē ,pig }

guinea worm [INVERTEBRATE ZOOLOGY] *Dracunculus medinensis.* A parasitic nematode that infects the subcutaneous tissues of humans and other mammals. { 'gin·ē ,wərm }

guitarfish [VERTEBRATE ZOOLOGY] The common name for fishes composing the family Rhinobatidae. { gə'tär,fish }

gular [ANATOMY] Of, pertaining to, or situated in the gula or upper throat. [VERTEBRATE ZOOLOGY] A horny shield on the plastron of turtles. { 'gyü·lər }

gulfweed [BOTANY] Brown algae of the genus *Sargassum*. { 'gəlf,wēd }

gull [VERTEBRATE ZOOLOGY] The common name for a number of long-winged swimming birds in the family Laridae having a stout build, a thick, somewhat hooked bill, a short tail, and webbed feet. { gəl }

gullet [ANATOMY] *See* esophagus. [INVERTEBRATE ZOOLOGY] A canal between the cytostome and reservoir that functions in food intake in ciliates. { 'gəl·ət }

gumbo [BOTANY] *See* okra. { 'gəm·bō }

gum vein [BOTANY] Local accumulation of resin occurring as a wide streak in certain hardwoods. { 'gəm ,vān }

Gunneraceae [BOTANY] A family of dicotyledonous terrestrial herbs in the order Haloragales, distinguished by two to four styles, a unilocular bitegmic ovule, large inflorescences with no petals, and drupaceous fruit. { ,gən·ə'rās·ē,ē }

gustation [PHYSIOLOGY] The act or the sensation of tasting. { gə'stā·shən }

gustatoreceptor [ANATOMY] A taste bud. [PHYSIOLOGY] Any sense organ that functions as a receptor for the sense of taste. { ¦gəs·tə·tō ri 'sep·tər }

gut [ANATOMY] The intestine. [EMBRYOLOGY] The embryonic digestive tube. { gət }

guttation [BOTANY] The discharge of water from a plant surface, especially from a hydathode. { ,gə'tā·shən }

Guttiferae [BOTANY] A family of dicotyledonous plants in the order Theales characterized by extipulate leaves and conspicuous secretory canals or cavities in all organs. { gə'tif·ə,rē }

Guttulinaceae [MICROBIOLOGY] A family of microorganisms in the Acrasiales characterized by simple fruiting structures with only slightly differentiated component cells containing little or no cellulose. { ,güd·əl·ə'nās·ē,ē }

Gymnarchidae [VERTEBRATE ZOOLOGY] A monotypic family of electrogenic fishes in the order Osteoglossiformes in which individuals lack pelvic, anal, and caudal fins. { ,jim'närk·ə,dē }

Gymnarthridae [PALEONTOLOGY] A family of extinct lepospondylous amphibians that have a skull with only a single bone representing the tabular and temporal elements of the primitive skull roof. { ,jim'närth·rə,dē }

Gymnoascaceae [MYCOLOGY] A family of ascomycetous fungi in the order Eurotiales including dermatophytes and forms that grow on dung, soil, and feathers. { ¦jim·nō·ə'skās·ē,ē }

Gymnoblastea [INVERTEBRATE ZOOLOGY] A suborder of cnidarians in the order Hydroida comprising hydroids without protective cups around the hydranths and gonozooids. { ,jim·nə'blas·tē·ə }

gymnoblastic [INVERTEBRATE ZOOLOGY] Having naked medusa buds, referring to anthomedusan hydroids. { ¦jim·nə¦bla·stik }

gymnocarpous [BOTANY] Having the hymenium uncovered on the surface of the thallus or fruiting body of lichens or fungi. { ¦jim·nə¦kär·pəs }

gymnocephalous cercaria [INVERTEBRATE ZOOLOGY] A type of digenetic trematode larva. { ¦jim·nə¦sef·ə·ləs sər'kar·ē·ə }

Gymnocerata [INVERTEBRATE ZOOLOGY] An equivalent name for Hydrocorisae. { ,jim·nō 'ser·əd·ə }

Gymnocodiaceae [PALEOBOTANY] A family of fossil red algae. { ¦jim·nō,kō·dē'as·ē,ē }

Gymnodinia [INVERTEBRATE ZOOLOGY] A suborder of flagellate protozoans in the order Dinoflagellida that are naked or have thin pellicles. { ,jim·nə'din·ē·ə }

gymnogynous [BOTANY] Having a naked ovary. { ,jim'näj·ə·nəs }

Gymnolaemata [INVERTEBRATE ZOOLOGY] A class of ectoproct bryozoans possessing lophophores which are circular in basal outline and zooecia which are short, wide, and vaselike or boxlike. { ˌjim·nə'lē·məd·ə }

Gymnonoti [VERTEBRATE ZOOLOGY] An equivalent name for Cypriniformes. { ˌjim·nə'nōd·ī }

Gymnophiona [VERTEBRATE ZOOLOGY] An equivalent name for Apoda. { ˌjim·nə'fī·ə·nə }

gymnoplast [BOTANY] In angiosperms, a cell without a cell wall. { 'jim·nəˌplast }

Gymnopleura [INVERTEBRATE ZOOLOGY] A subsection of brachyuran decapod crustaceans including the primitive burrowing crabs with trapezoidal or elongate carapaces, the first pereiopods subchelate, and some or all of the remaining pereiopods flattened and expanded. { ˌjim·nə'plu̇r·ə }

gymnosperm [BOTANY] The common name for members of the division Pinophyta; seed plants having naked ovules at the time of pollination. { 'jim·nəˌspərm }

Gymnospermae [BOTANY] The equivalent name for Pinophyta. { ˌjim·nə'spər·mē }

Gymnostomatida [INVERTEBRATE ZOOLOGY] An order of the protozoan subclass Holotrichia containing the most primitive ciliates, distinguished by the lack of ciliature in the oral area. { ˌjim·nə 'stō·məd·ə }

Gymnotidae [VERTEBRATE ZOOLOGY] The single family of the suborder Gymnotoidei; eel-shaped fishes having numerous vertebrae, and anus located far forward, and lacking pelvic and developed dorsal fins. { ˌjim'näd·əˌdē }

Gymnotoidei [VERTEBRATE ZOOLOGY] A monofamilial suborder of actinopterygian fishes in the order Cypriniformes. { ˌjim·nə'tȯid·ēˌī }

gynaecandrous [BOTANY] Having staminate and pistillate flowers on the same spike. { ¦gī·nə ¦kan·drəs }

gynander [BIOLOGY] A mosaic individual composed of diploid female portions derived from both parents and haploid male portions derived from an extra egg or sperm nucleus. { gīn'an·dər }

gynandromorph [BIOLOGY] An individual of a dioecious species made up of a mosaic of tissues of male and female genotypes. { gī'nan·drə ˌmȯrf }

gynandry [PHYSIOLOGY] A form of pseudohermaphroditism in which the external sexual characteristics are partly or wholly of the male aspect, but internal female genitalia are present. Also known as female pseudohermaphroditism; virilism. { gīn'an·drē }

gynobase [BOTANY] A gynoecium-bearing elongation of the receptacle in certain plants. { 'gīn·ōˌbās }

gynodioecious [BOTANY] Dioecious but with some perfect flowers on a plant bearing pistillate flowers. { ¦gīn·ō·dī'ē·shəs }

gynoecious [BOTANY] Pertaining to plants that have only female flowers. { gī'nē·shəs }

gynoecium [BOTANY] The aggregate of carpels in a flower. { gī'nē·sē·əm }

gynogenesis [EMBRYOLOGY] Development of a fertilized egg through the action of the egg nucleus, without participation of the sperm nucleus. { ¦gīn·ō'jen·ə·səs }

gynomerogony [EMBRYOLOGY] Development of a fragment of a fertilized egg containing the haploid egg nucleus. { ¦gīn·ō·mə'räg·ə·nē }

gynomonoecious [BOTANY] Having complete and pistillate flowers on the same plant. { ¦gīn·ō·mä'nē·shəs }

gynophore [BOTANY] **1.** A stalk that bears the gynoecium. **2.** An elongation of the receptacle between pistil and stamens. { 'gīn·əˌfȯr }

gynostemium [BOTANY] The column composed of the united gynoecia and androecium. { ¦gīn·ō 'stē·mē·əm }

gypsophilous [ECOLOGY] Flourishing on a gypsum-rich substratum. { ˌjip'säf·ə·ləs }

gypsy moth [INVERTEBRATE ZOOLOGY] *Porthetria dispar.* A large lepidopteran insect of the family Lymantriidae that was accidentally imported into New England from Europe in the late 19th century; larvae are economically important as pests of deciduous trees. { 'jip·sē ¦mȯth }

Gyracanthididae [PALEONTOLOGY] A family of extinct acanthodian fishes in the suborder Diplacanthoidei. { ˌjī·rəˌkan'thid·əˌdē }

Gyrinidae [INVERTEBRATE ZOOLOGY] The whirligig beetles, a family of large coleopteran insects in the suborder Adephaga. { jə'rin·əˌdē }

Gyrinocheilidae [VERTEBRATE ZOOLOGY] A monogeneric family of cypriniform fishes in the suborder Cyprinoidei. { ¦jir·ə·nō·kī'lī·əˌdē }

Gyrocotylidae [INVERTEBRATE ZOOLOGY] An order of tapeworms of the subclass Cestodaria; species are intestinal parasites of chimaeroid fishes and are characterized by an anterior eversible proboscis and a posterior ruffled adhesive organ. { ¦jī·rō·kä'til·əd·ē }

Gyrocotyloidea [INVERTEBRATE ZOOLOGY] A class of trematode worms according to some systems of classification. { ¦jī·rōˌkäd·əl'ȯid·ē·ə }

Gyrodactyloidea [INVERTEBRATE ZOOLOGY] A superfamily of ectoparasitic trematodes in the subclass Monogenea; the posterior holdfast is solid and is armed with central anchors and marginal hooks. { ¦jī·rōˌdak·tə'lȯid·ē·ə }

gyrogonite [PALEOBOTANY] A minute, ovoid body that is the residue of the calcareous encrustation about the female sex organs of a fossil stonewort. { jī'räg·əˌnīt }

Gyropidae [INVERTEBRATE ZOOLOGY] A family of biting lice in the order Mallophaga; members are ectoparasites of South American rodents. { jə 'räp·əˌdē }

gyrus [ANATOMY] One of the convolutions (ridges) on the surface of the cerebrum. { 'jī·rəs }

H

habenula [ANATOMY] **1.** Stalk of the pineal body. **2.** A ribbonlike structure. { hə'ben·yə·lə }

habenular commissure [ANATOMY] The commissure connecting the habenular ganglia in the roof of the diencephalon. { hə'ben·yə·lər 'käm·ə‚shu̇r }

habenular ganglia [ANATOMY] Olfactory centers anterior to the pineal body. { hə'ben·yə·lər 'gaŋ·glē·ə }

habenular nucleus [ANATOMY] Either of a pair of nerve centers that are located at the base of the pineal body on either side and serve as an olfactory correlation center. { hə'ben·yə·lər 'nü·klē·əs }

habitat [ECOLOGY] The part of the physical environment in which a plant or animal lives. { 'hab·ə‚tat }

habitus [BIOLOGY] General appearance or constitution of an organism. { 'hab·ə·təs }

hackberry [BOTANY] **1.** *Celtis occidentalis.* A tree of the eastern United States characterized by corky or warty bark, and by alternate, long-pointed serrate leaves unequal at the base; produces small, sweet, edible drupaceous fruit. **2.** Any of several other trees of the genus *Celtis.* { 'hak‚ber·ē }

hackmarack [BOTANY] *See* tamarack. { 'hak·mə ‚rak }

haddock [VERTEBRATE ZOOLOGY] *Melanogrammus aeglefinus* A fish of the family Gadidae characterized by a black lateral line and a dark spot behind the gills. { 'had·ək }

Hadromerina [INVERTEBRATE ZOOLOGY] A suborder of sponges in the class Clavaxinellida having monactinal megascleres, usually with a terminal knob at one end. { ‚had·rō·mə'rī·nə }

hadrosaur [PALEONTOLOGY] A duck-billed dinosaur. { 'had·rə‚sȯr }

Haeckel's law [BIOLOGY] *See* recapitulation theory. { 'hek·əlz ‚lȯ }

haematodocha [INVERTEBRATE ZOOLOGY] A sac in the palpus of male spiders that fills with hemolymph and becomes distended during pairing. { ‚hē·məd·ō'dō·kə }

Haemobartonella [MICROBIOLOGY] A genus of bacteria in the family Anaplasmataceae; parasites in or on red blood cells of many vertebrates. { ¦hē·mō‚bart·ən'el·ə }

Haemophilus [MICROBIOLOGY] A genus of gram-negative coccobacilli or rod-shaped bacteria of uncertain affiliation; cells may form threads and filaments and are aerobic or facultatively anaerobic; strictly blood parasites. { hē'mä·fə·ləs }

Haemosporina [INVERTEBRATE ZOOLOGY] A suborder of sporozoan protozoans in the subclass Coccidia; all are parasites of vertebrates, and human malarial parasites are included. { ‚hē·mō 'spȯr·ə·nə }

haerangium [MYCOLOGY] The fruiting body of *Fugascus* and *Ceratostomella.* { hē'ran·je·əm }

Hafnia [MICROBIOLOGY] A genus of bacteria in the family Enterobacteriaceae; motile rods that can utilize citrate as the only source of carbon. { 'haf·ne·ə }

Haftplatte [INVERTEBRATE ZOOLOGY] An adhesive plate or disk in some turbellarians; it is a glanduloepidermal organ. { 'häft‚pläd·ə }

Hageman factor [BIOCHEMISTRY] *See* factor XII. { 'häg·ə·män ‚fak·tər }

hagfish [VERTEBRATE ZOOLOGY] The common name for the jawless fishes composing the order Myxinoidea. { 'hag‚fish }

H agglutinin [IMMUNOLOGY] An antibody that is type-specific for the flagella of cells or microorganisms. { ¦āch ə'glüt·ən·ən }

hair [ZOOLOGY] **1.** A threadlike outgrowth of the epidermis of animals, especially a keratinized structure in mammalian skin. **2.** The hairy coat of a mammal, or of a part of the animal. { her }

hair ball [ECOLOGY] *See* lake ball. { 'her ‚bȯl }

hair cell [HISTOLOGY] The basic sensory unit of the inner ear of vertebrates; a columnar, polarized structure with specialized cilia on the free surface. { 'her ‚sel }

hair cycle [PHYSIOLOGY] The formation and growth of a new hair, followed by a resting stage, and ending with growth of another new hair from the same follicle. { 'her ‚sī·kəl }

hair follicle [ANATOMY] An epithelial ingrowth of the dermis that surrounds a hair. { 'her ‚fäl·ə·kəl }

hair gland [ANATOMY] Sebaceous gland associated with hair follicles. { 'her ‚gland }

hairpin loop [MOLECULAR BIOLOGY] Any double-stranded region of single-stranded deoxyribonucleic acid or ribonucleic acid formed by base-pairing between complementary base sequences on the same strand. { 'her‚pin ‚lüp }

hairworm [INVERTEBRATE ZOOLOGY] The common name for about 80 species of worms composing the class Nematomorpha. { 'her‚wərm }

Halacaridae [INVERTEBRATE ZOOLOGY] A family of marine arachnids in the order Acarina. { ˌhal·ə ˈkar·əˌdē }

Haldane's rule [GENETICS] The rule that if one sex in a first generation of hybrids is rare, absent, or sterile, then it is the heterogametic sex. { ˈhölˌdānz ˌrül }

Halecomorphi [VERTEBRATE ZOOLOGY] The equivalent name for Amiiformes. { ˌhal·ə·kō ˈmör·fī }

Halecostomi [VERTEBRATE ZOOLOGY] The equivalent name for Pholidophoriformes. { ˌhal·ə ˈkäst·ə·mē }

half-hardy plant [BOTANY] A plant that can withstand relatively low temperatures but cannot survive severe freezing in cold climates unless carefully protected. { ¦haf ˌhär·dē ˈplant }

halibut [VERTEBRATE ZOOLOGY] Either of two large species of flatfishes in the genus *Hippoglossus*; commonly known as a right-eye flounder. { ˈhal·ə·bət }

Halichondrida [INVERTEBRATE ZOOLOGY] A small order of sponges in the class Demospongiae with a skeleton of diactinal or monactinal, siliceous megascleres (or both), a skinlike dermis, and small amounts of spongin. { ˌhal·əˈkän·drē·də }

Halictidae [INVERTEBRATE ZOOLOGY] The halictid and sweat bees, a family of hymenopteran insects in the superfamily Apoidea. { həˈlik·tə ˌdē }

Halimeda [BOTANY] A genus of small, bushy green algae in the family Codiaceae composed of thick, leaflike segments; important as a fossil and as a limestone builder. { ˌhal·əˈmē·də }

Haliotis [INVERTEBRATE ZOOLOGY] A genus of gastropod mollusks commonly known as the abalones. { ˌhal·ēˈōd·əs }

Haliplidae [INVERTEBRATE ZOOLOGY] The crawling water beetles, a family of coleopteran insects in the suborder Adephaga. { həˈlip·ləˌdē }

Halitheriinae [PALEONTOLOGY] A subfamily of extinct sirenian mammals in the family Dugongidae. { həˌlith·əˈrī·əˌnē }

Haller's organ [INVERTEBRATE ZOOLOGY] A chemoreceptor on the tarsus of certain ticks. { ˈhal·ərz ˌör·gən }

hallux [ANATOMY] The first digit of the hindlimb; the big toe of a human. { ˈhal·əks }

halmophagous [ZOOLOGY] Pertaining to organisms which infest and eat stalks or culms of plants. { halˈmäf·ə·gəs }

Halobacteriaceae [MICROBIOLOGY] A family of gram-negative, aerobic rods and cocci which require high salt (sodium chloride) concentrations for maintenance and growth. { ˌhal·əˌbak·tir·ē ˈās·ēˌē }

Halobacterium [MICROBIOLOGY] A genus of bacteria in the family Halobacteriaceae; single, rod-shaped cells which may be pleomorphic when media are deficient. { ˌhal·ə·bakˈtir·ē·əm }

Halococcus [MICROBIOLOGY] A genus of bacteria in the family Halobacteriaceae; nonmotile cocci which occur in pairs, tetrads, or clusters of tetrads. { ˌhal·əˈkäk·əs }

Halocypridacea [INVERTEBRATE ZOOLOGY] A superfamily of ostracods in the suborder Myodocopa; individuals are straight-backed with a very thin, usually calcified carapace. { ˌhal·əˌsip·rə ˈdās·ē·ə }

halonate [MYCOLOGY] Pertaining to a spore surrounded by a colored circle. { ˈhal·əˌnāt }

halophile [BIOLOGY] An organism that requires high salt concentrations for growth and maintenance. { ˈhal·əˌfīl }

halophilism [BIOLOGY] The phenomenon of demand for high salt concentrations for growth and maintenance. { ¦hal·ə¦fil·iz·əm }

halophyte [ECOLOGY] A plant or microorganism that grows well in soils having a high salt content. { ˈhal·əˌfīt }

Haloragaceae [BOTANY] A family of dicotyledonous plants in the order Haloragales distinguished by an apical ovary of 2-4 loculi, small inflorescences, and small, alternate or opposite or whorled, exstipulate leaves. { ˌhal·ə·rəˈgās· ēˌē }

Haloragales [BOTANY] An order of dicotyledonous plants in the subclass Rosidae containing herbs with perfect or often unisexual, more or less reduced flowers, and a minute or vestigial perianth. { ˌhal·ə·rəˈgā·lēz }

Halosauridae [VERTEBRATE ZOOLOGY] A family of mostly extinct deep-sea teleost fishes in the order Notacanthiformes. { ˌhal·əˈsör·əˌdē }

halosere [ECOLOGY] The series of communities succeeding one another, from the pioneer stage to the climax, and commencing in salt water or on saline soil. { ˈhal·əˌsir }

haltere [INVERTEBRATE ZOOLOGY] Either of a pair of capitate filaments representing rudimentary hindwings in Diptera. Also known as balancer. { ˈhölˌtir }

Halysitidae [PALEONTOLOGY] A family of extinct Paleozoic corals of the order Tabulata. { ˌhal·ə ˈsid·əˌdē }

Hamamelidaceae [BOTANY] A family of dicotyledonous trees or shrubs in the order Hamamelidales characterized by united carpels, alternate leaves, perfect or unisexual flowers, and free filaments. { ˌha·məˌmel·əˈdās·ēˌē }

Hamamelidae [BOTANY] A small subclass of plants in the class Magnoliopsida having strongly reduced, often unisexual flowers with poorly developed or no perianth. { ˌha·məˈmel· əˌdē }

Hamamelidales [BOTANY] A small order of dicotyledonous plants in the subclass Hamamelidae characterized by vessels in the wood and a gynoecium consisting either of separate carpels or of united carpels that open at maturity. { ˌha·mə ˌmel·əˈdā·lēz }

hamate [BIOLOGY] Hook-shaped or hooked. { ˈhāˌmāt }

Hamoproteidae [INVERTEBRATE ZOOLOGY] A family of parasitic protozoans in the suborder Haemosporina; only the gametocytes occur in blood cells. { ˌham·ō·prəˈtē·əˌdē }

hamster [VERTEBRATE ZOOLOGY] The common name for any of 14 species of rodents in the family Cricetidae characterized by scent glands in the flanks, large cheek pouches, and a short tail. { 'ham·stər }

hamstring muscles [ANATOMY] The biceps femoris, semitendinosus, and semimembranosus collectively. { 'ham,striŋ ,məs·əlz }

hamulus [VERTEBRATE ZOOLOGY] A hooklike process, especially a small terminal hook on the barbicel of a feather. { 'ham·yə·ləs }

hamus [BIOLOGY] A hook or a curved process. { 'hā·məs }

hand [ANATOMY] The terminal part of the upper extremity modified for grasping. { hand }

hanging-drop preparation [MICROBIOLOGY] A technique used in microscopy in which a specimen is placed in a drop of a suitable fluid on a cover slip and the cover slip is inverted over a concavity on a slide. { 'haŋ·iŋ ,dräp ,prep·ə 'rā·shən }

H antigen [MICROBIOLOGY] A general term for microbial flagellar antigens; former designation for species-specific flagellar antigens of *Salmonella*. { 'āch ¦ant·i·jən }

haplobiont [BOTANY] A plant that produces only sexual haploid individuals. { ¦ha·plō¦bī,änt }

haplocaulescent [BOTANY] Having a simple axis with the reproductive organs on the principal axis. { ¦ha·plō·kȯ'les·ənt }

Haplodoci [VERTEBRATE ZOOLOGY] The equivalent name for Batrachoidiformes. { hə'pläd·ə ,sī }

haploid [GENETICS] Having half of the diploid or full complement of chromosomes, as in mature gametes. { 'ha,plȯid }

haploidization [MYCOLOGY] In certain fungi, the transformation of a diploid into a haploid cell by progressive loss of chromosomes due to nondisjunction. { ,hap,lȯid·i'zā·shən }

Haplolepidae [PALEONTOLOGY] A family of Carboniferous chondrostean fishes in the suborder Palaeoniscoidei having a reduced number of fin rays and a vertical jaw suspension. { ¦ha·plō ¦lep·ə,dē }

Haplomi [VERTEBRATE ZOOLOGY] An equivalent name for Salmoniformes. { ha'plō,mī }

haplomitosis [CELL BIOLOGY] Type of primitive mitosis in which the nuclear granules form into threadlike masses rather than clearly differentiated chromosomes. { ,ha·plō,mī'tō·səs }

haplont [BOTANY] A plant with only haploid somatic cells; the zygote is diploid. { 'ha,plänt }

haplophase [BIOLOGY] Haploid stage in the life cycle of an organism. { 'ha·plō,fāz }

haplopore [PALEONTOLOGY] Any randomly distributed pore on the surface of fossil cystoid echinoderms. { 'ha·plō,pȯr }

Haplosclerida [INVERTEBRATE ZOOLOGY] An order of sponges in the class Demospongiae including species with a skeleton made up of siliceous megascleres embedded in spongin fibers or spongin cement. { ,ha·plō'skler·ə·də }

haplosis [CELL BIOLOGY] Reduction of the chromosome number to half during meiosis. { ha 'plō·səs }

Haplosporea [INVERTEBRATE ZOOLOGY] A class of Protozoa in the subphylum Sporozoa distinguished by the production of spores lacking polar filaments. { ¦ha·plō¦spȯr·ē·ə }

Haplosporida [INVERTEBRATE ZOOLOGY] An order of Protozoa in the class Haplosporea distinguished by the production of uninucleate spores that lack both polar capsules and filaments. { ¦ha·plō¦spȯr·ə·də }

haplostele [BOTANY] A type of protostele with the core of xylem characterized by a smooth outline. { 'ha·plō,stēl }

haplo-sufficient gene [GENETICS] Any gene that allows the production of viable adults even when one copy of the gene in diploids is deleted from one of the homologous chromosomes. { ¦hap,lō sə,fish·ənt 'jēn }

hapten [IMMUNOLOGY] An incomplete antigen that cannot induce antibody formation by itself but can do so by coupling with a larger carrier molecule.

hapteron [BOTANY] A disklike holdfast on the stem of certain algae. { 'hap·tə,rän }

haptochlamydeous [BOTANY] Having the sporophylls protected by rudimentary perianth leaves. { ,hap·tō·klə'mid·ē·əs }

haptoglobin [BIOCHEMISTRY] An alpha globulin that constitutes 1-2% of normal blood serum; contains about 5% carbohydrate. { 'hap·tə,glō·bən }

Haptophyceae [BOTANY] A class of the phylum Chrysophyta that contains the Coccolithophorida. { ,hap·tə'fīs·ē,ē }

haptor [INVERTEBRATE ZOOLOGY] The posterior organ of attachment in certain monogenetic trematodes characterized by multiple suckers and the presence of hooks. { 'hap·tər }

haptotropism [BIOLOGY] Movement of sessile organisms in response to contact, especially in plants. { hap'tä·trə,piz·əm }

hardening [BOTANY] Treatment of plants designed to increase their resistance to extremes in temperature or drought. { 'hard·ən·iŋ }

Harderian gland [VERTEBRATE ZOOLOGY] An accessory lacrimal gland associated with lower eyelid structures in all vertebrates except land mammals. { här'dir·ē·ən ,gland }

hard fiber [BOTANY] A heavily lignified leaf fiber used in making cordage, twine, and textiles. { 'härd ,fī·bər }

hard palate [ANATOMY] The anterior portion of the roof of the mouth formed by paired palatine processes of the maxillary bones and by the horizontal part of each palate bone. { 'härd ¦pal·ət }

hardwood forest [ECOLOGY] **1.** An ecosystem having deciduous trees as the dominant form of vegetation. **2.** An ecosystem consisting principally of trees that yield hardwood. { 'härd,wu̇d ¦fär·əst }

hardy plant [BOTANY] A plant able to withstand low temperatures without artificial protection. { ¦här·dē ¦plant }

Hardy-Weinberg law [GENETICS] The concept that frequencies of both genes and genotypes will remain constant from generation to generation in an idealized population where mating is random and evolutionary forces are absent. { ¦här·dē 'wīn‚bərg ‚lȯ }

hare [VERTEBRATE ZOOLOGY] The common name for a number of lagomorphs in the family Leporidae; they differ from rabbits in being larger with longer ears, legs, and tails. { her }

Harpacticoida [INVERTEBRATE ZOOLOGY] An order of minute copepod crustaceans of variable form, but generally being linear and more or less cylindrical. { här‚pak·tə'kȯid·ə }

harpago [INVERTEBRATE ZOOLOGY] Part of the clasper on the copulatory organ of certain male insects. { 'här·pə‚gō }

Harpidae [INVERTEBRATE ZOOLOGY] A family of marine gastropod mollusks in the order Neogastropoda. { 'här·pə‚dē }

Hart-Park virus [VIROLOGY] A ribonucleic acid-containing animal virus of the rhabdovirus group. { 'härt 'pärk ‚vī·rəs }

Hassal's body [HISTOLOGY] *See* thymic corpuscle. { 'has·əlz ‚bäd·ē }

hastate [BIOLOGY] Shaped like an arrowhead with divergent barbs. { 'ha‚stāt }

Hatch-Slack pathway [BIOCHEMISTRY] A metabolic cycle involved in the non-light-requiring phase of photosynthesis in certain plants having specific metabolic and anatomical modifications in their mesophyll and bundle sheath cells which facilitate the temporary fixation of carbon dioxide (CO_2) into four-carbon organic acid; these acids are next broken down to three-carbon organic acids plus CO_2 in bundle sheath cells, where this freed CO_2 is then fixed into carbohydrates in a normal Calvin cycle pathway. { 'hach 'slak 'path ‚wā }

haustellum [INVERTEBRATE ZOOLOGY] A proboscis modified for sucking. { hȯ'stel·əm }

haustorium [BOTANY] **1.** An outgrowth of certain parasitic plants which serves to absorb food from the host. **2.** Food-absorbing cell of the embryo sac in nonparasitic plants. { hȯ'stȯr·ē·əm }

haustrum [ANATOMY] An outpocketing or pouch of the colon. { 'hȯ·strəm }

Haversian canal [HISTOLOGY] The central, longitudinal channel of an osteon containing blood vessels and connective tissue. { hə'vər·zhən kə‚nal }

Haversian lamella [HISTOLOGY] One of the concentric layers of bone composing a Haversian system. { hə'vər·zhən lə'mel·ə }

Haversian system [HISTOLOGY] *See* osteon. { hə'vər·zhən ‚sis·təm }

HA virus [VIROLOGY] *See* hemadsorption virus. { ¦āch¦ā ‚vī·rəs }

hawk [VERTEBRATE ZOOLOGY] Any of the various smaller diurnal birds of prey in the family Accipitridae; some species are used for hunting hare and partridge in India and other parts of Asia. { hȯk }

Hayflick limit [PHYSIOLOGY] The finite replicative capacity of normal somatic cells. { 'hā‚flik ‚lim·ət }

hazelnut [BOTANY] *See* filbert. { 'ha·zəl‚nət }

H-B virus [VIROLOGY] A subgroup-A picornavirus associated with diseases in rodents. { ¦āch ¦bē ‚vī·rəs }

HCG [BIOCHEMISTRY] *See* human chorionic gonadotropin.

H chain [IMMUNOLOGY] *See* heavy chain. { 'āch ‚chān }

head [ANATOMY] **1.** The region of the body consisting of the skull, its contents, and related structures. **2.** Proximal end of a long bone. [BOTANY] A dense cluster of nearly sessile flowers on a very short stem. { hed }

head bulb [INVERTEBRATE ZOOLOGY] A structure armed with spines behind the lips of spiruroid nematodes. { 'hed ‚bəlb }

head fold [EMBRYOLOGY] A ventral fold formed by rapid growth of the head of the embryo over the embryonic disk, resulting in the formation of the foregut accompanied by anteroposterior reversal of the anterior part of the embryonic disk. { 'hed ‚fōld }

head organ [INVERTEBRATE ZOOLOGY] One of the bulbous structures in the prohaptor of monogenetic trematodes which are openings for adhesive glands. { 'hed ‚ȯr·gən }

head process [EMBRYOLOGY] The notochord or notochordal plate formed as an axial outgrowth of the primitive node. { 'hed ‚prä·səs }

head shield [INVERTEBRATE ZOOLOGY] A conspicuous structure arching over the lips of certain nematodes. { 'hed ‚shēld }

hearing [PHYSIOLOGY] The general perceptual behavior and the specific responses made in relation to sound stimuli. { 'hir·iŋ }

heart [ANATOMY] The hollow muscular pumping organ of the cardiovascular system in vertebrates. { härt }

heartbeat [PHYSIOLOGY] Pulsation of the heart coincident with ventricular systole. { 'härt‚bet }

heart rate [PHYSIOLOGY] The number of heartbeats per minute. { 'härt ‚rāt }

heart valve [ANATOMY] Flaps of tissue that prevent reflux of blood from the ventricles to the atria or from the pulmonary arteries or aorta to the atria. { 'härt ‚valv }

heartwood [BOTANY] Xylem of an angiosperm. { 'härt‚wu̇d }

heart worm [INVERTEBRATE ZOOLOGY] *Dirofilaria immitis.* A filarial nematode parasitic on dogs and other carnivores. { 'härt ‚wərm }

heath [ECOLOGY] *See* temperate and cold scrub. { hēth }

heather [BOTANY] *Calluna vulgaris.* An evergreen heath of northern and alpine regions distinguished by racemes of small purple-pink flowers. { 'heth·ər }

heat shock protein [MOLECULAR BIOLOGY] Any of a group of proteins that are synthesized in the cytoplasm of cells as part of the heat shock response and act to protect the chromosomes from damage. { ¦hēt ‚shäk 'prō‚tēn }

heat shock response [MOLECULAR BIOLOGY] A cellular reaction to a stimulus such as elevated temperatures or abrupt environmental changes, in which there is cessation or slowdown of normal protein synthesis and activation of previously inactive genes, resulting in the production of heat shock proteins. { 'hēt ‚shäk ri‚späns }

heat stress index [PHYSIOLOGY] Relation of the

amount of evaporation or perspiration required for particular job conditions as related to the maximum evaporative capacity of an average person. Abbreviated HSI. { 'hēt ˌstres ¦inˌdeks }

heavy chain [IMMUNOLOGY] The heavier of the two types of polypeptide chains occurring in immunoglobulin molecules, its molecular weight range being 50,000-70,000. Also known as A chain; H chain. { 'hev·ē 'chān }

heavy meromyosin [BIOCHEMISTRY] The larger of two fragments obtained from the muscle protein myosin following limited proteolysis by trypsin or chymotrypsin. { 'hev·ē ¦mer·ə¦mī·əˌsin }

Hebrovellidae [INVERTEBRATE ZOOLOGY] A family of hemipteran insects in the subdivision Amphibicorisae. { ˌheb·rō'vel·əˌdē }

hectocotylus [INVERTEBRATE ZOOLOGY] A specialized appendage of male cephalopods adapted for the transference of sperm. { ˌhek·tō 'käd·əl·əs }

hedgehog [VERTEBRATE ZOOLOGY] The common name for members of the insectivorous family Erinaceidae characterized by spines on their back and sides. { 'hejˌhäg }

hedonic gland [VERTEBRATE ZOOLOGY] One of the mucus-secreting scent glands in many urodeles; functions in courtship. { hē'dän·ik ˌgland }

Hedwigiaceae [BOTANY] A family of mosses in the order Isobryales. { ˌhed·vig·ē'ās·ēˌē }

Heidelberg man [PALEONTOLOGY] An early type of European fossil man known from an isolated lower jaw; considered a variant of *Homo erectus* or an early stock of Neanderthal man. { 'hīd·əl·bərg ˌman }

hekistotherm [ECOLOGY] Plant adapted for conditions of minimal heat; can withstand long dark periods. { he'kis·tōˌthərm }

Helaletidae [PALEONTOLOGY] A family of extinct perissodactyl mammals in the superfamily Tapiroidea. { ˌhel·ə'led·əˌdē }

Helcionellacea [PALEONTOLOGY] A superfamily of extinct gastropod mollusks in the order Aspidobranchia. { ¦hel·sē·ō·nə'las·ē·ə }

Heleidae [INVERTEBRATE ZOOLOGY] The biting midges, a family of orthorrhaphous dipteran insects in the series Nematocera. { hə'lē·əˌdē }

Heliasteridae [INVERTEBRATE ZOOLOGY] A family of echinoderms in the subclass Asteroidea lacking pentameral symmetry but structurally resembling common asteroids. { ˌhēl·ē·ə'ster·əˌdē }

helical repeat [MOLECULAR BIOLOGY] The number of base pairs in one turn of a deoxyribonucleic acid helix. { 'hel·ə·kəl ri'pēt }

helicase [BIOCHEMISTRY] An enzyme that is capable of unwinding the deoxyribonucleic acid double helix at a replication fork. { 'hel·əˌkās }

Helicinidae [INVERTEBRATE ZOOLOGY] A family of gastropod mollusks in the order Archeogastropoda containing tropical terrestrial snails. { ˌhel·ə'sin·əˌdē }

helicoid [INVERTEBRATE ZOOLOGY] Of a gastropod shell, shaped like a flat coil or flattened spiral. { 'hel·əˌkȯid }

helicoid cyme [BOTANY] A type of determinate inflorescence having a coiled cluster, with flowers on only one side of the axis. { 'hel·əˌkȯid ¦sīm }

Heliconiaceae [BOTANY] A family of monocotyledonous plants in the order Zingiberales characterized by perfect flowers with a solitary ovule in each locule, schizocarpic fruit, and capitate stigma. { ˌhel·əˌkän·ē'ās·ēˌē }

Helicoplacoidea [PALEONTOLOGY] A class of free-living, spindle- or pear-shaped, plated echinozoans known only from the Lower Cambrian of California. { ˌhel·ə·kō·plə'kȯid·ē·ə }

Helicosporae [MYCOLOGY] A spore group of the Fungi Imperfecti characterized by spirally coiled, septate spores. { ˌhel·ə'kä·spəˌrē }

helicospore [INVERTEBRATE ZOOLOGY] Mature spore of the Helicosporida characterized by a peripheral spiral filament. { 'hel·ə·kəˌspȯr }

Helicosporida [INVERTEBRATE ZOOLOGY] An order of protozoans in the class Myxosporidea characterized by production of spores with a relatively thick, single intrasporal filament and three uninucleate sporoplasms. { ¦hel·ə·kō ¦spȯr·ə·də }

helicotrema [ANATOMY] The opening at the apex of the cochlea through which the scala tympani and the scala vestibuli communicate with each other. { ˌhel·ə·kō'tre·mə }

Heligmosomidae [INVERTEBRATE ZOOLOGY] A family of parasitic roundworms belonging to the Strongyloidea. { həˌlig·mō'sō·məˌdē }

Heliodinidae [INVERTEBRATE ZOOLOGY] A family of lepidopteran insects in the suborder Heteroneura. { ˌhē·lē·ə'din·əˌdē }

Heliolitidae [PALEONTOLOGY] A family of extinct corals in the order Tabulata. { ˌhē·lē·ō'lid·ə ˌdē }

heliophilous [ECOLOGY] Attracted by and adapted for a high intensity of sunlight. { ¦hē·lē ¦äf·ə·ləs }

heliophyte [ECOLOGY] A plant that thrives in full sunlight. { 'hē·lē·əˌfīt }

Heliornithidae [VERTEBRATE ZOOLOGY] The lobed-toed sun grebes, a family of pantropical birds in the order Gruiformes. { ¦hē·lēˌȯr'nith·ə ˌdē }

heliotaxis [BIOLOGY] Orientation movement of an organism in response to the stimulus of sunlight. { ¦hē·lē·ō¦tak·səs }

heliotrope [BOTANY] A plant whose flower or stem turns toward the sun. { 'hē·lē·əˌtrōp }

heliotropism [BIOLOGY] Growth or orientation movement of a sessile organism or part, such as a plant, in response to the stimulus of sunlight. { ˌhē·lē'ä·trəˌpiz·əm }

Heliozoia [INVERTEBRATE ZOOLOGY] A subclass of the protozoan class Actinopodea; individuals lack a central capsule and have either axopodia or filopodia. { ˌhē·lē·ə'zōi·ə }

heliozooid [BIOLOGY] Ameboid, but with distinct filamentous pseudopodia. { ˌhē·lē·ə'zōˌōid }

helix [MOLECULAR BIOLOGY] A spiral structure with a repeating pattern that characterizes many biological polymers, for example, double-stranded nucleic acids and proteins. { 'hēˌliks }

Helix [INVERTEBRATE ZOOLOGY] A genus of pulmonate land mollusks including many of the edible snails; individuals have a coiled shell with a low conical spire. { 'hē‚liks }

helix-destabilizing protein [BIOCHEMISTRY] Any of a group of proteins that bind to single-stranded regions of duplex deoxyribonucleic acid and cause unwinding of the helix. { ¦hē‚liks di 'stā·bə‚līz·iŋ ‚prō·tēn }

hellbender [VERTEBRATE ZOOLOGY] *Cryptobranchus alleganiensis.* A large amphibian of the order Urodela which is the most primitive of the living salamanders, retaining some larval characteristics. { 'hel¦ben·der }

Helmholtz theory [PHYSIOLOGY] *See* Young-Helmholtz theory. { 'helm‚hōlts ‚thē·ə·rē }

helminth [INVERTEBRATE ZOOLOGY] Any parasitic worm. { 'hel‚minth }

helminthoid [BIOLOGY] Resembling a helminth. { hel'min‚thȯid }

helminthologist [BIOLOGY] An individual who studies helminths. { ‚hel‚mən'thäl·ə·jəst }

helminthosporin [BIOCHEMISTRY] $C_{15}H_{10}O_5$ A maroon, crystalline pigment formed by certain fungi growing on a sugar substrate. { hel‚min·thə'spȯr·ən }

Helminthosporium [MYCOLOGY] A genus of parasitic fungi of the family Dematiaceae having conidiophores which are more or less irregular or bent and bear conidia successively on new growing tips. { hel‚min·thə'spȯr·ē·əm }

Helobiae [BOTANY] The equivalent name for Helobiales. { he'lō·bē‚ē }

Helobiales [BOTANY] An order embracing most of the Alismatidae in certain systems of classification. { he‚lō·bē'ā·lēz }

Heloderma [VERTEBRATE ZOOLOGY] The single genus in the reptilian family Helodermatidae; contains the only known poisonous lizards, the Gila monster (*H. suspectum*) and the beaded lizard (*H. horridum*). { ‚hē·lō'dər·mə }

Helodermatidae [VERTEBRATE ZOOLOGY] A family of lizards in the suborder Sauria. { ‚hē·lō·dər 'mad·ə‚dē }

Helodidae [INVERTEBRATE ZOOLOGY] The marsh beetles, a family of coleopteran insects in the superfamily Dascilloidea. { hə'lōd·ə‚dē }

Helodontidae [PALEONTOLOGY] A family of extinct ratfishes conditionally placed in the order Bradyodonti. { ‚he·lō'dänt·ə‚dē }

Helomyzidae [INVERTEBRATE ZOOLOGY] The sun flies, a family of myodarian cyclorrhaphous dipteran insects in the subsection Acalypteratae. { ‚he·lō'mīz·ə‚dē }

helophyte [ECOLOGY] A marsh plant; buds overwinter underwater. { 'he·lə‚fīt }

helophytia [ECOLOGY] Differences in ecological control by fluctuations in water level such as in marshes. { ‚he·lə'fī·shə }

Heloridae [INVERTEBRATE ZOOLOGY] A family of hymenopteran insects in the superfamily Proctotrupoidea. { hə'lȯr·ə‚dē }

Helotiales [MYCOLOGY] An order of fungi in the class Ascomycetes. { hə‚lō·shē'ā·lēz }

Helotidae [INVERTEBRATE ZOOLOGY] The metallic sap beetles, a family of coleopteran insects in the superfamily Cucujoidea. { hə'läd·ə‚dē }

helotism [ECOLOGY] Symbiosis in which one organism is a slave to the other, as between certain species of ants. { 'hel·ə‚tiz·əm }

Helotrephidae [INVERTEBRATE ZOOLOGY] A family of true aquatic, tropical hemipteran insects in the subdivision Hydrocorisae. { ‚he·lə'tref·ə ‚dē }

helper virus [VIROLOGY] A virus that, by its infection of a cell, enables a defective virus to multiply by supplying one or more functions that the defective virus lacks. { 'hel·pər ‚vī·rəs }

hem-, hema-, hemo-, haem- [HISTOLOGY] Combining form for blood. { hēm, 'hēm·ə, 'hēm·o, hēm }

hemadsorption virus [VIROLOGY] A descriptive term for myxoviruses that agglutinate red blood cells and cause the cells to adsorb to each other. Abbreviated HA virus. { ‚hēm·ad'sȯrp·shən ‚vī·rəs }

hemagglutination [IMMUNOLOGY] Agglutination of red blood cells. { ‚hē·mə‚glüd·ən'ā·shən }

hemagglutination-inhibition test [IMMUNOLOGY] A test to identify a virus antigen or to quantitate an antibody by adding virus-specific antibody to a mixture of agglutinating virus and red blood cells. { ‚hē·mə‚glüd·ən'ā·shən ‚in·ə'bish·ən ‚test }

hemagglutinin [IMMUNOLOGY] An erythrocyte-agglutinating antibody. { ‚hē·mə'glüd·ən·ən }

hemal arch [ANATOMY] **1.** A ventral loop on the body of vertebrate caudal vertebrae surrounding the blood vessels. **2.** In humans, the ventral vertebral process formed by the centrum together with the ribs. { 'hē·məl 'ärch }

hemal ring [INVERTEBRATE ZOOLOGY] A vessel in certain echinoderms, variously located, associated with the coelom and axial gland. { 'hē·məl 'riŋ }

hemal sinus [INVERTEBRATE ZOOLOGY] The two principal lacunae along the digestive tube in certain echinoderms. { 'hē·məl 'sī·nəs }

hemal tuft [INVERTEBRATE ZOOLOGY] Series of fine vessels in echioderms arising from the axial gland. { 'hē·məl 'təft }

hemapodium [INVERTEBRATE ZOOLOGY] The dorsal lobe of a parapodium. { ‚hē·mə'pō·dē·əm }

hematein [BIOCHEMISTRY] $C_{16}H_{12}O_6$ A brownish stain and chemical indicator obtained by oxidation of hematoxylin. { ‚hē·mə'tē·ən }

hematoblast [HISTOLOGY] An immature erythrocyte. { 'he·məd·ō‚blast }

hematochrome [BIOCHEMISTRY] A red pigment occurring in green algae, especially when plants are exposed to intense light on subaerial habitats. { hi'mad·ə‚krōm }

hematodocha [INVERTEBRATE ZOOLOGY] In some spiders, a thin sac on the male that is distended during copulation. { hi‚mad·ə'dō·kə }

hematogenous [PHYSIOLOGY] **1.** Pertaining to the production of blood or of its fractions. **2.** Carried by way of the bloodstream. **3.** Originating in blood. { ¦hēm·ə¦täj·ə·nəs }

hematophagous [ZOOLOGY] Feeding on blood. { ¦hē·mə¦täf·ə·gəs }

hematopoiesis [PHYSIOLOGY] The process by

which the cellular elements of the blood are formed. Also known as hemopoiesis. { ˌhe·məd·ō·pȯiʹē·səs }

hematopoietic system [ANATOMY] *See* reticuloendothelial system. { ¦he·məd·ō·pōi¦ed·ik ʹsis·təm }

hematopoietic tissue [HISTOLOGY] Blood-forming tissue, consisting of reticular fibers and cells. Also known as hemopoietic tissue. { ¦he·məd·ō·pōi¦ed·ik ʹtish·ü }

hematopoietin [BIOCHEMISTRY] A substance which is produced by the juxtaglomerular apparatus in the kidney and controls the rate of red cell production. Also known as hemopoietin. { ˌhe·məd·ōʹpȯi·ət·ən }

hematoporphyrin [BIOCHEMISTRY] $C_{34}H_{38}O_6N$ Iron-free heme, a porphyrin obtained by treating hemoglobin with sulfuric acid in vitro. Also known as hemoporphyrin. { ¦he·məd·ō¦pȯr·fə·rən }

hematoxylon [BOTANY] The heartwood of *Haematoxylon campechianum*. Also known as logwood. { ˌhē·məʹtäk·səˌlän }

heme [BIOCHEMISTRY] $C_{34}H_{32}O_4N_4Fe$ An iron-protoporphyrin complex associated with each polypeptide unit of hemoglobin. { hēm }

hemerythrin [BIOCHEMISTRY] *See* hemoerythrin. { ˌhem·əʹrith·rən }

heme synthetase [BIOCHEMISTRY] An enzyme which combines protoporphyrin IX, ferrous iron, and globin to form the intact hemoglobin molecule. { ˌhēm ʹsin·thəˌtās }

hemi- [BIOLOGY] **1.** Prefix for half. **2.** Prefix denoting one side of the body. { ʹhe·mē }

Hemiascomycetes [MYCOLOGY] The equivalent name for Hemiascomycetidae. { ¦he·mēˌas·kō ˌmīʹsēd·ēz }

Hemiascomycetidae [MYCOLOGY] A subclass of fungi in the class Ascomycetes. { ¦he·mēˌas·kō ˌmīʹsed·əˌdē }

hemiazygous vein [ANATOMY] A vein on the left side of the vertebral column which drains blood from the left ascending lumbar vein to the azygos vein. { ¦he·mēʹaz·ə·gəs ʹvān }

Hemibasidiomycetes [MYCOLOGY] The equivalent name for Heterobasidiomycetidae. { ¦he·mē·bəˌsid·ē·ōˌmīʹs ēd·ēz }

hemicellulose [BIOCHEMISTRY] $(C_6H_{10}O_5)_n$ A type of polysaccharide found in plant cell walls in association with cellulose and lignin; it is soluble in and extractable by dilute alkaline solutions. Also known as hexosan. { ¦he·mēʹsel·yə ˌlōs }

Hemichordata [SYSTEMATICS] A group of marine animals categorized as either a phylum of deuterostomes or a subphylum of chordates; includes the Enteropneusta, Pterobranchia, and Graptolithina. { ¦he·mē·kȯrʹdäd·ə }

Hemicidaridae [PALEONTOLOGY] A family of extinct Echinacea in the order Hemicidaroida distinguished by a stirodont lantern, and ambulacra abruptly widened at the ambitus. { ¦he·mē·si ʹdär·əˌdē }

Hemicidaroida [PALEONTOLOGY] An order of extinct echinoderms in the superorder Echinacea characterized by one very large tubercle on each interambulacral plate. { ¦he·mēˌsid·əʹrȯid·ə }

hemicryptophyte [ECOLOGY] A plant having buds at the soil surface and protected by scales, snow, or litter. { ¦he·mēʹkrip·təˌfīt }

hemicydic [BOTANY] Of flowers, having the floral leaves arranged partly in whorls and partly in spirals. { ʹhe·mēʹsī·dik }

hemidiaphragm [ANATOMY] A lateral half of a diaphragm. { ¦he·mēʹdī·əˌfram }

Hemidiscosa [INVERTEBRATE ZOOLOGY] An order of sponges in the subclass Amphidiscophora distinguished by birotulates that are hemidiscs with asymmetrical ends. { ¦he·mē·disʹkō·sə }

Hemileucinae [INVERTEBRATE ZOOLOGY] A subfamily of lepidopteran insects in the family Saturnidae consisting of the buck moths and relatives. { ¦he·mēʹlüs·ənˌē }

Hemimetabola [INVERTEBRATE ZOOLOGY] A division of the insect subclass Pterygota; members are characterized by hemimetabolous metamorphosis. { ¦he·mē·meʹtab·ə·lə }

hemimetabolous metamorphosis [INVERTEBRATE ZOOLOGY] An incomplete metamorphosis; gills are present in aquatic larvae, or naiads. { ¦he·mē·meʹtab·ə·ləs ˌmed·əʹmȯr·fə·səs }

hemin [BIOCHEMISTRY] $C_{34}H_{32}O_4N_4FeCl$ The crystalline salt of ferriheme, containing iron in the ferric state. { ʹhē·mən }

hemiparasite [ECOLOGY] A parasite capable of a saprophytic existence, especially certain parasitic plants containing some chlorophyll. Also known as semiparasite. { ¦he·mēʹpar·əˌsīt }

hemipelagic [ECOLOGY] Of the biogeographic environment of the hemipelagic region with both neritic and pelagic qualities. { ¦he·mē·pəʹlaj·ik }

hemipenis [VERTEBRATE ZOOLOGY] Either of a pair of nonerectile, evertible sacs that lie on the floor of the cloaca in snakes and lizards; used as intromittent organs. { ¦he·mēʹpē·nəs }

Hemipeplidae [INVERTEBRATE ZOOLOGY] An equivalent name for Cucujidae. { ¦he·mēʹpep·lə ˌdē }

Hemiprocnidae [VERTEBRATE ZOOLOGY] The crested swifts, a family comprising three species of perching birds found only in southeastern Asia. { ¦he·mēʹpräk·nəˌdē }

Hemiptera [INVERTEBRATE ZOOLOGY] The true bugs, an order of the class Insecta characterized by forewings differentiated into a basal area and a membranous apical region. { heʹmip·tə·rə }

Hemisphaeriales [MYCOLOGY] A group of ascomycetous fungi characterized by the wall of the fruit body being a stroma. { ¦he·mēˌsfir·ēʹā·lēz }

hemithorax [ANATOMY] One side of the chest. { ¦he·mēʹthȯrˌaks }

Hemizonida [PALEONTOLOGY] A Paleozoic order of echinoderms of the subclass Asteroidea having an ambulacral groove that is well defined by adambulacral ossicles, but with restricted or undeveloped marginal plates. { ¦he·mēʹzän·ə·də }

hemizygous [GENETICS] Pertaining to the condition or state of having a gene present in a single dose; for instance, in the X chromosome of male mammals. { ¦he·mē¦zī·gəs }

hemlock [BOTANY] The common name for members of the genus *Tsuga* in the pine family characterized by two white lines beneath the flattened, needlelike leaves. { 'hem,läk }

hemobilirubin [BIOCHEMISTRY] Bilirubin in normal blood serum before passage through the liver. { ¦hē·mō,bil·i'rü·bən }

hemoblast [HISTOLOGY] *See* hemocytoblast. { 'hē·mə,blast }

hemochorial placenta [EMBRYOLOGY] A type of placenta having the maternal blood in direct contact with the chorionic trophoblast. Also known as labyrinthine placenta. { ¦hē·mō'kȯr·ē·əl plə'sen·tə }

hemocoel [INVERTEBRATE ZOOLOGY] An expanded portion of the blood system in arthropods that replaces a portion of the coelom. { 'hē·mə,sēl }

hemoconia [BIOCHEMISTRY] Round or dumbbell-shaped, refractile, colorless particles found in blood plasma. { ,hē·mə'kō·nē·ə }

hemocyanin [BIOCHEMISTRY] A blue respiratory pigment found only in mollusks and in arthropods other than insects. { ,hē·mō'sī·ə·nən }

hemocyte [INVERTEBRATE ZOOLOGY] A cellular element of blood, especially in invertebrates. { 'hē·mə,sīt }

hemocytoblast [HISTOLOGY] A pluripotential blast cell thought to be capable of giving rise to all other blood cells. Also known as hemoblast; stem cell. { ,hē·mə'sīd·ə,blast }

hemocytolysis [PHYSIOLOGY] The dissolution of blood cells. { ,hē·mə,sī'täl·ə·səs }

hemodichorial placenta [EMBRYOLOGY] A placenta with a double trophoblastic layer. { ,hē·mə·də'kȯr·ē·əl plə'sen·tə }

hemodynamics [PHYSIOLOGY] A branch of physiology concerned with circulatory movements of the blood and the forces involved in circulation. { ¦hē·mō·dī'nam·iks }

hemoendothelial placenta [EMBRYOLOGY] A placenta having the endothelium of vessels of chorionic villi in direct contact with the maternal blood. { ¦hē·mō,en·də'thē·lē·əl plə'sen·tə }

hemoerythrin [BIOCHEMISTRY] A red respiratory pigment found in a few annelid and sipunculid worms and in the brachiopod *Lingula*. Also known as hemerythrin. { ,hē·mō·ə'rith·rən }

hemoflagellate [INVERTEBRATE ZOOLOGY] A parasitic, flagellate protozoan that lives in the blood of the host. { ,hē·mə'flaj·ə·lət }

hemoglobin [BIOCHEMISTRY] The iron-containing, oxygen-carrying molecule of the red blood cells of vertebrates comprising four polypeptide subunits in a heme group. { 'hē·mə,glō·bən }

hemoglobin A [BIOCHEMISTRY] The type of hemoglobin found in normal adults, which moves as a single component in an electrophoretic field, is rapidly denatured by highly alkaline solutions, and contains two titratable sulfhydryl groups per molecule. { 'hē·mə,glō·bən 'ā }

hemohistioblast [HISTOLOGY] The hypothetical reticuloendothelial cell from which all the cells of the blood are eventually differentiated. { ,hē·mō'his·tē·ō,blast }

hemolymph [INVERTEBRATE ZOOLOGY] The circulating fluid of the open circulatory systems of many invertebrates. { 'hē·mə,limf }

hemolysin [IMMUNOLOGY] A substance that lyses erythrocytes. { ,hē·mə'līs·ən }

hemolysis [PHYSIOLOGY] The lysis, or destruction, of erythrocytes with the release of hemoglobin. { hē'mäl·ə·səs }

hemomonochorial placenta [EMBRYOLOGY] A placenta with a single trophoblastic layer. { ¦hē·mō,män·ə'kȯr·ē·əl plə'sen·tə }

hemoparasite [INVERTEBRATE ZOOLOGY] A parasitic animal that lives in the blood of a vertebrate. { ¦hē·mō'par·ə,sīt }

hemopexin [BIOCHEMISTRY] A heme-binding protein in human plasma that may be a regulator of heme and drug metabolism, and a distributor of heme. { ,hē·mə'pek·sən }

hemophilic bacteria [MICROBIOLOGY] Bacteria of the genera *Hemophilus*, *Bordetella*, and *Moraxella*; all are small, gram-negative, nonmotile, parasitic rods, dependent upon blood factors for growth. { ,hē·mə'fil·ik bak'tir·ē·ə }

Hemophilus [MICROBIOLOGY] A genus of hemophilic bacteria in the family Brucellaceae requiring hemin and nicotinamide nucleoside for growth. { hə'mäf·ə·ləs }

hemopoiesis [PHYSIOLOGY] *See* hematopoiesis. { ,hē·mō,pȯi'ē·səs }

hemopoietic tissue [HISTOLOGY] *See* hematopoietic tissue. { ,hē·mō,pȯi'ed·ik 'tish·ü }

hemopoietin [BIOCHEMISTRY] *See* hematopoietin. { ,hē·mō,pȯi'ēt·ən }

hemoporphyrin [BIOCHEMISTRY] *See* hematoporphyrin. { ,hē·mō'pȯr·fə·rən }

hemorrhagic fever virus [VIROLOGY] Any of several arboviruses causing acute infectious human diseases characterized by fever, prostration, vomiting, and hemorrhage. { ,hem·ə'raj·ik ¦fē·vər ,vī·rəs }

hemorrhagic unit [BIOLOGY] A unit for the standardization of snake venom. { ,hem·ə'raj·ik 'yü·nət }

hemosiderin [BIOCHEMISTRY] An iron-containing glycoprotein found in most tissues and especially in liver. { ,hē·mō'sid·ə·rən }

hemosiderosis [PHYSIOLOGY] Deposition of hemosiderin in body tissues without tissue damage, reflecting an increase in body iron stores. { ,hē·mō,sid·ə'rō·səs }

hemotrichorial placenta [EMBRYOLOGY] A placenta with a triple trophoblastic layer. { ,hē·mō·trə'kȯr·ē·əl plə'sen·tə }

hemotrophe [BIOCHEMISTRY] The nutritive substance supplied via the placenta to embryos of viviparous animals. { hē'mä·trə·fē }

hen [VERTEBRATE ZOOLOGY] The female of several bird species, especially gallinaceous species. { hen }

henbane [BOTANY] *Hyoscyamus niger*. A poisonous herb containing the toxic alkaloids hyoscyamine and hyoscine; extracts have properties similar to belladonna. { 'hen,bān }

Henicocephalidae [INVERTEBRATE ZOOLOGY] A

family of homopteran insects of uncertain affinities. { ˌhen·ə·kō·sə'fal·əˌdē }

Henle's loop [ANATOMY] *See* loop of Henle. { 'hen·lēz ˌlüp }

henna [BOTANY] *Lawsonia inermis*. An Old World plant having small opposite leaves and axillary panicles of white flowers; a reddish-brown dye extracted from the leaves is used in hair dyes. Also known as Egyptian henna. { 'hen·ə }

Hensen's node [EMBRYOLOGY] Thickening formed by a group of cells at the anterior end of the primitive streak in vertebrate gastrulas. { 'hen·sənz ˌnōd }

heparin [BIOCHEMISTRY] An acid mucopolysaccharide acting as an antithrombin, antithromboplastin, and antiplatelet factor to prolong the clotting time of whole blood; occurs in a variety of tissues, most abundantly in liver. { 'hep·ə·rən }

Hepaticae [BOTANY] The equivalent name for Marchantiatae. { he'pad·əˌsē }

hepatic artery [ANATOMY] A branch of the celiac artery that carries blood to the stomach, pancreas, great omentum, liver, and gallbladder. { he'pad·ik 'ärd·ə·rē }

hepatic cecum [INVERTEBRATE ZOOLOGY] A hollow outpocketing of the foregut of *Branchiostoma*; receives veins from the intestine. { he'pad·ik 'sē·kəm }

hepatic duct [ANATOMY] The common duct draining the liver. Also known as common hepatic duct. { he'pad·ik 'dəkt }

hepatic duct system [ANATOMY] The biliary tract including the hepatic ducts, gallbladder, cystic duct, and common bile duct. { he'pad·ik 'dəkt ˌsis·təm }

hepatic plexus [ANATOMY] Nerve network accompanying the hepatic artery to the liver. { he 'pad·ik 'plek·səs }

hepatic portal system [ANATOMY] A system of veins in vertebrates which collect blood from the digestive tract and spleen and pass it through capillaries in the liver. { he'pad·ik 'pȯrd·əl ˌsis·təm }

hepatic vein [ANATOMY] A blood vessel that drains blood from the liver into the inferior vena cava. { he'pad·ik 'vān }

hepatitis virus [VIROLOGY] Any of several viruses causing hepatitis in humans and lower mammals. { 'hep·ə'tīd·əs 'vī·rəs }

hepatocyte [HISTOLOGY] An epithelial cell constituting the major cell type in the liver. { hə 'pad·əˌsīt }

hepatopancreas [INVERTEBRATE ZOOLOGY] A gland in crustaceans and certain other invertebrates that combines the digestive functions of the liver and pancreas of vertebrates. { ¦hep·əd·ō'paŋ·krē·əs }

Hepialidae [INVERTEBRATE ZOOLOGY] A family of lepidopteran insects in the superfamily Hepialoidea. { ˌhep·ē'al·əˌdē }

Hepialoidea [INVERTEBRATE ZOOLOGY] A superfamily of lepidopteran insects in the suborder Homoneura including medium- to large-sized moths which possess rudimentary mouthparts. { ˌhep·ē·ə'lȯid·ē·ə }

Hepsogastridae [INVERTEBRATE ZOOLOGY] A family of parasitic insects in the order Mallophaga. { ˌhep·sə'gas·trəˌdē }

heptose [BIOCHEMISTRY] Any member of the group of monosaccharides containing seven carbon atoms. { 'hepˌtōs }

heptulose [BIOCHEMISTRY] The generic term for a ketose formed from a seven-carbon monosaccharide. { 'hep·təˌlōs }

herb [BOTANY] **1.** A seed plant that lacks a persistent, woody stem above ground and dies at the end of the season. **2.** An aromatic plant or plant part used medicinally or for food flavoring. { hərb }

herbaceous [BOTANY] **1.** Resembling or pertaining to a herb. **2.** Pertaining to a stem with little or no woody tissue. { hər'bā·shəs }

herbaceous dicotyledon [BOTANY] A type of dicotyledon in which the primary vascular cylinder forms an ectophloic siphonostele with widely separated vascular strands. { hər'bā·shəs ˌdī ˌkäd·əl'ēd·ən }

herbaceous monocotyledon [BOTANY] A type of monocotyledon with a vascular system composed of widely spaced strands arranged in one of four ways. { hər'bā·shəs ˌmän·əˌkäd·əl'ēd·ən }

herbarium [BOTANY] **1.** A collection of plant specimens, pressed and mounted on paper or placed in liquid preservatives, and systematically arranged. **2.** A building where a herbarium is housed. { hər'ber·ē·əm }

herbicolous [ECOLOGY] Living on herbs. { hər 'bik·əl·əs }

herbivore [VERTEBRATE ZOOLOGY] An animal that eats only vegetation. { 'hər·bəˌvȯr }

herbivory [ECOLOGY] The consumption of plants without killing them. { hər'biv·ə·rē }

Herbst corpuscle [VERTEBRATE ZOOLOGY] A cutaneous sense organ found in the mucous membrane of the tongue of the duck. { 'hərbst ¦kȯr·pə·səl }

herd [VERTEBRATE ZOOLOGY] A number of one kind of wild, semidomesticated, or domesticated animals grouped or kept together under human control. { hərd }

herd immunity [IMMUNOLOGY] Immunity of a sufficient number of individuals in a population such that infection of one individual will not result in an epidemic. { 'hərd iˌmyü·nəd·ē }

hereditary [GENETICS] Of or pertaining to heredity or inheritance. { hə'red·əˌter·ē }

hereditary determinant [MOLECULAR BIOLOGY] A nuclear or extranuclear genetically functional unit that is replicated with conservation of specificity. { hə¦red·əˌter·ē di'tər·mə·nənt }

heredity [GENETICS] The sum of genetic endowment obtained from the parents. { hə'red·əd·ē }

Hering theory [PHYSIOLOGY] A theory of color vision which assumes that three qualitatively different processes are present in the visual system, and that each of the three is capable of responding in two opposite ways. { 'her·iŋ ˌthē·ə·rē }

heritability [GENETICS] In a population, the ratio

of the total genetic variance to the total phenotypic variance. { ˌher·əd·əˈbil·əd·ē }

heritable change [GENETICS] A nonlethal genetic change that is passed on to the descendants. { ¦her·əd·ə·bəl ˈchānj }

hermaphrodite [BIOLOGY] An individual animal or plant exhibiting hermaphroditism. { hərˈmaf·rəˌdīt }

hermaphroditic [ZOOLOGY] *See* monoecious. { hərˌma·frəˈdid·ik }

hermaphroditism [PHYSIOLOGY] An abnormal condition, especially in humans and other higher vertebrates, in which both male and female reproductive organs are present in the individual. { hərˈma·frə·dīdˌiz·əm }

hermatype [INVERTEBRATE ZOOLOGY] *See* hermatypic coral. { ˈhər·məˌtīp }

hermatypic coral [INVERTEBRATE ZOOLOGY] Reef-building coral characterized by the presence of symbiotic algae within their endodermal tissue. Also known as hermatype. { ˌhər·məˈtip·ik ˈkär·əl }

hermit crab [INVERTEBRATE ZOOLOGY] The common name for a number of marine decapod crustaceans of the families Paguridae and Parapaguridae; all lack right-sided appendages and have a large, soft, coiled abdomen. { ˈhər·mət ˌkrab }

heron [VERTEBRATE ZOOLOGY] The common name for wading birds composing the family Ardeidae characterized by long legs and neck, a long tapered bill, large wings, and soft plumage. { ˈher·ən }

herpes simplex virus [VIROLOGY] Either of two types of subgroup A herpesviruses that are specific for humans; given the binomial designation *Herpesvirus hominis*. { ¦hərˌpēz ˈsimˌpleks ˈvī·rəs }

herpesvirus [VIROLOGY] A major group of deoxyribonucleic acid-containing animal viruses, distinguished by a cubic capsid, enveloped virion, and affinity for the host nucleus as a site of maturation. { ˈhərˌpēzˌvī·rəs }

Herpetosiphon [MICROBIOLOGY] A genus of bacteria in the family Cytophagaceae; cells are unbranched, sheathed rods or filaments; unsheathed segments are motile; microcysts are not known. { ¦hər·pəd·ōˈsī·fən }

herring [VERTEBRATE ZOOLOGY] The common name for fishes composing the family Clupeidae; fins are soft-rayed and have no supporting spines, there are usually four gill clefts, and scales are on the body but absent on the head. { ˈher·iŋ }

Herring body [HISTOLOGY] Any of the distinct colloid masses in the vertebrate pituitary gland, possibly representing greatly dilated endings of nerve fibers. { ˈher·iŋ ˌbäd·ē }

Hesionidae [INVERTEBRATE ZOOLOGY] A family of small polychaete worms belonging to the Errantia. { ˈhes·ēˈän·əˌdē }

hesperidium [BOTANY] A modified berry, with few seeds, a leathery rind, and membranous extensions of the endocarp dividing the pulp into chambers; an example is the orange. { ˈhes·pə ˈrid·ē·əm }

Hesperiidae [INVERTEBRATE ZOOLOGY] The single family of the superfamily Hesperioidea comprising butterflies known as skippers because of their rapid, erratic flight. { hes·pəˈrī·əˌdē }

Hesperioidea [INVERTEBRATE ZOOLOGY] A monofamilial superfamily of lepidopteran insects in the suborder Heteroneura including heavy-bodied, mostly diurnal insects with clubbed antennae that are bent, curved, or reflexed at the tip. { heˌspir·ēˈȯid·ē·ə }

Hesperornithidae [PALEONTOLOGY] A family of extinct North American birds in the order Hesperornithiformes. { ˌhes·pərˌȯrˈnith·əˌdē }

Hesperornithiformes [PALEONTOLOGY] An order of ancient extinct birds; individuals were large, flightless, aquatic diving birds with the shoulder girdle and wings much reduced and the legs specialized for strong swimming. { ˌhes·pəˌrȯr ˌnith·əˈfȯrˌmēz }

Heteractinida [PALEONTOLOGY] A group of Paleozoic sponges with calcareous spicules; probably related to the Calcarea. { ˌhed·ə·rakˈtin·əd·ə }

Heterakidae [INVERTEBRATE ZOOLOGY] A group of nematodes assigned either to the suborder Oxyurina or the suborder Ascaridina. { ˌhed·əˈrak·əˌdē }

Heterakoidea [INVERTEBRATE ZOOLOGY] A superfamily of parasitic nematodes of the order Ascaridida, characterized by small, well-developed lips with paired sensilla in the labial region, an infundibular stoma, and a rarely cylindrical esophagus divided into three parts. { ˌhed·ə·rə ˈkȯid·ē·ə }

heterandrous [BOTANY] Having stamens differing from each other in length or form. { ¦hed·ə ¦ran·drəs }

heterauxesis [BIOLOGY] *See* allometry. { ¦hed·ər ˌȯgˈzē·səs }

heteroagglutinin [IMMUNOLOGY] An antibody in normal blood serum capable of agglutinating foreign particles and erythrocytes of other species. { ¦hed·ə·rō·əˈglüd·ən·ən }

heteroallele [GENETICS] One of two or more alternative forms of a gene that differ at nonidentical mutation sites. { ˌhed·ə·rō·əˈlēl }

heteroauxin [BIOCHEMISTRY] $C_{10}H_9O_2N$ A plant growth hormone with an indole skeleton. { ¦hed·ə·rōˈȯk·sən }

Heterobasidiomycetidae [MYCOLOGY] A class of fungi in which the basidium either is branched or is divided into cells by cross walls. { ¦hed·ə·rō·bəˌsid·ē·ōˌmī¦sed·əˌdē }

heteroblastic [EMBRYOLOGY] Arising from different tissues or germ layers, in referring to similar organs in different species. { ¦hed·ə·rō¦blas·tik }

Heterocapsina [BOTANY] An order of green algae in the class Xanthophyceae. [INVERTEBRATE ZOOLOGY] A suborder of yellow-green to green flagellate protozoans in the order Heterochlorida. { ¦hed·ə·rōˈkap·sə·nə }

heterocarpous [BOTANY] Producing two distinct types of fruit. { ˌhed·ə·rōˈkär·pəs }

heterocentric chromosome [GENETICS] A dicentric chromosome whose centromeres are of unequal strength. { ¦hed·ə·rəˌsen·trik ˈkrō·mə ˌsōm }

Heterocera [INVERTEBRATE ZOOLOGY] A formerly recognized suborder of Lepidoptera including all forms without clubbed antennae. { ˌhed·ə ′räs·ə·rə }

heterocercal [VERTEBRATE ZOOLOGY] Pertaining to the caudal fin of certain fishes and indicating that the upper lobe is larger, with the vertebral column terminating in this lobe. { ¦hed·ə·rō ¦sər·kəl }

Heteroceridae [INVERTEBRATE ZOOLOGY] The variegated mud-loving beetles, a family of coleopteran insects in the superfamily Dryopoidea. { ¦hed·ə·rō¦ser·əˌdē }

Heterocheilidae [INVERTEBRATE ZOOLOGY] A family of parasitic roundworms in the superfamily Ascaridoidea. { ¦hed·ə·rō¦kī·ləˌdē }

heterochlamydeous [BOTANY] Having the perianth differentiated into a distinct calyx and a corolla. { ¦hed·ə·rō·klə′mid·ē·əs }

Heterochlorida [INVERTEBRATE ZOOLOGY] An order of yellow-green to green flagellate organisms of the class Phytamastigophorea. { ¦hed·ə·rō′klȯr·ə·də }

heterochromatin [CELL BIOLOGY] Specialized chromosome material which remains tightly coiled even in the nondividing nucleus and stains darkly in interphase. { ¦hed·ə·rō′krō·məd·ən }

heterochromia [PHYSIOLOGY] A condition in which the two irises of an individual have different colors, or in which one iris has two colors. { ˌhed·ə·rō′krō·mē·ə }

heterochronic mutation [GENETICS] A mutation that perturbs the relative timing of events during postembryonic development. { ¦hed·ə·rəˌkrä·nik myü′tā·shən }

heterochronism [EMBRYOLOGY] Deviation from the normal sequence of organ formation; a factor in evolution. { ˌhed·ə′räk·rəˌniz·əm }

heterococcolith [BIOLOGY] A coccolith with crystals arranged into boat, trumpet, or basket shapes. { ˌhed·ə·rō′käk·əˌlith }

heterocoelous [ANATOMY] Pertaining to vertebrae with centra having saddle-shaped articulations. { ¦hed·ə·rō¦sē·ləs }

Heterocorallia [PALEONTOLOGY] An extinct small, monofamilial order of fossil corals with elongate skeletons; found in calcareous shales and in limestones. { ˌhed·ə·rō·kə′ral·ē·ə }

Heterocotylea [INVERTEBRATE ZOOLOGY] The equivalent name for Monogenea. { ˌhed·ə·rō·kə ′til·ē·ə }

heterocyst [BOTANY] Clear, thick-walled cell occurring at intervals along the filament of certain blue-green algae. { ′hed·ə·rəˌsist }

heterodactylous [VERTEBRATE ZOOLOGY] Having the first two toes turned backward. { ¦hed·ə·rō ¦dakt·əl·əs }

Heterodera [INVERTEBRATE ZOOLOGY] The cyst nematodes, a genus of phytoparasitic worms that live in the internal root systems of many plants. { ˌhed·ə′räd·ə·rə }

heterodimer [BIOCHEMISTRY] A protein made of paired polypeptides that differ in their amino acid sequences. { ˌhed·ə·rō′dī·mər }

heterodont [ANATOMY] Having teeth that are variable in shape and differentiated into incisors, canines, and molars. [INVERTEBRATE ZOOLOGY] In bivalves, having two types of teeth on one valve which fit into depressions on the other valve. { ′hed·ə·rəˌdänt }

Heterodonta [INVERTEBRATE ZOOLOGY] An order of bivalve mollusks in some systems of classification; hinge teeth are few in number and variable in form. { ˌhed·ə·rə′dän·tə }

Heterodontoidea [VERTEBRATE ZOOLOGY] A suborder of sharks in the order Selachii which is represented by the single living genus *Heterodontus*. { ¦hed·ə·rōˌdän′tȯid·ē·ə }

heteroduplex [GENETICS] A double-stranded deoxyribonucleic molecule comprising strands from different individuals. [MOLECULAR BIOLOGY] A double-stranded molecule of deoxyribonucleic acid in which the two strands show noncomplementary sections. { ˌhed·ə·rō′dü ˌpleks }

heteroecious [BIOLOGY] Pertaining to forms that pass through different stages of a life cycle in different hosts. { ˌhed·ə′rē·shəs }

heterogamete [BIOLOGY] A gamete that differs in size, appearance, structure, or sex chromosome content from the gamete of the opposite sex. Also known as anisogamete. { ¦hed·ə·rō′ga ˌmēt }

heterogametic sex [GENETICS] That sex of some species in which the two sex chromosomes are different in gene content or size and which therefore produces two or more different kinds of gametes. { ¦hed·ə·rō·gə′med·ik ′seks }

heterogamety [GENETICS] The production of different kinds of gametes by one sex of a species. { ¦hed·ə·rō′gam·əd·ē }

heterogamous [BIOLOGY] Of or pertaining to heterogamy. { ˌhed·ə′räg·ə·məs }

heterogamy [BIOLOGY] **1.** Alternation of a true sexual generation with a parthenogenetic generation. **2.** Sexual reproduction by fusion of unlike gametes. Also known as anisogamy. [BOTANY] Condition of producing two kinds of flowers. { ˌhed·ə′räg·ə·mē }

heterogeneity [BIOLOGY] The condition or state of being different in kind or nature. { ¦hed·ə·rə·jə′nē·əd·ē }

heterogeneous ribonucleic acid [MOLECULAR BIOLOGY] A large molecule of ribonucleic acid that is believed to be the precursor of messenger ribonucleic acid. Abbreviated H-RNA. { ˌhed·ə·rə ′jē·nē·əs ¦rī·bō·nü¦klē·ik ′as·əd }

Heterogeneratae [BOTANY] A class of brown algae distinguished by a heteromorphic alteration of generations. { ¦hed·ə·rō·ji′ner·əd·ē }

heterogenesis [BIOLOGY] Alternation of generations in a complete life cycle, especially the alternation of a dioecious generation with one or more parthenogenetic generations. { ¦hed·ə·rō ′jen·ə·səs }

heterogenetic antigen [IMMUNOLOGY] *See* heterophile antigen. { ¦hed·ə·rō·jə¦ned·ik ′ant·i·jən }

heterogenous [BIOLOGY] Not originating within the body of the organism. { ˌhed·ə′räj·ə·nəs }

heterogenous vaccine [IMMUNOLOGY] A vaccine

derived from a source other than the patient. { ˌhed·əˈräj·ə·nəs vakˈsēn }

Heterognathi [VERTEBRATE ZOOLOGY] An equivalent name for Cypriniformes. { ˌhed·əˈräg·nə ˌthī }

heterogony [BIOLOGY] **1.** Alteration of generations in a complete life cycle, especially of a dioecious and hermaphroditic generation. **2.** *See* allometry. [BOTANY] Having heteromorphic perfect flowers with respect to lengths of the stamens or styles. { ˌhed·əˈräg·ə·nē }

heterograft [IMMUNOLOGY] A tissue or organ obtained from an animal of one species and transplanted to the body of an animal of another species. Also known as heterologous graft. { ˈhed·ə·rōˌgraft }

heterohemolysin [IMMUNOLOGY] Hemolytic amboceptor against the erythrocytes of a species different from that used to obtain the amboceptor. { ¦hed·ə·rō·həˈmäl·ə·sən }

heterokaryon [MYCOLOGY] A bi- or multinucleate cell having genetically different kinds of nuclei. { ¦hed·ə·rōˈkar·ēˌän }

heterokaryosis [MYCOLOGY] The condition of a bi- or multinucleate cell having nuclei of genetically different kinds. { ˌhed·ə·rōˌkar·ēˈō·səs }

heterokaryotype [GENETICS] A karyotype that is heterozygous for a chromosome mutation. { ˌhed·ə·rəˈkar·ē·əˌtīp }

heterokont [BIOLOGY] An individual, especially among certain algae, having unequal flagella. { ˈhed·ə·rōˌkänt }

heterolactic fermentation [MICROBIOLOGY] A type of lactic acid fermentation by which small yields of lactic acid are produced and much of the sugar is converted to carbon dioxide and other products. { ¦hed·ə·rōˈlak·tik ˌfər·mənˈtā·shən }

heterolateral [ANATOMY] Of, pertaining to, or located on the opposite side. { ˌhed·ə·rōˈlad·ə·rəl }

heterolecithal [CELL BIOLOGY] Of an egg, having the yolk distributed unevenly throughout the cytoplasm. { ˌhed·ə·rōˈles·ə·thəl }

heterologous graft [IMMUNOLOGY] *See* heterograft. { ˌhed·əˈräl·ə·gəs ˈgraft }

heterologous stimulus [PHYSIOLOGY] A form of energy capable of exciting any sensory receptor or form of nervous tissue. { ˌhed·əˈräl·ə·gəs ˈstim·yə·ləs }

heteromedusoid [INVERTEBRATE ZOOLOGY] A styloid type of sessile gonophore. { ¦hed·ə·rō ˈmed·yəˌsȯid }

Heteromera [INVERTEBRATE ZOOLOGY] The equivalent name for Tenebrionoidea. { ¦hed·ə¦räm·ə·rə }

heteromerous [BOTANY] Of a flower, having one or more whorls made up of a different number of members than the remaining whorls. { ¦hed·ə¦räm·ə·rəs }

Heteromi [VERTEBRATE ZOOLOGY] An equivalent name for Notacanthiformes. { ˌhed·əˈrōˌmī }

heteromixis [MYCOLOGY] In Fungi, sexual reproduction which involves the fusion of genetically different nuclei, each from a different thallus. { ¦hed·ə·rōˈmik·səs }

heteromorphic [CELL BIOLOGY] Having synoptic or sex chromosomes that differ in size or form. [ZOOLOGY] Having a different form at each stage of the life history. { ¦hed·ə·rōˈmȯr·fik }

heteromorphosis [BIOLOGY] Regeneration of an organ or part that differs from the original structure at the site. [EMBRYOLOGY] Formation of an organ at an abnormal site. Also known as homoeosis. { ˌhed·ə·rōˈmȯr·fə·səs }

Heteromyidae [VERTEBRATE ZOOLOGY] A family of the mammalian order Rodentia containing the North American kangaroo mice and the pocket mice. { ˌhed·əˈräm·əˌdē }

Heteromyinae [VERTEBRATE ZOOLOGY] The spiny pocket mice, a subfamily of the rodent family Heteromyidae. { ˌhed·ə·rōˈmī·əˌnē }

Heteromyota [INVERTEBRATE ZOOLOGY] A monospecific order of wormlike animals in the phylum Echiurida. { ˌhed·ə·rōˈmī·ə·tə }

Heterondontidae [VERTEBRATE ZOOLOGY] The Port Jackson sharks, a family of aberrant modern elasmobranchs in the suborder Heterodontoidea. { ˌhed·ə·ränˈdänt·əˌdē }

Heteronemertini [INVERTEBRATE ZOOLOGY] An order of the class Anopla; individuals have a middorsal blood vessel and a body wall composed of three muscular layers. { ¦hed·ə·rōˌnem·ərˈtī·nī }

Heteroneura [INVERTEBRATE ZOOLOGY] A suborder of Lepidoptera; individuals are characterized by fore- and hindwings that differ in shape and venation and by sucking mouthparts. { ˌhed·ə·rōˈnu̇r·ə }

heteronuclear culture [CELL BIOLOGY] A cell culture showing a marked variation in chromosome complement among the cells. { ¦hed·ə·rəˌnü·klē·ər ¦kəl·chər }

heteropelmous [VERTEBRATE ZOOLOGY] Having bifid flexor tendons of the toes. { ¦hed·ə·rō ¦pel·məs }

heterophagic vacuole [CELL BIOLOGY] *See* food vacuole. { ˌhed·ə·rōˈfaj·ik ˈvak·yə·wōl }

heterophile antibody [IMMUNOLOGY] Substance that will react with heterophile antigen; found in the serum of patients with infectious mononucleosis. { ˈhed·ə·rəˌfīl ˈan·ti·bäd·ē }

heterophile antigen [IMMUNOLOGY] A substance that occurs in unrelated species of animals but has similar serologic properties among them. Also known as heterogenetic antigen. { ˈhed·ə·rəˌfīl ˈant·i·jən }

heterophile leukocyte [HISTOLOGY] A neutrophile of vertebrates other than humans. { ˈhed·ə·rəˌfīl ˈlü·kəˌsīt }

Heterophyllidae [PALEONTOLOGY] The single family of the extinct cnidarian order Heterocorallia. { ¦hed·ə·rōˈfil·əˌdē }

heterophyllous [BOTANY] Having more than one form of foliage leaves on the same plant or stem. { ˌhed·əˈräf·ə·ləs }

heterophyte [BOTANY] A plant that depends upon living or dead plants or their products for food materials. { ˈhed·ə·rəˌfīt }

Heteropiidae [INVERTEBRATE ZOOLOGY] A family of calcareous sponges in the order Sycettida. { ¦hed·ə·rōˈpī·əˌdē }

heteroplastic [BIOLOGY] Pertaining to transplantation between individuals of different species within the same genus. { ˌhed·ə·rō′plas·tik }

heteroplastidy [BOTANY] The condition of having two kinds of plastids, chloroplasts and leukoplasts. { ˌhed·ə·rō′plas·təd·ē }

heteroploidy [GENETICS] The condition of a chromosome complement in which one or more chromosomes, or parts of chromosomes, are present in number different from the numbers of the rest. { ¦hed·ə·rō¦plȯid·ē }

heteropolysaccharide [BIOCHEMISTRY] A polysaccharide which is a polymer consisting of two or more different monosaccharides. { ¦hed·ə·rō ˌpäl·ē′sak·əˌrīd }

Heteroporidae [INVERTEBRATE ZOOLOGY] A family of trepostomatous bryozoans in the order Cyclostomata. { ¦hed·ə·rō′pȯr·əˌdē }

Heteroptera [INVERTEBRATE ZOOLOGY] The equivalent name for Hemiptera. { ˌhed·ə′räp·tə·rə }

heteropycnosis [CELL BIOLOGY] Differential condensation of certain chromosomes, such as sex chromosomes, or chromosome parts. { ¦hed·ə·rō·pik′nō·səs }

heterosis [GENETICS] The increase in size, yield, and performance found in some hybrids, especially of inbred parents. Also known as hybrid vigor. { ˌhed·ə′rō·səs }

Heterosomata [VERTEBRATE ZOOLOGY] The equivalent name for Pleuronectiformes. { ˌhed·ə·rō′sō·məd·ə }

Heterosoricinae [PALEONTOLOGY] A subfamily of extinct insectivores in the family Soricidae distinguished by a short jaw and hedgehoglike teeth. { ˌhed·ə·rō·sə′ris·əˌnē }

Heterospionidae [INVERTEBRATE ZOOLOGY] A monogeneric family of spioniform worms found in shallow and abyssal depths of the Atlantic and Pacific oceans. { ¦hed·ə·rōˌspī′än·əˌdē }

heterospory [BOTANY] Development of more than one type of spores, especially relating to the microspores and megaspores in ferns and seed plants. { ˌhed·ə′räs·pə·rē }

heterostemony [BOTANY] Presence of two or more different types of stamens in the same flower. { ˌhed·ə·rō′stem·ə·nē }

Heterostraci [PALEONTOLOGY] An extinct group of ostracoderms, or armored, jawless vertebrates; armor consisted of bone lacking cavities for bone cells. { ˌhed·ə′räs·trəˌsī }

heterostyly [BOTANY] Condition or state of flowers having unequal styles. { ˌhed·ə′räst·əl·ē }

Heterotardigrada [INVERTEBRATE ZOOLOGY] An order of the tardigrades exhibiting wide morphologic variations. { ¦hed·ə·rōˌtär′dig·rə·də }

heterothallic [BOTANY] Pertaining to a mycelium with genetically incompatible hyphae, therefore requiring different hyphae to form a zygospore; refers to fungi and some algae. { ¦hed·ə·rō ¦thal·ik }

heterotopia [ECOLOGY] An abnormal habitat. { ˌhed·ə·rō′tō·pē·ə }

heterotopic [BIOLOGY] Pertaining to transplantation of tissue from one site to another on the same organism. { ¦hed·ə·rō¦täp·ik }

heterotopic transplantation [BIOLOGY] A graft transplanted to an abnormal anatomical location on the host. { ¦hed·ə·rō¦täp·ik ˌtranz·plən ′tā·shən }

Heterotrichida [INVERTEBRATE ZOOLOGY] A large order of large ciliates in the protozoan subclass Spirotrichia; buccal ciliature is well developed and some species are pigmented. { ˌhed·ə·rō ′trik·ə·də }

Heterotrichina [INVERTEBRATE ZOOLOGY] A suborder of the protozoan order Heterotrichida. { ˌhed·ə·rō′trik·ə·nə }

heterotrichous [BOTANY] In certain algae, a body that is divided into both prostrate and erect parts. { ¦hed·ə¦rä·trə·kəs }

heterotroph [BIOLOGY] An organism that obtains nourishment from the ingestion and breakdown of organic matter. { ′hed·ə·rōˌträf }

heterotrophic effect [BIOCHEMISTRY] The interaction between different ligands, such as the effect of an inhibitor or activator on the binding of a substrate by an enzyme. { ¦hed·ə·rō¦träf·ik i ′fekt }

heterotropic enzyme [BIOCHEMISTRY] A type of allosteric enzyme in which a small molecule other than the substrate serves as the allosteric reflector. { ¦hed·ə·rō¦träp·ik ′enˌzīm }

heteroxenous [BIOLOGY] Requiring more than one host to complete a life cycle. { ¦hed·ə¦räk·sə·nəs }

heterozooid [INVERTEBRATE ZOOLOGY] Any of the specialized, nonfeeding zooids in a bryozoan colony. { ˌhed·ə·rō′zōˌȯid }

heterozygote [GENETICS] An individual that has different alleles at one or more loci and therefore produces gametes of two or more different kinds. { ¦hed·ə·rō′zīˌgōt }

heterozygote advantage [GENETICS] *See* heterozygote superiority. { ˌhed·ə·rō¦zīˌgōt ad′van·tij }

heterozygote superiority [GENETICS] The greater fitness of an organism that is heterozygous at a given genetic locus as compared with either homozygote. Also known as heterozygote advantage. { ˌhed·ə·rō¦zīˌgōt süˌpir·ē′ȯr·əd·ē }

heterozygous [GENETICS] Of or pertaining to a heterozygote. { ¦hed·ə·rō¦zī·gəs }

Hevea [BOTANY] The rubber tree genus of the order Euphoriales from which the largest volumes of latex are harvested for use in the manufacture of natural rubber. { ′hē·vē·ə }

hexacanth [INVERTEBRATE ZOOLOGY] Having six hooks; refers specifically to the embryo of certain tapeworms. { ′hek·səˌkanth }

Hexacorallia [INVERTEBRATE ZOOLOGY] The equivalent name for Zoantharia. { ˌhek·sə·kə ′ral·ē·ə }

hexactin [INVERTEBRATE ZOOLOGY] A spicule, especially in Porifera, having six equal rays at right angles to each other. { hek′sak·tən }

Hexactinellida [INVERTEBRATE ZOOLOGY] A class of the phylum Porifera which includes sponges with a skeleton made up basically of hexactinal siliceous spicules. { hekˌsak·tə′nel·ə·də }

Hexactinosa [INVERTEBRATE ZOOLOGY] An order of sponges in the subclass Hexasterophora; parenchymal megascleres form a rigid framework and consist of simple hexactins. { hek‚sak·tə 'nō·sə }

Hexanchidae [VERTEBRATE ZOOLOGY] The six- and seven-gill sharks, a group of aberrant modern elasmobranchs in the suborder Notidanoidea. { ‚hek'saŋ·kə‚dē }

hexapetalous [BOTANY] Having or being a perianth comprising six petaloid divisions. { ¦hek·sə 'ped·əl·əs }

Hexapoda [INVERTEBRATE ZOOLOGY] An equivalent name for Insecta. { hek'säp·əd·ə }

hexaster [INVERTEBRATE ZOOLOGY] A type of hexactin with branching rays that form star-shaped figures. { 'hek‚sas·tər }

Hexasterophora [INVERTEBRATE ZOOLOGY] A subclass of sponges of the class Hexactinellida in which parenchymal microscleres are typically hexasters. { ‚hek‚sas·tə'räf·ə·rə }

hexokinase [BIOCHEMISTRY] Any enzyme that catalyzes the phosphorylation of hexoses. { ¦hek·sō'kī‚nās }

hexosamine [BIOCHEMISTRY] A primary amine derived from a hexose by replacing the hydroxyl with an amine group. { hek'säs·ə‚mēn }

hexosaminidase A [BIOCHEMISTRY] An enzyme which catalyzes the hydrolysis of the N-acetylgalactosamine residue from certain gangliosides. { hek‚säs·ə'min·ə‚dās 'ā }

hexosan [BIOCHEMISTRY] *See* hemicellulose. { 'hek·sō‚san }

hexose [BIOCHEMISTRY] Any monosaccharide that contains six carbon atoms in the molecule. { 'hek‚sōs }

hexose diphosphate pathway [BIOCHEMISTRY] *See* glycolytic pathway. { 'hek‚sōs dī'fäs‚fāt 'path ‚wā }

hexose monophosphate cycle [BIOCHEMISTRY] A pathway for carbohydrate metabolism in which one molecule of hexose monophosphate is completely oxidized. { 'hek‚sōs ¦män·ō¦fäs‚fāt 'sī·kəl }

hexose phosphate [BIOCHEMISTRY] Any one of the phosphoric acid esters of a hexose, notably glucose, formed during the metabolism of carbohydrates by living organisms. { 'hek‚sōs 'fäs ‚fāt }

hexulose [BIOCHEMISTRY] A ketose made from a six-carbon-chain monosaccharide. { 'heks·yə ‚lōs }

Hfr [MICROBIOLOGY] *See* high-frequency recombination.

hiascent [BIOLOGY] Gaping. { hī'ā·shənt }

hiatus [ANATOMY] A space or a passage through an organ. { hī'ād·əs }

hibernaculum [BIOLOGY] A winter shelter for plants or dormant animals. [BOTANY] A winter bud or other winter plant part. [INVERTEBRATE ZOOLOGY] A winter resting bud produced by a few fresh-water bryozoans which grows into a new colony in the spring. { ‚hī·bər'nak·yə·ləm }

hibernation [PHYSIOLOGY] **1.** Condition of dormancy and torpor found in cold-blooded vertebrates and invertebrates. **2.** *See* deep hibernation. { ‚hī·bər'nā·shən }

hickory [BOTANY] The common name for species of the genus *Carya* in the order Fagales; tall deciduous tree with pinnately compound leaves, solid pith, and terminal, scaly winter buds. { 'hik·ə·rē }

hide [VERTEBRATE ZOOLOGY] Outer covering of an animal. { hīd }

hidrosis [PHYSIOLOGY] The formation and excretion of sweat. { hī'drō·səs }

high-frequency recombination [MICROBIOLOGY] A bacterial cell type, especially *Escherichia coli*, having an integrated F factor and characterized by a high frequency of recombination. Abbreviated Hfr. { 'hī ¦frē·kwən·sē ‚rē‚käm·bə'nā·shən }

highmoor bog [ECOLOGY] A bog whose surface is covered by sphagnum mosses and is not dependent upon the water table. { 'hī‚mür 'bäg }

Hikojima serotype [IMMUNOLOGY] An immunologically distinct group of *Vibrio* somatic O antigens. { ‚hē·kō'jē·mə 'ser·ə‚tīp }

Hill plot [BIOCHEMISTRY] A graphic representation of the Hill reaction. { 'hil ‚plät }

Hill reaction [BIOCHEMISTRY] The release of molecular oxygen by isolated chloroplasts in the presence of a suitable electron receptor, such as ferricyanide. { 'hil rē‚ak·shən }

hilum [ANATOMY] *See* hilus. [BOTANY] Scar on a seed marking the point of detachment from the funiculus. { 'hī·ləm }

hilus [ANATOMY] An opening or recess in an organ, usually for passage of a vessel or duct. Also known as hilum. { 'hī·ləs }

Himantandraceae [BOTANY] A family of dicotyledonous plants in the order Magnoliales characterized by several, uniovulate carpels and laminar stamens. { hə¦mant·ən'drās·ē‚ē }

himantioid [MYCOLOGY] Pertaining to a mycelium arranged in spreading fanlike cords. { hə 'man·tē‚ȯid }

Himantopterinae [INVERTEBRATE ZOOLOGY] A subfamily of lepidopteran insects in the family Zygaenidae including small, brightly colored moths with narrow hindwings, ribbonlike tails, and long hairs covering the body and wings. { hə ¦man·tō'ter·ə‚nē }

hindbrain [EMBRYOLOGY] *See* rhombencephalon. { 'hīn‚brān }

hindgut [EMBRYOLOGY] The caudal portion of the embryonic digestive tube in vertebrates. { 'hīn ‚gət }

hinge joint [ANATOMY] *See* ginglymus. { 'hinj ‚jȯint }

hinge plate [INVERTEBRATE ZOOLOGY] **1.** In bivalve mollusks, the portion of a valve that supports the hinge teeth. **2.** The socket-bearing part of the dorsal valve in brachiopods. { 'hinj ‚plāt }

hinge tooth [INVERTEBRATE ZOOLOGY] A projection on a valve of a bivalve mollusk near the hinge line. { 'hinj ‚tüth }

Hiodontidae [VERTEBRATE ZOOLOGY] A family of tropical, fresh-water actinopterygian fishes in the order Osteoglossiformes containing the mooneyes of North America. { ‚hī·ə'dänt·ə‚dē }

hip [ANATOMY] **1.** The region of the junction of thigh and trunk. **2.** The hip joint, formed by articulation of the femur and hipbone. { hip }

hipbone [ANATOMY] A large broad bone consisting of three parts, the ilium, ischium, and pubis; makes up a lateral half of the pelvis in mammals. Also known as innominate. { 'hip,bōn }

Hippidea [INVERTEBRATE ZOOLOGY] A group of decapod crustaceans belonging to the Anomura and including cylindrical or squarish burrowing crustaceans in which the abdomen is symmetrical and bent under the thorax. { hi'pid·ē·ə }

Hippoboscidae [INVERTEBRATE ZOOLOGY] The louse flies, a family of cyclorrhaphous dipteran insects in the section Pupipara. { ,hip·ə'bäs·kə ,dē }

hippocampal sulcus [ANATOMY] A fissure on the brain situated between the para hippocampal gyrus and the fimbria hippocampi. Also known as dentate fissure. { ¦hip·ə¦kam·pəl 'səl·kəs }

hippocampus [ANATOMY] A ridge that extends over the floor of the descending horn of each lateral ventricle of the brain. { ,hip·ə'kam·pəs }

Hippocampus [VERTEBRATE ZOOLOGY] A genus of marine fishes in the order Gasterosteiformes which contains the sea horses. { ,hip·ə'kam· pəs }

Hippocrateaceae [BOTANY] A family of dicotyledonous plants in the order Celastrales distinguished by an extrastaminal disk, mostly opposite leaves, seeds without endosperm, and a well-developed latex system. { ,hip·ə,krād·ē 'ās·ē,ē }

hippocrepiform [BIOLOGY] Horseshoe-shaped. { ¦hip·ə¦krep·ə,fòrm }

Hippoglossidae [VERTEBRATE ZOOLOGY] A family of actinopterygian fishes in the order Pleuronectiformes composed of the flounders and plaice. { ,hip·ə'gläs·ə,dē }

Hippomorpha [VERTEBRATE ZOOLOGY] A suborder of the mammalian order Perissodactyla containing horses, zebras, and related forms. { ,hip·ə'mòr·fə }

Hippopotamidae [VERTEBRATE ZOOLOGY] The hippopotamuses, a family of palaeodont mammals in the superfamily Anthracotherioidea. { ,hip·ə·pəd'am·ə,dē }

hippopotamus [VERTEBRATE ZOOLOGY] The common name for two species of artiodactyl ungulates composing the family Hippopotamidae. { ,hip·ə'päd·ə·məs }

Hipposideridae [VERTEBRATE ZOOLOGY] The Old World leaf-nosed bats, a family of mammals in the order Chiroptera. { ,hi,pō·sə'der·ə,dē }

hirsute [BIOLOGY] Shaggy; hairy. { 'hər,süt }

Hirudinea [INVERTEBRATE ZOOLOGY] A class of parasitic or predatory annelid worms commonly known as leeches; all have 34 body segments and terminal suckers for attachment and locomotion. { ,hi·rə'din·ē·ə }

Hirudinidae [VERTEBRATE ZOOLOGY] The swallows, a family of passeriform birds in the suborder Oscines. { ,hi·rə'din·ə,dē }

his operon [GENETICS] A sequence of nine genes in the chromosomes of many bacteria which code for all the enzymes of histidine biosynthesis. { 'his 'äp·ə,rän }

hispid [BIOLOGY] Having a covering of bristles or minute spines. { 'his·pəd }

hispidulous [BIOLOGY] Hispid to a minute degree. { his'pij·ə·ləs }

histamine [BIOCHEMISTRY] $C_5H_9N_3$ An amine derivative of histadine which is widely distributed in human tissues. { 'his·tə,mēn }

Histeridae [INVERTEBRATE ZOOLOGY] The clown beetles, a large family of coleopteran insects in the superfamily Histeroidea. { hi'ster·ə,dē }

Histeroidea [INVERTEBRATE ZOOLOGY] A superfamily of coleopteran insects in the suborder Polyphaga. { ,his·tə'ròid·ē·ə }

histidase [BIOCHEMISTRY] An enzyme found in the liver of higher vertebrates that catalyzes the deamination of histidine to urocanic acid. { 'his·tə,dās }

histidine [BIOCHEMISTRY] $C_6H_9O_2N_3$ A crystalline basic amino acid present in large amounts in hemoglobin and resulting from the hydrolysis of most proteins. { 'his·tə,dēn }

histiocyte [HISTOLOGY] *See* macrophage. { 'his· tē·ə,sīt }

Histioteuthidae [INVERTEBRATE ZOOLOGY] A family of cephalopod mollusks containing several species of squids. { ,his·tē·ō'tü·thə,dē }

histochemistry [BIOCHEMISTRY] A science that deals with the distribution and activities of chemical components in tissues. { ¦hi·stō'kem· ə·strē }

histocompatibility [IMMUNOLOGY] The capacity to accept or reject a tissue graft. { ¦hi·stō·kəm 'pad·ə'bil·əd·ē }

histocompatibility gene [GENETICS] In mammals, any of the genes belonging to the major histocompatibility complex. { ,his·tə·kəm'pad·ə ,bil·əd·ē ,jēn }

histodifferentiation [EMBRYOLOGY] Differentiation of cell groups into tissues. { ¦hi·stō·dif·ə ,ren·chē'ā·shən }

histogen [BOTANY] A clearly delimited region or primary tissue from which the specific parts of a plant organ are thought to be produced. { 'his· tə·jən }

histogenesis [EMBRYOLOGY] The developmental process by which the definite cells and tissues which make up the body of an organism arise from embryonic cells. { ,his·tə'jen·ə·səs }

histoincompatibility [IMMUNOLOGY] The condition in which a recipient rejects a tissue graft. { ¦hi·stō,in·kəm,pad·ə'bil·əd·ē }

histologist [ANATOMY] An individual who specializes in histology. { hi'stäl·ə·jəst }

histology [ANATOMY] The study of the structure and chemical composition of animal and plant tissues as related to their function. { hi'stäl·ə· jē }

histone [BIOCHEMISTRY] Any of the strong, soluble basic proteins of cell nuclei that are precipitated by ammonium hydroxide. { 'hi,stōn }

histophysiology [PHYSIOLOGY] The science of tissue functions. { ,hi·stō,fiz·ē'äl·ə·jē }

Histoplasma [MYCOLOGY] A genus of parasitic fungi. { ,his·tə'plaz·mə }

Histoplasma capsulatum [MYCOLOGY] The parasitic fungus that causes histoplasmosis in humans. { ,his·tə'plaz·mə ,kap·sə'läd·əm }

histoplasmin test [IMMUNOLOGY] Skin test for

hypersensitivity reaction to *Histoplasma capsulatum* products in the diagnosis of histoplasmosis. { ˌhis·tə'plaz·mən ˌtest }

historadiography [BIOPHYSICS] A technique for taking x-ray pictures of cells, tissues, or sometimes whole small organisms. { ˌhis·təˌrād·ē'äg·rə·fē }

histotome [BIOLOGY] A microtome used to cut tissue sections for microscopic examination. { 'his·təˌtōm }

Histriobdellidae [INVERTEBRATE ZOOLOGY] A small family of errantian polychaete worms that live as ectoparasites on crayfishes. { ˌhis·trē·əb'del·əˌdē }

hitchhiking effect [GENETICS] The increase in frequency of a neutral allele closely linked to a selectively favored allele. { 'hichˌhīk·iŋ iˌfekt }

HIV [VIROLOGY] *See* human immunodeficiency virus.

hive [INVERTEBRATE ZOOLOGY] *See* beehive. { hīv }

HLA [IMMUNOLOGY] *See* human leukocyte antigen.

HLA complex [IMMUNOLOGY] The major histocompatibility complex of humans. { ¦āch¦el'ā ˌkämˌpleks }

hoarding behavior [VERTEBRATE ZOOLOGY] The carrying of food to the home nest for storage, in quantities exceeding daily need. { 'hȯrd·iŋ biˌhāv·yər }

hoarhound [BOTANY] *See* marrubium. { 'hȯrˌhau̇nd }

hoary [BOTANY] Having grayish or whitish color, referring to leaves. { hȯr·ē }

Hodotermitidae [INVERTEBRATE ZOOLOGY] A family of lower (primitive) termites in the order Isoptera. { ˌhäd·ō·tər'mid·əˌdē }

Hofbauer cell [HISTOLOGY] A large, possibly phagocytic cell found in chorionic villi. { 'hōf·bau̇r ˌsel }

holandric trait [GENETICS] Any trait appearing only in males. { hə¦lanˌdrik 'trāt }

holarctic zoogeographic region [ECOLOGY] A major unit of the earth's surface extending from the North Pole to 30-45°N latitude and characterized by faunal homogeneity. { hō'lärd·ik ˌzō·ōˌjē·ə¦graf·ik 'rē·jən }

Holasteridae [INVERTEBRATE ZOOLOGY] A family of exocyclic Euechinoidea in the order Holasteroida; individuals are oval or heart-shaped, with fully developed pore pairs. { ˌhäl·ə'ster·əˌdē }

Holasteroida [INVERTEBRATE ZOOLOGY] An order of exocyclic Euechinoidea in which the apical system is elongated along the anteroposterior axis and teeth occur only in juvenile stages. { ˌhäl·ə·stə'rȯid·ə }

holcodont [VERTEBRATE ZOOLOGY] Having the teeth fixed in a long, continuous groove. { 'häl·kəˌdänt }

holdfast [BOTANY] **1.** A suckerlike base which attaches the thallus of certain algae to the support. **2.** A disklike terminal structure on the tendrils of various plants used for attachment to a flat surface. [INVERTEBRATE ZOOLOGY] An organ by which parasites such as tapeworms attach themselves to the host. { 'hōlˌfast }

Holectypidae [PALEONTOLOGY] A family of extinct exocyclic Euechinoidea in the order Holectypoida; individuals are hemispherical. { hōˌlek'tip·əˌdē }

Holectypoida [INVERTEBRATE ZOOLOGY] An order of exocyclic Euechinoidea with keeled, flanged teeth, with distinct genital plates, and with the ambulacra narrower than the interambulacra on the adoral side. { hōˌlek·tə'pȯid·ə }

holism [BIOLOGY] The view that the whole of a complex system, such as a cell or organism, is functionally greater than the sum of its parts. Also known as organicism. { 'hōˌliz·əm }

Hollinacea [PALEONTOLOGY] A dimorphic superfamily of extinct ostracods in the suborder Beyrichicopina including forms with sulci, lobation, and some form of velar structure. { ˌhäl·ə'nās·ē ə }

Hollinidae [PALEONTOLOGY] An extinct family of ostracods in the superfamily Hollinacea distinguished by having a bulbous third lobe on the valve. { hə'lin·əˌdē }

holly [BOTANY] The common name for the trees and shrubs composing the genus *Ilex*; distinguished by spiny leaves and small berries. { 'häl·ē }

hollywood lignumvitae [BOTANY] *See* lignumvitae. { 'häl·ēˌwu̇d ˌlig·nəm'vī·dē }

Holobasidiomycetes [MYCOLOGY] An equivalent name for Homobasidiomycetidae. { ¦häl·ō·bəˌsid·ē·ō·mī' sēd·ēz }

holoblastic [EMBRYOLOGY] Pertaining to eggs that undergo total cleavage due to the absence of a mass of yolk. { ¦häl·ə¦blas·tik }

holobranch [VERTEBRATE ZOOLOGY] A gill with a row of filaments on each side of the branchial arch. { 'häl·əˌbraŋk }

holocarpic [BOTANY] **1.** Having the entire thallus developed into a fruiting body or sporangium. **2.** Lacking rhizoids and haustoria. { ¦häl·ō¦kär·pik }

holocellulose [BIOCHEMISTRY] The total polysaccharide fraction of wood, straw, and so on, that is composed of cellulose and all of the hemicelluloses and that is obtained when the extractives and the lignin are removed from the natural material. { ¦häl·ō'sel·yəˌlōs }

holocentric chromosome [GENETICS] A chromosome with nonlocalized centromeres. { ˌhō·ləˌsen·trik 'krō·məˌsōm }

Holocentridae [VERTEBRATE ZOOLOGY] A family of nocturnal beryciform fishes found in shallow tropical and subtropical reefs; contains the squirrelfishes and soldierfishes. { ˌhäl·ə'sen·trəˌdē }

Holocephali [VERTEBRATE ZOOLOGY] The chimaeras, a subclass of the Chondrichthyes, distinguished by four pairs of gills and gill arches, an erectile dorsal fin and spine, and naked skin. { ¦häl·ō¦sef·əˌlī }

holococcolith [BIOLOGY] A coccolith with simple rhombic or hexagonal crystals arranged like a mosaic. { ¦häl·ō'käk·əˌlith }

holocoenosis [ECOLOGY] The nature of the action of the environment on living organisms. { ¦häl·ō·sə¦nō·səs }

holocrine gland [PHYSIOLOGY] A structure whose cells undergo dissolution and are entirely extruded, together with the secretory product. { 'häl·ə·krən ˌgland }

holoechinate [INVERTEBRATE ZOOLOGY] Having the whole body covered with spines. { ¦hō·lō·ē 'kīˌnāt }

holoenzyme [BIOCHEMISTRY] A complex, fully active enzyme, containing an apoenzyme and a coenzyme. { ¦häl·ō'enˌzīm }

hologamy [BIOLOGY] Condition of having gametes similar in size and form to somatic cells. [BOTANY] Condition of having the whole thallus develop into a gametangium. { hə'läg·ə·mē }

hologony [INVERTEBRATE ZOOLOGY] Condition of having the germinal area extend the full length of a gonad; refers specifically to certain nematodes. { hə'läg·ə·nē }

hologynic trait [GENETICS] Any trait appearing only in females. { ˌhō·ləˌgī·nik 'trāt }

Holometabola [INVERTEBRATE ZOOLOGY] A division of the insect subclass Pterygota whose members undergo holometabolous metamorphosis during development. { ¦häl·ō·mə'tab·ə·lə }

holometabolous metamorphosis [INVERTEBRATE ZOOLOGY] Complete metamorphosis, during which there are four stages; the egg, larva, pupa, and imago or adult. { ¦häl·ō·mə'tab·ə·ləs ˌmed·ə'mȯr·fə·səs }

holomorphosis [BIOLOGY] Complete regeneration of a lost body structure. { ¦häl·ō'mȯr·fə·səs }

holomyarian [INVERTEBRATE ZOOLOGY] Having zones of muscle layers but no muscle cells; refers specifically to certain nematodes. { ¦häl·ō·mī ¦a·rē·ən }

holonephros [VERTEBRATE ZOOLOGY] Type of kidney having one nephron beside each somite along the entire length of the coelom; seen in larvae of myxinoid cyclostomes. { ¦häl·ō'ne ˌfrōs }

holophyte [BIOLOGY] An organism that obtains food in the manner of a green plant, that is, by synthesis of organic substances from inorganic substances using the energy of light. { 'häl·ə ˌfīt }

holoplankton [ZOOLOGY] Organisms that live their complete life cycle in the floating state. { ˌhäl·ō'plaŋk·tən }

Holoptychidae [PALEONTOLOGY] A family of extinct lobefin fishes in the order Osteolepiformes. { ˌhäl·əp·tə'kī·əˌdē }

holorhinal [VERTEBRATE ZOOLOGY] Among birds, having a rounded anterior margin on the nasal bones. { ¦häl·ō'rīn·əl }

Holostei [VERTEBRATE ZOOLOGY] An infraclass of fishes in the subclass Actinopterygii descended from the Chondrostei and ancestral to the Teleostei. { hə'läs·tēˌī }

holostome [INVERTEBRATE ZOOLOGY] A type of adult digenetic trematode having a portion of the ventral surface modified as a complex adhesive organ. { 'häl·əˌstōm }

Holothuriidae [VERTEBRATE ZOOLOGY] A family of aspidochirotacean echinoderms in the order Aspidochirotida possessing tentacular ampullae and only the left gonad. { ˌhäl·ō·thə'rī·əˌdē }

Holothuroidea [INVERTEBRATE ZOOLOGY] The sea cucumbers, a class of the subphylum Echinozoa characterized by a cylindrical body and smooth, leathery skin. { ˌhäl·ōˌthu̇·rē'ȯid·ē·ə }

Holothyridae [INVERTEBRATE ZOOLOGY] The single family of the acarine suborder Holothyrina. { ˌhäl·ō'thī·rəˌdē }

Holothyrina [INVERTEBRATE ZOOLOGY] A suborder of mites in the order Acarina which are large and hemispherical with a deep-brown, smooth, heavily sclerotized cuticle. { ˌhäl·ō'thī·rə·nə }

Holotrichia [INVERTEBRATE ZOOLOGY] A major subclass of the protozoan class Ciliatea; body ciliation is uniform with cilia arranged in longitudinal rows. { ˌhäl·ō'trik·ē·ə }

holotype [SYSTEMATICS] A nomenclatural type for the single specimen designated as "the type" by the original author at the time of publication of the original description. { 'häl·əˌtīp }

holozoic [ZOOLOGY] Obtaining food in the manner of most animals, by ingesting complex organic matter. { ¦häl·ə¦zō·ik }

Holuridae [PALEONTOLOGY] A group of extinct chondrostean fishes in the suborder Palaeonisocoidei distinguished in having lepidotrichia of all fins articulated but not bifurcated, fins without fulcra, and the tail not cleft. { hə'lu̇r·əˌdē }

Homacodontidae [PALEONTOLOGY] A family of extinct palaeodont mammals in the superfamily Dichobunoidea. { ˌhäm·ə·kō'dänt·əˌdē }

Homalopteridae [VERTEBRATE ZOOLOGY] A small family of cypriniform fishes in the suborder Cyprinoidei. { ¦häm·ə·läp'ter·əˌdē }

Homalorhagae [INVERTEBRATE ZOOLOGY] A suborder of the class Kinorhyncha having a single dorsal plate covering the neck and three ventral plates on the third zonite. { ˌhäm·ə'lȯr·əˌgē }

Homalozoa [INVERTEBRATE ZOOLOGY] A subphylum of echinoderms characterized by the complete absence of radial symmetry. { hə¦mal·ə ¦zō·ə }

Homaridae [INVERTEBRATE ZOOLOGY] A family of marine decapod crustaceans containing the lobsters. { hō'mar·əˌdē }

Homarus [INVERTEBRATE ZOOLOGY] A genus of the family Homaridae comprising most species of lobsters. { 'häm·ə·rəs }

homaxial [BIOLOGY] Having all axes equal. { hōm 'ak·sē·əl }

homeobox [MOLECULAR BIOLOGY] A highly conserved sequence of deoxyribonucleic acid (DNA) that occurs in the coding region of development-controlling regulatory genes and codes for a protein domain that is similar in structure to certain DNA-binding proteins and is thought to be involved in the control of gene expression during morphogenesis and development. { 'hō·mē·ə ˌbäks }

homeomorph [BIOLOGY] An organism which exhibits a superficial resemblance to another organism even though they have different ancestors. { 'hō·mē·əˌmȯrf }

homeosis [EMBRYOLOGY] *See* heteromorphosis. { ˌhō·mē'ō·səs }

homeostasis [BIOLOGY] In higher animals, the maintenance of an internal constancy and an independence of the environment. { ˌhō·mē·ō 'stā·səs }

homeotherm [PHYSIOLOGY] An endotherm that maintains a constant body temperature, as do most mammals and birds. { 'hō·mē·əˌthərm }

homeothermia [PHYSIOLOGY] The condition of being warm-blooded. { ¦hō·mē·ə¦thər·mē·ə }

homeotic gene [GENETICS] Any gene that determines the morphogenetic fate of groups of cells in terms of spatial orientation. { 'hō·mēˌäd·ik 'jēn }

homeotic mutation [GENETICS] A mutation that causes cells to switch from one developmental pattern to a different but homologous one. { 'hō·mēˌäd·ik myü'tā·shən }

home range [ECOLOGY] The physical area of an organism's normal activity. { 'hōm ¦rānj }

Hominidae [VERTEBRATE ZOOLOGY] A family of primates in the superfamily Hominoidea containing one living species, *Homo sapiens*. { hä 'min·əˌdē }

Hominoidea [VERTEBRATE ZOOLOGY] A superfamily of the order Primates comprising apes and humans. { hä·mə'nȯid·ē·ə }

Homo [VERTEBRATE ZOOLOGY] The genus of human beings, including modern humans and many extinct species. { 'hō·mō }

homoallele [GENETICS] One of two or more alternative forms of a gene that differ at identical mutation sites. { ˌhō·mō·ə'lēl }

Homobasidiomycetidae [MYCOLOGY] A subclass of basidiomycetous fungi in which the basidium is not divided by cross walls. { ¦hä·mō·bəˌsid·ē·ōˌmī'sed·əˌdē }

homocercal [VERTEBRATE ZOOLOGY] Pertaining to the caudal fin of certain fishes which has almost equal lobes, with the vertebral column terminating near the middle of the base. { ˌhäm·ə 'sər·kəl }

homochlamydeous [BIOLOGY] Having all members of the perianth similar or not differentiated into calyx or corolla. { ¦hä·mō·klə'mid·ē·əs }

homochromy [ZOOLOGY] A form of protective coloration whereby the individual blends into the background. { hə'mäk·rə·mē }

homocysteine [BIOCHEMISTRY] $C_4H_9O_2NS$ An amino acid formed in animals by demethylation of methionine. { ¦hä·mə'sis·tēn }

homodimer [BIOCHEMISTRY] A protein made of paired identical polypeptides. { ˌhō·mō'dī·mər }

homodont [VERTEBRATE ZOOLOGY] Having all teeth similar in form; characteristic of nonmammalian vertebrates. { 'hä·məˌdänt }

homoduplex [MOLECULAR BIOLOGY] A deoxyribonucleic acid duplex in which the nitrogenous bases of the two strands are precisely complementary. { ˌhō·mō'düˌpleks }

homodynamic [INVERTEBRATE ZOOLOGY] Developing through continuous successive generations without a diapause; applied to insects. { ˌhä·məˌdī'näm·ik }

homoecious [BIOLOGY] Having one host for all stages of the life cycle. { hō'mē·shəs }

homoeomerous [BOTANY] Having algae distributed uniformly throughout the thallus of a lichen. { ¦hō·mē¦äm·ə·rəs }

Homo erectus [PALEONTOLOGY] A type of fossil human from the Pleistocene of Java and China representing a specialized side branch in human evolution. { 'hō·mō ə'rek·təs }

homofermentative lactobacilli [MICROBIOLOGY] Bacteria that produce a single end product, lactic acid, from fermentation of carbohydrates. { ¦hō·mō·fər'men·tə·tiv ¦lak·tō·bə¦sil·ē }

homogametic sex [GENETICS] The sex of a species in which the paired sex chromosomes are of equal size and which therefore produces homogametes. { ¦hä·mō·gə¦med·ik 'seks }

homogamety [GENETICS] The production of homogametes by one sex of a species. { ¦hä·mō 'ga·məd·ē }

homogamous [BIOLOGY] Of or pertaining to homogamy. { hə'mäg·ə·məs }

homogamy [BIOLOGY] Inbreeding due to isolation. [BOTANY] Condition of having all flowers alike. { hə'mäg·ə·mē }

homogenate [BIOLOGY] A tissue that has been finely divided and mixed. { hə'mäj·ə·nət }

homogeneously staining region [CELL BIOLOGY] In human chromosomes, an extended chromosomal segment that has a banding pattern and represents a site of gene amplification; found mostly in cancer cells. { ˌhō·mə¦jēn·ē·əs·lē ¦stān·iŋ ˌrē·jən }

homogentisase [BIOCHEMISTRY] The enzyme that catalyzes the conversion of homogentisic acid to fumaryl acetoacetic acid. { ¦hä·mə¦jen·təˌsās }

homogentisic acid [BIOCHEMISTRY] $C_8H_8O_4$ An intermediate product in the metabolism of phenylalanine and tyrosine; found in excess in persons with phenylketonuria and alkaptonuria. { ¦hä·məˌjen¦tiz·ik 'as·əd }

homogony [BOTANY] Condition of having one type of flower, with stamens and pistil of uniform length. { hə'mäg·ə·nē }

homograft rejection [IMMUNOLOGY] An immunologic process by which an individual destroys and casts off a tissue transplanted from a donor of the same species. { 'hä·məˌgraft ri'jek·shən }

homoiochlamydeous [BOTANY] Having perianth leaves alike, not differentiated into sepals and petals. { hō¦mȯi·ō·klə'mid·ē·əs }

homoiogenetic [EMBRYOLOGY] Of a determined part of an embryo, capable of inducing formation of a similar part when grafted into an undetermined field. { hō¦mȯi·ō·jə'ned·ik }

homoiothermal [PHYSIOLOGY] Referring to an organism which maintains a constant internal temperature which is often higher than that of the environment; common among birds and mammals. Also known as warm-blooded. { hō ¦mȯi·ō¦thər·məl }

Homoistela [PALEONTOLOGY] A class of extinct echinoderms in the subphylum Homalozoa. { hō 'mȯi·stə·lə }

homokaryon [MYCOLOGY] A bi- or multinucleate

cell having nuclei all of the same kind. { ˌhä·mə ˈkar·ēˌän }

homokaryosis [MYCOLOGY] The condition of a bi- or multinucleate cell having nuclei all of the same kind. { ˌhä·məˌkar·ēˈō·səs }

homokaryotype [GENETICS] A karyotype that is homozygous for a chromosome mutation. { ˌhō·məˈkar·ē·əˌtīp }

homolecithal [CELL BIOLOGY] Referring to eggs having small amounts of evenly distributed yolk. Also known as isolecithal. { ˌhä·məˈles·ə·thəl }

homologous [BIOLOGY] Pertaining to a structural relation between parts of different organisms due to evolutionary development from the same or a corresponding part, such as the wing of a bird and the pectoral fin of a fish. { həˈmäl·ə·gəs }

homologous stimulus [PHYSIOLOGY] A form of energy to which a specific sensory receptor is most sensitive. { həˈmäl·ə·gəs ˈstim·yə·ləs }

homology [BIOLOGY] A fundamental similarity between structures or processes in different organisms that usually results from their having descended from a common ancestor. { həˈmäl·ə·jē }

homomorphism [BOTANY] Having perfect flowers consisting of only one type. { ˌhä·məˈmȯrˌfiz·əm }

Homoneura [INVERTEBRATE ZOOLOGY] A suborder of the Lepidoptera with mandibulate mouthparts, and fore- and hindwings that are similar in shape and venation. { ˌhä·məˈnu̇r·ə }

homopetalous [BOTANY] Having all petals identical. { ¦hä·mə¦ped·əl·əs }

homoplastic [BIOLOGY] Pertaining to transplantation between individuals of the same species. { ˌhō·mōˈplas·tik }

homoplastidy [BOTANY] The condition of having one kind of plastid. { ˌhō·mōˈplas·təd·ē }

homoplasy [BIOLOGY] Correspondence between organs or structures in different organisms acquired as a result of evolutionary convergence or of parallel evolution. { ˈhä·məˌplā·sē }

homopolymer tail [MOLECULAR BIOLOGY] A segment that contains only one sort of nucleotide at the 3′ end of a deoxyribonucleic acid or ribonucleic acid molecule. { ˌhō·mə¦päl·i·mər ˈtāl }

homopolysaccharide [BIOCHEMISTRY] A polysaccharide which is a polymer of one kind of monosaccharide. { ¦hä·mōˌpäl·ēˈsak·əˌrīd }

Homoptera [INVERTEBRATE ZOOLOGY] An order of the class Insecta including a large number of sucking insects of diverse forms. { hōˈmäp·tə·rə }

Homo sapiens [VERTEBRATE ZOOLOGY] Modern human species; a large, erect, omnivorous terrestrial biped of the primate family Hominidae. { ¦hō·mō ˈsap·ē·ənz }

Homosclerophorida [INVERTEBRATE ZOOLOGY] An order of primitive sponges of the class Demospongiae with a skeleton consisting of equirayed, tetraxonid, siliceous spicules. { ¦hä·mō ¦skler·əˈfȯr·ə·də }

homoserine [BIOCHEMISTRY] $C_4H_9O_3N$ An amino acid formed as an intermediate product in animals in the metabolic breakdown of cystathionine to cysteine. { həˈmäs·əˌrēn }

homosexual [BIOLOGY] Of, pertaining to, or being the same sex. { ¦hō·məˈsek·shə·wəl }

homospory [BOTANY] Production of only one kind of asexual spore. { həˈmäs·pə·rē }

homothallic [MYCOLOGY] Having genetically compatible hyphae and therefore forming zygospores from two branches of the same mycelium. { ˌhä·məˈthal·ik }

homotropic enzyme [BIOCHEMISTRY] A type of allosteric enzyme in which the substrate serves as the allosteric reflector. { ¦hä·mə¦träp·ik ˈen ˌzīm }

homotropous [BOTANY] Having the radicle directed toward the hilum. { hōˈmä·trə·pəs }

homotype [SYSTEMATICS] A taxonomic type for a specimen which has been compared with the holotype by another than the author of the species and determined by him to be conspecific with it. { ˈhä·məˌtīp }

homotypy [BIOLOGY] Protective device based on resemblance of shape to the background. { ¦hä·mə¦tī·pē }

homovanillic acid [BIOCHEMISTRY] $HOC_6H_3(OCH_3)CH_2CO_2H$ A major metabolite of 3-O-methyl dopa; used in enzyme determination. Abbreviated HVA. { ¦hä·mə·və¦nil·ik ˈas·əd }

homozygote [GENETICS] An individual that has identical alleles at one or more loci and therefore produces gametes which are all identical. { ¦hō·məˈzīˌgōt }

homozygous [GENETICS] Of or pertaining to a homozygote. { ˌhō·məˈzī·gəs }

honeybee [INVERTEBRATE ZOOLOGY] *Apis mellifera*. The bee kept for the commercial production of honey; a member of the dipterous family Apidae. { ˈhən·ēˌbē }

honeycomb [INVERTEBRATE ZOOLOGY] A mass of wax cells in the form of hexagonal prisms constructed by honeybees for their brood and honey. { ˈhən·ēˌkōm }

honeycomb coral [PALEONTOLOGY] The common name for members of the extinct order Tabulata; has prismatic sections arranged like the cells of a honeycomb. { ˈhən·ēˌkōm ¦kär·əl }

honeydew [INVERTEBRATE ZOOLOGY] The viscous secretion deposited on leaves by many aphids and scale insects; an attractant for ants. { ˈhən·ēˌdü }

Honey Dew melon [BOTANY] A variety of muskmelon (*Cucumis melo*) belonging to the Violales; fruit is large, oval, smooth, and creamy yellow to ivory, without surface markings. { ˈhən·ē ˌdü ˌmel·ən }

honey tube [INVERTEBRATE ZOOLOGY] Either of a pair of cornicles on the dorsal aspect of one abdominal segment in certain aphids. { ˈhən·ē ˌtüb }

hoof [VERTEBRATE ZOOLOGY] **1.** Horny covering for terminal portions of the digits of ungulate mammals. **2.** A hoofed foot, as of a horse. { hu̇f }

Hookeriales [BOTANY] An order of the mosses with irregularly branched stems and leaves that appear to be in one plane. { hu̇ˌkir·ēˈā·lēz }

hookworm [INVERTEBRATE ZOOLOGY] The common name for parasitic roundworms composing the family Ancylostomidae. { 'hu̇k,wərm }

hop [BOTANY] *Humulus lupulus*.A dioecious liana of the order Urticales distinguished by herbaceous vines produced from a perennial crown; the inflorescence, a catkin, of the female plant is used commercially for beer production. { häp }

hophornbeam [BOTANY] Any tree of the genus *Ostrya* in the birch family recognized by its very scaly bark and the fruit which closely resembles that of the hopvine. { ¦häp 'hȯrn,bēm }

Hoplocarida [INVERTEBRATE ZOOLOGY] A superorder of the class Crustacea with the single order Stomatopoda. { ,häp·lō'kar·ə·də }

Hoplonemertini [INVERTEBRATE ZOOLOGY] An order of unsegmented, ribbonlike worms in the class Enopla; all species have an armed proboscis. { ¦häp·lō,ne·mər'tī,nī }

Hoplophoridae [INVERTEBRATE ZOOLOGY] A family of prawns containing numerous bathypelagic representatives. { ,häp·lə'fȯr·ə,dē }

Hordeum [BOTANY] A genus of the order Cyperales containing all species of barley. { 'hȯr·dē·əm }

horehound [BOTANY] *See* marrubium. { 'hȯr ,hau̇nd }

horizontal evolution [EVOLUTION] *See* coincidental evolution. { ,hä·ri¦zänt·əl ,ev·ə'lü·shən }

horizontal plane [ANATOMY] A transverse plane at right angles to the longitudinal axis of the body. { ,här·ə'zänt·əl 'plān }

horizontal transmission [GENETICS] Passage of genetic information by invasive processes between individual organisms or cells of the same species. { ,här·ə¦zänt·əl tranz'mish·ən }

hormesis [BIOLOGY] Providing stimulus by nontoxic amounts of a toxic agent. { 'hȯr·mə·səs }

hormogonium [BOTANY] Portion of a filament between heterocysts in certain algae; detaches as a reproductive body. { ,hȯr·mə'gō·nē·əm }

hormone [BIOCHEMISTRY] **1.** A chemical messenger produced by endocrine glands and secreted directly into the bloodstream to exert a specific effect on a distant part of the body. **2.** An organic compound that is synthesized in minute quantities in one part of a plant and translocated to another part, where it influences a specific physiological process. { 'hȯr,mōn }

hormone-responsive element [MOLECULAR BIOLOGY] Specific regulatory nucleotide sequences in most hormonally regulated genes that are located near the promoter and mediate the action of steroid hormones. { 'hȯr,mōn ri ,spän·siv ,el·ə·mənt }

hornbeam [BOTANY] Any tree of the genus *Carpinus* in the birch family distinguished by doubly serrate leaves and by small, pointed, angular winter buds with scales in four rows. { 'hȯrn ,bēm }

horned dinosaur [PALEONTOLOGY] Common name for extinct reptiles of the suborder Ceratopsia. { ¦hȯrnd 'dīn·ə,sȯr }

horned liverwort [BOTANY] The common name for bryophytes of the class Anthocerotae. Also known as hornwort. { ¦hȯrnd 'liv·ər,wȯrt }

horned toad [VERTEBRATE ZOOLOGY] The common name for any of the lizards of the genus *Phrynosoma*; a reptile that resembles a toad but is less bulky. { ¦hȯrnd 'tōd }

horned-toad dinosaur [PALEONTOLOGY] The common name for extinct reptiles composing the suborder Ankylosauria. { ¦hȯrnd ,tōd 'dīn·ə ,sȯr }

hornet [INVERTEBRATE ZOOLOGY] The common name for a number of large wasps in the hymenopteran family Vespidae. { 'hȯr·nət }

hornwort [BOTANY] *See* horned liverwort. { 'hȯrn ,wȯrt }

horny coral [INVERTEBRATE ZOOLOGY] The common name for cnidarian members of the order Gorgonacea. { 'hȯr·nē 'kär·əl }

horse [VERTEBRATE ZOOLOGY] *Equus caballus*. A herbivorous mammal in the family Equidae; the feet are characterized by a single functional digit. { hȯrs }

horse chestnut [BOTANY] *Aesculus hippocastanum*. An ornamental buckeye tree in the order Sapindales, usually with seven leaflets per leaf and resinous buds. { 'hȯrs ¦ches·nət }

horseradish [BOTANY] *Armoracia rusticana*. A perennial crucifer belonging to the order Capparales and grown for its pungent roots, used as a condiment. { 'hȯrs¦rad·ish }

horse serum [IMMUNOLOGY] Immune serum obtained from the blood of horses. { 'hȯrs ,sir·əm }

horseshoe crab [INVERTEBRATE ZOOLOGY] The common name for arthropods composing the subclass Xiphosurida, especially the subgroup Limulida. { 'hȯr,shü ,krab }

horsetail [BOTANY] The common name for plants of the genus *Equisetum* composing the order Equisetales. Also known as scouring rush. { 'hȯrs,tāl }

horticulture [BOTANY] The art and science of growing plants. { 'hȯrd·ə,kəl·chər }

host [BIOLOGY] **1.** An organism on or in which a parasite lives. **2.** The dominant partner of a symbiotic or commensal pair. { hōst }

hot flash [PHYSIOLOGY] A sudden transitory sensation of heat, often involving the whole body, due to cessation of ovarian function; a symptom of the climacteric. { 'hät ,flash }

hot spot [MOLECULAR BIOLOGY] A site in a gene at which there is an unusually high frequency of mutation. { 'hät ,spät }

housefly [INVERTEBRATE ZOOLOGY] *Musca domestica*. A dipteran insect with lapping mouthparts commonly found near human habitations; a vector in the transmission of many disease pathogens. { 'hau̇s,flī }

housekeeping gene [GENETICS] *See* constitutive gene. { 'hau̇s,kēp·iŋ ,jēn }

Howship's lacuna [HISTOLOGY] Minute depressions in the surface of a bone undergoing resorption. { 'hau̇·shəps lə'kü·nə }

H-RNA [MOLECULAR BIOLOGY] *See* heterogeneous ribonucleic acid.

HSI [PHYSIOLOGY] *See* heat stress index.

H substance [BIOCHEMISTRY] An agent similar to

histamine and believed to play a role in local blood vessel response in tissue damage. { 'äch ˌsəb·stəns }

huarizo [VERTEBRATE ZOOLOGY] A hybrid offspring of a male llama and a female alpaca, bred for its fine fleece. { wä'rē·zō }

huckleberry [BOTANY] The common name for shrubs of the genus *Gaylussacia* in the family Ericaceae distinguished by an ovary with 10 locules and 10 ovules; the dark-blue berries are edible. { 'hək·əlˌber·ē }

Hudsonian life zone [ECOLOGY] A zone comprising the climate and biotic communities of the northern portions of North American coniferous forests and the peaks of high mountains. { ¦həd¦sō·nē·ən 'līf ˌzōn }

hull [BOTANY] The outer, usually hard, covering of a fruit or seed. { həl }

hulless buckwheat [BOTANY] *See* tartary buckwheat. { 'həl·əs 'bəkˌwēt }

human biogeography [ECOLOGY] The science concerned with the distribution of human populations on the earth. { 'hyü·mən ¦bī·ō·jē'äg·rə·fē }

human chorionic gonadotropin [BIOCHEMISTRY] A gonadotropic and luteotropic hormone secreted by the chorionic vesicle. Abbreviated HCG. Also known as chorionic gonadotropin. { 'hyü·mən kȯr·ē'än·ik gōˌnad·ə'trō·pən }

human community [ECOLOGY] That portion of a human ecosystem composed of human beings and associated plant and animal species. { 'hyü·mən kə'myün·əd·ē }

human ecology [ECOLOGY] The branch of ecology that considers the relations of individual persons and of human communities with their particular environment. { 'hyü·mən ē'käl·ə·jē }

human genetics [GENETICS] The study of human heredity and variation, including the physical basis, mechanisms of inheritance, inheritance patterns, and population analysis. { 'hyü·mən jə'ned·iks }

human immunodeficiency virus [VIROLOGY] The retrovirus that causes acquired immune deficiency syndrome. Abbreviated HIV. { 'hyü·mən ¦im·yə·nō·di'fish·ən·sē ˌvī·rəs }

human leukocyte antigen [IMMUNOLOGY] Any of a group of antigens present on the surface of nucleated body cells that are coded for by the major histocompatibility complex of humans and thus allow the immune system to distinguish self and nonself. Abbreviated HLA. { ¦yü·mən ¦lü·kəˌsīt 'ant·i·jən }

human measles immune serum [IMMUNOLOGY] Serum from the blood of a person who has recovered from measles. { 'hyü·mən ¦mē·zəlz i'myün ˌsir·əm }

human threadworm [INVERTEBRATE ZOOLOGY] *See* pinworm. { 'hyü·mən 'thredˌwərm }

human T-lymphotropic virus type 1 [VIROLOGY] A retrovirus associated with a rare form of T-cell leukemia occurring primarily in the Caribbean area, Africa, and Japan. Also known as human T-cell lymphotropic virus type 1. { 'hyü·mən 'tē ¦lim·fə¦träp·ik ˌvī·rəs ¦tīp 'wən }

human T-lymphotropic virus type 2 [VIROLOGY] A retrovirus associated with at least one form of leukemia. Also known as human T-cell lymphotropic virus type 2. { 'hyü·mən 'tē ¦lim·fə¦träp·ik ˌvī·rəs ¦tīp 'tü }

human T-lymphotropic virus type 3 [VIROLOGY] A former designation for the human immunodeficiency virus. Abbreviated HTLV-3. Also known as human T-cell lymphotropic virus type 3. { 'hyü·mən 'tē ¦lim·fə¦träp·ik ˌvī·rəs ¦tīp 'thrē }

humeral [ANATOMY] Of or pertaining to the humerus or the shoulder region. { 'hyüm·ə·rəl }

humeroglandular effector system [PHYSIOLOGY] Glandular effector system in which the activating agent is a blood-borne chemical. { ¦hyüm·ə·rō¦glan·jə·lər i'fek·tər ˌsis·təm }

humeromuscular effector system [PHYSIOLOGY] Muscular effector system in which the activating agent is a blood-borne chemical. { ¦hyüm·ə·rō¦məs·kyə·lər i'fek·tər ˌsis·təm }

humerus [ANATOMY] The proximal bone of the forelimb in vertebrates; the bone of the upper arm in humans, articulating with the glenoid fossa and the radius and ulna. { 'hyüm·ə·rəs }

humicolous [ECOLOGY] Of or pertaining to plant species inhabiting medium-dry ground. { hyü'mik·ə·ləs }

humid transition life zone [ECOLOGY] A zone comprising the climate and biotic communities of the northwest moist coniferous forest of the north-central United States. { 'hyü·məd tran¦zish·ən 'līf ˌzōn }

humifuse [BIOLOGY] Spread over the ground surface. { 'hyü·məˌfyüs }

Humiriaceae [BOTANY] A family of dicotyledonous plants in the order Linales characterized by exappendiculate petals, usually five petals, flowers with an intrastaminal disk, and leaves lacking stipules. { hyüˌmir·ē'ās·ēˌē }

humivore [ECOLOGY] An organism that feeds on humus. { 'hyü·məˌvȯr }

hummingbird [VERTEBRATE ZOOLOGY] The common name for members of the family Trochilidae; fast-flying, short-legged, weak-footed insectivorous birds with a tubular, pointed bill and a fringed tongue. { 'həm·iŋˌbərd }

hummock [ECOLOGY] A rounded or conical knoll frequently formed of earth and covered with vegetation. { 'həm·ək }

humor [PHYSIOLOGY] A fluid or semifluid part of the body. { 'hyü·mər }

humoral immunity [IMMUNOLOGY] Immunity in which immune responses are mediated by immunoglobulins. Also known as antibody-mediated immunity; immunoglobulin-mediated immunity. { ¦hyüm·ə·rəl i'myü·nəd·ē }

husk [BOTANY] The outer coat of certain seeds, particularly if it is a dry, membranous structure. { həsk }

Huxley's anastomosis [HISTOLOGY] Polyhedral cells forming the middle layer of a hair root sheath. { 'həks·lēz əˌnas·tə'mō·səs }

HVA [BIOCHEMISTRY] *See* homovanillic acid.

Hyaenidae [VERTEBRATE ZOOLOGY] A family of catlike carnivores in the superfamily Feloidea including the hyenas and aardwolf. { hī'e·nəˌdē }

Hyaenodontidae [PALEONTOLOGY] A family of extinct carnivorous mammals in the order Deltatheridia. { hī,ē·nə'dänt·ə,dē }

Hyalellidae [INVERTEBRATE ZOOLOGY] A family of amphipod crustaceans in the suborder Gammaridea. { ,hī·ə'lel·ə,dē }

hyaline [BIOCHEMISTRY] A clear, homogeneous, structureless material found in the matrix of cartilage, vitreous body, mucin, and glycogen. { 'hī·ə·lən }

hyaline cartilage [HISTOLOGY] A translucent connective tissue comprising about two-thirds clear, homogeneous matrix with few or no collagen fibrils. { 'hī·ə·lən 'kärt·lij }

hyaline membrane [HISTOLOGY] **1.** A basement membrane. **2.** A membrane of a hair follicle between the inner fibrous layer and the outer root sheath. { 'hī·ə·lən 'mem,brān }

hyaline test [INVERTEBRATE ZOOLOGY] A translucent wall or shell of certain foraminiferans composed of layers of calcite interspersed with separating membranes. { 'hī·ə·lən ,test }

Hyalodictyae [MYCOLOGY] A subdivision of the spore group Dictyosporae characterized by hyaline spores. { ,hi·ə·lō'dik·tē,ē }

Hyalodidymae [MYCOLOGY] A subdivision of the spore group Didymosporae characterized by hyaline spores. { ,hi·ə·lō'did·ə,mē }

Hyalohelicosporae [MYCOLOGY] A subdivision of the spore group Helicosporae characterized by hyaline spores. { ¦hī·ə·lō,hel·ə'käs·pə,rē }

hyaloid membrane [ANATOMY] The limiting membrane surrounding the vitreous body of the eyeball, and forming the suspensory ligament. { 'hī·ə,lòid'mem,brān }

Hyalophragmiae [MYCOLOGY] A subdivision of the spore group Phragmosporae characterized by hyaline spores. { ,hī·ə·lō'frag·mē,ē }

hyaloplasm [CELL BIOLOGY] The optically clear, viscous to gelatinous ground substance of cytoplasm in which formed bodies are suspended. { hī'al·ə,plaz·əm }

Hyaloscolecosporae [MYCOLOGY] A subdivision of the spore group Scalecosporae characterized by hyaline spores. { ¦hī·ə·lō,skäl·ə'käs·pə,rē }

Hyalospongia [PALEONTOLOGY] A class of extinct glass sponges, equivalent to the living Hexactinellida, having siliceous spicules made of opaline silica. { ,hi·ə·lō'spən·jē·ə }

Hyalosporae [MYCOLOGY] A subdivision of the spore group Amerosporae characterized by hyaline spores. { ,hī·ə'läs·pə·rē }

Hyalostaurosporae [MYCOLOGY] A subdivision of the spore group Staurosporae characterized by hyaline spores. { ¦hī·ə·lō·stó'räs·pə,rē }

hyaluronate [BIOCHEMISTRY] A salt or ester of hyaluronic acid. { ,hī·ə'lu̇r·ə,nāt }

hyaluronate lyase [BIOCHEMISTRY] *See* hyaluronidase. { ,hī·ə'lu̇r·ə,nāt 'lī,ās }

hyaluronic acid [BIOCHEMISTRY] A polysaccharide found as an integral part of the gellike substance of animal connective tissue. { ¦hī·ə·lu̇ ¦rän·ik 'as·əd }

hyaluronidase [BIOCHEMISTRY] Any one of a family of enzymes which catalyze the breakdown of hyaluronic acid. Also known as hyaluronate lyase; spreading factor. { ,hī·ə·lu̇'rän·ə,dās }

Hybodontoidea [PALEONTOLOGY] An ancient suborder of extinct fossil sharks in the order Selachii. { ,hī·bə,dän'tòid·ē·ə }

hybrid [GENETICS] The offspring of genetically dissimilar parents. { 'hī·brəd }

hybrid-arrested translation [MOLECULAR BIOLOGY] A method for identifying the proteins coded for by a cloned deoxyribonucleic acid (DNA) sequence by depending on the ability of the cloned DNA to form a base pair with its messenger ribonucleic acid and thereby inhibit its translation. { ¦hī·brəd ə,res·təd tranz'lā·shən }

hybrid dysgenesis [GENETICS] A syndrome of abnormal traits that appears in the hybrids between certain strains of the fruit fly *Drosophila melanogaster*, and includes such traits as partial sterility and greatly elevated genetic mutations and chromosome rearrangements. { 'hī,brid dis'jen·ə·səs }

hybrid enzyme [BIOCHEMISTRY] A form of polymeric enzyme occurring in heterozygous individuals that shows a hybrid molecular form made up of subunits differing in one or more amino acids. { 'hī·brəd 'en,zīm }

hybrid gene [MOLECULAR BIOLOGY] Any gene constructed by recombinant deoxyribonucleic acid technology that contains segments derived from different parents. { 'hī·brəd ,jēn }

hybridization [CELL BIOLOGY] The production of viable hybrid somatic cells following experimentally induced cell fusion. [GENETICS] The act or process of producing hybrids. [MOLECULAR BIOLOGY] **1.** Production of a hybrid by pairing complementary ribonucleic acid and deoxyribonucleic acid (DNA) strands. **2.** Production of a hybrid by pairing complementary DNA single strands. { ,hī·brəd·ə'zā·shən }

hybridization probe [MOLECULAR BIOLOGY] A small molecule of deoxyribonucleic acid or ribonucleic acid that is radioactively labeled and used to identify complementary nucleic acid sequences by hybridization. { ,hī·brəd·ə'zā·shən 'prōb }

hybrid merogony [EMBRYOLOGY] The fertilization of cytoplasmic fragments of the egg of one species by the sperm of a related species. { 'hī·brəd mə'räj·ə·nē }

hybrid molecule [BIOCHEMISTRY] A single molecule, usually protein, peculiar to heterozygotes and containing two structurally different polypeptide chains determined by two different alleles. { 'hī·brəd 'mäl·ə,kyül }

hybridoma [IMMUNOLOGY] A hybrid myeloma formed by fusing myeloma cells with lymphocytes that produce a specific antibody; the individual cells can be cloned, and each clone produces large amounts of identical (monoclonal) antibody. { ,hī·brə'dō·mə }

hybrid plasmid [MOLECULAR BIOLOGY] In recombinant deoxyribonucleic acid (DNA) technology, any plasmid chimera containing inserted DNA sequences. { 'plaz·mid }

hybrid sterility [GENETICS] Inability to form functional gametes in a hybrid due to disturbances in sex-cell development or in meiosis, caused by incompatible genetic constitution. { 'hī·brəd stə'ril·əd·ē }

hybrid swarm [GENETICS] A collection of hybrids produced when there is a breakdown of isolating barriers between two species whose areas of distribution overlap. { 'hī·brəd 'swórm }

hybrid vigor [GENETICS] *See* heterosis. { 'hī·brəd 'vig·ər }

hybrid zone [ECOLOGY] A geographic zone in which two populations hybridize after the breakdown of the geographic barrier that separated them. { ¦hī·brəd ¦zōn }

Hydatellales [BOTANY] An order of monocotyledonous flowering plants, division Magnoliophyta, of the class Liliopsida, characterized by small, submersed or partly submersed aquatic annuals with greatly simplified internal anatomy; consists of a single family with five species native to Australia, New Zealand, and Tasmania. { ˌhī·dad·ə'lā·lēz }

hydathode [BOTANY] An opening of the epidermis of higher plants specialized for exudation of water. { 'hīd·əˌthōd }

hydra [INVERTEBRATE ZOOLOGY] Any species of *Hydra* or related genera, consisting of a simple, tubular body with a mouth at one end surrounded by tentacles, and a foot at the other end for attachment. { 'hī·drə }

Hydra [INVERTEBRATE ZOOLOGY] A common genus of cnidarians in the suborder Anthomedusae. { 'hī·drə }

Hydrachnellae [INVERTEBRATE ZOOLOGY] A family of generally fresh-water predacious mites in the suborder Trombidiformes, including some parasitic forms. { ˌhīˌdrak'ne·lē }

Hydraenidae [INVERTEBRATE ZOOLOGY] The equivalent name for Limnebiidae. { hī'drē·nə ˌdē }

hydranth [INVERTEBRATE ZOOLOGY] Nutritive individual in a hydroid colony. { 'hīˌdranth }

hydrase [BIOCHEMISTRY] An enzyme that catalyzes removal or addition of water to a substrate without hydrolyzing it. { 'hīˌdrās }

hydric [ECOLOGY] Characterized by or thriving in abundance of moisture. { 'hī·drik }

Hydrobatidae [VERTEBRATE ZOOLOGY] The storm petrels, a family of oceanic birds in the order Procellariiformes. { 'hī·drō'bad·əˌdē }

hydrocaulus [INVERTEBRATE ZOOLOGY] The branched, upright stem of a hydroid colony. { 'hī·drō'kȯl·əs }

Hydrocharitaceae [BOTANY] The single family of the order Hydrocharitales, characterized by an inferior, compound ovary with laminar placentation. { ˌhī·drōˌkar·ə'tās·ēˌē }

Hydrocharitales [BOTANY] A monofamilial order of aquatic monocotyledonous plants in the subclass Alismatidae. { ˌhī·drōˌkar·ə'tā·lēz }

hydrochory [BIOLOGY] Dispersal of disseminules by water. { 'hī·drəˌkȯr·ē }

hydrocladium [INVERTEBRATE ZOOLOGY] The branchlet of a hydrocaulus. { ˌhī·drə'klād·ē·əm }

hydrocoele [INVERTEBRATE ZOOLOGY] **1.** Water vascular system in Echinodermata. **2.** Embryonic precursor of the system. { 'hī·drəˌsel }

Hydrocorallina [INVERTEBRATE ZOOLOGY] An order in some systems of classification set up to include the cnidarian groups Milleporina and Stylasterina. { ˌhī·drōˌkȯr·ə'lī·nə }

Hydrocorisae [INVERTEBRATE ZOOLOGY] A subdivision of the Hemiptera containing water bugs with concealed antennae and without a bulbus ejaculatorius in the male. { ˌhī·drō'kȯr·əˌsē }

hydrocortisone [BIOCHEMISTRY] $C_{21}H_{30}O_5$ The generic name for 17-hydroxycorticosterone; an adrenocortical steroid occurring naturally and prepared synthetically; its effects are similar to cortisone, but it is more active. Also known as cortisol. { ˌhī·drə'kȯrd·əˌzōn }

Hydrodamalinae [VERTEBRATE ZOOLOGY] A monogeneric subfamily of sirenian mammals in the family Dugongidae. { ˌhī·drō·də'mal·əˌnē }

hydroecium [INVERTEBRATE ZOOLOGY] The closed, funnel-shaped tube at the upper end of cnidarians belonging to the Siphonophora. { hī 'drē·shəm }

hydrofuge [ZOOLOGY] Of a structure, shedding water, as the hair on certain animals. { 'hī·drə ˌfyüj }

hydrogenase [BIOCHEMISTRY] Enzyme that catalyzes the oxidation of hydrogen. { hī'dräj·ə ˌnās }

hydrogen bacteria [MICROBIOLOGY] Bacteria capable of obtaining energy from the oxidation of molecular hydrogen. { 'hī·drə·jən bak'tir·ē·ə }

hydroid [INVERTEBRATE ZOOLOGY] **1.** The polyp form of a hydrozoan cnidarian. Also known as hydroid polyp; hydropolyp. **2.** Any member of the Hydroida. { 'hīˌdrȯid }

Hydroida [INVERTEBRATE ZOOLOGY] An order of cnidarians in the class Hydrozoa including usually colonial forms with a well-developed polyp stage. { hī'drȯi·də }

hydroid polyp [INVERTEBRATE ZOOLOGY] *See* hydroid. { 'hīˌdrȯid 'päl·əp }

hydrolase [BIOCHEMISTRY] Any of a class of enzymes which catalyze the hydrolysis of proteins, nucleic acids, starch, fats, phosphate esters, and other macromolecular substances. { 'hī·drə ˌlās }

hydrolytic enzyme [BIOCHEMISTRY] A catalyst that acts like a hydrolase. { ¦hī·drə¦lid·ik 'en ˌzīm }

Hydrometridae [INVERTEBRATE ZOOLOGY] The marsh treaders, a family of hemipteran insects in the subdivision Amphibicorisae. { ˌhī·drə 'me·trəˌdē }

Hydrophiidae [VERTEBRATE ZOOLOGY] A family of proglyphodont snakes in the suborder Serpentes found in Indian-Pacific oceans. { ˌhī·drō'fī·ə ˌdē }

Hydrophilidae [INVERTEBRATE ZOOLOGY] The water scavenger beetles, a large family of coleopteran insects in the superfamily Hydrophiloidea. { ˌhī·drə'fil·əˌdē }

Hydrophiloidea [INVERTEBRATE ZOOLOGY] A superfamily of coleopteran insects in the suborder Polyphaga. { ˌhī·drə·fə'lȯid·ē·ə }

hydrophilous [ECOLOGY] Inhabiting moist places. { hī'dräf·ə·ləs }

Hydrophyllaceae [BOTANY] A family of dicotyledonous plants in the order Polemoniales distinguished by two carpels, parietal placentation, and generally imbricate corolla lobes in the bud. { ,hī·drō·fə'lās·ē,ē }

hydrophyllium [INVERTEBRATE ZOOLOGY] A transparent body partly covering the spore sacs of siphonophoran cnidarians. { ,hī·drə'fil·ē·əm }

hydrophyte [BOTANY] **1.** A plant that grows in a moist habitat. **2.** A plant requiring large amounts of water for growth. Also known as hygrophyte. { 'hī·drə,fīt }

hydroplanula [INVERTEBRATE ZOOLOGY] A cnidarian larval stage between the planula and actinula stages. { ¦hī·drə'plan·yə·lə }

hydropolyp [INVERTEBRATE ZOOLOGY] *See* hydroid. { ¦hī·drō'päl·əp }

hydroponics [BOTANY] Growing of plants in a nutrient solution with the mechanical support of an inert medium such as sand. { ,hī·drə'pän·iks }

hydropore [INVERTEBRATE ZOOLOGY] In certain asteroids and echinoids, an opening on the aboral surface of a canal which extends from the ring canal in one of the interradii. { 'hī·drə ,pȯr }

hydrorhiza [INVERTEBRATE ZOOLOGY] Rootlike structure of a hydroid colony. { ,hī·drō'rī·zə }

Hydroscaphidae [INVERTEBRATE ZOOLOGY] The skiff beetles, a small family of coleopteran insects in the suborder Myxophaga. { ,hī·drə 'skaf·ə,dē }

hydrosere [ECOLOGY] Community in which the pioneer plants invade open water, eventually forming some kind of soil such as peat or muck. { 'hī·drə,sir }

hydroskeleton [INVERTEBRATE ZOOLOGY] Water contained within the coelenteron and serving a skeletal function in most cnidarian polyps. { ¦hī·drō'skel·ət·ən }

hydrospire [INVERTEBRATE ZOOLOGY] Either of a pair of flattened tubes composing part of the respiratory system in blastoids. { 'hī·drə,spīr }

hydrospore [INVERTEBRATE ZOOLOGY] Opening into the hydrocoele on the right side in echinoderm larvae. { 'hī·drə,spȯr }

hydrotheca [INVERTEBRATE ZOOLOGY] Cup-shaped portion of the perisarc in some cnidarians that serves to hold and protect a withdrawn hydranth. { ,hī·drō'thē·kə }

hydrotropism [BIOLOGY] Orientation involving growth or movement of a sessile organism or part, especially plant roots, in response to the presence of water. { hī'drä·trə,piz·əm }

β-hydroxybutyric dehydrogenase [BIOCHEMISTRY] The enzyme that catalyzes the conversion of L-β-hydroxybutyric acid to acetoacetic acid by dehydrogenation. { ¦bād·ə hī¦dräk·sē·byü'tir·ik dē·hī'dräj·ə·nās }

3-hydroxyflavone [BIOCHEMISTRY] *See* flavanol. { ¦thrē hī¦dräk·sē'fla,vōn }

hydroxylase [BIOCHEMISTRY] Any of several enzymes that catalyze certain hydroxylation reactions involving atomic oxygen. { hī'dräk·sə,lās }

hydroxyproline [BIOCHEMISTRY] $C_5H_9O_3N$ An amino acid that is essentially limited to structural proteins of the collagen type. { hī¦dräk·sə 'prō,lēn }

5-hydroxytryptamine [BIOCHEMISTRY] *See* serotonin. { ¦fīv hī¦dräk·sē'trip·tə,mēn }

5-hydroxytryptophan [BIOCHEMISTRY] $C_{11}H_{12}N_2O_3$ Minute rods or needlelike crystals; the biological precursor of serotonin. { ¦fīv hī¦dräk·sē'trip·tə,fan }

Hydrozoa [INVERTEBRATE ZOOLOGY] A class of the phylum Cnidaria which includes the freshwater hydras, the marine hydroids, many small jellyfish, a few corals, and the Portuguese man-of-war. { ,hī·drə'zō·ə }

hyena [VERTEBRATE ZOOLOGY] An African carnivore represented by three species of the family Hyaenidae that resemble dogs but are more closely related to cats. { hī'ē·nə }

Hyeniales [PALEOBOTANY] An order of Devonian plants characterized by small, dichotomously forked leaves borne in whorls. { ,hī·ə'nā·lēz }

Hyeniopsida [PALEOBOTANY] An extinct class of the division Equisetophyta. { ¦hī·ə·nē'äp·sə·də }

Hygrobiidae [INVERTEBRATE ZOOLOGY] The squeaker beetles, a small family of coleopteran insects in the suborder Adephaga. { hī·grə'bī·ə ,dē }

hygromycin [MICROBIOLOGY] $C_{25}H_{33}O_{12}N$ A weakly acidic, soluble antibiotic with a fairly broad spectrum, produced by a strain of *Streptomyces hygroscopicus*. { ,hī·grə'mīs·ən }

hygrophyte [BOTANY] *See* hydrophyte. { 'hī·grə ,fīt }

hygroscopic [BOTANY] Being sensitive to moisture, such as certain tissues. { ¦hī·grə¦skäp·ik }

hylaea [ECOLOGY] *See* tropical rainforest. { hī·lē·ə }

Hylidae [VERTEBRATE ZOOLOGY] The tree frogs, a large amphibian family in the suborder Procoela; many are adapted to arboreal life, having expanded digital disks. { 'hī·lə,dē }

Hylobatidae [VERTEBRATE ZOOLOGY] A family of anthropoid primates in the superfamily Hominoidea including the gibbon and the siamang of southeastern Asia. { ,hī·lō'bad·ə,dē }

hylophagous [ZOOLOGY] Feeding on wood, as termites. { hi¦läf·ə·gəs }

hylotomous [ZOOLOGY] Cutting wood, as wood-boring insects. { hī¦läd·ə·məs }

hymen [ANATOMY] A mucous membrane partly closing off the vaginal orifice. Also known as maidenhead. { 'hī·mən }

hymenium [MYCOLOGY] The outer, sporebearing layer of certain fungi or their fruiting bodies. { hī 'mē·nē·əm }

Hymenolepis [INVERTEBRATE ZOOLOGY] A genus of tapeworms parasitic in humans, birds, and mammals. { ,hī·mə'näl·ə·pəs }

Hymenomycetes [MYCOLOGY] A group of the Homobasidiomycetidae including forms such as mushrooms and pore fungi in which basidia are formed in an exposed layer (hymenium) and basidiospores are borne asymmetrically on slender stalks. { ¦hī·mə·nō,mī¦sēd·ēz }

hymenophore [MYCOLOGY] Portion of a sporophore that bears the hymenium. { hī'men·ə ˌfȯr }

hymenopodium [MYCOLOGY] **1.** Tissue beneath the hymenium in certain fungi. **2.** A genus of the Moniliales. { ˌhī·mə·nō'pōd·ē·əm }

Hymenoptera [INVERTEBRATE ZOOLOGY] A large order of insects including ants, wasps, bees, sawflies, and related forms; head, thorax and abdomen are clearly differentiated; wings, when present, and legs are attached to the thorax. { ˌhī·mə'näp·trə }

Hymenostomatida [INVERTEBRATE ZOOLOGY] An order of ciliated protozoans in the subclass Holotrichia having fairly uniform ciliation and a definite buccal ciliature. { ¦hī·mə·nō·stō'mad·ə·də }

Hynobiidae [VERTEBRATE ZOOLOGY] A family of salamanders in the suborder Cryptobranchoidea. { ˌhī·nō'bī·əˌdē }

hyobranchium [VERTEBRATE ZOOLOGY] A Y-shaped bone supporting the tongue and tongue muscles in a snake. { ¦hī·ō'braŋ·kē·əm }

Hyocephalidae [INVERTEBRATE ZOOLOGY] A monospecific family of hemipteran insects in the superfamily Pentatomorpha. { ˌhī·ō·sə'fal·ə ˌdē }

hyoglossus [ANATOMY] An extrinsic muscle of the tongue arising from the hyoid bone. { ˌhī·ō 'glä·səs }

hyoid [ANATOMY] **1.** A bone or complex of bones at the base of the tongue supporting the tongue and its muscles. **2.** Of or pertaining to structures derived from the hyoid arch. { 'hīˌȯid }

hyoid arch [EMBRYOLOGY] Either of the second pair of pharyngeal segments or gill arches in vertebrate embryos. { 'hīˌȯid ˌärch }

hyoid tooth [VERTEBRATE ZOOLOGY] One of a number of teeth on the tongue of fishes. { 'hīˌȯid ˌtüth }

hyomandibular [VERTEBRATE ZOOLOGY] The upper portion of the hyoid arch in fishes. { ¦hī·ō ˌman'dib·yə·lər }

hyomandibular cleft [EMBRYOLOGY] The space between the hyoid arch and the mandibular arch in the vertebrate embryo. { ¦hī·ōˌman'dib·yə·lər 'kleft }

hyomandibular pouch [EMBRYOLOGY] A portion of the endodermal lining of the pharyngeal cavity which separates the paired hyoid and mandibular arches in vertebrate embryos. { ¦hī·ōˌman 'dib·yə·lər 'pau̇ch }

Hyopssodontidae [PALEONTOLOGY] A family of extinct mammalian herbivores in the order Condylarthra. { ¦hī·äp·sə'dänt·əˌdē }

hyostylic [VERTEBRATE ZOOLOGY] Having the jaws and cranium connected by the hyomandibular, as certain fishes. { ¦hī·ō'stī·lik }

hypalgesia [PHYSIOLOGY] Diminished sensitivity to pain. { ¦hip·al'jē·zhə }

hypandrium [INVERTEBRATE ZOOLOGY] A plate covering the genitalia on the ninth abdominal segment of certain male insects. { hi'pan·drē·əm }

hypanthium [BOTANY] Expanded receptacle margin to which the sepals, petals, and stamens are attached in some flowers. { hi'pan·thē·əm }

hypantrum [VERTEBRATE ZOOLOGY] In reptiles, a notch on the anterior portion of the neural arch that articulates with the hyposphene. { hi'pan·trəm }

hypaxial musculature [ANATOMY] The ventral portion of the axial musculature of vertebrates including subvertebral flank and ventral abdominal muscle groups. { hi'pak·sē·əl 'məs·kyə·lə ˌchər }

hyperacid [PHYSIOLOGY] Containing more than the normal concentration of acid in the gastric juice. { ¦hī·pər'as·əd }

hyperactivity [PHYSIOLOGY] Excessive or pathologic activity. { ¦hī·pər·ak'tiv·əd·ē }

hyperalgesia [PHYSIOLOGY] Increased or heightened sensitivity to pain stimulation. { ¦hī·pər·al 'jē·zhə }

hyperchromatic [BIOLOGY] Staining more intensely than normal. { ¦hī·pər·krō'mad·ik }

hypercoracoid [VERTEBRATE ZOOLOGY] The upper of two bones at the base of the pectoral fin in teleosts. { ¦hī·pər'kȯr·əˌkȯid }

hyperergia [IMMUNOLOGY] An altered state of reactivity to antigenic materials, in which the response is more marked than usual; one form of allery or pathergy. { ˌhī·pər'ər·jē·ə }

hyperesthesia [PHYSIOLOGY] Increased sensitivity or sensation. { ˌhī·pər·əs'thē·zhə }

hyperglycemic factor [BIOCHEMISTRY] *See* glucagon. { ˌhī·pərˌglī'sē·mik 'fak·tər }

hyperglycemic glycogenolytic factor [BIOCHEMISTRY] *See* glucagon. { ˌhī·pərˌglī'sē·mik ¦glī·kō ˌjen·ə¦lid·ik 'fak·tər }

Hyperiidea [INVERTEBRATE ZOOLOGY] A suborder of amphipod crustaceans distinguished by large eyes which cover nearly the entire head. { ˌhī·pə·rī'id·ē·ə }

hyperimmune antibody [IMMUNOLOGY] An antibody having the characteristics of a blocking antibody. { ¦hī·pər·ə'myün 'ant·iˌbäd·ē }

hyperimmune serum [IMMUNOLOGY] An antiserum that provides a very high degree of immunity due to a high antibody titer. { ¦hī·pər·ə'myün 'sir·əm }

Hypermastigida [INVERTEBRATE ZOOLOGY] An order of the multiflagellate protozoans in the class Zoomastigophorea; all inhabit the alimentary canal of termites, cockroaches, and woodroaches. { ˌhī·pərˌma'stij·ə·də }

hypermetamorphism [INVERTEBRATE ZOOLOGY] Type of embryological development in certain insects in which one or more stages have been interpolated between the full-grown larva and the adult. { ¦hī·pərˌmed·ə'mȯrˌfiz·əm }

Hyperoartii [VERTEBRATE ZOOLOGY] A superorder in the subclass Monorhina distinguished by the single median dorsal nasal opening leading into a blind hypophyseal sac. { ˌhī·pə·rō'är·shēˌī }

Hyperotreti [VERTEBRATE ZOOLOGY] A suborder in the subclass Monorhina distinguished by the nasal opening which is located at the tip of the snout and communicates with the pharynx by a long duct. { ˌhī·pə·rō'trēdˌī }

hyperparasite [ECOLOGY] An organism that is parasitic on other parasites. { ¦hī·pər'par·ə ˌsīt }

hyperploid [GENETICS] Having one or more chromosomes or parts of chromosomes in excess of the haploid number, or of a whole multiple of the haploid number. { 'hī·pərˌplȯid }

hyperploidy [GENETICS] The condition or state of being hyperploid. { 'hī·pərˌplȯid·ē }

hypersensitivity [IMMUNOLOGY] The state of being abnormally sensitive, especially to allergens; responsible for allergic reactions. { ¦hī·pərˌsen·sə'tiv·əd·ē }

hypersensitization [IMMUNOLOGY] The process of producing hypersensitivity. { ¦hī·pərˌsen·sə·tə'zā·shən }

hypertelorism [ANATOMY] An unusually large distance between paired body parts or organs. { ¦hī·pər'tel·əˌriz·əm }

hypertely [EVOLUTION] An extreme overdevelopment of an organ or body part during evolution that is disadvantageous to the organism. [ZOOLOGY] An extreme degree of imitative coloration, beyond the aspect of utility. { hī'pərd·əl·ē }

hypertensin [BIOCHEMISTRY] *See* angiotensin. { ˌhī·pər'ten·sən }

hyperthermia [PHYSIOLOGY] A condition of elevated body temperature. { ˌhī·pər'thər·mē·ə }

hypertonic [PHYSIOLOGY] **1.** Excessive or above normal in tone or tension, as a muscle. **2.** Having an osmotic pressure greater than that of physiologic salt solution or of any other solution taken as a standard. { ˌhī·pər'tän·ik }

Hypertragulidae [PALEONTOLOGY] A family of extinct chevrotainlike pecoran ruminants in the superfamily Traguloidea. { ˌhī·pər·trə'gyül·əˌdē }

hypervariable minisatellite [GENETICS] One of a number of tandem repetitive regions of deoxyribonucleic acid dispersed throughout the human genome that display polymorphism associated with the allelic variation in the number of repetitive copies of each minisatellite. { ˌhī·pər 'ver·ē·ə·bəl ¦min·ē'sad·əlˌīt }

hypha [MYCOLOGY] One of the filaments composing the mycelium of a fungus. { 'hī·fə }

hyphidium [MYCOLOGY] A sterile hymenial structure of hyphal origin. { hī'fid·ē·əm }

Hyphochytriales [MYCOLOGY] An order of aquatic fungi in the class Phycomycetes having a saclike to limited hyphal thallus and zoospores with two flagella. { ˌhī·fō·kiˌtrī'ā·lēz }

Hyphochytridiomycetes [MYCOLOGY] A class of the true fungi; usually grouped with other classes under the general term Phycomycetes. { ¦hī·fō·ki¦trid·ē·ōˌmī'sēdˌēz }

hyphoid [MYCOLOGY] Hyphalike. { 'hīˌfȯid }

Hyphomicrobiaceae [MICROBIOLOGY] Formerly a family of bacteria in the order Hyphomicrobiales; cells occurring in free-floating groups with individual cells attached to each other by a slender filament. { ¦hī·fō·mīˌkrō·bē'ās·ēˌē }

Hyphomicrobiales [MICROBIOLOGY] Formerly an order of bacteria in the class Schizomycetes containing forms that multiply by budding. { ˌhī·fō·mīˌkrō·bē'ā·lēz }

Hyphomicrobium [MICROBIOLOGY] A genus of prosthecate bacteria; they reproduce by budding at hyphal tips; cells are rod-shaped with pointed ends, ovoid, or bean-shaped; hyphae are not septate. { ˌhī·fō·mī'krō·bē·əm }

Hyphomonas [MICROBIOLOGY] A genus of prosthecate bacteria; cells are oval or pear-shaped, and reproduction is by budding of the hyphae or by direct budding of cells. { ˌhi·fō'mō·nəs }

hyphopodium [MYCOLOGY] Hypha with a haustorium in certain ectoparasitic fungi. { ˌhī·fō 'pō·dē·əm }

Hypnineae [BOTANY] A suborder of the Hypnobryales characterized by complanate, glossy plants with ecostate or costate leaves and paraphyllia rarely present. { hip'nī·nēˌē }

Hypnobryales [BOTANY] An order of mosses composed of procumbent and pleurocumbent plants with usually symmetrical leaves arranged in more than two rows. { ¦hip·nōˌbrī'ā·lēz }

hypnotoxin [BIOCHEMISTRY] A supposed hormone produced by brain tissue and inducing sleep. { ¦hip·nə'täk·sən }

hypobasal [BOTANY] Located posterior to the basal wall. { ¦hī·pō¦bā·səl }

hypoblast [EMBRYOLOGY] *See* endoderm. { 'hī·pō ˌblast }

hypobranchial musculature [ANATOMY] The ventral musculature in vertebrates extending from the pectoral girdle forward to the hyoid arch, chin, and tongue. { ¦hī·po'braŋ·kē·əl 'məs·kyə·ləˌchər }

hypochil [BOTANY] Lower portion of the lip in orchids. Also known as hypochillium. { 'hī·pə ˌkil }

Hypochilidae [INVERTEBRATE ZOOLOGY] A family of true spiders in the order Araneida. { ˌhī·pə 'kil·əˌdē }

hypochillium [BOTANY] *See* hypochil. { ˌhī·pə 'kil·ē·əm }

Hypochilomorphae [INVERTEBRATE ZOOLOGY] A monofamilial suborder of spiders in the order Araneida. { ˌhī·pəˌkil·ə'mȯr·fē }

hypochnoid [MYCOLOGY] Having generally compacted hyphae. { hī'päkˌnȯid }

hypochondriac region [ANATOMY] The upper, lateral abdominal region just below the ribs on each side of the body. { ¦hī·pə¦kän·drēˌak ˌrē·jən }

hypocleidium [ANATOMY] The median, ventral bone between clavicles. [VERTEBRATE ZOOLOGY] Median process on the wishbone of birds. { ˌhī·pə'klīd·ē·əm }

hypocone [ANATOMY] The posterior inner cusp of an upper molar. [INVERTEBRATE ZOOLOGY] Region of a dinoflagellate posterior to the girdle. { 'hī·pəˌkōn }

Hypocopridae [INVERTEBRATE ZOOLOGY] A small family of coleopteran insects in the superfamily Cucujoidea. { ˌhī·pə'käp·rəˌdē }

hypocoracoid [VERTEBRATE ZOOLOGY] The lower of two bones at the base of the pectoral fin in teleosts. { ¦hī·pō'kȯr·əˌkȯid }

hypocotyl [BOTANY] The portion of the embryonic plant axis below the cotyledon. { 'hī·pə ˌkäd·əl }

hypocrateriform [BIOLOGY] Saucer-shaped. { ˌhī·pə·krə'ter·əˌfȯrm }

Hypocreales [MYCOLOGY] An order of fungi belonging to the Ascomycetes and including several entomophilic fungi. { ˌhī·pōˌkrē'ā·lēz }

Hypodermatidae [INVERTEBRATE ZOOLOGY] The warble flies, a family of myodarian cyclorrhaphous dipteran insects in the subsection Calypteratae. { ¦hī·pō·dər'mad·əˌdē }

hypodermis [BOTANY] The outermost cell layer of the cortex of plants. Also known as exodermis. [INVERTEBRATE ZOOLOGY] The layer of cells that underlies and secretes the cuticle in arthropods and some other invertebrates. { ˌhī·pə'dər·mis }

hypoergia [IMMUNOLOGY] A state of less than normal reactivity to antigenic materials, in which the response is less marked than usual; one form of allergy or pathergy. { ¦hī·pō¦er·jē·ə }

hypogeal [BIOLOGY] *See* hypogeous. { ¦hī·pə¦jē·əl }

hypogeous [BIOLOGY] Living or maturing below the surface of the ground. Also known as hypogeal. { ˌhī·pə'jē·əs }

hypoglossal nerve [ANATOMY] The twelfth cranial nerve; a paired motor nerve in tetrapod vertebrates innervating tongue muscles; corresponds to the hypobranchial nerve in fishes. { ¦hī·pə¦gläs·əl 'nərv }

hypoglossal nucleus [ANATOMY] A long nerve nucleus throughout most of the length of the medulla oblongata; cells give rise to the hypoglossal nerve fibers. { ¦hī·pə¦gläs·əl 'nü·klē·əs }

hypogynium [BOTANY] Structure that supports the ovary in plants such as sedges. { ˌhī·pəˌjīn·ē·əm }

hypogynous [BOTANY] Having all flower parts attached to the receptacle below the pistil and free from it. { ˌhī'päj·ə·nəs }

hypokalemia [PHYSIOLOGY] A reduction in the normal amount of potassium in the blood. { ˌhī·pō·kə'lē·mē·ə }

hypomere [EMBRYOLOGY] The lateral or lower mesodermal plate zone in vertebrate embryos. [INVERTEBRATE ZOOLOGY] The basal portion of certain sponges that contain no flagellated chambers. { 'hī·pəˌmir }

hypomorph [GENETICS] An allele having an effect similar to the normal allele, but being less active. { 'hī·pəˌmȯrf }

hypomorphic allele [GENETICS] An allele that has reduced levels of gene activity. { ¦hī·pə¦mȯr·fik ə'lēl }

hyponasty [BOTANY] A nastic movement involving inward and upward bending of a plant part. { 'hī·pəˌnas·tē }

hyponeural sinus [INVERTEBRATE ZOOLOGY] Tubular portion of the coelom containing hemal vessels and motor nerves in certain echinoderms. { ˌhī·pə'nu̇r·əl 'sī·nəs }

hyponychium [HISTOLOGY] The thickened stratum corneum of the epidermis, which lies under the free edge of the nail. { hī'päŋ·kē·əm }

hypopharynx [ANATOMY] *See* laryngopharynx. [INVERTEBRATE ZOOLOGY] A sensory, tonguelike structure on the floor of the mouth of many insects; sometimes modified for piercing. { ¦hī·pō 'far·iŋks }

hypophysis [ANATOMY] A small rounded endocrine gland which lies in the sella turcica of the sphenoid bone and is attached to the floor of the third ventricle of the brain in all craniate vertebrates. Also known as pituitary gland. { hī'päf·ə·səs }

hypoplankton [BIOLOGY] Forms of marine life whose swimming ability lies somewhere between that of the plankton and the nekton; includes some mysids, amphipods, and cumacids. { ¦hī·pō¦plaŋk·tən }

hypoplastron [VERTEBRATE ZOOLOGY] Either of the third pair of lateral bony plates in the plastron of most turtles. { ¦hī·pō¦plas·trən }

hypopleura [INVERTEBRATE ZOOLOGY] Sclerite above and in front of the hind coxa in Diptera. { ¦hī·pō¦plu̇r·ə }

hypoploid [GENETICS] Having one or more less chromosomes, or parts of chromosomes, than a whole multiple of the haploid number. { 'hī·pōˌplȯid }

hypoploidy [GENETICS] The condition or state of being hypoploid. { 'hī·pəˌplȯid·ē }

hypopus [INVERTEBRATE ZOOLOGY] The resting larval stage of certain mites. { 'hī·pə·pəs }

hypopygium [INVERTEBRATE ZOOLOGY] A modified ninth abdominal segment together with the copulatory apparatus in Diptera. { ˌhī·pə'pij·ē·əm }

hyposensitization [IMMUNOLOGY] *See* desensitization. { ¦hī·pōˌsen·səd·ə'zā·shən }

hypostatic [GENETICS] Subject to being suppressed, as a gene that can be suppressed by a nonallelic gene. { ¦hī·pō'stad·ik }

hypostome [INVERTEBRATE ZOOLOGY] **1.** Projection surrounding the oral aperture in many cnidarian polyps. **2.** Anteroventral part of the head in Diptera. **3.** Median ventral mouthpart in ticks. **4.** Raised area on the posterior oral margin in crustaceans. { 'hī·pəˌstōm }

hypostracum [INVERTEBRATE ZOOLOGY] The innermost layer of the cuticle of ticks lying above the hypodermis. { ¦hī·pō¦strak·əm }

hyposulculus [INVERTEBRATE ZOOLOGY] A groove of the siphonoglyph below the pharynx in anthozoans. { ˌhī·pō'səl·kyə·ləs }

hypotarsus [VERTEBRATE ZOOLOGY] A process on the tarsometatarsal bone in birds. { ¦hī·pō¦tär·səs }

hypothalamic center [ANATOMY] Any of the neural centers which regulate autonomic functions. { ¦hī·pō·thə'lam·ik ¦sen·tər }

hypothalamic releasing factor [BIOCHEMISTRY] Any of the hormones secreted by the hypothalamus which travel by way of nerve fibers to the anterior pituitary, where they cause selective release of specific pituitary hormones. { ¦hī·pō·thə'lam·ik ri'lēs·iŋ ˌfak·tər }

hypothalamoneurohypophyseal system [PHYSIOLOGY] The hormones and neurosecretory structures involved in the endocrine activity of the adenohypophysis, neurohypophysis, and hypothalamus. { ¦hī·pō·thə¦lam·ə¦nu̇r·ō,hī¦päf·ə'sē·əl 'sis·təm }

hypothalamoneurohypophyseal tract [ANATOMY] A bundle of nerve fibers connecting the supraoptic and paraventricular neurons of the hypothalamus with the infundibular stem and neurohypophysis. { ¦hī·pō·thə¦lam·ə¦nu̇r·ō,hī¦päf·ə'sē·əl 'trakt }

hypothalamus [ANATOMY] The floor of the third brain ventricle; site of production of several substances that act on the adenohypophysis. { ¦hī·pō'thal·ə·məs }

hypotheca [INVERTEBRATE ZOOLOGY] **1.** The lower valve of a diatom frustule. **2.** Covering on the hypocone in dinoflagellates. { ¦hī·pō'thē·kə }

hypothenar [ANATOMY] Of or pertaining to the prominent portion of the palm above the base of the little finger. { hī'päth·ə,när }

hypothermia [PHYSIOLOGY] Condition of reduced body temperature in homeotherms. { ,hī·pō'thər·mē·ə }

hypotonic [PHYSIOLOGY] **1.** Pertaining to subnormal muscle strength or tension. **2.** Referring to a solution with a lower osmotic pressure than physiological saline. { ¦hī·pə'tän·ik }

Hypotrichida [INVERTEBRATE ZOOLOGY] An order of highly specialized protozoans in the subclass Spirotrichia characterized by cirri on the ventral surface and a lack of ciliature on the dorsal surface. { ,hī·pə'trik·ə·də }

hypotype [SYSTEMATICS] A specimen of a species, which, though not a member of the original type series, is known from a published description or listing. { 'hī·pə,tīp }

hypoxanthine [BIOCHEMISTRY] $C_5H_4ON_4$ An intermediate product derived from adenine in the hydrolysis of nucleic acid. { ¦hī·pō'zan,thēn }

hypoxia [ECOLOGY] A condition characterized by a low level of dissolved oxygen in an aquatic environment. { hī'päk·sē·ə }

hypozygal [INVERTEBRATE ZOOLOGY] In comatulids, the proximal member of adjacent brachials in an articulation. { ,hī·pō'zīg·əl }

hypsodont [VERTEBRATE ZOOLOGY] Of teeth, having crowns that are high or deep and roots that are short. { 'hip·sə,dänt }

hypural [VERTEBRATE ZOOLOGY] Of or pertaining to the bony structure formed by fusion of the hemal spines of the last few vertebrae in most teleost fishes. { hī'pyu̇r·əl }

Hyracodontidae [PALEONTOLOGY] The running rhinoceroses, an extinct family of perissodactyl mammals in the superfamily Rhinoceratoidea. { ,hī·rə·ko'dänt·ə,dē }

Hyracoidea [VERTEBRATE ZOOLOGY] An order of ungulate mammals represented only by the conies of Africa, Arabia, and Syria. { ,hī·rə'kȯid·ē·ə }

hysteriaceous [MYCOLOGY] Of, belonging to, or resembling the Hysteriales. { hi,ster·ē'ā·shəs }

Hysteriales [BOTANY] An order of lichens in the class Ascolichenes including those species with an ascolocular structure. { hi,ster·ē'ā·lēz }

hysterochroic [MYCOLOGY] Having fruiting bodies which discolor progressively from base to apex with age. { ¦his·tə·rō¦krō·ik }

hysterosoma [INVERTEBRATE ZOOLOGY] A body division of an acarid mite composed of the metapodosoma and opisthosoma. { ¦his·tə·rə'sō·mə }

Hystrichospherida [PALEONTOLOGY] A group of protistan microfossils. { ,his·trə·kō'sfer·ə·də }

Hystricidae [VERTEBRATE ZOOLOGY] The Old World porcupines, a family of Rodentia ranging from southern Europe to Africa and eastern Asia and into the Philippines. { hi'stris·ə,dē }

Hystricomorpha [VERTEBRATE ZOOLOGY] A superorder of the class Rodentia. { ,his·trə·kō'mȯr·fə }

H zone [HISTOLOGY] The central portion of an A band in a sarcomere; characterized by the presence of myosin filaments. { 'āch ,zōn }

I

IAA [BIOCHEMISTRY] *See* indoleacetic acid.

Iballidae [INVERTEBRATE ZOOLOGY] A small family of hymenopteran insects in the superfamily Cynipoidea. { ī'bal·ə‚dē }

I band [HISTOLOGY] The band on either side of a Z line; encompasses portions of two adjacent sarcomeres and is characterized by the presence of actin filaments. { 'ī ‚band }

ibis [VERTEBRATE ZOOLOGY] The common name for wading birds making up the family Threskiornithidae and distinguished by a long, slender, downward-curving bill. { 'ī·bəs }

I blood group [IMMUNOLOGY] The erythrocyte antigens defined by reactions with anti-I and anti-i antibodies, which occur both in acquired hemolytic anemia and naturally in normal persons of the rare phenotype i. { ¦ī 'bləd ‚grüp }

IC_{50} [PHYSIOLOGY] *See* incapacitating concentration 50. { ¦ī¦sē 'fif·tē }

Icacinaceae [BOTANY] A family of dicotyledonous plants in the order Celastrales characterized by haplostemonous flowers, pendulous ovules, stipules wanting or vestigial, a polypetalous corolla, valvate petals, and usually one (sometimes three) locules. { i‚ka·sə‚nās·ē‚ē }

ichneumon [INVERTEBRATE ZOOLOGY] The common name for members of the family Ichneumonidae. { ik'nü·mən }

Ichneumonidae [INVERTEBRATE ZOOLOGY] The ichneumon flies, a large family of cosmopolitan, parasitic wasps in the superfamily Ichneumonoidea. { ik·nü'män·ə‚dē }

Ichneumonoidea [INVERTEBRATE ZOOLOGY] A superfamily of hymenopteran insects; members are parasites of other insects. { ik‚nü·mə'nȯid·ē·ə }

ichnite [PALEONTOLOGY] An ichnofossil of the footprint or track of an organism. Also known as ichnolite. { 'ik‚nīt }

ichnolite [PALEONTOLOGY] **1.** A rock containing a fossilized track or footprint. **2.** *See* ichnite. { 'ik·nə‚līt }

ichnology [PALEONTOLOGY] The study of ichnofossils, especially fossil footprints. { ik'näl·ə·jē }

Ichthyobdellidae [INVERTEBRATE ZOOLOGY] A family of leeches in the order Rhynchobdellae distinguished by cylindrical bodies with conspicuous, powerful suckers. { ‚ik·thē·äb'del·ə‚dē }

Ichthyodectidae [PALEONTOLOGY] A family of Cretaceous marine osteoglossiform fishes. { ‚ik·thē·ə'dek·tə‚dē }

ichthyology [VERTEBRATE ZOOLOGY] A branch of vertebrate zoology that deals with the study of fishes. { ‚ik·thē'äl·ə·jē }

ichthyophagous [ZOOLOGY] Subsisting on a diet of fish. { ‚ik·thē'äf·ə·gəs }

Ichthyopterygia [PALEONTOLOGY] A subclass of extinct Mesozoic reptiles composed of predatory fish-finned and sea-swimming forms with short necks and a porpoiselike body. { ‚ik·thē‚äp·tə'rij·ē·ə }

Ichthyornis [PALEONTOLOGY] The type genus of Ichthyornithidae. { ik·thē'ȯr·nəs }

Ichthyornithes [PALEONTOLOGY] A superorder of fossil birds of the order Ichthyornithiformes according to some systems of classification. { ‚ik·thē'ȯr·nə‚thēz }

Ichthyornithidae [PALEONTOLOGY] A family of extinct birds in the order Ichthyornithiformes. { ‚ik·thē‚ȯr'nith·ə‚dē }

Ichthyornithiformes [PALEONTOLOGY] An order of ancient fossil birds including strong flying species from the Upper Cretaceous that possessed all skeletal characteristics of modern birds. { ‚ik·thē‚ȯr·nə·thə'fȯr‚mēz }

Ichthyosauria [PALEONTOLOGY] The only order of the reptilian subclass Ichthyopterygia, comprising the extinct predacious fish-lizards; all were adapted to a sea life in having tail flukes, paddles, and dorsal fins. { ‚ik·thē·ə'sȯr·ē·ə }

Ichthyostegalia [PALEONTOLOGY] An extinct Devonian order of labyrinthodont amphibians, the oldest known representatives of the class. { ‚ik·thē·ō·stə'gal·ē·ə }

Ichthyotomidae [INVERTEBRATE ZOOLOGY] A monotypic family of errantian annelids in the superfamily Eunicea. { ‚ik·thē·ō'täm·ə‚dē }

icosahedral virus [VIROLOGY] A virion in the form of an icosahedron. { ī¦kä·sə¦hē·drəl 'vī·rəs }

Icosteidae [VERTEBRATE ZOOLOGY] The ragfishes, a family of perciform fishes in the suborder Stromateoidei found in high seas. { ‚ī‚kä'stē·ə‚dē }

icotype [SYSTEMATICS] A typical, accurately identified specimen of a species, but not the basis for a published description. { 'ī·kə‚tīp }

ICSH [BIOCHEMISTRY] *See* luteinizing hormone.

Icteridae [VERTEBRATE ZOOLOGY] The troupials, a family of New World perching birds in the suborder Oscines. { ik'ter·ə‚dē }

Ictidosauria [PALEONTOLOGY] An extinct order of

mammallike reptiles in the subclass Synapsida including small carnivorous and herbivorous terrestrial forms. { ik'tid·ə'sór·ē·ə }

ID_{50} [MICROBIOLOGY] *See* infective dose 50.

identical twins [BIOLOGY] *See* monozygotic twins. { ī'dent·ə·kəl 'twinz }

ideomotor [PHYSIOLOGY] **1.** Pertaining to involuntary movement resulting from or accompanying some mental activity, as moving the lips while reading. **2.** Pertaining to both ideation and motor activity. { 'id·ē·ə,mōd·ər }

ideotype [SYSTEMATICS] A specimen identified as belonging to a specific taxon but collected from other than the type locality. { 'id·ē·o,tīp }

idioblast [BOTANY] A plant cell that differs markedly in shape or function from neighboring cells within the same tissue. { 'īd·ē·ō,blast }

idiochromatin [CELL BIOLOGY] The portion of the nuclear chromatin thought to function as the physical carrier of genes. { ¦id·ē·ō'krō·mə·tən }

idiogram [GENETICS] A diagram of chromosome morphology that is used to compare karyotypes of different cells. { 'id·ē·ə,gram }

idiomuscular [PHYSIOLOGY] Pertaining to any phenomenon occurring in a muscle which is independent of outside stimuli. { ¦id·ē·ō'məs·kyə·lər }

idiosome [CELL BIOLOGY] **1.** A hypothetical unit of a cell, such as the region of modified cytoplasm surrounding the centriole or centrosome. **2.** A sex chromosome. { 'id·ē·ə,sōm }

Idiostolidae [INVERTEBRATE ZOOLOGY] A small family of hemipteran insects in the superfamily Pentatomorpha. { ,id·ē·ə'stäl·ə,dē }

Idoteidae [INVERTEBRATE ZOOLOGY] A family of isopod crustaceans in the suborder Valvifera having a flattened body and seven pairs of similar walking legs. { ,ī·dō'tē·ə,dē }

Ig [IMMUNOLOGY] *See* immunoglobulin.

iguana [VERTEBRATE ZOOLOGY] The common name for a number of species of herbivorous, arboreal reptiles in the family Iguanidae characterized by a dorsal crest of soft spines and a dewlap; there are only two species of true iguanas. { i'gwän·ə }

iguanid [VERTEBRATE ZOOLOGY] The common name for members of the reptilian family Iguanidae. { i'gwän·əd }

Iguanidae [VERTEBRATE ZOOLOGY] A family of reptiles in the order Squamata having teeth fixed to the inner edge of the jaws, a nonretractile tongue, a compressed body, five clawed toes, and a long but rarely prehensile tail. { i'gwän·ə,dē }

ileocecal valve [ANATOMY] A muscular structure at the junction of the ileum and cecum which prevents reflex of the cecal contents. { ¦il·ē·ō¦sē·kəl 'valv }

ileocolic artery [ANATOMY] A branch of the superior mesenteric artery that supplies blood to the terminal part of the ileum and the beginning of the colon. { ¦il·ē·ō'käl·ik 'ärd·ə·rē }

ileum [ANATOMY] The last portion of the small intestine, extending from the jejunum to the large intestine. { 'il·ē·əm }

iliac artery [ANATOMY] Either of the two large arteries arising by bifurcation of the abdominal aorta and supplying blood to the lower trunk and legs (or hind limbs in quadrupeds). Also known as common iliac artery. { 'il·ē,ak 'ärd·ə·rē }

iliac fascia [ANATOMY] The fascia covering the pelvic surface of the iliacus muscle. { 'il·ē,ak 'fā·shə }

iliac region [ANATOMY] *See* inguinal region. { 'il·ē,ak ,rē·jən }

iliacus [ANATOMY] The portion of the iliopsoas muscle arising from the iliac fossa and sacrum. { i'lī·ə·kəs }

iliac vein [ANATOMY] Any of the three veins on each side of the body which correspond to and accompany the iliac artery. { 'il·ē,ak ¦vān }

iliocostalis [ANATOMY] The lateral portion of the erector spinal muscle that extends the vertebral column and assists in lateral movements of the trunk. { ,il·ē·ō,kä'sta·ləs }

iliofemoral ligament [ANATOMY] A strong band of dense fibrous tissue extending from the anterior inferior iliac spine to the lesser trochanter and the intertrochanteric line. Also known as Y ligament. { ,il·ē·ō'fem·ə·rəl ,lig·ə·mənt }

iliolumbar ligament [ANATOMY] A fibrous band that radiates laterally from the transverse processes of the fourth and fifth lumbar vertebrae and attaches to the pelvis by two main bands. { ,il·ē·ō'ləm,bär ,lig·ə·mənt }

iliopsoas [ANATOMY] The combined iliacus and psoas muscles. { ,il·ē·ō'sō·əs }

iliotibial tract [ANATOMY] A thickened portion of the fascia lata extending from the lateral condyle of the tibia to the iliac crest. { ,il·ē·ō'tib·ē·əl ,trakt }

ilium [ANATOMY] Either of a pair of bones forming the superior portion of the pelvis bone in vertebrates. { 'il·ē·əm }

Illiciales [BOTANY] An order of dicotyledonous flowering plants, division Magnoliophyta, of the class Magnoliopsida, characterized by having woody plants with scattered spherical cells containing volatile oils. { i,lis·ē'ā·lēz }

imaginal disk [INVERTEBRATE ZOOLOGY] Any of the thickened areas within the sac of the body wall in holometabolous insects which give rise to specific organs in the adult. { ə'maj·ən·əl ,disk }

imago [INVERTEBRATE ZOOLOGY] The sexually mature, usually winged stage of insect development. { ə'mä·gō }

imbed [BIOLOGY] *See* embed. { im'bed }

imbricate [BIOLOGY] Having overlapping edges, such as scales, or the petals of a flower. { 'im·brə·kət }

immarginate [BIOLOGY] Lacking a clearly defined margin. { i'mär·jə,nāt }

immediate hypersensitivity [IMMUNOLOGY] A type of hypersensitivity in which the response rapidly occurs following exposure of a sensitized individual to the antigen. { i'mē·dē·ət ,hī·pər·sen·sə'tiv·əd·ē }

immigrant [ECOLOGY] An organism that settles in a zone where it was previously unknown. { 'im·ə·grənt }

immigration [ECOLOGY] The one-way inward

movement of individuals or their disseminules into a population or population area. [GENETICS] Gene flow from one population into another by interbreeding between members of the populations. { ˌim·əˈgrā·shən }

immune [IMMUNOLOGY] **1.** Safe from attack; protected against a disease by an innate or an acquired immunity. **2.** Pertaining to or conferring immunity. { iˈmyün }

immune body [IMMUNOLOGY] *See* antibody. { iˈmyün ˌbäd·ē }

immune hemolysin [IMMUNOLOGY] A substance formed in blood in response to an injection of erythrocytes from another species. { iˈmyün hē·məˈlīs·ən }

immune horse serum [IMMUNOLOGY] Serum obtained from the blood of an immunized horse. { iˈmyün ˈhȯrs ˌsir·əm }

immune lysin [IMMUNOLOGY] An antibody that will disrupt a particular type of cell in the presence of complement and cofactors, such as magnesium or calcium ions. { iˈmyün ˈlīs·ən }

immune opsonin [IMMUNOLOGY] A substance produced in blood in response to an infection or to inoculation with dead cells of the infecting species of bacteria. { iˈmyün ˈäp·sə·nən }

immune precipitation [IMMUNOLOGY] A method of isolating a protein from mixtures by using a specific antibody as the precipitating agent. { iˈmyün prəˌsip·əˈtā·shən }

immune protein [IMMUNOLOGY] Any antibody. { iˈmyün ˈprōˌtēn }

immune response [IMMUNOLOGY] The physiological responses stemming from activation of the immune system by antigens, consisting of a primary response in which the antigen is recognized as foreign and eliminated, and a secondary response to subsequent contact with the same antigen. { iˈmyün riˌspäns }

immune response gene [IMMUNOLOGY] Any of a group of genes in the major histocompatibility complex that determines the degree of immune response. Abbreviated IR gene. { i¦myün riˈspäns ˌjēn }

immune serum [IMMUNOLOGY] Blood serum obtained from an immunized individual and carrying antibodies. { iˈmyün ˌsir·əm }

immunity [IMMUNOLOGY] The condition of a living organism whereby it resists and overcomes an infection or a disease. { iˈmyü·nəd·ē }

immunization [IMMUNOLOGY] Rendering an organism immune to a specific communicable disease. { ˌim·yə·nəˈzā·shən }

immunochemistry [IMMUNOLOGY] A branch of science dealing with the chemical changes associated with immunity factors. { ¦im·yə·nō ˈkem·ə·strē }

immunodeficiency [IMMUNOLOGY] Any defect of antibody function or cell-mediated immunity. { ¦im·yəˌnō·dəˈfish·ən·sē }

immunodiffusion [IMMUNOLOGY] A serological procedure in which antigen and antibody solutions are permitted to diffuse toward each other through a gel matrix; interaction is manifested by a precipitin line for each system. { ¦im·yə·nō·dəˈfyü·zhən }

immunoelectrophoresis [IMMUNOLOGY] A serological procedure in which the components of an antigen are separated by electrophoretic migration and then made visible by immunodiffusion of specific antibodies. { ¦im·yə·nō·iˌlek·trə·fə ˈrē·səs }

immunofluorescence [IMMUNOLOGY] Fluorescence as the result of, or identifying, an immune response; a specifically stained antigen fluoresces in ultraviolet light and can thus be easily identified with a homologous antigen. { ¦im·yə·nō·fləˈres·əns }

immunogen [IMMUNOLOGY] A substance which stimulates production of specific antibody or of cellular immunity, and which can react with these products. { əˈmyü·nə·jən }

immunogenic [IMMUNOLOGY] Producing immunity. { ¦im·yə·nō¦jen·ik }

immunoglobulin [IMMUNOLOGY] Any of a set of serum glycoproteins which have the ability to bind other molecules with a high degree of specificity. Abbreviated Ig. { ¦im·yə·nōˈglä·byə·lən }

immunoglobulin-mediated immunity [IMMUNOLOGY] *See* humoral immunity. { ˌim·yə·nō¦gläb·yə·lən ¦mē·dēˌād·əd iˈmyü·nəd·ē }

immunological memory [IMMUNOLOGY] The capacity of the immune system to respond more rapidly and vigorously to the second contact with a specific antigen than to the primary contact. { ˌim·yə·nə¦läj·ə·kəl ˈmem·rē }

immunological paralysis [IMMUNOLOGY] *See* acquired immunological tolerance. { ¦im·yə·nə ¦läj·ə·kəl pəˈral·ə·səs }

immunologic suppression [IMMUNOLOGY] The use of x-irradiation, chemicals, corticosteroid hormones, or antilymphocyte antisera to suppress antibody production, particularly in graft transplants. Also known as immunosuppression. { ¦im·yə·nə¦läj·ik səˈpresh·ən }

immunologic tolerance [IMMUNOLOGY] **1.** A condition in which an animal will accept a homograft without rejection. **2.** A state of specific unresponsiveness to an antigen or antigens in adult life as a consequence of exposure to the antigen in utero or in the neonatal period. { ¦im·yə·nə ¦läj·ik ˈtäl·ə·rəns }

immunologist [IMMUNOLOGY] A person who specializes in immunology. { ˌim·yəˈnäl·ə·jəst }

immunology [BIOLOGY] A branch of biological science concerned with the native or acquired resistance of higher animal forms and humans to infection with microorganisms. { ˌim·yəˈnäl·ə·jē }

immunonephelometry [IMMUNOLOGY] The application of nephelometry to the quantification of antigen or antibody. { ¦im·yə·nōˌnef·əˈläm·ə·trē }

immunopotentiation [IMMUNOLOGY] Enhancement of an immune response by a variety of adjuvants. { ˌim·yə·nō·pəˌten·chēˈā·shən }

immunosuppression [IMMUNOLOGY] *See* immunologic suppression. { ¦im·yə·nō·səˈpresh·ən }

immunotoxin [IMMUNOLOGY] Conjugate of antibody and toxic protein such that the specificity of the antibody molecule is combined with the cytotoxic property of the toxin. { ˌim·yə·nō'täk·sən }

imparipinnate [BOTANY] *See* odd-pinnate. { ˌim·par·ə'piˌnāt }

Impennes [VERTEBRATE ZOOLOGY] A superorder of birds for the order Sphenisciformes in some systems of classification. { im'pen·ēz }

imperfect flower [BOTANY] A flower lacking either stamens or carpels. { im'pər·fikt 'flau̇·ər }

imperforate [BIOLOGY] Lacking a normal opening. { im'pər·fə·rət }

impermeable junction [CELL BIOLOGY] *See* tight junction. { im¦pər·mē·ə·bəl 'jəŋk·shən }

implexed [INVERTEBRATE ZOOLOGY] In insects, having the integument infolded for muscle attachment. { 'imˌplekst }

impunctate [BIOLOGY] Lacking pores. { im'pəŋkˌtāt }

IMVIC test [MICROBIOLOGY] A group of four cultural tests used to differentiate genera of bacteria in the family Enterobacteriaceae and to distinguish them from other bacteria; tests are indole, methyl red, Voges-Proskauer, and citrate. { 'imˌvik ˌtest }

inactivated vaccine [IMMUNOLOGY] *See* killed vaccine. { in¦ak·təˌvād·əd vakˌsēn }

Inadunata [PALEONTOLOGY] An extinct subclass of stalked Paleozoic Crinozoa characterized by branched or simple arms that were free and in no way incorporated into the calyx. { i¦nä·jə¦näd·ə }

inadunate [BIOLOGY] Not united. [INVERTEBRATE ZOOLOGY] In crinoids, having the arms free from the calyx. { i'näj·ə·nət }

inaperturate [BIOLOGY] Lacking apertures. { in'ap·əˌchu̇r·ət }

inarching [BOTANY] A kind of repair grafting in which two plants growing on their own roots are grafted together and one plant is severed from its roots after the graft union is established. { 'in'ärch·iŋ }

Inarticulata [INVERTEBRATE ZOOLOGY] A class of the phylum Brachiopoda; valves are typically not articulated and are held together only by soft tissue of the living animal. { ¦in·ärˌtik·yə'läd·ə }

inborn [BIOLOGY] Of or pertaining to a congenital or hereditary characteristic. { 'inˌbȯrn }

inbreeding [GENETICS] Breeding of closely related individuals; self-fertilization, as in some plants, is the most extreme form. { 'inˌbrēd·iŋ }

inbreeding coefficient [GENETICS] A measure of the rate of inbreeding or the degree to which an individual is inbred. Also known as Wright's inbreeding coefficient. { 'inˌbrēd·iŋ ˌkō·əˌfish·ənt }

inbreeding depression [GENETICS] A decrease in fitness and vigor as a result of inbreeding. { 'inˌbrēd·iŋ diˌpresh·ən }

incapacitating concentration 50 [PHYSIOLOGY] The concentration of gas or smoke that will incapacitate 50% of the test animals within a given time of exposure. Abbreviated IC_{50}. { ˌin·kə¦pas·əˌtād·iŋ ˌkans·ən¦trā·shən 'fif·tē }

incapsidation [VIROLOGY] The construction of a capsid around the genetic material of a virus. { inˌkap·sə'dā·shən }

incertae sedis [SYSTEMATICS] Placed in an uncertain taxonomic position. { iŋ¦kərˌtī 'sā·dəs }

incipient species [EVOLUTION] Populations that are in the process of diverging to the point of speciation but still have the potential to interbreed. { in'sip·ē·ənt ¦spē·shēz }

Incirrata [INVERTEBRATE ZOOLOGY] A suborder of cephalopod mollusks in the order Octopoda. { ¦in·sə¦räd·ə }

incised [BIOLOGY] Having a deeply and irregularly notched margin. { in'sīzd }

incisive canal [ANATOMY] The bifurcated bony passage from the floor of the nasal cavity to the incisive fossa. { in'sī·siv kə¦nal }

incisive foramen [ANATOMY] One of the two to four openings of the incisive canal on the floor of the incisive fossa. { in'sī·siv fə'rā·mən }

incisive fossa [ANATOMY] **1.** A bony pit behind the upper incisors into which the incisive canals open. **2.** A depression on the maxilla at the origin of the depressor muscle of the nose. **3.** A depression of the mandible at the origin of the mentalis muscle. { in'sī·siv 'fäs·ə }

incisor [ANATOMY] A tooth specialized for cutting, especially those in front of the canines on the upper jaw of mammals. { in'sīz·ər }

inclusion [CELL BIOLOGY] A visible product of cellular metabolism within the protoplasm. { in'klü·zhən }

inclusion body [VIROLOGY] Any of the abnormal structures appearing within the cell nucleus or the cytoplasm during the course of virus multiplication. { in'klü·zhən ˌbäd·ē }

inclusive fitness [GENETICS] The relative number of alleles that are passed on by an individual to the next generation. { iŋ'klü·siv 'fit·nəs }

incompatibility [IMMUNOLOGY] Genetic or antigenic differences between donor and recipient tissues that result in a rejection response. { ˌin·kəmˌpad·ə'bil·əd·ē }

incomplete dominance [GENETICS] In a heterozygote, the phenomenon in which the appearance of the phenotype resembles both alleles at a particular locus. { ¦iŋ·kəmˌplēt 'däm·ən·əns }

incomplete flower [BOTANY] A flower lacking one or more modified leaves, such as petals, sepals, pistils, or stamens. { ˌin·kəm'plēt 'flau̇·ər }

incrassate [BIOLOGY] State of being swollen or thickened. { in'kraˌsāt }

incretion [PHYSIOLOGY] An internal secretion. { in'krē·shən }

incross [GENETICS] Mating between individuals from the same inbred line. { 'inˌkrȯs }

incubation [VERTEBRATE ZOOLOGY] The act or process of brooding. { ˌiŋ·kyə'bā·shən }

incubation period [VERTEBRATE ZOOLOGY] The brooding period required to bring an egg to hatching. { ˌiŋ·kyə'bā·shən ˌpir·ē·əd }

incubator [MICROBIOLOGY] A laboratory cabinet with controlled temperature for the cultivation of bacteria, or for facilitating biologic tests. { 'iŋ·kyəˌbād·ər }

incubous [BOTANY] The juxtaposition of leaves such that the anterior margins of older leaves overlap the posterior margins of younger leaves. { 'iŋ·kyə·bəs }

incudate [BIOLOGY] Of, pertaining to, or having an incus. { 'iŋ·kyə‚dāt }

incumbent [BIOLOGY] Lying on or down. { in 'kəm·bənt }

incurrent canal [INVERTEBRATE ZOOLOGY] A canal through which water enters a sponge. { in'kər·ənt kə'nal }

incurrent siphon [INVERTEBRATE ZOOLOGY] *See* inhalant siphon. { in'kə·rənt 'sī·fən }

Incurvariidae [INVERTEBRATE ZOOLOGY] A family of lepidopteran insects in the superfamily Incurvarioidea; includes yucca moths and relatives. { ‚in‚kər·və'rī·ə‚dē }

Incurvarioidea [INVERTEBRATE ZOOLOGY] A monofamilial superfamily of lepidopteran insects in the suborder Heteroneura having wings covered with microscopic spines, a single genital opening in the female, and venation that is almost complete. { ‚in·kər‚var·ē'ȯid·ē·ə }

incus [ANATOMY] The middle one of three ossicles in the middle ear. Also known as anvil. { 'iŋ·kəs }

indeciduate placenta [EMBRYOLOGY] A placenta having the maternal and fetal elements associated but not fused. { ¦in·də¦sij·ə·wət plə 'sent·ə }

indehiscent [BOTANY] **1.** Remaining closed at maturity, as certain fruits. **2.** Not splitting along regular lines. { ¦in·də'his·ənt }

independent assortment [GENETICS] The random assortment of the alleles at two or more loci on different chromosome pairs or far apart on the same chromosome pair which occurs at meiosis; first discovered by G. Mendel. { ‚in·də'pen·dənt ə'sȯrt·mənt }

indeterminate cleavage [EMBRYOLOGY] Cleavage in which all the early cells have the same potencies with respect to development of the entire zygote. { ‚in·də'tərm·ə·nət 'klē·vij }

indeterminate growth [BOTANY] Growth of a plant in which the axis is not limited by development of a reproductive structure, and therefore growth continues indefinitely. { ‚in·də 'tərm·ə·nət 'grōth }

index fossil [PALEONTOLOGY] The ancient remains and traces of an organism that lived during a particular geologic time period and that geologically date the containing rocks. { 'in‚deks ‚fäs·əl }

indican [BIOCHEMISTRY] C_8H_6NOSOK The potassium salt of indoxylsulfate found in urine as a result of bacterial action on tryptophan in the bowel. { 'in·də‚kan }

indicator medium [MICROBIOLOGY] A usually solid culture medium containing substances capable of undergoing a color change in the vicinity of a colony which has effected a particular chemical change, such as fermenting a certain sugar. { 'in·də‚kād·ər ‚mē·dē·əm }

indicator plant [BOTANY] A plant used in geobotanical prospecting as an indicator of a certain geological phenomenon. { 'in·də‚kād·ər ‚plant }

indicator species [ECOLOGY] A species whose presence is directly related to a particular quality in its environment at a given location. { 'in·də ‚kād·ər ‚spē·shēz }

indirect Coombs test [IMMUNOLOGY] *See* Rh blocking test. { ‚in·də'rekt 'külmz ‚test }

indirect developing test [IMMUNOLOGY] *See* Rh blocking test. { ‚in·də'rekt di'vel·ə·piŋ ‚test }

indirect immunofluorescence [IMMUNOLOGY] The use of a labeled indicator antibody which reacts with an unlabeled detector antibody that has previously reacted with an antigen. { ‚in·də 'rekt ¦im·yə·nō·flu̇'res·əns }

indirect vision [PHYSIOLOGY] *See* peripheral vision. { ‚in·də'rekt 'vizh·ən }

individuation [EMBRYOLOGY] The process whereby, through induction, a spatially organized tissue, organ, or embryo develops. { ‚in·di ‚vij·ə'wā·shən }

indole [BIOCHEMISTRY] $C_6H_4(CHNH)CH$ A decomposition product of tryptophan formed in the intestine during putrefaction and by certain cultures of bacteria. Also known as 2,3-benzopyrrole. { 'in‚dōl }

indoleacetic acid [BIOCHEMISTRY] $C_{10}H_9O_2N$ A decomposition product of tryptophan produced by bacteria and occurring in urine and feces; used as a hormone to promote plant growth. Abbreviated IAA. { ¦in‚dōl·ə¦sēd·ik 'as·əd }

indole test [MICROBIOLOGY] A test for the production of indole from tryptophan by microorganisms; a solution of *para*-dimethylaminobenzaldehyde, amyl alcohol, and hydrochloric acid added to the incubated culture of bacteria shows a red color in the alcoholic layer if indole is present. { 'in‚dōl ‚test }

Indo-Pacific faunal region [ECOLOGY] A marine littoral faunal region extending eastward from the east coast of Africa, passing north of Australia and south of Japan, and ending in the east Pacific south of Alaska. { ¦in·dō·pə'sif·ik 'fȯn·əl ‚rē·jən }

Indriidae [VERTEBRATE ZOOLOGY] A family of Madagascan prosimians containing wholly arboreal vertical clingers and leapers. { in'drī·ə ‚dē }

inducer [EMBRYOLOGY] The cell group that functions as the acting system in embryonic induction by controlling the mode of development of the reacting system. Also known as inductor. { in 'dü·sər }

inducible enzyme [BIOCHEMISTRY] An enzyme which is present in trace quantities within a cell but whose concentration increases dramatically in the presence of substrate molecules. { in'dü·sə·bəl 'en‚zīm }

induction [EMBRYOLOGY] *See* embryonic induction. { in'dək·shən }

inductor [EMBRYOLOGY] *See* inducer. { in'dək·tər }

inductura [INVERTEBRATE ZOOLOGY] A layer of lamellar shell material along the inner lip of the aperture in gastropods. { in'dək·chə·rə }

indumentum [BOTANY] A covering, such as one

that is woolly. [MYCOLOGY] A covering of hairs. [VERTEBRATE ZOOLOGY] The plumage covering a bird. { ˌin·də′men·təm }

induplicate [BOTANY] Having the edges turned or rolled inward without twisting or overlapping; applied to the leaves of a bud. { in′dü·plə·kət }

induration [BIOLOGY] The process of hardening, especially by increasing the fibrous elements. { ˌin·də′rā·shən }

indusium [ANATOMY] A covering membrane such as the amnion. [BOTANY] An epidermal outgrowth covering the sori in many ferns. [MYCOLOGY] The annulus of certain fungi. { in′dü·zē·əm }

industrial microbiology [MICROBIOLOGY] The study, utilization, and manipulation of those microorganisms capable of economically producing desirable substances or changes in substances, and the control of undesirable microorganisms. { in′dəs·trē·əl ˌmī·krō·bī′äl·ə·jē }

industrial microorganism [MICROBIOLOGY] Any microorganism utilized for industrial microbiology. { in′dəs·trē·əl ˌmī·krō′ȯr·gəˌniz·əm }

industrial yeast [MICROBIOLOGY] Any yeast used for the production of fermented foods and beverages, for baking, or for the production of vitamins, proteins, alcohol, glycerol, and enzymes. { in′dəs·trē·əl ¦yēst }

inequilateral [BIOLOGY] Having the two sides or ends unequal, as the ends of a bivalve mollusk on either side of a line from umbo to gape. { in¦ē·kwə′lad·ə·rəl }

inermous [BIOLOGY] Lacking mechanisms for defense or offense, especially spines. { i′nər·məs }

infantile genitalia [ANATOMY] The genital organs of an infant. { ′in·fənˌtīl ˌjen·ə′tāl·yə }

infauna [ZOOLOGY] Aquatic animals which live in the bottom sediment of a body of water. { in′fȯn·ə }

infect [MICROBIOLOGY] To cause a phage infection of bacteria. { in′fekt }

infectious drug resistance [MICROBIOLOGY] A type of drug resistance that is transmissible from one bacterium to another by infectivelike agents referred to as resistance factors. { in′fek·shəs ′drəg riˌzis·təns }

infectious nucleic acid [VIROLOGY] Purified viral nucleic acid capable of infecting a host cell and causing the production of viral progeny. { in¦fek·shəs nü¦klē·ik ′as·əd }

infectious transfer [GENETICS] The rapid spread of extrachromosomal episomes from donor to recipient cells in a bacterial population. { in¦fek·shən ′tranzˌfər }

infectious unit [VIROLOGY] The smallest number of virus particles that will cause a lytic infection in a susceptible cell. { in′fek·shəs ˌyü·nət }

infective dose 50 [MICROBIOLOGY] The dose of microorganisms required to cause infection in 50% of the experimental animals; a special case of the median effective dose. Abbreviated ID_{50}. Also known as median infective dose. { in′fek·tiv ¦dōs ′fif·tē }

inferior [BIOLOGY] The lower of two structures. { in′fir·ē·ər }

inferior alveolar artery [ANATOMY] A branch of the internal maxillary artery supplying the mucous membrane of the mouth and teeth of the lower jaw. { in′fir·ē·ər al¦vē·ə·lər ′ärd·ə·rē }

inferior alveolar nerve [ANATOMY] A branch of the mandibular nerve that innervates the teeth of the lower jaw. { in′fir·ē·ər al¦vē·ə·lər ′nərv }

inferior cerebellar peduncle [ANATOMY] A large bundle of nerve fibers running from the medulla oblongata to the cerebellum. Also known as restiform body. { in′fir·ē·ər ¦ser·ə¦bel·ər ′pēˌdəŋ·kəl }

inferior colliculus [ANATOMY] One of the posterior pair of rounded eminences arising from the dorsal portion of the mesencephalon. { in′fir·ē·ər kə′lik·yə·ləs }

inferior ganglion [ANATOMY] **1.** The lower sensory ganglion in the glossopharyngeal nerve. **2.** The lower sensory ganglion on the vagus. { in′fir·ē·ər ′gaŋ·glē·ən }

inferior hypogastric plexus [ANATOMY] A network of nerves in the pelvic fascia containing autonomic nerve elements. { in′fir·ē·ər ˌhī·pə′gas·trik ′plek·səs }

inferior mesenteric ganglion [ANATOMY] A sympathetic ganglion within the inferior mesenteric plexus at the origin of the inferior mesenteric artery. { in′fir·ē·ər ˌmez·ən′ter·ik ′gaŋ·glē·ən }

inferior temporal gyrus [ANATOMY] A convolution on the temporal lobe of the cerebral hemispheres lying below the middle temporal sulcus and extending to the inferior sulcus. { in′fir·ē·ər ¦tem·pə·rəl ′jī·rəs }

inferior vena cava [ANATOMY] A large vein which drains blood from the iliac veins, lower extremities, and abdomen into the right atrium. { in′fir·ē·ər ˌvē·nə ′kā·və }

inferior vermis [ANATOMY] The inferior portion of the vermis of the brain. { in′fir·ē·ər ′vər·məs }

inferior vestibular nucleus [ANATOMY] The terminal nucleus for the spinal vestibular nerve tract. { in′fir·ē·ər və′stib·yə·lər ′nü·klē·əs }

inflammatory response [IMMUNOLOGY] A nonspecific defensive reaction of the body to invasion by a foreign substance or organism that involves phagocytosis by white blood cells and is often accompanied by accumulation of pus and an increase in the local temperature. { in¦flam·əˌtȯrē ri′späns }

inflated [BIOLOGY] **1.** Distended, applied to a hollow structure. **2.** Open and enlarged. { in′flād·əd }

inflected [BOTANY] Curved or bent sharply inward, downward, or toward the axis. Also known as inflexed. { in′flek·təd }

inflexed [BOTANY] *See* inflected. { in′flekst }

inflorescence [BOTANY] **1.** A flower cluster. **2.** The arrangement of flowers on a plant. { ˌin·flə′res·əns }

influent [ECOLOGY] An organism that disturbs the ecological balance of a community. { ′in ˌflü·ənt }

influenza vaccine [IMMUNOLOGY] A vaccine prepared from formaldehyde-attenuated mixtures of strains of influenza virus. { ˌin·flü'en·zə vak 'sēn }

influenza virus [VIROLOGY] Any of three immunological types, designated A, B, and C, belonging to the myxovirus group which cause influenza. { ˌin·flü'en·zə ˌvī·rəs }

informosome [CELL BIOLOGY] A type of cellular particle that is thought to be a complex of messenger ribonucleic acid with ribonucleoprotein. { in'fōr·məˌsōm }

infrabasal [BIOLOGY] Inferior to a basal structure. { ¦in·frə'bā·səl }

infrabranchial [VERTEBRATE ZOOLOGY] Situated below the gills. { ¦in·frə'braŋ·kē·əl }

infracentral [ANATOMY] Located below the centrum. { ¦in·frə'sen·trəl }

infracerebral gland [INVERTEBRATE ZOOLOGY] A structure lying ventral to the brain in annelids which is thought to produce a hormone that inhibits maturation of the gametes. { ¦in·frə·sə 'rē·brəl ¦gland }

infraciliature [INVERTEBRATE ZOOLOGY] The neuromotor apparatus, silverline system, or neuroneme system of ciliates. { ˌin·frə'sil·yə·chər }

infraclass [SYSTEMATICS] A subdivision of a subclass; equivalent to a superorder. { 'in·frəˌklas }

infraclavicle [VERTEBRATE ZOOLOGY] A bony element of the shoulder girdle located below the cleithrum in some ganoid and crossopterygian fishes. { ¦in·frə'klav·ə·kəl }

infrafoliar [BOTANY] Located below the leaves. { ˌin·frə'fō·lē·ər }

infraglenoid [ANATOMY] Below the glenoid cavity of the scapula. { ¦in·frə'gleˌnȯid }

infraglenoid tubercle [ANATOMY] A rough impression below the glenoid cavity, from which the long head of the triceps muscle arises. { ¦in·frə'gleˌnȯid 'tü·bər·kəl }

infraorbital [ANATOMY] Located beneath the orbit. { ¦in·frə'ȯr·bəd·əl }

infraspinous [ANATOMY] Below the spine of the scapula. { ¦in·frə'spī·nəs }

infraspinous fossa [ANATOMY] The recess on the posterior surface of the scapula occupied by the infraspinatus muscle. { ¦in·frə'spī·nəs 'fäs·ə }

infratemporal [ANATOMY] Situated below the temporal fossa. { ¦in·frə'tem·prəl }

infratemporal fossa [ANATOMY] An irregular space situated below and medial to the zygomatic arch, behind the maxilla and medial to the upper part of the ramus of the mandible. { ¦in·frə'tem·prəl 'fäs·ə }

infructescence [BOTANY] An inflorescence's fruiting stage. { ˌinˌfrək'tes·əns }

infundibular canal [INVERTEBRATE ZOOLOGY] A pathway from the mantle cavity through the funnel for water in cephalopods. { ˌin·fən'dib·yə·lər kə'nal }

infundibular ganglion [INVERTEBRATE ZOOLOGY] A branch of pedal ganglion which supplies the funnel in cephalopods. { ˌin·fən'dib·yə·lər 'gaŋ·glē·ən }

infundibular process [ANATOMY] The distal portion of the neural lobe of the pituitary. { ˌin·fən 'dib·yə·lər 'prä·səs }

infundibulum [ANATOMY] **1.** A funnel-shaped passage or part. **2.** The stalk of the neurohypophysis. { ˌin·fən'dib·yə·ləm }

infusoriform larva [INVERTEBRATE ZOOLOGY] The final larval stage, arising from germ cells within the infusorigen, in the life cycle of dicyemid mesozoans. { ¦in·fyə¦zȯr·əˌfȯrm 'lär·və }

infusorigen [INVERTEBRATE ZOOLOGY] An individual that gives rise to the infusoriform larva in dicyemid mesozoans. { ˌin·fyə'zȯr·ə·jən }

ingesta [BIOLOGY] Food and other substances taken into an animal body. { in'jes·tə }

ingestion [BIOLOGY] The act or process of taking food and other substances into the animal body. { in'jes·chən }

Ingolfiellidea [INVERTEBRATE ZOOLOGY] A suborder of amphipod crustaceans in which both abdomen and maxilliped are well developed and the head often bears a separate ocular lobe lacking eyes. { in¦gäl·fē·ə¦lid·ē·ə }

inguinal canal [ANATOMY] A short, narrow passage between the abdominal ring and the inguinal ring in which lies the spermatic cord in males and the round ligament in females. { 'iŋ·gwən·əl kə'nal }

inguinal fold [EMBRYOLOGY] A fold of embryonic tissue on the urogenital ridge in which the gubernaculum testis develops. { 'iŋ·gwən·əl 'fōld }

inguinal gland [ANATOMY] Any of the superficial lymphatic glands in the groin. { 'iŋ·gwən·əl ¦gland }

inguinal ligament [ANATOMY] The thickened lower portion of the aponeurosis of the external oblique muscle extending from the anterior superior spine of the ileum to the tubercle of the pubis and the pectineal line. Also known as Poupart's ligament. { 'iŋ·gwən·əl 'lig·ə·mənt }

inguinal region [ANATOMY] The abdominal region occurring on each side of the body as a depression between the abdomen and the thigh. Also known as iliac region. { 'iŋ·gwən·əl ˌrē·jən }

inhalant canal [INVERTEBRATE ZOOLOGY] The incurrent canal in sponges and mollusks. { in'hā·lənt kəˌnal }

inhalant siphon [INVERTEBRATE ZOOLOGY] A channel for water intake in the mantle of bivalve mollusks. Also known as incurrent siphon. { in 'hā·lənt 'sī·fən }

inhalation [PHYSIOLOGY] The process of breathing in. { ˌin·ə'lā·shən }

inheritance [GENETICS] **1.** The acquisition of characteristics by transmission of germ plasm from ancestor to descendant. **2.** The sum total of characteristics dependent upon the constitution of the fertilized ovum. { in'her·əd·əns }

inhibiting antibody [IMMUNOLOGY] A substance sometimes produced in the blood of immunized persons which is thought to prevent the expected antigen-reagin reaction. { in'hib·əd·iŋ 'ant·iˌbäd·ē }

inhibition index [BIOCHEMISTRY] The amount of antimetabolite that will overcome the biological effect of a unit weight of metabolite. { ˌin·ə 'bish·ən ˌinˌdeks }

Iniomi [VERTEBRATE ZOOLOGY] An equivalent name for Salmoniformes. { ˌin·ē'ōˌmī }

initiation codon [GENETICS] A codon that signals the first amino acid in a protein sequence; usually AUG, but sometimes GUG. Also known as start codon. { iˌnish·ē'ā·shən 'kōˌdän }

initiation complex [MOLECULAR BIOLOGY] An intermediate of protein synthesis consisting of messenger ribonucleic acid, initiator codons, initiation factors, and initiator transfer ribonucleic acid. { iˌnish·ē'ā·shən ˌkämˌpleks }

initiation factor [MOLECULAR BIOLOGY] Any protein required for the initiation of protein synthesis. { iˌnish·ē'ā·shən ˌfak·tər }

initiator codon [MOLECULAR BIOLOGY] A codon that acts as a start signal for the synthesis of a polypeptide. { i'nish·ēˌād·ər 'kōˌdän }

initiator ribonucleic acid [MOLECULAR BIOLOGY] An oligoribonucleotide that primes the initiation of Okazaki fragments during deoxyribonucleic acid synthesis. { iˌnish·ē¦ād·ər ˌrī·bō·nü¦klē·ik 'as·əd }

injection chimera [BIOLOGY] A chimera produced experimentally by inserting embryonic cells of different genetic makeup into the preimplantation blastocyst. { in'jek·shən kīˌmir·ə }

injury current [PHYSIOLOGY] *See* injury potential. { 'in·jə·rē ˌkə·rənt }

injury potential [PHYSIOLOGY] The potential difference observed between the injured and the noninjured regions of an injured tissue or cell. Also known as demarcation potential; injury current. { 'in·jə·rē pəˌten·chəl }

ink sac [INVERTEBRATE ZOOLOGY] An organ attached to the rectum in many cephalopods which secretes and ejects an inky fluid. { 'iŋk ˌsak }

inlet of the pelvis [ANATOMY] The space within the brim of the pelvis. { 'inˌlet əv thə 'pel·vəs }

innate [BIOLOGY] Pertaining to a natural or inborn character dependent on genetic constitution. [BOTANY] Positioned at the apex of a supporting structure. [MYCOLOGY] Embedded in, especially of an organ such as the fruiting body embedded in the thallus of some fungi. { i'nāt }

inner cell mass [EMBRYOLOGY] The cells at the animal pole of a blastocyst which give rise to the embryo and certain extraembryonic membranes. { ¦in·ər 'sel ˌmas }

inner ear [ANATOMY] The part of the vertebrate ear concerned with labyrinthine sense and sound reception; consists generally of a bony and a membranous labyrinth, made up of the vestibular apparatus, three semicircular canals, and the cochlea. Also known as internal ear. { ¦in·ər 'ir }

innervation [ANATOMY] The distribution of nerves to a part. [PHYSIOLOGY] The amount of nerve stimulation received by a part. { ˌin·ər 'vā·shən }

innominate [ANATOMY] *See* hip bone. { i¦näm·ə·nət }

innominate artery [ANATOMY] The first artery branching from the aortic arch; distributes blood to the head, neck, shoulder, and arm on the right side of the body. { i¦näm·ə·nət 'ärd·ə·rē }

inoculation [BIOLOGY] Introduction of a disease agent into an animal or plant to produce a mild form of disease and render the individual immune. [MICROBIOLOGY] Introduction of microorganisms onto or into a culture medium. { iˌnäk·yə'lā·shən }

inoculum [MICROBIOLOGY] A small amount of substance containing bacteria from a pure culture which is used to start a new culture or to infect an experimental animal. { i'näk·yə·ləm }

inoperculate [BIOLOGY] Lacking an operculum. { ¦in·ä'pər·kyə·lət }

inosine [BIOCHEMISTRY] $C_{10}H_{12}N_4O_5$ A compound occurring in muscle; a hydrolysis product of inosinic acid. { 'in·əˌsēn }

inosinic acid [BIOCHEMISTRY] $C_{10}H_{13}N_4O_8P$ A nucleotide constituent of muscle, formed by deamination of adenylic acid; on hydrolysis it yields hypoxanthine and D-ribose-5-phosphoric acid. { ¦in·ə¦sin·ik 'as·əd }

inquiline [ZOOLOGY] An animal that inhabits the nest of another species. { 'in·kwəˌlīn }

inrolling [BOTANY] Inward rolling of the corolla of a flower, a physical process associated with senescence. { 'inˌrōl·iŋ }

insect [INVERTEBRATE ZOOLOGY] **1.** A member of the Insecta. **2.** An invertebrate that resembles an insect, such as a spider, mite, or centipede. { 'inˌsekt }

Insecta [INVERTEBRATE ZOOLOGY] A class of the Arthropoda typically having a segmented body with an external, chitinous covering, a pair of compound eyes, a pair of antennae, three pairs of mouthparts, and two pairs of wings. { in'sek·tə }

insect control [ECOLOGY] Regulation of insect populations by biological or chemical means. { 'inˌsekt kənˌtrōl }

insectistasis [ECOLOGY] The use of pheromones to trap, confuse, or inhibit insects in order to hold populations below a level where they can cause significant economic damage. { in¦sek·tə ¦stā·səs }

Insectivora [VERTEBRATE ZOOLOGY] An order of mammals including hedgehogs, shrews, moles, and other forms, most of which have spines. { in ˌsek'tiv·ə·rə }

insectivorous [BIOLOGY] Feeding on a diet of insects. { inˌsek'tiv·ə·rəs }

insectivorous plant [BOTANY] A plant that captures and digests insects as a source of nutrients by using specialized leaves. Also known as carnivorous plant. { inˌsek'tiv·ə·rəs 'plant }

insect pathology [INVERTEBRATE ZOOLOGY] A biological discipline embracing the general principles of pathology as applied to insects. { 'in ˌsekt pə'thäl·ə·jē }

insect physiology [INVERTEBRATE ZOOLOGY] The study of the functional properties of insect tissues and organs. { 'inˌsekt ˌfiz·ē'äl·ə·je }

insemination [BIOLOGY] The planting of seed. [PHYSIOLOGY] **1.** The introduction of sperm into the vagina. **2.** Impregnation. { inˌsem·ə'nā·shən }

inserted [BIOLOGY] United or attached to the

supporting structure by natural growth. { in 'sərd·əd }

insertion [ANATOMY] The point at which a muscle is attached to a bone that moves when the muscle contracts; it is the distal end of the muscle. [MOLECULAR BIOLOGY] The addition of an extranumerary base pair to double-stranded deoxyribonucleic acid; causes errors in transcription. { in'sər·shən }

insertion mutagenesis [MOLECULAR BIOLOGY] Gene alteration due to insertion of unusual nucleotide sequences from sources such as transposons, viruses, or synthetic deoxyribonucleic acid. { in'sər·shən ˌmyüd·əˌjen·ə·səs }

insertion site [MOLECULAR BIOLOGY] **1.** In a cloning vector molecule of deoxyribonucleic acid (DNA), a restriction site into which foreign DNA can be inserted. **2.** The position at which a transposable genetic element is integrated. { in'sər·shən ˌsīt }

inspiration [PHYSIOLOGY] The drawing in of the breath. { ˌin·spə'rā·shən }

inspiratory capacity [PHYSIOLOGY] Commencing from expiratory standstill, the maximum volume of gas which can be drawn into the lungs. { in ·'spī·rəˌtȯr·ē kə'pas·əd·ē }

inspiratory reserve volume [PHYSIOLOGY] The amount of air that can be inhaled by forcible inspiration after completion of a normal inspiration. { in'spī·rəˌtȯr·ē ri¦zərv ˌväl·yəm }

instar [INVERTEBRATE ZOOLOGY] A stage between molts in the life of arthropods, especially insects. { 'inˌstär }

instep [ANATOMY] The arch on the medial side of the foot. { 'inˌstep }

instinct [ZOOLOGY] A precise form of behavior in which there is an invariable association of a particular series of responses with specific stimuli; an unconditioned compound reflex. { 'in ˌstiŋkt }

instinctive behavior [ZOOLOGY] Any species-typical pattern of responses not clearly acquired through training. { in'stiŋk·tiv bi'hā·vyər }

instinctual [ZOOLOGY] Of or pertaining to instincts. { in'stiŋk·chə·wəl }

insulin [BIOCHEMISTRY] A protein hormone produced by the beta cells of the islets of Langerhans which participates in carbohydrate and fat metabolism. { 'in·sə·lən }

insulinase [BIOCHEMISTRY] An enzyme produced by the liver which is able to inactivate insulin. { 'in·sə·ləˌnās }

integrase [BIOCHEMISTRY] An enzyme that facilitates prophage integration into or excision from a bacterial chromosome. { 'int·əˌgrās }

integration [GENETICS] Recombination involving insertion of a genetic element. { ˌint·ə'grā·shən }

integration efficiency [MOLECULAR BIOLOGY] The frequency with which a segment of foreign deoxyribonucleic acid is incorporated into a host bacterial genome. { int·ə'grā·shən iˌfish·ən·sē }

integument [ANATOMY] An outer covering, especially the skin, together with its various derivatives. { in'teg·yə·mənt }

integumentary musculature [VERTEBRATE ZOOLOGY] Superficial skeletal muscles which are spread out beneath the skin and are inserted into it in some terrestrial vertebrates. { in¦teg·yə¦men·trē 'məs·kyə·lə·chər }

integumentary pattern [ANATOMY] Any of the features of the skin and its derivatives that are arranged in designs, such as scales, epidermal ridges, feathers coloration, or hair. { in¦teg·yə¦men·trē 'pad·ərn }

integumentary system [ANATOMY] A system encompassing the integument and its derivatives. { in¦teg·yə¦men·trē 'sis·təm }

interambulacrum [INVERTEBRATE ZOOLOGY] In echinoderms, an area between two ambulacra. { ¦in·tərˌam·byə'la·krəm }

interarticular [ANATOMY] Situated between articulating surfaces. { ¦in·tərˌär'tik·yə·lər }

interatrial [ANATOMY] Located between the atria of the heart. { ¦in·tər'ā·trē·əl }

interatrial septum [ANATOMY] *See* atrial septum. { ¦in·tər'ā·trē·əl 'sep·təm }

interaxillary [BOTANY] Located within or between the axils of leaves. { ˌin·tər'ak·səˌler·ē }

intercalary [BOTANY] Referring to growth occurring between the apex and leaf. { in'tər·kə ˌler·ē }

intercalary meristem [BOTANY] A meristem that is forming between regions of permanent or mature meristem. { in'tər·kəˌler·ē 'mer·əˌstem }

intercalated disc [HISTOLOGY] A dense region at the junction of cellular units in cardiac muscle. { in'tər·kəˌlād·əd 'disk }

intercalated nucleus [ANATOMY] A nucleus of the medulla oblongata in the central gray matter of the ventricular floor located between the hypoglossal nucleus and the dorsal motor nucleus of the vagus. { in'tər·kəˌlād·əd 'nü·klē·əs }

intercalating agent [MOLECULAR BIOLOGY] A chemical substance that can insert itself between base pairs in a deoxyribonucleic acid molecule. { ˌin·tər'kaˌlād·iŋ ˌā·jənt }

intercapillary [ANATOMY] Located between capillaries. { ¦in·tər'kap·əˌler·ē }

intercarpal [ANATOMY] Located between the carpal bones. { ¦in·tər'kärp·əl }

intercavernous sinuses [ANATOMY] Venous sinuses located on the median line of the dura mater, connecting the cavernous sinuses of each side. { ¦in·tər'kav·ər·nəs 'sī·nə·səz }

intercellular [HISTOLOGY] Of or pertaining to the region between cells. { ¦in·tər'sel·yə·lər }

intercellular cement [HISTOLOGY] A substance bonding epithelial cells together. { ¦in·tər'sel·yə·lər si'ment }

intercellular junction [CELL BIOLOGY] Any specialized region of contact between the membranes of adjacent cells. { ¦in·tər'səl·yə·lər 'jəŋk·shən }

intercellular plexus [HISTOLOGY] A network of neuronal processes surrounding the cells in a sympathetic ganglion. { ¦in·tər'sel·yə·lər 'plek·səs }

intercellular space [HISTOLOGY] A space between adjacent cells. { ¦in·tər'sel·yə·lər ¦spās }

intercellular substance [HISTOLOGY] Tissue component that lies between cells. { ¦in·tər 'sel·yə·lər ¦səb·stəns }

intercentrum [VERTEBRATE ZOOLOGY] A type of crescentic intervertebral structure between successive centra in certain reptilian and mammalian tails. { ¦in·tər'sen·trəm }

interclavicle [VERTEBRATE ZOOLOGY] A membrane bone in front of the sternum and between the clavicles in monotremes and most reptiles. { ¦in·tər'klav·ə·kəl }

intercostal [ANATOMY] Situated or occurring between adjacent ribs. { ¦in·tər¦käst·əl }

intercostal muscles [ANATOMY] Voluntary muscles between adjacent ribs. { ¦in·tər¦käst·əl ¦məs·əlz }

intercostal nerve [ANATOMY] Any of the branches of the thoracic nerves in the intercostal spaces. { ¦in·tər¦käst·əl ¦nerv }

intercourse [ZOOLOGY] *See* coitus. { 'in·tər ˌkȯrs }

intercrescence [BIOLOGY] A growing together of tissues. { ˌin·tər'krēs·əns }

interfascicular cambium [BOTANY] The vascular cambium that develops between vascular bundles. { ¦in·tər·fa'sik·yə·lər 'kam·bē·əm }

interference phenomenon [VIROLOGY] Inhibition by a virus of the simultaneous infection of host cells by some other virus. { ˌin·tər'fir·əns fə 'näm·əˌnən }

interference range [GENETICS] The smallest genetic distance that is large enough for two crossing-over events not to interfere with each other. { ˌin·tər'fir·əns ˌrānj }

interferon [BIOCHEMISTRY] A protein produced by intact animal cells when infected with viruses; acts to inhibit viral reproduction and to induce resistance in host cells. { ˌin·tər'firˌän }

interferonogen [VIROLOGY] A preparation made of inactivated virus particles used as an inoculant to stimulate formation of interferon. { ¦in·tə·fə'rän·ə·jən }

interfoliaceous [BOTANY] Between a pair of leaves, such as between those which are opposite or verticillate. { ¦in·tərˌfō·lē'ā·shəs }

intergenic crossing-over [MOLECULAR BIOLOGY] Recombination between distinct genes or cistrons. { ¦in·tər¦jen·ik ¦krȯs·iŋ'ō·vər }

intergenic suppression [GENETICS] The restoration of a suppressed function or character by a second mutation that is located in a different gene than the original or first mutation. { ¦in·tər ¦jen·ik sə'presh·ən }

intergenote [MOLECULAR BIOLOGY] In hybrid bacteria, a chromosome with integrated deoxyribonucleic acid of foreign origin. { ˌin·tər'jēˌnōt }

interkinesis [CELL BIOLOGY] *See* interphase. { ¦in·tər·kə'nē·səs }

interlabium [INVERTEBRATE ZOOLOGY] A small lobe situated between the lips in certain nematodes. { ¦in·tər'lā·bē·əm }

interleukin [IMMUNOLOGY] Any of a class of proteins that are secreted mostly by macrophages and T lymphocytes and induce growth and differentiation of lymphocytes and hematopoietic stem cells. { ˌin·tər'lü·kən }

interleukin-1 [IMMUNOLOGY] A cytokine produced by macrophages, endothelial cells, lymphocytes, and epidermal cells that plays roles in the inflammatory process and in the immune response. { ˌin·tər¦lük·ən 'wən }

interleukin-2 [IMMUNOLOGY] A lymphokine secreted mostly by helper T lymphocytes that promotes the growth of T lymphocytes. { ˌin·tər ¦lük·ən 'tü }

intermediary metabolism [BIOCHEMISTRY] Intermediate steps in the chemical synthesis and breakdown of foodstuffs within body cells. { ¦in·tər¦mēd·ēˌer·ē me'tab·əˌliz·əm }

intermediate filament [CELL BIOLOGY] Any of several classes of cell-specific cytoplasmic filaments of 8-12 nanometers in diameter; protein composition varies from one cell type to another. { ˌint·ər'mēd·ē·ət 'fil·ə·mənt }

intermediate ganglion [ANATOMY] Any of certain small groups of nerve cells found along communicating branches of spinal nerves. { ˌin·tər 'mēd·ē·ət 'gaŋ·glē·ən }

intermediate host [BIOLOGY] The host in which a parasite multiplies asexually. { ˌin·tər'mēd·ē·ət 'hōst }

intermediate lobe [ANATOMY] The intermediate portion of the adenohypophysis. { ˌin·tər'mēd·ē·ət ¦lōb }

intermedin [BIOCHEMISTRY] A hormonal substance produced by the intermediate portion of the hypophysis of certain animal species which influences pigmentation; similar to melanocyte-stimulating hormone in humans. { ˌin·tər'mēd·ən }

intermembranous ossification [HISTOLOGY] Ossification within connective tissue with no prior formation of cartilage. { ¦in·tər'mem·brə·nəs ˌäs·ə·fə'kā·shən }

intermeningeal [ANATOMY] Between any two of the three meninges covering the brain and spinal cord. { ˌin·tər·men·ən 'jē·əl }

intermenstrual [PHYSIOLOGY] Between periods of menstruation. { ˌin·tər'men·strəl }

intermetameric [ANATOMY] Between adjacent metameres. { ¦in·tər¦med·ə¦mer·ik }

intermetatarsal [ANATOMY] Between adjacent bones of the metatarsus. { ˌin·tərˌmed·ə'tär·səl }

intermitotic [CELL BIOLOGY] Of or pertaining to a stage of the cell cycle between two successive mitoses. { ˌin·tərˌmī'täd·ik }

intermural [ANATOMY] Between the walls of an organ. { ¦in·tər'myu̇r·əl }

intermuscular [ANATOMY] Between muscles. { ¦in·tər'məs·kyə·lər }

intermuscular septum [ANATOMY] A connective-tissue partition between muscles. { ¦in·tər 'məs·kyə·lər 'sep·təm }

internal acoustic meatus [ANATOMY] An opening in the hard portion of the temporal bone for passage of the facial and acoustic nerves and internal auditory vessels. { in'tərn·əl ə'kü·stik mē 'ād·əs }

internal capsule [ANATOMY] A layer of nerve fibers on the outer side of the thalamus and caudate nucleus, which it separates from the lenticular nucleus. { in'tərn·əl 'kap·səl }

internal carotid [ANATOMY] A main division of the common carotid artery, distributing blood through three sets of branches to the cerebrum, eye, forehead, nose, internal ear, trigeminal nerve, dura mater, and hypophysis. Also known as internal carotid artery. { in'tərn·əl kə'räd·əd }

internal carotid artery [ANATOMY] *See* internal carotid. { in'tərn·əl kə'räd·əd 'ärd·ə·rē }

internal carotid nerve [ANATOMY] A sympathetic nerve which forms networks of branches around the internal carotid artery and its branches. { in'tərn·əl kə'räd·əd 'nərv }

internal ear [ANATOMY] *See* inner ear. { in'tərn·əl 'ir }

internal elastic membrane [HISTOLOGY] A sheet of elastin found between the tunica intima and the tunica media in medium- and small-caliber arteries. { in'tərn·əl i'las·tik 'mem,brān }

internal fertilization [PHYSIOLOGY] Fertilization of the egg within the body of the female. { in'tərn·əl ,fərd·əl·ə'zā·shən }

internal fistula [ANATOMY] A fistula which has no opening through the skin. { in'tərn·əl 'fis·chə·lə }

internal granular layer [HISTOLOGY] The fourth layer of the cerebral cortex. { in'tərn·əl 'gran·yə·lər ,lā·ər }

internal iliac artery [ANATOMY] The medial terminal division of the common iliac artery. { in'tərn·əl ¦il·ē,ak 'ärd·ə·rē }

internal respiration [PHYSIOLOGY] The gas exchange which occurs between the blood and tissues of an organism. { in'tərn·əl ,res·pə'rā·shən }

internal secretion [PHYSIOLOGY] A secreted substance that is absorbed directly into the blood. { in'tərn·əl si'krē·shən }

international unit [BIOLOGY] A quantity of a vitamin, hormone, antibiotic, or other biological that produces a specific internationally accepted biological effect. { ¦in·tər¦nash·ən·əl 'yü·nət }

internode [BIOLOGY] The interval between two nodes, as on a stem or along a nerve fiber. { 'in·tər,nōd }

internuncial neuron [HISTOLOGY] A neuron located in the spinal cord which connects motor and sensory neurons. { ¦in·tər¦nən·chəl 'nu̇r,än }

interoceptor [PHYSIOLOGY] A sense receptor located in visceral organs and yielding diffuse sensations. { ,in·tə·rō'sep·tər }

interocular distance [ANATOMY] The distance between the centers of rotation of the human eyes. { ¦in·tər'äk·yə·lər 'dis·təns }

interorbital [ANATOMY] Between the orbits of the eyes. { ¦in·tər'ȯr·bəd·əl }

interparietal [ANATOMY] Between the parietal bones. { ¦in·tər·pə'rī·əd·əl }

interphase [CELL BIOLOGY] Also known as interkinesis. **1.** The period between succeeding mitotic divisions. **2.** The period between the first and second meiotic divisions in those organisms where nuclei are reconstituted at the end of the first division. { 'in·tər,fāz }

interproglottid gland [INVERTEBRATE ZOOLOGY] Any of a number of cell clusters or glands arranged transversely along the posterior margin of the proglottids of certain tapeworms. { ¦in·tər,prō'gläd·əd ,gland }

interpterygoid [ZOOLOGY] A space between palatal plates in certain chordates. { ¦in·tər'ter·ə,gȯid }

interpulmonary [ANATOMY] Located between the lungs. { ¦in·tər'pu̇l·mə,ner·ē }

interradial canal [INVERTEBRATE ZOOLOGY] Any of the radially arranged gastrovascular canals in certain jellyfishes and ctenophores. { ¦in·tər'rād·ē·əl kə,nal }

interradius [INVERTEBRATE ZOOLOGY] The area between two adjacent arms in echinoderms. { ,in·tər'rād·ē·əs }

interray [INVERTEBRATE ZOOLOGY] A division of the radiate body of echinoderms. { ¦in·tər'rā }

interrenal [ANATOMY] Located between the kidneys. { ¦in·tər'rēn·əl }

interrupted gene [GENETICS] *See* split gene. { 'in·tə,rəp·təd 'jēn }

interrupted mating [GENETICS] A technique for mapping bacterial genes by determining the sequence of gene transfer between conjugating bacteria. { 'in·tə,rəp·təd 'mād·iŋ }

interscapular [ANATOMY] Between the shoulders or shoulder blades. { ¦in·tər'skap·yə·lər }

intersegmental [BIOLOGY] Situated between or involving segments. [EMBRYOLOGY] Situated between the primordial segments of the embryo. { ¦in·tər·seg'ment·əl }

intersegmental reflex [PHYSIOLOGY] An unconditioned reflex arc connecting input and output by means of afferent pathways in the dorsal spinal roots and efferent pathways in the ventral spinal roots. { ¦in·tər·seg'ment·əl 'rē,fleks }

intersex [PHYSIOLOGY] An individual who is intermediate in sexual constitution between male and female. { 'in·tər,seks }

interspace [ANATOMY] An interval between the ribs or the fibers or lobules of a tissue or organ. { 'in·tər,spās }

interspersion [ECOLOGY] **1.** An intermingling of different organisms within a community. **2.** The level or degree of intermingling of one kind of organism with others in the community. [MOLECULAR BIOLOGY] A regular pattern of alternating sequences of repetitious and nonrepetitious deoxyribonucleic acid in the genome of eukaryotes. { ,in·tər'spər·zhən }

interspinal [ANATOMY] Situated between or connecting spinous processes; interspinous. { ¦in·tər'spīn·əl }

intersternite [INVERTEBRATE ZOOLOGY] An intersegmental plate on the ventral surface of the abdomen in insects. { ¦in·tər'stər,nīt }

interstitial cell [HISTOLOGY] A cell that is not peculiar to or characteristic of a particular organ or tissue but which comprises fibrous tissue binding other cells and tissue elements; examples are neuroglial cells and Leydig cells. { ¦in·tər¦stish·əl ¦sel }

interstitial-cell-stimulating hormone [BIOCHEMISTRY] *See* luteinizing hormone. { ¦in·tər¦stish·əl ¦sel 'stim·yə,lād·iŋ ,hȯr,mōn }

interstitial gland [HISTOLOGY] **1.** Groups of Leydig cells which secrete angiogens. **2.** Groups of epithelioid cells in the ovarian medulla of some lower animals. { ¦in·tər¦stish·əl ¦gland }

interstitial lamella [HISTOLOGY] Any of the layers of bone between adjacent Haversian systems. { ¦in·tər¦stish·əl lə'mel·ə }

intertergite [INVERTEBRATE ZOOLOGY] One of the small plates between the tergites of certain insects. { ¦in·tər'tər,jīt }

intertubercular sulcus [ANATOMY] A deep groove on the anterior surface of the upper end of the humerus, separating the greater and lesser tubercles; contains the tendon of the long head of the biceps brachii muscle. Also known as bicipital groove. { ¦in·tər·tə'bər·kyə·lər 'səl·kəs }

intervallum [INVERTEBRATE ZOOLOGY] The space between the walls of pleosponges. { ¦in·tər¦val·əm }

intervascular [ANATOMY] Located between or surrounded by blood vessels. { ¦in·tər'vas·kyə·lər }

intervening sequence [GENETICS] A region of a gene that is transcribed but is not included in the final transcript of a ribonucleic acid. { ¦in·tər ¦vēn·iŋ 'sē·kwəns }

interventricular foramen [ANATOMY] Either one of the two foramens that connect the third ventricle of the brain with each lateral ventricle. Also known as foramen of Monro. { ¦in·tər·ven'trik·yə·lər fə'rā·mən }

interventricular septum [ANATOMY] The muscular wall between the heart ventricles. Also known as ventricular septum. { ¦in·tər·ven'trik·yə·lər 'sep·təm }

intervertebral [ANATOMY] Being or located between the vertebrae. { ¦in·tər'vərd·ə·bəl }

intervillous spaces [HISTOLOGY] Spaces in the placenta which communicate with the maternal blood vessels. { ¦in·tər'vil·əs 'spās·əz }

intestinal crura [INVERTEBRATE ZOOLOGY] The main intestinal branches in certain trematodes. { in'tes·tən·əl 'krůr·ə }

intestinal digestion [PHYSIOLOGY] Conversion of food to an assimilable form by the action of intestinal juices. { in'tes·tən·əl di'jes·chən }

intestinal hormone [BIOCHEMISTRY] Either of two hormones, secretin and cholecystokinin, secreted by the intestine. { in'tes·tən·əl 'hȯr ,mōn }

intestinal juice [PHYSIOLOGY] An alkaline fluid composed of the combined secretions of all intestinal glands. { in'tes·tən·əl ¦jüs }

intestinal villi [ANATOMY] Fingerlike projections of the small intestine, composed of a core of vascular tissue covered by epithelium and containing smooth muscle cells and an efferent lacteal end capillary. { in'tes·tən·əl 'vil,ī }

intestine [ANATOMY] The tubular portion of the vertebrate digestive tract, usually between the stomach and the cloaca or anus. { in'tes·tən }

intima [HISTOLOGY] The innermost coat of a blood vessel. Also known as tunica intima. { 'in·tə·mə }

intorsion [BIOLOGY] Inward rotation of a structure about a fixed point or axis. { in'tȯr·shən }

intraabdominal [ANATOMY] Occurring or being within the cavity of the abdomen. { ¦in·trə·ab 'däm·ən·əl }

intracartilaginous ossification [PHYSIOLOGY] *See* endochondral ossification. { ¦in·trə,kärd·əl'aj·ə·nəs ,äs·ə·fə'kā·shən }

intracellular [CELL BIOLOGY] Within a cell. { ¦in·trə'sel·yə·lər }

intracellular canaliculi [CELL BIOLOGY] A system of minute canals within certain gland cells which are thought to drain the glandular secretions. { ¦in·trə'sel·yə·lər ,kan·əl'ik·yə,lī }

intracellular digestion [PHYSIOLOGY] Digestion which takes place within the cytoplasm of the organism, as in many unicellular protozoans. { ¦in·trə'sel·yə·lər di'jes·chən }

intracellular enzyme [BIOCHEMISTRY] An enzyme that remains active only within the cell in which it is formed. Also known as organized ferment. { ¦in·trə'sel·yə·lər 'en,zīm }

intracellular symbiosis [CELL BIOLOGY] Existence of a self-duplicating unit within the cytoplasm of a cell, such as a kappa particle in *Paramecium*, which seems to be an infectious agent and may influence cell metabolism. { ¦in·trə 'sel·yə·lər ,sim·bē'ō·səs }

intracervical [ANATOMY] Located within the cervix of the uterus. { ¦in·trə'ser·və·kəl }

intracistron complementation [GENETICS] The process whereby two different mutant alleles, each of which determines in homozygotes an inactive enzyme, determine the formation of the active enzyme when present in the same nucleus. { ¦in·trə'sis·trən ,käm·plə·mən'tā·shən }

intracodon recombination [GENETICS] Genetic recombination between neighboring nucleotides within a codon. { ,in·trə'kō,dän rē,käm·bə¦nā·shən }

intracortical [ANATOMY] Occurring or located within the cortex. { ¦in·trə'kȯrd·ə·kəl }

intracranial [ANATOMY] Within the cranium. { ¦in·trə'krā·nē·əl }

intracytoplasmic [CELL BIOLOGY] Being or occurring within the cytoplasm of a cell. { ¦in·trə ,sīd·ə'plaz·mik }

intradermal [ANATOMY] Within the skin. { ¦in·trə'dər·məl }

intraductal [ANATOMY] Within a duct. { ¦in·trə 'dək·təl }

intradural [ANATOMY] Within the dura mater. { ¦in·trə'důr·əl }

intraembryonic [EMBRYOLOGY] Within the embryo. { ¦in·trə,em·brē'än·ik }

intraepidermal [ANATOMY] Within the epidermis. { ¦in·trə,ep·ə'dər·məl }

intraepithelial [ANATOMY] Within the epithelium. { ¦in·trə,ep·ə'thē·lē·əl }

intraesophageal [ANATOMY] Within the esophagus. { ¦in·trə·ə,säf·ə'jē·əl }

intrafascicular [BOTANY] Located or occurring within a vascular bundle. { ¦in·trə·fə'sik·yə·lər }

intrafusal fiber [HISTOLOGY] Any of the striated muscle fibers contained in a muscle spindle. { ¦in·trə'fyüz·əl 'fī·bər }

intragenic [MOLECULAR BIOLOGY] Within a gene, in referring to certain events. { ¦in·trə¦jen·ik }

intragenic recombination [MOLECULAR BIOLOGY] Recombination occurring between the mutons of one cistron. { ¦in·trə¦jen·ik rē,käm·bə'nā·shən }

intragenic suppression [GENETICS] The restoration of a suppressed function or character as a consequence of a second mutation located within the same gene as the original or first mutation. { ¦in·trə¦jen·ik sə'presh·ən }

intrahepatic [ANATOMY] Within the liver. { ¦in·trə·he'pad·ik }

intrajugular process [ANATOMY] **1.** A small, curved process on some occipital bones which partially or completely divides the jugular foramen into lateral and medial parts. **2.** A small process on the hard portion of the temporal bone which completely or partly separates the jugular foramen into medial and lateral parts. { ¦in·trə 'jəg·yə·lər ,prä·səs }

intralaminar nuclei [ANATOMY] A diffuse group of nuclei located in the internal medullary lamina of the thalamus. { ,in·trə'la·mən·ər 'nü·klē,ī }

intraluminal [ANATOMY] Within the lumen of a structure. { ,in·trə'lü·mən·əl }

intramarginal [BIOLOGY] Within a margin. { ,in·trə'mär· jən·əl }

intramedullary [ANATOMY] **1.** Within the tissues of the spinal cord or medulla oblongata. **2.** Within the bone marrow. **3.** Within the adrenal medulla. { ,in·trə·mə'dəl·ə·rē }

intramembranous [HISTOLOGY] Formed or occurring within a membrane. { ,in·trə'mem·brə·nəs }

intramembranous ossification [HISTOLOGY] Formation of bone tissue directly from connective tissue without a preliminary cartilage stage. { ,in·trə'mem·brə·nəs ,äs·ə·fə'kā·shən }

intramural [ANATOMY] Within the substance of the walls of an organ. { ¦in·trə'myůr·əl }

intramuscular [ANATOMY] Lying within or going into the substance of a muscle. { ¦in·trə'məs·kyə·lər }

intraocular [ANATOMY] Occurring within the globe of the eye. { ,in·trə'äk·yə·lər }

intraocular pressure [PHYSIOLOGY] The hydrostatic pressure within the eyeball. { ,in·trə'äk·yə·lər 'presh·ər }

intraparietal [ANATOMY] **1.** Within the wall of an organ or cavity. **2.** Within the parietal region of the cerebrum. **3.** Within the body wall. { ,in·trə·pə'rī·əd·əl }

intraperitoneal [ANATOMY] **1.** Within the peritoneum. **2.** Within the peritoneal cavity. { ¦in·trə ,per·ə·tə'nē·əl }

intrapetiolar [BOTANY] **1.** Enclosed by the base of the petiole. **2.** Between the petiole and the stem. { ¦in·trə¦ped·ē¦äl·yə·lər }

intrapulmonic [ANATOMY] Being or occurring within the lungs. { ¦in·trə·půl'män·ik }

intraspecific [BIOLOGY] Being within or occurring among the members of the same species. { ¦in·trə·spə·'sif·ik }

intrathecal [ANATOMY] Within the subarachnoid space. { ¦in·trə'thē·kəl }

intrathoracic [ANATOMY] Within the thoracic cavity. { ¦in·trə·thə'ras·ik }

intratracheal [ANATOMY] Being or occurring within the trachea. { ¦in·trə'trā·kē·əl }

intrauterine [ANATOMY] Being or occurring within the uterus. { ¦in·trə'yüd·ə·rən }

intravaginal [ANATOMY] **1.** Being or occurring within the vagina. **2.** Located within a tendon sheath. [BOTANY] Located within a sheath, referring to branches of grass. { ¦in·trə'vaj·ə·nəl }

intravenous [ANATOMY] Located within, or going into, the veins. { ,in·trə'vē·nəs }

intravesical [ANATOMY] Within the urinary bladder. { ,in·trə'ves·ə·kəl }

intravital [BIOLOGY] Occurring while the cell or organism is alive. { ,in·trə'vīd·əl }

intravital stain [BIOLOGY] A nontoxic dye injected into the body to selectively stain certain cells or tissues. { ,in·trə'vīd·əl 'stān }

intrinsic factor [BIOCHEMISTRY] A substance, produced by the stomach, which combines with the extrinsic factor (vitamin B_{12}) in food to yield an antianemic principle; lack of the intrinsic factor is believed to be a cause of pernicious anemia. Also known as Castle's intrinsic factor. { in 'trin·sik ¦fak·tər }

intrinsic nerve supply [ANATOMY] The nerves contained entirely within an organ or structure. { in'trin·sik 'nərv sə,plī }

introgressive hybridization [GENETICS] The spreading of genes of a species into the gene complex of another due to hybridization between numerically dissimilar populations, the extensive backcrossing preventing formation of a single stable population. { ¦in·trə¦gres·iv ,hī·brəd·ə'zā·shən }

introitus [ANATOMY] An opening or entryway, especially the opening into the vagina. { in'trō·ə·dəs }

intromission [ZOOLOGY] The act or process of inserting one body into another, specifically, of the penis into the vagina. { ,in·trə'mish·ən }

intromittent [ZOOLOGY] Adapted for intromission; applied to a copulatory organ. { ¦in·trə ¦mit·ənt }

intron [GENETICS] A noncoding segment of a gene. { 'in,trän }

introrse [BIOLOGY] Turned inward or toward the axis. { 'in,trȯrs }

introspective diplopia [PHYSIOLOGY] *See* physiologic diplopia. { ¦in·trə¦spek·tiv də'plō·pē·ə }

introvert [ZOOLOGY] **1.** A structure capable of introversion. **2.** To turn inward. { 'in·trə,vərt }

inulase [BIOCHEMISTRY] An enzyme produced by certain molds that catalyzes the conversion of inulin to levulose. { 'in·yə,lās }

inulin [BIOCHEMISTRY] A polysaccharide made up of polymerized fructofuranose units. { 'in·yə·lən }

in utero [EMBRYOLOGY] Within the uterus, referring to the fetus. { in 'yüd·ə,rō }

invagination [EMBRYOLOGY] The enfolding of a part of the wall of the blastula to form a gastrula. [PHYSIOLOGY] **1.** The act of ensheathing or becoming ensheathed. **2.** The process of burrowing or enfolding to form a hollow space within a previously solid structure, as the invagination of the nasal mucosa within a bone of the skull to form a paranasal sinus. { in,vaj·ə'nā·shən }

inversion [GENETICS] A type of chromosomal rearrangement in which two breaks take place in a chromosome and the fragment between breaks rotates 180° before rejoining. { in'vər·zhən }

inversion heterozygote [GENETICS] An organism in which one member of a pair of homologous chromosomes has an inverted gene sequence and the other has the normal gene sequence. { in 'vər·zhən ,hed·ə·rə'zī,gōt }

invertase [BIOCHEMISTRY] *See* saccharase. { 'in·vər,tās }

Invertebrata [INVERTEBRATE ZOOLOGY] A division of the animal kingdom including all animals which lack a spinal column; has no taxonomic status. { in,vərd·ə'bräd·ə }

invertebrate [INVERTEBRATE ZOOLOGY] An animal lacking a backbone and internal skeleton. { in 'vərd·ə,brət }

invertebrate pathology [INVERTEBRATE ZOOLOGY] All studies having to do with the principles of pathology as applied to invertebrates. { in'vərd·ə,brət pə'thäl·ə·jē }

invertebrate zoology [ZOOLOGY] A branch of biology that deals with the study of Invertebrata. { in'vərd·ə,brət zō'äl·ə·jē }

inverted repeats [GENETICS] Two copies of the same nucleotide sequence oriented in opposite directions on the same molecule. Also known as IR sequences. { in'vərd·əd ri'pēts }

inverted terminal repeats [MOLECULAR BIOLOGY] Related or identical sequences of deoxyribonucleic acid in inverted form occurring at opposite ends of some transposons. { in¦vərd·əd ¦tər·mə·nəl ri'pēts }

invertin [BIOCHEMISTRY] *See* saccharase. { in 'vərt·ən }

investing bone [ANATOMY] *See* dermal bone. { in 'vest·iŋ ,bōn }

in vitro [BIOLOGY] Pertaining to a biological reaction taking place in an artificial apparatus. { in 'vē·trō }

in vivo [BIOLOGY] Pertaining to a biological reaction taking place in a living cell or organism. { in 'vē·vō }

involucrate [BOTANY] Having an involucre. { ¦in·və¦lü·krət }

involucre [BOTANY] Bracts forming one or more whorls at the base of an inflorescence or fruit in certain plants. { 'in·və,lü·kər }

involucrum [ANATOMY] **1.** The covering of a part. **2.** New bone laid down by periosteum around a sequestrum in osteomyelitis. { ,in·və'lü·krəm }

involuntary fiber [HISTOLOGY] *See* smooth muscle fiber. { in'väl·ən,ter·ē 'fī·bər }

involuntary muscle [PHYSIOLOGY] Muscle not under the control of the will; usually consists of smooth muscle tissue and lies within a vescus, or is associated with skin. { in'väl·ən,ter·ē 'məs·əl }

involute [BIOLOGY] Being coiled, curled, or rolled in at the edge. { ¦in·və¦lüt }

involution [BIOLOGY] A turning or rolling in. [EMBRYOLOGY] Gastrulation by ingrowth of blastomeres around the dorsal lip. { ,in·və'lü·shən }

involution form [CELL BIOLOGY] A cell with a bizarre configuration caused by abnormal culture conditions. { ,in·və'lü·shən ,fórm }

involvucel [BOTANY] A secondary involucre. { in 'väl·və·səl }

involvucellate [BOTANY] Possessing an involvucel. { in¦väl·və¦sel,āt }

iodopsin [BIOCHEMISTRY] The visual pigment found in the retinal cones, consisting of retinene, combined with photopsin. { ,ī·ə'däp·sən }

ionophore [BIOCHEMISTRY] Any of a class of compounds, generally cyclic, having the ability to carry ions across lipid barriers due to the property of cation selectivity; examples are valinomycin and nonactin. { ī'än·ə,fór }

Iospilidae [INVERTEBRATE ZOOLOGY] A small family of pelagic polychaetes assigned to the Errantia. { ,ī·ə'spil·ə,dē }

ipecac [BOTANY] Any of several low, perennial, tropical South American shrubs or half shrubs in the genus *Cephaelis* of the family Rubiaceae; the dried rhizome and root, containing emetine, cephaeline, and other alkaloids, is used as an emetic and expectorant. { 'ip·ə,kak }

Ipidae [INVERTEBRATE ZOOLOGY] The equivalent name for Scolytidae. { 'ip·ə,dē }

IR gene [IMMUNOLOGY] *See* immune response gene. { ¦ī'är ,jēn }

Iridaceae [BOTANY] A family of monocotyledonous herbs in the order Liliales distinguished by three stamens and an inferior ovary. { ,ir·ə'dās·ē,ē }

iridocyte [HISTOLOGY] A specialized cell in the integument of certain animal species which is filled with iridescent crystals of guanine and a variety of lipophores. Also known as guanophore; iridophore. { i'rid·ə,sīt }

iridophore [HISTOLOGY] *See* iridocyte. { i'rid·ə ,fór }

iris [ANATOMY] A pigmented diaphragm perforated centrally by an adjustable pupil which regulates the amount of light reaching the retina in vertebrate eyes. [BOTANY] Any plant of the genus *Iris*, the type genus of the family Iridaceae, characterized by linear or sword-shaped leaves, erect stalks, and bright-colored flowers with the three inner perianth segments erect and the outer perianth segments drooping. { 'ī·rəs }

Irish moss [BOTANY] *See* carrageen. { 'ī,rish 'mós }

Irish potato [BOTANY] *See* potato. { 'ī·rish pə 'tād·ō }

iron bacteria [MICROBIOLOGY] The common name for bacteria capable of oxidizing ferrous iron to the ferric state. { 'ī·ərn bak'tir·ē·ə }

iron-binding protein [BIOCHEMISTRY] A serum

protein, such as hemoglobin, for the transport of iron ions. { 'ī·ərn ˌbīnd iŋ 'prōˌtēn }

iron metabolism [BIOCHEMISTRY] The chemical and physiological processes involved in absorption of iron from the intestine and in its role in erythrocytes. { 'ī·ərn mə'tab·əˌliz·əm }

Ironoidea [INVERTEBRATE ZOOLOGY] A superfamily of presumably carnivorous, fresh-water and soil-dwelling nematodes in the order Enoplida, having two circlets of cephalic sensory organs, a cylindrical, elongate stoma armed either anteriorly with three large teeth or posteriorly with small teeth, and cuticularized stomatal walls. { ˌi·rə'nȯid·ē·ə }

iron-porphyrin protein [BIOCHEMISTRY] Any protein containing iron and porphyrin; examples are hemoglobin, the cytochromes, and catalase. { 'ī·ərn'fȯr·fə·rən 'prōˌtēn }

ironwood [BOTANY] Any of a number of hardwood trees in the United States, including the American hornbeam, the buckwheat, and the eastern hophornbeam. { 'ī·ərn ˌwu̇d }

irradiation [BIOPHYSICS] Subjection of a biological system to sound waves of sufficient intensity to modify their structure or function. { iˌrād·ē 'ā·shən }

irregular [BOTANY] Lacking symmetry, as of a flower having petals unlike in size or shape. { i'reg·yə·lər }

irregular cleavage [EMBRYOLOGY] Division of a zygote into random masses of cells, as in certain cnidarians. { i'reg·yə·lər 'klē·vij }

irregular connective tissue [HISTOLOGY] A loose or dense connective tissue with fibers irregularly woven and irregularly distributed; collagen is the dominant fiber type. { i'reg·yə·lər kə'nek·tiv ˌtish·ü }

Irregularia [INVERTEBRATE ZOOLOGY] An artificial assemblage of echinoderms in which the anus and periproct lie outside the apical system, the ambulacral plates remain simple, the primary radioles are hollow, and the rigid test shows some degree of bilateral symmetry. { iˌreg·yə'lar·ē·ə }

irritability [PHYSIOLOGY] **1.** A condition or quality of being excitable; the power of responding to a stimulus. **2.** A condition of abnormal excitability of an organism, organ, or part, when it reacts excessively to a slight stimulation. { ˌir·əd·ə 'bil·əd·ē }

IR sequences [GENETICS] *See* inverted repeats. { ¦ī'är ˌsē·kwən·səs }

isanthous [BOTANY] Having regular flowers. { ī'san·thəs }

ischiopodite [INVERTEBRATE ZOOLOGY] The segment nearest the basipodite of walking legs in certain crustaceans. Also known as ischium. { ˌis·kē'äp·əˌdīt }

ischiorectal region [ANATOMY] The region between the ischium and the rectum. { ¦is·kē·ə 'rek·təl ˌrē·jən }

ischium [ANATOMY] Either of a pair of bones forming the dorsoposterior portion of the vertebrate pelvis; the inferior part of the human pelvis upon which the body rests in sitting. [INVERTEBRATE ZOOLOGY] *See* ischiopodite. { 'is·kē·əm }

ischium-pubis index [ANATOMY] The ratio (length of pubis × 100/length of ischium) by which the sex of an adult pelvis may usually be determined; the index is greater than 90 in females, and less than 90 in males. { 'is·kē·əm 'pyü·bəs ˌinˌdeks }

Ischnacanthidae [PALEONTOLOGY] The single family of the acanthodian order Ischnacanthiformes. { ˌisk·nə'kan·thəˌdē }

Ischnacanthiformes [PALEONTOLOGY] A monofamilial order of extinct fishes of the order Acanthodii; members were slender, lightly armored predators with sharp teeth, deeply inserted fin spines, and two dorsal fins. { ˌisk·nəˌkan·thə'fȯr ˌmēz }

Isectolophidae [PALEONTOLOGY] A family of extinct ceratomorph mammals in the superfamily Tapiroidea. { īˌsek·tə'läf·əˌdē }

island of Langerhans [HISTOLOGY] *See* islet of Langerhans. { 'ī·lənd əv 'läŋ·gərˌhänz }

island of Reil [ANATOMY] The insula of the cerebral hemisphere. { 'ī·lənd əv 'rīl }

islet of Langerhans [HISTOLOGY] A mass of cell cords in the pancreas that is of an endocrine nature, secreting insulin and a minor hormone like lipocaic. Also known as island of Langerhans; islet of the pancreas. { 'ī·lət əv 'läŋ·gər ˌhänz }

islet of the pancreas [HISTOLOGY] *See* islet of Langerhans. { 'ī·lət əv thə 'pan·krē·əs }

isoacceptor [MOLECULAR BIOLOGY] Any of several species of transfer ribonucleic acid that can accept the same amino acid. { ¦ī·sō·ak'sep·tər }

isoagglutinin [IMMUNOLOGY] An agglutinin which acts upon the red blood cells of members of the same species. Also known as isohemagglutinin. { ¦ī·sō·ə'glut·ən·ən }

isoallele [GENETICS] An allele that carries mutational alterations at the same site. { ¦ī·sō·ə 'lēl }

isoalloxazine mononucleotide [BIOCHEMISTRY] *See* riboflavin 5′-phosphate. { ˌī·sō·ə'läk·səˌzēn ˌmänō'nü·klē·əˌtīd }

isoantibody [IMMUNOLOGY] An antibody formed in response to immunization with tissue constituents derived from an individual of the same species. { ¦īˌsō'ant·iˌbäd·ē }

isoantigen [IMMUNOLOGY] An antigen in an individual capable of stimulating production of a specific antibody in another member of the same species. Also known as alloantigen. { ¦ī·sō'ant·ə·jən }

isobiochore [ECOLOGY] A boundary line on a map connecting world environments that have similar floral and faunal constituents. { ˌī·sə'bī·əˌkȯr }

Isobryales [BOTANY] An order of mosses in which the plants are slender to robust and up to 36 inches (90 centimeters) in length. { ¦ī·sō·brī'ā·lēz }

isocarpic [BOTANY] Having the same number of carpels and perianth divisions. { ˌī·sə'kär·pik }

isocercal [VERTEBRATE ZOOLOGY] Of the tail fin

of a fish, having the upper and lower lobes symmetrical and the vertebral column gradually tapering. { ¦ī·sə¦sər·kəl }

isochela [INVERTEBRATE ZOOLOGY] **1.** A chela having two equally developed parts. **2.** A chelate spicule with both ends identical. { ¦ī·sə'kē·lə }

isochromosome [CELL BIOLOGY] An abnormal chromosome with a medial centromere and identical arms formed as a result of transverse, rather than longitudinal, splitting of the centromere. { ¦ī·sō'krō·mə,sōm }

isocitric acid [BIOCHEMISTRY] $HOOCCH_2CH(COOH)CH(OH)COOH$ An isomer of citric acid that is involved in the Krebs tricarboxylic acid cycle in bacteria and plants. { ¦ī·sə¦si·trik 'as·əd }

isocoding mutation [GENETICS] A point mutation that changes a codon's nucleotide sequence but does not change the amino acid specified by the codon. { ¦ī·sə,kōd·iŋ myü'tā·shən }

Isocrinida [INVERTEBRATE ZOOLOGY] An order of stalked articulate echinoderms with nodal rings of cirri. { ¦i·sō'krī·nə·də }

isodiametric [BIOLOGY] Having equal diameters or dimensions. { ¦ī·sō,dī·ə'me·trik }

isodont [VERTEBRATE ZOOLOGY] **1.** Having all teeth alike. **2.** Of a snake, having the maxillary teeth of equal length. { 'ī·sə,dänt }

isodulcitol [BIOCHEMISTRY] *See* rhamnose. { ,ī·sə 'dəl·sə,tòl }

isoenzyme [BIOCHEMISTRY] Any of the electrophoretically distinct forms of an enzyme, representing different polymeric states but having the same function. Also known as isozyme. { ¦ī·sō 'en,zīm }

Isoetaceae [BOTANY] The single family assigned to the order Isoetales in some systems of classification. { ,ī·sō·ə'tās·ē,ē }

Isoetales [BOTANY] A monotypic order of the class Isoetopsida containing the single genus *Isoetes*, characterized by long, narrow leaves with a spoonlike base, spirally arranged on an underground cormlike structure. { ,ī·sō·ə'tā·lēz }

Isoetopsida [BOTANY] A class of the division Lycopodiophyta; members are heterosporous and have a distinctive appendage, the ligule, on the upper side of the leaf near the base. { ,ī·sō·ə 'täp·sə·də }

isoflavone [BIOCHEMISTRY] $C_{15}H_{10}O_2$ A colorless, crystalline ketone, occurring in many plants, generally in the form of a hydroxy derivative. { ¦ī·sō'fla,vōn }

isogamete [BIOLOGY] A reproductive cell that is morphologically similar in both male and female and cannot be distinguished on form alone. { ¦ī·sō'ga,mēt }

isogamy [BIOLOGY] Sexual reproduction by union of gametes or individuals of similar form or structure. { ī'säg·ə·mē }

isogeneic [GENETICS] Having the same origin and thus being genetically identical, as an inbred strain. Also known as isogenic; syngeneic. [IMMUNOLOGY] Referring to cells, tissues, or organs used in transplantation that originate in identical species. { ,ī·sə·jə'nē·ik }

Isogeneratae [BOTANY] A class of brown algae distinguished by having an isomorphic alternation of generations. { ,ī·sō,jen·ə'rä,tē }

isogony [BIOLOGY] Growth of parts at such a rate as to maintain relative size differences. { 'ī'säg·ə·nē }

isograft [BIOLOGY] A tissue transplant from one organism to another organism which is genetically identical. { 'ī·sə,graft }

isohemagglutinin [IMMUNOLOGY] *See* isoagglutinin. { ¦ī·sō,hē·mə'glüt·ən·ən }

isohemolysin [IMMUNOLOGY] A hemolysin produced by an individual injected with erythrocytes from another individual of the same species. { ¦ī·sō·hə'mäl·ə·sən }

isohemolysis [IMMUNOLOGY] Hemolysis induced by the action of an isohemolysin. { ¦ī·sō·hə 'mäl·ə·səs }

isoimmunization [IMMUNOLOGY] Immunization of an individual by the introduction of antigens from another individual of the same species. { ¦ī·sō,im·yə·nə'zā·shən }

isoinertial [BIOPHYSICS] Pertaining to the force of a human muscle that is applied to a constant mass in motion. { ¦ī·sō·i'nərsh·əl }

isokinetic [BIOPHYSICS] Pertaining to the force of a human muscle that is applied during constant velocity of motion. { ,i·sə·ki'ned·ik }

Isolaimoidea [INVERTEBRATE ZOOLOGY] A superfamily of rather large, free-living soil nematodes in the order Isolaimida, characterized by six hollow tubes around the oral opening, two circlets of six circumoral sensilla, the absence of amphids, and an elongated triradiate stoma with thickened anterior walls. { ,ī·sō·lə'mȯid·ē·ə }

isolate [GENETICS] A segment of a population within which assortative mating occurs. { 'ī·sə ,lāt }

isolating mechanism [GENETICS] A geographic barrier or biological difference that prevents mating or genetic exchange between individuals of different populations or species. { 'ī·sə,lād·iŋ mek·ə,niz·əm }

isolation [EVOLUTION] The restriction or limitation of gene flow between distinct populations due to barriers to interbreeding. [MICROBIOLOGY] Separation of an individual or strain from a natural, mixed population. [PHYSIOLOGY] Separation of a tissue, organ, system, or other part of the body for purposes of study. { ,ī·sə'lā·shən }

isolation gene [GENETICS] Any gene that causes reduction in viability or fertility when it is present in the heterozygous form. { ,ī·sə,lā·shən ,jēn }

isolecithal [CELL BIOLOGY] *See* homolecithal. { ¦ī·sō'les·ə·thəl }

isoleucine [BIOCHEMISTRY] $C_6H_{13}O_2$ An essential monocarboxylic amino acid occurring in most dietary proteins. { ¦ī·sō'lü,sēn }

isomerase [BIOCHEMISTRY] An enzyme that catalyzes isomerization reactions. { ī'säm·ə,rās }

isomerism [BIOLOGY] The condition of having two or more comparable parts made up of identical numbers of similar segments. { ī'säm·ə ,riz·əm }

isomerous [BIOLOGY] Characterized by isomerism. { ī'säm·ə·rəs }

Isometopidae [INVERTEBRATE ZOOLOGY] A family of hemipteran insects in the superfamily Cimicimorpha. { ¦ī·sō·mə'täp·ə,dē }

isometric contraction [PHYSIOLOGY] A contraction in which muscle tension is increased, but the muscle is not shortened because the resistance cannot be overcome. Also known as static contraction. { ¦ī·sə'me·trik kən'trak·shən }

isometric particle [VIROLOGY] A plant virus particle that appears at first sight to be spherical when viewed in the electron microscope, but which is actually an icosahedron, possessing 20 sides. { ¦ī·sə'me·trik 'pärd·ə·kəl }

isometric work [BIOPHYSICS] Physiologic work that is performed by muscles in terms of energy utilization and heat production and involves muscular contraction that is not accompanied by movement. Also known as static work. { ¦ī·sə 'me·trik 'wərk }

isotonic sodium chloride solution [PHYSIOLOGY] *See* normal saline. { ¦ī·sə¦tän·ik ¦sōd·ē·əm 'klȯr ,īd sə,lü·shən }

isonymous substitution [GENETICS] Any deoxyribonucleic acid base-pair substitution that does not result in change in the amino acid for which it codes. { ī¦sän·ə·məs ,səb·stə'tü·shən }

isophene [BIOLOGY] A line on a chart connecting those places within a given region where a particular biological phenomenon (as the flowering of a certain plant) occurs at the same time. { 'ī·sə,fēn }

isophyllous [BOTANY] Having foliage leaves of similar form on a plant or stem. { ¦ī·sə¦fil·əs }

Isopoda [INVERTEBRATE ZOOLOGY] An order of malacostracan crustaceans characterized by a cephalon bearing one pair of maxillipeds in addition to the antennae, mandibles, and maxillae. { ī'säp·ə·də }

isoprecipitin [IMMUNOLOGY] A precipitin effective only against the serum of individuals of the same species from which it is derived. { ¦ī·sō·prə'sip·ə·tən }

Isoptera [INVERTEBRATE ZOOLOGY] An order of Insecta containing morphologically primitive forms characterized by gradual metamorphosis, lack of true larval and pupal stages, biting and prognathous mouthparts, two pairs of subequal wings, and the abdomen joined broadly to the thorax. { ī'säp·tə·rə }

isopygous [INVERTEBRATE ZOOLOGY] Having a pygidium and a cephalon of equal size, as in certain trilobites. { ¦ī'säp·ə·gəs }

isoschizomer [BIOCHEMISTRY] One of two or more restriction endonucleases that cleave a deoxyribonucleic acid molecule at the same site. { ,ī·sə'siz·ə·mər }

Isospondyli [VERTEBRATE ZOOLOGY] A former equivalent name for Clupeiformes. { ,ī·sə'spän·də,lī }

isospore [BIOLOGY] A spore that does not display sexual dimorphism. { 'ī·sə,spȯr }

isostemonous [BOTANY] Having the number of stamens of a flower equal to the number of perianth divisions. { ¦ī·sə¦stē·mə·nəs }

isotonic [PHYSIOLOGY] **1.** Having uniform tension, as the fibers of a contracted muscle. **2.** Of a solution, having the same osmotic pressure as the fluid phase of a cell or tissue. { ¦ī·sə¦tän·ik }

isotope farm [BOTANY] A carbon-14 (^{14}C) growth chamber, or greenhouse, arranged as a closed system in which plants can be grown in an atmosphere of carbon dioxide (CO_2) containing ^{14}C and thus become labeled with ^{14}C; isotope farms also can be used with other materials, such as heavy water (D_2O), phosphorus-35 (^{35}P), and so forth, to produce biochemically labeled compounds. { 'ī·sə,tōp ,färm }

isotropic [BIOLOGY] Having a tendency for equal growth in all directions. [CELL BIOLOGY] An ovum lacking any predetermined axis. { ¦ī·sə ¦trä·pik }

isotypes [GENETICS] A series of antigens, for example, blood types, common to all members of a species but differentiating classes and subclasses within the species. { 'ī·sə,tīps }

isozyme [BIOCHEMISTRY] *See* isoenzyme. { 'ī·sə ,zīm }

isthmus [BIOLOGY] A passage or constricted part connecting two parts of an organ. { 'is·məs }

Istiophoridae [VERTEBRATE ZOOLOGY] The billfishes, a family of oceanic perciform fishes in the suborder Scombroidei. { ,is·tē·ə'fȯr·ə,dē }

Isuridae [VERTEBRATE ZOOLOGY] The mackerel sharks, a family of pelagic, predacious galeoids distinguished by a heavy body, nearly symmetrical tail, and sharp, awllike teeth. { I'sůr·ə,dē }

itch [PHYSIOLOGY] An irritating cutaneous sensation allied to pain. { ich }

iteroparity [BIOLOGY] Reproduction that occurs repeatedly over the life of the individual. { ,īd·ə·rə'par·əd·ē }

Ithomiinae [INVERTEBRATE ZOOLOGY] The glossywings, a subfamily of weak-flying lepidopteran insects having on the wings broad, transparent areas in which the scales are reduced to short hairs. { ,ith·ə'mī·ə,nē }

Itonididae [INVERTEBRATE ZOOLOGY] The gall midges, a family of orthorrhaphous dipteran insects in the series Nematocera; most are plant pests. { ,id·ə'nid·ə,dē }

Ixodides [INVERTEBRATE ZOOLOGY] The ticks, a suborder of the Acarina distinguished by spiracles behind the third or fourth pair of legs. { ,ik 'säd·ə,dēz }

J

Jacanidae [VERTEBRATE ZOOLOGY] The jacanas or lily-trotters, constituting the single family of the avian superfamily Jacanoidea. { jə'kan·ə,dē }

Jacanoidea [VERTEBRATE ZOOLOGY] A monofamilial superfamily of colorful marshbirds distinguished by greatly elongated toes and claws, long legs, a short tail, and a straight bill. { ,jak·ə'nȯid·ē·ə }

jackal [VERTEBRATE ZOOLOGY] **1.** *Canis aureus.* A wild dog found in southeastern Europe, southern Asia, and northern Africa. **2.** Any of various similar Old World wild dogs; they resemble wolves but are smaller and more yellowish. { 'jak·əl }

Jacobsoniidae [INVERTEBRATE ZOOLOGY] The false snout beetles, a small family of coleopteran insects in the superfamily Dermestoidea. { ,jā·kəb·sə'nī·ə,dē }

Jacobson's cartilage [ANATOMY] *See* vomeronasal cartilage. { 'jā·kəb·sənz ¦kärt·lij }

Jacobson's organ [VERTEBRATE ZOOLOGY] An olfactory canal in the nasal mucosa which ends in a blind pouch; it is highly developed in reptiles and vestigial in humans. { 'jā·kəb·sənz ¦ȯr·gən }

jaguar [VERTEBRATE ZOOLOGY] *Felis onca.* A large, wild cat indigenous to Central and South America; it is distinguished by a buff-colored coat with black spots, and has a relatively large head and short legs. { 'jag,wär }

Jamaica bayberry [BOTANY] *See* bayberry. { jə 'mā·kə 'bā,ber·ē }

Japygidae [INVERTEBRATE ZOOLOGY] A family of wingless insects in the order Diplura with forcepslike anal appendages; members attack and devour small soil arthropods. { jə'pij·ə,dē }

jaundice [INVERTEBRATE ZOOLOGY] *See* grasserie. { 'jȯn·dəs }

Java cotton [BOTANY] *See* kapok. { 'jäv·ə 'kat·ən }

Java man [PALEONTOLOGY] An overspecialized, apelike form of *Homo sapiens* from the middle Pleistocene having a small brain capacity, low cranial vault, and massive browridges. { ¦jäv·ə ¦man }

jaw [ANATOMY] Either of two bones forming the skeleton of the mouth of vertebrates: the upper jaw or maxilla, and the lower jaw or mandible. { jȯ }

jawless vertebrate [VERTEBRATE ZOOLOGY] The common name for members of the Agnatha. { 'jȯ·ləs 'vərd·ə·brət }

jejunum [ANATOMY] The middle portion of the small intestine, extending between the duodenum and the ileum. { jə'jü·nəm }

jellyfish [INVERTEBRATE ZOOLOGY] Any of various free-swimming marine cnidarians belonging to the Hydrozoa or Scyphozoa and having a bell- or bowl-shaped body. Also known as medusa. { 'jel·ē,fish }

jelly fungus [MYCOLOGY] The common name for many members of the Heterobasidiomycetidae, especially the orders Tremallales and Dacromycetales, distinguished by a jellylike appearance or consistency. { 'jel·ē ,fəŋ·gəs }

jenny [VERTEBRATE ZOOLOGY] **1.** A female animal, as a jenny wren. **2.** A female donkey. { 'jen·ē }

jerboa [VERTEBRATE ZOOLOGY] The common name for 25 species of rodents composing the family Dipodidae; all are adapted for jumping, having extremely long hindlegs and feet. { jər 'bō·ə }

jimsonweed [BOTANY] *Datura stramonium.* A tall, poisonous annual weed having large white or violet trumpet-shaped flowers and globose prickly fruits. Also known as apple of Peru. { 'jim·sən ,wēd }

jird [VERTEBRATE ZOOLOGY] Any one of the diminutive rodents composing related species of the genus *Meriones* which are inhabitants of northern Africa and southwestern Asia; they serve as experimental hosts for studies of schistosomiasis. { jərd }

joint [ANATOMY] A contact surface between two individual bones. Also known as articulation. { jȯint }

joint capsule [ANATOMY] A sheet of fibrous connective tissue enclosing a synovial joint. { 'jȯint ,kap·səl }

Joppeicidae [INVERTEBRATE ZOOLOGY] A monospecific family of hemipteran insects included in the Pentatomorpha; found in the Mediterranean regions. { ,jäp·ə'īs·ə,dē }

jordanon [ECOLOGY] *See* microspecies. { 'jȯrd·ən ,än }

Jordan's rule [EVOLUTION] The rule that organisms which are closely related tend to occupy adjacent rather than identical or distant ranges. [VERTEBRATE ZOOLOGY] The rule that fishes in areas of low temperatures tend to have more vertebrae than those in warmer waters. { 'jȯrd·ənz ,rül }

jugal [ANATOMY] Pertaining to the zygomatic

bone. [VERTEBRATE ZOOLOGY] In lower vertebrates, a bone lying below the orbit of the eye. { 'jü·gəl }

Jugatae [INVERTEBRATE ZOOLOGY] The equivalent name for Homoneura. { 'jü·gə,tē }

jugate [BIOLOGY] Structures which are joined together. { 'jü,gāt }

Juglandaceae [BOTANY] A family of dicotyledonous plants in the order Juglandales having unisexual flowers, a solitary basal ovule in a unilocular inferior ovary, and pinnately compound, exstipulate leaves. { ,jū,glan'dās·ē,ē }

Juglandales [BOTANY] An order of dicotyledonous plants in the subclass Hamamelidae distinguished by compound leaves; includes hickory, walnut, and butternut. { ,jü,glan'dā·lēz }

jugular [ANATOMY] Pertaining to the region of the neck above the clavicle. { 'jəg·yə·lər }

jugular foramen [ANATOMY] An opening in the cranium formed by the jugular notches of the occipital and temporal bones for passage of an internal jugular vein, the ninth, tenth, and eleventh cranial nerves, and the inferior petrosal sinus. { 'jəg·yə·lər fə'rā·mən }

jugular process [ANATOMY] A rough process external to the condyle of the occipital bone. { 'jəg·yə·lər ,prä·səs }

jugular vein [ANATOMY] The vein in the neck which drains the brain, face, and neck into the innominate. { 'jəg·yə·lər ,vān }

jugum [BOTANY] One pair of opposite leaflets of a pinnate leaf. [INVERTEBRATE ZOOLOGY] **1.** The most posterior and basal portion of the wing of an insect. **2.** A crossbar connecting the two arms of the brachidium in some brachiopods. { 'jü·gəm }

jumping gene [GENETICS] A mobile genetic entity, such as a transposon. { ¦jəmp·iŋ ¦jēn }

Juncaceae [BOTANY] A family of monocotyledonous plants in the order Juncales characterized by an inflorescence of diverse sorts, vascular bundles with abaxial phloem, and cells without silica bodies. { ,jəŋ'kās·ē,ē }

Juncales [BOTANY] An order of monocotyledonous plants in the subclass Commelinidae marked by reduced flowers and capsular fruits with one too many anatropous ovules per carpel. { ,jəŋ'kā·lēz }

junctional complex [CELL BIOLOGY] Any specialized area of intercellular adhesion. { 'jəŋk·shən·əl ¦käm,pleks }

junctional receptor [PHYSIOLOGY] An acetylcholine receptor which occurs in clusters in a muscle membrane at the nerve-muscle junction. { 'jəŋk·shən·əl ri'sep·tər }

junction sequence [MOLECULAR BIOLOGY] Either of the two terminal regions of the intron in ribonucleic acid precursors. { 'jəŋk·shən ,sē·kwəns }

Jungermanniales [BOTANY] The leafy liverworts, an order of bryophytes in the class Marchantiatae characterized by chlorophyll-containing, ribbonlike or leaflike bodies and an undifferentiated thallus. { ,jəŋ·gər,man·ē'ā·lēz }

Jungermanniidae [BOTANY] A subclass of liverworts of the class Hepticopsida, division Bryophyta, distinguished by little or no tissue differentiation or organized into erect or prostrate stems with leafy appendages, leaves generally one cell in thickness and mostly arranged in three rows, with the third row of underleaves commonly reduced or even lacking, and oil bodies usually present in all cells. { ,jəŋ·gər'man·ə ,dē }

jungle [ECOLOGY] An impenetrable thicket of second-growth vegetation replacing tropical rain forest that has been disturbed; lower growth layers are dense. { 'jəŋ·gəl }

junk deoxyribonucleic acid [MOLECULAR BIOLOGY] *See* selfish deoxyribonucleic acid. { ¦jəŋk dē¦äk·sē,rī·bō·nü¦klē·ik ,as·əd }

jute [BOTANY] Either of two Asiatic species of tall, slender, half-shrubby annual plants, *Corchorus capsularis* and *C. olitorius*, in the family Malvaceae, useful for their fiber. { jüt }

juvenile cell [HISTOLOGY] *See* metamyelocyte. { 'jü·vən·əl 'sel }

juvenile hormone [BIOCHEMISTRY] *See* neotenin. { 'jü·vən·əl 'hór,mōn }

Jynginae [VERTEBRATE ZOOLOGY] The wrynecks, a family of Old World birds in the order Piciformes; a subfamily of the Picidae in some systems of classification. { jin'jī,nē }

K

kaempferol [BIOCHEMISTRY] A flavonoid with a structure similar to that of quercetin but with only one hydroxyl in the B ring; acts as an enzyme cofactor and causes growth inhibition in plants. { 'kemp·fə,ròl }

kairomone [PHYSIOLOGY] A chemical produced by an organism that benefits the recipient, which is an individual of a different species. { 'kī·rə ,mōn }

Kaiserling's method [BIOLOGY] A method for preserving organ specimens and retaining their color by fixing in a solution of formalin, water, potassium nitrate, and potassium acetate, immersing in ethyl alcohol to restore color, and preserving in a solution of glycerin, aqueous arsenious acid, water, potassium acetate, and thymol. { 'kī·zər·liŋz ,meth·əd }

kaki [BOTANY] *Diospyros kaki*. The Japanese persimmon; it provides a type of ebony wood that is black with gray, yellow, and brown streaks, has a close, even grain, and is very hard. { 'kä·kē }

kale [BOTANY] Either of two biennial crucifers, *Brassica oleracea* var. *acephala* and *B. fimbriata*, in the order Capparales, grown for the nutritious curled green leaves. { kāl }

kallidin I [BIOCHEMISTRY] *See* bradykinin. { kə 'līd·ən 'wən }

Kalotermitidae [INVERTEBRATE ZOOLOGY] A family of relatively primitive, lower termites in the order Isoptera. { ,kal·ə,tər'mid·ə,dē }

Kamptozoa [INVERTEBRATE ZOOLOGY] An equivalent name for Entoprocta. { ,kam·tə'zō·ə }

kanamycin [MICROBIOLOGY] $C_{18}H_{36}O_{11}N_4$ A water-soluble, basic antibiotic produced by strains of *Streptomyces kanamyceticus*; the sulfate salt is effective in infections caused by gram-negative bacteria. { ,kan·ə'mīs·ən }

kangaroo [VERTEBRATE ZOOLOGY] Any of various Australian marsupials in the family Macropodidae generally characterized by a long, thick tail that is used as a balancing organ, short forelimbs, and enlarged hindlegs adapted for jumping. { ,kaŋ·gə'rü }

kapok [BOTANY] Silky fibers that surround the seeds of the kapok or ceiba tree. Also known as ceiba; Java cotton; silk cotton. { 'kā,päk }

kapok tree [BOTANY] *Ceiba pentandra*. A tree of the family Bombacaceae which produces pods containing seeds covered with silk cotton. Also known as silk cotton tree. { 'kā,päk ,trē }

kappa particle [CELL BIOLOGY] A self-duplicating nucleoprotein particle found in various strains of *Paramecium* and thought to function as an infectious agent; classed as an intracellular symbiont, occupying a position between the viruses and the bacteria and organelles. { 'kap·ə ,pard·ə·kəl }

Karumiidae [INVERTEBRATE ZOOLOGY] The termitelike beetles, a small family of coleopteran insects in the superfamily Cantharoidea distinguished by having a tenth tergum. { ,kar·ə'mī·ə ,dē }

karyocyte [HISTOLOGY] *See* normoblast. { 'kar·ē·ə,sīt }

karyodesma [CELL BIOLOGY] *See* nucleodesma. { ,kar·ē·ə'dez·mə }

karyogamy [CELL BIOLOGY] Fusion of gametic nuclei, as in fertilization. { ,kar·ē'äg·ə·mē }

karyokinesis [CELL BIOLOGY] Nuclear division characteristic of mitosis. { ¦kar·ē·ō·kə'nē·səs }

karyolymph [CELL BIOLOGY] The clear material composing the ground substance of a cell nucleus. { 'kar·ē·ə,limf }

karyolysis [CELL BIOLOGY] Dissolution of a cell nucleus. { ,kar·ē'äl·ə·səs }

karyomastigont [INVERTEBRATE ZOOLOGY] Pertaining to members of the protozoan order Oxymonadida; individuals can be uni- or multinucleate, and unattached forms give rise to two pairs of flagella. { ¦kar·ē·ō'mas·tə,gänt }

karyoplasm [CELL BIOLOGY] *See* nucleoplasm. { 'kar·ē·ə,plaz·əm }

karyoplasmic ratio [CELL BIOLOGY] *See* nucleocytoplasmic ratio. { ,kar·ē·ə'plaz·mik 'rā·shō }

karyorrhexis [CELL BIOLOGY] Fragmentation of a nucleus with scattering of the pieces in the cytoplasm. { ,kar·ē·ə'rek·səs }

karyosphere [CELL BIOLOGY] The fraction of nuclear volume to which the chromosomes are confined in nuclei that are rich in karyolymph. { 'kar·ē·ə,sfir }

karyotype [CELL BIOLOGY] The normal diploid or haploid complement of chromosomes, with respect to size, form, and number, characteristic of an individual, species, genus, or other grouping. { 'kar·ē·ə,tīp }

kasugamycin [MICROBIOLOGY] $C_{14}H_{28}ClN_3O_{10}$ A white, crystalline antibiotic used as a fungicide for rice crops. Also known as kasugamycin hydrochloride; kasumin. { kə,sü·gə'mīs·ən }

kasugamycin hydrochloride [MICROBIOLOGY] *See*

kasugamycin. { kə,sü·gə'mīs·ən ,hī·drə'klȯr ,īd }

kasumin [MICROBIOLOGY] *See* kasugamycin. { kə 'sü·mən }

Kathlaniidae [INVERTEBRATE ZOOLOGY] A family of nematodes assigned to the Ascaridina by some authorities and to the Oxyurina by others. { ,kath·lə'nī·ə,dē }

kb [GENETICS] *See* kilobase.

kbp [GENETICS] *See* kilobase.

keel [VERTEBRATE ZOOLOGY] The median ridge on the breastbone in certain birds. Also known as carina. { kēl }

Keilor skull [PALEONTOLOGY] An Australian fossil type specimen of *Homo sapiens* from the Pleistocene. { 'kē·lər ,skəl }

Kell blood group system [IMMUNOLOGY] A family of antigens found in erythrocytes and designated K, k, Kp^a, Kp^b, and Ku; antibodies to the K antigen, which occurs in about 10% of the population of England, have been associated with hemolytic transfusion reactions and with hemolytic disease. { 'kel 'bləd ,grüp ,sis·təm }

kelp [BOTANY] The common name for brown seaweed belonging to the Laminariales and Fucales. { kelp }

kenozooecium [INVERTEBRATE ZOOLOGY] The outer, nonliving, hardened portion of a kenozooid. { ¦kēn·ə·zō'ē·sē·əm }

kenozooid [INVERTEBRATE ZOOLOGY] A type of bryozoan heterozooid possessing a slender tubular or boxlike chamber, completely enclosed and lacking an aperture. { ¦kēn·ə¦zō,ȯid }

Kentucky coffee tree [BOTANY] *Gymnocladus dioica*. An extremely tall, dioecious tree of the order Rosales readily recognized when in fruit by its leguminous pods containing heavy seeds, once used as a coffee substitute. { kən'tək·ē 'kȯf·ē ,trē }

Kenyapithecus [PALEONTOLOGY] An early member of Hominidae from the Miocene. { ,ken·yə 'pith·ə·kəs }

keratin [BIOCHEMISTRY] Any of various albuminoids characteristic of epidermal derivatives, such as nails and feathers, which are insoluble in protein solvents, have a high sulfur content, and generally contain cystine and arginine as the predominating amino acids. { 'ker·əd·ən }

keratinized tissue [HISTOLOGY] Any tissue with a high keratin content, such as the epidermis or its derivatives. { 'ker·əd·ə,nīzd 'tish·ü }

keratinocyte [HISTOLOGY] A specialized epidermal cell that synthesizes keratin. { kə'rat·ən·ə ,sīt }

keratinous degeneration [CELL BIOLOGY] The occurrence of keratin granules in the cytoplasm of a cell, other than a keratinocyte. { kə'rat·ən·əs di,jen·ə'rā·shən }

keratohyalin [HISTOLOGY] Granules in the stratum granulosum of keratinized stratified squamous epithelium which become keratin. { ,ker·əd·ō'hī·ə·lən }

Kerguelen faunal region [ECOLOGY] A marine littoral faunal region comprising a large area surrounding Kerguelen Island in the southern Indian Ocean. { 'kər·gə·lən 'fȯn·əl ,rē·jən }

kernel [BOTANY] **1.** The inner portion of a seed. **2.** A whole grain or seed of a cereal plant, such as corn or barley. { 'kərn·əl }

ketoadipic acid [BIOCHEMISTRY] $C_6H_8O_5$ An intermediate product in the metabolism of lysine to glutaric acid. { ¦kēd·ō·ə¦dip·ik 'as·əd }

ketogenesis [BIOCHEMISTRY] Production of ketone bodies. { ¦kēd·ō'jen·ə·səs }

ketogenic hormone [BIOCHEMISTRY] A factor originally derived from crude anterior hypophysis extract which stimulated fatty-acid metabolism; now known as a combination of adrenocorticotropin and the growth hormone. Also known as fat-metabolizing hormone. { ¦kēd·ə ¦jen·ik 'hȯr,mōn }

ketogenic substance [BIOCHEMISTRY] Any foodstuff which provides a source of ketone bodies. { ¦kēd·ə¦jen·ik 'səb·stəns }

ketoglutaric acid [BIOCHEMISTRY] $C_5H_6O_5$ A dibasic keto acid occurring as an intermediate product in carbohydrate and protein metabolism. { ¦kēd·ō·glü¦tar·ik 'as·əd }

ketohexose [BIOCHEMISTRY] Any monosaccharide composed of a six-carbon chain and containing one ketone group. { ¦kēd·ō'hek,sōs }

ketolase [BIOCHEMISTRY] A type of enzyme that catalyzes cleavage of carbohydrates at the carbonyl carbon position. { 'kēd·əl,ās }

ketolysis [BIOCHEMISTRY] Dissolution of ketone bodies. { kē'täl·ə·səs }

ketone body [BIOCHEMISTRY] Any of various ketones which increase in blood and urine in certain conditions, such as diabetic acidosis, starvation, and pregnancy. Also known as acetone body. { 'kē,tōn ,bäd·ē }

ketose [BIOCHEMISTRY] A carbohydrate that has a ketone group. { 'kē,tōs }

ketosteroid [BIOCHEMISTRY] One of a group of neutral steroids possessing keto substitution, which produces a characteristic red color with *m*-dinitrobenzene in an alkaline solution; these compounds are principally metabolites of adrenal cortical and gonadal steroids. { ¦kēd·ō'stir ,ȯid }

key [SYSTEMATICS] An arrangement of the distinguishing features of a taxonomic group to serve as a guide for establishing relationships and names of unidentified members of the group. { kē }

Kidd blood group system [IMMUNOLOGY] The erythrocyte antigens defined by reactions to anti-Jk^a antibodies, originally found in the mother (Mrs. Kidd) of the erythroblastotic infant, and to anti-Jk^b antibodies. { 'kid 'bləd ,grüp ,sis·təm }

kidney [ANATOMY] Either of a pair of organs involved with the elimination of water and waste products from the body of vertebrates; in humans they are bean-shaped, about 5 inches (12.7 centimeters) long, and are located in the posterior part of the abdomen behind the peritoneum. { 'kid·nē }

killed vaccine [IMMUNOLOGY] A suspension of killed microorganisms used as antigens to produce immunity. Also known as inactivated vaccine. { 'kild vak'sēn }

killer whale [VERTEBRATE ZOOLOGY] *Orcinus orca*.

A predatory, cosmopolitan cetacean mammal, about 30 feet (9 meters) long, found only in cold waters. { 'kil·ər ‚wāl }

kilobase [GENETICS] Unit of length equal to 1000 base pairs in deoxyribonucleic acid or 1000 nitrogenous bases in ribonucleic acid. Abbreviated kb; kbp. { 'kil·ō‚bās }

kinase [BIOCHEMISTRY] Any enzyme that catalyzes phosphorylation reactions. { 'kī‚nās }

kinematic chain [ANATOMY] A group of body segments that are connected by joints so that the segments operate together to provide a wide range of motion for a limb. { ¦kin·ə‚mad·ik 'chān }

kinesiology [PHYSIOLOGY] The study of human motion through anatomical and mechanical principles. { kə‚nēz·ē'äl·ə·jē }

kinesis [PHYSIOLOGY] The general term for physical movement, including that induced by stimulation, for example, light. { ki'nē·səs }

kinesthesis [PHYSIOLOGY] The system of sensitivity present in the muscles and their attachments. { ‚kin·əs'thē·səs }

kinetid [CELL BIOLOGY] In eukaryotic cells, any locomotory structure, that is, a cilium or flagellum. { ki'ned·əd }

kinetin [BIOCHEMISTRY] $C_{10}H_9ON_5$ A cytokinin formed in many plants which has a stimulating effect on cell division. { 'kin·ə·tən }

kinetochore [CELL BIOLOGY] Within the centromere, the granule upon which the spindle fibers attach. { kə'ned·ə‚kór }

kinetoplast [CELL BIOLOGY] A genetically autonomous, membrane-bound organelle associated with the basal body at the base of flagella in certain flagellates, such as the trypanosomes. Also known as parabasal body. { kə'ned·ə‚plast }

Kinetoplastida [INVERTEBRATE ZOOLOGY] An order of colorless protozoans in the class Zoomastigophorea having pliable bodies and possessing one or two flagella in some stage of their life. { kə‚ned·ə'plas·tə·də }

kinetosome [CELL BIOLOGY] *See* basal body. { kə 'ned·ə‚sōm }

kingdom [SYSTEMATICS] One of the primary divisions that include all living organisms: most authorities recognize two, the animal kingdom and the plant kingdom, while others recognize three or more, such as Protista, Plantae, Animalia, and Mycota. { 'kiŋ·dəm }

kingfisher [VERTEBRATE ZOOLOGY] The common name for members of the avian family Alcedinidae; most are tropical Old World species characterized by short legs, long bills, bright plumage, and short wings. { 'kiŋ‚fish·ər }

kink [MOLECULAR BIOLOGY] A bend between two helical segments of deoxyribonucleic acid achieved by unstacking one base pair and twisting the polynucleotide backbone. { kiŋk }

kinocilium [CELL BIOLOGY] A type of cilium containing one central pair of microfibrils and nine peripheral pairs; they extend from the apex of hair cells in all vertebrate ears except mammals. { ¦kin·ə'sil·ē·əm }

kinomere [CELL BIOLOGY] *See* centromere. { 'kin·ə‚mir }

kinoplasm [CELL BIOLOGY] The substance of the protoplasm that is thought to form astral rays and spindle fibers. { 'kin·ə‚plaz·əm }

Kinorhyncha [INVERTEBRATE ZOOLOGY] A class of the phylum Aschelminthes consisting of superficially segmented microscopic marine animals lacking external ciliation. { ‚kin·ə'riŋ·kē·ə }

Kinosternidae [VERTEBRATE ZOOLOGY] The mud and musk turtles, a family of chelonian reptiles in the suborder Cryptodira found in North, Central, and South America. { ‚kin·ə'stər·nə‚dē }

Kirkbyacea [PALEONTOLOGY] A monomorphic superfamily of extinct ostracods in the suborder Beyrichicopina, all of which are reticulate. { ‚kərk·bē'ās·ē·ə }

Kirkbyidae [PALEONTOLOGY] A family of extinct ostracods in the superfamily Kirkbyacea in which the pit is reduced and lies below the middle of the valve. { kərk'bē·ə‚dē }

Kitasatoa [MICROBIOLOGY] A genus of bacteria in the family Actinoplanaceae; club-shaped sporangia, each containing a single chain of diplococcuslike uniflagellate spores. { kə¦ta·sə 'tō·ə }

kiwi [VERTEBRATE ZOOLOGY] The common name for three species of nocturnal ratites of New Zealand composing the family Apterygidae; all have small eyes, vestigial wings, a long slender bill, and short powerful legs. { 'kē‚wē }

Klebsiella [MICROBIOLOGY] A genus of bacteria in the family Enterobacteriaceae; nonmotile, encapsulated rods arranged singly, in pairs, or in chains; some species are human pathogens. { ‚kleb·zē'el·ə }

Klebsiella pneumoniae [MICROBIOLOGY] An encapsulated pathogenic bacterium that causes severe pneumonitis in humans.Formerly known as Friedlander's bacillus; pneumobacillus. { ‚kleb·zē'el·ə nə'mō·nē‚ī }

Klebs-Loeffler bacillus [MICROBIOLOGY] *See* Corynebacterium diphtheriae. { 'klāps 'lef·lər bə ‚sil·əs }

Kleinschmidt spread [CELL BIOLOGY] A visualization technique for electron microscopy in which molecules are mounted in a positively charged protein monolayer, which is spread on the surface of water, and then are transferred to a hydrophobic grid. { 'klīn‚shmit ‚spred }

klendusity [BOTANY] The tendency of a plant to resist disease due to a protective covering, such as a thick cuticle, that prevents inoculation. { klen'dü·səd·ē }

klinotaxis [BIOLOGY] Positive orientation movement of a motile organism induced by a stimulus. { ¦klī·nə'tak·səs }

Kloedenellacea [PALEONTOLOGY] A dimorphic superfamily of extinct ostracods in the suborder Kloedenellocopina having the posterior part of one dimorph longer and more inflated than the other dimorph. { ‚klēd·ən·ə'lās·ē·ə }

Kloedenellocopina [PALEONTOLOGY] A suborder of extinct ostracods in the order Paleocopa characterized by a relatively straight dorsal border with a gently curved or nearly straight ventral border. { ‚klēd·ən‚el·ə'käp·ə·nə }

knee [ANATOMY] **1.** The articulation between the femur and the tibia in humans. Also known as genu. **2.** The corresponding articulation in the hindlimb of a quadrupedal vertebrate. { nē }

kneecap [ANATOMY] *See* patella. { 'nē,kap }

Kneriidae [VERTEBRATE ZOOLOGY] A small family of tropical African fresh-water fishes in the order Gonorynchiformes. { nə'rī·ə,dē }

koala [VERTEBRATE ZOOLOGY] *Phascolarctos cinereus.* An arboreal marsupial mammal of the family Phalangeridae having large hairy ears, gray fir, and two clawed toes opposing three others on each limb. { kō'äl·ə }

Koch's postulates [MICROBIOLOGY] A set of laws elucidated by Robert Koch: the microorganism identified as the etiologic agent must be present in every case of the disease; the etiologic agent must be isolated and cultivated in pure culture; the organism must produce the disease when inoculated in pure culture into susceptible animals; a microorganism must be observed in and recovered from the experimentally diseased animal. Also known as law of specificity of bacteria. { 'kōks 'päs·chə·ləts }

kohlrabi [BOTANY] A biennial crucifer, designated *Brassica caulorapa* and *B. oleracea* var. *caulo-rapa*, of the order Capparales grown for its edible turniplike, enlarged stem. { kōl'rä·bē }

Komodo dragon [VERTEBRATE ZOOLOGY] *Varanus komodoensis.* A predatory reptile of the family Varanidae found only on the island of Komodo; it is the largest living lizard and may grow to 10 feet (3 meters). { kə'mō·dō 'drag·ən }

Koonungidae [INVERTEBRATE ZOOLOGY] A family of Australian crustaceans in the order Anaspidacea with sessile eyes and the first thoracic limb modified for digging. { kü'nən·jə,dē }

Kozak sequence [MOLECULAR BIOLOGY] A nucleotide sequence in the 5′ untranslated messenger ribonucleic acid region that allows ribosomes to recognize the initiator codon. { 'kō,zak ,sē·kwəns }

Krause's corpuscle [ANATOMY] One of the spheroid nerve-end organs resembling lamellar corpuscles, but having a more delicate capsule; found especially in the conjunctiva, the mucosa of the tongue, and the external genitalia; they are believed to be cold receptors. Also known as end bulb of Krause. { 'krau̇s·əz ,kȯr·pə·səl }

Krebs cycle [BIOCHEMISTRY] A sequence of enzymatic reactions involving oxidation of a two-carbon acetyl unit to carbon dioxide and water to provide energy for storage in the form of high-energy phosphate bonds. Also known as citric acid cycle; tricarboxylic acid cycle. { 'krebz ,sī·kəl }

Krebs-Henseleit cycle [BIOCHEMISTRY] A cyclic reaction pathway involving the breakdown of arginine to urea in the presence of arginase. { 'krebz 'hen·sə,līt ,sī·kəl }

Krebspest [INVERTEBRATE ZOOLOGY] A fatal fungus disease of crayfish caused by *Aphanomyces mystaci.* { 'krebs,pest }

krill [INVERTEBRATE ZOOLOGY] A name applied to planktonic crustaceans that constitute the diet of many whales, particularly whalebone whales. { kril }

kringle domain [BIOCHEMISTRY] *See* kringle region. { 'kriŋ·gəl dō,mān }

kringle region [BIOCHEMISTRY] A unique protein structural configuration composed of three disulfide bonds. Also known as kringle domain. { 'kriŋ·gəl ,rē·jən }

krummholz [ECOLOGY] Stunted alpine forest vegetation. Also known as elfinwood. { 'kru̇m ,hōlts }

K selection [ECOLOGY] Selection favoring species that reproduce slowly where a resource is constant but available in limited quantities; population is maintained at or near the carrying capacity (K) of the habitat. { 'kā si,lek·shən }

kudzu [BOTANY] Any of various perennial vine legumes of the genus *Pueraria* in the order Rosales cultivated principally as a forage crop. { 'ku̇d,zü }

Kuehneosauridae [PALEONTOLOGY] The gliding lizards, a family of Upper Triassic reptiles in the order Squamata including the earliest known aerial vertebrates. { ¦kyün,nē·ō'sȯr·ə,dē }

kumquat [BOTANY] A citrus shrub or tree of the genus *Fortunella* in the order Sapindales grown for its small, flame- to orange-colored edible fruit having three to five locules filled with an acid pulp, and a sweet, pulpy rind. { 'kəm,kwät }

Kupffer cell [HISTOLOGY] One of the fixed macrophages lining the hepatic sinusoids. { 'ku̇p·fər ,sel }

Kurthia [MICROBIOLOGY] A genus of gram-positive, aerobic, rod-shaped to coccoid bacteria in the coryneform group; metabolism is respiratory. { 'kər·thē·ə }

Kurtoidei [VERTEBRATE ZOOLOGY] A monogeneric suborder of perciform fishes having a unique ossification that encloses the upper part of the swim bladder, and an occipital hook in the male for holding eggs during brooding. { kər'tȯid·ē ,ī }

Kusnezovia [MICROBIOLOGY] A genus of bacteria of uncertain affiliation; coccoid, nonmotile cells attach to substrate and reproduce by budding. { ,küz·nə'zō·vē·ə }

Kutorginida [PALEONTOLOGY] An order of extinct brachiopod mollusks that is unplaced taxonomically. { ,küd·ər'jīn·ə·də }

K virus [VIROLOGY] A group 2 papovavirus affecting rats and mice. { 'kā ¦vī·rəs }

Kyasanur Forest virus [VIROLOGY] A group B arbovirus recognized as an agent that causes hemorrhagic fever. { kī'az·ə·nu̇r ¦fär·əst 'vī·rəs }

kynurenic acid [BIOCHEMISTRY] $C_{10}H_7O_3N$ A product of tryptophan metabolism found in the urine of mammals. { ¦kin·yə¦ren·ik 'as·əd }

kynurenine [BIOCHEMISTRY] $C_{10}H_{12}O_3N_2$ An intermediate product of tryptophan metabolism occurring in the urine of mammals. { ,kin·yə're ,nēn }

L

labellate [BIOLOGY] Having a labellum. { lə'be ˌlāt }

labellum [BOTANY] The median membrane of the corolla of an orchid often differing in size and morphology from the other two petals. [INVERTEBRATE ZOOLOGY] **1.** A prolongation of the labrum in certain beetles and true bugs. **2.** In Diptera, either of a pair of sensitive fleshy lobes consisting of the expanded end of the labium. { lə'bel·əm }

labial gland [ANATOMY] Any of the small, tubular mucous and serous glands underneath the mucous membrane of mammalian lips. [INVERTEBRATE ZOOLOGY] A salivary gland, or modification thereof, opening at the base of the labium in certain insects. { 'lā·bē·əl ¦gland }

labial palp [INVERTEBRATE ZOOLOGY] **1.** Either of a pair of fleshy appendages on either side of the mouth of certain bivalve mollusks. **2.** A jointed appendage attached to the labium of certain insects. { 'lā·bē·əl ¦palp }

labial papilla [INVERTEBRATE ZOOLOGY] Any of the sensory bristles around the mouth of many nematodes; they are jointed projections of the cuticle. { 'lā·bē·əl pə'pil·ə }

Labiatae [BOTANY] A large family of dicotyledonous plants in the order Lamiales; members are typically aromatic and usually herbaceous or merely shrubby. { ˌlā·bē'āˌtē }

labiate [ANATOMY] Having liplike margins that are thick and fleshy. [BIOLOGY] Having lips. [BOTANY] Having the limb of a tubular calyx or corolla divided into two unequal overlapping parts. { 'lā·bē·ət }

labile factor [BIOCHEMISTRY] *See* proaccelerin. { 'lāˌbīl ˌfak·tər }

labium [BIOLOGY] **1.** A liplike structure. **2.** The lower lip, as of a labiate corolla or of an insect. { 'lā·bē·əm }

labium majus [ANATOMY] Either of the two outer folds surrounding the vulva in the female. { 'lā·bē·əm 'mä·jəs }

labium minus [ANATOMY] Either of the two inner folds, at the inner surfaces of the labia majora, surrounding the vulva in the female. { 'lā·bē·əm 'mē·nəs }

Laboulbeniales [MYCOLOGY] An order of ascomycetous fungi made up of species that live primarily on the external surfaces of insects. { lə ˌbül·ben·ē'ā·lēz }

Labridae [VERTEBRATE ZOOLOGY] The wrasses, a family of perciform fishes in the suborder Percoidei. { 'lab·rəˌdē }

labrum [INVERTEBRATE ZOOLOGY] **1.** The upper lip of certain arthropods, lying in front of or above the mandibles. **2.** The outer edge of a gastropod shell. { 'lā·brəm }

labyrinth [ANATOMY] **1.** Any body structure full of intricate cavities and canals. **2.** The inner ear. { 'lab·əˌrinth }

labyrinthine placenta [EMBRYOLOGY] *See* hemochorial placenta. { ˌlab·ə'rinˌthēn plə'sen·tə }

labyrinthine reflex [PHYSIOLOGY] The involuntary response to stimulation of the vestibular apparatus in the inner ear. { ˌlab·ə'rinˌthēn 'rē ˌfleks }

Labyrinthodontia [PALEONTOLOGY] A subclass of fossil amphibians descended from crossopterygian fishes, ancestral to reptiles, and antecedent to at least part of other amphibian types. { ˌlab·əˌrin·thə'dän·chə }

Labyrinthulia [INVERTEBRATE ZOOLOGY] A subclass of the protozoan class Rhizopoda containing mostly marine, ovoid to spindle-shaped, uninucleate organisms that secrete a network of filaments (slime tubes) along which they glide. { ˌlab·ə·rən'thül·ē·ə }

Labyrinthulida [INVERTEBRATE ZOOLOGY] The single order of the protozoan subclass Labyrinthulia. { ˌlab·ə·rən'thül·ə·də }

laccal [BIOCHEMISTRY] $C_{17}H_{31}C_6H_3(OH)_2$ A phenol compound which is found in the sap of lacquer trees, and which can be isolated in crystalline form. { 'laˌkȯl }

laccase [BIOCHEMISTRY] Any of a class of plant oxidases which catalyze the oxidation of phenols. { 'laˌkās }

laccate [BIOLOGY] Having a lacquered appearance. { 'laˌkāt }

Lacciferinae [INVERTEBRATE ZOOLOGY] A subfamily of scale insects in the superfamily Coccoidea in which the male lacks compound eyes, the abdomen is without spiracles in all stages, and the apical abdominal segments of nymphs and females do not form a pygidium. { lak·sə 'fer·əˌnē }

lacerated [BIOLOGY] Having a deeply and irregularly incised margin or apex. { 'las·əˌrād·əd }

Lacertidae [VERTEBRATE ZOOLOGY] A family of reptiles in the suborder Sauria, including all typical lizards, characterized by movable eyelids, a fused lower jaw, homodont dentition, and epidermal scales. { lə'sərd·ə,dē }

Lachnospira [MICROBIOLOGY] A genus of weakly gram-positive, anaerobic bacteria of uncertain affiliation; motile, curved rods that ferment glucose and are found in the rumen of bovine animals. { lak'näs·pə·rə }

laciniate [BIOLOGY] **1.** Having a fringed border. **2.** Narrowly and deeply incised to form irregular lobes, which may be pointed. { lə'sin·ē,āt }

lac operon [GENETICS] A sequence of three linked genes which code for the enzymes involved in lactose utilization in many bacteria. { 'lak 'äp·ə,rän }

lacquer tree [BOTANY] *See* varnish tree. { 'lak·ər ,trē }

lacrimal [ANATOMY] Pertaining to tears, tear ducts, or tear-secreting organs. { 'lak·rə·məl }

lacrimal apparatus [ANATOMY] The functional and structural mechanisms for secreting and draining tears; includes the lacrimal gland, lake, puncta, canaliculi, sac, and nasolacrimal duct. { 'lak·rə·məl ,ap·ə'rad·əs }

lacrimal bone [ANATOMY] A small bone located in the anterior medial wall of the orbit, articulating with the frontal, ethmoid, maxilla, and inferior nasal concha. { 'lak·rə·məl ,bōn }

lacrimal canal [ANATOMY] *See* nasolacrimal canal. { 'lak·rə·məl kə,nal }

lacrimal canaliculus [ANATOMY] A small tube lined with stratified squamous epithelium which runs vertically a short distance from the punctum of each eyelid and then turns horizontally in the lacrimal part of the lid margin to the lacrimal sac. Also known as lacrimal duct. { 'lak·rə·məl ,kan·ə'lik·yə·ləs }

lacrimal duct [ANATOMY] *See* lacrimal canaliculus. { 'lak·rə·məl ,dəkt }

lacrimal gland [ANATOMY] A compound tubuloalveolar gland that secretes tears. Also known as tear gland. { 'lak·rə·məl ,gland }

lacrimal sac [ANATOMY] The dilation at the upper end of the nasolacrimal duct within the medial canthus of the eye. Also known as dacryocyst. { 'lak·rə·məl ,sak }

lacrimation [PHYSIOLOGY] **1.** Normal secretion of tears. **2.** Excessive secretion of tears, as in weeping. { ,lak·rə'mā·shən }

lactalbumin [BIOCHEMISTRY] A simple protein contained in milk which resembles serum albumin and is of high nutritional quality. { ,lak,tal 'byü·mən }

lactase [BIOCHEMISTRY] An enzyme that catalyzes the hydrolysis of lactose to dextrose and galactose. { 'lak,tās }

lactate [PHYSIOLOGY] To secrete milk. { 'lak,tāt }

lactate dehydrogenase [BIOCHEMISTRY] A zinc-containing enzyme which catalyzes the oxidation of several α-hydroxy acids to corresponding α-keto acids. { 'lak,tāt dē'hī·drə·jə,nās }

lactation [PHYSIOLOGY] Secretion of milk by the mammary glands. { lak'tā·shən }

lacteal [ANATOMY] One of the intestinal lymphatics that absorb chyle. [PHYSIOLOGY] Pertaining to or resembling milk. { 'lak·tē·əl }

lactescent [BIOLOGY] Having a milky appearance. [PHYSIOLOGY] Secreting milk or a milklike substance. { lak'tes·ənt }

lactic acid [BIOCHEMISTRY] $C_3H_6O_3$ A hygroscopic α-hydroxy acid, occurring in three optically isomeric forms: L form, in blood and muscle tissue as a product of glucose and glycogen metabolism; D form, obtained by fermentation of sucrose; and DL form, a racemic mixture present in foods prepared by bacterial fermentation, and also made synthetically. { 'lak·tik 'as·əd }

lactic dehydrogenase [BIOCHEMISTRY] An enzyme that catalyzes the dehydrogenation of L-lactic acid to pyruvic acid. Abbreviated LDH. { 'lak·tik dē'hī·drə·jə,nās }

lactic dehydrogenase virus [VIROLOGY] A virus of the rubella group which infects mice. { 'lak·tik dē'hī·drə·jə,nās ,vī·rəs }

lactiferous duct [BOTANY] A tubular channel consisting of latex vessels or latex cells; carries the latex produced by the plant. { lak'tif·ə·rəs 'dəkt }

lactin [BIOCHEMISTRY] *See* lactose. { 'lak·tən }

lactivorous [ZOOLOGY] Feeding on milk. { lak 'tiv·ə·rəs }

Lactobacillaceae [MICROBIOLOGY] The single family of gram-positive, asporogenous, rod-shaped bacteria; they are saccharoclastic, and produce lactate from carbohydrate metabolism. { ,lak·tō·bə,sil·ē'ās·ē,ē }

Lactobacilleae [MICROBIOLOGY] Formerly a tribe of rod-shaped bacteria in the family Lactobacillaceae. { ¦lak·tō·bə'sil·ē,ē }

Lactobacillus [MICROBIOLOGY] Lactic acid bacteria, the single genus of the family Lactobacillaceae; found in dairy products, meat products, fruits, beer, wine, and other food products. { ¦lak·tō·bə'sil·əs }

lactoferrin [BIOCHEMISTRY] An iron-binding protein found in milk, saliva, tears, and intestinal and respiratory secretions that interferes with the iron metabolism of bacteria; in conjunction with antibodies, it plays an important role in resistance to certain infectious diseases. { ¦lak·tō 'fer·ən }

lactoflavin [BIOCHEMISTRY] *See* riboflavin. { ¦lak·tō'flav·ən }

lactogenic hormone [BIOCHEMISTRY] *See* prolactin. { ¦lak·tə¦jen·ik 'hȯr,mōn }

lactoglobulin [BIOCHEMISTRY] A crystalline protein fraction of milk, which is soluble in half-saturated ammonium sulfate solution and insoluble in pure water. { ¦lak·tō'gläb·yə·lən }

lactonase [BIOCHEMISTRY] The enzyme that catalyzes the hydrolysis of 6-phosphoglucono-Δ-lactone to 6-phosphogluconic acid in the pentose phosphate pathway. { 'lak·tə,nās }

lactose [BIOCHEMISTRY] $C_{12}H_{22}O_{11}$ A disaccharide composed of D-glucose and D-galactose which occurs in milk. Also known as lactin; milk sugar. { 'lak,tōs }

lacuna [BIOLOGY] A small space or depression. [HISTOLOGY] A cavity in the matrix of bone or cartilage which is occupied by the cell body. { lə 'kü·nə }

lacunar system [INVERTEBRATE ZOOLOGY] A series of intercommunicating spaces branching from two longitudinal vessels in the hypodermis of many acanthocephalins. { lə'kü·nər ˌsis·təm }

Lacydonidae [INVERTEBRATE ZOOLOGY] A benthic family of pelagic errantian polychaetes. { ˌlas·ə 'dän·əˌdē }

Laemobothridae [INVERTEBRATE ZOOLOGY] A family of lice in the order Mallophaga including parasites of aquatic birds, especially geese and coots. { ˌlē·mə'bäth·rəˌdē }

Lagenidiales [MYCOLOGY] An order of aquatic fungi belonging to the class Phycomycetales characterized by a saclike to limited hyphal thallus and zoospores having two flagella. { ˌla·jə ˌnid·ē'ā·lēz }

lageniform [BIOLOGY] Flask-shaped. { lə'jen·ə ˌfȯrm }

lagging [CELL BIOLOGY] Pertaining to chromosomes that show little or no movement during metaphase and anaphase of meiosis or mitosis. { 'lag·iŋ }

lagging strand [MOLECULAR BIOLOGY] In deoxyribonucleic acid (DNA) replication, the 3′ to 5′ DNA strand that is discontinuously synthesized as a series of Okazaki fragments in the 5′ to 3′ direction. { 'lag·iŋ ˌstrand }

lag load [GENETICS] A type of genetic load that measures the rate of evolution of a species relative to changes in the environment. { 'lag ˌlōd }

Lagomorpha [VERTEBRATE ZOOLOGY] The order of mammals including rabbits, hares, and pikas; differentiated from rodents by two pairs of upper incisors covered by enamel, vertical or transverse jaw motion, three upper and two lower premolars, fused tibia and fibula, and a spiral valve in the cecum. { ˌlag·ə'mȯr·fə }

lag phase [MICROBIOLOGY] The period of physiological activity and diminished cell division following the addition of inoculum of bacteria to a new culture medium. { 'lag ˌfāz }

Lagriidae [INVERTEBRATE ZOOLOGY] The long-jointed bark beetles, a family of coleopteran insects in the superfamily Tenebrionoidea. { lə'grī·əˌdē }

Lagynacea [INVERTEBRATE ZOOLOGY] A superfamily of foraminiferan protozoans in the suborder Allogromiina having a free or attached test that has a membranous to tectinous wall and a single, ovoid, tubular, or irregular chamber. { ˌlag·ə'nās·ē·ə }

lake ball [ECOLOGY] A spherical mass of tangled, waterlogged fibers and other filamentous material of living or dead vegetation, produced mechanically along a lake bottom by wave action, and usually impregnated with sand and fine-grained mineral fragments. Also known as burr ball; hair ball. { 'lāk ˌbȯl }

Lamarckism [EVOLUTION] The theory that organic evolution takes place through the inheritance of modifications caused by the environment, and by the effects of use and disuse of organs. { lə'märˌkiz·əm }

lamb [VERTEBRATE ZOOLOGY] A young sheep. { lam }

lamb's wool [VERTEBRATE ZOOLOGY] The first fleece taken from a sheep up to 7 months old, having natural tapered fiber tip and spinning qualities superior to those of wool taken from previously shorn sheep. { 'lamz ˌwu̇l }

lamella [ANATOMY] A thin scale or plate. { lə 'mel·ə }

lamellar bone [HISTOLOGY] Any bone with a microscopic structure consisting of thin layers or plates. { lə'mel·ər ¦bōn }

lamellar chloroplast [CELL BIOLOGY] A type of chloroplast in which the layered structure extends more or less uniformly through the whole chloroplast body. { lə'mel·ər 'klȯr·əˌplast }

Lamellibranchiata [INVERTEBRATE ZOOLOGY] An equivalent name for Bivalvia. { ləˌmel·əˌbraŋ·kē 'äd·ə }

Lamellisabellidae [INVERTEBRATE ZOOLOGY] A family of marine animals in the order Thecanephria. { ləˌmel·ə·sə'bel·əˌdē }

Lamiaceae [BOTANY] An equivalent name for Labiatae. { ˌlā·mē'ās·ēˌē }

Lamiales [BOTANY] An order of dicotyledonous plants in the subclass Asteridae marked by its characteristic gynoecium, consisting of usually two biovulate carpels, with each carpel divided between the ovules by a false partition, or with the two halves of the carpel seemingly wholly separate. { ˌlā·mē'ā·lēz }

lamina [ANATOMY] A thin sheet or layer of tissue; a scalelike structure. [BOTANY] *See* blade. { 'lam·ə·nə }

lamina cribrosa [ANATOMY] **1.** The portion of the sclera which is perforated for the passage of the optic nerve. **2.** The fascia covering the saphenous opening in the thigh. **3.** The anterior or posterior perforated space of the brain. **4.** The perforated plates of bone through which pass branches of the cochlear part of the vestibulocochlear nerve. { 'lam·ə·nə krə'brō·sə }

laminal placentation [BOTANY] Condition in which the ovules occur on the inner surface of the carpels. { 'lam·ən·əl ˌplas·ən'tā·shən }

Laminariales [BOTANY] An order of brown, large, structurally complicated, often highly differentiated members, commonly called kelps, of the algal class Phaeophyceae; distinctive features include a life history in which microscopic, filamentous, dioecious gametophytes alternate with a massive, parenchymatous sporophyte, and a mature sporophyte typically consisting of a holdfast, stipe, and one or more blades. { ˌlam·əˌnar·ē'ā·lēz }

Laminariophyceae [BOTANY] A class of algae belonging to the division Phaeophyta. { ˌlam·iˌnar·ē·ō'fīs·ēˌē }

lamina terminalis [ANATOMY] The layer of gray matter in the brain connecting the optic chiasma and the anterior commissure where the latter becomes continuous with the rostral lamina. { 'lam·ə·nə ˌtər·mə'nāl·is }

lampbrush chromosome [CELL BIOLOGY] An exceptionally large chromosome characterized by fine lateral projections which are associated with active ribonucleic acid and protein synthesis. { 'lamp ˌbrəsh 'krō·məˌsōm }

lamprey [VERTEBRATE ZOOLOGY] The common

name for all members of the order Petromyzonida. { 'lam·prē }

Lampridiformes [VERTEBRATE ZOOLOGY] An order of teleost fishes characterized by a compressed, often ribbonlike body, fins composed of soft rays, a ductless swim bladder, and protractile maxillae among other distinguishing features. { ˌlam·prid·ə'fōrˌmēz }

Lamprocystis [MICROBIOLOGY] A genus of bacteria in the family Chromatiaceae; cells are spherical and motile, have gas vacuoles, and contain bacteriochlorophyll *a* on vesicular photosynthetic membranes. { ˌlam·prə'sis·təs }

Lampropedia [MICROBIOLOGY] A genus of gram-negative, obligately anaerobic cocci of uncertain affiliation; cells form pairs, tetrads, or flat squared tablets. { ˌlam·prə'pēd·ē·ə }

Lampyridae [INVERTEBRATE ZOOLOGY] The firefly beetles, a large cosmopolitan family of coleopteran insects in the superfamily Cantharoidea. { lam'pir·əˌdē }

lanatoside [BIOCHEMISTRY] Any of three natural glycosides from the leaves of *Digitalis lanata*; on hydrolysis with acid, it yields one molecule of D-glucose, three molecules of digitoxose, and one molecule of acetic acid; all three glycosides are cardioactive. { lə'nad·əˌsīd }

Lancefield groups [MICROBIOLOGY] Antigenically determined categories for classification of β-hemolytic streptococci. { 'lansˌfēld ˌgrüps }

lancelet [ZOOLOGY] The common name for members of the subphylum Cephalochordata. { 'lans·lət }

lanceolate [BIOLOGY] Shaped like the head of a lance. { 'lan·sē·əˌlāt }

Lanceolidae [INVERTEBRATE ZOOLOGY] A family of bathypelagic amphipod crustaceans in the suborder Hyperiidea. { ˌlan·sē'äl·əˌdē }

landmark [CELL BIOLOGY] Any distinctive feature that can be used to identify a chromosome. { 'lanˌmärk }

Langerhans cell [HISTOLOGY] **1.** A type of cytotrophoblast in the human chorionic vesicle which is thought to secrete chorionic gonadotropin. **2.** A highly branched dendritic cell of the mammalian epidermis showing a lobulated nucleus and a diagnostic organelle resembling a tennis racket. { 'läŋ·ərˌhänz ˌsel }

Lange's nerve [INVERTEBRATE ZOOLOGY] One of the paired cords of nervous tissue lying in the wall of the radial perihemal canal of asteroids. { 'läŋ·əz ˌnərv }

Lang's vesicle [INVERTEBRATE ZOOLOGY] A seminal bursa in many polyclad flatworms. { 'laŋz ¦ves·ə·kəl }

Languriidae [INVERTEBRATE ZOOLOGY] The lizard beetles, a cosmopolitan family of coleopteran insects in the superfamily Cucujoidea. { ˌlaŋ·gə'rī·əˌdē }

Laniatores [INVERTEBRATE ZOOLOGY] A suborder of arachnids in the order Phalangida having flattened, often colorful bodies and found chiefly in tropical areas. { ˌlan·ē·ə'tȯr·ēz }

lanosterol [BIOCHEMISTRY] $C_{30}H_{50}O$ An unsaturated sterol occurring in wool fat and yeast. { lə'näs·təˌrȯl }

lantern fish [VERTEBRATE ZOOLOGY] The common name for the deep-sea teleost fishes composing the family Myctophidae and distinguished by luminous glands that are widely distributed upon the body surface. { 'lan·tərn ˌfish }

Lanthonotidae [VERTEBRATE ZOOLOGY] A family of lizards (Sauria) belonging to the Anguimorpha line; restricted to North Borneo. { lan·thə'näd·əˌdē }

lanugo [ANATOMY] A downy covering of hair, especially that seen on the fetus or persisting on the adult body. { lə'nü·gō }

lapachoic acid [BIOCHEMISTRY] *See* lapachol. { lə'päch·ə·wik 'as·əd }

lapachol [BIOCHEMISTRY] $C_{15}H_{14}O_3$ A yellow crystalline compound obtained from lapacho, a hardwood in Argentina and Paraguay. Also known as lapachoic acid; targusic acid. { lə'päˌchȯl }

lapidicolous [ECOLOGY] Living under a stone. { ¦lap·ə¦dik·ə·ləs }

lappet [ZOOLOGY] A lobe or flaplike projection, such as on the margin of a jellyfish or the wattle of a bird. { 'lap·ət }

larch [BOTANY] The common name for members of the genus *Larix* of the pine family, having deciduous needles and short, spurlike branches which annually bear a crown of needles. { lärch }

large intestine [ANATOMY] *See* colon. { 'lärj in'tes·tən }

Largidae [INVERTEBRATE ZOOLOGY] A family of hemipteran insects in the superfamily Pentatomorpha. { 'lär·jəˌdē }

Laridae [VERTEBRATE ZOOLOGY] A family of birds in the order Charadriiformes composed of the gulls and terns. { 'lar·əˌdē }

Larinae [VERTEBRATE ZOOLOGY] A subfamily of birds in the family Laridae containing the gulls and characterized by a thick, slightly hooked beak, a square tail, and a stout white body, with shades of gray on the back and the upper wing surface. { 'lar·əˌnē }

larva [INVERTEBRATE ZOOLOGY] An independent, immature, often vermiform stage that develops from the fertilized egg and must usually undergo a series of form and size changes before assuming characteristic features of the parent. { 'lär·və }

Larvacea [INVERTEBRATE ZOOLOGY] A class of the subphylum Tunicata consisting of minute planktonic animals in which the tail, with dorsal nerve cord and notochord, persists throughout life. { lär'vā·shē·ə }

Larvaevoridae [INVERTEBRATE ZOOLOGY] The tachina flies, a large family of dipteran insects in the suborder Cyclorrhapha distinguished by a thick covering of bristles on the body; most are parasites of arthropods. { ˌlär·və'vȯr·əˌdē }

larva migrans [INVERTEBRATE ZOOLOGY] Fly larva, *Hypoderma* or *Gastrophilus*, that produces a creeping eruption in the dermis. { 'lär·və 'mīˌgranz }

larviporous [INVERTEBRATE ZOOLOGY] Feeding on larva, referring especially to insects. { lär'vip·ə·rəs }

laryngeal pouch [VERTEBRATE ZOOLOGY] A lateral

saclike expansion of the cavity of the larynx that is greatly developed in certain monkeys. { lə 'rin·jē·əl 'paúch }

laryngopharynx [ANATOMY] The lower portion of the pharynx, lying adjacent to the larynx. Also known as hypopharynx. { lə¦riŋ·gō'far·iŋks }

laryngotracheal groove [EMBRYOLOGY] A channel in the floor of the pharynx serving as the anlage of the respiratory system. { lə¦riŋ·gō'trā·kē·əl ˌgrüv }

laryngotracheobronchitis virus [VIROLOGY] *See* croup-associated virus. { lə¦riŋ·gō¦trā·kē·ō ˌbraŋ'kīd·əs 'vī·rəs }

larynx [ANATOMY] The complex of cartilages and related structures at the opening of the trachea into the pharynx in vertebrates; functions in protecting the entrance of the trachea, and in phonation in higher forms. { 'larˌiŋks }

Lasiocampidae [INVERTEBRATE ZOOLOGY] The tent caterpillars and lappet moths, a family of cosmopolitan (except New Zealand) lepidopteran insects in the suborder Heteroneura. { ˌlas·ē·ō'kam·pəˌdē }

lasso cell [INVERTEBRATE ZOOLOGY] *See* adhesive cell. { 'las·ō ˌsel }

late gene [GENETICS] Any gene expressed late in the life cycle. { 'lāt ˌjēn }

latency [PHYSIOLOGY] The period between the introduction of and the response to a stimulus. { 'lat·ən·sē }

latent bud [BOTANY] An axillary bud whose development is inhibited, sometimes for many years, due to the influence of apical and other buds. Also known as dormant bud. { 'lāt·ənt 'bəd }

latent period [PHYSIOLOGY] The period between the introduction of a stimulus and the response to it. [VIROLOGY] The initial period of phage growth after infection during which time virus nucleic acid is manufactured by the host cell. { 'lāt·ənt ¦pir·ē·əd }

latent virus [VIROLOGY] A virus that remains dormant within body cells but can be reactivated by conditions such as reduced host defenses, toxins, or irradiation, to cause disease. { ¦lāt·ənt 'vī·rəs }

latent-virus infection [VIROLOGY] A persistent viral infection in which there is little or no demonstrable presence of the virus and disease symptoms for a long time between episodes of recurrent outbreaks. { 'lāt·ənt ¦vī·rəs inˌfek·shən }

laterad [ANATOMY] Toward the lateral aspect. { 'lad·əˌrad }

lateral [ANATOMY] At, pertaining to, or in the direction of the side; on either side of the medial vertical plane. { 'lad·ə·rəl }

lateral bud [BOTANY] Any bud that develops on the side of a stem. { 'lad·ə·rəl 'bəd }

lateralia [INVERTEBRATE ZOOLOGY] Paired sensory bristles on the lateral aspect of the head of gnathostomulids. { ˌlad·ə'rāl·yə }

lateral lemniscus [ANATOMY] The secondary auditory pathway arising in the cochlear nuclei and terminating in the inferior colliculus and medial geniculate body. { 'lad·ə·rəl lem'nis·kəs }

lateral line [INVERTEBRATE ZOOLOGY] A longitudinal lateral line along the sides of certain oligochaetes consisting of cell bodies of the layer of circular muscle. [VERTEBRATE ZOOLOGY] A line along the sides of the body of most fishes, often distinguished by differently colored scales, which marks the lateral line organ. { 'lad·ə·rəl 'līn }

lateral-line organ [VERTEBRATE ZOOLOGY] A small, pear-shaped sense organ in the skin of many fishes and amphibians that is sensitive to pressure changes in the surrounding water. { 'lad·ə·rəl ˌlīn ˌȯr·gən }

lateral-line system [VERTEBRATE ZOOLOGY] The complex of lateral-line end organs and nerves in skin on the sides of many fishes and amphibians. { 'lad·ə·rəl ˌlīn ˌsis·təm }

lateral meristem [BOTANY] Strips or cylinders of dividing cells located parallel to the long axis of the organ in which they occur; the lateral meristem functions to increase the diameter of the organ. { 'lad·ə·rəl 'mer·əˌstem }

lateral root [BOTANY] A root branch arising from the main axis. { 'lad·ə·rəl 'rüt }

lateral ventricle [ANATOMY] The cavity of a cerebral hemisphere; communicates with the third ventricle by way of the interventricular foramen. { 'lad·ə·rəl 'ven·trə·kəl }

latewood [BOTANY] The portion of the annual ring that is formed after formation of earlywood has ceased. { 'lātˌwu̇d }

latex cell [BIOLOGY] A coenocytic cell of a lactiferous duct in a latex producing plant. { 'lāˌteks ˌsel }

latex vessel [BOTANY] An elongated cell joined end to end with other like cells to form a type of lactiferous duct. { 'lāˌteks ˌves·əl }

Lathridiidae [INVERTEBRATE ZOOLOGY] The minute brown scavenger beetles, a large cosmopolitan family of coleopteran insects in the superfamily Cucujoidea. { ˌlath·rə'dī·əˌdē }

laticifer [BOTANY] A latex duct found in the midcortex of certain plants. { lā'tis·ə·fər }

latiferous [BOTANY] Containing or secreting latex. { lā'tif·ə·rəs }

Latimeridae [VERTEBRATE ZOOLOGY] A family of deep-sea lobefin fishes (Coelacanthiformes) known from a single living species, *Latimeria chalumnae*. { ˌlad·ə'mer·əˌdē }

lattissimus dorsi [ANATOMY] The widest muscle of the back; a broad, flat muscle of the lower back that adducts and extends the humerus, is used to pull the body upward in climbing, and is an accessory muscle of respiration. { lə'tis·ə·məs 'dȯr·sē }

Laugiidae [PALEONTOLOGY] A family of Mesozoic fishes in the order Coelacanthiformes. { lau̇'jī·ə ˌdē }

Lauraceae [BOTANY] The laurel family of the order Magnoliales distinguished by definite stamens in series of three, a single pistil, and the lack of petals. { lȯ'rās·ēˌē }

Laurales [BOTANY] An order of dicotyledonous flowering, mostly woody plants of the class Magnoliopsida, division Magnoliophyta; commonly have scattered spherical cells containing volatile oils, leaves usually simple and mostly entire, and flowers often pollinated by beetles. { lȯ'rā·lēz }

Lauratonematidae [INVERTEBRATE ZOOLOGY] A family of marine nematodes of the superfamily Enoploidea; many females possess a cloaca. { ˌlȯr·ə·tō·nē'mad·əˌdē }

laurel forest [ECOLOGY] *See* temperate rainforest. { 'lȯr·əl 'fär·əst }

Laurer's canal [INVERTEBRATE ZOOLOGY] In certain flukes, a canal which passes from the oviduct to the ventral surface of the body. { 'lȯr·ərz kəˌnal }

laurisilva [ECOLOGY] *See* temperate rainforest. { ¦lȯr·ə¦sil·və }

Lauxaniidae [INVERTEBRATE ZOOLOGY] A family of myodarian cyclorrhaphous dipteran insects in the subsection Acalypteratae; larvae are leaf miners. { lȯk·sə'nī·əˌdē }

LAV [VIROLOGY] *See* lymphadenopathy-associated virus.

law of minimum [BIOLOGY] The law that those essential elements for which the ratio of supply to demand (A/N) reaches a minimum will be the first to be removed from the environment by life processes; it was proposed by J. von Liebig, who recognized phosphorus, nitrogen, and potassium as minimum in the soil; in the ocean the corresponding elements are phosphorus, nitrogen, and silicon. Also known as Liebig's law of minimum. { 'lȯ əv 'min·ə·məm }

law of specificity of bacteria [MICROBIOLOGY] *See* Koch's postulates. { 'lȯ əv ¦spes·ə¦fis·əd·ē əv bak'tir·ē·ə }

layering [BOTANY] A propagation method by which root formation is induced on a branch or a shoot attached to the parent stem by covering the part with soil. [ECOLOGY] A stratum of plant forms in a community, such as mosses, shrubs, or trees in a bog area. { 'lā·ə·riŋ }

LC_{50} [PHYSIOLOGY] *See* lethal concentration 50. { ¦el¦sē 'fif·tē }

L chain [IMMUNOLOGY] *See* light chain. { 'el ˌchān }

LDH [BIOCHEMISTRY] *See* lactic dehydrogenase.

leading strand [MOLECULAR BIOLOGY] In deoxyribonucleic acid (DNA) replication, the 5′ to 3′ DNA strand that is synthesized with few or no interruptions. { 'lēd·iŋ ˌstrand }

leaf [BOTANY] A modified aerial appendage which develops from a plant stem at a node, usually contains chlorophyll, and is the principal organ in which photosynthesis and transpiration occur. { lēf }

leaf bud [BOTANY] A bud that produces a leafy shoot. { 'lēf ˌbəd }

leaf cushion [BOTANY] The small part of the thickened leaf base that remains after abscission in various conifers, and also in some extinct plants. { 'lēf ˌku̇sh·ən }

leaf fiber [BOTANY] A long, multiple-celled fiber extracted from the leaves of many plants that is used for cordage, such as sisal for binder, and abaca for manila hemp. { 'lēf ˌfī·bər }

leaf gap [BOTANY] The place where the vascular bundle of the stem interrupts above a leaf trace as a result of the diversion of vascular tissue from the stem into a leaf, occurring in many vascular plants. { 'lēf ˌgap }

leafhopper [INVERTEBRATE ZOOLOGY] The common name for members of the homopteran family Cicadellidae. { 'lēfˌhäp·ər }

leaflet [BOTANY] **1.** A division of a compound leaf. **2.** A small or young foliage leaf. { 'lēf·lət }

leaf miner [INVERTEBRATE ZOOLOGY] Any of the larvae of various insects which burrow into and eat the parenchyma of leaves. { 'lēf ˌmīn·ər }

leaf-nosed [VERTEBRATE ZOOLOGY] Having a leaflike membrane on the nose, as certain bats. { 'lēf ˌnōzd }

leaf primordium [BOTANY] An immature leaf that arises as an emergence on the flanks of the apical meristem of the shoot tip. { 'lēf prī'mȯr·dē·əm }

leaf scar [BOTANY] A mark on a stem, formed by secretion of suberin and a gumlike substance, showing where a leaf has abscised. { 'lēf ˌskär }

leaf scorch [BOTANY] Any of several disorders and fungus diseases marked by a burned appearance of the leaves; for example, caused by the fungus *Diplocarpon earliana* in strawberry. { 'lēf ˌskȯrch }

leaf trace [BOTANY] A section of the vascular bundle that leads from the stele to the base of the leaf. { 'lēf ˌtrās }

leaky [GENETICS] Pertaining to a genetic block that is incomplete. [MOLECULAR BIOLOGY] Pertaining to a protein coded for by a mutant gene that shows subnormal activity. { 'lēk·ē }

leaky mutant gene [GENETICS] An allele with reduced activity relative to that of the normal allele. { 'lēk·ē ¦myüt·ənt 'jēn }

Lecanicephaloidea [INVERTEBRATE ZOOLOGY] An order of tapeworms of the subclass Cestoda distinguished by having the scolex divided into two portions; all species are intestinal parasites of elasmobranch fishes. { leˌkan·əˌsef·ə'lȯid·ē·ə }

Lecanoraceae [BOTANY] A temperate and boreal family of lichens in the order Lecanorales characterized by a crustose thallus and a distinct thalloid rim on the apothecia. { le·kə·nō'rās·ēˌē }

Lecanorales [BOTANY] An order of the Ascolichenes having open, discoid apothecia with a typical hymenium and hypothecium. { le·kə·nō'rā·lēz }

Lecideaceae [BOTANY] A temperate and boreal family of lichens in the order Lecanorales; members lack a thalloid rim around the apothecia. { ləˌsid·ē'ās·ēˌē }

lecithin [BIOCHEMISTRY] Any of a group of phospholipids having the general composition $CH_2OR_1 \cdot CHOR_2 \cdot CH_2OPO_2OHR_3$, in which R_1 and R_2 are fatty acids and R_3 is choline, and with emulsifying, wetting, and antioxidant properties. { 'les·ə·thən }

lecithinase [BIOCHEMISTRY] An enzyme that catalyzes the breakdown of a lecithin into its constituents. { 'les·ə·thəˌnās }

lecithinase A [BIOCHEMISTRY] An enzyme that catalyzes the removal of only one fatty acid from lecithin, yielding lipolecithin. { 'les·ə·thəˌnās 'ā }

lecithinase C [BIOCHEMISTRY] An enzyme that catalyzes the removal of the nitrogenous base of lecithin to produce the base and a phosphatidic acid. { 'les·ə·thə,nās 'sē }

lecithinase D [BIOCHEMISTRY] An enzyme that catalyzes the removal of the phosphorylated base from lecithins, producing α-β-diglyceride. { 'les·ə·thə,nās 'dē }

lectin [BIOCHEMISTRY] Any of various proteins that agglutinate erythrocytes and other types of cells and also have other properties, including mitogenesis, agglutination of tumor cells, and toxicity toward animals; found widely in plants, predominantly in legumes, and also occurring in bacteria, fish, and invertebrates. { 'lek·tən }

lectotype [SYSTEMATICS] A specimen selected as the type of a species or subspecies if the type was not designated by the author of the classification. { 'lek·tə,tīp }

Lecythidaceae [BOTANY] The single family of the order Lecythidales. { ¦les·ə·thə'dās·ē,ē }

Lecythidales [BOTANY] A monofamilial order of dicotyledonous tropical woody plants in the subclass Dilleniidae; distinguished by entire leaves, valvate sepals, separate petals, numerous centrifugal stamens, and a syncarpous, inferior ovary with axile placentation. { ¦les·ə·thə'dā·lēz }

Lederberg technique [MICROBIOLOGY] A method for rapid isolation of individual bacterial cells for demonstrating the spontaneous origin of bacterial mutants. { 'lā·də,berk tek,nēk }

Leeaceae [BOTANY] A family of dicotyledonous plants in the order Rhamnales distinguished by solitary ovules in each locule, simple to compound leaves, a small embryo, and hypogynous flowers. { lē'ās·ē,ē }

leech [INVERTEBRATE ZOOLOGY] The common name for members of the annelid class Hirudinea. { lēch }

leek [BOTANY] *Allium porrum*. A biennial herb known only by cultivation; grown for its mildly pungent succulent leaves and thick cylindrical stalk. { lēk }

left splicing junction [MOLECULAR BIOLOGY] *See* donor splicing site. { 'left ,splīs·iŋ ,jəŋk·shən }

left ventricular thrust [PHYSIOLOGY] *See* apex impulse. { 'left ven¦trik·yə·lər 'thrəst }

leg [ANATOMY] The lower extremity of a human limb, between the knee and the ankle. [ZOOLOGY] An appendage or limb used for support and locomotion. { leg }

legena [VERTEBRATE ZOOLOGY] An appendage of the sacculus containing sensory areas in the inner ear of tetrapods; termed the cochlea in humans. { lə'jē·nə }

legume [BOTANY] A dry, dehiscent fruit derived from a single simple pistil; common examples are alfalfa, beans, peanuts, and vetch. { lə 'gyüm }

Leguminosae [BOTANY] The legume family of the plant order Rosales characterized by stipulate, compound leaves, 10 or more stamens, and a single carpel; many members harbor symbiotic nitrogen-fixing bacteria in their roots. { lə ,gyüm·ə'nō·sē }

Leiodidae [INVERTEBRATE ZOOLOGY] The round carrion beetles, a cosmopolitan family of coleopteran insects in the superfamily Staphylinoidea; commonly found under decaying bark. { lī'äd·ə,dē }

leiosporous [MYCOLOGY] Having smooth spores. { lī'äs·pə·rəs }

Leishmania [INVERTEBRATE ZOOLOGY] A genus of flagellated protozoan parasites that are the etiologic agents of several diseases of humans, such as leishmaniasis. { lēsh'man·ē·ə }

Leishmania donovani [INVERTEBRATE ZOOLOGY] The protozoan parasite that causes kala-azar. { lēsh'man·ē·ə ¦dan·ō¦vän·ē }

Leishmania infantum [INVERTEBRATE ZOOLOGY] The protozoan parasite that causes infantile leishmaniasis. { lēsh'man·ē·ə in'fan·təm }

Leitneriales [BOTANY] A monofamilial order of flowering plants in the subclass Hamamelidae; members are simple-leaved, dioecious shrubs with flowers in catkins, and have a superior, pseudomonomerous ovary with a single ovule. { ,līt·nir·ē'ā·lēz }

lek [ZOOLOGY] A gathering place for courtship. Also known as arena. { lek }

Lelapiidae [INVERTEBRATE ZOOLOGY] A family of calcaronean sponges in the order Sycettida characterized by a rigid skeleton composed of tracts or bundles of modified triradiates. { le·lə'pī·ə ,dē }

lemma [BOTANY] Either of the pair of bracts that are borne above the glumes and enclose the flower of a grass spikelet. { 'lem·ə }

lemming [VERTEBRATE ZOOLOGY] The common name for the small burrowing rodents composing the subfamily Microtinae. { 'lem·iŋ }

Lemnaceae [BOTANY] The duckweeds, a family of monocotyledonous plants in the order Arales; members are small, free-floating, thalloid aquatics with much reduced flowers that form a miniature spadix. { lem'nās·ē,ē }

lemniscus [ANATOMY] A secondary sensory pathway of the central nervous system, usually terminating in the thalamus. { lem'nis·kəs }

lemon [BOTANY] *Citrus limon*. A small evergreen tree belonging to the order Sapindales cultivated for its acid citrus fruit which is a modified berry called a hesperidium. { 'lem·ən }

lemur [VERTEBRATE ZOOLOGY] The common name for members of the primate family Lemuridae; characterized by long tails, foxlike faces, and scent glands on the shoulder region and wrists. { 'lē·mər }

Lemuridae [VERTEBRATE ZOOLOGY] A family of prosimian primates of Madagascar belonging to the Lemuroidea; all members are arboreal forest dwellers. { lə'myůr·ə,dē }

Lemuroidea [VERTEBRATE ZOOLOGY] A suborder or superfamily of Primates including the lemurs, tarsiers, and lorises, or sometimes simply the lemurs. { ,lem·ə'ròid·ē·ə }

lengthening reaction [PHYSIOLOGY] Sudden inhibition of the stretch reflex when extensor muscles are subjected to an excessive degree of stretching by forceful flexion of a limb. { 'leŋk·thə,niŋ rē,ak·shən }

lenitic [ECOLOGY] *See* lentic. { lə'nid·ik }

lens [ANATOMY] A transparent, encapsulated, nearly spherical structure located behind the pupil of vertebrate eyes, and in the complex eyes of many invertebrates, that focuses light rays on the retina. Also known as crystalline lens. { lenz }

lens placode [EMBRYOLOGY] The ectodermal anlage of the lens of the eye; its formation is induced by the presence of the underlying optic vesicle. { 'lenz 'pla,kōd }

lentic [ECOLOGY] Of or pertaining to still waters such as lakes, reservoirs, ponds, and bogs. Also spelled lenitic. { 'len·tik }

lenticel [BOTANY] A loose-structured opening in the periderm beneath the stomata in the stem of many woody plants that facilitates gas transport. { 'len·tə·səl }

lentiginose [ANATOMY] Of or pertaining to pigment spots in the skin; freckled. { len'tij·ə ,nōs }

lentil [BOTANY] *Lens esculenta*. A semiviny annual legume having pinnately compound, vetchlike leaves; cultivated for its thin, lens-shaped, edible seed. { 'lent·əl }

leopard [VERTEBRATE ZOOLOGY] *Felis pardus*. A species of wildcat in the family Felidae found in Africa and Asia; the coat is characteristically buff-colored with black spots. { 'lep·ərd }

Leotichidae [INVERTEBRATE ZOOLOGY] A small Oriental family of hemipteran insects in the superfamily Leptopodoidea. { ,lē·ə'tik·ə,dē }

Lepadomorpha [INVERTEBRATE ZOOLOGY] A suborder of barnacles in the order Thoracica having a peduncle and a capitulum which is usually protected by calcareous plates. { ,lep·ə·də'mȯr·fə }

Leperditicopida [PALEONTOLOGY] An order of extinct ostracods characterized by very thick, straight-backed valves which show unique muscle scars and other markings. { ,le·pər,did·ə 'käp·ə·də }

Leperditillacea [PALEONTOLOGY] A superfamily of extinct paleocopan ostracods in the suborder Kloedenellocopina including the unisulcate, nondimorphic forms. { ,le·pər,did·ə'lās·ē·ə }

Lepiceridae [INVERTEBRATE ZOOLOGY] Horn's beetle, a family of Central American coleopteran insects composed of two species. { ,lep·ə'ser·ə ,dē }

Lepidocentroida [INVERTEBRATE ZOOLOGY] The name applied to a polyphyletic assemblage of echinoids that are now regarded as members of the Echinocystitoida and Echinothurioida. { ,lep·ə·dō,sen'trȯid·ə }

Lepidodendrales [PALEOBOTANY] The giant club mosses, an order of extinct lycopods (Lycopodiopsida) consisting primarily of arborescent forms characterized by dichotomous branching, small amounts of secondary vascular tissue, and heterospory. { ,lep·ə·dō,den'drā·lēz }

lepidophyllous [BOTANY] Having scaly leaves. { ¦lep·ə·dō¦fil·əs }

Lepidoptera [INVERTEBRATE ZOOLOGY] Large order of scaly-winged insects, including the butterflies, skippers, and moths; adults are characterized by two pairs of membranous wings and sucking mouthparts, featuring a prominent, coiled proboscis. { ,lep·ə'däp·tə·rə }

Lepidorsirenidae [VERTEBRATE ZOOLOGY] A family of slender, obligate air-breathing, eellike fishes in the order Dipteriformes having small thin scales, slender ribbonlike paired fins, and paired ventral lungs. { ,lep·ə·dō·sə'ren·ə,dē }

Lepidosaphinae [INVERTEBRATE ZOOLOGY] A family of homopteran insects in the superfamily Coccoidea having dark-colored, noncircular scales. { ,lep·ə·dō'saf·ə,nē }

Lepidosauria [VERTEBRATE ZOOLOGY] A subclass of reptiles in which the skull structure is characterized by two temporal openings on each side which have reduced bony arcades, and by the lack of an antorbital opening in front of the orbit. { ,lep·ə·dō'sȯr·ē·ə }

Lepidotrichidae [INVERTEBRATE ZOOLOGY] A family of wingless insects in the order Thysanura. { ,lep·ə·dō'trik·ə,dē }

Lepismatidae [INVERTEBRATE ZOOLOGY] A family of silverfish in the order Thysanura characterized by small or missing compound eyes. { ,lep·əz 'mad·ə,dē }

Lepisostei [VERTEBRATE ZOOLOGY] An equivalent name for Semionotiformes. { ,lep·ə'säs·tē,ī }

Lepisosteidae [VERTEBRATE ZOOLOGY] A family of fishes in the order Semionotiformes. { ,lep·ə·säs'tē·ə,dē }

Lepisosteiformes [VERTEBRATE ZOOLOGY] An equivalent name for Semionotiformes. { ,lep·ə ,säs·tē·ə'fȯr,mēz }

Leporidae [VERTEBRATE ZOOLOGY] A family of mammals in the order Lagomorpha including the rabbits and hares. { lə'pȯr·ə,dē }

Lepospondyli [PALEONTOLOGY] A subclass of extinct amphibians including all forms in which the vertebral centra are formed by ossification directly around the notochord. { ,lep·ə'spänd·əl,ī }

lepospondylous [VERTEBRATE ZOOLOGY] Having the notochord enclosed by cylindrical vertebrae shaped like an hourglass in longitudinal section. { ,lep·ə'spänd·əl·əs }

lepromin [IMMUNOLOGY] An emulsion of ground lepromata containing the leprosy bacillus; used for intradermal skin tests in Hansen's disease. { lə'prō·mən }

Leptaleinae [INVERTEBRATE ZOOLOGY] A subfamily of the Formicidae including largely arboreal ant forms which inhabit plants in tropical and subtropical regions. { ,lep·tə'lī,nē }

Leptictidae [PALEONTOLOGY] A family of extinct North American insectivoran mammals belonging to the Proteutheria which ranged from the Cretaceous to middle Oligocene. { ,lep'tik·tə ,dē }

Leptinidae [INVERTEBRATE ZOOLOGY] The mammal nest beetles, a small European and North American family of coleopteran insects in the superfamily Staphylinoidea. { lep'tin·ə,dē }

Leptocardii [ZOOLOGY] The equivalent name for Cephalochordata. { ,lep·tə'kärd·ē,ī }

leptocephalous larva [VERTEBRATE ZOOLOGY] The marine larva of the fresh-water European eel *Anguilla vulgaris*. { ,lep·tə'sef·ə·ləs 'lär·və }

leptocercal [VERTEBRATE ZOOLOGY] Of the tail of a fish, tapering to a long, slender point. { ¦lep·tə¦sər·kəl }

Leptochoeridae [PALEONTOLOGY] An extinct family of palaeodont artiodactyl mammals in the superfamily Dichobunoidea. { ˌlep·tə'kir·əˌdē }

Leptodactylidae [VERTEBRATE ZOOLOGY] A large family of frogs in the suborder Procoela found principally in the American tropics and Australia. { ˌlep·təˌdak'til·əˌdē }

leptodactylous [VERTEBRATE ZOOLOGY] Having slender toes, as certain birds. { ¦lep·tə¦dak·tə·ləs }

Leptodiridae [INVERTEBRATE ZOOLOGY] The small carrion beetles, a cosmopolitan family of coleopteran insects in the superfamily Staphylinoidea. { ˌlep·tə'dir·əˌdē }

Leptolepidae [PALEONTOLOGY] An extinct family of fishes in the order Leptolepiformes representing the first teleosts as defined on the basis of the advanced structure of the caudal skeleton. { ˌlep·tə'lep·əˌdē }

Leptolepiformes [PALEONTOLOGY] An extinct order of small, ray-finned teleost fishes characterized by a relatively strong, ossified axial skeleton, thin cycloid scales, and a preopercle with an elongated dorsal portion. { ˌlep·təˌlep·ə'fōr ˌmēz }

Leptomedusae [INVERTEBRATE ZOOLOGY] A suborder of hydrozoan cnidarians in the order Hydroida characterized by the presence of a hydrotheca. { ¦lep·tō·mə'dü·sē }

leptomeninges [ANATOMY] The pia mater and arachnoid considered together. { ¦lep·tō·mə 'nin·jēz }

Leptomitales [MYCOLOGY] An order of aquatic Phycomycetes characterized by a hyphal thallus, or basal rhizoids and terminal hyphae, and zoospores with two flagella. { ˌlep·tō·mī'tā·lēz }

leptophyll [ECOLOGY] A growth-form class of plants having a leaf surface area of 0.04 square inch (25 square millimeters) or less; common in alpine and desert habitats. { 'lep·təˌfil }

Leptopodidae [INVERTEBRATE ZOOLOGY] A tropical and subtropical family of hemipteran insects in the superfamily Leptopodoidea distinguished by the spiny body and appendages. { ˌlep·tə 'päd·əˌdē }

Leptopodoidea [INVERTEBRATE ZOOLOGY] A superfamily of hemipteran insects in the subdivision Geocorisae. { ˌlep·tə·pə'dȯid·ē·ə }

Leptosomatidae [VERTEBRATE ZOOLOGY] The cuckoo rollers, a family of Madagascan birds in the order Coraciiformes composed of a single species distinguished by the downy covering on the newly hatched young. { ˌlep·tə·sə'mad·ə ˌdē }

Leptospira [MICROBIOLOGY] A genus of bacteria in the family Spirochaetaceae; thin, helical cells with bent or hooked ends. { ˌlep·tə'spī·rə }

leptosporangium [BOTANY] A sporangium derived from a single actively dividing cell in a meristem. { ¦lep·tō·spə'ran·jē·əm }

Leptostraca [INVERTEBRATE ZOOLOGY] A primitive group of crustaceans considered as one of a series of Malacostraca distinguished by an additional abdominal somite that lacks appendages, and a telson bearing two movable articulated prongs. { lep'täs·trə·kə }

Leptostromataceae [MYCOLOGY] A family of fungi of the order Sphaeropsidales; pycnidia are black and shield-shaped, circular or oblong, and slightly asymmetrical; included are some fruit-tree pathogens. { ˌlep·təˌstrō·mə'tās·ēˌē }

leptotene [CELL BIOLOGY] The first stage of meiotic prophase, when the chromosomes appear as thin threads having well-defined chromomeres. { 'lep·təˌtēn }

Leptothrix [MICROBIOLOGY] A genus of sheathed bacteria; single cells are motile by means of polar or subpolar flagella, and sheaths are encrusted with iron or manganese oxides. { 'lep·tə ˌthriks }

Leptotrichia [MICROBIOLOGY] A genus of bacteria in the family Bacteroidaceae; straight or slightly curved rods with pointed or rounded ends arranged in filaments. { ˌlep·tə'trik·ē·ə }

Leptotyphlopidae [VERTEBRATE ZOOLOGY] A family of small, harmless, burrowing circumtropical snakes (Serpentes) in the order Squamata; teeth are present only on the lower jaw and are few in number. { ˌlep·tō·ti'fläp·əˌdē }

Lepus [VERTEBRATE ZOOLOGY] The type genus of the family Leporidae, comprising the typical hares. { 'lē·pəs }

Lernaeidae [INVERTEBRATE ZOOLOGY] A family of copepod crustaceans in the suborder Caligoida; all are fixed ectoparasites, that is, they penetrate the skin of fresh-water fish. { lər'nē·əˌdē }

Lernaeopodidae [INVERTEBRATE ZOOLOGY] A family of ectoparasitic crustaceans belonging to the Lernaeopodoida; individuals are attached to the walls of the fishes' gill chambers by modified second maxillae. { ˌlər·nē·ə'päd·əˌdē }

Lernaeopodoida [INVERTEBRATE ZOOLOGY] The fish maggots, a group of ectoparasitic crustaceans characterized by a modified postembryonic development reduced to two or three stages, a free-swimming larva, and the lack of external signs of physical maturity in adults. { ˌlər·nē·ə'pȯid·ē·ə }

lesion [BIOLOGY] A structural or functional alteration due to injury or disease. [MOLECULAR BIOLOGY] A damaged site in a gene, chromosome, or protein molecule. { 'lē·zhən }

Leskeineae [BOTANY] A suborder of mosses in the order Hypnobryales; plants are not complanate, paraphyllia are frequently present, leaves are costate, and alar cells are not generally differentiated. { ˌles'kī·nēˌē }

lespedeza [BOTANY] Any of various legumes of the genus *Lespedeza* having trifoliate leaves, small purple pea-shaped blossoms, and one seed per pod. { ˌles·pə'dē·zə }

lesser circulation [PHYSIOLOGY] *See* pulmonary circulation. { 'les·ər ˌsər·kyə'lā·shən }

lesser omentum [ANATOMY] A fold of the peritoneum extending from the lesser curvature of the stomach to the transverse hepatic fissure. { 'les·ər ō'men·təm }

Lestidae [INVERTEBRATE ZOOLOGY] A family of odonatan insects belonging to the Zygoptera;

distinguished by the wings being held in a V position while at rest. { 'les·tə‚dē }

Lestoniidae [INVERTEBRATE ZOOLOGY] A monospecific family of hemipteran insects in the superfamily Pentatomorpha found only in Australia. { ‚les·tə'nī·ə‚dē }

lethal concentration 50 [PHYSIOLOGY] In a fire, the concentration of a gas or smoke that will kill 50% of the test animals within a given time of exposure. Abbreviated LC_{50}. { ¦leth·əl ‚käns·ən ¦trā·shən 'fif·tē }

lethal equivalent value [GENETICS] The product of the mean number of deleterious genes carried by each member of a population and the mean probability that each gene will cause premature death when homozygous. { ¦lē·thəl i‚kwiv·ə·lənt 'val·yü }

lethal gene [GENETICS] A gene mutation that causes premature death in heterozygotes if dominant, and in homozygotes if recessive. Also known as lethal mutation. { 'lē·thəl 'jēn }

lethal mutation [GENETICS] *See* lethal gene. { 'lē·thəl myü'tā·shən }

lettuce [BOTANY] *Lactuca sativa*. An annual plant of the order Asterales cultivated for its succulent leaves; common varieties are head lettuce, leaf or curled lettuce, romaine lettuce, and iceberg lettuce. { 'led·əs }

Leucaltidae [INVERTEBRATE ZOOLOGY] A family of calcinean sponges in the order Leucettida having numerous small, interstitial, flagellated chambers. { lü'kal·tə‚dē }

Leucascidae [INVERTEBRATE ZOOLOGY] A family of calcinean sponges in the order Leucettida having a radiate arrangement of flagellated chambers. { lü'kas·ə‚dē }

Leucettida [INVERTEBRATE ZOOLOGY] An order of calcareous sponges in the subclass Calcinea having a leuconoid structure and a distinct dermal membrane or cortex. { lü'sed·ə·də }

leucine [BIOCHEMISTRY] $C_6H_{13}O_2N$ A monocarboxylic essential amino acid obtained by hydrolysis of protein-containing substances such as milk. { 'lü‚sēn }

leucine amino peptidase [BIOCHEMISTRY] An enzyme that acts on peptides to catalyze the release of terminal amino acids, especially leucine residues, having free α-amino groups. { 'lü‚sēn ə¦mē·nō 'pep·tə‚dās }

Leucodontineae [BOTANY] A family of mosses in the order Isobryales with foliated branches, often bearing catkins. { ‚lü·kə·dän'tin·ē‚ē }

leucon [INVERTEBRATE ZOOLOGY] A type of sponge having the choanocytes restricted to flagellated chambers inserted between the incurrent and excurrent canals, and a reduced or absent paragastric cavity. { 'lü‚kän }

Leuconostoc [MICROBIOLOGY] A genus of bacteria in the family Streptococcaceae; spherical or lenticular cells occurring in pairs or chains; ferment glucose with production of levorotatory lactic acid (heterofermentative), ethanol, and carbon dioxide. { ¦lü·kə'näs‚täk }

leucophore [HISTOLOGY] A white reflecting chromatophore. { 'lü·kə‚fór }

leucoplast [BOTANY] A nonpigmented plastid; capable of developing into a chromoplast. { 'lü·kə‚plast }

Leucosiidae [INVERTEBRATE ZOOLOGY] The purse crabs, a family of true crabs belonging to the Oxystomata. { ‚lü·kə'sī·ə‚dē }

leucosin [BIOCHEMISTRY] A simple protein of the albumin type found in wheat and other cereals. { 'lü·kə·sən }

Leucosoleniida [INVERTEBRATE ZOOLOGY] An order of calcareous sponges in the subclass Calcaronea characterized by an asconoid structure and the lack of a true dermal membrane or cortex. { ‚lü·kə‚sō·lə'nī·ə·də }

Leucospidae [INVERTEBRATE ZOOLOGY] A small family of hymenopteran insects in the superfamily Chalcidoidea distinguished by a longitudinal fold in the forewings. { lü'käs·pə‚dē }

leucosporous [MYCOLOGY] Having white spores. { lü'käs·pə·rəs }

Leucothoidae [INVERTEBRATE ZOOLOGY] A family of amphipod crustaceans in the suborder Gammaridea including semiparasitic and commensal species. { lü'kä·thȯi‚dē }

Leucothrix [MICROBIOLOGY] The type genus of the family Leucotrichaceae; cells do not form sulfur deposits. { 'lü·kə‚thriks }

Leucotrichaceae [MICROBIOLOGY] A family of bacteria in the order Cytophagales; long, colorless, unbranched filaments having conspicuous cross-walls and containing cylindrical or ovoid cells; filaments attach to substrate. { ‚lü·kə·tri'kās·ē‚ē }

leukemia virus [VIROLOGY] *See* leukovirus. { lü'kē·mē·ə ¦vī·rəs }

leukocidin [BIOCHEMISTRY] A toxic substance released by certain bacteria which destroys leukocytes. { ‚lü·kə'sīd·ən }

leukocyte [HISTOLOGY] A colorless, ameboid blood cell having a nucleus and granular or nongranular cytoplasm. Also known as white blood cell; white corpuscle. { 'lü·kə‚sīt }

leukocytolysin [BIOCHEMISTRY] A leukocyte-disintegrating lysin. { ‚lü·kə‚sī'täl·ə·sən }

leukocytopoiesis [PHYSIOLOGY] Formation of leukocytes. { ‚lü·kə‚sīd·ə‚pȯi'ē·səs }

leukotriene [BIOCHEMISTRY] Any of a family of oxidized metabolites of certain polyunsaturated fatty acids, predominantly arachidonic acid, that mediate responses in allergic reactions and inflammations, produced in specific cells upon stimulation. { ‚lü·kō'trī‚ēn }

leukovirus [VIROLOGY] A major group of animal viruses including those causing leukemia in birds, mice, and rats. Also known as leukemia virus. { ¦lü·kō¦vī·rəs }

levan [BIOCHEMISTRY] $(C_6H_{10}O_5)_n$ A polysaccharide consisting of repeating units of D-fructose and produced by a range of microorganisms, such as *Bacillus mesentericus*. { 'le‚van }

levator [PHYSIOLOGY] Any muscle that raises or elevates a part. { lə'vād·ər }

level above threshold [PHYSIOLOGY] Also known as sensation level. **1.** The pressure level of a sound in decibels above its threshold of audibility for the individual observer. **2.** In general, the level of any psychophysical stimulus, such as light, above its threshold of perception. { 'lev·əl ə,bəv 'thresh,hōld }

leveret [VERTEBRATE ZOOLOGY] A young hare. { 'lev·rət }

levigate [BOTANY] *See* glabrous. { 'lev·ə,gāt }

levotropic cleavage [EMBRYOLOGY] Spiral cleavage with the cells displaced counterclockwise. { ¦lē·və¦träp·ik 'klēv·ij }

levulose [BIOCHEMISTRY] Levorotatory D-fructose. { 'lev·yə,lōs }

Lewis blood group system [IMMUNOLOGY] An antigen, designated by Le^a, first recognized in a Mrs. Lewis, occurring in about 22% of the population, detected by anti-Le^a antibodies; primarily composed of soluble antigens of serum and body fluids like saliva, with secondary absorption by erythrocytes. { 'lü·əs 'bləd ,grüp ,sis·təm }

Leydig cell [HISTOLOGY] One of the interstitial cells of the testes; thought to produce androgen. { 'lī·dig ,sel }

L form [MICROBIOLOGY] A variant form of bacterial cells that has lost its cell wall because of the action of penicillin. { 'el ,fórm }

LH [BIOCHEMISTRY] *See* luteinizing hormone.

LH-RH [BIOCHEMISTRY] *See* luteinizing-hormone releasing hormone.

liana [BOTANY] A woody or herbaceous climbing plant with roots in the ground. { lē'än·ə }

Libellulidae [INVERTEBRATE ZOOLOGY] A large family of odonatan insects belonging to the Anisoptera and distinguished by a notch on the posterior margin of the eyes and a foot-shaped anal loop in the hindwing. { ,lī·bə'lü·lə,dē }

Libytheidae [INVERTEBRATE ZOOLOGY] The snout butterflies, a family of cosmopolitan lepidopteran insects in the suborder Heteroneura distinguished by long labial palps; represented in North America by a single species. { ,lib·ə'thē·ə,dē }

Liceaceae [MYCOLOGY] A family of plasmodial slime molds in the order Liceales. { ,lī·sē'ās·ē ,ē }

Liceales [MYCOLOGY] An order of plasmodial slime molds in the subclass Myxogastromycetidae. { ,lī·sē'ā·lēz }

lichen [BOTANY] The common name for members of the Lichenes. { 'lī·kən }

Lichenes [BOTANY] A group of organisms consisting of fungi and algae growing together symbiotically. { lī'kē·nēz }

Lichenes Imperfecti [BOTANY] A class of the Lichenes containing species with no known method of sexual reproduction. { lī'kē·nēz ,im·pər'fek·tī }

lichenology [BOTANY] The study of lichens. { ,lī·kə'näl·ə·jē }

lichenophagous [ZOOLOGY] Feeding on lichens. { ,lī·kə'näf·ə·gəs }

Lichnophorina [INVERTEBRATE ZOOLOGY] A suborder of ciliophoran protozoans belonging to the Heterotrichida. { ,līk·nə'fór·ə·nə }

licorice [BOTANY] *Glycyrrhiza glabra*. A perennial herb of the legume family (Leguminosae) cultivated for its roots, which when dried provide a product used as a flavoring in medicine, candy, and tobacco and in the manufacture of shoe polish. { 'lik·rəs }

Liebig's law of the minimum [BIOLOGY] *See* law of minimum. { 'lē·bigz ¦ló əv thə 'min·ə·məm }

Lieskeela [MICROBIOLOGY] A genus of sheathed bacteria; cells are nonmotile, and sheaths are not attached and may be encrusted with iron and manganese oxides. { ¦lēs·kē¦el·ə }

life cycle [BIOLOGY] The functional and morphological stages through which an organism passes between two successive primary stages. { 'līf ,sī·kəl }

life expectancy [BIOLOGY] The expected number of years that an organism will live based on statistical probability. { 'līf ik'spek·tən·sē }

life form [ECOLOGY] The form characteristically taken by a plant at maturity. { 'līf ,fórm }

life zone [ECOLOGY] A portion of the earth's land area having a generally uniform climate and soil, and a biota showing a high degree of uniformity in species composition and adaptation. { 'līf ,zōn }

lifting torque [BIOPHYSICS] A measure of the stress arising from the performance of a task that requires lifting; it is the product of the weight of the load and the load's distance from a point within the vertebral column that serves as a fulcrum. { 'lift·iŋ ,tórk }

ligament [HISTOLOGY] A flexible, dense white fibrous connective tissue joining, and sometimes encapsulating, the articular surfaces of bones. { 'lig·ə·mənt }

ligamentum nuchae [ANATOMY] *See* nuchal ligament. { ,lig·ə'men·təm 'nü,kē }

ligase [BIOCHEMISTRY] An enzyme that catalyzes the union of two molecules, involving the participation of a nucleoside triphosphate which is converted to a nucleoside diphosphate or monophosphate. Also known as synthetase. { 'lī ,gās }

ligation [MOLECULAR BIOLOGY] **1.** The process of joining two adjacent nitrogenous bases separated by a nick in one strand of a deoxyribonucleic acid duplex or of two linear nucleic acid molecules. **2.** Formation of a phosphodiester bond to join adjacent nucleotides in the same nucleic acid chain. { lī'gā·shən }

light adaptation [PHYSIOLOGY] The disappearance of dark adaptation; the chemical processes by which the eyes, after exposure to a dim environment, become accustomed to bright illumination, which initially is perceived as quite intense and uncomfortable. { 'līt ,ad,ap'tā·shən }

light chain [IMMUNOLOGY] The smaller of the two types of chains found in immunoglobulin molecules, molecular weight 23,500. Also known as B chain; L chain. { 'līt 'chān }

light meromyosin [BIOCHEMISTRY] The smaller of two fragments obtained from the muscle protein myosin following limited proteolysis by trypsin or chymotrypsin. { 'līt ,mer·ə'mī·ə·sən }

light reflex [PHYSIOLOGY] **1.** The postural orientation response of certain aquatic forms stimulated by the source of light; receptors may be on the ventral or dorsal surface. **2.** The response in which the pupil dilates when light levels are lowered, and constricts when light levels are raised. { 'līt ˌrēˌfleks }

Ligiidae [INVERTEBRATE ZOOLOGY] A family of primitive terrestrial isopods in the suborder Oniscoidea. { lə'jī·əˌdē }

ligneous [BIOLOGY] Of, pertaining to, or resembling wood. { 'lig·nē·əs }

lignify [BOTANY] To convert cell wall constituents into wood or woody tissue by chemical and physical changes. { 'lig·nəˌfī }

lignin [BIOCHEMISTRY] A substance that together with cellulose forms the woody cell walls of plants and cements them together. { 'lig·nən }

ligninase [BIOCHEMISTRY] Any of a group of enzymes that breaks down lignin. { 'lig·nəˌnās }

lignocellulose [BIOCHEMISTRY] Any of a group of substances in woody plant cells consisting of cellulose and lignin. { ¦lig·nō'sel·yəˌlōs }

lignosa [BOTANY] Woody vegetation. { lig'nō·sə }

lignumvitae [BOTANY] *Guaiacum sanctum*. A medium-sized evergreen tree in the order Sapindales that yields a resin or gum known as gum guaiac or resin of guaiac. Also known as hollywood lignumvitae. { 'lig·nəm'vīd·ē }

ligulate [BOTANY] **1.** Strap-shaped. **2.** Having ligules. { 'lig·yə·lət }

ligule [BOTANY] **1.** A small outgrowth in the axis of the leaves in Selaginellales. **2.** A thin outgrowth of a foliage leaf or leaf sheath. [INVERTEBRATE ZOOLOGY] A small lobe on the parapodium of certain polychaetes. { 'lig·yül }

Liliaceae [BOTANY] A family of the order Liliales distinguished by six stamens, typically narrow, parallel-veined leaves, and a superior ovary. { ˌlil·ē'ās·ēˌē }

Liliales [BOTANY] An order of monocotyledonous plants in the subclass Liliidae having the typical characteristics of the subclass. { ˌlil·ē'ā·lēz }

Liliidae [BOTANY] A subclass of the Liliopsida; all plants are syncarpous and have a six-membered perianth, with all members petaloid. { lə'lī·əˌdē }

Liliopsida [BOTANY] The monocotyledons, making up a class of the Magnoliophyta; characterized generally by a single cotyledon, parallel-veined leaves, and stems and roots lacking a well-defined pith and cortex. { ˌlil·ē'äp·səd·ə }

lily [BOTANY] **1.** Any of the perennial bulbous herbs with showy unscented flowers constituting the genus *Lilium*. **2.** Any of various other plants having similar flowers. { 'lil·ē }

Limacidae [INVERTEBRATE ZOOLOGY] A family of gastropod mollusks containing the slugs. { lī'mas·əˌdē }

limb [ANATOMY] An extremity or appendage used for locomotion or prehension, such as an arm or a leg. [BOTANY] A large primary tree branch. { limb }

limbate [BIOLOGY] Having a part of one color bordered with a different color. { 'limˌbāt }

limb bud [EMBRYOLOGY] A mound-shaped lateral proliferation of the embryonic trunk; the anlage of a limb. { 'lim ˌbəd }

limbic system [ANATOMY] The inner edge of the cerebral cortex in the medial and temporal regions of the cerebral hemispheres. { 'lim·bik ˌsis·təm }

limbus [BIOLOGY] A border clearly defined by its color or structure, as the margin of a bivalve shell or of the cornea of the eye. { 'lim·bəs }

lime [BOTANY] *Citrus aurantifolia*. A tropical tree with elliptic oblong leaves cultivated for its acid citrus fruit which is a hesperidium. { līm }

limicolous [ECOLOGY] Living in mud. { lī'mik·ə·ləs }

limited chromosome [CELL BIOLOGY] A chromosome that occurs only in germ cell nuclei. { ¦lim·əd·əd ¦krō·məˌsōm }

limivorous [ZOOLOGY] Feeding on mud, as certain annelids, for the organic matter it contains. { li'miv·ə·rəs }

Limnebiidae [INVERTEBRATE ZOOLOGY] The minute moss beetles, a family of coleopteran insects in the superfamily Hydrophiloidea. { ˌlim·nə'bī·əˌdē }

limnetic [ECOLOGY] Of, pertaining to, or inhabiting the pelagic region of a body of fresh water. { lim'ned·ik }

Limnichidae [INVERTEBRATE ZOOLOGY] The minute false water beetles, a cosmopolitan family of coleopteran insects in the superfamily Dryopoidea. { lim'nik·əˌdē }

Limnocharitaceae [BOTANY] A family of monocotyledonous plants in the order Alismatales characterized by schizogenous secretory canals, multiaperturate pollen, several or many ovules, and a horseshoe-shaped embryo. { ¦lim·nōˌkar·ə'tās·ēˌē }

limnology [ECOLOGY] The science of the life and conditions for life in lakes, ponds, and streams. { lim'näl·ə·jē }

Limnomedusae [INVERTEBRATE ZOOLOGY] A suborder of hydrozoan cnidarians in the order Hydroida characterized by naked hydroids. { ¦lim·nō·mə'dü·sē }

limnoplankton [BIOLOGY] Plankton found in fresh water, especially in lakes. { ¦lim·nō 'plaŋk·tən }

Limnoriidae [INVERTEBRATE ZOOLOGY] The gribbles, a family of isopod crustaceans in the suborder Flabellifera that burrow into submerged marine timbers. { ˌlim·nə'rī·əˌdē }

limoniform [BOTANY] Lemon-shaped. { li'män·əˌfȯrm }

limpet [INVERTEBRATE ZOOLOGY] Any of several species of marine gastropod mollusks composing the families Patellidae and Acmaeidae which have a conical and tentlike shell with ridges extending from the apex to the border. { 'lim·pət }

Limulacea [INVERTEBRATE ZOOLOGY] A group of horseshoe crabs belonging to the Limulida. { ˌlim·yə'lās·ē·ə }

Limulida [INVERTEBRATE ZOOLOGY] A subgroup of Xiphosurida including all living members of the subclass. { lə'myül·ə·də }

Limulodidae [INVERTEBRATE ZOOLOGY] The horseshoe crab beetles, a family of coleopteran insects in the superfamily Staphylinoidea. { ˌlim·myə'läd·əˌdē }

Limulus [INVERTEBRATE ZOOLOGY] The horseshoe crab; the type genus of the Limulacea. { 'lim·yə·ləs }

Linaceae [BOTANY] A family of herbaceous or shrubby dicotyledonous plants in the order Linales characterized by mostly capsular fruit, stipulate leaves, and exappendiculate petals. { lī 'nās·ē,ē }

Linales [BOTANY] An order of dicotyledonous plants in the subclass Orsidae containing simple-leaved herbs or woody plants with hypogynous, regular, syncarpous flowers having five to many stamens which are connate at the base. { lī'nā·lēz }

lincomycin [MICROBIOLOGY] $C_{18}H_{34}O_6N_2S \cdot HCl$ A monobasic crystalline antibiotic, produced by *Streptomyces lincolnensis*, that is active as lincomycin hydrochloride mainly toward gram-positive microorganisms. { ,liŋ·kə'mīs·ən }

linden [BOTANY] *See* basswood. { 'lin·dən }

line [BOTANY] A unit of length, equal to $^1/_{12}$ inch, or approximately 2.117 millimeters; it is most frequently used by botanists in describing the size of plants. { līn }

LINE [MOLECULAR BIOLOGY] *See* long interspersed nucleotide element.

linea alba [ANATOMY] A tendinous ridge extending in the median line of the abdomen from the pubis to the tiphoid process and formed by the blending of aponeuroses of the oblique and transverse muscles of the abdomen. { 'lin·ē·ə 'al·bə }

lineage [GENETICS] Descent from a common progenitor. { 'lin·ē·ij }

linearization [MOLECULAR BIOLOGY] Conversion of a circular deoxyribonucleic acid molecule into a linear molecule. { ,lin·ē·ər·ə'zā·shən }

Lineidae [INVERTEBRATE ZOOLOGY] A family of the Heteronemertini. { li'nē·ə,dē }

linellae [INVERTEBRATE ZOOLOGY] Thin organic threads in the tests of some xenophyophores. { lə'nel·ē }

Lineolaceae [MICROBIOLOGY] A family of bacteria in some systems of classification that includes coenocytic members (*Lineola*) of the Caryophanales. { ,lin·ē·ə'lās·ē,ē }

lineolate [BIOLOGY] Marked with fine lines. { 'lin·ē·ə,lāt }

lingual artery [ANATOMY] An artery originating in the external carotid and supplying the tongue. { 'liŋ·gwəl 'ärd·ə·rē }

lingual gland [ANATOMY] A serous, mucous, or mucoserous gland lying deep in the mucous membrane of the mammalian tongue. { 'liŋ·gwəl 'gland }

lingual nerve [ANATOMY] A branch of the mandibular nerve having somatic sensory components and innervating the mucosa of the floor of the mouth and the anterior two-thirds of the tongue. { 'liŋ·gwəl 'nərv }

lingual tonsil [ANATOMY] An aggregation of lymphoid tissue composed of 35-100 separate tonsillar units occupying the posterior part of the tongue surface. { 'liŋ·gwəl 'tän·səl }

Linguatulida [INVERTEBRATE ZOOLOGY] The equivalent name for Pentastomida. { ,liŋ·gwə 'tül·ə·də }

Linguatuloidea [INVERTEBRATE ZOOLOGY] A suborder of pentastomid arthropods in the order Porocephalida; characterized by an elongate, ventrally flattened, annulate, posteriorly attenuated body, simple hooks on the adult, and binate hooks in the larvae. { liŋ,gwach·ə'lōid·ē·ə }

lingula [ANATOMY] A tongue-shaped organ, structure, or part thereof. { 'liŋ·gyə·lə }

Lingulacea [INVERTEBRATE ZOOLOGY] A superfamily of inarticulate brachiopods in the order Lingulida characterized by an elongate, biconvex calcium phosphate shell, with the majority having a pedicle. { ,liŋ·gyə'lās·ē·ə }

lingulate [BIOLOGY] Tongue- or strap-shaped. { 'liŋ·gyə,lāt }

Lingulida [INVERTEBRATE ZOOLOGY] An order of inarticulate brachiopods represented by two living genera, *Lingula* and *Glottidia*. { liŋ'gyül·ə·də }

linin net [CELL BIOLOGY] The reticulum composed of chromatinic or oxyphilic substances in a cell nucleus. { 'lī·nən ¦net }

linkage [GENETICS] Failure of nonallelic genes to recombine at random as a result of their being located within the same chromosome or chromosome fragment. { 'liŋ·kij }

linkage group [GENETICS] The genes located on a single chromosome. { 'liŋ·kij ,grüp }

linkage map [GENETICS] A chromosome map showing relative positions of known genes in a linkage group. { 'liŋ·kij ,map }

linker [MOLECULAR BIOLOGY] *See* linker deoxyribonucleic acid. { 'liŋ·kər }

linker deoxyribonucleic acid [MOLECULAR BIOLOGY] **1.** A short, synthetic deoxyribonucleic acid (DNA) molecule that contains the recognition site for a specific restriction endonuclease. Also known as linker. **2.** A segment of DNA to which lysine-rich histone is bound and which connects the adjacent nucleosomes of a chromosome. { ¦liŋ·kər dē,äk·sē,rī·bō·nü¦klē·ik 'as·əd }

linoleic acid [BIOCHEMISTRY] $C_{17}H_{31}COOH$ A yellow unsaturated fatty acid, boiling at 229°C (14 mmHg), occurring as a glyceride in drying oils; obtained from linseed, safflower, and tall oils; a principal fatty acid in plants, and considered essential in animal nutrition; used in medicine, feeds, paints, and margarine. Also known as linolic acid; 9,12-octadecadienoic acid. { ¦lin·ə¦lē·ik 'as·əd }

linolenate [BIOCHEMISTRY] A salt or ester of linolenic acid. { ,lin·ə'lē,nāt }

linolenic acid [BIOCHEMISTRY] $C_{17}H_{29}COOH$ One of the principal unsaturated fatty acids in plants and an essential fatty acid in animal nutrition; a colorless liquid that boils at 230°C (17 mmHg or 2266 pascals), soluble in many organic solvents; used in medicine and drying oils. Also known as 9,12,15-octadecatrienoic acid. { ¦lin·ə¦lin·ik 'as·əd }

linolic acid [BIOCHEMISTRY] *See* linoleic acid. { lə 'nō·lik 'as·əd }

lion [VERTEBRATE ZOOLOGY] *Felis leo*. A large carnivorous mammal of the family Felidae distinguished by a tawny coat and blackish tufted tail, with a heavy blackish or dark-brown mane in the male. { 'lī·ən }

Liopteridae [INVERTEBRATE ZOOLOGY] A small family of hymenopteran insects in the superfamily Cynipoidea. { ˌlī·əp'ter·əˌdē }

lip [ANATOMY] A fleshy fold above and below the entrance to the mouth of mammals. { lip }

Liparidae [INVERTEBRATE ZOOLOGY] The equivalent name for Lymantriidae. { lə'par·əˌdē }

lipase [BIOCHEMISTRY] An enzyme that catalyzes the hydrolysis of fats or the breakdown of lipoproteins. { 'līˌpās }

Liphistiidae [INVERTEBRATE ZOOLOGY] A family of spiders in the suborder Liphistiomorphae in which the abdomen shows evidence of true segmentation by the presence of tergal and sternal plates. { ˌlif·ə'stī·əˌdē }

Liphistiomorphae [INVERTEBRATE ZOOLOGY] A suborder of arachnids in the order Araneida containing families with a primitively segmented abdomen. { ləˌfis·tē·ə'mȯr·fē }

lipid [BIOCHEMISTRY] One of a class of compounds which contain long-chain aliphatic hydrocarbons and their derivatives, such as fatty acids, alcohols, amines, amino alcohols, and aldehydes; includes waxes, fats, and derived compounds. Also known as lipin; lipoid. { 'lip·əd }

lipid metabolism [BIOCHEMISTRY] The physiologic and metabolic processes involved in the assimilation of dietary lipids and the synthesis and degradation of lipids. { 'lip·əd me'tab·əˌliz·əm }

lipin [BIOCHEMISTRY] **1.** A compound lipid, such as a cerebroside. **2.** *See* lipid. { 'lip·ən }

lipochrome [BIOCHEMISTRY] Any of various fat-soluble pigments, such as carotenoid, occurring in natural fats. Also known as chromolipid. { 'lip·əˌkrōm }

lipofuscin [BIOCHEMISTRY] Any of a group of lipid pigments found in cardiac and smooth muscle cells, in macrophages, and in parenchyma and interstitial cells; differential reactions include sudanophilia, Nile blue staining, fatty acid, glycol, and ethylene. { ˌlip·ə'fyüs·ən }

lipoic acid [BIOCHEMISTRY] $C_8H_{14}O_2S_2$ A compound which participates in the enzymatic oxidative decarboxylation of α-keto acids in a stage between thiamine pyrophosphate and coenzyme A. { lī'pō·ik 'as·əd }

lipoid [BIOCHEMISTRY] **1.** A fatlike substance. **2.** *See* lipid. { 'līˌpȯid }

Lipomycetoideae [MICROBIOLOGY] A subfamily of oxidative yeasts in the family Saccharomycetaceae characterized by budding cells and a saclike appendage which develops into an ascus. { ˌlip·əˌmī·sə'tȯid·ēˌē }

lipophore [HISTOLOGY] A chromatophore which contains lipochrome. { 'lip·əˌfȯr }

lipopolysaccharide [BIOCHEMISTRY] Any of a class of conjugated polysaccharides consisting of a polysaccharide combined with a lipid. { ˌlip·ōˌpäl·ē'sak·əˌrīd }

lipoprotein [BIOCHEMISTRY] Any of a class of conjugated proteins consisting of a protein combined with a lipid. { ˌlip·ə'prōˌtēn }

liposome [CELL BIOLOGY] One of the fatty droplets occurring in the cytoplasm, particularly of an egg. { 'lip·əˌsōm }

Lipostraca [PALEONTOLOGY] An order of the subclass Branchiopoda erected to include the single fossil species *Lepidocaris rhyniensis*. { li'päs·trə·kə }

lipotropic [BIOCHEMISTRY] Having an affinity for lipid compounds. { ˌlip·ə'trä·pik }

lipotropic hormone [BIOCHEMISTRY] Any hormone having lipolytic activity on adipose tissue. { ˌlip·ə'trä·pik 'hȯrˌmōn }

Lipotyphla [VERTEBRATE ZOOLOGY] A group of insectivoran mammals composed of insectivores which lack an intestinal cecum and in which the stapedial artery is the major blood supply to the brain. { ˌlip·ə'tī·fē·ə }

lipoxidase [BIOCHEMISTRY] An enzyme catalyzing the oxidation of the double bonds of an unsaturated fatty acid. { li'päk·səˌdās }

lirella [BOTANY] A long, narrow apothecium with a medial longitudinal furrow, occurring in certain lichens. { lə'rel·ə }

Liriopeidae [INVERTEBRATE ZOOLOGY] The phantom craneflies, a family of dipteran insects in the suborder Orthorrhapha distinguished by black and white banded legs. { ˌlir·ē·ə'pē·əˌdē }

Lissamphibia [VERTEBRATE ZOOLOGY] A subclass of Amphibia including all living amphibians; distinguished by pedicellate teeth and an operculum-plectrum complex of the middle ear. { ˌli ˌsam'fib·ē·ə }

Listeria [MICROBIOLOGY] A genus of small, gram-positive, motile coccoid rods of uncertain affiliation; found in animal and human feces. { li 'stir·ē·ə }

litchi [BOTANY] *See* lychee. { 'līˌchē }

Lithistida [PALEONTOLOGY] An order of fossil sponges in the class Demospongia having a reticulate skeleton composed of irregular and knobby siliceous spicules. { lə'this·tə·də }

Lithobiomorpha [INVERTEBRATE ZOOLOGY] An order of chilopods in the subclass Pleurostigmophora; members are anamorphic and have 15 leg-bearing trunk segments, and when eyes are present, they are ocellar. { ˌlith·ōˌbī·ə'mȯrˌfə }

lithocholic acid [BIOCHEMISTRY] $C_{24}H_{40}O_3$ A crystalline substance with a melting point of 184–186°C; soluble in hot alcohol; found in ox, human, and rabbit bile. { ¦lith·ə¦kä·lik 'as·əd }

lithocyst [BOTANY] Epidermal plant cell in which cytoliths are formed. [INVERTEBRATE ZOOLOGY] One of the minute sacs containing lithites in many invertebrates; thought to function in audition and orientation. { 'lith·əˌsist }

lithocyte [INVERTEBRATE ZOOLOGY] A special cell in anthomedusae containing a statolith. { 'lith·əˌsīt }

Lithodidae [INVERTEBRATE ZOOLOGY] The king crabs, a family of anomuran decapods in the superfamily Paguridea distinguished by reduced last pereiopods and by the asymmetrical arrangement of the abdominal plates in the female. { lə'thäd·əˌdē }

lithodomous [ZOOLOGY] Burrowing in rock. { lə 'thäd·ə·məs }

lithophagous [ZOOLOGY] Feeding on stone, as certain mollusks. { lə'thäf·ə·gəs }

lithophyte [ECOLOGY] A plant that grows on rock. { 'lith·ə,fīt }

lithosere [ECOLOGY] A succession of plant communities that originate on rock. { 'lith·ə,sir }

lithostyle [INVERTEBRATE ZOOLOGY] A static organ in Narcomedusae. Also known as tentaculocyst. { 'lith·ə,stīl }

Litopterna [PALEONTOLOGY] An order of hoofed, herbivorous mammals confined to the Cenozoic of South America; characterized by a skull without expansion of the temporal or squamosal sinuses, a postorbital bar, primitive dentition, and feet that were three-toed or reduced to a single digit. { ,lid·əp'tər·nə }

little-drop technique [BIOLOGY] A method for isolating single cells in which a drop of a cellular suspension containing a single cell, as determined by microscopic examination, is transferred with a capillary pipet to an appropriate culture medium. { 'lid·əl 'dräp ,tek,nēk }

littoral zone [ECOLOGY] Of or pertaining to the biogeographic zone between the high- and low-water marks. { 'lit·ə·rəl ,zōn }

Littorinacea [PALEONTOLOGY] An extinct superfamily of gastropod mollusks in the order Prosobranchia. { ,lid·ə·rə'nās·ē·ə }

Littorinidae [INVERTEBRATE ZOOLOGY] The periwinkles, a family of marine gastropod mollusks in the order Pectinibranchia distinguished by their spiral, globular shells. { ,lid·ə'rin·ə,dē }

lituate [BOTANY] Having a forked member or part with the ends turned slightly outward, as in certain fungi. { 'lich·ə·wət }

Lituolacea [INVERTEBRATE ZOOLOGY] A superfamily of benthic marine foraminiferans in the suborder Textulariina having a multilocular, rectilinear, enrolled or uncoiled test with a simple to labyrinthic wall. { ,lich·ə'lās·ē·ə }

liver [ANATOMY] A large vascular gland in the body of vertebrates, consisting of a continuous parenchymal mass covered by a capsule; secretes bile, manufactures certain blood proteins and enzymes, and removes toxins from the systemic circulation. { 'liv·ər }

liver fluke [INVERTEBRATE ZOOLOGY] Any trematode, especially *Clonorchis sinensis*, that lodges in the biliary passages within the liver. { 'liv·ər ,flük }

liver phosphorylase [BIOCHEMISTRY] An enzyme that catalyzes the breakdown of liver glycogen to glucose-1-phosphate. { 'liv·ər ,fäs'fȯr·ə,lās }

liverwort [BOTANY] The common name for members of the Marchantiatae. { 'liv·ər,wȯrt }

live-virus vaccine [IMMUNOLOGY] A suspension of attenuated live viruses injected to produce immunity. { 'līv ¦vī·rəs vak'sēn }

living fossil [BIOLOGY] A living species belonging to an ancient stock otherwise known only as fossils. { 'liv·iŋ 'fäs·əl }

lizard [VERTEBRATE ZOOLOGY] Any reptile of the suborder Sauria. { 'liz·ərd }

lizard-hipped dinosaur [PALEONTOLOGY] The name applied to members of the Saurichia because of the comparatively unspecialized three-pronged pelvis. { 'liz·ərd ¦hipt 'dī·nə,sȯr }

LK virus [VIROLOGY] A type of equine herpesvirus. { ¦el¦kā 'vī·rəs }

llama [VERTEBRATE ZOOLOGY] Any of three species of South American artiodactyl mammals of the genus *Lama* in the camel family; differs from the camel in being smaller and lacking a hump. { 'yäm·ə }

llano [ECOLOGY] A savannah of Spanish America and the southwestern United States generally having few trees. { 'yä·nō }

loach [VERTEBRATE ZOOLOGY] The common name for fishes composing the family Cobitidae; most are small and many are eel-shaped. { lōch }

Lobata [INVERTEBRATE ZOOLOGY] An order of the Ctenophora in which the body is helmet-shaped. { lō'bäd·ə }

lobate [BIOLOGY] Having lobes. [VERTEBRATE ZOOLOGY] Of a fish, having the skin of the fin extend onto the bases of the fin rays. { 'lō,bāt }

lobe [BIOLOGY] A rounded projection on an organ or body part. { lōb }

lobefin fish [VERTEBRATE ZOOLOGY] The common name for members composing the subclass Crossopterygii. { 'lōb,fin ,fish }

loblolly pine [BOTANY] *Pinus taeda*. A hard yellow pine of the central and southeastern United States having a reddish-brown fissured bark, needles in groups of three, and a full bushy top. { 'läb,läl·ē 'pīn }

lobopodia [INVERTEBRATE ZOOLOGY] Broad, thick pseudopodia. { ,lō·bə'päd·ē·ə }

Lobosia [INVERTEBRATE ZOOLOGY] A subclass of the protozoan class Rhizopodea generally characterized by lobopodia. { lō'bō·sē·ə }

lobster [INVERTEBRATE ZOOLOGY] The common name for several bottom-dwelling decapod crustaceans making up the family Homaridae which are commercially important as a food item. { 'läb·stər }

lobule [BIOLOGY] **1.** A small lobe. **2.** A division of a lobe. { 'läb·yül }

local immunity [IMMUNOLOGY] Immunity localized in a specific tissue or region of the body. { 'lō·kəl i'myü·nəd·ē }

locellus [BOTANY] **1.** In some legumes, a secondary compartment of a unilocular ovary that is formed by a false partition. **2.** One of the two cavities of a pollen sac. { lō'sel·əs }

lociation [ECOLOGY] One of the subunits of a faciation, distinguished by the relative abundance of a dominant species. { lō·sē'ā·shən }

locomotor system [ZOOLOGY] Appendages and associated parts, such as muscles, joints, and bones, concerned with motor activities and locomotion of the animal body. { ,lō·kə'mōd·ər ,sis·təm }

loco weed [BOTANY] Any species of *Astragalus* containing selenium taken up from the soil. { 'lō·kō ,wēd }

loculate [BIOLOGY] Having, or divided into, loculi. { 'läk·yə·lət }

locule [BOTANY] A small chamber in plant tissue within which specialized structures may develop, such as within an ovary or anther. { 'lä,kyül }

loculicidal [BOTANY] Pertaining to dehiscence that extends along the dorsal midline of a carpel. { ¦lak·yə¦lis·əd·əl }

loculus [BIOLOGY] A small cavity or chamber. { 'läk·yə·ləs }

locus [GENETICS] The fixed position of a gene in a chromosome. { 'lō·kəs }

locust [BOTANY] Either of two species of commercially important trees, black locust (*Robinia pseudoacacia*) and honey locust (*Gladitsia triacanthos*), in the family Leguminosae. [INVERTEBRATE ZOOLOGY] The common name for various migratory grasshoppers of the family Locustidae. { 'lō·kəst }

Locustidae [INVERTEBRATE ZOOLOGY] A family of insects in the order Orthoptera; antennae are usually less than half the body length, hindlegs are adapted for jumping, and the ovipositor is multipartite. { lō'kəs·tə,dē }

lodicule [BOTANY] One of the minute membranous bodies found at the base of the carpel in most flowering grasses; usually occurs in pairs. { 'läd·ə,kyül }

Loganiaceae [BOTANY] A family of mostly woody dicotyledonous plants in the order Gentianales; members lack a latex system and have fully united carpels and axile placentation. { lō,gān·ē'ā·sē,ē }

logarithmic growth [MICROBIOLOGY] *See* exponential growth. { 'läg·ə,rith·mik 'grōth }

logistic growth [BIOLOGY] Population growth in which the growth rate decreases with increasing number of individuals until it becomes zero when the population reaches a maximum. { lə 'jis·tik 'grōth }

logotype [SYSTEMATICS] The selection or designation of a genotype after the generic name was published. { 'lō·gə,tīp }

logwood [BOTANY] *See* hematoxylon. { 'läg ,wu̇d }

loment [BOTANY] A dry, indehiscent single-celled fruit that is formed from a single superior ovary; splits transversely in numerous segments at maturity. { 'lō,ment }

Lonchaeidae [INVERTEBRATE ZOOLOGY] A family of minute myodarian cyclorrhaphous dipteran insects in the subsection Acalyptreatae. { läŋ 'kē·ə,dē }

long bone [ANATOMY] A bone in which the length markedly exceeds the width, as the femur or the humerus. { 'lȯŋ ¦bōn }

long-day plant [BOTANY] A plant that flowers in response to a long photoperiod. { 'lȯŋ ¦dā ,plant }

longicollous [BIOLOGY] Having a long beak or neck. { ¦län·jə¦käl·əs }

Longidorinae [INVERTEBRATE ZOOLOGY] A subfamily of nematodes belonging to the Dorylaimoidea including economically important plant parasites. { ,län·jə'dȯr·ə,nē }

long interspersed nucleotide element [MOLECULAR BIOLOGY] In mammalian deoxyribonucleic acid, any of the 5-10-kilobase repeated sequences that are grouped with the nonviral retroposons. Abbreviated LINE. { ¦lȯŋ ¦in·tər ,spərst 'nü·klē·ə,tīd ,el·ə·mənt }

Longipennes [VERTEBRATE ZOOLOGY] An equivalent name for Charadriiformes. { ¦län·jə'pen·ēz }

loon [VERTEBRATE ZOOLOGY] The common name for birds composing the family Gaviidae, all of which are fish-eating diving birds. { lün }

loop of Henle [ANATOMY] The U-shaped portion of a renal tubule formed by a descending and an ascending tubule. Also known as Henle's loop. { 'lüp əv 'hen·lē }

loop pattern [ANATOMY] A type of fingerprint pattern in which one or more of the ridges enter on either side of the impression, recurve, touch, or pass an imaginary line from the delta to the core and terminate on the entering side. { 'lüp ,pad·ərn }

loose connective tissue [HISTOLOGY] A type of irregularly arranged connective tissue having a relatively large amount of ground substance. { 'lüs kə'nek·tiv 'tish·ü }

Lopadorrhynchidae [INVERTEBRATE ZOOLOGY] A small family of pelagic polychaete annelids belonging to the Errantia. { ,lō·pə·dō'riŋk·ə,dē }

Lophialetidae [PALEONTOLOGY] A family of extinct perissodactyl mammals in the superfamily Tapiroidea. { ,lä·fē·ə'led·ə,dē }

Lophiiformes [VERTEBRATE ZOOLOGY] A modified order of actinopterygian fishes distinguished by the reduction of the first dorsal fin to a few flexible rays, the first of which is on the head and bears a terminal bulb; includes anglerfish and allies. { ,lä·fē·ə'fȯr,mēz }

Lophiodontidae [PALEONTOLOGY] An extinct family of perissodactyl mammals in the superfamily Tapiroidea. { ,lä·fē·ə'dänt·ə,dē }

lophocercous [VERTEBRATE ZOOLOGY] Having a ridgelike caudal fin that lacks rays. { ¦lä·fə¦sər·kəs }

lophocyte [INVERTEBRATE ZOOLOGY] A specialized cell of uncertain function beneath the dermal membrane of certain Demospongiae which bears a process terminating in a tuft of fibrils. { 'läf·ə,sīt }

lophodont [VERTEBRATE ZOOLOGY] Having molar teeth whose grinding surfaces have transverse ridges. { 'läf·ə,dänt }

Lophogastrida [INVERTEBRATE ZOOLOGY] A suborder of free-swimming marine crustaceans in the order Mysidacea characterized by imperfect fusion of the sixth and seventh abdominal somites, seven pairs of gills and brood lamellae, and natatory, biramous pleopods. { ,läf·ə'gas·trə·də }

lophophore [INVERTEBRATE ZOOLOGY] A food-gathering organ consisting of a fleshy basal ridge or lobe, from which numerous ciliated tentacles arise; found in Bryozoa, Phoronida, and Brachiopoda. { 'läf·ə,fȯr }

lophotrichous [CELL BIOLOGY] Having a polar tuft of flagella. { lə'fä·trə·kəs }

Loranthaceae [BOTANY] A family of dicotyledonous plants in the order Santalales in which the ovules have no integument and are embedded in a large, central placenta. { 'lȯ,ran 'thās·ē,ē }

lorate [BIOLOGY] Strap-shaped. { 'lȯr,āt }

L organisms [MICROBIOLOGY] Pleomorphic

forms of bacteria occurring spontaneously, or favored by agents such as penicillin, which lack cell walls and grow in minute colonies; transition may be reversible under certain conditions. { 'el ˌȯr·gəˌniz·əmz }

lorica [INVERTEBRATE ZOOLOGY] A hard shell or case in certain invertebrates, as in many rotifers and protozoans; functions as an exoskeleton. { lə 'rī·kə }

Loricaridae [VERTEBRATE ZOOLOGY] A family of catfishes in the suborder Siluroidei found in the Andes. { ˌlȯr·ə·kə'rī·əˌdē }

Loricata [INVERTEBRATE ZOOLOGY] An equivalent name for Polyplacophora. { ˌlȯr·ə'käd·ə }

loricate [INVERTEBRATE ZOOLOGY] Of, pertaining to, or having a lorica. { 'lȯr·əˌkāt }

loris [VERTEBRATE ZOOLOGY] Either of two slow-moving, nocturnal, arboreal primates included in the family Lorisidae: the slender loris (*Loris tardigradus*) and slow loris (*Nycticebus coucang*). { 'lȯr·əs }

Lorisidae [VERTEBRATE ZOOLOGY] A family of prosimian primates comprising the lorises of Asia and the galagos of Africa. { lə'ris·əˌdē }

lotic [ECOLOGY] Of or pertaining to swiftly moving waters. { 'lōd·ik }

louping-ill virus [VIROLOGY] A group B arbovirus that is infectious in sheep, monkeys, mice, horses, and cattle. { 'lüp·iŋ ˌil ˌvī·rəs }

louse [INVERTEBRATE ZOOLOGY] The common name for the apterous ectoparasites composing the orders Anoplura and Mallophaga. { laủs }

Lower Austral life zone [ECOLOGY] A term used by C.H. Merriam to describe the southern portion of the Austral life zone, characterized by accumulated temperatures of 18,000°F (10,000°C). { 'lō·ər ¦ȯs·trəl 'līf ˌzōn }

lower motor neuron [ANATOMY] An efferent neuron which has its body located in the anterior gray column of the spinal cord or in the brainstem nuclei, and its axon passing by way of a peripheral nerve to skeletal muscle. Also known as final common pathway. { 'lō·ər 'mōd·ər ˌnü ˌrän }

Lower Sonoran life zone [ECOLOGY] A term used by C.H. Merriam to describe the climate and biotic communities of subtropical deserts and thorn savannas in the southwestern United States. { 'lō·ər sə'nȯr·ən 'līf ˌzōn }

low-zone tolerance [IMMUNOLOGY] Immunologic tolerance induced by repeated administration of very low doses of a protein antigen. { ¦lō ¦zōn 'täl·ə·rəns }

loxodont [VERTEBRATE ZOOLOGY] Having molar teeth with shallow hollows between the ridges. { 'läk·səˌdänt }

loxolophodont [VERTEBRATE ZOOLOGY] Having crests on the molar teeth that connect three of the tubercles and with the fourth or posterior inner tubercle being rudimentary or absent. { ¦läk·sə¦läf·əˌdänt }

Loxonematacea [PALEONTOLOGY] An extinct superfamily of gastropod mollusks in the order Prosobranchia. { ˌläk·səˌne·mə'tās·ē·ə }

Lucanidae [INVERTEBRATE ZOOLOGY] The stag beetles, a cosmopolitan family of coleopteran insects in the superfamily Scarabaeoidea. { lü 'kan·əˌdē }

lucerne [BOTANY] *See* alfalfa. { lü'sərn }

Lucibacterium [MICROBIOLOGY] A genus of light-emitting bacteria in the family Vibrionaceae; motile, asporogenous rods with peritrichous flagella. { ˌlü·siˌbak'tir·ē·əm }

luciferase [BIOCHEMISTRY] An enzyme that catalyzes the oxidation of luciferin. { lü'sif·əˌrās }

luciferin [BIOCHEMISTRY] A species-specific pigment in many luminous organisms that emits heatless light when combined with oxygen. { lü 'sif·ə·rən }

Luciocephalidae [VERTEBRATE ZOOLOGY] A family of fresh-water fishes in the suborder Anabantoidei. { ˌlü·sē·ō·sə'fal·əˌdē }

Luidiidae [INVERTEBRATE ZOOLOGY] A family of echinoderms in the suborder Paxillosina. { lü·ə 'dī·əˌdē }

lumbar artery [ANATOMY] Any of the four or five pairs of branches of the abdominal aorta opposite the lumbar region of the spine; supplies blood to loin muscles, skin on the sides of the abdomen, and the spinal cord. { 'ləmˌbär 'ärd·ə·rē }

lumbar nerve [ANATOMY] Any of five pairs of nerves arising from lumbar segments of the spinal cord; characterized by motor, visceral sensory, somatic sensory, and sympathetic components; they innervate the skin and deep muscles of the lower back and the lumbar plexus. { 'ləm ˌbär 'nərv }

lumbar vertebrae [ANATOMY] Those vertebrae located between the lowest ribs and the pelvic girdle in all vertebrates. { 'ləmˌbär 'vərd·əˌbrā }

lumbodorsal fascia [ANATOMY] The sheath of the erector spinae muscle alone, or the sheaths of the erector spinae and the quadratus lumborum muscles. { ¦ləm·bō'dȯr·səl 'fā·shə }

lumbosacral plexus [ANATOMY] A network formed by the anterior branches of lumbar, sacral, and coccygeal nerves which for descriptive purposes are divided into the lumbar, sacral, and pudendal plexuses. { ¦ləm·bō'sak·rəl 'plek·səs }

Lumbricidae [INVERTEBRATE ZOOLOGY] A family of annelid worms in the order Oligochaeta; includes the earthworm. { ləm'bris·əˌdē }

Lumbriclymeninae [INVERTEBRATE ZOOLOGY] A subfamily of mud-swallowing sedentary worms in the family Maldanidae. { ˌləm·bri·klī'men·ə ˌnē }

Lumbriculidae [INVERTEBRATE ZOOLOGY] A family of aquatic annelids in the order Oligochaeta. { ˌləm·bri'kyül·əˌdē }

Lumbricus [INVERTEBRATE ZOOLOGY] A genus of earthworms recognized as the type genus of the family Lumbricidae. { 'ləm·brə·kəs }

Lumbrineridae [INVERTEBRATE ZOOLOGY] A family of errantian polychaetes in the superfamily Eunicea. { ˌləm·bri'ner·əˌdē }

lumen [ANATOMY] The interior space within a tubular structure, such as within a blood vessel, a duct, or the intestine. { 'lü·mən }

lumper [SYSTEMATICS] A taxonomist who tends to recognize large taxa. { 'ləm·pər }

lunate [BIOLOGY] Crescent-shaped { 'lü,nāt }

lung [ANATOMY] Either of the paired air-filled sacs, usually in the anterior or anteroventral part of the trunk of most tetrapods, which function as organs of respiration. { ləŋ }

lung bud [EMBRYOLOGY] A primary outgrowth of the embryonic trachea; the anlage of a primary bronchus and all its branches. { 'ləŋ ,bəd }

lungfish [VERTEBRATE ZOOLOGY] The common name for members of the Dipnoi; all have lungs that arise from a ventral connection with the gut. { 'ləŋ,fish }

lungworm [INVERTEBRATE ZOOLOGY] Any of the nematodes that are parasites of terrestrial and marine nematodes, most commonly found in the respiratory tract, characterized by a reduced or absent stoma capsule, and an oral opening surrounded by six well-developed lips. { 'ləŋ ,wərm }

lunule [BIOLOGY] A crescent-shaped organ, structure, or mark. { 'lün,yül }

lupine [BOTANY] A leguminous plant of the genus *Lupinus* with an upright stem, leaves divided into several digitate leaflets, and terminal racemes of pea-shaped blossoms. { 'lü,pən }

lutein [BIOCHEMISTRY] **1.** A dried, powdered preparation of corpus luteum. **2.** *See* xanthophyll. { 'lüd·ē·ən }

luteinization [PHYSIOLOGY] Acquisition of characteristics of cells of the corpus luteum by ovarian follicle cells following ovulation. { ,lüd·ē·ə·nə'zā·shən }

luteinizing hormone [BIOCHEMISTRY] A glycoprotein hormone secreted by the adenohypophysis of vertebrates which stimulates hormone production by interstitial cells of gonads. Abbreviated LH. Also known as interstitial-cell-stimulating hormone (ICSH). { 'lüd·ē·ə,nīz·iŋ 'hör ,mōn }

luteinizing hormone-releasing hormone [BIOCHEMISTRY] A small peptide hormone released from the hypothalamus which acts on the pituitary gland to cause release of luteinizing hormone. Abbreviated LH-RH. { 'lüd·ē·ə,nīz·iŋ ¦hör,mōn ri'lēs·iŋ ¦hör,mōn }

luteotropic hormone [BIOCHEMISTRY] *See* prolactin. { ¦lüd·ē·ə¦träp·ik 'hör,mōn }

Lutheran blood group [IMMUNOLOGY] The erythrocyte antigens defined by reactions with an antibody designated anti-Lu[a], initially detected in the serum of a multiply transfused patient with lupus erythematosus, who developed antibodies against erythrocytes of a donor named Lutheran, and by anti-Lu[b]. { 'lüth·rən 'bləd ,grüp }

Lutjanidae [VERTEBRATE ZOOLOGY] The snappers, a family of perciform fishes in the suborder Percoidei. { lü'chan·ə,dē }

lyase [BIOCHEMISTRY] An enzyme that catalyzes the nonhydrolytic cleavage of its substrate with the formation of a double bond; examples are decarboxylases. { 'lī,ās }

Lycaenidae [INVERTEBRATE ZOOLOGY] A family of heteroneuran lepidopteran insects in the superfamily Papilionoidea including blue, gossamer, hairstreak, copper, and metalmark butterflies. { lī 'sēn·ə,dē }

Lycaeninae [INVERTEBRATE ZOOLOGY] A subfamily of the Lycaenidae distinguished by functional prothoracic legs in the male. { lī'sēn·ə,nē }

lychee [BOTANY] A tree of the genus *Litchi* in the family Sapindaceae, especially *L. chinensis* which is cultivated for its edible fruit, a one-seeded berry distinguished by the thin, leathery, rough pericarp that is red in most varieties. Also spelled litchi. { 'lī,chē }

lychnisc [INVERTEBRATE ZOOLOGY] A hexactin in which the central part of the spicule is surrounded by a system of 12 struts. { 'lik·nisk }

Lychniscosa [INVERTEBRATE ZOOLOGY] An order of sponges in the subclass Hexasterophora in which parenchymal megascleres form a rigid framework and are all or in part lychniscs. { ,lik·ni'skō·sə }

lycopene [BIOCHEMISTRY] $C_{40}H_{50}$ A red, crystalline hydrocarbon that is the coloring matter of certain fruits, such as tomatoes; it is isomeric with carotene. { 'lī·kə,pēn }

lycophore larva [INVERTEBRATE ZOOLOGY] A larva of certain cestodes characterized by cilia, large frontal glands, and 10 hooks. Also known as decanth larva. { 'lī·kə,fòr ,lär·və }

Lycopodiales [BOTANY] The type order of Lycopodiopsida. { ,lī·kə·pō·dī'ā·lēz }

Lycopodineae [BOTANY] The equivalent name for Lycopodiopsida. { ,lī·kə·pō'din·ē,ē }

Lycopodiophyta [BOTANY] A division of the subkingdom Embryobionta characterized by a dominant independent sporophyte, dichotomously branching roots and stems, a single vascular bundle, and small, simple, spirally arranged leaves. { ,lī·kə·pō·dī'äf·əd·ə }

Lycopodiopsida [BOTANY] The lycopods, the type class of Lycophodiophyta. { ,lī·kə·pō·dī 'äp·səd·ə }

Lycopsida [BOTANY] Former subphylum of the Embryophyta now designated as the division Lycopodiophyta. { lī'käp·sə·də }

Lycoriidae [INVERTEBRATE ZOOLOGY] A family of small, dark-winged dipteran insects in the suborder Orthorrhapha. { ,lī·kə'rī·ə,dē }

Lycosidae [INVERTEBRATE ZOOLOGY] A family of hunting spiders in the suborder Dipneumonomorphae that actively pursue their prey. { lī 'käs·ə,dē }

Lycoteuthidae [INVERTEBRATE ZOOLOGY] A family of squids. { ,lī·kə'tü·thə,dē }

Lyctidae [INVERTEBRATE ZOOLOGY] The large-winged beetles, a large cosmopolitan family of coleopteran insects in the superfamily Bostrichoidea. { 'lik·tə,de }

Lygaeidae [INVERTEBRATE ZOOLOGY] The lygaeid bugs, a large family of phytophagous hemipteran insects in the superfamily Lygaeoidea. { lī'jē·ə ,dē }

Lygaeoidea [INVERTEBRATE ZOOLOGY] A superfamily of pentatomorphan insects having four-segmented antennae and ocelli. { ,lī·jē'òid·ē·ə }

Lyginopteridaceae [PALEOBOTANY] An extinct family of the Lyginopteridales including monostelic pteridosperms having one or two vascular traces entering the base of the petiole. { ˌlī·jə·näpˌter·ə'dās·ēˌē }

Lyginopteridales [PALEOBOTANY] An order of the Pteridospermae. { ˌlī·jə·näpˌter·ə'dā·lēz }

Lyginopteridatae [PALEOBOTANY] The equivalent name for Pteridospermae. { ˌlī·jə·näp·fə'rid·ədˌē }

Lymantriidae [INVERTEBRATE ZOOLOGY] The tussock moths, a family of heteroneuran lepidopteran insects in the superfamily Noctuoidea; the antennae of males is broadly pectinate and there is a tuft of hairs on the end of the female abdomen. { ¦lī·mən'trī·əˌdē }

Lymexylonidae [INVERTEBRATE ZOOLOGY] The ship timber beetles composing the single family of the coleopteran superfamily Lymexylonoidea. { ləˌmek·sə'län·əˌdē }

Lymexylonoidea [INVERTEBRATE ZOOLOGY] A monofamilial superfamily of wood-boring coleopteran insects in the suborder Polyphaga characterized by a short neck and serrate antennae. { ləˌmek·sə·lə'nȯid·ē·ə }

lymph [HISTOLOGY] The colorless fluid which circulates through the vessels of the lymphatic system. { limf }

lymphadenopathy-associated virus [VIROLOGY] A former designation for the human immunodeficiency virus. Abbreviated LAV. { ˌlimˌfad·ən'äp·ə·thē ə¦sō·shēˌād·əd 'vī·rəs }

lymphatic [ANATOMY] *See* lymph vessel. { lim'fad·ik }

lymphatic system [ANATOMY] A system of vessels and nodes conveying lymph in the vertebrate body, beginning with capillaries in tissue spaces and eventually forming the thoracic ducts which empty into the subclavian veins. { lim'fad·ik ˌsis·təm }

lymphatic tissue [HISTOLOGY] Tissue consisting of networks of lymphocytes and reticular and collagenous fibers. Also known as lymphoid tissue. { lim'fad·ik ˌtish·ü }

lymph gland [ANATOMY] *See* lymph node. { 'limf ˌgland }

lymph heart [VERTEBRATE ZOOLOGY] A muscular expansion of a lymphatic vessel which contracts, driving lymph to the veins, as in amphibians. { 'limf ˌhärt }

lymph node [ANATOMY] An aggregation of lymphoid tissue surrounded by a fibrous capsule; found along the course of lymphatic vessels. Also known as lymph gland. { 'limf ˌnōd }

lymphoblast [HISTOLOGY] Precursor of a lymphocyte. { 'lim·fəˌblast }

lymphocyte [HISTOLOGY] An agranular leukocyte formed primarily in lymphoid tissue; occurs as the principal cell type of lymph and composes 20-30% of the blood leukocytes. { 'lim·fəˌsīt }

lymphocyte transformation [IMMUNOLOGY] *See* transformation. { ¦lim·fəˌsīt ˌtranz·fər'mā·shən }

lymphocytotropic [IMMUNOLOGY] Having an affinity for lymphocytes. { ˌlim·fōˌsīd·ə'träp·ik }

lymphoid cell [HISTOLOGY] A mononucleocyte that resembles a leukocyte. { 'limˌfȯid ˌsel }

lymphoid hemoblast of Pappenheim [HISTOLOGY] *See* pronormoblast. { 'limˌfȯid 'hē·mə ˌblast əv 'päp·ənˌhīm }

lymphoid tissue [HISTOLOGY] *See* lymphatic tissue. { 'limˌfȯid ˌtish·ü }

lymphokine [IMMUNOLOGY] A cytokine released from T lymphocytes after contact with an antigen. { 'lim·fəˌkīn }

lymphopoiesis [PHYSIOLOGY] The production of lymph. { ˌlim·fəˌpȯi'ē·səs }

lymphoproliferative virus group [VIROLOGY] *See* Gammaherpesvirinae. { ¦lim·fō·prō'lif·rəd·iv 'vī·rəs ˌgrüp }

lymph sinus [ANATOMY] One of the tracts of diffuse lymphatic tissue between the cords and nodules, and between the septa and capsule of a lymph node. { 'limf ˌsī·nəs }

lymph vessel [ANATOMY] A tubular passage for conveying lymph. Also known as lymphatic. { 'limf ˌves·əl }

lyngbyatoxin A [BIOCHEMISTRY] An indole alkaloid toxin produced by *Lyngbya majuscula.* { 'liŋ·bē·əˌtäk·sən 'ā }

lynx [VERTEBRATE ZOOLOGY] Any of several wildcats of the genus *Lynx* having long legs, short tails, and usually tufted ears; differs from other felids in having 28 instead of 30 teeth. { liŋks }

Lyomeri [VERTEBRATE ZOOLOGY] The equivalent name for Saccopharyngiformes. { lī'äm·əˌrī }

Lyon hypothesis [GENETICS] The assumption that mammalian females are X-chromosome mosaics as a result of the inactivation of one X chromosome in some embryonic cells and the other in the rest. { 'lī·ən hīˌpäth·ə·səs }

Lyon phenomenon [GENETICS] In the normal human female, the rendering of two X chromosomes inactive, or at least largely so, at an early stage in embryogenesis. { 'lī·ən fəˌnäm·əˌnän }

Lyopomi [VERTEBRATE ZOOLOGY] An equivalent name for Notacanthiformes. { lī'äp·əˌmī }

Lysaretidae [INVERTEBRATE ZOOLOGY] A family of errantian polychaete worms in the superfamily Eunicea. { ˌlī·sə'red·əˌdē }

lyse [CELL BIOLOGY] To undergo lysis. { līz }

Lysenkoism [BIOLOGY] A pseudoscientific theory that flourished in the Soviet Union from the early 1930s to the mid-1960s; advocated by T. D. Lysenko, who called it agrobiology, it was claimed to be a revolutionary fusion of agronomy and biological science, and it opposed traditional biology and the gene concept but supported the inheritance of acquired characteristics. { lī'seŋ·kōˌiz·əm }

Lysianassidae [INVERTEBRATE ZOOLOGY] A family of pelagic amphipod crustaceans in the suborder Gammaridea. { ˌlī·sē·ə'nas·əˌdē }

lysigenous [BIOLOGY] Of or pertaining to the space formed following lysis of cells. { 'lī'sij·ə·nəs }

lysin [IMMUNOLOGY] A substance, particularly antibodies, capable of lysing a cell. { 'līs·ən }

lysine [BIOCHEMISTRY] $C_6H_{14}O_2N_2$ An essential, basic amino acid obtained from many proteins by hydrolysis. { 'līˌsēn }

Lysiosquillidae [INVERTEBRATE ZOOLOGY] A family of crustaceans in the order Stomatopoda. { ˌlī·sē·ō'skwil·əˌdē }

lysis [CELL BIOLOGY] Dissolution of a cell or tissue by the action of a lysin. { 'lī·səs }

lysogeny [MICROBIOLOGY] Lysis of bacteria, with the liberation of bacteriophage particles. { lī'säj·ə·nē }

lysosome [CELL BIOLOGY] A specialized cell organelle surrounded by a single membrane and containing a mixture of hydrolytic (digestive) enzymes. { 'lī·səˌsōm }

lysozyme [BIOCHEMISTRY] An enzyme present in certain body secretions, principally tears, which acts to hydrolyze certain bacterial cell walls. { 'lī·səˌzīm }

lyssacine [INVERTEBRATE ZOOLOGY] An early stage of the skeletal network in hexactinellid sponges. { 'lī·səˌsēn }

Lyssacinosa [INVERTEBRATE ZOOLOGY] An order of sponges in the subclass Hexasterophora in which parenchymal megascleres are typically free and unconnected but are sometimes secondarily united. { ¦lī·sə·sə¦nō·sə }

lysyl oxidase [BIOCHEMISTRY] Any of a group of enzymes found in bone and connective tissue which oxidize terminal amino groups of lysine residues in tropocollagen molecules to aldehyde residues. { 'lī·səl 'äk·səˌdās }

lytic infection [MICROBIOLOGY] Penetration of a host cell by lytic phage. { 'lid·ik in'fek·shən }

lytic phage [VIROLOGY] Any phage that causes host cells to lyse. { 'lid·ik 'fāj }

lytic reaction [CELL BIOLOGY] A reaction that leads to lysis of a cell. { 'lid·ik rē'ak shən }

M

M [GENETICS] *See* Morgan.

macadamia nut [BOTANY] The hard-shelled seed obtained from the fruit of a tropical evergreen tree, *Macadamia ternifolia*. { ˌmak·əˈdā·mē·ə ˌnət }

macaque [VERTEBRATE ZOOLOGY] The common name for 12 species of Old World monkeys composing the genus *Macaca*, including the Barbary ape and the rhesus monkey. { məˈkak }

macaw [VERTEBRATE ZOOLOGY] The common name for large South and Central American parrots of the genus *Ara* and related genera; individuals are brilliantly colored with a long tail, a hooked bill, and a naked area around the eyes. { məˈkȯ }

Machaeridea [INVERTEBRATE ZOOLOGY] A class of homolozoan echinoderms in older systems of classification. { ˌmak·əˈrid·ē·ə }

Machilidae [INVERTEBRATE ZOOLOGY] A family of insects belonging to the Thysanura having large compound eyes and ocelli and a monocondylous mandible of the scraping type. { məˈkil·əˌdē }

machopolyp [INVERTEBRATE ZOOLOGY] *See* machozooid. { ¦ma·kōˈpäl·əp }

machozooid [INVERTEBRATE ZOOLOGY] A defensive polyp equipped with stinging organs in certain hydroid colonies. Also known as machopolyp. { ¦ma·kōˈzō·ȯid }

mackerel [VERTEBRATE ZOOLOGY] The common name for perciform fishes composing the subfamily Scombroidei of the family Scombridae, characterized by a long slender body, pointed head, and large mouth. { ˈmak·rəl }

mackerel shark [VERTEBRATE ZOOLOGY] The common name for isurid galeoid elasmobranchs making up the family Isuridae; heavy-bodied fish with sharp-edged, awllike teeth and a nearly symmetrical tail. { ˈmak·rəl ˈshärk }

macrandrous [BOTANY] Having both antheridia and oogonia on the same plant; used especially for certain green algae. { maˈkran·drəs }

Macraucheniidae [PALEONTOLOGY] A family of extinct herbivorous mammals in the order Litopterna; members were proportioned much as camels are, and eventually lost the vertebral arterial canal of the cervical vertebrae. { ˌmaˌkrȯ·kəˈnī·əˌdē }

Macristiidae [VERTEBRATE ZOOLOGY] A family of oceanic teleostean fishes assigned by some zoologists to the order Ctenothrissiformes. { ˌmak·rəˈstī·əˌdē }

macroblast of Naegeli [HISTOLOGY] *See* pronormoblast. { ˈmak·rəˌblast əv ˈneg·ə·lē }

macroconsumer [ECOLOGY] A large consumer which ingests other organisms or particulate organic matter. Also known as biophage. { ¦mak·rō·kənˈsü·mər }

macrocyclic [MYCOLOGY] Of a rust fungus, having binuclear spores as well as teliospores and sporidia, or having a life cycle that is long or complex. { ¦mak·rōˈsī·klik }

Macrocypracea [INVERTEBRATE ZOOLOGY] A superfamily of marine ostracods in the suborder Podocopa having all thoracic legs different from each other, greatly reduced furcae, and long, thin Zenker's organs. { ˈmak·rō·səˈprās·ē·ə }

macrocyte [HISTOLOGY] An erythrocyte whose diameter or mean corpuscular volume exceed that of the mean normal by more than two standard deviations. Also known as macronormocyte. { ˈmak·rəˌsīt }

Macrodasyoidea [INVERTEBRATE ZOOLOGY] An order of wormlike invertebrates of the class Gastrotricha characterized by distinctive, cylindrical adhesive tubes in the cuticle which are moved by delicate muscle strands. { ˌmak·rōˌda·sēˈȯid·ē·ə }

macroevolution [EVOLUTION] The larger course of evolution by which the categories of animal and plant classification above the species level have been evolved from each other and have differentiated into the forms within each. { ¦mak·rōˌev·əˈlü·shən }

macrofauna [ECOLOGY] **1.** Widely distributed fauna. **2.** Fauna of a macrohabitat. [ZOOLOGY] Animals visible to the naked eye. { ¦mak·rō ˈfȯn·ə }

macroflora [BOTANY] Plants which are visible to the naked eye. [ECOLOGY] **1.** Widely distributed flora. **2.** Flora of a macrohabitat. { ¦mak·rō ˈflȯr·ə }

macrofossil [PALEONTOLOGY] A fossil large enough to be observed with the naked eye. { ¦mak·rōˈfäs·əl }

macrogamete [BIOLOGY] The larger, usually female gamete produced by a heterogamous organism. { ¦mak·rōˈgaˌmēt }

macroglia [HISTOLOGY] That portion of the neuroglia composed of astrocytes. { məˈkräg·lē·ə }

macroglobulin [BIOCHEMISTRY] Any gamma globulin with a sedimentation constant of 19S. { ¦mak·rəˈgläb·yə·lən }

macrohabitat [ECOLOGY] An extensive habitat presenting considerable variation of the environment, containing a variety of ecological niches, and supporting a large number and variety of complex flora and fauna. { ¦mak·rō'hab·ə,tat }

Macrolepidoptera [INVERTEBRATE ZOOLOGY] A former division of Lepidoptera that included the larger moths and butterflies. { ¦mak·rō,lep·ə 'däp·trə }

macrolide antibiotic [MICROBIOLOGY] A basic antibiotic characterized by a macrocyclic ring structure. { 'mak·rə,līd ,ant·i,bī'äd·ik }

macrolymphocyte [HISTOLOGY] A large lymphocyte. { ,mak·rō'lim·fə,sīt }

macromere [EMBRYOLOGY] Any of the large blastomeres composing the vegetative hemisphere of telolecithal morulas and blastulas. { 'mak·rō ,mir }

Macromonas [MICROBIOLOGY] A genus of gram-negative, chemolithotrophic bacteria; large, motile, cylindrical to bean-shaped cells containing sulfur granules and calcium carbonate inclusions. { ¦mak·rə'mō·nəs }

macromutation [GENETICS] Any genetic change that leads to a pronounced phenotypic alteration. { ,mak·rə·myü'tā·shən }

macronormocyte [HISTOLOGY] *See* macrocyte. { ,mak·rə'nȯr·mə,sīt }

macronucleus [INVERTEBRATE ZOOLOGY] A large, densely staining nucleus of most ciliated protozoans, believed to influence nutritional activities of the cell. { ¦mak·rō'nü·klē·əs }

macronutrient [BIOCHEMISTRY] An element, such as potassium and nitrogen, essential in large quantities for plant growth. { ¦mak·rō'nü·trē·ənt }

macrophage [HISTOLOGY] A large phagocyte of the reticuloendothelial system. Also known as a histiocyte. { 'mak·rə,fāj }

macrophagy [BIOLOGY] Feeding on large particulate matter. { 'mak·rə,fā·jē }

macrophreate [INVERTEBRATE ZOOLOGY] A comatulid with a large, deep cavity in the calyx. { ¦mak·rō'frē,āt }

macrophyllous [BOTANY] Having large or long leaves. { ¦mak·rō'fil·əs }

macrophyte [ECOLOGY] A macroscopic plant, especially one in an aquatic habitat. { 'mak·rə ,fīt }

macropinocytosis [CELL BIOLOGY] A mechanism of endocytosis in which large droplets of fluid are trapped underneath extensions (ruffles) of the cell surface. { ,mak·rō,pin·ə,sī'tō·səs }

Macropodidae [VERTEBRATE ZOOLOGY] The kangaroos, a family of Australian herbivorous mammals in the order Marsupialia. { ,mak·rə'päd·ə ,dē }

macropodous [BOTANY] **1.** Having a large or long hypocotyl. **2.** Having a long stem or stalk. { mak 'kräp·ə·dəs }

macropterous [ZOOLOGY] Having large or long wings or fins. { ma'kräp·tə·rəs }

Macroscelidea [VERTEBRATE ZOOLOGY] A monofamilial order of mammals containing the elephant shrews and their allies. { ,mak·rō·sə'lid·ē·ə }

Macroscelididae [VERTEBRATE ZOOLOGY] The single, African family of the mammalian order Macroscelidea. { ,mak·rō·sə'lid·ə,dē }

macroseptum [INVERTEBRATE ZOOLOGY] A primary septum in certain anthozoans. { ¦mak·rō 'sep·təm }

macrosporangium [BOTANY] A spore case in which macrospores are produced. Also known as megasporangium. { ¦mak·rə·spə'ran·jē·əm }

macrospore [BOTANY] The larger of two spore types produced by heterosporous plants; the female gamete. Also known as megaspore. [INVERTEBRATE ZOOLOGY] The larger gamete produced by certain radiolarians; the female gamete. { 'mak·rə,spȯr }

macrosporogenesis [BOTANY] In angiosperms, the formation of macrospores and the production of the embryo sac from one or occasionally several cells of the subepidermal cell layer within the ovule of a closed ovary. Also known as megasporogenesis. { ¦mak·rō¦spȯr·ō'jen·ə·səs }

Macrostomida [INVERTEBRATE ZOOLOGY] An order of rhabdocoels having a simple pharynx, paired protonephridia, and a single pair of longitudinal nerves. { ,mak·rə'stäm·ə·də }

Macrostomidae [INVERTEBRATE ZOOLOGY] A family of rhabdocoels in the order Macrostomida; members are broad and flattened in shape and have paired sex organs. { ,mak·rə'stäm·ə,dē }

macrostylous [BOTANY] **1.** Having long styles. **2.** Having long styles and short stamens. { ¦mak·rō¦stī·ləs }

Macrotermitinae [INVERTEBRATE ZOOLOGY] A subfamily of termites in the family Termitidae. { ,mak·rō,tər'mid·ə,nē }

macrothermophyte [ECOLOGY] *See* megathermophyte. { ,mak·rə'thər·mə,fīt }

Macrouridae [VERTEBRATE ZOOLOGY] The grenadiers, a family of actinopterygian fishes in the order Gadiformes in which the body tapers to a point, and the dorsal, caudal, and anal fins are continuous. { mə'krůr·ə,dē }

Macroveliidae [INVERTEBRATE ZOOLOGY] A family of hemipteran insects in the subdivision Amphibicorisae. { ,mak·rō·və'lī·ə,dē }

Macrura [INVERTEBRATE ZOOLOGY] A group of decapod crustaceans in the suborder Reptantia including eryonids, spiny lobsters, and true lobsters; the abdomen is extended and bears a well-developed tail fan. { mə'krůr·ə }

macrurous [ZOOLOGY] Having a long tail. { mə 'krůr·əs }

macula [ANATOMY] Any anatomical structure having the form of a spot or stain. { 'mak·yə·lə }

macula lutea [ANATOMY] A yellow spot on the retina; the area of maximum visual acuity, being made up almost entirely of retinal cones. { 'mak·yə·lə 'lüd·ē·ə }

maculate [BOTANY] Marked with speckles or spots. { 'mak·yə·lət }

Madreporaria [INVERTEBRATE ZOOLOGY] The equivalent name for Scleractinia. { ,mad·rə·pə 'rar·ē·ə }

madreporite [INVERTEBRATE ZOOLOGY] A delicately perforated sieve plate at the distal end of the stone canal in echinoderms. { 'mad·rə‚pȯr ‚īt }

Magelonidae [INVERTEBRATE ZOOLOGY] A monogeneric family of spioniform annelid worms belonging to Sedentaria. { ‚maj·ə'län·ə‚dē }

magenstrasse [ANATOMY] Gastric canal. { 'mäg·ən‚shträ·sə }

maggot [INVERTEBRATE ZOOLOGY] Larva of a dipterous insect. { 'mag·ət }

magnocellular [CELL BIOLOGY] Having large cell bodies; said of various nuclei of the central nervous system. { ‚mag·nə'sel·yə·lər }

Magnolia [BOTANY] A genus of trees, the type genus of the Magnoliaceae, with large, chiefly white, yellow, or pinkish flowers, and simple, entire, usually large evergreen or deciduous alternate leaves. { mag'nōl·yə }

Magnoliaceae [BOTANY] A family of dicotyledonous plants of the order Magnoliales characterized by hypogynous flowers with few to numerous stamens, stipulate leaves, and uniaperturate pollen. { mag‚nō·lē'ās·ē‚ē }

Magnoliales [BOTANY] The type order of the subclass Magnoliidae; members are woody plants distinguished by the presence of spherical ethereal oil cells and by a well-developed perianth of separate tepals. { mag‚nō·lē'ā·lēz }

Magnoliidae [BOTANY] A primitive subclass of flowering plants in the class Magnoliopsida generally having a well-developed perianth, numerous centripetal stamens, and bitegmic, crassinucellate ovules. { ‚mag·nō'lī·ə‚dē }

Magnoliophyta [BOTANY] The angiosperms, a division of vascular seed plants having the ovules enclosed in an ovary and well-developed vessels in the xylem. { mag‚nō·lē'äf·əd·ə }

Magnoliopsida [BOTANY] The dicotyledons, a class of flowering plants in the division Magnoliophyta generally characterized by having two cotyledons and net-veined leaves, with vascular bundles borne in a ring enclosing a pith. { mag ‚nō·lē'äp·sə·də }

magnum [ANATOMY] Large, as in foramen magnum. { 'mag·nəm }

mahogany [BOTANY] Any of several tropical trees in the family Meliaceae of the Geraniales. { mə 'häg·ə·nē }

maidenhead [ANATOMY] *See* hymen. { 'mād·ən ‚hed }

Maillard reaction [BIOCHEMISTRY] A reaction in which the amino group in an amino acid tends to form condensation products with aldehydes; believed to cause the Browning reaction when an amino acid and a sugar coexist, evolving a characteristic flavor useful in food preparations. { mī 'yär rē‚ak·shən }

main-band deoxyribonucleic acid [MOLECULAR BIOLOGY] A peak band of deoxyribonucleic acid obtained by density gradient centrifugation. { ¦mān ¦band ‚dē‚äk·sē‚rī·bō·nü'klē·ik 'as·əd }

Maindroniidae [INVERTEBRATE ZOOLOGY] A family of wingless insects belonging to the Thysanura proper. { ‚mān·drō'nī·ə‚dē }

maize [BOTANY] *Zea mays.* Indian corn, a tall cereal grass characterized by large ears. { māz }

Majidae [INVERTEBRATE ZOOLOGY] The spider, or decorator, crabs, a family of decapod crustaceans included in the Brachyura; members are slow-moving animals that often conceal themselves by attaching seaweed and sessile animals to their carapace. { 'maj·ə‚dē }

major gene [GENETICS] Any gene individually associated with pronounced phenotypic effects. { 'mā·jər ‚jēn }

major histocompatibility complex [IMMUNOLOGY] In vertebrates, a family of genes that encode cell surface glycoproteins that regulate interactions among cells of the immune system, some components of the complement system, and perhaps other related functions connected with intercell recognition. Abbreviated MHC. { 'mā·jər ¦hi·stō·kəm'pad·ə'bil·əd·ē 'käm‚pleks }

major immunogene complex [IMMUNOLOGY] A genetic region containing loci that code for lymphocyte surface antigens, histocompatibility antigens, immune response gene products, and proteins of the complement system. Abbreviated MIC. { 'mā·jar ‚ə'myü·nə·jēn 'käm‚pleks }

Malachiidae [INVERTEBRATE ZOOLOGY] An equivalent name for Malyridae. { ‚mal·ə'kī·ə‚dē }

Malacobothridia [INVERTEBRATE ZOOLOGY] A subclass of worms in the class Trematoda; members typically have one or two soft, flexible suckers and are endoparasitic in vertebrates and invertebrates. { ‚mal·ə·kō·bə'thrid·ē·ə }

Malacocotylea [INVERTEBRATE ZOOLOGY] The equivalent name for Digenea. { ‚mal·ə·kō‚käd·əl 'ē·ə }

malacology [INVERTEBRATE ZOOLOGY] The study of mollusks. { ‚mal·ə'käl·ə·jē }

Malacopoda [INVERTEBRATE ZOOLOGY] A subphylum of invertebrates in the phylum Oncopoda. { ‚mal·ə'käp·ə·də }

Malacopterygii [VERTEBRATE ZOOLOGY] An equivalent name for Clupeiformes in older classifications. { ¦mal·ə·kō·tə¦rij·ē‚ī }

Malacostraca [INVERTEBRATE ZOOLOGY] A large, diversified subclass of Crustacea including shrimps, lobsters, crabs, sow bugs, and their allies; generally characterized by having a maximum of 19 pairs of appendages and trunk limbs which are sharply differentiated into thoracic and abdominal series. { ‚mal·ə'kä·strə·kə }

Malapteruridae [VERTEBRATE ZOOLOGY] A family of African catfishes in the suborder Siluroidei. { mə‚lap·tə'rür·ə‚dē }

malar [ANATOMY] Of or pertaining to the cheek or to the zygomatic bone. { 'mā·lər }

malar bone [ANATOMY] *See* zygomatic bone. { 'mā·lər ‚bōn }

malar stripe [VERTEBRATE ZOOLOGY] **1.** The area extending from the corner of the mouth backward and down in birds. **2.** Area on side of throat below the base of the lower mandible. { 'mā·lər ‚strīp }

malate dehydrogenase [BIOCHEMISTRY] *See* malic enzyme. { 'ma‚lāt dē'hī·drə·jə‚nās }

Malcidae [INVERTEBRATE ZOOLOGY] A small family of Ethiopian and Oriental hemipternan insects in the superfamily Pentatomorpha. { 'mal·sə,dē }

Malcodermata [INVERTEBRATE ZOOLOGY] The equivalent name for Cantharoidea. { ,mal·kō 'dər·məd·ə }

Maldanidae [INVERTEBRATE ZOOLOGY] The bamboo worms, a family of mud-swallowing annelids belonging to the Sedentaria. { mal'dan·ə,dē }

Maldaninae [INVERTEBRATE ZOOLOGY] A subfamily of the Maldanidae distinguished by cephalic and anal plaques with the anal aperture located dorsally. { mal'dan·ə,nē }

male [BOTANY] A flower lacking pistils. [ZOOLOGY] **1.** Of or pertaining to the sex that produces spermatozoa. **2.** An individual of this sex. { māl }

male climacteric [PHYSIOLOGY] A condition presumably due to loss of testicular function, associated with an elevated urinary excretion of gonadotropins and symptoms of loss of sexual desire and potency, hot flashes, and vasomotor instability. { 'māl klə'mak·tə·rik }

male heterogamety [GENETICS] Having paired sex chromosomes of unequal sizes in the male. { 'māl ¦hed·ə·rō¦gam·əd·ē }

male homogamety [GENETICS] Having the paired sex chromosomes of equal size in the male. { 'māl ¦hō·mō¦gam·əd·ē }

male sterility [PHYSIOLOGY] The condition in which male gametes are absent, deficient in number, or nonfunctional. { māl stə'ril·əd·ē }

malezal swamp [ECOLOGY] A swamp resulting from drainage of water over an extensive plain with a slight, almost imperceptible slope. { mə 'lēz·əl ,swämp }

malic acid [BIOCHEMISTRY] $COOH \cdot CH_2 \cdot CHOH \cdot COOH$ Hydroxysuccinic acid: a dibasic hydroxy acid existing in two optically active isomers and a racemic form; found in apples and many other fruits. { 'mal·ik 'as·əd }

malic dehydrogenase [BIOCHEMISTRY] An enzyme in the Krebs cycle that catalyzes the conversion of L-malic acid to oxaloacetic acid. { 'mal·ik dē'hī·drə·jə,nās }

malic enzyme [BIOCHEMISTRY] An enzyme which utilizes nicotinamide-adenine dinucleotide phosphate (NADP) to catalyze the oxidative decarboxylation of malic acid to pyruvic acid and carbon dioxide. Also known as malate dehydrogenase. { 'mal·ik 'en,zīm }

malignant [CELL BIOLOGY] Pertaining to cells that have undergone phenotypic transformation by oncogenes or protooncogenes. { mə'lig·nənt }

malleate trophus [INVERTEBRATE ZOOLOGY] A type of crushing masticatory apparatus in rotifers that are incidentally predatory, such as brachionids. { 'mal·ē,āt ,trō·fəs }

mallee [ECOLOGY] *See* tropical scrub. { 'mä·lē }

malleolus [ANATOMY] A projection on the distal end of the tibia and fibula at the ankle. { mə'lē·ə·ləs }

malleoramate trophus [INVERTEBRATE ZOOLOGY] An intermediate type of rotiferan masticatory apparatus having a looped manubrium and teeth on the incus (comprising the fulcrum and rami); developed for grinding. { ¦mal·ē·ə¦ra,māt ,trō·fəs }

malleus [ANATOMY] The outermost, hammer-shaped ossicle of the middle ear; attaches to the tympanic membrane and articulates with the incus. { 'mal·ē·əs }

Mallophaga [INVERTEBRATE ZOOLOGY] Biting lice, a comparatively small order of wingless insects characterized by five-segmented antennae, distinctly developed mandibles, one or two terminal claws on each leg, and a prothorax developed as a distinct segment. { mə'läf·ə·gə }

Malpighiaceae [BOTANY] A family of dicotyledonous plants in the order Polygalales distinguished by having three carpels, several fertile stamens, five petals that are commonly fringed or toothed, and indehiscent fruit. { mal,pig·ē'ās·ē,ē }

Malpighian corpuscle [ANATOMY] **1.** A lymph nodule of the spleen. **2.** *See* renal corpuscle. { mal 'pig·ē·ən 'kȯr·pə·səl }

Malpighian layer [HISTOLOGY] The germinative layer of the epidermis. { mal'pig·ē·ən ,lā·ər }

Malpighian pyramid [ANATOMY] *See* renal pyramid. { mal'pig·ē·ən 'pir·ə·məd }

Malpighian tubule [INVERTEBRATE ZOOLOGY] Any of the blind tubes that open into the posterior portion of the gut in most insects and certain other arthropods and excrete matter or secrete substances such as silk. { mal'pig·ē·ən 'tü ,byül }

maltase [BIOCHEMISTRY] An enzyme that catalyzes the conversion of maltose to dextrose. { 'mȯl,tās }

Malthusianism [BIOLOGY] The theory that population increases more rapidly than the food supply unless held in check by epidemics, wars, or similar phenomena. { mal'thü·zhə,niz·əm }

maltobiose [BIOCHEMISTRY] *See* maltose. { ¦mȯl·tō'bī,ōs }

maltose [BIOCHEMISTRY] $C_{12}H_{22}O_{11}$ A crystalline disaccharide that is a product of the enzymatic hydrolysis of starch, dextrin, and glycogen; does not appear to exist free in nature. Also known as maltobiose; malt sugar. { 'mȯl,tōs }

maltose phosphorylase [BIOCHEMISTRY] An enzyme which reacts maltose with inorganic phosphate to yield glucose and glucose-1-phosphate. { 'mȯl,tōs ,fäs'fȯr·ə,lās }

malt sugar [BIOCHEMISTRY] *See* maltose. { 'mȯlt 'shu̇g·ər }

Malvaceae [BOTANY] A family of herbaceous dicotyledons in the order Malvales characterized by imbricate or contorted petals, mostly unilocular anthers, and minutely spiny, multiporate pollen. { mal'vās·ē,ē }

Malvales [BOTANY] An order of flowering plants in the subclass Dilleniidae having hypogynous flowers with valvate calyx, mostly separate petals, numerous centrifugal stamens, and a syncarpous pistil. { mal'vā·lēz }

mamelon [BIOLOGY] Any dome-shaped protrusion or elevation. { 'mam·ə·lən }

mamma [ANATOMY] A milk-secreting organ characterizing all mammals. { 'mam·ə }

mammal [VERTEBRATE ZOOLOGY] A member of Mammalia. { 'mam·əl }

Mammalia [VERTEBRATE ZOOLOGY] A large class of warm-blooded vertebrates containing animals characterized by mammary glands, a body covering of hair, three ossicles in the middle ear, a muscular diaphragm separating the thoracic and abdominal cavities, red blood cells without nuclei, and embryonic development in the allantois and amnion. { mə'māl·yə }

mammary [ANATOMY] Of or pertaining to the mamma, or breast. { 'mam·ə·rē }

mammary gland [PHYSIOLOGY] A highly modified sebaceous gland that secretes milk; a unique anatomical feature of mammals. { 'mam·ə·rē ˌgland }

mammary lymphatic plexus [ANATOMY] A network of anastomosing lymphatic vessels in the walls of the ducts and between the lobules of the mamma; also functions to drain skin, areola, and nipple. { 'mam·ə·rē lim'fad·ik 'plek·səs }

mammary region [ANATOMY] The space on the anterior surface of the chest between a line drawn through the lower border of the third rib and one drawn through the upper border of the xiphoid cartilage. { 'mam·ə·rē ˌrē·jən }

mammary ridge [EMBRYOLOGY] An ectodermal thickening forming a longitudinal elevation on the chest between the limbs from which the mammary glands develop. { 'mam·ə·rē ˌrij }

mammary-stimulating hormone [BIOCHEMISTRY] **1.** Estrogen and progesterone considered together as the hormones which induce proliferation of the mammary ductile and acinous elements respectively. **2.** *See* prolactin. { 'mam·ə·rē ¦stim·yəˌlād·iŋ 'hȯrˌmōn }

mammary tumor agent [VIROLOGY] A milk-borne virus that produces mammary cancer in mice with the appropriate genotype. { 'mam·ə·rē ¦tü·mər ˌāj·ənt }

mammillary [ANATOMY] **1.** Of or pertaining to the nipple. **2.** Breast- or nipple-shaped. { 'ma·mə ˌler·ē }

mammillary body [ANATOMY] Either of two small, spherical masses of gray matter at the base of the brain in the space between the hypophysis and oculomotor nerve, which receive and relay olfactory impulses. { 'ma·məˌler·ē ¦bäd·ē }

mammillary line [ANATOMY] A vertical line passing through the center of the nipple. { 'ma·mə ˌler·ē ¦līn }

mammillary process [ANATOMY] One of the tubercles on the posterior part of the superior articular processes of the lumbar vertebrae. { 'ma məˌler ē 'präs əs }

mamm-, mammo- [ANATOMY] A combining form meaning breast. { 'mam·ō }

mammogen [BIOCHEMISTRY] *See* prolactin. { ˌmam·ə'jen }

mammogenic hormone [BIOCHEMISTRY] **1.** Any hormone that stimulates or induces development of the mammary gland. **2.** *See* prolactin. { ¦mam·ə¦jen·ik 'hȯrˌmōn }

mammoplasia [PHYSIOLOGY] Development of breast tissue. { ˌmam·ə'plā·zhə }

mammoth [PALEONTOLOGY] Any of various large Pleistocene elephants having long, upcurved tusks and a heavy coat of hair. { 'mam·əth }

mammotropin [BIOCHEMISTRY] *See* prolactin. { ¦mam·ə¦trä·pən }

Mammutinae [PALEONTOLOGY] A subfamily of extinct proboscidean mammals in the family Mastodontidae. { mə'myüt·ənˌē }

mandarin [BOTANY] A large and variable group of citrus fruits in the species *Citrus reticulata* and some of its hybrids; many varieties of the trees are compact with willowy twigs and small, narrow, pointed leaves; includes tangerines, King oranges, Temple oranges, and tangelos. { 'man·də·rən }

mandible [ANATOMY] **1.** The bone of the lower jaw. **2.** The lower jaw. [INVERTEBRATE ZOOLOGY] Any of various mouthparts in many invertebrates designed to hold or bite into food. { 'man·də·bəl }

mandibular arch [EMBRYOLOGY] The first visceral arch in vertebrates. { man'dib·yə·lər 'ärch }

mandibular cartilage [EMBRYOLOGY] The bar of cartilage supporting the mandibular arch. { man 'dib yə·lər 'kärt·lij }

mandibular fossa [ANATOMY] The depression in the temporal bone that articulates with the mandibular condyle. { man'dib·yə·lər 'fäs·ə }

mandibular gland [ANATOMY] *See* submandibular gland. { man'dib·yə·lər 'gland }

mandibular nerve [ANATOMY] A mixed nerve branch of the trigeminal nerve; innervates various structures of the lower jaw and face. { man 'dib·yə·lər 'nərv }

Mandibulata [INVERTEBRATE ZOOLOGY] A subphylum of Arthropoda; members possess a pair of mandibles which characterize the group. { man ˌdib·yə'läd·ə }

mandrill [VERTEBRATE ZOOLOGY] *Mandrillus sphinx*. An Old World cercopithecoid monkey found in west-central Africa and characterized by large red callosities near the ischium and by blue ridges on each side of the nose in males. { 'man·drəl }

mango [BOTANY] *Mangifera indica*. A large evergreen tree of the sumac family (Anacardiaceae), native to southeastern Asia, but now cultivated in Africa, tropical America, Florida, and California for its edible fruit, a thick-skinned, yellowish-red, fleshy drupe. { 'maŋ·gō }

mangrove [BOTANY] A tropical tree or shrub of the genus *Rhizophora* characterized by an extensive, impenetrable system of prop roots which contribute to land building. { 'maŋˌgrōv }

mangrove swamp [ECOLOGY] A tropical or subtropical marine swamp distinguished by the abundance of low to tall trees, especially mangrove trees. { 'maŋˌgrōv ˌswämp }

Manidae [VERTEBRATE ZOOLOGY] The pangolins, a family of mammals comprising the order Pholidota. { 'man·əˌdē }

manihot [BOTANY] *See* cassava. { 'man·əˌhät }

Manila hemp [BOTANY] *See* abaca. { mə'nil·ə 'hemp }

Manila maguey [BOTANY] *See* cantala. { mə'nil·ə mə'gā }

manioc [BOTANY] *See* cassava. { 'man·ēˌäk }

mannan [BIOCHEMISTRY] Any of a group of polysaccharides composed chiefly or entirely of D-mannose units. { 'ma,nan }

mannose [BIOCHEMISTRY] $C_6H_{12}O_6$ A fermentable monosaccharide obtained from manna. { 'ma,nōs }

Mantidae [INVERTEBRATE ZOOLOGY] A family of predacious orthopteran insects characterized by a long, slender prothorax bearing a pair of large, grasping legs, and a freely moving head with large eyes. { 'man·tə,dē }

mantis [INVERTEBRATE ZOOLOGY] The common name for insects composing the family Mantidae. { 'man·təs }

mantle [ANATOMY] Collectively, the convolutions, corpus callosum, and fornix of the brain. [BIOLOGY] An enveloping layer, as the external body wall lining the shell of many invertebrates, or the external meristematic layers in a stem apex. [VERTEBRATE ZOOLOGY] The back and wing plumage of a bird if distinguished from the rest of the plumage by a uniform color. { 'mant·əl }

mantle cavity [INVERTEBRATE ZOOLOGY] The space between mantle and body proper in bivalve mollusks. { 'mant·əl ,kav·əd·ē }

mantle lobe [INVERTEBRATE ZOOLOGY] Either of the flaps on the dorsal and ventral sides of the mantle in bivalve mollusks. { 'mant·əl ,lōb }

Mantodea [INVERTEBRATE ZOOLOGY] An order equivalent to the family Mantidae in some systems of classification. { man'tō·dē·ə }

Mantoux test [IMMUNOLOGY] An intradermal test for tuberculin sensitivity, that is, for past or present infection with tubercle bacilli. { man'tü ,test }

manubrium [ANATOMY] **1.** The triangular cephalic portion of the sternum in humans and certain other mammals. **2.** The median anterior portion of the sternum in birds. **3.** The process of the malleus. [BOTANY] A cylindrical cell that projects inward from the middle of each shield composing the antheridium in stoneworts. [INVERTEBRATE ZOOLOGY] The elevation bearing the mouth in hydrozoan polyps. { mə'nü·brē·əm }

manus [ANATOMY] The hand of a human or the forefoot of a quadruped. [INVERTEBRATE ZOOLOGY] The proximal enlargement of the propodus of the chela of arthropods. { 'mä·nəs }

manyplies [VERTEBRATE ZOOLOGY] *See* omasum. { 'men·ē,plīz }

map distance [GENETICS] The distance, in terms of percentage of crossing-over, between linked genes. { 'map ,dis·təns }

maple [BOTANY] Any of various broad-leaved, deciduous trees of the genus *Acer* in the order Sapindales characterized by simple, opposite, usually palmately lobed leaves and a fruit consisting of two long-winged samaras. { 'mā·pəl }

map unit [GENETICS] A measure of genetic distance corresponding to a recombination frequency of 1% or 1 centriMorgan (cM). { 'map ,yü·nət }

maquis [ECOLOGY] A type of vegetation composed of shrubs, or scrub, usually not exceeding 10 feet (3 meters) in height, the majority having small, hard, leathery, often spiny or needlelike drought-resistant leaves and occurring in areas with a Mediterranean climate. { mä'kē }

Marantaceae [BOTANY] A family of monocotyledonous plants in the order Zingiberales characterized by one functional stamen with a single functional pollen sac, solitary ovules in each locule, and mostly arillate seeds. { ,mar·ən'tās·ē ,ē }

Marattiaceae [BOTANY] A family of ferns coextensive with the order Marattiales. { mə,rad·ē 'ās·ē,ē }

Marattiales [BOTANY] An ancient order of ferns having massive eusporangiate sporangia in sori on the lower side of the circinate leaves. { mə ,rad·ē'ā·lēz }

Marburg virus [VIROLOGY] A large virus transmitted to humans by the grivet monkey (*Cercopithecus aethiops*). { 'mär,bu̇rg ,vī·rəs }

marcescent [BOTANY] Withering without falling off. { mär'ses·ənt }

Marcgraviaceae [BOTANY] A family of dicotyledonous shrubs or vines in the order Theales having exstipulate leaves with scanty or no endosperm, two integuments, and highly modified bracts. { märk,grā·vē'ās·ē,ē }

Marchantiales [BOTANY] The thallose liverworts, an order of the class Marchantiopsida having a flat body composed of several distinct tissue layers, smooth-walled and tuberculate-walled rhizoids, and male and female sex organs borne on stalks on separate plants. { mär,shan·tē'ā·lēz }

Marchantiidae [BOTANY] A subclass of liverworts of the class Hepaticopsida, having gametophytes with ribbonlike or rosette-shaped thalli and generally reduced sporophytes. { ,mär·shən'tī·ə ,dē }

Marchantiopsida [BOTANY] The liverworts, a class of lower green plants; the plant body is usually a thin, prostrate thallus with rhizoids on the lower surface. { mär,shan·tē'äp·sə·də }

mare [VERTEBRATE ZOOLOGY] A mature female horse or other equine. { 'mär·ā, mer }

Margaritiferidae [INVERTEBRATE ZOOLOGY] A family of gastropod mollusks with nacreous shells that provide an important source of commercial pearls. { mär¦gär·ə·də'fer·ə,dē }

Margarodinae [INVERTEBRATE ZOOLOGY] A subfamily of homopteran insects in the superfamily Coccoidea in which abdominal spiracles are present in all stages of development. { ,mär·gə 'räd·ən,ē }

marginal placentation [BOTANY] Arrangement of ovules near the margins of carpels. { 'mär·jən· əl ,plas·ən'tā·shən }

marginal sinus [ANATOMY] **1.** One of the small, bilateral sinuses of the dura mater which skirt the edge of the foramen magnum, usually uniting posteriorly to form the occipital sinus. **2.** *See* terminal sinus. [EMBRYOLOGY] An enlarged venous sinus incompletely encircling the margin of the placenta. { 'mär·jən·əl 'sī·nəs }

marginate [BOTANY] Having a distinct margin or border. { 'mär·jə,nāt }

maricolous [ECOLOGY] Living in the sea. { mə 'rik·ə·ləs }

marihuana [BOTANY] *See* marijuana. { ˌmar·ə ˈwän·ə }

marijuana [BOTANY] The Spanish name for the dried leaves and flowering tops of the hemp plant (*Cannabis sativa*), which have narcotic ingredients and are smoked in cigarettes. Also spelled marihuana. { ˌmar·əˈwän·ə }

marine biocycle [ECOLOGY] A major division of the biosphere composed of all biochores of the sea. { məˈrēn ˈbī·ōˌsī·kəl }

marine biology [BIOLOGY] A branch of biology that deals with those living organisms which inhabit the sea. { məˈrēn bīˈäl·ə·jē }

marine littoral faunal region [ECOLOGY] A geographically determined division of that portion of the zoosphere composed of marine animals. { məˈrēn ˈlit·ə·rəl ˈfón·əl ˌrē·jən }

marine marsh [ECOLOGY] A flat, savannalike land expanse at the edge of a sea; usually covered by water during high tide. { məˈrēn ˈmärsh }

marine microbiology [MICROBIOLOGY] The study of the microorganisms living in the sea. { məˈrēn ˌmī·krō·bīˈäl·ə·jē }

marita [INVERTEBRATE ZOOLOGY] An adult trematode. { məˈrīd·ə }

marjoram [BOTANY] Any of several perennial plants of the genera *Origanum* and *Majorana* in the mint family, Labiatae, the leaves are used as a food seasoning. { ˈmär·jə·rəm }

marker [GENETICS] A gene with a known location on a chromosome and a clear-cut phenotype. [IMMUNOLOGY] Any antigen that serves to distinguish cell types. { ˈmärk·ər }

marmoset [VERTEBRATE ZOOLOGY] Any of 10 species of South American primates belonging to the family Callithricidae; individuals are primitive in that they have claws rather than nails and a nonprehensile tail. { ˈmär·məˌset }

marmot [VERTEBRATE ZOOLOGY] Any of several species of stout-bodied, short-legged burrowing rodents of the genus *Marmota* in the squirrel family Sciuridae. { ˈmär·mət }

marrubium [BOTANY] *Marrubium vulgari*. An aromatic plant of the mint family, Labiatae; leaves have a bitter taste and are used as a tonic and anthelmintic. Also known as hoarhound; horehound. { məˈrü·bē·əm }

marsh [ECOLOGY] A transitional land-water area, covered at least part of the time by estuarine or coastal waters, and characterized by aquatic and grasslike vegetation, especially without peatlike accumulation. { märsh }

Marsileales [BOTANY] A small monofamilial order of heterosporous, leptosporangiate ferns (Polypodiophyta); leaves arise on long stalks from the rhizome, and sporangia are enclosed in modified folded leaves or leaf segments called sporocarps. { märˌsil·ēˈā·lēz }

Marsipobranchii [VERTEBRATE ZOOLOGY] An equivalent name for Cyclostomata. { ˌmär·sə·pō ˈbraŋ·kēˌī }

marsupial [VERTEBRATE ZOOLOGY] **1.** A member of the Marsupialia. **2.** Having a marsupium. **3.** Of, pertaining to, or constituting a marsupium. { märˈsü·pē·əl }

marsupial anteater [VERTEBRATE ZOOLOGY] *Myrmecobius fasciatus*. An anteater belonging to the Marsupialia. Also known as banded anteater. { märˈsü·pē·əl ˈantˌēd·ər }

marsupial frog [VERTEBRATE ZOOLOGY] Any of various South American tree frogs which carry the eggs in a pouch on the back. { märˈsü·pē·əl ˈfräg }

Marsupialia [VERTEBRATE ZOOLOGY] The single order of the mammalian infraclass Metatheria, characterized by the presence of a marsupium in the female. { märˌsü·pēˈā·lē·ə }

marsupial mole [VERTEBRATE ZOOLOGY] *Notoryctes typhlops*. A marsupial of Australia that resembles the euterian mole. { märˌsü·pē·əl ˈmōl }

Marsupicarnivora [VERTEBRATE ZOOLOGY] An order proposed to include the polydactylous and polyprotodont carnivorous superfamilies of Marsupialia. { märˌsü·pə·kärˈniv·ə·rə }

marsupium [VERTEBRATE ZOOLOGY] A fold of skin that forms a pouch enclosing the mammary glands on the abdomen of most marsupials. { märˈsü·pē·əm }

marten [VERTEBRATE ZOOLOGY] Any of seven species of carnivores of the genus *Martes* in the family Mustelidae which resemble the weasel but are larger and of a semiarboreal habit. { ˈmärt·ən }

masculine [BIOLOGY] Having an appearance or qualities distinctive for a male. { ˈmas·kyə·lən }

masculine pelvis [ANATOMY] A female pelvis similar to the normal male pelvis in having a deeper cavity and more conical shape. Also known as android pelvis. { ˈmas·kyə·lən ˈpel·vəs }

masculinize [PHYSIOLOGY] To cause a female or a sexually immature animal to take on male secondary sex characteristics. { ˈmas·kyə·ləˌnīz }

masked messenger ribonucleic acid [MOLECULAR BIOLOGY] *See* maternal messenger ribonucleic acid. { ¦maskd ˈmes·ən·jer ¦rī·bo·nüˈklē·ik ˈas·əd }

masseter [ANATOMY] The masticatory muscle, arising from the zygomatic arch and inserted into the lower jaw. { məˈsēd·ər }

mass extinction [EVOLUTION] *See* faunal extinction. { ˈmas ikˈstiŋk·shən }

massive [PALEONTOLOGY] Of corallum, composed of closely packed corallites. { ˈmas·iv }

mass reflex [PHYSIOLOGY] A spread of reflexes suggesting lack of control by higher cortical centers; seen in normal newborns, in persons under the influence of drugs or in severe emotional states, and in encephalopathy or high spinal cord transections. { ˈmas ˈrēˌfleks }

Mastacembeloidei [VERTEBRATE ZOOLOGY] The spiny eels, a suborder of perciform fishes that are eellike in shape and have the pectoral girdle suspended from the vertebral column. { ˌmas·tə ˌsem·bəˈlòid·ēˌī }

mastax [INVERTEBRATE ZOOLOGY] The muscular pharynx in rotifers. { ˈmaˌstaks }

mast cell [HISTOLOGY] A connective-tissue cell with numerous large, basophilic, metachromatic granules in the cytoplasm. { ˈmast ˌsel }

masticate [PHYSIOLOGY] To chew. { ˈmas·tə ˌkāt }

Mastigamoebidae [INVERTEBRATE ZOOLOGY] A

family of ameboid protozoans possessing one or two flagella, belonging to the order Rhizomastigida. { ˌmas·tə·gə'mē·bəˌdē }

mastigoneme [CELL BIOLOGY] Any of the fine hairlike appendages that extend from the shaft of the flagellum in certain motile cells. { ˌmas·tə'gōˌnēm }

Mastigophora [INVERTEBRATE ZOOLOGY] A superclass of the Protozoa characterized by possession of flagella. { ˌmas·tə'gäf·ə·rə }

mast-, masto- [ANATOMY] A combining form denoting breast; denoting mastoid. { mastˌ'mas·dō }

mastodon [PALEONTOLOGY] A member of the Mastodontidae, especially the genus *Mammut*. { 'mas·təˌdän }

Mastodontidae [PALEONTOLOGY] An extinct family of elephantoid proboscideans that had low-crowned teeth with simple ridges and without cement. { ˌmas·tə'dän·təˌdē }

mastoid [ANATOMY] **1.** Breast-shaped. **2.** The portion of the temporal bone where the mastoid process is located. { 'maˌstȯid }

mastoid air cell [ANATOMY] *See* mastoid cell. { 'maˌstȯid 'er ˌsel }

mastoid antrum [ANATOMY] An air-filled space between the upper portion of the middle ear and the mastoid cells. { 'maˌstȯid 'an·trəm }

mastoid canaliculus [ANATOMY] A small canal opening just above the stylomastoid foramen; gives passage to the auricular branch of the vagus nerve. { 'maˌstȯid ˌkan·ə'lik·yə·ləs }

mastoid cell [ANATOMY] One of the compartments in the mastoid portion of the temporal bone, connected with the mastoid antrum and lined with a mucous membrane. Also known as mastoid air cell; mastoid sinus. { 'maˌstȯid ˌsel }

mastoid foramen [ANATOMY] A small opening behind the mastoid process. { 'maˌstȯid fə'rā·mən }

mastoid fossa [ANATOMY] The depression behind the suprameatal spine on the lateral surface of the temporal bone. { 'maˌstȯid 'fäs·ə }

mastoid process [ANATOMY] A nipple-shaped, inferior projection of the mastoid portion of the temporal bone. { 'maˌstȯid 'prä·səs }

mastoid sinus [ANATOMY] *See* mastoid cell. { 'ma ˌstȯid 'sī·nəs }

Mastotermitidae [INVERTEBRATE ZOOLOGY] A family of lower termites in the order Isoptera with a single living species, in Australia. { ˌmas·tō·tər'mid·əˌdē }

match [IMMUNOLOGY] To select blood donors whose erythrocytes are compatible with those of the recipient. { mach }

mate [BIOLOGY] **1.** To pair for breeding. **2.** To copulate. { māt }

mate killer [VIROLOGY] Paramecia that contain the mu phage and cause their sensitive partners to die after conjugation. { 'māt ˌkil·ər }

maternal [BIOLOGY] Of, pertaining to, or related to a mother. { mə'tərn·əl }

maternal effect [GENETICS] Determination of characters of the progeny by the maternal parent; mediated by the genetic constitution of the mother. { mə'tərn·əl iˌfekt }

maternal inheritance [GENETICS] The acquisition of characters transmitted through the cytoplasm of the egg. { mə'tərn·əl in'her·əd·əns }

maternal messenger ribonucleic acid [MOLECULAR BIOLOGY] In certain oocytes, messenger ribonucleic acid that is stored during oogenesis for translation during early embryogenesis. Also known as masked messenger ribonucleic acid. { mə¦tərn·əl ¦mes·ən·jər ˌrī·bō·nü¦klē·ik 'as·əd }

maternal placenta [EMBRYOLOGY] The outer placental layer, developed from the decidua basalis. { mə'tərn·əl plə'sen·tə }

maternity [BIOLOGY] **1.** Motherhood. **2.** The state of being pregnant. { mə'tərn·əd·ē }

mathematical biology [BIOLOGY] A discipline that encompasses all applications of mathematics, computer technology, and quantitative theorizing to biological systems, and the underlying processes within the systems. { ¦math·ə¦mad·ə·kəl bī'äl·ə·jē }

mathematical biophysics [BIOLOGY] A discipline which attempts to utilize mathematics to explain biophysical processes. { ¦math·ə¦mad·ə·kəl ˌbī·ō'fiz·iks }

mathematical ecology [ECOLOGY] The application of mathematical theory and technique to ecology. { ¦math·ə¦mad·ə·kəl ē'käl·ə·jē }

matico [BOTANY] *Piper angustifolium*. An aromatic wild pepper found in tropical America whose leaves are rich in volatile oil, gums, and tannins; leaves were used medicinally as a stimulant and hemostatic. { mə'tē·kō }

mating [BIOLOGY] The meeting of individuals for sexual reproduction. { 'mād·iŋ }

mating type [MICROBIOLOGY] The genetically determined mating behavior characteristic of certain species of microorganisms; only different mating types can conjugate. { 'mad·iŋ ˌtīp }

Matricaria [BOTANY] A genus of weedy herbs having a strong odor and white and yellow disk flowers; chamomile oil is obtained from certain species. { ˌma·trə'kar·ē·ə }

matrix [HISTOLOGY] **1.** The intercellular substance of a tissue. Also known as ground substance. **2.** The epithelial tissue from which a toenail or fingernail develops. [MYCOLOGY] The substrate on or in which fungus grows. { 'mā·triks }

matrix association region [CELL BIOLOGY] In eukaryotic interphase nuclei, any of the specific sites to which the chromatin loop domains are anchored. { ¦ma·triks əˌsō·sē'ā·shən ˌrē·jən }

matroclinous inheritance [GENETICS] Inheritance in which the offspring more closely resemble the female parent than the male parent. { ¦ma·trəˌklī·nəs in'her·əd·əns }

matromycin [MICROBIOLOGY] *See* oleandomycin. { ¦ma·trə¦mīs·ən }

maturase [MOLECULAR BIOLOGY] Any enzyme encoded by self-splicing introns that catalyzes excision of the intron from its own primary transcript. { 'mach·əˌrās }

maturation [BIOLOGY] **1.** The process of coming

to full development. **2.** The final series of changes in the growth and formation of germ cells. [VIROLOGY] The process that leads to incorporation of viral genomes into capsids and complete virions. { ,mach·ə'rā·shən }

mature [BIOLOGY] **1.** Being fully grown and developed. **2.** Ripe. { mə'chür }

Mauritius hemp [BOTANY] A hard fiber obtained from the leaves of the cabuya, grown on the island of Mauritius; not a true hemp. { mò'rish·əs 'hemp }

maxilla [ANATOMY] **1.** The upper jawbone. **2.** The upper jaw. [INVERTEBRATE ZOOLOGY] Either of the first two pairs of mouthparts posterior to the mandibles in certain arthropods. { mak 'sil·ə }

maxillary air sinus [ANATOMY] *See* maxillary sinus. { 'mak·sə,ler·ē ¦er ,sī·nəs }

maxillary antrum [ANATOMY] *See* maxillary sinus. { 'mak·sə,ler·ē 'an·trəm }

maxillary arch [ANATOMY] *See* palatomaxillary arch. { 'mak·sə,ler·ē 'ärch }

maxillary artery [ANATOMY] A branch of the external carotid artery which supplies the deep structures of the face (internal maxillary) and the side of the face and nose (external maxillary). { 'mak·sə,ler·ē 'ärd·ə·rē }

maxillary hiatus [ANATOMY] An opening in the maxilla connecting the nasal cavity with the maxillary sinus. { 'mak·sə,ler·ē hī'ād·əs }

maxillary nerve [ANATOMY] A somatic sensory branch of the trigeminal nerve; innervates the meninges, the skin of the upper portion of the face, upper teeth, and mucosa of the nose, palate, and cheeks. { 'mak·sə,ler·ē 'nərv }

maxillary process of the embryo [EMBRYOLOGY] An outgrowth of the dorsal part of the mandibular arch that forms the lateral part of the upper lip, the upper cheek region, and the upper jaw except the premaxilla. { 'mak·sə,ler·ē ¦prä·səs əv thē 'em·brē·ō }

maxillary sinus [ANATOMY] A paranasal air cavity in the body of the maxilla. Also known as maxillary air sinus; maxillary antrum. { 'mak·sə,ler·ē 'sī·nəs }

maxilliped [INVERTEBRATE ZOOLOGY] One of the three pairs of crustacean appendages immediately posterior to the maxillae. { mak'sil·ə ,ped }

maximal breathing capacity [PHYSIOLOGY] The greatest respiratory minute volume which an individual can produce during a given period of extremely forceful breathing. { 'mak·sə·məl 'brēth·iŋ kə,pas·əd·ē }

maximum breathing capacity [PHYSIOLOGY] The greatest volume of air an individual can breathe voluntarily in 10-30 seconds; expressed as liters per minute. Abbreviated MBC. { 'mak·sə·məm 'breth·iŋ kə,pas·əd·ē }

mayfly [INVERTEBRATE ZOOLOGY] The common name for insects composing the order Ephemeroptera. { 'mā,flī }

mazaedium [BOTANY] The fruiting body of certain lichens, with the spores lying in a powdery mass in the capitulum. [MYCOLOGY] A slimy layer on the hymenial surface of some ascomycetous fungi. { mə'zē·dē·əm }

maz-, mazo- [EMBRYOLOGY] A combining form denoting placenta. { māz,'mā·zō }

Mazzoni's corpuscle [ANATOMY] A specialized encapsulated sensory nerve end organ on a tendon in series with muscle fibers. { mät'zō·nēz ,kòr·pə·səl }

MBC [PHYSIOLOGY] *See* maximum breathing capacity.

McBurney's point [ANATOMY] A point halfway between the umbilicus and the anterior superior iliac spine; a point of extreme tenderness in appendicitis. { mək'bər·nēz ,pòint }

meadow [ECOLOGY] A vegetation zone which is a low grassland, dense and continuous, variously interspersed with forbs but few if any shrubs. Also known as pelouse; Wiesen. { 'med·ō }

mealybug [INVERTEBRATE ZOOLOGY] Any of various scale insects of the family Pseudococcidae which have a powdery substance covering the dorsal surface; all are serious plant pests. { 'mē·lē,bəg }

Meantes [VERTEBRATE ZOOLOGY] The mud eels, a small suborder of the Urodela including three species of aquatic eellike salamanders with only anterior limbs. { mē'an,tēz }

measles immune globulin [IMMUNOLOGY] Sterile human globulin used to provide passive immunization against measles. { 'mē·zəlz i'myün 'gläb·yə·lən }

measles virus vaccine [IMMUNOLOGY] A suspension of live attenuated or inactivated measles virus used for active immunization against measles. { 'mē·zəlz 'vī·rəs ,vak,sēn }

meatal plate [EMBRYOLOGY] A mass of ectodermal cells on the bottom of the branchial groove in a 2-month embryo. { mē'ād·əl 'plāt }

meatus [ANATOMY] A natural opening or passage in the body. { mē'ād·əs }

mechanoreceptor [PHYSIOLOGY] A receptor that provides the organism with information about such mechanical changes in the environment as movement, tension, and pressure. { ¦mek·ə·nō·ri'sep·tər }

Meckel's cartilage [EMBRYOLOGY] The cartilaginous axis of the mandibular arch in the embryo and fetus. { 'mek·əlz ,kärt·lij }

Meckel's diverticulum [EMBRYOLOGY] The persistent blind end of the yolk stalk forming a tube connected with the lower ileum. { 'mek·əlz ,dī·vər'tik·yə·ləm }

meconium [EMBRYOLOGY] A greenish mass of mucous, desquamated epithelial cells, lanugo, and vernix caseosa that collects in the fetal intestine, becoming the first fecal discharge of the newborn. { mə'kō·nē·əm }

Mecoptera [INVERTEBRATE ZOOLOGY] The scorpion flies, a small order of insects; adults are distinguished by the peculiar prolongation of the head into a beak, which bears chewing mouthparts. { me'käp·tə·rə }

mecystasis [PHYSIOLOGY] Increase in muscle length with maintenance of the original degree of tension. { me'sis·tə·səs }

media [HISTOLOGY] The middle, muscular layer in the wall of a vein, artery, or lymph vessel. { 'mē·dē·ə }

mediad [ANATOMY] Toward the median line or plane of the body or of a part of the body. { 'mē·dē,ad }

medial [ANATOMY] **1.** Being internal as opposed to external (lateral). **2.** Toward the midline of the body. { 'mē·dē·əl }

medial lemniscus [ANATOMY] A lemniscus arising in the nucleus gracilis and nucleus cuneatus of the brain, crossing immediately as internal arcuate fibers, and terminating in the posterolateral ventral nucleus of the thalamus. { 'mē·dē·əl lem'nis·kəs }

median infective dose [MICROBIOLOGY] *See* infective dose 50. { 'mē·dē·ən in'fek·tiv 'dōs }

median lethal time [MICROBIOLOGY] The period of time required for 50% of a large group of organisms to die following a specific dose of an injurious agent, such as a drug or radiation. { 'mē·dē·ən 'lēth·əl ,tīm }

median nasal process [EMBRYOLOGY] The region below the frontonasal sulcus between the olfactory sacs; forms the bridge and mobile septum of the nose and various parts of the upper jaw and lip. { 'mē·dē·ən 'nāz·əl ,prä·səs }

mediastinum [ANATOMY] **1.** A partition separating adjacent parts. **2.** The space in the middle of the chest between the two pleurae. { ,mē·dē·ə 'stī·nəm }

medical genetics [GENETICS] A field of human genetics concerned with the relationship between heredity and disease. { 'med·ə·kəl jə 'ned·iks }

Mediterranean faunal region [ECOLOGY] A marine littoral faunal region including that offshore portion of the Atlantic Ocean from northern France to near the Equator. { ,med·ə·tə'rā·nē·ən 'fón·əl ,rē·jən }

medius [ANATOMY] The middle finger. { 'mē·dē·əs }

medulla [ANATOMY] **1.** The central part of certain organs and structures such as the adrenal glands and hair. **2.** Marrow, such as of bone or the spinal cord. **3.** *See* medulla oblongata. [BOTANY] **1.** Pith. **2.** The central spongy portion of some fungi. { mə'dəl·ə }

medulla oblongata [ANATOMY] The somewhat pyramidal, caudal portion of the vertebrate brain which extends from the pons to the spinal cord. Also known as medulla. { mə'dəl·ə ,äb,lóŋ 'gäd·ə }

medullary cord [ANATOMY] Dense lymphatic tissue separated by sinuses in the medulla of a lymph node. [EMBRYOLOGY] A primary invagination of the germinal epithelium of the embryonic gonad that differentiates into rete testis and seminiferous tubules or into rete ovarii. { mə 'dəl·ə·rē 'kórd }

medullary ray [BOTANY] An extension of pith between vascular bundles in the plant stem. Also known as pith ray. { me'dəl·ə·rē ,rā }

Medullosaceae [PALEOBOTANY] A family of seed ferns; these extinct plants all have large spirally arranged petioles with numerous vascular bundles. { mə'dəl·ō'sās·ē,ē }

medusa [INVERTEBRATE ZOOLOGY] *See* jellyfish. { mə'düs·ə }

megacanthopore [INVERTEBRATE ZOOLOGY] A large prominent tube, commonly projecting as a spine in a mature region of a bryozoan colony. { ,meg·ə'kan·thə,pór }

Megachilidae [INVERTEBRATE ZOOLOGY] The leaf-cutting bees, a family of hymenopteran insects in the superfamily Apoidea. { ,meg·ə'kil·ə,dē }

Megachiroptera [VERTEBRATE ZOOLOGY] The fruit bats, a group of Chiroptera restricted to the Old World; most species lack a tail, but when present it is free of the interfemoral membrane. { ¦meg·ə·kī'räp·tə·rə }

megacine [MICROBIOLOGY] The bacteriocidin produced by *Bacillus megaterium*. { 'meg·ə·sən }

Megadermatidae [VERTEBRATE ZOOLOGY] The false vampires, a family of tailless bats with large ears and a nose leaf; found in Africa, Australia, and the Malay Archipelago. { ,meg·ə·dər'mad·ə ,dē }

megagametophyte [BOTANY] The female gametophyte in plants having two types of spores. { ¦meg·ə·gə'mēd·ə,fīt }

megakaryocyte [HISTOLOGY] A giant bone-marrow cell characterized by a large, irregularly lobulated nucleus; precursor to blood platelets. { ¦meg·ə'kar·ē·ə,sīt }

megaloblast of Sabin [HISTOLOGY] *See* pronormoblast. { 'meg·ə·lō,blast əv 'sā·bən }

Megalodontoidea [INVERTEBRATE ZOOLOGY] A superfamily of hymenopteran insects in the suborder Symphyta. { ¦meg·ə·lō,dän'tóid·ē·ə }

Megalomycteroidei [VERTEBRATE ZOOLOGY] The mosaic-scaled fishes, a monofamilial suborder of the Cetomimiformes; members are rare species of small, elongate deep-sea fishes with degenerate eyes and irregularly disposed scales. { ,meg·ə·lō,mik·tə'róid·ē,ī }

megalops larva [INVERTEBRATE ZOOLOGY] A preimago stage of certain crabs having prominent eyes and chelae. { 'meg·ə,läps ,lär·və }

Megaloptera [INVERTEBRATE ZOOLOGY] A suborder included in the order Neuroptera by some authorities. { ,meg·ə'läp·tə·rə }

Megalopygidae [INVERTEBRATE ZOOLOGY] The flannel moths, a small family of lepidopteran insects in the suborder Heteroneura. { ¦meg·ə·lō 'pij·ə,dē }

megalosphere [INVERTEBRATE ZOOLOGY] The initial, large-chambered shell of sexual individuals of certain dimorphic species of Foraminifera. { 'meg·ə·lō,sfir }

Megamerinidae [INVERTEBRATE ZOOLOGY] A family of myodarian cyclorrhaphous dipteran insects in the subsection Acalypteratae. { ,meg·ə·mə 'rin·ə,dē }

megaphenic [GENETICS] Pertaining to genetic or environmental factors that are individually of large effect relative to the phenotypic standard deviation. { ,meg·ə'fē·nik }

megaphyllous [BOTANY] Having large leaves or leaflike extensions. { ¦meg·ə'fil·əs }

Megapodiidae [VERTEBRATE ZOOLOGY] The

mound birds and brush turkeys, a family of birds in the order Galliformes; distinguished by their method of incubating eggs in mounds of dirt or in decomposing vegetation. { ˌmeg·ə·pə'dī·əˌdē }

megasclere [INVERTEBRATE ZOOLOGY] A large sclerite. { 'meg·əˌsklir }

Megasphaera [MICROBIOLOGY] A genus of bacteria in the family Veillonellaceae; relatively large cells occurring in pairs arranged in chains. { mə'gas·fə·rə }

megasporangium [BOTANY] *See* macrosporangium. { ¦meg·ə·spə'ran·jē·əm }

megaspore [BOTANY] *See* macrospore. { 'meg·əˌspȯr }

megaspore mother cell [BOTANY] *See* megasporocyte. { 'meg·əˌspȯr 'məth·ər ˌsel }

megasporocyte [BOTANY] A diploid cell from which four megaspores are produced by meiosis. Also known as megaspore mother cell. { ¦meg·ə'spȯr·əˌsīt }

megasporogenesis [BOTANY] *See* macrosporogenesis. { ˌmeg·əˌspȯr·ə'jen·ə·səs }

megasporophyll [BOTANY] A leaf bearing megasporangia. { ˌmeg·ə'spȯr·əˌfil }

megathermophyte [ECOLOGY] A plant that requires great heat and abundant moisture for normal growth. Also known as macrothermophyte. { ˌmeg·ə'thər·məˌfīt }

Megathymiinae [INVERTEBRATE ZOOLOGY] The giant skippers, a subfamily of lepidopteran insects in the family Hesperiidae. { ˌmeg·ə·thə'mī·əˌnē }

Mehlis' gland [INVERTEBRATE ZOOLOGY] One of the large unicellular glands around the ootype of flatworms. { 'mā·ləs ˌgland }

Meibomian gland [ANATOMY] *See* tarsal gland. { mī'bō·mē·ən 'gland }

Meinertellidae [INVERTEBRATE ZOOLOGY] A family of wingless insects belonging to the Microcoryphia. { ˌmī·nər'tel·əˌdē }

meiocyte [CELL BIOLOGY] A cell undergoing meiotic division. { 'mī·əˌsīt }

meiofauna [ECOLOGY] Small benthic animals ranging in size between macrofauna and microfauna; includes interstitial animals. { ¦mī·ə'fȯn·ə }

meioflora [ECOLOGY] Small benthic plants ranging in size between macroflora and microflora; includes interstitial plants. { ¦mī·ə'flȯr·ə }

meiosis [CELL BIOLOGY] A type of cell division occurring in diploid or polyploid tissues that results in a reduction in chromosome number, usually by half. { mī'ō·səs }

meiospore [BIOLOGY] A spore produced as the result of meiosis. { 'mī·əˌspȯr }

meiotic drive [GENETICS] Any meiotic deviation from Mendelian segregation ratio in heterozygotes resulting from a disturbance in the meiotic mechanism. { mē¦äd·ik 'drīv }

Meissner's corpuscle [ANATOMY] An ovoid, encapsulated cutaneous sense organ presumed to function in touch sensation in hairless portions of the skin. { 'mīs·nərz 'kȯr·pə·səl }

Meissner's plexus [ANATOMY] *See* submucous plexus. { 'mīs·nərz 'plek·səs }

Melamphaidae [VERTEBRATE ZOOLOGY] A family of bathypelagic fishes in the order Beryciformes. { ˌmel·əm'fā·əˌdē }

Melampsoraceae [MYCOLOGY] A family of parasitic fungi in the order Uredinales in which the teleutospores are laterally united to form crusts or columns. { ˌmel·əm·sə'rās·ēˌē }

Melanconiaceae [MYCOLOGY] The single family of the order Melanconiales. { ˌmel·ənˌkō·nē'ās·ēˌē }

Melanconiales [MYCOLOGY] An order of the class Fungi Imperfecti including many plant pathogens commonly causing anthracnose; characterized by densely aggregated cnidophores on an acervulus. { ˌmel·ənˌkō·nē'ā·lēz }

Melandryidae [INVERTEBRATE ZOOLOGY] The false darkling beetles, a family of coleopteran insects in the superfamily Tenebrionoidea. { ˌmel·ən'drī·əˌdē }

melanin [BIOCHEMISTRY] Any of a group of brown or black pigments occurring in plants and animals. { 'mel·ə·nən }

melanoblast [HISTOLOGY] **1.** Precursor cell of melanocytes and melanophores. **2.** An immature pigment cell in certain vertebrates. **3.** A mature cell that elaborates melanin. { mə'lan·əˌblast }

melanocyte [HISTOLOGY] A cell containing dark pigments. { mə'lan·əˌsīt }

melanocyte-stimulating hormone [BIOCHEMISTRY] A protein substance secreted by the intermediate lobe of the pituitary of humans which causes dispersion of pigment granules in the skin; similar to intermedins in other vertebrates. Abbreviated MSH. Also known as melanophore-dilating principle; melanophore hormone. { mə'lan·əˌsīt ¦stim·yəˌlād·iŋ 'hȯrˌmōn }

melanogen [BIOCHEMISTRY] A colorless precursor of melanin. { mə'lan·ə·jən }

melanogenesis [BIOCHEMISTRY] The formation of melanin. { ˌmel·ə·nō'jen·ə·səs }

melanophage [HISTOLOGY] A phagocytic cell which engulfs and contains melanin. { mə'lan·əˌfāj }

melanophore [HISTOLOGY] A type of chromatophore containing melanin. { mə'lan·əˌfȯr }

melanophore hormone [BIOCHEMISTRY] *See* melanocyte-stimulating hormone. { mə'lan·əˌfȯr 'hȯrˌmōn }

melanophore-dilating principle [BIOCHEMISTRY] *See* melanocyte-stimulating hormone. { mə'lan·əˌfȯr 'dīˌlād·iŋ ˌprin·sə·pəl }

melanoprotein [BIOCHEMISTRY] A conjugated protein in which melanin is the associated chromagen. { ˌmel·ə·nō'prōˌtēn }

melanosome [CELL BIOLOGY] An organelle which contains melanin and in which tyrosinase activity is not demonstrable. { mə'lan·əˌsōm }

Melasidae [INVERTEBRATE ZOOLOGY] The equivalent name for Eucnemidae. { mə'las·əˌdē }

Melastomataceae [BOTANY] A large family of dicotyledonous plants in the order Myrtales characterized by an inferior ovary, axile placentation, up to twice as many stamens as petals (or sepals), anthers opening by terminal pores, and leaves with prominent, subparallel longitudinal ribs. { ˌmel·əˌstō·mə'tās·ēˌē }

Meleagrididae [VERTEBRATE ZOOLOGY] The turkeys, a family of birds in the order Galliformes characterized by a bare head and neck. { ˌmel·ē·əˈgrid·əˌdē }

Meliaceae [BOTANY] A family of dicotyledonous plants in the order Sapindales characterized by mostly exstipulate, alternate leaves, stamens mostly connate by their filaments, and syncarpous flowers. { ˌmel·ēˈās·ēˌē }

Melinae [VERTEBRATE ZOOLOGY] The badgers, a subfamily of carnivorous mammals in the family Mustelidae. { ˈmel·əˌnē }

Melinninae [INVERTEBRATE ZOOLOGY] A subfamily of sedentary annelids belonging to the Ampharetidae which have a conspicuous dorsal membrane, with or without dorsal spines. { mə ˈlin·əˌnē }

Meliolaceae [MYCOLOGY] The sooty molds, a family of ascomycetous fungi in the order Erysiphales, with dark mycelia and conidia. { ˌmel·ē·əˈlās·ēˌē }

melitose [BIOCHEMISTRY] *See* raffinose. { ˈmel·ə ˌtōs }

melitriose [BIOCHEMISTRY] *See* raffinose. { məˈlī·trēˌōs }

Melittidae [INVERTEBRATE ZOOLOGY] A family of hymenopteran insects in the superfamily Apoidea. { məˈlid·əˌdē }

Meloidae [INVERTEBRATE ZOOLOGY] The blister beetles, a large cosmopolitan family of coleopteran insects in the superfamily Meloidea; characterized by soft, flexible elytra and the strongly vesicant properties of the body fluids. { məˈlō·ə ˌdē }

Meloidea [INVERTEBRATE ZOOLOGY] A superfamily of coleopteran insects in the suborder Polyphaga. { məˈlȯid·ē·ə }

melon [BOTANY] Either of two soft-fleshed edible fruits, muskmelon or watermelon, or varieties of these. { ˈmel·ən }

melting profile [BIOCHEMISTRY] A plot of the degree of denaturation of the strands in a nucleic acid duplex in a specified time as a function of temperature. { ˈmelt·iŋ ˌprōˌfīl }

melting temperature [BIOCHEMISTRY] The temperature at which denaturing occurs for half of the double helices of deoxyribonucleic acid. { ˈmelt·iŋ ˌtem·prə·chər }

Melusinidae [INVERTEBRATE ZOOLOGY] A family of orthorrhaphous dipteran insects in the series Nematocera. { ˌmel·əˈsin·əˌdē }

Melyridae [INVERTEBRATE ZOOLOGY] The soft-winged flower beetles, a large family of cosmopolitan coleopteran insects in the superfamily Cleroidea. { məˈlir·əˌdē }

Membracidae [INVERTEBRATE ZOOLOGY] The treehoppers, a family of homopteran insects included in the series Auchenorrhyncha having a pronotum that extends backward over the abdomen, and a vertical upper portion of the head. { memˈbras·əd·ē }

membrane [HISTOLOGY] A thin layer of tissue surrounding a part of the body, separating adjacent cavities, lining cavities, or connecting adjacent structures. { ˈmemˌbrān }

membrane bone [ANATOMY] *See* dermal bone. { ˈmemˌbrān ˌbōn }

membrane carrier [CELL BIOLOGY] Any protein that facilitates the movement of small molecules across cell membranes. { ˈmemˌbrān ˌkar·ē·ər }

membrane potential [PHYSIOLOGY] A potential difference across a living cell membrane. { ˈmemˌbrān pəˌten·chəl }

membranous labyrinth [ANATOMY] The membranous portion of the inner ear of vertebrates. { ˈmem·brə·nəs ˈlab·əˌrinth }

membranous urethra [ANATOMY] The part of the urethra between the two facial layers of the urogenital diaphragm. { ˈmemˌbrə·nəs yu̇ˈrē·thrə }

memory trace [PHYSIOLOGY] *See* engram. { ˈmem·rē ˈtrās }

menacme [PHYSIOLOGY] The period of a woman's life during which menstruation persists. { mə ˈnak·mē }

menarche [PHYSIOLOGY] The onset of menstruation. { məˈnär·kē }

Mendelian genetics [GENETICS] Scientific study of heredity as related to or in accordance with Mendel's laws. { menˈdēl·yən jəˈned·iks }

Mendelian population [GENETICS] A group of individuals who interbreed according to a certain system of mating; the total genic content of the individuals of the group is called gene pool. { menˈdēl·yən ˌpäp·yəˈlā·shən }

Mendelian ratio [GENETICS] The ratio of occurrence of various phenotypes in any cross involving Mendelian characters. { menˈdēl·yən ˈrā·shō }

Mendelism [GENETICS] The basic laws of inheritance as formulated by Mendel. { ˈmen·dəˈlīz·əm }

Mendel's laws [GENETICS] Two basic principles of genetics formulated by Mendel: the law of segregation and law of independent assortment. { ˈmen·dəlz ˌlȯz }

meninges [ANATOMY] The membranes that cover the brain and spinal cord; there are three in mammals and one or two in submammalian forms. { məˈnin·jēz }

meningococcus [MICROBIOLOGY] Common name for *Neisseria meningitidis*. { məˌniŋ·gəˌkäk·əs }

meningothelium [HISTOLOGY] Epithelium of the arachnoid which envelops the brain. { məˌniŋ·gəˈthē·lē·əm }

meninx [ANATOMY] Any one of the three membranes covering the brain and spinal cord. { ˈmeˌniŋks }

meninx primitiva [VERTEBRATE ZOOLOGY] The single membrane covering the brain and spinal cord of certain submammalian vertebrates. { ˈmeˌniŋks ˌprim·əˈtī·və }

Meniscotheriidae [PALEONTOLOGY] A family of extinct mammals of the order Condylarthra possessing selenodont teeth and molarized premolars. { məˌnis·kō·thəˈrī·əˌdē }

meniscus [ANATOMY] A crescent-shaped body, especially an interarticular cartilage. { məˈnis·kəs }

Menispermaceae [BOTANY] A family of dicotyledonous woody vines in the order Ranunculales distinguished by mostly alternate, simple leaves, unisexual flowers, and a dioecious habit. { ˌmen·ə·spər′mās·ēˌē }

men-, meno- [PHYSIOLOGY] A combining form denoting menses. { men, ′men·ō }

menopause [PHYSIOLOGY] The natural physiologic cessation of menstruation, usually occurring in the last half of the fourth decade. Also known as climacteric. { ′men·əˌpȯz }

Menoponidae [INVERTEBRATE ZOOLOGY] A family of biting lice (Mallophaga) adapted to life only upon domestic and sea birds. { ˌmen·ə′pän·əˌdē }

menorrhea [PHYSIOLOGY] The normal flow of the menses. { ˌmen·ə′rē·ə }

menses [PHYSIOLOGY] *See* menstruation. { ′menˌsēz }

menstrual age [EMBRYOLOGY] The age of an embryo or fetus calculated from the first day of the mother's last normal menstruation preceding pregnancy. { ′men·strə·wəl ′āj }

menstrual cycle [PHYSIOLOGY] The periodic series of changes associated with menstruation and the intermenstrual cycle; menstrual bleeding indicates onset of the cycle. { ′men·strə·wəl ′sī·kəl }

menstrual period [PHYSIOLOGY] The time of menstruation. { ′men·strə·wəl ′pir·ē·əd }

menstruate [PHYSIOLOGY] To discharge the products of menstruation. { ′men·strəˌwāt }

menstruation [PHYSIOLOGY] The periodic discharge of sanguineous fluid and sloughing of the uterine lining in women from puberty to the menopause. Also known as menses. { ˌmen·strə′wā·shən }

mental [ANATOMY] Pertaining to the chin. Also known as genial. { ′men·təl }

Menthaceae [BOTANY] An equivalent name for Labiatae. { men′thās·ēˌē }

mentum [ANATOMY] The chin. [BOTANY] A projection formed by union of the sepals at the base of the column in some orchids. [INVERTEBRATE ZOOLOGY] **1.** A projection between the mouth and foot in certain gastropods. **2.** The median or basal portion of the labium in insects. { ′men·təm }

Menurae [VERTEBRATE ZOOLOGY] A small suborder of suboscine perching birds restricted to Australia, including the lyrebirds and scrubbirds. { mə′nyu̇r·ē }

Menuridae [VERTEBRATE ZOOLOGY] The lyrebirds, a family of birds in the suborder Menurae notable for their vocal mimicry. { mə′nyu̇r·əˌdē }

meraspis [PALEONTOLOGY] Advanced larva of a trilobite; stage in which the pygidium begins to form. { mə′rap·səs }

merganser [VERTEBRATE ZOOLOGY] Any of several species of diving water fowl composing a distinct subfamily of Anatidae and characterized by a serrate bill adapted for catching fish. { mər′gan·sər }

mericarp [BOTANY] An individual, one-seeded carpel of a schizocarp. { ′mer·əˌkärp }

mericlinal chimera [BIOLOGY] An organism or organ composed of two genetically different tissues, one of which partly surrounds the other. { ¦mer·əˌklīn·əl ′kim·ə·rə }

Meridosternata [INVERTEBRATE ZOOLOGY] A suborder of echinoderms including various deep-sea forms of sea urchins. { ¦mer·ə·dō·stər′näd·ə }

meristem [BOTANY] Formative plant tissue composed of undifferentiated cells capable of dividing and giving rise to other meristematic cells as well as to specialized cell types; found in growth areas. { ′mer·əˌstem }

meristic [BIOLOGY] Pertaining to a change in number or in geometric relation of parts of an organism. [ZOOLOGY] Of, pertaining to, or divided into segments. { mə′ris·tik }

Merkel's corpuscles [ANATOMY] Touch receptors consisting of flattened platelets at the tips of certain cutaneous nerves. { ′mər·kəlz ˌkər·pə·səlz }

Mermithidae [INVERTEBRATE ZOOLOGY] A family of filiform nematodes in the superfamily Mermithoidea; only juveniles are parasitic. { mər′mith·əˌdē }

Mermithoidea [INVERTEBRATE ZOOLOGY] A superfamily of nematodes composed of two families, both of which are invertebrate parasites. { mər·mə′thȯid·ē·ə }

meroblastic [EMBRYOLOGY] Of or pertaining to an ovum that undergoes incomplete cleavage due to large amounts of yolk. { ¦mer·ə¦blas·tik }

merocrine [PHYSIOLOGY] Pertaining to glands in which the secretory cells undergo cytological changes without loss of cytoplasm during secretion. { ′mer·ə·krən }

merogony [EMBRYOLOGY] The normal or abnormal development of a part of an egg following cutting, shaking, or centrifugation of the egg before or after fertilization. { mə′räj·ə·nē }

meromixis [GENETICS] Genetic exchange in bacteria involving a unidirectional transfer of a partial genome. { ˌmer·ə′mik·səs }

meromyarian [INVERTEBRATE ZOOLOGY] Having few muscle cells in each quadrant as seen in cross section; applied especially to nematodes. { ¦mer·ə·mī¦ar·ē·ən }

meromyosin [BIOCHEMISTRY] Protein fragments of a myosin molecule, produced by limited proteolysis. { ˌmer·ə′mī·ə·sən }

Meropidae [VERTEBRATE ZOOLOGY] The bee-eaters, a family of brightly colored Old World birds in the order Coraciiformes. { mə′räp·əˌdē }

meroplankton [BIOLOGY] Plankton composed of floating developmental stages (that is, eggs and larvae) of the benthos and nekton organisms. Also known as temporary plankton. { ¦mer·ə′plaŋk·tən }

merospermy [CELL BIOLOGY] Fusion of an egg with an anucleate sperm. { mi′räs·pər·mē }

Merostomata [INVERTEBRATE ZOOLOGY] A class of primitive arthropods of the subphylum Chelicerata distinguished by their aquatic mode of life and the possession of abdominal appendages which bear respiratory organs; only three living species are known. { ˌmer·ə′stō·mə·də }

-merous [BIOLOGY] Combining form that denotes having such parts or so many parts. { mər·əs }

Merozoa [INVERTEBRATE ZOOLOGY] The equivalent name for Cestoda. { ˌmer·əˈzō·ə }

merozoite [INVERTEBRATE ZOOLOGY] An ameboid trophozoite in some sporozoans produced from a schizont by schizogony. { ˌmer·əˈzōˌīt }

merozygote [MICROBIOLOGY] In bacteria, a zygote that has some diploid and some haploid genetic material because a chromosomal fragment was transferred by the F+ mate. { ˌmir·əˈzī ˌgōt }

Merycoidodontidae [PALEONTOLOGY] A family of extinct tylopod ruminants in the superfamily Merycoidodontoidea. { ˌmer·əˌkȯid·əˈdän·tə ˌdē }

Merycoidodontoidea [PALEONTOLOGY] A superfamily of extinct ruminant mammals in the infraorder Tylopoda which were exceptionally successful in North America. { ˈmer·əˌkȯid·əˌdän ˈtȯid·ē·ə }

Mesacanthidae [PALEONTOLOGY] An extinct family of primitive acanthodian fishes in the order Acanthodiformes distinguished by a pair of small intermediate spines, large scales, superficially placed fin spines, and a short branchial region. { ˌmes·əˈkan·thəˌdē }

mesappendix [ANATOMY] The mesentery of the vermiform appendix. { ˌmes·əˈpen·diks }

mesarch [BOTANY] Having metaxylem on both sides of the protoxylem in a siphonostele. [ECOLOGY] Originating in a mesic environment. { ˈmeˌzärk }

mescal buttons [BOTANY] The dried tops from the cactus *Lophophora williamsii*; capable of producing inebriation and hallucinations. { meˈskal ˌbət·ənz }

mesectoderm [EMBRYOLOGY] The portion of the mesenchyme arising from ectoderm. { məˈzek·tə ˌdərm }

mesencephalon [EMBRYOLOGY] The middle portion of the embryonic vertebrate brain; gives rise to the cerebral peduncles and the tectum. Also known as midbrain. { ¦mez·ənˈsef·əˌlän }

mesenchymal cell [HISTOLOGY] An undifferentiated cell found in mesenchyme and capable of differentiating into various specialized connective tissues. { ¦mez·ən¦kī·məl ˌsel }

mesenchymal epithelium [HISTOLOGY] A layer of squamous epithelial cells lining subdural, subarachnoid, and perilymphatic spaces, and the chambers of the eyeball. { ¦mez·ən¦kī·məl ˌep·əˈthē·lē·əm }

mesenchymal tissue [EMBRYOLOGY] Undifferentiated tissue composed of branching cells embedded in a coagulable fluid matrix. { ¦mez·ən ¦kī·məl ˈtish·ü }

mesenchyme [EMBRYOLOGY] That part of the mesoderm from which all connective tissues, blood vessels, blood, lymphatic system proper, and the heart are derived. { ˈmez·ənˌkīm }

mesendoderm [EMBRYOLOGY] Embryonic tissue which differentiates into mesoderm and endoderm. { məˈzen·dəˌdərm }

mesenteric [ANATOMY] Of or pertaining to the mesentery. { ¦mez·ən¦ter·ik }

mesenteric artery [ANATOMY] Either of two main arterial branches arising from the abdominal aorta: the inferior, supplying the descending colon and the rectum, and the superior, supplying the small intestine, the cecum, and the ascending and transverse colon. { ¦mez·ən¦ter·ik ˈärd·ə·rē }

mesenteron [EMBRYOLOGY] *See* midgut. [INVERTEBRATE ZOOLOGY] Central gastric cavity in an actinozoan. { meˈzen·təˌrän }

mesentery [ANATOMY] A fold of the peritoneum that connects the intestine with the posterior abdominal wall. { ˈmez·ənˌter·ē }

mesentoderm [EMBRYOLOGY] **1.** The entodermal portion of the mesoderm. **2.** Undifferentiated tissue from which entoderm and mesoderm are derived. **3.** That part of the mesoderm which gives rise to certain structures of the digestive tract. { məˈzen·təˌdərm }

mesethmoid [ANATOMY] A bone or cartilage in the center of the ethmoid region of the vertebrate skull; usually constitutes the greater portion of the nasal septum. { meˈzethˌmȯid }

mesic [ECOLOGY] **1.** Of or pertaining to a habitat characterized by a moderate amount of water. **2.** Of or pertaining to a mesophyte. { ˈme·zik }

mesoappendix [ANATOMY] The mesentery of the vermiform appendix. { ¦me·zō·ə¦pen·diks }

mesobilirubin [BIOCHEMISTRY] $C_{33}H_{40}O_6N_4$ Yellow, crystalline by-product of bilirubin reduction. { ¦me·zōˌbil·iˈrü·bən }

mesobilirubinogen [BIOCHEMISTRY] $C_{33}H_{44}O_6N_4$ Colorless, crystalline by-product of bilirubin reduction; may be converted to urobilin, stercobilinogen, or stercobilin. { ¦me·zōˌbil·i·rəˈbin·ə·jən }

mesobiliverdin [BIOCHEMISTRY] $C_{28}H_{38}O_6N_4$ A structural isomer of phycoerythrin and phycocyanobilin released by certain biliprotein by treatment with alkali. { ¦me·zōˌbil·iˈvərd·ən }

mesoblast [EMBRYOLOGY] Undifferentiated mesoderm of the middle layer of the embryo. { ˈme·zōˌblast }

mesoblastema [EMBRYOLOGY] *See* mesoderm. { ˌme·zō·blaˈstē·mə }

mesocardium [ANATOMY] Epicardium covering the blood vessels which enter and leave the heart. [EMBRYOLOGY] The mesentery supporting the embryonic heart. { ˌme·zōˈkärd·ē·əm }

mesocarp [BOTANY] The middle layer of the pericarp. { ˈmez·əˌkärp }

mesocercaria [INVERTEBRATE ZOOLOGY] The developmental stage in the second intermediate host of *Alaria*, a digenetic trematode. { ˌme·zō·sərˈkar·ē·ə }

mesocolon [ANATOMY] The part of the mesentery that is attached to the colon. { ¦me·zōˈkō·lən }

mesoderm [EMBRYOLOGY] The third germ layer, lying between the ectoderm and endoderm; gives rise to the connective tissues, muscles, urogenital system, vascular system, and the epithelial lining of the coelom. Also known as mesoblastema. { ˈmez·əˌdərm }

mesogaster [ANATOMY] The mesentery of the stomach. { ˈmez·əˌgas·tər }

Mesogastropoda [INVERTEBRATE ZOOLOGY] The equivalent name for Pectinibranchia. { ˌmez·ə·gaˈsträp·ə·də }

mesoglea [INVERTEBRATE ZOOLOGY] The gelatinous layer between the ectoderm and endoderm in cnidarians and certain sponges. { ¦me·zō 'glē·ə }

Mesohippus [PALEONTOLOGY] An early ancestor of the modern horse; occurred during the Oligocene. { ¦me·zō'hip·əs }

mesokaryotic [INVERTEBRATE ZOOLOGY] Pertaining to an organism that shares characteristics of both prokaryotic and eukaryotic organisms. { ˌmez·əˌkar·ē'äd·ik }

mesolamella [INVERTEBRATE ZOOLOGY] A thin gelatinous membrane between the epidermis and gastrodermis in hydrozoans. { ¦me·zō·lə 'mel·ə }

mesomere [EMBRYOLOGY] The muscle-plate region between the epimere and hypomere in vertebrates. { 'me·zōˌmir }

mesometrium [ANATOMY] The part of the broad ligament attached directly to the uterus. { ˌmez·ə'mē·trē·əm }

mesonephric duct [EMBRYOLOGY] The efferent duct of the mesonephros. Also known as Wolffian duct. { ¦me·zō¦nef·rik 'dəkt }

mesonephric fold [EMBRYOLOGY] *See* mesonephric ridge. { ¦me·zō¦nef·rik 'fōld }

mesonephric ridge [EMBRYOLOGY] A fold of the dorsal wall of the coelom lateral to the mesentery formed by development of the mesonephros. Also known as mesonephric fold. { ¦me·zō ¦nef·rik 'rij }

mesonephros [EMBRYOLOGY] One of the middle of three pairs of embryonic renal structures in vertebrates; persists in adult fish and is replaced by the metanephros in higher forms. { ¦mez·ə'ne ˌfrōs }

Mesonychidae [PALEONTOLOGY] A family of extinct mammals of the order Condylarthra. { ˌme ˌzäŋ'kid·əˌdē }

mesophile [BIOLOGY] An organism, as certain bacteria, that grows at moderate temperature. { 'mez·əˌfīl }

mesophily [ECOLOGY] Physiological response of organisms living in environments with moderate temperatures and a fairly high, constant amount of moisture. { 'mez·əˌfil·ē }

mesophyll [BOTANY] Parenchymatous tissue between the upper and lower epidermal layers in foliage leaves. { 'mez·əˌfil }

mesophyte [ECOLOGY] A plant requiring moderate amounts of moisture for optimum growth. { 'mez·əˌfīt }

mesopore [PALEONTOLOGY] A tube paralleling the autopore or chamber in fossil bryozoans. { 'mez·əˌpȯr }

mesopterygium [VERTEBRATE ZOOLOGY] The middle one of three basal cartilages in the pectoral fin of sharks and rays. { ¦meˌzäp·tə'rij·ē· əm }

mesoptic vision [PHYSIOLOGY] Vision in which the human eye's spectral sensitivity is changing from the photoptic state to the scotoptic state. { me'zäp·tik 'vizh·ən }

mesorchium [EMBRYOLOGY] The mesentery that supports the embryonic testis in vertebrates. { me'zȯr·kē·əm }

mesosalpinx [ANATOMY] The portion of the broad ligament forming the mesentery of the uterine tube. { ¦me·zō'salˌpiŋks }

Mesosauria [PALEONTOLOGY] An order of extinct aquatic reptiles which is known from a single genus, *Mesosaurus*, characterized by a long snout, numerous slender teeth, small forelimbs, and webbed hindfeet. { ˌme·zō'sȯr·ē·ə }

mesosere [ECOLOGY] A sere originating in a mesic habitat and characterized by mesophytes. { 'me·zōˌsir }

mesosoma [INVERTEBRATE ZOOLOGY] **1.** The anterior portion of the abdomen in certain arthropods. **2.** The middle of the body of some invertebrates, especially when the phylogenetic segmentation pattern cannot be determined. { ¦me·zō'sō·mə }

mesosome [MICROBIOLOGY] An extension of the cell membrane within a bacterial cell; possibly involved in cross-wall formation, cell division, and the attachment of daughter chromosomes following deoxyribonucleic acid replication. { 'mez·əˌsōm }

mesosternum [ANATOMY] The middle portion of the sternum in vertebrates. Also known as gladiolus. [INVERTEBRATE ZOOLOGY] The ventral portion of the mesothorax in insects. { ¦me·zō 'stər·nəm }

Mesostigmata [INVERTEBRATE ZOOLOGY] The mites, a suborder of the Acarina characterized by a single pair of breathing pores (stigmata) that are located laterally in the middle of the idiosoma between the second and third, or third and fourth, legs. { ¦me·zōˌstig'mäd·ə }

Mesosuchia [PALEONTOLOGY] A suborder of extinct crocodiles of the Late Jurassic and Early Cretaceous. { ˌme·zō'sü·kē·ə }

Mesotaeniaceae [BOTANY] The saccoderm desmids, a family of fresh-water algae in the order Conjugales; cells are oval, cylindrical, or rectangular and have simple, undecorated walls in one piece. { ˌme·zōˌtē·nē'ās·ēˌē }

Mesotardigrada [INVERTEBRATE ZOOLOGY] An order of tardigrades which combines certain echiniscoidean features with eutardigradan characters. { ¦me·zōˌtär'dig·rə·də }

mesotheca [INVERTEBRATE ZOOLOGY] The middle lamina of bifoliate bryozoan colonies. { ¦me·zō 'thē·kə }

mesothelium [ANATOMY] The simple squamous-cell epithelium lining the pleural, pericardial, peritoneal, and scrotal cavities. [EMBRYOLOGY] The lining of the wall of the primitive body cavity situated between the somatopleure and splanchnopleure. { ˌme·zō'thē·lē·əm }

mesotherm [ECOLOGY] A plant that grows successfully at moderate temperatures. { 'mez·ə ˌthərm }

mesothorax [INVERTEBRATE ZOOLOGY] The middle of three somites composing the thorax in insects. { ¦me·zō'thȯˌraks }

mesovarium [ANATOMY] A fold of the peritoneum that connects the ovary with the broad ligament. { ˌme·zō'var·ē·əm }

Mesoveliidae [INVERTEBRATE ZOOLOGY] The water treaders, a small family of hemipteran insects in the subdivision Amphibicorisae having well-developed ocelli. { ˌme·zō·və'lī·əˌdē }

mesoxyalyurea [BIOCHEMISTRY] *See* alloxan. { mə ˌzäk·sē·al'yu̇r·ē·ə }

Mesozoa [INVERTEBRATE ZOOLOGY] A division of the animal kingdom sometimes ranked intermediate between the Protozoa and the Metazoa; composed of two orders of small parasitic, wormlike organisms. { ˌmez·ə'zō·ə }

mesozooid [INVERTEBRATE ZOOLOGY] A type of bryozoan heterozooid that produces slender tubes (mesozooecia or mesopores), internally subdivided by many closely spaced diaphragms, that open as tiny polygonal apertures. { ¦mez·ə ¦zōˌȯid }

mesquite [BOTANY] Any plant of the genus *Prosopis*, especially *P. juliflora*, a spiny tree or shrub bearing sugar-rich pods; an important livestock feed. { mə'skēt }

messenger ribonucleic acid [BIOCHEMISTRY] A linear sequence of nucleotides which is transcribed from and complementary to a single strand of deoxyribonucleic acid and which carries the information for protein synthesis to the ribosomes. Abbreviated mRNA. { 'mes·ən·jər ¦rī·bō·nü'klē·ik 'as·əd }

metabiosis [ECOLOGY] An ecological association in which one organism precedes and prepares a suitable environment for a second organism. { ˌmed·ə·bī'ō·səs }

metabolic [PHYSIOLOGY] Of or pertaining to metabolism. { ˌmed·ə'bäl·ik }

metabolic block [BIOCHEMISTRY] A nonfunctional reaction in a metabolic pathway due to a defective enzyme whose normal counterpart catalyzes the reaction. { ˌmed·ə¦bäl·ik 'bläk }

metabolism [PHYSIOLOGY] The physical and chemical processes by which foodstuffs are synthesized into complex elements (assimilation, anabolism), complex substances are transformed into simple ones (disassimilation, catabolism), and energy is made available for use by an organism. { mə'tab·əˌliz·əm }

metabolite [BIOCHEMISTRY] A product of intermediary metabolism. { mə'tab·əˌlīt }

metabolize [PHYSIOLOGY] To transform by metabolism; to subject to metabolism. { mə'tab·ə ˌlīz }

metacarpus [ANATOMY] The portion of a hand or forefoot between the carpus and the phalanges. { ¦med·ə'kär·pəs }

metacentric [CELL BIOLOGY] Having the centromere near the middle of the chromosome. { ¦med·ə'sen·trik }

metacercaria [INVERTEBRATE ZOOLOGY] Encysted cercaria of digenetic trematodes; the infective form. { ¦med·ə·sər'kar·ē·ə }

metacestode [INVERTEBRATE ZOOLOGY] Encysted larva of a tapeworm; occurs in the intermediate host. { ¦med·ə'sesˌtōd }

Metachlamydeae [BOTANY] An artificial group of flowering plants, division Magnoliophyta, recognized in the Englerian system of classification; consists of families of dicotyledons in which petals are characteristically fused, forming a sympetalous corolla. { ˌmed·ə·klə'mid·ēˌē }

metachromatic granules [CELL BIOLOGY] Granules which assume a color different from that of the dye used to stain them. { ¦med·ə·krō'mad·ik 'gran·yülz }

metachrosis [VERTEBRATE ZOOLOGY] The ability of some animals to change color by the expansion and contraction of chromatophores. { ˌmed·ə'krō·səs }

metacneme [INVERTEBRATE ZOOLOGY] A secondary mesentery in many zoantharians. { 'meˌtak ˌnēm }

metacone [VERTEBRATE ZOOLOGY] **1.** The posterior of three cusps of primitive upper molars. **2.** The posteroexternal cusp of an upper molar in higher vertebrates, especially mammals. { 'med·əˌkōn }

metaconid [VERTEBRATE ZOOLOGY] The posteroexternal cusp of a lower molar in mammals; corresponds with the metacone. { ˌmed·ə'kō·nəd }

Metacopina [PALEONTOLOGY] An extinct suborder of ostracods in the order Podocopida. { ˌmed·ə'käp·ə·nə }

metagenesis [BIOLOGY] The phenomenon in which one generation of certain plants and animals reproduces asexually, followed by a sexually reproducing generation. Also known as alternation of generations. { ¦med·ə'jen·ə·səs }

metagranulocyte [HISTOLOGY] *See* metamyelocyte. { ˌmed·ə'gran·yə·ləˌsīt }

metakaryocyte [HISTOLOGY] *See* normoblast. { ˌmed·ə'kar·ē·əˌsīt }

Metallogenium [MICROBIOLOGY] A genus of bacteria of uncertain affiliation; coccoid cells that attach to substrate; they germinate directly or form groups of elementary bodies by budding, and filaments form from these bodies. { məˌtal·ə'jē·nē·əm }

metalloporphyrin [BIOCHEMISTRY] A compound, such as heme, consisting of a porphyrin combined with a metal such as iron, copper, silver, zinc, or magnesium. { məˌtal·ō'pȯr·fə·rən }

metalloprotein [BIOCHEMISTRY] A protein enzyme containing a metallic atom as an inherent portion of its molecule. { məˌtal·ō'prōˌtēn }

metallothionein [BIOCHEMISTRY] A group of vertebrate and invertebrate proteins that bind heavy metals; it may be involved in zinc homeostasis and resistance to heavy-metal toxicity. { məˌtal·ō'thī·əˌnēn }

metamere [ZOOLOGY] One of the linearly arranged similar segments of the body of metameric animals. Also known as somite. { 'med·ə ˌmir }

metamerism [ZOOLOGY] The condition of an animal body characterized by the repetition of similar segments (metameres), exhibited especially by arthropods, annelids, and vertebrates in early embryonic stages and in certain specialized adult structures. Also known as segmentation. { mə'tam·əˌriz·əm }

metamorphosis [BIOLOGY] **1.** A structural transformation. **2.** A marked structural change in an animal during postembryonic development. { ˌmed·ə'mȯr·fə·səs }

metamyelocyte [HISTOLOGY] A granulocytic cell

intermediate in development between the myelocyte and granular leukocyte; characterized by a full complement of cytoplasmic granules and a bean-shaped nucleus. Also known as juvenile cell; metagranulocyte. { ˌmed·əˈmī·ə·lōˌsīt }

metanauplius [INVERTEBRATE ZOOLOGY] A primitive larval stage of certain decapod crustaceans characterized by seven pairs of appendages; follows the nauplius stage. { ˌmed·əˈnȯ·plē·əs }

metanephridium [INVERTEBRATE ZOOLOGY] A type of nephridium consisting of a tubular structure lined with cilia which opens into the coelomic cavity. { ¦med·ə·neˈfrid·ē·əm }

metanephrine [BIOCHEMISTRY] An inactive metabolite of epinephrine (3-O-methylepinephrine) that is excreted in the urine; it is recovered and measured as a test for pheochromocytoma. { ˌmed·əˈneˌfrən }

metanephros [EMBRYOLOGY] One of the posterior of three pairs of vertebrate renal structures; persists as the definitive or permanent kidney in adult reptiles, birds, and mammals. { ˌmed·əˈne ˌfräs }

metanitricyte [HISTOLOGY] *See* normoblast. { ˌmed·əˈnī·trəˌsīt }

metaphase [CELL BIOLOGY] **1.** The phase of mitosis during which centromeres are arranged on the equator of the spindle. **2.** The phase of the first meiotic division when centromeric regions of homologous chromosomes come to lie equidistant on either side of the equator. { ˈmed·ə ˌfāz }

metaphloem [BOTANY] The primary phloem that forms after differentiation of the protophloem. { ¦med·əˈflō·əm }

metaphysis [ANATOMY] *See* epiphyseal plate. { məˈta·fə·səs }

Metaphyta [BIOLOGY] A kingdom set up to include mosses, ferns, and other plants in some systems of classification. { məˈtaf·əd·ə }

metaplasm [CELL BIOLOGY] The ergastic substance of protoplasm. { ˈmed·əˌplaz·əm }

metapodium [ANATOMY] **1.** The metatarsus in bipeds. **2.** The metatarsus and metacarpus in quadrupeds. [INVERTEBRATE ZOOLOGY] Posterior portion of the foot of a mollusk. { ¦med·ə ˈpō·dē·əm }

metapodosoma [INVERTEBRATE ZOOLOGY] Portion of the body bearing the third and fourth pairs of legs in Acarina. { ˌmed·əˌpäd·əˈsō·mə }

metapterygium [VERTEBRATE ZOOLOGY] The posterior one of three basal cartilages in the pectoral fin of sharks and rays. { mə¦tap·tə¦rij·ē·əm }

metarubricyte [HISTOLOGY] *See* normoblast. { ˌmed·əˈrü·brəˌsīt }

metascutellum [INVERTEBRATE ZOOLOGY] The scutellum of the metathorax in insects. { ¦med·ə·sküˈtel·əm }

metasicula [INVERTEBRATE ZOOLOGY] The succeeding part of the sicula or colonial tube of graptolites. { ¦med·əˈsik·yə·lə }

metasoma [INVERTEBRATE ZOOLOGY] The posterior region of the body of certain invertebrates, a term used especially when the phylogenetic segmentation pattern cannot be identified. { ˌmed·əˈsō·mə }

metasternum [INVERTEBRATE ZOOLOGY] The ventral portion of the metathorax in insects. { ¦med·əˈstər·nəm }

metastoma [INVERTEBRATE ZOOLOGY] Median plate posterior to the mouth in certain crustaceans and related arthropods. { məˈtas·tə·mə }

Metastrongylidae [INVERTEBRATE ZOOLOGY] A family of roundworms belonging to the Strongyloidea; species are parasitic in sheep, cattle, horses, dogs, and other domestic animals. { ¦med·əˌstränˈjil·əˌdē }

Metastrongyloidea [INVERTEBRATE ZOOLOGY] A superfamily of parasitic nematodes, characterized by a reduced or absent stoma capsule and an oral opening surrounded by six well-developed lips. { ¦med·əˌsträn·jəˈlȯid·ē·ə }

metatarsal [ANATOMY] Of or pertaining to the metatarsus. { ¦med·əˈtär·səl }

metatarsus [ANATOMY] The part of a foot or hindfoot between the tarsus and the phalanges. { ˌmed·əˈtär·səs }

Metatheria [VERTEBRATE ZOOLOGY] An infraclass of therian mammals including a single order, the Marsupialia; distinguished by a small braincase, a total of 50 teeth, the inflected angular process of the mandible, and a pair of marsupial bones articulating with the pelvis. { ˌmed·əˈthir·ē·ə }

metathorax [INVERTEBRATE ZOOLOGY] Posterior segment of the thorax in insects. { ¦med·əˈthȯr ˌaks }

metatroch [INVERTEBRATE ZOOLOGY] A segmented larval form following the trochophore in annelids. { ˈmed·əˌträk }

metaxylem [BOTANY] Primary xylem differentiated after and distinguished from protoxylem by thicker tracheids and vessels with pitted or reticulated walls. { ¦med·əˈzī·ləm }

Metazoa [ZOOLOGY] The multicellular animals that make up the major portion of the animal kingdom; cells are organized in layers or groups as specialized tissues or organ systems. { ˌmed·əˈzō·ə }

metazoea [INVERTEBRATE ZOOLOGY] The last zoea of certain decapod crustaceans; metamorphoses into a megalopa. { ˌmed·ə·zōˈē·ə }

metencephalon [EMBRYOLOGY] The cephalic portion of the rhombencephalon; gives rise to the cerebellum and pons. { ˌmed·inˈsef·əˌlän }

Meteoriaceae [BOTANY] A family of mosses in the order Isobryales in which the calyptra is frequently hairy. { ˌmēd·ēˌȯr·ēˈās·ēˌē }

metestrus [PHYSIOLOGY] The beginning of the luteal phase following estrus. { medˈes·trəs }

methane-oxidizing bacteria [MICROBIOLOGY] Bacteria that derive energy from oxidation of methane. { ˈmethˌān ¦äk·səˌdīz·iŋ bakˈtir·ē·ə }

Methanobacteriaceae [MICROBIOLOGY] The single family of methane-producing bacteria; anaerobes which obtain energy via formation of methane. { ¦meth·ə·nōˌbak·tir·ēˈās·ēˌē }

methanogen [BIOLOGY] An organism carrying out methanogenesis, requiring completely anaerobic conditions for growth; considered by some authorities to be distinct from bacteria. { mə 'than·ə·jən }

methanogenesis [BIOCHEMISTRY] The biosynthesis of the hydrocarbon methane; common in certain bacteria. Also known as bacterial methanogenesis. { ¦meth·ə·nō'jen·ə·səs }

Methanomonadaceae [MICROBIOLOGY] Formerly a family of bacteria in the suborder Pseudomonadineae; members identified as gram-negative rods are able to use carbon monoxide (*Carboxydomonas*), methane (*Methanomonas*), and hydrogen (*Hydrogenomonas*) as their sole source of energy for growth. { ¦meth·ə·nō‚män·ə'dās·ē‚ē }

methemoglobin [BIOCHEMISTRY] *See* ferrihemoglobin. { met‚hē·mə'glō·bən }

methionine [BIOCHEMISTRY] $C_5H_{11}O_2NS$ An essential amino acid; furnishes both labile methyl groups and sulfur necessary for normal metabolism. { mə'thī·ə‚nēn }

methylated cap [MOLECULAR BIOLOGY] A modified guanine nucleotide that terminates a messenger ribonucleic acid molecule. { ¦meth·ə‚lād·əd 'kap }

Methylomonadaceae [MICROBIOLOGY] A family of gram-negative, aerobic bacteria which utilize only one-carbon compounds as a source of carbon. { ‚meth·ə·lō‚män·ə'dās·ē‚ē }

methylotrophic bacteria [MICROBIOLOGY] Bacteria that are capable of growing on methane derivatives as their sole source of carbon and metabolic energy. { ¦meth·ə·lə¦trä·fik bak'tir·ē·ə }

methyl red test [MICROBIOLOGY] A cultural test for the ability of bacteria to ferment carbohydrate to form acid; uses methyl red as the indicator. { 'meth·əl ¦red ‚test }

methyl transferase [BIOCHEMISTRY] Any of a group of enzymes which catalyze the reaction of S-adenosyl methionine with a suitable acceptor to yield the methylated acceptor molecule and S-adenosyl homocysteine. { 'meth·əl 'tranz·fə ‚rās }

metraterm [INVERTEBRATE ZOOLOGY] The distal portion of the uterus in trematodes. { 'mē·trə ‚tərm }

Metridiidae [INVERTEBRATE ZOOLOGY] A family of zoantharian cnidarians in the order Actiniaria. { ‚me·trə'dī·ə‚dē }

Metzgeriales [BOTANY] An order of liverworts in the subclass Jungermannidae, class Hepaticopsida, distinguished by archegonia produced behind a growing apex, a flat elongated gametophyte with no tissue differentiation or surface pores and, less commonly, a stem with two rows of leaves. { ‚mets·gə·rē'ā·lēz }

Meyliidae [INVERTEBRATE ZOOLOGY] A family of free-living nematodes in the superfamily Desmoscolecoidea. { ma'lī·ə‚dē }

Meziridae [INVERTEBRATE ZOOLOGY] A family of hemipteran insects in the superfamily Aradoidea. { mə'zir·ə‚dē }

MHC [IMMUNOLOGY] *See* major histocompatibility complex.

Miacidae [PALEONTOLOGY] The single, extinct family of the carnivoran superfamily Miacoidea. { mī'as·ə‚dē }

Miacoidea [PALEONTOLOGY] A monofamilial superfamily of extinct carnivoran mammals; a stem group thought to represent the progenitors of the earliest member of modern carnivoran families. { ‚mī·ə'kȯid·ē·ə }

MIC [IMMUNOLOGY] *See* major immunogene complex.

micelle [MOLECULAR BIOLOGY] A submicroscopic structural unit of protoplasm built up from polymeric molecules. { mī'sel }

Michaelis constant [BIOCHEMISTRY] A constant K_m such that the initial rate of reaction V, produced by an enzyme when the substrate concentration is high enough to saturate the enzyme, is related to the rate of reaction v at a lower substrate concentration c by the formula $V = v(1 + K_m/c)$. { mi'kā·ləs ‚kän·stənt }

micracanthopore [INVERTEBRATE ZOOLOGY] Small, minute tubes projecting from the surface of bryozoan colonies. { ‚mī·krə'kan·thrə‚pȯr }

microaerophilic [MICROBIOLOGY] Pertaining to those microorganisms requiring free oxygen but in very low concentration for optimum growth. { ¦mī·krō¦er·ə¦fil·ik }

microanatomy [ANATOMY] Anatomical study of microscopic tissue structures. { ¦mī·krō·ə'nad·ə·mē }

microbe [MICROBIOLOGY] A microorganism, especially a bacterium of a pathogenic nature. { ¦mī‚krōb }

microbial insecticide [MICROBIOLOGY] Species-specific bacteria which are pathogenic for and used against injurious insects. { mī'krō·bē·əl in 'sek·tə‚sīd }

microbiology [BIOLOGY] The science and study of microorganisms, including protozoans, algae, fungi, bacteria, viruses, and rickettsiae. { ¦mī·krō·bī'äl·ə·jē }

microbody [CELL BIOLOGY] Any of three distinct classes (peroxisomes, glyoxysomes, and microperoxisomes) of cytoplasmic organelles that are bounded by a single membrane and contain a variety of enzymes. { 'mī·krə‚bäd·ē }

microcell [CELL BIOLOGY] A micronucleus within a layer of cytoplasm and a membrane. { 'mī·krə ‚sel }

microcentrum [CELL BIOLOGY] The centrosome, or a group of centrosomes, functioning as the dynamic center of a cell. { 'mī·krō‚sen·trəm }

microceratous [INVERTEBRATE ZOOLOGY] Having short antennae. { ‚mī·krō'ser·ə·təs }

microcercous cercaria [INVERTEBRATE ZOOLOGY] A cercaria with a very short broad tail. { ¦mi·krō ¦sər·kəs sər'kar·ē·ə }

microchemistry [BIOCHEMISTRY] The chemistry of individual cells and minute organisms. { ¦mī·krō'kem·ə·strē }

Microchiroptera [VERTEBRATE ZOOLOGY] A suborder of the mammalian order Chiroptera composed of the insectivorous bats. { ¦mī·krō·kī 'räp·tə·rə }

microcirculation [PHYSIOLOGY] The flow of blood or lymph in the vessels of the microcirculatory system. { ¦mī·krō‚sər·kyə'lā·shən }

microcirculatory system [ANATOMY] Those vessels of the blood and lymphatic systems which are visible only with a microscope. { ¦mī·krō 'sər·kyə·lə,tȯr·ē ,sis·təm }

microcneme [INVERTEBRATE ZOOLOGY] Microsepta in certain anemones. { 'mī·krə,nēm }

Micrococcaceae [MICROBIOLOGY] A family of gram-positive cocci; chemoorganotrophic organisms with respiratory or fermentative metabolism. { ¦mī·krō·käk'sās·ē,ē }

microconsumer [ECOLOGY] *See* decomposer. { ¦mī·krō·kən'sü·mər }

Microcotyloidea [INVERTEBRATE ZOOLOGY] A superfamily of ectoparasitic trematodes in the subclass Monogenea. { ¦mī·krō,käd·əl'ȯid·ē·ə }

Microcyprini [VERTEBRATE ZOOLOGY] The equivalent name for Cyprinodontiformes. { ,mī·krō·sə 'prē,nē }

microdissection [BIOLOGY] Dissection under a microscope. { ¦mī·krō·di'sek·shən }

Microdomatacea [PALEONTOLOGY] An extinct superfamily of gastropod mollusks in the order Aspidobranchia. { ,mī·krə,dō·mə'tās·ē·ə }

microenvironment [ECOLOGY] The specific environmental factors in a microhabitat. { ¦mī·krō·in'vī·ərn·mənt }

microevolution [EVOLUTION] **1.** Evolutionary processes resulting from the accumulation of minor changes over a relatively short period of time; evolutionary changes due to gene mutation. **2.** Evolution of species. { ¦mī·krō,ev·ə'lü·shən }

microfibril [MOLECULAR BIOLOGY] The submicroscopic unit of a microscopic cellular fiber. { ¦mī·krō'fi·brəl }

microfilament [CELL BIOLOGY] One of the cytoplasmic fibrous structures, about 5 nanometers in diameter, virtually identical to actin; thought to be important in the processes of phagocytosis and pinocytosis. { ¦mī·krō'fil·ə·mənt }

microfilaria [INVERTEBRATE ZOOLOGY] Slender, motile prelarval forms of filarial nematodes measuring 150-300 micrometers in length; adult filaria are mammalian parasites. { ¦mī·krō·fə'lar·ē·ə }

microflora [BOTANY] Microscopic plants. [ECOLOGY] The flora of a microhabitat. { ¦mī·krō 'flȯr·ə }

microfossil [PALEONTOLOGY] A small fossil which is studied and identified by means of the microscope. { ¦mī·krō'fäs·əl }

microgamete [BIOLOGY] The smaller, or male gamete produced by heterogametic species. { ¦mī·krō'ga,mēt }

microgametocyte [BIOLOGY] A cell that gives rise to microgametes. { ¦mī·krō·gə'mēd·ə,sīt }

microgametophyte [BOTANY] The male gametophyte in plants having two types of spores. { ¦mī·krō·gə'mēd·ə,fīt }

microgamy [BIOLOGY] Sexual reproduction by fusion of the small male and female gametes in certain protozoans and algae. { mī'kräg·ə·mē }

microgenesis [BIOLOGY] Abnormally small development of a part. { ¦mī·krō'jen·ə·səs }

microglia [HISTOLOGY] Small neuroglia cells of the central nervous system having long processes and exhibiting ameboid and phagocytic activity under certain pathologic conditions. { mī 'kräg·lē·ə }

microhabitat [ECOLOGY] A small, specialized, and effectively isolated location. { ¦mī·krō'hab·ə,tat }

Microhylidae [VERTEBRATE ZOOLOGY] A family of anuran amphibians in the suborder Diplasiocoela including many heavy-bodied forms with a pointed head and tiny mouth. { ,mī·krō'hī·lə ,dē }

microinjection [CELL BIOLOGY] Injection of cells with solutions by using a micropipet. { ¦mī·krō·in'jek·shən }

Microlepidoptera [INVERTEBRATE ZOOLOGY] A former division of Lepidoptera. { ¦mī·krō,lep·ə 'däp·tə·rə }

Micromalthidae [INVERTEBRATE ZOOLOGY] A family of coleopteran insects in the superfamily Cantharoidea; the single species is the telephone pole beetle. { ,mī·krō'mȯl·thə,dē }

micromanipulation [BIOLOGY] The techniques and practice of microdissection, microvivisection, microisolation, and microinjection. { ¦mī·krō·mə,nip·yə'lā·shən }

micromere [EMBRYOLOGY] A small blastomere of the upper or animal hemisphere in eggs that undergo uneven cleavage. { 'mī·krə,mir }

Micromonospora [MICROBIOLOGY] A genus of bacteria in the family Micromonosporaceae; the mycelium is well developed, branched, and septate; single spores are formed onhyphae.

Micromonosporaceae [MICROBIOLOGY] A family of bacteria in the order Actinomycetales; aerial hyphae are formed in all genera except *Micromonospora*; saprophytic soil organisms.

Micromonospora purpurea [MICROBIOLOGY] The bacterium that produces the antibiotic gentamycin.

micronekton [ECOLOGY] Active pelagic crustaceans and other forms intermediate between thrusting nekton and feebler-swimming plankton. { ,mī·krə'nek,tän }

micronucleus [INVERTEBRATE ZOOLOGY] The smaller, reproductive nucleus in multinucleate protozoans. { ¦mī·krō'nü·klē·əs }

micronutrient [BIOCHEMISTRY] *See* trace element. { ¦mī·krō'nü·trē·ənt }

microorganism [MICROBIOLOGY] A microscopic organism, including bacteria, protozoans, yeast, viruses, and algae. { ¦mī·krō'ȯr·gə,niz·əm }

micropaleontology [PALEONTOLOGY] A branch of paleontology that deals with the study of microfossils. { ¦mī·krō,pā·lē·ən'täl·ə·jē }

Micropezidae [INVERTEBRATE ZOOLOGY] A family of myodarian cyclorrhaphous dipteran insects in the subsection Acalypteratae. { ,mī·krō'pez·ə ,dē }

microphage [HISTOLOGY] A small phagocyte, especially a neutrophil. { 'mī·krə,fāj }

microphagy [BIOLOGY] Feeding on minute organisms or particles. { mī'kräf·ə·jē }

microphenic [GENETICS] Pertaining to genetic or environmental factors that are numerous but individually of small effect relative to the phenotypic standard deviation. { ,mī·krə'fen·ik }

microphyllous [BOTANY] **1.** Having small leaves.

2. Having leaves with a single, unbranched vein. { ¦mī·krō¦fil·əs }

Microphysidae [INVERTEBRATE ZOOLOGY] A palearctic family of hemipteran insects in the subfamily Cimicimorpha. { ˌmī·krə'fīs·əˌdē }

microphyte [ECOLOGY] 1. A microscopic plant. 2. A dwarfed plant due to unfavorable environmental conditions. { 'mī·krəˌfīt }

micropinocytosis [CELL BIOLOGY] A mechanism of endocytosis in which fluid droplets are internalized by indentations (caveolae) on the surface membrane which pinch off as tiny internal vesicles (micropinosomes). { ¦mī·krōˌpin·ə·sī'tō·səs }

micropinosome [CELL BIOLOGY] A very tiny vesicle that is pinched off from the plasma membrane of a cell during micropinocytosis. { ¦mī·krō'pin·əˌsōm }

Micropterygidae [INVERTEBRATE ZOOLOGY] The single family of the lepidopteran superfamily Micropterygoidea; members are minute moths possessing toothed, functional mandibles and lacking a proboscis. { mīˌkräp·tə'rij·əˌdē }

Micropterygoidea [INVERTEBRATE ZOOLOGY] A monofamilial superfamily of lepidopteran insects in the suborder Homoneura. { mīˌkräp·tə·rə'gȯid·ē·ə }

Micropygidae [INVERTEBRATE ZOOLOGY] A family of echinoderms in the order Diadematoida that includes only one genus, *Micropyga*, which has noncrenulate tubercles and umbrellalike outer tube feet. { ˌmī·krə'pij·əˌdē }

micropyle [BOTANY] A minute opening in the integument at the tip of an ovule through which the pollen tube commonly enters; persists in the seed as an opening or a scar between the hilum and point of radicle. { 'mī·krəˌpīl }

Microsauria [PALEONTOLOGY] An order of Carboniferous and early Permian lepospondylous amphibians. { ˌmī·krō'sȯr·ē·ə }

microsclere [INVERTEBRATE ZOOLOGY] A minute sclerite in Porifera. { 'mī·krəˌsklir }

microseptum [INVERTEBRATE ZOOLOGY] An incomplete or imperfect mesentery in zoantharians. { ¦mī·krō'sep·təm }

microsome [CELL BIOLOGY] 1. A fragment of the endoplasmic reticulum. 2. A minute granule of protoplasm. { 'mī·krəˌsōm }

microspecies [ECOLOGY] A small, localized species population that is clearly differentiated from related forms. Also known as jordanon. { ¦mī·krō 'spē·shēz }

microspike [CELL BIOLOGY] Any of the narrow cytoplasmic projections that extend or retract from the surface of a cell and may have a sensory function. { 'mi·krəˌspīk }

Microsporaceae [BOTANY] A monogeneric family of green algae in the suborder Ulotrichineae; the chloroplast is a parietal network. { ˌmī·krō·spə 'rās·ēˌē }

microsporangium [BOTANY] A sporangium bearing microspores. { ¦mī·krō·spə'ran·jē·əm }

microspore [BOTANY] The smaller spore of heterosporous plants; gives rise to the male gametophyte. { 'mī·krəˌspȯr }

microspore mother cell [BOTANY] *See* microsporocyte. { 'mī·krəˌspȯr 'məth·ər ˌsel }

Microsporida [INVERTEBRATE ZOOLOGY] The single order of the class Microsporidea. { ˌmī·krə 'spȯr·ə·də }

Microsporidae [INVERTEBRATE ZOOLOGY] The equivalent name for Sphaeriidae. { ˌmī·krə 'spȯr·əˌdē }

Microsporidea [INVERTEBRATE ZOOLOGY] A class of Cnidospora characterized by the production of minute spores with a single intrasporal filament or one or two intracapsular filaments and a single sporoplasm; mainly intracellular parasites of arthropods and fishes. { ˌmī·krə·spə'rid·ē·ə }

microsporocyte [BOTANY] A diploid cell from which four microspores are produced by meiosis. Also known as microspore mother cell. { ¦mī·krō 'spȯr·əˌsīt }

microsporogenesis [BOTANY] In angiosperms, formation of microspores and production of the male gametophyte. { ˌmī·krəˌspȯr·ə'jen·ə·səs }

microsporophyll [BOTANY] A sporophyll bearing microsporangia. { ¦mī·krō'spȯr·əˌfil }

microsurgery [BIOLOGY] Surgery on single cells by micromanipulation. { ¦mī·krō'sər·jə·rē }

Microtatobiotes [BIOLOGY] An artificial taxonomic category, comprising two unrelated groups of biological entities, the rickettsiae and the viruses. { mī¦kräd·ə·dōˌbī'ōd·ēz }

microtherm [ECOLOGY] A plant requiring a mean annual temperature range of 0-14°C for optimum growth. { 'mī·krəˌthərm }

Microtinae [VERTEBRATE ZOOLOGY] A subfamily of rodents in the family Muridae that includes lemmings and muskrats. { mī'krät·ənˌē }

microtomy [BIOLOGY] Cutting of thin sections of specimens with a microtome. { mī'kräd·ə·mē }

microtrabecular lattice [CELL BIOLOGY] A network of thin filaments that interconnect the cytoplasmic filaments. { ˌmī·krə·trə¦bek·yə·lər 'lad·əs }

Microtragulidae [PALEONTOLOGY] A group of saltatorial caenolistoid marsupials that appeared late in the Cenozoic and paralleled the small kangaroos of Australia. { ˌmī·krō·trə'gyül·ə ˌdē }

microtrichia [INVERTEBRATE ZOOLOGY] Small hairs on the integument of various insects, especially on the wings. { ˌmī·krō'trik·ē·ə }

microtubule [CELL BIOLOGY] One of the hollow tubelike filaments found in certain cell components, such as cilia and the mitotic spindle, and composed of repeating subunits of the protein tubulin. { ¦mī·krō'tüb·yül }

microvillus [CELL BIOLOGY] One of the filiform processes that form a brush border on the surfaces of certain specialized cells, such as intestinal epithelium. { ¦mī·krō'vil·əs }

mictic [BIOLOGY] 1. Requiring or produced by sexual reproduction. 2. Of or pertaining to eggs which without fertilization develop into males and with fertilization develop into amictic females, as occurs in rotifers. { 'mik·tik }

micturition [PHYSIOLOGY] *See* urination. { ˌmik·chə'ri·shən }

midaxillary line [ANATOMY] A perpendicular line

drawn downward from the apex of the axilla. { ˌmid'ak·səˌler·ē 'līn }

midbrain [ANATOMY] Those portions of the adult brain derived from the embryonic midbrain. [EMBRYOLOGY] The middle portion of the embryonic vertebrate brain. Also known as mesencephalon. { 'midˌbrān }

midclavicular line [ANATOMY] A vertical line parallel to and midway between the midsternal line and a vertical line drawn downward through the outer end of the clavicle. { 'mid·klə¦vik·yə·lər 'līn }

middle ear [ANATOMY] The middle portion of the ear in higher vertebrates; in mammals it contains three ossicles and is separated from the external ear by the tympanic membrane and from the inner ear by the oval and round windows. { 'mid·əl 'ir }

middle lamella [CELL BIOLOGY] The layer of a cell wall that is derived from the phragmoplast. { 'mid·əl lə'mel·ə }

midge [INVERTEBRATE ZOOLOGY] Any of various dipteran insects, principally of the families Ceratopogonidae, Cecidomyiidae, and Chironomidae; many are biting forms and are vectors of parasites of man and other vertebrates. { mij }

midgut [EMBRYOLOGY] The middle portion of the digestive tube in vertebrate embryos. Also known as mesenteron. [INVERTEBRATE ZOOLOGY] The mesodermal intermediate part of an invertebrate intestine. { 'midˌgət }

midrib [BOTANY] The large central vein of a leaf. { 'midˌrib }

migrant [ZOOLOGY] An animal that moves from one habitat to another. { 'mī·grənt }

migration [GENETICS] The transfer of genetic information among populations by the movement of individuals or groups of individuals from one population into another. [VERTEBRATE ZOOLOGY] Periodic movement of animals to new areas or habitats. { mī'grā·shən }

mildew [MYCOLOGY] **1.** A whitish growth on plants, organic matter, and other materials caused by a parasitic fungus. **2.** Any fungus producing such growth. { 'milˌdü }

Milichiidae [INVERTEBRATE ZOOLOGY] A family of myodarian cyclorrhaphous dipteran insects in the subsection Acalypteratae. { ˌmil·ə'kī·əˌdē }

milieu interieur [PHYSIOLOGY] The fundamental concept that the living organism exists in an aqueous internal environment which bathes all tissues and provides a medium for the elementary exchange of nutrients and waste. { mēl'yü in ¦tir·ē·ər }

Miliolacea [INVERTEBRATE ZOOLOGY] A superfamily of marine or brackish foraminiferans in the suborder Miliolina characterized by an imperforate test wall of tiny, disordered calcite rhombs. { ˌmil·ē·ə'lās·ē·ə }

Miliolidae [INVERTEBRATE ZOOLOGY] A family of foraminiferans in the superfamily Miliolacea. { ˌmil·ē'äl·əˌdē }

Miliolina [INVERTEBRATE ZOOLOGY] A suborder of the Foraminiferida characterized by a porcelaneous, imperforate calcite wall. { ˌmil·ē'äl·ə·nə }

milk [PHYSIOLOGY] **1.** The whitish fluid secreted by the mammary gland for the nourishment of the young; composed of carbohydrates, proteins, fats, mineral salts, vitamins, and antibodies. **2.** Any whitish fluid in nature resembling milk, as coconut milk. { milk }

milk factor [BIOCHEMISTRY] A filtrable, noncellular agent in the milk and tissues of certain strains of inbred mice; transmitted from the mother to the offspring by nursing. Also known as Bittner milk factor. { 'milk ˌfak·tər }

milk sugar [BIOCHEMISTRY] *See* lactose. { 'milk ˌshu̇g·ər }

milk teeth [ANATOMY] *See* deciduous teeth. { 'milk ˌtēth }

milkweed [BOTANY] Any of several latex-secreting plants of the genus *Asclepias* in the family Asclepiadaceae. { 'milkˌwēd }

milky disease [INVERTEBRATE ZOOLOGY] A bacterial disease of Japanese beetle larvae or related grubs caused by *Bacillus papilliae* and *B. lentimorbus* that penetrate the intestine and sporulate in the body cavity; blood of the grub eventually turns milky white. { 'mil·kē diˌzēz }

Milleporina [INVERTEBRATE ZOOLOGY] An order of the class Hydrozoa known as the stinging corals; they resemble true corals because of a calcareous exoskeleton. { mil·ə·pə'rī·nə }

millet [BOTANY] A common name applied to at least five related members of the grass family grown for their edible seeds. { 'mil·ət }

millipede [INVERTEBRATE ZOOLOGY] The common name for members of the arthropod class Diplopoda. { 'mil·əˌpēd }

mimetic [ZOOLOGY] Pertaining to or exhibiting mimicry. { mə'med·ik }

mimicry [ZOOLOGY] Assumption of color, form, or behavior patterns by one species of another species, for camouflage and protection. { 'mim·ə·krē }

Mimidae [VERTEBRATE ZOOLOGY] The mockingbirds, a family of the Oscines in the order Passeriformes. { 'mim·əˌdē }

Mimosoideae [BOTANY] A subfamily of the legume family, Leguminosae; members are largely woody and tropical or subtropical with regular flowers and usually numerous stamens. { ˌmim·ə'sȯid·ēˌē }

mineralocorticoid [BIOCHEMISTRY] A steroid hormone secreted by the adrenal cortex that regulates mineral metabolism and, secondarily, fluid balance. { ¦min·rə·lō'kȯrd·əˌkȯid }

minicell [MICROBIOLOGY] A small anucleate bacterial cell produced by abnormal and unequal division of a parent cell. { 'min·ēˌsel }

minichromosome [CELL BIOLOGY] A eukaryotic chromosome reduced in size by deletion of a segment of deoxyribonucleic acid. [VIROLOGY] Viral deoxyribonucleic acid combined with histone to form a chromatin-like structure. { ˌmin·ē'krō·məˌsōm }

minimal recognition length [MOLECULAR BIOLOGY] The shortest length of base pairs that will form a stable deoxyribonucleic acid duplex in genetic recombination. { ¦min·ə·məl ,rek·ig'nish·ən ,leŋkth }

miniplasmid [MOLECULAR BIOLOGY] Any plasmid that has been reduced in size by means of recombinant deoxyribonucleic acid technology. { ,min·ē'plaz·mid }

mink [VERTEBRATE ZOOLOGY] Any of three species of slender-bodied aquatic carnivorous mammals in the genus *Mustela* of the family Mustelidae. { miŋk }

minnow [VERTEBRATE ZOOLOGY] The common name for any fresh-water fish composing the family Cyprinidae, order Cypriniformes. { 'min·ō }

minus strand [MOLECULAR BIOLOGY] A polynucleotide strand that is complementary to, and formed by transcription from, another specific polynucleotide (plus) strand, with which it produces the double-stranded (double-helix) ribonucleic acid. { 'mī·nəs ,strand }

Miosireninae [PALEONTOLOGY] A subfamily of extinct sirenian mammals in the family Dugongidae. { 'mī·ō·sə'ren·ə,nē }

miosis [PHYSIOLOGY] Contraction of the pupil of the eye. { mī'ō·səs }

miotic [PHYSIOLOGY] Of or pertaining to miosis. { mī'äd·ik }

miracidium [INVERTEBRATE ZOOLOGY] The ciliated first larva of a digenetic trematode; forms a sporocyst after penetrating intermediate host tissues. { ,mī·rə'sid·ē·əm }

Mirapinnatoidei [VERTEBRATE ZOOLOGY] A suborder of tiny oceanic fishes in the order Cetomimiformes. { ,mir·ə,pin·ə'tȯid·ē,ī }

Miridae [INVERTEBRATE ZOOLOGY] The largest family of the Hemiptera; included in the Cimicomorpha, it contains herbivorous and predacious plant bugs which lack ocelli and have a cuneus and four-segmented antennae. { 'mir·ə ,dē }

Miripinnati [VERTEBRATE ZOOLOGY] The equivalent name for Marapinnatoidei. { ¦mir·ə·pə'näd·ē }

mispairing [MOLECULAR BIOLOGY] Pairing of a nucleotide in one chain of a deoxyribonucleic acid molecule that is not complementary to the nucleotide occupying the corresponding position in the other chain. { mis'per·iŋ }

misrepair [MOLECULAR BIOLOGY] Repair of deoxyribonucleic acid that gives rise to gene mutations or changes in chromosome structure. { 'mis·ri,per }

missense codon [GENETICS] A mutant codon that directs the incorporation of a different amino acid and results in the synthesis of a protein with a sequence in which one amino acid has been replaced by a different one; in some cases the mutant protein may be unstable or less active. { 'mis·əns 'kō,dän }

missense mutation [MOLECULAR BIOLOGY] A mutation that converts a codon coding for one amino acid to a codon coding for another amino acid. { 'mis·əns myü'tā·shən }

missense suppressor [MOLECULAR BIOLOGY] A suppressor that incorporates the correct amino acid at the site of a codon that has been altered because of a missense mutation. { 'mis·əns sə ,pres·ər }

mistletoe [BOTANY] **1.** *Viscum album*. The true, Old World mistletoe having dichotomously branching stems, thick leathery leaves, and waxy-white berries. **2.** Any of several species of green hemiparasitic plants of the family Loranthaceae. { 'mis·əl,tō }

mistranslation [MOLECULAR BIOLOGY] Incorporation of the wrong amino acid into a protein due to misreading of a codon. { ,mis·tranz'lā·shən }

mite [INVERTEBRATE ZOOLOGY] The common name for the acarine arthropods composing the diverse suborders Onychopalpida, Mesostigmata, Trombidiformes, and Sarcoptiformes. { mīt }

mitochondria [CELL BIOLOGY] Minute cytoplasmic organelles in the form of spherical granules, short rods, or long filaments found in almost all living cells; submicroscopic structure consists of an external membrane system. { ,mīd·ə'kän·drē·ə }

mitochondrial crest [CELL BIOLOGY] Any of the infoldings of the mitochondrial inner membrane that extend into the matrix. { ,mīd·ə¦kän·drē·əl 'kres }

mitochondrial deoxyribonucleic acid [BIOCHEMISTRY] The circular deoxyribonucleic acid duplex, generally 5 to 10 copies, contained within a mitochondrion and maternally inherited since only the egg cell contributes significant numbers of mitochondria to the zygote. Abbreviated mtDNA. Also known as mitochondrial genome. { ,mīd·ə ¦kän·drē·əl dē¦äk·sē,rī·bō·nü'klē·ik 'as·əd }

mitochondrial genome [BIOCHEMISTRY] *See* mitochondrial deoxyribonucleic acid. { ,mīd·ə¦kän·drē·əl 'je,nōm }

mitochondrial plasmid [CELL BIOLOGY] Plasmid-like deoxyribonucleic acid molecules in mitochondria of certain higher plants and some fungi. { ,mīd·ə¦kän·drē·əl 'plaz·mid }

mitogen [CELL BIOLOGY] A compound that stimulates cells to undergo mitosis. { 'mīd·ə,jen }

mitogenesis [CELL BIOLOGY] **1.** Induction of mitosis. **2.** Formation as a result of mitosis. { ,mīd·ə'jen·ə·səs }

mitomycin [MICROBIOLOGY] A complex of three antibiotics (mitomycin A, B, and C) produced by *Streptomyces caespitosus*. { ¦mīd·ə¦mīs·ən }

mitoplast [CELL BIOLOGY] **1.** A mitochondrion that has had its outer membrane removed. **2.** The cytoplast of a mitotic cell after the chromosomes are extruded. { 'mīd·ə,plast }

mitosis [CELL BIOLOGY] Nuclear division involving exact duplication and separation of the chromosome threads so that each of the two daughter nuclei carries a chromosome complement identical to that of the parent nucleus. { mī'tō·səs }

mitotic apparatus [CELL BIOLOGY] A transitory organelle-like formation that is seen during mitosis and meiosis and consists of the asters, the spindle, and the traction fibers. { mī¦täd·ik ,ap·ə'rad·əs }

mitotic center [CELL BIOLOGY] A structure that

defines the poles toward which chromosomes move during mitosis and meiosis. { mī¦täd·ik 'sen·tər }

mitotic index [CELL BIOLOGY] The number of cells undergoing mitosis per thousand cells. { mī'täd·ik 'in,deks }

mitotic inhibitor [CELL BIOLOGY] A compound that inhibits mitosis. { mī'täd·ik in'hib·əd·ər }

mitotic poison [CELL BIOLOGY] A compound that prevents or affects the completion of mitosis. { mī'täd·ik 'pȯiz·ən }

mitral valve [ANATOMY] The atrioventricular valve on the left side of the heart. { 'mī·trəl 'valv }

mitriform [BIOLOGY] Shaped like a miter. { 'mī·trə,fȯrm }

Mitteniales [BOTANY] An order of true mosses, class Bryopsida, characterized by branches of protonema consisting of spherical cells which reflect light from a backing of chloroplasts, thus providing a glow. { ,mit·ən'ā·lēz }

mixed bud [BOTANY] A bud that contains both rudimentary leaves and rudimentary flowers. { 'mikst 'bəd }

mixed cryoglobulin [BIOCHEMISTRY] A cryoglobulin with a monoclonal component made of immunoglobulin belonging to two different classes, one of which is monoclonal. { 'mikst ,krī·ō'gläb·yə·lən }

mixed gland [PHYSIOLOGY] A gland that secretes more than one substance, especially a gland containing both mucous and serous components. { 'mikst 'gland }

mixed nerve [PHYSIOLOGY] A nerve containing both sensory and motor components. { 'mikst 'nərv }

Mixodectidae [PALEONTOLOGY] A family of extinct insectivores assigned to the Proteutheria; a superficially rodentlike group confined to the Paleocene of North America. { ,mik·sə'dek·tə,dē }

mixoploidy [CELL BIOLOGY] The presence of cells having different chromosome numbers in the same cell population. { ¦mik·sə,plȯid·ē }

mixotrophic [BIOLOGY] Obtaining nutrition by combining autotrophic and heterotrophic mechanisms. { ¦mik·sə¦träf·ik }

Mnesarchaeidae [INVERTEBRATE ZOOLOGY] A family of lepidopteran insects in the suborder Homoneura; members are confined to New Zealand. { ,nē·sär'kē·ə,dē }

Mobilina [INVERTEBRATE ZOOLOGY] A suborder of ciliophoran protozoans in the order Peritrichida. { ,mō·bə'lī·nə }

Mobulidae [VERTEBRATE ZOOLOGY] The devil rays, a family of batoids that are surface feeders and live mostly on plankton. { mə'byül·ə,dē }

modal number [GENETICS] **1.** The typical chromosome number of a taxonomic group. **2.** The typical chromosome number of a tumor cell population. { 'mōd·əl ,nəm·bər }

modification [MOLECULAR BIOLOGY] In nucleic acid metabolism, any changes made to deoxyribonucleic acid or ribonucleic acid after their original incorporation into a polynucleotide chain. { ,mäd·ə·fə'kā·shən }

modified base [MOLECULAR BIOLOGY] A nucleotide that is an altered form of the usual four nucleic acid bases. { ¦mäd·ə,fīd 'bās }

modifier gene [GENETICS] A gene that alters the phenotypic expression of a nonallelic gene. { 'mäd·ə,fī·ər ,jēn }

modiolus [ANATOMY] The central axis of the cochlea. { mə'dī·ə·ləs }

modulating codon [MOLECULAR BIOLOGY] A codon that controls the frequency of transcription of a cistron. { ¦mäj·ə,lād·iŋ 'kō,dän }

Moeritheriidae [PALEONTOLOGY] The single family of the extinct order Moeritherioidea. { ,mir·ə·thə'rī·ə,dē }

Moeritherioidea [PALEONTOLOGY] A suborder of extinct sirenian mammals considered as primitive proboscideans by some authorities and as a sirenian offshoot by others. { ,mir·ə,thir·ē'ȯid·ē·ə }

molar [ANATOMY] **1.** A tooth adapted for grinding. **2.** Any of the three pairs of cheek teeth behind the premolars on each side of the jaws in humans. { 'mō·lər }

mold [MYCOLOGY] Any of various woolly fungus growths. [PALEONTOLOGY] An impression made in rock or earth material by an inner or outer surface of a fossil shell or other organic structure; a complete mold would be the hollow space. { mōld }

mole [VERTEBRATE ZOOLOGY] Any of 19 species of insectivorous mammals composing the family Talpidae; the body is stout and cylindrical, with a short neck, small or vestigial eyes and ears, a long naked muzzle, and forelimbs adapted for digging. { mōl }

molecular biology [BIOLOGY] That part of biology which attempts to interpret biological events in terms of the physicochemical properties of molecules in a cell. { mə'lek·yə·lərbī'äl·ə·jē }

molecular chaperon [MOLECULAR BIOLOGY] Any of a class of cellular proteins involved in correct folding of certain polypeptide chains and their assemblage into an oligomer. { mə'lək·yə·lər 'shap·ə,rän }

molecular genetics [MOLECULAR BIOLOGY] The approach which deals with the physics and chemistry of the processes of inheritance. { mə 'lek·yə·lər jə'ned·iks }

molecular mimicry [IMMUNOLOGY] The sharing, by two organisms closely related ecologically but not phylogenetically, of common macromolecular structures that are not attributable to evolutionary conservation of these structures. { mə 'lek·yə·lər 'mim·i,krē }

Molidae [VERTEBRATE ZOOLOGY] A family of marine fishes, including some species of sunfishes, in the order Perciformes. { 'mäl·ə,dē }

Mollicutes [MICROBIOLOGY] The mycoplasmas, a class of prokaryotic organisms lacking a true cell wall; cells are very small to submicroscopic. { mə 'lik·yə,dēz }

Mollusca [INVERTEBRATE ZOOLOGY] One of the divisions of phyla of the animal kingdom containing snails, slugs, octopuses, squids, clams, mussels, and oysters; characterized by a shell-secreting organ, the mantle, and a radula, a food-rasping organ located in the forward area of the mouth. { mə'ləs·kə }

mollusk [INVERTEBRATE ZOOLOGY] Any member of the Mollusca. { 'mäl·əsk }

Molossidae [VERTEBRATE ZOOLOGY] The free-tailed bats, a family of tropical and subtropical insectivorous mammals in the order Chiroptera. { mə'läs·ə,dē }

Molpadida [INVERTEBRATE ZOOLOGY] An order of sea cucumbers belonging to the Apodacea and characterized by a short, plump body bearing a taillike prolongation. { mäl'pā·də·də }

Molpadidae [INVERTEBRATE ZOOLOGY] The single family of the echinoderm order Molpadida. { mäl 'pā·də,dē }

molt [PHYSIOLOGY] To shed an outer covering as part of a periodic process of growth. { mōlt }

molting hormone [BIOCHEMISTRY] Any of several hormones which activate molting in arthropods. { 'mōl·tiŋ 'hór,mōn }

Momotidae [VERTEBRATE ZOOLOGY] The motmots, a family of colorful New World birds in the order Coraciiformes. { mə'mäd·ə,dē }

monactine [INVERTEBRATE ZOOLOGY] A single-rayed spicule in the sponges. { mä'nak·tən }

monadelphous [BOTANY] Having the filaments of the stamens united into one set. { ¦män·ə¦del·fəs }

Monadidae [INVERTEBRATE ZOOLOGY] A family of flagellated protozoans in the order Kinetoplastida having two flagella of uneven length. { mə 'nad·ə,dē }

monandrous [BOTANY] Having one stamen. { mə 'nan·drəs }

monaxon [INVERTEBRATE ZOOLOGY] A spicule formed by growth along a single axis. { mä'nak ,sän }

Monera [BIOLOGY] A kingdom that includes the bacteria and blue-green algae in some classification schemes. { mə'nir·ə }

monestrous [PHYSIOLOGY] Having a single estrous cycle per year. { män'es·trəs }

mongoose [VERTEBRATE ZOOLOGY] The common name for 39 species of carnivorous mammals which are members of the family Viveridae; they are plantigrade animals and have a long slender body, short legs, nonretractile claws, and scent glands. { 'mäŋ,güs }

Monhysterida [INVERTEBRATE ZOOLOGY] An order of aquatic nematodes in the subclass Chromadoria. { ,män·hi'ster·ə·də }

Monhysteroidea [INVERTEBRATE ZOOLOGY] A superfamily of free-living nematodes in the order Monhysterida characterized by single or paired outstretched ovaries, circular to cryptospiral amphids, and a stoma which is usually shallow and unarmed. { ,män ,hi·stə'ròid·ē·ə }

Moniliaceae [MYCOLOGY] A family of fungi in the order Moniliales; sporophores are usually lacking, but when present they are aggregated into fascicles, and hyphae and spores are hyaline or brightly colored. { mə,nil·ē'ās·ē,ē }

Moniliales [MYCOLOGY] An order of fungi of the Fungi Imperfecti containing many plant pathogens; asexual spores are always formed free on the surface of the material on which the organism is living, and never occur in either pycnidia or acervuli. { mə,nil·ē'ā·lēz }

moniliform [BIOLOGY] Constructed with contractions and expansions at regular alternating intervals, giving the appearance of a string of beads. { mə'nil·ə,fórm }

monitor [VERTEBRATE ZOOLOGY] Any of 27 carnivorous, voracious species of the reptilian family Varanidae characterized by a long, slender forked tongue and a dorsal covering of small, rounded scales containing pointed granules. { 'män·əd·ər }

monkey [VERTEBRATE ZOOLOGY] Any of several species of frugivorous and carnivorous primates which compose the families Cercopithecidae and Cebidae in the suborder Anthropoidea; the face is typically flattened and hairless, all species are pentadactyl, and the mammary glands are always in the pectoral region. { 'məŋ·kē }

monkshood [BOTANY] *See* aconite. { 'məŋks ,hùd }

monoamine oxidase [BIOCHEMISTRY] A mitochondrial enzyme which oxidatively deaminates intraneuronal biogenic amines, some of which are important neurotransmitters in the peripheral and central nervous system. { män·ō'am,ēn 'äk·sə,dās }

monoblast [HISTOLOGY] A motile cell of the spleen and bone marrow from which monocytes are derived. { 'män·ō,blast }

Monoblepharidales [MYCOLOGY] An order of aquatic fungi in the class Phycomycetes; distinguished by a mostly hyphal thallus and zoospores with one posterior flagellum. { ¦män·ō ,blef·ə·rə'dā·lēz }

Monobothrida [PALEONTOLOGY] An extinct order of monocyclic camerate crinoids. { ,män·ō 'bäth·rə·də }

monocardiogram [PHYSIOLOGY] *See* vectorcardiogram. { ¦män·ō'kär·dē·ə,gram }

monocarpic [BOTANY] Bearing fruit once and then dying. { ¦män·ō¦kär·pik }

monocarpous [BOTANY] Having a single ovary. { ¦män·ō¦kär·pəs }

monochlamydous [BOTANY] Referring to flowers having only one set of floral envelopes, that is, either a calyx or a corolla. { ¦män·ō'klam·ə·dəs }

Monochoidea [INVERTEBRATE ZOOLOGY] A superfamily of free-living, nonparasitic nematodes in the order Mononchida, characterized by angular, distinct lips bearing papilliform cephalic sensilla, an expanded lip region that is flattened anteriorly, and a heavily cuticularized, barrel or globular stoma, with one or more teeth or denticles. { ,män·ə'kòid·ē·ə }

Monochuloidea [INVERTEBRATE ZOOLOGY] A superfamily of nonparasitic nematodes in the order Mononchida, distinguished by papilliform cephalic sense organs, an inconspicuous, small, slitlike amphid aperture, and a stoma with a thick-walled, slightly tapered anterior and an elongate, thin-walled posterior. { ,män·ə·kə 'lòid·ē·ə }

monocistronic messenger [BIOCHEMISTRY] A messenger ribonucleic acid molecule that contains the amino acid sequence for a single polypeptide chain. { ˌmän·ə·sis¦trän·ik 'mes·ən·jər }

Monocleales [BOTANY] An order of liverworts of the subclass Marchantiidae consisting of a single genus, *Monoclea*, which has the largest gametophyte of all liverworts, and lobed spore mother cells. { ¦män·ō·klē'ā·lēz }

monoclimax [ECOLOGY] A climax community controlled primarily by one factor, as climate. { ¦män·ō¦klīˌmaks }

monoclinic [BOTANY] Having both stamens and pistils in the same flower. { ¦män·ə'klin·ik }

monoclonal antibody [IMMUNOLOGY] A highly specific antibody produced by hybridoma cells; the antibody binds with a single antigenic determinant. { ¦män·ə¦klō·nəl 'ant·iˌbäd·ē }

monoclonal cryoglobulin [BIOCHEMISTRY] A cryoglobulin composed of immunoglobin with only one class or subclass of heavy and light chain. { ¦män·ə¦klō·nəl ¦krī·ō'gläb·yə·lən }

monocolpate pollen [BOTANY] Pollen grains having a single furrow. { ¦män·ə'kōlˌpāt ˌpäl·ən }

monocotyledon [BOTANY] Any plant of the class Liliopsida; all have a single cotyledon. { ¦män·əˌkäd·əl'ēd·ən }

Monocotyledoneae [BOTANY] The equivalent name for Liliopsida. { ˌmän·əˌkäd·əl·ə'dō·nēˌē }

monocrepid [INVERTEBRATE ZOOLOGY] A desma formed by secondary deposits of silica on a monaxon. { ¦män·ə¦krep·əd }

Monocyathea [PALEONTOLOGY] A class of extinct parazoans in the phylum Archaeocyatha containing single-walled forms. { ˌmän·ō·sī'ā·thē·ə }

monocyte [HISTOLOGY] A large (about 12 micrometers), agranular leukocyte with a relatively small, eccentric, oval or kidney-shaped nucleus. { 'män·əˌsīt }

monodactylous [ZOOLOGY] Having a single digit or claw. { ¦män·ə'dak·tə·ləs }

Monodellidae [INVERTEBRATE ZOOLOGY] A monogeneric family of crustaceans in the order Thermosbaenacea distinguished by seven pairs of biramous pereiopods on thoracomeres 2-8, and by not having the telson united to the last pleonite. { ¦män·ə'del·əˌdē }

monodelphic [VERTEBRATE ZOOLOGY] **1.** Having a single genital tract, in the female. **2.** Having a single uterus. { ¦män·ō¦del·fik }

monoecious [BOTANY] **1.** Having both staminate and pistillate flowers on the same plant. **2.** Having archegonia and antheridia on different branches. [ZOOLOGY] Having male and female reproductive organs in the same individual. Also known as hermaphroditic. { mə'nē·shəs }

Monoedidae [INVERTEBRATE ZOOLOGY] An equivalent name for Colydiidae. { mə'nē·dəˌdē }

monogamous bivalent [IMMUNOLOGY] Antigen-antibody complex in which each bivalent antibody combines with two determinant groups on a single antigen molecule. { mə¦näg·ə·məs bī 'vā·lənt }

monogastric [VERTEBRATE ZOOLOGY] Having only one digestive cavity. { ¦män·ō'gas·trik }

Monogenea [INVERTEBRATE ZOOLOGY] A diverse subclass of the Trematoda which are principally ectoparasites of fishes; individuals have enlarged anterior and posterior holdfasts with paired suckers anteriorly and opisthaptors posteriorly. { ˌmän·ə'jē·nē·ə }

Monogenoidea [INVERTEBRATE ZOOLOGY] A class of the Trematoda in some systems of classification; equivalent to the Monogenea of other systems. { ˌmän·ə·jə'nȯid·ē·ə }

Monogonota [INVERTEBRATE ZOOLOGY] An order of the class Rotifera, characterized by the presence of a single gonad in both males and females. { ˌmän·ō·gō'näd·ə }

monogony [BIOLOGY] Asexual reproduction. { mə'näg·ə·mē }

monogynous [BOTANY] Having only one pistil. [VERTEBRATE ZOOLOGY] **1.** Having only one female in a colony. **2.** Consorting with only one female. { mə'näj·ə·nəs }

monohybrid [GENETICS] A hybrid individual heterozygous for one gene or a single character. { ¦män·ō'hī·brəd }

monokine [BIOCHEMISTRY] A cytokine released from macrophages. { 'män·əˌkīn }

Monommidae [INVERTEBRATE ZOOLOGY] A family of coleopteran insects in the superfamily Tenebrionoidea. { mə'nam·əˌdē }

monomorphic [BIOLOGY] Having or exhibiting only a single form. { ¦män·ə¦mȯr·fik }

mononuclear [CELL BIOLOGY] Having only one nucleus. { ¦män·ō'nü·klē·ər }

monophagous [ZOOLOGY] Subsisting on a single kind of food. Also known as monotrophic. { mə 'näf·ə·gəs }

Monophisthocotylea [INVERTEBRATE ZOOLOGY] An order of the Monogenea in which the posthaptor is without discrete multiple suckers or clamps. { ˌmän·əˌfis·thəˌkäd·əl'ē·ə }

Monophlebinae [INVERTEBRATE ZOOLOGY] A subfamily of the homopteran superfamily Coccoidea distinguished by a dorsal anus. { ˌmän·ə'fleb·əˌnē }

monophyletic [EVOLUTION] Pertaining to any form evolved from a single interbreeding population. { ˌmän·ə·fə'led·ik }

monophyodont [VERTEBRATE ZOOLOGY] Having only one set of teeth throughout life. { ¦män·ō 'fī·əˌdänt }

Monopisthocotylea [INVERTEBRATE ZOOLOGY] An order of trematode worms in the subclass Pectobothridia. { ˌmän·əˌfis·thəˌkäd·əl'ē·ə }

Monoplacophora [INVERTEBRATE ZOOLOGY] A group of shell-bearing mollusks represented by few living forms; considered to be a sixth class of mollusks. { ˌmän·ō·plə'käf·ə·rə }

monoploid [GENETICS] **1.** Having only one set of chromosomes. **2.** Having the haploid number of chromosomes. { 'män·əˌplȯid }

monopodial [BOTANY] Stem branching in which there are lateral shoots on a primary axis. { ˌmän·ə'pōd·ē·əl }

monopodium [BOTANY] A primary axis that continues to grow while giving off successive lateral branches. { ˌmän·ə'pōd·ē·əm }

Monoposthioidea [INVERTEBRATE ZOOLOGY] A superfamily of chiefly marine nematodes in the order Desmodorida, represented by the single family Monoposthiidae; distingushed by an annulate cuticle with spikelike ornamentation and a stoma that may or may not possess a well-developed tooth opposed by small subventral teeth. { ˌmän·ōˌpäs·thē′ōid·ē·ə }

Monopylina [INVERTEBRATE ZOOLOGY] A suborder of radiolarian protozoans in the order Oculosida in which pores lie at one pole of a single-layered capsule. { ˌmän·ō·pə′lī·nə }

monorchid [ANATOMY] **1.** Having one testis. **2.** Having one testis descended into the scrotum. { mä′nȯr·kəd }

Monorhina [VERTEBRATE ZOOLOGY] The subclass of Agnatha that includes the jawless vertebrates with a single median nostril. { ˌmän·ə′rī·nə }

monorhinal [ANATOMY] Having only one nostril. { ¦män·ə¦rī·nəl }

monosaccharide [BIOCHEMISTRY] A carbohydrate which cannot be hydrolyzed to a simpler carbohydrate; a polyhedric alcohol having reducing properties associated with an actual or potential aldehyde or ketone group; classified on the basis of the number of carbon atoms, as triose (3C), tetrose (4C), pentose (5C), and so on. { ¦män·ō¦sak·əˌrīd }

Monosigales [BOTANY] A botanical order equivalent to the Choanoflagellida in some systems of classification. { ˌmän·ō·si′gā·lēz }

monosiphonous [BIOLOGY] Having a single central tube, as in the thallus of certain filamentous algae or the hydrocaulus of some hydrozoans. { ¦män·ō′sī·fə·nəs }

monosome [CELL BIOLOGY] **1.** A single ribosome attached to messenger ribonucleic acid. **2.** A chromosome in the diploid chromosome complement that lacks a homolog. { ′män·əˌsōm }

monosomy [GENETICS] The condition in which one chromosome of a pair is missing. { ′män·əˌsōm·ē }

monospermous [BOTANY] Having or producing one seed. { ¦män·ō¦spər·məs }

monosporangium [BOTANY] A sporangium producing monospores. { ¦män·ō·spə′ran·jē·əm }

monospore [BOTANY] A simple or undivided nonmotile asexual spore; produced by the diploid generation of some algae. { ′män·əˌspȯr }

monostome [INVERTEBRATE ZOOLOGY] A cercaria having only one mouth or sucker. { ′män·əˌstōm }

Monostylifera [INVERTEBRATE ZOOLOGY] A suborder of the Hoplonemertini characterized by a single stylet. { ˌmän·ō·stī′līf·ə·rə }

Monotomidae [INVERTEBRATE ZOOLOGY] The equivalent name for Rhizophagidae. { ˌmän·ə′täm·əˌdē }

Monotremata [VERTEBRATE ZOOLOGY] The single order of the mammalian subclass Prototheria containing unusual mammallike reptiles, or quasi-mammals. { ¦män·ō·trə′mad·ə }

monotrichous [MICROBIOLOGY] Of bacteria, having an individual flagellum at one pole. { mə′nä·trə·kəs }

Monotropaceae [BOTANY] A family of dicotyledonous herbs or half shrubs in the order Ericales distinguished by a small, scarcely differentiated embryo without cotyledons, lack of chlorophyll, leaves reduced to scales, and anthers opening by longitudinal slits. { ˌmän·ō·trə′pās·ēˌē }

monotrophic [ZOOLOGY] *See* monophagous. { ¦män·ə¦träf·ik }

monotype [BIOLOGY] A single type of organism that constitutes a species or genus. { ′män·əˌtīp }

monotypic [SYSTEMATICS] Pertaining to a taxon that constains only one immediately subordinate taxon. { ¦män·ə¦tip·ik }

monozygotic twins [BIOLOGY] Twins which develop from a single fertilized ovum. Also known as identical twins. { ¦män·ōˌzī′gäd·ik ′twinz }

mons [ANATOMY] An eminence. { mänz }

monsoon forest [ECOLOGY] A tropical forest occurring in regions where a marked dry season is followed by torrential rain; characterized by vegetation adapted to withstand drought. { ¦mänˌsün ′fär·əst }

mons pubis [ANATOMY] The eminence of the lower anterior abdominal wall above the superior rami of the pubic bones. { ′mänz ′pyü·bəs }

Monstrilloida [INVERTEBRATE ZOOLOGY] A suborder or order of microscopic crustaceans in the subclass Copepoda; adults lack a second antenna and mouthparts, and the digestive tract is vestigial. { ˌmän·strə′lȯid·ə }

mons veneris [ANATOMY] The mons pubis of the female. { ′mänz ′ven·ə·rəs }

montane [ECOLOGY] Of, pertaining to, or being the biogeographic zone composed of moist, cool slopes below the timberline and having evergreen trees as the dominant life-form. { män ′tān }

Montgomery's tubercles [ANATOMY] Elevations in the areola of the nipple due to apocrine sweat glands; most prominent during pregnancy and lactation. { mənt′gəm·rēz ′tü·bər·kəlz }

Monticellidae [INVERTEBRATE ZOOLOGY] A family of tapeworms in the order Proteocephaloidea, in which some or all of the organs are in the cortical mesenchyme; catfish parasites. { ˌmän·tə′sel·əˌdē }

monticulus [ANATOMY] The median dorsal portion of the cerebellum. { män′tik·yə·ləs }

moor [ECOLOGY] *See* bog. { mu̇r }

moose [VERTEBRATE ZOOLOGY] An even-toed ungulate of the genus *Alces* in the family Cervidae; characterized by spatulate antlers, long legs, a short tail, and a large head with prominent overhanging snout. { müs }

Moraceae [BOTANY] A family of dicotyledonous woody plants in the order Urticales characterized by two styles or style branches, anthers inflexed in the bud, and secretion of a milky juice. { mə′rās·ēˌē }

Mordellidae [INVERTEBRATE ZOOLOGY] The tumbling flower beetles, a family of coleopteran insects in the superfamily Meloidea. { mȯr′del·əˌdē }

morel [MYCOLOGY] Any fungus belonging to the genus *Morchella*, distinguished by a large, pitted, spongelike cap; it is a highly prized food, but may be poisonous when taken with alcohol. { mə′rel }

mores [ECOLOGY] Groups of organisms preferring the same physical environment and having the same reproductive season. { 'mȯr‚āz }

Morgan [GENETICS] A unit that expresses the relative distance between the genes on a chromosome, equal to a crossover value of 100%. Abbreviated M. { 'mȯr·gən }

moribund [BIOLOGY] **1.** In a dying or deathlike state. **2.** In a state of suspended life functions; dormant. { 'mȯr·ə·bənd }

Moridae [VERTEBRATE ZOOLOGY] A family of actinopterygian fishes in the order Gadiformes. { 'mȯr·ə‚dē }

Morinae [VERTEBRATE ZOOLOGY] The deep-sea cods, a subfamily of the Moridae. { 'mȯr·ə‚nē }

Mormyridae [VERTEBRATE ZOOLOGY] A large family of electrogenic fishes belonging to the Osteoglossiformes; African river and lake fishes characterized by small eyes, a slim caudal peduncle, and approximately equal dorsal and anal fins in most. { mȯr'mir·ə‚dē }

Mormyriformes [VERTEBRATE ZOOLOGY] Formerly an order of fishes which are now assigned to the Osteoglossiformes. { ‚mȯr·mə·rə'fȯr‚mēz }

Moro reflex [PHYSIOLOGY] The startle reflex observed in normal infants from birth through the first few months, consisting of abduction and extension of all extremities, followed by flexion and abduction of the extremities. { 'mȯr·ō ‚rē ‚fleks }

morph [GENETICS] An individual variant in a polymorphic population. { mȯrf }

morphallaxis [PHYSIOLOGY] Regeneration whereby one part is transformed into another by reorganization of tissue fragments rather than by cell proliferation. { ‚mȯr·fə'lak·səs }

Morphinae [INVERTEBRATE ZOOLOGY] A subfamily of large tropical butterflies in the family Nymphalidae. { 'mȯr·fə‚nē }

morphogen [BIOCHEMISTRY] Any compound that exerts a morphogenetic effect at low concentrations. { 'mȯr·fə·jən }

morphogene [GENETICS] Any gene involved directly or indirectly in the control of growth and morphogenesis. { 'mȯr·fə‚jēn }

morphogenesis [EMBRYOLOGY] The transformation involved in the growth and differentiation of cells and tissue. Also known as topogenesis. { ‚mȯr·fə'jen·ə·səs }

morphogenetic movement [EMBRYOLOGY] Any movement of or within a cell that changes the shape of differentiating cells or tissues. { ‚mȯr·fə·jə¦ned·ik müv·mənt }

morphogenetic stimulus [EMBRYOLOGY] A stimulus exerted by one part of the developing embryo on another, leading to morphogenesis in the reacting part. { ‚mȯr·fə·jə¦ned·ik 'stim·yə·ləs }

morphology [BIOLOGY] A branch of biology that deals with structure and form of an organism at any stage of its life history. { mȯr'fäl·ə·jē }

morphospecies [SYSTEMATICS] A typological species distinguished solely on the basis of morphology. { ¦mȯr·fō'spē‚shēz }

morula [EMBRYOLOGY] A solid mass of blastomeres formed by cleavage of the eggs of many animals; precedes the blastula or gastrula, depending on the type of egg. [INVERTEBRATE ZOOLOGY] A cluster of immature male gametes in which differentiation occurs outside the gonad; common in certain annelids. { 'mȯr·ə·lə }

Moruloidea [INVERTEBRATE ZOOLOGY] The only class of the phylum Mesozoa; embryonic development in the organisms proceeds as far as the morula or stereoblastula stage. { ‚mȯr·ə'lȯid·ē·ə }

mosaic [BIOLOGY] An organism or part made up of tissues or cells exhibiting mosaicism. [EMBRYOLOGY] An egg in which the cytoplasm of early cleavage cells is of the type which determines its later fate. { mō'zā·ik }

mosaic evolution [EVOLUTION] The tendency of one or more characters to undergo evolutionary change at different rates than other characters in a lineage. { mō¦zā·ik ‚ev·ə'lü·shən }

mosaicism [GENETICS] The coexistence in an individual of somatic cells of genetically different types; it is caused by gene or chromosome mutations, especially nondisjunction, after fertilization, by double fertilization, or by fusion of embryos. { mō'zā·ə‚siz·əm }

mosasaur [PALEONTOLOGY] Any reptile of the genus *Mosasaurus*; large, aquatic, fish-eating lizards from the Cretaceous which are related to the monitors but had paddle-shaped limbs. { 'mō·sə‚sȯr }

mosquito [INVERTEBRATE ZOOLOGY] Any member of the dipterous subfamily Culicinae; a slender fragile insect, with long legs, a long slender abdomen, and narrow wings. { mə'skēd·ō }

moss [BOTANY] Any plant of the class Bryatae, occurring in nearly all damp habitats except the ocean. { mȯs }

moss forest [ECOLOGY] *See* temperate rainforest. { 'mȯs 'fär·əst }

moss land [ECOLOGY] An area which contains abundant moss but is not wet enough to be a bog. { 'mȯs ‚land }

Motacillidae [VERTEBRATE ZOOLOGY] The pipits, a family of passeriform birds in the suborder Oscines. { ‚mōd·ə'sil·ə‚dē }

moth [INVERTEBRATE ZOOLOGY] Any of various nocturnal or crepuscular insects belonging to the lepidopteran suborder Heteroneura; typically they differ from butterflies in having the antennae feathery and rarely clubbed, a stouter body, less brilliant coloration, and proportionately smaller wings. { mȯth }

mother-of-pearl [INVERTEBRATE ZOOLOGY] The pearly iridescent internal layer of the shell of various pearl-bearing bivalve mollusks. { 'məth·ər əv 'pərl }

motile [BIOLOGY] Being capable of spontaneous movement. { mōd·əl }

motility symbiosis [ECOLOGY] A symbiotic relationship in which motility is conferred upon an organism by its symbiont. { mō¦til·əd·ē ‚sim·bē'ō·səs }

motoneuron [PHYSIOLOGY] *See* motor neuron. { ¦mōd·ə'nu̇r‚än }

motor [PHYSIOLOGY] **1.** That which causes action

or movement. **2.** Pertaining to efferent nerves which innervate muscles and glands. { 'mōd·ər }

motor area [ANATOMY] The ascending frontal gyrus containing nerve centers for voluntary movement; characterized by the presence of Betz cells. Also known as Broadman's area 4; motor cortex; pyramidal area. { 'mōd·ər ˌer·ē·ə }

motor cell [BOTANY] *See* bulliform cell. [PHYSIOLOGY] An efferent nerve cell in the anterior horn of the spinal cord. { 'mōd·ər ˌsel }

motor cortex [ANATOMY] *See* motor area. { 'mōd·ər 'kȯrˌteks }

motor end plate [ANATOMY] A specialized area beneath the sarcolemma where functional contact is made between motor nerve fibers and muscle fibers. { 'mōd·ər 'end ˌplāt }

motor nerve [PHYSIOLOGY] A nerve composed wholly or principally of motor fibers. { 'mōd·ər ˌnərv }

motor neuron [PHYSIOLOGY] An efferent nerve cell. Also known as motoneuron. { 'mōd·ər 'nu̇r ˌän }

motor speech area [ANATOMY] The cortical area located in the triangular and opercular portions of the inferior frontal gyrus; in right-handed people it is more developed on the left side. { 'mōd·ər 'spēch ˌer·ē·ə }

motor system [PHYSIOLOGY] Any portion of the nervous system that regulates and controls the contractile activity of muscle and the secretory activity of glands. { 'mōd·ər ˌsis·təm }

motor unit [ANATOMY] The axon of an anterior horn cell, or the motor fiber of a cranial nerve, together with the striated muscle fibers innervated by its terminal branches. { 'mōd·ər ˌyü·nət }

mountain lion [VERTEBRATE ZOOLOGY] *See* puma. { 'mau̇nt·ən ˌlī·ən }

mouse [VERTEBRATE ZOOLOGY] Any of various rodents which are members of the families Muridae, Heteromyidae, Cricetidae, and Zapodidae; characterized by a pointed snout, short ears, and an elongated body with a long, slender, sparsely haired tail. { mau̇s }

mousebane [BOTANY] *See* aconite. { 'mau̇s ˌbān }

mouse deer [VERTEBRATE ZOOLOGY] *See* chevrotain. { 'mau̇s ˌdir }

mouth [ANATOMY] The oral or buccal cavity and its related structures. { mau̇th }

MSH [BIOCHEMISTRY] *See* melanocyte-stimulating hormone.

mtDNA [BIOCHEMISTRY] *See* mitochondrial deoxyribonucleic acid.

muc-, muci-, muco- [ZOOLOGY] A combining form denoting pertaining to mucus, mucin, mucosa. { myük,'myü·sē, 'myü·kō }

Mucedinaceae [MYCOLOGY] The equivalent name for Moniliaceae. { myüˌsed·ən'ās·ēˌē }

mucigel [BIOCHEMISTRY] A complex polysaccharide material that is composed of root mucilage and bacterial slime and acts to control aggregation of soil particles in the rhizosphere in the vicinity of older portions of plant roots. { 'myü·səˌjel }

mucigen [BIOCHEMISTRY] A substance from which mucin is derived; contained in mucus-secreting epithelial cells. { 'myü·sə·jən }

mucin [BIOCHEMISTRY] A glycoprotein constituent of mucus and various other secretions of humans and lower animals. { 'myü·sən }

mucocutaneous [ANATOMY] Pertaining to a mucous membrane and the skin, and to the line where these join. { ¦myü·kō·kyü'tā·nē·əs }

mucoid [BIOCHEMISTRY] **1.** Any of various glycoproteins, similar to mucins but differing in solubilities and precipitation properties and found in cartilage, in the crystalline lens, and in white of egg. **2.** Resembling mucus. [MICROBIOLOGY] Pertaining to large colonies of bacteria characterized by being moist and sticky. { 'myüˌkȯid }

mucolytic [BIOCHEMISTRY] Effecting the dissolution, liquefaction, or dispersion of mucus and mucopolysaccharides. { ˌmyü·kə'lid·ik }

mucopolysaccharide [BIOCHEMISTRY] Any of a group of polysaccharides containing an amino sugar and uronic acid; a constituent of mucoproteins, glycoproteins, and blood-group substances. { ¦myü·kōˌpäl·ē'sak·əˌrīd }

mucoprotein [BIOCHEMISTRY] Any of a group of glycoproteins containing a sugar, usually chondroitinsulfuric or mucoitinsulfuric acid, combined with amino acids or polypeptides. { ¦myü·kō'prōˌtēn }

Mucorales [MYCOLOGY] An order of terrestrial fungi in the class Phycomycetes, characterized by a hyphal thallus and nonmotile sporangiospores, or conidiospores. { ˌmyü·kə'rā·lēz }

mucosa [HISTOLOGY] A mucous membrane. { myü'kō·sə }

mucous [PHYSIOLOGY] Of or pertaining to mucus; secreting mucus. { 'myü·kəs }

mucous cell [PHYSIOLOGY] A mucus-secreting cell. { 'myü·kəs ˌsel }

mucous connective tissue [HISTOLOGY] A type of loose connective tissue in which the ground substance is especially prominent and soft; occurs in the umbilical cord. { 'myü·kəs kə'nek·tiv 'tish·ü }

mucous epithelium [EMBRYOLOGY] The epidermis of an embryo, excluding the epitrichium. [HISTOLOGY] The germinative layer of a stratified squamous epithelium. { 'myü·kəs ˌep·ə'thē·lē·əm }

mucous gland [PHYSIOLOGY] A gland that secretes mucus. { 'myü·kəs ˌgland }

mucous membrane [HISTOLOGY] The type of membrane lining cavities and canals which have communication with air; it is kept moist by glandular secretions. Also known as tunica mucosa. { 'myü·kəs 'memˌbrān }

mucro [BIOLOGY] An abrupt, sharp terminal tip or process. { 'myü·krō }

mucronate [BIOLOGY] Terminated abruptly by a sharp terminal tip or process. { 'myü·krəˌnāt }

mucus [PHYSIOLOGY] A viscid fluid secreted by mucous glands, consisting of mucin, water, inorganic salts, epithelial cells, and leukocytes held in suspension. { 'myü·kəs }

mudfish [VERTEBRATE ZOOLOGY] *See* bowfin. { 'mədˌfish }

mud puppy [VERTEBRATE ZOOLOGY] Any of several American salamanders of the genera *Necturus* and *Proteus* making up the family Proteidae; distinguished by having both lungs and gills as an adult. { 'məd ˌpəp·ē }

Mugilidae [VERTEBRATE ZOOLOGY] The mullets, a family of perciform fishes in the suborder Mugiloidei. { myü'jil·əˌdē }

Mugiloidei [VERTEBRATE ZOOLOGY] A suborder of fishes in the order Perciformes; individuals are rather elongate, terete fishes with a short spinous dorsal fin that is well separated from the soft dorsal fin. { ˌmyü·jə'lȯid·ēˌī }

mulberry [BOTANY] Any of various trees of the genus *Morus* (family Moraceae), characterized by milky sap and simple, often lobed alternate leaves. { 'məlˌber·ē }

mule [VERTEBRATE ZOOLOGY] The sterile hybrid offspring of the male ass and the mare, or female horse. { myül }

Müllerian duct [EMBRYOLOGY] *See* paramesonephric duct. { mi'ler·ē·ən 'dəkt }

Müllerian mimicry [ZOOLOGY] Mimicry between two aposematic species. { mi'ler·ē·ən 'mim·ə·krē }

Müller's larva [INVERTEBRATE ZOOLOGY] The ciliated larva characteristic of various members of the Polycladida; resembles a modified ctenophore. { 'mil·ərz ˌlär·və }

multicellular [BIOLOGY] Consisting of many cells. { ¦məl·tē'sel·yə·lər }

multicipital [BIOLOGY] Having many heads or branches arising from one point. { ˌməl·tə'sip·əd·əl }

multicompartmental genome [VIROLOGY] In certain viruses, separation of genetic information into encapsulated nucleic acid molecules. { ˌməl·tə·kəmˌpärt'ment·əl 'jēˌnōm }

multifid [BIOLOGY] Divided into many lobes. { 'məl·təˌfid }

multigene family [GENETICS] A set of genes that arose from duplication and variation of a single ancestral gene. { 'məl·təˌjēn ˌfam·lē }

multiglandular [ANATOMY] Of or pertaining to several glands. { ˌməl·tə'glan·jə·lər }

Multillidae [INVERTEBRATE ZOOLOGY] An economically important family of Hymenoptera; includes the cow killer, a parasite of bumblebee pupae. { məl'til·əˌdē }

multilocular [BIOLOGY] Having many small chambers or vesicles. { ¦məl·tē'äk·yə·lər }

multimer [BIOCHEMISTRY] A protein molecule composed of two or more monomers. { 'məl·tə·mər }

multiple epidermis [BOTANY] Epidermis that is several layers thick, occurring in many species of *Ficus*, *Begonia*, and *Peperomia*. { 'məl·tə·pəl ˌep·ə'dər·məs }

multiple fruit [BOTANY] Any fruit derived from the ovaries and accessory structures of several flowers consolidated into one mass, such as a pineapple and mulberry. { 'məl·tə·pəl 'früt }

Multituberculata [PALEONTOLOGY] The single order of the nominally mammalian suborder Allotheria; multituberculates had enlarged incisors, the coracoid bones were fused to the scapula, and the lower jaw consisted of the dentary bone alone. { ˌməl·tē·təˌbər·kyə'läd·ə }

multituberculate [VERTEBRATE ZOOLOGY] Of teeth, having several or many simple conical cusps. { ˌməl·tē·təˌbər·kyə'lāt }

mu phage [VIROLOGY] A temperate phage with properties similar to those of transposable genetic elements. { 'myü ˌfāj }

muramic acid [BIOCHEMISTRY] An organic acid found in the mucopeptide (murein) in the cell walls of bacteria and blue-green algae. { myu̇'ram·ik 'as·əd }

muramidase [BIOCHEMISTRY] Lysozyme when acting as an enzyme on the hydrolysis of the muramic acid-containing mucopeptide in the cell walls of some bacteria. { myu̇'ram·əˌdās }

Murchisoniacea [PALEONTOLOGY] An extinct superfamily of gastropod mollusks in the order Prosobranchia. { ¦mər·chəˌsən·ē'ā·shē·ə }

murein [BIOCHEMISTRY] The peptidoglycan of bacterial cell walls. { myu̇r·ē·ən }

Muricacea [INVERTEBRATE ZOOLOGY] A superfamily of gastropod mollusks in the order Prosobranchia. { ˌmyu̇r·ə'ka·shē·ə }

muricate [ZOOLOGY] Covered with sharp, hard points. { 'myu̇r·əˌkāt }

Muricidae [INVERTEBRATE ZOOLOGY] A family of predatory gastropod mollusks in the order Neogastropoda; contains the rock snails. { ˌmyu̇'ris·əˌdē }

Muridae [VERTEBRATE ZOOLOGY] A large diverse family of relatively small cosmopolitan rodents; distinguished from closely related forms by the absence of cheek pouches. { 'myu̇r·əˌdē }

muriform [BIOLOGY] **1.** Resembling the arrangement of courses in a brick wall, especially having both horizontal and vertical septa. **2.** Pertaining to or resembling a rat or mouse. { 'myu̇r·əˌfȯrm }

Murinae [VERTEBRATE ZOOLOGY] A subfamily of the Muridae which contains such forms as the striped mouse, house mouse, harvest mouse, and field mouse. { myu̇'rīˌnē }

Musaceae [BOTANY] A family of monocotyledonous plants in the order Zingiberales characterized by five functional stamens, unisexual flowers, spirally arranged leaves and bracts, and fleshy, indehiscent fruit. { myü'zā·sēˌē }

Musci [BOTANY] *See* Bryopsida. { 'məˌsī }

Muscicapidae [VERTEBRATE ZOOLOGY] A family of passeriform birds assigned to the Oscines; includes the Old World flycatchers or fantails. { ˌməs·ə'kap·əˌdē }

Muscidae [INVERTEBRATE ZOOLOGY] A family of myodarian cyclorrhaphous dipteran insects in the subsection Calypteratae; includes the houseflies, stable flies, and allies. { 'məs·əˌdē }

muscle [ANATOMY] A contractile organ composed of muscle tissue that changes in length and effects movement when stimulated. [HISTOLOGY] A tissue composed of cells containing contractile fibers; three types are smooth, cardiac, and skeletal. { 'məs·əl }

muscle fiber [HISTOLOGY] The contractile cell or unit of which muscle is composed. { 'məs·əl ˌfī·bər }

muscle hemoglobin [BIOCHEMISTRY] *See* myoglobin. { 'məs·əl 'hē·mə,glō·bən }

muscle tone [PHYSIOLOGY] *See* tonus. { 'məs·əl ,tōn }

muscul-, musculo- [ZOOLOGY] A combining form denoting muscle, muscular. { 'məs·kə·lō }

muscularis externa [HISTOLOGY] The layer of the digestive tube consisting of smooth muscles. { ,məs·kyə'lar·əs ek'stər·nə }

muscularis mucosae [HISTOLOGY] Thin, deep layer of smooth muscle in some mucous membranes, as in the digestive tract. { ,məs·kyə'lar·əs myü'kō·sē }

muscular system [ANATOMY] The muscle cells, tissues, and organs that effect movement in all vertebrates. { 'məs·kyə·lər ,sis·təm }

musculoaponeurotic [HISTOLOGY] Composed of muscle and of fibrous connective tissue in the form of a membrane. { ¦məs·kyə·lō¦ap·ə·nü¦räd·ik }

musculocutaneous [ANATOMY] Of or pertaining to muscles and skin. { ¦məs·kyə·lō·kyü'tā·nē·əs }

musculocutaneous nerve [ANATOMY] A branch of the brachial plexus with both motor and somatic sensory components; innervates flexor muscles of the upper arm, and skin of the lateral aspect of the forearm. { ¦məs·kyə·lō·kyü'tā·nē·əs 'nərv }

musculoskeletal system [ANATOMY] The muscular and skeletal elements of vertebrates, considered as a functional unit. { ¦məs·kyə·lō'skel·ə·təl ,sis·təm }

mushroom [MYCOLOGY] **1.** A fungus belonging to the basidiomycetous order Agaricales. **2.** The fruiting body (basidiocarp) of such a fungus. { 'məsh,rüm }

Musidoridae [INVERTEBRATE ZOOLOGY] A family of orthorrhaphous dipteran insects in the series Brachycera distinguished by spear-shaped wings. { ,myü·zə'dȯr·ə,dē }

musk [PHYSIOLOGY] Any of various strong-smelling substances obtained from the musk glands of musk deer or similar animals; used in the form of a tincture as a fixative for perfume. { məsk }

musk bag [VERTEBRATE ZOOLOGY] *See* musk gland. { 'məsk ,bag }

muskeg [ECOLOGY] A peat bog or tussock meadow, with variably woody vegetation. { 'mə ,skeg }

musk gland [VERTEBRATE ZOOLOGY] A large preputial scent gland of the musk deer and various other animals, including skunk and musk-ox. Also known as musk bag. { 'məsk ,gland }

muskmelon [BOTANY] *Cucumis melo.* The edible, fleshy, globular to long-tapered fruit of a trailing annual plant of the order Violales; surface is uniform to broadly sutured to wrinkled, and smooth to heavily netted, and flesh is pale green to orange; varieties include cantaloupe, Honey Dew, Casaba, and Persian melons. { 'məsk,mel·ən }

musk-ox [VERTEBRATE ZOOLOGY] *Ovibos moschatus.* An even-toed ungulate which is a member of the mammalian family Bovidae; a heavy-set animal with a shag pilage, splayed feet, and flattened horns set low on the head. { 'məs,käks }

muskrat [VERTEBRATE ZOOLOGY] *Ondatra zibethica.* The largest member of the rodent subfamily Microtinae; essentially a water rat with a laterally flattened, long, naked tail, a broad blunt head with short ears, and short limbs. { 'mə,skrat }

Musophagidae [VERTEBRATE ZOOLOGY] The turacos, an African family of birds of uncertain affinities usually included in the order Cuculiformes; resemble the cuckoos anatomically but have two unique pigments, turacin and turacoverdin. { ,myü·zə'faj·ə,dē }

Muspiceoidea [INVERTEBRATE ZOOLOGY] A superfamily of parasitic nematodes in the order Dioctophymatida, distinguished by a greatly reduced neurosensory structure, the absence of (except in one species) amphids and cephalic papillae, and, in females, a reduced digestive tube; males have never been reported. { myüs ,pī·sē'ȯid·ē·ə }

mustard [BOTANY] Any of several annual crucifers belonging to the genus *Brassica* of the order Capparales; leaves are lyrately lobed, flowers are yellow, and pods have linear beaks; the mustards are cultivated for their pungent seed and edible foliage, and the seeds of *B. niger* are used as a condiment, prepared as a powder, paste, or oil. { 'məs·tərd }

Mustilidae [VERTEBRATE ZOOLOGY] A large, diverse family of low-slung, long-bodied carnivorous mammals including minks, weasels, and badgers; distinguished by having only one molar in each upper jaw, and two at the most in the lower jaw. { mə'stil·ə,dē }

mutable gene [GENETICS] Any of a class of unstable genes that spontaneously mutate at a sufficiently high rate to produce mosaicism. { ¦myüd·ə·bəl 'jēn }

mutagen [GENETICS] An agent that raises the frequency of mutation above the spontaneous rate. { 'myüd·ə·jən }

mutagen persistence [GENETICS] The stability of a mutagen in the environment or in the human body. { ¦myüd·ə·jən pər'sis·təns }

mutagen specificity [GENETICS] The tendency of a mutagen to induce only one type of mutation. { ¦myüd·ə·jən ,spes·ə'fis·əd·ē }

mutant [GENETICS] An individual bearing an allele that has undergone mutation and is expressed in the phenotype. { 'myüt·ənt }

mutase [BIOCHEMISTRY] An enzyme able to catalyze a dismutation or a molecular rearrangement. { 'myü,tās }

mutation [GENETICS] An abrupt change in the genotype of an organism, not resulting from recombination; genetic material may undergo qualitative or quantitative alteration, or rearrangement. { myü'tā·shən }

mutation fixation [MOLECULAR BIOLOGY] The condition of changing a premutational deoxyribonucleic acid lesion to mutation. { myü'tā·shən fik,sā·shən }

mutation hot spot [MOLECULAR BIOLOGY] Any locus in the deoxyribonucleic acid sequence or on a chromosome where mutations or aberrations occur preferentially. { myü¦tā·shən 'hät ,spöt }

Mutillidae [INVERTEBRATE ZOOLOGY] The velvet ants, a family of hymenopteran insects in the superfamily Scolioidea. { myü'til·ə,dē }

muton [MOLECULAR BIOLOGY] The smallest unit of genetic material capable of undergoing mutation. { 'myü,tän }

mutualism [ECOLOGY] Mutual interactions between two species that are beneficial to both species. { 'myü·chə·wə,liz·əm }

muzzle [VERTEBRATE ZOOLOGY] The snout of an animal, as a dog or horse. { 'məz·əl }

Mycelia Sterilia [MYCOLOGY] An order of fungi of the class Fungi Imperfecti distinguished by the lack of spores; certain members are plant pathogens. { mī'sel·yə stə'ril·yə }

mycelium [BIOLOGY] A mass of filaments, or hyphae, composing the vegetative body of many fungi and some bacteria. { mī'sē·lē·əm }

Mycetaeidae [INVERTEBRATE ZOOLOGY] The equivalent name for Endomychidae. { ,mī·sə 'tē·ə,dē }

mycetocyte [INVERTEBRATE ZOOLOGY] **1.** One of the cells clustered together to form a mycetome. **2.** An individual cell functioning like a mycetome. { mī'sēd·ə,sīt }

mycetome [INVERTEBRATE ZOOLOGY] One of the specialized structures in the body of certain insects for holding endosymbionts. { 'mī·sə ,tōm }

Mycetophagidae [INVERTEBRATE ZOOLOGY] The hairy fungus beetles, a cosmopolitan family of coleopteran insects in the superfamily Cucujoidea. { mī,sēd·ə'faj·ə,dē }

Mycetozoa [BIOLOGY] A zoological designation for organisms that exhibit both plant and animal characters during their life history (Myxomycetes); equivalent to the botanical Myxomycophyta. { mī,sēd·ə'zō·ə }

Mycetozoia [INVERTEBRATE ZOOLOGY] A subclass of the protozoan class Rhizopodea. { mī,sēd·ə 'zöi·ə }

Mycobacteriaceae [MICROBIOLOGY] A family of bacteria in the order Actinomycetales; acid-fast, aerobic rods form a filamentous or myceliumlike growth. { ,mī·kō,bak·tir·ē'ās·ē,ē }

mycobactin [BIOCHEMISTRY] Any compound produced by some strains, and required for growth by other strains, of *Mycobacteria*. { ¦mī·kō¦bak·tən }

mycobiont [BOTANY] The fungal component of a lichen, commonly an ascomycete. { ¦mī·kə'bī ,änt }

mycology [BOTANY] The branch of botany that deals with the study of fungi. { mī'käl·ə·jē }

mycomycin [MICROBIOLOGY] $C_{13}H_{10}O_2$ An antibiotic produced by *Nocardia acidophilus* and a species of *Actinomyces*; characterized as a highly unsaturated aliphatic acid that shows strong activity against *Mycobacterium tuberculosis*. { ,mīk·ə 'mīs·ən }

mycophagous [ZOOLOGY] Feeding on fungi. { mī 'käf·ə·gəs }

Mycophiformes [VERTEBRATE ZOOLOGY] An equivalent name for Salmoniformes. { mī,käf·ə 'fór·mēz }

Mycoplasmataceae [MICROBIOLOGY] A family of the order Mycoplasmatales; distinguished by sterol requirement for growth. { ,mī·kō,plaz·mə 'tās·ē,ē }

Mycoplasmatales [MICROBIOLOGY] The single order of the class Mollicutes; organisms are gram-negative, generally nonmotile, nonsporing bacteria which lack a true cell wall. { ,mī·kō,plaz·mə 'tā·lēz }

mycorrhiza [BOTANY] A mutual association in which the mycelium of a fungus invades the roots of a seed plant. { ,mīk·ə'rīz·ə }

Mycota [MYCOLOGY] An equivalent name for Eumycetes. { mī'käd·ə }

mycotoxin [MYCOLOGY] A toxin produced by a fungus. { 'mī·kə,täk·sən }

Myctophidae [VERTEBRATE ZOOLOGY] The lantern fishes, a family of deep-sea forms of the suborder Myctophoidei. { mik'täf·ə,dē }

Myctophoidei [VERTEBRATE ZOOLOGY] A large suborder of marine salmoniform fishes characterized by having the upper jaw bordered only by premaxillae, and lacking a mesocoracoid arch in the pectoral girdle. { mik·tə'föid·ē,ī }

Mydaidae [INVERTEBRATE ZOOLOGY] The mydas flies, a family of orthorrhaphous dipteran insects in the series Brachycera. { mī'dā·ə,dē }

myel-, myelo- [ANATOMY] A combining form indicating relationship to marrow, often in specific reference to the spinal cord. { 'mī·əl, 'mī·ə·lō }

myelencephalon [EMBRYOLOGY] The caudal portion of the hindbrain; gives rise to the medulla oblongata. { ,mī·ə·lən'sef·ə,län }

myelin [BIOCHEMISTRY] A soft, white fatty substance that forms a sheath around certain nerve fibers. { 'mī·ə·lən }

myelin sheath [HISTOLOGY] An investing cover of myelin around the axis cylinder of certain nerve fibers. { 'mī·ə·lən 'shēth }

myeloblast [HISTOLOGY] The youngest precursor cell for blood granulocytes, having a nucleus with finely granular chromatin and nucleoli and intensely basophilic cytoplasm. { 'mī·ə·lō ,blast }

myeloblastic [HISTOLOGY] Of, pertaining to, or characterized by the presence of myeloblasts. { ,mī·ə·lə'blas·tik }

myelocele [ANATOMY] The canal of the spinal cord. { 'mī·ə·lō,sēl }

myelocyte [HISTOLOGY] A motile precursor cell of blood granulocytes found in bone marrow. { 'mī·ə·lə,sīt }

myeloid [ANATOMY] **1.** Of or pertaining to bone marrow. **2.** Of or pertaining to the spinal cord. { 'mī·ə,lóid }

myeloid tissue [HISTOLOGY] Red bone marrow attached to argyrophile fibers which form wide meshes containing scattered fat cells, erythroblasts, myelocytes, and mature myeloid elements. { 'mī·ə,lóid ,tish·ü }

myelomonocyte [HISTOLOGY] **1.** A monocyte developing in bone marrow. **2.** A blood cell intermediate between monocytes and granulocytes. { ¦mī·ə·lō'män·ə,sīt }

myelopoiesis [PHYSIOLOGY] The process by which blood cells form in the bone marrow. { ˌmī·ə·lōˌpȯi'ē·səs }

myenteric [HISTOLOGY] Of or pertaining to the muscular coat of the intestine. { ¦mī·ən¦ter·ik }

myenteric plexus [ANATOMY] A network of nerves between the circular and longitudinal layers of the muscular coat of the digestive tract. Also known as Auerbach's plexus. { ¦mī·ən¦ter·ik 'plek·səs }

myenteron [HISTOLOGY] The muscular coat of the intestine. { mī'ent·əˌrän }

Mygalomorphae [INVERTEBRATE ZOOLOGY] A suborder of spiders (Araneida) including American tarantulas, trap-door spiders, and purse-web spiders; the tarantulas may attain a leg span of 10 inches (25 centimeters). { ˌmig·ə·lō'mȯrˌfē }

Mylabridae [INVERTEBRATE ZOOLOGY] The equivalent name for Bruchidae. { mə'lab·rəˌdē }

Myliogatidae [VERTEBRATE ZOOLOGY] The eagle rays, a family of batoids which may reach a length of 15 feet (4.6 meters). { ˌmil·ē·ə'gad·ə ˌdē }

Mymaridae [INVERTEBRATE ZOOLOGY] The fairy flies, a family of hymenopteran insects in the superfamily Chalcidoidea. { mī'mar·əˌdē }

myoblast [EMBRYOLOGY] A precursor cell of a muscle fiber. { 'mī·əˌblast }

myocardium [HISTOLOGY] The muscular tissue of the heart wall. { ˌmī·ə'kärd·ē·əm }

myocoel [EMBRYOLOGY] Portion of the coelom enclosed in a myotome. { 'mī·əˌsēl }

myocomma [HISTOLOGY] A ligamentous connection between successive myomeres. Also known as myoseptum. { ˌmī·ə'käm·ə }

myocyte [HISTOLOGY] **1.** A contractile cell. **2.** A muscle cell. { 'mī·əˌsīt }

Myodaria [INVERTEBRATE ZOOLOGY] A section of the Schizophora series of cyclorrhaphous dipterans; in this group adult antennae consist of three segments, and all families except the Conopidae have the second cubitus and the second anal veins united for almost their entire length. { ˌmī·ə'dar·ē·ə }

Myodocopa [INVERTEBRATE ZOOLOGY] A suborder of the order Myodocopida; includes exclusively marine ostracods distinguished by possession of a heart. { ˌmī·ə'däk·ə·pə }

Myodocopida [INVERTEBRATE ZOOLOGY] An order of the subclass Ostracoda. { ¦mī·ə·də¦käp·ə·də }

Myodopina [INVERTEBRATE ZOOLOGY] The equivalent name for Myodocopa. { ˌmī·ə'däp·ə·nə }

myoelastic fiber [HISTOLOGY] An elastic fiber associated with the smooth muscles in bronchi and bronchioles. { ¦mī·ō·i'las·tik 'fī·bər }

myoelectric potential [PHYSIOLOGY] The electrical potential created by muscle action. { ¦mī·ō·i'lek·trik pə'ten·chəl }

myofibril [CELL BIOLOGY] A contractile fibril in a muscle cell. [INVERTEBRATE ZOOLOGY] *See* myoneme. { ¦mī·ō'fī·brəl }

myofilament [CELL BIOLOGY] The structural unit of muscle proteins in a muscle cell. { ¦mī·ō'fil·ə·mənt }

myofrisk [INVERTEBRATE ZOOLOGY] A contractile structure surrounding the spines of certain radiolarians. { 'mī·ōˌfrisk }

myoglobin [BIOCHEMISTRY] A hemoglobinlike iron-containing protein pigment occurring in muscle fibers. Also known as muscle hemoglobin; myohemoglobin. { ¦mī·ə¦glō·bən }

myohematin [BIOCHEMISTRY] A cytochrome respiratory enzyme allied to hematin. { ¦mī·ō'hē·mə·tən }

myohemoglobin [BIOCHEMISTRY] *See* myoglobin. { ¦mī·ō'hē·məˌglō·bən }

myoinositol [BIOCHEMISTRY] The commonest isomer of inositol. Also known as mesionositol. { ¦mī·ō·i'näs·əˌtȯl }

myokinase [BIOCHEMISTRY] An enzyme that catalyzes the reversible transfer of phosphate groups in adenosinediphosphate; occurs in muscle and other tissues. { ¦mī·ō'kīˌnās }

myomere [EMBRYOLOGY] A muscle segment differentiated from the myotome, which divides to form the epimere and hypomere. { 'mī·əˌmir }

myometrium [HISTOLOGY] The muscular tissue of the uterus. { ˌmī·ə'mē·trē·əm }

Myomorpha [VERTEBRATE ZOOLOGY] A suborder of rodents recognized in some systems of classification. { ˌmī·ə'mȯr·fə }

myoneme [INVERTEBRATE ZOOLOGY] A contractile fibril in a protozoan. Also known as myofibril. { 'mī·əˌnēm }

myoneural junction [ANATOMY] The point of junction of a motor nerve with the muscle which it innervates. Also known as neuromuscular junction. { ¦mī·ō'nu̇r·əl 'jəŋk·shən }

Myopsida [INVERTEBRATE ZOOLOGY] A natural assemblage of cephalopod mollusks considered as a suborder in the order Teuthoida according to some systems of classification, and a group of the Decapoda according to other systems; the eye is covered by the skin of the head in all species. { mī'äp·sə·də }

myoseptum [HISTOLOGY] *See* myocomma. { ¦mī·ō'sep·təm }

myosin [BIOCHEMISTRY] A muscle protein, comprising up to 50% of the total muscle proteins; combines with actin to form actomycin. { 'mī·ə·sən }

myostatic reflex [PHYSIOLOGY] *See* stretch reflex. { ¦mī·ə¦stad·ik 'rēˌfleks }

myotasis [PHYSIOLOGY] Stretching of a muscle. { mī'äd·ə·səs }

myotome [ANATOMY] A group of muscles innervated by a single spinal nerve. [EMBRYOLOGY] The muscle plate that differentiates into myomeres. { 'mī·əˌtōm }

Myriangiales [MYCOLOGY] An order of parasitic fungi of the class Ascomycetes which produce asci at various levels in uniascal locules within stromata. { ˌmir·ēˌan·jē'ā·lēz }

Myriapoda [INVERTEBRATE ZOOLOGY] Informal designation for those mandibulate arthropods having two body tagmata, one pair of antennae, and more than three pairs of adult ambulatory appendages. { ˌmir·ē'äp·ə·də }

Myricaceae [BOTANY] The single family of the plant order Myricales. { ˌmir·ə'kās·ēˌē }

Myricales [BOTANY] An order of dicotyledonous

plants in the subclass Hamamelidae, marked by its simple, resinous-dotted, aromatic leaves, and a unilocular ovary with two styles and a single ovule. { ˌmir·əˈkā·lēz }

Myrientomata [INVERTEBRATE ZOOLOGY] The equivalent name for the Protura. { ˌmir·ē·ən ˈtäm·əd·ə }

Myriotrochidae [INVERTEBRATE ZOOLOGY] A family of holothurian echinoderms in the order Apodida, distinguished by eight or more spokes in each wheel-shaped spicule. { ˌmir·ē·əˈtrō·kə ˌdē }

Myrmecophagidae [VERTEBRATE ZOOLOGY] A small family of arboreal anteaters in the order Edentata. { ˌmər·mə·kōˈfaj·əˌdē }

myrmecophagous [ZOOLOGY] Feeding on ants. { ˌmər·məˈkäf·ə·gəs }

myrmecophile [ECOLOGY] An organism, usually a beetle, that habitually inhabits the nest of ants. { mərˈmek·əˌfīl }

myrmecophyte [ECOLOGY] A plant that houses and benefits from the habitation of ants. { mər ˈmek·əˌfīt }

Myrmeleontidae [INVERTEBRATE ZOOLOGY] The ant lions, a family of insects in the order Neuroptera; larvae are commonly known as doodlebugs. { ˌmər·mə·lēˈän·təˌdē }

Myrmicinae [INVERTEBRATE ZOOLOGY] A large diverse subfamily of ants (Formicidae); some members are inquilines and have no worker caste. { mərˈmis·əˌnē }

Myrsinaceae [BOTANY] A family of mostly woody dicotyledonous plants in the order Primulales characterized by flowers without staminodes, a schizogenous secretory system, and gland-dotted leaves. { ˌmər·səˈnās·ēˌē }

Myrtaceae [BOTANY] A family of dicotyledonous plants in the order Myrtales characterized by an inferior ovary, numerous stamens, anthers usually opening by slits, and fruit in the form of a berry, drupe, or capsule. { mərˈtās·ēˌē }

Myrtales [BOTANY] An order of dicotyledonous plants in the subclass Rosidae characterized by opposite, simple, entire leaves and perigynous to epigynous flowers with a compound pistil. { mərˈtā·lēz }

Mysida [INVERTEBRATE ZOOLOGY] A suborder of the crustacean order Mysidacea characterized by fusion of the sixth and seventh abdominal somites in the adult, lack of gills, and other specializations. { ˈmī·sə·də }

Mysidacea [INVERTEBRATE ZOOLOGY] An order of free-swimming Crustacea included in the division Pericarida; adult consists of 19 somites, each bearing one pair of functionally modified, biramous appendages, and the carapace envelops most of the thorax and is fused dorsally with up to four of the anterior thoracic segments. { ˌmī·səˈdās·ē·ə }

mysis [INVERTEBRATE ZOOLOGY] A larva of certain higher crustaceans, characterized by biramous thoracic appendages. { ˈmī·səs }

Mystacinidae [VERTEBRATE ZOOLOGY] A monospecific family of insectivorous bats (Chiroptera) containing the New Zealand short-tailed bat; hindlegs and body are stout, and fur is thick. { ˌmis·təˈsin·əˌdē }

Mystacocarida [INVERTEBRATE ZOOLOGY] An order of primitive Crustacea, the body is wormlike and the cephalothorax bears first and second antennae, mandibles, and first and second maxillae. { ˌmis·tə·kōˈkar·ə·də }

Mysticeti [VERTEBRATE ZOOLOGY] The whalebone whales, a suborder of the mammalian order Cetacea, distinguished by horny filter plates of suspended from the upper jaws. { ˌmis·təˈsēˌtī }

Mytilacea [INVERTEBRATE ZOOLOGY] A suborder of bivalve mollusks in the order Filibranchia. { ˌmid·əlˈā·shē·ə }

Mytilidae [INVERTEBRATE ZOOLOGY] A family of mussels in the bivalve order Anisomyaria. { mī ˈtil·əˌdē }

myx-, myxo- [ZOOLOGY] A combining form denoting mucus, mucous, mucin, mucinous. { ˈmik·sō }

myxameba [BIOLOGY] An independent ameboid cell of the vegetative phase of Acrasiales. { ˌmiks·əˈmē·bə }

Myxicolinae [INVERTEBRATE ZOOLOGY] A subfamily of sedentary polychaete annelids in the family Sabellidae. { ˌmik·səˈkäl·əˌnē }

Myxiniformes [VERTEBRATE ZOOLOGY] The equivalent name for the Myxinoidea. { ˌmik·sə·nə ˈfȯr·mēz }

Myxinoidea [VERTEBRATE ZOOLOGY] The hagfishes, an order of eellike, jawless vertebrates (Agnatha) distinguished by having the nasal opening at the tip of the snout and leading to the pharynx, with barbels around the mouth and 6-15 pairs of gill pouches. { ˌmik·səˈnȯid·ē·ə }

Myxobacterales [MICROBIOLOGY] An order of gliding bacteria; unicellular, gram-negative rods embedded in a layer of slime and capable of gliding movement; form fruiting bodies containing resting cells (myxospores) under certain environmental conditions. { ˌmik·səˌbak·təˈrā·lēz }

Myxococcaceae [MICROBIOLOGY] A family of the order Myxobacterales; vegetative cells are straight to slightly tapered, and spherical to ovoid microcysts (myxospores) are produced. { ˌmik·səˌkäkˈsās·ēˌē }

Myxogastromycetidae [MYCOLOGY] A large subclass of plasmodial slime molds (Myxomycetes). { ¦mik·səˌgas·trōˌmīˈsed·əˌdē }

Myxomycetes [BIOLOGY] Plasmodial (acellular or true) slime molds, a class of microorganisms of the division Mycota; they are on the borderline of the plant and animal kingdoms and have a noncellular, multinucleate, jellylike, creeping, assimilative stage (the plasmodium) which alternates with a myxameba stage. { ˌmik·səˌmī ˈsēd·ēz }

Myxomycophyta [BOTANY] An order of microorganisms, equivalent to the Mycetozoia of zoological classification. { ˌmik·səˌmīˈkäf·əd·ə }

Myxophaga [INVERTEBRATE ZOOLOGY] A suborder of the Coleoptera. { mikˈsäf·ə·gə }

Myxophyceae [BOTANY] An equivalent name for the Cyanophyceae. { ˌmik·səˈfī·sēˌē }

Myxosporida [INVERTEBRATE ZOOLOGY] An order of the protozoan class Myxosporidea characterized by the production of spores with one or more valves and polar capsules, and by possession of a single sporoplasm. { ˌmik·səˈspȯr·ə·də }

Myxosporidea [INVERTEBRATE ZOOLOGY] A class of the protozoan subphylum Cnidospora; members are parasitic in some fish, a few amphibians, and certain invertebrates. { ˌmik·sə·spəˈrid·ē·ə }

myxovirus [VIROLOGY] A group of ribonucleic-acid animal viruses characterized by hemagglutination and hemadsorption; includes influenza and fowl plague viruses and the paramyxoviruses. { ˈmik·səˌvī·rəs }

Myzopodidae [VERTEBRATE ZOOLOGY] A monospecific order of insectivorous bats (Chiroptera) containing the Old World disk-winged bat of Madagascar; characterized by long ears and by a vestigial thumb with a monostalked sucking disk. { ˌmī·zəˈpäd·əˌdē }

myzorhynchus [INVERTEBRATE ZOOLOGY] An apical sucker on the scolex of certain tapeworms. { ˌmī·zəˈriŋ·kəs }

Myzostomaria [INVERTEBRATE ZOOLOGY] An aberrant group of Polychaeta; most are greatly depressed, broad, and very small, and true segmentation is delayed or absent in the adult; all are parasites of echinoderms. { ˌmī·zə·stəˈmar·ē·ə }

Myzostomidae [INVERTEBRATE ZOOLOGY] A monogeneric family of the Myzostomaria. { ˌmī·zəˈstäm·əˌdē }

N

Nabidae [INVERTEBRATE ZOOLOGY] The damsel bugs, a family of hemipteran insects in the superfamily Cimicimorpha. { 'nab·ə‚dē }

Nabothian glands [ANATOMY] Mucous glands of the uterine cervix. { nə'bō·thē·ən 'glanz }

nacre [INVERTEBRATE ZOOLOGY] An iridescent inner layer of many mollusk shells. { 'nā·kər }

NAD [BIOCHEMISTRY] *See* diphosphopyridine nucleotide.

nail [ANATOMY] The horny epidermal derivative covering the dorsal aspect of the terminal phalanx of each finger and toe. { nāl }

nail-patella syndrome [GENETICS] A genetic disorder inherited as an autosomal dominant that is part of a linkage group with ABO blood group genes; characterized by defects in the nails and abnormalities of the elbows and other bones, skin, eyes, and kidneys. { 'nāl pə'tel·ə ‚sin ‚drōm }

Najadaceae [BOTANY] A family of monocotyledonous, submerged aquatic plants in the order Najadales distinguished by branching stems and opposite or whorled leaves. { ‚nāj·ə'dās·ē‚ē }

Najadales [BOTANY] An order of aquatic and semiaquatic flowering plants in the subclass Alismatidae; the perianth, when present, is not differentiated into sepals and petals, and the flowers are usually not individually subtended by bracts. { ‚nāj·ə'dā·lēz }

naked bud [BOTANY] A bud covered only by rudimentary foliage leaves. { 'nā·kəd 'bəd }

Namanereinae [INVERTEBRATE ZOOLOGY] A subfamily of largely fresh-water errantian annelids in the family Nereidae. { ‚nā·mə‚nə'rē·ə‚nē }

nannandrous [BOTANY] Pertaining to species of plants in which male members are markedly smaller than females, such as in some algal species of *Oedogonium* that have antheridia produced in special dwarf filaments. { na'nan·drəs }

nannoplankton [BIOLOGY] Minute plankton; the smallest plankton, including algae, bacteria, and protozoans. { ¦nan·ō'plaŋk·tən }

nano- [BIOLOGY] A prefix meaning dwarfed. { 'nan·ō }

nanophanerophyte [ECOLOGY] A shrub not exceeding 6.6 feet (2 meters) in height. { ¦nan·ō 'fan·ə·rə‚fīt }

nanozooid [INVERTEBRATE ZOOLOGY] Dwarf zooid; bryozoan heterozooid possessing only a single tentacle. { ¦nan·ō¦zō‚òid }

nape [ANATOMY] The back of the neck. { nāp }

napex [ANATOMY] That portion of the scalp just below the occipital protuberance. { 'nā‚peks }

napiform [BOTANY] Turnip-shaped, referring to roots. { 'nāp·ə‚fȯrm }

Narcomedusae [INVERTEBRATE ZOOLOGY] A suborder of hydrozoan cnidarians in the order Trachylina; the hydroid generation is represented by an actinula larva. { ‚när·kə·mə'dü‚sē }

naris [ANATOMY] *See* nostril. { 'nar·əs }

narrow-spectrum antibiotic [MICROBIOLOGY] An antibiotic effective against a limited number of microorganisms. { 'nar·ō ¦spek·trəm ‚ant·i‚bī 'äd·ik }

narwhal [VERTEBRATE ZOOLOGY] *Monodon monoceros.* An arctic whale characterized by lack of a dorsal fin, and by possession in the male of a long, twisted, pointed tusk (or rarely, two tusks) which is a source of ivory. { 'när‚wäl }

nasal [ANATOMY] Of or pertaining to the nose. { 'nā·zəl }

nasal bone [ANATOMY] Either of two rectangular bone plates forming the bridge of the nose; they articulate with the frontal, ethmoid, and maxilla bones. { 'nā·zəl ‚bōn }

nasal cavity [ANATOMY] Either of a pair of cavities separated by a septum and located between the nasopharynx and anterior nares. { 'nā·zəl ¦kav·əd·ē }

nasal crest [ANATOMY] **1.** The linear prominence on the medial border of the palatal process of the maxilla. **2.** The linear prominence on the medial border of the palatine bone. **3.** The linear prominence on the internal border of the nasal bone and forming part of the nasal septum. { 'nā·zəl ¦krest }

nasal pit [EMBRYOLOGY] *See* olfactory pit. { 'nā·zəl ‚pit }

nasal process of the frontal bone [ANATOMY] The downward projection of the nasal part of the frontal bone which terminates as the nasal spine. { 'nā·zəl ¦prä·səs əv thə 'frənt·əl ‚bōn }

nasal process of the maxilla [ANATOMY] Frontal process of the maxilla. { 'nā·zəl ¦prä·səs əv thə mak'sil·ə }

nasal septum [ANATOMY] The partition separating the two nasal cavities. { 'nā·zəl ¦sep·təm }

nascent ribonucleic acid [MOLECULAR BIOLOGY] **1.** A ribonucleic acid (RNA) molecule in the process of being synthesized. **2.** A complete, newly synthesized RNA molecule before any alterations have been made. { ¦nās·ənt ¦rī·bō·nü ˌklē·ik 'as·əd }

Nasellina [INVERTEBRATE ZOOLOGY] The equivalent name for Monopylina. { ˌnas·ə'lī·nə }

nasolacrimal canal [ANATOMY] The bony canal that lodges the nasolacrimal duct. Also known as lacrimal canal. { ¦nā·zō'lak·rə·məl kə'nal }

nasolacrimal duct [ANATOMY] The membranous duct lodged within the nasolacrimal canal; it gives passage to the tears from the lacrimal sac to the inferior meatus of the nose. { ¦nā·zō'lak·rə·məl 'dəkt }

nasolacrimal groove [EMBRYOLOGY] The furrow, the maxillary, and the lateral nasal processes of the embryo. { ¦nā·zō'lak·rə·məl 'grüv }

nasopalatine duct [EMBRYOLOGY] A canal between the oral and nasal cavities of the embryo at the point of fusion of the maxillary and palatine processes. { ¦nā·zō'pal·əˌtēn 'dəkt }

nasopharynx [ANATOMY] The space behind the posterior nasal orifices, above a horizontal plane through the lower margin of the palate. { ¦nā·zō 'far·iŋks }

nastic movement [BOTANY] Movement of a flat plant part, oriented relative to the plant body and produced by diffuse stimuli causing disproportionate growth or increased turgor pressure in the tissues of one surface. { 'nas·tik 'müv·mənt }

Nasutitermitinae [INVERTEBRATE ZOOLOGY] A subfamily of termites in the family Termitidae, characterized by having the cephalic glands open at the tip of an elongated tube which projects anteriorly. { nə¦süd·ə·tər'mit·ənˌē }

Natalidae [VERTEBRATE ZOOLOGY] The funnel-eared bats, a monogeneric family of small, tropical American insectivorous bats (Chiroptera) with large, funnellike ears. { nə'tal·əˌdē }

Natantia [INVERTEBRATE ZOOLOGY] A suborder of decapod crustaceans comprising shrimp and related forms characterized by a long rostrum and a ventrally flexed abdomen. { nə'tan·chə }

Naticacea [INVERTEBRATE ZOOLOGY] A superfamily of gastropod mollusks in the order Prosobranchia. { ˌnad·ə'kā·shē·ə }

Naticidae [INVERTEBRATE ZOOLOGY] A family of gastropod mollusks in the order Pectinibranchia comprising the moon-shell snails. { nə'tis·ə ˌdē }

native [BIOLOGY] Grown, produced or originating in a specific region or country. { 'nād·iv }

natriuresis [PHYSIOLOGY] Excretion of sodium in the urine. { ¦na·trē·yü'rē·səs }

natural immunity [IMMUNOLOGY] Native immunity possessed by the individuals of a race, strain, or species. { 'nach·rəl i'myü·nəd·ē }

naturalized [ECOLOGY] Of a species, having become permanently established after being introduced. { 'nach·rəˌlīzd }

natural killer cell [HISTOLOGY] A large, granular lymphocyte that can lyse a variety of target cells when it is activated by interferon. Abbreviated NK. { ¦nach·rəl 'kil·ər ˌsel }

natural selection [EVOLUTION] Darwin's theory of evolution, according to which organisms tend to produce progeny far above the means of subsistence; in the struggle for existence that ensues, only those progeny with favorable variations survive; the favorable variations accumulate through subsequent generations, and descendants diverge from their ancestors. { 'nach·rəl si'lek·shən }

Naucoridae [INVERTEBRATE ZOOLOGY] A family of hemipteran insects in the superfamily Naucoroidea. { nȯ'kȯr·əˌdē }

Naucoroidea [INVERTEBRATE ZOOLOGY] The creeping water bugs, a superfamily of hemipteran insects in the subdivision Hydrocorisae; they are suboval in form, with chelate front legs. { nȯ·kə'rȯid·ē·ə }

nauplius [INVERTEBRATE ZOOLOGY] A larval stage characteristic of many groups of Crustacea; the oval, unsegmented body has three pairs of appendages: uniramous antennules, biramous antennae, and mandibles. { 'nȯ·plē·əs }

Nautilidae [INVERTEBRATE ZOOLOGY] A monogeneric family of cephalopod mollusks in the order Nautiloidea; *Nautilus pompilius* is the only well-known living species. { nȯ'til·əˌdē }

Nautiloidea [INVERTEBRATE ZOOLOGY] A primitive order of tetrabranchiate cephalopods; shells are external and smooth, being straight or coiled and chambered with curved transverse septa. { ˌnȯd·əl'ȯid·ē·ə }

Navarroan [PALEONTOLOGY] A North American (Gulf Coast) stage of Upper Cretaceous geologic time, above the Tayloran and below the Midwayan of the Tertiary. { ˌnav·ə'rō·ən }

navel [ANATOMY] The umbilicus. { 'nā·vəl }

navicular [ANATOMY] A boat-shaped bone, especially the lateral bone on the radial side of the proximal row of the carpus. [BIOLOGY] Resembling or having the shape of a boat. { nə'vik·yə·lər }

neallotype [SYSTEMATICS] A type specimen that, compared with the holotype, is of the opposite sex, and was collected and described later. { nē 'al·əˌtīp }

Neanderthal man [PALEONTOLOGY] A type of fossil human that is a subspecies of *Homo sapiens* and is distinguished by a low broad braincase, continuous arched browridges, projecting occipital region, short limbs, and large joints. { nē 'an·dərˌtäl 'man }

Nearctic fauna [ECOLOGY] The indigenous animal communities of the Nearctic zoogeographic region. { nē'ärd·ik 'fȯn·ə }

Nearctic zoogeographic region [ECOLOGY] The zoogeographic region that includes all of North America to the edge of the Mexican Plateau. { nē 'ärd·ik ¦zō·ōˌjē·ə'graf·ik ˌrē·jən }

near point [PHYSIOLOGY] The smallest distance from the eye at which a small object can be seen without blurring. { 'nir ˌpȯint }

Nebaliacea [INVERTEBRATE ZOOLOGY] A small, marine order of Crustacea in the subclass Leptostraca distinguished by a large bivalve shell, without a definite hinge line, an anterior articulated rostrum, eight thoracic and seven abdominal somites, a pair of articulated furcal rami, and the telson. { nəˌbā·lē'ā·shə }

neck [ANATOMY] The usually constricted communicating column between the head and trunk of the vertebrate body. { nek }

Neckeraceae [BOTANY] A family of mosses in the order Isobryales distinguished by undulate leaves. { ˌnek·əˈrās·ēˌē }

Necrolestidae [PALEONTOLOGY] An extinct family of insectivorous marsupials. { ¦ne·krōˈles·tə ˌdē }

necrophagous [ZOOLOGY] Feeding on dead bodies. { neˈkräf·ə·gəs }

nectar [BOTANY] A sugar-containing liquid secretion of the nectaries of many flowers. { ˈnek·tər }

nectarine [BOTANY] A smooth-skinned, fuzzless fruit originating as a spontaneous somatic mutation of the peach, *Prunus persica* and P. *persica* var. *nectarina*. { ¦nek·tə¦rēn }

nectary [BOTANY] A secretory organ or surface modification of a floral organ in many flowers, occurring on the receptacle, in and around ovaries, on stamens, or on the perianth; secretes nectar. { ˈnek·tə·rē }

nectocalyx [INVERTEBRATE ZOOLOGY] A swimming bell of a siphonophore. Also known as nectophore. { ¦nek·tōˈkā·liks }

Nectonematoidea [INVERTEBRATE ZOOLOGY] A monogeneric order of worms belonging to the class Nematomorpha, characterized by dorsal and ventral epidermal chords, a pseudocoele, and dorsal and ventral rows of bristles; adults are parasites of true crabs and hermit crabs. { ¦nek·tōˌne·məˈtȯid·ē·ə }

nectophore [INVERTEBRATE ZOOLOGY] *See* nectocalyx. { ˈnek·təˌfȯr }

nectosome [INVERTEBRATE ZOOLOGY] The part of a complex siphonophore that bears swimming bells. { ˈnek·təˌsōm }

Nectridea [PALEONTOLOGY] An order of extinct lepospondylous amphibians characterized by vertebrae in which large fan-shaped hemal arches grow directly downward from the middle of each caudal centrum. { nekˈtrid·ē·ə }

Nectrioidaceae [MYCOLOGY] The equivalent name for Zythiaceae. { ¦nek·trēˌȯid·ēˈās·ēˌē }

needle [BOTANY] A slender-pointed leaf, as of the firs and other evergreens. { ˈnēd·əl }

neencephalon [ANATOMY] The neopallium and the phylogenetically new acquisitions of the cerebellum and thalamus collectively. Also spelled neoencephalon. { ¦nē·inˈsef·əˌlän }

negative afterimage [PHYSIOLOGY] An afterimage that is seen on a bright background and is complementary in color to the initial stimulus. { ˈneg·əd·iv ¦af·tər¦im·ij }

negative gene control [MOLECULAR BIOLOGY] Prevention of gene expression by the binding of specific repressor molecules to operator sites. { ¦neg·əd·iv ˈjēn kənˌtrōl }

negative interference [GENETICS] A crossover exchange between homologous chromosomes which increases the likelihood of another in the same vicinity. { ˈneg·əd·iv ˌin·tərˈfir·əns }

negative phase [IMMUNOLOGY] The temporary quantitative reduction of serum antibodies immediately following a second inoculation of antigen. { ˈneg·əd·iv ˈfāz }

negative regulator [GENETICS] Any regulator that acts to prevent transcription or translation. { ˈneg·əd·iv ˌreg·yəˈlād·ər }

negative staining [BIOLOGY] A method in microscopy for demonstrating the form of cells, bacteria, and other small objects by staining the ground rather than the objects. { ˈneg·əd·iv ˈstān·iŋ }

Neididae [INVERTEBRATE ZOOLOGY] A small family of thread-legged hemipteran insects in the superfamily Lygaeoidea. { nēˈid·əˌdē }

Neisseriaceae [MICROBIOLOGY] The single family of gram-negative aerobic cocci and coccobacilli; some species are human parasites and pathogens. { ˈnī·sər·ēˈās·ēˌē }

nektobenthos [ECOLOGY] Those forms of marine life that exist just above the ocean bottom and occasionally rest on it. { ¦nek·təˈbenˌthȯs }

nekton [INVERTEBRATE ZOOLOGY] Free-swimming aquatic animals, essentially independent of water movements. { ˈnek·tən }

Nelumbonaceae [BOTANY] A family of flowering aquatic herbs in the order Nymphaeales characterized by having roots, perfect flowers, alternate leaves, and triaperturate pollen. { nəˌləm·bəˈnās·ēˌē }

Nemata [INVERTEBRATE ZOOLOGY] An equivalent name for Nematoda. { nəˈmad·ə }

Nemataceae [BOTANY] A family of mosses in the order Hookeriales distinguished by having perichaetial leaves only. { ˌnem·əˈtās·ēˌē }

Nemathelminthes [INVERTEBRATE ZOOLOGY] A subdivision of the Amera which comprised the classes Rotatoria, Gastrotrichia, Kinorhyncha, Nematoda, Nematomorpha, and Acanthocephala. { ˌnem·əˌthelˈmin·thēz }

Nematocera [INVERTEBRATE ZOOLOGY] A series of dipteran insects in the suborder Orthorrhapha; adults have antennae that are usually longer than the head, and the flagellum consists of 10-65 similar segments. { ˌnem·əˈtäs·ə·rə }

nematocyst [INVERTEBRATE ZOOLOGY] An intracellular effector organelle in the form of a coiled tube which may be rapidly everted in food gathering or defense by cnidarians. { nəˈmad·ə ˌsist }

Nematoda [INVERTEBRATE ZOOLOGY] A group of unsegmented worms which have been variously recognized as an order, class, and phylum. { ˌnem·əˈtō·də }

nematode [INVERTEBRATE ZOOLOGY] **1.** Any member of the Nematoda. **2.** Of or pertaining to the Nematoda. { ˈnem·əˌtōd }

Nematodonteae [BOTANY] A group of mosses included in the subclass Eubrya in which there may be faint transverse bars on the peristome teeth. { nə¦mad·əˈdänt·ēˌē }

nematogen [INVERTEBRATE ZOOLOGY] A reproductive phase of the Dicyemida during which vermiform larvae are formed asexually from the germ cells in the axial cells. { nəˈmad·ə·jən }

Nematognathi [VERTEBRATE ZOOLOGY] The equivalent name for Siluriformes. { ˌnem·əˈtäg·nəˌthī }

Nematoidea [INVERTEBRATE ZOOLOGY] An equivalent name for Nematoda. { ˌnem·əˈtȯid·ē·ə }

nematology [INVERTEBRATE ZOOLOGY] The study of nematodes. { ˌnem·ə'täl·ə·jē }

Nematomorpha [INVERTEBRATE ZOOLOGY] A group of the Aschelminthes or a separate phylum that includes the horsehair worms. { ˌnem·əd·ə'mȯr·fə }

Nematophytales [PALEOBOTANY] A group of fossil plants from the Silurian and Devonian periods that bear some resemblance to the brown seaweeds (Phaeophyta). { ¦nem·əd·ō·fī'tā·lēz }

Nematosporoideae [BOTANY] A subfamily of the Saccharomycetaceae containing parasitic yeasts; two genera have been studied in culture: *Nematospora* with asci that contain eight spindle-shaped ascospores, and *Metschnikowia* whose asci contain one or two needle-shaped ascospores. { ¦nem·əd·ō·spə'rȯid·ēˌē }

nematozooid [INVERTEBRATE ZOOLOGY] A zooid bearing organs of defense, in hydroids and siphonophores. { ¦nem·əd·ə'zōˌȯid }

Nemertea [INVERTEBRATE ZOOLOGY] An equivalent name for Rhynchocoela. { nə'mərd·ē·ə }

Nemertina [INVERTEBRATE ZOOLOGY] An equivalent name for Rhynchocoela. { ˌne·mər'tī·nə }

Nemertinea [INVERTEBRATE ZOOLOGY] An equivalent name for Rhynchocoela. { ˌne·mər'tin·ē·ə }

Nemestrinidae [INVERTEBRATE ZOOLOGY] The hairy flies, a family of dipteran insects in the series Brachycera of the suborder Orthorrhapha. { ¦nem·ə¦strin·əˌdē }

Nemichthyidae [VERTEBRATE ZOOLOGY] A family of bathypelagic, eellike amphibians in the order Apoda. { ˌnem·ik'thī·əˌdē }

Nemognathinae [INVERTEBRATE ZOOLOGY] A subfamily of the coleopteran family Meloidae; members have greatly elongate maxillae that form a poorly fitted tube. { ˌnem·əg'nath·əˌnē }

nemoral [ECOLOGY] Pertaining to or inhabiting a grove or wooded area. { 'nem·rəl }

Neoanthropinae [PALEONTOLOGY] A subfamily of the Hominidae in some systems of classification, set up to include *Homo sapiens* and direct ancestors of *H. sapiens*. { ¦nē·ō·an'thräp·əˌnē }

neoblast [INVERTEBRATE ZOOLOGY] Any of various undifferentiated cells in annelids which migrate to and proliferate at sites of repair and regeneration. { 'nē·əˌblast }

Neocathartidae [PALEONTOLOGY] An extinct family of vulturelike diurnal birds of prey (Falconiformes) from the Upper Eocene. { ¦nē·ō·kə'thärd·əˌdē }

neocentric activity [CELL BIOLOGY] In plants, an aberrant behavior during meiosis in which specific chromosome regions act as secondary sites of attachment for spindle fibers. { ˌnē·ə'sen·trik ak¦tiv·əd·ē }

neocerebellum [ANATOMY] Phylogenetically, the most recent part of the cerebellum; receives cerebral cortex impulses via the corticopontocerebellar tract. { ¦nē·ōˌser·ə'bel·əm }

neocortex [ANATOMY] Phylogenetically the most recent part of the cerebral cortex; includes all but the olfactory, hippocampal, and piriform regions of the cortex. { ¦nē·ō'kȯrˌteks }

neoencephalon [ANATOMY] *See* neencephalon. { ¦nē·ō·in'sef·əˌlän }

Neogastropoda [INVERTEBRATE ZOOLOGY] An order of gastropods which contains the most highly developed snails; respiration is by means of ctenidia, the nervous system is concentrated, an operculum is present, and the sexes are separate. { ¦nē·ō·ga'sträp·ə·də }

Neognathae [VERTEBRATE ZOOLOGY] A superorder of the avian subclass Neornithes, characterized as flying birds with fully developed wings and sternum with a keel, fused caudal vertebrae, and absence of teeth. { nē'äg·nəˌthē }

Neogregarinida [INVERTEBRATE ZOOLOGY] An order of sporozoan protozoans in the subclass Gregarinia which are insect parasites. { ¦nē·ōˌgreg·ə'rin·ə·də }

neomorph [GENETICS] A mutant allele that produces an effect different from that produced by the normal allele. { 'nē·əˌmȯrf }

neomycin [MICROBIOLOGY] The collective name for several colorless antibiotics produced by a strain of *Streptomyces fradiae*; the commercial fraction ($C_{23}H_{46}N_6O_{13}$) has a broad spectrum of activity. { ¦nē·ə¦mīs·ən }

neonatal line [ANATOMY] A prominent incremental line formed in the neonatal period in the enamel and dentin of a deciduous tooth or of a first permanent molar. { ¦nē·ə'nād·əl 'līn }

neopallium [ANATOMY] Phylogenetically, the new part of the cerebral cortex; formed from the region between the pyriform lobe and the hippocampus, it comprises the nonolfactory region. { ¦nē·ō'pal·ē·əm }

neopalynology [BOTANY] A field of palynology concerned with extant microorganisms and disassociated microscopic parts of megaorganisms. { ¦nē·ōˌpal·ə'näl·ə·jē }

neoplastoid [CELL BIOLOGY] Pertaining to immortal mammalian cell lines that may behave like neoplasms. { ˌnē·ə'plasˌtȯid }

Neopseustidae [INVERTEBRATE ZOOLOGY] A family of Lepidoptera in the superfamily Eriocranioidea. { ˌnē·əp'sü·stəˌdē }

Neoptera [INVERTEBRATE ZOOLOGY] A section of the insect subclass Pterygota; members have a muscular and articular mechanism allowing the wings to be flexed over the abdomen when at rest. { nē'äp·tə·rə }

Neopterygii [VERTEBRATE ZOOLOGY] An equivalent name for Actinopterygii. { nēˌäp·tə'rij·ēˌī }

Neorhabdocoela [INVERTEBRATE ZOOLOGY] A group of the Rhabdocoela comprising fresh-water, marine, or terrestrial forms, with a bulbous pharynx, paired protonephridia, sexual reproduction, and ventral gonopores. { ˌnē·ōˌrab·də'sē·lə }

Neornithes [VERTEBRATE ZOOLOGY] A subclass of the class Aves containing all known birds except the fossil *Archaeopteryx*. { nē'ȯr·nəˌthēz }

neossoptile [VERTEBRATE ZOOLOGY] A downy feather on most newly hatched birds. { ˌnē·ə'säp·təl }

neotenin [BIOCHEMISTRY] A hormone secreted by cells of the corpus allatum in arthropod larvae and nymphs; inhibits the development of adult characters. Also known as juvenile hormone. { nē 'ät·ən·ən }

neoteny [VERTEBRATE ZOOLOGY] A phenomenon peculiar to some salamanders, in which large larvae become sexually mature while still retaining gills and other larval features. { ·'nē·ə‚tē·nē *or* nē 'ät·ən·ē }

Neotropical zoogeographic region [ECOLOGY] A zoogeographic region that includes Mexico south of the Mexican Plateau, the West Indies, Central America, and South America. { ¦nē·ō 'träp·ə·kəl ¦zō·ō‚jē·ə'graf·ik 'rē·jən }

neotype [SYSTEMATICS] A specimen selected as type subsequent to the original description when the primary types are known to be destroyed; a nomenclatural type. { 'nē·ə‚tīp }

neounitarian theory of hematopoiesis [HISTOLOGY] A theory that under certain conditions, such as in tissue culture or in pathologic states, lymphocytes or cells resembling lymphocytes can become multipotent. { ¦nē·ō·yü·nə'tar·ē·ən ¦thē·ə·re əv ‚hem·əd·ō·pȯi'ē·səs }

Nepenthaceae [BOTANY] A family of dicotyledonous plants in the order Sarraceniales; includes many of the pitcher plants. { ‚nep·ən'thās·ē‚ē }

nephr-, nephro- [ANATOMY] Combining form denoting kidney. { 'nef·rō }

nephric tubule [ANATOMY] *See* uriniferous tubule. { 'nef·rik 'tüb‚yül }

nephridioblast [INVERTEBRATE ZOOLOGY] An ecodermal precursor cell of a nephridium in certain animals. { nə'frid·ē·ə‚blast }

nephridioduct [INVERTEBRATE ZOOLOGY] The duct of a nephridium, sometimes serving as a common excretory and genital outlet. { nə'frid·ē·ə‚dəkt }

nephridiopore [INVERTEBRATE ZOOLOGY] The external opening of a nephridium. { nə'frid·ē·ə ‚pȯr }

nephridium [INVERTEBRATE ZOOLOGY] Any of various paired excretory structures present in the Platyhelminthes, Rotifera, Rhynchocoela, Acanthocephala, Priapuloidea, Entoprocta, Gastrotricha, Kinorhyncha, Cephalochorda, and some Archiannelida and Polychaeta. { nə'frid·ē·əm }

nephroabdominal [ANATOMY] Of or pertaining to the kidneys and abdomen. { ¦nef·rō·ab'däm·ə·nəl }

nephrocoel [ANATOMY] The cavity of a nephrotome. { ¦nef·rə‚sēl }

nephrogenic [EMBRYOLOGY] **1.** Having the potential to develop into kidney tissue. **2.** Of renal origin. { ¦nef·rə¦jen·ik }

nephrogenic cord [EMBRYOLOGY] The longitudinal cordlike mass of mesenchyme derived from the mesomere or nephrostomal plate of the mesoderm, from which develop the functional parts of the pronephros, mesonephros, and metanephros. { ¦nef·rə¦jen·ik 'kȯrd }

nephrogenic tissue [EMBRYOLOGY] The tissue of the nephrogenic cord derived from the nephrotome plate that forms the blastema or primordium from which the embryonic and definitive kidneys develop. { ¦nef·rə¦jen·ik 'tish·ü }

nephrolysin [BIOCHEMISTRY] A toxic substance capable of disintegrating kidney cells. { nə'fräl·ə·sən }

nephromixium [INVERTEBRATE ZOOLOGY] A compound nephridium composed of flame cells and the coelomic funnel; functions as both an excretory organ and a genital duct. { ¦nef·rō'mik·sē·əm }

nephron [ANATOMY] The functional unit of a kidney, consisting of the glomerulus with its capsule and attached uriniferous tubule. { 'nef ‚rän }

Nephropidae [INVERTEBRATE ZOOLOGY] The true lobsters, a family of decapod crustaceans in the superfamily Nephropidea. { nə'fräp·ə‚dē }

Nephropidea [INVERTEBRATE ZOOLOGY] A superfamily of the decapod section Macrura including the true lobsters and crayfishes, characterized by a rostrum and by chelae on the first three pairs of pereiopods, with the first pair being noticeably larger. { ‚nef·rə'pid·ē·ə }

nephros [ANATOMY] The kidney. { 'nef·rəs }

nephrostome [INVERTEBRATE ZOOLOGY] The funnel-shaped opening of a nephridium into the coelom. { 'nef rə‚stōm }

nephrotome [EMBRYOLOGY] The narrow mass of embryonic mesoderm connecting somites and lateral mesoderm, from which the pronephros, mesonephros, metanephros, and their ducts develop. { 'nef·rə‚tōm }

Nephtyidae [INVERTEBRATE ZOOLOGY] A family of errantian annelids of highly opalescent colors, distinguished by an eversible pharynx. { nef'tī·ə ‚dē }

Nepidae [INVERTEBRATE ZOOLOGY] The water scorpions, a family of hemipteran insects in the superfamily Nepoidea, characterized by a long breathing tube at the tip of the abdomen, chelate front legs, and a short stout beak. { 'nep·ə‚dē }

Nepoidea [INVERTEBRATE ZOOLOGY] A superfamily of hemipteran insects in the subdivision Hydrocorisae. { nə'pȯid·ē·ə }

Nepticulidae [INVERTEBRATE ZOOLOGY] The single family of the lepidopteran superfamily Nepticuloidea. { ‚nep·tə'kyül·ə‚dē }

Nepticuloidea [INVERTEBRATE ZOOLOGY] A monofamilial superfamily of heteroneuran Lepidoptera; members are tiny moths with wing spines, and the females have a single genital opening. { ‚nep·tə·kyə'lȯid·ē·ə }

Nereidae [INVERTEBRATE ZOOLOGY] A large family of mostly marine errantian annelids that have a well-defined head, elongated body with many segments, and large complex parapodia on most segments. { nə'rē·ə‚dē }

Nerillidae [INVERTEBRATE ZOOLOGY] A family of archiannelids characterized by well-developed parapodia and setae. { nə'ril·ə‚dē }

Neritacea [INVERTEBRATE ZOOLOGY] A superfamily of gastropod mollusks in the order Aspidobranchia. { ‚ner·ə'tās·ē·ə }

Neritidae [INVERTEBRATE ZOOLOGY] A family of

primitive marine, fresh-water, and terrestrial snails in the order Archaeogastropoda. { nə'rid·ə,dē }

nerve [ANATOMY] A bundle of nerve fibers or processes held together by connective tissue. { nərv }

nerve block [PHYSIOLOGY] Interruption of impulse transmission through a nerve. { 'nərv ,bläk }

nerve cord [INVERTEBRATE ZOOLOGY] Paired, ventral cords of nervous tissue in certain invertebrates, such as insects or the earthworm. [ZOOLOGY] Dorsal, hollow tubular cord of nervous tissue in chordates. { 'nərv ,kȯrd }

nerve ending [ANATOMY] **1.** The structure on the distal end of an axon. **2.** The termination of a nerve. { 'nərv ,end·iŋ }

nerve fiber [CELL BIOLOGY] The long process of a neuron, usually the axon. { 'nərv ,fī·bər }

nerve impulse [PHYSIOLOGY] The transient physicochemical change in the membrane of a nerve fiber which sweeps rapidly along the fiber to its termination, where it causes excitation of other nerves, muscle, or gland cells, depending on the connections and functions of the nerve. { 'nərv ,im,pəls }

nerve net [INVERTEBRATE ZOOLOGY] A network of continuous nerve cells characterized by diffuse spread of excitation, local and equipotential autonomy, spatial attenuation of conduction, and facilitation; occurs in cnidarians and certain other invertebrates. { 'nərv ,net }

nerve tract [ANATOMY] A bundle of nerve fibers having the same general origin and destination. { 'nərv ,trakt }

nervous [BIOLOGY] **1.** Of or pertaining to nerves. **2.** Originating in or affected by nerves. **3.** Affecting or involving nerves. { 'nər·vəs }

nervous system [ANATOMY] A coordinating and integrating system which functions in the adaptation of an organism to its environment; in vertebrates, the system consists of the brain, brainstem, spinal cord, cranial and peripheral nerves, and ganglia. { 'nər·vəs ,sis·təm }

nervous tissue [HISTOLOGY] The nerve cells and neuroglia of the nervous system. { 'nər·vəs ¦tish·ü }

Nesiotinidae [INVERTEBRATE ZOOLOGY] A family of bird-infesting biting lice (Mallophaga) that are restricted to penguins. { ¦nes·ē·ō'tin·ə,dē }

Nesophontidae [PALEONTOLOGY] An extinct family of large, shrewlike lipotyphlans from the Cenozoic found in the West Indies. { ,nes·ə'fän·tə ,dē }

nest [VERTEBRATE ZOOLOGY] A bed, receptacle, or location in which the eggs of animals are laid and hatched. { nest }

net aerial production [ECOLOGY] The biomass or biocontent which is incorporated into the aerial parts, that is, the leaf, stem, seed, and associated organs, of a plant community. { 'net 'er·ē·əl prə 'dək·shən }

net plankton [ECOLOGY] Plankton that can be removed from sea water by the process of filtration through a fine net. { 'net 'plaŋk·tən }

net primary production [ECOLOGY] Over a specified period of time, the biomass or biocontent which is incorporated into a plant community. { 'net 'prīm·ə·rē prə'dək·shən }

net production rate [ECOLOGY] The assimilation rate (gross production rate) minus the amount of matter lost through predation, respiration, and decomposition. { 'net prə'dək·shən ,rāt }

nettle [BOTANY] A prickly or stinging plant of the family Urticaceae, especially in the genus *Urtica*. { 'ned·əl }

nettle cell [INVERTEBRATE ZOOLOGY] *See* cnidoblast. { 'ned·əl ,sel }

net-veined [BIOLOGY] Having a network of veins, as a leaf or an insect wing. { 'net ,vānd }

neural arc [PHYSIOLOGY] A nerve circuit consisting of effector and receptor with intercalated neurons between them. { 'nu̇r·əl 'ärk }

neural arch [ANATOMY] *See* vertebral arch. { 'nu̇r·əl 'ärch }

neural canal [EMBRYOLOGY] The embryonic vertebral canal. { 'nu̇r·əl kə'nal }

neural crest [EMBRYOLOGY] Ectoderm composing the primordium of the cranial, spinal, and autonomic ganglia and adrenal medulla, located on either side of the neural tube. { 'nu̇r·əl 'krest }

neural ectoderm [EMBRYOLOGY] Embryonic ectoderm which will form the neural tube and neural crest. { 'nu̇r·əl 'ek·tə,dərm }

neural fold [EMBRYOLOGY] Either of a pair of dorsal longitudinal folds of the neural plate which unite along the midline, forming the neural tube. { 'nu̇r·əl 'fōld }

neural groove [EMBRYOLOGY] A longitudinal groove between the neural folds of the vertebrate embryo before the neural tube is completed. { 'nu̇r·əl 'grüv }

neural plate [EMBRYOLOGY] The thickened dorsal plate of ectoderm that differentiates into the neural tube. { 'nu̇r·əl 'plāt }

neural spine [ANATOMY] The spinous process of a vertebra. { 'nu̇r·əl 'spīn }

neural tube [EMBRYOLOGY] The embryonic tube that differentiates into brain and spinal cord. { 'nu̇r·əl 'tüb }

neuraminic acid [BIOCHEMISTRY] $C_9H_{17}NO_8$ An amino acid, the aldol condensation product of pyruvic acid and N-acetyl-D-mannosamine, regarded as the parent acid of a family of widely distributed acyl derivatives known as sialic acids. { ¦nu̇r·ə¦min·ik 'as·əd }

neuraminidase [MICROBIOLOGY] A bacterial enzyme that acts to split salic acid from neuraminic acid glycosides. { ,nu̇r·ə'min·ə,dās }

neurapophysis [EMBRYOLOGY] Either of two projections on each embryonic vertebra which unite to form the neural arch. { ¦nu̇r·ə'päf·ə·səs }

neurenteric canal [EMBRYOLOGY] A temporary duct connecting the neural tube and primitive gut in certain vertebrate and tunicate embryos. { ¦nu̇r·ən'ter·ik kə'nal }

neurilemma [HISTOLOGY] A thin tissue covering the axon directly, or covering the myelin sheath when present, of peripheral nerve fibers. { ,nu̇r·ə'lem·ə }

neurine [BIOCHEMISTRY] $CH_2{=}CHN(CH_3)_3OH$ A

very poisonous, syrupy liquid with fishy aroma; soluble in water and alcohol; a product of putrefaction of choline in brain tissue and bile, and in cadavers. Also known as trimethylvinylammonium hydroxide. { 'nü,rēn }

neurite [ANATOMY] *See* axon. { 'nu̇,rīt }

neuroanatomy [ANATOMY] The study of the anatomy of the nervous system and nerve tissue. { ¦nu̇r·ō·ə'nad·ə·mē }

neurobiotaxis [EVOLUTION] Hypothetical migration of nerve cells and ganglia toward regions of maximum stimulation during phylogenetic development. { ¦nu̇r·ō,bī·ō'tak·səs }

neuroblast [EMBRYOLOGY] Embryonic, undifferentiated neuron, derived from neural plate ectoderm. { 'nu̇r·ō,blast }

neurochemistry [BIOCHEMISTRY] Chemistry of the nervous system. { ¦nu̇r·ō'kem·ə·strē }

neurocirculatory [ANATOMY] Pertaining to both the nervous and the vascular systems. { nu̇r·ō 'sər·kyə·lə,tȯr·ē }

neurocoele [ANATOMY] The system of cavities and ventricles in the brain and spinal cord. { 'nu̇r·ō,sēl }

neurocranium [ANATOMY] The portion of the cranium which forms the braincase. { ¦nu̇r·ə'krā·nē·əm }

neurocutaneous [ANATOMY] **1.** Concerned with both the nerves and skin. **2.** Pertaining to innervation of the skin. { ¦nu̇r·ō·kyə'tā·nē·əs }

neurocyte [CELL BIOLOGY] The body of a nerve cell. { 'nu̇r·ə,sīt }

Neurodontiformes [PALEONTOLOGY] A suborder of Conodontophoridia having a lamellar internal structure. { ¦nu̇r·ō,dänt·ə'fōr·mēz }

neuroelectricity [PHYSIOLOGY] A current or voltage generated in the nervous system. { ¦nu̇r·ō,i ,lek'tris·əd·ē }

neuroendocrine [BIOLOGY] Pertaining to both the nervous and endocrine systems, structurally and functionally. { ¦nu̇r ō'en·də·krən }

neuroendocrinology [BIOLOGY] The study of the structural and functional interrelationships between the nervous and endocrine systems. { ¦nu̇r·ō,en·də·krə'näl·ə·jē }

neuroepidermal [BIOLOGY] Pertaining to both the nerves and epidermis, structurally and functionally. { ¦nu̇r·ō,ep·ə'dər·məl }

neurofibril [CELL BIOLOGY] A fibril of a neuron, usually extending from the processes and traversing the cell body. { ¦nu̇r·ō'fī·brəl }

neurogenesis [EMBRYOLOGY] The formation of nerves. { ¦nu̇r·ō'jen·ə·səs }

neurogenic [BIOLOGY] **1.** Originating in nervous tissue. **2.** Innervated by nerves. { ¦nu̇r·ō¦jen·ik }

neuroglia [HISTOLOGY] The nonnervous, supporting elements of the nervous system. { nü 'räg·lē·ə }

neurohemal organ [ZOOLOGY] Any of various structures in vertebrates and some invertebrates that consist of clusters of bulbous, secretion-filled axon terminals of neurosecretory cells which function as storage-and-release centers for neurohormones. { ¦nu̇r·ō'hē·məl ,ȯr·gən }

neurohormone [BIOCHEMISTRY] A hormone produced by nervous tissue. { ¦nu̇r·ō'hȯr,mōn }

neurohumor [BIOCHEMISTRY] A hormonal transmitter substance, such as acetylcholine, released by nerve endings in the transmission of impulses. { ¦nu̇r·ō'hyü·mər }

neurohypophysis [ANATOMY] The neural portion or posterior lobe of the hypophysis. { ¦nu̇r·ō·hī 'päf·ə·səs }

neuromast [VERTEBRATE ZOOLOGY] A lateral-line sensory organ in fishes and other lower vertebrates consisting of a cluster of receptor cells connected with nerve fibers. { 'nu̇r·ō,mast }

neuromere [EMBRYOLOGY] An embryonic segment of the central nervous system in vertebrates. { 'nu̇r·ō,mir } .

neuromodulator [PHYSIOLOGY] A chemical agent that is released by a neurosecretory cell and acts on other neurons in a local region of the central nervous system by modulating their response to neurotransmitters. { ,nu̇r·ō'mäj·ə,lād·ər }

neuromuscular [BIOLOGY] Pertaining to both nerves and muscles, functionally and structurally. { ¦nu̇r·ō'məs·kyə·lər }

neuromuscular junction [ANATOMY] *See* myoneural junction. { ¦nu̇r·ō'məs·kyə·lər 'jəŋk·shən }

neuron [HISTOLOGY] A nerve cell, including the cell body, axon, and dendrites. { 'nu̇,rän }

neuron doctrine [BIOLOGY] A doctrine that the neuron is the basic structural and functional unit of the nervous system, and that it acts upon another neuron through the synapse. { 'nu̇,rän ,däk·trən }

neurophysiology [PHYSIOLOGY] The study of the functions of the nervous system. { ¦nu̇r·ō,fiz·ē 'äl·ə·jē }

neuropil [HISTOLOGY] Nervous tissue consisting of a fibrous network of nonmyelinated nerve fibers; gray matter with few nerve cell bodies; usually a region of synapses between axons and dendrites. { 'nu̇r·ō,pil }

neuroplasm [CELL BIOLOGY] Protoplasm of nerve cells. { 'nu̇r·ō,plaz·əm }

neuropodium [CELL BIOLOGY] A terminal branch of an axon. { ,nu̇r·ō'pō·dē·əm }

neuropore [EMBRYOLOGY] A terminal aperture of the neural tube before complete closure at the 20-25 somite stage. { 'nu̇r·ō,pȯr }

Neuroptera [INVERTEBRATE ZOOLOGY] An order of delicate insects having endopterygote development, chewing mouthparts, and soft bodies. { nü 'räp·tə·rə }

neurosecretion [PHYSIOLOGY] The synthesis and release of hormones by nerve cells. { ¦nu̇r·ō·si 'krē·shən }

neurosecretory cell [HISTOLOGY] A neuron that releases one or more hormones into the circulatory system. { ,nu̇r·ō·si'krēd·ə·rē ,sel }

neurotoxin [BIOCHEMISTRY] A poisonous substance in snake venom that acts as a nervous system depressant by blocking neuromuscular transmission by binding acetylcholine receptors on motor end plates, or on the innervated face of an electroplax. { ¦nu̇r·ō'täk·sən }

neurotransmitter [PHYSIOLOGY] A chemical

agent that is released by a neuron at a synapse, diffuses across the synapse, and acts upon a postsynaptic neuron, a muscle, or a gland cell. { ˌnu̇r·ōˌtranz′mid·ər }

neurotropic [BIOLOGY] Having an affinity for nerve tissue. { ¦nu̇r·ō¦träp·ik }

neurovascular [BIOLOGY] Pertaining structurally and functionally to both the nervous and vascular structures. { ¦nu̇r·ō′vas·kyə·lər }

neurulation [EMBRYOLOGY] Differentiation of nerve tissue and formation of the neural tube. { ˌnu̇r·ə′lā·shən }

neuston [BIOLOGY] Minute organisms that float or swim on surface water or on a surface film of water. { ′nüˌstän }

neutralism [ECOLOGY] A neutral interaction between two species, that is, one having no evident effect on either species. { ′nü·trəˌliz·əm }

neutralizing antibody [IMMUNOLOGY] An antibody that reduces or abolishes some biological activity of a soluble antigen or of a living microorganism. { ′nü·trəˌlīz·iŋ ′ant·iˌbäd·ē }

neutral mutation [GENETICS] A mutation that has no phenotypic effect or adaptive significance. { ′nü·trəl myüˌtā·shən }

neutrophil [HISTOLOGY] A large granular leukocyte with a highly variable nucleus, consisting of three to five lobes, and cytoplasmic granules which stain with neutral dyes and eosin. { ′nü·trəˌfil }

neutrophilia [BIOLOGY] Affinity for neutral dyes. { ˌnü·trə′fil·ē·ə }

neutrophilous [BIOLOGY] Preferring an environment free of excess acid or base. { nü′träf·ə·ləs }

Newcastle virus [VIROLOGY] A ribonucleic acid hemagglutinating myxovirus responsible for Newcastle disease. { ′nüˌkas·əl ˌvī·rəs }

newt [VERTEBRATE ZOOLOGY] Any of the small, semiaquatic salamanders of the genus *Triturus* in the family Salamandridae; all have an aquatic larval stage. { nüt }

nexus [PHYSIOLOGY] *See* gap junction. { ′nek·səs }

niacin [BIOCHEMISTRY] *See* nicotinic acid. { ′nī·ə·sən }

niacinamide [BIOCHEMISTRY] *See* nicotinamide. { ˌnī·ə′sin·ə·mīd }

niche [ECOLOGY] The unique role or way of life of a plant or animal species. { nich }

nick [BIOCHEMISTRY] The absence of a phosphodiester bond between adjacent nucleotides in one strand of duplex deoxyribonucleic acid. { nik }

nickase [BIOCHEMISTRY] An enzyme that causes single-stranded breaks in duplex deoxyribonucleic acid, allowing it to unwind. { ′niˌkās }

Nicoletiidae [INVERTEBRATE ZOOLOGY] A family of the insect order Thysanura proper. { ˌnik·ə·lə ′tī·əˌdē }

Nicomachinae [INVERTEBRATE ZOOLOGY] A subfamily of the limnivorous sedentary annelids in the family Maldanidae. { ˌnik·ə′mak·əˌnē }

nicotinamide [BIOCHEMISTRY] $C_6H_6ON_2$ Crystalline basic amide of the vitamin B complex that is interconvertible with nicotinic acid in the living organism; the amide of nicotinic acid. Also known as niacinamide. { ˌnik·ə′tin·əˌmīd }

nicotinamide adenine dinucleotide [BIOCHEMISTRY] *See* diphosphopyridine nucleotide. { ˌnik·ə ′tin·əˌmīd ′ad·ənˌēn dī′nü·klē·əˌtīd }

nicotinic acid [BIOCHEMISTRY] $C_6H_5NO_2$ A component of the vitamin B complex; a white, water-soluble powder stable to heat, acid, and alkali; used for the treatment of pellagra. Also known as niacin. { ¦nik·ə¦tin·ik ′as·əd }

nictitating membrane [VERTEBRATE ZOOLOGY] A membrane of the inner angle of the eye or below the eyelid in many vertebrates, and capable of extending over the eyeball. { ′nik·əˌtād·iŋ ′memˌbrān }

nidamental gland [ZOOLOGY] Any of various structures that secrete covering material for eggs or egg masses. { ¦nīd·ə¦ment·əl ′gland }

nidicolous [ZOOLOGY] **1.** Spending a short time in the nest after hatching. **2.** Sharing the nest of another species. { nī′dik·ə·ləs }

nidus [ZOOLOGY] A nest or breeding place. { ′nīd·əs }

night ape [VERTEBRATE ZOOLOGY] *See* bushbaby. { ′nīt ˌāp }

night vision [PHYSIOLOGY] *See* scotopic vision. { ′nīt ˌvizh·ən }

nigrescent [BIOLOGY] Blackish. { nī′gres·ənt }

Nilionidae [INVERTEBRATE ZOOLOGY] The false ladybird beetles, a family of coleopteran insects in the superfamily Tenebrionoidea. { ˌnil·ē′än·ə ˌdē }

nipple [ANATOMY] The conical projection in the center of the mamma, containing the outlets of the milk ducts. { ′nip·əl }

Nippotaeniidea [INVERTEBRATE ZOOLOGY] An order of tapeworms of the subclass Cestoda including some internal parasites of certain freshwater fishes; the head bears a single terminal sucker. { ˌnip·ōˌtē·nē′ī·dē·ə }

Nissl bodies [CELL BIOLOGY] Chromophil granules of nerve cells which ultrastructurally are composed of large ribosomes. { ′nis·əl ˌbäd·ēz }

Nitelleae [BOTANY] A tribe of stoneworts, order Charales, characterized by 10 cells in two tiers of five each composing the apical crown. { ni′tel·ēˌē }

Nitidulidae [INVERTEBRATE ZOOLOGY] The sap-feeding beetles, a large family of coleopteran insects in the superfamily Cucujoidea; individuals have five-jointed tarsi and antennae with a terminal three-jointed clavate expansion. { ˌnid·ə ′dyül·əˌdē }

nitrification [MICROBIOLOGY] Formation of nitrous and nitric acids or salts by oxidation of the nitrogen in ammonia; specifically, oxidation of ammonium salts to nitrites and oxidation of nitrites to nitrates by certain bacteria. { ˌnī·trə·fə ′kā·shən }

nitrifying bacteria [MICROBIOLOGY] Members of the family Nitrobacteraceae. { ′nī·trəˌfī·iŋ bak ′tir·ē·ə }

Nitrobacteraceae [MICROBIOLOGY] The nitrifying bacteria, a family of gram-negative, chemolithotrophic bacteria; autotrophs which derive energy from nitrification of ammonia or nitrite, and obtain carbon for growth by fixation of carbon dioxide. { ¦nī·trō,bak·tə'rās·ē,ē }

nitrocellulose filter [MOLECULAR BIOLOGY] A very thin filter whose fibers selectively bind single-stranded deoxyribonucleic acid (DNA) but not double-stranded DNA or ribonucleic acid. { ¦nī·trō'sel·yə,lōs ,fil·tər }

nitrogenase [BIOCHEMISTRY] An enzyme that catalyzes a six-electron reduction of N_2 in the process of nitrogen fixation. { nī'trä·jə,nās }

nitrogen balance [PHYSIOLOGY] The difference between nitrogen intake (as protein) and total nitrogen excretion for an individual. { 'nī·trə·jən ,bal·əns }

nitrogen fixation [MICROBIOLOGY] Assimilation of atmospheric nitrogen by heterotrophic bacteria. Also known as dinitrogen fixation. { 'nī·trə·jən ,fik¦sā·shən }

nitrogenous base [BIOCHEMISTRY] A purine or a pyrimidine derivative which is one of the three components of a nucleotide of nucleic acids. Also known as base. { nī'trä·jə·nəs 'bās }

nitrophyte [BOTANY] A plant that requires nitrogen-rich soil for growth. { ¦nī·trə,fīt }

nival [ECOLOGY] **1.** Characterized by or living in or under the snow. **2.** Of or pertaining to a snowy environment. { 'nī·vəl }

niveous [BIOLOGY] *See* niveus. { 'niv·ē·əs }

niveus [BIOLOGY] Snow-white in color. Also spelled niveous. { 'niv·ē·əs }

NK [HISTOLOGY] *See* natural killer cell.

Nocardiaceae [MICROBIOLOGY] A family of aerobic bacteria in the order Actinomycetales; mycelium and spore production is variable. { nō ,kär·dē'ās·ē,ē }

nocioceptive reflex [PHYSIOLOGY] *See* flexion reflex. { ¦nō·sē·ō·ri'sep·tiv 'rē,fleks }

Noctilionidae [VERTEBRATE ZOOLOGY] The fish-eating bats, a tropical American monogeneric family of the Chiroptera having small eyes and long, narrow wings. { näk,til·ē'än·ə,dē }

Noctuidae [INVERTEBRATE ZOOLOGY] A large family of dull-colored, medium-sized moths in the superfamily Noctuoidea; larva are mostly exposed foliage feeders, representing an important group of agricultural pests. { näk'tü·ə,dē }

Noctuoidea [INVERTEBRATE ZOOLOGY] A large superfamily of lepidopteran insects in the suborder Heteroneura; most are moderately large moths with reduced maxillary palpi. { ,näk·tü'óid·ē·ə }

nocturnal emission [PHYSIOLOGY] Normal, involuntary seminal discharge occurring during sleep in males after puberty. { näk'tərn·əl i 'mish·ən }

nodal rhythm [PHYSIOLOGY] A cardiac rhythm characterized by pacemaker function originating in the atrioventricular node, with a heart rate of 40-70 per minute. { 'nōd·əl 'rith·əm }

nodal tissue [HISTOLOGY] **1.** Tissue from the sinoatrial node, and the atrioventricular node and bundle and its branches, composed of a dense network of Purkinje fibers. **2.** Tissue from a lymph node. { 'nōd·əl ,tish·ü }

node [ANATOMY] **1.** A knob or protuberance. **2.** A small, rounded mass of tissue, such as a lymph node. **3.** A point of constriction along a nerve. [BOTANY] A site on a plant stem at which leaves and axillary buds arise. { nōd }

node of Ranvier [ANATOMY] The region of a local constriction in a myelinated nerve; formed at the junction of two Schwann cells. { 'nōd əv rän 'vyā }

Nodosariacea [INVERTEBRATE ZOOLOGY] A superfamily of Foraminiferida in the suborder Rotaliina characterized by a radial calcite test wall with monolamellar septa, and a test that is coiled, uncoiled, or spiral about the long axis. { ,nōd·ə ,sar·ē'ās·ē·ə }

nodose [BIOLOGY] Having many or noticeable protuberances; knobby. { 'nō,dōs }

nodule [ANATOMY] **1.** A small node. **2.** A small aggregation of cells. [BOTANY] A bulbous enlargement found on roots of legumes and certain other plants, whose formation is stimulated by symbiotic, nitrogen-fixing bacteria that colonize the roots. { 'näj·ül }

nodules of the semilunar valves [ANATOMY] Small nodes in the midregion of the pulmonary and aortic semilunar valves. { 'näj·ülz əv thə 'sem·i,lü·nər 'valvz }

nodulose [BIOLOGY] Having minute nodules or fine knobs. { 'näj·ə,lōs }

Noeggerathiales [PALEOBOTANY] A poorly defined group of fossil plants whose geologic range extends from Upper Carboniferous to Triassic. { ,neg·ə,rath·ē'ā·lēz }

nogalamycin [MICROBIOLOGY] $C_{39}H_{45}NO_{16}$ An antineoplastic antibiotic produced by *Streptomyces nogalaster*. { nō,gal·ə'mīs·ən }

nomen dubium [SYSTEMATICS] A proposed taxonomic name invalid because it is not accompanied by a definition or description of the taxon to which it applies. { 'nō·mən 'dü·bē·əm }

nomen nudum [SYSTEMATICS] A proposed taxonomic name invalid because the accompanying definition or description of the taxon cannot be interpreted satisfactorily. { 'nō·mən 'nü·dəm }

noncompetitive inhibition [BIOCHEMISTRY] Enzyme inhibition in which the inhibitor can combine with either the free enzyme or the enzyme-substrate complex so that the inhibitor does not compete with the substrate for the enzyme. { ¦nän·kəm'ped·əd·iv ,in·ə'bish·ən }

noncongression [CELL BIOLOGY] The failure of pairing chromosomes, in certain stages of mitosis and meiosis, to orient in an orderly arrangement on the spindle equator. { ,nän·kən 'gre·shən }

nonconjugative plasmid [GENETICS] Any plasmid that prevents conjugation of its bacterial host. { nän¦kän·jə,gād·iv 'plaz·mid }

nondisjunction mosaic [GENETICS] A population of cells with different chromosome numbers produced when one chromosome is lost during mitosis or when both members of a pair of chromosomes are included in the same daughter nucleus; can occur during embryogenesis or adulthood. { ¦nän·dis'jəŋk·shən mō'zā·ik }

nonessential amino acid [BIOCHEMISTRY] An amino acid which can be synthesized by an organism and thus need not be supplied in the diet. { ¦nän·i'sen·chəl ə'mē·nō 'as·əd }

nongranular leukocyte [HISTOLOGY] A white blood cell, such as a lymphocyte or monocyte, with clear homogeneous cytoplasm. { ¦nän 'gran·yə·lər 'lü·kə,sīt }

nonhistone protein [BIOCHEMISTRY] A class of acidic proteins in the cell nucleus associated with deoxyribonucleic acid. { nän¦hi,stōn 'prō ,tēn }

Nonionacea [INVERTEBRATE ZOOLOGY] A superfamily of Foraminiferida in the suborder Orbitoidacea, characterized by a granular calcite test wall with monolamellar septa, and a planispiral to trochospiral test. { ,nō·nē·ə'nās·ē·ə }

nonpermissive cell [VIROLOGY] A cell that does not support replication of a virus. { ,nän·pər ¦mis·iv 'sel }

nonpositive tactile stimulus [PHYSIOLOGY] A cessation of a feedback signal from tactile sensors to the brain, such as when an object held in the hand falls. { ¦nän¦päz·əd·iv ¦tak,tīl 'stim·yə·ləs }

nonproductive infection [VIROLOGY] *See* abortive infection. { ¦nän·prə'dək·tiv in'fek·shən }

nonprotein nitrogen [BIOCHEMISTRY] The nitrogen fraction in the body tissues, excretions, and secretions, not precipitated by protein precipitants. { ¦nän¦prō,tēn 'nī·trə·jən }

nonsaccharine sorghum [BOTANY] *See* grain sorghum. { ¦nän'sak·ə·rən 'sȯr·gəm }

nonselective medium [MICROBIOLOGY] A culture medium that supports the growth of all genotypes. { ¦nän·si,lek·tiv 'mē·dē·əm }

nonsense mutation [MOLECULAR BIOLOGY] A mutation that changes a codon that codes for one amino acid into a codon that does not specify any amino acid (a nonsense codon). { 'nän ,səns myü,tā·shən }

nonsense suppression [GENETICS] Suppression of the termination effect of a nonsense codon on a growing polypeptide chain, allowing return to the normal phenotype. { 'nän,sens sə,presh·ən }

nonspecific immunity [IMMUNOLOGY] Resistance attributable to factors other than specific antibodies, including genetic, age, or hormonal factors. { ¦nän·spə¦sif·ik i'myün·əd·ē }

noosphere [ECOLOGY] *See* anthroposphere. { 'nō·ə,sfir }

noradrenaline [BIOCHEMISTRY] *See* norepinephrine. { ,nȯr·ə'dren·ə·lən }

noradrenergic system [BIOLOGY] A system of neurons that is responsible for the synthesis, storage, and release of the neurotransmitter norepinephrine. { nȯr¦ad·rə¦nər·jik 'sis·təm }

norepinephrine [BIOCHEMISTRY] $C_8H_{11}O_3N$ A hormone produced by chromaffin cells of the adrenal medulla; acts as a vasoconstrictor and mediates transmission of sympathetic nerve impulses. Also known as noradrenaline. { ,nȯr·ep·ə'ne·frən }

normalizing selection [GENETICS] Removal of alleles that produce divergence from the average phenotype in a population by selecting against deviant individuals. { 'nȯr·mə,līz·iŋ si,lek·shən }

normal saline [PHYSIOLOGY] U.S. Pharmacopoeia title for a sterile solution of sodium chloride in purified water, containing 0.9 gram of sodium chloride in 100 milliliters; isotonic with body fluids. Also known as isotonic sodium chloride solution; normal salt solution; physiological saline; physiological salt solution; physiological sodium chloride solution; sodium chloride solution. { 'nȯr·məl 'sā,lēn }

normal salt solution [BIOCHEMISTRY] *See* normal saline. { 'nȯr·məl 'sȯlt sə,lü·shən }

normoblast [HISTOLOGY] The smallest of the nucleated precursors of the erythrocyte; slightly larger than a mature adult erythrocyte. Also known as acidophilic erythroblast; arthochromatic erythroblast; eosinophilic erythroblast; karyocyte; metakaryocyte; metanitricyte; metarubricyte. { 'nȯr·mə,blast }

normochromatic [CELL BIOLOGY] Pertaining to cells of the erythrocytic series which have a normal staining color; attributed to the presence of a full complement of hemoglobin. { ¦nȯr·mə·krə 'mad·ik }

normochromic [CELL BIOLOGY] Pertaining to erythrocytes which have a mean corpuscular hemoglobin (MCH), or color, index and a mean corpuscular hemoglobin concentration (MCHC), or saturation, index within two standard deviations above or below the mean normal. { ¦nȯr·mə 'krō·mik }

normocyte [HISTOLOGY] An erythrocyte having both a diameter and a mean corpuscular volume (MCV) within two standard deviations above or below the mean normal. { 'nȯr·mə,sīt }

normothermia [PHYSIOLOGY] A state of normal body temperature. { ,nȯr·mə'thər·mē·ə }

nose [ANATOMY] The nasal cavities and the structures surrounding and associated with them in all vertebrates. { nōz }

nose leaf [VERTEBRATE ZOOLOGY] A leaflike expansion of skin on the nose of certain bats; believed to have a tactile function. { 'nōz ,lēf }

Nosodendridae [INVERTEBRATE ZOOLOGY] The wounded-tree beetles, a small family of coleopteran insects in the superfamily Dermestoidea. { ,näz·ə'den·drə,dē }

nostril [ANATOMY] One of the external orifices of the nose. Also known as naris. { 'näs·trəl }

Notacanthidae [VERTEBRATE ZOOLOGY] A family of benthic, deep-sea teleosts in the order Notacanthiformes, including the spiny eel. { ,nōd·ə 'kan·thə,dē }

Notacanthiformes [VERTEBRATE ZOOLOGY] An order of actinopterygian fishes whose body is elongated, tapers posteriorly, and has no caudal fin. { ,nōd·ə,kan·thə'fȯr·mēz }

notch graft [BOTANY] A plant graft in which the scion is inserted in a narrow slit in the stock. { 'näch ,graft }

Noteridae [INVERTEBRATE ZOOLOGY] The burrowing water beetles, a small family of coleopteran insects in the suborder Adephaga. { nō'ter·ə ,dē }

Nothosauria [PALEONTOLOGY] A suborder of

chiefly marine Triassic reptiles in the order Sauropterygia. { ˌnäth·ə'sȯr·ē·ə }

Notidanoidea [VERTEBRATE ZOOLOGY] A suborder of rare sharks in the order Selachii; all retain the primitive jaw suspension of the order. { ˌnōd·ə·də'nȯid·ē·ə }

Notiomastodontinae [PALEONTOLOGY] A subfamily of extinct elephantoid proboscidean mammals in the family Gomphotheriidae. { ˌnōd·ē·ōˌmas·tə'dän·təˌnē }

Notioprogonia [PALEONTOLOGY] A suborder of extinct mammals comprising a diversified archaic stock of Notoungulata. { ˌnōd·ē·ō·prə'gō·nē·ə }

notochord [VERTEBRATE ZOOLOGY] An elongated dorsal cord of cells which is the primitive axial skeleton in all chordates; persists in adults in the lowest forms (*Branchiostoma* and lampreys) and as the nuclei pulposi of the intervertebral disks in adult vertebrates. { 'nōd·əˌkȯrd }

notochordal canal [EMBRYOLOGY] A canal formed by a continuation of the primary pit into the head process of mammalian embryos; provides a temporary connection between the yolk sac and amnion. { ¦nōd·ə¦kȯrd·əl kə'nal }

notochordal plate [EMBRYOLOGY] A plate of cells representing the root of the head process of the embryo after the embryo becomes vesiculated. { ¦nōd·ə¦kȯrd·əl 'plāt }

Notodelphyidiformes [INVERTEBRATE ZOOLOGY] A tribe of the Gnathostoma in some systems of classification. { ¦nōd·ə·delˌfid·ə'fȯr·mēz }

Notodelphyoida [INVERTEBRATE ZOOLOGY] A small group of crustaceans bearing a superficial resemblance to many insect larvae as a result of uniform segmentation, comparatively small trunk appendages, and crowding of inconspicuous oral appendages into the anterior portion of the head. { ˌnōd·əˌdel·fē'ȯid·ə }

Notodontidae [INVERTEBRATE ZOOLOGY] The puss moths, a family of lepidopteran insects in the superfamily Noctuoidea, distinguished by the apparently three-branched cubitus. { ˌnōd·ə'dän·təˌdē }

Notogaean [ECOLOGY] Pertaining to or being a biogeographic region including Australia, New Zealand, and the southwestern Pacific islands. { ¦nōd·ə¦jē·ən }

Notommatidae [INVERTEBRATE ZOOLOGY] A family of rotifers in the order Monogonota including forms with a cylindrical body covered by a nonchitinous cuticle and with a slender posterior foot. { ˌnōd·ə'mat·əˌdē }

Notomyotina [INVERTEBRATE ZOOLOGY] A suborder of echinoderms in the order Phanerozonida in which the upper marginals alternate in position with the lower marginals, and each tube foot has a terminal sucking disk. { ˌnōd·ə·mī'ät·ən·ə }

Notonectidae [INVERTEBRATE ZOOLOGY] The backswimmers, a family of aquatic, carnivorous hemipteran insects in the superfamily Notonectoidea; individuals swim ventral side up, aided in breathing by an air bubble. { ˌnōd·ə'nek·təˌdē }

Notonectoidea [INVERTEBRATE ZOOLOGY] A superfamily of Hemiptera in the subdivision Hydrocorisae. { ˌnod·ə·nek'tȯid·ē·ə }

notopodium [INVERTEBRATE ZOOLOGY] The dorsal branch of a parapodium in certain annelids. { ˌnōd·ə'pō·dē·əm }

Notopteridae [VERTEBRATE ZOOLOGY] The featherbacks, a family of actinopterygian fishes in the order Osteoglossiformes; bodies are tapered and compressed, with long anal fins that are continuous with the caudal fin. { ˌnōˌtäp'ter·əˌdē }

Notoryctidae [PALEONTOLOGY] An extinct family of Australian insectivorous mammals in the order Marsupialia. { ˌnōd·ə'rik·təˌdē }

Notostigmata [INVERTEBRATE ZOOLOGY] The single suborder of the Opilioacriformes, an order of mites. { ˌnōd·ə·stig'mäd·ə }

Notostigmophora [INVERTEBRATE ZOOLOGY] A subclass or suborder of the Chilopoda, including those centipedes embodying primitive as well as highly advanced characters, distinguished by dorsal respiratory openings. { ˌnōd·ə·stig'mäf·ə·rə }

Notostraca [INVERTEBRATE ZOOLOGY] The tadpole shrimps, an order of crustaceans generally referred to the Branchiopoda, having a cylindrical trunk that consists of 25-44 segments, a dorsoventrally flattened dorsal shield, and two narrow, cylindrical cercopods on the telson. { nə'täs·trə·kə }

Nototheniidae [VERTEBRATE ZOOLOGY] A family of perciform fishes in the suborder Blennioidei, including most of the fishes of the permanently frigid waters surrounding Antarctica. { ˌnōd·ə·thə'nī·əˌdē }

Notoungulata [PALEONTOLOGY] An extinct order of hoofed herbivorous mammals, characterized by a skull with an expanded temporal region, primitive dentition, and primitive feet with five toes, the weight borne mainly by the third digit. { ˌnōd·ōˌəŋ·gyə'läd·ə }

novobiocin [MICROBIOLOGY] $C_{30}H_{36}O_{11}N_2$ A moderately broad-spectrum antibiotic produced by strains of *Streptomyces niveus* and S. *spheroides*; it is a dibasic acid and is converted either to the monosodium salt or to the calcium acid salt for pharmaceutical use. { ˌnō·və'bī·ə·sən }

nucellus [BOTANY] The oval central mass of tissue in the ovule; contains the embryo sac. { ˌnü'sel·əs }

nucha [ANATOMY] The nape of the neck. { 'nü·kə }

nuchal ligament [ANATOMY] An elastic ligament extending from the external occipital protuberance and middle nuchal line to the spinous process of the seventh cervical vertebra. Also known as ligamentum nuchae. { 'nü·kəl 'lig·ə·mənt }

nuchal organ [INVERTEBRATE ZOOLOGY] Any of various sense organs on the prostomium of many annelids, which are sensitive to changes in the immediate environment of the individual. { 'nü·kəl ˌȯr·gən }

nuchal tentacle [INVERTEBRATE ZOOLOGY] Any of various filiform or thick, fleshy tactoreceptors on anterior segments of many annelids. { 'nü·kəl ˌten·tə·kəl }

nuclear dimorphism [INVERTEBRATE ZOOLOGY] In

ciliated protozoas, the occurrence of two types of nuclei in a cell, each with different genetic functions. { ¦nü·klē·ər dī'mȯr‚fiz·əm }

nuclear envelope [CELL BIOLOGY] A structure consisting of two membranes that surrounds the nucleus; the outermost membrane is continuous with the rough endoplasmic reticulum. { 'nü·klē·ər 'en·və‚lōp }

nuclear lamina [CELL BIOLOGY] A protein meshwork lining the inner surface of the nuclear envelope. { 'nü·klē·ər 'lam·ə·nə }

nuclear membrane [CELL BIOLOGY] The envelope surrounding the cell nucleus, separating the nucleoplasm from the cytoplasm; composed of two membranes and contains numerous pores. { 'nü·klē·ər 'mem‚brān }

nuclear plaque [MYCOLOGY] In yeast, any region on the nuclear envelope from which the spindle originates. { 'nü·klē·ər 'plak }

nuclear pore complex [CELL BIOLOGY] Any of the nonrandomly distributed, octagonal orifices in the nuclear envelope. { ¦nü·klē·ər ¦pȯr 'käm ‚pleks }

nuclear transfer [CELL BIOLOGY] Insertion of a diploid somatic nucleus into an egg from which the nucleus has been removed. { 'nü·klē·ər 'tranz·fər }

nuclease [BIOCHEMISTRY] An enzyme that catalyzes the splitting of nucleic acids to nucleotides, nucleosides, or the components of the latter. { 'nü·klē‚ās }

nucleic acid [BIOCHEMISTRY] A large, acidic, chainlike molecule containing phosphoric acid, sugar, and purine and pyrimidine bases; two types are ribonucleic acid and deoxyribonucleic acid. { nü'klē·ik 'as·əd }

nuclein [BIOCHEMISTRY] Any of a poorly defined group of nucleic acid protein complexes occurring in cell nuclei. { 'nü·klē·ən }

nucleocapsid [VIROLOGY] The nucleic acid of a virus and its surrounding capsid. { ¦nü·klē·ō 'kap·səd }

nucleocytoplasmic ratio [CELL BIOLOGY] The ratio between the measured cross-sectional area or estimated volume of the nucleus of a cell to the volume of its cytoplasm. Also known as karyoplasmic ratio. { ¦nü·klē·ō‚sīd·ə'plaz·mik 'rā·shō }

nucleodesma [CELL BIOLOGY] A connection composed of fibrils between the nucleus and cytoplasm. Also known as karyodesma. { ‚nü·klē·ə 'dez·mə }

nucleoid [CELL BIOLOGY] A discrete region within mitochondria, chloroplasts, and prokaryotes that contain molecules of deoxyribonucleic acid. [VIROLOGY] The ribonucleic acid (RNA) core that is enveloped by a protein capsid in RNA tumor viruses. { 'nü·klē‚ȯid }

nucleolar [CELL BIOLOGY] Of or pertaining to the nucleolus. { nü'klē·ə·lər }

nucleolus [CELL BIOLOGY] A small, spherical body composed principally of protein and located in the metabolic nucleus. Also known as plasmosome. { nü'klē·ə·ləs }

nucleoplasm [CELL BIOLOGY] The protoplasm of a nucleus. Also known as karyoplasm. { 'nü·klē·ə‚plaz·əm }

nucleoprotein [BIOCHEMISTRY] Any member of a class of conjugated proteins in which molecules of nucleic acid are closely associated with molecules of protein. { ¦nü·klē·ō'prō‚tēn }

nucleoreticulum [CELL BIOLOGY] Any type of network found within a nucleus. { ¦nü·klē·ō·rə'tik·yə·ləm }

nucleosidase [BIOCHEMISTRY] Any enzyme involved in splitting a nucleoside into its component base and pentose. { ‚nü·klē·ə'sī‚dās }

nucleoside [BIOCHEMISTRY] The glycoside resulting from removal of the phosphate group from a nucleotide; consists of a pentose sugar linked to a purine or pyrimidine base. { 'nü·klē·ə‚sīd }

nucleosome [CELL BIOLOGY] A morphologically repeating unit of deoxyribonucleic acid (DNA) containing 190 base pairs of DNA folded together with eight histone molecules. Also known as v-body. { 'nü·klē·ə‚sōm }

nucleospindle [CELL BIOLOGY] A mitotic spindle derived from nuclear material. { ¦nü·klē·ō 'spind·əl }

nucleotidase [BIOCHEMISTRY] Any of a group of enzymes which split phosphoric acid from nucleotides, leaving nucleosides. { ‚nü·klē·ə'tī‚dās }

nucleotide [BIOCHEMISTRY] An ester of a nucleoside and phosphoric acid; the structural unit of a nucleic acid. { 'nü·klē·ə‚tīd }

nucleus [CELL BIOLOGY] A small mass of differentiated protoplasm rich in nucleoproteins and surrounded by a membrane; found in most animal and plant cells, contains chromosomes, and functions in metabolism, growth, and reproduction. [HISTOLOGY] A mass of nerve cells in the central nervous system. { 'nü·klē·əs }

nucleus pulposus [ANATOMY] The soft, fibrocartilaginous central portion of the intervertebral disk. { ¦nü·klē·əs pəl'pō·səs }

Nuda [INVERTEBRATE ZOOLOGY] A class of the phylum Ctenophora distinguished by the lack of tentacles. { 'nüd·ə }

Nudechiniscidae [INVERTEBRATE ZOOLOGY] A family of heterotardigrades in the suborder Echiniscoidea characterized by a uniform cuticle. { ‚nüd·ə·ki'nis·kə‚dē }

Nudibranchia [INVERTEBRATE ZOOLOGY] A suborder of the Opisthobranchia containing the sea slugs; these mollusks lack a shell and a mantle cavity, and the gills are variable in size and shape. { ‚nüd·ə'braŋ·kē·ə }

null allele [GENETICS] An allele that does not produce a functional product and behaves as a recessive. { 'nəl ə‚lēl }

null cell [IMMUNOLOGY] A lymphocyte without T- or B-cell markers on its surface. { 'nəl ‚sel }

numerical taxonomy [SYSTEMATICS] The numerical evaluation of the affinity or similarity between taxonomic units and the ordering of these units into taxa on the basis of their affinities. { nü 'mer·i·kəl tak'sän·ə·mē }

Numididae [VERTEBRATE ZOOLOGY] A family of birds in the order Galliformes commonly known as guinea fowl; there are few if any feathers on the neck or head, but there may be a crest of feathers and various fleshy appendages. { nü 'mid·ə,dē }

nurse cell [HISTOLOGY] A cell type of the ovary of many animals which nourishes the developing egg cell. { 'nərs ,sel }

nurse graft [BOTANY] A plant graft in which the scion remains united with the stock only until roots develop on the scion. { 'nərs ,graft }

nut [BOTANY] **1.** A fruit which has at maturity a hard, dry shell enclosing a kernel consisting of an embryo and nutritive tissue. **2.** An indehiscent, one-celled, one-seeded, hard fruit derived from a single, simple, or compound ovary. { nət }

nutation [BOTANY] The rhythmic change in the position of growing plant organs caused by variation in the growth rates on different sides of the growing apex. { nü'tā·shən }

nutmeg [BOTANY] *Myristica fragrans.* A dark-leafed evergreen tree of the family Myristicaceae cultivated for the golden-yellow fruits which resemble apricots; a delicately flavored spice is obtained from the kernels inside the seeds. { 'nət,meg }

nutrient [BIOLOGY] Providing nourishment. { 'nü·trē·ənt }

nutrient biopurification [ECOLOGY] A process taking place within a nutrient cycle that maintains the pools of nutrient substances at optimum concentrations, to the exclusion of nonnutrient substances. { 'nü·trē·ənt ¦bī·ō,pyür·ə·fə'kā·shən }

nutrient foramen [ANATOMY] The opening into the canal which gives passage to the blood vessels of the medullary cavity of a bone. { 'nü·trē·ənt fə'rā·mən }

nutrition [BIOLOGY] The science of nourishment, including the study of nutrients that each organism must obtain from its environment in order to maintain life and reproduce. { nü'trish·ən }

Nuttalliellidae [INVERTEBRATE ZOOLOGY] A family of ticks (Ixodides) containing one rare African species, *Nuttalliella namaqua*, morphologically intermediate between the families Argasidae and Ixodidae. { nə,tal·ē'el·ə,dē }

nux vomica [BOTANY] The seed of *Strychnos nux-vomica*, an Indian tree of the family Loganiaceae; contains the alkaloid strychnine, and was formerly used in medicine. { 'nəks 'väm·ə·kə }

Nyctaginaceae [BOTANY] A family of dicotyledonous plants in the order Caryophyllales characterized by an apocarpous, monocarpous, or syncarpous gynoecium, sepals joined to a tube, a single carpel, and a cymose inflorescence. { ,nik·tə·jə'nās·ē,ē }

Nycteridae [VERTEBRATE ZOOLOGY] The slit-faced bats, a monogeneric family of insectivorous chiropterans having a simple, well-developed nose leaf, and large ears joined together across the forehead. { nik'ter·ə,dē }

Nyctibiidae [VERTEBRATE ZOOLOGY] A family of birds in the order Caprimulgiformes including the neotropical potoos. { ,nik·tə'bī·ə,dē }

nyctinasty [BOTANY] A nastic movement in higher plants associated with diurnal light and temperature changes. { 'nik·tə,nas·tē }

Nyctribiidae [INVERTEBRATE ZOOLOGY] The bat tick flies, a family of myodarian cyclorrhaphous dipteran insects in the subsection Acalyptratae. { ,nik·trə'bī·ə,dē }

nygmata [INVERTEBRATE ZOOLOGY] Sensory spots on wings of certain insects, such as some neuropterans. { nig'mäd·ə }

Nygolaimoidea [INVERTEBRATE ZOOLOGY] A superfamily of predaceous nematodes of the order Dorylaimida, distinguished by an eversible stoma with a protrusible subventral mural tooth, and a bottle-shaped esophagus. { ,nī·gō·lə 'mȯid·ē·ə }

nymph [INVERTEBRATE ZOOLOGY] Any immature larval stage of various hemimetabolic insects. { nimf }

Nymphaeaceae [BOTANY] A family of dicotyledonous plants in the order Nymphaeales distinguished by the presence of roots, perfect flowers, alternate leaves, and uniaperturate pollen. { ,nim·fē'ās·ē,ē }

Nymphaeales [BOTANY] An order of flowering aquatic herbs in the subclass Magnoliidae; all lack cambium and vessels and have laminar placentation. { ,nim·fē'ā·lēz }

Nymphalidae [INVERTEBRATE ZOOLOGY] The four-footed butterflies, a family of lepidopteran insects in the superfamily Papilionoidea; prothoracic legs are atrophied, and the well-developed patagia are heavily sclerotized. { nim'fal·ə,dē }

Nymphalinae [INVERTEBRATE ZOOLOGY] A subfamily of the lepidopteran family Nymphalidae. { nim'fal·ə,nē }

Nymphonidae [INVERTEBRATE ZOOLOGY] A family of marine arthropods in the subphylum Pycnogonida; members have chelifores, five-jointed palpi, and ten-jointed ovigers. { nim'fän·ə,dē }

Nymphulinae [INVERTEBRATE ZOOLOGY] A subfamily of the lepidopteran family Pyralididae which is notable because some species are aquatic. { nim'fyül·ə,nē }

Nysmyth's membrane [ANATOMY] The primary enamel cuticle which is the transitory remnants of the enamel organ and oral epithelium covering the enamel of a tooth after eruption. { 'nī ,smiths ,mem,brān }

Nyssaceae [BOTANY] A family of dicotyledonous plants in the order Cornales characterized by perfect or unisexual flowers with imbricate petals, a solitary ovule in each locule, a unilocular ovary, and more stamens than petals. { nə'sās·ē,ē }

nystatin [MICROBIOLOGY] $C_{46}H_{77}NO_{19}$ An antifungal antibiotic produced by *Streptomyces noursei*; used for the treatment of infections caused by *Candida* (*Monilia*) *albicans*. { 'nis·təd·ən }

O

oak [BOTANY] Any tree of the genus *Quercus* in the order Fagales, characterized by simple, usually lobed leaves, scaly winter buds, a star-shaped pith, and its fruit, the acorn, which is a nut; the wood is tough, hard, and durable, generally having a distinct pattern. { ōk }

O antigen [MICROBIOLOGY] A somatic antigen of certain flagellated microorganisms. { 'ō 'ant·i·jən }

oat [BOTANY] Any plant of the genus *Avena* in the family Gramineae, cultivated as an agricultural crop for its seed, a cereal grain, and for straw. { ōt }

obclavate [BIOLOGY] Inversely clavate. { äb'kla‚vāt }

obcordate [BOTANY] Referring to a leaf, heart-shaped with the notch apical. { äb'kȯr‚dāt }

obdiplastemonous [BOTANY] Having the stamens arranged in two whorls, with members of the outer whorl positioned opposite the petals. { äb¦dip·lō¦stē·mə·nəs }

obese [ANATOMY] Extremely fat. { ō'bēs }

obligate [BIOLOGY] Restricted to a specified condition of life, as an obligate parasite. { 'äb·lə·gət }

oblique [ANATOMY] Referring to a muscle, positioned obliquely and having one end that is not attached to bone. [BOTANY] Referring to a leaf, having the two sides of a blade unequal. { ə 'blēk }

Obolellida [PALEONTOLOGY] A small order of Early and Middle Cambrian inarticulate brachiopods, distinguished by a shell of calcium carbonate. { ‚äb·ə'lel·ə·də }

obovate [BIOLOGY] Inversely ovate. { 'äb·ə‚vāt }

obsolete [BIOLOGY] A part of an organism that is imperfect or indistinct, compared with a corresponding part of similar organisms. { ‚äb·sə 'lēt }

obturator [ANATOMY] **1.** Pertaining to that which closes or stops up, as an obturator membrane. **2.** Either of two muscles, originating at the pubis and ischium, which rotate the femur laterally. { 'äb·tə‚rād·ər }

obturator artery [ANATOMY] A branch of the internal iliac; it branches into the pubic and acetabular arteries. { 'äb·tə‚rād·ər ¦ärd·ə·rē }

obturator foramen [ANATOMY] A large opening in the pelvis, between the ischium and the pubis, that gives passage to vessels and nerves; it is partly closed by a fibrous obturator membrane. { 'äb·tə‚rād·ər fə'rā·mən }

obturator membrane [ANATOMY] **1.** A fibrous membrane closing the obturator foramen of the pelvis. **2.** A thin membrane between the crura and foot plates of the stapes. { 'äb·tə‚rād·ər 'mem‚brān }

obturator nerve [ANATOMY] A mixed nerve arising in the lumbar plexus; innervates the adductor, gracilis, and obturator externus muscles, and the skin of the medial aspect of the thigh, hip, and knee joints. { 'äb·tə‚rād·ər ¦nərv }

obtuse [BOTANY] Of a leaf, having a blunt or rounded free end. { äb'tüs }

obvallate [BIOLOGY] Surrounded by or as if by a wall. { äb'va‚lāt }

obvolute [BIOLOGY] Overlapping. { 'äb·və‚lüt }

occipital arch [INVERTEBRATE ZOOLOGY] A part of an insect cranium lying between the occipital suture and postoccipital suture. { äk'sip·əd·əl ¦ärch }

occipital artery [ANATOMY] A branch of the external carotid which branches into the mastoid, auricular, sternocleidomastoid, and meningeal arteries. { äk'sip·əd·əl ¦ärd·ə·rē }

occipital bone [ANATOMY] The bone which forms the posterior portion of the skull, surrounding the foramen magnum. { äk'sip·əd·əl ¦bōn }

occipital condyle [ANATOMY] An articular surface on the occipital bone which articulates with the atlas. [INVERTEBRATE ZOOLOGY] A projection on the posterior border of an insect head which articulates with the lateral neck plates. { äk'sip·əd·əl 'kän‚dīl }

occipital crest [ANATOMY] Either of two transverse ridges connecting the occipital protuberances with the foramen magnum. { äk'sip·əd·əl ¦krest }

occipital ganglion [INVERTEBRATE ZOOLOGY] One of a pair of ganglia located just posterior to the brain in insects. { äk'sip·əd·əl 'gaŋ·glē‚än }

occipitalia [INVERTEBRATE ZOOLOGY] An unpaired row of dorsal cilia on the head of gnathostomulids. { äk‚sip·ə'tal·ē·ə }

occipital lobe [ANATOMY] The posterior lobe of the cerebrum having the form of a three-sided pyramid. { äk'sip·əd·əl ¦lōb }

occipital pole [ANATOMY] The tip of the occipital lobe of the brain. { äk'sip·əd·əl ¦pōl }

occipital protuberance [ANATOMY] A prominence on the surface of the occipital bone to which the ligamentum nuchae is attached. { äk 'sip·əd·əl prə'tü·bə·rəns }

occipitofrontalis [ANATOMY] A muscle in two parts, the frontal (inserting in the skin of the forehead) and the occipital (inserting in the galea sponeurotica). { äk¦sip·əd·ō,frən'tal·is }

occiput [ZOOLOGY] The back of the head of an insect or vertebrate. { 'äk·sə,pu̇t }

occluding junction [CELL BIOLOGY] *See* tight junction. { ə'klüd·iŋ ,jəŋk·shən }

occlusion [ANATOMY] The relationship of the masticatory surfaces of the maxillary teeth to the masticatory surfaces of the mandibular teeth when the jaws are closed. [PHYSIOLOGY] The deficit in muscular tension when two afferent nerves that share certain motor neurons in the central nervous system are stimulated simultaneously, as compared to the sum of tensions when the two nerves are stimulated separately. { ə'klü·zhən }

occult virus [VIROLOGY] A virus whose presence is assumed but which cannot be recovered. { ə 'kəlt 'vī·rəs }

Oceanian [ECOLOGY] Of or pertaining to the zoogeographic region that includes the archipelagos and islands of the central and south Pacific. { ,ō·shē'an·ē·ən }

oceanodromous [VERTEBRATE ZOOLOGY] Of a fish, migratory in salt water. { ¦ō·shə¦nä·drə·məs }

ocellus [INVERTEBRATE ZOOLOGY] A small, simple invertebrate eye composed of photoreceptor cells and pigment cells. { ō'sel·əs }

ocelot [VERTEBRATE ZOOLOGY] *Felis pardalis.* A small arboreal wild cat, of the family Felidae, characterized by a golden head and back, silvery flanks, and rows of somewhat metallic spots on the body. { 'äs·ə,lät }

Ochnaceae [BOTANY] A family of dicotyledonous plants in the order Theales, characterized by simple, stipulate leaves, a mostly gynobasic style, and anthers that generally open by terminal pores. { äk'nās·ē,ē }

Ochotonidae [VERTEBRATE ZOOLOGY] A family of the mammalian order Lagomorpha; members are relatively small, and all four legs are about equally long. { ,äk·ə'tän·ə,dē }

ochratoxin A [BIOCHEMISTRY] $C_{20}H_{18}ClNO_6$ A toxic metabolite from *Aspergillus ochraceus*; a crystalline compound which exhibits green fluorescence; melting point is 169°C; inhibits phosphorylase and mitochondrial respiration in rat liver. { ¦äk·rə¦täk·sən 'ā }

ochre mutation [GENETICS] Alteration of a codon to UAA, a codon that results in premature termination of the polypeptide chain in bacteria. { 'ō·kər myü'tā·shən }

Ochrobium [MICROBIOLOGY] A genus of bacteria in the family Siderocapsaceae; cells are ellipsoidal to rod-shaped and are surrounded by a delicate sheath, resembling a horseshoe, that is heavily embedded with iron oxides. { ō'krō·bē·əm }

ochroleucous [BIOLOGY] Pale ocher or buff colored. { ,ō·krə'lü·kəs }

Ochteridae [INVERTEBRATE ZOOLOGY] The velvety shorebugs, the single family of the hemipteran superfamily Ochteroidea. { äk'ter·ə,dē }

Ochteroidea [INVERTEBRATE ZOOLOGY] A monofamilial tropical and subtropical superfamily of hemipteran insects in the subdivision Hydrocorisae; individuals are black with a silky sheen, and the antennae are visible from above. { äk·tə 'ròid·ē·ə }

ocrea [BOTANY] A tubular stipule or pair of coherent stipules. { 'ä·krē·ə }

Octocorallia [INVERTEBRATE ZOOLOGY] The equivalent name for Alcyonaria. { ,äk·tō·kə'ril·yə }

Octopoda [INVERTEBRATE ZOOLOGY] An order of the dibranchiate cephalopods, characterized by having eight arms equipped with one to three rows of suckers. { äk'täp·ə·də }

Octopodidae [INVERTEBRATE ZOOLOGY] The octopuses, in family of cephalopod mollusks in the order Octopoda. { äk·tə'päd·ə,dē }

octopus [INVERTEBRATE ZOOLOGY] Any member of the genus *Octopus* in the family Octopodidae; the body is round with a large head and eight partially webbed arms, each bearing two rows of suckers, and there is no shell. { 'äk·tə,pu̇s }

ocular [BIOLOGY] Of or pertaining to the eye. { 'äk·yə·lər }

ocular skeleton [VERTEBRATE ZOOLOGY] A rigid structure in most submammalian vertebrates consisting of a cup of hyaline cartilage enclosing the posterior part of the eye, and a thin-walled ring of intramembranous bones in the edge of the sclera at its junction with the cornea. { 'äk·yə·lər 'skel·ə·tən }

oculomotor [PHYSIOLOGY] **1.** Pertaining to eye movement. **2.** Pertaining to the oculomotor nerve. { ¦äk·yə·lō'mōd·ər }

oculomotor nerve [ANATOMY] The third cranial nerve; a paired somatic motor nerve arising in the floor of the midbrain, which innervates all extrinsic eye muscles except the lateral rectus and superior oblique, and furnishes autonomic fibers to the ciliary and pupillary sphincter muscles within the eye. { ¦äk·yə·lō'mōd·ər 'nərv }

oculomotor nucleus [ANATOMY] A nucleus in the floor of the midbrain that gives rise to motor fibers of the oculomotor nerve. { ¦äk·yə·lō'mōd·ər 'nü·klē·əs }

Oculosida [INVERTEBRATE ZOOLOGY] An order of the protozoan subclass Radiolaria; pores are restricted to certain areas in the central capsule, and an olive-colored material is always present near the astropyle. { ,äk·yə'läs·ə·də }

odd-pinnate [BOTANY] Of a compound leaf, having a single leaflet at the tip of the petiole with leaflets on both sides of the petiole. Also known as imparipinnate. { 'äd 'pi,nāt }

Odiniidae [INVERTEBRATE ZOOLOGY] A family of cyclorrhaphous myodarian dipteran insects in the subsection Acalyptratae. { ,ōd·ən'ī·ə,dē }

Odobenidae [VERTEBRATE ZOOLOGY] A family of carnivorous mammals in the suborder Pinnipedia; contains a single species, the walrus (*Odobenus rosmarus*). { ,ō·dō'ben·ə,dē }

Odonata [INVERTEBRATE ZOOLOGY] The dragonflies, an order of the class Insecta, characterized by a head with large compound eyes, and wings with clear or transparent membranes traversed by networks of veins. { ˌōd·ənˈad·ə }

odontoblast [HISTOLOGY] One of the elongated, dentin-forming cells covering the dental papilla. { ōˈdänt·əˌblast }

Odontoceti [VERTEBRATE ZOOLOGY] The toothed whales, a suborder of cetacean mammals distinguished by a single blowhole. { ōˌdänt·əˈsēˌtī }

odontoclast [HISTOLOGY] A multinuclear cell concerned with resorption of the roots of milk teeth. { ōˈdänt·əˌklast }

odont-, odonto- [VERTEBRATE ZOOLOGY] A combining form meaning tooth. { ˌōˌdän·tō }

odontogenesis [EMBRYOLOGY] Formation of teeth. { ōˌdänt·əˈjen·ə·səs }

odontogenic [HISTOLOGY] **1.** Pertaining to the origin and development of teeth. **2.** Originating in tissues associated with teeth. { ō¦dänt·ə¦jen·ik }

Odontognathae [PALEONTOLOGY] An extinct superorder of the avian subclass Neornithes, including all large, flightless aquatic forms and other members of the single order Hesperornithiformes. { ōˌdänˈtäg·nəˌthē }

odontoid [BIOLOGY] Toothlike. { ōˈdänˌtȯid }

odontoid process [ANATOMY] A toothlike projection on the anterior surface of the axis vertebra with which the atlas articulates. { ōˈdänˌtȯid ˌprä·səs }

odontology [VERTEBRATE ZOOLOGY] A branch of science that deals with the formation, development, and abnormalities of teeth. { ˌōˌdänˈtäl·ə·jē }

Odontostomatida [INVERTEBRATE ZOOLOGY] An order of the protozoan subclass Spirotrichia; individuals are compressed laterally and possess very little ciliature. { ō¦dänt·ə·stōˈmad·ə·də }

Oecophoridae [INVERTEBRATE ZOOLOGY] A family of small to moderately small moths in the lepidopteran superfamily Tineoidea, characterized by a comb of bristles, the pecten, on the scape of the antennae. { ˌēk·əˈfȯr·əˌdē }

Oedemeridae [INVERTEBRATE ZOOLOGY] The false blister beetles, a large family of coleopteran insects in the superfamily Tenebrionoidea. { ˌēd·əˈmer·əˌdē }

Oedogoniales [BOTANY] An order of fresh-water algae in the division Chlorophyta; characterized as branched or unbranched microscopic filaments with a basal holdfast cell. { ˌēd·əˌgō·nēˈā·lēz }

Oegophiurida [INVERTEBRATE ZOOLOGY] An order of echinoderms in the subclass Ophiuroidea, represented by a single living genus; members have few external skeletal plates and lack genital bursae, dorsal and ventral arm plates, and certain jaw plates. { ˌēg·ə·fiˈyu̇r·ə·də }

Oegopsida [INVERTEBRATE ZOOLOGY] A suborder of cephalopod mollusks in the order Decapoda of one classification system, and in the order Teuthoidea of another system. { ēˈgäp·sə·də }

Oepikellacea [PALEONTOLOGY] A dimorphic superfamily of extinct ostracods in the order Paleocopa, distinguished by convex valves and the absence of any trace of a major sulcus in the external configuration. { ēˌpik·əˈlās·ē·ə }

Oestridae [INVERTEBRATE ZOOLOGY] A family of cyclorrhaphous myodarian dipteran insects in the subsection Calypteratae. { ˈes·trəˌdē }

oidiophore [MYCOLOGY] A hypha that produces oidia. { ōˈid·əˌfȯr }

oidium [MYCOLOGY] One of the small, thin-walled spores with flat ends produced by autofragmentation of the vegetative hyphae in certain Eumycetes. { ōˈid·ē·əm }

Oikomonadidae [INVERTEBRATE ZOOLOGY] A family of protozoans in the order Kinetoplastida containing organisms that have a single flagellum. { ˌȯik·ə·məˈnad·əˌdē }

oil gland [VERTEBRATE ZOOLOGY] *See* uropygial gland. { ˈȯil ˌgland }

okapi [VERTEBRATE ZOOLOGY] *Okapia johnstoni.* An artiodactylous mammal in the family Giraffidae; has a hazel coat with striped hindquarters, and the head shape, lips, and tongue are the same as those of the giraffe, but the neck is not elongate. { ōˈkä·pē }

Okazaki fragment [MOLECULAR BIOLOGY] In deoxyribonucleic acid replication, a discontinuous segment in which the lagging strand is synthesized. { ˌō·kə¦zä·kē ˌfrag·mənt }

okra [BOTANY] *Hibiscus esculentus.* A tall annual plant grown for its edible immature pods. Also known as gumbo. { ˈō·krə }

Olacaceae [BOTANY] A family of dicotyledonous plants in the order Santalales characterized by dry or fleshy indehiscent fruit, the presence of petals, stamens, and chlorophyll, and a 2-5-celled ovary. { ˌō·ləˈsās·ēˌē }

Oldhaminidina [PALEONTOLOGY] A suborder of extinct articulate brachiopods in the order Strophomenida distinguished by a highly lobate brachial valve seated within an irregular convex pedicle valve. { ˌōl·də·məˈnī·də·nə }

Oleaceae [BOTANY] A family of dicotyledonous plants in the order Scrophulariales characterized generally by perfect flowers, two stamens, axile to parietal or apical placentation, a four-lobed corolla, and two ovules in each locule. { ˌō·lēˈās·ēˌē }

oleandomycin [MICROBIOLOGY] $C_{35}H_{61}O_{12}N$ A macrolide antibotic produced by *Streptomyces antibioticus;* active mainly against gram-positive microorganisms. Also known as matromycin. { ˌō·lēˌan·dəˈmīs·ən }

olecranon [ANATOMY] The large process at the distal end of the ulna that forms the bony prominence of the elbow and receives the insertion of the triceps muscle. { ōˈlek·rəˌnän }

Olenellidae [PALEONTOLOGY] A family of extinct arthropods in the class Trilobita. { ˌō·ləˈnel·əˌdē }

Olethreutidae [INVERTEBRATE ZOOLOGY] A family of moths in the superfamily Tortricoidea whose hindwings usually have a fringe of long hairs along the basal part of the cubitus. { ˌō·ləˈthrüd·əˌdē }

olfaction [PHYSIOLOGY] **1.** The function of smelling. **2.** The sense of smell. { älˈfak·shən }

olfactoreceptor [PHYSIOLOGY] A structure which is a receptor for the sense of smell. { äl¦fak·tō·ri¦sep·tər }

olfactory bulb [VERTEBRATE ZOOLOGY] The bulbous distal end of the olfactory tract located beneath each anterior lobe of the cerebrum; well developed in lower vertebrates. { äl′fak·trē ,bəlb }

olfactory cell [PHYSIOLOGY] One of the sensory nerve cells in the olfactory epithelium. { äl′fak·trē ,sel }

olfactory foramen [ANATOMY] Any of the openings in the cribriform plate of the ethmoid bone through which pass the fila olfactoria of the olfactory nerves. { äl′fak·trē fə′rā·mən }

olfactory gland [PHYSIOLOGY] A type of serous gland in the nasal mucous membrane. { äl′fak·trē ,gland }

olfactory lobe [VERTEBRATE ZOOLOGY] A lobe projecting forward from the inferior surface of the frontal lobe of each cerebral hemisphere, including the olfactory bulb, tracts, and trigone; well developed in most vertebrates, but reduced in humans. { äl′fak·trē ,lōb }

olfactory nerve [ANATOMY] The first cranial nerve; a paired sensory nerve with its origin in the olfactory lobe and formed by processes of the olfactory cells which lie in the nasal mucosa; greatly reduced in humans. { äl′fak·trē ,nərv }

olfactory organ [PHYSIOLOGY] Any of the small chemoreceptors in the mucous membrane lining the upper part of the nasal gravity which receive stimuli interpreted as odors. { äl′fak·trē ,ȯr·gən }

olfactory pit [EMBRYOLOGY] A depression near the olfactory placode in the embryo that develops into part of the nasal cavity. Also known as nasal pit. { äl′fak·trē ,pit }

olfactory region [ANATOMY] The area on and above the superior conchae and on the adjoining nasal septum where the mucous membrane has olfactory epithelium and olfactory glands. { äl′fak·trē ,rē·jən }

olfactory stalk [ANATOMY] The structure that connects the olfactory bulb to the cerebrum of the vertebrate brain. { äl′fak·trē ,stȯk }

olfactory tract [ANATOMY] A narrow tract of white nerve fibers originating in the olfactory bulb and extending posteriorly to the anterior perforated substance, where it enlarges to form a lateral root (olfactory trigone). { äl′fak·trē ,trakt }

Oligobrachiidae [INVERTEBRATE ZOOLOGY] A monotypic family of the order Athecanephria. { ,äl·ə·gō·brə′kī·ə,dē }

Oligochaeta [INVERTEBRATE ZOOLOGY] A class of the phylum Annelida including worms that exhibit both external and internal segmentation, and setae which are not borne on parapodia. { ,äl·ə·gō′kēd·ə }

oligodendroglia [HISTOLOGY] Small neuroglial cells with spheroidal or ovoid nuclei and fine cytoplasmic processes with secondary divisions. { ,äl·ə·gō·den′dräg·lē·ə }

oligodynamic action [MICROBIOLOGY] The inhibiting or killing of microorganisms by use of very small amounts of a chemical substance. { ¦äl·ə·gō·dī′nam·ik ′ak·shən }

Oligomera [INVERTEBRATE ZOOLOGY] A subphylum of the phylum Vermes comprising groups with two or three coelomic divisions. { ,äl·ə ′gäm·ə·rə }

oligomeric protein [BIOCHEMISTRY] A protein composed of two or more polypeptide chains. { ə,lig·ə′mer·ik ′prō,tēn }

oligomerous [BOTANY] Having one or more whorls with fewer members than other whorls of the flower. { ,äl·ə′gäm·ə·rəs }

oligomycin [MICROBIOLOGY] Any of a group of antifungal antibiotics produced by an actinomycete resembling *Streptomyces diastachromogenes*; the colorless, hexagonal crystals are soluble in many organic solvents. { ə,lig·ə′mīs·ən }

oligonucleotide [BIOCHEMISTRY] A polynucleotide of low molecular weight, consisting of less than 20 nucleotide polymers. { ,äl·ə·gō′nü·klē·ə,tīd }

oligophagous [ZOOLOGY] Eating only a limited variety of foods. { ¦äl·ə¦gäf·ə·gəs }

Oligopygidae [PALEONTOLOGY] An extinct family of exocyclic Euechinoidia in the order Holectypoida which were small ovoid forms of the Early Tertiary. { ¦äl·ə·gō¦pij·ə,dē }

oligosaccharide [BIOCHEMISTRY] A sugar composed of two to eight monosaccharide units joined by glycosidic bonds. Also known as compound sugar. { ,äl·ə·gō′sak·ə,rīd }

Oligotrichida [INVERTEBRATE ZOOLOGY] A minor order of the Spirotrichia; the body is round in cross section, and the adoral zone of membranelles is often highly developed at the oral end of the organism. { ¦äl·ə·gō′trik·ə·də }

olivaceous [BIOLOGY] **1.** Resembling an olive. **2.** Olive colored. { ,äl·ə′vā·shəs }

olivary nucleus [ANATOMY] A prominent, convoluted gray band that opens medially and occupies the upper two-thirds of the medulla oblongata. { ′äl·ə,ver·ē ′nü·klē·əs }

olive [BOTANY] Any plant of the genus *Olea* in the order Schrophulariales, especially the evergreen olive tree (*O. europea*) cultivated for its drupaceous fruit, which is eaten ripe (black olives) and unripe (green), and is of high oil content. { ′äl·əv }

Olividae [INVERTEBRATE ZOOLOGY] A family of snails in the gastropod order Neogastropoda. { ō ′liv·ə,dē }

omasum [VERTEBRATE ZOOLOGY] The third chamber of the ruminant stomach where the contents are mixed to a more or less homogeneous state. Also known as manyplies; psalterium. { ō′mā·səm }

ombrophilous [ECOLOGY] Able to thrive in areas of abundant rainfall. { äm′bräf·ə·ləs }

ombrophobous [ECOLOGY] Unable to live in the presence of long, continuous rain. { äm′bräf·ə·bəs }

omentum [ANATOMY] A fold of the peritoneum connecting or supporting abdominal viscera. { ō ′ment·əm }

ommatidium [INVERTEBRATE ZOOLOGY] The structural unit of a compound eye, composed of a cornea, a crystalline cone, and a receptor element connected to the optic nerve. { ˌäm·ə'tid·ē·əm }

ommatophore [INVERTEBRATE ZOOLOGY] A movable peduncle that bears an eye, as in snails. { ə 'mad·əˌfȯr }

omnivore [ZOOLOGY] An organism that eats both animal and vegetable matter. { 'äm·nəˌvȯr }

omohyoid [ANATOMY] **1.** Pertaining conjointly to the scapula and the hyoid bone. **2.** A muscle attached to the scapula and the hyoid bone. { ¦ō·mō'hīˌȯid }

Omophronidae [INVERTEBRATE ZOOLOGY] The savage beetles, a small family of coleopteran insects in the suborder Adephaga. { ˌo·mə'frän·əˌdē }

omphalomesenteric artery [EMBRYOLOGY] *See* vitelline artery. { ¦äm·fə·lōˌmez·ən'ter·ik 'ärd·ə·rē }

omphalomesenteric duct [EMBRYOLOGY] *See* vitelline duct. { ¦äm·fə·lōˌmez·ən'ter·ik 'dəkt }

omphalomesenteric vein [EMBRYOLOGY] *See* vitelline vein. { ¦äm·fə·lōˌmez·ən'ter·ik 'vān }

Omphralidae [INVERTEBRATE ZOOLOGY] A family of orthorrhaphous dipteran insects in the series Nematocera. { äm'fral·əˌdē }

Onagraceae [BOTANY] A family of dicotyledonous plants in the order Myrtales characterized generally by an inferior ovary, axile placentation, twice as many stamens as petals, a four-nucleate embryo sac, and many ovules. { ˌän·ə'grās·ēˌē }

Onchidiidae [INVERTEBRATE ZOOLOGY] An intertidal family of sluglike pulmonate mollusks of the order Systellommatophora in which the body is oval or lengthened, with the convex dorsal integument lacking a mantle cavity or shell. { ˌäŋ·kə'dī·əˌdē }

Oncholaimoidea [INVERTEBRATE ZOOLOGY] A superfamily of nematodes in the order Enoplida, characterized by a stoma armed with one dorsal tooth and two subventral teeth and sometimes with transverse rows of small denticles from its walls, and two whorls of cephalic sensilla. { ˌäŋ·kō·lə'mȯid·ē·ə }

onchosphere [INVERTEBRATE ZOOLOGY] The hexacanth embryo identified as the earliest differentiated stage of cyclophyllidean tapeworms. { 'äŋ·kōˌsfir }

oncocyte [HISTOLOGY] A columnar-shaped cell with finely granular eosinophilic cytoplasm, found in salivary and certain endocrine glands, nasal mucosa, and other locations. { 'äŋ·kōˌsīt }

oncogene [GENETICS] A gene that causes cancer in an animal; the gene specifies the structure of an enzyme that catalyzes events that can induce cancerous growth. { 'äŋ·kōˌjēn }

oncogenic virus [VIROLOGY] A virus that transforms the infected cells so that they undergo uncontrolled proliferation. { ¦aŋ·kəˌjen·ik 'vī·rəs }

oncomouse [BIOLOGY] A laboratory mouse that carries activated human cancer genes. { 'äŋk·əˌmau̇s }

Oncopoda [INVERTEBRATE ZOOLOGY] A phylum of the superphylum Articulata. { äŋ'käp·ə·də }

oncotic pressure [PHYSIOLOGY] Also known as colloidal osmotic pressure. **1.** The osmotic pressure exerted by colloids in a solution. **2.** The pressure exerted by plasma proteins. { äŋ'käd·ik 'presh·ər }

onion [BOTANY] **1.** *Allium cepa*. A biennial plant in the order Liliales cultivated for its edible bulb. **2.** Any plant of the genus *Allium*. { 'ən·yən }

onisciform [INVERTEBRATE ZOOLOGY] Ovate and slightly flattened. { ä'nis·əˌfȯrm }

Oniscoidea [INVERTEBRATE ZOOLOGY] A terrestrial suborder of the Isopoda; the body is either dorsoventrally flattened or highly vaulted, and the head, thorax, and abdomen are broadly joined. { ˌän·ə'skȯid·ē·ə }

ontogeny [EMBRYOLOGY] The origin and development of an organism from conception to adulthood. { än'täj·ə·nē }

Onuphidae [INVERTEBRATE ZOOLOGY] A family of tubicolous, herbivorous, scavenging errantian annelids in the superfamily Eunicea. { än'yüf·əˌdē }

Onychodontidae [PALEONTOLOGY] A family of Lower Devonian lobefin fishes in the order Osteolepiformes. { ¦än·ə·kō'dänt·əˌdē }

onych-, onycho- [ZOOLOGY] A combining form denoting claw or nail. { 'än·ə·kō }

Onychopalpida [INVERTEBRATE ZOOLOGY] A suborder of mites in the order Acarina. { ¦an·ə·kō 'pal·pə·də }

Onychophora [INVERTEBRATE ZOOLOGY] A phylum of wormlike animals that combine features of both the annelids and the arthropods. { ˌän·ə 'käf·ə·rə }

Onygenaceae [MYCOLOGY] A family of ascomycetous fungi in the order Eurotiales comprising forms that inhabit various animal substrata, such as horns and hoofs. { ˌän·ə·jə'nās·ēˌē }

oocyst [INVERTEBRATE ZOOLOGY] The encysted zygote of some Sporozoa. { 'ō·əˌsist }

oocyte [HISTOLOGY] An egg before the completion of maturation. { 'ō·əˌsīt }

oogamete [BIOLOGY] A large, nonmotile female gamete containing reserve materials. { ¦ō·ə'ga ˌmēt }

oogamous [BIOLOGY] Of sexual reproduction, characterized by fusion of a motile sperm with an oogamete. { ō'äg·ə·məs }

oogenesis [PHYSIOLOGY] Processes involved in the growth and maturation of the ovum in preparation for fertilization. { ˌō·ə'jen·ə·səs }

oogonium [BOTANY] The unisexual female sex organ in oogamous algae and fungi. [HISTOLOGY] A descendant of a primary germ cell which develops into an oocyte. { ˌō·ə'gō·nē·əm }

ookinete [INVERTEBRATE ZOOLOGY] The elongated, mobile zygote of certain Sporozoa, as that of the malaria parasite. { ˌō·ə'kīˌnēt }

oolemma [HISTOLOGY] *See* zona pellucida. { ¦ō·ə 'lem·ə }

oology [VERTEBRATE ZOOLOGY] A branch of zoology concerned with the study of eggs, especially bird eggs. { ō'äl·ə·jē }

Oomycetes [MYCOLOGY] A class of the Phycomycetes comprising the biflagellate water molds and downy mildews. { ˌō·ə·mī'sēd·ēz }

oophagous [ZOOLOGY] Feeding or living on eggs. { ō′äf·ə·gəs }

ooplasm [CELL BIOLOGY] Cytoplasm of an egg. { ′ō·ə,plaz·əm }

oospore [BOTANY] A spore which is produced by heterogamous fertilization and from which the sporophytic generation develops. { ′ō·ə,spòr }

oostegite [INVERTEBRATE ZOOLOGY] In many crustaceans, a platelike expansion of the basal segment of a thoracic appendage that aids in forming an egg receptacle. { ō′äs·tə,jīt }

ootype [INVERTEBRATE ZOOLOGY] In trematodes and tapeworms, a thickening of the oviduct near the ovaries. { ′ō·ə,tīp }

Opalinata [INVERTEBRATE ZOOLOGY] A superclass of the subphylum Sarcomastigophora containing highly specialized forms which resemble ciliates. { o¦pal·ə¦näd·ə }

Opegraphaceae [BOTANY] A family of the Hysteriales characterized by elongated ascocarps; members are crustose on bark and rocks. { ,ō·pə·grə′fās·ə,ē }

open bundle [BOTANY] A vascular bundle containing cambium. { ′ō·pən ′bən·dəl }

open-circle deoxyribonucleic acid [BIOCHEMISTRY] *See* relaxed circular deoxyribonucleic acid. { ′ō·pən ,sər·kəl dē,äk·sē,rī·bō·nü′klē·ik′as·əd }

open community [ECOLOGY] A community which other organisms readily colonize because some niches are unoccupied. { ′ō·pən kə′myü·nəd·ē }

open reading frame [MOLECULAR BIOLOGY] A stretch of triplets contained between an initator codon and a terminator codon. Abbreviated ORF. { ¦ō·pən ¦rēd·iŋ ¦frām }

open water [ECOLOGY] Lake water that is free from emergent vegetation, artificial obstructions, or tangled masses of underwater vegetation at very shallow depths. { ′ō·pən ′wòd·ər }

operator [GENETICS] A sequence at one end of an operon on which a repressor acts, thus regulating the transcription of the operon. { ′äp·ə ,rād·ər }

operculum [ANATOMY] **1.** The soft tissue partially covering the crown of an erupting tooth. **2.** That part of the cerebrum which borders the lateral fissure. [BIOLOGY] **1.** A lid, flap, or valve. **2.** A lidlike body process. { ō′pər·kyə·ləm }

operon [GENETICS] A functional unit composed of a number of adjacent cistrons on the chromosome; its transcription is regulated by a receptor sequence, the operator, and a repressor. { ′äp·ə,rän }

operon network [GENETICS] A group of operons and their associated regulator genes that interact such that the products of one operon activate or suppress another operon. { ′äp·ə,rän ,net ,wərk }

Opheliidae [INVERTEBRATE ZOOLOGY] A family of limivorous worms belonging to the annelid group Sedentaria. { ,äf·ə′lī·ə,dē }

Ophiacodonta [VERTEBRATE ZOOLOGY] A suborder of extinct reptiles in the order Pelycosauria, including primitive, partially aquatic carnivores. { ¦af·ē·ə·kə′dän·tə }

Ophidiidae [VERTEBRATE ZOOLOGY] A family of small actinopterygian fishes in the order Gadiformes, comprising the cusk eels and brotulas. { ,äf·ə′dī·ə,dē }

Ophiocanopidae [INVERTEBRATE ZOOLOGY] A family of asterozoan echinoderms in the subclass Ophiuroidea. { ,äf·ē·ō·kə′näp·ə,dē }

Ophiocistioidea [PALEONTOLOGY] A small class of extinct Echinozoa in which the domed aboral surface of the test was roofed by polygonal plates and carried an anal pyramid. { ,äf·ē·ō,sis·tē ′òid·ē·ə }

Ophioglossales [BOTANY] An order of ferns in the subclass Ophioglossidae. { ,äf·ē·ō·glä′sā·lēz }

Ophioglossidae [BOTANY] The adder's-tongue ferns, a small subclass of the class Polypodiopsida; the plants are homosporous and eusporangiate and are distinguished by the arrangement of the sporogenous tissue in the characteristic fertile spike of the sporophyte. { ,äf·ē·ō′gläs·ə,dē }

Ophiomyxidae [INVERTEBRATE ZOOLOGY] The single family of the echinoderm suborder Ophiomyxina distinguished by a soft, unprotected integument. { ,äf·ē·ō′mik·sə,dē }

Ophiomyxina [INVERTEBRATE ZOOLOGY] A monofamilial suborder of ophiuroid echinoderms in the order Phrynophiurida. { ,äf·ē·ō ′mik·sə·nə }

ophiopluteus [INVERTEBRATE ZOOLOGY] The pluteus larva of brittle stars. { ¦äf·ē·ō′plüd·ē·əs }

Ophiurida [INVERTEBRATE ZOOLOGY] An order of echinoderms in the subclass Ophiuroidea in which the vertebrae articulate by means of ball-and-socket joints, and the arms, which do not branch, move mainly from side to side. { ,äf·ə ′yúr·ə·də }

Ophiuroidea [INVERTEBRATE ZOOLOGY] The brittle stars, a subclass of the Asterozoa in which the arms are usually clearly demarcated from the central disk and perform whiplike locomotor movements. { äf·ə·yə′ròid·ē·ə }

ophthalmic [ANATOMY] Of or pertaining to the eye. { äf′thal·mik }

ophthalmic nerve [ANATOMY] A sensory branch of the trigeminal nerve which supplies the lacrimal glands, upper eyelids, skin of the forehead, and anterior portion of the scalp, meninges, nasal mucosa, and frontal, ethmoid, and sphenoid air sinuses. { äf′thal·mik ′nərv }

Opilioacaridae [INVERTEBRATE ZOOLOGY] The single family of moderately large mites of the suborder Notostigmata which comprises the Opiliocariformes. { ō¦pil·ē·ō′kar·ə,dē }

Opilioacariformes [INVERTEBRATE ZOOLOGY] A small monofamilial order of the Acari comprising large mites characterized by long legs and by the possession of a pretarsus on the pedipalp, with prominent claws. { ō¦pil·ē·ō,kar·ə′fòr·mēz }

opine [BIOCHEMISTRY] A type of amino acid usually not found in nature, such as that secreted by a crown gall. { ′ō,pīn }

opisthaptor [INVERTEBRATE ZOOLOGY] A posterior adhesive organ in monogenetic trematodes. { ¦äp·əs¦thap·tər }

Opisthobranchia [INVERTEBRATE ZOOLOGY] A subclass of the class Gastropoda containing the sea hares, sea butterflies, and sea slugs; generally characterized by having gills, a small external or internal shell, and two pairs of tentacles. { ə ˌpis·thə'braŋ·kē·ə }

Opisthocoela [VERTEBRATE ZOOLOGY] A suborder of the order Anura; members have opisthocoelous trunk vertebrae, and the adults typically have free ribs. { əˌpis·thə'sē·lə }

opisthocoelous [ANATOMY] Of, related to, or being a vertebra with the centrum convex anteriorly and concave posteriorly. { ə¦pis·thə¦sē·ləs }

Opisthocomidae [VERTEBRATE ZOOLOGY] A family of birds in the order Galliformes, including the hoatzins. { əˌpis·thə'käm·əˌdē }

opisthognathous [INVERTEBRATE ZOOLOGY] Having the mouthparts ventral and posterior to the cranium. [VERTEBRATE ZOOLOGY] Having retreating jaws. { ¦ä·pəs¦thäg·nə·thəs }

opisthonephros [VERTEBRATE ZOOLOGY] The fundamental adult kidney in amphibians and fishes. { əˌpis·thə'neˌfräs }

Opisthopora [INVERTEBRATE ZOOLOGY] An order of the class Oligochaeta distinguished by meganephridiostomal, male pores opening posteriorly to the last testicular segment. { ¦ä·pəs ¦thäp·ə·rə }

opisthotic [ANATOMY] Of, relating to, or being the posterior and inferior portions of the bony elements in the inner ear capsule. { ¦ä·pəs ¦thäd·ik }

Opomyzidae [INVERTEBRATE ZOOLOGY] A family of cyclorrhaphous myodarian dipteran insects in the subsection Acalypteratae. { ˌäp·ə'mī·zə ˌdē }

opossum [VERTEBRATE ZOOLOGY] Any member of the family Didelphidae in the order Marsupialia; these mammals are arboreal and mainly omnivorous, and have many incisors, with all teeth pointed and sharp. { ə'päs·əm }

opportunistic microorganism [MICROBIOLOGY] A normally harmless endogenous microorganism that produces disease due to fortuitous events that affect the host. { ¦äp·ərˌtü¦nis·tik ¦mī·krō 'ór·gəˌniz·əm }

opportunistic species [ECOLOGY] Species characterized by high reproduction rates, rapid development, early reproduction, small body size, and uncertain adult survival. { ¦äp·ərˌtü¦nis·tik 'spē·shēz }

opposite [BOTANY] **1.** Located side by side. **2.** Of leaves, being in pairs on an axis with each member separated from the other of the pair by half the circumference of the axis. { 'äp·ə·zət }

opsonic action [IMMUNOLOGY] The effect produced upon susceptible microorganisms and other cells by opsonins, which renders them vulnerable to phagocytes. { äp'sän·ik 'ak·shən }

opsonic index [IMMUNOLOGY] A numerical measure of the opsonic activity of sera, expressed as the ratio of the average number of bacteria engulfed per phagocytic cell in immune serum compared with the corresponding value for normal serum. { äp'sän·ik 'inˌdeks }

opsonin [IMMUNOLOGY] A substance in blood serum that renders bacteria more susceptible to phagocytosis by leukocytes. { 'äp·sə·nən }

opsonize [IMMUNOLOGY] To render microorganisms susceptible to phagocytosis. { 'äp·sə ˌnīz }

optic [BIOLOGY] Pertaining to the eye. { 'äp·tik }

optical axis [ANATOMY] An imaginary straight line passing through the midpoint of the cornea (anterior pole) and the midpoint of the retina (posterior pole). { 'äp·tə·kəl 'ak·səs }

optical-righting reflex [PHYSIOLOGY] *See* visual-righting reflex. { 'äp·tə·kəl ¦rīd·iŋ ˌrēˌfleks }

optic canal [ANATOMY] The channel at the apex of the orbit, the anterior termination of the optic groove, just beneath the lesser wing of the sphenoid bone; it gives passage to the optic nerve and ophthalmic artery. { 'äp·tik kə'nal }

optic capsule [VERTEBRATE ZOOLOGY] A cartilaginous capsule that develops around the eye in elasmobranchs and higher vertebrate embryos. { 'äp·tik 'kap·səl }

optic chiasma [ANATOMY] The partial decussation of the optic nerves on the undersurface of the hypothalamus. { 'äp·tik kī'az·mə }

optic cup [EMBRYOLOGY] A two-layered depression formed by invagination of the optic vesicle from which the pigmented and sensory layers of the retina will develop. { 'äp·tik ˌkəp }

optic disk [ANATOMY] The circular area in the retina that is the site of the convergence of fibers from the ganglion cells of the retina to form the optic nerve. { 'äp·tik ˌdisk }

optic gland [INVERTEBRATE ZOOLOGY] Either of a pair of endocrine glands in the octopus and squid which are found near the brain and produce a substance which causes gonadal maturation. { 'äp·tik ˌgland }

optic lobe [ANATOMY] One of the anterior pair of colliculi of the mammalian corpora quadrigemina. [INVERTEBRATE ZOOLOGY] A lateral lobe of the forebrain in certain arthropods. [VERTEBRATE ZOOLOGY] Either of the corpora bigemina of lower vertebrates. { 'äp·tik ˌlōb }

optic nerve [ANATOMY] The second cranial nerve; a paired sensory nerve technically consisting of three layers of special nerve cells in the retina of the eye; fibers converge to form the optic tracts. { 'äp·tik ˌnərv }

optic stalk [EMBRYOLOGY] The constriction of the optic vesicle which connects the embryonic eye and forebrain in vertebrates. { 'äp·tik ˌstók }

optic tectum [VERTEBRATE ZOOLOGY] The roof of the mesencephalon constituting a major visual center and association area of the brain of premature vertebrates. { 'äp·tik 'tek·təm }

optic tract [ANATOMY] The band of optic nerve fibers running from the optic chiasma to the lateral geniculate body and midbrain. { 'äp·tik ˌtrakt }

optic vesicle [EMBRYOLOGY] An evagination of the lateral wall of the forebrain in vertebrate embryos which precedes formation of the optic cup. { 'äp·tik 'ves·ə·kəl }

oral [ANATOMY] Of or pertaining to the mouth. { 'ór·əl }

oral arm [INVERTEBRATE ZOOLOGY] In a jellyfish,

any of the prolongations of the distal end of the manubrium. { 'ȯr·əl ¦ärm }

oral cavity [ANATOMY] The cavity of the mouth. { 'ȯr·əl 'kav·əd·ē }

oral disc [INVERTEBRATE ZOOLOGY] The flattened upper or free end of the body of a polyp that has the mouth in the center and tentacles around the margin. { 'ȯr·əl ¦disk }

oral groove [INVERTEBRATE ZOOLOGY] A depressed, groovelike peristome. { 'ȯr·əl ¦grüv }

orange [BOTANY] Any of various evergreen trees of the genus *Citrus*, cultivated for the edible fruit, a berry with an aromatic, leathery rind containing numerous oil glands. { 'är·inj }

orangeophile [HISTOLOGY] A type of acidophile cell of the anterior lobe of the adenohypophysis, presumed to elaborate growth hormone. { ə'ran·jē·ə,fīl }

orangutan [VERTEBRATE ZOOLOGY] *Pongo pygmaeus*.The largest of the great apes, a long-armed primate distinguished by long sparse reddish-brown hair, naked face and hands and feet, and a large laryngeal cavity which appears as a pouch below the chin. { ə'raŋ·ü,tan }

Orbiniidae [INVERTEBRATE ZOOLOGY] A family of polychaete annelids belonging to the Sedentaria; the prostomium is exposed, and the thorax and abdomen are weakly separated. { ,ȯr·bə'nī·ə,dē }

Orbiniinae [INVERTEBRATE ZOOLOGY] A subfamily of sedentary polychaete annelids in the family Orbiniidae. { ȯr·bə'nī·ə,nē }

orbit [ANATOMY] The bony cavity in the lateral front of the skull beneath the frontal bone which contains the eyeball. Also known as eye socket. { 'ȯr·bət }

orbital fossa [INVERTEBRATE ZOOLOGY] A depression from which the eyestalk arises on the front of the carapace of crustaceans. { 'ȯr·bəd·əl 'fäs·ə }

Orbitoidacea [INVERTEBRATE ZOOLOGY] A superfamily of foraminiferan protozoans in the suborder Rotaliina characterized by a low trochospire or a planispiral, uncoiled or branching test composed of radial calcite with bilamellar septa. { ,ȯr·bə·tȯi'dās·ē,ə }

orch-, orchi-, orchid-, orchido-, orchio- [ZOOLOGY] A combining form denoting testis. { 'ȯr·kē, 'ȯr·kəd·ō, 'ȯr·kē·ō }

orchid [BOTANY] Any member of the family Orchidaceae; plants have complex, specialized irregular flowers usually with only one or two stamens. { 'ȯr·kəd }

Orchidaceae [BOTANY] A family of monocotyledonous plants in the order Orchidales characterized by irregular flowers with only one or two stamens which are adnate to the style, and pollen grains which cohere in large masses called pollinia. { ,ȯr·kə'dās·ē,ē }

Orchidales [BOTANY] An order of monocotyledonous plants in the subclass Liliidae; plants are mycotropic and sometimes nongreen with numerous tiny seeds that have an undifferentiated embryo and little or no endosperm. { ,ȯr·kə'dā·lēz }

order [SYSTEMATICS] A taxonomic category ranked below the class and above the family, made up either of families, subfamilies, or suborders. { 'ȯrd·ər }

ordered octad [MYCOLOGY] A linear sequence of pairs of each of four haploid cells produced by a postmeiotic division within a fungal ascus. { 'ȯrd·ərd 'äk,tad }

ordered tetrad [MYCOLOGY] A linear sequence of four haploid meiotic cells within a fungal ascus. { 'ȯrd·ərd 'te,trad }

Orectolobidae [VERTEBRATE ZOOLOGY] An ancient isurid family of galeoid sharks, including the carpet and nurse sharks, which are primarily bottom feeders with small teeth and a blunt rostrum with barbels near the mouth. { ȯ,rek·tə'läb·ə,dē }

oreodont [PALEONTOLOGY] Any member of the family Merycoidodontidae. { 'ȯr·ē·ō,dänt }

ORF [MOLECULAR BIOLOGY] *See* open reading frame. { ȯrf *or* ¦ō¦är'ef }

organ [ANATOMY] A differentiated structure of an organism composed of various cells or tissues and adapted for a specific function. { 'ȯr·gən }

organelle [CELL BIOLOGY] A specialized subcellular structure, such as a mitochondrion, having a special function; a condensed system showing a high degree of internal order and definite limits of size and shape. { ¦ȯr·gə¦nel }

organic evolution [EVOLUTION] The processes of change in organisms by which descendants come to differ from their ancestors, and a history of the sequence of such changes. { ȯr'gan·ik ,ev·ə'lü·shən }

organicism [BIOLOGY] *See* holism. { ȯr'gan·ə,siz·əm }

organism [BIOLOGY] An individual constituted to carry out all life functions. { 'ȯr·gə,niz·əm }

organized ferment [BIOCHEMISTRY] *See* intracellular enzyme. { ,ȯr·gə·nə'zā·shən 'fər·ment }

organizer [EMBRYOLOGY] Any part of the embryo which exerts a morphogenetic stimulus on an adjacent part or parts, as in the induction of the medullary plate by the dorsal lip of the blastopore. { 'ȯr·gə,niz·ər }

organ of Corti [ANATOMY] A specialized structure located on the basilar membrane of the mammalian cochlea, which contains rods of Corti and hair cells connected to ganglia of the cochlear nerve. Also known as spiral organ. { 'ȯr·gən əv 'kȯrd·ē }

organ of Leydig [VERTEBRATE ZOOLOGY] Two large accumulations of lymphoid tissues which run longitudinally the length of the esophagus in selachian fishes. { 'ȯr·gən əv 'lī,dig }

organogenesis [EMBRYOLOGY] The formation of an organ. { ȯr,gan·ə'jen·ə·səs }

organoleptic [PHYSIOLOGY] Having an effect or making an impression on sense organs; usually used in connection with subjective testing of food and drug products. { ȯr¦gan·ə¦lep·tik }

organotropic [MICROBIOLOGY] Of microorganisms, localizing in or entering the body by way of the viscera or, occasionally, somatic tissue. { ȯr¦gan·ə¦träp·ik }

orgasm [PHYSIOLOGY] The intense, diffuse, and subjectively pleasurable sensation experienced during sexual intercourse or genital manipulation, culminating in the male with seminal ejaculation and in the female with uterine contractions, warm suffusion, and pelvic throbbing sensations. { 'ȯr‚gaz·əm }

Oribatei [INVERTEBRATE ZOOLOGY] A heavily sclerotized group of free-living mites in the suborder Sarcoptiformes which serve as intermediate hosts of tapeworms. { ‚ȯr·ə'bad·ē‚ī }

Oribatulidae [INVERTEBRATE ZOOLOGY] A family of oribatid mites in the suborder Sarcoptiformes. { ‚ȯr·ə·bə'tül·ə‚dē }

Oriental zoogeographic region [ECOLOGY] A zoogeographic region which encompasses tropical Asia from the Iranian Peninsula eastward through the East Indies to, and including, Borneo and the Philippines. { ‚ȯr·ē'ent·əl ‚zō·ə‚jē·ə'graf·ik ‚rē·jən }

origin [ANATOMY] The point at which the nonmoving end of a muscle is attached to a bone; it is at the proximal end of the muscle. { 'är·ə·jən }

ormer [INVERTEBRATE ZOOLOGY] *See* abalone. { 'ȯr·mər }

Ormyridae [INVERTEBRATE ZOOLOGY] A small family of hemipteran insects in the superfamily Chalcidoidea. { ȯr'mī·rə‚dē }

Orneodidae [INVERTEBRATE ZOOLOGY] A small family of lepidopteran insects in the superfamily Tineoidea; adults have each wing divided into six featherlike plumes. { ‚ȯr·nē'äd·ə‚dē }

ornithine [BIOCHEMISTRY] $C_5H_{12}O_2N_2$ An amino acid occurring in the urine of some birds, but not found in native proteins. { 'ȯr·nə‚thēn }

ornithine cycle [BIOCHEMISTRY] A sequence of cyclic reactions in which potentially toxic products of protein catabolism are converted to nontoxic urea. { 'ȯr·nə‚thīn ‚sī·kəl }

Ornithischia [PALEONTOLOGY] An order of extinct terrestrial reptiles, popularly known as dinosaurs; distinguished by a four-pronged pelvis, and a median, toothless predentary bone at the front of the lower jaw. { ‚ȯr·nə'this·kē·ə }

ornithology [VERTEBRATE ZOOLOGY] The study of birds. { ‚ȯr·nə'thäl·ə·jē }

Ornithopoda [PALEONTOLOGY] A suborder of extinct reptiles in the order Ornithischia including all bipedal forms in the order. { ‚ȯr·nə'thäp·ə·də }

Ornithorhynchidae [VERTEBRATE ZOOLOGY] A monospecific order of monotremes containing the semiaquatic platypus; characterized by a duck-billed snout, horny plates instead of teeth in the adult, and a flattened, well-developed tail. { ‚ȯr·nə·thō'riŋ·kə‚dē }

Oromericidae [VERTEBRATE ZOOLOGY] An extinct family of camellike tylopod runinants in the superfamily Cameloidea. { ‚ȯr·ə·mə'ris·ə‚dē }

oropharynx [ANATOMY] The oral pharynx, located between the lower border of the soft palate and the larynx. { ¦ȯr·ō'far‚iŋks }

orophyte [ECOLOGY] Any plant that grows in the subalpine region. { 'ȯr·ə‚fīt }

orotic acid [BIOCHEMISTRY] $C_4H_4O_4N_2$ A crystalline acid which is a growth factor for certain bacteria and is also a pyrimidine precursor. { ə'räd·ik 'as·əd }

orphan virus [VIROLOGY] Any nonpathogenic virus found in the human digestive and respiratory tracts. { ¦ȯr·fən 'vī·rəs }

Orthacea [PALEONTOLOGY] An extinct group of articulate brachiopods in the suborder Orthidina in which the delthyrium is open. { ȯr'thās·ē·ə }

Ortheziinae [INVERTEBRATE ZOOLOGY] A subfamily of homopteran insects in the superfamily Coccoidea having abdominal spiracles present in all stages and a flat anal ring bearing pores and setae in immature forms and adult females. { ‚ȯr·thə'zī·ə‚nē }

Orthida [PALEONTOLOGY] An order of extinct articulate brachiopods which includes the oldest known representatives of the class. { 'ȯr·thə·də }

Orthidina [PALEONTOLOGY] The principal suborder of the extinct Orthida, including those articulate brachiopods characterized by biconvex, finely ribbed shells with a straight hinge line and well-developed interareas on both valves. { ȯr'thid·ən·ə }

orthoceratite [INVERTEBRATE ZOOLOGY] Any nautiloid belonging to the genus *Orthoceras*, characterized by the presence of three longitudinal furrows on the body chamber. { ‚ȯr·thə'ser·ə‚tīt }

orthochromatic [BIOLOGY] Having normal staining characteristics. { ‚ȯr·thə·krə'mad·ik }

orthogenesis [EVOLUTION] A unidirectional evolutionary change among a related group of animals. { ‚ȯr·thə'jen·ə·səs }

orthokinesis [BIOLOGY] Random movement of a motile cell or organism in response to a stimulus. { ¦ȯr·thə·ki'nē·səs }

orthologous locus [GENETICS] A gene that evolved without diverging from an ancestral locus. { ȯr'thäl·ə·gəs ‚lō·kəs }

Orthonectida [INVERTEBRATE ZOOLOGY] An order of Mesozoa; orthonectids parasitize various marine invertebrates as multinucleate plasmodia, and sexually mature forms are ciliated organisms. { ‚ȯr·thə'nek·tə·də }

Orthoperidae [INVERTEBRATE ZOOLOGY] The minute fungus beetles, a family of coleopteran insects in the superfamily Cucujoidea. { ‚ȯr·thə'per·ə‚dē }

Orthopsida [INVERTEBRATE ZOOLOGY] An order of echinoderms in the subclass Euechinoidea. { ȯr'thäp·sə·də }

Orthopsidae [PALEONTOLOGY] A family of extinct echinoderms in the order Hemicidaroida distinguished by a camarodont lantern. { ȯr'thäp·sə‚dē }

Orthoptera [INVERTEBRATE ZOOLOGY] A heterogeneous order of generalized insects with gradual metamorphosis, chewing mouthparts, and four wings. { ȯr'thäp·tə·rə }

Orthorrhapha [INVERTEBRATE ZOOLOGY] A suborder of the Diptera; in this group of flies, the adult escapes from the puparium through a T-shaped opening. { or'thȯr·ə·fə }

Orthotrichales [BOTANY] An order of true mosses in the subclass Bryidae, characterized by dull, tuft- or mat-forming plants that are probably heterogeneous, making a generalized description difficult. { ˌȯr·thō·trəˈkā·lēz }

orthotropism [BOTANY] The tendency of a plant to grow with the longer axis oriented vertically. { ȯrˈthä·trəˌpiz·əm }

orthotropous [BOTANY] Having a straight ovule with the micropyle at the end opposite the stalk. { ȯrˈthä·trə·pəs }

Orussidae [INVERTEBRATE ZOOLOGY] A small family of hymenopteran insects in the superfamily Siricoidea. { ōˈrüs·əˌdē }

oryctocoenosis [PALEONTOLOGY] The part of a thanatocoenosis that has been preserved as a fossil. { əˌrik·tə·səˈnō·səs }

osage orange [BOTANY] *Maclura pomifera*. A tree in the mulberry family of the Urticales characterized by yellowish bark, milky sap, simple entire leaves, strong axillary thorns, and aggregate green fruit about the size and shape of an orange. { ˈōˌsāj ˈär·inj }

osazone [BIOCHEMISTRY] Any of the compounds that contain two phenylhydrazine residues and are produced by a reaction between a reducing sugar and phenylhydrazine. { ˈō·səˌzōn }

Oscillatoriales [BOTANY] An order of blue-green algae (Cyanophyceae) which are filamentous and truly multicellular. { ¦äs·ə·ləˌtȯr·ēˈā·lēz }

Oscillospiraceae [MICROBIOLOGY] Formerly a family of large, gram-negative, motile bacteria of the order Caryophanales which lose motility on exposure to oxygen. { ˌäs·ə·lō·spəˈrās·ēˌē }

Oscines [VERTEBRATE ZOOLOGY] The songbirds, a suborder of the order Passeriformes. { ˈäs·ə ˌnēz }

osculum [INVERTEBRATE ZOOLOGY] An excurrent orifice in Porifera. { ˈäs·kyə·ləm }

osmophile [MICROBIOLOGY] A microorganism adapted to media with high osmotic pressure. { ˈäz·məˌfīl }

osmoreceptor [PHYSIOLOGY] One of a group of structures in the hypothalamus which respond to changes in osmotic pressure of the blood by regulating the secretion of the neurohypophyseal antidiuretic hormone. { ¦äz·mō·riˈsep·tər }

osmoregulatory mechanism [PHYSIOLOGY] Any physiological mechanism for the maintenance of an optimal and constant level of osmotic activity of the fluid in and around the cells. { ¦äz·mō ˈreg·yə·ləˌtȯr·ē ˈmek·əˌniz·əm }

osmotic fragility [PHYSIOLOGY] Susceptibility of red blood cells to lyses when placed in dilute (hypotonic) salt solutions. { äzˈmäd·ik frəˈjil·əd·ē }

osmotic shock [PHYSIOLOGY] The bursting of cells suspended in a dilute salt solution. { äz ˈmäd·ik ˈshäk }

osphradium [INVERTEBRATE ZOOLOGY] *See* asphradium. { äsˈfrād·ē·əm }

os priapi [VERTEBRATE ZOOLOGY] *See* baculum. { ˈäs prīˈä·pē }

osseous [ANATOMY] Bony; composed of or resembling bone. { ˈäs·ē·əs }

osseous system [ANATOMY] The skeletal system of the body. { ˈäs·ē·əs ¦sis·təm }

osseous tissue [HISTOLOGY] Bone tissue. { ˈäs·ē·əs ¦tish·ü }

ossicle [ANATOMY] Any of certain small bones, as those of the middle ear. [INVERTEBRATE ZOOLOGY] Any of various calcareous bodies. { ˈäs·ə·kəl }

ossify [PHYSIOLOGY] To form or turn into bone. { ˈäs·əˌfī }

Ostariophysi [VERTEBRATE ZOOLOGY] A superorder of actinopterygian fishes distinguished by the structure of the anterior four or five vertebrae which are modified as an encasement for the bony ossicles connecting the inner ear and swim bladder. { ä¦stär·ē·ōˈfīˌsī }

Osteichthyes [VERTEBRATE ZOOLOGY] The bony fishes, a class of fishlike vertebrates distinguished by having a bony skeleton, a swim bladder, a true gill cover, and mesodermal ganoid, cycloid, or ctenoid scales. { ˌä·stēˈik·thēˌēz }

osteoblast [HISTOLOGY] A bone-forming cell of mesenchymal origin. { ˈäs·tē·əˌblast }

osteoclasis [PHYSIOLOGY] **1.** Destruction of bony tissue. **2.** Bone resorption. { ˌäs·tēˈäk·lə·səs }

osteoclast [HISTOLOGY] A large multinuclear cell associated with bone resorption. { ˈäs·tē·ə ˌklast }

osteocyte [HISTOLOGY] A bone cell. { ˈäs·tē·ə ˌsīt }

osteogenesis [PHYSIOLOGY] Formation or histogenesis of bone. { ¦äs·tē·ōˈjen·ə·səs }

Osteoglossidae [VERTEBRATE ZOOLOGY] The bony tongues, a family of actinopterygian fishes in the order Osteoglossiformes. { ˌäs·tē·ōˈgläs·əˌdē }

Osteoglossiformes [VERTEBRATE ZOOLOGY] An order of soft-rayed, actinopterygian fishes distinguished by paired, usually bony rods at the base of the second gill arch, a single dorsal fin, no adipose fin, and a usually abdominal pelvic fin. { ˌäs·tē·ōˌgläs·əˈfȯr·mēz }

osteoid [HISTOLOGY] The young hyaline matrix of true bone in which the calcium salts are deposited. { ˈäs·tēˌȯid }

Osteolepidae [PALEONTOLOGY] A family of extinct fishes in the order Osteolepiformes. { ˌäs·tē·ōˈlep·əˌdē }

Osteolepiformes [PALEONTOLOGY] A primitive order of fusiform lobefin fishes, subclass Crossopterygii, generally characterized by rhombic bony scales, two dorsal fins placed well back on the body, and a well-ossified head covered with large dermal plating bones. { ˌäs·tē·ōˌlep·ə ˈfȯr·mēz }

osteolith [PALEONTOLOGY] A fossil bone. { ˈäs·tē·əˌlith }

osteology [ANATOMY] The study of anatomy and structure of bone. { ˌäs·tēˈäl·ə·jē }

osteolysis [PHYSIOLOGY] Resorption of bone. { ˌäs·tēˈäl·ə·səs }

osteometry [ANATOMY] The study of the size and proportions of the osseous system. { ˌäs·tē ˈäm·ə·trē }

osteon [HISTOLOGY] A microscopic unit of mature bone composed of layers of osteocytes and bone surrounding a central canal. Also known as Haversian system. { ˈäs·tēˌän }

osteophony [PHYSIOLOGY] Conduction of sound by bone. { ˌäs·tēˈäf·ə·nē }

Osteostraci [PALEONTOLOGY] An order of extinct jawless vertebrates; they were mostly small, with the head and part of the body encased in a solid armor of bone, and the posterior part of the body and the tail covered with thick scales. { ˌäs·tē 'äs·trəˌsī }

ostiole [BIOLOGY] A small orifice or pore. { 'äs·tēˌōl }

ostium [BIOLOGY] A mouth, entrance, or aperture. { 'äs·tē·əm }

Ostomidae [INVERTEBRATE ZOOLOGY] The bark-gnawing beetles, a family of coleopteran insects in the superfamily Cleroidea. { ä'stäm·əˌdē }

ost-, oste-, osteo- [ANATOMY] A combining form meaning bone. { äst, 'ä·stē, 'ä·stē·ō }

Ostracoda [INVERTEBRATE ZOOLOGY] A subclass of the class Crustacea containing small, bivalved aquatic forms; the body is unsegmented and there is no true abdominal region. { ä'sträk·ə·də }

ostracoderm [PALEONTOLOGY] Any of various extinct jawless vertebrates covered with an external skeleton of bone which together with the Cyclostomata make up the class Agnatha. { 'ä·strə·kō ˌdərm }

Ostreidae [INVERTEBRATE ZOOLOGY] A family of bivalve mollusks in the order Anisomyaria containing the oysters. { ä'strē·əˌdē }

ostrich [VERTEBRATE ZOOLOGY] *Struthio camelus.* A large running bird with soft plumage, naked head, neck and legs, small wings, and thick powerful legs with two toes on each leg; the only living species of the Struthioniformes. { 'ö ˌstrich }

ot-, oto- [ANATOMY] A combining form meaning ear. { ōt, 'ō·dō }

Otariidae [VERTEBRATE ZOOLOGY] The sea lions, a family of carnivorous mammals in the superfamily Canoidea. { ˌōd·ə'rī·əˌdē }

Othniidae [INVERTEBRATE ZOOLOGY] The false tiger beetles, a small family of coleopteran insects in the superfamily Tenebrionoidea. { ˌōth'nī·ə ˌdē }

otic [ANATOMY] Of or pertaining to the ear or a part thereof. { 'ōd·ik }

otic capsule [EMBRYOLOGY] A cartilaginous capsule surrounding the auditory vesicle during development, later fusing with the spheroid and occipital cartilages. { 'ōd·ik 'kap·səl }

otic ganglion [ANATOMY] The nerve ganglion located immediately below the foramen ovale of the sphenoid bone. { 'ōd·ik 'gaŋ·glēˌän }

Otitidae [INVERTEBRATE ZOOLOGY] A family of cyclorrhaphous myodarian dipteran insects in the subsection Acalyptratae. { ō'tid·əˌdē }

otocyst [EMBRYOLOGY] The auditory vesicle of vertebrate embryos. [INVERTEBRATE ZOOLOGY] An auditory vesicle, otocell, or otidium in some invertebrates. { 'ōd·əˌsist }

otolith [ANATOMY] A calcareous concretion on the end of a sensory hair cell in the vertebrate ear and in some invertebrates. { 'ōd·əˌlith }

Otopheidomenidae [INVERTEBRATE ZOOLOGY] A family of parasitic mites in the suborder Mesostigmata. { ˌōd·əˌfē·dō'men·əˌdē }

otter [VERTEBRATE ZOOLOGY] Any of various members of the family Mustelidae, having a long thin body, short legs, a somewhat flattened head, webbed toes, and a broad flattened tail; all are adapted to aquatic life. { 'äd·ər }

outbreed [BIOLOGY] *See* crossbreed. { 'au̇t ˌbrēd }

outbreeding [GENETICS] *See* exogamy. { 'au̇t ˌbrēd·iŋ }

outgroup [SYSTEMATICS] A monophyletic taxon that is used in a phylogenetic study to resolve which of two homologous character states are apomorphous. { 'au̇tˌgrüp }

ovalbumin [BIOCHEMISTRY] The major, conjugated protein of eggwhite. { ¦ov·al'byü·mən }

oval window [ANATOMY] The membrane-covered opening into the inner ear of tetrapods, to which the ossicles of the middle ear are connected. { 'ō·vəl 'win·dō }

ovarian [ANATOMY] Of or pertaining to the ovaries. { ō'ver·ē·ən }

ovarian follicle [HISTOLOGY] An ovum and its surrounding follicular cells, found in the ovarian cortex. { ō'ver·ē·ən 'fäl·ə·kəl }

ovariole [INVERTEBRATE ZOOLOGY] The tubular structural unit of an insect ovary. { ō'var·ēˌōl }

ovary [ANATOMY] A glandular organ that produces hormones and gives rise to ova in female vertebrates. [BOTANY] The enlarged basal portion of a pistil that bears the ovules in angiosperms. { 'ōv·ə·rē }

overdominance [GENETICS] Monohybrid heterosis, that is, the phenomenon of the phenotype being more pronounced in the heterozygote than in either homozygote with respect to a single specified pair of alleles. { ¦ō·vər'däm·ə·nəns }

overlapping genes [GENETICS] Genes having nucleotide sequences that may overlap in a way that involves control genes or structural genes. { ¦ō·vərˌlap·iŋ 'jēnz }

overwinding [MOLECULAR BIOLOGY] Supercoiling of a deoxyribonucleic acid molecule in the same direction as that of the winding of the double helix, resulting in increased tension in the two strands of the molecule. { ¦ō·vər¦wīn·diŋ }

ovicell [INVERTEBRATE ZOOLOGY] A broad chamber in certain bryozoans. { 'ō·vəˌsel }

ovicyst [INVERTEBRATE ZOOLOGY] The pouch of a tunicate in which the eggs develop. { 'ō·və ˌsist }

oviduct [ANATOMY] A tube that serves to conduct ova from the ovary to the exterior or to an intermediate organ such as the uterus. Also known in mammals as Fallopian tube; uterine tube. { 'ō·vəˌdəkt }

oviger [INVERTEBRATE ZOOLOGY] A modified leg used for carrying eggs in some pycnogonids. { 'ō·və·jər }

oviparous [VERTEBRATE ZOOLOGY] Producing eggs that develop and hatch externally. { ō'vip·ə·rəs }

oviposit [ZOOLOGY] To lay or deposit eggs, especially by means of a specialized organ, as found in certain insects and fishes. { 'ō·vəˌpäz·ət }

ovipositor [INVERTEBRATE ZOOLOGY] A specialized structure in many insects for depositing eggs. [VERTEBRATE ZOOLOGY] A tubular extension of the genital orifice in most fishes. { 'ō·və ˌpäz·əd·ər }

ovoviviparous [VERTEBRATE ZOOLOGY] Producing eggs that develop internally and hatch before or soon after extrusion. { ¦ō·vōˌvī'vip·ə·rəs }

ovulation [PHYSIOLOGY] Discharge of an ovum or ovule from the ovary. { ˌäv·yə'lā·shən }

ovule [BOTANY] A structure in the ovary of a seed plant that develops into a seed following fertilization. { 'ävˌyül }

ovum [CELL BIOLOGY] A female gamete. Also known as egg. { 'ō·vəm }

Oweniidae [INVERTEBRATE ZOOLOGY] A family of limivorous polychaete annelids of the Sedentaria. { ˌō·wə'nī·əˌdē }

owl [VERTEBRATE ZOOLOGY] Any of a number of diurnal and nocturnal birds of prey composing the order Strigiformes; characterized by a large head, more or less forward-directed large eyes, a short hooked bill, and strong talons. { aủl }

Oxalidaceae [BOTANY] A family of dicotyledonous plants in the order Geraniales, generally characterized by regular flowers, two or three times as many stamens as sepals or petals, a style which is not gynobasic, and the fruit which is a beakless, loculicidal capsule. { äkˌsal·ə 'dās·ēˌē }

oxidase [BIOCHEMISTRY] An enzyme that catalyzes oxidation reactions by the utilization of molecular oxygen as an electron acceptor. { 'äk·sə ˌdās }

oxidative phosphorylation [BIOCHEMISTRY] Conversion of inorganic phosphate to the energy-rich phosphate of adenosinetriphosphatase by reactions associated with the electron transfer system. { ˌäk·səˌdād·iv ˌfäs·fə·rə'lā·shən }

oxidoreductase [BIOCHEMISTRY] An enzyme catalyzing a reaction in which two molecules of a compound interact so that one molecule is oxidized and the other reduced, with a molecule of water entering the reaction. { ¦äk·sə·dō·ri'dək ˌtās }

oximetry [PHYSIOLOGY] Optical measurement of the degree of oxygen saturation of the blood hemoglobin by determining the variation in the color of the blood. { äk'sim·ə·trē }

Oxyaenidae [PALEONTOLOGY] An extinct family of mammals in the order Deltatheridea; members were short-faced carnivores with powerful jaws. { ˌäk·sē'en·əˌdē }

oxygenase [BIOCHEMISTRY] An oxidoreductase that catalyzes the direct incorporation of oxygen into its substrate. { 'äk·sə·jəˌnās }

oxygen debt [PHYSIOLOGY] **1.** A bodily condition in which oxygen demand is greater than oxygen supply. **2.** The amount of oxygen needed to restore the body to a steady state after a muscular exertion. { 'äk·sə·jən ˌdet }

oxygen toxicity [PHYSIOLOGY] **1.** Harmful effects of breathing oxygen at pressures greater than atmospheric. **2.** A toxic effect in a living organism caused by a species of oxygen-containing reactive intermediate produced during the reduction of dioxygen. { 'äk·sə·jən täk'sis·əd·ē }

oxyhemoglobin [BIOCHEMISTRY] The red crystalline pigment formed in blood by the combination of oxygen and hemoglobin, without the oxidation of iron. { ¦äk·sē'hē·məˌglō·bən }

oxylophyte [ECOLOGY] A plant that thrives in or is restricted to acid soil. { äk'sil·əˌfīt }

Oxymonadida [INVERTEBRATE ZOOLOGY] An order of xylophagous protozoans in the class Zoomastigophorea; colorless flagellate symbionts in the digestive tract of the roach *Cryptocercus* and of certain termites. { ˌäk·sē·mə 'näd·ə·də }

oxyntic cell [HISTOLOGY] *See* parietal cell. { äk 'sin·tik 'sel }

oxypetalous [BOTANY] Having sharp-pointed petals. { ¦äk·sē'ped·əl·əs }

oxyphytia [ECOLOGY] Discordant habitat control due to an excessively acidic substratum. { ˌäk·sə 'fīd·ē·ə }

oxyreductase [BIOCHEMISTRY] Any of a class of enzymes that catalyze electron-transfer reactions. { ˌäk·sē·ri'dəkˌtās }

Oxystomata [INVERTEBRATE ZOOLOGY] A subsection of the Brachyura, including those true crabs in which the first pair of pereiopods is chelate, and the mouth frame is triangular and forward. { ˌäk·sē'stō·mə·də }

Oxystomatidae [INVERTEBRATE ZOOLOGY] A family of free-living marine nematodes in the superfamily Enoploidea, distinguished by amphids that are elongated longitudinally. { ˌäk·sē·stō 'mad·əˌdē }

oxytetracycline [MICROBIOLOGY] $C_{22}H_{24}O_9N_2$ A crystalline, amphoteric, broad-spectrum antibiotic produced by *Streptomyces rimosus*; produced commercially by fermentation. { ¦äk·sēˌte·trə'sī ˌklēn }

oxytocin [BIOCHEMISTRY] $C_{43}H_{66}O_{12}N_{12}S_2$ A polypeptide hormone secreted by the neurohypophysis that stimulates contraction of the uterine muscles. { ¦äk·sē¦tō·sən }

Oxyurata [INVERTEBRATE ZOOLOGY] The equivalent name for Oxyurina. { ˌäk·sē·yủ'räd·ə }

Oxyuridae [INVERTEBRATE ZOOLOGY] A family of the nematode superfamily Oxyuroidea. { ˌäk·sē 'yủr·əˌdē }

Oxyurina [INVERTEBRATE ZOOLOGY] A suborder of nematodes in the order Ascaridida. { ˌäk·sē'yủ 'rī·nə }

Oxyuroidea [INVERTEBRATE ZOOLOGY] A superfamily of marine nematodes in the order Enoplida; contains species that maintain the most ancestral characters known in the phylum, such as a stoma composed entirely of esophastome, which is the ancestral primary blastocoel invagination. { ˌäk·sē·yủ'rȯid·ē·ə }

oyster [INVERTEBRATE ZOOLOGY] Any of various bivalve mollusks of the family Ostreidae; the irregular shell is closed by a single adductor muscle, the foot is small or absent, and there is no siphon. { 'ȯi·stər }

Ozawainellidae [PALEONTOLOGY] A family of extinct protozoans in the superfamily Fusulinacea. { ō¦zä·wə·i'nel·əˌdē }

P

P_1 [GENETICS] The parental generation; parents of the F_1 generation.

PABA [BIOCHEMISTRY] *See para*-aminobenzoic acid.

paca [VERTEBRATE ZOOLOGY] Any of several rodents of the genus *Cuniculus*, especially *C. paca*, with a white-spotted brown coat, found in South and Central America. { 'päk·ə }

Pacchionian bodies [ANATOMY] *See* arachnoidal granulations. { ¦pak·ē¦ō·nē·ən ˌbäd·ēz }

pachycephalosaur [PALEONTOLOGY] A boneheaded dinosaur, composing the family Pachycephalosauridae. { ˌpak·ə'sef·ə·ləˌsȯr }

Pachycephalosauridae [PALEONTOLOGY] A family of ornithischian dinosaurs characterized by a skull with a solid rounded mass of bone 4 inches (10 centimeters) thick above the minute brain cavity. { ˌpak·əˌsef·ə·lə'sȯr·əˌdē }

pachyderm [VERTEBRATE ZOOLOGY] Any of various nonruminant hooved mammals characterized by thick skin, including the elephants, hippopotamuses, rhinoceroses, and others. { 'pak·əˌdərm }

pachyglossal [VERTEBRATE ZOOLOGY] Of lizards, having a thick tongue. { ¦pak·ə'gläs·əl }

pachynema [CELL BIOLOGY] *See* pachytene. { pə'kin·ə·mə }

pachytene [CELL BIOLOGY] The third stage of meiotic prophase during which paired chromosomes thicken, each chromosome splits into chromatids, and breakage and crossing over between nonsister chromatids occur. Also known as pachynema. { 'pak·əˌtēn }

Pacific faunal region [ECOLOGY] A marine littoral faunal region including offshore waters west of Central America, running from the coast of South America at about 5° south latitude to the southern tip of California. { pə'sif·ik 'fȯn·əl ˌrē·jən }

Pacific temperate faunal region [ECOLOGY] A marine littoral faunal region including a narrow zone in the North Pacific Ocean, from Indochina to Alaska and along the west coast of the United States to about 40° north latitude. { pə'sif·ik 'tem·prət 'fȯn·əl ˌrē·jən }

Pacinian corpuscle [ANATOMY] An encapsulated lamellar sensory nerve ending that functions as a kinesthetic receptor. { pə'chin·ē·ən 'kȯr·pə·səl }

packet [BIOLOGY] A cluster of organisms in the form of a cube resulting from cell division in three planes. { 'pak·ət }

packet gland [INVERTEBRATE ZOOLOGY] A cluster of gland cells opening through the epidermis of nemertines. { 'pak·ət ˌgland }

pactamycin [MICROBIOLOGY] An antitumor and antibacterial antibiotic produced by *Streptomyces pactum* var. *pactum*. { ˌpak·tə'mīs·ən }

pad [ANATOMY] A small circumscribed mass of fatty tissue, as in terminal phalanges of the fingers or the underside of the toes of an animal, such as a dog. { pad }

paedogamy [INVERTEBRATE ZOOLOGY] A type of autogamy in certain protozoans whereby there is mutual fertilization of gametes derived from a single cell. { pē'däg·ə·mē }

paedomorphosis [EVOLUTION] Phylogenetic change in which adults retain juvenile characters, accompanied by an increased capacity for further change; indicates potential for further evolution. { ˌpēd·ə'mȯr·fə·səs }

Paenungulata [VERTEBRATE ZOOLOGY] A superorder of mammals, including proboscideans, xenungulates, and others. { pēnˌəŋ·gyə'läd·ə }

Paeoniaceae [BOTANY] A monogeneric family of dicotyledonous plants in the order Dilleniales; members are mesophyllic shrubs characterized by cleft leaves, flowers with an intrastaminal disk, and seeds having copious endosperm. { ˌpē·ə·nē'ās·ēˌē }

PAF [IMMUNOLOGY] *See* platelet activating factor.

Paguridae [INVERTEBRATE ZOOLOGY] The hermit crabs, a family of decapod crustaceans belonging to the Paguridea. { pə'gyůr·əˌdē }

Paguridea [INVERTEBRATE ZOOLOGY] A group of anomuran decapod crustaceans in which the abdomen is nearly always asymmetrical, being either soft and twisted or bent under the thorax. { ˌpag·yə'rid·ē·ə }

pain [PHYSIOLOGY] Patterns of somesthetic sensation, generally unpleasant, or causing suffering or distress. { pān }

pain spot [PHYSIOLOGY] Any of the small areas of skin overlying the endings of either very small myelinated (delta) or unmyelinated (C) nerve fibers whose stimulation, depending on the intensity and duration, results in the sensation of either pain or itching. { 'pān ˌspät }

pain threshold [PHYSIOLOGY] The lowest limit for the perception of pain sensations. { 'pān ˌthreshˌhōld }

Palaeacanthaspidoidei [PALEONTOLOGY] A suborder of extinct, placoderm fishes in the order

Rhenanida; members were primitive, arthrodire-like species. { ¦pāl·ē·ə‚kan·thə·spi′dȯid·ē‚ī }

Palaeacanthocephala [INVERTEBRATE ZOOLOGY] An order of the Acanthocephala including parasitic worms characterized by fragmented nuclei in the hypodermis, lateral placement of the chief lacunar vessels, and proboscis hooks arranged in long rows. { ¦pāl·ē·ə‚kan·thə′sef·ə·lə }

Palaechinoida [PALEONTOLOGY] An extinct order of echinoderms in the subclass Perischoechinoidea with a rigid test in which the ambulacra bevel over the adjoining interambulacra. { ¦pāl·ē·kī′nȯid·ē·ə }

Palaemonidae [INVERTEBRATE ZOOLOGY] A family of decapod crustaceans in the group Caridea. { ¦pāl·ē′män·ə‚dē }

Palaeocaridacea [INVERTEBRATE ZOOLOGY] An order of crustaceans in the superorder Syncarida. { ¦pāl·ē·ō‚kar·ə′dās·ē·ə }

Palaeocaridae [INVERTEBRATE ZOOLOGY] A family of the crustacean order Palaeocaridacea. { ¦pāl·ē·ō′kar·ə‚dē }

Palaeoconcha [PALEONTOLOGY] An extinct order of simple, smooth-hinged bivalve mollusks. { ¦pāl·ē·ō′kaŋ·kə }

Palaeocopida [PALEONTOLOGY] An extinct order of crustaceans in the subclass Ostracoda characterized by a straight hinge and by the anterior location for greatest height of the valve. { ¦pāl·ē·ō′käp·ə·də }

Palaeodonta [VERTEBRATE ZOOLOGY] A suborder of artiodactylous mammals including piglike forms such as the extinct "giant pigs" and the hippopotami. { ‚pāl·ē·ə′dän·tə }

Palaeognathae [VERTEBRATE ZOOLOGY] The ratites, making up a superorder of birds in the subclass Neornithes; merged with the Neognathae in some systems of classification. { ‚pāl·ē′äg·nə‚thē }

Palaeoisopus [PALEONTOLOGY] A singular, monospecific, extinct arthropod genus related to the pycnogonida, but distinguished by flattened anterior appendages. { ¦pāl·ē·ō′ī·sə·pəs }

Palaeomastodontinae [PALEONTOLOGY] An extinct subfamily of elaphantoid proboscidean mammals in the family Mastodontidae. { ¦pāl·ē·ō‚mas·tə′dänt·ən‚ē }

Palaeomerycidae [PALEONTOLOGY] An extinct family of pecoran ruminants in the superfamily Cervoidea. { ¦pāl·ē·ō·mə′ris·ə‚dē }

Palaeonemertini [INVERTEBRATE ZOOLOGY] A family of the class Anopla distinguished by the two- or three-layered nature of the body-wall musculature. { ¦pāl·ē·ō‚ne·mər′tī‚nī }

Palaeonisciformes [PALEONTOLOGY] A large extinct order of chondrostean fishes including the earliest known and most primitive ray-finned forms. { ¦pāl·ē·ō‚nis·ə′fȯr·mēz }

Palaeoniscoidei [PALEONTOLOGY] A suborder of extinct fusiform fishes in the order Palaeonisciformes with a heavily ossified exoskeleton and thick rhombic scales on the body surface. { ¦pāl·ē·ō·nis′kȯid·ē‚ī }

Palaeopantopoda [PALEONTOLOGY] A monogeneric order of extinct marine arthropods in the subphylum Pycnogonida. { ¦pāl·ē·ō·pan′täp·ə·də }

Palaeopneustidae [INVERTEBRATE ZOOLOGY] A family of deep-sea echinoderms in the order Spatangoida characterized by an oval test, long spines, and weakly developed fascioles and petals. { ¦pāl·ē·ō′nü·stə‚dē }

Palaeopterygii [VERTEBRATE ZOOLOGY] An equivalent name for the Actinopterygii. { ‚pāl·ē‚äp·tə′rij·ē‚ī }

Palaeoryctidae [PALEONTOLOGY] A family of extinct insectivorous mammals in the order Deltatheridia. { ¦pāl·ē·ō′rik·tə‚dē }

Palaeospondyloidea [PALEONTOLOGY] An ordinal name assigned to the single, tiny fish *Palaeospondylus*, known only from Middle Devonian shales in Carthness, Scotland. { ¦pāl·ē·ō‚spän·də′lōid·ē·ɔ }

Palaeotheriidae [PALEONTOLOGY] An extinct family of perissodactylous mammals in the superfamily Equoidea. { ¦pāl·ē·ō·thə′rī·ə‚dē }

palaeotheriodont [VERTEBRATE ZOOLOGY] Being or having lophodont teeth with longitudinal external tubercles that are connected with inner tubercles by transverse oblique crests. { ¦pāl·ē·ō¦ther·ē·ə‚dänt }

palaeotropical [ECOLOGY] *See* paleotropical. { ¦pāl·ē·ō′träp·ə·kəl }

palama [VERTEBRATE ZOOLOGY] The membranous web on the feet of aquatic birds. { ′pal·ə·mə }

palate [ANATOMY] The roof of the mouth. { ′pal·ət }

palatine bone [ANATOMY] Either of a pair of irregularly L-shaped bones forming portions of the hard palate, orbits, and nasal cavities. { ′pal·ə‚tīn ′bōn }

palatine canal [ANATOMY] One of the canals in the palatine bone, giving passage to branches of the descending palatine nerve and artery. { ′pal·ə‚tīn kə′nal }

palatine gland [ANATOMY] Any of numerous small oral glands on the palate of mammals. { ′pal·ə‚tīn ′gland }

palatine process [ANATOMY] A thick process that projects horizontally mediad from the medial aspect of the maxilla. [EMBRYOLOGY] An outgrowth on the ventromedial aspect of the maxillary process that develops into the definite palate. { ′pal·ə‚tīn ′prä·səs }

palatine suture [ANATOMY] The median suture joining the bones of the palate. { ′pal·ə‚tīn ′sü·chər }

palatine tonsil [ANATOMY] Either of a pair of almond-shaped aggregations of lymphoid tissue embedded between folds of tissue connecting the pharynx and posterior part of the tongue with the soft palate. Also known as faucial tonsil; tonsil. { ′pal·ə‚tīn ′tän·səl }

palatomaxillary arch [ANATOMY] An arch formed by the palatine, maxillary, and premaxillary bones. Also known as maxillary arch. { ¦pal·ə·dō ′mak·sə‚ler·ē ′ärch }

palatoquadrate [VERTEBRATE ZOOLOGY] A series of bones or a cartilaginous rod constituting part of the roof of the mouth or upper jaw of most nonmammalian vertebrates. { ¦pal·ə·dō'kwä ˌdrāt }

palea [BOTANY] **1.** The upper, enclosing bract of a grass flower. **2.** A chaffy scale found on the receptacle of the disk flowers of some composite plants. [INVERTEBRATE ZOOLOGY] One of the enlarged flattened setae forming the operculum of the tube of certain polychaete worms. { 'pā·lē·ə }

Palearctic [ECOLOGY] Pertaining to a biogeographic region including Europe, northern Asia and Arabia, and Africa north of the Sahara. { ¦pāl·ē'ärd·ik }

paleate [BOTANY] Having a covering of chaffy scales, as some rhizomes. { 'pā·lēˌāt }

paleoagrostology [PALEOBOTANY] The study of fossil grasses. { ¦pāl·ē·ōˌag·rə'stäl·ə·jē }

paleoalgology [PALEOBOTANY] The study of fossil algae. Also known as paleophycology. { ¦pāl·ē·ō·al'gäl·ə·jē }

paleobiochemistry [PALEONTOLOGY] The study of chemical processes used by organisms that lived in the geologic past. { ¦pāl·ē·ōˌbī·ō'kem·ə·strē }

paleobioclimatology [PALEONTOLOGY] The study of climatological events affecting living organisms for millennia or longer. { ¦pāl·ē·ōˌbī·ō ˌklī·mə'täl·ə·jē }

paleobiocoenosis [PALEONTOLOGY] An assemblage of organisms that lived together in the geologic past as an interrelated community. Also known as paleocoenosis. { ¦pāl·ē·ōˌbī·ō·sə'nō·səs }

paleobotany [PALEONTOLOGY] The branch of paleontology concerned with the study of ancient and fossil plants and vegetation of the geologic past. { ¦pāl·ē·ō'bät·ən·ē }

Paleocharaceae [PALEOBOTANY] An extinct group of fossil plants belonging to the Charophyta distinguished by sinistrally spiraled gyrogonites. { ¦pāl·ē·ō·kə'rās·ēˌē }

paleocoenosis [PALEONTOLOGY] *See* paleobiocoenosis. { ¦pāl·ē·ō·sə'nō·səs }

Paleocopa [PALEONTOLOGY] An order of extinct ostracods distinguished by a long, straight hinge. { ˌpāl·ē'äk·ə·pə }

paleodepth [PALEONTOLOGY] The water level at which an ancient organism or group of organisms flourished. { 'pāl·ē·ōˌdepth }

paleoecology [PALEONTOLOGY] The ecology of the geologic past. { ¦pāl·ē·ō·i'käl·ə·jē }

paleoherpetology [PALEONTOLOGY] The study of fossil reptiles. { ¦pāl·ē·ōˌhər·pə'täl·ə·jē }

paleoichnology [PALEONTOLOGY] The study of trace fossils in the fossil state. Also spelled palichnology. { ¦pāl·ē·ō·ik'näl·ə·jē }

paleomalacology [PALEONTOLOGY] A branch of paleontology concerned with the study of mollusks. { ¦pāl·ē·ōˌmal·ə'käl·ə·jē }

paleomorphology [PALEONTOLOGY] The study of the form and structure of fossil remains in order to describe the original anatomy of an organism. { ¦pāl·ē·ō·mȯr'fäl·ə·jē }

paleomycology [PALEOBOTANY] The study of fossil fungi. { ¦pāl·ē·ō·mī'käl·ə·jē }

Paleonthropinae [PALEONTOLOGY] A former subfamily of fossil man in the family Hominidae; set up to include the Neanderthalers together with Rhodesian man. { ˌpāl·ē·ən'thräp·əˌnē }

paleontology [BIOLOGY] The study of life of the past as recorded by fossil remains. { ˌpāl·ē·ən 'täl·ə·jē }

paleopalynology [PALEONTOLOGY] A field of palynology concerned with fossils of microorganisms and of dissociated microscopic parts of megaorganisms. { ¦pāl·ē·ōˌpal·ə'näl·ə·jē }

Paleoparadoxidae [PALEONTOLOGY] A family of extinct hippopotamuslike animals in the order Desmostylia. { ¦pāl·ē·ōˌpar·ə'däk·səˌdē }

paleophycology [PALEOBOTANY] *See* paleoalgology. { ¦pāl·ē·ō·fī'käl·ə·jē }

Paleophytic [PALEOBOTANY] A paleobotanic division of geologic time, signifying that period during which the pteridophytes flourished, sometime between the evolution of the algae and the appearance of the first gymnosperms. Also known as Pteridophytic. { ¦pāl·ē·ə¦fid·ik }

Paleoptera [INVERTEBRATE ZOOLOGY] A section of the insect subclass Pterygota including primitive forms that are unable to flex their wings over the abdomen when at rest. { ˌpāl·ē'äp·tə·rə }

paleosere [ECOLOGY] A series of ecologic communities that have led to a climax community. { 'pāl·ē·əˌsir }

paleospecies [PALEONTOLOGY] The species that are given ancestor and descendant status in a phyletic lineage, depending on the geological strata in which they are found. { 'pē·lē·ōˌspē ˌshēz }

paleotropical [ECOLOGY] Of or pertaining to a biogeographic region that includes the Oriental and Ethiopian regions. Also spelled palaeotropical. { ¦pāl·ē·ō'träp·ə·kəl }

paleozoology [PALEONTOLOGY] The branch of paleontology concerned with the study of ancient animals as recorded by fossil remains. { ¦pāl·ē·ō·zō'äl·ə·jē }

palichnology [PALEONTOLOGY] *See* paleoichnology. { ˌpal·ik'näl·ə·jē }

palindrome [GENETICS] A nucleic acid sequence that is self-complementary. { 'pal·ənˌdrōm }

palingenesis [EMBRYOLOGY] Unaltered recapitulation of ancestral features by the developing stages of an organism. { ˌpal·ən'jen·ə·səs }

Palinuridae [INVERTEBRATE ZOOLOGY] The spiny lobsters or langoustes, a family of macruran decapod crustaceans belonging to the Scyllaridea. { ˌpal·ə'nyu̇r·əˌdē }

palisade cell [BOTANY] One of the columnar cells of the palisade mesophyll which contain numerous chloroplasts. { ˌpal·ə'sād ˌsel }

palisade mesophyll [BOTANY] A tissue system of the chlorenchyma in well-differentiated broad leaves composed of closely spaced palisade cells oriented parallel to one another, but with their long axes perpendicular to the surface of the blade. { ˌpal·ə'sād 'mez·əˌfil }

pallet [INVERTEBRATE ZOOLOGY] One of a pair of plates on the siphon tubes of certain Bivalvia. { 'pal·ət }

pallial artery [INVERTEBRATE ZOOLOGY] The artery

that supplies blood to the mantle of a mollusk. { 'pal·ē·əl ˌärd·ə·rē }

pallial chamber [INVERTEBRATE ZOOLOGY] The mantle cavity in mollusks. { 'pal·ē·əl ˌchām·bər }

pallial line [INVERTEBRATE ZOOLOGY] A mark on the inner surface of a bivalve shell caused by attachment of the mantle. { 'pal·ē·əl ˌlīn }

pallial nerve [INVERTEBRATE ZOOLOGY] One of the pair of dorsal nerves that innervate the mantle in mollusks. { 'pal·ē·əl ˌnərv }

pallial sinus [INVERTEBRATE ZOOLOGY] An inward bend in the posterior portion of the pallial line in bivalve mollusks. { 'pal·ē·əl ˌsī·nəs }

pallium [ANATOMY] The cerebral cortex. [INVERTEBRATE ZOOLOGY] The mantle of a mollusk or brachiopod. { 'pal·ē·əm }

Pallopteridae [INVERTEBRATE ZOOLOGY] A family of myodarian cyclorrhaphous dipteran insects in the subsection Acalypteratae. { ˌpal·əp'ter·ə ˌdē }

palm [ANATOMY] The flexor or volar surface of the hand. [BOTANY] Any member of the monocotyledonous family Arecaceae; most are trees with a slender, unbranched trunk and a terminal crown of large leaves that are folded between the veins. { päm }

Palmales [BOTANY] An equivalent name for Arecales. { pä'mā·lēz }

palmar [ANATOMY] Of or pertaining to the palm of the hand. { 'päm·ər }

palmar aponeurosis [ANATOMY] Bundles of fibrous connective tissue which radiate from the tendons of the deep fascia of the forearm toward the proximal ends of the fingers. { 'päm·ər ˌap·ə·nə'rōs·əs }

palmar arch [ANATOMY] *See* deep palmar arch; superficial palmar arch. { 'päm·ər 'ärch }

palmar reflex [PHYSIOLOGY] Flexion of the fingers when the palm of the hand is irritated. { 'päm·ər 'rēˌfleks }

palmate [BOTANY] Having lobes, such as on leaves, that radiate from a common point. [VERTEBRATE ZOOLOGY] Having webbed toes. [ZOOLOGY] Having the distal portion broad and lobed, resembling a hand with the fingers spread. { 'päˌmāt }

palmella stage [BOTANY] A stage in the life history of some unicellular flagellate algae in which the cells lose their flagella and form a gelatinous aggregation. { päl'mel·ə ˌstāj }

palmelloid [BOTANY] Pertaining to a colony of cells that aggregates in a gelatinous matrix, as is characteristic of blue-green algae. { päl'me ˌlȯid }

palmetto fiber [BOTANY] Brush or broom fiber obtained from young leafstalks of the cabbage palm tree (*Sabal palmetto*). { päl'med·ō ˌfī·bər }

palm nut [BOTANY] The edible seed of the African oil palm (*Elaeis guineensis*). { 'päm ˌnət }

Palmyridae [INVERTEBRATE ZOOLOGY] A mongeneric family of errantian polychaete annelids. { pal'mir·əˌdē }

palp [INVERTEBRATE ZOOLOGY] Any of various sensory, usually fleshy appendages near the oral aperture of certain invertebrates. { palp }

palpal organ [INVERTEBRATE ZOOLOGY] An organ on the terminal joint of each pedipalp of a male spider which functions to convey sperm to the female genital orifice. { 'pal·pəl ˌȯr·gən }

Palpatores [INVERTEBRATE ZOOLOGY] A suborder of long-legged arachnids in the order Phalangida. { ˌpal·pə'tȯr·ēz }

palpebra [ANATOMY] The eyelid. { 'pal·pə·brə }

palpebral disk [VERTEBRATE ZOOLOGY] A scale, often transparent, covering the eyelid of certain lizards. { 'pal·pə·brəl 'disk }

palpebral fissure [ANATOMY] The opening between the eyelids. { 'pal·pə·brəl 'fish·ər }

palpebral fold [ANATOMY] A fold formed by the reflection of the conjunctiva from the eyelid onto the eye. { 'pal·pə·brəl 'fōld }

Palpicornia [INVERTEBRATE ZOOLOGY] The equivalent name for Hydrophiloidea. { ˌpal·pə'kȯr·nē·ə }

palpiger [INVERTEBRATE ZOOLOGY] The palpi-bearing portion of an insect labium. { 'pal·pə·jər }

Palpigradida [INVERTEBRATE ZOOLOGY] An order of rare tropical and warm-temperate arachnids; all are minute, whitish, eyeless animals with an elongate body that terminates in a slender, multisegmented flagellum set with setae. { ˌpal·pə 'grad·əd·ə }

palpocil [INVERTEBRATE ZOOLOGY] A fine, filamentous tactile hair. { 'pal·pəˌsil }

palpus [INVERTEBRATE ZOOLOGY] **1.** A process on a mouthpart of an arthropod that has a tactile or gustatory function. **2.** Any similar process on other invertebrates. { 'pal·pəs }

paludal [ECOLOGY] Relating to swamps or marshes and to material that is deposited in a swamp environment. { pə'lüd·əl }

paludification [ECOLOGY] Bog expansion resulting from the gradual rising of the water table as accumulation of peat impedes water drainage. { pəˌlüd·ə·fə'kā·shən }

palustrine [ECOLOGY] Being, living, or thriving in a marsh. { pə'ləs·trən }

palynofacies [PALEONTOLOGY] An assemblage of palynomorphs in a portion of a sediment, representing local environmental conditions, but not representing the regional palynoflora. { ˌpal·ə·nō'fā·shēz }

palynology [PALEONTOLOGY] The study of spores, pollen, microorganisms, and microscopic fragments of megaorganisms that occur in sediments. { ˌpal·ə'näl·ə·jē }

palynomorph [PALEONTOLOGY] A microscopic feature such as a spore or pollen that is of interest in palynological studies. { pə'lin·əˌmȯrf }

palynostratigraphy [PALEONTOLOGY] The stratigraphic application of palynologic methods. { ¦pal·ə·nō·strə'tig·rə·fē }

palytoxin [BIOCHEMISTRY] A water-soluble toxin produced by several species of *Palythoa*; considered to be one of the most poisonous substances known. { 'pal·əˌtäk·sən }

pampa [ECOLOGY] An extensive plain in South America, usually covered with grass. { 'päm·pə }

Pamphiliidae [INVERTEBRATE ZOOLOGY] The web-spinning sawflies, a family of hymenopteran insects in the superfamily Megalodontoidea. { ˌpam·fəˈlī·əˌdē }

pampiniform [ANATOMY] Of the network of veins in the spermatic cord and in the broad ligament, having the form of a tendril. { pamˈpin·əˌfȯrm }

pampiniform plexus [ANATOMY] A venous network in the spermatic cord in the male, and in the broad ligament in the female. { pamˈpin·ə ˌfȯrm ˈplek·səs }

pamprodactylous [VERTEBRATE ZOOLOGY] Having the toes turned forward, as of certain birds. { ˌpam·prəˈdak·təl·əs }

panagglutinin [IMMUNOLOGY] An agglutinin lacking specificity, which agglutinates erythrocytes of various types. { ˌpan·əˈglüt·ən·ən }

Panagrolaimoidea [INVERTEBRATE ZOOLOGY] A superfamily of free-living nematodes in the order Rhabditida, characterized by a broad, open, thick-walled stoma that forms a chamber as long as its breadth. { pəˌna·grō·ləˈmȯid·ē·ə }

Pancarida [INVERTEBRATE ZOOLOGY] A superorder of the subclass Malacostraca; the cylindrical, cruciform body lacks an external division between the thorax and pleon and has the cephalon united with the first thoracomere. { panˈkar·ə·də }

panclimax [ECOLOGY] Two or more related climax communities or formations having similar climate, life forms, and genera or dominants. Also known as panformation. { panˈklīˌmaks }

pancreas [ANATOMY] A composite gland in most vertebrates that produces and secretes digestive enzymes, as well as at least two hormones, insulin and glucagon. { ˈpan·krē·əs }

pancreatic diverticulum [EMBRYOLOGY] One of two diverticula (dorsal and ventral) from the embryonic duodenum or hepatic diverticulum that form the pancreas or its ducts. { ¦pan·krē¦ad·ik ˌdī·vərˈtik·yə·ləm }

pancreatic duct [ANATOMY] The main duct of the pancreas formed from the dorsal and ventral pancreatic ducts of the embryo. { ¦pan·krē¦ad·ik ˌdəkt }

pancreatic juice [PHYSIOLOGY] The thick, transparent, colorless secretion of the pancreas. { ¦pan·krē¦ad·ik ˈjüs }

pancreatic lipase [BIOCHEMISTRY] *See* steapsin. { ¦pan·krē¦ad·ik ˈlīˌpās }

pancreatin [BIOCHEMISTRY] A cream-colored, amorphous powder obtained from the fresh pancreas of a hog; contains amylopsin, trypsin, steapsin, and other enzymes. { panˈkrē·əd·ən }

pancreozymin [BIOCHEMISTRY] A crude extract of the intestinal mucosa that stimulates secretion of pancreatic juice. { ˌpan·krē·ōˈzī·mən }

panda [VERTEBRATE ZOOLOGY] Either of two Asian species of carnivores in the family Procyonidae; the red panda (*Ailurus fulgens*) has long, thick, red fur, with black legs; the giant panda (*Ailuropoda melanoleuca*) is white, with black legs and black patches around the eyes. { ˈpan·də }

Pandanaceae [BOTANY] The single, pantropical family of the plant order Pandanales. { ˌpan·də ˈnās·ēˌē }

Pandanales [BOTANY] A monofamilial order of monocotyledonous plants; members are more or less arborescent and sparingly branched, with numerous long, firm, narrow, parallel-veined leaves that usually have spiny margins. { ˌpan· dəˈnā·lēz }

Pandaridae [INVERTEBRATE ZOOLOGY] A family of dimorphic crustaceans in the suborder Caligoida; members are external parasites of sharks. { panˈdar·əˌdē }

Pandionidae [VERTEBRATE ZOOLOGY] A monospecific family of birds in the order Falconiformes; includes the osprey (*Pandion haliaetus*), characterized by a reversible hindtoe, well-developed claws, and spicules on the scales of the feet. { pan·dēˈän·əˌdē }

pandurate [BOTANY] Of a leaf, having the outline of a fiddle. { ˈpan·dyu̇r·ət }

Paneth cells [HISTOLOGY] *See* cells of Paneth. { ˈpan·əth ˌselz }

panformation [ECOLOGY] *See* panclimax. { ˈpan· fərˌmā·shən }

pangene [CELL BIOLOGY] A hypothetical heredity-controlling protoplasmic particle proposed by Darwin. { ˈpanˌjēn }

pangenesis [BIOLOGY] Darwin's comprehensive theory of heredity and development, according to which all parts of the body give off gemmules which aggregate in the germ cells; during development, they are sorted out from one another and give rise to parts similar to those of their origin. { panˈjen·ə·səs }

pangolin [VERTEBRATE ZOOLOGY] Any of seven species composing the mammalian family Manidae; the entire dorsal surface of the body is covered with broad, horny scales, the small head is elongate, and the mouth is terminal in the snout. { ˈpaŋ·gə·lən }

panicle [BOTANY] A branched or compound raceme in which the secondary branches are often racemose as well. { ˈpan·ə·kəl }

panmixis [BIOLOGY] Random mating within a breeding population; in a closed population this results in a high degree of uniformity. { pan ˈmik·səs }

Panmycin [MICROBIOLOGY] A trade name for tetracycline. { panˈmīs·ən }

panniculus [ANATOMY] A membrane or layer. { pəˈnik·yə·ləs }

pannose [BIOLOGY] Having a felty or woolly texture. { ˈpaˌnōs }

panspermia [BIOLOGY] A 19th-century theory in opposition to the theory of spontaneous generation and proposing that all reproductive bodies are universal, developing wherever there is a favorable environment. { panˈspər·mē·ə }

pansporoblast [INVERTEBRATE ZOOLOGY] A sporont of cnidosporan protozoans that contains two sporoblasts. { panˈspȯr·əˌblast }

Pantodonta [PALEONTOLOGY] An extinct order of mammals which included the first large land animals of the Tertiary. { ˌpan·təˈdän·tə }

Pantodontidae [VERTEBRATE ZOOLOGY] A family of fishes in the order Osteoglossiformes; the single, small species is known as African butterflyfish because of its expansive pectoral fins. { ˌpan·təˈdän·təˌdē }

Pantolambdidae [PALEONTOLOGY] A family of middle to late Paleocene mammals of North America in the superfamily Pantolambdoidea. { ˌpan·təˈlam·dəˌdē }

Pantolambdodontidae [PALEONTOLOGY] A family of late Eocene mammals of Asia in the superfamily Pantolambdoidea. { ˌpan·təˌlam·dəˈdän·tə ˌdē }

Pantolambdoidea [PALEONTOLOGY] A superfamily of extinct mammals in the order Pantodonta. { ˌpan·tə·lamˈdȯid·ē·ə }

Pantolestidae [PALEONTOLOGY] An extinct family of large aquatic insectivores referred to the Proteutheria. { ˌpan·təˈles·təˌdē }

pantophagous [ZOOLOGY] Feeding on a variety of foods. { panˈtäf·ə·gəs }

Pantophthalmidae [INVERTEBRATE ZOOLOGY] The wood-boring flies, a family of orthorrhaphous dipteran insects in the series Brachycera. { ˌpan ˌtäfˈthal·məˌdē }

Pantopoda [INVERTEBRATE ZOOLOGY] The equivalent name for Pycnogonida. { panˈtäp·ə·də }

pantothenate [BIOCHEMISTRY] A salt or ester of pantothenic acid. { ˌpan·təˈtheˌnāt }

pantothenic acid [BIOCHEMISTRY] $C_9H_{17}O_5N$ A member of the vitamin B complex that is essential for nutrition of some animal species. Also known as vitamin B_3. { ¦pan·tə¦then·ik ˈas·əd }

Pantotheria [PALEONTOLOGY] An infraclass of carnivorous and insectivorous Jurassic mammals; early members retained many reptilian features of the jaws. { ˌpan·təˈthir·ē·ə }

papain [BIOCHEMISTRY] An enzyme preparation obtained from the juice of the fruit and leaves of the papaya (*Carica papaya*); contains proteolytic enzymes. { pəˈpī·ən }

Papaveraceae [BOTANY] A family of dicotyledonous plants in the order Papaverales, with regular flowers, numerous stamens, and a well-developed latex system. { pəˌpav·əˈrās·ēˌē }

Papaverales [BOTANY] An order of dicotyledonous plants in the subclass Magnoliidae, marked by a syncarpous gynoecium, parietal placentation, and only two sepals. { pəˌpav·əˈrā·lēz }

Papilionidae [INVERTEBRATE ZOOLOGY] A family of lepidopteran insects in the superfamily Papilionoidea; members are the only butterflies with fully developed forelegs bearing an epiphysis. { pəˌpil·ēˈän·əˌdē }

Papilionoidea [INVERTEBRATE ZOOLOGY] A superfamily of diurnal butterflies (Lepidoptera) with clubbed antennae, which are rounded at the tip, and forewings that always have two or more veins. { pəˌpil·ē·əˈnȯid·ē·ə }

Papilionoideae [BOTANY] A subfamily of the family Leguminosae with characteristic irregular flowers that have a banner, two wing petals, and two lower petals united to form a boat-shaped keel. { pəˌpil·ē·əˈnȯid·ēˌē }

papilla [BIOLOGY] A small, nipplelike eminence. { pəˈpil·ə }

papillary muscle [ANATOMY] Any of the muscular eminences in the ventricles of the heart from which the chordae tendineae arise. { ˈpap·ə ˌler·ē ˈməs·əl }

papillate [BIOLOGY] **1.** Having or covered with papillae. **2.** Resembling a papilla. Also known as papillose. { ˈpap·əˌlāt }

papillose [BIOLOGY] *See* papillate. { ˈpap·əˌlōs }

papovavirus [VIROLOGY] A deoxyribonucleic acid-containing group of animal viruses, including papilloma and vacuolating viruses. { ¦pap·ə·vəˈvī·rəs }

Pappotheriidae [PALEONTOLOGY] A family of primitive, tenreclike Cretaceous insectivores assigned to the Proteutheria. { ˌpap·ə·thəˈrī·ə ˌdē }

pappus [BOTANY] An appendage or group of appendages consisting of a modified perianth on the ovary or fruit of various seed plants; adapted to dispersal by wind and other means. { ˈpap·əs }

paprika [BOTANY] *Capsicum annuum.* A type of pepper with nonpungent flesh, grown for its long red fruit from which a dried, ground condiment is prepared. { pəˈprē·kə }

papula [BIOLOGY] A small papilla. { ˈpap·yə·lə }

paraaortic body [ANATOMY] One of the small masses of chromaffin tissue lying along the abdominal aorta. Also known as glomus aorticum. { ¦par·ə·ā¦ȯrd·ik ˈbäd·ē }

parabasal body [CELL BIOLOGY] *See* kinetoplast. { ¦par·ə¦bā·səl ˈbäd·ē }

parabiosis [BIOLOGY] Experimental joining of two individuals to study the effects of one partner upon the other. { ¦par·ə·bīˈō·səs }

Paracanthopterygii [VERTEBRATE ZOOLOGY] A superorder of teleost fishes, including the codfishes and allied groups. { ˌpar·əˌkanˌthäp·tə ˈrij·ēˌī }

paracentric inversion [GENETICS] A type of chromosomal alteration that occurs within one arm of a chromosome and does not span the centromere. { ¦par·ə¦sen·trik inˈvər·zhən }

paracolon bacteria [MICROBIOLOGY] A group of bacteria intermediate between the *Escherichia-Aerobacter* genera and the *Salmonella-Shigella* group. { ¦par·əˈkō·lən bakˈtir·ē·ə }

paracondyloid [VERTEBRATE ZOOLOGY] A process on the outer side of each condyle of the occipital bone in the skull of certain mammals. { ¦par·ə ˈkänd·əlˌȯid }

paracone [VERTEBRATE ZOOLOGY] **1.** The anterior cusp of a primitive tricuspid upper molar. **2.** The principal anterior, external cusp of an upper molar in higher forms. { ˈpar·əˌkōn }

paraconid [VERTEBRATE ZOOLOGY] **1.** The cusp of a primitive lower molar corresponding to the paracone. **2.** The anterior, internal cusp of a lower molar in higher forms. { ˌpar·əˈkän·əd }

paracrine signaling [PHYSIOLOGY] Signaling in which the target cell is close to the signaling cell and the signal molecule affects only adjacent target cells. { ˈpar·əˌkrēn ˌsig·nəl·iŋ }

Paracrinoidea [PALEONTOLOGY] A class of extinct Crinozoa characterized by the numerous, irregularly arranged plates, uniserial armlike appendages, and no clear distinction between adoral and aboral surfaces. { ¦par·ə·krəˈnȯid·ē·ə }

Paracucumidae [INVERTEBRATE ZOOLOGY] A family of holothurian echinoderms in the order Dendrochirotida; the body is invested with plates and has a simplified calcareous ring. { ¦par·ə·kə 'kyüm·ə,dē }

paradidymis [ANATOMY] Atrophic remains of the paragenital tubules of the mesonephros, occurring near the convolutions of the epididymal duct. { ¦par·ə¦did·ə·məs }

paraganglion [ANATOMY] Any of various isolated chromaffin bodies associated with structures such as the abdominal aorta, heart, kidney, and gonads. Also known as chromaffin body. { ¦par·ə'gaŋ·glē,än }

paragastric [ANATOMY] Located near the stomach. [INVERTEBRATE ZOOLOGY] A cavity in Porifera into which radial canals open, and which opens to the outside through the cloaca. { ¦par·ə¦gas·trik }

paragenetic [GENETICS] Pertaining to chromosome changes that alter gene expression but not makeup. { ,par·ə·jə'ned·ik }

paragglutination [IMMUNOLOGY] Agglutination of colon bacteria with the serum of a patient infected, or recovering from an infection, with dysentery bacilli. { ,par·ə,glüt·ən'ā·shən }

paragnath [INVERTEBRATE ZOOLOGY] **1.** One of the paired leaflike lobes of the metastoma situated behind the mandibles in most crustaceans. **2.** One of the paired lobes of the hypopharynx in certain insects. **3.** One of the small, sharp and hard jaws of certain annelids. { 'par·əg,nath }

parainfluenza [MICROBIOLOGY] An organism exhibiting growth characteristics of *Hemophilus influenzae*. { ¦par·ə,in·flü'en·zə }

parakeet [VERTEBRATE ZOOLOGY] Any of various small, slender species of parrots with long tails in the family Psittacidae. { 'par·ə,kēt }

parallel evolution [EVOLUTION] Evolution of similar characteristics in different groups of organisms. { 'par·ə,lel ,ev·ə'lü·shən }

parallel muscle [ANATOMY] Any muscle having the long fibers arranged parallel to each other. { 'par·ə,lel 'məs·əl }

parallel-veined [BOTANY] Of a leaf, having the veins parallel, or nearly parallel, to each other. { 'par·ə,lel ¦vānd }

paralogous locus [GENETICS] A gene that arose by duplication and later diverged from the parent gene. { pə'ral·ə·gəs ,lō·kəs }

paralutein cells [HISTOLOGY] Epithelioid cells of the corpus luteum. { ¦par·ə'lüd·ē·ən ,selz }

paralytic secretion [PHYSIOLOGY] Glandular secretion occurring in a denervated gland. { ¦par·ə¦lid·ik si'krē·shən }

Parameciidae [INVERTEBRATE ZOOLOGY] A family of ciliated protozoans in the order Holotrichia; the body has differentiated anterior and posterior ends and is bounded by a hard but elastic pellicle. { ,par·ə·mə'sī·ə,dē }

paramecium [INVERTEBRATE ZOOLOGY] A single-celled protozoan belonging to the family Parameciidae. { ,par·ə'mē·sē·əm }

Paramecium [INVERTEBRATE ZOOLOGY] The genus of protozoans composing the family Parameciidae. { ,par·ə'mē·sē·əm }

Paramelina [VERTEBRATE ZOOLOGY] An order of marsupials that includes the bandicoots in some systems of classification. { ,par·ə'mel·ə·nə }

paramere [BIOLOGY] One half of a bilaterally symmetrical animal or somite. [INVERTEBRATE ZOOLOGY] Any of several paired structures of an insect, especially those on the ninth abdominal segment. { 'par·ə,mir }

paramesonephric duct [EMBRYOLOGY] An embryonic genital duct; in the female, it is the anlage of the oviducts, uterus, and vagina; in the male, it degenerates, leaving the appendix testes. Also known as Müllerian duct. { ¦par·ə¦me·zə¦nef·rik 'dəkt }

paramo [ECOLOGY] A biological community, essentially a grassland, covering extensive high areas in equatorial mountains of the Western Hemisphere. { 'pär·ə,mō }

paramutation [GENETICS] A mutation in which one member of a heterozygous pair of alleles permanently changes its partner allele. { ¦par·ə·myü'tā·shən }

paramylum [BIOCHEMISTRY] A reserve, starchlike carbohydrate of various protozoans and algae. { pə'ram·ə·ləm }

paramyosin [BIOCHEMISTRY] A type of fibrous protein found in the adductor muscles of bivalves and thought to form the core of a filament with myosin molecules at the surface. { ,par·ə 'mī·ə·sən }

paramyxovirus [VIROLOGY] A subgroup of myxoviruses, including the viruses of mumps, measles, parainfluenza, and Newcastle disease; all are ribonucleic acid-containing viruses and possess an ether-sensitive lipoprotein envelope. { ¦par·ə,mik·sō'vī·rəs }

paranasal sinus [ANATOMY] Any of the paired sinus cavities of the human face; includes the frontal, ethmoid, and sphenoid sinuses. { ¦par·ə¦nā·zəl 'sī·nəs }

paranthropophytia [ECOLOGY] Discrepant control of regions or areas due to immediate and continuous or periodic interference, as by certain cultivation practices. { ,par·an,thräp·ə'fī·shə }

Paranyrocidae [PALEONTOLOGY] An extinct family of birds in the order Anseriformes, restricted to the Miocene of South Dakota. { pə,ran·ə'räs·ə,dē }

Paraonidae [INVERTEBRATE ZOOLOGY] A family of small, slender polychaete annelids belonging to the Sedenteria. { ¦par·ə'än·ə,dē }

Paraparchitacea [PALEONTOLOGY] A superfamily of extinct ostracods in the suborder Kloedenellocopina including nonsulcate, nondimorphic forms. { ¦par·ə,pär·kə'tās·ē·ə }

parapatric [ECOLOGY] Referring to populations or species that occupy nonoverlapping but adjacent geographical areas without interbreeding. { ¦par·ə¦pa·trik }

parapatric speciation [EVOLUTION] Gradual speciation whereby new species are created from populations that maintain overlapping geographic zones of genetic contact. { ¦par·ə¦pa·trik ,spe·shē'ā·shən }

paraphysis [BOTANY] A sterile filament borne among the sporogenous or gametogenous organs in many cryptogams. [VERTEBRATE ZOOL-

OGY] A median evagination of the roof of the telencephalon of some lower vertebrates. { pə 'raf·ə·səs }

parapodium [INVERTEBRATE ZOOLOGY] **1.** One of the short, paired processes on the sides of the body segments in certain annelids. **2.** A lateral expansion of the foot in gastropod mollusks. { ,par·ə'pōd·ē·əm }

parapolar cell [INVERTEBRATE ZOOLOGY] Either of the first two trunk cells in the development of certain Mesozoa. { ¦par·ə¦pō·lər 'sel }

Parasaleniidae [INVERTEBRATE ZOOLOGY] A family of echinacean echinoderms in the order Echinoida composed of oblong forms with trigeminate ambulacral plates. { ¦par·ə,sal·ə'nī·ə,dē }

Paraselloidea [INVERTEBRATE ZOOLOGY] A group of the Asellota that contains forms in which the first pleopods of the male are coupled along the midline, and are lacking in the female. { ,par·ə·sə'lȯid·ē·ə }

Paraseminotidae [PALEONTOLOGY] A family of Lower Triassic fishes in the order Palaeonisciformes. { ,par·ə,sem·ə'näd·ə,dē }

parasexual cycle [GENETICS] A series of events leading to genetic recombination in vegetative or somatic cells; it was first described in filamentous fungi; there are three essential steps: heterokaryosis; fusion of unlike haploid nuclei in the heterokaryon to yield heterozygous diploid nuclei; and recombination and segregation at mitosis by two independent processes, mitotic crossing-over and loss of chromosomes. { ¦par·ə'sek·shə·wəl 'sī·kəl }

parasite [BIOLOGY] An organism that lives in or on another organism of different species from which it derives nutrients and shelter. { 'par·ə,sīt }

Parasitica [INVERTEBRATE ZOOLOGY] A group of hymenopteran insects that includes four superfamilies of the Apocrita: Ichneumonoidea, Chalcidoidea, Cynipoidea, and Proctotrupoidea; some are phytophagous, while others are parasites of other insects. { ,par·ə'sid·ə·kə }

parasitic castration [BIOLOGY] Destruction of the reproductive organs by parasites. { ¦par·ə¦sid·ik ka'strā·shən }

parasitism [ECOLOGY] A symbiotic relationship in which the host is harmed, but not killed immediately, and the species feeding on it is benefited. { 'par·ə·sə,tiz·əm }

parasitoidism [BIOLOGY] Systematic feeding by an insect larva on living host tissues so that the host will live until completion of larval development. { ¦par·ə·sə'tȯid,iz·əm }

parasitology [BIOLOGY] A branch of biology which deals with those organisms, plant or animal, which have become dependent on other living creatures. { ,par·ə·sə'täl·ə·jē }

parasphenoid [VERTEBRATE ZOOLOGY] A bone in the base of the skull of many vertebrates. { ¦par·ə'sfē,nȯid }

parastacidae [INVERTEBRATE ZOOLOGY] A family of crayfishes assigned to the Nephropoidea. { ,par·ə'stās·ə,dē }

parastyle [VERTEBRATE ZOOLOGY] A small cusp anterior to the paracone of an upper molar. { 'par·ə,stīl }

Parasuchia [PALEONTOLOGY] The equivalent name for Phytosauria. { ,par·ə'sü·kē·ə }

parasympathetic [ANATOMY] Of or pertaining to the craniosacral division of the autonomic nervous system. { ¦par·ə,sim·pə'thed·ik }

parasympathetic nervous system [ANATOMY] The craniosacral portion of the autonomic nervous system, consisting of preganglionic nerve fibers in certain sacral and cranial nerves, outlying ganglia, and postganglionic fibers. { ¦par·ə,sim·pə'thed·ik 'nər·vəs ,sis·təm }

parathormone [BIOCHEMISTRY] A polypeptide hormone that functions in regulating calcium and phosphate metabolism. Also known as parathyroid hormone. { ,par·ə'thȯr,mōn }

Parathuramminacea [PALEONTOLOGY] An extinct superfamily of foraminiferans in the suborder Fusulinina, with a test having a globular or tubular chamber and a simple, undifferentiated wall. { ,par·ə·thə,ram·ə'nās·ē·ə }

parathyroid gland [ANATOMY] A paired endocrine organ located within, on, or near the thyroid gland in the neck region of all vertebrates except fishes. { ¦par·ə'thī,rȯid ,gland }

parathyroid hormone [BIOCHEMISTRY] *See* parathormone. { ¦par·ə'thī,rȯid 'hȯr,mōn }

paratonic movement [BOTANY] The movement of the whole or parts of a plant due to the influence of an external stimulus, such as gravity, chemicals, heat, light, or electricity. { ¦par·ə¦tän·ik 'müv·mənt }

paratroch [INVERTEBRATE ZOOLOGY] The ciliated band encircling the anus in certain trichophore larvae. { 'par·ə,trōk }

paratype [SYSTEMATICS] A specimen other than the holotype which is before the author at the time of original description and which is designated as such or is clearly indicated as being one of the specimens upon which the original description was based. { 'par·ə,tīp }

parauterine organ [INVERTEBRATE ZOOLOGY] In certain tapeworms, a pouchlike sac which receives and retains the embryos. { ¦par·ə'yüd·ə·rən ,ȯr·gən }

paraxonic [VERTEBRATE ZOOLOGY] Pertaining to a state or condition wherein the axis of the foot lies between the third and fourth digits. { ¦par·ak'sän·ik }

Parazoa [INVERTEBRATE ZOOLOGY] A name proposed for a subkingdom of animals which includes the sponges (Porifera). { 'par·ə'zō·ə }

Pareiasauridae [PALEONTOLOGY] A family of large, heavy-boned terrestrial reptiles of the late Permian, assigned to the order Cotylosauria. { pə 'rī·ə'sȯr·ə'dē }

parenchyma [BOTANY] A tissue of higher plants consisting of living cells with thin walls that are agents of photosynthesis and storage; abundant in leaves, roots, and the pulp of fruit, and found also in leaves and stems. [HISTOLOGY] The specialized epithelial portion of an organ, as contrasted with the supporting connective tissue and nutritive framework. { pə'reŋ·kə·mə }

parenchymella [INVERTEBRATE ZOOLOGY] *See* diploblastula. { pə¦reŋ·kə'mel·ə }

parenchymula [INVERTEBRATE ZOOLOGY] The flagellate larva of calcinean sponges in which there is a cavity filled with gelatinous connective tissue. { ˌpar·əŋ'kim·yə·lə }

parental imprinting [GENETICS] The condition whereby the extent of gene expression depends upon the sex of the parent that transmits the gene. Also known as genomic imprinting. { pə ¦ren·təl 'imˌprint·iŋ }

Pareulepidae [INVERTEBRATE ZOOLOGY] A monogeneric family of errantian polychaete annelids. { ˌpar·yü'lep·əˌdē }

parietal [ANATOMY] Of or situated on the wall of an organ or other body structure. [BOTANY] Of a plant part, having a peripheral location or orientation; in particular, attached to the main wall of an ovary. { pə'rī·əd·əl }

parietal bone [ANATOMY] The bone that forms the side and roof of the cranium. { pə'rī·əd·əl ˌbōn }

parietal cell [HISTOLOGY] One of the peripheral, hydrochloric acid-secreting cells in the gastric fundic glands. Also known as acid cell; delomorphous cell; oxyntic cell. { pə'rī·əd·əl ˌsel }

Parietales [BOTANY] An order of plants in the Englerian system; families are placed in the order Violales in other systems. { pəˌrī·ə'tā·lēz }

parietal lobe [ANATOMY] The cerebral lobe of the brain above the lateral cerebral sulcus and behind the central sulcus. { pə'rī·əd·əl ˌlōb }

parietal peritoneum [ANATOMY] The portion of the peritoneum lining the interior of the body wall. { pə'rī·əd·əl ˌper·ə·tə'nē·əm }

parietal pleura [ANATOMY] The pleura lining the inner surface of the thoracic cavity. { pə'rī·əd·əl 'plu̇r·ə }

Parkinje effect [PHYSIOLOGY] *See* Purkinje effect. { pər'kin·jē iˌfekt }

parkland [ECOLOGY] *See* temperate woodland; tropical woodland. { 'pärkˌland }

Parmeliaceae [BOTANY] The foliose shield lichens, a family of the order Lecanorales. { ˌpär·mel·ē'ās·ēˌē }

Parnidae [INVERTEBRATE ZOOLOGY] The equivalent name for Dryopidae. { 'par·nəˌdē }

paromomycin [MICROBIOLOGY] A broad-spectrum antibiotic produced by *Streptomyces rimosus forma paromomycinus*; it is effective in the treatment of intestinal amebiasis in humans. { ˌpar·ə·me'mīs·ən }

paronychium [ANATOMY] *See* perionychium. { ˌpar·ə'nik·ē·əm }

parotid duct [ANATOMY] The duct of the parotid gland. Also known as Stensen's duct. { pə'räd·əd 'dəkt }

parotid gland [ANATOMY] The salivary gland in front of and below the external ear; the largest salivary gland in humans; a compound racemose serous gland that communicates with the mouth by Steno's duct. { pə'räd·əd 'gland }

parovarium [ANATOMY] *See* epoophoron. { ˌpar·ə 'var·ē·əm }

parrot [VERTEBRATE ZOOLOGY] Any member of the avian family Psittacidae, distinguished by the short, stout, strongly hooked beak. { 'par·ət }

pars anterior [ANATOMY] The major secretory portion of the anterior lobe of the adenohypophysis. Also known as pars distalis. { ¦pärzˌan 'tir·ē·ər }

pars distalis [ANATOMY] *See* pars anterior. { ¦pärz 'dis·tə·ləs }

pars intermedia [ANATOMY] The intermediate lobe of the adenohypophysis. { ¦pärz ˌin·tər 'mē·dē·ə }

parsley [BOTANY] *Petroselinum crispum*. A biennial herb of European origin belonging to the order Umbellales; grown for its edible foliage. { 'pär·slē }

pars nervosa [ANATOMY] The inferior subdivision of the neurohypophysis. Also known as pars neuralis. { ¦pärz nər'vō·sə }

pars neuralis [ANATOMY] *See* pars nervosa. { ¦pärz nu̇'räl·əs }

parsnip [BOTANY] *Pastinaca sativa*. A biennial herb of Mediterranean origin belonging to the order Umbellales; grown for its edible thickened taproot. { 'pär·snəp }

pars tuberalis [ANATOMY] A pair of processes that grow forward or upward along the stalk of the adenohypophysis. { ¦pärz ˌtü·bə'ral·əs }

parthenita [INVERTEBRATE ZOOLOGY] A stage, such as the sporocyst, redia, or cercaria, in the development of a fluke which reproduces parthenogenetically. { pär'then·əd·ə }

parthenocarpy [BOTANY] Production of fruit without fertilization. { 'pär·thə·nōˌkär·pē }

parthenogenesis [INVERTEBRATE ZOOLOGY] A special type of sexual reproduction in which an egg develops without entrance of a sperm; common among rotifers, aphids, thrips, ants, bees, and wasps. { ¦pär·thə·nō'jen·ə·səs }

parthenomerogony [EMBRYOLOGY] Development of a nucleated fragment of an unfertilized egg following parthenogenetic stimulation. { ¦pär·thə·nō·mə'räg·ə·nē }

parthenospore [MYCOLOGY] *See* azygospore. { 'pär·thə·nəˌspȯr }

partial cleavage [EMBRYOLOGY] Cleavage in which only part of the egg divides into blastomeres. { 'pär·shəl 'klē·vij }

partridge [VERTEBRATE ZOOLOGY] Any of the game birds comprising the genera *Alectoris* and *Perdix* in the family Phasianidae. { 'pär·trij }

parvovirus [VIROLOGY] The equivalent name for picodnavirus. { ¦pär·vō'vī·rəs }

passage number [MICROBIOLOGY] The number of times that a culture has been subcultured. { 'pas·ij ˌnəm·bər }

Passalidae [INVERTEBRATE ZOOLOGY] The peg beetles, a family of tropical coleopteran insects in the superfamily Scarabaeoidea. { pə'sal·ə ˌdē }

passenger [MOLECULAR BIOLOGY] A deoxyribonucleic acid segment that will be spliced into a plasmid or bacteriophage for subsequent cloning. { 'pas·ən·jər }

Passeres [VERTEBRATE ZOOLOGY] The equivalent name for Oscines. { 'pas·əˌrēz }

Passeriformes [VERTEBRATE ZOOLOGY] A large

order of perching birds comprising two major divisions: Suboscines and Oscines. { ˌpas·ə·rə'fór ˌmēz }

Passifloraceae [BOTANY] A family of dicotyledonous, often climbing plants in the order Violales; flowers are polypetalous and hypogynous with a corona, and seeds are arillate with an oily endosperm. { ˌpas·ə·flə'rās·ēˌē }

passive anaphylaxis [IMMUNOLOGY] Anaphylaxis elicited by temporary sensitization with antibodies followed by injection of the corresponding sensitizing antigen. { 'pas·iv ˌan·ə·fə'lak·səs }

passive cutaneous anaphylaxis [IMMUNOLOGY] The vascular reaction at the site of intradermally injected antibody when, 3 hours later, the specific antigen, usually mixed with Evans blue dye, is injected intravenously. { 'pas·iv kyü'tā·nē·əs ˌan·ə·fə'lak·səs }

passive immunity [IMMUNOLOGY] **1.** Immunity acquired by injection of antibodies in another individual or in an animal. **2.** Immunity acquired by the fetus by the transfer of maternal antibodies through the placenta. { 'pas·iv i'myün·əd·ē }

Pasteur-Chamberland filter [MICROBIOLOGY] A porcelain filter with small pores used in filtration sterilization. { pa'stər 'chām·bər·lənd ˌfil·tər }

Pasteur effect [MICROBIOLOGY] Inhibition of fermentation by supplying an abundance of oxygen to replace anerobic conditions. { pa'stər iˌfekt }

Pasteuriaceae [MICROBIOLOGY] Formerly a family of stalked bacteria in the order Hyphomicrobiales. { ˌpa·stə·rēˌās·ēˌē }

patagium [VERTEBRATE ZOOLOGY] **1.** A membrane or fold of skin extending between the forelimbs and hindlimbs of flying squirrels, flying lizards, and other arboreal animals. **2.** A membrane or fold of skin on a bird's wing anterior to the humeral and radioulnar bones. { pə'tā·jē·əm }

patch budding [BOTANY] Budding in which a small rectangular patch of bark bearing the scion (a bud) is fitted into a corresponding opening in the bark of the stock. { 'pach ˌbəd·iŋ }

patch test [IMMUNOLOGY] A test in which material is applied and left in contact with intact skin surfaces for 48 hours in order to demonstrate tissue sensitivity. { 'pach ˌtest }

patella [ANATOMY] A sesamoid bone in front of the knee, developed in the tendon of the quadriceps femoris muscle. Also known as kneecap. { pə'tel·ə }

Patellacea [PALEONTOLOGY] An extinct superfamily of gastropod mollusks in the order Aspidobranchia which developed a cap-shaped shell and were specialized for clinging to rock. { ˌpad·əl'ās·ē·ə }

Patellidae [INVERTEBRATE ZOOLOGY] The true limpets, a family of gastropod mollusks in the order Archeogastropoda. { pə'tel·əˌdē }

Paterinida [PALEONTOLOGY] A small extinct order of inarticulated brachiopods, characterized by a thin shell of calcium phosphate and convex valves. { ˌpad·ə'rin·əd·ə }

paternity test [IMMUNOLOGY] Identification of the blood groups of a mother, her child, and a putative father in order to establish the probability of paternity or nonpaternity; actually, only nonpaternity can be established. { pə'tər·nəd·ē ˌtest }

pathergy [IMMUNOLOGY] Either a subnormal response to an allergen or an unusually intense one in which the individual becomes sensitive not only to the specific substance but to others. { 'path·ər·jē }

pathovar [MICROBIOLOGY] A pathological variant of a nonpathological bacterial species. Abbreviated pv. { 'path·əˌvär }

patroclinous inheritance [GENETICS] Inheritance in which the offspring more closely resembles the male parent than the female parent. { ¦pa·trəˌklī·nəs in'her·əd·əns }

pattern formation [EMBRYOLOGY] The embryogenic process in which the spatial differentiation of cells is specified in a structure that initially is largely homogeneous. { 'pad·ərn fərˌmā·shən }

pattern gene [GENETICS] A gene involved in the establishment of a particular pattern during cell differentiation. { 'pad·ərn ˌjēn }

paua [INVERTEBRATE ZOOLOGY] *See* abalone. { paů·ə }

Paucituberculata [VERTEBRATE ZOOLOGY] An order of marsupial mammals in some systems of classification, including the opossum, rats, and polydolopids. { ¦pós·ē·təˌbər·kyə'läd·ə }

paunch [ANATOMY] In colloquial usage, the cavity of the abdomen and its contents. [VERTEBRATE ZOOLOGY] *See* rumen. { pónch }

paurometabolous metamorphosis [INVERTEBRATE ZOOLOGY] A simple, gradual, direct metamorphosis in which immature forms resemble the adult except in size and are referred to as nymphs. { ˌpór·ē·mə'tab·ə·ləs ˌmed·ə'mór·fə·səs }

Pauropoda [INVERTEBRATE ZOOLOGY] A class of the Myriapoda distinguished by bifurcate antennae, 12 trunk segments with 9 pairs of functional legs, and the lack of eyes, spiracles, tracheae, and a circulatory system. { pó'räp·ə·də }

Paussidae [INVERTEBRATE ZOOLOGY] The flat-horned beetles, a family of coleopteran insects in the suborder Adephaga. { 'pós·əˌdē }

Pavlov's pouch [PHYSIOLOGY] A small portion of stomach, completely separated from the main stomach, but retaining its vagal nerve branches, which communicates with the exterior; used in the long-term investigation of gastric secretion, and particularly in the study of conditioned reflexes. { 'pavˌläfs ˌpaůch }

paw [VERTEBRATE ZOOLOGY] The foot of an animal, especially a quadruped having claws. { pó }

paxilla [INVERTEBRATE ZOOLOGY] A pillarlike spine in certain starfishes that sometimes has a flattened summit covered with spinules. { pak'sil·ə }

Paxillosida [INVERTEBRATE ZOOLOGY] An order of the Asteroidea in some systems of classification, equivalent to the Paxillosina. { ˌpak·sə'läs·əd·ə }

Paxillosina [INVERTEBRATE ZOOLOGY] A suborder

of the Phanerozonida with pointed tube feet which lack suckers, and with paxillae covering the upper body surface. { ˌpak·sə'läs·ə·nə }

PBI [BIOCHEMISTRY] *See* protein-bound iodine.

P blood group [IMMUNOLOGY] A system of immunologically distinct, genetically determined erythrocyte antigens first defined by their reaction with anti-P, and immune rabbit antiserum, and later broadened to include related antigens. { 'pē 'bləd ˌgrüp }

PC [BIOCHEMISTRY] *See* phosphocreatine.

PCR [MOLECULAR BIOLOGY] *See* polymerase chain reaction.

pea [BOTANY] **1.** *Pisum sativum*. The garden pea, an annual leafy leguminous vine cultivated for its smooth or wrinkled, round edible seeds which are borne in dehiscent pods. **2.** Any of several related or similar cultivated plants. { pē }

peach [BOTANY] *Prunus persica*. A low, spreading, freely branching tree of the order Rosales, cultivated in less rigorous parts of the temperate zone for its edible fruit, a juicy drupe with a single large seed, a pulpy yellow or white mesocarp, and a thin firm epicarp. { pēch }

peak zone [PALEONTOLOGY] An informal biostratigraphic zone consisting of a body of strata characterized by the exceptional abundance of some taxon (or taxa) or representing the maximum development of some taxon. { 'pēk ˌzōn }

peanut [BOTANY] *Arachis hypogaea*. A low, branching, self-pollinated annual legume cultivated for its edible seed, which is a one-loculed legume formed beneath the soil in a pod. { 'pē·nət }

pear [BOTANY] Any of several tree species of the genus *Pyrus* in the order Rosales, cultivated for their fruit, a pome that is wider at the apical end and has stone cells throughout the flesh. { per }

pearl moss [BOTANY] *See* carrageen. { 'pərl ˌmȯs }

peat ball [ECOLOGY] A lake ball containing an abundance of peaty fragments. { 'pēt ˌbȯl }

peat flow [ECOLOGY] A mudflow of peat produced in a peat bog by a bog burst. { 'pēt ˌflō }

peat moss [ECOLOGY] Moss, especially sphagnum moss, from which peat has been produced. { 'pēt ˌmȯs }

pébrine [INVERTEBRATE ZOOLOGY] A contagious protozoan disease of silkworms and other caterpillars caused by *Nosema bombycis*. { pā'brēn }

pecan [BOTANY] *Carya illinoensis*. A large deciduous hickory tree in the order Fagales which produces an edible, oblong, thin-shelled nut. { pi 'kän }

peccary [VERTEBRATE ZOOLOGY] Either of two species of small piglike mammals in the genus *Tayassu*, composing the family Tayassuidae. { 'pek·ə·rē }

pecking order [VERTEBRATE ZOOLOGY] A hierarchy of social dominance within a flock of poultry where each bird is allowed to peck another lower in the scale and must submit to pecking by one of higher rank. { 'pek·iŋ ˌȯr·dər }

Pecora [VERTEBRATE ZOOLOGY] An infraorder of the Artiodactyla; includes those ruminants with a reduced ulna and usually with antlers, horns, or deciduous horns. { 'pek·ə·rə }

pecten [ZOOLOGY] Any of various comblike structures possessed by animals. { 'pek·tən }

Pectenidae [INVERTEBRATE ZOOLOGY] A family of bivalve mollusks in the order Anisomyaria; contains the scallops. { pek'ten·əˌdē }

pectic acid [BIOCHEMISTRY] A complex acid, partially demethylated, obtained from the pectin of fruits. { 'pek·tik 'as·əd }

pectin [BIOCHEMISTRY] A purified carbohydrate obtained from the inner portion of the rind of citrus fruits, or from apple pomace; consists chiefly of partially methoxylated polygalacturonic acids. { 'pek·tən }

Pectinariidae [INVERTEBRATE ZOOLOGY] The cone worms, a family of polychaete annelids belonging to the Sedentaria. { ˌpek·tə·nə'rī·əˌdē }

pectinase [BIOCHEMISTRY] An enzyme that catalyzes the transformation of pectin into sugars and galacturonic acid. { 'pek·təˌnās }

pectinesterase [BIOCHEMISTRY] An enzyme that catalyzes the hydrolytic breakdown of pectins to pectic acids. { ¦pek·tən'es·təˌrās }

pectineus [ANATOMY] A muscle arising from the pubis and inserted on the femur. { pek'tin·ē·əs }

Pectinibranchia [INVERTEBRATE ZOOLOGY] An order of gastropod mollusks which contains many families of snails; respiration is by means of ctenidia, the nervous system is not concentrated, and sexes are separate. { ˌpek·tə·nə'braŋ·kē·ə }

Pectobothridia [INVERTEBRATE ZOOLOGY] A subclass of parasitic worms in the class Trematoda, characterized by caudal hooks or hard posterior suckers or both. { ˌpek·tə·bä'thrid·ē·ə }

pectoral fin [VERTEBRATE ZOOLOGY] One of the pair of fins of fishes corresponding to forelimbs of a quadruped. { 'pek·tə·rəl 'fin }

pectoral girdle [ANATOMY] The system of bones supporting the upper or anterior limbs in vertebrates. Also known as shoulder girdle. { 'pek·tə·rəl 'gərd·əl }

pectoralis major [ANATOMY] The large muscle connecting the anterior aspect of the chest with the shoulder and upper arm. { ˌpek·tə'ral·əs 'mā·jər }

pectoralis minor [ANATOMY] The small, deep muscle connecting the third to fifth ribs with the scapula. { ˌpek·tə'ral·əs 'mīn·ər }

pedal [BIOLOGY] Of or pertaining to the foot. { 'ped·əl }

pedal disk [INVERTEBRATE ZOOLOGY] The broad, flat base of many sea anemones, used for attachment to a substrate. { 'ped·əl ˌdisk }

pedal ganglion [INVERTEBRATE ZOOLOGY] One of the paired ganglia supplying nerves to the foot muscles in most mollusks. { 'ped·əl 'gaŋ·glē·ən }

pedal gland [INVERTEBRATE ZOOLOGY] *See* foot gland. { 'ped·əl ˌgland }

pedate [BIOLOGY] **1.** Having toelike parts. **2.** Having a foot. **3.** Having tube feet. { 'pe ˌdāt }

pedicel [BOTANY] **1.** The stem of a fruiting or

sporebearing organ. **2.** The stem of a single flower. [ZOOLOGY] A short stalk in an animal body. { 'ped·ə,sel }

pedicellaria [INVERTEBRATE ZOOLOGY] In echinoids and starfishes, any of various small grasping organs in the form of a beak carried on a stalk. { ,ped·ə·sə'ler·ē·ə }

pedicellate [BIOLOGY] Having a pedicel. { pe'dis·ə,lāt }

Pedicellinea [INVERTEBRATE ZOOLOGY] The single order of the class Calyssozoa, including all entoproct bryozoans. { ,ped·ə·sə'lin·ē·ə }

pedicle [ANATOMY] A slender process acting as a foot or stalk (as the base of a tumor), or the basal portion of an organ that is continuous with other structures. { 'ped·ə·kəl }

pedigree [GENETICS] The ancestral line of an individual. { 'ped·ə,grē }

pedigree selection [GENETICS] Artificial selection of an individual for mating based on attributes of its ancestors. { 'ped·ə·grē sə,lek·shən }

Pedilidae [INVERTEBRATE ZOOLOGY] The false ant-loving flower beetles, a family of coleopteran insects in the superfamily Tenebrionoidea. { pə'dil·ə,dē }

Pedinidae [INVERTEBRATE ZOOLOGY] The single family of the order Pedinoida. { pe'din·ə,dē }

Pedinoida [INVERTEBRATE ZOOLOGY] An order of Diadematacea making up those forms of echinoderms which possess solid spines and a rigid test. { ,ped·ən'òid·ə }

Pedionomidae [VERTEBRATE ZOOLOGY] A family of quaillike birds in the order Gruiformes. { ,ped·ē·ə'näm·ə,dē }

Pedipalpida [INVERTEBRATE ZOOLOGY] Former order of the Arachnida; these animals are now placed in the orders Uropygi and Amblypygi. { ,ped·ə'pal·pəd·ə }

pedipalpus [INVERTEBRATE ZOOLOGY] One of the second pair of appendages of an arachnid. { ¦ped·ə¦pal·pəs }

peduncle [ANATOMY] A band of white fibers joining different portions of the brain. [BOTANY] **1.** A flower-bearing stalk. **2.** A stalk supporting the fruiting body of certain thallophytes. [INVERTEBRATE ZOOLOGY] The stalk supporting the whole or a large part of the body of certain crinoids, brachiopods, and barnacles. { 'pē,dəŋ·kəl }

pedunculate [BIOLOGY] **1.** Having or growing on a peduncle. **2.** Being attached to a peduncle. { pē'dəŋ·kyə·lət }

Pegasidae [VERTEBRATE ZOOLOGY] The single family of the order Pegasiformes. { pə'gas·ə,dē }

Pegasiformes [VERTEBRATE ZOOLOGY] The sea moths or sea dragons, a small order of actinopterygian fishes; the anterior of the body is encased in bone, and the nasal bones are enlarged to form a rostrum that projects well forward of the mouth. { pə,gas·ə'fòr,mēz }

peg graft [BOTANY] A graft made by driving a scion of leafless dormant wood with wedge-shaped base into an opening in the stock and sealing with wax or other material. { 'peg ,graft }

Peisidicidae [INVERTEBRATE ZOOLOGY] A monogeneric family of polychaete annelids belonging to the Errantia. { ,pī·sə'dis·ə,dē }

Peking man [PALEONTOLOGY] *Sinanthropus pekinensis.* An extinct human type; the braincase was thick, with a massive basal and occipital torus structure and heavy browridges. { 'pē,kiŋ 'man }

pelargonidin [BIOCHEMISTRY] An anthocyanidin pigment obtained by hydrolysis of pelargonin in the form of its red-brown crystalline chloride, $C_{15}H_{11}ClO_5$. { ¦pe,lär'gän·əd·ən }

pelargonin [BIOCHEMISTRY] An anthocyanin obtained from the dried petals of red pelargoniums or blue cornflowers in the form of its red crystalline chloride, $C_{27}H_{31}ClO_{15}$. { ,pe,lär'gō·nən }

Pelecanidae [VERTEBRATE ZOOLOGY] The pelicans, a family of aquatic birds in the order Pelecaniformes. { ,pel·ə'kan·ə,dē }

Pelecaniformes [VERTEBRATE ZOOLOGY] An order of aquatic, fish-eating birds characterized by having all four toes joined by webs. { ,pel·ə,kan·ə'fòr,mēz }

Pelecanoididae [VERTEBRATE ZOOLOGY] The diving petrels, a family of oceanic birds in the order Procellariiformes. { ,pel·ə·kə'nòid·ə,dē }

Pelecinidae [INVERTEBRATE ZOOLOGY] The pelecinid wasps, a monospecific family of hymenopteran insects in the superfamily Proctotrupoidea. { ,pel·ə'sin·ə,dē }

Pelecypoda [INVERTEBRATE ZOOLOGY] The equivalent name for Bivalvia. { ,pel·ə'si·pə·də }

pelican [VERTEBRATE ZOOLOGY] Any of several species of birds composing the family Pelecanidae, distinguished by the extremely large bill which has a distensible pouch under the lower mandible. { 'pel·ə·kən }

pellet [VERTEBRATE ZOOLOGY] A mass of undigestible material regurgitated by a carnivorous bird. { 'pel·ət }

pellicle [CELL BIOLOGY] A plasma membrane. [INVERTEBRATE ZOOLOGY] A thin protective membrane, as on certain protozoans. { 'pel·ə·kəl }

Pelmatozoa [INVERTEBRATE ZOOLOGY] A division of the Echinodermata made up of those forms which are anchored to the substrate during at least part of their life history. { pel,mad·ə'zō·ə }

Pelobatidae [VERTEBRATE ZOOLOGY] A family of frogs in the suborder Anomocoela, including the spadefoot toads. { ,pel·ō'bad·ə,dē }

Pelodytidae [VERTEBRATE ZOOLOGY] A family of frogs in the suborder Anomocoela. { ,pel·ə'did·ə,dē }

Pelogonidae [INVERTEBRATE ZOOLOGY] The equivalent name for Ochteridae. { ,pel·ə'gän·ə,dē }

Pelomedusidae [VERTEBRATE ZOOLOGY] The side-necked or hidden-necked turtles, a family of the order Chelonia. { ,pel·ō·mə'düs·ə,dē }

Pelonemataceae [MICROBIOLOGY] A family of gliding bacteria of uncertain affiliation; straight, flexuous, or spiral, unbranched filaments containing colorless, cylindrical cells. { ¦pel·ō,nem·ə'tās·ē,ē }

Pelopidae [INVERTEBRATE ZOOLOGY] A family of oribatid mites, order Sarcoptiformes. { pə'läp·ə ˌdē }

Peloplocaceae [MICROBIOLOGY] Formerly a family in the order Chlamydobacteriales; long, unbranched trichomes in a delicate sheath. { ˌpel·ō·plə'kās·ēˌē }

Peloridiidae [INVERTEBRATE ZOOLOGY] The single family of the homopteran series Coleorrhyncha. { ˌpel·ō·rə'dī·əˌdē }

pelouse [ECOLOGY] *See* meadow. { pə'lüz }

peltate [BOTANY] Of leaves, having the petiole attached to the lower surface instead of the base. { 'pelˌtāt }

pelvic cavity [ANATOMY] *See* pelvis. { 'pel·vik ¦kav·əd·ē }

pelvic fin [VERTEBRATE ZOOLOGY] One of the pair of fins of fishes corresponding to the hindlimbs of a quadruped. { 'pel·vik ¦fin }

pelvic girdle [ANATOMY] The system of bones supporting the lower limbs, or the hindlimbs, of vertebrates. { 'pel·vik ¦gərd·əl }

pelvic index [ANATOMY] The ratio of the anteroposterior diameter to the transverse diameter of the pelvis. { 'pel·vik ¦inˌdeks }

pelvis [ANATOMY] **1.** The main, basin-shaped cavity of the kidney into which urine is discharged by nephrons. **2.** The basin-shaped structure formed by the hipbones together with the sacrum and coccyx, or caudal vertebrae. **3.** The cavity of the bony pelvis. Also known as pelvic cavity. { 'pel·vəs }

Pelycosauria [PALEONTOLOGY] An extinct order of primitive, mammallike reptiles of the subclass Synapsida, characterized by a temporal fossa that lies low on the side of the skull. { ˌpel·ə·kə 'sȯr·ē·ə }

pen [INVERTEBRATE ZOOLOGY] The inner horny, feather-shaped, chitnous shell of a squid. Also known as gladius. { pen }

Penaeidea [INVERTEBRATE ZOOLOGY] A primitive section of the Decapoda in the suborder Natantia; in these forms, the pleurae of the first abdominal somite overlap those of the second, the third legs are chelate, and the gills are dendrobranchiate. { ˌpen·ē'id· ē·ə }

penesaline [ECOLOGY] Referring to an environment intermediate between normal marine and saline, characterized by evaporitic carbonates often interbedded with gypsum or anhydrite, and by a salinity high enough to be toxic to normal marine organisms. { ˌpēn·ə'sāˌlēn }

penetrance [GENETICS] The proportion of individuals carrying a dominant gene in the heterozygous condition or a recessive gene in the homozygous condition in which the specific phenotypic effect is apparent. Also known as gene penetrance. { 'pen·ə·trəns }

penetrant [INVERTEBRATE ZOOLOGY] A large barbed nematocyst that pierces the body of the prey and injects a paralyzing agent. { 'pen·ə·trənt }

penetration gland [INVERTEBRATE ZOOLOGY] A gland at the anterior end of certain cercariae that secretes a histolytic substance. { ˌpen·ə'trā·shən ˌgland }

penguin [VERTEBRATE ZOOLOGY] Any member of the avian order Sphenisciformes; structurally modified wings do not fold and they function like flippers, the tail is short, feet are short and webbed, and the legs are set far back on the body. { 'peŋ·gwən }

penicillate [BIOLOGY] Having a tuft of fine hairs. { ¦pen·ə¦sil·ət }

penicillin [MICROBIOLOGY] **1.** The collective name for salts of a series of antibiotic organic acids produced by a number of *Penicillium* and *Aspergillus* species; active against most gram-positive bacteria and some gram-negative cocci. **2.** *See* benzyl penicillin sodium. { ˌpen·ə'sil·ən }

penicillinase [BIOCHEMISTRY] A bacterial enzyme that hydrolyzes and inactivates penicillin. { ˌpen·ə'sil·əˌnās }

Peniculina [INVERTEBRATE ZOOLOGY] A suborder of the Hymenostomatida. { peˌnik·yə'lī·nə }

penis [ANATOMY] The male organ of copulation in humans and certain other vertebrates. Also known as phallus. { 'pē·nəs }

penna [VERTEBRATE ZOOLOGY] *See* contour feather. { 'pen·ə }

Pennales [BOTANY] An order of diatoms (Bacillariophyceae) in which the form is often circular, and the markings on the valves are radial. { pə 'nā·lēz }

pennate [BIOLOGY] **1.** Wing-shaped. **2.** Having wings. **3.** Having feathers. { 'peˌnāt }

Pennatulacea [INVERTEBRATE ZOOLOGY] The sea pens, an order of the subclass Alcyonaria; individuals lack stolons and live with their bases embedded in the soft substratum of the sea. { pə ˌnach·ə'lā·shə }

Pennellidae [INVERTEBRATE ZOOLOGY] A family of copepod crustaceans in the suborder Caligoida; skin-penetrating external parasites of various marine fishes and whales. { ˌpen·ə'lī·əˌdē }

penniculus [ANATOMY] A tuft of arterioles in the spleen. [BIOLOGY] A brush-shaped structure. { pə'nik·yə·ləs }

pentacrinoid [INVERTEBRATE ZOOLOGY] The larva of a feather star. { pen'tak·rəˌnȯid }

pentactinal [ZOOLOGY] Having five rays or branches. { pen'tak·tə·nəl }

pentacula [INVERTEBRATE ZOOLOGY] The five-tentacled stage in the life history of echinoderms. { pen'tak·yə·lə }

pentadactyl [VERTEBRATE ZOOLOGY] Having five digits on the hand or foot. { ¦pen·tə¦dakt·əl }

pentadelphous [BOTANY] Having the stamens in five sets with the filaments more or less united within each set. { ¦pen·tə¦del·fəs }

pentahydroxyhexoic acid [BIOCHEMISTRY] *See* galactonic acid. { ˌpen·tə·hī¦dräk·sē·hek'sō·ik 'as·əd }

Pentamerida [PALEONTOLOGY] An extinct order of articulate brachiopods. { ˌpen·tə'mer·ə·də }

Pentameridina [PALEONTOLOGY] A suborder of extinct brachiopods in the order Pentamerida; dental plates associated with the brachiophores were well developed, and their bases enclosed the dorsal adductor muscle field. { ˌpen·tə·mə 'rid·ən·ə }

pentamerous [BOTANY] Having each whorl of the

flower consisting of five members, or a multiple of five. { pen'tam·ə·rəs }

pentandrous [BOTANY] Having five stamens. { pen'tan·drəs }

Pentastomida [INVERTEBRATE ZOOLOGY] A class of bloodsucking parasitic arthropods; the adult is vermiform, and there are two pairs of hooklike, retractile claws on the cephalothorax. { ˌpen·tə'stäm·ə·də }

Pentatomidae [INVERTEBRATE ZOOLOGY] The true stink bugs, a family of hemipteran insects in the superfamily Pentatomoidea. { ˌpen·tə'täm·əˌdē }

Pentatomoidea [INVERTEBRATE ZOOLOGY] A subfamily of the hemipteran group Pentatomorpha distinguished by marginal trichobothria and by antennae which are usually five-segmented. { ˌpen·tə·tə'mȯid·ē·ə }

Pentatomorpha [INVERTEBRATE ZOOLOGY] A large group of hemipteran insects in the subdivision Geocorisae in which the eggs are nonoperculate, a median spermatheca is present, accessory salivary glands are tubular, and the abdomen has trichobothria. { ˌpen·tə·də'mȯr·fə }

pentosan [BIOCHEMISTRY] A hemicellulose present in cereal, straws, brans, and other woody plant tissues; yields five-carbon-atom sugars. { 'pen·təˌsan }

pentose [BIOCHEMISTRY] Any one of a class of carbohydrates containing five atoms of carbon. { 'penˌtōs }

pentose phosphate pathway [BIOCHEMISTRY] A pathway by which glucose is metabolized or transformed in plants and microorganisms; glucose-6-phosphate is oxidized to 6-phosphogluconic acid, which then undergoes oxidative decarboxylation to form ribulose-5-phosphate, which is ultimately transformed to fructose-6-phosphate. { 'penˌtōs 'fäsˌfāt 'pathˌwā }

pepo [BOTANY] A fleshy indehiscent berry with many seeds and a hard rind; characteristic of the Cucurbitaceae. { 'pēˌpō }

pepper [BOTANY] Any of several warm-season perennials of the genus *Capsicum* in the order Polemoniales, especially *C. annum* which is cultivated for its fruit, a many-seeded berry with a thickened integument. { 'pep·ər }

peppermint [BOTANY] Any of various aromatic herbs of the genus *Mentha* in the family Labiatae, especially *M. piperita*. { 'pep·ərˌmint }

pepsin [BIOCHEMISTRY] A proteolytic enzyme found in the gastric juice of mammals, birds, reptiles, and fishes. { 'pep·sən }

pepsinogen [BIOCHEMISTRY] The precursor of pepsin, found in the stomach mucosa. { pep'sin·ə·jən }

peptic [PHYSIOLOGY] **1.** Of or pertaining to pepsin. **2.** Of or pertaining to digestion. { 'pep·tik }

peptidase [BIOCHEMISTRY] An enzyme that catalyzes the hydrolysis of peptides to amino acids. { 'pep·təˌdās }

peptide [BIOCHEMISTRY] A compound of two or more amino acids joined by peptide bonds. { 'pepˌtīd }

peptide synthetase [BIOCHEMISTRY] A ribosomal synthetase that catalyzes the formation of peptide bonds during protein synthesis. { 'pepˌtīd 'sin·thəˌtās }

Peptococcaceae [MICROBIOLOGY] A family of gram-positive cocci; organisms can use either amino acids or carbohydrates for growth and energy. { ˌpep·tə·käk'sās·ēˌē }

peptone [BIOCHEMISTRY] A water-soluble mixture of proteoses and amino acids derived from albumin, meat, or milk; used as a nutrient and to prepare nutrient media for bacteriology. { 'pepˌtōn }

Peracarida [INVERTEBRATE ZOOLOGY] A superorder of the Eumalacostraca; these crustaceans have the first thoracic segment united with the head, the cephalothorax usually larger than the abdomen, and some thoracic segments free from the carapace. { ˌper·ə'kar·ə·də }

Peramelidae [VERTEBRATE ZOOLOGY] The bandicoots, a family of insectivorous mammals in the order Marsupialia. { ˌpər·ə'mel·əˌdē }

perception [PHYSIOLOGY] Recognition in response to sensory stimuli; the act or process by which the memory of certain qualities of an object is associated with other qualities impressing the senses, thereby making possible recognition of the object. { pər'sep·shən }

perceptual overload [PHYSIOLOGY] Saturation of the nervous system by an input of excess sensory information, resulting in an absence of response. { pər¦sep·chə·wəl 'ō·vərˌlōd }

perch [VERTEBRATE ZOOLOGY] **1.** Any member of the family Percidae. **2.** The common name for a number of unrelated species of fish belonging to the Centrarchidae, Anabantoidei, and Percopsiformes. { pərch }

Percidae [VERTEBRATE ZOOLOGY] A family of fresh-water actinopterygian fishes in the suborder Percoidei; comprises the true perches. { 'pər·səˌdē }

Perciformes [VERTEBRATE ZOOLOGY] The typical spiny-rayed fishes, comprising the largest order of vertebrates; characterized by fin spines, a swim bladder without a duct, usually ctenoid scales, and 17 or fewer caudal fin rays. { ˌpər·sə'fȯrˌmēz }

Percoidei [VERTEBRATE ZOOLOGY] A large, typical suborder of the order Perciformes; includes over 50% of the species in this order. { pər'kȯid·ēˌī }

Percomorphi [VERTEBRATE ZOOLOGY] An equivalent, ordinal name for the Perciformes. { ˌpər·kə'mȯrˌfī }

Percopsidae [VERTEBRATE ZOOLOGY] A family of fishes in the order Percopsiformes. { pər'käp·səˌdē }

Percopsiformes [VERTEBRATE ZOOLOGY] A small order of actinopterygian fishes characterized by single, ray-supported dorsal and anal fins and a subabdominal pelvic fin with three to eight soft rays. { pərˌkäp·sə'fȯrˌmēz }

perennial [BOTANY] A plant that lives for an indefinite period, dying back seasonally and producing new growth from a perennating part. { pə'ren·ē·əl }

perfect flower [BOTANY] A flower having both stamens and pistils. { 'pər·fikt 'flau̇·ər }

perfoliate [BOTANY] Pertaining to the form of a leaf having its base united around the stem. [INVERTEBRATE ZOOLOGY] Pertaining to the form of certain insect antennae having the terminal joints expanded and flattened to form plates which encircle the stalk. { pər'fōl·ē,āt }

perforatorium [CELL BIOLOGY] *See* acrosome. { ,pər·fə·rə'tór·ē·əm }

perforin [IMMUNOLOGY] An enzyme that is secreted by natural killer cells and cytotoxic T cells and destroys foreign cells by puncturing their membranes, causing leakage of the cell contents. { 'pər·fər·ən }

perfusion [PHYSIOLOGY] The pumping of a fluid through a tissue or organ by way of an artery. { pər'fyü·zhən }

Pergidae [INVERTEBRATE ZOOLOGY] A small family of hymenopteran insects in the superfamily Tenthredinoidea. { 'pər·jə,dē }

perianal [ANATOMY] Situated or occurring around the anus. { ¦per·ē'ān·əl }

perianth [BOTANY] The calyx and corolla considered together. { 'per·ē,anth }

periblast [EMBRYOLOGY] The nucleated layer of cytoplasm that surrounds the blastodisk of an egg undergoing discoidal cleavage. { 'per·ə ,blast }

periblastula [EMBRYOLOGY] The blastula of a centrolecithal egg, formed by superficial segmentation. { ¦per·ə'blas·tyə·lə }

periblem [BOTANY] A layer of primary meristem which produces the cortical cells. { 'per·ə ,blem }

pericardial [ANATOMY] **1.** Of or pertaining to the pericardium. **2.** Located around the heart. { ,per·ə'kärd·ē·əl }

pericardial cavity [ANATOMY] A potential space between the inner layer of the pericardium and the epicardium of the heart. { ,per·ə'kärd·ē·əl 'kav·əd·ē }

pericardial fluid [PHYSIOLOGY] The fluid in the pericardial cavity. { ,per·ə'kärd·ē·əl 'flü·əd }

pericardial organ [INVERTEBRATE ZOOLOGY] One of the neurohemal organs associated with the pericardial cavity in crustaceans. { ,per·ə'kärd·ē·əl 'ór·gən }

pericardium [ANATOMY] The membranous sac that envelops the heart; it contains 5-20 grams of clear serous fluid. { ,per·ə'kärd·ē·əm }

pericarp [BOTANY] The wall of a fruit, developed by ripening and modification of the ovarian wall. { 'per·ə,kärp }

pericentric inversion [GENETICS] A type of chromosome aberration in which chromosome material involving both arms of the chromosome is inverted, thus spanning the centromere. { ,per·ə 'sen·trik in'vər·zhən }

perichondrium [ANATOMY] The fibrous connective tissue covering cartilage, except at joints. { ,per·ə'kän·drē·əm }

periclinal [BOTANY] Pertaining to a cell layer that is parallel to the surface of a plant part. { ¦per·ə ¦klīn·əl }

periclinal chimera [GENETICS] A plant carrying a mixture of cells of two distinct species. { ¦per·ə ¦klīn·əl kī'mir·ə }

pericranium [ANATOMY] The periosteum on the outer surface of the cranial bones. { ¦per·ə'krā·nē·əm }

pericycle [BOTANY] The outer boundary of the stele of plants; may not be present as a distinct layer of cells. { 'per·ə,sī·kəl }

pericystium [ANATOMY] The tissues surrounding a bladder. { ,per·ə'sis·tē·əm }

pericyte [HISTOLOGY] A mesenchymal cell found around a capillary; it may or may not be contractile. { 'per·ə,sīt }

periderm [BOTANY] A group of secondary tissues forming a protective layer which replaces the epidermis of many plant stems, roots, and other parts; composed of cork cambium, phelloderm, and cork. [EMBRYOLOGY] The superficial transient layer of epithelial cells of the embryonic epidermis. { 'per·ə,dərm }

peridium [BOTANY] The outer investment of the sporophore of many fungi. { pə'rid·ē·əm }

perifollicular [HISTOLOGY] Surrounding a follicle. { ¦per·ə·fə'lik·yə·lər }

perigonium [BOTANY] The perianth of a liverwort. [INVERTEBRATE ZOOLOGY] The sac containing the generative bodies in the gonophore of a hydroid. { ,per·ə'gō·nē·əm }

perigynium [BOTANY] A fleshy cup- or tubelike structure surrounding the archegonium of various bryophytes. { ,per·ə'jin·ē·əm }

perigynous [BOTANY] Bearing the floral organs on the rim of an expanded saucer- or cup-shaped receptacle or hypanthium. { pə'rij·ə·nəs }

perikaryon [CELL BIOLOGY] **1.** The body of a nerve cell, containing the nucleus. **2.** A cytoplasmic mass surrounding a nucleus. { ¦per·ə'kar·ē,än }

Perilampidae [INVERTEBRATE ZOOLOGY] A family of hymenopteran insects in the superfamily Chalcidoidea. { ,per·ə'lam·pə,dē }

perilymph [PHYSIOLOGY] The fluid separating the membranous from the osseous labyrinth of the internal ear. { 'per·ə,limpf }

perimetrium [ANATOMY] The serous covering of the uterus. { ,per·ə'mē·trē·əm }

perimysium [ANATOMY] The connective tissue sheath enveloping a muscle or a bundle of muscle fibers. { ,per·ə'mī·sē·əm }

perineum [ANATOMY] **1.** The portion of the body included in the outlet of the pelvis, bounded in front by the pubic arch, behind by the coccyx and sacrotuberous ligaments, and at the sides by the tuberosities of the ischium. **2.** The region between the anus and the scrotum in the male, between the anus and the posterior commissure of the vulva in the female. { ,per·ə'nē·əm }

perineural [ANATOMY] Situated around nervous tissue or a nerve. { ,per·ə'nùr·əl }

periocular [ANATOMY] Surrounding the eye. { ¦per·ē'äk·yə·lər }

periodicity [MOLECULAR BIOLOGY] The number of base pairs in one turn of the deoxyribonucleic acid duplex. { ,pir·ē·ə'dis·əd·ē }

periodontal [ANATOMY] **1.** Surrounding a tooth. **2.** Of or pertaining to the periodontium. { ¦per·ē·ō¦dänt·əl }

periodontium [ANATOMY] The tissues surrounding a tooth. { ¦per·ē·ō'dan·chəm }

perionychium [ANATOMY] The border of epidermis surrounding an entire nail. Also known as paronychium. { ˌper·ē·ō′nik·ē·əm }

periosteum [ANATOMY] The fibrous membrane enveloping bones, except at joints and the points of tendonous and ligamentous attachment. { ˌper·ē′äs·tē·əm }

periostracum [INVERTEBRATE ZOOLOGY] A protective layer of chitin covering the outer portion of the shell in many mollusks, especially freshwater forms. { ˌper·ē′äs·trə·kəm }

periotic [ANATOMY] **1.** Situated about the ear. **2.** Of or pertaining to the parts immediately about the internal ear. { ¦per·ē¦äd·ik }

peripheral [ANATOMY] Pertaining to or located at or near the surface of a body or an organ. { pə′rif·ə·rəl }

peripheral hemodynamics [PHYSIOLOGY] A division of hematology concerned with blood flow in regions of the body that are close to the surface, such as the extremities. { pə¦rif·ə·rəl ˌhē·mō·dī′nam·iks }

peripheral nervous system [ANATOMY] The autonomic nervous system, the cranial nerves, and the spinal nerves including their associated sensory receptors. { pə′rif·ə·rəl ′nər·vəs ˌsis·təm }

peripheral vision [PHYSIOLOGY] The act of seeing images that fall upon parts of the retina outside the macula lutea. Also known as indirect vision. { pə′rif·ə·rəl ′vizh·ən }

periphyton [ECOLOGY] Sessile biotal components of a fresh-water ecosystem. { pə′rif·əˌtän }

periplast [CELL BIOLOGY] **1.** A cell membrane. **2.** A pellicle covering ectoplasm. [HISTOLOGY] The stroma of an animal organ. [INVERTEBRATE ZOOLOGY] The ectoplasm of a flagellate. { ′per·əˌplast }

periproct [INVERTEBRATE ZOOLOGY] The area surrounding the anus of echinoids. { ′per·əˌpräkt }

Periptychidae [PALEONTOLOGY] A family of extinct herbivorous mammals in the order Condylarthra distinguished by specialized, fluted teeth. { ˌper·əp′tik·əˌdē }

Peripylina [INVERTEBRATE ZOOLOGY] An equivalent name for Porulosida. { ¦per·ə′plī·nə }

perisarc [INVERTEBRATE ZOOLOGY] The outer integument of a hydroid. { ′per·əˌsärk }

Periscelidae [INVERTEBRATE ZOOLOGY] A family of myodarian cyclorrhaphous dipteran insects in the subsection Acalypteratae. { ˌper·ə′sel·əˌdē }

Perischoechinoidea [INVERTEBRATE ZOOLOGY] A subclass of principally extinct echinoderms belonging to the Echinoidea and lacking stability in the number of columns of plates that make up the ambulacra and interambulacra. { pə¦ris·kōˌek·ə′nȯid·ē·ə }

perisperm [BOTANY] In a seed, the nutritive tissue that is derived from the nucellus and deposited on the outside of the embryo sac. { ′per·əˌspərm }

Perissodactyla [VERTEBRATE ZOOLOGY] An order of exclusively herbivorous mammals distinguished by an odd number of toes and mesaxonic feet, that is, with the axis going through the third toe. { pəˌris·ō′dak·tə·lə }

peristalsis [PHYSIOLOGY] The rhythmic progressive wave of muscular contraction in tubes, such as the intestine, provided with both longitudinal and transverse muscular fibers. { ˌper·ə′stäl·səs }

peristome [BOTANY] The fringe around the opening of a moss capsule. [INVERTEBRATE ZOOLOGY] The area surrounding the mouth of various invertebrates. { ′per·əˌstōm }

perithecium [MYCOLOGY] A spherical, cylindrical, or oval ascocarp which usually opens by a terminal slit or pore. { ˌper·ə′thē·shəm }

peritoneal cavity [ANATOMY] The potential space between the visceral and parietal layers of the peritoneum. { ¦per·ə·tə¦nē·əl ′kav·əd·ē }

peritoneum [ANATOMY] The serous membrane enveloping the abdominal viscera and lining the abdominal cavity. { ˌper·ə·tə′nē·əm }

Peritrichia [INVERTEBRATE ZOOLOGY] A specialized subclass of the class Ciliatea comprising both sessile and mobile forms. { ˌper·ə′trik·ē·ə }

Peritrichida [INVERTEBRATE ZOOLOGY] The single order of the protozoan subclass Peritrichia. { ˌper·ə′trik·ə·də }

peritrichous [INVERTEBRATE ZOOLOGY] Of certain protozoans, having spirally arranged cilia around the oral disk. [MICROBIOLOGY] Of bacteria, having a uniform distribution of flagella on the body surface. { pə′ri·trə·kəs }

perivitelline space [CELL BIOLOGY] In mammalian ova, the space formed between the ovum and the zona pellucida at the time of maturation, into which the polar bodies are given off. { ¦per·ə′vid·əlˌēn ˌspās }

permanent teeth [ANATOMY] The second set of teeth of a mammal, following the milk teeth; in humans, the set of 32 teeth consists of 8 incisors, 4 canines, 8 premolars, and 12 molars. { ′pər·mə·nənt ¦tēth }

permease [BIOCHEMISTRY] Any of a group of enzymes which mediate the phenomenon of active transport. { ′pər·mēˌās }

permissive cell [VIROLOGY] A cell that supports replication of a virus. { pər′mis·iv ′sel }

permissive condition [GENETICS] An environmental condition under which a conditional lethal mutant can survive and show the wild phenotype. { pər′mis·iv kənˌdish·ən }

Perognathinae [VERTEBRATE ZOOLOGY] A subfamily of the rodent family Heteromyidae, including the pocket and kangaroo mice. { ˌper·əg′nath·əˌnē }

Peronosporales [MYCOLOGY] An order of aquatic and terrestrial phycomycetous fungi with a hyphal thallus and zoospores with two flagella. { ˌper·əˌnäs·pə′rā·lēz }

Perothopidae [INVERTEBRATE ZOOLOGY] A small family of coleopteran insects in the superfamily Elateroidea found only in the United States. { ˌper·ə′thäp·əˌdē }

peroxidase [BIOCHEMISTRY] An enzyme that catalyzes reactions in which hydrogen peroxide is an electron acceptor. { pə′räk·səˌdās }

peroxisome [CELL BIOLOGY] Any of a subclass of microbodies that contain at least four enzymes involved in the metabolism of hydrogen peroxide. { pə'räk·sə,sōm }

perpetuation [VIROLOGY] Maintenance of a viral genome within bacterial host cells without killing them due to weakening of viral virulence. { pər,pech·ə'wā·shən }

Persian melon [BOTANY] A variety of muskmelon (*Cucumis melo*) in the order Violales; the fruit is globular and without sutures, and has dark-green skin, thin abundant netting, and firm, thick, orange flesh. { 'pər·zhən 'mel·ən }

persistence of vision [PHYSIOLOGY] The ability of the eye to retain the impression of an image for a short time after the image has disappeared. { pər'sis·təns əv 'vizh·ən }

persistent [BOTANY] Of a leaf, withering but remaining attached to the plant during the winter. { pər'sis·tənt }

persistent virus infection [VIROLOGY] A covert viral infection in which a degree of equilibrium is established between the virus and the host's immune system, resulting in an infection of long duration. { pər'sis·tənt 'vī·rəs in,fek·shən }

perspiration [PHYSIOLOGY] **1.** The secretion of sweat. **2.** *See* sweat. { ,pər·spə'rā·shən }

pessulus [VERTEBRATE ZOOLOGY] A bar composed of cartilage or bone that crosses the windpipe of a bird at its division into bronchi. { 'pes·yə·ləs }

petal [BOTANY] One of the sterile, leaf-shaped flower parts that make up the corolla. { 'ped·əl }

Petalichthyida [PALEONTOLOGY] A small order of extinct dorsoventrally flattened fishes belonging to the class Placodermi; the external armor is in two shields of large plates. { ,ped·ə·lik'thē·ə·də }

Petalodontidae [PALEONTOLOGY] A family of extinct cartilaginous fishes in the order Bradyodonti distinguished by teeth with deep roots and flattened diamond-shaped crowns. { ,ped·əl·ə'dänt·ə,dē }

Petaluridae [INVERTEBRATE ZOOLOGY] A family of dragonflies in the suborder Anisoptera. { ,ped·ə'lu̇r·ə,dē }

petasma [INVERTEBRATE ZOOLOGY] A modified endopodite of the first abdominal appendage in a male decapod crustacean. { pə'taz·mə }

petiole [BOTANY] The stem which supports the blade of a leaf. { 'ped·ē,ōl }

petrel [VERTEBRATE ZOOLOGY] A sea bird of the families Procellariidae and Hydrobatidae, generally small to medium-sized with long wings and dark plumage with white areas near the rump. { 'pe·trəl }

petri dish [MICROBIOLOGY] A shallow glass or plastic dish with a loosely fitting overlapping cover used for bacterial plate cultures and plant and animal tissue cultures. { 'pē·trē ,dish }

Petriidae [INVERTEBRATE ZOOLOGY] A small family of coleopteran insects in the superfamily Tenebrionoidea. { pə'trī·ə,dē }

petroleum microbiology [MICROBIOLOGY] Those aspects of microbiological science and engineering of interest to the petroleum industry, including the role of microbes in petroleum formation, and the exploration, production, manufacturing, storage, and food synthesis from petroleum. { pə'trō·lē·əm ,mī·krō·bī'äl·ə·jē }

Petromyzonida [VERTEBRATE ZOOLOGY] The lampreys, an order of eellike, jawless vertebrates (Agnatha) distinguished by a single, dorsal nasal opening, and the mouth surrounded by an oral disk and provided with a rasping tongue. { ,pe·trō·mī'zän·ə·də }

Petromyzontiformes [VERTEBRATE ZOOLOGY] The equivalent name for Petromyzonida. { ,pe·trō·mī ,zänt·ə'fór,mēz }

petrosal nerve [ANATOMY] Any of several small nerves passing through the petrous part of the temporal bone and usually attached to the geniculate ganglion. { pə'trō·səl ,nərv }

petrosal process [ANATOMY] A sharp process of the sphenoid bone located below the notch for the passage of the abducens nerve, which articulates with the apex of the petrous portion of the temporal bone and forms the medial boundary of the foramen lacerum. { pə'trō·səl ,prä·səs }

Petrosaviaceae [BOTANY] A small family of monocotyledonous plants in the order Triuridales characterized by perfect flowers, three carpels, and numerous seeds. { ,pe·trō,sav·ē'ās·ē,ē }

Peyer's patches [HISTOLOGY] Aggregates of lymph nodules beneath the epithelium of the ileum. { 'pī·ərz ,pach·əz }

PGA [BIOCHEMISTRY] *See* folic acid.

phacella [INVERTEBRATE ZOOLOGY] *See* gastric filament. { fə'sel·ə }

Phaenocephalidae [INVERTEBRATE ZOOLOGY] A monospecific family of coleopteran insects in the superfamily Cucujoidea, found only in Japan. { ,fē·nō·sə'fal·ə,dē }

Phaenothontidae [VERTEBRATE ZOOLOGY] The tropic birds, a family of fish-eating aquatic forms in the order Pelecaniformes. { ,fē·nō'thänt·ə ,de }

Phaeocoleosporae [MYCOLOGY] A spore group of the Fungi Imperfecti with dark filiform spores. { fē,kō·lē'äs·pə,rē }

Phaeodictyosporae [MYCOLOGY] A spore group of the Fungi Imperfecti with dark muriform spores. { ,fē·ō,dik·tē'äs·pə,rē }

Phaeodidymae [MYCOLOGY] A spore group of the Fungi Imperfecti with dark two-celled spores. { ,fē·ō'did·ə,mē }

Phaeodorina [INVERTEBRATE ZOOLOGY] The equivalent name for Tripylina. { ,fē·ə·də'rī·nə }

Phaeohelicosporae [MYCOLOGY] A spore group of the Fungi Imperfecti with dark, spirally coiled, septate spores. { ,fē·ō,hel·ə'käs·pə,rē }

Phaeophragmiae [MYCOLOGY] A spore group of the Fungi Imperfecti with dark three- to many-celled spores. { ,fē·ō'frag·mē,ē }

Phaeophyta [BOTANY] The brown algae, constituting a division of plants; the plant body is multicellular, varying from a simple filamentous form to a complex, sometimes branched body having a basal attachment. { fē'äf·əd·ə }

Phaeosporae [MYCOLOGY] A spore group of

Fungi Imperfecti characterized by dark one-celled, nonfiliform spores. { fē'äs·pə,rē }

Phaeostaurosporae [MYCOLOGY] A spore group of the Fungi Imperfecti with dark star-shaped or forked spores. { ,fē·ō·stȯ'räs·pə,rē }

phage [VIROLOGY] *See* bacteriophage. { fāj }

phage cross [VIROLOGY] Multiple infection of a single bacterium by phages that differ at one or more genetic sites, leading to the production of recombinant progeny phage. { 'fāj ,krȯs }

phage induction [VIROLOGY] Prophage stimulation by a variety of means that induce the vegetative state. { 'fāj in,dək·shən }

phage restriction [VIROLOGY] The inability of a phage to replicate due to an enzyme mechanism for degrading foreign deoxyribonucleic acid that enter the bacterial host cell. { 'fāj ri,strik·shən }

phagocyte [CELL BIOLOGY] An ameboid cell that engulfs foreign material. { 'fag·ə,sīt }

phagocytic vacuole [CELL BIOLOGY] *See* food vacuole. { ¦fag·ə¦sid·ik 'vak·yə,wōl }

phagocytin [BIOCHEMISTRY] A type of bactericidal agent present within phagocytic cells. { ¦fag·ə¦sīt·ən }

phagocytosis [CELL BIOLOGY] A specialized form of macropinocytosis in which cells engulf large solid objects such as bacteria and deliver the internalized objects to special digesting vacuoles; exists in certain cell types, such as macrophages and neutrophils. { ,fag·ə,sī'tō·səs }

phagolysosome [CELL BIOLOGY] An intracellular vesicle formed by fusion of a lysosome with a phagosome. { ,fag·ə'lī·sə,sōm }

phagosome [CELL BIOLOGY] A closed intracellular vesicle containing material captured by phagocytosis. { 'fag·ə,sōm }

phagotroph [INVERTEBRATE ZOOLOGY] An organism that ingests nutrients by phagocytosis. { 'fag·ə,träf }

Phalacridae [INVERTEBRATE ZOOLOGY] The shining flower beetles, a family of coleopteran insects in the superfamily Cucujoidea. { fə'lak·rə ,dē }

Phalacrocoracidae [VERTEBRATE ZOOLOGY] The cormorants, a family of aquatic birds in the order Pelecaniformes. { ,fal·ə,krō·kə'ras·ə,dē }

Phalaenidae [INVERTEBRATE ZOOLOGY] The equivalent name for Noctuidae. { fə'len·ə,dē }

Phalangeridae [VERTEBRATE ZOOLOGY] A family of marsupial mammals in which the marsupium is well developed and opens anteriorly, the hindfeet are syndactylous, and the hallux is opposable and lacks a claw. { ,fal·ən'jer·ə,dē }

Phalangida [INVERTEBRATE ZOOLOGY] An order of the class Arachnida characterized by an unsegmented cephalothorax broadly joined to a segmented abdomen, paired chelate chelicerae, and paired palpi. { fə'lan·jə·də }

phalanx [ANATOMY] One of the bones of the fingers or toes. { 'fā,laŋks }

Phalaropodidae [VERTEBRATE ZOOLOGY] The phalaropes, a family of migratory shore birds characterized by lobate toes and by reversal of the sex roles with respect to dimorphism and care of the young. { fə,ler·ə'päd·ə,de }

Phallostethidae [VERTEBRATE ZOOLOGY] A family of actinopterygian fishes in the order Atheriniformes. { ,fal·ō'steth·ə,dē }

Phallostethiformes [VERTEBRATE ZOOLOGY] An equivalent name for Atheriniformes. { ,fal·ō 'steth·ə'fȯr,mēz }

phallotoxin [BIOCHEMISTRY] One of a group of toxic peptides produced by the mushroom *Amanita phalloides*. { ¦fal·ō¦täk·sən }

phallus [ANATOMY] *See* penis. [EMBRYOLOGY] An undifferentiated embryonic structure derived from the genital tubercle that differentiates into the penis in males and the clitoris in females. { 'fal·əs }

phanerophyte [ECOLOGY] A perennial tree or shrub with dormant buds borne on aerial shoots. { 'fan·ə·rō,fīt }

Phanerorhynchidae [PALEONTOLOGY] A family of extinct chondrostean fishes in the order Palaeonisciformes having vertical jaw suspension. { ¦fan·ə·rō'riŋ·kə,dē }

Phanerozonida [INVERTEBRATE ZOOLOGY] An order of the Asteroidea in which the body margins are defined by two conspicuous series of plates and in which pentamerous symmetry is generally constant. { ,fan·ə·rō'zän·ə·də }

Phanodermatidae [INVERTEBRATE ZOOLOGY] A family of free-living nematodes in the superfamily Enoploidea. { ,fa·nō·dər'mad·ə,dē }

Pharetronida [INVERTEBRATE ZOOLOGY] An order of calcareous sponges in the subclass Calcinea characterized by a leuconoid structure. { ¦far·ə 'trän·ə·də }

pharmacogenetics [GENETICS] The science of genetically determined variations in drug responses. { ¦fär·mə·kō·jə'ned·iks }

pharyngeal aponeurosis [ANATOMY] The fibrous submucous layer of the pharynx. { fə'rin·jē·əl ,ap·ō·nu̇'rō·səs }

pharyngeal bursa [EMBRYOLOGY] A small pit caudal to the pharyngeal tonsil, resulting from the ingrowth of epithelium along the course of the degenerating tip of the notochord of the vertebrate embryo. { fə'rin·jē·əl 'bər·sə }

pharyngeal cleft [EMBRYOLOGY] One of the paired open clefts on the sides of the embryonic pharynx between successive visceral arches in vertebrates. { fə'rin·jē·əl 'kleft }

pharyngeal plexus [ANATOMY] **1.** A nerve plexus innervating the pharynx. **2.** A plexus of veins situated at the side of the pharynx. { fə'rin·jē·əl 'plek·səs }

pharyngeal pouch [EMBRYOLOGY] One of the five paired sacculations in the lateral aspect of the pharynx in vertebrate embryos. Also known as visceral pouch. { fə'rin·jē·əl 'pau̇ch }

pharyngeal tonsil [ANATOMY] *See* adenoid. { fə 'rin·jē·əl 'tän·səl }

pharyngeal tooth [VERTEBRATE ZOOLOGY] A tooth developed on the pharyngeal bone in many fishes. { fə'rin·jē·əl 'tüth }

Pharyngobdellae [INVERTEBRATE ZOOLOGY] A family of leeches in the order Arhynchobdellae that is distinguished by the lack of jaws. { fə,riŋ ,gäb'del·ə,dē }

pharynx [ANATOMY] A chamber at the oral end of

the vertebrate alimentary canal, leading to the esophagus. { 'far·iŋks }

phase-shift mutation [GENETICS] *See* frameshift mutation. { 'fāz ¦shift myü,tā·shən }

Phasianidae [VERTEBRATE ZOOLOGY] A family of game birds in the order Galliformes; typically, members are ground feeders, have bare tarsi and copious plumage, and lack feathers around the nostrils. { ,fāz·ē'an·ə,dē }

phasmid [INVERTEBRATE ZOOLOGY] One of a pair of lateral caudal pores which function as chemoreceptors in certain nematodes. { 'faz·məd }

Phasmidae [INVERTEBRATE ZOOLOGY] A family of the insect order Orthoptera including the walking sticks and leaf insects. { 'faz·mə,dē }

Phasmidea [INVERTEBRATE ZOOLOGY] An equivalent name for Secernentea. { faz'mid·ē·ə }

Phasmidia [INVERTEBRATE ZOOLOGY] An equivalent name for Secernentea. { faz'mid·ē·ə }

pheasant [VERTEBRATE ZOOLOGY] Any of various large sedentary game birds with long tails in the family Phasianidae; sexual dimorphism is typical of the group. { 'fez·ənt }

phellem [BOTANY] Cork; the outer tissue layer of the periderm. { 'fel·əm }

phelloderm [BOTANY] Layers of parenchymatous cells formed as inward derivatives of the phellogen. { 'fel·ə,dərm }

phellogen [BOTANY] The meristematic portion of the periderm, consisting of one layer of cells that initiate formation of the cork and secondary cortex tissue. { 'fel·ə·jən }

Phenacodontidae [PALEONTOLOGY] An extinct family of large herbivorous mammals in the order Condylarthra. { fə,näk·ə'dänt·ə,dē }

Phengodidae [INVERTEBRATE ZOOLOGY] The fire beetles, a New World family of coleopteran insects in the superfamily Cantharoidea. { fen 'gäd·ə,dē }

phenocopy [GENETICS] The nonhereditary alteration of a phenotype to a form imitating a mutant trait; caused by external conditions during development. { ¦fēn·ə¦käp·ē }

phenocritical period [EMBRYOLOGY] During development, that period during which a gene's effect can most readily be influenced by external factors. { ¦fē·nə,krid·ə·kəl 'pir·ē·əd }

phenogenetics [GENETICS] The study of the phenotypic effects of the genetic material. Also known as physiological genetics. { ¦fēn·ə·jə 'ned·iks }

phenotype [GENETICS] The observable characters of an organism. { 'fē·nə,tīp }

phenotypic lag [GENETICS] Delay in the expression of a newly acquired character. { ¦fē·nə¦tip·ik 'lag }

phenotypic masking [MICROBIOLOGY] Masking of the phenotype in strains of bacteria that are drug-dependent. { ¦fē·nə,tip·ik 'mask·iŋ }

phenotypic mixing [VIROLOGY] The production of virus particles having different structural components in the protein coats, each synthesized under the direction of a different genome. { ¦fē·nə,tip·ik 'mik·siŋ }

phenotypic plasticity [GENETICS] The extent of genotype expression in different environments. { ¦fē·nə,tip·ik plas'tis·əd·ē }

phenotypic sex determination [BIOLOGY] Control of the development of gonads by environmental stimuli, such as temperature. { ¦fē·nə ,tip·ik 'seks di,tər·mə,nā·shən }

phenotypic suppression [BIOLOGY] Prevention of mutant phenotype expression. { ¦fē·nə,tip·ik sə'presh·ən }

phenylalanine [BIOCHEMISTRY] $C_9H_{11}O_2N$ An essential amino acid, obtained in the levo form by hydrolysis of proteins (as lactalbumin); converted to tyrosine in the normal body. Also known as α-aminohydrocinnamic acid; α-amino-β-phenylpropionic acid; β-phenylalanine. { ¦fen·əl'al·ə,nēn }

phenylpyruvic acid [BIOCHEMISTRY] $C_6H_5CH_2 \cdot CO \cdot COOH$ A keto acid, occurring as a metabolic product of phenylalanine. { ¦fen·əl·pī'rü·vik 'as·əd }

pheochromoblast [HISTOLOGY] A precursor of a pheochromocyte. { ,fē·ō'krō·mə,blast }

pheoplast [CELL BIOLOGY] A plastid containing brown pigment and found in diatoms, dinoflagellates, and brown algae. { 'fē·ə,plast }

pheromone [PHYSIOLOGY] Any substance secreted by an animal which influences the behavior of other individuals of the same species. { 'fer·ə,mōn }

phialospore [MYCOLOGY] One of a chain of spores produced successively on phialides. { 'fī·ə·lə,spȯr }

Philomycidae [INVERTEBRATE ZOOLOGY] A family of pulmonate gastropods composed of slugs. { ,fil·ə'mīs·ə,dē }

philopatry [ECOLOGY] A dispersal method in which reproductive particles remain near their point of origin. { fə'läp·ə·trē }

Philopteridae [INVERTEBRATE ZOOLOGY] A family of biting lice (Mallophaga) that are parasitic on most land birds and water birds. { ,fil·əp'ter·ə ,dē }

phlebotaxis [BIOLOGY] Movement of a simple motile organism in response to the presence of blood. { 'fle·bə,tak·səs }

phlegm [PHYSIOLOGY] A viscid, stringy mucus, secreted by the mucosa of the air passages. { flem }

phleomycin [MICROBIOLOGY] An antibacterial antibiotic produced by *Streptomyces verticillatus*; antitumor activity has also been demonstrated. { 'flē·ə,mīs·ən }

Phloeidae [INVERTEBRATE ZOOLOGY] The bark bugs, a small neotropical family of hemipteran insects in the superfamily Pentatomoidea. { 'flē·ə,dē }

phloem [BOTANY] A complex, food-conducting vascular tissue in higher plants; principal conducting cells are sieve elements. Also known as bast; sieve tissue. { 'flō·əm }

Phocaenidae [VERTEBRATE ZOOLOGY] The porpoises, a family of marine mammals in the order Cetacea. { fō'sē·nə,dē }

Phocidae [VERTEBRATE ZOOLOGY] The seals, a pinniped family of carnivoran mammals in the superfamily Canoidea. { 'fō·sə,dē }

Phodilidae [VERTEBRATE ZOOLOGY] A family of birds in the order Strigiformes; the bay owl (*Pholidus bodius*) is the single species. { fō'dil·ə‚dē }

Phoenicopteridae [VERTEBRATE ZOOLOGY] The flamingos, a family of long-legged, long-necked birds in the order Ciconiiformes. { ‚fēn·ə·käp 'ter·ə‚dē }

Phoenicopteriformes [VERTEBRATE ZOOLOGY] An order comprising the flamingos in some systems of classification. { ‚fē·nə‚käp·tə·rə'fōr‚mēz }

Phoeniculidae [VERTEBRATE ZOOLOGY] The African wood hoopoes, a family of birds in the order Coraciiformes. { ‚fēn·ə'kyü·lə‚dē }

Pholadidae [INVERTEBRATE ZOOLOGY] A family of bivalve mollusks in the subclass Eulamellibranchia; individuals may have one or more dorsal accessory plates, and the visceral mass is attached to the valves in the dorsal portion of the body. { fō'lad·ə‚dē }

Pholidophoridae [PALEONTOLOGY] A generalized family of extinct fishes belonging to the Pholidophoriformes. { fə‚lid·ə'fȯr·ə‚dē }

Pholidophoriformes [PALEONTOLOGY] An extinct actinopterygian group composed of mostly small fusiform marine and fresh-water fishes of an advanced holostean level. { fə‚lid·ə‚fȯr·ə'fȯr ‚mēz }

Pholidota [VERTEBRATE ZOOLOGY] An order of mammals comprising the living pangolins and their fossil predecessors; characterized by an elongate tubular skull with no teeth, a long protrusive tongue, strong legs, and five-toed feet with large claws. { ‚fäl·ə'dōd·ə }

Phomaceae [MYCOLOGY] The equivalent name for Sphaerioidaceae. { fō'mās·ē‚ē }

Phomales [MYCOLOGY] The equivalent name for Sphaeropsidales. { fō'mā·lēz }

phonation [PHYSIOLOGY] Production of speech sounds. { fō'nā·shən }

phonoreception [PHYSIOLOGY] The perception of sound through specialized sense organs. { ¦fō·nō·ri'sep·shən }

phoresy [ECOLOGY] A type of commensalism which involves the transporation of one organism (the guest) by a larger organism (the host) of a different species. { 'fȯr·ə·sē }

Phoridae [INVERTEBRATE ZOOLOGY] The humpbacked flies, a family of cyclorrhaphous dipteran insects in the series Aschiza. { 'fȯr·ə‚dē }

Phoronida [INVERTEBRATE ZOOLOGY] A small, homogeneous group, or phylum, of animals having an elongate body, a crown of tentacles surrounding the mouth, and the anus occurring at the level of the mouth. { fə'rän·ə·də }

phosphatase [BIOCHEMISTRY] An enzyme that catalyzes the hydrolysis and synthesis of phosphoric acid esters and the transfer of phosphate groups from phosphoric acid to other compounds. { 'fäs·fə‚tās }

phosphatide [BIOCHEMISTRY] *See* phospholipid. { 'fäs·fə‚tīd }

phosphocreatine [BIOCHEMISTRY] $C_4H_{10}N_3O_5P$ Creatine phosphate, a phosphoric acid derivative of creatine which contains an energy-rich phosphate bond; it is present in muscle and other tissues, and during the anaerobic phase of muscular contraction it hydrolyzes to creatine and phosphate and makes energy available. Abbreviated PC. { ¦fäs·fō'krē·ə‚tēn }

phosphoenolpyruvic acid [BIOCHEMISTRY] $CH_2{=\!=}O(OPO_3H_2)COOH$ A high-energy phosphate formed by dehydration of 2-phosphoglyceric acid; it reacts with adenosinediphosphate to form adenosinetriphosphate and enolpyruvic acid. { ¦fäs·fō¦ē‚nȯl·pī'rü·vik 'as·əd }

phosphoglucoisomerase [BIOCHEMISTRY] An enzyme that catalyzes the conversion of galactose-1-phosphate to glucose-1-phosphate. { ¦fäs·fō¦glü·kō·ī'säm·ə‚rās }

phosphoglucomutase [BIOCHEMISTRY] An enzyme that catalyzes the conversion of glucose-1-phosphate to glucose-6-phosphate. { ¦fäs·fō ¦glü·kō'myü‚tās }

phospholipase [BIOCHEMISTRY] An enzyme that catalyzes a hydrolysis of a phospholipid, especially a lecithinase that acts in this manner on a lecithin. { ¦fäs·fō'lī‚pās }

phospholipid [BIOCHEMISTRY] Any of a class of esters of phosphoric acid containing one or two molecules of fatty acid, an alcohol, and a nitrogenous base. Also known as phosphatide. { ¦fäs·fō'lip·əd }

phosphomonoesterase [BIOCHEMISTRY] An enzyme catalyzing hydrolysis of phosphoric acid esters containing one ester linkage. { ¦fäs·fō ‚män·ō'es·tə‚rās }

phosphorolysis [BIOCHEMISTRY] A reaction by which elements of phosphoric acid are incorporated into the molecule of a compound. { ‚fäs·fə 'räl·ə·səs }

phosphorylase [BIOCHEMISTRY] An enzyme that catalyzes the formation of glucose-1-phosphate (Cori ester) from glycogen and inorganic phosphate; it is widely distributed in animals, plants, and microorganisms. { fäs'fȯr·ə‚lās }

phosphotransacetylase [BIOCHEMISTRY] An enzyme that catalyzes the reversible transfer of an acetyl group from acetyl coenzyme A to a phosphate, with formation of acetyl phosphate. { ¦fä·sfō‚tranz·ə'sed·əl‚ās }

photic zone [ECOLOGY] The uppermost layer of a body of water (approximately the upper 330 feet or 100 meters) that receives enough sunlight to permit the occurrence of photosynthesis. { 'fōd·ik }

Photidae [INVERTEBRATE ZOOLOGY] A family of amphipod crustaceans in the suborder Gammaridea. { 'fäd·ə‚dē }

photoautotrophic [BIOLOGY] Pertaining to organisms which derive energy from light and manufacture their own food. { ¦fōd·ō‚ȯd·ō'träf·ik }

photobiont [ECOLOGY] A photosynthetic partner of a symbiotic pair, such as the algal component of the fungal-algal association in lichens. { ¦fōd·ō'bī‚änt }

photoinhibition [BOTANY] Damage to the light-harvesting reactions of the photosynthetic apparatus caused by excess light energy trapped by the chloroplast. { ¦fōd·ō‚in·ə'bish·ən }

photomorphogenesis [BOTANY] The control exerted by light over growth, development, and differentiation of plants that is independent of photosynthesis. { ¦fōd·ō‚mȯr·fō'jen·ə·səs }

photoperiodism [PHYSIOLOGY] The physiological responses of an organism to the length of night or day or both. { ¦fōd·ō'pir·ē·ə‚diz·əm }

photophilic [BIOLOGY] Thriving in full light. { ¦fōd·ō¦fil·ik }

photophobic [BIOLOGY] **1.** Avoiding light. **2.** Exhibiting negative phototropism. { ‚fōd·ə'fō·bik }

photophore gland [VERTEBRATE ZOOLOGY] A highly modified integumentary gland which develops into a luminous organ composed of a lens and a light-emitting gland; occurs in deep-sea teleosts and elasmobranchs. { 'fōd·ə‚fȯr ‚gland }

photophosphorylase [BIOCHEMISTRY] An enzyme that is associated with the surface of a thylakoid membrane and is involved in the final stages of adenosinetriphosphate production by photosynthetic phosphorylation. { ¦fōd·ō‚fä'sfȯr·ə‚lās }

photophosphorylation [BIOCHEMISTRY] Phosphorylation induced by light energy in photosynthesis. { ¦fōd·ō‚fä·sfə·rə'lā·shən }

photophygous [BIOLOGY] Thriving in shade. { fə'täf·ə·gəs }

photopic vision [PHYSIOLOGY] *See* foveal vision. { fō'täp·ik 'vizh·ən }

photopigment [BIOCHEMISTRY] A pigment that is unstable in the presence of light of appropriate wavelengths, such as the chromophore pigment which combines with opsins to form rhodopsin in the rods and cones of the vertebrate eye. { ¦fōd·ō¦pig·mənt }

photoreactive chlorophyll [BIOCHEMISTRY] Chlorophyll molecules which receive light quanta from antenna chlorophyll and constitute a photoreaction center where light energy conversion occurs. { ¦fōd·ō·rē'ak·tiv 'klȯr·ə‚fil }

photoreception [PHYSIOLOGY] The process of absorption of light energy by plants and animals and its utilization for biological functions, such as photosynthesis and vision. { ¦fōd·ō·ri'sep·shən }

photoreceptor [PHYSIOLOGY] A highly specialized, light-sensitive cell or group of cells containing photopigments. { ¦fōd·ō·ri'sep·tər }

photorespiration [BIOCHEMISTRY] Respiratory activity taking place in plants during the light period; CO_2 is released and O_2 is taken up, but no useful form of energy, such as adenosinetriphosphate, is derived. { ¦fōd·ō‚res·pə'rā·shən }

photosynthesis [BIOCHEMISTRY] Synthesis of chemical compounds in light, especially the manufacture of organic compounds (primarily carbohydrates) from carbon dioxide and a hydrogen source (such as water), with simultaneous liberation of oxygen, by chlorophyll-containing plant cells. { ¦fōd·ō'sin·thə·səs }

photosystem I [BIOCHEMISTRY] One of two reaction sequences of the light phase of photosynthesis in green plants that involves a pigment system which is excited by wavelengths shorter than 700 nanometers and which transfers this energy to energy carriers such as NADPH that are subsequently utilized in carbon dioxide fixation. { 'fōd·ō‚sis·təm 'wən }

photosystem II [BIOCHEMISTRY] One of two reaction sequences of the light phase of photosynthesis in green plants which involves a pigment system excited by wavelengths shorter than 685 nanometers and which is directly involved in the splitting or photolysis of water. { 'fōd·ō‚sis·təm 'tü }

phototaxis [BIOLOGY] Movement of a motile organism or free plant part in response to light stimulation. { ¦fōd·ə¦tak·səs }

phototroph [BIOLOGY] An organism that utilizes light as a source of metabolic energy. { 'fōd·ə‚träf }

phototrophic bacteria [MICROBIOLOGY] Primarily aquatic bacteria comprising two principal groups: purple bacteria and green sulfur bacteria; all contain bacteriochlorophylls. { ¦fōd·ə¦träf·ik bak'tir·ē·ə }

phototropism [BOTANY] A growth-mediated response of a plant to stimulation by visible light. { fō'tä·trə‚piz·əm }

Phoxichilidiidae [INVERTEBRATE ZOOLOGY] A family of marine arthropods in the subphylum Pycnogonida; typically, chelifores are present, palpi are lacking, and ovigers have five to nine joints in males only. { ‚fäk·sə‚kil·ə'dī·ə‚dē }

Phoxocephalidae [INVERTEBRATE ZOOLOGY] A family of amphipod crustaceans in the suborder Gammaridea. { ‚fäk·sō·sə'fal·ə‚dē }

Phractolaemidae [VERTEBRATE ZOOLOGY] A family of tropical African fresh-water fishes in the order Gonorynchiformes. { ‚frak·tə'lē·mə‚dē }

phragmacone [INVERTEBRATE ZOOLOGY] *See* phragmocone. { 'frag·mə‚kōn }

Phragmobasidiomycetes [MYCOLOGY] An equivalent name for Heterobasidiomycetidae. { ¦frag·mō·bə¦sid·ē·ō·mī'sēd·ēz }

phragmocone [INVERTEBRATE ZOOLOGY] The siphuncular tube of the chambered part of the shell of certain mollusks. Also spelled phragmacone. { 'frag·mə‚kōn }

phragmoid [BOTANY] Having septae perpendicular to the long axis, as the conidia of certain fungi. { 'frag‚mȯid }

phragmoplast [CELL BIOLOGY] A thin barrier which is formed across the spindle equator in late cytokinesis in plant cells and within which the cell wall is laid down. { 'frag·mə‚plast }

phragmosome [CELL BIOLOGY] A differentiated cytoplasmic partition in which the phragmoplast and cell plate develop during cell division in plant cells. { 'frag·mə‚sōm }

Phragmosporae [MYCOLOGY] A spore group of the Fungi Imperfecti with three- to many-celled spores. { frag'mäs·pə‚rē }

Phreatoicidae [INVERTEBRATE ZOOLOGY] A family of isopod crustaceans in the suborder Phreatoicoidea in which only the left mandible retains a lacinia mobilis. { frē‚ad·ō'ī·sə‚dē }

Phreatoicoidea [INVERTEBRATE ZOOLOGY] A suborder of the Isopoda having a subcylindrical body that appears laterally compressed, antennules shorter than the antennae, and the first thoracic segment fused with the head. { frē,ad·ō·i'kȯid·ē·ə }

phreatophyte [ECOLOGY] A plant with a deep root system which obtains water from the groundwater or the capillary fringe above the water table. { frē'ad·ə,fīt }

phrenic nerve [ANATOMY] A nerve, arising from the third, fourth, and fifth cervical (cervical plexus) segments of the spinal cord; innervates the diaphragm. { 'fren·ik ¦nərv }

Phrynophiurida [INVERTEBRATE ZOOLOGY] An order of the Ophiuroidea in which the vertebrae usually articulate by means of hourglass-shaped surfaces, and the arms are able to coil upward or downward in the vertical plane. { ,frī·nə'fyu̇r·ə·də }

Phycitinae [INVERTEBRATE ZOOLOGY] A large subfamily of moths in the family Pyralididae in which the frenulum of the female is a simple spine rather than a bundle of bristles. { fī'sīt·ən,ē }

phycobilin [BIOCHEMISTRY] Any of various protein-bound pigments which are open-chain tetrapyrroles and occur in some groups of algae. { ,fī·kō'bī·lən }

phycobiont [BOTANY] The algal component of a lichen, commonly the green unicell of the genus *Trebouxia*. { ,fī·kō'bī,änt }

phycocyanin [BIOCHEMISTRY] A blue phycobilin. { ¦fī·kō'sī·ə·nən }

phycocyanobilin [BIOCHEMISTRY] $C_{31}H_{38}O_2N_4$ Phycobilin with an ethylidene side chain ($=CH-CH_3$) and only one asymmetric carbon atom (C_1). { ¦fī·kō,sī·ə·nō'bī·lən }

phycoerythrin [BIOCHEMISTRY] A red phycobilin. { ¦fī·kō·ə'rith·rən }

phycoerythrobilin [BIOCHEMISTRY] $C_{31}H_{38}O_2N_4$ Phycobilin with seven conjugated double bonds, an ethylidine side chain ($=CH-CH_3$), and two asymmetric carbon atoms (C_1 and C_7). { ¦fī·kō·ə ,rith·rə'bī·lən }

phycology [BOTANY] *See* algology. { fī'käl·ə·jē }

Phycomycetes [MYCOLOGY] A primitive class of true fungi belonging to the Eumycetes; they lack regularly spaced septa in the actively growing portions of the plant body, and have the sporangiospore, produced in the sporangium by cleavage, as the fundamental, asexual reproductive unit. { ¦fī·kō,mī'sēd·ēz }

Phycosecidae [INVERTEBRATE ZOOLOGY] A small family of coleopteran insects of the superfamily Cucujoidea, including five species found in New Zealand, Australia, and Egypt. { ¦fī·kō'sē·sə ,dē }

Phylactolaemata [INVERTEBRATE ZOOLOGY] A class of fresh-water ectoproct bryozoans; individuals have lophophores which are U-shaped in basal outline, and relatively short, wide zooecia. { fə,lak·tō'lē·məd·ə }

phyletic evolution [EVOLUTION] The gradual evolution of population without separation into isolated parts. { fī'led·ik ,ev·ə'lü·shən }

phyletic gradualism [EVOLUTION] *See* gradualism. { fī'led·ik 'gra·jə·wə,liz·əm }

phyllary [BOTANY] A bract of the involucre of a composite plant. { 'fil·ə·rē }

Phyllobothrioidea [INVERTEBRATE ZOOLOGY] The equivalent name for Tetraphyllidea. { ,fil·ō ,bäth·rē'ȯid·ē·ə }

phyllobranchiate gill [INVERTEBRATE ZOOLOGY] A type of decapod crustacean gill with flattened branches, or lamellae usually arranged in two opposite series. { ,fil·ō'braŋ·kē·ət 'gil }

phylloclade [BOTANY] A flattened stem that fulfills the same functions as a leaf. { 'fil·ə,klād }

phyllode [BOTANY] A broad, flat petiole that replaces the blade of a foliage leaf. [INVERTEBRATE ZOOLOGY] A petal-shaped group of ambulacra near the mouth of certain echinoderms. { 'fi ,lōd }

Phyllodocidae [INVERTEBRATE ZOOLOGY] A leaf-bearing family of errantian annelids in which the species are often brilliantly iridescent and are highly motile. { ,fil·ə'däs·ə,dē }

Phyllogoniaceae [BOTANY] A family of mosses in the order Isobryales in which the leaves are equitant. { ,fil·ə,gō·nē'ās·ē,ē }

Phyllolepida [PALEONTOLOGY] A monogeneric order of placoderms from the late Upper Devonian in which the armor is broad and low with a characteristic ornament of concentric and transverse ridges on the component plates. { ,fil·ə 'lep·ə·də }

Phyllophoridae [INVERTEBRATE ZOOLOGY] A family of dendrochirotacean holothurians in the order Dendrochirotida having a rather naked skin and a complex calcareous ring. { ,fil·ə'fȯr·ə ,dē }

phyllosoma [INVERTEBRATE ZOOLOGY] A flat, transparent, long-legged larval stage of various spiny lobsters. { ,fil·ə'sō·mə }

phyllospondylous [VERTEBRATE ZOOLOGY] Of vertebrae, having a hypocentrum but no pleurocentra; the neural arch extends ventrad to enclose the notochord and form transverse processes which articulate with the ribs. { ,fil·ə 'spän·də·ləs }

Phyllostictales [MYCOLOGY] An equivalent name for Sphaeropsidales. { ,fil·ə,stik'tā·lēz }

Phyllostomatidae [VERTEBRATE ZOOLOGY] The New World leaf-nosed bats (Chiroptera), a large tropical and subtropical family of insect- and fruit-eating forms with narrow, pointed ears. { ,fil·ə·stō'mad·ə,dē }

phyllotaxy [BOTANY] The arrangement of leaves on a stem. { ¦fil·ə¦tak·sē }

Phylloxerinae [INVERTEBRATE ZOOLOGY] A subfamily of homopteran insects in the family Chermidae in which the sexual forms lack mouthparts, and the parthenogenetic females have a beak but the digestive system is closed, and no honeydew is produced. { ,fi,läk'ser·ə,nē }

phylogenetic tree [EVOLUTION] *See* evolutionary tree. { ,fī·lō·jə¦ned·ik 'trē }

phylogeny [EVOLUTION] The evolutionary or ancestral history of organisms. { fə'läj·ə·nē }

phylum [SYSTEMATICS] A major taxonomic category in classifying animals (and plants in some systems), composed of groups of related classes. { 'fī·ləm }

Phymatidae [INVERTEBRATE ZOOLOGY] A family of

carnivorous hemipteran insects characterized by strong, thick forelegs. { fī′mad·ə‚dē }

Phymosomatidae [INVERTEBRATE ZOOLOGY] A family of echinacean echinoderms in the order Phymosomatoida with imperforate crenulate tubercles; one surviving genus is known. { ‚fī·mə·sō′mad·ə‚dē }

Phymosomatoida [INVERTEBRATE ZOOLOGY] An order of Echinacea with a stirodont lantern and diademoid ambulacral plates. { ‚fī·mə·sə·mə ′tȯid·ē·ə }

physa [INVERTEBRATE ZOOLOGY] The rounded basal portion of the body of certain sea anemones. { ′fī·sə }

Physalopteridae [INVERTEBRATE ZOOLOGY] A family of parasitic nematodes in the superfamily Spiruroidea. { ‚fī·sə·läp′ter·ə‚dē }

Physalopteroidea [INVERTEBRATE ZOOLOGY] A superfamily of parasitic nematodes in the order Spirurida, characterized by two large lateral lips generally provided with teeth on their inner surfaces, reduced stoma, and a reduced or absent inner whorl of circumoral sensilla and an external circle with four fused sensilla. { ′fī·sə‚läp·tə ′rȯid·ē·ə }

Physaraceae [MYCOLOGY] A family of slime molds in the order Physarales. { ‚fī·sə′rās·ē‚ē }

Physarales [MYCOLOGY] An order of Myxomycetes in the subclass Myxogastromycetidae. { ‚fī·sə′rā·lēz }

physiological biophysics [BIOPHYSICS] An area of biophysics concerned with the use of physical mechanisms to explain the behavior and the functioning of living organisms or parts thereof, and with the response of living organisms to physical forces. { ‚fiz·ē·ə′läj·ə·kəl ‚bī·ō′fiz·iks }

physiological dead space [PHYSIOLOGY] *See* dead space. { ‚fiz·ē·ə′läj·ə·kəl ′ded ‚spās }

physiological ecology [ECOLOGY] The study of biophysical, biochemical, and physiological processes used by animals to cope with factors of their physical environment, or employed during ecological interactions with other organisms. { ‚fiz·ē·ə′läj·ə·kəl ē′käl·ə·jē }

physiological genetics [GENETICS] *See* phenogenetics. { ‚fiz·ē·ə′läj·ə·kəl jə′ned·iks }

physiological saline [PHYSIOLOGY] *See* normal saline. { ‚fiz·ē·ə′läj·ə·kəl ′sā‚lēn }

physiological salt solution [PHYSIOLOGY] *See* normal saline. { ‚fiz·ē·ə′läj·ə·kəl ′sȯlt sə‚lü·shən }

physiological sodium chloride solution [PHYSIOLOGY] *See* normal saline. { ‚fiz·ē·ə′läj·ə·kəl ′sōd·ē·əm ′klȯr‚īd sə‚lü·shən }

physiologic diplopia [PHYSIOLOGY] A normal phenomenon in which there is formation of images in noncorresponding retinal points, giving a perception of depth. Also known as introspective diplopia. { ‚fiz·ē·ə′läj·ik di′plō·pē·ə }

physiologic tremor [PHYSIOLOGY] A tremor in normal individuals, caused by fatigue, apprehension, or overexposure to cold. { ‚fiz·ē·ə′läj·ik ′trem·ər }

physiology [BIOLOGY] The study of the basic activities that occur in cells and tissues of living organisms by using physical and chemical methods. { ‚fiz·ē′äl·ə·jē }

Physopoda [INVERTEBRATE ZOOLOGY] The equivalent name for Thysanoptera. { fī′säp·ə·də }

Physosomata [INVERTEBRATE ZOOLOGY] A superfamily of amphipod crustaceans in the suborder Hyperiidea; the eyes are small or rarely absent, and the inner plates of the maxillipeds are free at the apex. { ‚fī·sə′säm·əd·ə }

Phytalmiidae [INVERTEBRATE ZOOLOGY] A family of myodarian cyclorrhaphous dipteran insects in the subsection Acalypteratae. { ‚fīd·əl′mī·ə‚dē }

phytal zone [ECOLOGY] The part of a lake bottom covered by water shallow enough to permit the growth of rooted plants. { ′fīd·əl ‚zōn }

Phytamastigophorea [INVERTEBRATE ZOOLOGY] A class of the subphylum Sarcomastigophora, including green and colorless phytoflagellates. { ¦fīd·ə‚ma·stə¦gäf·ə′rē·ə }

phytase [BIOCHEMISTRY] An enzyme occurring in plants, especially cereals, which catalyzes hydrolysis of phytic acid to inositol and phosphoric acid. { ′fī‚tās }

phytoalexin [BIOCHEMISTRY] A natural substance that is toxic to fungi and is synthesized by a plant as a response to fungal infection. { ′fīd·ō·ə′lek·sən }

phytochemistry [BOTANY] The study of the chemistry of plants, plant products, and processes taking place within plants. { ′fīd·ō‚kem·ə·strē }

phytochorology [BOTANY] *See* plant geography. { ¦fīd·ō·kō′räl·ə·jē }

phytochrome [BIOCHEMISTRY] A protein plant pigment which serves to direct the course of plant growth and development in response variously to the presence or absence of light, to photoperiod, and to light quality. { ′fīd·ə ‚krōm }

phytocoenosis [ECOLOGY] The entire plant population of a particular habitat. { ¦fīd·ō·sē′nō·səs }

phytogenic dam [ECOLOGY] A natural dam consisting of plants and plant remains. { ¦fīd·ə ¦jen·ik ′dam }

phytogenic dune [ECOLOGY] Any dune in which the growth of vegetation influences the form of the dune, for example, by arresting the drifting of sand. { ¦fīd·ə¦jen·ik ′dün }

phytogeography [BOTANY] *See* plant geography. { ¦fīd·ō·jē′äg·rə·fē }

phytohemagglutinin [BIOCHEMISTRY] *See* phytolectin. { ¦fīd·ō‚hē·mə′glüt·ən·ən }

phytohormone [BIOCHEMISTRY] *See* plant hormone. { ¦fīd·ō′hȯr‚mōn }

phytolectin [BIOCHEMISTRY] A lectin found in plants. Also known as phytohemagglutinin. { ‚fīd·ə′lek·tən }

phytolith [PALEONTOLOGY] A fossilized part of a living plant that secreted mineral matter. { ′fīd·ə‚lith }

Phytomastigina [INVERTEBRATE ZOOLOGY] The equivalent name for Phytamastigophorea. { ‚fīd·ō·mas·tə′jī·nə }

Phytomonadida [INVERTEBRATE ZOOLOGY] The equivalent name for Volvocida. { ˌfīd·ō·mō 'näd·ə·də }

phytopathogen [ECOLOGY] An organism that causes a disease in a plant. { ¦fīd·ō'path·ə·jən }

phytophagous [ZOOLOGY] Feeding on plants. { fī 'täf·ə·gəs }

phytoplankton [ECOLOGY] Planktonic plant life. { ¦fīd·ə'plaŋk·tən }

Phytosauria [PALEONTOLOGY] A suborder of Late Triassic long-snouted aquatic thecodonts resembling crocodiles but with posteriorly located external nostrils, absence of a secondary palate, and a different structure of the pelvic and pectoral girdles. { ˌfīd·ə'sȯr·ē·ə }

Phytoseiidae [INVERTEBRATE ZOOLOGY] A family of the suborder Mesostigmata. { ˌfīd·ō·sē'ī·ə ˌdē }

phytosociology [ECOLOGY] A broad study of plants that includes the study of all phenomena affecting their lives as social units. { ¦fīd·ōˌsō·sē 'äl·ə·jē }

phytosterol [BIOCHEMISTRY] Any of various sterols obtained from plants, including ergosterol and stigmasterol. { fī'täs·təˌrȯl }

phytotoxin [BIOCHEMISTRY] **1.** A substance toxic to plants. **2.** A toxin produced by plants. { ¦fīd·ə 'täk·sən }

phytotron [BOTANY] A research tool used to study whole plants; contains a large number of individually controlled environments that provide the means of studying the effect of each environmental factor, such as temperature or light, at many levels simultaneously. { 'fīd·əˌträn }

pia arachnoid [VERTEBRATE ZOOLOGY] The outer meninx of certain submammalian forms having two membranes covering the brain and spinal cord. { 'pī·ə ə'rakˌnȯid }

pia mater [ANATOMY] The vascular membrane covering the surface of the brain and spinal cord. { 'pē·ə ˌmād·ər }

Picidae [VERTEBRATE ZOOLOGY] The woodpeckers, a large family of birds in the order Piciformes; adaptive modifications include a long tongue and hyoid mechanism, and stiffened tail feathers. { 'pis·əˌdē }

Piciformes [VERTEBRATE ZOOLOGY] An order of birds characterized by the peculiar arrangement of the tendons of the toes. { ˌpis·ə'fȯrˌmēz }

Picinae [VERTEBRATE ZOOLOGY] The true woodpeckers, a subfamily of the Picidae. { 'pis·ə ˌnē }

picodnavirus [VIROLOGY] A group of deoxyribonucleic acid-containing animal viruses including the adeno-satellite viruses. { pē'käd·nəˌvī·rəs }

picornavirus [VIROLOGY] A viral group made up of small (18-30 nanometers), ether-sensitive viruses that lack an envelope and have a ribonucleic acid genome; among subgroups included are enteroviruses and rhinoviruses, both of human origin. { pē'kȯr·nəˌvī·rəs }

Picrodendraceae [BOTANY] A small family of dicotyledonous plants in the order Juglandales characterized by unisexual flowers borne in catkins, four apical ovules in a superior ovary, and trifoliate leaves. { ˌpik·rō·den'drās·ēˌē }

picrotoxin [BIOCHEMISTRY] $C_{30}H_{34}O_{13}$ A poisonous, crystalline plant alkaloid found primarily in *Cocculus indicus*; used as a stimulant and convulsant drug. Also known as cocculin. { ˌpik·rə'täk·sən }

Picumninae [VERTEBRATE ZOOLOGY] The piculets, a subfamily of the avian family Picidae. { pi 'kyüm·nəˌnē }

piece-root grafting [BOTANY] Grafting in which each piece of a cut seedling root is used as a stock. { 'pēs ˌrüt ˌgraft·iŋ }

Pieridae [INVERTEBRATE ZOOLOGY] A family of lepidopteran insects in the superfamily Papilionoidea including white, sulfur, and orange-tip butterflies; characterized by the lack of a prespiracular bar at the base of the abdomen. { pī'er·ə ˌdē }

Piesmatidae [INVERTEBRATE ZOOLOGY] The ash-gray leaf bugs, a family of hemipteran insects belonging to the Pentatomorpha. { pēz'mad·ə ˌdē }

Piesmidae [INVERTEBRATE ZOOLOGY] A small family of hemipteran insects in the superfamily Lygaeoidea. { 'pez·məˌdē }

pig [VERTEBRATE ZOOLOGY] Any wild or domestic mammal of the superfamily Suoidea in the order Artiodactyla; toes terminate in nails which are modified into hooves, the tail is short, and the body is covered sparsely with hair which is frequently bristlelike. { pig }

pigeon [VERTEBRATE ZOOLOGY] Any of various stout-bodied birds in the family Columbidae having short legs, a bill with a horny tip, and a soft cere. { 'pij·ən }

pigeon milk [PHYSIOLOGY] A milky glandular secretion of the crop of pigeons that is regurgitated to feed newly hatched young. { 'pij·ən ˌmilk }

pigment [BIOCHEMISTRY] Any coloring matter in plant or animal cells. { 'pig·mənt }

pigmentation [PHYSIOLOGY] The normal color of the body and its organs, resulting from a summation of the natural color of the tissue, the pigments deposited therein, and the pigments carried through the blood bathing the tissue. { ˌpig·mən'tā·shən }

pigment cell [CELL BIOLOGY] Any cell containing deposits of pigment. { 'pig·mənt ˌsel }

pika [VERTEBRATE ZOOLOGY] Any member of the family Ochotonidae, which includes 14 species of lagomorphs resembling rabbits but having a vestigial tail and short, rounded ears. { 'pī·kə }

pike [VERTEBRATE ZOOLOGY] Any of about five species of predatory fish which compose the family Esocidae in the order Clupeiformes; the body is cylindrical and compressed, with cycloid scales that have deeply scalloped edges. { pīk }

Pilacraceae [MYCOLOGY] A family of Basidiomycetes. { ˌpil·ə'krās·ēˌē }

Pilargidae [INVERTEBRATE ZOOLOGY] A family of small, short, depressed errantian polychaete annelids. { pə'lär·jəˌdē }

pileum [VERTEBRATE ZOOLOGY] The top of a bird's head, from the nape to the bill. { 'pil·ē·əm }

pileus [BIOLOGY] The umbrella-shaped upper cap of mushrooms and other basidiomycetous fungi. { 'pil·ē·əs }

Pilidae [INVERTEBRATE ZOOLOGY] A family of fresh-water snails in the order Pectinibranchia. { 'pil·ə,dē }

Pilifera [VERTEBRATE ZOOLOGY] Collective designation for animals with hair, that is, mammals. { pī'lif·ə·rə }

pillotina [MICROBIOLOGY] A large spirochete that contains microtubules and lives symbiotically in the hind gut of termites. { ,pil·ə'tē·nə }

pilomotor nerve [ANATOMY] A nerve causing contraction of one of the arrectoris pilorum muscles. { ¦pī·lō'mōd·ər ,nərv }

pilomotor reflex [PHYSIOLOGY] Erection of the hairs of the skin (gooseflesh) in response to chilling or irritation of the skin or to an emotional stimulus. { ¦pī·lō'mōd·ər 'rē,fleks }

pilosebaceous [ANATOMY] Pertaining to the hair follicles and sebaceous glands, as the pilosebaceous apparatus, comprising the hair follicle and its attached gland. { ¦pī·lō·sə'bā·shəs }

Piltdown man [PALEONTOLOGY] An alleged fossil man based on fragments of a skull and mandible that were eventually discovered to constitute a skillful hoax. { 'pilt,daůn ,man }

pilus [ANATOMY] A hair. [BIOLOGY] A fine, slender, hairlike body. [MICROBIOLOGY] Any filamentous appendage other than flagella on certain gram-negative bacteria. Also known as fimbria. { 'pī·ləs }

pimento [BOTANY] *Capsicum annuum*. A type of pepper in the order Polemoniales grown for its thick, sweet-fleshed red fruit. { pə'ment·ō }

piña [BOTANY] A fiber obtained from the large leaves of the pineapple plant. Also known as pineapple fiber. { 'pēn·yə }

pinacocyte [INVERTEBRATE ZOOLOGY] A flattened polygonal cell occurring in the dermal epithelium of sponges, and lining the exhalant canals. { 'pin·ə·kō,sīt }

Pinales [BOTANY] An order of gymnospermous woody trees and shrubs in the class Pinopsida, including pine, spruce, fir, cypress, yew, and redwood; the largest plants are the conifers. { pī'nā·lēz }

pincer [INVERTEBRATE ZOOLOGY] A grasping apparatus, as on the anterior legs of a lobster, consisting of two hinged jaws. { 'pin·sər }

pine [BOTANY] Any of the cone-bearing trees composing the genus *Pinus*; characterized by evergreen leaves (needles), usually in tight clusters of two to five. { 'pīn }

pineal body [ANATOMY] An unpaired, elongated, club-shaped, knoblike or threadlike organ attached by a stalk to the roof of the vertebrate forebrain. Also known as conarium; epiphysis. { 'pin·ē·əl ,bäd·ē }

pineapple [BOTANY] *Ananas sativus*. A perennial plant of the order Bromeliales with long, swordlike, usually rough-edged leaves and a dense head of small abortive flowers; the fruit is a sorosis that develops from the fleshy inflorescence and ripens into a solid mass, covered by the persistent bracts and crowned by a tuft of leaves. { 'pī,nap·əl }

pineapple fiber [BOTANY] *See* piña. { 'pī,nap·əl ,fī·bər }

pine nut [BOTANY] The edible seed borne in the cone of various species of pine (*Pinus*), such as stone pine (*P. pinea*) and piñon pine (*P. cembroides* var. *edulis*). { 'pīn ,nət }

pinfeather [VERTEBRATE ZOOLOGY] A young, underdeveloped feather, especially one still enclosed in a cylindrical horny sheath which is afterward cast off. { 'pin,feth·ər }

pinguecula [ANATOMY] A small patch of yellowish-white connective tissue located on the conjunctiva, between the cornea and the canthus of the eye. { piŋ'gwek·yə·lə }

Pinicae [BOTANY] A large subdivision of the Pinophyta, comprising woody plants with a simple trunk and excurrent branches, simple, usually alternative, needlelike or scalelike leaves, and wood that lacks vessels and usually has resin canals. { 'pī·nə,sē }

pinion [VERTEBRATE ZOOLOGY] The distal portion of a bird's wing. { 'pin·yən }

pinna [ANATOMY] The cartilaginous, projecting flap of the external ear of vertebrates. Also known as auricle. { 'pin·ə }

pinnate [BOTANY] Having parts arranged like a feather, branching from a central axis. { 'pi ,nāt }

pinnate muscle [ANATOMY] A muscle having a central tendon onto which many short, diagonal muscle fibers attach at rather acute angles. { 'pi,nāt ¦məs·əl }

Pinnipedia [VERTEBRATE ZOOLOGY] A suborder of aquatic mammals in the order Carnivora, including walruses and seals. { ,pin·ə'pē·dē·ə }

Pinnotheridae [INVERTEBRATE ZOOLOGY] The pea crabs, a family of decapod crustaceans belonging to the Brachygnatha. { ,pin·ə'ther·ə,dē }

pinnulate [BIOLOGY] Having pinnules. { 'pin·yə·lāt }

pinnule [BIOLOGY] The secondary branch of a plumelike or pinnate organ. { 'pin,yül }

pinocytosis [CELL BIOLOGY] Deprecated term formerly used to describe the process of uptake or internalization of particles, macromolecules, and fluid droplets by living cells; the process is now termed endocytosis. { ¦pin·ō·sī'tō·səs }

Pinophyta [BOTANY] The gymnosperms, a division of seed plants characterized as vascular plants with roots, stems, and leaves, and with seeds that are not enclosed in an ovary but are borne on cone scales or exposed at the end of a stalk. { pə'näf·əd·ə }

Pinopsida [BOTANY] A class of gymnospermous plants in the subdivision Pinicae characterized by entire-margined or slightly toothed, narrow leaves. { pə'näp·səd·ə }

pinosome [CELL BIOLOGY] A closed intracellular vesicle containing material captured by pinocytosis. { 'pin·ə,sōm }

pinulus [INVERTEBRATE ZOOLOGY] A sponge spicule, usually with five rays, one of which develops numerous small spines. { 'pin·yə·ləs }

pinworm [INVERTEBRATE ZOOLOGY] *Enterobius vermicularis.* A phasmid nematode of the superfamily Oxyuroidea; causes enterobiasis. Also known as human threadworm; seatworm. { 'pin,wərm }

pioneer [ECOLOGY] An organism that is able to establish itself in a barren area and begin an ecological cycle. { ,pī·ə'nir }

Piophilidae [INVERTEBRATE ZOOLOGY] The skipper flies, a family of myodarian cyclorrhaphous dipteran insects in the subsection Acalypteratae. { ,pī·ə'fil·ə,dē }

Piperaceae [BOTANY] A family of dicotyledonous plants in the order Piperales characterized by alternate leaves, a solitary ovule, copious perisperm, and scanty endosperm. { ,pip·ə'rās·ē,ē }

Piperales [BOTANY] An order of dicotyledonous herbaceous plants marked by ethereal oil cells, uniaperturate pollen, and reduced crowded flowers with orthotropous ovules. { 'pip·ə'rā·lēz }

Pipidea [VERTEBRATE ZOOLOGY] A family of frogs sometimes included in the suborder Opisthocoela, but more commonly placed in its own suborder, Aglossa; a definitive tongue is lacking, and free ribs are present in the tadpole but they fuse to the vertebrae in the adult. { pə'pid·ē·ə }

Pipridae [VERTEBRATE ZOOLOGY] The manakins, a family of colorful, neotropical suboscine birds in the order Passeriformes. { 'pip·rə,dē }

piptoblast [INVERTEBRATE ZOOLOGY] A statoblast that is free but has no float. { 'pip·tə,blast }

piriformis [ANATOMY] A muscle arising from the front of the sacrum and inserted into the greater trochanter of the femur. { ,pir·ə'fȯr·məs }

Piroplasmea [INVERTEBRATE ZOOLOGY] A class of parasitic protozoans in the superclass Sarcodina; includes the single genus *Babesia.* { ,pir·ə'plaz·mē·ə }

Pisces [VERTEBRATE ZOOLOGY] The fish and fishlike vertebrates, including the classes Agnatha, Placodermi, Chondrichthyes, and Osteichthyes. { 'pī·sēz }

piscivorous [ZOOLOGY] Feeding on fishes. { pə'siv·ə·rəs }

Pisionidae [INVERTEBRATE ZOOLOGY] A small family of errantian polychaete annelids; allies of the scale bearers. { ,pī·sē'än·ə,dē }

pistachio [BOTANY] *Pistacia vera.* A small, spreading dioecious evergreen tree with leaves that have three to five broad leaflets, and with large drupaceous fruit; the edible seed consists of a single green kernel covered by a brown coat and enclosed in a tough shell. { pə'stash·ē,ō }

pistil [BOTANY] The ovule-bearing organ of angiosperms; consists of an ovary, a style, and a stigma. { 'pist·əl }

pistillate [BOTANY] **1.** Having a pistil. **2.** Having pistils but no stamens. { 'pist·əl,āt }

pit [BOTANY] **1.** A cavity in the secondary wall of a plant cell, formed where secondary deposition has failed to occur, and the primary wall remains uncovered; two main types are simple pits and bordered pits. **2.** The stone of a drupaceous fruit. { pit }

pitch [MOLECULAR BIOLOGY] The distance between two adjacent turns of double-stranded deoxyribonucleic acid. { pich }

pitcher plant [BOTANY] Any of various insectivorous plants of the families Sarraceniaceae and Nepenthaceae; the leaves form deep pitchers in which water collects and insects are drowned and digested. { 'pich·ər ,plant }

pith [BOTANY] A central zone of parenchymatous tissue that occurs in most vascular plants and is surrounded by vascular tissue. { pith }

pith ray [BOTANY] *See* medullary ray. { 'pith ,rā }

Pittidae [VERTEBRATE ZOOLOGY] The pittas, a homogeneous family of brightly colored suboscine birds with an erectile crown of feathers, in the suborder Tyranni. { 'pid·ə,dē }

pituicyte [HISTOLOGY] The characteristic cell of the neurohypophysis; these cells are pigmented and fusiform and are probably derived from neuroglial cells. { pə'tü·ə,sīt }

pituitary [ANATOMY] Of or pertaining to the hypophysis. [PHYSIOLOGY] Secreting phlegm or mucus (archaic usage). { pə'tü·ə,ter·ē }

pituitary gland [ANATOMY] *See* hypophysis. { pə'tü·ə,ter·ē ,gland }

Pityaceae [PALEOBOTANY] A family of fossil plants in the order Cordaitales known only as petrifactions of branches and wood. { ,pid·ē'ās·ē,ē }

pivot joint [ANATOMY] A diarthrosis that permits a rotation of one bone around another; an example is the articulation of the atlas with the axis. Also known as trochoid. { 'piv·ət ,jȯint }

placenta [BOTANY] A plant surface bearing a sporangium. [EMBRYOLOGY] A vascular organ that unites the fetus to the wall of the uterus in all mammals except marsupials and monotremes. { plə'sent·ə }

placental barrier [EMBRYOLOGY] The tissues intervening between the maternal and the fetal blood of the placenta, which prevent or hinder certain substances or organisms from passing from mother to fetus. { plə'sent·əl 'bar·ē·ər }

placentation [BOTANY] The attachment of ovules along the inner ovarian wall by means of the placenta. [EMBRYOLOGY] The formation and fusion of the placenta to the uterine wall. { plas·ən'tā·shən }

placode [EMBRYOLOGY] A platelike epithelial thickening, frequently marking, in the embryo, the anlage of an organ or part. { 'pla,kōd }

Placodermi [PALEONTOLOGY] A large and varied class of Paleozoic fishes characterized by a complex bony armor covering the head and the front portion of the trunk. { ,pla·kə'dər·mē }

Placodontia [PALEONTOLOGY] A small order of Triassic marine reptiles of the subclass Euryapsida characterized by flat-crowned teeth in both the upper and lower jaws and on the palate. { ,plā·kə'dän·chə }

Placothuriidae [INVERTEBRATE ZOOLOGY] A family of holothurian echinoderms in the order Dendrochirotida; individuals are invested in plates and have a complex calcareous ring mechanism. { ,pla·kə·thə'rī·ə,dē }

Plagiaulacida [PALEONTOLOGY] A primitive, monofamilial suborder of multituberculate mammals distinguished by their dentition (dental formula I 3/0 C 0/0 Pm 5/4 M 2/2), having cutting premolars and two rows of cusps on the upper molars. { ¦plā·jē·ə·yü'läs·ə·də }

Plagiaulacidae [PALEONTOLOGY] The single family of the extinct mammalian suborder Plagiaulacida. { ¦plā·jē·ə·yü'läs·ə‚dē }

plagioclimax [ECOLOGY] A plant community which is in equilibrium under present conditions, but which has not reached its natural climax, or has regressed from it, due to biotic factors such as human intervention. { ¦plā·jē·ō'klī‚maks }

plagiodont [VERTEBRATE ZOOLOGY] Of a snake, having obliquely set, or two converging series of, palatal teeth. { 'plā·jē·ə‚dänt }

plagiogravitropism [BOTANY] A response of root and shoot branches to gravity where growth is at different angles from the vertical. { ¦plā·jē·ō‚grav·ə'trō·piz·əm }

Plagiosauria [PALEONTOLOGY] An aberrant Triassic group of labyrinthodont amphibians. { ¦plā·jē·ə'sȯr·ē·ə }

plagiosere [ECOLOGY] A plant succession deflected from its normal course by biotic factors. { 'plā·jē·ə‚sir }

planaria [INVERTEBRATE ZOOLOGY] Any flatworm of the turbellarian order Tricladida; the body is broad and dorsoventrally flattened, with anterior lateral projections, the auricles, and a pair of eyespots on the anterior dorsal surface. { plə'ner·ē·ə }

Planctomyces [MICROBIOLOGY] A genus of appendaged bacteria; spherical, oblong, or pear-shaped cells with long, slender stalks; reproduce by budding. { ‚plaŋk·tə'mī·sēz }

planidium [INVERTEBRATE ZOOLOGY] A first-stage legless larva of various insects in the orders Diptera and Hymenoptera. { plə'nid·ē·əm }

Planipennia [INVERTEBRATE ZOOLOGY] A suborder of insects in the order Neuroptera in which the larval mandibles are modified for piercing and for sucking. { ‚plan·ə'pen·ē·ə }

plankton [ECOLOGY] Passively floating or weakly motile aquatic plants and animals. { 'plaŋk·tən }

planoblast [INVERTEBRATE ZOOLOGY] The medusa form of a hydrozoan. { 'plan·ə‚blast }

planomycin [MICROBIOLOGY] *See* fervenulin. { ‚plan·ə'mīs·ən }

planospiral [INVERTEBRATE ZOOLOGY] Having the shell coiled in one plane, used particularly of foraminiferans and mollusks. { ¦plā·nō'spī·rəl }

plant [BOTANY] Any organism belonging to the kingdom Plantae, generally distinguished by the presence of chlorophyll, a rigid cell wall, and abundant, persistent, active embryonic tissue, and by the absence of the power of locomotion. { plant }

Plantae [BOTANY] The plant kingdom. { 'plan‚tē }

Plantaginaceae [BOTANY] The single family of the plant order Plantaginales. { ‚plan·tə·jə'nās·ē‚ē }

Plantaginales [BOTANY] An order of dicotyledonous herbaceous plants in the subclass Asteridae, marked by small hypogynous flowers with a persistent regular corolla and four petals. { ‚plan·tə·jə'nā·lēz }

plantar [ANATOMY] Of or relating to the sole of the foot. { 'plan·tər }

plantaris [ANATOMY] A small muscle of the calf of the leg; origin is the lateral condyle of the femur, and insertion is the calcaneus; flexes the knee joint. { plan'tar·əs }

plantar reflex [PHYSIOLOGY] Flexion of the toes in response to stroking of the outer surface of the sole, from heel to little toe. { 'plan·tər 'rē‚fleks }

plant fermentation [BIOCHEMISTRY] A form of plant metabolism in which carbohydrates are partially degraded without the consumption of molecular oxygen. { 'plant ‚fər·mən'tā·shən }

plant geography [BOTANY] A major division of botany, concerned with all aspects of the spatial distribution of plants. Also known as geographical botany; phytochorology; phytogeography. { 'plant jē‚äg·rə·fē }

plant hormone [BIOCHEMISTRY] An organic compound that is synthesized in minute quantities by one part of a plant and translocated to another part, where it influences physiological processes. Also known as phytohormone. { 'plant ‚hȯr‚mōn }

plantigrade [VERTEBRATE ZOOLOGY] Pertaining to walking with the whole sole of the foot touching the ground. { 'plan·tə‚grād }

plant key [BOTANY] An analytical guide to the identification of plants, based on the use of contrasting characters to subdivide a group under study into branches. { 'plant ‚kē }

plant kingdom [BOTANY] The worldwide array of plant life constituting a major division of living organisms. { 'plant ‚kiŋ·dəm }

plant pathology [BOTANY] The branch of botany concerned with diseases of plants. { 'plant pə'thäl·ə·jē }

plant physiology [BOTANY] The branch of botany concerned with the processes which occur in plants. { 'plant ‚fiz·ē'äl·ə·jē }

plant societies [ECOLOGY] Assemblages of plants which constitute structural parts of plant communities. { 'plant sə‚sī·əd·ēz }

plantula [INVERTEBRATE ZOOLOGY] A small, cushionlike structure on the ventral surface of the segments of insect tarsi. { 'plan·chə·lə }

plant virus [VIROLOGY] A virus that replicates only within plant cells. { 'plant ‚vī·rəs }

planula [INVERTEBRATE ZOOLOGY] The ciliated, free-swimming larva of coelenterates. { 'plan·yə·lə }

Planuloidea [INVERTEBRATE ZOOLOGY] The equivalent name for Moruloidea. { ‚plan·yə'lȯid·ē·ə }

plaque [VIROLOGY] A clear area representing a colony of viruses on a plate culture formed by lysis of the host cell. { plak }

plasma [HISTOLOGY] The fluid portion of blood or lymph. { 'plaz·mə }

plasma cell [HISTOLOGY] *See* plasmacyte. { 'plaz·mə ‚sel }

plasmacyte [HISTOLOGY] A fairly large, generally

ovoid cell with a small, eccentrically placed nucleus; the chromatin material is adherent to the nuclear membrane and the cytoplasm is agranular and deeply basophilic everywhere except for a clear area adjacent to the nucleus in the area of the cytocentrum. Also known as plasma cell. { 'plaz·mə,sīt }

plasmagel [CELL BIOLOGY] The outer, gelated zone of protoplasm in a pseudopodium. { 'plaz·mə,jel }

plasmagene [CELL BIOLOGY] A cytoplasmic particle or substance, which may be present in bodies such as plastids or mitochondria, and which can reproduce and pass on inherited qualities to daughter cells. { 'plaz·mə,jēn }

plasma membrane [CELL BIOLOGY] *See* cell membrane. { 'plaz·mə 'mem,brān }

plasmasol [CELL BIOLOGY] The inner, solated zone of protoplasm in a pseudopodium. { 'plaz·mə,sól }

plasma thromboplastin antecedent [BIOCHEMISTRY] *See* factor XI. { 'plaz·mə ¦thräm·bō¦plas·tən 'ant·i,sēd·ənt }

plasma thromboplastin component [BIOCHEMISTRY] *See* Christmas factor. { 'plaz·mə ¦thräm·bō ¦plas·tən kəm'pō·nənt }

plasmalemma [CELL BIOLOGY] *See* cell membrane. { ¦plaz·mə'lem·ə }

plasmid [GENETICS] An extrachromosomal genetic element found among various strains of *Escherichia coli* and other bacteria. { 'plaz·məd }

plasmid cloning vector [GENETICS] A plasmid that accepts foreign deoxyribonucleic acid (DNA) and is therefore used in recombinant DNA experiments. { 'plaz·mid ,klōn·iŋ ,vek·tər }

plasmid donation [GENETICS] The transfer of a nonconjugative plasmid from a donor cell to a recipient cell by way of a contact function provided by a conjugative plasmid. { 'plaz·mid dō ,nā·shən }

plasmin [BIOCHEMISTRY] A proteolytic enzyme in plasma which can digest many proteins through the process of hydrolysis. Also known as fibrinolysin. { 'plaz·mən }

plasminogen [BIOCHEMISTRY] The inert precursor, or zymogen, of plasmin. Also known as profibrinolysin. { plaz'min·ə·jən }

plasmodesma [CELL BIOLOGY] An intercellular bridge, thought to be strands of cytoplasm connecting two cells. { ,plaz·mə'dez·mə }

Plasmodiidae [INVERTEBRATE ZOOLOGY] A family of parasitic protozoans in the suborder Haemosporina inhabiting the erythrocytes of the vertebrate host. { ,plaz·mə'dī·ə,dē }

plasmoditrophoblast [CELL BIOLOGY] *See* syncytiotrophoblast. { ,plaz·mō·dī'träf·ə,blast }

plasmosome [CELL BIOLOGY] *See* nucleolus. { 'plaz·mə,sōm }

Plasmodiophorida [INVERTEBRATE ZOOLOGY] An order of the protozoan subclass Mycetozoia occurring as endoparasites of plants. { ,plaz·mō ,dī·ə'fór·ə·də }

Plasmodiophoromycetes [MYCOLOGY] A class of the Fungi. { ,plaz·mō·dī,äf·ə·rō·mī'sē·dēz }

plasmodium [MICROBIOLOGY] The noncellular, multinucleate, jellylike, ameboid, assimilative stage of the Myxomycetes. { plaz'mō·dē·əm }

Plasmodroma [INVERTEBRATE ZOOLOGY] A subphylum of the Protozoa, including Mastigophora, Sarcodina, and Sporozoa, in some taxonomic systems. { plaz'mä·drə·mə }

plasmogamy [INVERTEBRATE ZOOLOGY] Fusion of protoplasts, without nuclear fusion, to form a multinucleate mass; occurs in certain protozoans. { plaz'mäg·ə·mē }

plasmolysis [PHYSIOLOGY] Shrinking of the cytoplasm away from the cell wall due to exosmosis by immersion of a plant cell in a solution of higher osmotic activity. { plaz'mäl·ə·səs }

plasmon [GENETICS] The cytoplasmic genetic system in eukaryotes consisting primarily of mitochondrial deoxyribonucleic acid (DNA) and chloroplast DNA. { 'plaz,män }

plasmotomy [INVERTEBRATE ZOOLOGY] Subdivision of a plasmodium into two or more parts. { plas'mäd·ə·mē }

plastid [CELL BIOLOGY] One of the specialized cell organelles containing pigments or protein materials, often serving as centers of special metabolic activities; examples are chloroplasts and leukoplasts. { 'plas·təd }

plastogene [CELL BIOLOGY] A cytoplasmic factor, controlled by or interacting with the nucleus, which determines differentiation of a plastid. { 'plas·tə,jēn }

plastoquinone [BIOCHEMISTRY] Any of a group of quinones that are involved in electron transport in chloroplasts during photosynthesis. { ,plas·tə·kwə'nōn }

plastron [INVERTEBRATE ZOOLOGY] The ventral plate of the cephalothorax of spiders. [VERTEBRATE ZOOLOGY] The ventral portion of the shell of tortoises and turtles. { 'plas,trän }

Platanaceae [BOTANY] A small family of monoecious dicotyledonous plants in which flowers have several carpels which are separate, three or four stamens, and more or less orthotropous ovules, and leaves are stipulate. { ,plat·ən'ās·ē,ē }

Plataspidae [INVERTEBRATE ZOOLOGY] A family of shining, oval hemipteran insects in the superfamily Pentatomoidea. { plə'tas·pə,dē }

plate budding [BOTANY] Plant budding by inserting a rectangular scion with a bud under a flap of bark on the stock in such a manner that the exposed wood on the stock is covered. { 'plāt ,bəd·iŋ }

plate count [MICROBIOLOGY] The number of bacterial colonies that develop on a medium in a petri dish seeded with a known amount of inoculum. { 'plāt ,kaúnt }

platelet [HISTOLOGY] *See* thrombocyte. { 'plāt·lət }

platelet-activating factor [IMMUNOLOGY] A phospholipid released by leukocytes that causes aggregation of platelets and other effects, such as an increase in vascular permeability and bronchoconstriction. Abbreviated PAF. { ¦plāt·lət 'ak·tə,vād·iŋ ,fak·tər }

Platyasterida [INVERTEBRATE ZOOLOGY] An order of Asteroidea in which traces of metapinnules persist, the ossicles of the arm skeleton being arranged in two growth gradient systems. { ¦plad·ē·a'ster·ə·də }

Platybelondoninae [PALEONTOLOGY] A subfamily of extinct elephantoid mammals in the family Gomphotheriidae consisting of species with digging specializations of the lower tusks. { ˌplad·ē ˌbel·ən'dän·əˌnē }

platycelous [VERTEBRATE ZOOLOGY] Of a vertebra, having a flat or concave ventral surface and a convex dorsal surface. { ¦plad·ē¦sē·ləs }

Platycephalidae [VERTEBRATE ZOOLOGY] The flatheads, a family of perciform fishes in the suborder Cottoidei. { ˌplad·ē·sə'fal·əˌdē }

Platyceratacea [PALEONTOLOGY] A specialized superfamily of extinct gastropod mollusks which adapted to a coprophagous life on crinoid calices. { ˌplad·ēˌser·ə'tās·ē·ə }

Platycopa [INVERTEBRATE ZOOLOGY] A suborder of ostracod crustaceans in the order Podocopida including marine forms with two pairs of thoracic legs. { plə'tik·ə·pə }

Platycopina [INVERTEBRATE ZOOLOGY] The equivalent name for Platycopa. { ˌplad·ə'käp·ə·nə }

Platyctenea [INVERTEBRATE ZOOLOGY] An order of the ctenophores whose members are sedentary or parasitic; adults often lack ribs and are flattened due to shortening of the main axis. { ˌplad·ik'tē·nē·ə }

Platygasteridae [INVERTEBRATE ZOOLOGY] A family of hymenopteran insects in the superfamily Proctotrupoidea. { ˌplad·ē·ga'ster·əˌdē }

Platyhelminthes [INVERTEBRATE ZOOLOGY] A phylum of invertebrates composed of bilaterally symmetrical, nonsegmented, dorsoventrally flattened worms characterized by lack of coelom, anus, circulatory and respiratory systems, and skeleton. { ¦plad·ē·hel'min·thēz }

platymyarian [INVERTEBRATE ZOOLOGY] In nematodes, pertaining to flat muscle cells with the fibrillar region limited to a basal zone. { ¦plad·ē ˌmī¦a·rē·ən }

Platypodidae [INVERTEBRATE ZOOLOGY] The ambrosia beetles, a family of coleopteran insects in the superfamily Curculionoidea. { ˌplad·ə'päd·ə ˌdē }

Platypsyllidae [INVERTEBRATE ZOOLOGY] The equivalent name for Leptinidae. { ˌplad·ə'sil·ə ˌdē }

platypus [VERTEBRATE ZOOLOGY] *Ornithorhynchus anatinus.* A monotreme, making up the family Ornithorhynchidae, which lays and incubates eggs in a manner similar to birds, and retains some reptilian characters; the female lacks a marsupium. Also known as duckbill platypus. { 'plad· əˌpu̇s }

platysma [ANATOMY] A subcutaneous muscle of the neck, extending from the face to the clavicle; muscle of facial expression. { plə'tiz·mə }

Platysomidae [PALEONTOLOGY] A family of extinct palaeonisciform fishes in the suborder Platysomoidei; typically, the body is laterally compressed and rhombic-shaped, with long dorsal and anal fins. { ˌplad·ē'säm·əˌdē }

Platysomoidei [PALEONTOLOGY] A suborder of extinct deep-bodied marine and fresh-water fishes in the order Palaeonisciformes. { ˌplad·ē· sə'mȯid·ēˌī }

Platysternidae [VERTEBRATE ZOOLOGY] The big-headed turtles, a family of Asiatic fresh-water Chelonia with a single species (*Platysternon megacephalum*), characterized by a large head, hooked mandibles, and a long tail. { ˌplad·ē 'stər·nəˌdē }

Plecoptera [INVERTEBRATE ZOOLOGY] The stoneflies, an order of primitive insects in which adults differ only slightly from immature stages, except for wings and tracheal gills. { plə'käp·tə·rə }

plectane [INVERTEBRATE ZOOLOGY] A cuticular plate supporting papillae in some nematodes. { 'plekˌtān }

Plectascales [MYCOLOGY] An equivalent name for Eurotiales. { ˌplek·tə'skā·lēz }

Plectognathi [VERTEBRATE ZOOLOGY] The equivalent name for Tetraodontiformes. { plek'täg·nə ˌthī }

Plectoidea [INVERTEBRATE ZOOLOGY] A superfamily of small, free-living nematodes characterized by simple spiral amphids or variants thereof, elongate cylindroconoid stoma, and reflexed ovaries. { plek'tȯid·ē·ə }

plectostele [BOTANY] A protostele that has the xylem divided into plates. { 'plek·təˌstēl }

Pleidae [INVERTEBRATE ZOOLOGY] A family of hemipteran insects in the superfamily Pleoidea. { 'plē·əˌdē }

pleiomorphism [GENETICS] The occurrence of variable phenotypes in a group of organisms with the same genotype. { ˌplē·ə'mȯrˌfiz·əm }

pleiotropy [GENETICS] The quality of a gene having more than one phenotypic effect. { plī'ä· trə·pē }

pleodont [VERTEBRATE ZOOLOGY] Having solid teeth. { 'plē·əˌdänt }

Pleoidea [INVERTEBRATE ZOOLOGY] A superfamily of suboval hemipteran insects belonging to the subdivision Hydrocoriseae. { plē'ȯid·ē·ə }

pleomorphism [BIOLOGY] The occurrence of more than one distinct form of an organism in a single life cycle. { ˌplē·ō'mȯrˌfiz·əm }

pleopod [INVERTEBRATE ZOOLOGY] An abdominal appendage in certain crustaceans that is modified for swimming. { 'plē·əˌpäd }

Pleosporales [BOTANY] The equivalent name for the lichenized Pseudophaeriales. { ˌplē·ə·spə 'rā·lēz }

plerocercoid [INVERTEBRATE ZOOLOGY] The infective metacestode of certain cyclophyllidean tapeworms; distinguished by a solid body. { ˌplir·ə'sərˌkȯid }

plerome [BOTANY] Central core of primary meristem which gives rise to all cells of the stele from the pericycle inward. { 'pliˌrōm }

plesioaster [INVERTEBRATE ZOOLOGY] A type of poriferan microscleric monaxonic spicule. { ˌplē·sē·ō'as·tər }

Plesiocidaroida [PALEONTOLOGY] An extinct order of echinoderms assigned to the Euechinoidea. { ˌplē·sē·ōˌsik·ə'rȯid·ə }

plesiomorph [EVOLUTION] The original character

of a branching phyletic lineage, found in the ancestral forms. { 'plē·sē·ə,mȯrf }

Plesiosauria [PALEONTOLOGY] A group of extinct reptiles in the order Sauropterygia constituting a highly specialized offshoot of the nothosaurs. { ,plē·sē·ə'sȯr·ē·ə }

plesiotype [SYSTEMATICS] A specimen or specimens on which subsequent descriptions are based. { 'plē·sē·ə,tīp }

Plethodontidae [VERTEBRATE ZOOLOGY] A large family of salamanders in the suborder Salamandroidea characterized by the absence of lungs and the presence of a fine groove from nostril to upper lip. { ,pleth·ə'dänt·ə,dē }

pleura [ANATOMY] The serous membrane covering the lung and lining the thoracic cavity. { plu̇r·ə }

Pleuracanthodii [PALEONTOLOGY] An order of Paleozoic sharklike fishes distinguished by two-pronged teeth, a long spine projecting from the posterior braincase, and direct backward extension of the tail. { plü,rak·ən'thō·dē,ī }

pleural cavity [ANATOMY] The potential space included between the parietal and visceral layers of the pleura. { 'plu̇r·əl ¦kav·əd·ē }

pleural rib [VERTEBRATE ZOOLOGY] *See* ventral rib. { 'plu̇r·əl ¦rib }

pleurapophysis [ANATOMY] One of the lateral processes of a vertebra, corresponding morphologically to a rib. { ,plu̇r·ə'päf·ə·səs }

pleurobranchia [INVERTEBRATE ZOOLOGY] A gill that arises from the lateral wall of the thorax in certain arthropods. { ,plu̇r·ə'braŋ·kē·ə }

pleurocarpous [BOTANY] Having the sporophyte in leaf axils along the side of the stem or on lateral branches; refers specifically to mosses. { ¦plu̇r·ə¦kär·pəs }

Pleuroceridae [INVERTEBRATE ZOOLOGY] A family of fresh-water snails in the order Pectinibranchia. { ,plu̇r·ə'ser·ə,dē }

Pleurocoelea [INVERTEBRATE ZOOLOGY] An extinct superfamily of gastropod mollusks of the order Opisthobranchia in which the shell, mantle cavity, and gills were present. { ,plu̇r·ə'sē·lē·ə }

Pleurodira [VERTEBRATE ZOOLOGY] A suborder of turtles (Chelonia) distinguished by spines on the posterior cervical vertebrae so that the head is retractile laterally. { ,plu̇r·ə'dī·rə }

pleurodontia [VERTEBRATE ZOOLOGY] Attachment of the teeth to the inner surface of the jawbone. { ,plu̇r·ə'dän·chə }

pleurolophocercous cercaria [INVERTEBRATE ZOOLOGY] A larval digenetic trematode distinguished by a long, powerful tail with a pair of fin folds, a protrusible oral sucker, and pigmented dorsal eyespots. { ¦plu̇r·ə¦läf·ə¦sər·kəs sər'kar·ē·ə }

Pleuromeiaceae [PALEOBOTANY] A family of plants in the order Pleuromiales, but often included in the Isoetales due to a phylogenetic link. { ,plu̇r·ō·mē'ās·ē,ē }

Pleuromeiales [PALEOBOTANY] An order of Early Triassic lycopods consisting of the genus *Pleuromeia*; the upright branched stem had grasslike leaves and a single terminal strobilus. { ,plu̇r·ō·mē'ā·lēz }

pleuron [INVERTEBRATE ZOOLOGY] The lateral portion of a single thoracic segment in arthropods. { 'plu̇r,än }

Pleuronectiformes [VERTEBRATE ZOOLOGY] The flatfishes, an order of actinopterygian fishes distinguished by the loss of bilateral symmetry. { ,plu̇r·ō,nek·tə'fȯr·mēz }

Pleuronematina [INVERTEBRATE ZOOLOGY] A suborder of the Hymenostomatida. { ,plu̇r·ō,nem·ə'tī·nə }

pleuroperitoneal cavity [VERTEBRATE ZOOLOGY] The body cavity containing both the lungs and the abdominal viscera in all pulmonate vertebrates except mammals. { ,plu̇r·ō¦per·ə·tə¦nē·əl 'kav·əd·ē }

pleuropneumonialike organism [MICROBIOLOGY] Any of a poorly defined group of microorganisms classified in the order Mycoplasmatales, including the smallest organisms capable of independent life, and comparable in size to the large filterable viruses. Abbreviated PPLO. { ,plu̇r·ō·nü¦mō·nyə,lik 'ȯr·gə,niz·əm }

Pleurostigmophora [INVERTEBRATE ZOOLOGY] A subclass of the centipedes, in some taxonomic systems, distinguished by lateral spiracles. { ,plu̇r·ə·stig'mäf·ə·rə }

Pleurotomariacea [PALEONTOLOGY] An extinct superfamily of gastropod mollusks in the order Aspidobranchia. { ,plu̇r·əd·ə,mar·ē'ās·ē·ə }

plexus [ANATOMY] A network of interlacing nerves or anastomosing vessels. { 'plek·səs }

plica [BIOLOGY] A fold, as of skin or a leaf. { 'plī·kə }

Pliohyracinae [PALEONTOLOGY] An extinct subfamily of ungulate mammals in the family Procaviidae. { ,plī·ō·hī'ras·ə,nē }

ploidy [GENETICS] Number of complete chromosome sets in a nucleus. { 'plȯid·ē }

Plokiophilidae [INVERTEBRATE ZOOLOGY] A small family of predacious hemipteran insects in the superfamily Cimicoidea; individuals live in the webs of spiders and embiids. { ,pläk·ē·ō'fil·ə,dē }

Plotosidae [VERTEBRATE ZOOLOGY] A family of Indo-Pacific salt-water catfishes (Siluriformes). { plə'täs·ə,dē }

plum [BOTANY] Any of various shrubs or small trees of the genus *Prunus* that bear smooth-skinned, globular to oval, drupaceous stone fruit. { pləm }

plumage [VERTEBRATE ZOOLOGY] The entire covering of feathers of a bird. { 'plü·mij }

Plumatellina [INVERTEBRATE ZOOLOGY] The single order of the ectoproct bryozoan class Phylactolaemata. { ,plü·mə·tə'lī·nə }

Plumbaginaceae [BOTANY] The leadworts, the single family of the order Plumbaginales. { ,pləm·bə·jə'nās·ē,ē }

Plumbaginales [BOTANY] An order of dicotyledonous plants in the subclass Caryophyllidae; flowers are pentamerous with fused petals, trinucleate pollen, and a compound ovary containing a single basal ovule. { ,pləm·bə·jə'nā·lēz }

plumicome [BIOLOGY] A spicule with plumelike tufts. { 'plü·mə,kōm }

plumose [VERTEBRATE ZOOLOGY] Having feathers or plumes. { 'plü,mōs }

plumule [BOTANY] The primary bud of a plant embryo. [VERTEBRATE ZOOLOGY] A down feather. { 'plü·myül }

plurilocular sporangium [BOTANY] A multicelled, compartmentalized sporangium, such as is found in some brown algae. { ,plůr·ə'läk·yə·lər spə'ran·jē·əm }

plus strand [MOLECULAR BIOLOGY] A parental ribonucleic acid (RNA) strand in RNA bacteriophages that is used as a template for a complementary RNA strand (minus strand) produced with the formation of a double-stranded (double-helix) RNA. { 'pləs ,strand }

pluteus [INVERTEBRATE ZOOLOGY] The free-swimming, bilaterally symmetrical, easel-shaped larva of ophiuroids and echinoids. { 'plüd·ē·əs }

pluviilignosa [ECOLOGY] A tropical rain forest. { ¦plü·vē,il·əg'nō·sə }

pneumatocodon [INVERTEBRATE ZOOLOGY] Exumbrellar surface of the float or pneumatophore of siphonophorans. { ,nü·məd·ō'kō,dän }

pneumatophore [BOTANY] **1.** An air bladder in marsh plants. **2.** A submerged or exposed erect root that functions in the respiration of certain marsh plants. [INVERTEBRATE ZOOLOGY] The air sac of a siphonophore. { 'nü·məd·ə,fȯr }

pneumatosaccus [INVERTEBRATE ZOOLOGY] Subumbrellar surface of the float or pneumatophore of siphonophorans. { ,nü·mə·dō'sak·əs }

pneumobacillus [MICROBIOLOGY] *See* Klebsiella pneumoniae. { ¦nü·mō·bə'sil·əs }

pneumostome [INVERTEBRATE ZOOLOGY] The respiratory aperture of gastropod mollusks. { 'nü·mə,stōm }

Poaceae [BOTANY] The equivalent name for Gramineae. { pō'ās·ē,ē }

Poales [BOTANY] The equivalent name for Cyperales. { pō'ā·lēz }

pocket gopher [VERTEBRATE ZOOLOGY] *See* gopher. { 'päk·ət ,gō·fər }

pod [BOTANY] A dry dehiscent fruit; a legume. { päd }

Podargidae [VERTEBRATE ZOOLOGY] The heavy-billed frogmouths, a family of Asian and Australian birds in the order Caprimulgiformes. { pə'där·jə,dē }

Podicipedide [VERTEBRATE ZOOLOGY] The single family of the avian order Podicipediformes. { ,päd·ə·sə'ped·ə,dē }

Podicipediformes [VERTEBRATE ZOOLOGY] The grebes, an order of swimming and diving birds distinguished by dense, silky plumage, a rudimentary tail, and toes that are individually broadened and lobed. { ,päd·ə·sə,ped·ə'fȯr ,mēz }

Podicipitiformes [VERTEBRATE ZOOLOGY] The equivalent name for Podicipediformes. { ,päd·ə·sə,pid·ə'fȯr,mēz }

podite [INVERTEBRATE ZOOLOGY] A segment of a limb of an arthropod. { 'pä,dīt }

podium [INVERTEBRATE ZOOLOGY] The terminal portion of a body wall appendage in certain echinoderms. { 'pō·dē·əm }

podobranch [INVERTEBRATE ZOOLOGY] A gill of a crustacean attached to the basal segment of a thoracic limb. { 'päd·ə,braŋk }

Podocopa [INVERTEBRATE ZOOLOGY] A suborder of fresh-water ostracod crustaceans in the order Podocopida in which the inner lamella has a calcified rim joining the outer lamella along a chitinous zone of concrescence, and the two valves fit together firmly. { pə'däk·ə·pə }

Podocopida [INVERTEBRATE ZOOLOGY] An order of the Ostracoda; contains all fresh-water ostracods and is divided into the suborders Podocopa, Metacopina, and Platycopina. { ,päd·ə 'käp·ə·də }

Podocopina [INVERTEBRATE ZOOLOGY] The equivalent name for Podocopa. { ,päd·ə·kə'pī·nə }

podocyst [INVERTEBRATE ZOOLOGY] A sinus in the foot of certain gastropod mollusks. { 'päd·ə ,sist }

Podogona [INVERTEBRATE ZOOLOGY] The equivalent name for Ricinuleida. { pə'däg·ə·nə }

Podostemaceae [BOTANY] The single family of the order Podostemales. { pə,däs·tə'mās·ē,ē }

Podostemales [BOTANY] An order of dicotyledonous plants in the subclass Rosidae; plants are submerged aquatics with modified, branching shoots and small, perfect flowers having a superior ovary and united carpels. { ,päd·ə·stə 'mā·lēz }

Poeciliidae [VERTEBRATE ZOOLOGY] A family of fishes in the order Atheriniformes including the live-bearers, such as guppies, swordtails, and mollies. { ,pē·sə'lī·ə,dē }

Poecilosclerida [INVERTEBRATE ZOOLOGY] An order of sponges of the class Demospongiae in which the skeleton includes two or more types of megascleres. { ,pē·sə·lō'skler·ə·də }

Poeobiidae [INVERTEBRATE ZOOLOGY] A monotypic family of spioniform worms (*Poeobius meseres*) belonging to the Sedentaria and found in the North Pacific Ocean. { ,pē·ə'bī·ə,dē }

pogonochore [BOTANY] A type of plant that produces plumed disseminules. { pə'gän·ə,chȯr }

Pogonophora [INVERTEBRATE ZOOLOGY] The single class of the phylum Brachiata; the elongate body consists of three segments, each with a separate coelom; there is no mouth, anus, or digestive canal, and sexes are separate. { ,pō·gə 'näf·ə·rə }

poikilotherm [ZOOLOGY] An animal, such as reptiles, fishes, and invertebrates, whose body temperature varies with and is usually higher than the temperature of the environment; a cold-blooded animal. { pȯi'kil·ə,thərm }

point mutation [GENETICS] Mutation of a single gene due to addition, loss, replacement, or change of sequence in one or more base pairs of the deoxyribonucleic acid of that gene. { 'pȯint myü'tā·shən }

poison gland [VERTEBRATE ZOOLOGY] Any of various specialized glands in certain fishes and amphibians which secrete poisonous mucuslike substances. { 'pȯiz·ən ,gland }

poison hemlock [BOTANY] *Conium maculatum.* A branching biennial poisonous herb that contains a volatile alkaloid, coniine, in its fruits and leaves. { 'pȯiz·ən 'hem,läk }

poison ivy [BOTANY] Any of several climbing, shrubby, or arborescent plants of the genus *Rhus* in the sumac family (Anacardiaceae); characterized by ternate leaves, greenish flowers, and white berries that produce an irritating oil. { 'pȯiz·ən 'ī·vē }

poison oak [BOTANY] Any of several bushy poison ivy plants or shrubby poison sumacs. { 'pȯiz·ən 'ōk }

poisonous plant [BOTANY] Any of about 400 species of vascular plants containing principles which initiate pathological conditions in man and animals. { 'pȯiz·ən·əs 'plant }

poison sumac [BOTANY] *Rhus vernix.* A tall bush of the sumac family (Anacardiaceae) bearing pinnately compound leaves with 7-13 entire leaflets, and drooping, axillary clusters of white fruits that produce an irritating oil. { 'pȯiz·ən 'sü,mak }

polar bear [VERTEBRATE ZOOLOGY] *Thalarctos maritimus.* A large aquatic carnivore found in the polar regions of the Northern Hemisphere. { 'pō·lər ,ber }

polar body [CELL BIOLOGY] One of the small bodies cast off by the oocyte during maturation. { 'pō·lər ,bäd·ē }

polarity [MOLECULAR BIOLOGY] The orientation of a strand of polynucleotide with respect to its partner, expressed in terms of nucleotide linkages. { pə'lar·əd·ē }

polar mutations [GENETICS] A class of mutations in the genes of an operon that affect the expression not only of the gene in which the mutation resides, but also of the genes located to one side of the mutated gene. { 'pō·lər myü'tā·shənz }

polar nucleus [BOTANY] One of the two nuclei in the center of the embryo sac of a seed plant which fuse to form the endosperm nucleus. { 'pō·lər 'nü·klē·əs }

Polemoniaceae [BOTANY] A family of autotrophic dicotyledonous plants in the order Polemoniales distinguished by lack of internal phloem, corolla lobes that are convolute in the bud, three carpels, and axile placentation. { ,päl·ə,mō·nē'ās·ē,ē }

Polemoniales [BOTANY] An order of dicotyledonous plants in the subclass Asteridae, characterized by sympetalous flowers, a regular, usually five-lobed corolla, and stamens equal in number and alternate with the petals. { ,päl·ə,mō·nē'ā·lēz }

polian vesicle [INVERTEBRATE ZOOLOGY] Interradial reservoirs connecting with the ring vessel in most asteroids and holothuroids. { 'pō·lē·ən 'ves·ə·kəl }

poliovirus vaccine [IMMUNOLOGY] A vaccine prepared from one or all three types of polioviruses in a live or attenuated state. { ¦pō·lē·ō'vī·rəs vak'sēn }

pollen [BOTANY] The small male reproductive bodies produced in pollen sacs of the seed plants. { 'päl·ən }

pollen count [BOTANY] The number of grains of pollen that collect on a specified area (often taken as 1 square centimeter) in a specified time. { 'päl·ən ,kau̇nt }

pollen sac [BOTANY] In the anther of angiosperms and gymnosperms, a cavity that contains microspores. { 'päl·ən ,sak }

pollen tube [BOTANY] The tube produced by the wall of a pollen grain which enters the embryo sac and provides a passage through which the male nuclei reach the female nuclei. { 'päl·ən ,tüb }

pollination [BOTANY] The transfer of pollen from a stamen to a pistil; fertilization in flowering plants. { ,päl·ə'nā·shən }

pollution [ECOLOGY] Destruction or impairment of the purity of the environment. [PHYSIOLOGY] Emission of semen at times other than during coitus. { pə'lü·shən }

polyadelphous [BOTANY] Pertaining to stamens that are united by their filaments into several sets or bundles. { ¦päl·ē·ə¦del·fəs }

polyadenylation [MOLECULAR BIOLOGY] The addition of adenine nucleotides to the 3′ end of messenger ribonucleic acid molecules during posttranscriptional modification. { ,päl·ē·ə,den·ə'lā·shən }

Polyangiaceae [MICROBIOLOGY] A family of bacteria in the order Myxobacterales; vegetative cells and myxospores are cylindrical with blunt, rounded ends; the slime capsule is lacking; sporangia are sessile or stalked. { ,päl·ē,an·jē'ās·ē,ē }

polyaxon [INVERTEBRATE ZOOLOGY] A spicule that is laid down along several axes. { ¦päl·ē'ak,sän }

Polybrachiidae [INVERTEBRATE ZOOLOGY] A family of sedentary marine animals in the order Thecanephria. { ,päl·i·bra'kī·ə,dē }

Polychaeta [INVERTEBRATE ZOOLOGY] The largest class of the phylum Annelida, distinguished by paired, lateral, fleshy appendages (parapodia) provided with setae, on most segments. { ,päl·i'kēd·ə }

Polycirrinae [INVERTEBRATE ZOOLOGY] A subfamily of polychaete annelids in the family Terebellidae. { ,päl·i'sir·ə,nē }

polycistronic messenger [BIOCHEMISTRY] In ribonucleic acid viruses, messenger ribonucleic acid that contains the amino acid sequence for several proteins. { ¦päl·i·sis'trän·ik 'mes·ən·jər }

Polycladida [INVERTEBRATE ZOOLOGY] A class of marine Turbellaria whose leaflike bodies have a central intestine with radiating branches, many eyes, and tentacles in most species. { ,päl·i'klad·əd·ə }

polyclimax [ECOLOGY] A climax community under the controlling influence of many environmental factors, including soils, topography, fire, and animal interactions. { ,päl·i'klī,maks }

polyclonal [IMMUNOLOGY] Pertaining to cells or molecules that arise from more than one clone. { ,päl·i'klō·nəl }

polyclonal mixed cryoglobulin [BIOCHEMISTRY] A cryoglobulin made of heterogeneous immunoglobin molecules belonging to two or more different classes, and sometimes additional serum proteins. { ¦päl·i¦klōn əl 'mikst ¦krī·ə 'gläb·yə·lən }

Polycopidae [INVERTEBRATE ZOOLOGY] The single family of the suborder Cladocopa. { ,päl·i 'käp·ə,dē }

Polyctenidae [INVERTEBRATE ZOOLOGY] A family of hemipteran insects in the superfamily Cimicoidea; the individuals are bat ectoparasites which resemble bedbugs but lack eyes and have ctenidia and strong claws. { ,päl·ək'ten·ə,dē }

Polydolopidae [PALEONTOLOGY] A Cenozoic family of rodentlike marsupial mammals. { ,päl·i·də 'läp·ə,dē }

polyembryony [ZOOLOGY] A form of sexual reproduction in which two or more offspring are derived from a single egg. { ¦päl·ē·im'brī·ə ,nē }

polyestrous [PHYSIOLOGY] Having several periods of estrus in a year. { ¦päl·ē'es·trəs }

Polygalaceae [BOTANY] A family of dicotyledonous plants in the order Polygalales distinguished by having a bicarpellate pistil and monadelphous stamens. { ,päl·i·gə'lās·ē,ē }

polygalacturonase [BIOCHEMISTRY] An enzyme that catalyzes the hydrolysis of glycosidic linkage of polymerized galacturonic acids. { ,päl·i,ga,lak 'tůr·ə,nās }

Polygalales [BOTANY] An order of dicotyledonous plants in the subclass Rosidae characterized by its simple leaves and usually irregular, hypogynous flowers. { ,päl·i·gə'lā·lēz }

polygamous [BOTANY] Having both perfect and imperfect flowers on the same plant. [VERTEBRATE ZOOLOGY] Having more than one mate at one time. { pə'lig·ə·məs }

polygene [GENETICS] One of a group of nonallelic genes that collectively control a quantitative character. { 'päl·i,jēn }

Polygnathidae [PALEONTOLOGY] A family of Middle Silurian to Cretaceous conodonts in the suborder Conodontiformes, having platforms with small pitlike attachment scars. { ,pal·ig'nath·ə ,dē }

Polygonaceae [BOTANY] The single family of the order Polygonales. { pə,lig·ə'nās·ē,ē }

Polygonales [BOTANY] An order of dicotyledonous plants in the subclass Caryophyllidae characterized by well-developed endosperm, a unilocular ovary, and often trimerous flowers. { pə ,lig·ə'nā·lēz }

polyhedral disease [INVERTEBRATE ZOOLOGY] *See* polyhedrosis. { ¦päl·i¦hē·drəl di'zēz }

polyhedrosis [INVERTEBRATE ZOOLOGY] Any of several virus diseases of insect larvae characterized by the breakdown of tissues and presence of polyhedral granules. Also known as polyhedral disease. { ,päl·i·hē'drō·səs }

polykaryocyte [CELL BIOLOGY] *See* syncytium. { ¦päl·i'kar·ē·ə,sīt }

Polymera [INVERTEBRATE ZOOLOGY] Formerly a subphylum of the Vermes; equivalent to the phylum Annelida. { pə'lim·ə·rə }

polymerase [BIOCHEMISTRY] An enzyme that links nucleotides together to form polynucleotide chains. { pə'lim·ə,rās }

polymerase chain reaction [MOLECULAR BIOLOGY] A technique for copying and amplifying the complementary strands of a target deoxyribonucleic acid molecule. Abbreviated PCR. { pə ¦lim·ə,rās 'chān rē,ak·shən }

polymorph [BIOLOGY] An organism that exhibits polymorphism. [HISTOLOGY] *See* granulocyte. { 'päl·i,mórf }

polymorphism [BIOLOGY] **1.** Occurrence of different forms of individual in a single species. **2.** Occurrence of different structural forms in a single individual at different periods in the life cycle. [GENETICS] The coexistence of genetically determined distinct forms in the same population, even the rarest of them being too common to be maintained solely by mutation; human blood groups are an example. { ,päl·i 'mór,fiz·əm }

polymorphonuclear leukocyte [HISTOLOGY] *See* granulocyte. { ¦päl·i,mór·fō'nü·klē·ər 'lü·kə ,sīt }

polymyarian [INVERTEBRATE ZOOLOGY] Referring to the cross-sectional appearance of muscle cells in a nematode, having many cells in each quadrant. { ¦päl·i,mī¦ar·ē·ən }

polymyxin [MICROBIOLOGY] Any of the basic polypeptide antibiotics produced by certain strains of *Bacillus polymyxa*. { ,päl·i'mik·sən }

Polynemidae [VERTEBRATE ZOOLOGY] A family of perciform shore fishes in the suborder Mugiloidei. { ,päl·i'nem·ə,dē }

Polynoidae [INVERTEBRATE ZOOLOGY] The largest family of polychaetes, included in the Errantia and having a body of varying size and shape that is covered with elytra. { ,päl·ə'nói,dē }

polynucleotide [BIOCHEMISTRY] A linear sequence of nucleotides. { ¦päl·ə'nü·klē·ə,tīd }

Polyodontidae [INVERTEBRATE ZOOLOGY] A family of tubicolous, often large-bodied errantian polychaetes with characteristic cephalic and parapodial structures. { ,päl·ē·ō'dänt·ə,dē }

polyoma virus [VIROLOGY] A small deoxyribonucleic acid virus normally causing inapparent infection in mice, but experimentally capable of producing parotid tumors and a wide variety of other tumors. { ,päl·ē'ō·mə 'vī·rəs }

Polyopisthocotylea [INVERTEBRATE ZOOLOGY] An order of the trematode subclass Monogenea having a solid posterior holdfast bearing suckers or clamps. { ¦päl·ē·ō,pis·thə,kad·əl'ē·ə }

polyp [INVERTEBRATE ZOOLOGY] A sessile cnidarian individual having a hollow, somewhat cylindrical body, attached at one end, with a mouth surrounded by tentacles at the free end; may be solitary (hydra) or colonial (coral). { 'päl·əp }

polypeptide [BIOCHEMISTRY] A chain of amino acids linked together by peptide bonds but with a lower molecular weight than a protein; obtained by synthesis, or by partial hydrolysis of protein. { ¦päl·i'pep,tīd }

polypetalous [BOTANY] Having distinct petals, in reference to a flower or a corolla. Also known as choripetalous. { ¦päl·i'ped·əl·əs }

Polyphaga [INVERTEBRATE ZOOLOGY] A suborder of the order Coleoptera; members are distinguished by not having the hind coxae fused to the metasternum and by lacking notopleural sutures. { pə'lif·ə·gə }

polyphagous [ZOOLOGY] Feeding on many different kinds of plants or animals. { pə'lif·ə·gəs }

polyphenol oxidase [BIOCHEMISTRY] A copper-containing enzyme that catalyzes the oxidation of phenol derivatives to quinones. { ¦päl·i'fē,nȯl 'äk·sə,dās }

polyphyodont [VERTEBRATE ZOOLOGY] Having teeth which may be constantly replaced. { ¦päl·i ¦fī·ə,dänt }

polypide [INVERTEBRATE ZOOLOGY] The internal contents of an ectoproct bryozoan zooid. { 'päl·i,pīd }

Polyplacophora [INVERTEBRATE ZOOLOGY] The chitons, an order of mollusks in the class Amphineura distinguished by an elliptical body with a dorsal shell that comprises eight calcareous plates overlapping posteriorly. { ,päl·i·pla'käf·ə·rə }

polyploidy [GENETICS] The occurrence of related forms possessing chromosome numbers which are three or more times the haploid number. { 'päl·i,plȯid·ē }

Polypodiales [BOTANY] The true ferns; the largest order of modern ferns, distinguished by being leptosporangiate and by having small sporangia with a definite number of spores. { ,päl·i,päd·ē 'ā·lēz }

Polypodiophyta [BOTANY] The ferns, a division of the plant kingdom having well-developed roots, stems, and leaves that contain xylem and phloem and show well-developed alternation of generations. { ,päl·i,päd·ē'äf·əd·ē }

Polypodiopsida [BOTANY] A class of the division Polypodiophyta; stems of these ferns bear several large, spirally arranged, compound leaves with sporangia grouped in sori on their undermargins. { ,päl·i,päd·ē'äp·səd·ə }

polypore [MYCOLOGY] Any member of the Basidiomycetes having basidia that line the numerous tubes or pores of the basidiocarp. { 'päl·ē,pȯr }

Polypteridae [VERTEBRATE ZOOLOGY] The single family of the order Polypteriformes. { ,päl·əp 'ter·ə,dē }

Polypteriformes [VERTEBRATE ZOOLOGY] An ancient order of actinopterygian fishes distinguished by thick, rhombic, ganoid scales with an enamellike covering, a slitlike spiracle behind the eye, a symmetrical caudal fin, and a dorsal series of free, spinelike finlets. { ,päl·əp,ter·ə'fȯr ,mēz }

polyribosome [CELL BIOLOGY] *See* polysome. { ,päl·i'rī·bə,sōm }

polysaccharide [BIOCHEMISTRY] A carbohydrate composed of many monosaccharides. { ¦päl·i 'sak·ə,rīd }

polysaprobic [ECOLOGY] Referring to a body of water in which organic matter is decomposing rapidly and free oxygen either is exhausted or is present in very low concentrations. { ¦päl·ə·sə 'prō·bik }

polysepalous [BOTANY] Having separate sepals. Also known as chorisepalous. { ¦päl·ə'sep·ə·ləs }

polysome [CELL BIOLOGY] A complex of ribosomes bound together by a single messenger ribonucleic acid molecule. Also known as polyribosome. { 'päl·i,sōm }

polysomy [GENETICS] The occurrence in a nucleus of one or more individual chromosomes in a number higher than that of the remainder. { 'päl·ē,sō·mē }

polyspermy [PHYSIOLOGY] Penetration of the egg by more than one sperm. { 'päl·ē,spər·mē }

polyspore [BOTANY] In certain red algae, an asexual spore, of which there are 12 to 16. { 'päl·ē ,spȯr }

polystele [BOTANY] A stele consisting of vascular units in the parenchyma. { 'päl·i,stēl }

Polystomatoidea [INVERTEBRATE ZOOLOGY] A superfamily of monogeneid trematodes characterized by strong suckers and hooks on the posterior end. { ,päl·i,stō·mə'tȯid·ē·ə }

Polystylifera [INVERTEBRATE ZOOLOGY] A suborder of the Hoplonemertini distinguished by many stylets. { ,päl·i·stə'lif·ə·rə }

polytene chromosome [GENETICS] A giant, multistranded, cablelike chromosome composed of many identical chromosomes having their chromomeres in register and produced by polyteny. Also known as Balbiani chromosome. { 'päl·i ,tēn 'krō·mə,sōm }

Polytrichales [BOTANY] An order of ascocarpous perennial mosses; rigid, simple stems are highly developed and arise from a prostrate subterranean rhizome. { pə,li·trə'kā·lēz }

polytypic [SYSTEMATICS] A taxon that contains two or more taxa in the immediately subordinate category. { ¦päl·i¦tip·ik }

polyvalent [IMMUNOLOGY] **1.** Of antigens, having many combining sites or determinants. **2.** Pertaining to vaccines composed of mixtures of different organisms, and to the resulting mixed antiserum. { ¦päl·i'vā·lənt }

Polyzoa [INVERTEBRATE ZOOLOGY] The equivalent name for Bryozoa. { ,päl·ə'zō·ə }

Pomacentridae [VERTEBRATE ZOOLOGY] The damselfishes, a family of perciform fishes in the suborder Percoidei. { ,pō·mə'sen·trə,dē }

Pomadasyidae [VERTEBRATE ZOOLOGY] The grunts and sweetlips, a family of perciform fishes in the suborder Percoidei. { ,pō·mə·də'sī·ə ,dē }

Pomatiasidae [INVERTEBRATE ZOOLOGY] A family of land snails in the order Pectinibranchia. { ,pō·mə·tī'as·ə,dē }

Pomatomidae [VERTEBRATE ZOOLOGY] A monotypic family of the Perciformes containing the bluefish (*Pomatomus saltatrix*). { ,pō·mə'täm·ə ,dē }

pomegranate [BOTANY] *Punica granatum*. A small, deciduous ornamental tree of the order Myrtales cultivated for its fruit, which is a reddish, pomelike berry containing numerous seeds embedded in crimson pulp. { 'päm·ə,gran·ət }

Pompilidae [INVERTEBRATE ZOOLOGY] The spider wasps, the single family of the superfamily Pompiloidea. { pam'pil·ə,dē }

Pompiloidea [INVERTEBRATE ZOOLOGY] A monofamilial superfamily of hymenopteran insects in the suborder Apocrita with oval abdomen and strong spinose legs. { ˌpäm·pəˈlȯid·ē·ə }

ponderosa pine [BOTANY] *Pinus ponderosa*. A hard pine tree of western North America; attains a height of 150-225 feet (46-69 meters) and has long, dark-green leaves in bundles of two to five and tawny, yellowish bark. { ˌpän·dəˌrō·sə ˈpīn }

Ponerinae [INVERTEBRATE ZOOLOGY] A subfamily of tropical carnivorous ants (Formicidae) in which pupae characteristically form in cocoons. { pōˈner·əˌnē }

Pongidae [VERTEBRATE ZOOLOGY] A family of anthropoid primates in the superfamily Hominoidea; includes the chimpanzee, gorilla, and orangutan. { ˈpän·jəˌdē }

pons [ANATOMY] **1.** A process or bridge of tissue connecting two parts of an organ. **2.** A convex white eminence located at the base of the brain; consists of fibers receiving impulses from the cerebral cortex and sending fibers to the contralateral side of the cerebellum. { pänz }

pontine flexure [EMBRYOLOGY] A flexure in the embryonic brain concave dorsally, occurring in the region of the myelencephalon. { ˈpänˌtēn ˈflek·shər }

Pontodoridae [INVERTEBRATE ZOOLOGY] A monotypic family of pelagic polychaetes assigned to the Errantia. { ˌpän·təˈdȯr·əˌdē }

poplar [BOTANY] Any tree of the genus *Populus*, family Salicaceae, marked by simple, alternate leaves, scaly buds, bitter bark, and flowers and fruits in catkins. { ˈpäp·lər }

popliteal artery [ANATOMY] A continuation of the femoral artery in the posterior portion of the thigh above the popliteal space and below the buttock. { päpˈlid·ē·əl ˈärd·ə·rē }

popliteal nerve [ANATOMY] Either of two branches of the sciatic nerve in the lower part of the thigh; the larger branch continues as the tibial nerve, and the smaller branch continues as the peroneal nerve. { päpˈlid·ē·əl ˈnərv }

popliteal space [ANATOMY] A diamond-shaped area behind the knee joint. { päpˈlid·ē·əl ˈspās }

popliteal vein [ANATOMY] A vein passing through the popliteal space, formed by merging of the tibial veins and continuing to become the femoral vein. { päpˈlid·ē·əl ˈvān }

popliteus [ANATOMY] **1.** The ham or hinder part of the knee joint. **2.** A muscle on the back of the knee joint. { päpˈlid·ē·əs }

poppy [BOTANY] Any of various ornamental herbs of the genus *Papaver*, family Papaveraceae, with large, showy flowers; opium is obtained from the fruits of the opium poppy (*P. somniferum*). { ˈpäp·ē }

population [BIOLOGY] A group of organisms occupying a specific geographic area or biome. { ˌpäp·yəˈlā·shən }

population bottleneck [EVOLUTION] Genetic drift that occurs as a result of a drastic reduction in population by an event having little to do with the forces of natural selection. { ˌpäp·yəˈlā·shən ˈbäd·əlˌnek }

population density [ECOLOGY] The size of the population within a particular unit of space. { ˌpäp·yəˈlā·shən ˈden·səd·ē }

population dispersal [BIOLOGY] The process by which groups of living organisms expand the space or range within which they live. { ˌpäp·yə ˈlā·shən diˈspər·səl }

population dispersion [BIOLOGY] The spatial distribution at any particular moment of the individuals of a species of plant or animal. { ˌpäp·yəˈlā·shən diˈspər·zhən }

population dynamics [BIOLOGY] The aggregate of processes that determine the size and composition of any population. { ˌpäp·yəˈlā·shən dīˈnam·iks }

population genetics [GENETICS] The study of both experimental and theoretical consequences of Mendelian heredity on the population level; includes studies of gene frequencies, genotypes, phenotypes, and mating systems. { ˌpäp·yəˈlā·shən jəˈned·iks }

Porcellanasteridae [INVERTEBRATE ZOOLOGY] A family of essentially deep-water forms in the suborder Paxillosina. { pȯrˌsel·ə·nəˈster·əˌdē }

Porcellanidae [INVERTEBRATE ZOOLOGY] The rock sliders, a family of decapod crustaceans of the group Anomura which resemble true crabs but are distinguished by the reduced, chelate fifth pereiopods and the well-developed tail fan. { ˌpȯr·səˈlan·əˌdē }

porcupine [VERTEBRATE ZOOLOGY] Any of about 26 species of rodents in two families (Hystricidae and Erethizontidae) which have spines or quills in addition to regular hair. { ˈpȯr·kyəˌpīn }

pore [BIOLOGY] Any minute opening by which matter passes through a wall or membrane. { pȯr }

pore fungus [MYCOLOGY] The common name for members of the families Boletaceae and Polyporaceae in the group Hymenomycetes; spore-bearing surfaces are characteristically within tubes or pores. { ˈpȯr ˌfəŋ·gəs }

Porifera [INVERTEBRATE ZOOLOGY] The sponges, a phylum of the animal kingdom characterized by the presence of canal systems and chambers through which water is drawn in and released; tissues and organs are absent. { pəˈrif·ə·rə }

porocyte [INVERTEBRATE ZOOLOGY] One of the perforated, tubular cells which constitute the wall of the incurrent canals in certain Porifera. { ˈpȯr·əˌsīt }

porogamy [BOTANY] Passage of the pollen tube through the micropyle of an ovule in a seed plant. { pȯˈräg·ə·mē }

Poroxylaceae [PALEOBOTANY] A monogeneric family of extinct plants included in the Cordaitales. { pəˌräk·səˈlās·ēˌē }

porphin [BIOCHEMISTRY] A heterocyclic ring consisting of four pyrrole rings linked by methine (—CH=) bridges; the basic structure of chlorophyll, hemoglobin, the cytochromes, and certain other related substances. { ˈpȯr·fən }

porphobilinogen [BIOCHEMISTRY] $C_{10}H_{14}O_4N_2$ Dicarboxylic acid derived from pyrrole; a product of hemoglobin breakdown that gives the urine a Burgundy-red color. { ¦pȯr·fō·bəˈlin·ə·jən }

porphyrin [BIOCHEMISTRY] A class of red-pigmented compounds with a cyclic tetrapyrrolic structure in which the four pyrrole rings are joined through their α-carbon atoms by four methene bridges (=C—); the porphyrins form the active nucleus of chlorophylls and hemoglobin. { 'pȯr·fə·rən }

porpoise [VERTEBRATE ZOOLOGY] Any of several species of marine mammals of the family Phocaenidae which have small flippers, a highly developed sonar system, and smooth, thick, hairless skin. { 'pȯr·pəs }

porta hepatis [ANATOMY] The transverse fissure of the liver through which the portal vein and hepatic artery enter the liver and the hepatic ducts leave. { 'pȯrd·ə he'pad·əs }

portal [ANATOMY] **1.** Of or pertaining to the porta hepatis. **2.** Pertaining to the portal vein or system. { 'pȯrd·əl }

portal circulation [PHYSIOLOGY] The passage of venous blood through a portal system. { 'pȯrd·əl ˌsər·kyə'lā·shən }

portal system [ANATOMY] A system of veins that break into a capillary network before returning the blood to the heart. { 'pȯrd·əl ˌsis·təm }

portal vein [ANATOMY] Any vein that terminates in a network of capillaries. { 'pȯrd·əl ˌvān }

Portuguese man-of-war [INVERTEBRATE ZOOLOGY] Any of several brilliantly colored tropical siphonophores in the genus *Physalia* which possess a large float and extremely long tentacles. { ˌpȯr·chə'gēz ¦man əv 'wȯr }

Portulacaceae [BOTANY] A family of dicotyledonous plants in the order Caryophyllales distinguished by a syncarpous gynoecium, few, cyclic tepals and stamens, two sepals, and two to many ovules. { ˌpȯr·chə·lə'kās·ēˌē }

Portunidae [INVERTEBRATE ZOOLOGY] The swimming crabs, a family of the Brachyura having the last pereiopods modified as swimming paddles. { pȯr'tü·nəˌdē }

Porulosida [INVERTEBRATE ZOOLOGY] An order of the protozoan subclass Radiolaria in which the central capsule shows many pores. { ˌpȯr·yə'lä·səd·ə }

position effect [GENETICS] **1.** Change in expressivity of a gene associated with chromosome aberrations. **2.** Inherent gene expression as influenced by neighboring genes. { pə'zish·ən i ˌfekt }

positive afterimage [PHYSIOLOGY] An afterimage persisting after the eyes are closed or turned toward a dark background, and of the same color as the stimulating light. { 'päz·əd·iv 'af·tərˌim·ij }

positive gene control [MOLECULAR BIOLOGY] Enhancement of gene expression through binding of specific expressor molecules to promoter sites. { ¦päz·əd·iv 'jēn kənˌtrōl }

positive interference [GENETICS] The reduction, by one crossover exchange, of the likelihood of another crossover in its vicinity. { 'päz·əd·iv ˌin·tər'fir·əns }

postcentral gyrus [ANATOMY] The cerebral convolution that lies immediately posterior to the central sulcus and extends from the longitudinal fissure above the posterior ramus of the lateral sulcus. { pōst'sen·trəl 'jī·rəs }

postcentral sulcus [ANATOMY] The first sulcus of the parietal lobe of the cerebrum, lying behind and roughly parallel to the central sulcus. { pōst 'sen·trəl 'səl·kəs }

posterior [ZOOLOGY] **1.** The hind end of an organism. **2.** Toward the back, or hinder end, of the body. { pä'stir·ē·ər }

posterior chamber [ANATOMY] The space in the eye between the posterior surface of the iris and the ciliary body, and the lens. { pä'stir·ē·ər 'chām·bər }

postmeiotic fusion [CELL BIOLOGY] The union of two identical haploid nuclei produced by mitotic division of the egg nucleus, resulting in restoration of the diploid state. { ¦pōst·mēˌäd·ik 'fyü·zhən }

posttranslational modification [MOLECULAR BIOLOGY] Any polypeptide alteration that occurs after synthesis of the chain. { ˌpōs·tranz¦lā·shən·əl ˌmäd·ə·fə'kā·shən }

Potamogalinae [VERTEBRATE ZOOLOGY] An aberrant subfamily of West African tenrecs (Tenrecidae). { ˌpäd·ə·mō'gal·əˌnē }

Potamogetonaceae [BOTANY] A large family of monocotyledonous plants in the order Najadales characterized by a solitary, apical or lateral ovule, usually two or more carpels, flowers in spikes or racemes, and four each of tepals and stamens. { ˌpäd·ə·mōˌjēd·ə'nās·ēˌē }

Potamogetonales [BOTANY] The equivalent name for Najadales. { ˌpäd·ə·mōˌjēd·ə'nā·lēz }

Potamonidae [INVERTEBRATE ZOOLOGY] A family of fresh-water crabs included in the Brachyura. { ˌpäd·ə'män·əˌdē }

potamoplankton [BIOLOGY] Plankton found in rivers. { ¦päd·ə·mō¦plaŋk·tən }

potato [BOTANY] *Solanum tuberosum.* An erect herbaceous annual that has a round or angular aerial stem, underground lateral stems, pinnately compound leaves, and white, pink, yellow, or purple flowers occurring in cymose inflorescences; produces an edible tuber which is a shortened, thickened underground stem having nodes (eyes) and internodes. Also known as Irish potato; white potato. { pə'tā·dō }

Pottiales [BOTANY] An order of mosses distinguished by erect stems, lanceolate to broadly ovate or obovate leaves, a strong, mostly percurrent or excurrent costa, and a cucullate calyptra. { ˌpäd·ē'ā·lēz }

Poupart's ligament [ANATOMY] *See* inguinal ligament. { pü'pärz ˌlig·ə·mənt }

pour-plate culture [MICROBIOLOGY] A technique for pure-culture isolation of bacteria; liquid, cooled agar in a test tube is inoculated with one loopful of bacterial suspension and mixed by rolling the tube between the hands; subsequent transfers are made from this to a second test tube, and from the second to a third; contents of each tube are poured into separate petri dishes; pure cultures can be isolated from isolated colonies appearing on the plates after incubation. { 'pȯr ˌplāt ˌkəl·chər }

Pourtalesiidae [INVERTEBRATE ZOOLOGY] A family of exocyclic Euechinoidea in the order Holasteroida, including those forms with a bottle-shaped test. { ˌpȯrd·əl·ə′sī·əˌdē }

powdery mildew [MYCOLOGY] A fungus characterized by production of abundant powdery conidia on the host; a member of the family Erysiphaceae or the genus *Oidium*. { ′pau̇d·ə·rē ′milˌdü }

poxvirus [VIROLOGY] A deoxyribonucleic acid-containing animal virus group including the viruses of smallpox, molluscum contagiosum, and various animal pox and fibromas. { ′päksˌvī·rəs }

PPLO [MICROBIOLOGY] *See* pleuropneumonialike organism.

prairie dog [VERTEBRATE ZOOLOGY] The common name for three species of stout, fossorial rodents belonging to the genus *Cynomys* in the family Sciuridae; all have a short, flat tail, small ears, and short limbs terminating in long claws. { ′prer·ē ˌdȯg }

prairie wolf [VERTEBRATE ZOOLOGY] *See* coyote. { ′prer·ē ˌwu̇lf }

Prasinovolvocales [BOTANY] An order of green algae in which there are lateral appendages in the flagellum. { ˌpräz·ən·ōˌväl·və′kā·lēz }

pratincolous [ECOLOGY] **1.** Living in meadows. **2.** Living in low grass. { prə′tiŋ·kə·ləs }

preadaptation [EVOLUTION] Possession by an organism or group of organisms, specialized to one mode of life, of characters which favor easy adaptation to a new environment. { ¦prēˌad·əp′tā·shən }

precentral gyrus [ANATOMY] The cerebral convolution that lies between the precentral sulcus and the central sulcus and extends from the superomedial border of the hemisphere to the posterior ramus of the lateral sulcus. { ′prē¦sen·trəl ′jī·rəs }

precipitation [IMMUNOLOGY] Aggregation of soluble antigen by an antibody. { prəˌsip·ə′tā·shən }

precipitin test [IMMUNOLOGY] An immunologic test in which a specific reaction between antigen and antibody results in a visible precipitate. { prə′sip·ə·tən ˌtest }

precollagenous fiber [HISTOLOGY] *See* reticular fiber. { ˌprē·kə′laj·ə·nəs ′fī·bər }

predation [BIOLOGY] The killing and eating of an individual of one species by an individual of another species. { prə′dā·shən }

predator [ECOLOGY] An animal that preys on other animals as a source of food. { ′pred·əd·ər }

prednisolone [BIOCHEMISTRY] $C_{21}H_{28}O_5$ A glucocorticoid that is a dehydrogenated analog of hydrocortisone. { pred′nis·əˌlōn }

preen gland [VERTEBRATE ZOOLOGY] *See* uropygial gland. { ′prēn ˌgland }

prefoliation [BOTANY] *See* vernation. { ′prēˌfō·lē′ā·shən }

prefrontal [ANATOMY] Situated in the anterior part of the frontal lobe of the brain. [VERTEBRATE ZOOLOGY] **1.** Of or pertaining to a bone of some vertebrate skulls, located anterior and lateral to the frontal bone. **2.** Of, pertaining to, or being a scale or plate in front of the frontal scale on the head of some reptiles and fishes. { prē′frənt·əl }

pregnanediol [BIOCHEMISTRY] $C_{21}H_{36}O_2$ A metabolite of progesterone, present in urine during the progestational phase of the menstrual cycle and also during pregnancy. { pregˌnan·ə′dīˌȯl }

pregnenolone [BIOCHEMISTRY] $C_{21}H_{32}O_2$ A steroid ketone that is formed as an oxidation product of cholesterol, stigmasterol, and certain other steroids. { preg′nen·əˌlōn }

prehensile [VERTEBRATE ZOOLOGY] Adapted for seizing, grasping, or plucking, especially by wrapping around some object. { prē′hen·səl }

prehension [PHYSIOLOGY] A movement that involves holding, seizing, or grasping. { prē′hen·shən }

premaxilla [ANATOMY] Either of two bones of the upper jaw of vertebrates located in front of and between the maxillae. { ¦prē·mak′sil·ə }

premessenger ribonucleic acid [MOLECULAR BIOLOGY] A giant molecule of ribonucleic acid that is transcribed from a cistron. { ˌprē′mes·ən·jər ¦rī·bō¦nü¦klē·ik ′as·əd }

premitosis [CELL BIOLOGY] In certain protozoans, a kind of mitosis in which there is an intranuclear centriole and the mitotic apparatus is located within the nuclear membrane. { ˌprē·mī′tō·səs }

premolar [ANATOMY] In each quadrant of the permanent human dentition, one of the two teeth between the canine and the first molar; a bicuspid. { prē′mō·lər }

prepattern [EMBRYOLOGY] The organization in a developing organism before a definite organizational pattern is established. { ′prēˌpad·ərn }

prepuce [ANATOMY] **1.** The foreskin of the penis, a fold of skin covering the glans penis. **2.** A similar fold over the glans clitoridis. { ′prēp·əs }

preputial gland [ANATOMY] *See* Tyson's gland. { prē′pyü·shəl ˌgland }

prescutum [INVERTEBRATE ZOOLOGY] The anterior part of the tergum of a segment of the thorax in insects. { prē′skyüd·əm }

presoma [INVERTEBRATE ZOOLOGY] The anterior portion of an invertebrate that lacks a definitive head structure. { prē′sō·mə }

pressure point [PHYSIOLOGY] A point of marked sensibility to pressure or weight, arranged like the temperature spots, and showing a specific end apparatus arranged in a punctate manner and connected with the pressure sense. { ′presh·ər ˌpȯint }

pretransfer ribonucleic acid [MOLECULAR BIOLOGY] The primary transcription product of transfer ribonucleic acid-encoding genes. { ˌprē¦tranz·fər ˌrī·bō·nü¦klē·ik ′as·əd }

prevalence [GENETICS] In human genetics, the frequency with which a genetically based medical condition is found in the population. { ′prev·ə·ləns }

Priapulida [INVERTEBRATE ZOOLOGY] A minor phylum of wormlike marine animals; the body is made up of three distinct portions (proboscis, trunk, and caudal appendage) and is often covered with spines and tubercles, and the mouth is surrounded by concentric rows of teeth. { ˌprī·ə'pyül·əd·ə }

Priapuloidea [INVERTEBRATE ZOOLOGY] An equivalent name for Priapulida. { prīˌap·ə'lȯid·ē·ə }

Pribnow box [MOLECULAR BIOLOGY] In prokaryotes, a highly conserved sequence element located upstream from the transcriptional start site to which binds the sigma subunit of the ribonucleic acid polymerase. { 'pribˌnō ˌbäks }

primary [VERTEBRATE ZOOLOGY] Of or pertaining to quills on the distal joint of a bird wing. { 'prī ˌmer·ē }

primary constriction [CELL BIOLOGY] *See* centromere. { 'prīˌmer·ē kən'strik·shən }

primary culture [BIOLOGY] A tissue culture started from cells, tissues, or organs taken directly from the organism. { ¦prīˌmer·ē 'kəl·chər }

primary growth [BOTANY] Plant growth that originates in apical meristematic tissue of shoots and roots, giving rise to primary tissue. { 'prī ˌmer·ē 'grōth }

primary meristem [BOTANY] Meristem which is derived directly from embryonic tissue and which gives rise to epidermis, vascular tissue, and the cortex. { 'prīˌmer·ē 'mer·əˌstem }

primary phloem [BOTANY] Phloem derived from apical meristem. { 'prīˌmer·ē 'flō·əm }

primary production [ECOLOGY] The total amount of new organic matter produced by photosynthesis. { 'prīˌmer·ē prə'dək·shən }

primary root [BOTANY] The first plant root to develop; derived from the radicle. { 'prīˌmer·ē 'rüt }

primary structure [BIOCHEMISTRY] The sequence of amino acids in the molecule of a protein or a peptide. { 'prīˌmer·ē 'strək·chər }

primary succession [ECOLOGY] *See* prisere. { 'prī ˌmer·ē sək'sesh·ən }

primary tissue [BOTANY] Plant tissue formed during primary growth. [HISTOLOGY] Any of the four fundamental tissues composing the vertebrate body. { 'prīˌmer·ē 'tish·ü }

primary transcript [MOLECULAR BIOLOGY] The initial transcription of a ribonucleic acid molecule from deoxyribonucleic acid. { 'prīˌmer·ē 'tranˌskript }

primary xylem [BOTANY] Xylem derived from apical meristem. { 'prīˌmer·ē 'zī·ləm }

primase [MOLECULAR BIOLOGY] Any of a class of enzymes involved in initiation of deoxyribonucleic acid replication. { 'prīˌmās }

Primates [VERTEBRATE ZOOLOGY] The order of mammals to which man belongs; characterized in terms of evolutionary trends by retention of a generalized limb structure and dentition, increasing digital mobility, replacement of claws by flat nails, development of stereoscopic vision, and progressive development of the cerebral cortex. { prī'mād·ēz }

primed lymphocyte [IMMUNOLOGY] A lymphocyte that is from an immunized individual or that has been exposed to antigen in cell culture and is therefore sensitized. { 'prīmd 'lim·fəˌsīt }

prime mover [ANATOMY] A muscle that produces a specific motion or maintains a specific posture. { 'prīm 'müv·ər }

primer [MOLECULAR BIOLOGY] **1.** A short ribonucleic acid (RNA) sequence that is complementary to a sequence of deoxyribonucleic acid (DNA) and has a 3'-OH terminus at which a DNA polymerase begins synthesis of a DNA chain. **2.** A short sequence of DNA that is complementary to a messenger RNA (mRNA) sequence and enables reverse transcriptase to begin copying the neighboring sequences of mRNA. **3.** A transfer RNA whose elongation starts RNA-directed DNA synthesis in retroviruses. { 'prīm·ər }

primibrach [INVERTEBRATE ZOOLOGY] In crinoids, the brachials of the unbranched arm. { 'prī·mə ˌbrak }

Primitiopsacea [PALEONTOLOGY] A small dimorphic superfamily of extinct ostracods in the suborder Beyrichicopina; the velum of the male was narrow and uniform, but that of the female was greatly expanded posteriorly. { prīˌmid·ē·äp'sā·shə }

primitive gut [EMBRYOLOGY] The tubular structure in embryos which differentiates into the alimentary canal. { 'prim·əd·iv 'gət }

primitive streak [EMBRYOLOGY] A dense, opaque band of ectoderm in the bilaminar blastoderm associated with the morphogenetic movements and proliferation of the mesoderm and notochord; indicates the first trace of the vertebrate embryo. { 'prim·əd·iv 'strēk }

primordial gut [EMBRYOLOGY] *See* archenteron. { prī'mȯrd·ē·əl 'gət }

primordium [EMBRYOLOGY] *See* anlage. { prī 'mȯrd·ē·əm }

Primulaceae [BOTANY] A family of dicotyledonous plants in the order Primulales characterized by a herbaceous habit and capsular fruit with two to many seeds. { ˌprim·yə'lās·ēˌē }

Primulales [BOTANY] An order of dicotyledonous plants in the subclass Dilleniidae distinguished by sympetalous flowers, stamens located opposite the corolla lobes, and a compound ovary with a single style. { ˌprim·yə'lā·lēz }

prion [BIOCHEMISTRY] Any of a group of infectious proteins that cause fatal neurodegenerative diseases in humans and animals, including scrapie and bovine spongiform encephalopathy in animals and Creutzefeldt-Jakob disease and GerstmannStraussler-Scheinker disease in humans. { 'prīˌän }

Prioniodidae [PALEONTOLOGY] A family of conodonts in the suborder Conodontiformes having denticulated bars with a large denticle at one end. { ˌprī·əˌnī'äd·əˌdē }

Prioniodinidae [PALEONTOLOGY] A family of conodonts in the suborder Conodontiformes characterized by denticulated bars or blades with a large denticle in the middle third of the specimen. { ˌprī·əˌnī·ə'din·əˌdē }

prionodont [VERTEBRATE ZOOLOGY] Having many simple, similar teeth set in a row like sawteeth. { prī'än·əˌdänt }

prisere [ECOLOGY] The ecological succession of vegetation that occurs in passing from barren earth or water to a climax community. Also known as primary succession. { 'prī,sir }

Pristidae [VERTEBRATE ZOOLOGY] The sawfishes, a family of modern sharks belonging to the batoid group. { 'pris·tə,dē }

Pristiophoridae [VERTEBRATE ZOOLOGY] The saw sharks, a family of modern sharks often grouped with the squaloids which have a greatly extended rostrum with enlarged denticles along the margins. { ,pris·tē·ə'fór·ə,dē }

proaccelerin [BIOCHEMISTRY] A labile procoagulant in normal plasma but deficient in the blood of patients with parahemophilia; essential for rapid conversion of prothrombin to thrombin. Also known as factor V; labile factor. { ¦prō·ak 'sel·ə·rən }

proamnion [EMBRYOLOGY] The part of the embryonic area at the sides and in front of the head of the developing amniote embryo, which remains without mesoderm for a considerable period. { prō'am·nē,än }

Proanura [PALEONTOLOGY] Triassic forerunners of the Anura. { prō'an·yə·rə }

proband [GENETICS] The clinically affected member of a family through which the study of a particular character usually begins in a pedigree. Also known as propositus. { 'prō,band }

probe [BIOLOGY] A biochemical substance labeled with a radioactive isotope or tagged in some other way and used to identify or isolate a gene, a gene product, or a protein. { prōb }

probend [GENETICS] The clinically affected individual who is being studied with respect to a pedigree of interest to human genetics. { 'präb·ənd }

Proboscidea [VERTEBRATE ZOOLOGY] An order of herbivorous placental mammals characterized by having a proboscis, incisors enlarged to become tusks, and pillarlike legs with five toes bound together on a broad pad. { ,prō·bə'sid·ē·ə }

proboscis [INVERTEBRATE ZOOLOGY] A tubular organ of varying form and function on a large number of invertebrates, such as insects, annelids, and tapeworms. [VERTEBRATE ZOOLOGY] The flexible, elongated snout of certain mammals. { prə'bäs·kəs }

procambium [BOTANY] The part of the apical meristematic tissue from which primary vascular tissues are derived. { prō'käm·bē·əm }

Procampodeidae [INVERTEBRATE ZOOLOGY] A family of the insect order Diplura. { prō,kam·pə 'dē·ə,dē }

procarp [BOTANY] The reproductive structure of the female gametophyte that is found in certain red algae. { 'prō,kärp }

Procaviidae [VERTEBRATE ZOOLOGY] A family of mammals in the order Hyracoidea including the hyraxes. { ,prō·kə'vī·ə,dē }

Procaviinae [VERTEBRATE ZOOLOGY] A subfamily of ungulate mammals in the family Procaviidae. { ,prō·kə'vī·ə,nē }

Procellariidae [VERTEBRATE ZOOLOGY] A family of birds in the order Procellariiformes comprising the petrels, fulmars, and shearwaters. { ,prō·sə·lə'rī·ə,dē }

Procellariiformes [VERTEBRATE ZOOLOGY] An order of oceanic birds characterized by tubelike nostril openings, webbed feet, dense plumage, compound horny sheath of the bill, and, often, a peculiar musky odor. { ,prō·sə·lə,rī·ə'fór,mēz }

procephalon [INVERTEBRATE ZOOLOGY] The part of an insect's head that lies anteriorly to the segment in which the mandibles are located. { prō 'sef·ə,län }

procercoid [INVERTEBRATE ZOOLOGY] The solid parasitic larva of certain eucestodes, such as pseudophyllideans, that develops in the body of the intermediate host. { prō'sər,kȯid }

process [ANATOMY] A projection from the central mass of an organism. { 'prä,ses }

procoagulant [BIOCHEMISTRY] Any of blood clotting factors V to VIII; accelerates the conversion of prothrombin to thrombin in the presence of thromboplastin and calcium. { ¦prō·kō'ag·yə·lənt }

Procoela [VERTEBRATE ZOOLOGY] A suborder of the Anura characterized by a procoelous vertebral column and a free coccyx articulating with a double condyle. { prō'sēl·ə }

procoelous [VERTEBRATE ZOOLOGY] The form of a vertebra that is concave anteriorly and convex posteriorly. { prō'sel·əs }

procollagen [BIOCHEMISTRY] A high-molecular-weight form of collagen that is found in intracellular spaces and is believed to be the precursor of collagen. { prō'käl·ə·jən }

Procolophonia [PALEONTOLOGY] A subclass of extinct cotylosaurian reptiles. { ,präk·ə·lə'fō·nē·ə }

proctiger [INVERTEBRATE ZOOLOGY] The cone-shaped, reduced terminal segment of the abdomen of an insect which contains the anus. { 'präk·tə·jər }

proctodone [INVERTEBRATE ZOOLOGY] An insect hormone that causes diapause to end. { 'präk·tə ,dōn }

Proctotrupidae [INVERTEBRATE ZOOLOGY] A family of hymenopteran insects in the superfamily Proctotrupoidea. { ,präk·tə'trü·pə,dē }

Proctotrupoidea [INVERTEBRATE ZOOLOGY] A superfamily of parasitic Hymenoptera in the suborder Apocrita. { ,präk·tə·trə'pȯid·ē·ə }

procumbent [BOTANY] Having stems that lie flat on the ground but do not root at the nodes. { prō 'kəm·bənt }

Procyonidae [VERTEBRATE ZOOLOGY] A family of carnivoran mammals in the superfamily Canoidea, including raccoons and their allies. { ,prō·sē'än·ə,dē }

Prodinoceratinae [PALEONTOLOGY] A subfamily of extinct herbivorous mammals in the family Untatheriidae; animals possessed a carnivorelike body of moderate size. { ¦präd·ən·ō·sə'rat·ən,ē }

producer [ECOLOGY] An autotrophic organism of the ecosystem; any of the green plants. { prə 'dü·sər }

Productinida [PALEONTOLOGY] A suborder of extinct articulate brachiopods in the order Strophomenida characterized by the development of spines. { ,prä·dək'tin·ə·də }

proenzyme [BIOCHEMISTRY] *See* zymogen. { prō 'en,zīm }

proerythroblast of Ferrata [HISTOLOGY] *See* pronormoblast. { ¦prō·ə'rith·rə,blast əv fe'räd·ə }

proestrus [PHYSIOLOGY] The beginning of the follicular phase of estrus. { prō'es·trəs }

profibrinolysin [BIOCHEMISTRY] *See* plasminogen. { ¦prō,fī·brə'näl·ə·sən }

profunda [ANATOMY] Deep-seated; applied to certain arteries. { prō'fən·də }

profundal zone [ECOLOGY] The region occurring below the limnetic zone and extending to the bottom in lakes deep enough to develop temperature stratification. { prō'fənd·əl ,zōn }

Proganosauria [PALEONTOLOGY] The equivalent name for Mesosauria. { prō,gan·ə'sȯr·ē·ə }

progeny [BIOLOGY] Offspring; descendants. { 'präj·ə·nē }

progeny test [GENETICS] The assessment of parental genotype by study of its progeny under controlled conditions. { 'präj·ə·nē ,test }

progestational hormone [BIOCHEMISTRY] **1.** The natural hormone progesterone, which induces progestational changes of the uterine mucosa. **2.** Any derivative or modification or progesterone having similar actions. { ¦prō,je'stā·shən·əl 'hȯr,mōn }

progesterone [BIOCHEMISTRY] $C_{21}H_{30}O_2$ A steroid hormone produced in the corpus luteum, placenta, testes, and adrenals; plays an important physiological role in the luteal phase of the menstrual cycle and in the maintenance of pregnancy; it is an intermediate in the biosynthesis of androgens, estrogens, and the corticoids. { prō'jes·tə,rōn }

proglottid [INVERTEBRATE ZOOLOGY] One of the segments of a tapeworm. { prō'gläd·əd }

Progymnospermopsida [PALEONTOLOGY] A class of plants intermediate between ferns and gymnosperms; comprises the Devonian genus *Archaeopteris*. { prō¦jim·nō,spər'mäp·səd·ə }

prohaptor [INVERTEBRATE ZOOLOGY] The anterior attachment organ of a typical monogenetic trematode. { prō'hap·tər }

Projapygidae [INVERTEBRATE ZOOLOGY] A family of wingless insects in the order Diplura. { ¦prä·jə'pij·əd·ē }

projection area [ANATOMY] An area of the cortex connected with lower centers of the brain by projection fibers. { prə'jek·shən ,er·ē·ə }

projection fibers [ANATOMY] Fibers joining the cerebral cortex to lower centers of the brain, and vice versa. { prə'jek·shən ,fī·bərz }

Prokaryotae [BIOLOGY] A superkingdom of predominantly unicellular microorganisms lacking a membrane-bound nucleus containing chromosomes and having asexual reproduction by binary fission; it includes the kingdom Monera, and viruses, which are acellular, are included by some. { ¦prō·kar·ē'ō,dē }

prokaryote [CELL BIOLOGY] **1.** A primitive nucleus, where the deoxyribonucleic acid-containing region lacks a limiting membrane. **2.** Any cell containing such a nucleus, such as the bacteria and the blue-green algae. { prō'kar·ē,ōt }

Prolacertiformes [PALEONTOLOGY] A suborder of extinct terrestrial reptiles in the order Eosuchia distinguished by reduction of the lower temporal arcade. { prō¦las·ər·də'fȯr,mēz }

prolactin [BIOCHEMISTRY] A protein hormone produced by the adenohypophysis; stimulates lactation and promotes functional activity of the corpus luteum. Also known as lactogenic hormone; luteotropic hormone; mammary-stimulating hormone; mammogen; mammogenic hormone; mammotropin. { prō 'lak·tən }

prolamellar body [BOTANY] An accumulation of vesicles formed by the invagination of the proplastid membrane during etiolation. { ,prō·lə ¦mel·ər 'bäd·ē }

prolamin [BIOCHEMISTRY] Any of the simple proteins, such as zein, found in plants; soluble in strong alcohol, insoluble in absolute alcohol and water. { prō'lam·ən }

proline [BIOCHEMISTRY] $C_5H_9O_2$ A heterocyclic amino acid occurring in essentially all proteins, and as a major constituent in collagen proteins. { 'prō,lēn }

prometaphase [CELL BIOLOGY] A stage between prophase and metaphase in mitosis in which the nuclear membrane disappears and the spindle forms. { prō'med·ə,fāz }

promiscuous deoxyribonucleic acid [GENETICS] Segments of deoxyribonucleic acid that have been transferred between organelles, such as mitochondria and chloroplasts, as a result of transpositional events that occurred millions of years ago. { prə¦mis·kyə·wəs dē¦rī·bō·nü,klē·ik 'as·əd }

promoter [GENETICS] The site on deoxyribonucleic acid to which ribonucleic acid polymerase binds preparatory to initiating transcription of an operon. { prə'mōd·ər }

promyelocyte [HISTOLOGY] The earliest myelocyte stage derived from the myeloblast. { prō 'mī·ə·lə,sīt }

pronate [ANATOMY] **1.** To turn the forearm so that the palm of the hand is down or toward the back. **2.** To turn the sole of the foot outward with the lateral margin of the foot elevated; to evert. { 'prō,nāt }

pronator [PHYSIOLOGY] A muscle which pronates, as the muscles of the forearm attached to the ulna and radius. { 'prō,nād·ər }

pronephros [EMBRYOLOGY] One of the anterior pair of renal organs in higher vertebrate embryos; the pair initiates formation of the archinephric duct. { prō'ne·frəs }

pronghorn [VERTEBRATE ZOOLOGY] *Antilocapra americana*. An antelopelike artiodactyl composing the family Antilocapridae; the only hollow-horned ungulate with branched horns present in both sexes. { 'präŋ,hȯrn }

pronormoblast [HISTOLOGY] A nucleated erythrocyte precursor with scanty basophilic cytoplasm without hemoglobin. Also known as lymphoid hemoblast of Pappenheim; macroblast of Naegeli; megaloblast of Sabin; proerythroblast of Ferrata; prorubricyte; rubriblast; rubricyte. { prō'nȯr·mə,blast }

pronucleus [CELL BIOLOGY] One of the two nuclear bodies of a newly fertilized ovum, the male pronucleus and the female pronucleus, the fusion of which results in the formation of the germinal (cleavage) nucleus. { prō'nü·klē·əs }

proofreading [MOLECULAR BIOLOGY] Any mechanism for correcting errors in replication, transcription, or translation that involves monitoring of individual units after they have been added to the chain. Also known as editing. { 'prüf‚rēd·iŋ }

propagule [BOTANY] **1.** A reproductive structure of brown algae. **2.** A propagable shoot. { 'präp·ə‚gyül }

Propalticidae [INVERTEBRATE ZOOLOGY] A family of coleopteran insects of the superfamily Cucujoidea found in Old World tropics and Pacific islands. { ‚prō·pəl'sid·əd·ē }

properdin [IMMUNOLOGY] A macroglobin of normal plasma capable of killing various bacteria and viruses in the presence of complement and magnesium ions. { 'prō·pər·dən }

prophage [VIROLOGY] Integrated unit formed by union of the provirus into the bacterial genome. { 'prō‚fāj }

prophase [CELL BIOLOGY] The initial stage of mitotic or meiotic cell division in which chromosomes are condensed from the nuclear material and split logitudinally to form pairs. { 'prō‚fāz }

Propionibacteriaceae [MICROBIOLOGY] A family of bacteria related to the actinomycetes; gram-positive, anaerobic to aerotolerant rods or filaments; ferment carbohydrates, with propionic acid as the principal product. { ¦prō·pē‚än·ə‚bak·tir·ē'ās·ē‚ē }

proplastid [BOTANY] Precursor body of a cell plastid. { prō'plas·təd }

propleuron [INVERTEBRATE ZOOLOGY] A pleuron of the prothorax in insects. { prō'plu̇r‚än }

propodite [INVERTEBRATE ZOOLOGY] The sixth leg joint of certain crustaceans. Also known as propodus. { 'präp·ə‚dīt }

propodus [INVERTEBRATE ZOOLOGY] *See* propodite. { 'präp·əd·əs }

propositus [GENETICS] *See* proband. { prə'päz·əd·əs }

proprioception [PHYSIOLOGY] The reception of internal stimuli. { ‚prō·prē·ə'sep·shən }

proprioceptor [PHYSIOLOGY] A sense receptor that signals spatial position and movements of the body and its members, as well as muscular tension. { ‚prō·prē·ə'sep·tər }

prop root [BOTANY] A root that serves to support or brace the plant. Also known as brace root. { 'präp ‚rüt }

propterygium [VERTEBRATE ZOOLOGY] The anterior of the three principal basal cartilages forming a support of one of the paired fins of sharks, rays, and certain other fishes. { ‚präp·tə'rij·ē·əm }

Prorastominae [PALEONTOLOGY] A subfamily of extinct dugongs (Dugongidae) which occur in the Eocene of Jamaica. { ‚pro̊r·ə'stäm·ə‚nē }

prorennin [BIOCHEMISTRY] *See* renninogen. { prō'ren·ən }

Prorhynchidae [INVERTEBRATE ZOOLOGY] A family of turbellarians in the order Alloeocoela. { prō'riŋ·kə‚dē }

prorubricyte [HISTOLOGY] *See* pronormoblast. { prō'rü·brə‚sīt }

Prosauropoda [PALEONTOLOGY] A division of the extinct reptilian suborder Sauropodomorpha; they possessed blunt teeth, long forelimbs, and extremely large claws on the first finger of the forefoot. { ‚prä·so̊'räp·əd·ə }

Prosobranchia [INVERTEBRATE ZOOLOGY] The largest subclass of the Gastropoda; generally, respiration is by means of ctenidia, an operculum is present, there is one pair of tentacles, and the sexes are separate. { ‚prä·sə'braŋ·kē·ə }

prosodus [INVERTEBRATE ZOOLOGY] A canal leading from an incurrent canal to a flagellated chamber in Porifera. { 'präs·əd·əs }

prosoma [INVERTEBRATE ZOOLOGY] The anterior part of the body of mollusks and other invertebrates; primitive segmentation is not apparent. { prō'sō·mə }

Prosopora [INVERTEBRATE ZOOLOGY] An order of the class Oligochaeta comprising mesonephridiostomal forms in which there are male pores in the segment of the posterior testes. { prə'säp ə rə }

prosopyle [INVERTEBRATE ZOOLOGY] The opening into a flagellated chamber from an inhalant canal in sponges. { 'präs·ə‚pīl }

prostaglandin [BIOCHEMISTRY] Any of various physiologically active compounds containing 20 carbon atoms and formed from essential fatty acids; found in highest concentrations in normal human semen; activities affect the nervous system, circulation, female reproductive organs, and metabolism. { ‚präs·tə'glan·dən }

prostate [ANATOMY] A glandular organ that surrounds the urethra at the neck of the urinary bladder in the male. { 'prä‚stāt }

prosthecae [MICROBIOLOGY] Appendages that are part of the wall in bacteria in the genus *Caulobacter*. { präs'thē·sē }

prosthecate bacteria [MICROBIOLOGY] Single-celled microorganisms that differ from typical unicellular bacteria in having one or more appendages which extend from the cell surface; the best-known genus is *Caulobacter*. { 'präs·thə‚kāt bak'tir·ē·ə }

prosthetic group [BIOCHEMISTRY] A characteristic nonamino acid substance that is strongly bound to a protein and necessary for the protein portion of an enzyme to function; often used to describe the function, as in hemeprotein for hemoglobin. { präs'thed·ik 'grüp }

Prostigmata [INVERTEBRATE ZOOLOGY] The equivalent name for Trombidiformes. { ‚prō‚stig'mäd·ə }

prostomium [INVERTEBRATE ZOOLOGY] The portion of the head anterior to the mouth in annelids and mollusks. { prō'stō·mē·əm }

protamine [BIOCHEMISTRY] Any of the simple proteins that are combined with nucleic acid in the sperm of certain fish, and that upon hydrolysis yield basic amino acids; used in medicine to control hemorrhage, and in the preparation of an insulin form to control diabetes. { 'prōd·ə ˌmēn }

protandry [PHYSIOLOGY] That condition in which an animal is first a male and then becomes a female. { prō'tan·drē }

Proteaceae [BOTANY] A large family of dicotyledonous plants in the order Proteales, notable for often having a large cluster of small or reduced flowers. { prō'tās·ēˌē }

Proteales [BOTANY] An order of dicotyledonous plants in the subclass Rosidae marked by its strongly perigynous flowers, a four-lobed, often corollalike calyx, and reduced or absent true petals. { ˌprōd·ē'ā·lēz }

protease [BIOCHEMISTRY] An enzyme that digests proteins. { 'prōd·ēˌās }

protective coloration [ZOOLOGY] A color pattern that blends with the environment and increases the animal's probability of survival. { prə'tek·tiv ˌkəl·ə'rā·shən }

Proteeae [MICROBIOLOGY] Formerly a tribe of the Enterobacteriaceae comprising the genus *Proteus*; included organisms which were characteristically motile, fermented dextrose with gas production, and produced urease. { ˌprō'tē·ēˌē }

Proteida [VERTEBRATE ZOOLOGY] A suborder coextensive with Proteidae in some classification systems. { prō'tē·əd·ə }

Proteidae [VERTEBRATE ZOOLOGY] A family of the amphibian suborder Salamandroidea; includes the neotenic, aquatic *Necturus* and *Proteus* species. { prō'tē·əˌdē }

protein [BIOCHEMISTRY] Any of a class of high-molecular-weight polymer compounds composed of a variety of α-amino acids joined by peptide linkages. { 'prōˌtēn }

proteinase [BIOCHEMISTRY] A type of protease which acts directly on native proteins in the first step of their conversion to simpler substances. { 'pröt·ənˌās }

protein-bound iodine [BIOCHEMISTRY] Iodine bound to blood protein. Abbreviated PBI. { 'prō ˌtēn ˌbaůnd 'ī·əˌdīn }

protein coat [INVERTEBRATE ZOOLOGY] *See* capsid. { 'prōˌtēn ˌkōt }

protein engineering [MOLECULAR BIOLOGY] The design and construction of new proteins or enzymes with novel or desired functions by modifying amino acid sequences by using recombinant deoxyribonucleic acid technology. { 'prō ˌtēn ˌen·jə'nir·iŋ }

protein kinase [BIOCHEMISTRY] An enzyme that exerts regulatory effects on growth and malignant transformation by phosphorylating proteins. { ¦prōˌtēn 'kīˌnās }

Proteocephalidae [INVERTEBRATE ZOOLOGY] A family of tapeworms in the order Proteocephaloidea in which the reproductive organs are within the central mesenchyme of the segment. { ˌprōd·ē·ō·sə'fal·əˌdē }

Proteocephaloidea [INVERTEBRATE ZOOLOGY] An order of tapeworms of the subclass Cestoda in which the holdfast organ bears four suckers and, frequently, a suckerlike apical organ. { ˌprōd·ē·ōˌsef·ə'lȯid·ē·ə }

proteoglycan [BIOCHEMISTRY] A high-molecular-weight polyanionic substance covalently linked by numerous heteropolysaccharide side chains to a polypeptide chain backbone. { ˌprōd·ē·ō 'glī·kən }

proteolysin [BIOCHEMISTRY] A lysin that produces proteolysis. { ˌprōd·ē'äl·ə·sən }

proteolysis [BIOCHEMISTRY] Fragmentation of a protein molecule by addition of water to the peptide bonds. { ˌprōd·ē'äl·ə·səs }

proteolytic enzyme [BIOCHEMISTRY] Any enzyme that catalyzes the breakdown of protein. { ¦prōd·ē·ə¦lid·ik 'enˌzīm }

Proteomyxida [INVERTEBRATE ZOOLOGY] The single order of the Proteomyxidia. { ¦prōd·ē·ə 'mik·səd·ə }

Proteomyxidia [INVERTEBRATE ZOOLOGY] A subclass of Actinopodea including protozoan organisms which lack protective coverings or skeletal elements and have reticulopodia, or filopodia. { ¦prōd·ē·ə·mik'sid·ē·ə }

proteoplast [CELL BIOLOGY] A type of cell plastid containing crystalline, fibrillar, or amorphous masses of protein. { 'prōd·ē·əˌplast }

proteose [BIOCHEMISTRY] One of a group of derived proteins intermediate between native proteins and peptones; soluble in water, not coagulable by heat, but precipitated by saturation with ammonium or zinc sulfate. { 'prōd·ēˌōs }

Proterostomia [ZOOLOGY] That part of the animal kingdom in which cleavage of the egg is of the determinate type; includes all bilateral phyla except Echinodermata, Chaetognatha, Pogonophora, Hemichordata, and Chordata. { ˌpräd·ə· rō'stō·mē·ə }

Proterosuchia [PALEONTOLOGY] A suborder of moderate-sized thecodont reptiles with lightly built triangular skulls, downturned snouts, and palatal teeth. { ˌpräd·ə·rō'sü·kē·ə }

Proterotheriidae [PALEONTOLOGY] A group of extinct herbivorous mammals in the order Litopterna which displayed an evolutionary convergence with the horses in their dentition and in reduction of the lateral digits of their feet. { ˌpräd·ə·rō·thə'rī·əˌdē }

Proteutheria [VERTEBRATE ZOOLOGY] A group of primatelike insectivores that contains the living tree shrews. { ˌprōd·ē·yü'thir·ē·ə }

prothallium [BOTANY] The gametophyte of a pteridophyte in the form of a flat green thallus with thizoids. { prō'thal·ē·əm }

prothoracic gland [INVERTEBRATE ZOOLOGY] One of the paired glands in the prothorax of insects which produce ecdysone. { ¦prō·thə'ras·ik 'gland }

prothorax [INVERTEBRATE ZOOLOGY] The first thoracic segment of an insect; bears the first pair of legs. { prō'thȯrˌaks }

prothrombin [BIOCHEMISTRY] An inactive plasma protein precursor of thrombin. Also known as factor II; thrombinogen. { prō'thräm·bən }

prothrombin factor [BIOCHEMISTRY] *See* vitamin K. { prō'thräm·bən ˌfak·tər }

proticity [BIOCHEMISTRY] In oxidative phosphorylation, the flowing of protons in the proton circuit from high to low protic potential. { prō'tis·əd·ē }

Protista [BIOLOGY] A proposed kingdom to include all unicellular organisms lacking a definite cellular arrangement, such as bacteria, algae, diatoms, and fungi. { prə'tis·tə }

Protoariciinae [INVERTEBRATE ZOOLOGY] A subfamily of polychaete annelids in the family Orbiniidae. { ˌprōd·ōˌar·ə'sī·əˌnē }

Protobranchia [INVERTEBRATE ZOOLOGY] A small and primitive order in the class Bivalvia; the hinge is taxodont in all but one family, there is a central ligament pit, and the anterior and posterior adductor muscles are nearly equal in size. { ˌprōd·ō'braŋ·kē·ə }

Protoceratidae [PALEONTOLOGY] An extinct family of pecoran ruminants in the superfamily Traguloidea. { ¦prōd·ō·sə'rad·əˌdē }

Protochordata [INVERTEBRATE ZOOLOGY] The equivalent name for Hemichordata. { ¦prōd·ō·kȯr'dad·ə }

Protococcaceae [BOTANY] A monogeneric family of green algae in the suborder Ulotrichineae in which reproduction is entirely vegetative. { ˌprōd·ō·käk'sās·ēˌē }

Protococcida [INVERTEBRATE ZOOLOGY] A small order of the protozoan subclass Coccidia; all are invertebrate parasites, and only sexual reproduction is known. { ˌprōd·ō'käk·səd·ə }

Protocucujidae [INVERTEBRATE ZOOLOGY] A small family of coleopteran insects in the superfamily Cucujoidea found in Chile and Australia. { ˌprōd·ō·kə'kü·yəˌdē }

protoderm [BOTANY] *See* dermatogen. { 'prōd·əˌdərm }

Protodonata [PALEONTOLOGY] An extinct order of huge dragonflylike insects found in Permian rocks. { ¦prōd·ō·də'näd·ə }

Protodrilidae [INVERTEBRATE ZOOLOGY] A family of annelids belonging to the Archiannelida. { ˌprōd·ō'dril·əˌdē }

Protoeumalacostraca [PALEONTOLOGY] The stem group of the crustacean series Eumalacostraca. { ˌprōd·ōˌyü·mə·lə'käs·trə·kə }

protogyny [PHYSIOLOGY] A condition in hermaphroditic or dioecious organisms in which the female reproductive structures mature before the male structures. { prō'täj·ə·nē }

Protomastigida [INVERTEBRATE ZOOLOGY] The equivalent name for Kinetoplastida. { ˌprōd·ō·ma'stij·ə·də }

protomer [BIOCHEMISTRY] One of the polypeptide chains composing an oligomeric protein. Also known as subunit. { 'prōd·ə·mər }

Protomonadina [INVERTEBRATE ZOOLOGY] An order of flagellates, subclass Mastigophora, with one or two flagella, including many species showing protoplasmic collars ringing the base of the flagellum. { ˌprōd·ōˌmän·ə'dī·nə }

Protomonida [INVERTEBRATE ZOOLOGY] The equivalent name for Protomonadina. { ˌprōd·ō'män·ə·də }

Protomyzostomidae [INVERTEBRATE ZOOLOGY] A family of parasitic polychaetes belonging to the Myzostomaria and known for three species from Japan and the Murman Sea. { ˌprōd·ōˌmī·zə'stäm·əˌdē }

protonema [BOTANY] A green, filamentous structure that originates from an asexual spore of mosses and some liverworts and that gives rise by budding to a mature plant. { ˌprōt·ən'ē·mə }

protonephridium [INVERTEBRATE ZOOLOGY] **1.** A primitive excretory tube in many invertebrates. **2.** The duct of a flame cell. { ˌprōd·ō·nə'frid·ē·əm }

protophloem [BOTANY] The initial primary phloem developed from the procambium. { ¦prōd·ə'flō·əm }

Protophyta [BOTANY] A division of the plant kingdom, according to one system of classification, set up to include the bacteria, the blue-green algae, and the viruses. { prə'täf·əd·ə }

protoplasm [CELL BIOLOGY] The colloidal complex of protein that composes the living material of a cell. { 'prōd·əˌplaz·əm }

protoplast [CELL BIOLOGY] The living portion of a cell considered as a unit; includes the cytoplasm, the nucleus, and the plasma membrane. { 'prōd·əˌplast }

protoplast fusion [GENETICS] A technique by which two protoplasts are joined or a protoplast is joined with a component of another cell in order to accomplish genetic transformation. { 'prōd·əˌplast ˌfyü·zhən }

protopodite [INVERTEBRATE ZOOLOGY] The basal segment of a crustacean limb bearing an endopodite or exopodite, or both, at its distal extremity. { prə'täp·əˌdīt }

Protopteridales [PALEOBOTANY] An extinct order of ferns, class Polypodiatae. { ˌprōd·ōˌter·ə'dā·lēz }

Protosireninae [PALEONTOLOGY] An extinct superfamily of sirenian mammals in the family Dugongidae found in the middle Eocene of Egypt. { ˌprōd·ō·sə'ren·əˌnē }

Protospondyli [VERTEBRATE ZOOLOGY] An equivalent name for Semionotiformes. { ˌprōd·ō'spän·dəˌlī }

protostele [BOTANY] A stele consisting of a solid rod of xylem surrounded by phloem. { 'prōd·əˌstēl }

Protostomia [INVERTEBRATE ZOOLOGY] A major division of bilateral animals; includes most worms, arthropods, and mollusks. { ˌprōd·ə'stō·mē·ə }

Protosuchia [PALEONTOLOGY] A suborder of extinct crocodilians from the Late Triassic and Early Jurassic. { ¦prōd·ō'sü·kē·ə }

Prototheria [VERTEBRATE ZOOLOGY] A small subclass of Mammalia represented by a single order, the Monotremata. { ¦prōd·ō'thir·ē·ə }

prototroch [INVERTEBRATE ZOOLOGY] The band of cilia characteristic of a trochophore larva. { 'prōd·əˌträk }

prototrophic [MICROBIOLOGY] Pertaining to bacteria with the nutritional properties of the wild type, or the strains found in nature. { ¦prōd·ō¦träf·ik }

Prototrupoidea [INVERTEBRATE ZOOLOGY] A superfamily of the Hymenoptera. { ˌprōd·ō·trə ˈpöid·ē·ə }

protoxylem [BOTANY] The part of the primary xylem that differentiates from the procambium and is formed during elongation of an embryonic plant organ. { ¦prōd·ōˈzī·ləm }

Protozoa [INVERTEBRATE ZOOLOGY] A diverse phylum of eukaryotic microorganisms; the structure varies from a simple uninucleate protoplast to colonial forms, the body is either naked or covered by a test, locomotion is by means of pseudopodia or cilia or flagella, there is a tendency toward universal symmetry in floating species and radial symmetry in sessile types, and nutrition may be phagotrophic or autotrophic or saprozoic. { ¦prōd·ə¦zō·ə }

protozoology [INVERTEBRATE ZOOLOGY] That branch of biology which deals with the Protozoa. { ¦prōd·ō·zōˈäl·ə·jē }

Protrachaeta [INVERTEBRATE ZOOLOGY] The equivalent name for Onychophora. { prō·trə ˈkēd·ə }

protrypsin [BIOCHEMISTRY] *See* trypsinogen. { prō ˈtrip·sən }

Protura [INVERTEBRATE ZOOLOGY] An order of primitive wingless insects belonging to the subclass Apterygota; individuals are elongate and eyeless, lack antennae, and are from pale amber to white in color; anamorphosis is characteristic of the group. { prəˈtyu̇r·ə }

proventriculus [INVERTEBRATE ZOOLOGY] **1.** A sac anterior to the gizzard in earthworms. **2.** A dilation of the foregut anterior to the midgut of Mandibulata. [VERTEBRATE ZOOLOGY] The true stomach of a bird, usually separated from the gizzard by a constriction. { ˌprō·vənˈtrik·yə·ləs }

provirus [VIROLOGY] The phage genome. { prō ˈvī·rəs }

provitamin [BIOCHEMISTRY] A vitamin precursor; assumes vitamin activity upon activation or chemical change. { prōˈvīd·ə·mən }

proximal [ANATOMY] Near the body or the median line of the body. { ˈpräk·sə·məl }

proximal convoluted tubule [ANATOMY] The convoluted portion of the vertebrate nephron lying between Bowman's capsule and the loop of Henle; functions in the resorption of sugar, sodium and chloride ions, and water. { ˈpräk·sə·məl ˌkän·vəˈlüd·əd ˈtübˌyül }

proximoceptor [PHYSIOLOGY] An exteroceptor involved in taste or cutaneous sensations. { ¦präk·sə·mō¦sep·tər }

psalterium [VERTEBRATE ZOOLOGY] *See* omasum. { sölˈtir·ē·əm }

Psammettidae [INVERTEBRATE ZOOLOGY] A family of Psamminida, with a strongly built test, haphazardly arranged xenophyae, no specialized surface layer, and no large openings in the test. { sə ˈmed·əˌdē }

Psamminida [INVERTEBRATE ZOOLOGY] An order of Xenophyophorea distinguished by the absence of linellae in the test and, in general, rigidity of the body. { ˌsam·əˈnī·də }

Psamminidae [INVERTEBRATE ZOOLOGY] A family of Psamminida, with a solid, sometimes fragile test and external xenophyae arranged in a distinct surface layer. { səˈmin·əˌdē }

Psammodontidae [PALEONTOLOGY] A family of extinct cartilaginous fishes in the order Bradyodonti in which the upper and lower dentitions consisted of a few large quadrilateral plates arranged in two rows meeting in the midline. { ˌsam·əˈdänt·əˌdē }

Psammodrilidae [INVERTEBRATE ZOOLOGY] A small family of spioniform worms belonging to the Sedentaria. { ˌsam·əˈdril·əˌdē }

psammon [ECOLOGY] **1.** In a body of fresh water, that part of the environment composed of a sandy beach and bottom lakeward from the water line. **2.** Organisms which inhabit the interstitial water in the sands on a lake shore. { ˈsa ˌmän }

psammophilic [ECOLOGY] Pertaining to an organism found in sand. { ¦sam·ə¦fil·ik }

psammophyte [ECOLOGY] Thriving (as a plant) on sandy soil. { ˈsam·əˌfīt }

psammosere [ECOLOGY] Stages in plant succession which begin in sandy soil. { ˈsam·əˌsir }

Pselaphidae [INVERTEBRATE ZOOLOGY] The ant-loving beetles, a large family of coleopteran insects in the superfamily Staphylinoidea. { sə ˈlaf·əˌdē }

Psephenidae [INVERTEBRATE ZOOLOGY] The water penny beetles, a small family of coleopteran insects in the superfamily Dryopoidea. { səˈfen·əˌdē }

Pseudaliidae [INVERTEBRATE ZOOLOGY] A family of roundworms belonging to the Strongyloidea which occur as parasites of whales and porpoises. { ˌsüd·ə·dəˈlī·əˌdē }

pseudaposematic [ECOLOGY] Pertaining to an imitation in coloration or form by an organism of another organism that possesses dangerous or disagreeable characteristics. { ¦süd·əˌpōz·ə ˈmad·ik }

pseudoalleles [GENETICS] Closely linked genes that behave as alleles and can be separated by crossing over. { ˌsü·dō·əˈlēlz }

pseudoautosomal [GENETICS] Pertaining to segments of the X and Y chromosomes that are affected by obligatory crossing-over of these chromosomes so that they show a pattern of inheritance that does not appear to be sex-linked. { ¦süd·ōˌöd·ōˈsō·məl }

Pseudoborniales [PALEOBOTANY] An order of fossil plants found in Middle and Upper Devonian rocks. { ¦sü·dōˌbör·nēˈā·lēz }

pseudocentrum [VERTEBRATE ZOOLOGY] A centrum formed by fusion of the dorsal or dorsal and ventral arcualia, as in tailed amphibians. { ¦sü·dōˈsen·trəm }

pseudocercus [INVERTEBRATE ZOOLOGY] *See* urogomphus. { ¦sü·dōˈsər·kəs }

pseudocoele [INVERTEBRATE ZOOLOGY] A space between the body wall and internal organs that is not formed by gastrulation and lacks a cellular lining. { ˈsüd·əˌsēl }

pseudocoelocyte [INVERTEBRATE ZOOLOGY] In nematodes, a mesenchymal cell in the pseudocoelom. { ¦sü·dōˈsel·əˌsīt }

Pseudocoelomata [INVERTEBRATE ZOOLOGY] A

group comprising the animal phyla Entoprocta, Aschelminthes, and Acanthocephala; characterized by a pseudocoelom. { ¦sü·dō,sē·lə 'mäd·ə }

pseudocolumella [INVERTEBRATE ZOOLOGY] In anthozoans, a type of axial structure. { ¦sü·dō ,kal·yə'mel·ə }

Pseudocycnidae [INVERTEBRATE ZOOLOGY] A family of the Caligoida which comprises external parasites on the gills of various fishes. { ¦sü·dō 'sik·nə,dē }

Pseudodiadematidae [INVERTEBRATE ZOOLOGY] A family of Jurassic and Cretaceous echinoderms in the order Phymosomatoida which had perforate crenulate tubercles. { ¦sü·dō,dī·ə·də'mad·ə,dē }

pseudogamy [BIOLOGY] *See* pseudomixis. { sü 'däg·ə·mē }

pseudogene [GENETICS] A sequence of deoxyribonucleic acid resembling but not functioning like a gene. { 'sü·dō,jēn }

pseudohermaphroditism [PHYSIOLOGY] A condition in humans which simulates hermaphroditism, with gynandry in females and androgyny in males. { ¦sü·dō·hər'maf·rə·də,tiz·əm }

pseudomixis [BIOLOGY] Formation of an embryo from the fusion of vegetative cells instead of gametes. Also known as pseudogamy; somatogamy. { ¦süd·ō'mik·səs }

Pseudomonadaceae [MICROBIOLOGY] A family of gram-negative, aerobic, rod-shaped bacteria; cells are straight or curved and motile by polar flagella. { sü,däm·ə·nə'dās·ē,ē }

Pseudomonadales [MICROBIOLOGY] Formerly an order of ovoid, rod-shaped, comma-shaped, or spiral bacteria in the class Schizomycetes; cells characterized as rigid and motile by means of polar flagella. { sü,däm·ə·nə'dā·lēz }

Pseudomonadineae [MICROBIOLOGY] Formerly a suborder of bacteria of the order Pseudomonadales including those families whose cells lacked photosynthetic pigments. { sü,däm·ə·nə'dī·nē,ē }

Pseudophoracea [INVERTEBRATE ZOOLOGY] An extinct superfamily of gastropod mollusks in the order Aspidobranchia. { ,süd·ə·fə'rā·shə }

Pseudophyllidea [INVERTEBRATE ZOOLOGY] An order of tapeworms of the subclass Cestoda, parasitic principally in the intestine of cold-blooded vertebrates. { ,süd·ə·fə'lid·ē·ə }

pseudoplasmodium [INVERTEBRATE ZOOLOGY] An aggregate of amebas resembling a plasmodium. { ¦sü·dō·plaz'mō·dē·əm }

pseudoplastic [BIOLOGY] Referring to an organism which lacks the capacity for major modification or for evolutionary differentiation. { ¦sü·dō'plas·tik }

pseudopodium [BOTANY] A slender, leafless branch of the gametophyte in certain Bryatae. [CELL BIOLOGY] Temporary projection of the protoplast of ameboid cells in which cytoplasm streams actively during extension and withdrawal. [INVERTEBRATE ZOOLOGY] Foot of a rotifer. { ,süd·ə'pōd·ē·əm }

Pseudoscorpionida [INVERTEBRATE ZOOLOGY] An order of terrestrial Arachnida having the general appearance of miniature scorpions without the postabdomen and sting. { ¦sü·dō,skȯr·pē 'än·ə·də }

Pseudosphaeriales [BOTANY] An order of the class Ascolichenes, shared by the class Ascomycetes; the ascocarp is flask-shaped and lined with a layer of interwoven, branched pseudoparaphyses. { ¦sü·dō,sfir·ē'ā·lēz }

Pseudosporidae [INVERTEBRATE ZOOLOGY] A family of the protozoan subclass Proteomyxidia; flagellated stages invade Volvocidae and filamentous algae and become amebas. { ¦sü·dō 'spȯr·ə,dē }

pseudostem [BOTANY] A false stem composed of concentric rolled or folded blades and sheaths that surround the growing point. { 'süd·ō ,stem }

pseudostratified epithelium [HISTOLOGY] A type of epithelium in which all cells reach to the basement membrane but some extend toward the surface only part way, while others reach the surface. { ¦sü·dō'strad·ə,fīd ,ep·ə'thē·lē·əm }

Pseudosuchia [PALEONTOLOGY] A suborder of extinct reptiles of the order Thecodontia comprising bipedal, unarmored or feebly armored forms which resemble dinosaurs in many skull features but retain a primitive pelvis. { ,sü·dō 'sü·kē·ə }

Pseudothelphusidae [INVERTEBRATE ZOOLOGY] A family of fresh-water crabs belonging to the Brachyura. { ¦sü·dō·thel'fyüz·ə,dē }

Pseudotriakidae [VERTEBRATE ZOOLOGY] The false catsharks, a family of galeoids in the carcharinid line. { ¦sü·dō·trī'ak·ə,dē }

pseudo-wild type [GENETICS] A wild phenotype due to a second (suppressor) mutation of a mutant. { ¦sü·dō'wīld ,tīp }

psilate [BOTANY] Lacking ornamentation; generally applied to pollen. { 'sī,lāt }

Psilidae [INVERTEBRATE ZOOLOGY] The rust flies, a family of myodarian cyclorrhaphous dipteran insects in the subsection Acalyptreaatae. { 'sil·ə ,dē }

Psilophytales [PALEOBOTANY] A group formerly recognized as an order of fossil plants. { ,sī·lō,fī 'tā·lēz }

Psilophytineae [PALEONTOLOGY] The equivalent name for Rhyniopsida. { ,sī·lō,fī'tin·ē,ē }

Psilopsida [BOTANY] A subdivision of the Tracheophyta. { sī'läp·səd·ə }

Psilorhynchidae [VERTEBRATE ZOOLOGY] A small family of actinopterygian fishes belonging to the Cyprinoidei. { ,sī·lō'riŋ·kə,dē }

Psilotales [BOTANY] The equivalent name for Psilotophyta. { ,sī·lō'tā·lēz }

Psilotatae [BOTANY] A class of the Psilotophyta. { sī'läd·ə,dē }

Psilotophyta [BOTANY] A division of the plant kingdom represented by three living species; the life cycle is typical of the vascular cryptogams. { ,sī·lō'täf·əd·ə }

Psittacidae [VERTEBRATE ZOOLOGY] The single family of the Psittaciformes. { sə'tas·ə,dē }

Psittaciformes [VERTEBRATE ZOOLOGY] The parrots, a monofamilial order of birds that exhibit zygodactylism and have a strong hooked bill. { sə ˌtas·ə'fȯrˌmēz }

psoas [ANATOMY] Either of two muscles: psoas major which arises from the bodies and transverse processes of the lumbar vertebrae and is inserted into the lesser trochanter of the femur, and psoas minor which arises from the bodies and transverse processes of the lumbar vertebrae and is inserted on the pubis. { 'sō·əs }

Psocoptera [INVERTEBRATE ZOOLOGY] An order of small insects in which wings may be present or absent, tarsi are two- or three-segmented, cerci are absent, and metamorphosis is gradual. { sō 'käp·tə·rə }

Psolidae [INVERTEBRATE ZOOLOGY] A family of echinoderms in the order Dendrochirotida characterized by a ventral adhesive sucker and a U-shaped gut, with the mouth and anus opening upward on the adoral surface. { 'säl·əˌdē }

Psophiidae [VERTEBRATE ZOOLOGY] The trumpeters, a family of birds in the order Gruiformes. { sō 'fil·əˌdē }

psychiatric genetics [GENETICS] The study of the genetic causes and modes of inheritence that underlie the generally recognized mental illnesses. { ˌsī·kē'a·trik jə'ned·iks }

Psychidae [INVERTEBRATE ZOOLOGY] The bagworms, a family of lepidopteran insects in the superfamily Tineoidea; males are large, hairy moths, but females are degenerate, wingless, and legless and live in bag-shaped cases. { 'sī·kəˌdē }

Psychodidae [INVERTEBRATE ZOOLOGY] The moth flies, a family of orthorrhaphous dipteran insects in the series Nematocera. { sī'käd·ə ˌdē }

psychogalvanic reflex [PHYSIOLOGY] A variation in the electric conductivity of the skin in response to emotional stimuli, which cause changes in blood circulation, secretion of sweat, and skin temperature. { ¦sī·kō·gal'van·ik 'rē ˌfleks }

psychology [BIOLOGY] **1.** The science that deals with the functions of the mind and the behavior of an organism in relation to its environment. **2.** The mental activity characteristic of a person or a situation. { sī'käl·ə·jē }

psychrophile [BIOLOGY] An organism that thrives at low temperatures. { 'sī·krəˌfīl }

psychrophobia [PHYSIOLOGY] Abnormal sensitivity to cold. { ˌsī·krə'fō·bē·ə }

psychrophyte [ECOLOGY] A plant adapted to the climatic conditions of the arctic or alpine regions. { 'sī·krəˌfīt }

Psyllidae [INVERTEBRATE ZOOLOGY] The jumping plant lice, a family of the Homoptera in the series Sternorrhyncha in which adults have a transverse head with protuberant eyes and three ocelli, 6- to 10-segmented antennae, and wings with reduced but conspicuous venation. { 'sil·əˌdē }

PTA [BIOCHEMISTRY] *See* factor XI.

ptarmigan [VERTEBRATE ZOOLOGY] Any of various birds of the genus *Lagopus* in the family Tetraonidae; during the winter, plumage is white and hairlike feathers cover the feet. { 'tar·mə·gən }

PTC [BIOCHEMISTRY] *See* Christmas factor.

Pteraspidomorphi [VERTEBRATE ZOOLOGY] The equivalent name for Diplorhina. { təˌras·pə·də 'mȯrˌfī }

Pterasteridae [INVERTEBRATE ZOOLOGY] A family of deep-water echinoderms in the order Spinulosida distinguished by having webbed spine fins. { ˌter·ə'ster·əˌdē }

Pteridophyta [BOTANY] The equivalent name for Polypodiophyta. { ˌter·ə'däf·əd·ə }

Pteridophytic [PALEOBOTANY] *See* Paleophytic. { tə¦rid·ə¦fid·ik }

Pteridospermae [PALEOBOTANY] Seed ferns, a class of the Cycadicae comprising extinct plants characterized by naked seeds borne on large fernlike fronds. { ˌter·ə·dō'spərˌmē }

Pteridospermophyta [PALEOBOTANY] The equivalent name for Pteridospermae. { ˌter·ə·dō·spər 'mäf·əd·ē }

Pteriidae [INVERTEBRATE ZOOLOGY] Pearl oysters, a family of bivalve mollusks which have nacreous shells. { tə'rī·əˌdē }

pterinophore [CELL BIOLOGY] A yellow to orange chromatophore that contains pterine pigment. { tə'rin·əˌfōr }

Pterobranchia [INVERTEBRATE ZOOLOGY] A group of small or microscopic marine animals regarded as a class of the Hemichordata; all are sessile, tubicolous organisms with a U-shaped gut and three body segments. { ˌter·ə'braŋ·kē·ə }

pterochore [BOTANY] A type of plant that produces winged disseminules. { 'ter·əˌkȯr }

Pteroclidae [VERTEBRATE ZOOLOGY] The sandgrouse, a family of gramnivorous birds in the order Columbiformes; mainly an Afro-Asian group resembling pigeons and characterized by cryptic coloration, usually corresponding with the soil color of the habitat. { tə'räk·ləˌdē }

pterodactyl [PALEONTOLOGY] The common name for members of the extinct reptilian order Pterosauria. { ¦ter·ə¦dak·təl }

Pterodactyloidea [PALEONTOLOGY] A suborder of Late Jurassic and Cretaceous reptiles in the order Pterosauria distinguished by lacking tails and having increased functional wing length due to elongation of the metacarpels. { ¦ter·ə¦dak·tə 'lȯid·ē·ə }

pteroic acid [BIOCHEMISTRY] $C_{14}H_{12}N_6O_3$ A crystalline amino acid formed by hydrolysis of folic acid or other pteroylglutamic acids. { tə'rō·ik 'as·əd }

Pteromalidae [INVERTEBRATE ZOOLOGY] A family of hymenopteran insects in the superfamily Chalcidoidea. { ˌter·ə'mal·əˌdē }

Pteromedusae [INVERTEBRATE ZOOLOGY] A suborder of hydrozoan cnidarians in the order Trachylina characterized by a modified, bipyramidal medusae. { ˌter·ə·mə'düˌsē }

Pterophoridae [INVERTEBRATE ZOOLOGY] The plume moths, a family of the lepidopteran superfamily Pyralidoidea in which the wings are divided into featherlike plumes, maxillary palpi are lacking, and the legs are long. { ˌter·ə'fȯr·ə ˌdē }

Pteropidae [INVERTEBRATE ZOOLOGY] The fruit

bats, a large family of the Chiroptera found in Asia, Australia, and Africa. { tə'räp·ə,dē }

Pteropoda [INVERTEBRATE ZOOLOGY] The sea butterflies, an order of pelagic gastropod mollusks in the subclass Opisthobranchia in which the foot is modified into a pair of large fins and the shell, when present, is thin and glasslike. { tə 'räp·ə·də }

Pteropodidae [VERTEBRATE ZOOLOGY] A family of fruit-eating bats in the suborder Megachiroptera, characterized by primitive ears and by shoulder joints. { ,ter·ə'päd·ə,dē }

Pteropsida [BOTANY] A large group of vascular plants characterized by having parenchymatous leaf gaps in the stele and by having leaves which are thought to have originated in the distant past as branched stem systems. { tə'räp·səd·ə }

Pterosauria [PALEONTOLOGY] An extinct order of flying reptiles of the Mesozoic era belonging to the subclass Archosauria; the wing resembled that of a bat, and a large heeled sternum supported strong wing muscles. { ,ter·ə'sȯr·ē·ə }

pterostigma [INVERTEBRATE ZOOLOGY] Opaque thickened spot occurring on the costal margin of an insect wing. { ,ter·ə'stig·mə }

pteroylglutamic acid [BIOCHEMISTRY] *See* folic acid. { ¦ter·ə·wəl·glü'tam·ik 'as·əd }

pterygium [VERTEBRATE ZOOLOGY] A generalized vertebrate limb. { tə'rij·ē·əm }

pterygoid bone [VERTEBRATE ZOOLOGY] A rodlike bone or group of bones forming a portion of the palatoquadrate arch in lower vertebrates. { 'ter·ə,gȯid }

pterygopalatine fossa [ANATOMY] The gap between the pterygoid process of the sphenoid bone and the maxilla and palatine bone. { ¦ter·ə·gō'pal·ə,tēn 'fäs·ə }

pterygoquadrate [EMBRYOLOGY] Of, pertaining to, or being the first branchial arch in lower vertebrate embryos; gives rise to most of the upper jaw. { ¦ter·ə·gō'kwä,drāt }

Ptiliidae [INVERTEBRATE ZOOLOGY] The feather-winged beetles, a family of coleopteran insects in the superfamily Staphylinoidea. { 'til·ə,dē }

Ptilodactylidae [INVERTEBRATE ZOOLOGY] The toed-winged beetles, a family of the Coleoptera in the superfamily Dryopoidea. { ¦til·ō·dak'til·ə ,dē }

Ptilodontoidea [PALEONTOLOGY] A suborder of extinct mammals in the order Multituberculata. { ¦til·ō·dän'tȯid·ē·ə }

Ptinidae [INVERTEBRATE ZOOLOGY] The spider beetles, a family of coleopteran insects in the superfamily Bostrichoidea. { 'tin·ə,dē }

ptyalase [BIOCHEMISTRY] *See* ptyalin. { 'tī·ə,lās }

ptyalin [BIOCHEMISTRY] A diastatic enzyme found in saliva which catalyzes the hydrolysis of starch to dextrin, maltose, and glucose, and the hydrolysis of sucrose to glucose and fructose. Also known as ptyalase; salivary amylase; salivary diastase. { 'tī·ə·lən }

Ptychodactiaria [INVERTEBRATE ZOOLOGY] An order of the zoantharian anthozoans of the phylum Cnidaria known only from two genera, *Ptychodactis* and *Dactylanthus*. { ¦tī·kō,dak·tē'ar·ē·ə }

Ptychomniaceae [BOTANY] A family of mosses in the order Isobryales distinguished by an eight-ribbed capsule. { tī,käm·nē'ās·ē,ē }

Ptyctodontida [PALEONTOLOGY] An order of Middle and Upper Devonian fishes of the class Placodermi in which both the head and trunk shields are present, and the joint between them is a well-differentiated and variable structure. { ,tik·tə'dänt·əd·ə }

puberty [PHYSIOLOGY] The period at which the generative organs become capable of exercising the function of reproduction; signalized in the boy by a change of voice and discharge of semen, in the girl by the appearance of the menses. { 'pyü·bərd·ē }

puberulic acid [BIOCHEMISTRY] $(HO)_3(C_7H_2O)$-COOH A keto acid formed as a metabolic product of certain species of *Penicillium*; has some germicidal activity against gram-positive bacteria. { pyü'bər·yə·lik 'as·əd }

pubic arch [ANATOMY] The arch formed by the conjoined rami of the pubis and ischium. { 'pyü·bik 'ärch }

pubic crest [ANATOMY] The crest extending from the pubic tubercle to the medial extremity of the pubis. { 'pyü·bik 'krest }

pubic symphysis [ANATOMY] The fibrocartilaginous union of the pubic bones. Also known as symphysis pubis. { 'pyü·bik 'sim·fə·səs }

pubis [ANATOMY] The pubic bone, the portion of the hipbone forming the front of the pelvis. { 'pyü·bəs }

puff ball [BOTANY] A spherical basidiocarp that retains spores until fully mature and, when disturbed, releases them as puffs of fine dust. { 'pəf ,bȯl }

pulmonary artery [ANATOMY] A large artery that conducts venous blood from the heart to the lungs of tetrapods. { 'pu̇l·mə,ner·ē 'ärd·ə·rē }

pulmonary circulation [PHYSIOLOGY] The circulation of blood through the lungs for the purpose of oxygenation and the release of carbon dioxide. Also known as lesser circulation. { 'pu̇l·mə,ner·ē ,sər·kyə'lā·shən }

pulmonary plexus [ANATOMY] A nerve plexus composed chiefly of vagal fibers situated on the anterior and posterior aspects of the bronchi and accompanying them into the substance of the lung. { 'pu̇l·mə,ner·ē 'plek·səs }

pulmonary valve [ANATOMY] A valve consisting of three semilunar cusps situated between the right ventricle and the pulmonary trunk. { 'pu̇l·mə,ner·ē 'valv }

pulmonary vein [ANATOMY] A large vein that conducts oxygenated blood from the lungs to the heart in tetrapods. { 'pu̇l·mə,ner·ē 'vān }

pulmonary ventilation [PHYSIOLOGY] The volume of gas entering and exiting the lungs per unit time of respiration. { 'pu̇l·mə,ner·ē ,vent·əl'ā·shən }

Pulmonata [INVERTEBRATE ZOOLOGY] A subclass of the gastropod mollusks which contains the "lung"-bearing snails; the gills have been lost and in their place the mantle cavity has become a pulmonary sac. { ,pu̇l·mə'näd·ə }

pulp [ANATOMY] A mass of soft spongy tissue in

the interior of an organ. [BOTANY] The soft succulent portion of a fruit. { pəlp }

pulp cavity [ANATOMY] The space within the central part of a tooth containing the dermal pulp and made up of the pulp chamber and a root canal. { 'pəlp ˌkav·əd·ē }

pulp chamber [ANATOMY] The coronal portion of the central cavity of a tooth. { 'pəlp ˌchām·bər }

pulsation [PHYSIOLOGY] A beating or throbbing, usually rhythmic, as of the heart or an artery. { pəl'sā·shən }

pulse [PHYSIOLOGY] **1.** The regular, recurrent, palpable wave of arterial distention due to the pressure of the blood ejected with each contraction of the heart. **2.** A single wave. { pəls }

pulse pressure [PHYSIOLOGY] The difference between the systolic and diastolic blood pressure. { 'pəls ˌpresh·ər }

pulse rate [PHYSIOLOGY] The number of pulsations of an artery per minute. { 'pəls ˌrāt }

pulse wave [PHYSIOLOGY] A wave of increased pressure over the arterial system, started by contraction of septum and valves in the heart. { 'pəls ˌwāv }

pulvillus [INVERTEBRATE ZOOLOGY] A small cushion or cushionlike pad, often covered with short hairs, on an insect's foot between the claws of the last segment. { ˌpəl'vil·əs }

pulvinus [BOTANY] A cushionlike enlargement of the base of a petiole which functions in turgor movements of leaves. { ˌpəl'vī·nəs }

puma [VERTEBRATE ZOOLOGY] *Felis concolor.* A large, tawny brown wild cat (family Felidae) once widespread over most of the Americas. Also known as American lion; catamount; cougar; mountain lion. { 'pü·mə }

pumpkin [BOTANY] Any of several prickly vines with large lobed leaves and yellow flowers in the genus Cucurbita of the order Violales; the fruit is orange-colored and large, with a firm rind. { 'pəm·kən }

puna [ECOLOGY] An alpine biological community in the central portion of the Andes Mountains of South America characterized by low-growing, widely spaced plants that lack much green color most of the year. { 'pü·nə }

punctate [BIOLOGY] Dotted; full of minute points. { 'pəŋkˌtāt }

punctuated equilibrium [EVOLUTION] A model of evolution that proposes long periods without change punctuated by periods of rapid speciation, with natural selection acting on species as well as on individuals. Also known as punctuated evolution. { ¦pəŋk·chəˌwād·əd ˌē·kwə'lib·rē·əm }

punctuated evolution [EVOLUTION] *See* punctuated equilibrium. { ¦pəŋk·chəˌwād·əd ˌev·ə'lü·shən }

pupa [INVERTEBRATE ZOOLOGY] The quiescent, intermediate form assumed by an insect that undergoes complete metamorphosis; it follows the larva and precedes the adult stages and is enclosed in a hardened cuticle or a cocoon. { 'pyü·pə }

pupate [INVERTEBRATE ZOOLOGY] **1.** To develop into a pupa. **2.** To pass through a pupal stage. { 'pyüˌpāt }

pupil [ANATOMY] The contractile opening in the iris of the vertebrate eye. { 'pyü·pəl }

pupillary reflex [PHYSIOLOGY] **1.** Contraction of the pupil in response to stimulation of the retina by light. Also known as Whytt's reflex. **2.** Contraction of the pupil on accommodation for close vision, and dilation of the pupil on accommodation for distant vision. **3.** Contraction of the pupil on attempted closure of the eye. Also known as Westphal-Pilcz reflex; Westphal's pupillary reflex. { 'pyü·pəˌler·ē 'rēˌfleks }

Pupipara [INVERTEBRATE ZOOLOGY] A section of cyclorrhaphous dipteran insects in the Schizophora series in which the young are born as mature maggots ready to become pupae. { pyü'pip·ə·rə }

pure culture [MICROBIOLOGY] A culture that contains cells of one kind, all progeny of a single cell. { 'pyůr ¦kəl·chər }

purine [BIOCHEMISTRY] A heterocyclic compound containing fused pyrimidine and imidazole rings; adenine and guanine are the purine components of nucleic acids and coenzymes. { 'pyůˌrēn }

Purkinje cell [HISTOLOGY] Any of the cells of the cerebral cortex with large, flask-shaped bodies forming a single cell layer between the molecular and granular layers. { pər'kin·jē ˌsel }

Purkinje effect [PHYSIOLOGY] When illumination is reduced to a low level, slowly enough to allow adaptation by the eye, the sensation produced by the longer-wave stimuli (red, orange) decreases more rapidly than that produced by shorter-wave stimuli (violet, blue). Also spelled Parkinje effect. { pər'kin·jē iˌfekt }

Purkinje fibers [HISTOLOGY] Modified cardiac muscle fibers composing the terminal portion of the conducting system of the heart. { pər'kin·jē ˌfī·bərz }

puromycin [MICROBIOLOGY] $C_{22}H_{29}O_5N_7$ A colorless, crystalline broad-spectrum antibiotic produced by a strain of *Streptomyces.* { ¦pyůr·ə¦mīs·ən }

purple bacteria [MICROBIOLOGY] Any of various photosynthetic bacteria that contain bacteriochlorophyll, distinguished by purplish or reddish-brown pigments. { 'pər·pəl bak'tir·ē·ə }

purple nonsulfur bacteria [MICROBIOLOGY] Any of various purple photosynthetic bacteria, especially members of the family Athiorhodaceae, that utilize organic hydrogen donor compounds. { 'pər·pəl ¦nän¦səl·fər bak'tir·ē·ə }

purple sulfur bacteria [MICROBIOLOGY] Any of various anaerobic photosynthetic purple bacteria, especially in the family Thiorhodaceae, that utilize H_2S and other inorganic sulfur compounds as a source of hydrogen, while the carbon source can be carbon monoxide. { 'pər·pəl 'səl·fər bakˌtir·ē·ə }

Pustulosa [PALEONTOLOGY] An extinct suborder of echinoderms in the order Phanerozonida found in the Paleozoic. { ˌpəs·chə'lō·sə }

pusule [INVERTEBRATE ZOOLOGY] A noncontractile fluid-filled vacuole emptied by means of a duct; found in dinoflagellates. { 'pəs·yül }

putrefaction [BIOCHEMISTRY] Decomposition of organic matter, particularly the anaerobic breakdown of proteins by bacteria, with the production of foul-smelling compounds. { ˌpyü·trə'fak·shən }

pv [MICROBIOLOGY] *See* pathovar.

pycnidiospore [MYCOLOGY] A conidium produced by a pycnidium. { pik'nid·ē·ōˌspȯr }

pycnidium [MYCOLOGY] A cavity that bears pycnidiospores in certain fungi. { pik'nid·ē·əm }

pycniospore [MYCOLOGY] A haploid spore of a rust fungus that fuses with a haploid hypha of opposite sex to produce dikaryotic aeciospores. { 'pik·nē·əˌspȯr }

pycnium [MYCOLOGY] A flask-shaped fruit body of a rust fungus formed in clusters just beneath the surface of a host tissue. { 'pik·nē·əm }

Pycnodontiformes [PALEONTOLOGY] An extinct order of specialized fishes characterized by a laterally compressed, disk-shaped body, long dorsal and anal fins, and an externally symmetrical tail. { ˌpik·nəˌdänt·ə'fȯrˌmēz }

Pycnogonida [INVERTEBRATE ZOOLOGY] The sea spiders, a subphylum of marine arthropods in which the body is reduced to a series of cylindrical trunk somites supporting the appendage. { ˌpik·nə'gän·əd·ə }

Pycnogonidae [INVERTEBRATE ZOOLOGY] A family of the Pycnogonida lacking both chelifores and palpi and having six to nine jointed ovigers in the male only. { ˌpik·nə'gän·əˌdē }

Pygasteridae [PALEONTOLOGY] The single family of the extinct order Pygasteroida. { ˌpī·gə'ster·əˌdē }

Pygasteroida [PALEONTOLOGY] An order of extinct echinoderms in the superorder Diadematacea having four genital pores, noncrenulate tubercles, and simple ambulacral plates. { ˌpī·gə·stə'rȯid·ə }

pygidium [INVERTEBRATE ZOOLOGY] **1.** A caudal shield on the abdomen of some Arthropoda. **2.** The terminal body segment of many invertebrates. { pī'jid·ē·əm }

Pygopodidae [VERTEBRATE ZOOLOGY] The flap-footed lizards, a family of the suborder Sauria. { ˌpī·gə'päd·əˌdē }

pylome [INVERTEBRATE ZOOLOGY] An aperture for emission of pseudopodia and intake of food in some Sarcodina. { 'pīˌlōm }

pyloric caecum [INVERTEBRATE ZOOLOGY] **1.**One of the tubular pouches that open into the ventriculus of an insect. **2.** One of the paired tubes having lateral glandular diverticula in each ray of a starfish. [VERTEBRATE ZOOLOGY] One of the tubular pouches that open from the pyloric end of the stomach into the alimentary canal of most fishes. { pə'lȯr·ik 'sē·kəm }

pyloric sphincter [ANATOMY] The thickened ring of circular smooth muscle at the lower end of the pyloric canal of the stomach. { pə'lȯr·ik 'sfiŋk·tər }

pylorus [ANATOMY] The orifice of the stomach communicating with the small intestine. { pə'lȯr·əs }

pyocyanin [MICROBIOLOGY] $C_{13}H_{10}N_{20}$ An antibiotic substance forming blue crystals, produced by *Pseudomonas aeruginosa*; active against many bacteria and fungi. { ˌpī·ō'sī·ə·nən }

Pyralidae [INVERTEBRATE ZOOLOGY] The equivalent name for Pyralididae. { pə'ral·əˌdē }

Pyralididae [INVERTEBRATE ZOOLOGY] A large family of moths in the lepidopteran superfamily Pyralidoidea; the labial palpi are well developed, and the legs are usually long and slender. { ˌpir·ə'lid·əˌdē }

Pyralidinae [INVERTEBRATE ZOOLOGY] A subfamily of the Pyralididae. { ˌpir·ə'lin·əˌnē }

Pyralidoidea [INVERTEBRATE ZOOLOGY] A superfamily of the Lepidoptera belonging to the Heteroneura and including long-legged, slender-bodied moths with well-developed maxillary palpi. { ˌpir·ə·lə'dȯid·ē·ə }

pyramidal area [ANATOMY] *See* motor area. { ¦pir·ə¦mid·əl *or* pə'ram·ə·dəl 'er·ē·ə }

pyramidal system [ANATOMY] The corticospinal and corticobulbar tracts. { ¦pir·ə¦mid·əl 'sis·təm }

Pyramidellidae [INVERTEBRATE ZOOLOGY] A family of gastropod mollusks in the order Tectibranchia; the operculum is present in this group. { pə ˌram·ə'del·əˌdē }

pyramid of numbers [ECOLOGY] The concept that an organism making up the base of a food chain is numerically abundant while each succeeding member of the chain is represented by successively fewer individuals; uses feeding relationship as a basis for the quantitative analysis of an ecological system. { 'pir·əˌmid əv 'nəm·bərz }

Pyraustinae [INVERTEBRATE ZOOLOGY] A large subfamily of the Pyralididae containing relatively large, economically important moths. { pə'rȯs·təˌnē }

pyrenoid [BOTANY] A colorless body found within the chromatophore of certain algae; a center for starch formation and storage. { 'pir·əˌnȯid }

Pyrenolichenes [BOTANY] The equivalent name for Pyrenulales. { ˌpī·rə·nō·lī'kē·nēz }

Pyrenulaceae [BOTANY] A family of the Pyrenulales; all species are crustose and most common on tree bark in the tropics. { pīˌren·yə'lās·ēˌē }

Pyrenulales [BOTANY] An order of the class Ascolichenes including only those lichens with perithecia that contain true paraphyses and unitunicate asci. { pīˌren·yə'lā·lēz }

Pyrgotidae [INVERTEBRATE ZOOLOGY] A family of myodarian cyclorrhaphous dipteran insects in the subsection Acalyptratae. { pər'gäd·əˌdē }

2,3-pyridinedicarboxylic acid [BIOCHEMISTRY] $C_7H_5NO_4$ An odorless, crystalline compound with a melting point of 190°C; soluble in water; inhibits glucose synthesis. Also known as quinolinic acid. { ¦tü ¦thrē ¦pir·əˌdēnˌkär·bäk'sil·ik 'as·əd }

pyridoxal hydrochloride [BIOCHEMISTRY] *See* pyridoxine hydrochloride. { ˌpir·ə'däk·səl ˌhī·drə 'klȯrˌīd }

pyridoxal phosphate [BIOCHEMISTRY] *See* codecarboxylase. { ˌpir·ə'däk·səl 'fäˌsfāt }

pyridoxine hydrochloride [BIOCHEMISTRY] C_8H_{11}-

$NO_3 \cdot HCl$ A crystalline compound, decomposing at about 208°C; used in medicine in vitamin therapy. Also known as pyridoxal hydrochloride; vitamin B_6 hydrochloride. { ˌpir·ə'däkˌsēn ˌhī·drə'klȯrˌīd }

pyrimidine [BIOCHEMISTRY] $C_4H_4N_2$ A heterocyclic organic compound containing nitrogen atoms at positions 1 and 3; naturally occurring derivatives are components of nucleic acids and coenzymes. { pə'rim·əˌdēn }

pyrogen [BIOCHEMISTRY] A group of substances thought to be polysaccharides of microbial origin that produce an increase in body temperature when injected into humans and some animals. { 'pī·rəˌjən }

pyrophosphatase [BIOCHEMISTRY] An enzyme catalyzing hydrolysis of esters containing two or more molecules of phosphoric acid to form a simpler phosphate ester. { ˌpī·rō'fä· sfəˌtās }

Pyrosomida [INVERTEBRATE ZOOLOGY] An order of pelagic tunicates in the class Thaliacea in which species form tubular swimming colonies and are often highly luminescent. { ˌpī·rə'säm·əd·ə }

Pyrotheria [PALEONTOLOGY] An extinct monofamilial order of primitive, mastodonlike, herbivorous, hoofed mammals restricted to the Eocene and Oligocene deposits of South America. { ˌpī·rō'thir·ē·ə }

Pyrotheriidae [PALEONTOLOGY] The single family of the Pyrotheria. { ˌpī·rō·thə'rī·əˌdē }

Pyrrhocoridae [INVERTEBRATE ZOOLOGY] A family of hemipteran insects belonging to the superfamily Pyrrhocoroidea. { ˌpir·ə'kȯr·əˌdē }

Pyrrhocoroidea [INVERTEBRATE ZOOLOGY] A superfamily of the Pentatomorpha. { ˌpir·ə·kə'rȯid·ē·ə }

Pyrrhophyta [BOTANY] A small division of motile, generally unicellular flagellate algae characterized by the presence of yellowish-green to golden-brown plastids and by the general absence of cell walls. { pə'räf·əd·ə }

pyruvate [BIOCHEMISTRY] Salt of pyruvic acid, such as sodium pyruvate, $NaOOCCOCH_3$. { pī'rüˌvāt }

pyruvic acid [BIOCHEMISTRY] Important intermediate in protein and carbohydrate metabolism; liquid with acetic-acid aroma; melts at 11.8°C; miscible with alcohol, ether, and water; used in biochemical research. { pī'rü·vik 'as·əd }

Pythidae [INVERTEBRATE ZOOLOGY] An equivalent name for Salpingidae. { 'pith·əˌdē }

python [VERTEBRATE ZOOLOGY] The common name for members of the reptilian subfamily Pythoninae. { 'pīˌthän }

Pythoninae [VERTEBRATE ZOOLOGY] A subfamily of the reptilian family Boidae distinguished anatomically by the skull structure and the presence of a pair of vestigial hindlegs in the form of stout, movable spurs. { pī'thän·əˌnē }

pyxidium [BOTANY] A capsular fruiting body dehiscing around its circumference, thus causing the upper part to fall off. { pik'sid·ē·əm }

pyxis [BOTANY] A capsule that dehisces by a transverse fissure around the circumference. { 'pik·səs }

Q

quadrant [ANATOMY] One of the four regions into which the abdomen may be divided for purposes of physical diagnosis. [PHYSIOLOGY] A sector of one-fourth of the field of vision of one or both eyes. { 'kwä·drənt }

quadrate bone [VERTEBRATE ZOOLOGY] A small element forming part of the upper jaw joint on each side of the head in vertebrates below mammals. { 'kwä,drāt ,bōn }

quadratojugal [VERTEBRATE ZOOLOGY] A small bone connecting the quadrate and jugal bones on each side of the skull in many lower vertebrates. { kwä¦drā·dō'jü·gəl }

quadriceps [ANATOMY] Four-headed, as a muscle. { 'kwä·drə,seps }

quadriceps femoris [ANATOMY] The large extensor muscle of the thigh, combining the rectus femoris and vastus muscles. { 'kwä·drə,seps 'fem·ə·rəs }

quadrigeminal body [ANATOMY] *See* corpora quadrigemina. { ¦kwä·drə¦jem·ə·nəl 'bäd·ē }

Quadrijugatoridae [PALEONTOLOGY] A monomorphic family of extinct ostracods in the superfamily Hollinacea. { ¦kwä·drə,jü·gə'tór·ə ,dē }

quadruped [VERTEBRATE ZOOLOGY] An animal that has four legs. { 'kwä·drə,ped }

quagmire [ECOLOGY] *See* bog. { 'kwäg,mīr }

quail [VERTEBRATE ZOOLOGY] Any of several migratory game birds in the family Phasianidae. { kwāl }

quantasome [CELL BIOLOGY] One of the highly ordered array of units that has a "cobblestone" appearance in electron micrographs of the lamella of chloroplasts, and thought to be the most probable site of the light reaction in photosynthesis. { 'kwän·tə,sōm }

quantitative genetics [GENETICS] The study of continuously varying traits, such as those of the intellect and personality, which cannot be categorized as dichotomies. { 'kwän·ə·tād·iv jə 'ned·iks }

quantitative inheritance [GENETICS] The acquisition of characteristics, such as height, weight, and intelligence, which show a quantitative and continuous type of variation. { 'kwän·ə·tād·iv in 'her·əd·əns }

quantitative structure-activity relationships [BIOCHEMISTRY] The establishment of statistical correlations between the potencies of a series of structurally related compounds and one or more quantitative structural parameters, such as lipophilicity, polarity, and molecular size, by using multilinear regression analysis. { 'kwän·ə ,tād·iv ¦strək·chər ak¦tiv·əd·ē ri'lā·shən ,ships }

quantum evolution [EVOLUTION] A special but extreme case of phyletic evolution; the rapid evolution that takes place when relatively sudden and drastic change occurs in the environment or when organisms spread into new habitats where conditions differ from those to which they are adapted; the organisms must then adapt quickly to the new conditions if they are to survive. { 'kwän·təm ,ev·ə'lü·shən }

quebracho [BOTANY] Any of a number of South American trees in different genera in the order Sapindales, but all being a valuable source of wood, bark, and tannin. { kā'bra·chō }

quebracho bark [BOTANY] Bark of the white quebracho tree *Aspidosperma quebracho* of Chile and Argentina; main components are aspidospermine, tannin, and quebrachine; used in medicine and tanning. { kā'bra·chō ,bärk }

queen [INVERTEBRATE ZOOLOGY] A mature, fertile female in a colony of ants, bees, or termites, whose function is to lay eggs. { kwēn }

quellung reaction [MICROBIOLOGY] Swelling of the capsule of a bacterial cell, caused by contact with serum containing antibodies capable of reacting with polysaccharide material in the capsule; applicable to *Pneumococcus*, *Klebsiella*, and *Hemophilus*. { 'kwel·əŋ rē,ak·shən }

quenching [IMMUNOLOGY] An adaptation of immunofluorescence that uses two fluorochromes, one of which absorbs light emitted by the other; one fluorochrome labels that antigen, another the antibody, and the antigen-antibody complexes retain both; the initially emitted light is absorbed and so quenched by the second compound. { 'kwench·iŋ }

quercetin [BIOCHEMISTRY] $C_{15}H_5O_2(OH)_5$ A yellow, crystalline flavonol obtained from oak bark and Douglas-fir bark; used as an antioxidant and absorber of ultraviolet rays, and in rubber, plastics, and vegetable oils. { 'kwer·sə·tən }

quill [VERTEBRATE ZOOLOGY] The hollow, horny shaft of a large stiff wing or tail feather. { kwil }

quillwort [BOTANY] The common name for plants of the genus *Isoetes*. { 'kwil,wórt }

quince [BOTANY] *Cydonia oblonga*. A deciduous tree of the order Rosales characterized by crooked branching, leaves that are densely hairy on the underside and solitary white or pale-pink flowers; fruit is an edible pear- or apple-shaped tomentose pome. { kwins }

quinoa [BOTANY] *Chenopodium quinoa*. An annual herb of the family Chenopodiaceae grown at high altitudes in South America for the highly nutritious seeds. { kwi'nō·ə }

quinoprotein [BIOCHEMISTRY] A member of a class of proteins that uses pyrroloquinoline quinone as a cofactor. { ˌkwin·ə'prōˌtēn }

quinquefoliate [BOTANY] Of a leaf, having five leaflets. { ˌkwin·kə'fō·lēˌāt }

quintuplet [BIOLOGY] One of five children who have been born at one birth. { kwin'təp·lət }

R

rabbit [VERTEBRATE ZOOLOGY] Any of a large number of burrowing mammals in the family Leporidae. { 'rab·ət }

raccoon [VERTEBRATE ZOOLOGY] Any of 16 species of carnivorous nocturnal mammals belonging to the family Procyonidae; all are arboreal or semiarboreal and have a bushy, long ringed tail. { ra'kün }

race [BIOLOGY] **1.** An infraspecific taxonomic group of organisms, such as subspecies or microspecies. **2.** A fixed variety or breed. { rās }

racemase [BIOCHEMISTRY] Any of group of enzymes that catalyze racemization reactions. { 'ras·ə,mās }

raceme [BOTANY] An inflorescence on which flowers are borne on stalks of equal length on an unbranched main stalk that continues to grow during flowering. { rā'sēm }

racemose [ANATOMY] Of a gland, compound and shaped like a bunch of grapes, with freely branching ducts that terminate in acini. [BOTANY] Bearing, or occurring in the form of, a raceme. { 'ras·ə,mōs }

rachiglossate radula [INVERTEBRATE ZOOLOGY] A radula of certain gastropod mollusks which has one or three longitudinal series of teeth, each of which may bear many cusps. { ¦rā·kə'glä,sāt 'raj·ə·lə }

rachilla [BOTANY] The axis of a grass spikelet. { rə 'kil·ə }

rachis [ANATOMY] The vertebral column. [BIOLOGY] An axial structure such as the axis of an inflorescence, the central petiole of a compound leaf, or the central cord of an ovary in Nematoda. { 'rā·kəs }

radial artery [ANATOMY] A branch of the brachial artery in the forearm; principal branches are the radial recurrent and the main artery of the thumb. { 'rād·ē·əl 'ärd·ə·rē }

radial canal [INVERTEBRATE ZOOLOGY] **1.** One of the numerous canals that radiate from the spongocoel in certain Porifera. **2.** Any of the canals extending from the coelenteron to the circular canal in the margin of the umbrella in jellyfishes. **3.** A canal radiating from the circumoral canal along each ambulacral area in many echinoderms. { 'rād·ē·əl kə'nal }

radial cleavage [EMBRYOLOGY] A cleavage pattern characterized by formation of a mass of cells that show radial symmetry. { 'rād·ē·əl 'klē·vij }

radial deviation [BIOPHYSICS] A position of the human hand in which the wrist is bent toward the thumb. { ¦rād·ē·əl ,dē,·vē'ā·shən }

radial nerve [ANATOMY] A large nerve that arises in the brachial plexus and branches to enervate the extensor muscles and skin of the posterior aspect of the arm, forearm, and hand. { 'rād·ē·əl 'nərv }

Radiata [INVERTEBRATE ZOOLOGY] Members of the Eumetazoa which have a primary radial symmetry; includes the Cnidaria and Ctenophora. { ,rād·ē'äd·ə }

radiation biochemistry [BIOCHEMISTRY] The study of the response of the constituents of living matter to radiation. { ,rād·ē'ā·shən ¦bī·ō 'kem·ə·strē }

radiation biology [BIOLOGY] *See* radiobiology. { ,rād·ē'ā·shən bī'äl·ə·jē }

radiation biophysics [BIOPHYSICS] The study of the response of organisms to ionizing radiations and to ultraviolet light. { ,rād·ē'ā·shən ¦bī·ō 'fiz·iks }

radiation cytology [CELL BIOLOGY] An aspect of biology that deals with the effects of radiations on living cells. { ,rād·ē'ā·shən sī'täl·ə·jē }

radiation effects [BIOLOGY] The harmful effects of ionizing radiation on humans and other animals, such as production of cancers, cataracts, and radiation ulcers, loss of hair, reddening of skin, sterilization, nausea, vomiting, mucous or bloody diarrhea, purpura, epilation, and agranulocytic infections. { ,rād·ē'ā·shən i,feks }

radiation genetics [GENETICS] The study of the genetic effects of radiation. Also known as radiogenetics. { ,rād·ē'ā·shən jə,ned·iks }

radiation microbiology [MICROBIOLOGY] A field of basic and applied radiobiology concerned chiefly with the damaging effects of radiation on microorganisms. { ,rād·ē'ā·shən ¦mī·krō·bī'äl·ə·jē }

radical [BOTANY] **1.** Of, pertaining to, or proceeding from the root. **2.** Arising from the base of a stem or from an underground stem. { 'rad·ə·kəl }

radicle [BOTANY] The embryonic root of a flowering plant. { 'rad·ə·kəl }

radiobiology [BIOLOGY] Study of the scientific principles, mechanisms, and effects of the interaction of ionizing radiation with living matter. Also known as radiation biology. { 'rād·ē·ō·bī 'äl·ə·jē }

radioecology [ECOLOGY] The interdisciplinary

study of organisms, radionuclides, ionizing radiation, and the environment. { ¦rād·ē·ō·ē'käl·ə·jē }

radiogenetics [GENETICS] *See* radiation genetics. { ¦rād·ē·ō·jə'ned·iks }

radiohumeral joint [ANATOMY] The joint in the elbow between the radius and the humerus bones. { ¦rād·ē·ō'hyüm·ə·rəl 'jȯint }

radioimmunoassay [IMMUNOLOGY] A sensitive method for determining the concentration of an antigenic substance in a sample by comparing its inhibitory effect on the binding of a radioactivity-labeled antigen to a limited amount of a specific antibody with the inhibitory effect of known standards. { ¦rād·ē·ō¦im·yə·nō'a,sā }

Radiolaria [INVERTEBRATE ZOOLOGY] A subclass of the protozoan class Actinopodea whose members are noted for their siliceous skeletons and characterized by a membranous capsule which separates the outer from the inner cytoplasm. { ,rād·ē·ō'lar·ē·ə }

radiole [INVERTEBRATE ZOOLOGY] A spine on a sea urchin. { 'rād·ē·ōl }

radiomimetic activity [BIOLOGY] The radiation-like effects of certain chemicals, such as nitrogen mustard, urethane, and fluorinated pyrimidines. { ¦rād·ē·ō·mi'med·ik ak'tiv·əd·ē }

radiomutation [GENETICS] A mutation which is the result of exposure of living-tissue chromosomes to ionizing radiation. { ¦rād·ē·ō·myü'tā·shən }

radioresistance [BIOLOGY] The resistance of organisms or tissues to the harmful effects of various radiations. { ¦rād·ē·ō·ri'zis·təns }

radish [BOTANY] *Raphanus sativus*. **1.** An annual or biennial crucifer belonging to the order Capparales. **2.** The edible, thickened hypocotyl of the plant. { 'rad·ish }

radius [ANATOMY] The outer of the two bones of the human forearm or of the corresponding part in vertebrates other than fish. { 'rād·ē·əs }

radula [INVERTEBRATE ZOOLOGY] A filelike ribbon studded with horny or chitinous toothlike structures, found in the mouth of all classes of mollusks except Bivalvia. { 'raj·ə·lə }

raffinase [BIOCHEMISTRY] An enzyme that hydrolyzes raffinose, yielding fructose in the reaction. { 'raf·ə,nās }

raffinose [BIOCHEMISTRY] $C_{18}H_{32}O_{16} \cdot 5H_2O$ A white, crystalline trisaccharide found in sugar-beets, cottonseed meal, and molasses; yields glucose, fructose, and galactose on complete hydrolysis. Also known as gossypose; melitose; melitriose. { 'raf·ə,nōs }

Rafflesiales [BOTANY] A small order of dicotyledonous plants; members are highly specialized, nongreen, rootless parasites which grow from the roots of the host. { re,flē·zhē'ā·lēz }

Raillietiellidae [INVERTEBRATE ZOOLOGY] A small family of parasitic arthropods in the order Cephalobaenida. { ,rāl·yə'tyel·ə,dē }

rain desert [ECOLOGY] A desert in which rainfall is sufficient to maintain a sparse general vegetation. { 'rān ,dez·ərt }

Rainey's corpuscle [INVERTEBRATE ZOOLOGY] The sickle-shaped spore of an encysted sarcosporidian. { 'rā·nēz ,kȯr·pə·səl }

rainforest [ECOLOGY] A forest of broad-leaved, mainly evergreen, trees found in continually moist climates in the tropics, subtropics, and some parts of the temperate zones. { 'rān,fär·əst }

raised bog [ECOLOGY] An area of acid, peaty soil, especially that developed from moss, in which the center is relatively higher than the margins. { 'rāzd 'bäg }

Rajidae [VERTEBRATE ZOOLOGY] The skates, a family of elasmobranchs included in the batoid group. { 'raj·ə,dē }

Rajiformes [VERTEBRATE ZOOLOGY] The equivalent name for Batoidea. { ,raj·ə'fȯr,mēz }

Rallidae [VERTEBRATE ZOOLOGY] A large family of birds in the order Gruiformes comprising rails, gallinules, and coots. { 'ral·ə,dē }

ram [VERTEBRATE ZOOLOGY] A male sheep or goat. { ram }

Ramapithecinae [PALEONTOLOGY] A subfamily of Hominidae including the protohominids of the Miocene and Pliocene. { ,räm·ə·pə'thes·ən,ē }

Ramapithecus [PALEONTOLOGY] The genus name given to a fossilized upper jaw fragment found in the Siwalik hills, India; closely related to the human family. { ,räm·ə'pith·ə·kəs }

ramate [BIOLOGY] Having branches. { 'ra,māt }

ramentum [BOTANY] A thin brownish scale consisting of a single layer of cells and occurring on the leaves and young shoots of many ferns. { rə 'men·təm }

ramicolous [BOTANY] Living on twigs. { rə'mik·ə·ləs }

ramie [BOTANY] *Boehmeria nivea*. A shrub or half-shrub of the nettle family (Urticaceae) cultivated as a source of a tough, strong, durable, lustrous natural woody fiber resembling flax, obtained from the phloem of the plant; used for high-quality papers and fabrics. Also known as China grass; rhea. { 'rā·mē }

Ramon flocculation test [IMMUNOLOGY] A method of standardizing antitoxins; a toxin-antitoxin flocculation that is a precipitin reaction in which the end point is the zone of optimal proportion; that is, the zone in which there is no uncombined antigen or antibody. { rə'mōn ,fläk·yə'lā·shən ,test }

ramose [BIOLOGY] Having lateral divisions or branches. { 'rā,mōs }

Ramphastidae [VERTEBRATE ZOOLOGY] The toucans, a family of birds with large, often colorful bills in the order Piciformes. { ,ram'fas·tə,dē }

ramus [ANATOMY] A slender bone process branching from a large bone. [VERTEBRATE ZOOLOGY] The barb of a feather. [ZOOLOGY] The branch of a structure such as a blood vessel, nerve, arthropod appendage, and so on. { 'rā·məs }

random mating [GENETICS] A mating system in which there is an equal opportunity for all male and female geametes to join in fertilization. { 'ran·dəm ,mād·iŋ }

range [ECOLOGY] The area or region over which a species is distributed. { rānj }

range of motion [BIOPHYSICS] The degree of movement that can occur in a joint. { 'rānj əv 'mō·shən }

Ranidae [VERTEBRATE ZOOLOGY] A family of frogs in the suborder Diplasiocoela including the large, widespread genus *Rana*. { 'ran·ə,dē }

Ranunculaceae [BOTANY] A family of dicotyledonous herbs in the order Ranunculales distinguished by alternate leaves with net venation, two or more distinct carpels, and numerous stamens. { rə,nəŋ·kyə'lās·ē,ē }

Ranunculales [BOTANY] An order of dicotyledons in the subclass Magnoliidae characterized by its mostly separate carpels, triaperturate pollen, herbaceous or only secondarily woody habit, and frequently numerous stamens. { rə,nəŋ·kyə'lā·lēz }

rape [BOTANY] *Brassica napus.* A plant of the cabbage family in the order Capparales; the plant does not form a compact head, the leaves are bluish-green, deeply lobed, and curled, and the small flowers produce black seeds; grown for forage. { rāp }

raphe [ANATOMY] A broad seamlike junction between two lateral halves of an organ or other body part. [BOTANY] **1.** The part of the funiculus attached along its full length to the integument of an anatropous ovule, between the chalaza and the attachment to the placenta. **2.** The longitudinal median line or slit on a diatom valve. { 'rā·fē }

Raphidae [VERTEBRATE ZOOLOGY] A family of birds in the order Columbiformes that included the dodo (*Raphus calcullatus*); completely extirpated during the 17th and early 18th centuries. { 'raf·ə,dē }

raphide [BOTANY] One of the long, needle-shaped crystals, usually consisting of calcium oxalate, occurring as a metabolic by-product in certain plant cells. { 'rāf,īd }

raptorial [ZOOLOGY] **1.** Living on prey. **2.** Adapted for snatching or seizing prey, as birds of prey. { rap'tȯr·ē·əl }

rasorial [ZOOLOGY] Adapted for scratching for food; applied to birds. { rə'sȯr·ē·əl }

raspberry [BOTANY] Any of several species of upright shrubs of the genus *Rubus*, with perennial roots and prickly biennial stems, in the order Rosales; the edible black or red juicy berries are aggregate fruits, and when ripe they are easily separated from the fleshy receptacle. { 'raz ,ber·ē }

rat [VERTEBRATE ZOOLOGY] The name applied to over 650 species of mammals in several families of the order Rodentia; they differ from mice in being larger and in having teeth modified for gnawing. { rat }

rataria larva [INVERTEBRATE ZOOLOGY] The second, hourglass-shaped, free-swimming larva of the siphonophore *Velella*. { rə'tar·ē·ə ,lär·və }

ratfish [VERTEBRATE ZOOLOGY] The common name for members of the chondrichthyan order Chimaeriformes. { 'rat,fish }

Rathke's pouch [EMBRYOLOGY] *See* craniobuccal pouch. { 'rät,kēz ,paùch }

ratites [VERTEBRATE ZOOLOGY] A group of flightless, mostly large, running birds comprising several orders and including the emus, cassowaries, kiwis, and ostriches. { 'ra,tīts }

rattan [BOTANY] Any of several long-stemmed, climbing palms, especially of the genera *Calanius* and *Daemonothops*; stem material is used to make walking sticks, wickerwork, and cordage. { ra 'tan }

rattlesnake [VERTEBRATE ZOOLOGY] Any of a number of species of the genera *Sistrurus* or *Crotalus* distinguished by the characteristic rattle on the end of the tail. { 'rad·əl,snāk }

Raunkiaer system [BOTANY] A classification system for plant life-forms based on the position of perennating buds in relation to the soil surface. { 'raùn·kē·ir ,sis·təm }

Rauwolfia [BOTANY] A genus of mostly poisonous, tropical trees and shrubs of the dogbane family (Apocynaceae); certain species yield substances used as emetics and cathartics, while *R. serpentina* is a source of alkaloids used as tranquilizers. { raù'wùl·fē·ə }

ray [VERTEBRATE ZOOLOGY] Any of about 350 species of the elasmobranch order Batoidea having flattened bodies with large pectoral fins attached to the side of the head, ventral gill slits, and long, spikelike tails. { rā }

rayfin fish [VERTEBRATE ZOOLOGY] The common name for members of the Actinopterygii. { 'rā,fin ,fish }

ray flower [BOTANY] One of the small flowers with a strap-shaped corolla radiating from the margin of the head of a capitulum. { 'rā ,flaù·ər }

ray initial [BOTANY] One of the cells of the cambium which divide to produce new phloem and xylem ray cells. { 'rā i,nish·əl }

ray shake [BOTANY] A radial crack in wood caused by wounds in a tree along the barrier zone. { 'rā ,shāk }

reaction time [PHYSIOLOGY] The interval between application of a stimulus and the beginning of the response. { rē'ak·shən ,tīm }

reaction wood [BOTANY] An abnormal development of a tree and therefore its wood as the result of unusual forces acting on it, such as an atypical gravitational pull. { rē'ak·shən ,wùd }

reading [MOLECULAR BIOLOGY] A linear process by which amino acid sequences are recognized by the protein-synthesizing system of a cell from messenger ribonucleic codes. { 'rēd·iŋ }

reading frame [MOLECULAR BIOLOGY] A nucleotide sequence that starts with an initiation codon, partitions the subsequent nucleotides into a series of amino acid-encoding triplets, and ends with a termination codon. { 'rēd·iŋ ,frām }

reading mistake [MOLECULAR BIOLOGY] The incorrect placement of one or more amino acid residues in a polypeptide chain during genetic translation. { 'rēd·iŋ mə'stāk }

readthrough [GENETICS] Transcription beyond a termination sequence due to failure of ribonucleic acid polymerase to recognize the termination codon. { 'rēd,thrü }

reagin [IMMUNOLOGY] **1.** An antibody which occurs in human atopy, such as hay fever and asthma, and which readily sensitizes the skin. **2.** An antibody which reacts in various serologic tests for syphilis. { rē'ā·jən }

reassortant virus [VIROLOGY] A virion containing deoxyribonucleic acid from one virus species and a protein coat from another. { ˌrē·ə¦sȯrt·ənt 'vī·rəs }

recapitulation theory [BIOLOGY] The biological theory that an organism passes through developmental stages resembling various stages in the phylogeny of its group; ontogeny recapitulates phylogeny. Also known as biogenetic law; Haeckel's law. { ˌrē·kəˌpich·ə'lā·shən ˌthē·ə·rē }

recent [EVOLUTION] Referring to taxa which still exist; the antonym of fossil. { 'rē·sənt }

receptacle [BOTANY] The pointed end of a pedicel or peduncle from which the flower parts grow. { ri'sep·tə·kəl }

receptor [BIOCHEMISTRY] A site or structure in a cell which combines with a drug or other biological to produce a specific alteration of cell function. [PHYSIOLOGY] A sense organ. { ri'sep·tər }

recessive [GENETICS] **1.** An allele that is not expressed phenotypically when present in the heterozygous condition. **2.** An organism homozygous for a recessive gene. { ri'ses·iv }

reciprocal inhibition [PHYSIOLOGY] In muscular movement, the simultaneous relaxation of one muscle and the contraction of its antagonist. { ri'sip·rə·kəl ˌin·ə'bish·ən }

reciprocal recombination [GENETICS] In dihybrid gametes, the generation of linkage arrangements unlike those of the maternal and paternal homologues. { ri'sip·rə·kəl rēˌkäm·bə'nā·shən }

reciprocal translocation [CELL BIOLOGY] The special case of translocation in which two segments exchange positions. { ri'sip·rə·kəl ˌtranz·lō'kā·shən }

reclinate [BOTANY] Vernation in which the upper part of the leaf is bent down on the lower part. { 'rek·ləˌnāt }

recognition sequence [MOLECULAR BIOLOGY] A specific sequence of nucleotides at which a restriction endonuclease cleaves a deoxyribonucleic acid molecule. { ˌrek·ig'nish·ən ˌsē·kwəns }

recognition site [MOLECULAR BIOLOGY] The nucleotide sequence in duplex deoxyribonucleic acid (DNA) to which a restriction endonuclease binds initially and within which the endonuclease cuts the DNA. { ˌrek·ig'nish·ən ˌsīt }

recombinant [GENETICS] Any new cell, individual, or molecule that is produced in the laboratory by recombinant deoxyribonucleic acid technology or that arises naturally as a result of recombination. { rē'käm·bə·nənt }

recombinant technology [GENETICS] **1.** In genetic engineering, laboratory techniques used to join deoxyribonucleic acid (DNA) from different sources to produce novel DNA. Also known as gene splicing. **2.** In genetic engineering, laboratory techniques used to join ribonucleic acid (RNA) from different sources to produce novel RNA. { ri¦käm·bə·nənt tek'näl·ə·jē }

recombination [GENETICS] **1.** The occurrence of gene combinations in the progeny that differ from those of the parents as a result of independent assortment, linkage, and crossing-over. **2.** The production of genetic information in which there are elements of one line of descent replaced by those of another line, or additional elements. { ˌrēˌkäm·bə'nā·shən }

recombination frequency [GENETICS] The number of recombinants divided by the total number of progeny. { rē'käm·bə'nā·shən ˌfrē·kwən·sē }

recombination mosaic [GENETICS] A mosaic produced as the result of somatic crossing-over. { ˌrēˌkäm·bə'nā·shən mōˌzā·ik }

recombination repair [MOLECULAR BIOLOGY] A repair mechanism involving exchange of correct for incorrect segments between two damaged deoxyribonucleic acid molecules. { rēˌkäm·bə'nā·shən riˌper }

recombinator [MOLECULAR BIOLOGY] Any nucleotide sequence that stimulates genetic recombination at neighboring sites. { rē'käm·bəˌnād·ər }

recon [GENETICS] The smallest deoxyribonucleic acid unit capable of recombination. { 'rēˌkän }

recruitment [PHYSIOLOGY] A serial discharge from neurons innervating groups of muscle fibers. { ri'krüt·mənt }

Recticornia [INVERTEBRATE ZOOLOGY] A family of amphipod crustaceans in the superfamily Genuina containing forms in which the first antennae are straight, arise from the anterior margin of the head, and have few flagellar segments. { ˌrek·tə'kȯr·nē·ə }

rectrix [VERTEBRATE ZOOLOGY] One of the stiff tail feathers used by birds to control direction of flight. { 'rek·triks }

rectum [ANATOMY] The portion of the large intestine between the sigmoid flexure and the anus. { 'rek·təm }

rectus [ANATOMY] Having a straight course, as certain muscles. { 'rek·təs }

rectus abdominis [ANATOMY] The long flat muscle of the anterior abdominal wall which, as vertical fibers, arises from the pubic crest and symphysis, and is inserted into the cartilages of the fifth, sixth, and seventh ribs. { 'rek·təs ab'däm·ə·nəs }

rectus femoris [ANATOMY] A division of the quadriceps femoris inserting in the patella and ultimately into the tubercle of the tibia. { 'rek·təs 'fem·ə·rəs }

rectus oculi [ANATOMY] Any of four muscles (superior, inferior, lateral, and medial) of the eyeball, running forward from the optic foramen and inserted into the sclerotic coat. { 'rek·təs 'äk·yə·lī }

recumbent [BOTANY] Of or pertaining to a plant or plant part that tends to rest on the surface of the soil. { ri'kəm·bənt }

recurrent backcrossing [GENETICS] Repetitive sexual crossing of hybrids to one parent, used to eliminate all but the desired characteristics of the donor parent. { ri'kər·ənt 'bakˌkrȯs·iŋ }

recurrent parent [GENETICS] In recurrent backcrossing, the parent that is crossed with the first and the subsequent generations. Also known as backcross parent. { ri'kər·ənt 'per·ənt }

red algae [BOTANY] The common name for members of the phylum Rhodophyta. { 'red 'al·jē }

red blood cell [HISTOLOGY] *See* erythrocyte. { 'red 'bləd ,sel }

redia [INVERTEBRATE ZOOLOGY] A larva produced within the miracidial sporocyst of certain digenetic trematodes which may give rise to daughter rediae or to cercariae. { 'rē·dē·ə }

redifferentiation [PHYSIOLOGY] The return to a position of greater specialization in actual and potential functions, or the developing of new characteristics. { rē,dif·ə,ren·chē'ā·shən }

red nucleus [HISTOLOGY] A mass of reticular fibers in the gray matter of the tegmentum of the mesencephalon of higher vertebrates; it receives fibers from the cerebellum of the opposite side and gives rise to rubrospinal tract fibers of the opposite side. { 'red 'nü·klē·əs }

red tide [BIOLOGY] A reddish discoloration of coastal surface waters due to concentrations of certain toxin-producing dinoflagellates. Also known as red water. { 'red 'tīd }

redtop grass [BOTANY] One of the bent grasses, *Agrostis alba* and its relatives, which grow on a wide variety of soils; it is a perennial, spreads slowly by rootstocks, and has top growth 2-3 feet (60-90 centimeters) tall. { 'red,täp ,gras }

reducer [BIOLOGY] *See* decomposer. { ri'dü·sər }

redundancy [GENETICS] **1.** Repetition of a specified deoxyribonucleic acid sequence in a nucleus. **2.** Multiplicity of codons for individual amino acids. { ri'dən·dən·sē }

Reduviidae [INVERTEBRATE ZOOLOGY] The single family of the hemipteran group Reduvioidea; nearly all have a stridulatory furrow on the prosternum, ocelli are generally present, and the beak is three-segmented. { ,rej·ə'vī·ə,dē }

Reduvioidea [INVERTEBRATE ZOOLOGY] The assassin bugs or conenose bugs, a monofamilial group of hemipteran insects in the subdivision Geocorisae. { ,rej·ə,vē'òid·ē·ə }

red water [BIOLOGY] *See* red tide. { 'red 'wòd·ər }

redwood [BOTANY] *Sequoia sempervirens.* An evergreen tree of the pine family; it is the tallest tree in the Americas, attaining 350 feet (107 meters); its soft heartwood is a valuable building material. { 'red,wùd }

reed [BOTANY] Any tall grass characterized by a slender jointed stem. { rēd }

reflex [PHYSIOLOGY] An automatic response mediated by the nervous system. { 'rē,fleks }

reflex arc [ANATOMY] A chain of neurons composing the anatomical substrate or pathway of the unconditioned reflex. { 'rē,fleks ,ärk }

reflexed [BOTANY] Turned abruptly backward. { 'rē,flekst }

reflex time [PHYSIOLOGY] The time required for the nerve impulse to travel in a reflex action. { 'rē,fleks ,tīm }

refractory period [PHYSIOLOGY] A brief period of time following the stimulation of a nerve during which the nerve will not respond to a second stimulus. { ri'frak·trē ,pir·ē·əd }

refugium [ECOLOGY] An area which has escaped the great changes which occurred in the region as a whole, often providing conditions in which relic colonies can survive; for example, a driftless area which has escaped the effects of glaciation because it projected above the ice. { rə'fyü·jē·əm }

regeneration [BIOLOGY] The replacement by an organism of tissues or organs which have been lost or severely injured. { rē,jen·ə'rā·shən }

regional anatomy [ANATOMY] The detailed study of the anatomy of a part or region of the body of an animal. { 'rēj·ən·əl ə'nad·ə·mē }

regular [BOTANY] Having radial symmetry, referring to a flower. { 'reg·yə·lər }

regular connective tissue [HISTOLOGY] Connective tissue in which the fibers are arranged in definite patterns. { 'reg·yə·lər kə'nek·tiv 'tish·ü }

Regularia [INVERTEBRATE ZOOLOGY] An assemblage of echinoids in which the anus and periproct lie within the apical system; not considered a valid taxon. { ,reg·yə'lar·ē·ə }

regulative egg [EMBRYOLOGY] An egg in which unfertilized fragments develop as complete, normal individuals. { 'reg·yə,lād·iv 'eg }

regulator gene [GENETICS] A gene that controls the rate of transcription of one or more other genes. { 'reg·yə,lād·ər ,jēn }

regulon [GENETICS] In bacteria, a system of genes, formed by one or more operons, which regulate enzyme induction and whose activity is controlled by a single repressor substance. { 'reg·yə,län }

regurgitation [PHYSIOLOGY] Bringing back into the mouth undigested food from the stomach. { ri,gər·jə'tā·shən }

Reichert's cartilage [EMBRYOLOGY] The cartilage of the hyoid arch in a human embryo. { 'rī·kərts ,kärt·lij }

Reighardiidae [INVERTEBRATE ZOOLOGY] A monotypic family of arthropods in the order Cephalobaenida; the posterior end of the organism is rounded, without lobes, and the cuticula is covered with minute spines. { ,rī·gär'dī·ə,dē }

reindeer [VERTEBRATE ZOOLOGY] *Rangifer tarandus.* A migratory ruminant of the deer family (Cervidae) which inhabits the Arctic region and has a circumpolar distribution; characteristically, both sexes have antlers and are brown with yellow-white areas on the neck and chest. { 'rān,dir }

Reissner's membrane [ANATOMY] The anterior wall of the cochlear duct, which separates the cochlear duct from the scala vestibuli. Also known as vestibular membrane of Reissner. { 'rīs·nərz ,mem,brān }

rejection [IMMUNOLOGY] Destruction of a graft by the immune system of the recipient. { ri'jek·shən }

relative refractory period [PHYSIOLOGY] A period of a few milliseconds following the absolute refractory period during which the excitation threshold of neural tissue is raised and a stronger-than-normal stimulus is required to initiate an action potential. { 'rel·əd·iv ri'frak·trē ˌpir·ē·əd }

relaxed circular deoxyribonucleic acid [BIOCHEMISTRY] A form of circular deoxyribonucleic acid in which the circle of one strand is broken. Also known as open-circle deoxyribonucleic acid. { ri'lakst 'sər·kyə·lər dē¦äk·sēˌrī·bō·nü'klē·ik 'as·əd }

relaxin [BIOCHEMISTRY] A hormone found in the serum of humans and certain other animals during pregnancy; probably acting with progesterone and estrogen, it causes relaxation of pelvic ligaments in the guinea pig. { ri'lak·sən }

release factor [MOLECULAR BIOLOGY] Any protein that responds to termination codons in messenger ribonucleic acid and causes the release of the finished polypeptide. { ri'lēs ˌfak·tər }

releaser stimulus [ZOOLOGY] A stimulus which affects an animal by initiating an instinctual behavior pattern. { ri'lēs·ər ˌstim·yə·ləs }

relict [BIOLOGY] A persistent, isolated remnant of a once-abundant species. { 'rel·ikt }

Remak's ganglion [ANATOMY] A ganglion near the junction of the coronary sinus and the right atrium. { 'rāˌmäks ˌgaŋ·glēˌän }

remex [VERTEBRATE ZOOLOGY] *See* flight feather. { 'rēˌmeks }

remineralization [PHYSIOLOGY] The continual reforming of tooth mineral that occurs at the surface of teeth, chiefly from constituents of saliva. { rēˌmin·rə·lə'zā·shən }

renal artery [ANATOMY] A branch of the abdominal or ventral aorta supplying the kidneys in vertebrates. { 'rēn·əl 'ärd·ə·rē }

renal corpuscle [ANATOMY] The glomerulus together with its Bowman's capsule in the renal cortex. Also known as Malpighian corpuscle. { 'rēn·əl 'kȯr·pə·səl }

renal papilla [ANATOMY] A fingerlike projection into the renal pelvis through which the collecting tubules discharge. { 'rēn·əl pə'pil·ə }

renal pyramid [ANATOMY] Any of the conical masses composing the medullary substance of the kidney. Also known as Malpighian pyramid. { 'rēn·əl 'pir·ə·məd }

renal threshold [PHYSIOLOGY] A concentration of a substance within the blood which, when reached, causes the substance to appear in the urine. { 'rēn·əl 'threshˌhōld }

renal tubule [ANATOMY] One of the glandular tubules which elaborate urine in the kidneys. { 'rēn·əl 'tü·byül }

renal vein [ANATOMY] A vein which returns blood from the kidney to the vena cava. { 'rēn·əl 'vān }

renaturation [BIOCHEMISTRY] The process of restoring denatured proteins to their original condition. { rēˌnach·ə'rā·shən }

renette [INVERTEBRATE ZOOLOGY] An excretory cell found in certain nematodes. { re'net }

renin [BIOCHEMISTRY] A proteolytic enzyme produced in the afferent glomerular arteriole which reacts with the plasma component hypertensinogen to produce angiotensin II. { 'rēn·ən }

rennet [VERTEBRATE ZOOLOGY] The lining of the stomach of certain animals, especially the fourth stomach in ruminants. { 'ren·ət }

rennin [BIOCHEMISTRY] An enzyme found in the gastric juice of the fourth stomach of calfs; used for coagulating milk casein in cheesemaking. Also known as chymosin. { 'ren·ən }

renninogen [BIOCHEMISTRY] The zymogen of rennin. Also known as prorennin. { rə'nin·ə·jən }

reovirus [VIROLOGY] A group of ribonucleic acid-containing animal viruses, including agents of encephalitis and phlebotomus fever. { 'rē·ōˌvī·rəs }

repair synthesis [MOLECULAR BIOLOGY] Enzymatic excision and replacement of regions of damaged deoxyribonucleic acid, as in repair of thymine dimers by ultraviolet irradiation. { ri'per ˌsin·thə·səs }

repand [BOTANY] Having a margin that undulates slightly, referring to a leaf. { rə'pand }

repent [BOTANY] Of a stem, creeping along the ground and rooting at the nodes. { 'rē·pent }

repetition frequency [MOLECULAR BIOLOGY] The number of copies of a given nucleotide sequence present in the haploid genome. { ˌrep·ə'tish·ən ˌfrē·kwən·sē }

repetitious deoxyribonucleic acid [MOLECULAR BIOLOGY] Nucleotide sequences occurring repeatedly in chromosomal deoxyribonucleic acid. { rep·ə¦tish·əs dēˌäk·sē¦rī·bō·nüˌklē·ik 'as·əd }

replacement [PALEONTOLOGY] Substitution of inorganic matter for the original organic constituents of an organism during fossilization. { ri 'plās·mənt }

replica plating [MICROBIOLOGY] A method for the isolation of nutritional mutants in microorganisms; colonies are grown from a microorganism suspension previously exposed to a mutagenic agent, on a complete medium in a petri dish; a velour surface is used to transfer the impression of all these colonies to a petri dish containing a minimal medium; colonies that do not grow on the minimal medium are the mutants. { 'rep·lə·kə ˌplād·iŋ }

replicating fork [MOLECULAR BIOLOGY] The Y-shaped region of a chromosome that is a growing point in the replication of deoxyribonucleic acid. { 'rep·ləˌkād·iŋ ¦fȯrk }

replication [MOLECULAR BIOLOGY] Duplication, as of a nucleic acid, by copying from a molecular template. [VIROLOGY] Multiplication of phage in a bacterial cell. { ˌrep·lə'kā·shən }

replication bubble [MOLECULAR BIOLOGY] *See* replication eye. { ˌrep·lə'kā·shən ˌbəb·əl }

replication eye [MOLECULAR BIOLOGY] A replicated region of deoxyribonucleic acid contained within a longer, unreplicated region and presented in the shape of an eye. Also known as replication bubble. { ˌrep·lə'kā·shən ī }

replicon [GENETICS] A genetic element characterized by possession of the structural gene that controls a specific initiator and receptor locus (replicator) for its action. { 'rep·ləˌkän }

replum [BOTANY] A thin wall separating the two valves or chambers of certain fruits. { 'rep·ləm }

reporter gene [BIOCHEMISTRY] A gene coding sequence that is used experimentally to replace a normal sequence, and codes for an enzymatic marker or a protein which confers resistance to an antibiotic or cell toxin. { ri'pórd·ər ˌjēn }

repressing [BIOCHEMISTRY] The termination of enzyme synthesis when the products of the reaction catalyzed by the enzyme reach a critical concentration. [MOLECULAR BIOLOGY] Inhibition of transcription or translation due to binding of a repressor to an operator on a deoxyribonucleic acid molecule or to a specific messenger ribonucleic acid site. { ri 'pres·iŋ }

repression [BIOCHEMISTRY] The termination of enzyme synthesis when the products of the reaction catalyzed by the enzyme reach a critical concentration. [MOLECULAR BIOLOGY] Inhibition of transcription or translation due to binding of a repressor to an operator on a deoxyribonucleic acid molecule or to a specific messenger ribonucleic acid site. { ri'presh·ən }

repressor [BIOCHEMISTRY] An end product of metabolism which represses the synthesis of enzymes in the metabolic pathway. [GENETICS] The product of a regulator gene that acts to repress the transcription of another gene. { ri 'pres·ər }

reproduction [BIOLOGY] The mechanisms by which organisms give rise to other organisms of the same kind. { ¦rē·prə¦dək·shən }

reproductive behavior [ZOOLOGY] The behavior patterns in different types of animals by means of which the sperm is brought to the egg and the parental care of the resulting young insured. { ¦rē·prə¦dək·tiv bi'hā·vyər }

reproductive distribution [ECOLOGY] The range of areas where conditions are favorable to maturation, spawning, and early development of marine animals. { ¦rē·prə¦dək·tiv ˌdis·trə'byü·shən }

reproductive system [ANATOMY] The structures concerned with the production of sex cells and perpetuation of the species. { ¦rē·prə¦dək·tiv ˌsis·təm }

Reptantia [INVERTEBRATE ZOOLOGY] A suborder of the crustacean order Decapoda including all decapods other than shrimp. { rep'tan·chē·ə }

reptile [VERTEBRATE ZOOLOGY] Any member of the class Reptilia. { 'repˌtīl }

Reptilia [VERTEBRATE ZOOLOGY] A class of terrestrial vertebrates composed of turtles, tuatara, lizards, snakes, and crocodileans; characteristically they lack hair, feathers, and mammary glands, the skin is covered with scales, they have a three-chambered heart, and the pleural and peritoneal cavities are continuous. { rep'til·yə }

RES [ANATOMY] *See* reticuloendothelial system.

Resedaceae [BOTANY] A family of dicotyledonous herbs in the order Capparales having irregular, hypogynous flowers. { ˌres·ə'dās·ēˌē }

reserve cell [HISTOLOGY] **1.** One of the small, undifferentiated epithelial cells at the base of the stratified columnar lining of the bronchial tree. **2.** A chromophobe cell. { ri'zərv 'sel }

residual air [PHYSIOLOGY] *See* residual volume. { rə'zij·ə·wəl 'er }

residual volume [PHYSIOLOGY] Air remaining in the lungs after the most complete expiration possible; it is elevated in diffuse obstructive emphysema and during an attack of asthma. Also known as residual air. { rə'zij·ə·wəl 'väl·yəm }

resin duct [BOTANY] A canal (intercellular space) lined with secretory cells that release resins into the canal; common in gymnosperms. { 'rez·ən ˌdəkt }

resistance factor [GENETICS] *See* R factor. { ri 'zis·təns ˌfak·tər }

resistance transfer factor [GENETICS] A carrier of genetic information in bacteria which is considered to control the ability of self-replication and conjugal transfer of R factors. Abbreviated RTF. { ri'zis·təns 'tranz·fər ˌfak·tər }

respiration [PHYSIOLOGY] **1.** The processes by which tissues and organisms exchange gases with their environment. **2.** The act of breathing with the lungs, consisting of inspiration and expiration. { ˌres·pə'rā·shən }

respiratory center [PHYSIOLOGY] A large area of the brain involved in regulation of respiration. { 'res·prəˌtór·ē ¦sen·tər }

respiratory dead space [PHYSIOLOGY] That part of the respiratory system which has no alveoli and in which little or no exchange of gas between air and blood takes place. { 'res·prəˌtór·ē 'ded ˌspās }

respiratory epithelium [HISTOLOGY] The ciliated pseudostratified epithelium lining the respiratory tract. { 'res·prəˌtór·ē ˌep·ə'thē·lē·əm }

respiratory minute volume [PHYSIOLOGY] The total amount of air which moves in and out of the lungs in a minute. { 'res·prəˌtór·ē 'min·ət 'väl·yəm }

respiratory pigment [BIOCHEMISTRY] Any of various conjugated proteins that function in living organisms to transfer oxygen in cellular respiration. { 'res·prəˌtór·ē ¦pig·mənt }

respiratory quotient [PHYSIOLOGY] The ratio of volumes of carbon dioxide evolved and oxygen consumed during a given period of respiration. Abbreviated RQ. { 'res·prəˌtor·ē ¦kwō·shənt }

respiratory syncytial virus [VIROLOGY] An enveloped, single-stranded RNA animal virus belonging to the Paramyxoviridae genus *Pneumovirus*; associated with a large proportion of respiratory illnesses in very young children, particularly bronchiolitis and pneumonia. { 'res·prəˌtór·ē sin'sish·əl 'vī·rəs }

respiratory system [ANATOMY] The structures and passages involved with the intake, expulsion, and exchange of oxygen and carbon dioxide in the vertebrate body. { 'res·prəˌtór·ē ¦sis·təm }

respiratory tree [ANATOMY] The trachea, bronchi, and bronchioles. [INVERTEBRATE ZOOLOGY] Either of a pair of branched tubular appendages of the cloaca in certain holothurians that is thought to have a respiratory function. { 'res·prəˌtor·ē ˌtrē }

restiform body [ANATOMY] *See* inferior cerebellar peduncle. { 'res·təˌfórm ˌbäd·ē }

resting cell [CELL BIOLOGY] An interphase cell. { 'rest·iŋ ˌsel }

resting metabolism [PHYSIOLOGY] The metabolism of a person at rest while seated or standing in a normal position. { 'rest·iŋ me,tab·ə,liz·əm }

resting potential [PHYSIOLOGY] The potential difference between the interior cytoplasm and the external aqueous medium of the living cell. { 'rest·iŋ pə,ten·chəl }

resting spore [BIOLOGY] A spore that remains dormant for long periods before germination, withstanding adverse conditions; usually invested in a thickened cell wall. { 'rest·iŋ ,spór }

Restionaceae [BOTANY] A large family of monocotyledonous plants in the order Restionales characterized by unisexual flowers, wholly cauline leaves, unilocular anthers, and a more or less open inflorescence. { ,res·tē·ə'nās·ē,ē }

Restionales [BOTANY] An order of monocotyledonous plants in the subclass Commelinidae having reduced flowers and a single, pendulous, orthotropous ovule in each of the one to three locules of the ovary. { ,res·tē·ə'nā·lēz }

restoration [ECOLOGY] A conservation measure involving the correction of past abuses that have impaired the productivity of the resources base. { ,res·tə'rā·shən }

restriction [BIOCHEMISTRY] The degradation of foreign deoxyribonucleic acid by restriction endonucleases capable of recognizing particular patterns of specificity. { ri¦strik·shən }

restriction endonuclease [BIOCHEMISTRY] Any of the specific endonucleases that recognizes a short specific sequence within a deoxyribonucleic acid molecule and then catalyzes double-stranded cleavage of that molecule. Also known as endodeoxyribonuclease. { ri'strik·shən ¦en·dō'nü·klē,ās }

restriction fragment [BIOCHEMISTRY] Any of the individual polynucleotide sequences produced by digestion of deoxyribonucleic acid with a restriction endonuclease. { ri'strik·shən ,frag·mənt }

restriction fragment length polymorphism [MOLECULAR BIOLOGY] Variations in the length of restriction fragments resulting from action by a specific endonuclease. { ri¦strik·shən ¦frag·mənt ,leŋkth ,päl·i'mór,fiz·əm }

restriction map [MOLECULAR BIOLOGY] A diagram of a deoxyribonucleic acid molecule showing sites at which restriction endonucleases produce cleavage. { ri'strik·shən ,map }

restriction site [MOLECULAR BIOLOGY] A sequence in a deoxyribonucleic acid molecule that can be cleaved with a specific restriction endonuclease. { ri'strik·shən ,sīt }

restrictive condition [GENETICS] An environmental condition under which a conditional lethal mutant either cannot grow or shows the mutant phenotype. { ri¦strik·tiv kən¦dish·ən }

resupinate [BOTANY] Inverted, usually through 180°, so as to appear upside down or reversed. { rē'sü·pə,nāt }

rete cord [EMBRYOLOGY] One of the deep, anastomosing strands of cells of the medullary cords of the vertebrate embryo that form the rete testis or the rete ovarii. { 'rēd·ē ,kórd }

rete mirabile [VERTEBRATE ZOOLOGY] A network of small blood vessels that are formed by the branching of a large vessel and that usually reunite into a single trunk; believed to have an oxygen-storing function in certain aquatic fauna. { 'rēd·ē mi'räb·ə·lē }

rete ovarii [ANATOMY] Vestigial tubules or cords of cells near the hilus of the ovary, corresponding with the rete testis, but not connected with the mesonephric duct. { 'rēd·ē ō'var·ē,ī }

rete testis [ANATOMY] The network of anastomosing tubules in the mediastinum testis. { 'rēd·ē 'tes·təs }

reticular cell [HISTOLOGY] *See* reticulocyte. { re 'tik·yə·lər 'sel }

reticular fiber [HISTOLOGY] Any of the delicate, branching argentophile fibers conspicuous in the connective tissue of lymphatic tissue, myeloid tissue, the red pulp of the spleen, and most basement membranes. Also known as argentaffin fiber; argyrophil lattice fiber; precollagenous fiber. { re'tik·yə·lər 'fī·bər }

reticular formation [ANATOMY] The portion of the central nervous system which consists of small islands of gray matter separated by fine bundles of nerve fibers running in every direction. { re 'tik·yə·lər fór¦mā·shən }

Reticulariaceae [MYCOLOGY] A family of plasmodial slime molds in the order Liceales. { rə ¦tik·yə,lar·ē'ās·ē,ē }

reticular system [ANATOMY] *See* reticuloendothelial system. { re'tik·yə·lər ,sis·təm }

reticular tissue [HISTOLOGY] Connective tissue having reticular fibers as the principal element. { re'tik·yə·lər ,tish·ü }

reticulate [BIOLOGY] Having or resembling a network of fibers, veins, or lines. [GENETICS] Of or relating to evolutionary change resulting from genetic recombination between strains in an interbreeding population. { rə'tik·yə·lət }

reticulin [BIOCHEMISTRY] A protein isolated from reticular fibers. { rə'tik·yə·lən }

reticulocyte [HISTOLOGY] Also known as reticular cell. **1.** A large, immature red blood cell, having a reticular appearance when stained due to retention of portions of the nucleus. **2.** A cell of reticular tissue. { rə'tik·yə·lə,sīt }

reticuloendothelial system [ANATOMY] The macrophage system, including all phagocytic cells such as histiocytes, macrophages, reticular cells, monocytes, and microglia, except the granular white blood cells. Abbreviated RES. Also known as hematopoietic system; reticular system. { rə ¦tik·yə·lō,en·dō'thē·lē·əl ,sis·təm }

reticuloendothelium [HISTOLOGY] The cells making up the reticuloendothelial system. { rə,tik·yə·lō,en·də'thē·lē·əm }

reticulopodia [INVERTEBRATE ZOOLOGY] Pseudopodia in the form of a branching network. { rə ¦tik·yə·lō'päd·ē·ə }

Reticulosa [PALEONTOLOGY] An order of Paleozoic hexactinellid sponges with a branching form in the subclass Hexasterophora. { rə,tik·yə'lō·sə }

reticulospinal tract [ANATOMY] Nerve fibers descending from large cells of the reticular formation of the pons and medulla into the spinal cord. { rə¦tik·yə·lō'spīn·əl 'trakt }

reticulum [BIOLOGY] A fine network. [VERTEBRATE ZOOLOGY] The second stomach in ruminants. { rə'tik·yə·ləm }

retina [ANATOMY] The photoreceptive layer and terminal expansion of the optic nerve in the dorsal aspect of the vertebrate eye. { 'ret·ən·ə }

retinaculum [INVERTEBRATE ZOOLOGY] **1.** A clasp on the forewing of certain moths for retaining the frenulum of the hindwing. **2.** An appendage on the third abdominal somite of springtails that articulates with the furcula. { ˌret·ən'ak·yə·ləm }

retinal [BIOCHEMISTRY] A carotenoid, produced as an intermediate in the bleaching of rhodopsin and decomposition to vitamin A. Also known as vitamin A aldehyde. { 'ret·ən·əl }

retinal pigment [BIOCHEMISTRY] *See* rhodopsin. { 'ret·ən·əl 'pig·mənt }

retinene [BIOCHEMISTRY] A pigment extracted from the retina, which turns yellow by the action of light; the chief carotenoid of the retina. { 'ret·ənˌēn }

retinoid [BIOCHEMISTRY] The set of molecules composing vitamin A and its synthetic analogs, such as retinal or retinyl acetate. { 'ret·ənˌȯid }

retinol [BIOCHEMISTRY] *See* vitamin A. { 'ret·ən ˌȯl }

retinula [INVERTEBRATE ZOOLOGY] The receptor element at the inner end of the ommatidium in a compound eye. { rə'tin·yə·lə }

Retortamonadida [INVERTEBRATE ZOOLOGY] An order of parasitic flagellate protozoans belonging to the class Zoomastigophorea, having two or four flagella and a complex blepharoplast-centrosome-axostyle apparatus. { ri'tȯr·də·mə 'näd·ə·də }

retractor [ANATOMY] A muscle that draws a limb or other body part toward the body. { ri'trak·tər }

retrocerebral gland [INVERTEBRATE ZOOLOGY] Any of various endocrine glands located behind the brain in insects which function in postembryonic development and metamorphosis. { ¦re·trō·sə'rē·brəl 'gland }

retroflexion [ANATOMY] The state of being bent backward. { ¦re·trə'flek·shən }

retroposon [GENETICS] A mobile genetic element that transposes by reverse transcription from a ribonucleic acid intermediate. { ˌre·trə'pōˌzän }

retrorse [BIOLOGY] Bent downward or backward. { ri'trȯrs }

retrostalsis [PHYSIOLOGY] Reverse peristalsis. { ¦re·trō'stäl·səs }

retroversion [ANATOMY] A turning back. { ˌre·trə 'vər·zhən }

retrovirus [VIROLOGY] A family of ribonucleic acid viruses distinguished by virions which possess reverse transcriptase and which have two proteinaceous structures, a dense core, and an envelope that surrounds the core. { 're·trōˌvī·rəs }

retuse [BOTANY] Having a rounded apex with a slight, central notch. { rə'tüs }

reverse graft [BOTANY] A plant graft made by inserting the scion in an inverted position. { ri'vərs 'graft }

reverse mutation [GENETICS] A mutation in a mutant allele which makes it capable of producing the nonmutant phenotype; may actually restore the original deoxyribonucleic-acid sequence of the gene or produce a new one which has a similar effect. Also known as back mutation. { ri'vərs myü'tā·shən }

reverse passive anaphylaxis [IMMUNOLOGY] Hypersensitivity produced when the antigen is injected first, then followed in several hours by the specific antibody, causing shock. { ri'vərs 'pas·iv ˌan·ə·fə'lak·səs }

reverse pinocytosis [CELL BIOLOGY] *See* emiocytosis. { ri'vərs ¦pī·nō·sī'tō·səs }

reverse transcript [MOLECULAR BIOLOGY] A deoxyribonucleic acid sequence obtained from a ribonucleic acid sequence by means of reverse transcription. { ri¦vərs 'tranˌskript }

reverse transcriptase [GENETICS] A polymerase that mediates deoxyribonucleic acid synthesis by using a ribonucleic acid template. { ri'vərs tran 'skripˌtās }

reverse transcription [GENETICS] The synthesis of deoxyribonucleic acid from a ribonucleic acid template. { ri'vərs tran'skrip·shən }

revolute [BOTANY] Rolled backward and downward. { 'rev·əˌlüt }

R factor [GENETICS] A self-replicating, infection-like agent that carries genetic information and transmits drug resistance from bacterium to bacterium by conjugation of cell. Also known as resistance factor. { 'är ˌfak·tər }

Rhabdiasoidea [INVERTEBRATE ZOOLOGY] An order or superfamily of parasitic nematodes. { ˌrab·dē·ə'sȯid·ē·ə }

rhabdion [INVERTEBRATE ZOOLOGY] One of the sclerotized segments lining the buccal cavity of nematodes. { 'rab·dēˌän }

rhabdite [INVERTEBRATE ZOOLOGY] A small rod-like or fusiform body secreted by epidermal or parenchymal cells of certain turbellarians and trematodes. { 'rabˌdīt }

Rhabditia [INVERTEBRATE ZOOLOGY] A subclass of nematodes in the class Secernentea. { rab 'dish·ə }

Rhabditidia [INVERTEBRATE ZOOLOGY] An order of nematodes in the subclass Rhabditia including parasites of humans and domestic animals. { ˌrab·də'tid·ē·ə }

Rhabditoidea [INVERTEBRATE ZOOLOGY] A superfamily of small to moderate-sized nematodes in the order Rhabditidia with small, porelike, anteriorly located amphids, and esophagus with corpus, isthmus, and valvulated basal bulb. { ˌrab·də'tȯid·ē·ə }

Rhabdocoela [INVERTEBRATE ZOOLOGY] Formerly an order of the Turbellaria, and now divided into three orders, Catenulida, Macrostomida, and Neorhabdocoela. { ˌrab·də'sē·lə }

rhabdoglyph [PALEONTOLOGY] A trace fossil consisting of a presumable worm trail appearing on the undersurface of flysch beds (sandstones) as a nearly straight bulge with little or no branching. { 'rab·də,glif }

rhabdolith [BOTANY] A minute coccolith having a shield surmounted by a long stem and found at all depths in the ocean, from the surface to the bottom. { 'rab·də,lith }

rhabdome [INVERTEBRATE ZOOLOGY] The central translucent cylinder in the retinula of a compound eye. { 'rab,dōm }

Rhabdophorina [INVERTEBRATE ZOOLOGY] A suborder of ciliates in the order Gymnostomatida. { ,rab·dō·fə'rī·nə }

rhabdosome [INVERTEBRATE ZOOLOGY] A colonial graptolite that develops from a single individual. { 'rab·də,sōm }

rhabdovirus [VIROLOGY] A group of ribonucleic acid-containing animal viruses, including rabies virus and certain infective agents of fish and insects. { ¦rab·dō'vī·rəs }

rhabdus [INVERTEBRATE ZOOLOGY] A uniaxial sponge spicule. { 'rab·dəs }

Rhachitomi [PALEONTOLOGY] A group of extinct amphibians in the order Temnospondyli in which pleurocentra were retained. { rə'kid·ə,mī }

rhachitomous [VERTEBRATE ZOOLOGY] Being, having, or pertaining to vertebrae with centra whose parts do not fuse. { rə'kid·ə·məs }

Rhacophoridae [VERTEBRATE ZOOLOGY] A family of arboreal frogs in the suborder Diplasiocoela. { ,rak·ō'fór·ə,dē }

Rhacopilaceae [BOTANY] A family of mosses in the order Isobryales generally having dimorphous leaves with smaller dorsal leaves and a capsule that is plicate when dry. { ,rak·ō·pə 'lās·ē,ē }

Rhagionidae [INVERTEBRATE ZOOLOGY] The snipe flies, a family of predatory orthorrhaphous dipteran insects in the series Brachycera that are brownish or gray with spotted wings. { ,rag·ē 'än·ə,dē }

rhagon [INVERTEBRATE ZOOLOGY] A pyramid-shaped, colonial sponge having an osculum at the apex and flagellated chambers in the upper wall only. { 'rā,gän }

Rhamnaceae [BOTANY] A family of dicotyledonous plants in the order Rhamnales characterized by a solitary ovule in each locule, free stamens, simple leaves, and flowers that are hypogynous to perigynous or epigynous. { ram'nās·ē,ē }

Rhamnales [BOTANY] An order of dicotyledonous plants in the subclass Rosidae having a single set of stamens, opposite the petals, usually a well-developed intrastamenal disk, and two or more locules in the ovary. { ram'nā·lēz }

rhamnose [BIOCHEMISTRY] $C_6H_{12}O_5$ A deoxysugar occurring free in poison sumac, and in glycoside combination in many plants. Also known as isodulcitol. { 'ram,nōs }

rhamphoid [BIOLOGY] Beak-shaped. { 'ram ,fóid }

Rhamphorhynchoidea [PALEONTOLOGY] A Jurassic suborder of the Pterosauria characterized by long, slender tails with an expanded tip. { ,ram·fə·riŋ'kóid·ē·ə }

rhampotheca [VERTEBRATE ZOOLOGY] The horny sheath covering a bird's beak. { ,ram·fə'thē·kə }

Rh antigen [IMMUNOLOGY] *See* Rh factor. { ¦är ¦āch 'ant·i·jən }

Rh blocking serum [IMMUNOLOGY] A serum that reacts with Rh-positive blood without causing agglutination, but which blocks the action of anti-Rh serums that are subsequently introduced. { ¦är¦āch ¦bläk·iŋ ,sir·əm }

Rh blocking test [IMMUNOLOGY] A test for the detection of Rh antibody in plasma wherein erythrocytes having the Rh antigen are incubated in the patient's serum so that the antibodies may be adsorbed on these cells, which are then employed in the antiglobulin test. Also known as indirect Coombs test; indirect developing test. { ¦är¦āch ¦bläk·iŋ ,test }

Rh blood group [IMMUNOLOGY] The extensive, genetically determined system of red blood cell antigens defined by the immune serum of rabbits injected with rhesus monkey erythrocytes, or by human antisera. Also known as rhesus blood group. { ¦är¦āch 'bləd ,grüp }

rhea [BOTANY] *See* ramie. [VERTEBRATE ZOOLOGY] The common name for members of the avian order Rheiformes. { 'rē·ə }

Rheidae [VERTEBRATE ZOOLOGY] The single family of the avian order Rheiformes. { 'rē·ə,dē }

Rheiformes [VERTEBRATE ZOOLOGY] The rheas, an order of South American running birds; called American ostriches, they differ from the true ostrich in their smaller size, feathered head and neck, three-toed feet, and other features. { ,rē·ə 'fór,mēz }

Rhenanida [PALEONTOLOGY] An order of extinct marine fishes in the class Placodermi distinguished by mosaics of small bones between the large plates in the head shield. { re'nan·ə·də }

rheobase [PHYSIOLOGY] The intensity of the steady current just sufficient to excite a tissue when suddenly applied. { 'rē·ō,bās }

rheophile [ECOLOGY] Living or thriving in running water. { 'rē·ə,fīl }

rheophilous bog [ECOLOGY] A bog which draws its source of water from drainage. { rē'äf·ə·ləs 'bäg }

rheoplankton [ECOLOGY] Plankton found in flowing water. { 'rē·ō,plaŋk·tən }

rheotaxis [BIOLOGY] Movement of a motile cell or organism in response to the direction of water currents. { ¦rē·ə¦tak·səs }

rheotropism [BIOLOGY] Orientation response of an organism to the stimulus of a flowing fluid, as water. { rē'ä·trə,piz·əm }

rhesus factor [IMMUNOLOGY] *See* Rh factor. { 'rē·səs ,fak·tər }

rhesus macaque [VERTEBRATE ZOOLOGY] *See* rhesus monkey. { 'rē·səs mə'kak }

rhesus monkey [VERTEBRATE ZOOLOGY] *Macaque mulatta.* An agile, gregarious primate found in southern Asia and having a short tail, short limbs of almost equal length, and a stocky build. Also known as rhesus macaque. { 'rē·səs ,məŋ·kē }

rheumatoid factor [IMMUNOLOGY] The immunoglobulin in the class IgM that is detected in the synovial fluid of individuals with rheumatoid arthritis. { 'rü·mə‚tȯid ‚fak·tər }

Rh factor [IMMUNOLOGY] Any of several red blood cell antigens originally identified in the blood of rhesus monkeys. Also known as Rh antigen; rhesus factor. { ¦är'āch ‚fak·tər }

Rhigonematidae [INVERTEBRATE ZOOLOGY] A family of nematodes in the superfamily Oxyuroidea. { ‚rig·ō·nə'mad·ə‚dē }

Rhincodontidae [VERTEBRATE ZOOLOGY] The whale sharks, a family of essentially tropical galeoid elasmobranchs in the isurid line. { ‚riŋ·kə 'dänt·ə‚dē }

rhinencephalon [ANATOMY] The anterior olfactory portion of the vertebrate brain. { ¦rīn·in 'sef·ə‚län }

Rhinobatidae [VERTEBRATE ZOOLOGY] The guitarfishes, a family of elasmobranchs in the batoid group. { ‚rī·nō'bad·ə‚dē }

Rhinoceratidae [VERTEBRATE ZOOLOGY] A family of perissodactyl mammals in the superfamily Rhinoceratoidea, comprising the living rhinoceroses. { rī‚näs·ə'räd·ə‚dē }

Rhinoceratoidea [VERTEBRATE ZOOLOGY] A superfamily of perissodactyl mammals in the suborder Ceratomorpha including living and extinct rhinoceroses. { rī‚näs·ə·rə'tȯid·ē·ə }

rhinoceros [VERTEBRATE ZOOLOGY] The common name for the odd-toed ungulates composing the family Rhinoceratidae, characterized by massive, thick-skinned limbs and bodies, and one or two horns which are composed of a solid mass of hairs attached to the bony prominence of the skull. { rī'näs·ə·rəs }

Rhinochimaeridae [VERTEBRATE ZOOLOGY] A family of ratfishes, order Chimaeriformes, distinguished by an extremely elongate rostrum. { ‚rīn·ə·ki'mer·ə‚dē }

Rhinocryptidae [VERTEBRATE ZOOLOGY] The tapaculos, a family of ground-inhabiting suboscine birds in the suborder Tyranni characterized by a large, movable flap which covers the nostrils. { ‚rīn·ə'krip·tə‚dē }

Rhinolophidae [VERTEBRATE ZOOLOGY] The horseshoe bats, a family of insect-eating chiropterans widely distributed in the Eastern Hemisphere and distinguished by extremely complex, horseshoe-shaped nose leaves. { ‚rīn·ə'läf·ə ‚dē }

rhinophore [INVERTEBRATE ZOOLOGY] An olfactoreceptor of certain land mollusks, usually borne on a tentacle. { 'rīn·ə‚fȯr }

Rhinopomatidae [VERTEBRATE ZOOLOGY] The mouse-tailed bats, a small family of insectivorous chiropterans found chiefly in arid regions of northern Africa and southern Asia and characterized by long, wirelike tails and rudimentary nose leaves. { ‚rīn·ə·pə'mad·ə‚dē }

Rhinopteridae [VERTEBRATE ZOOLOGY] The cow-nosed rays, a family of batoid sharks having a fleshy pad at the front end of the head and a well-developed poison spine. { ‚rī·näp'ter·ə‚dē }

Rhinotermitidae [INVERTEBRATE ZOOLOGY] A family of lower termites of the order Isoptera. { ¦rī·nō·tər'mad·ə‚dē }

rhinotheca [VERTEBRATE ZOOLOGY] The horny sheath on the upper part of a bird's bill. { ¦rīn·ə 'thē·kə }

rhinovirus [VIROLOGY] A subgroup of the picornavirus group including small, ribonucleic acid-containing forms which are not inactivated by ether. { ¦rīn·ə'vī·rəs }

Rhipiceridae [INVERTEBRATE ZOOLOGY] The cedar beetles, a family of coleopteran insects in the superfamily Elateroidea. { ‚rip·ə'ser·ə‚dē }

Rhipidistia [VERTEBRATE ZOOLOGY] The equivalent name for Osteolepiformes. { ‚rip·ə'dis· tē·ə }

rhipidium [BOTANY] A fan-shaped inflorescence with cymose branching in which branches lie in the same plane and are suppressed alternately on each side. { ri'pid·ē·əm }

Rhipiphoridae [INVERTEBRATE ZOOLOGY] The wedge-shaped beetles, a family of coleopteran insects in the superfamily Meloidea. { ‚rip·ə 'fȯr·ə‚dē }

rhizanthous [BOTANY] Producing flowers directly from the root. { rī'zan·thəs }

rhizautoicous [BOTANY] Of mosses, having the antheridial branch and the archegonial branch connected by rhizoids. { ¦rīz‚ȯ¦tȯi·kəs }

rhizine [BOTANY] The rhizoid of a lichen. { 'rī ‚zēn }

Rhizobiaceae [MICROBIOLOGY] A family of gram-negative, motile, aerobic rods; utilize carbohydrates and produce slime on carbohydrate media. { rī‚zō·bē'ās·ē‚ē }

rhizocarpous [BOTANY] Pertaining to perennial herbs having perennating underground parts from which stems and foliage arise annually. { ¦rī·zō¦kär·pəs }

Rhizocephala [INVERTEBRATE ZOOLOGY] An order of crustaceans which parasitize other crustaceans; adults have a thin-walled sac enclosing the visceral mass and show no trace of segmentation, appendages, or sense organs. { ‚rī·zō 'sef·ə·lə }

Rhizochloridina [INVERTEBRATE ZOOLOGY] A suborder of flagellate protozoans in the order Heterochlorida. { ‚rī·zō‚klȯr·ə'dī·nə }

Rhizodontidae [PALEONTOLOGY] An extinct family of lobefin fishes in the order Osteolepiformes. { ‚rī·zō'dänt·ə‚dē }

rhizoid [BOTANY] A rootlike structure which helps to hold the plant to a substrate; found on fungi, liverworts, lichens, mosses, and ferns. { 'rī ‚zȯid }

Rhizomastigida [INVERTEBRATE ZOOLOGY] An order of the protozoan class Zoomastigophorea; all species are microscopic and ameboid, and have one or two flagella. { ¦rī·zō·mas'tij·əd·ə }

Rhizomastigina [INVERTEBRATE ZOOLOGY] The equivalent name for Rhizomastigida. { 'rī·zō· mas·tə'jī·nə }

rhizome [BOTANY] An underground horizontal stem, often thickened and tuber-shaped, and possessing buds, nodes, and scalelike leaves. { 'rī‚zōm }

rhizomorph [BOTANY] A rootlike structure, characteristic of many basidiomycetes, consisting of a mass of densely packed and intertwined hyphae. { 'rī·zə,mȯrf }

Rhizophagidae [INVERTEBRATE ZOOLOGY] The root-eating beetles, a family of minute coleopteran insects in the superfamily Cucujoidea. { ,rī·zō'fā·jə,dē }

Rhizophoraceae [BOTANY] A family of dicolyledonous plants in the order Cornales distinguished by opposite, stipulate leaves, two ovules per locule, folded or convolute bud petals, and a berry fruit. { rī,zäf·ə'rās·ē,ē }

Rhizophorales [BOTANY] An order of dicotyledonous flowering plants, class Magnoliopsida; mostly tanniferous trees and shrubs with leaves opposite, simple, and entire, and flowers regular, mostly perfect, and variously perigynous or epigynous. { ¦rīz·ə·fō'rā·lēz }

rhizophore [BOTANY] A leafless, downward-growing dichotomous *Selaginella* shoot that has tufts of adventitious roots at the apex. { 'rī·zə,fȯr }

rhizoplast [CELL BIOLOGY] A delicate fiber or thread running between the nucleus and the blepharoplast in cells bearing flagella. { 'rī·zə ,plast }

rhizopod [INVERTEBRATE ZOOLOGY] An anastomosing rootlike pseudopodium. { 'rī·zə,päd }

Rhizopodea [INVERTEBRATE ZOOLOGY] A class of the protozoan superclass Sarcodina in which pseudopodia may be filopodia, lobopodia, or reticulopodia, or may be absent. { ,rī·zə'pō·dē·ə }

Rhizostomeae [INVERTEBRATE ZOOLOGY] An order of the class Scyphozoa having the umbrella generally higher than it is wide with the margin divided into many lappets but not provided with tentacles. { ,rī·zə'stō·mē,ē }

rhizotron [BOTANY] An underground laboratory system designed for examining plant root growth; contains enclosed columns of soil with transparent plastic windows which permit viewing, measuring, and photographing. { 'rīz·ə ,trän }

L-rhodeose [BIOCHEMISTRY] *See* L-fucose. { ¦el 'rōd·ē,ōs }

Rhodesian man [PALEONTOLOGY] A type of fossil man inhabiting southern and central Africa during the late Pleistocene; the skull was large and low, marked by massive browridges, with a cranial capacity of 1300 cubic centimeters or less. { rō'dē·zhən 'man }

Rhodininae [INVERTEBRATE ZOOLOGY] A subfamily of limivorous worms in the family Maldanidae. { rō'din·ə,nē }

Rhodobacteriineae [MICROBIOLOGY] Formerly a suborder of the order Pseudomonadales comprising all of the photosynthetic, or phototrophic, bacteria except those of the genus *Rhodomicrobium*. { ¦rō·dō·bak,tir·ē'ī·nē,ē }

Rhodophyceae [BOTANY] A class of algae belonging to the division or subphylum Rhodophyta. { ,rōd·ə'fīs·ē,ē }

Rhodophyta [BOTANY] The red algae, a large diverse phylum or division of plants distinguished by having an abundance of the pigment phycoerythrin. { rō'däf·əd·ə }

rhodoplast [BOTANY] A reddish chromatophore occurring in red algae. { 'rōd·ə,plast }

rhodopsin [BIOCHEMISTRY] A deep-red photosensitive pigment contained in the rods of the retina of marine fishes and most higher vertebrates. Also known as retinal pigment; visual purple. { rō'däp·sən }

Rhodospirillaceae [MICROBIOLOGY] A family of bacteria in the suborder Rhodospirillineae; cells are motile by flagella, multiplication is by budding or binary fission, and photosynthetic membranes are continuous with the cytoplasmic membrane. { ¦rō·dō,spī·rə'lās·ē,ē }

Rhodospirillales [MICROBIOLOGY] The single order of the phototrophic bacteria; cells are spherical, rod-shaped, spiral, or vibrio-shaped, and all contain bacteriochlorophylls and carotenoid pigments. { ¦rō·dō,spī·rə'lā·lēz }

Rhodospirillineae [MICROBIOLOGY] The purple bacteria, a suborder of the order Rhodospirillales; contain bacteriochlorophyll *a* or *b*, located on internal membranes. { ¦rō·dō,spī·rə'lin·ē,ē }

rhodoxanthin [BIOCHEMISTRY] $C_{40}H_{50}O_2$ A xanthophyll carotenoid pigment. { ¦rō·dō'zan·thən }

rhohelos [ECOLOGY] A stream-crossed, nonalluvial marsh typical of filled lake areas. { rō'hē ,lōs }

Rhoipteleaceae [BOTANY] A monotypic family of dicotyledonous plants in the order Juglandales having pinnately compound leaves, and flowers in triplets with four sepals and six stamens, and the lateral flowers female but sterile. { ,rȯip,tē·lē'ās·ē,ē }

rhombencephalon [EMBRYOLOGY] The most caudal of the primary brain vesicles in the vertebrate embryo. Also known as hindbrain. { ¦räm,ben 'sef·ə,län }

Rhombifera [PALEONTOLOGY] An extinct order of Cystoidea in which the thecal canals crossed the sutures at the edges of the plates, so that one-half of any canal lay in one plate and the other half on an adjoining plate. { räm'bif·ə·rə }

rhombogen [INVERTEBRATE ZOOLOGY] A form of reproductive individual of the mesozoan order Dicyemida found in the sexually mature host which arises from nematogens and gives rise to free-swimming infusorigens. { 'räm·bə·jən }

rhomboporoid cryptostome [PALEONTOLOGY] Any of a group of extinct bryozoans in the order Cryptostomata that built twiglike colonies with zooecia opening out in all directions from the central axis of each branch. { ¦räm·bō¦pȯr,ȯid 'krip·tə,sōm }

Rhopalidae [INVERTEBRATE ZOOLOGY] A family of pentatomorphan hemipteran insects in the superfamily Coreoidea. { rō'päl·ə,dē }

rhopalium [INVERTEBRATE ZOOLOGY] A sense organ found on the margin of a discomedusan. { rō 'päl·ē·əm }

Rhopalocera [INVERTEBRATE ZOOLOGY] Formerly a suborder of Lepidoptera comprising those forms with clubbed antennae. { ,rō·pə'läs·ə·rə }

rhopalocercous cercaria [INVERTEBRATE ZOOL-

OGY] A free-swimming digenetic trematode larva distinguished by a very wide tail. { ¦rō·pə·lō ¦sər·kəs sər'kar·ē·ə }

Rhopalodinidae [INVERTEBRATE ZOOLOGY] A family of holothurian echinoderms in the order Dactylochirotida in which the body is flask-shaped, the mouth and anus lying together. { ¦rō·pə·lō 'din·ə,dē }

Rhopalosomatidae [INVERTEBRATE ZOOLOGY] A family of hymenopteran insects in the superfamily Scolioidea. { ¦rō·pə·lō·sō'mad·ə,dē }

rhubarb [BOTANY] *Rheum rhaponticum*. A herbaceous perennial of the order Polygoniales grown for its thick, edible petioles. { 'rü,bärb }

Rhynchobdellae [INVERTEBRATE ZOOLOGY] An order of the class Hirudinea comprising leeches that possess an eversible proboscis and lack hemoglobin in the blood. { ,riŋ,käb'de·lē }

Rhynchocephalia [VERTEBRATE ZOOLOGY] An order of lepidosaurian reptiles represented by a single living species, *Sphenodon punctatus*, and characterized by a diapsid skull, teeth fused to the edges of the jaws, and an overhanging beak formed by the upper jaw. { ,riŋ·kō·sə'fāl·yə }

rhynchocoel [INVERTEBRATE ZOOLOGY] A cavity that holds the inverted proboscis in nemertinean worms. { 'riŋ·kō,sēl }

Rhynchocoela [INVERTEBRATE ZOOLOGY] A phylum of bilaterally symmetrical, unsegmented, ribbonlike worms having an eversible proboscis and a complete digestive tract with an anus. { ,riŋ·kō'sē·lə }

rhynchodaeum [INVERTEBRATE ZOOLOGY] The part of the proboscis lying anterior to the brain in nemertinean worms. { riŋ'kō·dē·əm }

Rhynchodina [INVERTEBRATE ZOOLOGY] A suborder of ciliate protozoans in the order Thigmotrichida. { ,riŋ·kə'dī·nə }

Rhynchonellida [INVERTEBRATE ZOOLOGY] An order of articulate brachiopods; typical forms are dorsibiconvex, the posterior margin is curved, the dorsal interarea is absent, and the ventral one greatly reduced. { ,riŋ·kə'nel·əd·ə }

rhynchophorous [ZOOLOGY] Having a beak. { riŋ 'käf·ə·rəs }

Rhynchosauridae [PALEONTOLOGY] An extinct family of generally large, stout, herbivorous lepidosaurian reptiles in the order Rhynchocephalea. { ,riŋ·kə'sȯr·ə,dē }

Rhynchotheriinae [PALEONTOLOGY] A subfamily of extinct elaphantoid mammals in the family Gomphotheriidae comprising the beak-jawed mastodonts. { ,riŋ·kə·thə'rī·ə,nē }

Rhyniophyta [PALEOBOTANY] A subkingdom of the Embryobionta including the relatively simple, uppermost Silurian-Devonian vascular plants. { ,rī·nē'äf·əd·ə }

Rhyniopsida [PALEOBOTANY] A class of extinct plants in the subkingdom Rhyniophyta characterized by leafless, usually dichotomously branched stems that bore terminal sporangia. { ,rī·nē'äp·səd·ə }

Rhynochetidae [VERTEBRATE ZOOLOGY] A monotypic family of gruiform birds containing only the kagu of New Caledonia. { ,rī·nə'kēd·ə ,dē }

Rhysodidae [INVERTEBRATE ZOOLOGY] The wrinkled bark beetles, a family of coleopteran insects in the suborder Adephaga. { rī'säd·ə,dē }

rib [ANATOMY] One of the long curved bones forming the wall of the thorax in vertebrates. [BOTANY] A primary vein in a leaf. { rib }

ribitol [BIOCHEMISTRY] *See* adonitol. { 'rī·bə,tȯl }

riboflavin [BIOCHEMISTRY] $C_{17}H_{20}N_4O_6$ A water-soluble, yellow orange fluorescent pigment that is essential to human nutrition as a component of the coenzymes flavin mononucleotide and flavin adenine dinucleotide. Also known as lactoflavin; vitamin B_2; vitamin G. { 'rī·bə,flā·vən }

riboflavin 5′-phosphate [BIOCHEMISTRY] $C_{17}H_{21}N_4O_9P$ The phosphoric acid ester of riboflavin. Also known as flavin phosphate; flavin mononucleotide; FMN; isoalloxazine mononucleotide; vitamin B_2 phosphate. { 'rī·bə,flā·vən ¦fīv,prīm 'fä,sfāt }

ribonuclease [BIOCHEMISTRY] $C_{587}H_{909}N_{171}O_{197}S_{12}$ An enzyme that catalyzes the depolymerization of ribonucleic acid. { ,rī·bō'nü·klē,ās }

ribonucleic acid [BIOCHEMISTRY] A long-chain, usually single-stranded nucleic acid consisting of repeating nucleotide units containing four kinds of heterocyclic, organic bases: adenine, cytosine, quanine, and uracil; they are conjugated to the pentose sugar ribose and held in sequence by phosphodiester bonds; involved intracellularly in protein synthesis. Abbreviated RNA. { ¦rī·bō¦nü¦klē·ik 'as·əd }

ribonucleic acid polymerase [BIOCHEMISTRY] An enzyme that transcribes a ribonucleic acid (RNA) molecule from one strand of a deoxyribonucleic acid (DNA) molecule. Also known as transcriptase. { ¦rī·bō¦nü¦klē·ik 'as·əd pə'lim·ə,rās }

ribonucleoprotein [BIOCHEMISTRY] Any of a large group of conjugated proteins in which molecules of ribonucleic acid are closely associated with molecules of protein. { ¦rī·bō¦nü·klē·ō'prō ,tēn }

ribonucleotide [BIOCHEMISTRY] A ribose-containing nucleotide, the structural unit of ribonucleic acid. { ¦rī·bō'nü·klē·ə,tīd }

ribose [BIOCHEMISTRY] $C_5H_{10}O_5$ A pentose sugar occurring as a component of various nucleotides, including ribonucleic acid. { 'rī,bōs }

riboside [BIOCHEMISTRY] Any glycoside containing ribose as the sugar component. { 'rī·bə ,sīd }

ribosomal ribonucleic acid [BIOCHEMISTRY] Any of three large types of ribonucleic acid found in ribosomes: 5S RNA, with molecular weight 40,000; 14-16S RNA, with molecular weight 600,000; and 18-22S RNA with molecular weight 1,200,000. Abbreviated r-RNA. { ¦rī·bə¦sō·məl ¦rī·bō¦nü¦klē·ik 'as·əd }

ribosome [CELL BIOLOGY] One of the small, complex particles composed of various proteins and three molecules of ribonucleic acid which synthesize proteins within the living cell. { 'rī·bə ,sōm }

ribozyme [BIOCHEMISTRY] A ribonucleic acid molecule that can catalyze, or lower the activation energy for, specific biochemical reactions. { 'rīb·ə,zīm }

ribulose |BIOCHEMISTRY| $C_5H_{10}O_5$ A pentose sugar that exists only as a syrup; synthesized from arabinose by isomerization with pyridine; important in carbohydrate metabolism. Also known as D-erythropentose; D-riboketose. { 'rī·byə‚lōs }

ribulose diphosphate |BIOCHEMISTRY| $C_5H_{12}O_{11}P_2$ The phosphate ester of ribulose. { 'rī·byə‚lōs dī'fä‚sfāt }

rice |BOTANY| *Oryza sativa.* An annual cereal grass plant of the order Cyperales, cultivated as a source of human food for its carbohydrate-rich grain. { rīs }

Ricinidae |INVERTEBRATE ZOOLOGY| A family of bird lice, order Mallophaga, which occur on numerous land and water birds. { rə'sin·ə‚dē }

Ricinuleida |INVERTEBRATE ZOOLOGY| An order of rare, ticklike arachnids in which the two anterior pairs of appendages are chelate, and the terminal segments of the third legs of the male are modified as copulatory structures. { ‚ris·ən·yü 'lē·ə·də }

Rickettsiaceae |MICROBIOLOGY| A family of the order Rickettsiales; small, rod-shaped, coccoid, or diplococcoid cells often found in arthropods; includes human and animal parasites and pathogens. { ri‚ket·sē'ās·ē‚ē }

Rickettsiales |MICROBIOLOGY| An order of prokaryotic microorganisms; gram-negative, obligate, intracellular animal parasites (may be grown in tissue cultures); many cause disease in humans and animals. { ri‚ket·sē'ā·lēz }

Rickettsieae |MICROBIOLOGY| A tribe of the family Rickettsiaceae; cells are occasionally filamentous; infect arthropods and some vertebrates and are pathogenic for humans, most frequently an incidental host. { ri'ket·sē‚ē }

rictus |VERTEBRATE ZOOLOGY| The mouth aperture in birds. { 'rik·təs }

rifampicin |MICROBIOLOGY| An antibacterial and antiviral antibiotic; action depends upon its preferential inhibition of bacterial ribonucleic acid polymerase over animal-cell RNA polymerase. { rə'fam·pə·sən }

rind |BOTANY| **1.** The bark of a tree. **2.** The thick outer covering of certain fruits. { rīnd }

ring canal |INVERTEBRATE ZOOLOGY| In echinoderms, the circular tube of the water-vascular system that surrounds the esophagus. { 'riŋ kə ‚nal }

ring deoxyribonucleic acid |BIOCHEMISTRY| *See* circular deoxyribonucleicacid. { 'riŋ dē¦äk·sē¦rī·bō¦nü¦klē·ik 'as·əd }

ringent |BOTANY| Having widely separated, gaping lips. |ZOOLOGY| Gaping irregularly. { 'rin·jənt }

ringtail |VERTEBRATE ZOOLOGY| *See* cacomistle. { 'riŋ‚tāl }

ring test |IMMUNOLOGY| The simplest of the precipitin tests for antigen-antibody reaction; the solution containing antigen is layered on a solution containing antibody; a white disk or precipitate forms at the point where the two solutions diffuse until optimum concentration for precipitation is reached. { 'riŋ ‚test }

ring vessel |INVERTEBRATE ZOOLOGY| A part of the water-vascular system in echinoderms; it is the circular canal around the mouth into which the stone canal empties, and from which a radial water vessel traverses to each of five radii. { 'riŋ ‚ves·əl }

Riodininae |INVERTEBRATE ZOOLOGY| A subfamily of the lepidopteran family Lycaenidae in which prothoracic legs are nonfunctional in the male. { ‚rī·ə'din·ə‚dē }

riparian |BIOLOGY| Living or located on a riverbank. { rə'per·ē·ən }

ripe |BOTANY| Of fruit, fully developed, having mature seed and so usable as food. { rīp }

Rissoacea |PALEONTOLOGY| An extinct superfamily of gastropod mollusks. { ‚ris·ə'wās·ē·ə }

river-delta marsh |EVOLUTION| A brackish or fresh-water marsh bordering the mouth of a distributary stream. { 'riv·ər ¦del·tə ‚märsh }

rivulose |BOTANY| Marked by irregular, narrow lines. { 'riv·yə‚lōs }

RNA |BIOCHEMISTRY| *See* ribonucleic acid.

roach |INVERTEBRATE ZOOLOGY| An insect of the family Blattidae; the body is wide and flat, the anterior part of the thorax projects over the head, and antennae are long and filiform, with many segments. Also known as cockroach. { rōch }

Robertinacea |INVERTEBRATE ZOOLOGY| A superfamily of marine, benthic foraminiferans in the suborder Rotaliina characterized by a trochospiral or modified test with a wall of radial aragonite, and having bilamellar septa. { rä‚bərd·ə 'nās·ē·ə }

Roccilaceae |BOTANY| A family of fruticose species of Hysteriales that grow profusely on trees and rocks along the coastlines of Portugal, California, and western South America. { ‚räs·ə 'lās·ē‚ē }

rocket electrophoresis |IMMUNOLOGY| A variant of crossed electrophoresis in which the medium contains only one antibody; test substances are driven directly into the medium that contains the antibody, forming rocket-shaped (inverted V) trails of precipitation. { 'räk·ət i‚lek·trō·fə'rē·səs }

rock shell |INVERTEBRATE ZOOLOGY| The common name for a large number of gastropod mollusks composing the family Muricidae and characterized by having conical shells with various sculpturing. { 'räk ‚shel }

rod |HISTOLOGY| One of the rod-shaped sensory bodies in the retina which are sensitive to dim light. { räd }

rodent |VERTEBRATE ZOOLOGY| The common name for members of the order Rodentia. { 'rōd·ənt }

Rodentia |VERTEBRATE ZOOLOGY| An order of mammals characterized by a single pair of ever-growing upper and lower incisors, a maximum of five upper and four lower cheek teeth on each side, and free movement of the lower jaw in an anteroposterior direction. { rō'den·chə }

roll-tube technique |MICROBIOLOGY| A pure-culture technique, employed chiefly in tissue culture, in which, during incubation, the test tubes are held in a wheellike instrument at an angle of about 15° from the horizontal and the wheel is rotated vertically about once every 2 minutes. { 'rōl ˌtüb tekˌnēk }

rookery [ZOOLOGY] A location used by birds for breeding and nesting. { 'ru̇k·ə·rē }

rooster [VERTEBRATE ZOOLOGY] An adult male of certain birds and fowl, such as pheasants and ptarmigans. { 'rüs·tər }

root [BOTANY] The absorbing and anchoring organ of a vascular plant; it bears neither leaves nor flowers and is usually subterranean. { rüt }

root canal [ANATOMY] The cavity within the root of a tooth, occupied by pulp, nerves, and vessels. { 'rüt kəˌnal }

root cap [BOTANY] A thick, protective mass of parenchymal cells covering the meristematic tip of the root. { 'rüt ˌkap }

root hair [BOTANY] One of the hairlike outgrowths of the root epidermis that function in absorption. { 'rüt ˌher }

rootstock [BOTANY] A root or part of a root used as the stock for grafting. { 'rütˌstäk }

rootworm [INVERTEBRATE ZOOLOGY] **1.** An insect larva that feeds on plant roots. **2.** A nematode that infests the roots of plants. { 'rütˌwərm }

Roproniidae [INVERTEBRATE ZOOLOGY] A small family of hymenopteran insects in the superfamily Proctotrupoidea. { ˌräp·rə'nī·əˌdē }

Rosaceae [BOTANY] A family of dicotyledonous plants in the order Rosales typically having stipulate leaves and hypogynous, slightly perigynous, or epigynous flowers, numerous stamens, and several or many separate carpels. { rō'zās·ēˌē }

Rosales [BOTANY] A morphologically diffuse order of dicotyledonous plants in the subclass Rosidae. { rō'zā·lēz }

rose [BOTANY] A member of the genus *Rosa* in the rose family (Rosaceae); plants are erect, climbing, or trailing shrubs, generally prickly stemmed, and bear alternate, odd-pinnate single leaves. { rōz }

rose hip [BOTANY] The ripened false fruit of a rose plant. { 'rōz ˌhip }

rosemary [BOTANY] *Rosmarinus officinalis.* A fragrant evergreen of the mint family from France, Spain, and Portugal; leaves have a pungent bitter taste and are used as an herb and in perfumes. { 'rōzˌmer·ē }

Rosenmueller's organ [ANATOMY] *See* epoophoron. { 'rōz·ənˌmu̇l·ərz ˌȯr·gən }

rosette [BIOLOGY] Any structure or marking resembling a rose. { rō'zet }

Rosidae [BOTANY] A large subclass of the class Magnoliatae; most have a well-developed corolla with petals separate from each other, binucleate pollen, and ovules usually with two integuments. { 'rōz·əˌdē }

rostellum [BIOLOGY] The anterior, flattened region of the scolex of armed tapeworms. { rä 'stel·əm }

Rostratulidae [VERTEBRATE ZOOLOGY] A small family of birds in the order Charadriiformes containing the painted snipe; females are more brightly colored than males. { ˌrä·strə'tyü·lə ˌdē }

rostrum [BIOLOGY] A beak or beaklike process. { 'rä·strəm }

Rotaliacea [INVERTEBRATE ZOOLOGY] A superfamily of foraminiferans in the suborder Rotaliina characterized by a planispiral or trochospiral test having apertural pores and composed of radial calcite, with secondarily bilamellar septa. { rōˌtal·ē'ā·shə }

rotate [BOTANY] Of a sympetalous corolla, having a short tube and petals radiating like the spokes of a wheel. { 'rōˌtāt }

rotator [ANATOMY] A muscle that partially rotates a part of the body on the part's axis. { 'rō ˌtād·ər }

Rotatoria [INVERTEBRATE ZOOLOGY] The equivalent name for Rotifera. { ˌrōd·ə'tȯr·ē·ə }

Rotifera [INVERTEBRATE ZOOLOGY] A class of the phylum Aschelminthes distinguished by the corona, a retractile trochal disk provided with several groups of cilia and located on the head. { rō 'tif·ə·rə }

rough colony [MICROBIOLOGY] A flattened, irregular, and wrinkled colony of bacteria indicative of decreased capsule formation and virulence. { 'rəf 'käl·ə·nē }

round ligament [ANATOMY] **1.** A flattened band extending from the fovea on the head of the femur to attach on either side of the acetabular notch between which it blends with the transverse ligament. **2.** A fibrous cord running from the umbilicus to the notch in the anterior border of the liver; represents the remains of the obliterated umbilical vein. { 'rau̇nd 'lig·ə·mənt }

round window [ANATOMY] A membrane-covered opening between the middle and inner ears in amphibians and mammals through which energy is dissipated after traveling in the membranous labyrinth. { 'rau̇nd 'win·dō }

roundworm [INVERTEBRATE ZOOLOGY] The name applied to nematodes. { 'rau̇ndˌwərm }

RQ [PHYSIOLOGY] *See* respiratory quotient.

r-RNA [BIOCHEMISTRY] *See* ribosomal ribonucleic acid.

r selection [ECOLOGY] Selection that favors rapid population growth (r represents the intrinsic rate of increase). { 'är siˌlek·shən }

RTF [GENETICS] *See* resistance transfer factor.

rubber tree [BOTANY] *Hevea brasiliensis.* A tall tree of the spurge family (Euphorbiaceae) from which latex is collected and coagulated to produce rubber. { 'rəb·ər ˌtrē }

Rubiaceae [BOTANY] The single family of the plant order Rubiales. { ˌrü·bē'ās·ēˌē }

Rubiales [BOTANY] An order of dicotyledonous plants marked by their inferior ovary, regular or nearly regular corolla, and opposite leaves with interpetiolar stipules or whorled leaves without stipules. { ˌrü·bē'ā·lēz }

rubredoxin [BIOCHEMISTRY] A class of iron-sulfur proteins that contains one iron coordinated to the sulfur atom of four cysteine residues. { ¦rü·brə'däk·sən }

rubriblast [HISTOLOGY] *See* pronormoblast. { 'rü·brəˌblast }

ruddy turnstone [VERTEBRATE ZOOLOGY] *Arenaria interpes.* A member of the avian order Charadriiformes that perform transpacific flights during their migration. { 'rəd·ē 'tərn,stōn }

ruderal [ECOLOGY] **1.** Growing on rubbish, or waste or disturbed places. **2.** A plant that thrives in such a habitat. { 'rüd·ə·rəl }

rudistids [PALEONTOLOGY] Fossil sessile bivalves that formed reefs during the Cretaceous in the southern Mediterranean or the Tethyan belt. { rü 'dis·tədz }

Ruffini cylinder [ANATOMY] A cutaneous nerve ending suspected as the mediator of warmth. { rü'fē·nē ,sil·ən·dər }

rufous [BOTANY] Having a reddish-brown color. { 'rü·fəs }

Rugosa [PALEONTOLOGY] An order of extinct corals having either simple or compound skeletons with internal skeletal structures consisting mainly of three elements, the septa, tabulae, and dissepiments. { ,rü'gō·sə }

rugose [BIOLOGY] Having a wrinkled surface. { 'rü,gōs }

rumen [VERTEBRATE ZOOLOGY] The first chamber of the ruminant stomach. Also known as paunch. { 'rü·mən }

ruminant [PHYSIOLOGY] Characterized by the act of regurgitation and rechewing of food. [VERTEBRATE ZOOLOGY] A mammal belonging to the Ruminantia. { 'rü·mə·nənt }

Ruminantia [VERTEBRATE ZOOLOGY] A suborder of the Artiodactyla including sheep, goats, camels, and other forms which have a complex stomach and ruminate their food. { ,rü·mə'nan·chə }

rumination [PHYSIOLOGY] Regurgitation and remastication of food in preparation for true digestion in ruminants. { ,rü·mə'nā·shən }

runcinate [BOTANY] Pinnately cut with downward-pointing lobes. { 'rən·sə·nət }

runner [BOTANY] A horizontally growing, sympodial stem system; adventitious roots form near the apex, and a new runner emerges from the axil of a reduced leaf. Also known as stolon. { 'rən·ər }

running bird [VERTEBRATE ZOOLOGY] Any of the large, flightless, heavy birds usually categorized as ratites. { 'rən·iŋ ,bərd }

runoff desert [ECOLOGY] An arid region in which local rain is insufficient to support any perennial vegetation except in drainage or runoff channels. { 'rən,óf ,dez·ərt }

rupicolous [ECOLOGY] Living among or growing on rocks. { rü'pik·ə·ləs }

Russell's viper [VERTEBRATE ZOOLOGY] *See* tic polonga. { 'rəs·əlz 'vī·pər }

rut [PHYSIOLOGY] The period during which the male animal has a heightened mating drive. { rət }

rutabaga [BOTANY] *Brassica napobrassica*. A biennial crucifer of the order Capparales probably resulting from the natural crossing of cabbage and turnip and characterized by a large, edible, yellowish fleshy root. { ¦rüd·ə¦bā·gə }

Rutaceae [BOTANY] A family of dicotyledonous plants in the order Sapindales distinguished by mostly free stamens and glandular-punctate leaves. { rü'tās·ē,ē }

rye [BOTANY] *Secale cereale*. A cereal plant of the order Cyperales cultivated for its grain, which contains the most desirable gluten, next to wheat. { rī }

rye buckwheat [BOTANY] *See* tartary buckwheat. { 'rī 'bək,wēt }

Rynchopidae [VERTEBRATE ZOOLOGY] The skimmers, a family of birds in the order Charadriiformes with a knifelike lower beak that is longer and narrower than the upper one. { ,riŋ'käp·ə,dē }

Rytiodinae [VERTEBRATE ZOOLOGY] A subfamily of trichechiform sirenians in the family Dugongidae. { ,rid·ē'äd·ən,ē }

S

saba [BOTANY] A plant (*Musa sapientum* var. *compressa*) that is common in the Philippines; the fruit is a cooking banana. { sə'bä }

Sabellariidae [INVERTEBRATE ZOOLOGY] The sand-cementing worms, a family of polychaete annelids belonging to the Sedentaria and characterized by a compact operculum formed of setae of the first several segments. { sə,bel·ə'rī·ə ,dē }

Sabellidae [INVERTEBRATE ZOOLOGY] A family of sedentary polychaete annelids often occurring in intertidal depths but descending to great abyssal depths; one of two families that make up the feather-duster worms. { sə'bel·ə,dē }

Sabellinae [INVERTEBRATE ZOOLOGY] A subfamily of the Sabellidae including the most numerous and largest members. { sə'bel·ə,nē }

Sabin vaccine [IMMUNOLOGY] A live-poliovirus vaccine that is administered orally. { 'sā·bən vak 'sēn }

sable [VERTEBRATE ZOOLOGY] *Martes zibellina*. A carnivore of the family Mustelidae; a valuable fur-bearing animal, quite similar to the American marten. { 'sā·bəl }

Sabouraud's agar [MICROBIOLOGY] A peptone-maltose agar used as a culture medium for pathogenic fungi, especially the dermatophytes. { sa·bu̇'rōz 'ag·ər }

sac [BIOLOGY] A soft-walled cavity within a plant or animal, often containing a special fluid and usually having a narrow opening or none at all. { sak }

saccadic movement [PHYSIOLOGY] Rapid eye movement that transfers the gaze from one fixation point to another. { sə¦kad·ik 'müv·mənt }

saccate [BOTANY] Having a saclike or pouchlike form. { 'sa,kāt }

saccharase [BIOCHEMISTRY] An enzyme that catalyzes the hydrolysis of disaccharide to monosaccharides, specifically of sucrose to dextrose and levulose. Also known as invertase; invertin; sucrase. { 'sak·ə,rās }

Saccharomycetaceae [MYCOLOGY] The single family of the order Saccharomycetales. { ¦sak·ə·rō,mī·sə'tās·ē,ē }

Saccharomycetales [MYCOLOGY] An order of the subclass Hemiascomycetidae comprising typical yeasts, characterized by the presence of naked asci in which spores are formed by free cells. { ¦sak·ə·rō,mī·sə'tā·lēz }

Saccharomycetoideae [MYCOLOGY] A subfamily of Saccharomycetacae in which spores may be hat-, sickle-, or kidney-shaped, or round or oval. { ¦sak·ə·rō,mī·sə'tȯid·ē,ē }

Saccoglossa [INVERTEBRATE ZOOLOGY] An order of gastropod mollusks belonging to the Opisthobranchia. { ,sak·ə'gläs·ə }

Saccopharyngiformes [VERTEBRATE ZOOLOGY] Formerly an order of actinopterygian fishes, the gulpers, now included in the Anguilliformes. { ,sak·ō·fə,rin·jə'fȯr,mēz }

Saccopharyngoidei [VERTEBRATE ZOOLOGY] The gulpers, a suborder of actinopterygian fishes in the order Anguilliformes having degenerative adaptations, including loss of swim bladder, opercle, branchiostegal ray, caudal fin, scales, and ribs. { ,sak·ō,far·əŋ'gȯid·ē,ī }

sacculus [ANATOMY] The smaller, lower saclike chamber of the membranous labyrinth of the vertebrate ear. { 'sak·yə·ləs }

sac fungus [MYCOLOGY] The common name for members of the class Ascomycetes. { 'sak ,fəŋ·gəs }

sacral nerve [ANATOMY] Any of five pairs of spinal nerves in the sacral region which innervate muscles and skin of the lower back, lower extremities, and perineum, and branches to the hypogastric and pelvic plexuses. { 'sak·rəl 'nərv }

sacral vertebrae [ANATOMY] Three to five fused vertebrae that form the sacrum in most mammals; amphibians have one sacral vertebra, reptiles usually have two, and birds have 10-23 fused in the synsacrum. { 'sak·rəl 'vərd·ə,brā }

sacrococcygeus [ANATOMY] One of two inconstant thin muscles extending from the lower sacral vertebrae to the coccyx. { ¦sa·krō·käk'sij·ē·əs }

sacroiliac [ANATOMY] Pertaining to the sacrum and the ilium. { ¦sa·krō¦il·ē,ak }

sacrospinous [ANATOMY] Pertaining to the sacrum and the spine of the ischium. { ¦sa·krō 'spī·nəs }

sacrum [ANATOMY] A triangular bone, consisting in humans of five fused vertebrae, located below the last lumbar vertebra, above the coccyx, and between the hipbones. { 'sak·rəm }

Saefftigen's pouch [INVERTEBRATE ZOOLOGY] An elongated pouch inside the genital sheath in many acanthocephalans. { 'zef·ti·gənz ,pau̇ch }

safflower [BOTANY] *Carthamus tinctorius*. An annual thistlelike herb belonging to the composite family (Compositae); the leaves are edible, flowers yield dye, and seeds yield a cooking oil. { 'sa,flau̇·ər }

saffron [BOTANY] *Crocus sativus.* A crocus of the iris family (Iridaceae); the source of a yellow dye used for coloring food and medicine. { 'saf·rən }

Sagartiidae [INVERTEBRATE ZOOLOGY] A family of zoantharians in the order Actiniaria. { ,sag·ər'tī·ə,dē }

sage [BOTANY] *Salvia officinalis.* A half-shrub of the mint family (Labiatae); the leaves are used as a spice. { sāj }

sagebrush [BOTANY] Any of various hoary undershrubs of the genus *Artemisia* found on the alkaline plains of the western United States. { 'sāj,brəsh }

Sagenocrinida [PALEONTOLOGY] A large order of extinct, flexible crinoids that occurred from the Silurian to the Permian. { ¦saj·ə·nō'krī·nə·də }

Saghathiinae [PALEONTOLOGY] An extinct subfamily of hyracoids in the family Procaviidae. { ,sag·ə'thī·ə,nē }

sagitta [VERTEBRATE ZOOLOGY] The larger of two otoliths in the ear of most fishes. { sə'jid·ə }

sagittal [ZOOLOGY] In the median longitudinal plane of the body, or parallel to it. { 'saj·əd·əl }

Sagittariidae [VERTEBRATE ZOOLOGY] A family of birds in the order Falconiformes comprising a single species, the secretary bird, noted for its nuchal plumes resembling quill pens stuck behind an ear. { ,saj·ə·tə'rī·ə,dē }

sagittate [BOTANY] Shaped like an arrowhead, especially referring to leaves. { 'saj·ə,tāt }

sagittocyst [INVERTEBRATE ZOOLOGY] A cyst in the epidermis of certain turbellarians containing a single spindle-shaped needle. { sə'jid·ə,sist }

sahel [ECOLOGY] A region having characteristics of a savanna or a steppe and bordering on a desert. { sə'hel }

sailfish [VERTEBRATE ZOOLOGY] Any of several large fishes of the genus *Istiophorus* characterized by a very large dorsal fin that is highest behind its middle. { 'sāl,fish }

salamander [VERTEBRATE ZOOLOGY] The common name for members of the order Urodela. { 'sal·ə,man·dər }

Salamandridae [VERTEBRATE ZOOLOGY] A family of urodele amphibians in the suborder Salamandroidea characterized by a long row of prevomerine teeth. { ,sal·ə'man·drə,dē }

Salamandroidea [VERTEBRATE ZOOLOGY] The largest suborder of the Urodela characterized by teeth on the roof of the mouth posterior to the openings of the nostrils. { ,sal·ə,man'dröid·ē·ə }

Salangidae [VERTEBRATE ZOOLOGY] A family of soft-rayed fishes, in the suborder Galaxioidei, which live in estuaries of eastern Asia. { sə'lan·jə,dē }

Saldidae [INVERTEBRATE ZOOLOGY] The shore bugs, a family of predacious hemipteran insects in the superfamily Saldoidea. { 'sal·də,dē }

Saldoidea [INVERTEBRATE ZOOLOGY] A superfamily of the hemipteran group Leptopodoidea. { sal'dȯi·dē·ə }

Saleniidae [INVERTEBRATE ZOOLOGY] A family of echinoderms in the order Salenioida distinguished by imperforate tubercles. { ,sa·lə'nī·ə·dē }

Salenioida [INVERTEBRATE ZOOLOGY] An order of the Echinacea in which the apical system includes one or several large angular plates covering the periproct. { sə,lē·nē'ȯi·də }

Salicaceae [BOTANY] The single family of the order Salicales. { ,sal·ə'kās·ē,ē }

Salicales [BOTANY] A monofamilial order of dicotyledonous plants in the subclass Dilleniidae; members are dioecious, woody plants, with alternate, simple, stipulate leaves and plumose-hairy mature seeds. { ,sal·ə'kā·lēz }

Salientia [VERTEBRATE ZOOLOGY] The equivalent name for Anura. { ,sā·lē'en·chə }

saliva [PHYSIOLOGY] The opalescent, tasteless secretions of the oral glands. { sə'lī·və }

salivary amylase [BIOCHEMISTRY] *See* ptyalin. { 'sal·ə,ver·ē 'am·ə,lās }

salivary diastase [BIOCHEMISTRY] *See* ptyalin. { 'sal·ə,ver·ē 'dī·ə,stās }

salivary gland [PHYSIOLOGY] A gland that secretes saliva, such as the sublingual or parotid. { 'sal·ə,ver·ē ,gland }

salivary gland chromosomes [CELL BIOLOGY] Polytene chromosomes found in the interphase nuclei of salivary glands in the larvae of Diptera; chromosomes in the larva undergo complete somatic pairing to form two homologous polytene chromosomes fused side by side. { 'sal·ə,ver·ē ,gland 'krō·mə,sōmz }

salivation [PHYSIOLOGY] Excessive secretion of saliva. { ,sal·ə'vā·shən }

Salk vaccine [IMMUNOLOGY] A killed-virus vaccine administered for active immunization against poliomyelitis. { 'sȯk vak,sēn }

salmon [VERTEBRATE ZOOLOGY] The common name for a number of fish in the family Salmonidae which live in coastal waters of the North Atlantic and North Pacific and breed in rivers tributary to the oceans. { 'sam·ən }

Salmonelleae [MICROBIOLOGY] Formerly a tribe of the Enterobacteriaceae comprising the pathogenic genera *Salmonella* and *Shigella*. { ,sal·mə'nel·ē,ē }

Salmonidae [VERTEBRATE ZOOLOGY] A family of soft-rayed fishes in the suborder Salmonoidei including the trouts, salmons, whitefishes, and graylings. { sal'män·ə,dē }

Salmoniformes [VERTEBRATE ZOOLOGY] An order of soft-rayed fishes comprising salmon and their allies; the stem group from which most higher teleostean fishes evolved. { ,sal,män·ə'fȯr,mēz }

Salmonoidei [VERTEBRATE ZOOLOGY] A suborder of the Salmoniformes comprising forms having an adipose fin. { sal·mə'nȯid·ē,ī }

Salmopercae [VERTEBRATE ZOOLOGY] An equivalent name for Percopsiformes. { ,sal·mō'pər,sē }

Salpida [INVERTEBRATE ZOOLOGY] An order of tunicates in the class Thaliacea including transparent forms ringed by muscular bands. { 'sal·pə·də }

Salpingidae [INVERTEBRATE ZOOLOGY] The narrow-waisted bark beetles, a family of coleopteran insects in the superfamily Tenebrionoidea. { sal 'pin·jə,dē }

saltatorial [ZOOLOGY] Adapted for leaping. { ¦sal·tə¦tór·ē·əl }

salt gland [VERTEBRATE ZOOLOGY] A compound tubular gland, located around the eyes and nasal passages in certain marine turtles, snakes, and birds, which copiously secretes a watery fluid containing a high percentage of salt. { 'sólt ,gland }

Salticidea [INVERTEBRATE ZOOLOGY] The jumping spiders, a family of predacious arachnids in the suborder Dipneumonomorphae having keen vision and rapid movements. { ,sal·tə'sid·ē·ə }

saltmarsh [ECOLOGY] A maritime habitat found in temperate regions, but typically associated with tropical and subtropical mangrove swamps, in which excess sodium chloride is the predominant environmental feature. { 'sólt,märsh }

saltmarsh plain [ECOLOGY] A salt marsh that has been raised above the level of the highest tide and has become dry land. { 'sólt,märsh ,plān }

salt-spray climax [ECOLOGY] A climax community along exposed Atlantic and Gulf seacoasts composed of plants able to tolerate the harmful effects of salt picked up and carried by onshore winds from seawater. { 'sólt ¦sprā 'klī,maks }

Salviniales [BOTANY] A small order of heterosporous, leptosporangiate ferns (division Polypodiophyta) which float on the surface of the water. { ,sal,vin·ē'ā,lēz }

samara [BOTANY] A dry, indehiscent, winged fruit usually containing a single seed, such as sugar maple (*Acer saccharum*). { sə'mar·ə }

Sambonidae [INVERTEBRATE ZOOLOGY] A family of pentastomid arthropods in the suborder Porocephaloidea of the order Porocephalida. { sam 'bän·ə,dē }

Samythinae [INVERTEBRATE ZOOLOGY] A subfamily of sedentary polychaete annelids in the family Ampharetidae having a conspicuous dorsal membrane. { sə'mith·ə,nē }

Sandalidae [INVERTEBRATE ZOOLOGY] The equivalent name for Rhipiceridae. { san'dal·ə,dē }

sandalwood [BOTANY] **1.** Any species of the genus *Santalum* of the sandalwood family (Santalaceae) characterized by a fragrant wood. **2.** *S. album*. A parasitic tree with hard, close-grained, aromatic heartwood used in ornamental carving and cabinetwork. { 'san·dəl,wu̇d }

sandblow [ECOLOGY] A patch of coarse, sandy soil denuded of vegetation by wind action. { 'san,blō }

sand dollar [INVERTEBRATE ZOOLOGY] The common name for the flat, disk-shaped echinoderms belonging to the order Clypeasteroida. { 'san ,däl·ər }

sandfly [INVERTEBRATE ZOOLOGY] Any of various small biting Diptera, especially of the genus *Phlebotomus*, which are vectors for phlebotomus (sandfly) fever. { 'san,flī }

sand hopper [INVERTEBRATE ZOOLOGY] The common name for gammaridean crustaceans found on beaches. { 'san ,häp·ər }

sandpiper [VERTEBRATE ZOOLOGY] Any of various small birds that are related to plovers and that frequent sandy and muddy shores in temperate latitudes; bill is moderately long with a soft, sensitive tip, legs and neck are moderately long, and plumage is streaked brown, gray, or black above and is white below. { 'san,pī·pər }

sand shark [VERTEBRATE ZOOLOGY] Any of various shallow-water predatory elasmobranchs of the family Carchariidae. Also known as tiger shark. { 'san ,shark }

sanguivorous [ZOOLOGY] Feeding on blood. { saŋ'gwiv·ə·rəs }

sanidaster [INVERTEBRATE ZOOLOGY] A rod-shaped spicule having spines at intervals along its length. { ¦san·ə¦das·tər }

SA node [ANATOMY] *See* sinoauricular node. { ¦es ¦ā 'nōd }

Santalaceae [BOTANY] A family of parasitic dicotyledonous plants in the order Santalales characterized by dry or fleshy indehiscent fruit, plants with chlorophyll, petals absent, and ovules without integument. { ,san·tə'lās·ē,ē }

Santalales [BOTANY] An order of dicotyledonous plants in the subclass Rosidae characterized by progressive adaptation to parasitism, accompanied by progressive simplification of the ovules. { ,san·tə'lā·lēz }

sap [BOTANY] The fluid part of a plant which circulates through the vascular system and is composed of water, gases, salts, and organic products of metabolism. { sap }

saphenous nerve [ANATOMY] A somatic sensory nerve arising from the femoral nerve and innervating the skin of the medial aspect of the leg, foot, and knee joint. { sə'fē·nəs ,nərv }

Sapindaceae [BOTANY] A family of dicotyledonous plants in the order Sapindales distinguished by mostly alternate leaves, usually one and less often two ovules per locule, and seeds lacking endosperm. { ,sap·ən'dās·ē,ē }

Sapindales [BOTANY] An order of mostly woody dicotyledonous plants in the subclass Rosidae with compound or lobed leaves and polypetalous, hypogynous to perigynous flowers with one or two sets of stamens. { ,sap·ən'dā·lēz }

sapling [BOTANY] A young tree with a trunk less than 4 inches (10 centimeters) in diameter at a point approximately 4 feet (1.2 meters) above the ground. { 'sap·liŋ }

Sapotaceae [BOTANY] A family of dicotyledonous plants in the order Ebenales characterized by a well-developed latex system. { ,sap·ə'tās·ē,ē }

saprobic [BOTANY] Living on decaying organic matter; applied to plants and microorganisms. { sə'prō·bik }

saprogen [BIOLOGY] An organism that lives on nonliving organic matter. { 'sap·rə·jən }

Saprolegniales [MYCOLOGY] An order of aquatic fungi belonging to the class Phycomycetes, having a mostly hyphal thallus and zoospores with two flagella. { ,sap·rə,leg·nē'ā·lēz }

saprophage [BIOLOGY] An organism that lives on decaying organic matter. { 'sap·rə,fāj }

saprophyte [BOTANY] A plant that lives on decaying organic matter. { 'sap·rə,fīt }

saprovore [ZOOLOGY] A detritus-eating animal. { 'sap·rə,vȯr }

saprozoic [ZOOLOGY] Feeding on decaying organic matter; applied to animals. { ¦sap·rə¦zō·ik }

sapwood [BOTANY] The younger, softer, outer layers of a woody stem, between the cambium and heartwood. Also known as alburnum. { 'sap ,wu̇d }

Sapygidae [INVERTEBRATE ZOOLOGY] A family of hymenopteran insects in the superfamily Scolioidea. { sə'pij·ə,dē }

Sarcina [MICROBIOLOGY] A genus of strictly anaerobic bacteria in the family Peptococcaceae; spherical cells occur in packets; ferment carbohydrates. { 'sär·sə·nə }

sarcochore [BOTANY] A plant dispersing minute, light disseminules. { 'sär·kə,kȯr }

Sarcodina [INVERTEBRATE ZOOLOGY] A superclass of Protozoa in the subphylum Sarcomastigophora in which movement involves protoplasmic flow, often with recognizable pseudopodia. { ,sär·kə'dī·nə }

sarcoglia [CELL BIOLOGY] The protoplasm occurring at a myoneural junction. { sär'käg·lē·ə }

sarcolemma [HISTOLOGY] The thin connective tissue sheath enveloping a muscle fiber. { ,sär·kə'lem·ə }

Sarcomastigophora [INVERTEBRATE ZOOLOGY] A subphylum of Protozoa comprising forms that possess flagella or pseudopodia or both. { ,sar·kə,mas·tə'gäf·ə·rə }

sarcomere [HISTOLOGY] One of the segments defined by Z disks in a skeletal muscle fibril. { 'sär·kə,mir }

Sarcophagidae [INVERTEBRATE ZOOLOGY] A family of the myodarian orthorrhaphous dipteran insects in the subsection Calypteratae comprising flesh flies, blowflies, and scavenger flies. { ,sär·kə'fa·jə,dē }

sarcoplasm [HISTOLOGY] Hyaline, semifluid interfibrillar substance of striated muscle tissue. { 'sär·kə,plaz·əm }

sarcoplasmic reticulum [CELL BIOLOGY] Collectively, the cysternae of a single muscle fiber. { ¦sar·kə¦plaz·mik rə'tik·yə·ləm }

Sarcopterygii [VERTEBRATE ZOOLOGY] A subclass of Osteichthyes, including Crossopterygii and Dipnoi in some systems of classification. { sär ,käp·tə'rij·ē,ī }

Sarcoptiformes [INVERTEBRATE ZOOLOGY] A suborder of the Acarina including minute globular mites without stigmata. { sär,käp·tə'fȯr,mēz }

sarcosoma [INVERTEBRATE ZOOLOGY] The fleshy portion of an anthozoan. { ,sär·kə'sō·mə }

Sarcosporida [INVERTEBRATE ZOOLOGY] An order of Protozoa of the class Haplosporea which comprises parasites in skeletal and cardiac muscle of vertebrates. { ,sär·kə'spȯr·əd·ə }

sarcostyle [INVERTEBRATE ZOOLOGY] A fibril or column of muscular tissue. { 'sär·kə,stīl }

sarcotubule [CELL BIOLOGY] A tubular invagination of a muscle fiber. { ¦sär·kō'tü,byül }

sardine [VERTEBRATE ZOOLOGY] **1.** *Sardina pilchardus*. The young of the pilchard, a herringlike fish in the family Clupeidae found in the Atlantic along the European coasts. **2.** The young of any of various similar and related forms which are processed and eaten as sardines. { sär'dēn }

sarkomycin [MICROBIOLOGY] $C_7H_8O_3$ An antibiotic produced by an actinomycete which acts as a carcinolytic agent. { ,sär·kə'mīs·ən }

sarmentocymarin [BIOCHEMISTRY] A cardioactive, steroid glycoside from the seeds of *Strophanthus sarmentosus*; on hydrolysis it yields sarmentogenin and sarmentose. { sär¦men·tō'sī·mə·rən }

sarmentogenin [BIOCHEMISTRY] $C_{23}H_{34}O_5$ The steroid aglycon of sarmentocymarin; isometric with digitoxigenin, and characterized by a hydroxyl group at carbon number 11. { sär,men·tō 'jen·ən }

sarmentose [BOTANY] Producing slender, prostrate stems or runners. { sär'men,tōs }

Sarothriidae [INVERTEBRATE ZOOLOGY] The equivalent name for Jacobsoniidae. { ,sar·ə 'thrī·ə,dē }

Sarraceniaceae [BOTANY] A small family of dicotyledonous plants in the order Sarraceniales in which leaves are modified to form pitchers, placentation is axile, and flowers are perfect with distinct filaments. { ,sar·ə,sē·nē'ās·ē,ē }

Sarraceniales [BOTANY] An order of dicotyledonous herbs or shrubs in the subclass Dilleniidae; plants characteristically have alternate, simple leaves that are modified for catching insects, and grow in waterlogged soils. { ,sar·ə,sē·nē'ā·lēz }

sarsaparilla [BOTANY] Any of various tropical American vines of the genus *Smilax* (family Liliaceae) found in dense, moist jungles; a flavoring material used in medicine and soft drinks is obtained from the dried roots of at least four species. { ,sas·pə'ril·ə }

sartorius [ANATOMY] A large muscle originating in the anterior superior iliac spine and inserting in the tibia; flexes the hip and knee joints, and rotates the femur laterally. { sär'tȯr·ē·əs }

sassafras [BOTANY] *Sassafras albidum*. A medium-sized tree of the order Magnoliales recognized by the bright-green color and aromatic odor of the leaves and twigs. { 'sas·ə,fras }

satellite [CELL BIOLOGY] A chromosome segment distant from but attached to the rest of the chromosome by an achromatic filament. { 'sad·əl ,īt }

satellite band [MOLECULAR BIOLOGY] A fraction of the deoxyribonucleic acid (DNA) of an organism which has a different density from the rest and is therefore separable as a band in density gradient centrifugation; these bands are usually made up of highly repetitive sequences of DNA. { 'sad·əl,īt ,band }

satellite cell [HISTOLOGY] One of the neurilemmal cells surrounding nerve cells in the peripheral nervous system. { 'sad·əl,īt ,sel }

satellite deoxyribonucleic acid [MOLECULAR BIOLOGY] Any fraction, usually highly repetitious, of chromosomal deoxyribonucleic acid that differs significantly in its base composition from the majority of other fractions. { ¦sad·ə,līt dē ,äk·sē'rī·bō·nü,klē·ik 'as·əd }

saturation diving [PHYSIOLOGY] Diving in which

the tissues exposed to high pressure at great ocean depths for 24 hours become saturated with gases, especially inert gases, thereby reaching a new equilibrium state. { ˌsach·əˈrā·shən ˈdīv·iŋ }

Saturniidae [INVERTEBRATE ZOOLOGY] A family of medium- to large-sized moths in the superfamily Saturnioidea including the giant silkworms, characterized by reduced, often vestigial, mouthparts and strongly bipectinate antennae. { ˌsad·ərˈnī·əˌdē }

Saturnioidea [INVERTEBRATE ZOOLOGY] A superfamily of medium- to very-large-sized moths in the suborder Heteroneura having the frenulum reduced or absent, reduced mouthparts, no tympanum, and pectinate antennae. { ˌsad·ər·nē ˈoid·ē·ə }

Satyrinae [INVERTEBRATE ZOOLOGY] A large, cosmopolitan subfamily of lepidopterans in the family Nymphalidae, containing the wood nymphs, meadow browns, graylings, and arctics, characterized by bladderlike swellings at the bases of the forewing veins. { səˈtir·əˌnē }

Sauria [VERTEBRATE ZOOLOGY] The lizards, a suborder of the Squamata, characterized generally by two or four limbs but sometimes none, movable eyelids, external ear openings, and a pectoral girdle. { ˈsȯr·ē·ə }

Saurichthyidae [PALEONTOLOGY] A family of extinct chondrostean fishes bearing a superficial resemblance to the Aspidorhynchiformes. { ˌsȯr·əkˈthī·əˌdē }

Saurischia [PALEONTOLOGY] The lizard-hipped dinosaurs, an order of extinct reptiles in the subclass Archosauria characterized by an unspecialized, three-pronged pelvis. { sȯˈris·kē·ə }

Sauropoda [PALEONTOLOGY] A group of fully quadrupedal, seemingly herbivorous dinosaurs from the Jurassic and Cretaceous periods in the suborder Sauropodomorpha; members had small heads, spoon-shaped teeth, long necks and tails, and columnar legs. { sȯˈräp·əd·ə }

Sauropodomorpha [PALEONTOLOGY] A suborder of extinct reptiles in the order Saurischia, including large, solid-limbed forms. { sȯ¦räp·əd·ə ˈmȯr·fə }

Sauropterygia [PALEONTOLOGY] An order of Mesozoic marine reptiles in the subclass Euryapsida. { sȯˌräp·təˈrij·ē·ə }

Saururaceae [BOTANY] A family of dicotyledonous plants in the order Piperales distinguished by mostly alternate leaves, two to ten ovules per carpel, and carpels distinct or united into a compound ovary. { ˌsȯ·rəˈrās·ēˌē }

savane armée [ECOLOGY] *See* thornbush. { saˈvän ärˈmā }

savane épineuse [ECOLOGY] *See* thornbush. { sa ˈvän ā·pəˈnüz }

savanna [ECOLOGY] Any of a variety of physiognomically or environmentally similar vegetation types in tropical and extratropical regions; all contain grasses and one or more species of trees of the families Leguminosae, Bombacaceae, Bignoniaceae, or Dilleniaceae. { sə ˈvan·ə }

savanna-woodland [ECOLOGY] *See* tropical woodland. { səˈvan·ə ˈwu̇d·lənd }

savory [BOTANY] A herb of the mint family in the genus *Satureia*; of the more than 100 species, only summer savory (*S. hortensis*) and winter savory (*S. montana*) are grown for flavoring purposes. { ˈsav·ə·rē }

sawfish [VERTEBRATE ZOOLOGY] Any of several elongate viviparous fishes of the family Pristidae distinguished by a dorsoventrally flattened elongated snout with stout toothlike projections along each edge. { ˈsȯˌfish }

saxicolous [ECOLOGY] Living or growing among rocks. { sakˈsik·ə·ləs }

Saxifragaceae [BOTANY] A family of dicotyledonous plants in the order Rosales which are scarcely or not at all succulent and have two to five carpels usually more or less united, and leaves not modified into pitchers. { ˌsak·sə·frə ˈgās·ēˌē }

saxitoxin [BIOCHEMISTRY] A nonprotein toxin produced by the dinoflagellate *Gonyaulax catenella*. { ¦sak·sə¦täk·sən }

scabrous [BIOLOGY] Having a rough surface covered with stiff hairs or scales. { ˈskab·rəs }

scala media [ANATOMY] The middle channel of the cochlea, filled with endolymph and bounded above by Reissner's membrane and below by the basilar membrane. Also known as cochlear duct. { ˈskā·lə ˈmē·dē·ə }

scalariform [BIOLOGY] Resembling a ladder; having transverse markings or bars. { skəˈlar·ə ˌfȯrm }

scala tympani [ANATOMY] The lowest channel in the cochlea of the ear; filled with perilymph. { ˈskā·lə timˈpan·ē }

scala vestibuli [ANATOMY] The uppermost channel of the cochlea; filled with perilymph. { ˈskā· lə veˈstib·yə·lē }

scale [BOTANY] The bract of a catkin. [VERTEBRATE ZOOLOGY] A flat calcified or cornified platelike structure on the skin of most fishes and of some tetrapods. { skāl }

scale insect [INVERTEBRATE ZOOLOGY] Any of various small, structurally degenerate homopteran insects in the superfamily Coccoidea which resemble scales on the surface of a host plant. { ˈskāl ˈinˌsekt }

scalenus [ANATOMY] One of three muscles in the neck, arising from the transverse processes of the cervical vertebrae, and inserted on the first two ribs. { skāˈlē·nəs }

scale scar [BOTANY] A mark left on a stem after bud scales have fallen off. { ˈskāl ˌskär }

Scalibregmidae [INVERTEBRATE ZOOLOGY] A family of mud-swallowing worms belonging to the Sedentaria and found chiefly in sublittoral and great depths. { ˌskal·əˈbreg·məˌdē }

scaling [BIOLOGY] The removing of scales from fishes. { ˈskāl·iŋ }

scallion [BOTANY] *See* shallot. { ˈskal·yən }

scallop [INVERTEBRATE ZOOLOGY] Any of various bivalve mollusks in the family Pectinidae distinguished by radially ribbed valves with undulated margins. { ˈskäl·əp }

Scalpellidae [INVERTEBRATE ZOOLOGY] A primitive family of barnacles in the suborder Lepadomorpha having more than five plates. { skal 'pel·ə‚dē }

scandent [BOTANY] Climbing by stem-roots or tendrils. { 'skan·dənt }

scansorial [BOTANY] Adapted for climbing. { skan'sȯr·ē·əl }

Scapanorhychidae [VERTEBRATE ZOOLOGY] The goblin sharks, a family of deep-sea galeoids in the isurid line having long, sharp teeth and a long, pointed rostrum. { ‚skap·ə·nō'rik·ə‚dē }

scapha [ANATOMY] The furrow of the auricle between the helix and the antihelix. { 'skaf·ə }

Scaphidiidae [INVERTEBRATE ZOOLOGY] The shining fungus beetles, a family of coleopteran insects in the superfamily Staphylinoidea. { ‚skaf·ə'dī·ə‚dē }

scaphoid [ANATOMY] A boat-shaped bone of the carpus or of the tarsus. { 'skaf‚ȯid }

Scaphopoda [INVERTEBRATE ZOOLOGY] A class of the phylum Mollusca in which the soft body fits the external, curved and tapering, nonchambered, aragonitic shell which is open at both ends. { skə'fäp·əd·ə }

scapula [ANATOMY] The large, flat, triangular bone forming the back of the shoulder. Also known as shoulder blade. { 'skap·yə·lə }

scapulet [INVERTEBRATE ZOOLOGY] In some medusae, fringed outgrowths on the outer surfaces of the arms near the bell. { 'skap·yə‚let }

scapulus [INVERTEBRATE ZOOLOGY] A modified submarginal region in some sea anemones. { 'skap·yə·ləs }

scapus [BIOLOGY] The stem, shaft, or column of a structure. { 'skā·pəs }

Scarabaeidae [INVERTEBRATE ZOOLOGY] The lamellicorn beetles, a large cosmopolitan family of coleopteran insects in the superfamily Scarabaeoidea including the Japanese beetle and other agricultural pests. { ‚skar·ə'bē·ə‚dē }

Scarabaeoidea [INVERTEBRATE ZOOLOGY] A superfamily of Coleoptera belonging to the suborder Polyphaga. { ‚skar·ə·bē'ȯid·ē·ə }

Scaridae [VERTEBRATE ZOOLOGY] The parrotfishes, a family of perciform fishes in the suborder Percoidei which have the teeth of the jaw generally coalescent. { 'skar·ə‚dē }

scarious [BOTANY] Having a thin, membranous texture. { 'skar·ē·əs }

scarlet fever streptococcus antitoxin [IMMUNOLOGY] A sterile aqueous solution of antitoxins obtained from the blood of animals immunized against group A beta hemolytic streptococci toxin; formerly used in the treatment of, and to produce immunity against, scarlet fever. { 'skär·lət 'fē·vər ¦strep·tə¦käk·əs ‚ant·i'täk·sən }

scarlet fever streptococcus toxin [IMMUNOLOGY] Toxic filtrate of cultures of *Streptococcus pyogenes* responsible for the characteristic rash of scarlet fever; the toxin is used in the Dick test. { 'skär·lət 'fē·vər ¦strep·tə¦käk·əs 'täk·sən }

Scarpa's fascia [ANATOMY] The deep, membranous layer of the superficial fascia of the lower abdomen. { 'skär·pəz ‚fā·shə }

Scatopsidae [INVERTEBRATE ZOOLOGY] The minute black scavenger flies, a family of orthorrhaphous dipteran insects in the series Nematocera. { skə'täp·sə‚dē }

scavenger [ECOLOGY] An organism that feeds on carrion, refuse, and similar matter. { 'skav·ən·jər }

Scelionidae [INVERTEBRATE ZOOLOGY] A family of small, shining wasps in the superfamily Proctotrupoidea, characterized by elbowed, 11- or 12-segmented antennae. { ‚sel·ē'än·ə‚dē }

scent gland [VERTEBRATE ZOOLOGY] A specialized skin gland of the tubuloalveolar or acinous variety which produces substances having peculiar odors; found in many mammals. { 'sent ‚gland }

Schardinger dextrin [BIOCHEMISTRY] *See* cycloamylose. { 'shärd·ən·jər 'deks·trən }

schemochrome [ZOOLOGY] A feather color that originates within the feather structures, through refraction of light independent of pigments. { 'skē·mə‚krōm }

Schick test [IMMUNOLOGY] A skin test for determining susceptibility to diphtheria performed by the intradermal injection of diluted diphtheria toxin; a positive reaction, showing edema and scaling after 5 to 7 days, indicates lack of immunity. { 'shik ‚test }

Schindleriidae [VERTEBRATE ZOOLOGY] The single family of the order Schindlerioidei. { ‚shind·lə'rī·ə‚dē }

Schindlerioidei [VERTEBRATE ZOOLOGY] A suborder of fishes in the order Perciformes composed of one monogeneric family comprising two tiny oceanic species that are transparent and neotenic. { ‚shind·lə·rē'ȯid·ē‚ī }

schindylesis [ANATOMY] A synarthrosis in which a plate of one bone is fixed in a fissure of another. { ‚skin·də'lē·səs }

Schistostegiales [BOTANY] A monospecific order of mosses; the small, slender, glaucous plants are distinguished by the luminous protonema. { ¦shis·tə‚stej·ē'ā·lēz }

schizaxon [ANATOMY] An axon that divides, in its course, into equal or nearly equal branches. { skiz'ak‚sän }

schizocarp [BOTANY] A dry fruit that separates at maturity into single-seeded indehiscent carpels. { 'skiz·ə‚kärp }

Schizocoela [INVERTEBRATE ZOOLOGY] A group of animal phyla, including Bryozoa, Brachiopoda, Phoronida, Sipunculoidea, Echiuroidea, Priapuloidea, Mollusca, Annelida, and Arthropoda, all characterized by the appearance of the coelom as a space in the embryonic mesoderm. { ¦skiz·ə¦sē·lə }

schizodont [INVERTEBRATE ZOOLOGY] A multinucleate trophozoite that segments into merozoites. { 'skiz·ə‚dänt }

schizogamy [BIOLOGY] A form of reproduction involving division of an organism into a sexual and an asexual individual. { ski'zäg·ə·mē }

schizogenesis [BIOLOGY] Reproduction by fission. { ‚ski·zō'jen·ə·səs }

schizognathous [VERTEBRATE ZOOLOGY] Descriptive of birds having a palate in which the vomer is small and pointed, the maxillopalatines are not united with each other or with the vomer, and the palatines articulate posteriorly with the rostrum. { ski'zäg·nə·thəs }

Schizogoniales [BOTANY] A small order of the Chlorophyta containing algae that are either submicroscopic filaments or macroscopic ribbons or sheets a few centimeters wide and attached by rhizoids to rocks. { ,skiz·ə,gō·nē'ā·lēz }

schizogony [INVERTEBRATE ZOOLOGY] Asexual reproduction by multiple fission of a trophozoite; a characteristic of certain Sporozoa. { ski'zäg·ə·nē }

Schizomeridaceae [BOTANY] A family of green algae in the order Ulvales. { ¦skiz·ə,mer·ə'dās·ē,ē }

Schizomycetes [MICROBIOLOGY] Formerly a class of the division Protophyta which included the bacteria. { ,skiz·ə·mī'sēd·ēz }

Schizomycophyta [BOTANY] The designation for bacteria in those taxonomic systems that consider bacteria as plants. { ,skiz·ə·mī'käf·əd·ə }

schizont [INVERTEBRATE ZOOLOGY] A multinucleate cell in certain members of the Sporozoa that is produced from a trophozoite in a cell of the host, and that segments into merozoites. { 'skī ,zänt }

Schizopathidae [INVERTEBRATE ZOOLOGY] A family of dimorphic zoanthartans in the order Antipatharia. { ,skiz·ə'path·ə,dē }

schizopelmous [VERTEBRATE ZOOLOGY] Having the two flexor tendons of the toes separate, as in certain birds. { ¦ski·zō¦pel·məs }

Schizophora [INVERTEBRATE ZOOLOGY] A series of the dipteran suborder Cyclorrhapha in which adults possess a frontal suture through which a distensible sac, or ptilinum, is pushed to help the organism escape from its pupal case. { ski 'zäf·ə·rə }

Schizophyceae [MICROBIOLOGY] The blue-green algae, a class of the division Protophyta. { ,skiz·ə'fī·sē,ē }

Schizophyta [BOTANY] The prokaryotes, a division of the plant subkingdom Thallobionta; includes the bacteria and blue-green algae. { ski 'zäf·əd·ē }

schizopod [INVERTEBRATE ZOOLOGY] **1.** Having the limbs split so that each has an endopodite and an exopodite, as in certain crustaceans. **2.** A biramous appendage. { 'skiz·ə,päd }

Schizopteridae [INVERTEBRATE ZOOLOGY] A family of minute ground-inhabiting hemipterans in the group Dipsocoeoidea; individuals characteristically live in leaf mold. { ,ski·zäp'ter·ə,dē }

schizorhinal [VERTEBRATE ZOOLOGY] Having a deep cleft on the posterior margin of the osseous external nares, as in certain birds. { ¦skiz·ə¦rīn·əl }

schizothecal [VERTEBRATE ZOOLOGY] Having the horny envelope of the tarsus divided into scalelike plates; refers to most birds. { ¦skiz·ə'thē·kəl }

schizothoracic [INVERTEBRATE ZOOLOGY] Having a prothorax that is large and loosely articulated with the thorax. { ¦ski·zō·thə'ras·ik }

Schlemm's canal [ANATOMY] A space or series of spaces at the junction of the sclera and cornea in the eye; drains aqueous humor from the anterior chamber. { 'shlemz kə,nal }

Schneiderian membrane [ANATOMY] The mucosa lining the nasal cavities and paranasal sinuses. { shnī'dir·ē·ən 'mem,brān }

Schoenbiinae [INVERTEBRATE ZOOLOGY] A subfamily of moths in the family Pyralididae, including the genus *Acentropus*, the most completely aquatic Lepidoptera. { shən'bī·ə,nē }

Schubertellidae [PALEONTOLOGY] An extinct family of marine protozoans in the superfamily Fusulinacea. { ,shü·bər'tel·ə,dē }

Schultz-Charlton test [IMMUNOLOGY] A skin test for the diagnosis of scarlet fever, performed by the intradermal injection of human scarlet fever immune serum; a positive reaction consists of blanching of the rash in the area surrounding the point of injection. { 'shu̇lts 'chärlt·ən ,test }

Schultz-Dale reaction [IMMUNOLOGY] A method for demonstrating anaphylactic hypersensitivity outside the body by suspending an excised intestinal loop or uterine strip from a sensitized animal in an oxygenated, physiological salt solution; addition of the proper allergen causes contraction of the smooth muscle. { 'shu̇lts 'dāl rē,ak·shən }

Schwagerinidae [PALEONTOLOGY] A family of fusulinacean protozoans that flourished during the Early and Middle Pennsylvanian and became extinct during the Late Permian. { ,shwäg·ə'rin·ə ,dē }

Schwann cell [HISTOLOGY] One of the cells that surround peripheral axons forming sheaths of the neurilemma. { 'shwän ,sel }

Sciaenidae [VERTEBRATE ZOOLOGY] A family of perciform fishes in the suborder Percoidei, which includes the drums. { sī'ēn·ə,dē }

sciatic nerve [ANATOMY] Either of a pair of long nerves that originate in the lower spinal cord and send fibers to the upper thigh muscles and the joints, skin, and muscles of the leg. { sī'ad·ik 'nərv }

Scincidae [VERTEBRATE ZOOLOGY] The skinks, a family of the reptilian suborder Sauria which have reduced limbs and snakelike bodies. { 'skiŋ·kə,dē }

Scinidae [INVERTEBRATE ZOOLOGY] A family of bathypelagic, amphipod crustaceans in the suborder Hyperiidea. { 'skin·ə,dē }

Sciomyzidae [INVERTEBRATE ZOOLOGY] A family of myodarian cyclorrhaphous dipteran insects in the subsection Acalypteratae. { ,sī·ə'miz·ə,dē }

scion [BOTANY] A section of a plant, usually a stem or bud, which is attached to the stock in grafting. { 'sī·ən }

sciophilous [ECOLOGY] Capable of thriving in shade. { sī'äf·ə·ləs }

sciophyte [BOTANY] A plant that thrives at lowered light intensity. { 'sī·ə,fīt }

Scitaminales [BOTANY] An equivalent name for Zingiberales. { ,sīd·ə·mə'nā·lēz }

Scitamineae [BOTANY] An equivalent name for Zingiberales. { ,sīd·ə'min·ē,ē }

Sciuridae [VERTEBRATE ZOOLOGY] A family of rodents including squirrels, chipmunks, marmots, and related forms. { sī'yu̇r·ə,dē }

Sciuromorpha [VERTEBRATE ZOOLOGY] A suborder of Rodentia according to the classical arrangement of the order. { sī,yůr·ə'mòr·fə }

sclera [ANATOMY] The hard outer coat of the eye, continuous with the cornea in front and with the sheath of the optic nerve behind. { 'skler·ə }

Scleractinia [INVERTEBRATE ZOOLOGY] An order of the subclass Zoantharia which comprises the true, or stony, corals; these are solitary or colonial anthozoans which attach to a firm substrate. { ,skler·ək'tin·ē·ə }

Scleraxonia [INVERTEBRATE ZOOLOGY] A suborder of cnidarians in the order Gorgonacea in which the axial skeleton has calcareous spicules. { ,skleer·ək'sō·nē·ə }

sclereid [BOTANY] A thick-walled, lignified plant cell typically found in sclerenchyma. { 'sklir·ē·əd }

sclerenchyma [BOTANY] A supporting plant tissue composed principally of sclereids whose walls are often mineralized. { sklə'reŋ·kə·mə }

sclerite [INVERTEBRATE ZOOLOGY] One of the sclerotized plates of the integument of an arthropod. { 'skle,rīt }

scleroblast [INVERTEBRATE ZOOLOGY] A spicule-secreting cell in Porifera. { 'skler·ə,blast }

scleroblastema [EMBRYOLOGY] Embryonic tissue from which bones are formed. { ¦skler·ō·bla'stē·mə }

sclerocaulous [BOTANY] Having a hard, dry stem because of exceptional development of sclerenchyma. { ¦skler·ō'kòl·əs }

sclerochore [BOTANY] A plant that disperses disseminules without apparent morphological adaptations. { 'skler·ə,kòr }

Sclerodactylidae [INVERTEBRATE ZOOLOGY] A family of echinoderms in the order Dendrochirotida having a complex calcareous ring. { ,skler·ō,dak'til·ə,dē }

sclerodermatous [INVERTEBRATE ZOOLOGY] Having a skeleton that is composed of scleroderm, as certain corals. [VERTEBRATE ZOOLOGY] Having a hard outer covering, for example, hard plate or horny scale. { ¦sler·ō¦dər·məd·əs }

Sclerogibbidae [INVERTEBRATE ZOOLOGY] A monospecific family of the hymenopteran superfamily Bethyloidea. { ,skler·ō'jib·ə,dē }

sclerophyllous [BOTANY] Characterized by thick, hard foliage due to well-developed sclerenchymatous tissue. { ¦skler·ə¦fil·əs }

scleroprotein [BIOCHEMISTRY] Any one of a class of proteins, such as keratin, fibroin, and the collagens, which occur in hard parts of the animal body and serve to support or protect. Also known as albuminoid. { ¦skler·ō'prō,tēn }

scleroseptum [INVERTEBRATE ZOOLOGY] A calcareous radial septum in certain corals. { ¦skler·ō'sep·təm }

sclerotesta [BOTANY] The middle hard layer of the testa in various seeds. { ,skler·ə'tes·tə }

sclerotic [ANATOMY] Pertaining to the sclera. { sklə'räd·ik }

sclerotium [MICROBIOLOGY] The hardened, resting or encysted condition of the plasmodium of Myxomycetes. [MYCOLOGY] A hardened, resting mass of hyphae, usually black on the outside, from which fructifications may develop. { sklə'rō·shəm }

sclerotome [EMBRYOLOGY] The part of a mesodermal somite which enters into the formation of the vertebrae. [VERTEBRATE ZOOLOGY] The fibrous tissue separating successive myotomes in certain lower vertebrates. { 'skler·ə,tōm }

scobiform [BOTANY] Resembling sawdust. { 'skäb·ə,fòrm }

scolecodont [PALEONTOLOGY] Any of the paired, pincerlike jaws occurring as fossils of annelid worms. { skō'lē·kə,dänt }

Scolecosporae [MYCOLOGY] A spore group of Fungi Imperfecti characterized by filiform spores. { ,skō·lə'käs·pə,rē }

scolex [INVERTEBRATE ZOOLOGY] The head of certain tapeworms, typically having a muscular pad with hooks, and two pairs of lateral suckers. { 'skō,leks }

Scoliidae [INVERTEBRATE ZOOLOGY] A family of the Hymenoptera in the superfamily Scolioidea. { skō'lī·ə,dē }

Scolioidea [INVERTEBRATE ZOOLOGY] A superfamily of hymenopteran insects in the suborder Apocrita. { ,skō·lē'òid·ē·ə }

scolop [INVERTEBRATE ZOOLOGY] The thickened, distal tip of a vibration-sensitive organ in insects. { 'skäl·əp }

Scolopacidea [VERTEBRATE ZOOLOGY] A large, cosmopolitan family of birds of the order Charadriiformes including snipes, sandpipers, curlews, and godwits. { ,skäl·ə·pə'sīd·ē·ə }

Scolopendridae [INVERTEBRATE ZOOLOGY] A family of centipedes in the order Scolopendromorpha which characteristically possess eyes. { ,skäl·ə'pen·drə,dē }

Scolopendromorpha [INVERTEBRATE ZOOLOGY] An order of the chilopod subclass Pleurostigmophora containing the dominant tropical forms, and also the largest of the centipedes. { ¦skäl·ə,pen·drə'mòr·fə }

scolophore [INVERTEBRATE ZOOLOGY] *See* scolopophore. { 'skäl·ə,fòr }

scolopophore [INVERTEBRATE ZOOLOGY] A spindle-shaped, bipolar nerve ending in the integument of insects, believed to be auditory in function. Also known as scolophore. { skə'läp·ə,fòr }

Scolytidae [INVERTEBRATE ZOOLOGY] The bark beetles, a large family of coleopteran insects in the superfamily Curculionoidea characterized by a short beak and clubbed antennae. { skə'lid·ə,dē }

Scombridae [VERTEBRATE ZOOLOGY] A family of perciform fishes in the suborder Scombroidei including the mackerels and tunas. { 'skäm·brə,dē }

Scombroidei [VERTEBRATE ZOOLOGY] A suborder of fishes in the order Perciformes; all are moderate- to large-sized shore and oceanic fishes having fixed premaxillae. { skäm'bròi·dē,ī }

scopa [INVERTEBRATE ZOOLOGY] A brushlike arrangement of short stiff hairs on the body surface of certain insects. { 'skō·pə }

Scopeumatidae [INVERTEBRATE ZOOLOGY] The

dung flies, a family of myodarian cyclorrhaphous dipteran insects in the subsection Calypteratae. { ˌskä·pyü'mad·əˌdē }

Scopidae [VERTEBRATE ZOOLOGY] A family of birds in the order Ciconiiformes containing a single species, the hammerhead (*Scopus umbretta*) of tropical Africa. { 'skäp·əˌdē }

scopula [ZOOLOGY] A tuft of hair, as on the feet and chelicerae of certain spiders. { 'skäp·yə·lə }

Scorpaenidae [VERTEBRATE ZOOLOGY] The scorpion fishes, a family of Perciformes in the suborder Cottoidei, including many tropical shorefishes, some of which are venomous. { skȯr'pē·nəˌdē }

Scorpaeniformes [VERTEBRATE ZOOLOGY] An order of fishes coextensive with the perciform suborder Cottoidei in some systems of classification. { skȯrˌpē·nə'fōrˌmēz }

scorpioid cyme [BOTANY] A cyme with a curved axis and flowers arising two-ranked on alternate sides of the axis. { 'skȯr·pēˌȯid 'sīm }

scorpion [INVERTEBRATE ZOOLOGY] The common name for arachnids constituting the order Scorpionida. { 'skȯr·pē·ən }

Scorpionida [VERTEBRATE ZOOLOGY] The scorpions, an order of arachnids characterized by a shieldlike carapace covering the cephalothorax and by large pedipalps armed with chelae. { ˌskȯr·pē'än·əd·ə }

Scotch pine [BOTANY] *Pinus sylvestris*. A hard pine of North America having two short, bluish needles in a cluster. { 'skäch 'pīn }

scotochromogen [MICROBIOLOGY] **1.** Any microorganism which produces pigment when grown without light as well as with light. **2.** A member of group II of the atypical mycobacteria. { ˌskäd·ə'krō·mə·jən }

scotopic vision [PHYSIOLOGY] Vision that is due to the activity of the rods of the retina only; it is the type of vision that occurs at very low levels of illumination, and it can detect differences of brightness but not of hue. Also known as night vision. { skə'täp·ik·'vizh·ən }

scouring rush [BOTANY] *See* horsetail. { 'skau̇r·iŋ ˌrəsh }

Scraptidae [INVERTEBRATE ZOOLOGY] An equivalent name for Melandryidae. { 'skrap·təˌdē }

scrod [VERTEBRATE ZOOLOGY] A young fish, especially cod. { skräd }

Scrophulariaceae [BOTANY] A large family of dicotyledonous plants in the order Scrophulariales, characterized by a usually herbaceous habit, irregular flowers, axile placentation, and dry, dehiscent fruit. { ˌskräf·yəˌlar·ē'ās·ēˌē }

Scrophulariales [BOTANY] An order of flowering plants in the subclass Asteridae distinguished by a usually superior ovary and, generally, either by an irregular corolla or by fewer stamens than corolla lobes, or commonly both. { ˌskräf·yəˌlar·ē'ā·lēz }

scrotum [ANATOMY] The pouch containing the testes. { 'skrōd·əm }

scrub [ECOLOGY] A tract of land covered with a generally thick growth of dwarf or stunted trees and shrubs and a poor soil. { skrəb }

sculpin [VERTEBRATE ZOOLOGY] Any of several species of small fishes in the family Cottidae characterized by a large head that sometimes has spines, spiny fins, broad mouth, and smooth, scaleless skin. { 'skəl·pən }

scute [INVERTEBRATE ZOOLOGY] A cornified, epithelial, scalelike structure in lizards and snakes. { skyüt }

Scutechiniscidae [INVERTEBRATE ZOOLOGY] A family of heterotardigrades in the suborder Echiniscoidea, with segmental and intersegmental thickenings of cuticle. { sküˌtek·ə'nis·əˌdē }

Scutelleridae [INVERTEBRATE ZOOLOGY] The shield bugs, a family of Hemiptera in the superfamily Pentatomoidea. { ˌsküd·əl'er·əˌdē }

scutellum [BOTANY] **1.** A rounded apothecium with an elevated rim found in certain lichens. **2.** The flattened cotyledon of a monocotyledonous plant embryo, such as a grass. [INVERTEBRATE ZOOLOGY] The third of four pieces forming the upper part of the thoracic segment in certain insects. [VERTEBRATE ZOOLOGY] One of the scales on the tarsi and toes of birds. { skü'tel·əm }

Scutigeromorpha [INVERTEBRATE ZOOLOGY] The single order of notostigmophorous centipedes; members are distinguished by a dorsal respiratory opening, compound-type eyes, long flagellate multisegmental antennae, and long thin legs with multisegmental tarsi. { sküˌtij·ə·rə 'mȯr·fə }

scutum [INVERTEBRATE ZOOLOGY] **1.** A bony, horny, or chitinous plate. **2.** The second of four pieces forming the upper part of the thoracic segment in certain insects. **3.** One or two lower opercular valves in certain barnacles. { 'sküd·əm }

Scydmaenidae [INVERTEBRATE ZOOLOGY] The antlike stone beetles, a large cosmopolitan family of the Coleoptera in the superfamily Staphylinoidea. { sid'mē·nəˌdē }

Scyllaridae [INVERTEBRATE ZOOLOGY] The Spanish, or shovel-nosed, lobsters, a family of the Scyllaridea. { si'lar·əˌdē }

Scyllaridea [INVERTEBRATE ZOOLOGY] A superfamily of decapod crustaceans in the section Macrura including the heavily armored spiny lobsters and the Spanish lobsters, distinguished by the absence of a rostrum and chelae. { ˌsil·ə 'rid·ē·ə }

Scylliorhinidae [VERTEBRATE ZOOLOGY] The catsharks, a family of the carcharinid group of galeoids; members exhibit the most exotic color patterns of all sharks. { ˌsil·ē·ō'rin·əˌdē }

scyphistoma [INVERTEBRATE ZOOLOGY] A sessile, polyploid larva of many Scyphozoa which may produce either more scyphistomae or free-swimming medusae. { sī'fis·tə·mə }

scyphomedusa [INVERTEBRATE ZOOLOGY] A medusa of the scyphozoans. { ¦sī·fō·mə'dü·sə }

Scyphomedusae [INVERTEBRATE ZOOLOGY] A subclass of the class Scyphozoa characterized by reduced marginal tentacles, tetramerous medusae, and medusalike polyploids. { ¦sī·fō·mə'dü ˌsē }

scyphopolyp [INVERTEBRATE ZOOLOGY] A polyp of the scyphozoans. { ¦sī·fō'päl·əp }

Scyphozoa [INVERTEBRATE ZOOLOGY] A class of the phylum Cnidaria; all members are marine and are characterized by large, well-developed medusae and by small, fairly well-organized polyps. { ˌsī·fə'zō·ə }

sea anemone [INVERTEBRATE ZOOLOGY] Any of the 1000 marine cnidarians that constitute the order Actiniaria; the adult is a cylindrical polyp or hydroid stage with the free end bearing tentacles that surround the mouth. { 'sē əˌnem·ə·nē }

sea cucumber [INVERTEBRATE ZOOLOGY] The common name for the echinoderms that make up the class Holothuroidea. { 'sē kyüˌkəm·bər }

sea fan [INVERTEBRATE ZOOLOGY] A form of horny coral that branches like a fan. { 'sē ˌfan }

sea grass [BOTANY] Marine plants which are found in shallow brackish or marine waters, are more highly organized than algae, are seed-bearing, and attain lengths of up to 8 feet (2.4 meters). { 'sē ˌgras }

sea horse [INVERTEBRATE ZOOLOGY] Any of about 50 species of tropical and subtropical marine fishes constituting the genus *Hippocampus* in the family Syngnathidae; the body is compressed, the head is bent ventrally and has a tubiform snout, and the tail is tapering and prehensile. { 'sē ˌhȯrs }

seal [VERTEBRATE ZOOLOGY] Any of various carnivorous mammals of the suborder Pinnipedia, especially the families Phoridae, containing true seals, and Otariidae, containing the eared and fur seals. { sēl }

sea lily [INVERTEBRATE ZOOLOGY] The common name for those crinoids in which the body is flower-shaped and is carried at the tip of an anchored stem. { 'sē ˌlil·ē }

sea lion [VERTEBRATE ZOOLOGY] Any of several large, eared seals of the Pacific Ocean; related to fur seals but lack a valuable coat. { 'sē ˌlī·ən }

sea marsh [ECOLOGY] A salt marsh periodically overflowed or flooded by the sea. Also known as sea meadow. { 'sē ˌmärsh }

sea meadow [ECOLOGY] *See* sea marsh. { 'sē ˌmed·ō }

sea otter [VERTEBRATE ZOOLOGY] *Enhydra lutris*. A large marine otter found close to the shoreline in the North Pacific; these animals are diurnally active and live in herds. { 'sē ˌäd·ər }

sea pen [INVERTEBRATE ZOOLOGY] The common name for cnidarians constituting the order Pennatulacea. { 'sē ˌpen }

sea-run [VERTEBRATE ZOOLOGY] Having the habit of ascending a river from the sea, especially to spawn, as salmon and brook trout. { 'sēˌrən }

seashell [INVERTEBRATE ZOOLOGY] The shell of a marine invertebrate, especially a mollusk. { 'sē ˌshel }

sea slug [INVERTEBRATE ZOOLOGY] The common name for the naked gastropods composing the suborder Nudibranchia. { 'sē ˌsləg }

sea spider [INVERTEBRATE ZOOLOGY] The common name for arthropods in the subphylum Pycnogonida. { 'sē ˌspī·dər }

sea squirt [INVERTEBRATE ZOOLOGY] A sessile, marine tunicate of the class Ascidiacea; it squirts water from two openings in the unattached end when touched or disturbed. { 'sē ˌskwərt }

sea turtle [VERTEBRATE ZOOLOGY] Any of various marine turtles, principally of the families Cheloniidae and Dermochelidae, having paddle-shaped feet. { 'sē ˌtərd·əl }

seatworm [INVERTEBRATE ZOOLOGY] *See* pinworm. { 'sētˌwərm }

sea urchin [INVERTEBRATE ZOOLOGY] A marine echinoderm of the class Echinoidea; the soft internal organs are enclosed in and protected by a test or shell consisting of a number of close-fitting plates beneath the skin. { 'sē ˌər·chən }

seaweed [BOTANY] A marine plant, especially algae. { 'sēˌwēd }

sebaceous gland [PHYSIOLOGY] A gland, arising in association with a hair follicle, which produces and liberates sebum. { si'bā·shəs 'gland }

Sebekidae [INVERTEBRATE ZOOLOGY] A family of pentastomid arthropods in the suborder Porocephaloidea. { si'bek·əˌdē }

sebum [PHYSIOLOGY] The secretion of sebaceous glands, composed of fat, cellular debris, and keratin. { 'sē·bəm }

secalose [BIOCHEMISTRY] A polysaccharide consisting of fructose units; occurs in green rye and oats, and in rye flour. { 'sek·əˌlōs }

Secernentea [INVERTEBRATE ZOOLOGY] A class of the phylum Nematoda in which the primary excretory system consists of intracellular tubular canals joined anteriorly and ventrally in an excretory sinus, into which two ventral excretory gland cells may also open. { ˌse·sər'nen·chə }

secodont [VERTEBRATE ZOOLOGY] Having teeth adapted for cutting. { 'sek·əˌdänt }

secondary cambium [BOTANY] One of the tissue layers formed after the initial cambial layers in certain plant roots, and that produce a ring of tissue. { 'sek·ənˌder·ē 'kam·bē·əm }

secondary meristem [BOTANY] Meristem developed from differentiated living tissue. { 'sek·ən ˌder·ē 'mer·əˌstem }

secondary metabolite [BOTANY] A natural chemical product of plants not normally involved in primary metabolic processes such as photosynthesis and cell respiration. Also known as secondary plant product. { 'sek·ənˌder·ē mə'tab·ə ˌlīt }

secondary periderm [BOTANY] Any layer of the periderm except the first and outermost layer. { 'sek·ənˌder·ē 'per·iˌdərm }

secondary phloem [BOTANY] Phloem produced by the cambium, consisting of two interpenetrating systems, the vertical or axial and the horizontal or ray. { 'sek·ənˌder·ē 'flō·əm }

secondary plant product [BOTANY] *See* secondary metabolite. { 'sek·ənˌder·ē 'plant ˌpräd·əkt }

secondary root [BOTANY] A root arising from a primary root. { 'sek·ənˌder·ē 'rüt }

secondary structure [BIOCHEMISTRY] The conformation of a protein or peptide molecule with respect to nearest-neighbor amino acids. [PALEONTOLOGY] A coarse structure usually between the thin sheets in the protective wall of a tintinnid. { 'sek·ənˌder·ē 'strək·chər }

secondary tissue [BOTANY] Tissue that develops

from the vascular cambium or from differentiated tissues. { 'sek·ən,der·ē 'tish·ü }

secondary tympanic membrane [ANATOMY] The membrane closing the fenestra cochleae. { 'sek·ən,der·ē tim'pan·ik 'mem,brān }

secondary wall [BOTANY] The portion of a plant cell wall produced internal to and following deposition of the primary wall; usually consists of several anisotropic layers, and often has prominent internal rings, spirals, bars, or reticulations. { 'sek·ən,der·ē 'wȯl }

secondary xylem [BOTANY] Xylem produced by cambium, composed of two interpenetrating systems, the horizontal (ray) and vertical (axial). { 'sek·ən,der·ē 'zī·ləm }

second generation [GENETICS] *See* F. { 'sek·ənd ,jen·ə'rā·shən }

second messenger [MOLECULAR BIOLOGY] Any small molecule or ion that occurs in the cytoplasm of a cell, is generated in response to a hormone binding to a cell-surface receptor, and activates various kinases that regulate the activities of other enzymes. { 'sek·ənd 'mes·ən·jər }

secretin [BIOCHEMISTRY] A basic polypeptide hormone produced by the duodenum in response to the presence of acid; acts to excite the pancreas to activity. { si'krēt·ən }

secretion [PHYSIOLOGY] **1.** The act or process of producing a substance which is specialized to perform a certain function within the organism or is excreted from the body. **2.** The material produced by such a process. { si'krē·shən }

secretor gene [GENETICS] A dominant autosomal gene in man which controls secretion of A and B antigenic material in saliva, urine, plasma, and other body fluids; it is not linked to the ABO genes. { si'krēd·ər ,jēn }

secretory granules [CELL BIOLOGY] Accumulations of material produced within a cell for secretion outside the cell. { si'krēd·ə·rē ,gran ,yülz }

secretory structure [BOTANY] Plant cells or organizations of plant cells which produce a variety of secretions. { si'krēd·ə·rē ,strək·chər }

secund [BOTANY] Having lateral members arranged on one side only. { 'sē,kənd }

Sedentaria [INVERTEBRATE ZOOLOGY] A group of families of polychaete annelids in which the anterior, or cephalic, region is more or less completely concealed by overhanging peristomial structures, or the body is divided into an anterior thoracic and a posterior abdominal region. { ,sed·ən'tar·ē·ə }

sedoheptulose [BIOCHEMISTRY] A seven carbon ketose sugar widely distributed in plants of the Crassulaceae family; a significant intermediary compound in the cyclic regeneration of D-ribulose. { ¦sē·dō'hep·tə,lōs }

seed [BOTANY] A fertilized ovule containing an embryo which forms a new plant upon germination. { sēd }

seed coat [BOTANY] The envelope which encloses the seed except for a tiny pore, the micropyle. { 'sēd ,kōt }

seed fern [PALEOBOTANY] The common name for the extinct plants classified as Pteridospermae, characterized by naked seeds borne on large, fernlike fronds. { 'sēd ,fərn }

seedling [BOTANY] **1.** A plant grown from seed. **2.** A tree younger and smaller than a sapling. **3.** A tree grown from a seed. { 'sēd·liŋ }

segmental reflex [PHYSIOLOGY] A reflex arc having afferent inputs by way of the spinal dorsal roots, and efferent outputs over spinal ventral roots of the same or adjacent segments. { seg 'ment·əl 'rē,fleks }

segmentation [ZOOLOGY] *See* metamerism. { ,seg·mən'tā·shən }

segmented genome [GENETICS] In some ribonucleic acid viruses, two or more nonidentical ribonucleic acid molecules with different genetic information. { 'seg,men·təd 'jē,nōm }

segregation [GENETICS] The separation of alleles and homologous chromosomes during meiosis in the formation of gametes. { ,seg·rə'gā·shən }

segregation distorter [GENETICS] An abnormality of meiosis which produces a distortion of the 1:1 segregation ratio in a heterozygote. { ,seg·rə 'gā·shən dis,tȯr·dər }

Seisonacea [INVERTEBRATE ZOOLOGY] A monofamiliar order of the class Rotifera characterized by an elongated jointed body with a small head, a long slender neck region, a thick fusiform trunk, and an elongated foot terminating in a perforated disk. { ,sī·sə'nā·shə }

Seisonidea [INVERTEBRATE ZOOLOGY] The equivalent name for Seisonacea. { ,sī·sə'nīd·ē·ə }

Seitz filter [MICROBIOLOGY] A bacterial filter made of asbestos and used to sterilize solutions without the use of heat. { 'zītz ,fil·tər }

Selachii [VERTEBRATE ZOOLOGY] An order of elasmobranchs including all fossil sharks, except Cladoselachii and Pleuracanthodii. { sə'lāk·ē,ī }

Selaginellales [BOTANY] The plant order of small club mosses, containing one living genus, *Selaginella*; distinguished from other lycopods in being heterosporous and in having a ligule borne on the upper base of the leaf. { sə,laj·ə·nə'lā·lēz }

selection [GENETICS] Any natural or artificial process which favors the survival and propagation of individuals of a given genotype in a population. { si'lek·shən }

selection coefficient [GENETICS] A measure of the rate of transmission through successive generations of a given allele compared to the rate of transmission of another (usually the wild-type) allele. { si'lek·shən ,kō·i,fish·ənt }

selection pressure [EVOLUTION] Those factors that influence the direction of natural selection. { si'lek·shən ,presh·ər }

selective breeding [BIOLOGY] Breeding of animals or plants having desirable characters. { si 'lek·tiv 'brēd·iŋ }

selective medium [MICROBIOLOGY] A bacterial culture medium containing an individual organic compound as the sole source of carbon, nitrogen, or sulfur for growth of an organism. { si'lek·tiv 'mē·dē·əm }

selenodont [VERTEBRATE ZOOLOGY] **1.** Being or pertaining to molars having crescentic ridges on the crown. **2.** A mammal with selenodont dentition. { sə'lē·nə,dänt }

self-differentiation [PHYSIOLOGY] The differentiation of a tissue, even when isolated, solely as a result of intrinsic factors after determination. { ¦self ,dif·ə,ren·chē¦ā·shən }

self-incompatibility [BOTANY] Pertaining to an individual flower that cannot complete fertilization with its own pollen. { ¦self ,in·kəm,pad·ə 'bil·əd·ē }

selfish deoxyribonucleic acid [MOLECULAR BIOLOGY] Any tandemly repeated or dispersed repetitive deoxyribonucleic acid sequence that has no obvious function but can spread and accumulate in the species because of its rapid replication. Also known as junk deoxyribonucleic acid. { ¦sel·fish dē,äk·sē'rī·bō·nü,klē·ik 'as·əd }

self-pollination [BOTANY] Transfer of pollen from the anther to the stigma of the same flower or of another flower on the same plant. { ¦self ,päl·ə ¦nā·shən }

Seligeriales [BOTANY] An order of true mosses in the class Bryopsida; members grow on rocks and may be exceedingly small to moderate size and tufted; the double structure of the peristome is distinctive. { ,sel·ə,jir·ē'ā·lēz }

sella turcica [ANATOMY] A depression in the upper surface of the sphenoid bone in which the pituitary gland rests in vertebrates. { ,sel·ə 'tər·kə·kə }

selva [ECOLOGY] *See* tropical rainforest. { 'sel·və }

Semaeostomeae [INVERTEBRATE ZOOLOGY] An order of the class Scyphozoa including most of the common medusae, characterized by a flat, domelike umbrella whose margin is divided into many lappets. { sə,mē·ə'stō·mē,ē }

sematic [ECOLOGY] *See* aposematic. { si'mat·ik }

semelparity [BIOLOGY] Reproduction that occurs only one time during the life of an individual. { 'sem·əl,par·əd·ē }

semen [PHYSIOLOGY] The fluid that carries the male germ cells. Also known as seminal fluid. { 'sē·mən }

semicircular canal [ANATOMY] Any of three loop-shaped tubular structures of the vertebrate labyrinth; they are arranged in three different spatial planes at right angles to each other, and function in the maintenance of body equilibrium. { ¦sem·i'sər·kyə·lər kə'nal }

semiconservative replication [MOLECULAR BIOLOGY] Replication of deoxyribonucleic acid by longitudinal separation of the two complementary strands of the molecule, each being conserved and acting as a template for synthesis of a new complementary strand. { ¦sem·i·kən'sər·vəd·iv rep·li'kā·shən }

semidesert [ECOLOGY] An area intermediate in character and often located between a desert and a grassland or woodland. { ¦sem·i'dez·ərt }

semidormancy [BOTANY] Decrease in plant growth rate; may be seasonal or associated with unfavorable environmental conditions. { ¦sem·i 'dór·mən·sē }

semidouble [BOTANY] Pertaining to a flower that has more than the usual number of petals or disk florets while it retains some pollen-bearing stamens or some perfect disk florets. { ¦sem·i'dəb·əl }

semilate [BOTANY] Pertaining to a plant whose growing season is intermediate between midseason forms and late forms. { ¦sem·i'lāt }

semilethal gene [GENETICS] A mutant causing the death of some of the individuals of the relevant genotype, but never 100%. Also known as sublethal gene. { ¦sem·i'lēth·əl 'jēn }

semilunar cartilage [ANATOMY] One of the two interarticular knee cartilages. { ¦sem·i'lü·nər 'kärt·lij }

semilunar ganglion [ANATOMY] *See* Gasserian ganglion. { ¦sem·i'lü·nər 'gaŋ·glē·ən }

semilunar valve [ANATOMY] Either of two tricuspid valves in the heart, one at the orifice of the pulmonary artery and the other at the orifice of the aorta. { ¦sem·i'lü·nər 'valv }

semimembranosus [ANATOMY] One of the hamstring muscles, arising from the ischial tuber, and inserted into the tibia. { ¦sem·i,mem·brə 'nō·səs }

seminal bursa [INVERTEBRATE ZOOLOGY] A sac which retains sperm for a period of time in turbellarians. { 'sem·ən·əl 'bər·sə }

seminal fluid [PHYSIOLOGY] *See* semen. { 'sem·ə·nəl ,flü·əd }

seminal fructose [PHYSIOLOGY] Fructose that is normally produced in the seminal vesicles. { ¦sem·ən·əl 'frük,tōs }

seminal groove [ZOOLOGY] A passage in many animals providing a pathway for sperm. { 'sem·ən·əl 'grüv }

seminal receptacle [ZOOLOGY] *See* spermatheca. { 'sem·ən·əl ri'sep·tə·kəl }

seminal vesicle [ANATOMY] A saclike, glandular diverticulum on each ductus deferens in male vertebrates; it is united with the excretory duct and serves for temporary storage of semen. { 'sem·ən·əl 'ves·i·kəl }

seminiferous tubule [ANATOMY] Any of the tubercles of the testes which produce spermatozoa. { ¦sem·ə¦nif·rəs 'tü,byül }

seminivorous [ZOOLOGY] Feeding on seeds. { ¦sem·ə¦niv·ə·rəs }

semiochemical [PHYSIOLOGY] Any of a class of substances produced by organisms, especially insects, that participate in regulation of their behavior in such activities as aggregation of both sexes, sexual stimulation, and trail following. { ¦sem·ē·ə'kem·ə·kəl }

Semionotiformes [VERTEBRATE ZOOLOGY] An order of actinopterygian fishes represented by the single living genus *Lepisosteus*, the gars; the body is characteristically encased in a heavy armor of interlocking ganoid scales. { ,sem·ē·ə,nō·də'fór ,mēz }

semipalmate [VERTEBRATE ZOOLOGY] Having a web halfway down the toes. { ¦sem·i'päl,māt }

semiparasite [ECOLOGY] *See* hemiparasite. { ¦sem·i'par·ə,sīt }

semispecies [SYSTEMATICS] **1.** The species that compose a superspecies. **2.** Populations that have acquired some attributes of species rank.

3. Organisms that are borderline between species and subspecies. { ¦sem·i'spē·shēz }

semispinalis [ANATOMY] One of the deep longitudinal muscles of the back, attached to the vertebrae. { ,sem·i,spī'nal·əs }

semitendinosus [ANATOMY] One of the hamstring muscles, arising from the ischium and inserted into the tibia. { ,sem·ē,ten·də'nō·səs }

Semper's larva [INVERTEBRATE ZOOLOGY] A cylindrical larva in the life history of certain zoanthid corals, characterized by a hole at each end and an annular or longitudinal band of long cilia. { 'sem·pərz ,lär·və }

senescence [BIOLOGY] The study of the biological changes related to aging, with special emphasis on plant, animal, and clinical observations which may apply to humans. { si'nes·əns }

sensation [PHYSIOLOGY] The subjective experience that results from the stimulation of a sense organ. { sen'sā·shən }

sensation level [PHYSIOLOGY] *See* level above threshold. { sen'sā·shən ,lev·əl }

sense organ [PHYSIOLOGY] A structure which is a receptor for external or internal stimulation. { 'sens ,ȯr·gən }

sense strand [MOLECULAR BIOLOGY] The strand of a double-stranded deoxyribonucleic acid molecule that is complementary to the ribonucleic acid formed by transcription. Also known as coding strand. { 'sens ,strand }

sensillum [ZOOLOGY] A simple, epithelial sense organ composed of one cell or of a few cells. { sen'sil·əm }

sensitivity [PHYSIOLOGY] The capacity for receiving sensory impressions from the environment. { ,sen·sə'tiv·əd·ē }

sensitization [IMMUNOLOGY] The alteration of a body's responsiveness to a foreign antigen, usually an allergen, such that upon subsequent exposures to the allergen there is a heightened immune response. { ,sen·səd·ə'zā·shən }

sensorium [PHYSIOLOGY] **1.** A center, especially in the brain, for receiving and integrating sensations. **2.** The entire sensory apparatus of an individual. { sen'sȯr·ē·əm }

sensory area [PHYSIOLOGY] Any area of the cerebral cortex associated with the perception of sensations. { 'sen·sə·rē ,er·ē·ə }

sensory cell [PHYSIOLOGY] **1.** A neuron having its terminal processes connected with sensory nerve endings. **2.** A modified epithelial or connective tissue cell adapted for the reception and transmission of sensations. { 'sen·sə·rē ,sel }

sensory nerve [PHYSIOLOGY] A nerve that conducts afferent impulses from the periphery to the central nervous system. { 'sen·sə·rē ,nərv }

sepal [BOTANY] One of the leaves composing the calyx. { sēp·əl }

separation disk [BOTANY] A layer of gelatinous material between two adjacent negative cells in some blue-green algae; associated with hormogonium formation. { ,sep·ə'rā·shən ,disk }

separation layer [BOTANY] A structurally distinct layer of the abscission zone of a plant containing abundant starch and dense cytoplasm. { ,sep·ə'rā·shən ,lā ər }

Sepioidea [INVERTEBRATE ZOOLOGY] An order of the molluscan subclass Coleoidea having a well-developed eye, an internal shell, fins separated posteriorly, and chromatophores in the dermis. { ,sē·pē'ȯid·ē·ə }

Sepsidae [INVERTEBRATE ZOOLOGY] The spiny-legged flies, a family of myodarian cyclorrhaphous dipteran insects in the subsection Acalypteratae; development takes place in decaying organic matter. { 'sep·sə,dē }

septal filament [INVERTEBRATE ZOOLOGY] In anthozoans, the free edges of the septum containing gland cells and nematocysts. { 'sep·təl 'fil·ə·mənt }

septal ostium [INVERTEBRATE ZOOLOGY] Any of the openings in septa of anthozoans. { 'sep·təl 'äs·tē·əm }

septate [BIOLOGY] Having a septum. { 'sep,tāt }

Septibranchia [INVERTEBRATE ZOOLOGY] A small order of bivalve mollusks in which the anterior and posterior abductor muscles are about equal in size, the foot is long and slender, and the gills have been transformed into a muscular septum. { ,sep·tə'braŋ·kē·ə }

septicidal [BOTANY] A type of dehiscence exhibited by some fruit in which splitting of the pericarp occurs along the junction of component carpels. { ¦sep·tə¦sīd·əl }

septulum [ANATOMY] A small septum. { 'sep·tə·ləm }

septum [BIOLOGY] A partition or dividing wall between two cavities. { 'sep·təm }

septum pellucidum [ANATOMY] A thin translucent septum forming the internal boundary of the lateral ventricles of the brain and enclosing between its two laminas the so-called fifth ventricle. { 'sep·təm pə'lü·səd·əm }

septum primum [EMBRYOLOGY] The first incomplete interatrial septum of the embryo. { 'sep·təm 'prē·məm }

septum secundum [EMBRYOLOGY] The second incomplete interatrial septum of the embryo, containing the foramen ovale; it develops to the right of the septum primum and fuses with it to form the adult interatrial septum. { 'sep·təm si'kən·dəm }

Sequoia [BOTANY] A genus of conifers having overlapping, scalelike evergreen leaves and vertical grooves in the trunk; the giant sequoia (*Sequoia gigantea*) is the largest and oldest of all living things. { si'kwȯi·yə }

sere [ECOLOGY] A temporary community which occurs during a successional sequence on a given site. { sir }

Sergestidae [INVERTEBRATE ZOOLOGY] A family of decapod crustaceans including several species of prawns. { sər'jes·tə,dē }

serial homology [ZOOLOGY] The similarity between the members of a single series of structures, such as vertebrae, in an organism. { 'sir·ē·əl hə'mäl·ə·jē }

sericeous [BOTANY] Of, pertaining to, or consisting of silk. { ,sir·ē'ā·shəs }

serine [BIOCHEMISTRY] $C_3H_7O_3N$ An amino acid

obtained by hydrolysis of many proteins; a biosynthetic precursor of several metabolites, including cysteine, glycine, and choline. { 'se ˌrēn }

seritinous [ECOLOGY] Of, pertaining to, or occurring during the latter, drier half of the summer. { ¦ser·ə'tī·nəs }

serofibrinous [PHYSIOLOGY] Composed of serum and fibrin. { ¦si·rō'fī·brə·nəs }

Serolidae [INVERTEBRATE ZOOLOGY] A family of isopod crustaceans which contains greatly flattened forms that live partially buried on sandy bottoms. { sə'räl·əˌdē }

serology [BIOLOGY] The branch of science dealing with the properties and reactions of blood sera. { sə'räl·ə·jē }

serosa [ANATOMY] The serous membrane lining the pleural, peritoneal, and pericardial cavities. [EMBRYOLOGY] The chorion of reptile and bird embryos. { sə'rō·sə }

serotinous [BOTANY] Of plants, flowering or developing late in a season. { sə'rät·ən·əs }

serotonin [BIOCHEMISTRY] $C_{10}H_{12}ON_2$ A compound derived from tryptophan which functions as a local vasoconstrictor, plays a role in neurotransmission, and has pharmacologic properties. Also known as 5-hydroxytryptamine. { ˌsir·ə'tō·nən }

serotype [MICROBIOLOGY] A serological type of intimately related microorganisms, distinguished on the basis of antigenic composition. { 'sir·əˌtīp }

serous gland [PHYSIOLOGY] A structure that secretes a watery, albuminous fluid. { 'sir·əs ˌgland }

serous membrane [HISTOLOGY] A delicate membrane covered with flat, mesothelial cells lining closed cavities of the body. { 'sir·əs 'mem ˌbrān }

Serpentes [VERTEBRATE ZOOLOGY] The snakes, a suborder of the Squamata characterized by the lack of limbs and pectoral girdle and external ear openings, immovable eyelids, and a braincase that is completely bony anteriorly. { sər'pen ˌtēz }

serpentine locomotion [VERTEBRATE ZOOLOGY] The wavelike or undulating movements characteristic of snakes. { 'sər·pənˌtēn ˌlō·kə'mō·shən }

Serpulidae [INVERTEBRATE ZOOLOGY] A family of polychaete annelids belonging to the Sedentaria including many of the feather-duster worms which construct calcareous tubes in the earth, sometimes in such abundance as to clog drains and waterways. { sər'pyü·ləˌdē }

Serranidae [VERTEBRATE ZOOLOGY] A family of perciform fishes in the suborder Percoidei including the sea basses and groupers. { sə'ran·ə ˌdē }

serrate [BIOLOGY] Possessing a notched or toothed edge. { 'seˌrāt }

Serratieae [MICROBIOLOGY] Formerly a tribe of the Enterobacteriaceae containing the genus *Serratia*, with soil and water forms characterized by the production of a bright-orange to deep-red pigment, prodigiosin. { sə'rāsh·ēˌē }

Serridentinae [PALEONTOLOGY] An extinct subfamily of elephantoids in the family Gomphotheriidae. { ˌser·ə'dent·ənˌē }

Serritermitidae [INVERTEBRATE ZOOLOGY] A family of the Isoptera which contains the single monotypic genus *Serritermes*. { ˌser·ə·tər'mid·ə ˌdē }

Serropalpidae [INVERTEBRATE ZOOLOGY] An equivalent name for Melandryidae. { ˌser·ə'pal·pəˌdē }

serrulate [BIOLOGY] Finely serrate. { 'ser·ə·lət }

Sertoli cell [HISTOLOGY] One of the sustentacular cells of the seminiferous tubules. { sər'tō·lē ˌsel }

serum [PHYSIOLOGY] The liquid portion that remains when blood clots spontaneously and the formed and clotting elements are removed by centrifugation; it differs from plasma by the absence of fibrinogen. { 'sir·əm }

serum accident [IMMUNOLOGY] A serious allergic reaction which immediately follows the introduction of a foreign serum into a hypersensitive individual; dyspnea and flushing occur, soon followed by shock and occasionally by fatal termination. { 'sir·əm ˌak·sə·dənt }

serum albumin [BIOCHEMISTRY] The principal protein fraction of blood serum and serous fluids. { 'sir·əm al'byü·mən }

serum globulin [BIOCHEMISTRY] The globulin fraction of blood serum. { 'sir·əm 'glä·byə·lən }

sessile [BOTANY] Attached directly to a branch or stem without an intervening stalk. [ZOOLOGY] Permanently attached to the substrate. { 'ses·əl }

Sessilina [INVERTEBRATE ZOOLOGY] A suborder of ciliates in the order Peritrichida. { ˌses·ə'lī·nə }

sessoblast [INVERTEBRATE ZOOLOGY] A statoblast that attaches to zooecial tubes or to the substratum. { 'ses·əˌblast }

seta [BIOLOGY] **1.** A slender, usually rigid bristle or hair. Also known as chaeta. **2.** In mosses and liverworts, the stalk of the sporophyte supporting the capsule. { 'sed·ə }

setigerous [INVERTEBRATE ZOOLOGY] Referring to a segment with setae. { sə'tij·ə·rəs }

sex [BIOLOGY] **1.** The state of condition of an organism which comes to expression in the production of germ cells. **2.** To determine the sex of. { seks }

sex cell [BIOLOGY] *See* gamete. { 'seks ˌsel }

sex chromatin [CELL BIOLOGY] *See* Barr body. { 'seks 'krō·mə·tən }

sex chromosome [GENETICS] Either member of a pair of chromosomes responsible for sex determination of an organism. { 'seks ˌkrō·mə ˌsōm }

sex cords [EMBRYOLOGY] Cordlike masses of epithelial tissue that invaginate from germinal epithelium of the gonad and give rise to seminiferous tubules and rete testes in the male, and primary ovarian follicles and rete ovarii in the female. { 'seks ˌkȯrdz }

sex determination [GENETICS] The mechanisms by which sex is determined in a species. { 'seks diˌtər·məˌnā·shən }

sex factor [GENETICS] *See* fertility factor. { 'seks ˌfak·tər }

sex hormone [BIOCHEMISTRY] Any hormone secreted by a gonad, but also found in other tissues. { 'seks ˌhȯrˌmōn }

sex-influenced inheritance [GENETICS] That part of the inheritance pattern on which sex differences operate to promote character differences. { 'seks in¦flü·ənst in'her·ət·əns }

sex-limited inheritance [GENETICS] Expression of a phenotype in only one sex; may be due to either a sex-linked or autosomal gene. { 'seks ¦lim·əd·əd in'her·ət·əns }

sex-linked inheritance [GENETICS] The transmission to successive generations of differences that are due to genes located in the sex chromosomes. { 'seks ¦liŋkt in'her·ət·əns }

sex organs [ANATOMY] The organs pertaining entirely to the sex of an individual, both physiologically and anatomically. { 'seks ˌȯr·gənz }

sex ratio [BIOLOGY] The relative proportion of males and females in a population. { 'seks ˌrā·shō }

sexual cycle [PHYSIOLOGY] A cycle of physiological and structural changes associated with sex; examples are the estrous cycle and the menstrual cycle. { 'sek·shə·wəl 'sī·kəl }

sexual dimorphism [BIOLOGY] Diagnostic morphological differences between the sexes. { 'sek·shə·wəl dī'mȯrˌfiz·əm }

sexuality [BIOLOGY] **1.** The sum of a person's sexual attributes, behavior, and tendencies. **2.** The psychological and physiological sexual impulses whose satisfaction affords pleasure. { ˌsek·shə'wal·əd·ē }

sexual reproduction [BIOLOGY] Reproduction involving the paired union of special cells (gametes) from two individuals. { 'sek·shə·wəl ˌrē·prə'dək·shən }

sexual selection [EVOLUTION] A special form of natural selection responsible for the evolution of traits that promote success in competition for mates. { ¦seksh·ə·wəl si'lek·shən }

sexual spore [BIOLOGY] A spore resulting from conjugation of gametes or nuclei of opposite sex. { 'sek·shə·wəl 'spȯr }

Seymouriamorpha [PALEONTOLOGY] An extinct group of labyrinthodont Amphibia of the Upper Carboniferous and Permian in which the intercentra were reduced. { sēˌmȯr·ē·ə'mȯr·fə }

shadscale [BOTANY] *Atriplex confertifolia.* A small shiny shrub found in the Great Basin Desert. { 'shadˌskāl }

shake culture [MICROBIOLOGY] **1.** A method for isolating anaerobic bacteria by shaking a deep liquid culture of an agar or gelatin to distribute the inoculum before solidification of the medium. **2.** A liquid medium in a flask that has been inoculated with an aerobic microorganism and placed on a shaking machine; action of the machine continually aerates the culture. { 'shāk ˌkəl·chər }

shallot [BOTANY] *Allium ascalonicum.* A bulbous onionlike herb. Also known as scallion. { 'shal·ət }

shark [VERTEBRATE ZOOLOGY] Any of about 225 species of carnivorous elasmobranchs which occur principally in tropical and subtropical oceans; the body is fusiform with a heterocercal tail and a tough, usually gray, skin roughened by tubercles, and the snout extends beyond the mouth. { shärk }

shearwater [VERTEBRATE ZOOLOGY] Any of various species of oceanic birds of the genus *Puffinus* having tubular nostrils and long wings. { 'shir ˌwȯd·ər }

sheathed bacteria [MICROBIOLOGY] Chains of bacterial cells, usually rod-shaped, enclosed in a hyaline envelope or sheath. { 'shēthd bak'tir·ē·ə }

sheep [VERTEBRATE ZOOLOGY] Any of various mammals of the genus *Ovis* in the family Bovidae characterized by a stocky build and horns, when present, which tend to curl in a spiral. { shēp }

shell [ZOOLOGY] **1.** A hard, usually calcareous, outer covering on an animal body, as of bivalves and turtles. **2.** The hard covering of an egg. **3.** Chitinous exoskeleton of certain arthropods. { shel }

shellfish [INVERTEBRATE ZOOLOGY] An aquatic invertebrate, such as a mollusk or crustacean, that has a shell or exoskeleton. { 'shelˌfish }

shell gland [INVERTEBRATE ZOOLOGY] An organ that secretes the embryonic shell in many mollusks. [VERTEBRATE ZOOLOGY] A specialized structure attached to the oviduct in certain animals that secretes the eggshell material. { 'shel ˌgland }

shell membrane [CELL BIOLOGY] Either of a pair of membranes lining the inner surface of an egg shell; they allow free entry of oxygen but prevent rapid evaporation of moisture. { 'shel 'mem ˌbrān }

shelterbelt [ECOLOGY] A natural or planned barrier of trees or shrubs to reduce erosion and provide shelter from wind and storm activity. { 'shel·tərˌbelt }

Shetland sheep [VERTEBRATE ZOOLOGY] A breed of sheep raised in the Shetland Isles of Scotland. { 'shet·lənd 'shēp }

shikimic acid [BIOCHEMISTRY] $C_7H_{10}O_5$ A crystalline acid that is a plant constituent, and an intermediate in the biochemical pathway from phosphoenolpyruvic acid to tyrosine. { shə'kim·ik 'as·əd }

shipworm [INVERTEBRATE ZOOLOGY] Any of several bivalve mollusk species belonging to the family Teredinidae and which superficially resemble earthworms because the two valves are reduced to a pair of plates at the anterior of the animal or are used for boring into wood. { 'ship ˌwərm }

shock organ [IMMUNOLOGY] The organ or tissue that exhibits the most marked response to the antigen-antibody interaction in hypersensitivity, as the lungs in allergic asthma or the skin in allergic contact dermatitis. { 'shäk ˌȯr·gən }

shoot [BOTANY] **1.** The aerial portion of a plant, including stem, branches, and leaves. **2.** A new, immature growth on a plant. { shüt }

shore bird [VERTEBRATE ZOOLOGY] A general term applied to a large number of birds in 12 families of the suborder Charadrii which are always found near water, although the habitat and morphology is varied. Also known as wader. { 'shȯr ˌbərd }

short-term exposure limit [PHYSIOLOGY] The maximum amount of harmful gas or dust to which a person may be exposed for a brief period (usually 15 minutes) without being physically harmed. Abbreviated STEL. { ¦shȯrt ¦tərm ik 'spō·zhər ˌlim·ət }

shoulder [ANATOMY] **1.** The area of union between the upper limb and the trunk in humans. **2.** The corresponding region in other vertebrates. { 'shōl·dər }

shoulder blade [ANATOMY] *See* scapula. { 'shōl·dər ˌblād }

shoulder girdle [ANATOMY] *See* pectoral girdle. { 'shōl·dər ˌgərd·əl }

shrew [VERTEBRATE ZOOLOGY] Any of more than 250 species of insectivorous mammals of the family Soricidae; individuals are small with a moderately long tail, minute eyes, a sharp-pointed snout, and small ears. { shrü }

shrimp [INVERTEBRATE ZOOLOGY] The common name for a number of crustaceans, principally in the decapod suborder Natantia, characterized by having well-developed pleopods and by having the abdomen sharply bent in most species, producing a humped appearance. { shrimp }

shrub [BOTANY] A low woody plant with several stems. { shrəb }

shuttle vector [MOLECULAR BIOLOGY] A deoxyribonucleic acid vector able to replicate in two different organisms, and therefore able to shuttle foreign nucleic acids between two different hosts. Also known as bifunctional vector. { 'shəd·əl ˌvek·tər }

Shwartzman phenomenon [IMMUNOLOGY] A type of local tissue reactivity in the skin in which a preparatory injection of the endotoxin is followed by an intravenous injection of the same or another endotoxin 24 hours later, producing immediate neutropenia and thrombopenia with the development of leukocyte-platelet thrombi with subsequent hemorrhage. { 'shwȯrts·mən fə ˌnäm·əˌnän }

sialic acid [BIOCHEMISTRY] Any of a family of amino sugars, containing nine or more carbon atoms, that are nitrogen- and oxygen-substituted acyl derivatives of neuraminic acid; as components of lipids, polysaccharides, and mucoproteins, they are widely distributed in bacteria and in animal tissues. { sī'al·ik 'as·əd }

sialomucin [BIOCHEMISTRY] An acid mucopolysaccharide containing sialic acid as the acid component. { sī¦al·ə'myüs·ən }

Siboglinidae [INVERTEBRATE ZOOLOGY] A family of pogonophores in the order Athecanephria. { ˌsī·bə'glī·nəˌdē }

sicula [INVERTEBRATE ZOOLOGY] The cone-shaped chitinous skeleton of the first zooid of a graptolite colony. { 'sik·yə·lə }

Siderocapsaceae [MICROBIOLOGY] A family of gram-negative, chemolithotrophic bacteria; characterized by the ability to deposit iron or manganese compounds on or around the cells. { ˌsid·ə·rəˌkap'sās·ēˌē }

siderocyte [CELL BIOLOGY] An erythrocyte which contains granules staining blue with the Prussian blue reaction. { 'sid·ə·rəˌsīt }

siderophore [BIOCHEMISTRY] A molecular receptor that binds and transports iron. { 'sid·ə·rə ˌfȯr }

Sierolomorphidae [INVERTEBRATE ZOOLOGY] A small family of hymenopteran insects in the superfamily Scolioidea. { sē¦er·ə·lō'mȯr·fəˌdē }

sieve area [BOTANY] An area in the wall of a sieve-tube element, sieve cell, or parenchyma cell characterized by clusters of pores through which strands of cytoplasm pass to adjoining cells. { 'siv ˌer·ē·ə }

sieve cell [BOTANY] A long, tapering cell that is characteristic of phloem in gymnosperms and lower vascular plants, in which all the sieve areas are of equal specialization. { 'siv ˌsel }

sieve plate [BOTANY] A perforated section of the wall of a component member of a sieve tube. { 'siv ˌplāt }

sieve tissue [BOTANY] *See* phloem. { 'siv ˌtish·ü }

sieve tube [BOTANY] A phloem element consisting of a series of thin-walled cells arranged end to end, in which some sieve areas are more specialized than others. { 'siv ˌtüb }

Sigalionidae [INVERTEBRATE ZOOLOGY] A family of scale-bearing polychaete annelids belonging to the Errantia. { ˌsig·ə·lē'än·əˌdē }

Siganidae [VERTEBRATE ZOOLOGY] A small family of herbivorous perciform fishes in the suborder Acanthuroidei having minute concealed scales embedded in the skin and strong, sharp fin spines. { si'gan·əˌdē }

sight [PHYSIOLOGY] *See* vision. { sīt }

sigma [INVERTEBRATE ZOOLOGY] A C-shaped spicule. { 'sig·mə }

sigmaspire [INVERTEBRATE ZOOLOGY] An S-shaped sponge spicule. { 'sig·məˌspīr }

sigmoid [BIOLOGY] S-shaped. { 'sigˌmȯid }

sigmoid colon [ANATOMY] The S-shaped portion of the colon between the descending colon and the rectum. { 'sigˌmȯid 'kō·lən }

signaling cell [PHYSIOLOGY] A cell whose products induce a specific response in target cells. { 'sig·nə·liŋ ˌsel }

signal molecule [BIOCHEMISTRY] A molecule produced by a signaling cell. { 'sig·nəl ˌmäl·ə ˌkyül }

signet-ring cell [HISTOLOGY] A cell with a large fat- or carbohydrate-filled vacuole that pushes the nucleus against the cell membrane. { 'sig·nət ˌriŋ 'sel }

silent mutation [GENETICS] A mutation that does not result in amino acid sequence change. { 'sī·lənt myü'tā·shən }

silicle [BOTANY] A many-seeded capsule formed from two united carpels, usually of equal length and width, and divided on the inside by a replum. { 'sil·ə·kəl }

silicoblast [INVERTEBRATE ZOOLOGY] Poriferan

amebocytes involved in formation of siliceous spicules. { 'sil·ə·kə,blast }

Silicoflagellata [BOTANY] A class of unicellular flagellates of the plant division Chrysophyta represented by a single living genus, *Dictyocha*. { ¦sil·ə·kō,flaj·ə'läd·ə }

Silicoflagellida [INVERTEBRATE ZOOLOGY] An order of marine flagellates in the class Phytamastigophorea which have an internal, siliceous, tubular skeleton, numerous yellow chromatophores, and a single flagellum. { ¦sil·ə·kō·flə 'jel·əd·ə }

silique [BOTANY] A silicle-like capsule, but usually at least four times as long as it is wide, which opens by sutures at either margin and has parietal placentation. { si'lēk }

silk [INVERTEBRATE ZOOLOGY] A continuous protein fiber consisting principally of fibroin and secreted by various insects and arachnids, especially the silkworm, for use in spinning cocoons, webs, egg cases, and other structures. { silk }

silk cotton [BOTANY] *See* kapok. { 'silk ¦kät·ən }

silk gland [INVERTEBRATE ZOOLOGY] A gland in certain insects which secretes a viscous fluid in the form of filaments known as silk; it is a salivary gland in insects and an abdominal gland in spiders. { 'silk ,gland }

silkworm [INVERTEBRATE ZOOLOGY] The larva of various moths, especially *Bombyx mori*, that produces a large amount of silk for building its cocoon. { 'silk,wərm }

Silphidae [INVERTEBRATE ZOOLOGY] The carrion beetles, a family of coleopteran insects in the superfamily Staphylinoidea. { 'sil·fə,dē }

Siluridae [VERTEBRATE ZOOLOGY] A family of European catfish in the suborder Siluroidei in which the adipose dorsal fin is rudimentary or lacking. { si'lu̇r·ə,dē }

Siluriformes [VERTEBRATE ZOOLOGY] The catfishes, a distinctive order of actinopterygian fishes in the superorder Ostariophysi, distinguished by a complex Weberian apparatus that involves the fifth vertebrae and one to four pair of barbels. { si,lu̇r·ə'fȯr,mēz }

Siluroidei [VERTEBRATE ZOOLOGY] A suborder of the Siluriformes. { ,sil·yə'rȯid·ē,ī }

Silvanidae [INVERTEBRATE ZOOLOGY] An equivalent name for Cucujidae. { sil'van·ə,dē }

silverfish [INVERTEBRATE ZOOLOGY] Any of over 350 species of insects of the order Thysanura; they are small, wingless forms with biting mouthparts. { 'sil·vər,fish }

silverline system [INVERTEBRATE ZOOLOGY] A series of superficial argentophilic lines in many protozoans, especially ciliates. { 'sil·vər,līn ,sis·təm }

similarity coefficient [SYSTEMATICS] In numerical taxonomy, a factor S used to calculate the similarity between organisms, according to the formula $S = n_s/(n_s + n_d)$, where n_s represents the number of positive features shared by two strains, and n_d represents the number of features positive for one strain and negative for the other. { ,sim·ə'lar·əd·ē ,kō·i,fish·ənt }

Simonsiellaceae [MICROBIOLOGY] A family of bacteria in the order Cytophagales; cells are arranged to form flat filaments capable of gliding motility when the flat surface is in contact with the substrate. { sə,män·sē·ə'lās·ē,ē }

simple [BIOLOGY] **1.** Made up of one piece. **2.** Unbranched. **3.** Consisting of identical units, as a simple tissue. { 'sim·pəl }

simple branched tubular gland [ANATOMY] A structure consisting of two or more unbranched, tubular secreting units joining a common outlet duct. { 'sim·pəl 'brancht 'tüb·yə·lər 'gland }

simple fruit [BOTANY] A fruit that has developed from a single carpel or several united carpels. { 'sim·pəl 'früt }

simple gland [ANATOMY] A gland having a single duct. { 'sim·pəl 'gland }

simple leaf [BOTANY] A leaf having one blade, or a lobed leaf in which the separate parts do not reach down to the midrib. { 'sim·pəl 'lēf }

simple pistil [BOTANY] A pistil that consists of a single carpel. { 'sim·pəl 'pis·təl }

simple pit [BOTANY] A pit that lacks a border. { 'sim·pəl 'pit }

simple protein [BIOCHEMISTRY] One of a group of proteins which, upon hydrolysis, yield exclusively amino acids; included are globulins, glutelins, histones, prolamines, and protamines. { 'sim·pəl 'prō,tēn }

simple stomach [ANATOMY] A stomach consisting of a single dilation of the alimentary canal, as found in humans, dogs, and many higher and lower vertebrates. { 'sim·pəl 'stəm·ək }

simple tubular gland [ANATOMY] A gland consisting of a single, tubular secreting unit. { 'sim·pəl 'tü·byə·lər 'gland }

simplex uterus [ANATOMY] A uterus consisting of a single cavity, representing the greatest degree of fusion of the Müllerian ducts; found in man and apes. { 'sim,pleks ¦yüd·ə·rəs }

Simuliidae [INVERTEBRATE ZOOLOGY] The black flies, a family of orthorrhaphous dipteran insects in the series Nematocera. { ,sim·yə'lī·ə,dē }

sinistrorse [BIOLOGY] Twisting or coiling counterclockwise. { ¦sin·ə¦strȯrs }

sinoatrial node [ANATOMY] A bundle of Purkinje fibers located near the junction of the superior vena cava with the right atrium which acts as a pacemaker for cardiac excitation. Abbreviated SA node. Also known as sinoauricular node. { ¦sī·nō 'ā·trē·əl 'nōd }

sinoauricular node [ANATOMY] *See* sinoatrial node. { ¦sī·nō·ȯ'rik·yə·lər 'nōd }

sinuate [BOTANY] Having a wavy margin with strong indentations. { 'sin yə,wət }

sinus [BIOLOGY] A cavity, recess, or depression in an organ, tissue, or other part of an animal body. { 'sī·nəs }

sinus gland [INVERTEBRATE ZOOLOGY] An endocrine gland in higher crustaceans, lying in the eyestalk in most stalk-eyed species, which is the site of storage and release of a molt-inhibiting hormone. { 'sī·nəs ,gland }

sinus hairs [VERTEBRATE ZOOLOGY] *See* vibrissae. { 'sī·nəs ,herz }

sinus of Morgagni [ANATOMY] The space between the upper border of the levator veli palatini muscle and the base of the skull. { 'sī·nəs əv mȯr'gän·yē }

sinusoid [ANATOMY] Any of the relatively large spaces comprising part of the venous circulation in certain organs, such as the liver. { 'sī·nə ˌsȯid }

sinus venosus [EMBRYOLOGY] The vessel in the transverse septum of the embryonic mammalian heart into which open the vitelline, allantoic, and common cardinal veins. [VERTEBRATE ZOOLOGY] The chamber of the lower vertebrate heart to which the veins return blood from the body. { 'sī·nəs və'nō·səs }

Siphinodentallidae [INVERTEBRATE ZOOLOGY] A family of mollusks in the class Scaphopoda characterized by a subterminal epipodial ridge which is not slit dorsally and which terminates with a crenulated disk. { ¦sī·fə·nō·den'tal·əˌdē }

siphon [BOTANY] A tubular element in various algae. [INVERTEBRATE ZOOLOGY] **1.** A tubular structure for intake or output of water in bivalves and other mollusks. **2.** The sucking-type of proboscis in many arthropods. { 'sī·fən }

Siphonales [BOTANY] A large order of green algae (Chlorophyta) which are coenocytic, nonseptate, and mostly marine. { ˌsī·fə'nā·lez }

Siphonaptera [INVERTEBRATE ZOOLOGY] The fleas, an order of insects characterized by a small, laterally compressed, oval body armed with spines and setae, three pairs of legs modified for jumping, and sucking mouthparts. { ˌsī·fə'näp·trə }

Siphonocladaceae [BOTANY] A family of green algae in the order Siphonocladales. { ¦sī·fə·nō·klə'dās·ēˌē }

Siphonocladales [BOTANY] An order of green algae in the division Chlorophyta including marine, mostly tropical forms. { ¦sī·fə·nō·klə'dā·lēz }

siphonogamous [BOTANY] In plants, especially seed plants, the accomplishment of fertilization by means of a pollen tube. { ˌsī·fə'näg·ə·məs }

siphonoglyph [INVERTEBRATE ZOOLOGY] A ciliated groove leading from the mouth to the gullet in certain anthozoans. { sī'fän·əˌglif }

Siphonolaimidae [INVERTEBRATE ZOOLOGY] A family of nematodes in the superfamily Monhysteroidea in which the stoma is modified into a narrow, elongate, hollow, spearlike apparatus. { ¦sī·fə·nōˌlā'im·əˌdē }

Siphonolaimoidea [INVERTEBRATE ZOOLOGY] A superfamily of marine nematodes in the order Monhysterida, having a stoma in the form of a very narrow tube or a spear, and very large amphids. { ¦sī·fə·nō·lə'mȯid·ē·ə }

Siphonophora [INVERTEBRATE ZOOLOGY] An order of the cnidarian class Hydrozoa characterized by the complex organization of components which may be connected by a stemlike region or may be more closely united into a compact organism. { ˌsī·fə'näf·rə }

siphonosome [INVERTEBRATE ZOOLOGY] The lower part of a siphonophore colony, bearing the nutritive and reproductive zooids. { sī'fän·ə ˌsōm }

siphonostele [BOTANY] A type of stele consisting of pith surrounded by xylem and phloem. { sī 'fän·əˌstēl }

Siphonotretacea [PALEONTOLOGY] A superfamily of extinct, inarticulate brachiopods in the suborder Acrotretidina of the order Acrotretida having an enlarged, tear-shaped, apical pedicle valve. { ¦sī·fə·nō·trə'tās·ē·ə }

siphonozooid [INVERTEBRATE ZOOLOGY] A zooid of certain alcyonarians that lacks tentacles and gonads. { ¦sī·fə·nə'zōˌȯid }

siphuncle [INVERTEBRATE ZOOLOGY] **1.** A honeydew-secreting tube (cornicle) in aphids. **2.** A tubular extension of the mantle extending through all the chambers to the apex of a shelled cephalopod. { 'sīˌfəŋ·kəl }

Siphunculata [INVERTEBRATE ZOOLOGY] The equivalent name for Anoplura. { siˌfəŋ·kyə 'läd·ə }

Sipunculida [INVERTEBRATE ZOOLOGY] A phylum of marine worms which dwell in burrows, secreted tubes, or adopted shells; the mouth and anus occur close together at one end of the elongated body, and the jawless mouth, surrounded by tentacles, is situated in an eversible proboscis. { ˌsīˌpəŋ'kyü·lə·də }

Sipunculoidea [INVERTEBRATE ZOOLOGY] An equivalent name for Sipunculida. { sīˌpəŋ·kyə 'lȯid·ē·ə }

Sirenia [VERTEBRATE ZOOLOGY] An order of aquatic placental mammals which include the living manatees and dugongs; these are nearly hairless, thick-skinned mammals without hind limbs and with paddlelike forelimbs. { sī'rē·nē·ə }

Siricidae [INVERTEBRATE ZOOLOGY] The horntails, a family of the Hymenoptera in the superfamily Siricoidea; females use a stout, hornlike ovipositor to deposit eggs in wood. { sə'ris·əˌdē }

Siricoidea [INVERTEBRATE ZOOLOGY] A superfamily of wasps of the suborder Symphala in the order Hymenoptera. { ˌsir·ə'kȯid·ē·ə }

sisal [BOTANY] *Agave sisalina*. An agave of the family Amaryllidaceae indigenous to Mexico and Central America; a coarse, stiff yellow fiber produced from the leaves is used for making twine and brush bristles. { 'sī·səl }

sister chromatids [CELL BIOLOGY] The two daughter strands of a chromosome after it has duplicated. { 'sis·tər 'krō·məˌtədz }

skate [VERTEBRATE ZOOLOGY] Any of various batoid elasmobranchs in the family Rajidae which have flat bodies with winglike pectoral fins and a slender tail with two small dorsal fins. { skāt }

skeletal muscle [ANATOMY] A striated, voluntary muscle attached to a bone and concerned with body movements. { 'skel·əd·əl ˌməs·əl }

skeletal system [ANATOMY] Structures composed of bone or cartilage or a combination of both which provide a framework for the vertebrate body and serve as attachment for muscles. { 'skel·əd·əl ˌsis·təm }

skimmer [VERTEBRATE ZOOLOGY] Any of various ternlike birds, members of the Rhynchopidae, that inhabit tropical waters around the world and are unique in having the knifelike lower mandible substantially longer than the wider upper mandible. { 'skim·ər }

skin [ANATOMY] The external covering of the vertebrate body, consisting of two layers, the outer epidermis and the inner dermis. { skin }

skink [VERTEBRATE ZOOLOGY] Any of numerous small- to medium-sized lizards comprising the family Scincidae with a cylindrical body; short, sometimes vestigial, legs; cores of bone in the body scales; and pleurodont dentition. { skiŋk }

skin test [IMMUNOLOGY] A procedure for evaluating immunity status involving the introduction of a reagent into or under the skin. { 'skin ˌtest }

skipjack [VERTEBRATE ZOOLOGY] *See* bluefish. { 'skipˌjak }

skull [ANATOMY] The bones and cartilages of the vertebrate head which form the cranium and the face. { skəl }

skunk [VERTEBRATE ZOOLOGY] Any one of a group of carnivores in the family Mustelidae characterized by a glossy black and white coat and two musk glands at the base of the tail. { skəŋk }

slant culture [MICROBIOLOGY] A method for maintaining bacteria in which the inoculum is streaked on the surface of agar that has solidified in inclined glass tubes. { 'slant ˌkəl·chər }

slavery [INVERTEBRATE ZOOLOGY] An interspecific association among ants in which members of one species bring pupae of another species to their nest, which, when adult, become slave workers in the colony. { 'slav·ə·rē }

sleep [PHYSIOLOGY] A state of rest in which consciousness and activity are diminished. { 'slēp }

slime bacteria [MICROBIOLOGY] The common name for bacteria in the order Myxobacterales, so named for the layer of slime deposited behind cells as they glide on a surface. { 'slīm bakˌtir·ē·ə }

slime fungus [MYCOLOGY] *See* slime mold. { 'slīm ˌfəŋ·gəs }

slime gland [ZOOLOGY] A glandular structure in many animals producing a mucous material. { 'slīm ˌgland }

slime mold [MYCOLOGY] The common name for members of the Myxomycetes. Also known as slime fungus. { 'slīm ˌmōld }

slop culture [BOTANY] A method of growing plants in which surplus nutrient fluid is allowed to run through the sand or other medium in which the plants are growing. { 'släp ˌkəl·chər }

sloth [VERTEBRATE ZOOLOGY] Any of several edentate mammals in the family Bradypodidae found exclusively in Central and South America; all are slow-moving, arboreal species that cling to branches upside down by means of long, curved claws. { släth }

slow virus [VIROLOGY] Any member of a group of animal viruses characterized by prolonged periods of incubation and an extended clinical course lasting months or years. { 'slō 'vī·rəs }

slow virus infection [VIROLOGY] A persistent viral infection characterized by a long preclinical period extending for months or years from the time of exposure. { 'slō ¦vir·əs inˌfek·shən }

slug [INVERTEBRATE ZOOLOGY] Any of a number of pulmonate gastropods which have a rudimentary shell and the body elevated toward the middle and front end where the mantle covers the lung region. { sləg }

small intestine [ANATOMY] The anterior portion of the intestine in humans and other mammals; it is divided into three parts, the duodenum, the jejunum, and the ileum. { 'smȯl in'tes·tən }

smallpox vaccine [IMMUNOLOGY] A vaccine prepared from a glycerinated suspension of the exudate from cowpox vesicles obtained from healthy vaccinated calves or sheep. Also known as antismallpox vaccine; glycerinated vaccine virus; Jennerian vaccine; virus vaccinium. { 'smȯl ˌpäks vakˌsēn }

smear [BIOLOGY] A preparation for microscopic examination made by spreading a drop of fluid, such as blood, across a slide and using the edge of another slide to leave a uniform film. { smir }

smegma [PHYSIOLOGY] The sebaceous secretion that accumulates around the glans penis and the clitoris. { 'smeg·mə }

smell [PHYSIOLOGY] To perceive by olfaction. { smel }

Smilacaceae [BOTANY] A family of monocotyledonous plants in the order Liliales; members are usually climbing, leafy-stemmed plants with tendrils, trimerous flowers, and a superior ovary. { ˌsmī·lə'kās·ēˌē }

Sminthuridae [INVERTEBRATE ZOOLOGY] A family of insects in the order Collembola which have simple tracheal systems. { smin'thyu̇r·əˌdē }

smooth muscle [ANATOMY] The involuntary muscle tissue found in the walls of viscera and blood vessels, consisting of smooth muscle fibers. { 'smüth 'məs·əl }

smooth muscle fiber [HISTOLOGY] Any of the elongated, nucleated, spindle-shaped cells comprising smooth muscles. Also known as involuntary fiber; nonstriated fiber; unstriated fiber. { 'smüth 'məs·əl ˌfī·bər }

smut fungus [MYCOLOGY] The common name for members of the Ustilaginales. { 'smət ˌfəŋ·gəs }

snail [INVERTEBRATE ZOOLOGY] Any of a large number of gastropod mollusks distinguished by a spiral shell that encloses the body, a head, a foot, and a mantle. { snāl }

snake [VERTEBRATE ZOOLOGY] Any of about 3000 species of reptiles which belong to the 13 living families composing the suborder Serpentes in the order Squamata. { snāk }

sneeze [PHYSIOLOGY] A sudden, noisy, spasmodic expiration through the mouth and nose. { snēz }

snout [VERTEBRATE ZOOLOGY] The elongated nose of various mammals. { snau̇t }

sobole [BOTANY] An underground creeping stem. { 'sä·bəˌlē }

social animal [ZOOLOGY] An animal that exhibits social behavior. { 'sō·shəl 'an·ə·məl }

social behavior [ZOOLOGY] Any behavior on the part of an organism stimulated by, or acting upon, another member of the same species. { 'sō·shəl bi'hā·vyər }

social hierarchy [VERTEBRATE ZOOLOGY] The establishment of a dominance-subordination relationship among higher animal societies. { 'sō·shəl 'hī·ər,är·kē }

social parasitism [VERTEBRATE ZOOLOGY] An aberrant type of parasitism occurring in some birds, in which the female of one species lays her eggs in the nests of other species and permits the foster parents to raise the young. { 'sō·shəl 'par·ə ,sə,diz·əm }

social releaser [ZOOLOGY] A releaser stimulus which an animal receives from a member of its species. { 'sō·shəl ri'lē·sər }

society [ECOLOGY] A secondary or minor plant community forming part of a community. [ZOOLOGY] An organization of individuals of the same species in which there are divisions of resources and of labor as well as mutual dependence. { sə'sī·əd·ē }

sodium chloride solution [PHYSIOLOGY] *See* normal saline. { 'sōd·ē·əm 'klȯr,īd sə,lü·shən }

soft coral [INVERTEBRATE ZOOLOGY] The common name for cnidarians composing the order Alcyonacea; the colony is supple and leathery. { 'sȯft 'kär·əl }

soft palate [ANATOMY] The posterior part of the palate which consists of an aggregation of muscles, the tensor veli palatini, levator veli palatini, azygos uvulae, palatoglossus, and palatopharyngeus, and their covering mucous membrane. { 'sȯft 'pal·ət }

soft-shell disease [INVERTEBRATE ZOOLOGY] A disease of lobsters caused by a chitinous bacterium which extracts chitin from the exoskeleton. { 'sȯft ¦shel di,zēz }

soil conservation [ECOLOGY] Management of soil to prevent or reduce soil erosion and depletion by wind and water. { ¦sȯil ,kän·sər,vā·shən }

soil ecology [ECOLOGY] The study of interactions among soil organisms and interactions between biotic and abiotic aspects of the soil environment. { 'sȯil i,käl·ə·jē }

soil microbiology [MICROBIOLOGY] A study of the microorganisms in soil, their functions, and the effect of their activities on the character of the soil and the growth and health of plant life. { ¦sȯil ¦mī·krə·bī'äl·ə·jē }

Solanaceae [BOTANY] A family of dicotyledonous plants in the order Polemoniales having internal phloem, mostly numerous ovules and seeds on axile placentae, and mostly cellular endosperm. { ,sō·lə'nās·ē,ē }

solar propagation [BOTANY] A method of rooting plant cuttings involving the use of a modified hotbed; bottom heat is provided by radiation of stored solar heat from bricks or stones in the bottom of the hotbed frame. { 'sō·lər ,präp·ə 'gā·shən }

Solasteridae [INVERTEBRATE ZOOLOGY] The sun stars, a family of asteroid echinoderms in the order Spinulosida. { ,säl·ə'ster·ə,dē }

Solemyidae [INVERTEBRATE ZOOLOGY] A family of bivalve mollusks in the order Protobranchia. { ,säl·ə'mī·ə,dē }

Solenichthyes [VERTEBRATE ZOOLOGY] An equivalent name for Gasterosteiformes. { ,säl·ə'nik·thē,ēz }

solenium [INVERTEBRATE ZOOLOGY] A diverticulum of the enteron in certain hydroids. { sō'lē·nē·əm }

solenocyte [INVERTEBRATE ZOOLOGY] Any of various hollow, flagellated cells in the nephridia of the larvae of certain annelids, mollusks, rotifers, and lancelets. { sō'lē·nə,sīt }

solenodon [VERTEBRATE ZOOLOGY] Either of two species of insectivorous mammals comprising the family Solenodontidae; the almique (*Atopogale cubana*) is found only in Cuba, while the white agouta (*Solenodon paradoxus*) is confined to Haiti. { sō'lē·nə,dän }

Solenodontidae [VERTEBRATE ZOOLOGY] The solenodons, a family of insectivores belonging to the group Lipotyphla. { sō,lē·nə'dänt·ə,dē }

Solenogastres [INVERTEBRATE ZOOLOGY] The equivalent name for Aplacophora. { sō,lē·nə'ga ,strēz }

Solenopora [PALEOBOTANY] A genus of extinct calcareous red algae in the family Solenoporaceae that appeared in the Late Cambrian and lasted until the Early Tertiary. { ,säl·ə'näp·rə }

Solenoporaceae [PALEOBOTANY] A family of extinct red algae having compact tissue and the ability to deposit calcium carbonate within and between the cell walls. { sō,lē·nə·pə'rās·ē,ē }

soleus [ANATOMY] A flat muscle of the calf; origin is the fibula, popliteal fascia, and tibia, and insertion is the calcaneus; plantar-flexes the foot. { 'sō·lē·əs }

Solo man [PALEONTOLOGY] A relative but primitive form of fossil man from Java; this form had a small brain, heavy horizontal browridges, and a massive cranial base. { 'sō·lō 'man }

Solpugida [INVERTEBRATE ZOOLOGY] The sun spiders, an order of nonvenomous, spiderlike, predatory arachnids having large chelicerae for holding and crushing prey. { säl'pyü·jəd·ə }

solute compartmentation [BOTANY] The sequestering of a plant cell's salt in a vacuole so that the salt does not poison the cell. { 'säl·yüt kəm ,pärt·mən'tā·shən }

soma [BIOLOGY] The whole of the body of an individual, excluding the germ tract. { 'sō·mə }

Somasteroidea [INVERTEBRATE ZOOLOGY] A subclass of Asterozoa comprising sea stars of generalized structure, the jaws often only partly developed, and the skeletal elements of the arm arranged in a double series of transverse rows termed metapinnules. { ,sō·mə·stə'rȯid·ē·ə }

somatic aneuploidy [CELL BIOLOGY] An irregular variation in number of one or more individual chromosomes in the cells of a tissue. { sō'mad·ik 'a·nyü,plȯid·ē }

somatic cell [BIOLOGY] Any cell of the body of an organism except the germ cells. { sō'mad·ik 'sel }

somatic copulation [MYCOLOGY] A form of reproduction in ascomycetes and basidiomycetes involving sexual fusion of undifferentiated vegetative cells. { sō'mad·ik ,käp·yə'lā·shən }

somatic crossing-over [CELL BIOLOGY] Crossing-over during mitosis in somatic or vegetative cells. { sō'mad·ik ¦krȯs·iŋ 'ō·vər }

somatic death [BIOLOGY] The cessation of characteristic life functions. { sō'mad·ik 'deth }

somatic embryogenesis [BOTANY] The production of embryoids from sporophytic or somatic plant cells. { sə¦mad·ik ˌem·brē·ə'jen·ə·səs }

somatic mesoderm [EMBRYOLOGY] The external layer of the lateral mesoderm associated with the ectoderm after formation of the coelom. { sō 'mad·ik 'mez·əˌdərm }

somatic nervous system [PHYSIOLOGY] The portion of the nervous system concerned with the control of voluntary muscle and relating the organism with its environment. { sō'mad·ik 'nər·vəs ˌsis·təm }

somatic pairing [CELL BIOLOGY] The pairing of homologous chromosomes at mitosis in somatic cells; occurs in Diptera. { sō'mad·ik 'per·iŋ }

somatic reflex system [PHYSIOLOGY] An involuntary control system characterized by a control loop which includes skeletal muscles. { sō 'mad·ik 'rēˌfleks ˌsis·təm }

somatoblast [INVERTEBRATE ZOOLOGY] **1.** An undifferentiated cleavage cell that gives rise to somatic cells in annelids. **2.** The outer cell layer of the nematogen in Dicyemida. { 'sō·məd·ə ˌblast }

somatochrome [CELL BIOLOGY] A nerve cell possessing a well-defined body completely surrounding the nucleus on all sides, the cytoplasm having a distinct contour, and readily taking a stain. { 'sō·məd·əˌkrōm }

somatoclonal variation [GENETICS] The appearance of new traits in plants that regenerate from a callus in tissue culture. { səˌmad·ə¦klōn·əl ˌver·ē'ā·shən }

somatocyst [INVERTEBRATE ZOOLOGY] A cavity filled with air in the float of certain Siphonophora. { 'sō·məd·əˌsist }

somatogamy [BIOLOGY] *See* pseudomixis. { ˌsō·mə'täg·ə·mē }

somatometry [ANATOMY] Measurement of the human body with the soft parts intact. { ˌsō·mə 'täm·ə·trē }

somatophyte [BOTANY] A plant composed of distinct somatic cells that develop especially into mature or adult tissue. { 'sō·məd·əˌfīt }

somatopleure [EMBRYOLOGY] A complex layer of tissue consisting of the somatic layer of the mesoblast together with the epiblast, forming the body wall in craniate vertebrates and the amnion and chorion in amniotes. { 'sō·məd·ə ˌplu̇r }

somatostatin [BIOCHEMISTRY] A peptide secreted by the hypothalamus which acts primarily to inhibit the release of growth hormone from the anterior pituitary. { ¦sō·məd·ə'stat·ən }

somatotropin [BIOCHEMISTRY] The growth hormone of the pituitary gland. { ˌsō·mə'tä·trə·pən }

somesthesis [PHYSIOLOGY] The general name for all systems of sensitivity present in the skin, muscles and their attachments, visceral organs, and nonauditory labyrinth of the ear. { ¦sōm·es 'thē·səs }

somite [ZOOLOGY] *See* metamere. { 'sōˌmīt }

somnambulism [PHYSIOLOGY] **1.** Sleepwalking. **2.** The performance of any fairly complex act while in a sleeplike state or trance. { säm'näm·byəˌliz·əm }

sooty mold [MYCOLOGY] Ascomycetous fungi of the family Capnodiaceae, with dark mycelium and conidia. { 'su̇d·ē 'mōld }

sorbin [BIOCHEMISTRY] *See* sorbose. { 'sȯr·bən }

sorbose [BIOCHEMISTRY] $C_6H_{12}O_6$ A carbohydrate prepared by fermentation; produced as water-soluble crystals that melt at 165°C; used in the production of vitamin C. Also known as sorbin. { 'sȯrˌbōs }

soredium [BOTANY] A structure comprising algal cells wrapped in the hyphal tissue of lichens, as in certain Lecanorales; when separated from the thallus, it grows into a new thallus. { sȯ'rē·dē·əm }

sorghum [BOTANY] Any of a variety of widely cultivated grasses, especially *Sorghum bicolor* in the United States, grown for grain and herbage; growth habit and stem form are similar to Indian corn, but leaf margins are serrate and spikelets occur in pairs on a hairy rachis. { 'sȯr·gəm }

Soricidae [VERTEBRATE ZOOLOGY] The shrews, a family of insectivorous mammals belonging to the Lipotyphla. { sə'ris·əˌdē }

sorrel tree [BOTANY] *See* sourwood. { 'sär·əl ˌtrē }

sorus [BOTANY] **1.** A cluster of sporangia on the lower surface of a fertile fern leaf. **2.** A clump of reproductive bodies or spores in lower plants. { 'sȯr·əs }

sourwood [BOTANY] *Oxydendrum arboreum.* A deciduous tree of the heath family (Ericaceae) indigenous along the Alleghenies and having long, simple, finely toothed, long-pointed leaves that have an acid taste, and white, urn-shaped flowers. Also known as sorrel tree. { 'sau̇rˌwu̇d }

South Australian faunal region [ECOLOGY] A marine littoral region along the southwestern coast of Australia. { 'sau̇th ȯ'strāl·yən 'fȯn·əl ˌrē·jən }

sow [VERTEBRATE ZOOLOGY] An adult female swine. { sau̇ }

soybean [BOTANY] *Glycine max.* An erect annual legume native to China and Manchuria and widely cultivated for forage and for its seed. { 'sȯiˌbēn }

space biology [BIOLOGY] A term for the various biological sciences and disciplines that are concerned with the study of living things in the space environment. { 'spās bīˌäl·ə·jē }

space perception [PHYSIOLOGY] The awareness of the spatial properties and relations of an object, or of one's own body, in space; especially, the sensory appreciation of position, size, form, distance, and direction of an object, or of the observer himself, in space. { 'spās pərˌsep·shən }

spacer deoxyribonucleic acid [MOLECULAR BI-

OLOGY] Untranscribed deoxyribonucleic acid (DNA) segments, usually containing repetitious DNA, of eukaryotic and some viral genomes flanking functional genetic regions (cistrons). { ¦spās·ər dē¦äk·sē,rī·bō·nü,klē·ik 'as·əd }

spadix [BOTANY] A fleshy spike that is enclosed in a leaflike spathe and is the characteristic inflorescence of palms and arums. [INVERTEBRATE ZOOLOGY] A cone-shaped structure in male Nautiloidea formed of four modified tentacles, and believed to be homologous with the hectocotylus in male squids. { 'spā·diks }

Sparganiaceae [BOTANY] A family of monocotyledonous plants in the order Typhales distinguished by the inflorescence of globose heads, a vestigial perianth, and achenes that are sessile or nearly sessile. { spär,gā·nē'ās·ē,ē }

sparganum [INVERTEBRATE ZOOLOGY] The plerocercoid larva of a tapeworm. { 'spär·gə·nəm }

Sparidae [VERTEBRATE ZOOLOGY] A family of perciform fishes in the suborder Percoidei, including the porgies. { 'spar·ə,dē }

Spatangoida [INVERTEBRATE ZOOLOGY] An order of exocyclic Euechinoidea in which the posterior ambulacral plates form a shield-shaped area behind the mouth. { ,spat·ən'gȯid·ə }

spathe [BOTANY] A large, usually colored bract or pair of bracts enclosing an inflorescence, especially a spadix, on the same axis. { spāth }

spatulate [BIOLOGY] Shaped like a spoon. { 'spach·ə·lət }

spawn [BOTANY] Mycelium used for initiating mushroom propagation. [ZOOLOGY] **1.** The collection of eggs deposited by aquatic animals, such as fish. **2.** To produce or deposit eggs or discharge sperm; applied to aquatic animals. { spȯn }

spearmint [BOTANY] *Mentha spicata.* An aromatic plant of the mint family, Labiatae; the leaves are used as a flavoring in foods. { 'spir,mint }

speciation [EVOLUTION] The evolution of species. { ,spē·sē'ā·shən }

species [SYSTEMATICS] A taxonomic category ranking immediately below a genus and including closely related, morphologically similar individuals which actually or potentially interbreed. { 'spē·shēz }

species concept [EVOLUTION] The idea that the diversity of nature is divisible into a finite number of definable species. { 'spē·shēz ,kän,sept }

species population [ECOLOGY] A group of similar organisms residing in a defined space at a certain time. { 'spē·shēz ,päp·yə'lā·shən }

specific locus test [GENETICS] A technique used to detect recessive induced mutations in diploid organisms; a strain which carries several known recessive mutants in a homozygous condition is crossed with a nonmutant strain treated to induce mutations in its germ cells; induced recessive mutations allelic with those of the test strain will be expressed in the progeny. { spə'sif·ik 'lō·kəs ,test }

spectacle [ZOOLOGY] A colored marking in the form of rings around the eyes, as in certain birds, reptiles, and mammals (as the raccoon). { 'spek·tə·kəl }

speech [PHYSIOLOGY] A complex process in which the eating and breathing mechanisms are used to generate patterns of sounds that form words and sentences to express thoughts. { spēch }

Spelaeogriphacea [INVERTEBRATE ZOOLOGY] A peracaridan order of the Malacostraca comprised of the single species *Spelaeogriphus lepidops*, a small, blind, transparent, shrimplike crustacean with a short carapace that coalesces dorsally with the first thoracic somite. { ¦spē·lē·ō·gri'fās·ē·ə }

sperm [HISTOLOGY] *See* spermatozoon. { spərm }

spermatheca [ZOOLOGY] A sac in the female for receiving and storing sperm until fertilization; found in many invertebrates and certain vertebrates. Also known as seminal receptacle. { ¦spər·mə'thē·kə }

spermatic cord [ANATOMY] The cord consisting of the ductus deferens, epididymal and testicular nerves and blood vessels, and connective tissue that extends from the testis to the deep inguinal ring. { spər'mad·ik 'kȯrd }

spermatid [HISTOLOGY] A male germ cell immediately before assuming its final typical form. { 'spər·məd·əd }

spermatin [BIOCHEMISTRY] An albuminoid material occurring in semen. { 'spər·məd·ən }

spermatocyte [HISTOLOGY] A cell of the last or next to the last generation of male germ cells which differentiates to form spermatozoa. { spər 'mad·ə,sīt }

spermatogenesis [PHYSIOLOGY] The process by which spermatogonia undergo meiosis and transform into spermatozoa. { spər,mad·ə'jen·ə·səs }

spermatogonium [HISTOLOGY] A primitive male germ cell, the last generation of which gives rise to spermatocytes. { spər,mad·ə'gō·nē·əm }

spermatophore [ZOOLOGY] A bundle or packet of sperm produced by certain animals, such as annelids, arthropods, and some vertebrates. { spər 'mad·ə,fȯr }

spermatophyte [BOTANY] Any one of the seed-bearing vascular plants. { spər'mad·ə,fīt }

spermatozoon [HISTOLOGY] A mature male germ cell. Also known as sperm. { spər,mad·ə'zō·ən }

spermidine [BIOCHEMISTRY] $H_2N(CH_2)_3NH(CH_2)_4NH_2$ The triamine found in semen and other animal tissues. { 'spər·mə,dēn }

spermine [BIOCHEMISTRY] $C_{10}H_{26}N_4$ A tetramine found in semen, blood serum, and other body tissues. { 'spər,mēn }

spermiogenesis [CELL BIOLOGY] Nuclear and cytoplasmic transformation of spermatids into spermatozoa. { ,spər·mē·ō'jen·ə·səs }

sperm nucleus [BOTANY] One of the two nuclei in a pollen grain that function in double fertilization in seed plants. { 'spərm ,nü·klē·əs }

sperm whale [VERTEBRATE ZOOLOGY] *Physeter catadon.* An aggressive toothed whale belonging to the group Odontoceti of the order Cetacea; it produces ambergris and contains a mixture of spermaceti and oil in a cavity of the nasal passage. { 'spərm ,wāl }

Sphaeractinoidea [PALEONTOLOGY] An extinct group of fossil marine hydrozoans distinguished in part by the relative prominence of either vertical or horizontal trabeculae and by the presence of long, tabulate tubes called autotubes. { sfir,ak·tə'nȯid·ē·ə }

Sphaeriales [MYCOLOGY] An order of fungi in the subclass Euascomycetes characterized by hard, dark perithecia with definite ostioles. { ,sfir·ē'ā·lēz }

Sphaeriidae [INVERTEBRATE ZOOLOGY] The minute bog beetles, a small family of coleopteran insects in the suborder Myxophaga. { sfə'rī·ə,dē }

Sphaerioidaceae [MYCOLOGY] A family of fungi of the order Sphaeropsidales in which the pycnidia are black or dark-colored and are flask-, cone-, or lens-shaped with thin walls and a round, relatively small pore. { ,sfir·ē,ȯi'dās·ē,ē }

Sphaerocarpales [BOTANY] An order of liverworts in the subclass Marchantiidae, characterized by an envelope surrounding each antheridium and archegonium, absence of elaters, poor development of seta, and absence of thickenings in the unilayered wall of an indehiscent capsule. { ,sfir·o·kar'pa·lez }

Sphaeroceridae [INVERTEBRATE ZOOLOGY] A family of myodarian cyclorrhaphous dipteran insects in the subsection Acalypteratae. { ,sfir·ō'ser·ə,dē }

Sphaerodoridae [INVERTEBRATE ZOOLOGY] A family of polychaete annelids belonging to the Errantia in which species are characterized by small bodies, and are usually papillated. { ,sfir·ō'dȯr·ə,dē }

Sphaerolaimidae [INVERTEBRATE ZOOLOGY] A family of free-living nematodes in the superfamily Monhysteroidea characterized by a spacious and deep stoma. { ,sfir·ō'lī·mə,dē }

Sphaeromatidae [INVERTEBRATE ZOOLOGY] A family of isopod crustaceans in the suborder Flabellifera in which the body is broad and oval and the inner branch of the uropod is immovable. { ,sfir·ō'mad·ə,dē }

Sphaerophoraceae [BOTANY] A family of the Ascolichenes in the order Caliciales which are fruticose with a solid thallus. { sfə,räf·ə'rās·ē,ē }

Sphaeropleineae [BOTANY] A suborder of green algae in the order Ulotrichales distinguished by long, coenocytic cells, numerous bandlike chloroplasts, and heterogametes produced in undifferentiated vegetative cells. { ,sfir·ō'plān·ē,ē }

Sphaeropsidaceae [MYCOLOGY] An equivalent name for Sphaerioidaceae. { sfə,räp·sə'dās·ē,ē }

Sphaeropsidales [MYCOLOGY] An order of fungi of the class Fungi Imperfecti in which asexual spores are formed in pycnidia, which may be separate or joined to vegetative hyphae, conidiophores are short or absent, and conidia are always slime spores. { sfə,räp·sə'dā·lēz }

Sphagnaceae [BOTANY] The single monogeneric family of the order Sphagnales. { sfag'nās·ē,ē }

Sphagnales [BOTANY] The single order of mosses in the subclass Sphagnobrya containing the single family Sphagnaceae. { sfag'nā·lēz }

Sphagnobrya [BOTANY] A subclass of the Bryopsida; plants are grayish-green with numerous, spirally arranged branches and grow in deep tufts or mats, commonly in bogs and in other wet habitats. { sfag'näb·rē·ə }

sphagnum bog [ECOLOGY] A bog composed principally of mosses of the genus *Sphagnum* (Sphagnales) but also of other plants, especially acid-tolerant species, which tend to form peat. { 'sfag·nəm 'bäg }

Sphecidae [INVERTEBRATE ZOOLOGY] A large family of hymenopteran insects in the superfamily Sphecoidea. { 'sfes·ə,dē }

Sphecoidea [INVERTEBRATE ZOOLOGY] A superfamily of wasps belonging to the suborder Apocrita. { sfə'kȯid·ē·ə }

Sphenacodontia [PALEONTOLOGY] A suborder of extinct reptiles in the order Pelycosauria which were advanced, active carnivores. { sfə,näk·ə'dän·chə }

sphenethmoid [VERTEBRATE ZOOLOGY] A bone that surrounds the anterior portion of the brain in many amphibians. { sfēn·'eth,mȯid }

Spheniscidae [VERTEBRATE ZOOLOGY] The single family of the avian order Sphenisciformes. { sfə'nis·ə,dē }

Sphenisciformes [VERTEBRATE ZOOLOGY] The penguins, an order of aquatic birds found only in the Southern Hemisphere and characterized by paddlelike wings, erect posture, and scalelike feathers. { sfə,nis·ə'fȯr,mēz }

Sphenodontidae [VERTEBRATE ZOOLOGY] A family of lepidosaurian reptiles in the order Rhynchocephalia represented by a single living species, *Sphenodon punctatus*, a lizardlike form distinguished by lack of a penis. { ,sfē·nə'dänt·ə,dē }

sphenoid bone [ANATOMY] The butterfly-shaped bone forming the anterior part of the base of the skull and portions of the cranial, orbital, and nasal cavities. { 'sfē,nȯid ¦bōn }

sphenoid sinus [ANATOMY] Either of a pair of paranasal sinuses located centrally between and behind the eyes, below the ethymoid sinus. { 'sfē,nȯid ¦sī·nəs }

sphenopalatine [ANATOMY] Of or pertaining to the region of or surrounding the sphenoid and palatine bones. { ,sfē·nō'pal·ə,tēn }

sphenopalatine foramen [ANATOMY] The space between the sphenoid and orbital processes of the palatine bone; it opens into the nasal cavity and gives passage to branches from the pterygopalatine ganglion and the sphenopalatine branch of the maxillary artery. { ,sfē·nō'pal·ə,tēn fə'rā·mən }

Sphenopsida [BOTANY] A group of vascular cryptogams characterized by whorled, often very small leaves and by the absence of true leaf gaps in the stele; essentially equivalent to the division Equisetophyta. { sfə'näp·səd·ə }

Sphenyllopsida [PALEOBOTANY] An extinct class of embryophytes in the division Equisetophyta. { ,sfēn·əl'äp·səd·ə }

spheroplast [CELL BIOLOGY] A plant cell which

possesses only a partial or modified cell wall. [MICROBIOLOGY] A bacterial cell that assumes a spherical shape due to partial or complete absence of the wall. { 'sfir·ə‚plast }

sphincter [ANATOMY] A muscle that surrounds and functions to close an orifice. { 'sfiŋk·tər }

sphincter of Oddi [ANATOMY] Sphincter of the hepatopancreatic ampulla. { 'sfiŋk·tər əv 'äd·ē }

Sphinctozoa [PALEONTOLOGY] A group of fossil sponges in the class Calcarea which have a skeleton of massive calcium carbonate organized in the form of hollow chambers. { ‚sfiŋk·tə'zō·ə }

Sphindidae [INVERTEBRATE ZOOLOGY] The dry fungus beetles, a family of coleopteran insects in the superfamily Cucujoidea. { 'sfin·də‚dē }

Sphingidae [INVERTEBRATE ZOOLOGY] The single family of the lepidopteran superfamily Sphingoidea. { 'sfin·jə‚dē }

Sphingoidea [INVERTEBRATE ZOOLOGY] A superfamily of Lepidoptera in the suborder Heteroneura consisting of the sphinx, hawk, or hummingbird moths; these are heavy-bodied forms with antennae that are thickened with a pointed apex, a well-developed proboscis, and narrow wings. { sfiŋ'gȯid·ē·ə }

sphingolipid [BIOCHEMISTRY] Any lipid, such as a sphingomyelin, that yields sphingosine or one of its derivatives as a product of hydrolysis. { ¦sfiŋ·gō'lip·əd }

sphingomyelin [BIOCHEMISTRY] A phospholipid consisting of choline, sphingosine, phosphoric acid, and a fatty acid. { ¦sfiŋ·gō'mī·ə·lən }

sphingosine [BIOCHEMISTRY] $C_{18}H_{37}O_2N$ A moiety of sphingomyelin, cerebrosides, and certain other phosphatides. { 'sfiŋ·gə‚sēn }

Sphyraenidae [VERTEBRATE ZOOLOGY] A family of shore fishes in the suborder Mugiloidei of the order Perciformes comprising the barracudas. { sfə'rē·nə‚dē }

Sphyriidae [INVERTEBRATE ZOOLOGY] A family of ectoparasitic Crustacea belonging to the group Lernaeopodoida; the parasite embeds its head and part of its thorax into the host. { sfə'rī·ə‚dē }

spicule [BOTANY] An empty diatom shell. [INVERTEBRATE ZOOLOGY] A calcareous or siliceous, usually spikelike supporting structure in many invertebrates, particularly in sponges and alcyonarians. { 'spik·yül }

spiculin [BIOCHEMISTRY] An organic material making up a portion of a spicule. { 'spik·yə·lən }

spiculum [INVERTEBRATE ZOOLOGY] A bristlelike copulatory organ in certain nematodes. Also known as copulatory spicule. { 'spik·yə·ləm }

spider [INVERTEBRATE ZOOLOGY] The common name for arachnids comprising the order Araneida. { 'spīd·ər }

spike [BOTANY] An indeterminate inflorescence with sessile flowers. { spīk }

spikelet [BOTANY] The compound inflorescence of a grass consisting of one or several bracteate spikes. { 'spīk·lət }

spinacene [BIOCHEMISTRY] *See* squalene. { 'spin·ə‚sēn }

spinach [BOTANY] *Spinacia oleracea.* An annual potherb of Asiatic origin belonging to the order Caryophyllales and grown for its edible foliage. { 'spin·ich }

spinal column [ANATOMY] *See* spine. { 'spīn·əl ‚käl·əm }

spinal cord [ANATOMY] The cordlike posterior portion of the central nervous system contained within the spinal canal of the vertebral column of all vertebrates. { 'spīn·əl ‚kȯrd }

spinal foramen [ANATOMY] Central canal of the spinal cord. { 'spīn·əl fə'rā·mən }

spinal ganglion [ANATOMY] Any one of the sensory ganglions, each associated with the dorsal root of a spinal nerve. { 'spīn·əl 'gaŋ·glē·ən }

spinal nerve [ANATOMY] Any of the paired nerves arising from the spinal cord. { 'spīn·əl ¦nərv }

spinal reflex [PHYSIOLOGY] A reflex mediated through the spinal cord without the participation of the more cephalad structures of the brain or spinal cord. { 'spīn·əl 'rē‚fleks }

spindle [CELL BIOLOGY] A structure formed of fiberlike elements just before metaphase that extends between the poles of the achromatic figure and is attached to the centromeric regions of the chromatid pairs. { 'spin·dəl }

spindle fiber [CELL BIOLOGY] One of the fiberlike elements of the spindle; an aggregation of microtubules resulting from the polymerization of a series of small protein fibrils by primary —S—S— linkages. { 'spin·dəl ‚fī·bər }

spine [ANATOMY] An articulated series of vertebrae forming the axial skeleton of the trunk and tail, and being a characteristic structure of vertebrates. Also known as backbone; spinal column; vertebral column. [BOTANY] A rigid sharp-pointed process in plants; many are modified leaves. [INVERTEBRATE ZOOLOGY] One of the processes covering the surface of a sea urchin. [VERTEBRATE ZOOLOGY] **1.** One of the spiny rays supporting the fins of most fishes. **2.** A sharp-pointed modified hair on certain mammals, such as the porcupine. { spīn }

spiniger [INVERTEBRATE ZOOLOGY] Seta that tapers to a fine point, most frequently used in connection with compound seta (thus, compound spiniger). { 'spin·ə·jər }

spinneret [INVERTEBRATE ZOOLOGY] An organ that spins fiber from the secretion of silk glands. { ‚spin·ə'ret }

spinney [ECOLOGY] A small grove of trees or a thicket with undergrowth. { 'spin·ē }

spinoblast [INVERTEBRATE ZOOLOGY] A statoblast having a float of air cells and barbs or hooks on the surface. { 'spī·nə‚blast }

spinochrome [BIOCHEMISTRY] A type of echinochrome; an organic pigment that is known only from sea urchins and certain homopteran insects. { 'spī·nə‚krōm }

spinous process [ANATOMY] Any slender, sharp-pointed projection on a bone. { 'spī·nəs ‚prä·səs }

Spintheridae [INVERTEBRATE ZOOLOGY] An amphinomorphan family of small polychaete annelids included in the Errantia. { spin'ther·ə‚dē }

Spinulosida [INVERTEBRATE ZOOLOGY] An order of Asteroidea in which pedicellariae rarely occur, marginal plates bounding the arms and disk are small and inconspicuous, and spines occur in groups on the upper surface. { ˌspin·yəˈläs·əd·ē }

spiny-rayed fish [VERTEBRATE ZOOLOGY] The common designation for actinopterygian fishes, so named for the presence of stiff, unbranched, pointed fin rays, known as spiny rays. { ˈspī·nē ˈrād ˈfish }

Spionidae [INVERTEBRATE ZOOLOGY] A family of spioniform annelid worms belonging to the Sedentaria. { spīˈän·əˌdē }

spioniform worm [INVERTEBRATE ZOOLOGY] A polychaete annelid characterized by the presence of a pair of short to long, grooved palpi near the mouth. { spīˈän·əˌfórm ˈwərm }

spiracle [INVERTEBRATE ZOOLOGY] An external breathing orifice of the tracheal system in insects and certain arachnids. [VERTEBRATE ZOOLOGY] **1.** The external respiratory orifice in cetaceous and amphibian larvae. **2.** The first visceral cleft in fishes. { ˈspir·ə·kəl }

spiral cleavage [EMBRYOLOGY] A cleavage pattern characterized by formation of a cell mass showing spiral symmetry; occurs in mollusks. { ˈspī·rəl ˈklē·vij }

spiral ganglion [ANATOMY] The ganglion of the cochlear part of the vestibulocochlear nerve embedded in the spiral canal of the modiolus. { ˈspī·rəl ˈgaŋ·glē·ən }

spiral ligament [ANATOMY] The reticular connective tissue connecting the basilar membrane to the outer cochlear wall in the ear of mammals. { ˈspī·rəl ˈlig·ə·mənt }

spiral organ [ANATOMY] *See* organ of Corti. { ˈspī·rəl ¦ór·gən }

spiral valve [VERTEBRATE ZOOLOGY] A spiral fold of mucous membrane in the small intestine of elasmobranchs and some primitive fishes which increases the surface area for absorption. { ˈspī·rəl ¦valv }

spiramycin [MICROBIOLOGY] A complex of related antibiotics, which resemble erythromycin structurally and in antibacterial spectrum, produced by *Streptomyces ambofaciens*. { ¦spī·rəˈmīs·ən }

spiraster [INVERTEBRATE ZOOLOGY] A spiral spicule bearing rays in Porifera. { spīˈras·tər }

spire [BOTANY] A narrow, tapering blade or stalk. { spīr }

spiricle [BOTANY] Any of the coiled threads in certain seed coats which uncoil when moistened. { ˈspir·ə·kəl }

Spiriferida [PALEONTOLOGY] An order of fossil articulate brachiopods distinguished by the spiralium, a pair of spirally coiled ribbons of calcite supported by the crura. { ˌspī·rəˈfer·əd·ə }

Spiriferidina [PALEONTOLOGY] A suborder of the extinct brachiopod order Spiriferida including mainly ribbed forms having laterally or ventrally directed spires, well-developed interareas, and a straight hinge line. { spīˌrif·ə·rəˈdī·nə }

Spirillaceae [MICROBIOLOGY] A family of bacteria; motile, helically curved rods that move with a characteristic corkscrewlike motion. { ˈspī·rə ˈlās·ēˌē }

Spirillinacea [INVERTEBRATE ZOOLOGY] A superfamily of foraminiferan protozoans in the suborder Rotaliina characterized by a planispiral or low conical test with a wall composed of radial calcite. { spəˌril·əˈnās·ē·ə }

Spirobrachiidae [INVERTEBRATE ZOOLOGY] A family of the Brachiata in the order Thecanephria. { ˌspī·rə·brəˈkī·əˌdē }

Spirochaetaceae [MICROBIOLOGY] The single family of the order Spirochaetales. { ˌspī·rə·kē ˈtās·ēˌē }

Spirochaetales [MICROBIOLOGY] An order of bacteria characterized by slender, helically coiled cells sometimes occurring in chains. { ˌspī·rə·kē ˈtā·lēz }

spirochete [MICROBIOLOGY] The common name for any member of the order Spirochaetales. { ˈspī·rəˌkēt }

spirocyst [INVERTEBRATE ZOOLOGY] A thin-walled capsule that contains a long, unarmed, eversible, spirally coiled thread of uniform diameter; found in cnidarians. { ˈspī·rəˌsist }

spirometry [PHYSIOLOGY] The measurement, by a form of gas meter (spirometer), of volumes of air that can be moved in or out of the lungs. { spī ˈräm·ə·trē }

Spirotrichia [INVERTEBRATE ZOOLOGY] A subclass of the protozoan class Ciliatea which contains those ciliates characterized by conspicuous, compound ciliary structures, known as cirri, and buccal organelles. { ˌspī·rōˈtrik·ē·ə }

Spirulidae [INVERTEBRATE ZOOLOGY] A family of cephalopod mollusks containing several species of squids. { spīˈrül·əˌdē }

Spiruria [INVERTEBRATE ZOOLOGY] A subclass of nematodes in the class Secernentea. { spīˈrůr·ē·ə }

Spirurida [INVERTEBRATE ZOOLOGY] An order of phasmid nematodes in the subclass Spiruria. { spīˈrůr·əd·ə }

Spiruroidea [INVERTEBRATE ZOOLOGY] A superfamily of spirurid nematodes which are parasitic in the respiratory and digestive systems of vertebrates. { ˌspī·rəˈròid·ē·ə }

splanchnic mesoderm [EMBRYOLOGY] The internal layer of the lateral mesoderm that is associated with the entoderm after the formation of the coelom. { ˈsplaŋk·nik ˈmez·əˌdərm }

splanchnic nerve [ANATOMY] A nerve carrying nerve fibers from the lower thoracic paravertebral ganglions to the collateral ganglions. { ˈsplaŋk·nik ¦nərv }

splanchnocranium [ANATOMY] Portions of the skull derived from the primitive skeleton of the gill apparatus. { ¦splaŋk·nəˈkrā·nē·əm }

splanchnopleure [EMBRYOLOGY] The inner layer of the mesoblast from which part of the wall of the alimentary canal and portions of the visceral organs are derived in coelomates. { ˈsplaŋk·nə ˌplůr }

spleen [ANATOMY] A blood-forming lymphoid organ of the circulatory system, present in most vertebrates. { splēn }

splenic flexure [ANATOMY] An abrupt turn of the colon beneath the lower end of the spleen, connecting the descending with the transverse colon. { 'splen·ik 'flek·shər }

splenium [ANATOMY] The rounded posterior extremity of the corpus callosum. { 'splē·nē·əm }

split gene [GENETICS] A eukaryotic gene in which the coding sequences (exons) are interrupted by a number of usually noncoding regions (introns). Also known as interrupted gene. { 'split 'jēn }

splitter [SYSTEMATICS] A taxonomist who divides taxa very finely. { 'splid·ər }

sponge [INVERTEBRATE ZOOLOGY] The common name for members of the phylum Porifera. { spənj }

Spongiidae [INVERTEBRATE ZOOLOGY] A family of sponges of the order Dictyoceratida; members are encrusting, massive, or branching in form and have small spherical flagellated chambers which characteristically join the exhalant canals by way of narrow channels. { spən'jī·ə,dē }

Spongillidae [INVERTEBRATE ZOOLOGY] A family of fresh- and brackish water sponges in the order Haplosclerida which are chiefly gray, brown, or white in color, and encrusting, massive, or branching in form. { spən'jil·ə,dē }

spongin [BIOCHEMISTRY] A scleroprotein, occurring as the principal component of skeletal fibers in many sponges. { 'spən·jən }

spongioblast [EMBRYOLOGY] A primordial cell arising from the ectoderm of the embryonic neural tube which differentiates to form the neuroglia, the ependymal cells, the neurolemma sheath cells, the satellite cells of ganglions, and Müller's fibers of the retina. { 'spən·jē·ō,blast }

spongiocyte [HISTOLOGY] **1.** A neuroglia cell. **2.** A cell of the adrenal cortex which has a spongy appearance due to the solution of lipids during tissue preparation for microscopical examination. { 'spən·jē·ō,sīt }

Spongiomorphida [PALEONTOLOGY] A small, extinct Mesozoic order of fossil colonial Hydrozoa in which the skeleton is a reticulum composed of perforate lamellae parallel to the upper surface and of regularly spaced vertical elements in the form of pillars. { ,spən·jē·ō'mȯr·fə·də }

Spongiomorphidae [PALEONTOLOGY] The single family of extinct hydrozoans comprising the order Spongiomorphida. { ,spən·jē·ō'mȯr·fə,dē }

spongocoel [INVERTEBRATE ZOOLOGY] The branching, internal cavity of a sponge, connected to the outside by way of the osculum. { 'späŋ·gə,sēl }

spongy mesophyll [BOTANY] A system of loosely and irregularly arranged parenchymal cells with numerous intercellular spaces found near the lower surface in well-differentiated broad leaves. Also known as spongy parenchyma. { 'spən·jē 'mē·zō,fil }

spongy parenchyma [BOTANY] *See* spongy mesophyll. { 'spən·jē pə'reŋ·kə·mə }

spontaneous generation [BIOLOGY] *See* abiogenesis. { spän'tā·nē·əs ,jen·ə'rā·shən }

spontaneous mutation [GENETICS] A mutation that occurs naturally. { spän'tā·nē·əs myü'tā·shən }

sporangiophore [BOTANY] A stalk or filament on which sporangia are borne. { spə'ran·jē·ə,fȯr }

sporangiospore [BOTANY] A spore that forms in a sporangium. { spə'ran·jē·ə,spȯr }

sporangium [BOTANY] A case in which asexual spores are formed and borne. { spə'ran·jē·əm }

spore [BIOLOGY] A uni- or multicellular, asexual, reproductive or resting body that is resistant to unfavorable environmental conditions and produces a new vegetative individual when the environment is favorable. { spȯr }

spore mother cell [BOTANY] One of the cells of the archespore of a spore-bearing plant from which a spore, but usually a tetrad of spores, is produced. Also known as sporocyte. { 'spȯr 'məth·ər ,sel }

sporidium [MYCOLOGY] A small spore, especially one formed on a promycelium. { spə'rid·ē·əm }

sporoblast [INVERTEBRATE ZOOLOGY] A sporozoan cell from which sporozoites arise. { 'spȯr·ə ,blast }

Sporobolomycetaceae [MYCOLOGY] The single family of the order Sporobolomycetales. { spə ,räb·ə·lō,mī·sə'tās·ē,ē }

Sporobolomycetales [MYCOLOGY] An order of yeastlike and moldlike fungi assigned to the class Basidiomycetes characterized by the formation of sterigmata, upon which the asexual ballistospores are formed. { spə,räb·ə·lō,mī·sə 'tā·lēz }

sporocarp [BOTANY] Any multicellular structure in or on which spores are formed. { 'spȯr·ə ,kärp }

sporocyst [BOTANY] A unicellular resting body from which asexual spores arise. [INVERTEBRATE ZOOLOGY] **1.** A resistant envelope containing an encysted sporozoan. **2.** An encysted sporozoan. **3.** The first reproductive form of a digenetic trematode in which rediae develop. { 'spȯr·ə,sist }

sporocyte [BOTANY] *See* spore mother cell. { 'spȯr·ə,sīt }

sporogenesis [BIOLOGY] **1.** Reproduction by means of spores. **2.** Formation of spores. { ¦spȯr·ə'jen·ə·səs }

sporogony [BIOLOGY] Reproduction by means of spores. [INVERTEBRATE ZOOLOGY] Propagative reproduction involving formation, by sexual processes, and subsequent division of a zygote. { spə'räg·ə·nē }

sporont [INVERTEBRATE ZOOLOGY] A stage in the life history of sporozoans which gives rise to spores. { 'spȯr,änt }

sporophore [MYCOLOGY] A structure on the thallus of fungi which produces spores. { 'spȯr·ə ,fȯr }

sporophyll [BOTANY] A modified leaf that develops sporangia. { 'spȯr·ə,fil }

sporophyte [BOTANY] **1.** An individual of the spore-bearing generation in plants exhibiting alternation of generation. **2.** The spore-producing generation. **3.** The diplophase in a plant life cycle. { 'spȯr·ə,fīt }

sporopollenin [BIOCHEMISTRY] A substance related to suberin and cutin but more resistant to decay that is found in the exine of pollen grains. { ¦spȯr·ō'päl·ə·nən }

sporosac [INVERTEBRATE ZOOLOGY] A degenerate gonophore in certain hydroid cnidarians. { 'spȯr·ə‚sak }

Sporozoa [INVERTEBRATE ZOOLOGY] A subphylum of parasitic Protozoa, typically producing spores during the asexual stages of the life cycle. { ‚spȯr·ə'zō·ə }

sporozoite [INVERTEBRATE ZOOLOGY] A motile, infective stage of certain sporozoans, which is the result of sexual reproduction and which gives rise to an asexual cycle in the new host. { ‚spȯr·ə'zō‚īt }

sporulation [BIOLOGY] The act and process of spore formation. { ‚spȯr·yə'lā·shən }

spout hole [VERTEBRATE ZOOLOGY] **1.** A blowhole of a cetacean mammal. **2.** A nostril of a walrus or seal. { 'spau̇t ‚hōl }

spreading factor [BIOCHEMISTRY] *See* hyaluronidase. { 'spred·iŋ ‚fak·tər }

Sprigginidae [INVERTEBRATE ZOOLOGY] An extinct family of annelid worms distinguished by a horseshoe-shaped head. { spri'gin·ə‚dē }

springwood [BOTANY] The portion of an annual ring that is formed principally during the growing season; it is softer, more porous, and lighter than summerwood because of its higher proportion of large, thin-walled cells. { 'spriŋ‚wu̇d }

spruce [BOTANY] An evergreen tree belonging to the genus *Picea* characterized by single, four-sided needles borne on small peglike projections, pendulous cones, and resinous wood. { sprüs }

Spumellina [INVERTEBRATE ZOOLOGY] The equivalent name for Peripylina. { spyü'mel·ə·nə }

spur [BOTANY] **1.** A hollow process at the base of a petal or sepal. **2.** A short fruit-bearing tree branch. **3.** A short projecting root. [ZOOLOGY] A stiff, sharp outgrowth, as on the legs of certain birds and insects. { spər }

sputum [PHYSIOLOGY] Material discharged from the surface of the respiratory passages, mouth, or throat; may contain saliva, mucus, pus, microorganisms, blood, or inhaled particulate matter in any combination. { 'spyüd·əm }

squalene [BIOCHEMISTRY] $C_{30}H_{50}$ A liquid triterpene which is found in large quantities in shark liver oil, and which appears to play a role in the biosynthesis of sterols and polycyclic terpenes; used as a bactericide and as an intermediate in the synthesis of pharmaceuticals. Also known as spinacene. { 'skwā‚lēn }

Squalidae [VERTEBRATE ZOOLOGY] The spiny dogfishes, a family of squaloid elasmobranchs recognized by their well-developed fin spines. { 'skwā·lə‚dē }

Squamata [VERTEBRATE ZOOLOGY] An order of reptiles, composed of the lizards and snakes, distinguished by a highly modified skull that has only a single temporal opening, or none, by the lack of shells or secondary palates, and by possession of paired penes on the males. { skwə 'mäd·ə }

squamodisk [INVERTEBRATE ZOOLOGY] In monogenetic trematodes, a disk bearing concentric circles of spines, scales, or ridges, and located on the opisthaptor. { 'skwā·mō‚disk }

squamosal bone [ANATOMY] The part of the temporal bone in man corresponding with the squamosal bone in lower vertebrates. [VERTEBRATE ZOOLOGY] A membrane bone lying external and dorsal to the auditory capsule of many vertebrate skulls. { skwə'mō·səl ‚bōn }

squamous [BIOLOGY] Covered with or composed of scales. { 'skwā·məs }

squamous epithelium [HISTOLOGY] A single-layered epithelium consisting of thin, flat cells. { 'skwā·məs ‚ep·ə'thē·lē·əm }

squamulose [BIOLOGY] Covered with or composed of minute scales. { 'skwäm·yə‚lōs }

squarrose [BOTANY] Having stiff divergent bracts, or other processes. { 'skwä‚rōs }

squarrulose [BOTANY] Mildly squarrose. { 'skwä·rə‚lōs }

squash [BOTANY] Either of two plants of the genus *Cucurbita*, order Violales, cultivated for its fruit; some types are known as pumpkins. { skwäsh }

Squatinidae [VERTEBRATE ZOOLOGY] A group of squaloid elasmobranchs of uncertain affinity characterized by a greatly extended rostrum with enlarged denticles along the margins; maximum length is under 4 feet (1.2 meters). { skwə'tin·ə ‚dē }

squid [INVERTEBRATE ZOOLOGY] Any of a number of marine cephalopod mollusks characterized by a reduced internal shell, ten tentacles, an ink sac, and chromatophores. { skwid }

Squillidae [INVERTEBRATE ZOOLOGY] The single family of the eumalacostracan order Stomatopoda, the mantis shrimp. { 'skwil·ə‚dē }

squirrel [VERTEBRATE ZOOLOGY] Any of over 200 species of arboreal rodents of the families Sciuridae and Anomaluridae having a bushy tail and long, strong hind limbs. { 'skwərl }

stab culture [MICROBIOLOGY] A culture of anaerobic bacteria made by piercing a solid agar medium in a test tube with an inoculating needle covered with the bacterial inoculum. { 'stab ‚kəl·chər }

stable factor [BIOCHEMISTRY] *See* factor VII. { 'stā·bəl 'fak·tər }

Staffellidae [PALEONTOLOGY] An extinct family of marine protozoans (superfamily Fusulinacea) that persisted during the Pennsylvanian and Early Permian. { sta'fel·ə‚dē }

stage theory [PHYSIOLOGY] A theory of color vision which proposes that there are three or more types of cone receptors whose responses are conducted to higher visual centers, and that interactions occur at some stage along the conducting paths so that strong activity in one type of response inhibits that of other response paths. Also known as zone theory. { 'stāj ‚thē·ə·rē }

stalked barnacle [INVERTEBRATE ZOOLOGY] The common name for crustaceans composing the suborder Lepadomorpha. { 'stȯkt 'bär·nə·kəl }

stallion [VERTEBRATE ZOOLOGY] **1.** A mature male equine mammal. **2.** A male horse not castrated. { 'stal·yən }

stamen [BOTANY] The male reproductive structure of a flower, consisting of an anther and a filament. { 'stā·mən }

staminate flower [BOTANY] A flower having stamens but lacking functional carpels. { 'stam·ə·nət 'flaů·ər }

staminode [BOTANY] A stamen with no functional anther. { 'stā·mə,nōd }

stand [ECOLOGY] A group of plants, distinguishable from adjacent vegetation, which is generally uniform in species composition, age, and condition. { stand }

standing crop [ECOLOGY] The number of individuals or total biomass present in a community at one particular time. { 'stand·iŋ 'kräp }

stapedius muscle [ANATOMY] The muscle which attaches to and controls the stapes in the inner ear. { stə'pēd·ē·əs ,məs·əl }

stapes [ANATOMY] The stirrup-shaped middle-ear ossicle, articulating with the incus and the oval window. Also known as columella. { 'stā·pēz }

Staphylinidae [INVERTEBRATE ZOOLOGY] The rove beetles, a very large family of coleopteran insects in the superfamily Staphylinoidea. { ,staf·ə'lin·ə,dē }

Staphylinoidea [INVERTEBRATE ZOOLOGY] A superfamily of Coleoptera in the suborder Polyphaga. { ,staf·ə·lə'nȯid·ē·ə }

staphylomycin [MICROBIOLOGY] An antibiotic composed of three active components produced by a strain of *Actinomyces* that inhibits growth of gram-positive microorganisms and acid-fast bacilli. { ,staf·ə·lō'mīs·ən }

staphylotoxin [BIOCHEMISTRY] Any of the various toxins elaborated by strains of *Staphylococcus aureus*, including hemolysins, enterotoxins, and leukocidin. { ,staf·ə·lō'täk·sən }

starch [BIOCHEMISTRY] Any one of a group of carbohydrates or polysaccharides, of the general composition $(C_6H_{10}O_5)_n$, occurring as organized or structural granules of varying size and markings in many plant cells; it hydrolyzes to several forms of dextrin and glucose; its chemical structure is not completely known, but the granules consist of concentric shells containing at least two fractions: an inner portion called amylose, and an outer portion called amylopectin. { stärch }

starfish [INVERTEBRATE ZOOLOGY] The common name for echinoderms belonging to the subclass Asteroidea. { 'stär,fish }

Starling's law of the heart [PHYSIOLOGY] The energy associated with cardiac contraction is proportional to the length of the myocardial fibers in diastole. { 'stär·liŋz 'lȯ əv thə 'härt }

start codon [GENETICS] *See* initiation codon. { ¦stärt 'kō,dän }

starter [MICROBIOLOGY] A culture of microorganisms, either pure or mixed, used to commence a process, for example, cheese manufacture. { 'stär·dər }

startle response [PHYSIOLOGY] The complex, involuntary, usually spasmodic psychophysiological response movement of an organism to a sudden unexpected stimulus. { 'stärd·əl ri,späns }

startpoint [MOLECULAR BIOLOGY] The deoxyribonucleic acid base pair that corresponds to the first nucleotide incorporated into the primary ribonucleic acid (RNA) transcript by RNA polymerase. { 'stärt,pȯint }

static contraction [PHYSIOLOGY] *See* isometric contraction. { 'stad·ik kən'trak·shən }

static reflex [PHYSIOLOGY] Any one of a series of reflexes which are involved in the establishment of muscular tone for postural purpose. { 'stad·ik 'rē,fleks }

static work [BIOPHYSICS] *See* isometric work. { 'stad·ik 'wərk }

stationary phase [MICROBIOLOGY] The period following termination of exponential growth in a bacterial culture when the number of viable microorganisms remains relatively constant for a time. { 'stā·shə,ner·ē 'fāz }

stationary population [ECOLOGY] A population containing a basically even distribution of age groups. { 'stā·shə,ner·ē ,päp·yə'lā·shən }

statoblast [INVERTEBRATE ZOOLOGY] A chitin-encapsulated body which serves as a special means of asexual reproduction in the Phylactolaemata. { 'stad·ə,blast }

statocone [INVERTEBRATE ZOOLOGY] One of the minute calcareous granules found in the statocyst of certain animals. { 'stad·ə,kōn }

statocyst [BOTANY] A cell containing statoliths in a fluid medium. Also known as statocyte. [INVERTEBRATE ZOOLOGY] A sensory vesicle containing statoliths and which functions in the perception of the position of the body in space. { 'stad·ə,sist }

statocyte [BOTANY] *See* statocyst. { 'stad·ə,sīt }

statokinetic [PHYSIOLOGY] Pertaining to the balance and posture of the body or its parts during movement, as in walking. { ¦stad·ō·ki'ned·ik }

statolith [BOTANY] A sand grain or other solid inclusion which moves readily in the fluid contents of a statocyst, comes to rest on the lower surface of the cell, and is believed to function in gravity perception. [INVERTEBRATE ZOOLOGY] A secreted calcareous body, a sand grain, or other solid inclusion contained in a statocyst. { 'stad·ə,lith }

statoreceptor [PHYSIOLOGY] A sense organ concerned primarily with equilibrium. { ¦stād·ō·ri 'sep·tər }

statospore [BOTANY] In certain algae, an internally formed spore in its resting stage. { 'stad·ə ,spȯr }

stauractine [INVERTEBRATE ZOOLOGY] A sponge spicule in which the four rays lie in one plane. { stȯ'rak,tēn }

Stauromedusae [INVERTEBRATE ZOOLOGY] An order of the class Scyphozoa in which the medusa is composed of a cuplike bell called a calyx and a stem that terminates in a pedal disk. { ¦stȯ·rō·mi'dü·sē }

Staurosporae [MYCOLOGY] A spore group of the Fungi Imperfecti characterized by star-shaped or forked spores. { stȯ'räs·pə,rē }

steapsin [BIOCHEMISTRY] An enzyme in pancreatic juice that catalyzes the hydrolysis of fats. Also known as pancreatic lipase. { stē'ap·sən }

Steatornithidae [VERTEBRATE ZOOLOGY] A family

of birds in the order Caprimulgiformes which contains a single, South American species, the oilbird or guacharo (*Steatornis caripensis*). { 'stē·ə ˌtȯr'nith·əˌdē }

Steganopodes [VERTEBRATE ZOOLOGY] Formerly, an order of birds that included the totipalmate swimming birds. { ˌsteg·ə'näp·əˌdēz }

Stegodontinae [PALEONTOLOGY] An extinct subfamily of elephantoid proboscideans in the family Elephantidae. { ˌsteg·ə'dänt·əˌnē }

Stegosauria [PALEONTOLOGY] A suborder of extinct reptiles of the order Ornithischia comprising the plated dinosaurs of the Jurassic which had tiny heads, great triangular plates arranged on the back in two alternating rows, and long spikes near the end of the tail. { ˌsteg·ə'sȯr·ē·ə }

Steinheim man [PALEONTOLOGY] A prehistoric man represented by a skull, without mandible, found near Stuttgart, Germany; the browridges are massive, the face is relatively small, and the braincase is similar in shape to that of *Homo sapiens*. { 'shtīnˌhīm ˌman }

STEL [PHYSIOLOGY] *See* short-term exposure limit. { stel *or* ¦es¦tē¦ē'el }

stele [BOTANY] The part of a plant stem including all tissues and regions of plants from the cortex inward, including the pericycle, phloem, cambium, xylem, and pith. { 'stēl }

Stelenchopidae [INVERTEBRATE ZOOLOGY] A family of polychaete annelids belonging to the Myzostomaria, represented by a single species from Crozet Island in the Antarctic Ocean. { ˌstel·ən 'käp·əˌdē }

stellate ganglion [ANATOMY] The ganglion formed by the fusion of the inferior cervical and the first thoracic sympathetic ganglions. { 'ste ˌlāt 'gaŋ·glē·ən }

stellate reticulum [HISTOLOGY] The part of the epithelial dental organ of a developing tooth which lies between the inner and the outer dental epithelium; composed of stellate cells with long, anastomosing processes in a mucoid fluid in the interstitial spaces. { 'steˌlāt rə'tik·yə·ləm }

Stelleroidea [INVERTEBRATE ZOOLOGY] The single class of echinoderms in the subphylum Asterozoa; characters coincide with those of the subphylum. { ˌstel·ə'rȯid·ē·ə }

stem [BOTANY] The organ of vascular plants that usually develops branches and bears leaves and flowers. { stem }

stem cell [EMBRYOLOGY] A formative cell. [HISTOLOGY] *See* hemocytoblast. { 'stem ˌsel }

Stemonitaceae [MYCOLOGY] The single family of the order Stemonitales. { stēˌmän·ə'tās·ēˌē }

Stemonitales [MYCOLOGY] An order of fungi in the subclass Myxogastromycetidae of the class Myxomycetes. { stēˌmän·ə'tā·lēz }

Stenetrioidea [INVERTEBRATE ZOOLOGY] A group of isopod crustaceans in the suborder Asellota consisting mostly of tropical marine forms in which the first pleopods are fused. { stəˌne·trē 'ȯid·ē·ə }

Stenocephalidae [INVERTEBRATE ZOOLOGY] A family of Old World, neotropical Hemiptera included in the Pentatomorpha. { ˌsten·ə·sə'fal·ə ˌdē }

Stenoglossa [INVERTEBRATE ZOOLOGY] The equivalent name for Neogastropoda. { ˌsten·ə 'gläs·ə }

stenohaline [ECOLOGY] In marine organisms, indicating the ability to tolerate only a narrow range of salinities. { ¦sten·ə¦haˌlīn }

Stenolaemata [INVERTEBRATE ZOOLOGY] A class of marine ectoproct bryozoans having lophophores which are circular in basal outline and zooecia which are long, slender, tubular or prismatic, and gradually tapering to their proximal ends. { ˌsten·ə·lə'mäd·ə }

Stenomasteridae [PALEONTOLOGY] An extinct family of Euechinoidea, order Holasteroida, comprising oval and heart-shaped forms with fully developed pore pairs. { ˌsten·ə·mas'ter·ə ˌdē }

stenoplastic [BIOLOGY] Relating to an organism which exhibits a limited capacity for modification or adaptation to a new environment. { ¦sten·ə¦plas·tik }

Stenopodidea [INVERTEBRATE ZOOLOGY] A section of decapod crustaceans in the suborder Natantia which includes shrimps having the third pereiopods chelate and much longer and stouter than the first pair. { ˌsten·ə·pə'did·ē·ə }

Stenostomata [INVERTEBRATE ZOOLOGY] The equivalent name for Cyclostomata. { ˌsten·ə·stə 'mäd·ə }

stenotherm [BIOLOGY] An organism able to tolerate only a small variation of temperature in the environment. { 'sten·əˌthərm }

stenothermic [BIOLOGY] Indicating the ability to tolerate only a limited range of temperatures. { ¦sten·ə¦thər·mik }

Stenothoidae [INVERTEBRATE ZOOLOGY] A family of amphipod crustaceans in the suborder Gammaridea containing semiparasitic and commensal species. { ˌsten·ə'thȯiˌdē }

stenotopic [ECOLOGY] Referring to an organism with a restricted distribution. { ¦sten·ə¦täp·ik }

Stensen's duct [ANATOMY] *See* parotid duct. { 'sten·sənz ˌdəkt }

Stensioellidae [PALEONTOLOGY] A family of Lower Devonian placoderms of the order Petalichthyida having large pectoral fins and a broad subterminal mouth. { ˌsten·shō'el·əˌdē }

Stenurida [PALEONTOLOGY] An order of Ophiuroidea, comprising the most primitive brittlestars, known only from Paleozoic sediments. { stə 'nu̇r·əd·ə }

Stephanidae [INVERTEBRATE ZOOLOGY] A small family of the Hymenoptera in the superfamily Ichneumonoidea characterized by many-segmented filamentous antennae. { stə'fan·əˌdē }

stepping reflex [PHYSIOLOGY] A reflex response of the newborn and young infant, characterized by alternating stepping movements with the legs, as in walking, elicited when the infant is held upright so that both soles touch a flat surface while the infant is moved forward to accompany any step taken. { 'step·iŋ ˌrēˌfleks }

stercobilin [BIOCHEMISTRY] Urobilin as a component of the brown fecal pigment. { ˌstər·kō ˈbī·lən }

stercobilinogen [BIOCHEMISTRY] A colorless reduction product of stercobilin found in feces. { ˌstər·kō·bīˈlin·ə·jən }

Stercorariidae [VERTEBRATE ZOOLOGY] A family of predatory birds of the order Charadriiformes including the skuas and jaegers. { ˌstər·kə·rəˈrī·əˌdē }

Sterculiaceae [BOTANY] A family of dicotyledonous trees and shrubs of the order Malvales distinguished by imbricate or contorted petals, bilocular anthers, and ten to numerous stamens arranged in two or more whorls. { stərˌkyü·lē ˈās·ē }

stereoblastula [EMBRYOLOGY] A blastula that lacks a cavity, making it unable to gastrulate. { ¦ster·ē·əˈblas·chə·lə }

stereocilia [CELL BIOLOGY] **1.** Nonmotile tufts of secretory microvilli on the free surface of cells of the male reproductive tract. **2.** Homogeneous cilia within simple membrane coverings; found on the free-surface hair cells. { ¦ster·ē·əˈsil·ē·ə }

stereogastrula [EMBRYOLOGY] A gastrula that lacks a cavity. { ¦ster·ē·əˈgas·trə·lə }

stereopsis [PHYSIOLOGY] *See* stereoscopy. { ˌster·ēˈäp·səs }

stereoscopic radius [PHYSIOLOGY] The greatest distance at which there is a sensation of depth in vision due to the fact that the two eyes do not perceive exactly the same view. { ¦ster·ē·ə ¦skäp·ik ˈrād·ē·əs }

stereoscopic vision [PHYSIOLOGY] *See* stereoscopy. { ¦ster·ē·ə¦skäp·ik ˈvizh·ən }

stereoscopy [PHYSIOLOGY] The phenomenon of simultaneous vision with two eyes in which there is a vivid perception of the distances of objects from the viewer; it is present because the two eyes view objects in space from two points, so that the retinal image patterns of the same object are slightly different in the two eyes. Also known as stereopsis; stereoscopic vision. { ˌster·ēˈäs·kə·pē }

Stereospondyli [PALEONTOLOGY] A group of labyrinthodont amphibians from the Triassic characterized by a flat body without pleurocentra and with highly developed intercentra. { ˌster·ē·ə ˈspän·dəˌlī }

stereotaxis [BIOLOGY] An orientation movement in response to stimulation by contact with a solid body. Also known as thigmotaxis. { ˌster·ē·əˈtak·səs }

stereotropism [BIOLOGY] Growth or orientation of a sessile organism or part of an organism in response to the stimulus of a solid body. Also known as thigmotropism. { ˌster·ēˈä·trəˌpiz·əm }

sterigma [BOTANY] A peg-shaped structure to which needles are attached in certain conifers. [MYCOLOGY] A slender stalk arising from the basidium of some fungi, from the top of which basidiospores are formed by abstriction. { stəˈrig·mə }

sterile distribution [ECOLOGY] A range of areas in which marine animals may live and spawn, but in which eggs do not hatch and larvae do not survive. { ˈster·əl ˌdis·trəˈbyü·shən }

sterility [PHYSIOLOGY] The inability to reproduce because of congenital or acquired reproductive system disorders involving lack of gamete production or production of abnormal gametes. { stəˈril·əd·ē }

sterilization [MICROBIOLOGY] An act or process of destroying all forms of microbial life on and in an object. { ˌster·ə·ləˈzā·shən }

Sternaspidae [INVERTEBRATE ZOOLOGY] A monogeneric family of polychaete annelids belonging to the Sedentaria. { stərˈnas·pəˌdē }

sternebra [VERTEBRATE ZOOLOGY] A segment of the sternum in vertebrates. { ˈstər·nə·brə }

Sterninae [VERTEBRATE ZOOLOGY] A subfamily of birds in the family Laridae, including the Arctic tern. { ˈstər·nəˌnē }

sternite [INVERTEBRATE ZOOLOGY] **1.** The ventral part of an arthropod somite. **2.** The chitinous plate on the ventral surface of an abdominal segment of an insect. { ˈstərˌnīt }

sternocleidomastoid [ANATOMY] A muscle of the neck that flexes the head; origin is the manubrium of the sternum and clavicle, and insertion is the mastoid process. { ¦stər·nō¦klīd·əˈma ˌstȯid }

sternocostal [ANATOMY] Pertaining to the sternum and the ribs. { ¦stər·nəˈkäst·əl }

sternohyoid [ANATOMY] A muscle arising from the manubrium of the sternum and inserted into the hyoid bone. { ˌstər·nōˈhīˌȯid }

Sternorrhyncha [INVERTEBRATE ZOOLOGY] A series of the insect order Homoptera in which the beak appears to arise either between or behind the fore coxae, and the antennae are long and filamentous with no well-differentiated terminal setae. { ¦stər·nəˈriŋ·kə }

sternothyroid [ANATOMY] Pertaining to the sternum and thyroid cartilage. { ¦stər·nəˈthīˌrȯid }

Sternoxia [INVERTEBRATE ZOOLOGY] The equivalent name for Elateroidea. { stərˈnäk·sē·ə }

sternum [ANATOMY] The bone, cartilage, or series of bony or cartilaginous segments in the median line of the anteroventral part of the body of vertebrates above fishes, connecting with the ribs or pectoral girdle. { ˈstər·nəm }

steroid [BIOCHEMISTRY] A member of a group of compounds, occurring in plants and animals, that are considered to be derivatives of a fused, reduced ring system, cyclopenta[α]-phenanthrene, which consists of three fused cyclohexane rings in a nonlinear or phenanthrene arrangement. { ˈstiˌrȯid }

sterol [BIOCHEMISTRY] Any of the natural products derived from the steroid nucleus; all are waxy, colorless solids soluble in most organic solvents but not in water, and contain one alcohol functional group. { ˈstiˌrȯl }

sterone [BIOCHEMISTRY] A ketone derived from a steroid. { ˈstiˌrōn }

Sthenurinae [PALEONTOLOGY] An extinct subfamily of marsupials of the family Diprotodontidae, including the giant kangaroos. { sthəˈnu̇r·ə ˌnē }

Stichaeidae [VERTEBRATE ZOOLOGY] The pricklebacks, a family of perciform fishes in the suborder Blennioidei. { stə'kē·ə,dē }

Stichocotylidae [INVERTEBRATE ZOOLOGY] A family of trematodes in the subclass Aspidogastrea in which adults are elongate and have a single row of alveoli. { ¦stik·ə·kə'til·ə,dē }

Stichopodidae [INVERTEBRATE ZOOLOGY] A family of the echinoderm order Aspidochirotida characterized by tentacle ampullae and by left and right gonads. { ¦stik·ə'päd·ə,dē }

Stickland reaction [BIOCHEMISTRY] An amino acid fermentation involving the coupled decomposition of two or more substrates. { 'stik·lənd rē,ak·shən }

stickleback [VERTEBRATE ZOOLOGY] Any fish which is a member of the family Gasterosteidae, so named for the variable number of free spines in front of the dorsal fin. { 'stik·əl,bak }

sticky end [BIOCHEMISTRY] Any of the single-stranded complementary ends of a deoxyribonucleic acid molecule. Also known as cohesive end. { 'stik·ē ,end }

stigma [BOTANY] The rough or sticky apical surface of the pistil for reception of the pollen. [INVERTEBRATE ZOOLOGY] **1.** The eyespot of certain protozoans, such as *Euglena*. **2.** The spiracle of an insect or arthropod. **3.** A colored spot on many lepidopteran wings. { 'stig·mə }

stigmatism [PHYSIOLOGY] A condition of the refractive media of the eye in which rays of light from a point are accurately brought to a focus on the retina. { 'stig·mə,tiz·əm }

Stilbaceae [MYCOLOGY] The equivalent name for Stilbellaceae. { stil'bās·ē,ē }

Stilbellaceae [MYCOLOGY] A family of fungi of the order Moniliales in which conidiophores are aggregated in long bundles or fascicles, forming synnemata or coremia, generally having the conidia in a head at the top. { ,stil·bə'lās·ē,ē }

stilbesterol [BIOCHEMISTRY] *See* diethylstilbesterol. { stil'bes·tə,ról }

stilt root [BOTANY] A prop root of a mangrove tree. { 'stilt ,rüt }

stimulus [PHYSIOLOGY] An agent that produces a temporary change in physiological activity in an organism or in any of its parts. { 'stim·yə·ləs }

stinger [ZOOLOGY] A sharp piercing organ, as of a bee, stingray, or wasp, usually connected with a poison gland. { 'stiŋ·ər }

stinging cell [INVERTEBRATE ZOOLOGY] *See* cnidoblast. { 'stiŋ·iŋ ,sel }

stingray [VERTEBRATE ZOOLOGY] Any of various rays having a whiplike tail armed with a long serrated spine, at the base of which is a poison gland. { 'stiŋ,rā }

stipe [BOTANY] **1.** The petiole of a fern frond. **2.** The stemlike portion of the thallus in certain algae. [MYCOLOGY] The short stalk or stem of the fruit body of a fungus, such as a mushroom. { stīp }

stipule [BOTANY] Either of a pair of appendages that are often present at the base of the petiole of a leaf. { 'stip·yül }

Stirodonta [INVERTEBRATE ZOOLOGY] Formerly, an order of Euechinoidea that included forms with stirodont dentition. { ,stir·ə'dän·tə }

stirodont dentition [INVERTEBRATE ZOOLOGY] In Echinoidea, the condition in which the teeth are keeled within and the foramen magnum is open. { 'stir·ə,dänt den'tish·ən }

stolon [BOTANY] *See* runner. [INVERTEBRATE ZOOLOGY] An elongated projection of the body wall from which buds are formed giving rise to new zooids in Anthozoa, Hydrozoa, Bryozoa, and Ascidiacea. [MYCOLOGY] A hypha produced above the surface and connecting a group of conidiophores. { 'stō·lən }

Stolonifera [INVERTEBRATE ZOOLOGY] An order of the Alcyonaria, lacking a coenenchyme; they form either simple (*Clavularia*) or rather complex colonies (*Tubipora*). { ,stäl·ə'nif·rə }

stoma [BIOLOGY] A small opening or pore in a surface. [BOTANY] One of the minute openings in the epidermis of higher plants which are regulated by guard cells and through which gases and water vapor are exchanged between internal spaces and the external atmosphere. { 'stō·mə }

stomach [ANATOMY] The tubular or saccular organ of the vertebrate digestive system located between the esophagus and the intestine and adapted for temporary food storage and for the preliminary stages of food breakdown. { 'stəm·ək }

stomatoblastula [INVERTEBRATE ZOOLOGY] A blastula stage in some sponges capable of engulfing maternal amebocytes for nutrition. { ¦stō·məd·ō'blas·chə·lə }

Stomatopoda [INVERTEBRATE ZOOLOGY] The single order of the Eumalacostraca in the superorder Hoplocarida distinguished by raptorial arms, especially the second pair of maxillipeds. { ,stō·mə'täp·əd·ə }

Stomiatoidei [VERTEBRATE ZOOLOGY] A suborder of fishes of the order Salmoniformes including the lightfishes and allies, which are of small size and often grotesque form and are equipped with photophores. { ,stō·mē·ə'tóid·ē,ī }

stomium [BOTANY] A region along a sporangium or pollen sac where dehiscence takes place. { 'stō·mē·əm }

stomocnidae nematocyst [INVERTEBRATE ZOOLOGY] A nematocyst which has an open-ended thread. { stə'mäk·nə,dē nə'mad·ə,sist }

stomodaeum [EMBRYOLOGY] The anterior part of the embryonic alimentary tract formed as an invagination of the ectoderm. { ,stō·mə'dē·əm }

stone canal [INVERTEBRATE ZOOLOGY] A canal in many echinoderms that has a more or less calcified wall and that leads from the madreporite to the ring vessel. { 'stōn kə,nal }

stone cell [BOTANY] *See* brachysclereid. { 'stōn ,sel }

stone fruit [BOTANY] *See* drupe. { 'stōn ,früt }

stonewort [BOTANY] The common name for algae comprising the class Charophyceae, so named because most species are lime-encrusted. { 'stōn,wórt }

Stonnomida [INVERTEBRATE ZOOLOGY] An order

of Xenophyophorea distinguished by the presence of linellae in the test and a flexible body. { stə'näm·əd·ə }

Stonnomidae [INVERTEBRATE ZOOLOGY] A family coextensive with the order Stonnomida. { stə 'näm·ə,dē }

stony coral [INVERTEBRATE ZOOLOGY] Any coral characterized by a calcareous skeleton. { 'stō·nē 'kär·əl }

stop codon [GENETICS] *See* terminator codon. { 'stäp 'kō,dän }

stork [VERTEBRATE ZOOLOGY] Any of several species of long-legged wading birds in the family Ciconiidae. { stȯrk }

straight sinus [ANATOMY] A sinus of the dura mater running from the inferior sagittal sinus along the junction of the falx cerebri and tentorium to the transverse sinus. { 'strāt ¦sī·nəs }

strain [BIOLOGY] An intraspecific group of organisms that possess only one or a few distinctive traits and are maintained as an artificial breeding group. [CELL BIOLOGY] A population of cells derived either from a primary culture or from a cell line by the selection or cloning of cells having specific properties or markers. { strān }

strain propagation [PHYSIOLOGY] Transmission of a response to external stress within the body by mechanical or biological processes. { 'strān ,präp·ə,gā·shən }

strategy [ECOLOGY] A group of related traits that evolved under the influence of natural selection and solve particular problems encountered by organisms. { 'strad·ə·jē }

stratified squamous epithelium [HISTOLOGY] A multiple-layered epithelium composed of thin, flat superficial cells and cuboidal and columnar deeper cells. { 'strad·ə,fīd 'skwā·məs ,ep·ə 'thē·lē·əm }

Stratiomyidae [INVERTEBRATE ZOOLOGY] The soldier flies, a family of orthorrhaphous dipteran insects in the series Brachycera. { ,strad·ē·ō'mī·ə ,dē }

stratum corneum [HISTOLOGY] The outer layer of flattened keratinized cells of the epidermis. { 'strad·əm 'kȯr·nē·əm }

stratum disjunctum [HISTOLOGY] The outermost layer of desquamating keratinized cells of the stratum corneum. { 'strad·əm dis'jəŋk·təm }

stratum germinativum [HISTOLOGY] The innermost germinative layer of the epidermis. { 'strad·əm ,jər·mə·nə'tī·vəm }

stratum granulosum [HISTOLOGY] A layer of granular cells interposed between the stratum corneum and the stratum germinativum in the thick skin of the palms and soles. { 'strad·əm ,gran·yə'lō·səm }

stratum lucidum [HISTOLOGY] A layer of irregular transparent epidermal cells with traces of nuclei interposed between the stratum corneum and stratum germinativum in the thick skin of the palms and soles. { 'strad·əm 'lü·səd·əm }

Strauss reaction [IMMUNOLOGY] The exudative swelling of the scrotum in male hamsters and guinea pigs upon subcutaneous or intraperitoneal inoculation of *Pseudomonas mallei*, the causative agent of glanders. { 'straủs rē,ak·shən }

straw [BOTANY] A stem of grain, such as wheat or oats. { strȯ }

strawberry [BOTANY] A low-growing perennial of the genus *Fragaria*, order Rosales, that spreads by stolons; the juicy, usually red, edible fruit consists of a fleshy receptacle with numerous seeds in pits or nearly superficial on the receptacle. { 'strȯ,ber·ē }

streak plate [MICROBIOLOGY] A method of culturing aerobic bacteria by streaking the surface of a solid medium in a petri dish with an inoculating wire or glass rod in a continuous movement so that most of the surface is covered; used to isolate majority members of a mixed population. { 'strēk ,plāt }

Streblidae [INVERTEBRATE ZOOLOGY] The bat flies, a family of cyclorrhaphous dipteran insects in the section Pupipara; adults are ectoparasites on bats. { 'streb·lə,dē }

Strelitziaceae [BOTANY] A family of monocotyledonous plants in the order Zingiberales distinguished by perfect flowers with five functional stamens and without an evident hypanthium, penniveined leaves, and symmetrical guard cells. { strə,lit·sē'ās·ē,ē }

strepaster [INVERTEBRATE ZOOLOGY] A short, spiny microscleric, monaxonic spicule. { stre 'pas·tər }

strepogenin [BIOCHEMISTRY] A factor, possibly a peptide derivative of glutamic acid, reported to exist in certain proteins, acting as a growth stimulant in bacteria and mice in the presence of completely hydrolyzed protein. Also known as streptogenin. { strə'päj·ə·nən }

Strepsiptera [INVERTEBRATE ZOOLOGY] An order of the Coleoptera that is coextensive with the family Stylopidae. { strep'sip·tə·rə }

streptobiosamine [BIOCHEMISTRY] $C_{13}H_{23}NO_9$ A nitrogen-containing disaccharide, obtained when streptomycin undergoes acid hydrolysis; in the streptomycin molecule it is glycosidally linked to streptidine. { ¦strep·tō·bī'ä·sə,mēn }

Streptococcaceae [MICROBIOLOGY] A family of gram-positive cocci; chemoorganotrophs with fermentative metabolism. { ¦strep·tə·käk'sās·ē,ē }

Streptococceae [MICROBIOLOGY] Formerly a tribe of the family Lactobacillaceae including cocci that occur in pairs, short chains, or tetrads and which generally obtain energy by fermentation of carbohydrates or related compounds. { ¦strep·tə'käk·sē,ē }

streptogenin [BIOCHEMISTRY] *See* strepogenin. { strep'tä·jə·nən }

streptokinase [BIOCHEMISTRY] An enzyme occurring as a component of fibrinolysin in cultures of certain hemolytic streptococci. { ¦strep·tō'kī ,nās }

streptolysin [BIOCHEMISTRY] Any of a group of hemolysins elaborated by *Streptococcus pyogenes*. { ¦strep·tō'līs·ən }

Streptomycetaceae [MICROBIOLOGY] A family of soil-inhabiting bacteria in the order Actinomycetales; branched mycelia are produced by vegetative hyphae; spores are produced on aerial hyphae. { ,strep·tō,mī·sə'tās·ē,ē }

streptomycin [MICROBIOLOGY] $C_{21}H_{39}O_{12}N_7$ Water-soluble antibiotic obtained from *Streptomyces griseus* that is used principally in the treatment of tuberculosis. { ˌstrep·təˈmīs·ən }

streptothricin [MICROBIOLOGY] $C_{19}H_{34}O_7N_8$ An antibiotic produced by *Streptomyces lavendulae*; active against various gram-negative and gram-positive microorganisms. { ˌstrep·təˈthrīs·ən }

stress [BIOLOGY] A stimulus or succession of stimuli of such magnitude as to tend to disrupt the homeostasis of the organism. { stres }

stretch reflex [PHYSIOLOGY] Contraction of a muscle in response to a sudden, brisk, longitudinal stretching of the same muscle. Also known as myostatic reflex. { ˈstrech ˌrēˌfleks }

stria [BIOLOGY] A minute line, band, groove, or channel. { ˈstrī·ə }

striated muscle [HISTOLOGY] Muscle tissue consisting of muscle fibers having cross striations. { ˈstrīˌād·əd ˈməs·əl }

stridulation [INVERTEBRATE ZOOLOGY] Creaking and other audible sounds made by certain insects, produced by rubbing various parts of the body together. { ˌstrij·əˈlā·shən }

Strigidae [VERTEBRATE ZOOLOGY] A family of birds of the order Strigiformes containing the true owls. { ˈstrij·əˌdē }

Strigiformes [VERTEBRATE ZOOLOGY] The order of birds containing the owls. { ˌstrij·əˈfórˌmēz }

strigose [BIOLOGY] Covered with stiff, pointed, hairlike scales or bristles. { ˈstrīˌgōs }

Strigulaceae [BOTANY] A family of Ascolichenes in the order Pyrenulales comprising crustose species confined to tropical evergreens, and which form extensive crusts on or under the cuticle of leaves. { ˌstrig·yəˈlās·ēˌē }

strobilation [INVERTEBRATE ZOOLOGY] Asexual reproduction by segmentation of the body into zooids, proglottids, or separate individuals. { ˌsträb·əˈlā·shən }

strobilocercus [INVERTEBRATE ZOOLOGY] A larval tapeworm that has undergone strobilation. { ˌsträb·ə·lōˈsər·kəs }

strobilus [BOTANY] **1.** A conelike structure made up of sporophylls, or spore-bearing leaves, as in Equisetales. **2.** The cone of members of the Pinophyta. { ˈsträb·ə·ləs }

stroke volume [PHYSIOLOGY] The amount of blood pumped during each cardiac contraction; quantitatively, the diastolic volume of the left ventricle minus the volume of blood in the ventricle at the end of systole. { ˈstrōk ˌväl·yəm }

stroma [ANATOMY] The supporting tissues of an organ, including connective and nervous tissues and blood vessels. { ˈstrō·mə }

Stromateidae [INVERTEBRATE ZOOLOGY] A family of perciform fishes in the suborder Stromateoidei containing the butterfishes. { ˈstrō·məˈtē·ə ˌdē }

Stromateoidei [VERTEBRATE ZOOLOGY] A suborder of fishes of the order Perciformes in which most species have teeth in pockets behind the pharyngeal bone. { ˈstrō·mə·tēˈóid·ēˌī }

Stromatoporoidea [PALEONTOLOGY] An extinct order of fossil colonial organisms thought to belong to the class Hydrozoa; the skeleton is a coenosteum. { strəˌmad·ə·pəˌróid·ē·ə }

Strombacea [PALEONTOLOGY] An extinct superfamily of gastropod mollusks in the order Prosobranchia. { strämˈbās·ē·ə }

Strombidae [INVERTEBRATE ZOOLOGY] A family of gastopod mollusks comprising tropical conchs. { ˈsträm·bəˌdē }

strongyle [INVERTEBRATE ZOOLOGY] A monaxonic spicule rounded at each end. { ˈstränˌjīl }

Strongyloidea [INVERTEBRATE ZOOLOGY] The hookworms, an order or superfamily of roundworms which, as adults, are endoparasites of most vertebrates, including humans. { ˌsträn·jəˈlóid·ē·ə }

strongylote [INVERTEBRATE ZOOLOGY] Rounded at one end, referring to sponge spicules. { ˈsträn·jəˌlōt }

strophiole [BOTANY] A crestlike excrescence around the hilum in certain seeds. { ˈsträ·fē ˌōl }

Strophomenida [PALEONTOLOGY] A large diverse order of articulate brachiopods which first appeared in Lower Ordovician times and became extinct in the Late Triassic. { ˌsträ·fəˈmen·əd·ə }

Strophomenidina [PALEONTOLOGY] A suborder of extinct, articulate brachiopods in the order Strophomenida characterized by a concavo-convex shell, the pseudodeltidium and socket plates disposed subparallel to the hinge. { ˌsträ·fə ˌmen·əˈdī·nə }

structural gene [MOLECULAR BIOLOGY] *See* cistron. { ˈstrək·chə·rəl ˈgēn }

Struthionidae [VERTEBRATE ZOOLOGY] The single family of the avian order Struthioniformes. { ˌstrü·thēˈän·əˌdē }

Struthioniformes [VERTEBRATE ZOOLOGY] A monofamilial order of ratite birds containing the single living species of ostrich (*Struthio camelus*). { ˌstrü·thēˌän·əˈfórˌmēz }

Stuart factor [BIOCHEMISTRY] A procoagulant in normal plasma but deficient in the blood of patients with a hereditary bleeding disorder; may be closely related to prothrombin since both are formed in the liver by action of vitamin K. Also known as factor X; Stuart-Power factor. { ˈstü·ərt ˌfak·tər }

Stuart-Power factor [BIOCHEMISTRY] *See* Stuart factor. { ˈstü·ərt ˈpaů·ər ˌfak·tər }

sturgeon [VERTEBRATE ZOOLOGY] Any of 10 species of large bottom-living fish which comprise the family Acipenseridae; the body has five rows of bony plates, and the snout is elongate with four barbels on its lower surface. { ˈstər·jən }

Styginae [INVERTEBRATE ZOOLOGY] A subfamily of butterflies in the family Lycaenidae in which the prothoracic legs in the male are nonfunctional. { ˈstij·ə·nē }

Stygocaridacea [INVERTEBRATE ZOOLOGY] An order of crustaceans in the superorder Syncarida characterized by having a furca. { ˌstig·əˌkar·əˈdās·ē·ə }

Stylasterina [INVERTEBRATE ZOOLOGY] An order of the class Hydrozoa, including several brightly colored branching or encrusting corallike cnidarians of warm seas. { stə‚las·tə'rī·nə }

style [BOTANY] The portion of a pistil connecting the stigma and ovary. [ZOOLOGY] A slender elongated process on an animal. { stīl }

stylet [INVERTEBRATE ZOOLOGY] A slender, rigid, elongated appendage. { 'stī·lət }

styloglossus [ANATOMY] A muscle arising from the styloid process of the temporal bone, and inserted into the tongue. { ‚stī·lō'gläs·əs }

stylohyoid [ANATOMY] Pertaining to the styloid process of the temporal bone and the hyoid bone. { ¦stī·lō¦hī‚ȯid }

styloid [ZOOLOGY] Resembling a style. { 'stī‚lȯid }

stylomastoid [ANATOMY] Relating to the styloid and the mastoid processes of the temporal bone. { ¦stī·lō'ma‚stȯid }

Stylommatophora [INVERTEBRATE ZOOLOGY] A large order of the molluscan subclass Pulmonata characterized by having two pairs of retractile tentacles with eyes located on the tips of the large tentacles. { sti‚lam·ə'täf·ə·rə }

stylopodium [BOTANY] A conical or disk-shaped enlargement at the base of the style in plants of the family Umbelliferae. { ‚stī·lə'pōd·ē·əm }

Stypocapitellidae [INVERTEBRATE ZOOLOGY] A family of polychaete annelids belonging to the Sedentaria and consisting of a monotypic genus found in western Germany. { ‚stī·pō‚kap·ə'tel·ə‚dē }

subalpine [ECOLOGY] *See* alpestrine. { ¦səb'al‚pīn }

subarachnoid space [ANATOMY] The space between the pia mater and the arachnoid of the brain. { ¦səb·ə'rak‚nȯid ¦spās }

subboreal [ECOLOGY] A biogeographic zone whose climatic condition approaches that of the boreal. { ¦səb'bȯr·ē·əl }

subcardinal vein [VERTEBRATE ZOOLOGY] Either of a pair of longitudinal veins of the mammalian embryo or the adult of some lower vertebrates which partly replace the postcardinals in the abdominal region, ventromedial to the mesonephros. { ¦səb'kärd·nəl 'vān }

subclavian artery [ANATOMY] The proximal part of the principal artery in the arm or forelimb. { 'səb¦klā·vē·ən 'ärd·ə·rē }

subclavian vein [ANATOMY] The proximal part of the principal vein in the arm or forelimb. { 'səb¦klā·vē·ən 'vān }

subclavius [ANATOMY] A small muscle attached to the clavicle and the first rib. { ‚səb'klā·vē·əs }

subclimax [ECOLOGY] A community immediately preceding a climax in an ecological succession. { ¦səb'klī‚maks }

subcollateral [ANATOMY] Ventrad of the collateral sulcus of the brain. { ¦səb·kə'lad·ə·rəl }

subcutaneous connective tissue [HISTOLOGY] The layer of loose connective tissue beneath the dermis. { ¦səb·kyü'tā·nē·əs kə'nek·tiv ‚tish·ü }

subdominant [ECOLOGY] A species which may appear more abundant at particular times of the year than the true dominant in a climax; for example, in a savannah trees and shrubs are more conspicuous than the grasses, which are the true dominants. { ¦səb'däm·ə·nənt }

subendothelial layer [HISTOLOGY] The middle layer of the tunica intima of veins and of medium and larger arteries, consisting of collagenous and elastic fibers and a few fibroblasts. { ¦səb‚en·də'thē·lē·əl 'lā·ər }

suberin [BIOCHEMISTRY] A fatty substance found in many plant cell walls, especially cork. { 'sü·bə·rən }

suberization [BOTANY] Infiltration of plant cell walls by suberin resulting in the formation of corky tissue that is impervious to water. { ‚sü·bə·rə'zā·shən }

suberose [BOTANY] Having a texture like cork due to or resembling that due to suberization. { 'sü·bə‚rōs }

sublethal gene [GENETICS] *See* semilethal gene. { ¦səb'lē·thəl 'jēn }

subliminal [PHYSIOLOGY] Below the threshold of responsiveness, consciousness, or sensation to a stimulus. { sə'blim·ə·nəl }

sublingual gland [ANATOMY] A complex of salivary glands located in the sublingual fold on each side of the floor of the mouth. { ¦səb'liŋ·gwəl 'gland }

submandibular duct [ANATOMY] The duct of the submandibular gland which empties into the mouth on the side of the frenulum of the tongue. { ¦səb·man'dib·yə·lər 'dəkt }

submandibular gland [ANATOMY] A large seromucous or mixed salivary gland located below the mandible on each side of the jaw. Also known as mandibular gland; submaxillary gland. { ¦səb·man'dib·yə·lər 'gland }

submaxillary gland [ANATOMY] *See* submandibular gland. { ¦səb'mak·sə‚ler·ē ‚gland }

submerged culture [MICROBIOLOGY] A method for growing pure cultures of aerobic bacteria in which microorganisms are incubated in a liquid medium subjected to continuous, vigorous agitation. { səb'mərjd 'kəl·chər }

submerged fermentation [MICROBIOLOGY] Industrial production of antibiotics, enzymes, and other substances by growing the microorganisms that produce the product in a submerged culture. { səb'mərjd ‚fər·mən'tā·shən }

submucosa [HISTOLOGY] The layer of fibrous connective tissue that attaches a mucous membrane to its subadjacent parts. { ¦səb·myü'kō·sə }

submucous plexus [ANATOMY] A visceral nerve network lying in the submucosa of the digestive tube. Also known as Meissner's plexus. { ¦səb'myü·kəs 'plek·səs }

Suboscines [VERTEBRATE ZOOLOGY] A major division of the order Passeriformes, usually divided into the suborders Eurylaimi, Tyranni, and Menurae. { ¦səb'äs·ə‚nēz }

subscapularis [ANATOMY] A muscle arising from the costal surface of the scapula and inserted on the lesser tubercle of the humerus. { ‚səb‚skap·yə'lar·əs }

subsere [ECOLOGY] A secondary community that succeeds an interrupted climax. { 'səb‚sir }

subspecies [SYSTEMATICS] A geographically defined grouping of local populations which differs taxonomically from similar subdivisions of species. { ¦səb'spē·shēz }

substance P [BIOCHEMISTRY] An undecapeptide widely distributed in the central nervous system and found in highest concentrations in superficial layers of the dorsal horn of the spinal cord, in the trigeminal nerve nucleus, and in the substantia nigra; acts as a neurotransmitter. { 'səb·stəns 'pē }

substrain [CELL BIOLOGY] A strain derived by isolation of a single cell or group of cells having properties or markers not shared by the other cells of the cell strain. { 'səb·ˌstrān }

substrate [BIOCHEMISTRY] The substance with which an enzyme reacts. [ECOLOGY] The foundation to which a sessile organism is attached. { 'səbˌstrāt }

subtend [BOTANY] To lie adjacent to and below another structure, often enclosing it. { səb 'tend }

subtilin [MICROBIOLOGY] An antibiotic substance obtained from *Bacillus subtilis*, active against gram-positive bacteria. { 'səb·tə·lən }

Subtriquetridae [INVERTEBRATE ZOOLOGY] A family of arthropods in the suborder Porocephaloidea. { ˌsəb·trə'ke·trəˌdē }

subtropical forest [ECOLOGY] *See* temperate rainforest. { ˌsəb'träp·ə·kəl 'fär·əst }

subulate [BOTANY] Linear, delicate, and tapering to a sharp point. { 'səb·yə·lət }

Subulitacea [PALEONTOLOGY] An extinct superfamily of gastropod mollusks in the order Prosobranchia which possessed a basal fold but lacked an apertural sinus. { ˌsəb·yə·lə'tās·ē·ə }

Subuluridae [INVERTEBRATE ZOOLOGY] The equivalent name for Heterakidae. { ¦səb·yə'lu̇r·əˌdē }

Subuluroidea [INVERTEBRATE ZOOLOGY] A superfamily of parasitic nematodes in the order Ascaridida characterized by weakly developed lips with sensilla and a thick-walled stoma that is armed with three teeth. { ˌsəb·yə·lə'ro̊id·ē·ə }

subumbrella [INVERTEBRATE ZOOLOGY] The concave undersurface of the body of a jellyfish. { ¦səb·əm'brel·ə }

subunit [BIOCHEMISTRY] *See* protomer. { 'səbˌyü·nət }

succession [ECOLOGY] A gradual process brought about by the change in the number of individuals of each species of a community and by the establishment of new species populations which may gradually replace the original inhabitants. { sək'sesh·ən }

succinamide [BIOCHEMISTRY] $H_2NCOCH_2CONH_2$ The amide of succinic acid. { sək'sin·ə·mīd }

succinic acid dehydrogenase [BIOCHEMISTRY] An enzyme that catalyzes the dehydrogenation of succinic acid to fumaric acid in the presence of a hydrogen acceptor. Also known as succinic dehydrogenase. { sək'sin·ik 'as·əd dē'hī·drə·jə ˌnās }

succinic dehydrogenase [BIOCHEMISTRY] *See* succinic acid dehydrogenase. { sək'sin·ik dē'hī·drə·jəˌnās }

succinoxidase [BIOCHEMISTRY] A complex enzyme system containing succinic dehydrogenase and cytochromes that catalyzes the conversion of succinate ion and molecular oxygen to fumarate ion. { ˌsək·sən'äk·səˌdās }

succulent [BOTANY] Describing a plant having juicy, fleshy tissue. { 'sək·yə·lənt }

succus entericus [PHYSIOLOGY] The intestinal juice secreted by the glands of the intestinal mucous membrane; it is thin, opalescent, alkaline, and has a specific gravity of 1.011. { 'sək·əs in 'ter·ə·kəs }

sucker [BOTANY] A shoot that develops rapidly from the lower portion of a plant, and usually at the expense of the plant. [ZOOLOGY] A disk-shaped organ in various animals for adhering to or holding onto an individual, usually of another species. { 'sək·ər }

sucking louse [INVERTEBRATE ZOOLOGY] The common name for insects of the order Anoplura, so named for the slender, tubular mouthparts. { 'sək·iŋ ˌlau̇s }

sucrase [BIOCHEMISTRY] *See* saccharase. { 'sü ˌkrās }

Suctoria [INVERTEBRATE ZOOLOGY] A small subclass of the protozoan class Ciliatea, distinguished by having tentacles which serve as mouths. { sək'tor·ē·ə }

Suctorida [INVERTEBRATE ZOOLOGY] The single order of the protozoan subclass Suctoria. { sək 'to̊r·əd·ə }

sudomotor [PHYSIOLOGY] Pertaining to the efferent nerves that control the activity of sweat glands. { ¦süd·ə'mōd·ər }

suffrutescent [BOTANY] Of or pertaining to a stem intermediate between herbaceous and shrubby, becoming partly woody and perennial at the base. { ¦səˌfrü'tes·ənt }

suffruticose [BOTANY] Low stems which are woody, grading into herbaceous at the top. { sə 'früd·əˌkōs }

sugar [BIOCHEMISTRY] A generic term for a class of carbohydrates usually crystalline, sweet, and water soluble; examples are glucose and fructose. { 'shu̇g·ər }

sugarbeet [BOTANY] *Beta vulgaris*. A beet characterized by a white root and cultivated for the high sugar content of the roots. { 'shu̇g·ərˌbēt }

sugarcane [BOTANY] *Saccharum officinarum*. A stout, perennial grass plant characterized by two-ranked leaves, and a many-jointed stalk with a terminal inflorescence in the form of a silky panicle; the source of more than 50% of the world's annual sugar production. { 'shu̇g·ər ˌkān }

sugar maple [BOTANY] *Acer saccharum*. A commercially important species of maple tree recognized by its gray furrowed bark, sharp-pointed scaly winter buds, and symmetrical oval outline of the crown. { 'shu̇g·ər 'mā·pəl }

suicide [IMMUNOLOGY] Death of cells that have selectively taken up heavily radioactively labeled antigen. { 'sü·əˌsīd }

suicide inhibitor [BIOCHEMISTRY] A compound which resembles the normal substrate for an enzyme, but which interacts with the enzyme to form a covalent bond and thus inactivates the enzyme. { 'sü·ə‚sīd in'hib·əd·ər }

Suidae [VERTEBRATE ZOOLOGY] A family of paleodont artiodactyls in the superfamily Suoidae including wild and domestic pigs. { 'sü·ə‚dē }

sulcate [ZOOLOGY] Having furrows or grooves on the surface. { 'səl‚kāt }

sulculus [ZOOLOGY] A small sulcus. { 'səl·kyə·ləs }

sulcus [ZOOLOGY] A furrow or groove, especially one on the surface of the cerebrum. { 'səl·kəs }

sulfatase [BIOCHEMISTRY] Any of a group of esterases that catalyze the hydrolysis of sulfuric esters. { 'səl·fə‚tās }

sulf-heme protein [BIOCHEMISTRY] A heme protein that has reacted with sulfur to yield a new structure. { 'səlf ‚hēm 'prō‚tēn }

sulfhemoglobin [BIOCHEMISTRY] A greenish substance derived from hemoglobin by the action of hydrogen sulfide; it may appear in the blood following the ingestion of sulfanilamide and other substances. { ¦səlf'hē·mə‚glō·bən }

sulfur bacteria [MICROBIOLOGY] Any of various bacteria having the ability to oxidize sulfur compounds. { 'səl·fər bak‚tir·ē·ə }

Sulidae [VERTEBRATE ZOOLOGY] A family of aquatic birds in the order Pelecaniformes including the gannets and boobies. { 'sü·lə‚dē }

summation wave [PHYSIOLOGY] A sustained contraction of muscles, caused by the rapid firing of nerve impulses. { sə'mā·shən ‚wāv }

summerwood [BOTANY] The less porous, usually harder portion of an annual ring that forms in the latter part of the growing season. { 'səm·ər ‚wu̇d }

sundew [BOTANY] Any plant of the genus *Drosera* of the family Droseraceae; the genus comprises small, herbaceous, insectivorous plants that grow on all continents, especially Australia. { 'sən‚dü }

sunfish [VERTEBRATE ZOOLOGY] Any of several species of marine and freshwater fishes in the families Centrarchidae and Molidae characterized by brilliant metallic skin coloration. { 'sən ‚fish }

sunflower [BOTANY] *Helianthus annuus*. An annual plant native to the United States characterized by broad, ovate leaves growing from a single, usually long (3-20 feet or 1-6 meters) stem, and large, composite flowers with yellow petals. { 'sən‚flau̇·ər }

Suoidea [VERTEBRATE ZOOLOGY] A superfamily of artiodactyls of the suborder Paleodonta which comprises the pigs and peccaries. { sü'ȯid·ē·ə }

supercilium [ANATOMY] The eyebrow. { ‚sü·pər 'sil·ē·əm }

supercoiling [MOLECULAR BIOLOGY] Winding of the deoxyribonucleic acid duplex on itself so that it crosses its own axis; may be in the same (positive) direction as, or opposite (negative) direction to, the turns of the double helix. { 'sü·pər ‚kȯil·iŋ }

superfecundation [PHYSIOLOGY] Multiple, simultaneous fertilization by a number of sperm of many eggs released at ovulation. { ¦sü·pər‚fē ‚kən'dā·shən }

superfemale [GENETICS] A female with three X chromosomes and two sets of autosomes resulting in sterility and generally early death. { ¦sü·pər'fē‚māl }

superficial cleavage [EMBRYOLOGY] Meroblastic cleavage restricted to the peripheral cytoplasm, as in the centrolecithal insect ovum. { ¦sü·pər ¦fish·əl 'klē·vij }

superficial palmar arch [ANATOMY] The arterial anastomosis formed by the ulnar artery in the palm with a branch from the radial artery. Also known as palmar arch. { ¦sü·pər¦fish·əl 'päm·ər 'ärch }

supergene [GENETICS] A chromosomal segment protected from crossing over and therefore transmitted from generation to generation as if it were a single recon. { 'sü·pər‚jēn }

superhelix [BIOCHEMISTRY] A macromolecular structure consisting of a number of alpha-helical polypeptide strands which are twisted together. { ¦sü·pər'hē‚liks }

superinfection [VIROLOGY] An attack on a bacterial cell by several phages due to the introduction of large numbers of viruses into the bacterial culture. { ¦sü·pər·in'fek·shən }

superior [BOTANY] **1.** Positioned above another organ or structure. **2.** Referring to a calyx that is attached to the ovary. **3.** Referring to an ovary that is above the insertion of the floral parts. { sə 'pir·ē·ər }

superior alveolar canals [ANATOMY] The alveolar canals of the maxilla. { sə'pir·ē·ər al'vē ə lər kə 'nalz }

superior ganglion [ANATOMY] **1.** The upper sensory ganglion of the glossopharyngeal nerve, located in the upper part of the jugular foramen; it is inconstant. **2.** The upper sensory ganglion of the vagus nerve, located in the jugular foramen. { sə'pir·ē·ər 'gaŋ·glē·ən }

superior mesenteric artery [ANATOMY] A major branch of the abdominal aorta with branches to the pancreas and intestine. { sə'pir·ē·ər ‚mez·ən'ter·ik 'ärd·ə·rē }

superior vena cava [ANATOMY] The principal vein collecting blood from the head, chest wall, and upper extremities and draining into the right atrium. { sə'pir·ē·ər ¦vē·nə 'kä·və }

supermale [GENETICS] A male with one X chromosome and three or more sets of autosomes, resulting in sterility and generally early death. { 'sü·pər‚māl }

supernumerary bud [BOTANY] *See* accessory bud. { ¦sü·pər'nü·mə‚rer·ē 'bəd }

supernumerary chromosome [CELL BIOLOGY] A chromosome present in addition to the normal chromosome complement. Also known as accessory chromosome. { ¦sü·pər'nü·mə‚rer·ē 'krō·mə‚sōm }

superposed [BOTANY] **1.** Growing vertically over another part. **2.** Of or pertaining to floral parts that are opposite each other. { ¦sü·pər'pōzd }

superposition eye [INVERTEBRATE ZOOLOGY] A compound eye in which a given rhabdome receives light from a number of facets; visual acuity is reduced in this type of eye. { ˌsü·pər·pə'zish·ən 'ī }

supination [ANATOMY] **1.** Turning the palm upward. **2.** Inversion of the foot. { ˌsü·pə'nā·shən }

suppressor [GENETICS] A secondary mutation that reverses the effect of a mutation. { sə'pres·ər }

suppressor gene [GENETICS] A gene that reverses the effect of a mutation in another gene. { sə'pres·ər ˌjēn }

suppressor mutation [GENETICS] A mutation that restores functional loss of a primary mutation and is located at a different genetic site from the primary mutation. { sə'pres·ər myüˌtā·shən }

supracardinal veins [VERTEBRATE ZOOLOGY] Paired longitudinal veins in the mammalian embryo and various adult lower vertebrates in the thoracic and abdominal regions, dorsolateral to and on the sides of the descending aorta; they replace the postcardinal and subcardinal veins. { ¦sü·prə'kärd·nəl ˌvānz }

suprahyoid muscles [ANATOMY] The muscles attached to the upper margin of the hyoid bone. { ¦sü·prə'hīˌȯid ˌməs·əlz }

supraliminal [PHYSIOLOGY] Above, or in excess of, a threshold. { ¦sü·prə'lim·ə·nəl }

supranuclear [ANATOMY] In the nervous system, central to a nucleus. { ¦sü·prə'nü·klē·ər }

supraoccipital [ANATOMY] Situated above the occipital bone. { ¦sü·prə·äk'sip·əd·əl }

supraoptic [ANATOMY] Situated above the optic tract. { ¦sü·prə¦äp·tik }

suprarenal gland [ANATOMY] *See* adrenal gland. { ¦sü·prə'rēn·əl ˌgland }

suprascapula [ANATOMY] An anomalous bone sometimes found between the superior border of the scapula and the spines of the lower cervical or first thoracic vertebrae. { ¦sü·prə'skap·yə·lə }

suprasegmental reflex [PHYSIOLOGY] A reflex employing complex multineuronal channels to integrate the body and limb musculature with fixed positions or movements of the head. { ¦sü·prə·seg'ment·əl 'rēˌfleks }

suprasternal notch [ANATOMY] Jugular notch of the sternum. { ¦sü·prə'stərn·əl 'näch }

suprasternal space [ANATOMY] The triangular space above the manubrium, enclosed by the layers of the deep cervical fascia which are attached to the front and back of this bone. { ¦sü·prə'stərn·əl 'spās }

supravital [BIOLOGY] Pertaining to the staining of living cells after removal from a living animal or of still living cells from a recently killed animal. { ¦sü·prə'vīd·əl }

survival ratio [BIOLOGY] The number of organisms surviving irradiation by ionizing radiation divided by the number of organisms before irradiation. { sər'vī·vəl ˌrā·shō }

suspension feeder [ZOOLOGY] An animal that feeds on small particles suspended in water; particles may be minute living plants or animals, or products of excretion or decay from these or larger organisms. { sə'spen·shən ¦fēd·ər }

suspensor [BOTANY] A mass of cells in higher plants that pushes the embryo down into the embryo sac and into contact with the nutritive tissue. [MYCOLOGY] A hypha which bears an apical gametangium in fungi of the Mucorales. { sə'spen·sər }

sustained yield [BIOLOGY] In a biological resource such as timber or grain, the replacement of a harvest yield by growth or reproduction before another harvest occurs. { sə'stānd 'yēld }

sustentacular cell [HISTOLOGY] One of the supporting cells of an epithelium as contrasted with other cells with special function, as the nonnervous cells of the olfactory epithelium or the Sertoli cells of the seminiferous tubules. { ¦səs·tən¦tak·yə·lər ˌsel }

suture [BIOLOGY] A distinguishable line of union between two closely united parts. { 'sü·chər }

swamp [ECOLOGY] A waterlogged land supporting a natural vegetation predominantly of shrubs and trees. { swämp }

swan [VERTEBRATE ZOOLOGY] Any of several species of large waterfowl comprising the subfamily Anatinae; they are herbivorous stout-bodied forms with long necks and spatulate bills. { swän }

Swanscombe man [PALEONTOLOGY] A partial skull recovered in Swanscombe, Kent, England, which represents an early stage of *Homo sapiens* but differing in having a vertical temporal region and a rounded occipital profile. { 'swanz·kəm ¦man }

swarmer cell [MICROBIOLOGY] The daughter cell which separates from the stalked mother cell in bacteria in the genus *Caulobacter*. { 'swȯr·mər ˌsel }

sweat [PHYSIOLOGY] The secretion of the sweat glands. Also known as perspiration. { swet }

sweat gland [PHYSIOLOGY] A coiled tubular gland of the skin which secretes sweat. { 'swet ˌgland }

sweepstakes route [ECOLOGY] A means that allows chance migration across a sea on natural rafts, so that oceanic islands can be colonized. { 'swēpˌstāks ˌrüt }

sweetgum [BOTANY] *Liquidambar styraciflua*. A deciduous tree of the order Hamamelidales found in the southeastern United States, and distinguished by its five-lobed, or star-shaped, leaves, and by the corky ridges developed on the twigs. { 'swētˌgəm }

sweet potato [BOTANY] *Ipomoea batatas*. A tropical vine having variously shaped leaves, purplish flowers, and a sweet, fleshy, edible tuberous root. { 'swēt pəˌtād·ō }

swim bladder [VERTEBRATE ZOOLOGY] A gas-filled cavity found in the body cavities of most bony fishes, has various functions in different fishes, acting as a float, a lung, a hearing aid, and a sound-producing organ. { 'swim ˌblad·ər }

swimmeret [INVERTEBRATE ZOOLOGY] Any of a series of paired biramous appendages under the abdomen of many crustaceans, used for swimming and egg carrying. { ˌswim·əˌret }

swimming bird [VERTEBRATE ZOOLOGY] Any bird belonging to the orders Charadriiformes and Pelacaniformes. { 'swim·iŋ ˌbird }

swine [VERTEBRATE ZOOLOGY] Any of various species comprising the Suidae. { swīn }

switch gene [GENETICS] A gene that causes the epigenotype to switch to a different developmental pathway. { 'swich ˌjēn }

switching site [GENETICS] A point at which gene segments combine during gene rearrangement. { 'swich·iŋ ˌsīt }

sycamore [BOTANY] **1.** Any of several species of deciduous trees of the genus *Platanus*, especially *P. occidentalis* of eastern and central North America, distinguished by simple, large, three-to five-lobed leaves and spherical fruit heads. **2.** The Eurasian maple (*Acer pseudoplatanus*). { 'sik·ə ˌmȯr }

Sycettida [INVERTEBRATE ZOOLOGY] An order of calcareous sponges of the subclass Calcaronea in which choanocytes occur in flagellated chambers, and the spongocoel is not lined with these cells. { sə'sed·əd·ə }

Sycettidae [INVERTEBRATE ZOOLOGY] A family of sponges in the order Sycettida. { sə'sed·əˌdē }

Sycidales [PALEONTOLOGY] A group of fossil aquatic plants assigned to the Charophyta, characterized by vertically ribbed gyrogonites. { ˌsis·ə'dā·lēz }

sycon [INVERTEBRATE ZOOLOGY] A canal system in sponges in which the flagellated layer is confined to outpocketings of the paragaster that are indirectly connected to the incurrent canals. { 'sīˌkän }

syconium [BOTANY] A fleshy fruit, as a fig, with an enlarged pulpy receptacle internally lined with minute flowers. { sī'kō·nē·əm }

Syllidae [INVERTEBRATE ZOOLOGY] A large family of polychaete annelids belonging to the Errantia; identified by their long, linear, translucent bodies with articulated cirri; size ranges from minute *Exogone* to *Trypanosyllis*, which may be 4 inches (100 millimeters) long. { 'sil·əˌdē }

Syllinae [INVERTEBRATE ZOOLOGY] A subfamily of polychaete annelids of the family Syllidae. { 'sil·əˌnē }

Sylonidae [INVERTEBRATE ZOOLOGY] A family of parasitic crustaceans in the order Rhizocephala. { sə'län·əˌdē }

Sylopidae [INVERTEBRATE ZOOLOGY] A family of coleopteran insects in the superfamily Meloidea in which the elytra in males are reduced to small leathery flaps while the hindwings are large and fan-shaped. { sə'läp·əˌdē }

Sylvicolidae [INVERTEBRATE ZOOLOGY] A family of orthorrhaphous dipteran insects in the series Nematocera. { ˌsil·və'käl·əˌdē }

symbiont [ECOLOGY] A member of a symbiotic pair. { 'sim·bēˌänt }

symbiosis [ECOLOGY] **1.** An interrelationship between two different species. **2.** An interrelationship between two different organisms in which the effects of that relationship is expressed as being harmful or beneficial. Also known as consortism. { ˌsim·bē'ō·səs }

Symmetrodonta [PALEONTOLOGY] An order of the extinct mammalian infraclass Pantotheria distinguished by the central high cusp, flanked by two smaller cusps and several low minor cusps, on the upper and lower molars. { ˌsim·ə·trə'dänt·ə }

symmetry [BIOLOGY] The disposition of organs and other constituent parts of the body of living organisms with respect to imaginary axes. { 'sim·əˌtrē }

sympathetic nervous system [ANATOMY] The portion of the autonomic nervous system, innervating smooth muscle and glands of the body, which upon stimulation produces a functional state of preparation for flight or combat. { ˌsim·pə'thed·ik 'nər·vəs ˌsis·təm }

sympathicotropic cell [HISTOLOGY] Any of various cells possessing special affinity for the sympathetic nervous system. { sim¦path·ə·kō¦träp·ik 'sel }

sympathochromaffin cell [HISTOLOGY] One of the precursors of sympathetic and medullary cells in the adrenal medulla. { ¦sim·pə·thō·krō 'maf·ən ˌsel }

sympatric [ECOLOGY] Of a species, occupying the same range as another species but maintaining identity by not interbreeding. { sim'pa·trik }

sympatric speciation [EVOLUTION] Speciation that occurs without geographic isolation of a population. { sim¦pa·trik ˌspē·shē'ā·shən }

sympetalous [BOTANY] *See* gamopetalous. { sim 'ped·əl·əs }

symphile [ECOLOGY] An organism, usually a beetle, living as a guest in the nest of a social insect, such as an ant, where it is reared and bred in exchange for its exudates. { 'simˌfīl }

Symphyla [INVERTEBRATE ZOOLOGY] A class of the Myriapoda comprising tiny, pale, centipede-like creatures which inhabit humus or soil. { 'sim·fə·lə }

symphysis [ANATOMY] An immovable articulation of bones connected by fibrocartilaginous pads. { 'sim·fə·səs }

symphysis pubis [ANATOMY] *See* pubic symphysis. { 'sim·fə·səs 'pyü·bəs }

Symphyta [INVERTEBRATE ZOOLOGY] A suborder of the Hymenoptera including the sawflies and horntails characterized by a broad base attaching the abdomen to the thorax. { 'sim·fəd·ə }

sympodium [BOTANY] A branching system in trees in which the main axis is comprised of successive secondary branches, each representing the dominant fork of a dichotomy. { sim'pōd·ē·əm }

Synallactidae [INVERTEBRATE ZOOLOGY] A family of echinoderms of the order Aspidochirotida comprising mainly deep-sea forms which lack tentacle ampullae. { ˌsin·ə'lak·təˌdē }

synandrous [BOTANY] Having several united stamens. { sə'nan·drəs }

synangium [BOTANY] A compound sorus made up of united sporangia. [VERTEBRATE ZOOLOGY]

In lower vertebrates, a peripheral arterial trunk from which branches arise. { sə'nan·jē·əm }

Synanthae [BOTANY] An equivalent name for Cyclanthales. { sə'nan,thē }

Synanthales [BOTANY] An equivalent name for Cyclanthales. { ,sin·ən'thā·lēz }

synapse [ANATOMY] A site where the axon of one neuron comes into contact with and influences the dendrites of another neuron or a cell body. { 'si,naps }

synapsis [CELL BIOLOGY] Pairing of homologous chromosomes during the zygotene stage of meiosis. { sə'nap·səs }

synaptic transmission [PHYSIOLOGY] The mechanisms by which a presynaptic neuron influences the activity of an anatomically adjacent postsynaptic neuron. { sə'nap·tik tranz'mish·ən }

synapticulum [INVERTEBRATE ZOOLOGY] A conical or cylindrical supporting process, as those extending between septa in some corals, or connecting gill bars in *Branchiostoma*. { ,sin·ap'tik·yə·ləm }

Synaptidae [INVERTEBRATE ZOOLOGY] A family of large sea cucumbers of the order Apodida lacking a respiratory tree and having a reduced water-vascular system. { sə'nap·tə,dē }

synaptinemal complex [CELL BIOLOGY] Ribbonlike structures that extend the length of synapsing chromosomes and are believed to function in exchange pairing. { sə,nap·tə'nē·məl 'käm ,pleks }

synarthrosis [ANATOMY] An articulation in which the connecting material (fibrous connective tissue) is continuous, immovably binding the bones. { ¦sin·är'thrō·səs }

Synbranchiformes [VERTEBRATE ZOOLOGY] An order of eellike actinopterygian fishes that, unlike true eels, have the premaxillae present as distinct bones. { sin,braŋ·kə'fór,mēz }

Synbranchii [VERTEBRATE ZOOLOGY] The equivalent name for Synbranchiformes. { sin'braŋ·kē,ī }

Syncarida [INVERTEBRATE ZOOLOGY] A superorder of crustaceans of the subclass Melacostraca lacking a carapace and oostegites and having exopodites on all thoracic limbs. { siŋ'kar·əd·ə }

syncarp [BOTANY] A compound fleshy fruit. { 'sin,kärp }

syncarpous [BOTANY] Descriptive of a gynecium having the carpels united in a compound ovary. { sin'kär·pəs }

synchondrosis [ANATOMY] A type of synarthrosis in which the bone surfaces are connected by cartilage. { ,sin·kän'drō·səs }

synchorology [ECOLOGY] A study which involves the distribution ranges of plant communities, phytosociological zones, vegetation and geographical complexes, dissemination spectra, and current plant migration patterns. { ,sin·kə'räl·ə·jē }

synchronous growth [MICROBIOLOGY] A population of bacteria in which all cells divide at approximately the same time. { 'siŋ·krə·nəs 'grōth }

syncytial trophoblast [CELL BIOLOGY] *See* syncytiotrophoblast. { sin'sish·əl 'träf·ə,blast }

syncytiotrophoblast [CELL BIOLOGY] An irregular sheet or net of deeply staining cytoplasm in which nuclei are irregularly scattered. Also known as plasmoditrophoblast; syncytial trophoblast. { sin¦sish·ē·ō'träf·ə,blast }

syncytium [CELL BIOLOGY] A mass of multinucleated cytoplasm without division into separate cells. Also known as polykaryocyte. [INVERTEBRATE ZOOLOGY] Multinucleated cell or gland. { sin'sish·ē·əm }

syndactyly [ANATOMY] The condition characterized by union of two or more digits, as in certain birds and mammals; it is a familial anomaly in humans. { sin'dakt·əl·ē }

syndesmosis [ANATOMY] An articulation in which the bones are joined by collagen fibers. { ,sin,dez'mō·səs }

syndynamics [ECOLOGY] The study of the causes of and trends in successional changes within a plant community. { ¦sin·dī'nam·iks }

synecology [ECOLOGY] The study of environmental relations of groups of organisms, such as communities. { ¦sin·i'käl·ə·jē }

Synentognathi [VERTEBRATE ZOOLOGY] The equivalent name for Beloniformes. { ,sin·ən 'täg·nə·thē }

synergid [BOTANY] Either of two small cells lying in the embryo sac in seed plants adjacent to the egg cell toward the micropylar end. { sə'nər·jəd }

synergism [ECOLOGY] An ecological association in which the physiological processes or behavior of an individual are enhanced by the nearby presence of another organism. { 'sin·ər,jiz·əm }

synergist [ANATOMY] A muscle that assists a prime mover muscle in performing a specific action. { 'sin·ər,jist }

Syngamidae [INVERTEBRATE ZOOLOGY] A family of roundworms belonging to the Strongyloidea and including parasites of birds and mammals. { siŋ'gam·ə,dē }

syngamy [BIOLOGY] Sexual reproduction involving union of gametes. { 'siŋ·gə·mē }

syngeneic [GENETICS] *See* isogeneic. { ,sin·jə'nē·ik }

syngenesious [BOTANY] Pertaining to an aggregate of stamens fused at the anthers. { ¦sin·jə¦nē·zhəs }

Syngnathidae [VERTEBRATE ZOOLOGY] A family of fishes in the order Gasterosteiformes including the seahorses and pipefishes. { siŋ'nath·ə,dē }

synkinesia [PHYSIOLOGY] Involuntary movement coincident with purposeful movements carried out by a distant part of the body, such as swinging the arms while walking. Also known as accessory movement; associated automatic movement. { ¦sin,kī¦nē·zhə }

synonym [SYSTEMATICS] A taxonomic name that is rejected as being incorrectly applied, or incorrect in form, or not representative of a natural genetic grouping. { 'sin·ə,nim }

synopsis [SYSTEMATICS] In taxonomy, a brief summary of current knowledge about a taxon. { sə'näp·səs }

synostosis [ANATOMY] A type of synarthrosis in which the bones are continuous. { si,nä'stō·səs }

synovia [PHYSIOLOGY] *See* synovial fluid. { sə'nō·vē·ə }

synovial fluid [PHYSIOLOGY] A transparent viscid fluid secreted by synovial membranes. Also known as synovia. { sə'nō·vē·əl 'flü·əd }

synovial membrane [HISTOLOGY] A layer of connective tissue which lines sheaths of tendons at freely moving articulations, ligamentous surfaces of articular capsules, and bursae. { sə'nō·vē·əl 'mem,brān }

synpelmous [VERTEBRATE ZOOLOGY] Having the two main flexor tendons of the toes united beyond the branches that go to each digit, as in certain birds. { sin'pel·məs }

synphylogeny [ECOLOGY] The study of the trends and changes in plant communities through historical and evolutionary perspectives. { ¦sin·fə'läj·ə·nē }

synphysiology [ECOLOGY] The study of the metabolic processes of plant communities or species which constantly compete with each other, by investigating water needs, transpiration, assimilation and production or organic matter, physiological effects of light, temperature, root exudates, and various other ecological factors. { ¦sin,fiz·ē'äl·ə·jē }

synsepalous [BOTANY] *See* gamosepalous. { sin'sep·ə·ləs }

Synteliidae [INVERTEBRATE ZOOLOGY] The sap-flow beetles, a small family of coleopteran insects in the superfamily Histeroidea. { ,sint·əl'ī·ə,dē }

syntenic group [GENETICS] The loci on the same chromosome pair, irrespective of whether or not they are known to show linkage in heredity. { sin'ten·ik ,grüp }

Syntexidae [INVERTEBRATE ZOOLOGY] A family of the Hymenoptera in the superfamily Siricoidea. { sin'tek·sə,dē }

synthetase [BIOCHEMISTRY] *See* ligase. { 'sin·thə,tās }

Syntrophiidina [PALEONTOLOGY] A suborder of extinct articulate brachiopods of the order Pentamerida characterized by a strong dorsal median fold. { sin,träf·ē·ə'dī·nə }

syntrophism [BIOLOGY] Mutual dependence of cells for nutritional needs, especially between strains of bacteria. { 'sin·trə,fiz·əm }

syntrophoblast [EMBRYOLOGY] The outer syncytial layer of the trophoblast that forms the outermost fetal element of the placenta. { sin'träf·ə,blast }

syntype [SYSTEMATICS] Any specimen of a series when no specimen is designated as the holotype. Also known as cotype. { 'sin,tīp }

synusia [ECOLOGY] A structural unit of a community characterized by uniformity of life-form or of height. { sə'nü·zhə }

Synxiphosura [PALEONTOLOGY] An extinct heteorgeneous order of arthropods in the subclass Merostomata possibly representing an explosive proliferation of aberrant, terminal, and apparently blind forms. { ¦sin,zif·ə'sůr·ə }

Syringamminidae [INVERTEBRATE ZOOLOGY] A family of Psamminida, with a fragile test constructed of tubes of xenophyae tightly cemented together. { ,sir·iŋ·gə'min·ə,dē }

Syringophyllidae [PALEONTOLOGY] A family of extinct corals in the order Tabulata. { sə,riŋ·gō'fil·ə,dē }

syrinx [PALEONTOLOGY] A tube surrounding the pedicle in certain fossil brachiopods. [VERTEBRATE ZOOLOGY] The vocal organ in birds. { 'sir·iŋks }

Syrphidae [INVERTEBRATE ZOOLOGY] The flower flies, a family of cyclorrhaphous dipteran insects in the series Aschiza. { 'sər·fə,dē }

Systellomatophora [INVERTEBRATE ZOOLOGY] An order of the subclass Pulmonata in which the eyes are contractile but stalks are not retractile, the body is sluglike, oval, or lengthened, and the lung is posterior. { ¦sis·tə·lō·mə'täf·ə·rə }

systematics [BIOLOGY] The science of animal and plant classification. { ,sis·tə'mad·iks }

systemic circulation [PHYSIOLOGY] The general circulation, as distinct from the pulmonary circulation. { si'stem·ik ,sər·kyə'lā·shən }

systems ecology [ECOLOGY] The combined approaches of systems analysis and the ecology of whole ecosystems and subsystems. { 'sis·təmz i'käl·ə·jē }

systole [PHYSIOLOGY] The contraction phase of the heart cycle. { 'sis·tə·lē }

syzygy [INVERTEBRATE ZOOLOGY] End-to-end union of the sporonts of certain gregarine protozoans. { 'siz·ə·jē }

T

Tabanidae [INVERTEBRATE ZOOLOGY] The deer and horse flies, a family of orthorrhaphous dipteran insects in the series Brachycera. { tə'ban·ə,dē }

tabled whelk [INVERTEBRATE ZOOLOGY] *Neptunea tabulata.* A marine gastropod mollusk about 5 inches (13 centimeters) in length and 2 inches (5 centimeters) in diameter, found at depths of 150-200 feet (45-60 meters), off the west coast of Canada and the United States. { 'tā·bəld 'welk }

tabula [PALEONTOLOGY] A transverse septum that closes off the lower part of the polyp cavity in certain extinct corals and hydroids. { 'tab·yə·lə }

Tabulata [PALEONTOLOGY] An extinct Paleozoic order of corals of the subclass Zoantharia characterized by an exclusively colonial mode of growth and by secretion of a calcareous exoskeleton of slender tubes. { ,tab·yə'läd·ə }

TAB vaccine [IMMUNOLOGY] A vaccine containing killed typhoid bacilli and the paratyphoid organisms (*Salmonella paratyphi* A and B) most frequently involved in paratyphoid fever. { 'tab vak 'sēn }

Tachinidae [INVERTEBRATE ZOOLOGY] The tachina flies, a family of bristly, grayish or black Diptera whose larvae are parasitic in caterpillars and other insects. { tə'kin·ə,dē }

Tachyglossidae [VERTEBRATE ZOOLOGY] A family of monotreme mammals having relatively large brains with convoluted cerebral hemispheres; comprises the echidnas or spiny anteaters. { ,tak·ə'gläs·ə,dē }

Tachyniscidae [INVERTEBRATE ZOOLOGY] A family of myodarian cyclorrhaphous dipteran insects in the subsection Acalypteratae. { ,tak·ə'nis·ə ,dē }

tachyphylaxis [IMMUNOLOGY] Rapid desensitization against doses of organ extracts or serum by the previous inoculation of small subtoxic doses of the same preparation. { ,tak·ə·fə'lak·səs }

tachysterol [BIOCHEMISTRY] The precursor of calciferol in the irradiation of ergosterol; an isomer of ergosterol. { tə'kis·tə,ròl }

tachytely [EVOLUTION] Evolution at a rapid rate resulting in differential selection and fixation of new types. { ¦tak·ə¦tel·ē }

tactile [PHYSIOLOGY] Pertaining to the sense of touch. { 'tak·təl }

tactile hairs [VERTEBRATE ZOOLOGY] *See* vibrissae. { 'tak·təl 'herz }

tactile receptor [PHYSIOLOGY] *See* tactoreceptor. { 'tak·təl ri'sep·tər }

tactoid [BIOCHEMISTRY] A particle that appears as a spindle-shaped body under the polarizing microscope and occurs in mosaic virus, fibrin, and myosin. { 'tak,tòid }

tactoreceptor [PHYSIOLOGY] A sense organ that responds to touch. Also known as tactile receptor. { ¦tak·tō·ri'sep·tər }

tadpole [VERTEBRATE ZOOLOGY] The larva of a frog or toad; at hatching it has a rounded body with a long fin-bordered tail, and the gills are external but shortly become enclosed. { 'tad ,pōl }

tadpole shrimp [INVERTEBRATE ZOOLOGY] Any of the phyllopod crustaceans that are members of the genus *Lepidurus.* { 'tad,pōl ,shrimp }

taenia [ANATOMY] A ribbon-shaped band of nerve fibers or muscle. { 'tē·nē·ə }

Taeniodidea [INVERTEBRATE ZOOLOGY] An equivalent name for Cyclophyllidea. { ,tē·nē·ə'did·ē·ə }

Taeniodonta [PALEONTOLOGY] An order of extinct quadrupedal land mammals, known from early Cenozoic deposits in North America. { ,tē·nē·ə'dänt·ə }

Taenioidea [INVERTEBRATE ZOOLOGY] An equivalent name for Cyclophyllidea. { ,tē·nē'òid·ē·ə }

Taeniolabidoidea [PALEONTOLOGY] An advanced suborder of the extinct mammalian order Multituberculata having incisors that were self-sharpening in a limited way. { ,tē·nē·ō,lab·ə 'dòid·ē·ə }

taenoglossate radula [INVERTEBRATE ZOOLOGY] A long, narrow radula with seven teeth in each transverse row, found in certain pectinibranch bivalves. { ,tē·nə'glä,sāt 'raj·ə·lə }

tagma [INVERTEBRATE ZOOLOGY] A compound body section of a metameric animal that results from embryonic fusion of two or more somites. { 'tag·mə }

tagmosis [INVERTEBRATE ZOOLOGY] The formation of groups of metameres into body regions with functional differences. { tag'mō·səs }

tagua palm [BOTANY] *Phytelephas macrocarpa.* A palm tree of tropical America; the endosperm of the seed is used as an ivory substitute. { 'täg·wə ,päm }

taiga [ECOLOGY] A zone of forest vegetation encircling the Northern Hemisphere between the arctic-subarctic tundras in the north and the steppes, hardwood forests, and prairies in the south. Also known as boreal forest. { 'tī·gə }

tail [VERTEBRATE ZOOLOGY] **1.** The usually slender appendage that arises immediately above the anus in many vertebrates and contains the caudal vertebrae. **2.** The uropygium, and its feathers, of a bird. **3.** The caudal fin of a fish or aquatic mammal. { tāl }

Talitridae [INVERTEBRATE ZOOLOGY] A family of terrestrial amphipod crustaceans in the suborder Gammaridea. { tə'li·trə,dē }

talon [VERTEBRATE ZOOLOGY] A sharply hooked claw on the foot of a bird of prey. { 'tal·ən }

Talpidae [VERTEBRATE ZOOLOGY] The moles, a family of insectivoran mammals; distinguished by the forelimbs which are adapted for digging, having powerful muscles and a spadelike bony structure. { 'tal·pə,dē }

talus [ANATOMY] *See* astragalus. { 'tal·əs }

tamarack [BOTANY] *Larix laricina.* A larch and a member of the pine family; it has an erect narrowly pyramidal habit, and grows in wet and moist soils in the northeastern United States, west to the Lake States and across Canada to Alaska; used for railroad ties, posts, sills, and boats. Also known as hackmarack. { 'tam·ə ,rak }

Tanaidacea [INVERTEBRATE ZOOLOGY] An order of eumalacostracans of the crustacean superorder Peracarida; the body is linear, more or less cylindrical or dorsoventrally depressed, and the first and second thoracic segments are fused with the head, forming a carapace. { ,tan·ē·ə'dā·shə }

Tanaostigmatidae [INVERTEBRATE ZOOLOGY] A small family of hymenopteran insects in the superfamily Chalcidoidea. { tə,nā·ō·stig'mad·ə ,dē }

tandem duplication [CELL BIOLOGY] The occurrence of two identical sequences, one following the other, in a chromosome segment. { 'tan·dəm ,dü·plə'kā·shən }

tangelo [BOTANY] A tree that is hybrid between a tangerine or other mandarin and a grapefruit or shaddock; produces an edible fruit. { 'tan·jə· lō }

tangerine [BOTANY] Any of several trees of the species *Citrus reticulata*; the fruit is a loose-skinned mandarin with a deep-orange or scarlet rind. { 'tan·jə,rēn }

tangoreceptor [PHYSIOLOGY] A sense organ in the skin that responds to touch and pressure. { ¦taŋ·go·ri'sep·tər }

tannase [BIOCHEMISTRY] An enzyme that catalyzes the hydrolysis of tannic acid to gallic acid; found in cultures of *Aspergillus* and *Penicillium*. { 'ta,nās }

Tanyderidae [INVERTEBRATE ZOOLOGY] The primitive crane flies, a family of orthorrhaphous dipteran insects in the series Nematocera. { ,tan·ə 'der·ə,dē }

Tanypezidae [INVERTEBRATE ZOOLOGY] A family of myodarian cyclorrhaphous dipteran insects in the subsection Acalyptreatae. { ,tan·ə'pez·ə ,dē }

tape grass [BOTANY] *Vallisnerida spiralis.* An aquatic flowering plant belonging to the family Hydrocharitaceae. Also known as eel grass. { 'tāp ,gras }

tapetum [ANATOMY] **1.** A reflecting layer in the choroid coat behind the neural retina, chiefly in the eyes of nocturnal mammals. **2.** A tract of nerve fibers forming part of the roof of each lateral ventricle in the vertebrate brain. [BOTANY] A layer of nutritive cells surrounding the spore mother cells in the sporangium in higher plants; it is broken down to provide nourishment for developing spores. { tə'pēd·əm }

tapeworm [INVERTEBRATE ZOOLOGY] Any member of the class Cestoidea; all are vertebrate endoparasites, characterized by a ribbonlike body divided into proglottids, and the anterior end modified into a holdfast organ. { 'tāp,wərm }

taphocoenosis [PALEONTOLOGY] *See* thanatocoenosis. { ¦taf·ō·sē'nō·səs }

taphonomy [PALEONTOLOGY] The study of fossil preservation, including all events during the transition of organisms from the biosphere to the lithosphere. { tə'fän·ə·mē }

tapir [VERTEBRATE ZOOLOGY] Any of several large odd-toed ungulates of the family Tapiridae that have a heavy, sparsely hairy body, stout legs, a prehensile muzzle, a short tail, and small eyes. { 'tā·pər }

Tapiridae [VERTEBRATE ZOOLOGY] The tapirs, a family of perissodactyl mammals in the superfamily Tapiroidea. { tə'pir·ə,dē }

Tapiroidea [VERTEBRATE ZOOLOGY] A superfamily of the mammalian order Perissodactyla in the suborder Ceratomorpha. { ,tap·ə'ròid·ē·ə }

taproot [BOTANY] A root system in which the primary root forms a dominant central axis that penetrates vertically and rather deeply into the soil; it is generally larger in diameter than its branches. { 'tap,rüt }

tarantula [INVERTEBRATE ZOOLOGY] **1.** Any of various large hairy spiders of the araneid suborder Mygalomorphae. **2.** Any of the wolf spiders comprising the family Lycosidae. { tə'ran·chə·lə }

Tardigrada [INVERTEBRATE ZOOLOGY] A class of microscopic, bilaterally symmetrical invertebrates in the subphylum Malacopoda; the body consists of an anterior prostomium and five segments surrounded by a soft, nonchitinous cuticle, with four pairs of ventrolateral legs. { tär 'dig·rə·də }

target cell [PHYSIOLOGY] A cell that has receptors for the product of a signaling cell. { 'tär·gət ,sel }

tarpon [VERTEBRATE ZOOLOGY] *Megalops atlantica.* A herringlike fish of the family Elopidae weighing up to 300 pounds (136 kilograms) and reaching a length of 8 feet (2.4 meters); it has a single soft, rayed dorsal fin, strong jaws, a bony plate under the mouth, numerous small teeth, and coarse, bony flesh covered with large scales. { 'tär· pən }

tarsal gland [ANATOMY] Any of the sebaceous

glands in the tarsal plates of the eyelids. Also known as Meibomian gland. { 'tär·səl ˌgland }

tarsier [VERTEBRATE ZOOLOGY] Any of several species of primates comprising the genus *Tarsius* of the family Tarsiidae characterized by a round skull, a flattened face, and large eyes that are separated from the temporal fossae in the orbital depression, and by adhesive pads on the expanded ends of the fingers and toes. { 'tär·sēˌā }

Tarsiidae [VERTEBRATE ZOOLOGY] The tarsiers, a family of prosimian primates distinguished by incomplete postorbital closure and a greatly elongated ankle region. { tär'sī·əˌdē }

Tarsonemidae [INVERTEBRATE ZOOLOGY] A small family of phytophagous mites in the suborder Trombidiformes. { ˌtär·sə'nem·əˌdē }

tarsus [ANATOMY] **1.** The instep of the foot consisting of the calcaneus, talus, cuboid, navicular, medial, intermediate, and lateral cuneiform bones. **2.** The dense connective tissues supporting an eyelid. { 'tär·səs }

tartary buckwheat [BOTANY] One of three buckwheat species grown commercially; the leaves are narrower than the other two species and arrow-shaped, and the flowers are smaller with inconspicuous greenish-white sepals. Also known as duck wheat; hulless buckwheat; rye buckwheat. { 'tärd·ə·rē 'bəkˌwēt }

tassel [BOTANY] The male inflorescence of corn and certain other plants. { 'tas·əl }

taste [PHYSIOLOGY] A chemical sense by which flavors are perceived depending on taste, tactile, and warm and cold receptors in the mouth, as well as smell receptors in the nose. { tāst }

taste bud [ANATOMY] An end organ consisting of goblet-shaped clusters of elongate cells with microvilli on the distal end to mediate the sense of taste. { 'tāst ˌbəd }

taurocholic acid [BIOCHEMISTRY] $C_{26}H_{45}NO_7S$ A common bile acid with a five-carbon chain; it is the product of the conjugation of taurine with cholic acid; crystallizes from an alcohol ether solution, and decomposes at about 125°C. Also known as cholaic acid; cholytaurine. { ¦tȯr·ə ¦kōl·ik 'as·əd }

taurodont [ANATOMY] Of teeth, having a large pulp cavity and reduced roots. { 'tȯr·əˌdänt }

Taxales [BOTANY] A small order of gymnosperms in the class Pinatae; members are trees or shrubs with evergreen, often needlelike leaves, with a well-developed fleshy covering surrounding the individual seeds, which are terminal or subterminal on short shoots. { tak'sā·lēz }

taxis [PHYSIOLOGY] A mechanism of orientation by means of which an animal moves in a direction related to a source of stimulation. { 'tak·səs }

Taxocrinida [PALEONTOLOGY] An order of flexible crinoids distributed from Ordovician to Mississippian. { ˌtak·sə'krī·nəd·ə }

Taxodonta [INVERTEBRATE ZOOLOGY] A subclass of pelecypod mollusks in which the hinge is of the taxodontal type, that is, the dentition is a series of similar alternating teeth and sockets along the hinge margin. { ˌtak·sə'dänt·ə }

taxon [SYSTEMATICS] A taxonomic group or entity. { 'takˌsän }

taxonomic category [SYSTEMATICS] One of a hierarchy of levels in the biological classification of organisms; the seven major categories are kingdom, phylum, class, order, family, genus, species. { ¦tak·sə¦näm·ik 'kad·əˌgȯr·ē }

taxonomy [SYSTEMATICS] A study aimed at producing a hierarchical system of classification of organisms which best reflects the totality of similarities and differences. { tak'sän·ə·mē }

Tayassuidae [VERTEBRATE ZOOLOGY] The peccaries, a family of artiodactyl mammals in the superfamily Suoidae. { ˌtā·yə'sü·əˌdē }

T cell [IMMUNOLOGY] One of a heterogeneous population of thymus-derived lymphocytes which participates in the immune responses. Also known as T lymphocyte. { 'tē ˌsel }

T-cell receptor [IMMUNOLOGY] Protein on the surface of T lymphocytes that specifically recognizes molecules of the major histocompatibility complex, either alone or in association with foreign antigens. Abbreviated TCR. { 'tē ˌsel ri ˌsep·tər }

TCR [IMMUNOLOGY] *See* T-cell receptor.

tea [BOTANY] *Thea sinensis*. A small tree of the family Theaceae having lanceolate leaves and fragrant white flowers; a caffeine beverage is made from the leaves of the plant. { tē }

tear gland [ANATOMY] *See* lacrimal gland. { 'tir ˌgland }

technosphere [ECOLOGY] The part of the physical environment affected through building or modification by humans. { 'tek·nəˌsfir }

Tectibranchia [INVERTEBRATE ZOOLOGY] An order of mollusks in the subclass Opisthobranchia containing the sea hares and the bubble shells; the shell may be present, rudimentary, or absent. { ˌtek·tə'braŋ·kē·ə }

tectorial membrane [ANATOMY] **1.** A jellylike membrane covering the organ of Corti in the ear. **2.** A strong sheet of connective tissue running from the basilar part of the occipital bone to the dorsal surface of the bodies of the axis and third cervical vertebra. { tek'tȯr·ē·əl 'memˌbrān }

tectum [ANATOMY] A rooflike structure of the body, especially the roof of the midbrain including the corpora quadrigemina. { 'tek·təm }

tegmen [BIOLOGY] An integument or covering. [BOTANY] The inner layer of a seed coat. [INVERTEBRATE ZOOLOGY] A thickened forewing of Orthoptera, Coleoptera, and certain other insects. { 'teg·mən }

tegmentum [ANATOMY] A mass of white fibers with gray matter in the cerebral peduncles of higher vertebrates. [BOTANY] The outer layer, or scales, of a leaf bud. [INVERTEBRATE ZOOLOGY] The upper layer of a shell plate in Amphineura. { teg'men·təm }

teichoic acid [BIOCHEMISTRY] A polymer of ribitol or glycerol phosphate with additional compounds such as glucose linked to the backbone of the polymer; found in the cell walls of some bacteria. { tā'kō·ik 'as·əd }

Teiidae [VERTEBRATE ZOOLOGY] The tegus lizards, a diverse family of the suborder Sauria that is especially abundant and widespread in South America. { 'tē·ə,dē }

T-E index [ECOLOGY] *See* temperature-efficiency index. { ¦tē'ē ,in,deks }

telamon [INVERTEBRATE ZOOLOGY] A curved chitinous outgrowth of the cloacal wall in various male nematodes. { 'tel·ə,män }

teleceptor [PHYSIOLOGY] A sense receptor which transmits information about portions of the external environment which are not necessarily in direct contact with the organism, such as the receptors of the ear, eye, and nose. { 'tel·ə,sep·tər }

Telegeusidae [INVERTEBRATE ZOOLOGY] The long-lipped beetles, a small family of colepteran insects in the superfamily Cantharoidea confined to the western United States. { ,tel·ə'gyüs·ə ,dē }

telencephalon [EMBRYOLOGY] The anterior subdivision of the forebrain in a vertebrate embryo; gives rise to the olfactory lobes, cerebral cortex, and corpora striata. { ¦tel·en'sef·ə,län }

Teleosauridae [PALEONTOLOGY] A family of Jurassic reptiles in the order Crocodylia characterized by a long snout and heavy armor. { ¦tel·ē·ə 'sȯr·ə,dē }

Teleostei [VERTEBRATE ZOOLOGY] An infraclass of the subclass Actinopterygii, or rayfin fishes; distinguished by paired bracing bones in the supporting skeleton of the caudal fin, a homocercal caudal fin, thin cycloid scales, and a swim bladder with a hydrostatic function. { ,tel·ē'äs·tē,ī }

telephone theory [PHYSIOLOGY] *See* frequency theory. { 'tel·ə,fōn ,thē·ə·rē }

Telestacea [INVERTEBRATE ZOOLOGY] An order of the subclass Alcyonaria comprised of individuals which form erect branching colonies by lateral budding from the body wall of an elongated axial polyp. { ,tel·ə'stā·shē·ə }

teleutospore [MYCOLOGY] *See* teliospore. { tə 'lüd·ə,spȯr }

teliospore [MYCOLOGY] A thick-walled spore of the terminal stage of Uredinales and Ustilaginales which is a probosidium or a group of probosidia. Also known as teleutospore. { 'tē·lē·ə ,spȯr }

telium [BOTANY] A sorus that bears teliospores. { 'tē·lē·əm }

teloblast [INVERTEBRATE ZOOLOGY] A large cell that produces many smaller cells at the growing end of many embryos, especially in annelids and mollusks. { 'tel·ə,blast }

telocentric [CELL BIOLOGY] Pertaining to a chromosome with a terminal centromere. { ¦tel·ə ¦sen·trik }

telocoel [EMBRYOLOGY] A cavity of the telencephalon. { 'tel·ə,sēl }

telodendrion [ANATOMY] The terminal branching of an axon. Also known as telodendron. { ¦tel·ə 'den·drē,än }

telodendron [ANATOMY] *See* telodendrion. { ¦tel·ə'den·drən }

telogen [PHYSIOLOGY] A quiescent phase in the cycle of hair growth when the hair is retained in the hair follicle as a dead or "club" hair. { 'tel·ə·jən }

telolecithal [CELL BIOLOGY] Of an ovum, having a large, evenly dispersed volume of yolk and a small amount of cytoplasm concentrated at one pole. { ¦tel·ō'les·ə·thəl }

telomerase [BIOCHEMISTRY] A deoxyribonucleic acid polymerase that elongates telomeres. { tə 'läm·ə,rās }

telomere [CELL BIOLOGY] A centromere in the terminal position on a chromosome. { 'tel·ə,mir }

telophase [CELL BIOLOGY] The phase of meiosis or mitosis at which the chromosomes, having reached the poles, reorganize into interphase nuclei with the disappearance of the spindle and the reappearance of the nuclear membrane; in many organisms telophase does not occur at the end of the first meiotic division. { 'tel·ə,fāz }

Telosporea [INVERTEBRATE ZOOLOGY] A class of the protozoan subphylum Sporozoa in which the spores lack a polar capsule and develop from an oocyst. { ,tel·ə'spȯr·ē·ə }

telotaxis [BIOLOGY] Tactic movement of an organism by the orientation of one or the other of two bilaterally symmetrical receptors toward the stimulus source. { ¦tel·ō'tak·səs }

telotroch [INVERTEBRATE ZOOLOGY] A preanal tuft of cilia in a trochophore larva. { 'tel·ə,träk }

telson [INVERTEBRATE ZOOLOGY] The postabdominal segment in lobsters, amphipods, and certain other invertebrates. { 'tel·sən }

Temnocephalida [INVERTEBRATE ZOOLOGY] A group of rhabdocoeles sometimes considered a distinct order but usually classified under the Neorhabdocoela; members are characterized by the possession of tentacles and adhesive organs. { ,tem·nō·sə'fal·əd·ə }

Temnochilidae [INVERTEBRATE ZOOLOGY] The equivalent name for Ostomidae. { ,tem·nō'kil·ə ,dē }

Temnopleuridae [INVERTEBRATE ZOOLOGY] A family of echinoderms in the order Temnopleuroida whose tubercles are imperforate, though usually crenulate. { ,tem·nō'plu̇r·ə,dē }

Temnopleuroida [INVERTEBRATE ZOOLOGY] An order of echinoderms in the superorder Echinacea with a camarodont lantern, smooth or sculptured test, imperforate or perforate tubercles, and bronchial slits which are usually shallow. { ,tem·nō·plə'rȯid·ē·ə }

Temnospondyli [PALEONTOLOGY] An order of extinct amphibians in the subclass Labyrinthodontia having vertebrae with reduced pleurocentra and large intercentra. { ,tem·nō'spän·də,lī }

temperate and cold savannah [ECOLOGY] A regional vegetation zone, very extensively represented in North America and in Eurasia at high altitudes; consists of scattered or clumped trees (very often conifers and mostly needle-leaved evergreens) and a shrub layer of varying coverage; mosses and, even more abundantly, lichens form an almost continuous carpet. { ¦tem·prət ən ¦kōld sə'van·ə }

temperate and cold scrub [ECOLOGY] Regional vegetation zone whose density and periodicity vary a good deal; requires a considerable amount of moisture in the soil, whether from mist, seasonal downpour, or snowmelt; shrubs may be evergreen or deciduous; and undergrowth of ferns and other large-leaved herbs are quite frequent, especially at subalpine level; wind shearing and very cold winters prevent tree growth. Also known as bosque; fourré; heath. { ¦tem·prət ən ¦kōld 'skrəb }

temperate mixed forest [ECOLOGY] A forest of the North Temperate Zone containing a high proportion of conifers with a few broad-leafed species. { 'tem·prət 'mikst ¦fär·əst }

temperate phage [VIROLOGY] A deoxyribonucleic acid phage, the genome of which can under certain circumstances become integrated with the genome of the host. { 'tem·prət 'fāj }

temperate rainforest [ECOLOGY] A vegetation class in temperate areas of high and evenly distributed rainfall characterized by comparatively few species with large populations of each species; evergreens are somewhat short with small leaves, and there is an abundance of large tree ferns. Also known as cloud forest; laurel forest; laurisilva; moss forest; subtropical forest. { 'tem·prət 'rān,fär·əst }

temperate woodland [ECOLOGY] A vegetation class similar to tropical woodland in spacing, height, and stratification, but it can be either deciduous or evergreen, broad-leaved or needle-leaved. Also known as parkland; woodland. { 'tem·prət 'wu̇d·lənd }

temperature-efficiency index [ECOLOGY] For a given location, a measure of the long-range effectiveness of temperature (thermal efficiency) in promoting plant growth. Abbreviated T-E index. Also known as thermal-efficiency index. { 'tem·prə·chər i'fish·ən·sē ,in,deks }

temperature-efficiency ratio [ECOLOGY] For a given location and month, a measure of thermal efficiency; it is equal to the departure, in degrees Fahrenheit, of the normal monthly temperature above 32°F (0°C) divided by 4: $(T - 32)/4$. Abbreviated T-E ratio. Also known as thermal-efficiency ratio. { 'tem·prə·chər i'fish·ən·sē ,rā·shō }

temperature-sensitive mutant [GENETICS] A mutant gene that is functional at high (low) temperature but is inactivated by lowering (elevating) the temperature. { 'tem·prə·chər ¦sen·səd·iv 'myüt·ənt }

template [MOLECULAR BIOLOGY] The macromolecular model for the synthesis of another macromolecule. { 'tem·plət }

temples [INVERTEBRATE ZOOLOGY] The posterolateral angles of the head, in lice. { 'tem·pəlz }

temporal bone [ANATOMY] The bone forming a portion of the lateral aspect of the skull and part of the base of the cranium in vertebrates. { 'tem·prəl ,bōn }

temporary plankton [BIOLOGY] *See* meroplankton. { 'tem·pə,rer·ē 'plaŋk·tən }

tender plant [BOTANY] A plant that is incapable of resisting cold. { ¦ten·dər ¦plant }

Tendipedidae [INVERTEBRATE ZOOLOGY] The midges, a family of orthorrhaphous dipteran insects in the series Nematocera whose larvae occupy intertidal wave-swept rocks on the seacoasts. { ,ten·də'ped·ə,dē }

tendon [ANATOMY] A white, glistening, fibrous cord which joins a muscle to some movable structure such as a bone or cartilage; tendons permit concentration of muscle force into a small area and allow the muscle to act at a distance. { 'ten·dən }

tendon sheath [ANATOMY] The synovial membrane surrounding a tendon. { 'ten·dən ,shēth }

tendril [BOTANY] A stem modification in the form of a slender coiling structure capable of twining about a support to which the plant is then attached. { 'ten·drəl }

Tenebrionidae [INVERTEBRATE ZOOLOGY] The darkling beetles, a large cosmopolitan family of coleopteran insects in the superfamily Tenebrionoidea; members are common pests of grains, dried fruits, beans, and other food products. { tə,neb·rē'än·ə,dē }

Tenebrionoidea [INVERTEBRATE ZOOLOGY] A superfamily of the Coleoptera in the suborder Polyphaga. { tə,neb·rē·ə'nȯid·ē·ə }

teniae coli [HISTOLOGY] The three bands comprising the longitudinal layer of the tunica muscularis of the colon: the tenia libera, tenia mesocolica, and tenia omentalis. { 'tē·nē,ē 'kō·lī }

tenrec [VERTEBRATE ZOOLOGY] Any of about 30 species of unspecialized, insectivorous mammals which are indigenous to Madagascar and have poor vision and clawed digits. { 'ten,rek }

Tenrecidae [VERTEBRATE ZOOLOGY] The tenrecs, a family of insectivores in the group Lipotyphla. { ten'res·ə,dē }

tension wood [BOTANY] In some hardwood trees, wood characterized by the presence of gelatinous fibers and excessive longitudinal shrinkage; causes trees to lean. { 'ten·chən ,wu̇d }

tensor muscle [PHYSIOLOGY] A muscle that stretches a part or makes it tense. { 'ten·sər ,məs·əl }

tentacle [INVERTEBRATE ZOOLOGY] Any of various elongate, flexible processes with tactile, prehensile, and sometimes other functions, and which are borne on the head or about the mouth of many animals. { 'ten·tə·kəl }

Tentaculata [INVERTEBRATE ZOOLOGY] A class of the phylum Ctenophora whose members are characterized by having variously modified tentacles. { ten,tak·yə'läd·ə }

tentaculocyst [INVERTEBRATE ZOOLOGY] A sense organ located at the margin of the umbrella in some cnidarian medusoids, consisting of a modified tentacle with a cavity that often contains lithites. { ten'tak·yə·lō,sist }

tentaculozoid [INVERTEBRATE ZOOLOGY] A slender tentacular individual of a hydrozoan colony. { ten¦tak·yə·lō'zō,ȯid }

tented arch [ANATOMY] A fingerprint pattern which possesses either an angle, an upthrust, or two of the three basic characteristics of a loop. { 'ten·təd 'ärch }

Tenthredinidae [INVERTEBRATE ZOOLOGY] A family of hymenopteran insects in the superfamily Tenthredinoidea including economically important species whose larvae are plant pests. { ,ten·thrə'din·ə,dē }

Tenthredinoidea [INVERTEBRATE ZOOLOGY] A superfamily of Hymenoptera in the suborder Symphyla. { ˌten·thrə·də′nȯid·ē·ə }

tentillum [INVERTEBRATE ZOOLOGY] A contractile branch of a tentacle containing many nematocysts in certain siphonophores. { ten′til·əm }

Tenuipalpidae [INVERTEBRATE ZOOLOGY] A small family of mites in the suborder Trombidiformes. { ˌten·yə·wə′pal·pəˌdē }

tepary bean [BOTANY] *Phaseolus acutifolius* var. *latifolius.* One of the four species of beans of greatest economic importance in the United States. { ′tep·ə·rē ˌbēn }

Tephritidae [INVERTEBRATE ZOOLOGY] The fruit flies, a family of myodarian cyclorrhaphous dipteran insects in the subsection Acalyptratae. { tə ′frid·əˌdē }

T-E ratio [ECOLOGY] *See* temperature-efficiency ratio. { ¦tē′ē ′rā·shō }

Teratornithidae [PALEONTOLOGY] An extinct family of vulturelike birds of the Pleistocene of western North America included in the order Falconiformes. { ˌter·ə·tȯr′nith·əˌdē }

Terebellidae [INVERTEBRATE ZOOLOGY] A family of polychaete annelids belonging to the Sedentaria which are chiefly large, thick-bodied, tubicolous forms with the anterior end covered by a matted mass of tentacular cirri. { ˌter·ə′bel·ə ˌdē }

Terebratellidina [PALEONTOLOGY] An extinct suborder of articulate brachiopods in the order Terebratulida in which the loop is long and offers substantial support to the side arms of the lophophore. { ˌter·ə·brəˌtel·ə′dīn·ə }

Terebratulida [INVERTEBRATE ZOOLOGY] An order of articulate brachiopods that has a punctate shell structure and is characterized by the possession of a loop extending anteriorly from the crural bases, providing some degree of support for the lophophore. { ˌter·ə·brə′tül·əd·ə }

Terebratulidina [INVERTEBRATE ZOOLOGY] A suborder of articulate brachiopods in the order Terebratulida distinguished by a short V- or W-shaped loop. { ˌter·ə·brach·ə·lə′dīn·ə }

Teredinidae [INVERTEBRATE ZOOLOGY] The pileworms or shipworms, a family of bivalve mollusks in the subclass Eulamellibranchia distinguished by having the two valves reduced to a pair of small plates at the anterior end of the animal. { ˌter·ə′din·əˌdē }

terete [BOTANY] Of a stem, cylindrical in section, but tapering at both ends. { tə′rēt }

tergite [INVERTEBRATE ZOOLOGY] The dorsal plate covering a somite in arthropods and certain other articulate animals. { ′tərˌjīt }

tergum [INVERTEBRATE ZOOLOGY] A dorsal plate of the operculum in barnacles. { ′tər·gəm }

terminal bar [CELL BIOLOGY] One of the structures formed in certain epithelial cells by the combination of local modifications of contiguous surfaces and intervening intercellular substances; they become visible with the light microscope after suitable staining and appear to close the spaces between the epithelial cells of the intestine at their free surfaces. { ′tər·mən·əl ˌbär }

terminal bud [BOTANY] A bud that develops at the apex of a stem. Also known as apical bud. { ′tər·mən·əl ˌbəd }

terminal hair [ANATOMY] One of three types of hair in man based on hair size, time of appearance, and structural variations; the larger, coarser hair in the adult that replaces the vellus hair. { ′tər·mən·əl ¦her }

terminal nerve [ANATOMY] Either of a pair of small cranial nerves that run from the nasal area to the forebrain, present in most vertebrates; the function is not known. { ′tər·mən·əl ¦nərv }

terminal sinus [EMBRYOLOGY] The vascular sinus bounding the area vasculosa of the blastoderm of a meroblastic ovum. Also known as marginal sinus. { ′tər·mən·əl ¦sī·nəs }

terminator codon [GENETICS] A codon that acts as a stopping point in the code sequence for protein synthesis. Also known as stop codon. { ′tər·məˌnād·ər ′kōˌdän }

Termitaphididae [INVERTEBRATE ZOOLOGY] The termite bugs, a small family of Hemiptera in the superfamily Aradoidea. { terˌmīd·ə′fid·əˌdē }

termitarium [INVERTEBRATE ZOOLOGY] A termites' nest. { ˌtər·mə′ter·ē·əm }

termite [INVERTEBRATE ZOOLOGY] A soft-bodied insect of the order Isoptera; individuals feed on cellulose and live in colonies with a caste system comprising three types of functional individuals: sterile workers and soldiers, and the reproductives. Also known as white ant. { ′tərˌmīt }

termiticole [ECOLOGY] An organism that lives in a termites' nest. { tər′mīd·əˌkōl }

Termitidae [INVERTEBRATE ZOOLOGY] A large family of the order Isoptera which contains the higher termites, representing 80% of the species. { tər′mid·əˌdē }

termitophile [ECOLOGY] An organism that lives in a termites' nest in a symbiotic association with the termites. { tər′mīd·əˌfīl }

Termopsidae [INVERTEBRATE ZOOLOGY] A family of insects in the order Isoptera composed of damp wood-dwelling forms. { tər′mäp·səˌdē }

ternate [BOTANY] Composed of three subdivisions, as a leaf with three leaflets. { ′tərˌnāt }

Ternifine man [PALEONTOLOGY] The name for a fossil human type, represented by three lower jaws and a parietal bone discovered in France and thought to be from the upper part of the middle Pleistocene. { ′tər·nəˌfēn ′man }

terrapin [VERTEBRATE ZOOLOGY] Any of several North American tortoises in the family Testudinidae, especially the diamondback terrapin. { ′ter·ə·pən }

territoriality [ZOOLOGY] A pattern of behavior in which one or more animals occupy and defend a definite area or territory. { ter·əˌtȯr·ē′al·əd·ē }

tertiary structure [BIOCHEMISTRY] The characteristic three-dimensional folding of the polypeptide chains in a protein molecule. { ′tər·shēˌer·ē ′strək·chər }

Tessaratomidae [INVERTEBRATE ZOOLOGY] A family of large tropical Hemiptera in the superfamily Pentatomoidea. { ˌtes·ə·rə′täm·əˌdē }

tessellate [BOTANY] Marked by a pattern of small squares resembling a tiled pavement. { 'tes·ə ˌlāt }

test [INVERTEBRATE ZOOLOGY] A hard external covering or shell that is calcareous, siliceous, chitinous, fibrous, or membranous. { test }

testa [BOTANY] A seed coat. Also known as episperm. { 'tes·tə }

Testacellidae [INVERTEBRATE ZOOLOGY] A family of pulmonate gastropods that includes some species of slugs. { ˌtes·tə'sel·əˌdē }

testicular hormone [BIOCHEMISTRY] Any of various hormones secreted by the testes. { te'stik·yə·lər 'hȯrˌmōn }

testis [ANATOMY] One of a pair of male reproductive glands in vertebrates; after sexual maturity, the source of sperm and hormones. { 'tes·təs }

testosterone [BIOCHEMISTRY] $C_{19}H_{28}O_2$ The principal androgenic hormone released by the human testis; may be synthesized from cholesterol and certain other sterols. { tes'täs·təˌrōn }

Testudinata [VERTEBRATE ZOOLOGY] The equivalent name for Chelonia. { teˌstyüd·ən'äd·ə }

Testudinellidae [INVERTEBRATE ZOOLOGY] A family of free-swimming rotifers in the suborder Flosculariacea. { teˌstyüd·ən'el·əˌdē }

Testudinidae [VERTEBRATE ZOOLOGY] A family of tortoises in the suborder Cryptodira; there are about 30 species found on all continents except Australia. { ˌtest·yü'din·əˌdē }

tetanolysin [BIOCHEMISTRY] A hemolysin produced by *Clostridium tetani*. { ¦tet·ən·ō'līs·ən }

tetanospasmin [BIOCHEMISTRY] A neurotoxin elaborated by the bacterium *Clostridium tetani* and which is responsible for the manifestations of tetanus. { ˌtet·ən·ō'spaz·mən }

tetanus antitoxin [IMMUNOLOGY] A serum containing antibodies that neutralize tetanus toxin. { 'tet·ən·əs 'ant·iˌtäk·sən }

tetanus toxoid [IMMUNOLOGY] Detoxified tetanus toxin used to produce active immunity against tetanus. { 'tet·ən·əs 'täkˌsȯid }

Tethinidae [INVERTEBRATE ZOOLOGY] A family of myodarian cyclorrhaphous dipteran insects in the subsection Acalyptratae. { tə'thin·əˌdē }

Tetrabranchia [INVERTEBRATE ZOOLOGY] A subclass of primitive mollusks of the class Cephalopoda; *Nautilus* is the only living form and is characterized by having four gills. { ˌte·trə 'braŋ·kē·ə }

Tetracentraceae [BOTANY] A family of dicotyledonous trees in the order Trochodendrales distinguished by possession of a perianth, four stamens, palmately veined leaves, and secretory idioblasts. { ˌte·trəˌsen'trās·ēˌē }

Tetracorallia [PALEONTOLOGY] The equivalent name for Rugosa. { ˌte·trə·kə'ral·yə }

Tetractinomorpha [INVERTEBRATE ZOOLOGY] A heterogeneous subclass of Porifera in the class Demospongiae. { tə¦trak·tə·nə'mȯr·fə }

tetracycline [MICROBIOLOGY] **1.** Any of a group of broad-spectrum antibiotics produced biosynthetically by fermentation with a strain of *Streptomyces aureofaciens* and certain other species or chemically by hydrogenolysis of chlortetracycline. **2.** $C_{22}H_{24}O_8N_2$ A broad-spectrum antibiotic belonging to the tetracycline group of antibiotics; useful because of broad antimicrobial action, with low toxicity, in the therapy of infections caused by gram-positive and gram-negative bacteria as well as rickettsiae and large viruses such as psittacosis-lymphogranuloma viruses. { ˌte·trə'sīˌklēn }

tetrad [CELL BIOLOGY] A group of four chromatids lying parallel to each other as a result of the longitudinal division of each of a pair of homologous chromosomes during the pachytene and later stages of the prophase of meiosis. { 'te ˌtrad }

tetradactylous [VERTEBRATE ZOOLOGY] Having four digits on a limb. { ¦te·trə¦dakt·əl·əs }

tetrad analysis [GENETICS] A method of genetic analysis possible in fungi, algae, bryophytes, and orchids in which the four products of an individual cell which has gone through meiosis are recovered as a group; it provides more direct and complete information regarding segregation and recombination mechanisms than is possible to obtain from meiotic products collected at random. { 'teˌtrad əˌnal·ə·səs }

4,6,3′,4′-tetrahydroxyaurone [BIOCHEMISTRY] *See* aureusidin. { ¦fȯr ¦siks ¦thrēˌprīm ¦fȯrˌprīm ¦te·trə·hī¦dräk·sē'ȯˌrōn }

Tetrahymenina [INVERTEBRATE ZOOLOGY] A suborder of ciliated protozoans in the order Hymenostomatida. { ¦te·trəˌhī·mə'nīn·ə }

Tetralophodontinae [PALEONTOLOGY] An extinct subfamily of proboscidean mammals in the family Gomphotheridae. { ¦te·trəˌläf·ə'dänt·əˌdē }

tetramerous [BIOLOGY] Characterized by or having four parts. { te'tram·ə·rəs }

Tetranychidae [INVERTEBRATE ZOOLOGY] The spider mites, a family of phytophagous trombidiform mites. { ˌte·trə'nik·əˌdē }

Tetraodontiformes [VERTEBRATE ZOOLOGY] An order of specialized teleost fishes that includes the triggerfishes, puffers, trunkfishes, and ocean sunfishes. { ¦te·trə·ōˌdänt·ə'fȯrˌmēz }

Tetraonidae [VERTEBRATE ZOOLOGY] The ptarmigans and grouse, a family of upland game birds in the order Galliformes characterized by rounded tails and wings and feathered nostrils. { ˌte·trə'än·əˌdē }

Tetraphidaceae [BOTANY] The single family of the plant order Tetraphidales. { ˌte·trə·fə'dās·ēˌē }

Tetraphidales [BOTANY] A monofamilial order of mosses distinguished by scalelike protonema and the peristomes of four rigid, nonsegmented teeth. { ˌte·trə·fə'dā·lēz }

Tetraphyllidea [INVERTEBRATE ZOOLOGY] An order of small tapeworms of the subclass Cestoda characterized by the variation in the structure of the scolex; all species are intestinal parasites of elasmobranch fishes. { ˌte·trə·fə'līd·ē·ə }

tetraploidy [CELL BIOLOGY] The occurrence of related forms possessing in the somatic cells chromosome numbers four times the haploid number. { 'te·trəˌplȯid·ē }

tetrapod [VERTEBRATE ZOOLOGY] A four-footed animal. { 'te·trəˌpäd }

Thecideidina [PALEONTOLOGY] An extinct suborder of articulate brachiopods doubtfully included in the order Terebratulida. { thə‚sid·ē·ə 'dīn·ə }

Thecodontia [PALEONTOLOGY] An order of archosaurian reptiles, confined to the Triassic and distinguished by the absence of a supratemporal bone, parietal foramen, and palatal teeth, and by the presence of an antorbital fenestra. { ‚thek·ə 'dän·chə }

Thelastomidae [INVERTEBRATE ZOOLOGY] A family of nematode worms in the superfamily Oxyuroidea. { ‚thel·ə'stäm·ə‚dē }

thenar [ANATOMY] The ball of the thumb. { 'thē ‚när }

Theophrastaceae [BOTANY] A family of tropical and subtropical dicotyledonous woody plants in the order Primulales characterized by flowers having staminodes alternate with the corolla lobes. { ‚thē·ə·fras'tās·ē‚ē }

Therapsida [PALEONTOLOGY] An order of mammallike reptiles of the subclass Synapsida which first appeared in mid-Permian times and persisted until the end of the Triassic. { thə'rap·səd·ə }

Therevidae [INVERTEBRATE ZOOLOGY] The stiletto flies, a family of orthorrhaphous dipteran insects in the series Brachycera. { thə'rev·ə‚dē }

Theria [VERTEBRATE ZOOLOGY] A subclass of the class Mammalia including all living mammals except the monotremes. { 'ther· ē·ə }

Theridiidae [INVERTEBRATE ZOOLOGY] The comb-footed spiders, a family of the suborder Dipneumonomorphae. { ‚ther·ə'dī·ə‚de }

thermal belt [ECOLOGY] Any one of several possible horizontal belts of a vegetation type found in mountainous terrain, resulting primarily from vertical temperature variation. Also known as thermal zone. { 'thər·məl ‚belt }

thermal ecology [ECOLOGY] Study of the independent and interactive biotic and abiotic components of naturally heated environments. { 'thər·məl i'käl·ə·jē }

thermal-efficiency index [ECOLOGY] *See* temperature-efficiency index. { 'thər·məl i¦fish·ən·sē ‚in‚deks }

thermal-efficiency ratio [ECOLOGY] *See* temperature-efficiency ratio. { 'thər·məl i¦fish·ən·sē ‚rā·shō }

thermal pollution [ECOLOGY] The discharge of heated effluent into natural waters that causes a rise in temperature sufficient to upset the ecological balance of the waterway. { 'thər·məl pə 'lü·shən }

thermal zone [ECOLOGY] *See* thermal belt. { 'thər·məl 'zōn }

thermoacidophile [BIOLOGY] An organism that grows under extremely acidic conditions and at very high temperatures. { ¦thər·mō·a'sid·ə‚fīl }

thermoduric bacteria [MICROBIOLOGY] Bacteria which survive pasteurization, but do not grow at temperatures used in a pasteurizing process. { ¦thər·mō'dür·ik bak'tir·ē·ə }

thermometer bird [VERTEBRATE ZOOLOGY] The name applied to the brush turkey, native to Australia, because it lays its eggs in holes in mounds of earth and vegetation, with the heat from the decaying vegetation serving to incubate the eggs. { thər'mäm·əd·ər ‚bərd }

thermoperiodicity [BOTANY] The totality of responses of a plant to appropriately fluctuating temperatures. { ¦thər·mō‚pir·ē·ə'dis·əd·ē }

thermophile [BIOLOGY] An organism that thrives at high temperatures. { 'thər·mə‚fīl }

thermoreception [PHYSIOLOGY] The process by which environmental temperature affects specialized sense organs (thermoreceptors). { ¦thər·mō·ri'sep·shən }

thermoreceptor [PHYSIOLOGY] A sense receptor that responds to stimulation by heat and cold. { ¦thər·mō·ri'sep·tər }

thermoregulation [PHYSIOLOGY] A mechanism by which mammals and birds attempt to balance heat gain and heat loss in order to maintain a constant body temperature when exposed to variations in cooling power of the external medium. { ¦thər·mō‚reg·yə'lā·shən }

Thermosbaenacea [INVERTEBRATE ZOOLOGY] An order of small crustaceans in the superorder Pancarida. { ‚thər·məs·bə'nās·ē·ə }

Thermosbaenidae [INVERTEBRATE ZOOLOGY] A family of the crustacean order Thermosbaenacea. { ‚thər·məs'bē·nə‚dē }

thermotaxis [BIOLOGY] Orientation movement of a motile organism in response to the stimulus of a temperature gradient. { ¦thər·mō¦tak·səs }

therophyte [ECOLOGY] An annual plant whose seed is the only overwintering structure. { 'ther·ə‚fīt }

Theropoda [PALEONTOLOGY] A suborder of carnivorous bipedal saurischian reptiles which first appeared in the Upper Triassic and culminated in the uppermost Cretaceous. { thi'räp·əd·ə }

Tetrapoda [VERTEBRATE ZOOLOGY] The superclass of the subphylum Vertebrata whose members typically possess four limbs; includes all forms above fishes. { te'träp·əd·ə }

Tetrarhynchoidea [INVERTEBRATE ZOOLOGY] The equivalent name for Trypanorhyncha. { ‚te·trə·riŋ'kȯid·ē·ə }

Tetrasporales [BOTANY] A heterogeneous and artificial assemblage of colonial fresh-water and marine algae in the division Chlorophyta. { ‚te·trə·spə'rā·lēz }

tetraspore [BOTANY] One of the haploid asexual spores of the red algae formed in groups of four. { 'te·trə‚spȯr }

tetraxon [INVERTEBRATE ZOOLOGY] A type of sponge spicule with four axes. { tə'trak‚sän }

Tetrigidae [INVERTEBRATE ZOOLOGY] The grouse locusts or pygmy grasshoppers in the order Orthoptera in which the front wings are reduced to small scalelike structures. { te'trij·ə‚dē }

tetrodotoxin [BIOCHEMISTRY] $C_{11}H_{17}N_3O_8$ A toxin that blocks the action potential in the nerve impulse. { ¦te·trə·dō'täk·sən }

tetrose [BIOCHEMISTRY] Any of a group of monosaccharides that have a four-carbon chain; an example is erythrose, $CH_2OH \cdot (CHOH)_2 \cdot CHO$. { 'te‚trōs }

Tettigoniidae [INVERTEBRATE ZOOLOGY] A family of insects in the order Orthoptera which have long antennae, hindlegs fitted for jumping, and an elongate, vertically flattened ovipositor; consists of the longhorn or green grasshopper. { ˌted·ə·gə'nī·əˌdē }

Teuthidae [VERTEBRATE ZOOLOGY] The rabbitfishes, a family of perciform fishes in the suborder Acanthuroidei. { tü'thid·əˌdē }

Teuthoidea [INVERTEBRATE ZOOLOGY] An order of the molluscan subclass Coleoidea in which the rostrum is not developed, the proostracum is represented by the elongated pen or gladus, and ten arms are present. { tü'thȯid·ē·ə }

textile microbiology [MICROBIOLOGY] That branch of microbiology concerned with textile materials; deals with microorganisms that are harmful either to the fibers or to the consumer, and microorganisms that are useful, as in the retting process. { 'tek·stəl ¦mī·krō·bī'äl·ə·jē }

Textulariina [INVERTEBRATE ZOOLOGY] A suborder of foraminiferan protozoans characterized by an agglutinated wall. { ˌtek·styə·lə'rī·ən·ə }

thalamus [ANATOMY] Either one of two masses of gray matter located on the sides of the third ventricle and forming part of the lateral wall of that cavity. { 'thal·ə·məs }

Thalassinidea [INVERTEBRATE ZOOLOGY] The mud shrimps, a group of thin-shelled, burrowing decapod crustaceans belonging to the Macrura; individuals have large chelate or subchelate first pereiopods, and no chelae on the third pereiopods. { thəˌlas·ə'nid·ē·ə }

Thalattosauria [PALEONTOLOGY] A suborder of extinct reptiles in the order Eosuchia from the Middle Triassic. { thəˌlad·ə'sȯr·ē·ə }

Thaliacea [INVERTEBRATE ZOOLOGY] A small class of pelagic Tunicata in which oral and atrial apertures occur at opposite ends of the body. { ˌthā·lē'ā·shē·ə }

Thallobionta [BOTANY] One of the two subkingdoms of plants, characterized by the absence of specialized tissues or organs and multicellular sex organs. { ¦thal·ō·bī'änt·ə }

Thallophyta [BOTANY] The equivalent name for Thallobionta. { thə'läf·əd·ə }

thallospore [BOTANY] A spore that develops by budding of hyphal cells. { 'thal·əˌspȯr }

thallus [BOTANY] A plant body that is not differentiated into special tissue systems or organs and may vary from a single cell to a complex, branching multicellular structure. { 'thal·əs }

thanatocoenosis [PALEONTOLOGY] The assemblage of dead organisms or fossils that occurred together in a given area at a given moment of geologic time. Also known as death assemblage; taphocoenosis. { ˌthan·ə·tō·sə'nō·səs }

Thaumaleidae [INVERTEBRATE ZOOLOGY] A family of orthorrhaphous dipteran insects in the series Nematocera. { ˌthȯ·mə'lē·əˌdē }

Thaumastellidae [INVERTEBRATE ZOOLOGY] A monospecific family of the Hemiptera assigned to the Pentatomorpha found only in Ethiopia. { ˌthȯ·mə'stel·əˌdē }

Thaumastocoridae [INVERTEBRATE ZOOLOGY] The single family of the hemipteran superfamily Thaumastocoroidea. { thəˌmas·tə'kȯr·əˌdē }

Thaumastocoroidea [INVERTEBRATE ZOOLOGY] A monofamilial superfamily of the Hemiptera in the subdivision Geocorisae which occurs in Australia and the New World tropics. { thəˌmas·tə·kə'rȯid·ē·ə }

Thaumatoxenidae [INVERTEBRATE ZOOLOGY] A family of cyclorrhaphous dipteran insects in the series Aschiza. { ˌthȯ·mə·täk'sen·əˌdē }

Theaceae [BOTANY] A family of dicotyledonous erect trees or shrubs in the order Theales characterized by alternate, exstipulate leaves, usually five petals, and mostly numerous stamens. { thē'ās·ēˌē }

Theales [BOTANY] An order of dicotyledonous mostly woody plants in the subclass Dilleniidae with simple or occasionally compound leaves, petals usually separate, numerous stamens, and the calyx arranged in a tight spiral. { thē'ā·lēz }

theca [ANATOMY] The sheath of dura mater which covers the spinal cord. [BOTANY] **1.** A moss capsule. **2.** A pollen sac. [HISTOLOGY] The layer of stroma surrounding a Graafian follicle. [INVERTEBRATE ZOOLOGY] The test of a testate protozoan or a rotifer. { 'thē·kə }

theca folliculi [HISTOLOGY] The capsule surrounding a developing or mature Graafian follicle; consists of two layers, theca interna and theca externa. { 'thē·kə fə'lik·yə·lī }

Thecanephria [INVERTEBRATE ZOOLOGY] An order of the phylum Brachiata containing a group of elongate, tube-dwelling tentaculate, deep-sea animals of bizarre structure. { ˌthē·kə'nef·rē·ə }

thecate [BIOLOGY] Having a theca. { 'thēˌkāt }

Theropsida [PALEONTOLOGY] An order of extinct mammallike reptiles in the subclass Synapsida. { thi'räp·səd·ə }

Thesium [BOTANY] The hymenium of the apothecium in lichens. { 'thē·sē·əm }

thesocyte [INVERTEBRATE ZOOLOGY] An amebocyte in Porifera containing ergastic cytoplasmic inclusions. { 'thes·əˌsīt }

theta antigen [IMMUNOLOGY] A cell membrane constituent which distinguishes T cells from other lymphocytes. { 'thād·ə 'ant·i·jən }

thiamine [BIOCHEMISTRY] $C_{12}H_{17}ClN_4OS$ A member of the vitamin B complex that occurs in many natural sources, frequently in the form of cocarboxylase. Also known as aneurine; vitamin B_1. { 'thī·ə·mən }

thiamine pyrophosphate [BIOCHEMISTRY] The coenzyme or prosthetic component of carboxylase; catalyzes decarboxylation of various α-keto acids. Also known as cocarboxylase. { 'thī·ə·mən ¦pī·rō'fäˌsfāt }

Thiaridae [INVERTEBRATE ZOOLOGY] A family of freshwater gastropod mollusks in the order Pectinibranchia. { ˌthī'ar·əˌdē }

thicket [ECOLOGY] *See* tropical scrub. { 'thik·ət }

thick-tailed bushbaby [VERTEBRATE ZOOLOGY] *Galago crassicaudatus.* A primate animal in the family Lorisidae; one of the six species of bushbaby, the thick-tailed bushbaby is more aggressive and solitary than the other species, and grows to over 1 foot (30 centimeters) in length with an equally long tail. Also known as great galago. { 'thik ¦tāld 'bu̇shˌbā·bē }

thigh [ANATOMY] The upper part of the leg, from the pelvis to the knee. { thī }

thigmotaxis [BIOLOGY] *See* stereotaxis. { ¦thig·mə¦tak·səs }

Thigmotrichida [INVERTEBRATE ZOOLOGY] An order of ciliated protozoans in the subclass Holotrichia. { ¦thig·mō'trik·əd·ə }

thigmotropism [BIOLOGY] *See* stereotropism. { thig'mä·trə,piz·əm }

Thinocoridae [VERTEBRATE ZOOLOGY] The seed snipes, family of South American birds in the order Charadriiformes. { ,thin·ə'kȯr·ə,dē }

Thiobacteriaceae [MICROBIOLOGY] Formerly a family of nonfilamentous, gram-negative bacteria of the suborder Pseudomonadineae characterized by the ability to oxidize hydrogen sulfide, free sulfur, and inorganic sulfur compounds to sulfuric acid. { ¦thī·ō·bak,tir·ē'ās·ē,ē }

Thiorhodaceae [MICROBIOLOGY] Formerly a family of bacteria in the suborder Rhodobacteriineae composed of the purple, red, orange, and brown sulfur bacteria; characterized as strict anaerobes which oxidize hydrogen sulfide and store sulfur globules internally. { ,thī·ə·rō'dās·ē,ē }

thiostreptone [MICROBIOLOGY] A polypeptide antibiotic produced by a species of *Streptomyces* that crystallizes from a chloroform methanol solution; used in veterinary medicine. { ¦thī·ō'strep ,tōn }

thirst [PHYSIOLOGY] A sensation, as of dryness in the mouth and throat, resulting from water deprivation. { 'thərst }

thistle [BOTANY] Any of the various prickly plants comprising the family Compositae. { 'this·əl }

Thlipsuridae [PALEONTOLOGY] A Paleozoic family of ostracod crustaceans in the suborder Platycopa. { thlip'sür·ə,dē }

Thoracica [INVERTEBRATE ZOOLOGY] An order of the subclass Cirripedia; individuals are permanently attached in the adult stage, the mantle is usually protected by calcareous plates, and six pairs of biramous thoracic appendages are present. { thə'ras·ə·kə }

thoracic duct [ANATOMY] The common lymph trunk beginning in the crura of the diaphragm at the level of the last thoracic vertebra, passing upward, and emptying into the left subclavian vein at its junction with the left internal jugular vein. { thə'ras·ik 'dəkt }

thoracic cavity [ANATOMY] *See* thorax. { thə'ras·ik 'kav·əd·ē }

thoracic vertebrae [ANATOMY] The vertebrae associated with the chest and ribs in vertebrates; there are 12 in humans. { thə'ras·ik 'vərd·ə ,brā }

thoracoabdominal breathing [PHYSIOLOGY] The process of air breathing in reptiles, birds, and mammals that depends upon aspiration or sucking inspiration, and involves trunk musculature to supply pulmonary ventilation. { ¦thȯr·ə·kō·ab'däm·ə·nəl 'brēth·iŋ }

Thoracostomopsidae [INVERTEBRATE ZOOLOGY] A family of marine nematodes in the superfamily Enoploidea, which have the stomatal armature modified to form a hollow tube. { ¦thȯr·ə·kō·stō 'mäp·sə,dē }

thorax [ANATOMY] The chest; the cavity of the mammalian body between the neck and the diaphragm, containing the heart, lungs, and mediastinal structures. Also known as thoracic cavity. [INVERTEBRATE ZOOLOGY] The middle of three principal divisions of the body of certain classes of arthropods. { 'thȯr,aks }

Thorictidae [INVERTEBRATE ZOOLOGY] The ant blood beetles, a family of coleopteran insects in the superfamily Dermestoidea. { thə'rik·tə,dē }

thorn [BOTANY] A short, sharp, rigid, leafless branch on a plant. [ZOOLOGY] Any of various sharp spinose structures on an animal. { thȯrn }

thornback [VERTEBRATE ZOOLOGY] *Raja clavata.* A ray found in European waters and characterized by spines on its back. { 'thȯrn,bak }

thornbush [ECOLOGY] A vegetation class that is dominated by tall succulents and profusely branching smooth-barked deciduous hardwoods which vary in density from mesquite bush in the Caribbean to the open spurge thicket in Central Africa; the climate is that of a warm desert, except for a rather short intense rainy season. Also known as Dorngeholz; Dorngestrauch; dornveld; savane armée; savane épineuse; thorn scrub. { 'thȯrn,bu̇sh }

thorn forest [ECOLOGY] A type of forest formation, mostly tropical and subtropical, intermediate between desert and steppe; dominated by small trees and shrubs, many armed with thorns and spines; leaves are absent, succulent, or deciduous during long dry periods, which may also be cool; an example is the caatinga of northeastern Brazil. { 'thȯrn 'fär·əst }

thorn scrub [ECOLOGY] *See* thornbush. { 'thȯrn 'skrəb }

threadfin [VERTEBRATE ZOOLOGY] Common name for any of the fishes in the family Polynemidae. { 'thred,fin }

threonine [BIOCHEMISTRY] $CH_3CHOHCH(NH_2)$-COOH A crystalline α-amino acid considered essential for normal growth of animals; it is biosynthesized from aspartic acid and is a precursor of isoleucine in microorganisms. { 'thrē·ə,nēn }

thresher shark [VERTEBRATE ZOOLOGY] Common name for fishes in the family Alopiidae; pelagic predacious sharks of generally wide distribution that have an extremely long, whiplike tail with which they thrash the water, destroying schools of small fishes. { 'thresh·ər ,shärk }

threshold [PHYSIOLOGY] The minimum level of a stimulus that will evoke a response in an irritable tissue. { 'thresh,hōld }

threshold of audibility [PHYSIOLOGY] The minimum effective sound pressure of a specified signal that is capable of evoking an auditory sensation in a specified fraction of the trials; the threshold may be expressed in decibels relative to 0.0002 microbar (2×10^{-5} pascal) or 1 microbar (0.1 pascal). Also known as threshold of detectability; threshold of hearing. { 'thresh ,hōld əv ,ȯd·ə'bil·əd·ē }

threshold of detectability [PHYSIOLOGY] *See* threshold of audibility. { 'thresh,hōld əv di,tek·tə'bil·əd·ē }

threshold of feelings [PHYSIOLOGY] The minimum effective sound pressure of a specified signal that, in a specified fraction of trials, will stimulate the ear to a point at which there is a sensation of feeling, discomfort, tickle, or pain; normally expressed in decibels relative to 0.0002 microbar (0.00002 pascal) or 1 microbar (0.1 pascal). { 'thresh,hōld əv 'fēl·iŋz }

threshold of hearing [PHYSIOLOGY] *See* threshold of audibility. { 'thresh,hōld əv 'hir·iŋ }

Threskiornithidae [VERTEBRATE ZOOLOGY] The ibises, a family of long-legged birds in the order Ciconiiformes. { ,thres·kē·ȯr'nith·ə,dē }

thrip [INVERTEBRATE ZOOLOGY] A small, slender-bodied phytophagous insect of the order Thysanoptera with suctorial mouthparts, a stout proboscis, a vestigial right mandible, and a fully developed left mandible, while wings may be present or absent. { thrip }

Thripidae [INVERTEBRATE ZOOLOGY] A large family of thrips, order Thysanoptera, which includes the most common species. { 'thrip·ə,dē }

throat [ANATOMY] The region of the vertebrate body that includes the pharynx, the larynx, and related structures. [BOTANY] The upper, spreading part of the tube of a gamopetalous calyx or corolla. { thrōt }

thrombin [BIOCHEMISTRY] An enzyme elaborated from prothrombin in shed blood which induces clotting by converting fibrinogen to fibrin. { 'thräm·bən }

thrombinogen [BIOCHEMISTRY] *See* prothrombin. { thräm'bin·ə·jən }

thrombocyte [HISTOLOGY] One of the minute protoplasmic disks found in vertebrate blood; thought to be fragments of megakaryocytes. Also known as blood platelet; platelet. { 'thräm·bə ,sīt }

thrombokinase [BIOCHEMISTRY] A proteolytic enzyme in blood plasma that, together with thromboplastin, calcium, and factor V, converts prothrombin to thrombin. { ¦thräm·bō'kī,nās }

thromboplastin [BIOCHEMISTRY] Any of a group of lipid and protein complexes in blood that accelerate the conversion of prothrombin to thrombin. Also known as factor III; plasma thromboplastin component (PTC). { ,thräm·bō 'plas·tən }

thromboplastinogen [BIOCHEMISTRY] *See* antihemophilic factor. { ¦thräm·bō,plas'tin·ə·jən }

thromboxane [BIOCHEMISTRY] Any member of a group of 20-carbon fatty acids related to the prostaglandins and derived mainly from arachidonic acid. { ,thräm'bäk,sān }

Throscidae [INVERTEBRATE ZOOLOGY] The false metallic wood-boring beetles, a cosmopolitan family of the Coleoptera in the superfamily Elateroidea. { 'thräs·kə,dē }

Thunburg technique [BIOCHEMISTRY] A technique used to study oxidation of a substrate occurring by dehydrogenation reactions; methylene blue, a reversibly oxidizable indicator, substitutes for molecular oxygen as the ultimate hydrogen acceptor (oxidant), becoming reduced to the colorless leuco form. { 'thən,bərg tek ,nēk }

Thunnidae [VERTEBRATE ZOOLOGY] The tunas, a family of perciform fishes; there are no scales on the posterior part of the body, and those on the anterior are fused to form an armored covering, the body is streamlined, and the tail is crescent-shaped. { 'thən·ə,dē }

Thurniaceae [BOTANY] A small family of monocotyledonous plants in the order Juncales distinguished by an inflorescence of one or more dense heads, vascular bundles of the leaf in vertical pairs, and silica bodies in the leaf epidermis. { ,thər·nē'ās·ē,ē }

Thylacinidae [VERTEBRATE ZOOLOGY] A family of Australian carnivorous marsupials in the superfamily Dasyuroidea. { ,thī·lə'sīn·ə,dē }

Thylacoleonidae [PALEONTOLOGY] An extinct family of carnivorous marsupials in the superfamily Phalangeroidea. { ,thī·lə,kō·lē'än·ə,dē }

thylakoid [CELL BIOLOGY] An internal membrane system which occupies the main body of a plastid; particularly well developed in chloroplasts. { 'thī·lə,kȯid }

thyme [BOTANY] A perennial mint plant of the genus *Thymus*; pungent aromatic herb is made from the leaves. { tīm }

Thymelaeceae [BOTANY] A family of dicotyledonous woody plants in the order Myrtales characterized by a superior ovary with a solitary ovule, and petals, if present, are scalelike. { ,thī·mə·lē·ə'ās·ē,ē }

thymic corpuscle [HISTOLOGY] A characteristic, rounded, acidophil body in the medulla of the thymus; composed of hyalinized cells concentrically arranged about a core which is occasionally calcified. Also known as Hassal's body. { 'thī·mik 'kȯr·pə·səl }

thymidine [BIOCHEMISTRY] $C_{10}H_{14}N_2O_5$ A nucleoside derived from deoxyribonucleic acid; essential growth factor for certain microorganisms in mediums lacking vitamin B_{12} and folic acid. { 'thī·mə,dēn }

thymidylic acid [BIOCHEMISTRY] $C_{10}H_{15}N_2O_8P$ A mononucleotide component of deoxyribonucleic acid which yields thymine, D-ribose, and phosphoric acid on complete hydrolysis. { ¦thī·mə ¦dil·ik 'as·əd }

thymine [BIOCHEMISTRY] $C_5H_6N_2O_2$ A pyrimidine component of nucleic acid, first isolated from the thymus. { 'thī,mēn }

thymocyte [HISTOLOGY] A lymphocyte formed in the thymus. { 'thī·mə,sīt }

thymopharyngeal duct [EMBRYOLOGY] The third pharyngobranchial duct; it may elongate between the pharynx and thymus. { ¦thī·mō·fə 'rin·jē·əl 'dəkt }

thymosin [IMMUNOLOGY] Any of a group of hormones secreted by the thymus gland that stimulate lymphocyte production within the thymus and confer on lymphocytes elsewhere in the body the capacity to respond to antigenic stimulation. { 'thī·mə·sən }

thymus gland [ANATOMY] A lymphoid organ in the neck or upper thorax of all vertebrates; it is most prominent in early life and is essential for normal development of the circulating pool of lymphocytes. { 'thī·məs ,gland }

Thyrididae [INVERTEBRATE ZOOLOGY] The window-winged moths, a small tropical family of lepidopteran insects in the suborder Heteroneura. { thī'rid·ə,dē }

thyrocalcitonin [BIOCHEMISTRY] *See* calcitonin. { ¦thī·rō,kal·sə'tō·nən }

thyroglobulin [BIOCHEMISTRY] An iodinated protein found as the storage form of the iodinated hormones in the thyroid follicular lumen and epithelial cells. { ¦thī·rō'gläb·yə·lən }

thyroglossal duct [EMBRYOLOGY] A narrow temporary channel connecting the anlage of the thyroid with the surface of the tongue. { thī·rō 'gläs·əl 'dəkt }

thyroid cartilage [ANATOMY] The largest of the laryngeal cartilages in humans and most other mammals, located anterior to the cricoid; in humans, it forms the Adam's apple. { 'thī,ròid ,kärt·lij }

thyroid gland [ANATOMY] An endocrine gland found in all vertebrates that produces, stores, and secretes the thyroid hormones. { 'thī,ròid ,gland }

thyroid hormone [BIOCHEMISTRY] Commonly, thyroxine or triiodothyronine, or both; a metabolically active compound formed and stored in the thyroid gland which functions to regulate the rate of metabolism. { 'thī,ròid 'hòr,mōn }

thyroid-stimulating hormone [BIOCHEMISTRY] *See* thyrotropic hormone. { 'thī,ròid ¦stim·yə ,lād·iŋ 'hòr,mōn }

thyroprotein [BIOCHEMISTRY] A protein secreted in the thyroid gland, such as thyroxine. { ¦thī·rō 'prō,tēn }

Thyropteridae [VERTEBRATE ZOOLOGY] The New-World disk-winged bats, a family of the Chiroptera found in Central and South America, characterized by a stalked sucking disk and a well-developed claw on the thumb. { ,thī,räp 'ter·ə,dē }

thyrotropic hormone [BIOCHEMISTRY] A hormone produced by the adenohypophysis which regulates thyroid gland function. Also known as thyroid-stimulating hormone (TSH). { ¦thī·rə ¦träp·ik 'hòr,mōn }

thyrotropin [BIOCHEMISTRY] A thyroid-stimulating hormone produced by the adenohypophysis. { thī'rä·trə·pən }

thyroxine [BIOCHEMISTRY] $C_{15}H_{11}I_4NO_4$ The active physiologic principle of the thyroid gland; used in the form of the sodium salt for replacement therapy in states of hypothyroidism or absent thyroid function. { thī'räk,sēn }

thyrse [BOTANY] An inflorescence with a racemose primary axis and cymose secondary and later axes. { thərs }

Thysanidae [INVERTEBRATE ZOOLOGY] A family of hymenopteran insects in the superfamily Chalcidoidea. { thī'san·ə,dē }

Thysanoptera [INVERTEBRATE ZOOLOGY] The thrips, an order of small, slender insects having exopterygote development, sucking mouthparts, and exceptionally narrow wings with few or no veins and bordered by long hairs. { ,thī·sə'näp·tə·rə }

Thysanura [INVERTEBRATE ZOOLOGY] The silverfish, machilids, and allies, an order of primarily wingless insects with soft, fusiform bodies. { ,thī·sə'nu̇r·ə }

tibia [ANATOMY] The larger of the two leg bones, articulating with the femur, fibula, and talus. { 'tib·ē·ə }

tibialis [ANATOMY] **1.** A muscle of the leg arising from the proximal end of the tibia and inserted into the first cuneiform and first metatarsal bones. **2.** A deep muscle of the leg arising proximally from the tibia and fibula and inserted into the navicular and first cuneiform bones. { ,tib·ē 'al·əs }

tick [INVERTEBRATE ZOOLOGY] Any arachnid comprising Ixodoidea; a bloodsucking parasite and important vector of various infectious diseases of humans and lower animals. { tik }

tickle [PHYSIOLOGY] A tingling sensation of the skin or a mucous membrane following light, tactile stimulation. { 'tik·əl }

tic polonga [VERTEBRATE ZOOLOGY] *Vipera russellii.* A member of the Viperidae; one of the most deadly and most common snakes in India; it may reach a length of 5 feet (1.5 meters), is nocturnal in its habits, and pursues rodents into houses. Also known as Russell's viper. { ,tik pə'lòŋ·gə }

tidal air [PHYSIOLOGY] That air which is inspired and expired during normal breathing. { 'tīd·əl 'er }

tidal volume [PHYSIOLOGY] The volume of air moved in and out of the lungs during a single normal respiratory cycle. { 'tīd·əl 'väl·yəm }

Tiedenmann's body [INVERTEBRATE ZOOLOGY] One of the small glands opening into the ring vessel in many echinoderms in which amebocytes are produced. { 'tēd·ən,mänz ,bäd·ē }

tiger [VERTEBRATE ZOOLOGY] *Felis tigris.* An Asiatic carnivorous mammal in the family Felidae characterized by a tawny coat with transverse black stripes and white underparts. { 'tī·gər }

tiger beetle [INVERTEBRATE ZOOLOGY] The common name for any of the bright-colored beetles in the family Cicindelidae; there are about 1300 species distributed all over the world. { 'tī·gər ,bēd·əl }

tiger salamander [VERTEBRATE ZOOLOGY] *Ambystoma tigrinum.* A salamander in the family Ambystomatidae, found in a variety of subspecific forms from Canada to Mexico and over most of the United States; lives in arid and humid regions, and is the only salamander in much of the Great Plains and Rocky Mountains. { 'tī·gər ¦sal·ə,man·dər }

tiger shark [VERTEBRATE ZOOLOGY] *See* sand shark. { 'tī·gər ,shärk }

tight junction [CELL BIOLOGY] An intercellular junction composed of a series of fusions of the junctional membrane, forming a continuous seal; serves as a selective barrier to small molecules and as a total barrier to large molecules. Also known as impermeable junction; occluding junction; zonula occludens. { ¦tīt 'jəŋk·shən }

tiller [BOTANY] A shoot that develops from an axillary or adventitious bud at the base of a stem. { 'til·ər }

Tilletiaceae [MYCOLOGY] A family of fungi in the

order Ustilaginales in which basidiospores form at the tip of the apibasidium. { tə‚le·she'as·ē‚ē }

Tillodontia [PALEONTOLOGY] An order of extinct quadrupedal land mammals known from early Cenozoic deposits in the Northern Hemisphere and distinguished by large, rodentlike incisors, blunt-cuspid cheek teeth, and five clawed toes. { ‚til·ə'dän·chə }

timberline [ECOLOGY] The elevation or latitudinal limits for arboreal growth. Also known as tree line. { 'tim·bər‚līn }

timothy [BOTANY] *Phleum pratense.* A perennial hay grass of the order Cyperales characterized by moderately leafy stems and a dense cylindrical inflorescence. { 'tim·ə·thē }

Tinamidae [VERTEBRATE ZOOLOGY] The single family of the avian order Tinamiformes. { ti 'nam·ə‚dē }

Tinamiformes [VERTEBRATE ZOOLOGY] The tinamous, an order of South and Central American birds which are superficially fowllike but have fully developed wings and are weak fliers. { ti ‚nam·ə'fȯr‚mēz }

Tineidae [INVERTEBRATE ZOOLOGY] A family of small moths in the superfamily Tineoidea distinguished by an erect, bristling vestiture on the head. { ti'nē·ə‚dē }

Tineoidea [INVERTEBRATE ZOOLOGY] A superfamily of heteroneuran Lepidoptera which includes small moths that usually have well-developed maxillary palpi. { ‚tin·ē'ȯid·ē·ə }

Tingidae [INVERTEBRATE ZOOLOGY] The lace bugs, the single family of the hemipteran superfamily Tingoidea. { 'tin·jə‚dē }

Tingoidea [INVERTEBRATE ZOOLOGY] A superfamily of the Hemiptera in the subdivision Geocorisae characterized by the wings with many lacelike areolae. { tiŋ'gȯid·ē·ə }

Tintinnida [INVERTEBRATE ZOOLOGY] An order of ciliated protozoans in the subclass Spirotrichia whose members are conical or trumpet-shaped pelagic forms bearing shells. { tin'tin·əd·ə }

Tiphiidae [INVERTEBRATE ZOOLOGY] A family of the Hymenoptera in the superfamily Scolioidea. { tə'fī·ə‚dē }

Tipulidae [INVERTEBRATE ZOOLOGY] The crane flies, a family of orthorrhaphous dipteran insects in the series Nematocera. { tə'pyül·ə‚dē }

tissue [HISTOLOGY] An aggregation of cells more or less similar morphologically and functionally. { 'tish·ü }

tissue culture [CELL BIOLOGY] Growth of tissue cells in artificial media. { 'tish·ü ‚kəl·chər }

Titanoideidae [PALEONTOLOGY] A family of extinct land mammals in the order Pantodonta. { ‚tīt·ən·ȯi'dē·ə‚dē }

titanothere [PALEONTOLOGY] Any member of the family Brontotheriidae. { tī'tan·ə‚thir }

T lymphocyte [IMMUNOLOGY] *See* T cell. { 'tē 'lim·fə‚sīt }

T method [BOTANY] A budding method in which a T-shaped cut is made through the bark at the internode of the stock, the bark of the scion is separated from the xylem along the cambium and removed, and the scion is forced into the incision on the stock. { 'tē ‚meth·əd }

toad [VERTEBRATE ZOOLOGY] Any of several species of the amphibian order Anura, especially in the family Bufonidae; glandular structures in the skin secrete acrid, irritating substances of varying toxicity. { tōd }

toadstool [MYCOLOGY] Any of various fleshy, poisonous or inedible fungi with a large umbrella-shaped fruiting body. { 'tōd‚stül }

tobacco [BOTANY] **1.** Any plant of the genus *Nicotinia* cultivated for its leaves, which contain 1-3% of the alkaloid nicotine. **2.** The dried leaves of the plant. { tə'bak·ō }

Todidae [VERTEBRATE ZOOLOGY] The todies, a family of birds in the order Coraciiformes found in the West Indies. { 'tō·də‚dē }

toe [ANATOMY] One of the digits on the foot of humans and other vertebrates. { tō }

toleragen [IMMUNOLOGY] A substance which, in appropriate dosages, produces a state of specific immunological tolerance in humans or animals. { 'täl·ə·rə·jən }

tomato [BOTANY] A plant of the genus *Lycopersicon*, especially *L. esculentum*, in the family Solanaceae cultivated for its fleshy edible fruit, which is red, pink, orange, yellow, white, or green, with fleshy placentas containing many small, oval seeds with short hairs and covered with a gelatinous matrix. { tə 'mād·ō }

tomentose [BOTANY] Covered with densely matted hairs. { tə'men‚tōs }

tomentum [ANATOMY] The deep layer of the pia mater composed principally of minute blood vessels. [BIOLOGY] Pubescence consisting of densely matted wooly hairs. { tə'men·təm }

tomium [VERTEBRATE ZOOLOGY] The cutting edge of a bird's beak. { 'tō·mē·əm }

Tomopteridae [INVERTEBRATE ZOOLOGY] The glass worms, a family of pelagic polychaete annelids belonging to the group Errantia. { tə‚mäp 'ter·ə‚dē }

tongue [ANATOMY] A muscular organ located on the floor of the mouth in humans and most vertebrates which may serve various functions, such as taking and swallowing food or tasting or as a tactile organ or sometimes a prehensile organ. { təŋ }

tongue worm [INVERTEBRATE ZOOLOGY] *See* acorn worm. { 'təŋ ‚wərm }

tonic labyrinthine reflexes [PHYSIOLOGY] Rotation or deviation of the head causes extension of the limbs on the same side as the chin, and flexion of the opposite extremities: dorsiflexion of the head produces increased extensor tonus of the upper extremities and relaxation of the lower limbs, while ventroflexion of the head produces the reverse; seen in the young infant and patients with a lesion at the midbrain level or above. { 'tän·ik ‚lab·ə'rin‚thēn 'rē‚flek·səz }

tonic neck reflexes [PHYSIOLOGY] Reflexes in which rotation or deviation of the head causes extension of the limbs on the same side as the chin, and flexion of the opposite extremities; dorsiflexion of the head produces increased extension tonus of the upper extremities and relaxation of the lower limbs, and ventroflexion of the head, the reverse; seen normally in incomplete forms in the very young infant, and thereafter in patients with a lesion at the midbrain level or above. { 'tän·ik 'nek 'rē,flek·səz }

tonofibril [CELL BIOLOGY] Any of the fibrils converging on desmosomes in epithelial cells. { ¦tä·nō'fī·brəl }

tonoplast [BOTANY] The membrane surrounding a plant-cell vacuole. { 'tän·ə,plast }

tonsil [ANATOMY] **1.** Localized aggregation of diffuse and nodular lymphoid tissue found in the throat where the nasal and oral cavities open into the pharynx. **2.** *See* palatine tonsil. { 'tän·səl }

tonus [PHYSIOLOGY] The degree of muscular contraction when not undergoing shortening. Also known as muscle tone. { 'tō·nəs }

tooth [ANATOMY] One of the hard bony structures supported by the jaws in mammals and by other bones of the mouth and pharynx in lower vertebrates serving principally for prehension and mastication. [INVERTEBRATE ZOOLOGY] Any of various sharp, horny, chitinous, or calcareous processes on or about any part of an invertebrate that functions like or resembles vertebrate jaws. { 'tüth }

tooth shell [INVERTEBRATE ZOOLOGY] A mollusk of the class Scaphopoda characterized by the elongate, tube-shaped, or cylindrical shell which is open at both ends and slightly curved. { 'tüth ,shel }

top grafting [BOTANY] Grafting a scion of one variety of tree onto the main branch of another. { ¦täp ,graft·iŋ }

topocline [ECOLOGY] A graded series of characters exhibited by a species or other closely related organisms along a geographical axis. { 'täp·ə,klīn }

topogenesis [EMBRYOLOGY] *See* morphogenesis. { ¦täp·ə'jen·ə·səs }

topographic anatomy [ANATOMY] The use of bony and soft tissue landmarks on the surface of the body to indicate the known location of deeper structures. { ¦täp·ə¦graf·ik ə'nad·ə·mē }

topographic climax [ECOLOGY] A climax plant community under a uniform macroclimate over which minor topographic features such as hills, rivers, valleys, or undrained depressions exert a controlling influence. { ¦täp·ə¦graf·ik 'klī ,maks }

topoisomerases [BIOCHEMISTRY] Any of a group of enzymes capable of relaxing, unwinding, unpackaging, or changing the degree of supercoiling of deoxyribonucleic acid fiber. { ¦tä·pō·ī 'säm·ə,rās·əz }

topotaxis [INVERTEBRATE ZOOLOGY] *See* tropism. { ¦täp·ə¦tak·səs }

topotype [SYSTEMATICS] A specimen of a species not of the original type series collected at the type locality. { 'täp·ə,tīp }

top shell [INVERTEBRATE ZOOLOGY] Any of the marine snails of the family Trochidae characterized by a spiral conical shell with a flat base. { ¦täp ,shel }

topwork [BOTANY] A procedure employed to propagate seedless varieties of fruit and hybrids, to change the variety of fruit, and to correct pollination problems, using any of three methods: root grafting, crown grafting, and top grafting. { 'täp,wərk }

tornaria [INVERTEBRATE ZOOLOGY] The larva of some acorn worms (Enteropneusta) which is large and marked by complex bands of cilia. { tör'nar·ē·ə }

tornote [INVERTEBRATE ZOOLOGY] A monaxon spicule in certain Porifera having both ends terminating abruptly in points. { 'tör,nōt }

torose [INVERTEBRATE ZOOLOGY] **1.** Having knobby prominences on the surface. **2.** Cylindrical with alternate swellings and contractions. { 'tö,rōs }

Torpedinidae [VERTEBRATE ZOOLOGY] The electric rays or torpedoes, a family of batoid sharks. { ,tör·pə'din·ə,dē }

torpor [PHYSIOLOGY] The condition in hibernating poikilotherms during winter when body temperature drops in a parallel relation to ambient environmental temperatures. { 'tör·pər }

Torridincolidae [INVERTEBRATE ZOOLOGY] A small family of coleopteran insects in the suborder Myxophaga found only in Africa and Brazil. { tə,rid·ən'käl·ə,dē }

tortoise [VERTEBRATE ZOOLOGY] Any of various large terrestrial reptiles in the order Chelonia, especially the family Testudinidae. { 'tórd·əs }

Tortricidae [INVERTEBRATE ZOOLOGY] A family of phytophagous moths in the superfamily Tortricoidea which have a stout body, lightly fringed wings, and threadlike antennae. { tör'tris·ə,dē }

Tortricoidea [INVERTEBRATE ZOOLOGY] A superfamily of small wide-winged moths in the suborder Heteroneura. { ,tör·trə'köid·ē·ə }

Torulopsidales [MYCOLOGY] The equivalent name for Cryptococcales. { ,tör·ə,läp·sə'dā·lēz }

torus [ANATOMY] A rounded protuberance occurring on a body part. [BOTANY] The thickened membrane closing a bordered pit. { 'tör·əs }

Torymidae [INVERTEBRATE ZOOLOGY] A family of hymenopteran insects in the superfamily Chalcidoidea. { tö'rim·ə,dē }

total lung capacity [PHYSIOLOGY] The volume of gas contained within the lungs at the end of a maximum inspiration. { 'tōd·əl 'ləŋ kə,pas·əd·ē }

totipalmate [VERTEBRATE ZOOLOGY] Having all four toes connected by webs, as in the Pelecaniformes. { ,tōd·ə'pä,māt }

totipotence [EMBRYOLOGY] Capacity of a blastomere to develop into a fully formed embryo. { tō 'tip·əd·əns }

toucan [VERTEBRATE ZOOLOGY] Any of numerous fruit-eating birds, of the family Ramphastidae, noted for their large and colorful bills. { 'tü ,kan }

touch [PHYSIOLOGY] The array of sensations arising from pressure sensitivity of the skin. { təch }

toxa [INVERTEBRATE ZOOLOGY] A curved sponge spicule. { 'täk·sə }

Toxasteridae [PALEONTOLOGY] A family of Cretaceous echinoderms in the order Spatangoida which lacked fascioles and petals. { ˌtäk·sə ˈster·əˌdē }

toxicyst [INVERTEBRATE ZOOLOGY] A type of trichocyst in Protozoa which may, upon contact, induce paralysis or lysis of the prey. { ˈtäk·sə ˌsist }

toxigenicity [MICROBIOLOGY] A microorganism's capability for producing toxic substances. { ˌtäk·sə·jəˈnis·əd·ē }

toxin [BIOCHEMISTRY] Any of various poisonous substances produced by certain plant and animal cells, including bacterial toxins, phytotoxins, and zootoxins. { ˈtäk·sən }

Toxodontia [PALEONTOLOGY] An extinct suborder of mammals representing a central stock of the order Notoungulata. { ˌtäk·səˈdän·chə }

Toxoglossa [INVERTEBRATE ZOOLOGY] A group of carnivorous marine gastropod mollusks distinguished by a highly modified radula (toxoglossate). { ˌtäk·səˈgläs·ə }

toxoglossate radula [INVERTEBRATE ZOOLOGY] A radula in certain carnivorous gastropods having elongated, spearlike teeth often perforated by the ducts of large poison glands. { ¦täk·sə¦glä ˌsāt ˈraj·ə·lə }

toxoid [IMMUNOLOGY] Detoxified toxin, but with antigenic properties intact; toxoids of tetanus and diphtheria are used for immunization. { ˈtäkˌsȯid }

Toxoplasmea [INVERTEBRATE ZOOLOGY] A class of the protozoan subphylum Sporozoa composed of small, crescent-shaped organisms that move by body flexion or gliding and are characterized by a two-layered pellicle with underlying microtubules, micropyle, paired organelles, and micronemes. { ¦täk·səˈplaz·mē·ə }

Toxoplasmida [INVERTEBRATE ZOOLOGY] An order of the class Toxoplasmea; members are parasites of vertebrates. { ¦täk·səˈplaz· məd·ə }

toxoplasmin [BIOCHEMISTRY] The *Toxoplasma* antigen; used in a skin test to demonstrate delayed hypersensitivity to toxoplasmosis. { ¦täk·sə ˈplaz·mən }

Toxopneustidae [INVERTEBRATE ZOOLOGY] A family of Tertiary and extant echinoderms of the order Temnopleuroida where the branchial slits are deep and the test tends to be absent. { ˌtäk·səˈnyü·stəˌdē }

Toxotidae [VERTEBRATE ZOOLOGY] The archerfishes, a family of small fresh-water forms in the order Perciformes. { täkˈsäd·əˌdē }

T phage [VIROLOGY] Any of a series (T1-T7) of deoxyribonucleic acid phages which lyse strains of the gram-negative bacterium *Escherichia coli* and its relatives. { ˈtē ˌfāj }

TPN [BIOCHEMISTRY] *See* triphosphopyridine nucleotide.

trabecula [ANATOMY] A band of fibrous or muscular tissue extending from the capsule or wall into the interior of an organ. { trəˈbek·yə·lə }

trace element [BIOCHEMISTRY] A chemical element that is needed in minute quantities for the proper growth, development, and physiology of the organism. Also known as micronutrient. { ˈtrās ˌel·ə·mənt }

trachea [ANATOMY] The cartilaginous and membranous tube by which air passes to and from the lungs in humans and many vertebrates. [BOTANY] A xylem vessel resembling the trachea of vertebrates. [INVERTEBRATE ZOOLOGY] One of the anastomosing air-conveying tubules composing the respiratory system in most insects. { ˈtrā·kē·ə }

tracheid [BOTANY] An elongate, spindle-shaped xylem cell, lacking protoplasm at maturity, and having secondary walls laid in various thicknesses and patterns over the primary wall. { ˈtrā·kē·əd }

Tracheophyta [BOTANY] A large group of plants characterized by the presence of specialized conducting tissues (xylem and phloem) in the roots, stems, and leaves. { ˌtrā·kēˈäf·əd·ē }

Trachylina [INVERTEBRATE ZOOLOGY] An order of moderate-sized jellyfish of the class Hydrozoa distinguished by having balancing organs and either a small polyp stage or none. { ˌtrak·ə ˈlīn·ə }

Trachymedusae [INVERTEBRATE ZOOLOGY] A group of marine jellyfish, recognized as a separate order or as belonging to the order Trachylina whose tentacles have a solid core consisting of a single row of endodermal cells. { ¦tra·kē·mə ˈdüˌsē }

Trachypsammiacea [INVERTEBRATE ZOOLOGY] An order of colonial anthozoan cnidarians characterized by a dendroid skeleton. { ¦tra·kēˌsam· ēˈäs·ē·ə }

Trachystomata [VERTEBRATE ZOOLOGY] The name given to the Meantes when the group is considered to be an order. { ¦tra·kēˈstō· məd·ē }

Tragulidae [VERTEBRATE ZOOLOGY] The chevrotains, a family of pecoran ruminants in the superfamily Traguloidea. { trəˈgyül·əˌdē }

Traguloidea [VERTEBRATE ZOOLOGY] A superfamily of pecoran ruminants, composed of the most primitive forms with large canines; the chevrotain is the only extant member. { ˌtra·gyəˈlȯid· ē·ə }

tragus [ANATOMY] **1.** The prominence in front of the opening of the external ear. **2.** One of the hairs in the external ear canal. { ˈtrā·gəs }

trama [MYCOLOGY] The loosely woven hyphal tissue between adjacent hymenia in basidiomycetes. { ˈtrā·mə }

transaminase [BIOCHEMISTRY] One of a group of enzymes that catalyze the transfer of the amino group of an amino acid to a keto acid to form another amino acid. Also known as aminotransferase. { ¦tranzˈam·əˌnās }

transcapsidation [VIROLOGY] Change in the capsid of PARA (particle aiding replication of adenovirus) from one type of adenovirus to another. { ¦tranzˌkap·səˈdā·shən }

transcriptase [BIOCHEMISTRY] *See* ribonucleic acid polymerase. { tranˈskripˌtās }

transcription [MOLECULAR BIOLOGY] The process by which ribonucleic acid is formed from deoxyribonucleic acid. { tranzˈkrip·shən }

transcription unit [MOLECULAR BIOLOGY] The segment of deoxyribonucleic acid between the sites of initiation and termination of transcription by ribonucleic acid polymerase. { tranz 'krip·shən ˌyü·nət }

transduction [MICROBIOLOGY] Transfer of genetic material between bacterial cells by bacteriophages. { tranz'dək·shən }

transfection [GENETICS] Infection of a cell with viral deoxyribonucleic acid or ribonucleic acid. { trans'fek·shən }

transferase [BIOCHEMISTRY] Any of various enzymes that catalyze the transfer of a chemical group from one molecule to another. { 'tranz·fə ˌrās }

transfer immunity [IMMUNOLOGY] *See* adoptive immunity. { 'tranzˌfər iˌmyün·əd·ē }

transfer ribonucleic acid [MOLECULAR BIOLOGY] The smallest ribonucleic acid molecule found in cells; its structure is complementary to messenger ribonucleic acid and it functions by transferring amino acids from the free state to the polymeric form of growing polypeptide chains. Abbreviated t-RNA. { 'tranz·fər ¦rī·bō·nü¦klē·ik 'as·əd }

transferrin [BIOCHEMISTRY] Any of various beta globulins in blood serum which bind and transport iron to the bone marrow and storage areas. { 'tranz'fer·ən }

transformation [GENETICS] **1.** Transfer and incorporation of foreign deoxyribonucleic acid (DNA) into a cell and subsequent recombination of part or all of that DNA into the cell's genome. Also known as bacterial transformation; genetic transformation. **2.** Conversion of a normal cell to a neoplastic cell by a cascade of events under the control of different classes of oncogenes. Also known as cellular transformation. [IMMUNOLOGY] Change in a lymphocyte from a small, resting lymphocyte into a large lymphocyte following stimulation by antigens or lectin, or viral infection. Also known as lymphocyte transformation. { ˌtranz·fər'mā·shən }

transforming principle [MICROBIOLOGY] Deoxyribonucleic acid which effects transformation in bacterial cells. { tranz'fȯrm·iŋ ˌprin·sə·pəl }

transgenic organism [GENETICS] An organism into which genetic material from another organism has been experimentally transferred, so that the host acquires the genetic traits of the transferred genes in its chromosomal composition. { tranz'jen·ik 'ȯr·gəˌniz·əm }

transition [MOLECULAR BIOLOGY] A mutation resulting from the substitution in deoxyribonucleic acid or ribonucleic acid of one purine or pyrimidine for another. { tran'zish·ən }

transitional epithelium [HISTOLOGY] A form of stratified epithelium found in the urinary bladder; cells vary between squamous, when the tissue is stretched, and columnar, when not stretched. { tran'zish·ən·əl ˌep·ə'thē·lē·əm }

transketolase [BIOCHEMISTRY] An enzyme that cleaves a substrate at the position of the carbonyl carbon and transports a two-carbon fragment to an acceptor compound to form a new compound. { ¦tranz'kēd·əˌlās }

translation [MOLECULAR BIOLOGY] The process by which the linear sequence of nucleotides in a molecule of messenger ribonucleic acid directs the specific linear sequence of amino acids, as during protein synthesis. { tran'slā·shən }

translocation [BOTANY] Movement of water, mineral salts, and organic substances from one part of a plant to another. [CELL BIOLOGY] The transfer of a chromosome segment from its usual position to a new position in the same or in a different chromosome. { ¦tranz·lō'kā·shən }

transmethylase [BIOCHEMISTRY] A transferase enzyme involved in catalyzing chemical reactions in which methyl groups are transferred from a substrate to a new compound. { ¦tranz'meth·ə ˌlās }

transmethylation [BIOCHEMISTRY] A metabolic reaction in which a methyl group is transferred from one compound to another; methionine and choline are important donors of methyl groups. { ¦tranzˌmeth·ə'lā·shən }

transpiration [BIOLOGY] The passage of a gas or liquid (in the form of vapor) through the skin, a membrane, or other tissue. { ˌtranz·pə'rā·shən }

transplantation [BIOLOGY] **1.** The artificial removal of part of an organism and its replacement in the body of the same or of a different individual. **2.** To remove a plant from one location and replant it in another place. { ˌtranz·plan'tā·shən }

transplantation antigen [IMMUNOLOGY] An antigen in a cell which induces a histocompatibility reaction when the cell is transplanted into an organism not having that antigen. { ˌtranz·plan 'tā·shən 'ant·i·jən }

transposon [GENETICS] A kind of translocatable genetic element which comprises large discrete segments of deoxyribonucleic acid that are capable of moving from one chromosomal site to another in the same organism or in a different organism. { ¦tranz'pōˌzän }

transverse colon [ANATOMY] The portion of the colon between the right and left colic flexures, extending transversely across the upper abdomen. { trans¦vərs 'kō·lən }

transversion [MOLECULAR BIOLOGY] A mutation resulting from the substitution in deoxyribonucleic acid or ribonucleic acid of a purine for a pyrimidine or a pyrimidine for a purine. { trans 'vər·zhən }

trapeziform [BIOLOGY] Having the form of a trapezium. { trə'pē·zəˌfȯrm }

traumatotropism [BIOLOGY] Orientation response of an organ of a sessile organism in response to a wound. { ˌtraů·mə'tä·trəˌpiz·əm }

Trebidae [INVERTEBRATE ZOOLOGY] A family of copepod crustaceans of the order Caligoida which are external parasites on selachians. { 'treb·əˌdē }

tree [BOTANY] A perennial woody plant at least 20 feet (6 meters) in height at maturity, having an erect stem or trunk and a well-developed crown or leaf canopy. { trē }

tree fern [BOTANY] The common name for plants belonging to the families Cyatheaceae and Dicksoniaceae; all are ferns that exhibit an arborescent habit. { 'trē ˌfərn }

tree frog [VERTEBRATE ZOOLOGY] Any of the arboreal frogs comprising the family Hylidae characterized by expanded digital adhesive disks. { 'trē ˌfräg }

tree line [ECOLOGY] *See* timberline. { 'trē ˌlīn }

Trematoda [INVERTEBRATE ZOOLOGY] A loose grouping of acoelomate, parasitic flatworms of the phylum Platyhelminthes; they exhibit cephalization, bilateral symmetry, and well-developed holdfast structures. { trem·ə'tōd·ə }

Trematosauria [PALEONTOLOGY] A group of Triassic amphibians in the order Temnospondyli. { ˌtrem·əd·ə'sȯr·ē·ə }

Tremellales [MYCOLOGY] An order of basidiomycetous fungi in the subclass Heterobasidiomycetidae in which basidia have longitudinal walls. { ˌtrem·ə'lā·lēz }

Trentepohliaceae [BOTANY] A family of green algae belonging to the Ulotrichales having thick walls, bandlike or reticulate chloroplasts, and zoospores or isogametes produced in enlarged, specialized cells. { ˌtren·təˌpō·lē'ās·ēˌē }

Treponemataceae [MICROBIOLOGY] Formerly a family of the bacterial order Spirochaetales including the spirochetes less than 20 micrometers long and less than 5 micrometers in diameter; most species are parasitic. { ˌtrep·əˌnē·mə'tās·ēˌē }

Trepostomata [PALEONTOLOGY] An extinct order of ectoproct bryozoans in the class Stenolaemata characterized by delicate to massive colonies composed of tightly packed zooecia with solid calcareous zooecial walls. { ˌtrep·ə'stō·məd·ə }

Treroninae [VERTEBRATE ZOOLOGY] The fruit pigeons, a subfamily of the avian family Columbidae distinguished by the gaudy coloration of the feathers. { trə'rän·əˌnē }

Tretothoracidae [INVERTEBRATE ZOOLOGY] A family of the Coleoptera in the superfamily Tenebrionoidea which contains a single species found in Queensland, Australia. { ˌtred·ə·thə'ras·əˌdē }

triacetyloleandomycin [MICROBIOLOGY] An antibiotic produced by *Streptomyces antibioticus* and used clinically in the treatment of pneumonia, osteomyelitis, furuncles, and carbuncles. { trī·ə¦sēd·əlˌō·lēˌan·də'mīs·ən }

triaene [INVERTEBRATE ZOOLOGY] An elongated spicule in certain Porifera with three rays diverging from one end. { 'trīˌēn }

Triakidae [VERTEBRATE ZOOLOGY] A family of galeoid sharks in the carcharinid line. { trī'ak·əˌdē }

triandrous [BOTANY] Possessing three stamens. { trī'an·drəs }

triangular ligament [ANATOMY] *See* urogenital diaphragm. { trī'aŋ·gyə·lər 'lig·ə·mənt }

Triatominae [INVERTEBRATE ZOOLOGY] The kissing bugs, a subfamily of hemipteran insects in the family Reduviidae, distinguished by a long, slender rostrum. { trī·ə'täm·əˌnē }

triaxon [INVERTEBRATE ZOOLOGY] A spicule in Porifera having three axes which cross each other at right angles. { trī'akˌsän }

tricarboxylic acid cycle [BIOCHEMISTRY] *See* Krebs cycle. { trīˌkär·bäk'sil·ik 'as·əd ˌsī·kəl }

Trichechidae [VERTEBRATE ZOOLOGY] The manatees, a family of nocturnal, solitary sirenian mammals in the suborder Trichechiformes. { trə'kek·əˌdē }

Trichechiformes [VERTEBRATE ZOOLOGY] A suborder of mammals in the order Sirenia which contains the manatees and dugongids. { trəˌkek·ə'fȯrˌmēz }

trichesthesia [PHYSIOLOGY] A form of tactile sensibility in hair-covered regions of the body. { ˌtrik·əs'thēzh·ə }

Trichiaceae [MYCOLOGY] A family of slime molds in the order Trichiales. { ˌtrik·ē'ās·ēˌē }

Trichiales [MYCOLOGY] An order of Myxomycetes in the subclass Myxogastromycetidae. { ˌtrik·ē'ā·lēz }

Trichiuridae [VERTEBRATE ZOOLOGY] The cutlassfishes, a family of the suborder Scombroidei. { ˌtrik·ē'yu̇r·əˌdē }

trichobothrium [INVERTEBRATE ZOOLOGY] An erect, bristlelike sensory hair found on certain arthropods, insects, and other invertebrates. { ¦trik·ə'bäth·rē·əm }

trichobranchiate gill [INVERTEBRATE ZOOLOGY] A gill with filamentous branches arranged in several series around the axis; found in some decapod crustaceans. { ¦trik·ə'braŋ·kēˌāt ¦gil }

Trichobranchidae [INVERTEBRATE ZOOLOGY] A family of polychaete annelids belonging to the Sedentaria; most members are rare and live at great ocean depths. { ˌtrik·ə'braŋ·kəˌdē }

trichocercous cercaria [INVERTEBRATE ZOOLOGY] A trematode larva distinguished by a spiny tail. { ¦trik·ə¦sər·kəs sər'kar·ē·ə }

Trichocomaceae [MYCOLOGY] A small tropical family of ascomycetous fungi in the order Eurotiales with ascocarps from which a tuft of capillitial threads extrudes, releasing the ascospores after dissolution of the asci. { ˌtrik·ə·kə'mās·ēˌē }

trichocyst [INVERTEBRATE ZOOLOGY] A minute structure in the cortex of certain protozoans that releases filamentous or fibrillar threads when discharged. { 'trik·əˌsist }

Trichodactylidae [INVERTEBRATE ZOOLOGY] A family of fresh-water crabs in the section Brachyura, found mainly in tropical regions. { ˌtrik·ə·dak'til·əˌdē }

Trichogrammatidae [INVERTEBRATE ZOOLOGY] A family of the Hymenoptera in the superfamily Chalcidoidea whose larvae are parasitic in the eggs of other insects. { ˌtrik·ə·grə'mad·əˌdē }

trichogyne [BOTANY] A terminal portion of a procarp or archicarp which receives a spermatium. { 'trik·əˌjīn }

trichome [BOTANY] An appendage derived from the protoderm in plants, including hairs and scales. [INVERTEBRATE ZOOLOGY] A brightly colored tuft of hairs on the body of a myrmecophile that releases an aromatic substance attractive to ants. { 'trīˌkōm }

Trichomonadida [INVERTEBRATE ZOOLOGY] An order of the protozoan class Zoomastigophorea which contains four families of uninucleate species. { ˌtrik·ə·mə'näd·əd·ə }

Trichomonadidae [INVERTEBRATE ZOOLOGY] A family of flagellate protozoans in the order Trichomonadida. { ˌtrik·ə·mə'näd·əˌdē }

Trichomycetes [MYCOLOGY] A class of true fungi, division Fungi. { ¦trik·əˌmī'sēd·ēs }

trichomycin [MICROBIOLOGY] An antibiotic produced by *Streptomyces hachijoensis* and *S. abikoensis*; a water-soluble yellow powder that inhibits yeasts and fungi. { ˌtrik·ə'mīs·ən }

Trichoniscidae [INVERTEBRATE ZOOLOGY] A primitive family of isopod crustaceans in the suborder Oniscoidea found in damp littoral, halophilic, or riparian habitats. { ˌtrik·ə'nis·əˌdē }

Trichophilopteridae [INVERTEBRATE ZOOLOGY] A family of lice in the order Mallophaga adapted to life upon the lemurs of Madagascar. { ˌtrik·ə ˌfil·əp'ter·əˌdē }

trichophytin [IMMUNOLOGY] A group antigen obtained from filtrates of *Trichophyton mentagrophytes*; used in a skin test to ascertain past or present infection with the dermatophytes. { ˌtrik·ə'fīt·ən }

Trichoptera [INVERTEBRATE ZOOLOGY] The caddis flies, an aquatic order of the class Insecta; larvae are wormlike and adults have two pairs of well-veined hairy wings, long antennae, and mouthparts capable of lapping only liquids. { trə'käp·tə·rə }

Trichopterygidae [INVERTEBRATE ZOOLOGY] The equivalent name for Ptiliidae. { trəˌkäp·tə'rij·ə ˌdē }

Trichostomatida [INVERTEBRATE ZOOLOGY] An order of ciliated protozoans in the subclass Holotrichia in which no true buccal ciliature is present but there is a vestibulum. { ˌtrik·ə·stō 'mad·əd·ə }

Trichostrongylidae [INVERTEBRATE ZOOLOGY] A family of parasitic roundworms belonging to the Strongyloidea; hosts are cattle, sheep, goats, swine, and cats. { ˌtrik·ə·strän'jil·əˌdē }

Trichuroidea [INVERTEBRATE ZOOLOGY] A group of nematodes parasitic in various vertebrates and characterized by a slender body sometimes having a thickened posterior portion. { ˌtrik·yə 'ròid·ē·ə }

Tricladida [INVERTEBRATE ZOOLOGY] The planarians, an order of the Turbellaria distinguished by diverticulated intestines with a single anterior branch and two posterior branches separated by the pharynx. { trī'klad·əd·ə }

triconodont [VERTEBRATE ZOOLOGY] **1.** A tooth with three main conical cusps. **2.** Having such teeth. { trī'kän·əˌdänt }

Triconodonta [PALEONTOLOGY] An extinct mammalian order of small flesh-eating creatures of the Mesozoic era having no angle or a pseudoangle on the lower jaw and triconodont molars. { trīˌkän·ə'dänt·ə }

Trictenotomidae [INVERTEBRATE ZOOLOGY] A small family of Indian and Malaysian beetles in the superfamily Tenebrionoidea. { ˌtrik·tə·nō 'täm·əˌdē }

tricuspid valve [ANATOMY] A valve consisting of three flaps located between the right atrium and right ventricle of the heart. { trī'kəs·pəd 'valv }

Tridacnidae [INVERTEBRATE ZOOLOGY] A family of bivalve mollusks in the subclass Eulamellibranchia which contains the giant clams of the tropical Pacific. { trī'dak·nəˌdē }

Tridactylidae [INVERTEBRATE ZOOLOGY] The pygmy mole crickets, a family of insects in the order Orthoptera, highly specialized for fossorial existence. { ˌtrī·dak'til·əˌdē }

trifid [BIOLOGY] Divided into three lobes separated by narrow sinuses partway to the base. { 'trīˌfid }

trifoliate [BOTANY] Having three leaves or leaflets. { trī'fō·lēˌāt }

trigeminal nerve [ANATOMY] The fifth cranial nerve in vertebrates; either of a pair of composite nerves rising from the side of the medulla, and with three great branches: the ophthalmic, maxillary, and mandibular nerves. { trī'jem·ə·nəl 'nərv }

Triglidae [VERTEBRATE ZOOLOGY] The searobins, a family of perciform fishes in the suborder Cottoidei. { 'trig·ləˌdē }

Trigonalidae [INVERTEBRATE ZOOLOGY] A small family of hymenopteran insects in the superfamily Proctotrupoidea. { ˌtrig·ə'nal·əˌdē }

trigone [ANATOMY] A triangular area inside the bladder limited by the openings of the ureters and urethra. [BOTANY] A thickening of plant cell walls formed when three or more cells adjoin. { 'trīˌgōn }

Trigonostylopoidea [PALEONTOLOGY] A suborder of Paleocene-Eocene ungulate mammals in the order Astrapotheria. { ˌtrig·ə·nōˌstil·ə 'pòid·ē·ə }

trigonous [BIOLOGY] **1.** Having three corners. **2.** Having a triangular cross section. { 'trig·ə·nəs }

Trilobita [PALEONTOLOGY] The trilobites, a class of extinct Cambrian-Permian arthropods characterized by an exoskeleton covering the dorsal surface, delicate biramous appendages, body segments divided by furrows on the dorsal surface, and a pygidium composed of fused segments. { ˌtrī·lə'bīd·ə }

Trilobitoidea [PALEONTOLOGY] A class of Cambrian arthropods that are closely related to the Trilobita. { ˌtrī·lō·bə'tòid·ē·ə }

Trilobitomorpha [INVERTEBRATE ZOOLOGY] A subphylum of the Arthropoda including Trilobita. { trīˌläb·əd·ə'mòr·fə }

trilocular [BIOLOGY] Having three cavities or cells. { trī'äk·yə·lər }

Trimenoponidae [INVERTEBRATE ZOOLOGY] A family of lice in the order Mallophaga occurring as parasites on South American rodents. { ˌtrī·mə·nə'pän·əˌdē }

Trimerellacea [PALEONTOLOGY] A superfamily of extinct inarticulate brachiopods in the order Lingulida; they have valves, usually consisting of calcium carbonate. { trəˌmer·ə'lās·ē·ə }

Trimerophytopsida [PALEOBOTANY] A group of extinct land vascular plants with leafless, dichotomously branched stems that bear terminal sporangia. { trəˌmer·ə·fə'täp·səd·ə }

trimerous [BOTANY] Having parts in sets of three. [INVERTEBRATE ZOOLOGY] In insects, having the tarsus divided or apparently divided into three segments. { 'trim·ə·rəs }

trimethylvinylammonium hydroxide [BIOCHEMISTRY] *See* neurine. { trī¦meth·əl,vīn·əl·ə'mō·nē·əm hī'dräk,sīd }

trimorphous [BIOLOGY] Characterized by occurring in three distinct forms, as an organ or whole organism. { trī'mȯr·fəs }

trinomial [SYSTEMATICS] A nomenclatural designation for an organism composed of three terms: genus, species, and subspecies or variety. { trī 'nō·mē·əl }

trinomial nomenclature [SYSTEMATICS] The designation of subspecies by a three-word name. { trī'nō·mē·əl 'nō·mən,klā·chər }

Trionychidae [VERTEBRATE ZOOLOGY] The soft-shelled turtles, a family of reptiles in the order Chelonia. { trī·ə'nik·ə,dē }

triose [BIOCHEMISTRY] A group of monsaccharide compounds that have a three-carbon chain length. { 'trī,ōs }

triphosphopyridine dinucleotide [BIOCHEMISTRY] *See* triphosphopyridine nucleotide. { trī¦fä·sfō 'pir·ə,dēn dī'nü·klē·ə,tīd }

triphosphopyridine nucleotide [BIOCHEMISTRY] $C_{12}H_{28}N_7O_{17}P_3$ A grayish-white powder, soluble in methanol and in water; a coenzyme and an important component of enzymatic systems concerned with biological oxidation-reduction systems. Abbreviated TPN. Also known as codehydrogenase II; coenzyme II; triphosphopyridine dinucleotide. { trī¦fä·sfō'pir·ə,dēn 'nü·klē·ə,tīd }

tripinnate [BIOLOGY] Being bipinnate and having each division pinnate. { trī'pi,nāt }

triple response [PHYSIOLOGY] The three stages of vasomotor reaction consisting of reddening, flushing of adjacent skin, and development of wheals, when a pointed instrument is drawn heavily across the skin. { 'trip·əl ri'späns }

triplet [GENETICS] A three-base unit in deoxyribonucleic or ribonucleic acid which codes for a particular amino acid in a protein chain. { 'trip·lət }

Triploblastica [ZOOLOGY] Animals that develop from three germ layers. { ,trip·ō'blas·tə·kə }

triploidy [CELL BIOLOGY] The occurrence of related forms possessing chromosome numbers three times the haploid number. { 'tri,plȯid·ē }

Tripylina [INVERTEBRATE ZOOLOGY] A subdivision of the protozoan order Oculosida in which the major opening (astropyle) usually contains a perforated plate. { ,trip·ə'lī·nə }

trisaccate pollen [BOTANY] A three-pored pollen grain, often having a triangular outline in cross section. { trī'sa,kāt 'päl·ən }

trisaccharide [BIOCHEMISTRY] A carbohydrate which, on hydrolysis, yields three molecules of monosaccharides. { trī'sak·ə,rīd }

trisomy [CELL BIOLOGY] The presence in triplicate of one of the chromosomes of the complement. { 'trī,sō·mē }

tritonymph [INVERTEBRATE ZOOLOGY] The third stage of development in certain acarids. { 'trīd·ō,nimf }

Triuridaceae [BOTANY] A family of monocotyledonous plants in the order Triuridales distinguished by unisexual flowers and several carpels with one seed per carpel. { trī,yu̇r·ə'dās·ē,ē }

Triuridales [BOTANY] A small order of terrestrial, mycotrophic monocots in the subclass Alismatidae without chlorophyll, and with separate carpels, trinucleate pollen, and a well-developed endosperm. { trī,yu̇r·ə'dā·lēz }

trivium [INVERTEBRATE ZOOLOGY] The three rays opposite the madreporite in starfish. { 'triv·ē·əm }

t-RNA [MOLECULAR BIOLOGY] *See* transfer ribonucleic acid.

Trochacea [PALEONTOLOGY] A recent subfamily of primitive gastropod mollusks in the order Aspidobranchia. { trō'kāsh·ē·ə }

trochal disk [INVERTEBRATE ZOOLOGY] A flat or funnel-shaped ciliated disk at the anterior end of a rotifer that functions in locomotion and food ingestion. { 'trō·kəl ,disk }

trochanter [ANATOMY] A process on the proximal end of the femur in many vertebrates, which serves for muscle attachment and, in birds, for articulation with the ilium. [INVERTEBRATE ZOOLOGY] The second segment of an insect leg, counting from the base. { trō'kan·tər }

Trochidae [INVERTEBRATE ZOOLOGY] A family of gastropod mollusks in the order Aspidobranchia, including many of the top shells. { 'träk·ə,dē }

Trochili [VERTEBRATE ZOOLOGY] A suborder of the avian order Apodiformes. { 'träk·ə,lī }

Trochilidae [VERTEBRATE ZOOLOGY] The hummingbirds, a tropical New World family of the suborder Trochili with tubular tongues modified for nectar feeding; slender bills and the ability to hover are further feeding adaptations. { trä'kil·ə ,dē }

Trochiliscales [PALEOBOTANY] A group of extinct plants belonging to the Charophyta in which the gyrogonites are dextrally spiraled. { trə,kil·ə 'skā·lēz }

trochlea [ANATOMY] A pulleylike anatomical structure. { 'träk·lē·ə }

trochlear nerve [ANATOMY] The fourth cranial nerve; either of a pair of somatic motor nerves which innervate the superior oblique muscle of the eyeball. { 'träk·lē·ər ,nərv }

trochoblast [INVERTEBRATE ZOOLOGY] A cell bearing cilia on a trochophore. { 'träk·ə,blast }

Trochodendraceae [BOTANY] A family of dicotyledonous trees in the order Trochodendrales distinguished by the absence of a perianth and stipules, numerous stamens, and pinnately veined leaves. { ,träk·ō·den'drās·ē,ē }

Trochodendrales [BOTANY] An order of dicotyledonous trees in the subclass Hamamelidae characterized by primitively vesselless wood and unique, elongate, often branched idioblasts in the leaves. { ,träk·ō·den'drā·lēz }

trochoid [ANATOMY] *See* pivot joint. { 'trō,kȯid }

trochophore [INVERTEBRATE ZOOLOGY] A generalized but distinct free-swimming larva found in several invertebrate groups, having a pear-shaped form with an external circlet of cilia, apical ciliary tufts, a complete functional digestive tract, and paired nephridia with excretory tubules. Also known as trochosphere. { 'träk·ə ˌfȯr }

trochosphere [INVERTEBRATE ZOOLOGY] *See* trochophore. { 'träk·əˌsfir }

trochus [INVERTEBRATE ZOOLOGY] The inner band of cilia on a trochal disk. { 'trō·kəs }

Troglodytidae [VERTEBRATE ZOOLOGY] The wrens, a family of songbirds in the order Passeriformes. { ˌträg·lə'did·əˌdē }

Trogonidae [VERTEBRATE ZOOLOGY] The trogons, the single, pantropical family of the avian order Trogoniformes. { trō'gän·əˌdē }

Trogoniformes [VERTEBRATE ZOOLOGY] An order of brightly colored, slow-moving birds characterized by a unique foot structure with the first and second toes directed backward. { trōˌgän·ə'fȯr ˌmēz }

Troland and Fletcher theories [PHYSIOLOGY] Theories of hearing according to which the time nature of a sound stimulation affects the sensation of pitch. { 'trō·lənd ən 'flech·ər ˌthē·ə·rēz }

Trombiculidae [INVERTEBRATE ZOOLOGY] The chiggers, or red bugs, a family of mites in the suborder Trombidiformes whose larvae are parasites of vertebrates. { ˌträm·bə'kyül·əˌdē }

Trombidiformes [INVERTEBRATE ZOOLOGY] The trombidiform mites, a suborder of the Acarina distinguished by the presence of a respiratory system opening at or near the base of the chelicerae. { ˌträm·bə·də'fȯrˌmēz }

Tropaeolaceae [BOTANY] A family of dicotyledonous plants in the order Geraniales characterized by strongly irregular flowers, simple peltate leaves, eight stamens, and schizocarpous fruit. { ˌtrō·pē·ō'lās·ēˌē }

trophallaxis [ECOLOGY] Exchange of food between organisms, not only of the same species but between different species, especially among social insects. { ˌträf·ə'lak·səs }

trophic [BIOLOGY] Pertaining to or functioning in nutrition. { 'träf·ik }

trophic level [ECOLOGY] Any of the feeding levels through which the passage of energy through an ecosystem proceeds; examples are photosynthetic plants, herbivorous animals, and microorganisms of decay. { 'träf·ik ˌlev·əl }

trophobiosis [ECOLOGY] A nutritional relationship associated only with certain species of ants in which alien insects supply food to the ants and are milked by the ants for their secretions. { ˌträf·ōˌbī'ō·səs }

trophoblast [EMBRYOLOGY] A layer of ectodermal epithelium covering the outer surface of the chorion and chorionic villi of many mammals. { 'träf·əˌblast }

trophocyte [INVERTEBRATE ZOOLOGY] A nutritive cell of the ovary or testis of an insect. { 'träf·ə ˌsīt }

trophogenic [ECOLOGY] Originating from nutritional differences rather than resulting from genetic determinants, such as various castes of social insects. { ¦träf·ə¦jen·ik }

tropholytic [ECOLOGY] Pertaining to the deep zone in a lake where dissimilation of organic matter predominates. { ¦traf·ə¦lid·ik }

trophosome [INVERTEBRATE ZOOLOGY] The nutritional zooids of a hydroid colony. { 'träf·ə ˌsōm }

trophotaeniae [VERTEBRATE ZOOLOGY] Vascular rectal processes which establish placental relationships with the ovarian tissue in live-bearing fishes. { ˌträf·ō'tē·nēˌī }

trophozoite [INVERTEBRATE ZOOLOGY] A vegetative protozoan; used especially of a parasite. { ˌträf·ə'zōˌīt }

trophus [INVERTEBRATE ZOOLOGY] Masticatory apparatus in Rotifera. { 'träf·əs }

tropical life zone [ECOLOGY] A subdivision of the eastern division of Merriam's life zones; an example is southern Florida, where the vegetation is the broadleaf evergreen forest; typical and important plants are palms and mangroves; typical and important animals are the armadillo and alligator; typical and important crops are citrus fruits, avocado, and banana. { 'träp·ə·kəl 'līf ˌzōn }

tropical rainforest [ECOLOGY] A vegetation class consisting of tall, close-growing trees, their columnar trunks more or less unbranched in the lower two-thirds, and forming a spreading and frequently flat crown; occurs in areas of high temperature and high rainfall. Also known as hylaea; selva. { 'träp·ə·kəl 'rānˌfär·əst }

tropical savanna [ECOLOGY] *See* tropical woodland. { 'träp·ə·kəlsə'van·ə }

tropical scrub [ECOLOGY] A class of vegetation composed of low woody plants (shrubs), sometimes growing quite close together, but more often separated by large patches of bare ground, with clumps of herbs scattered throughout; an example is the Ghanaian evergreen coastal thicket. Also known as brush; bush; fourré; mallee; thicket. { 'träp·ə·kəl 'skrəb }

tropical woodland [ECOLOGY] A vegetation class similar to a forest but with wider spacing between trees and sparse lower strata characterized by evergreen shrubs and seasonal graminoids; the climate is warm and moist. Also known as parkland; savanna-woodland; tropical savanna. { 'träp·ə·kəl 'wu̇d·lənd }

Tropiometridae [INVERTEBRATE ZOOLOGY] A family of feather stars in the class Crinoidea which are bottom crawlers. { ˌträp·ē·ō'me·trəˌdē }

tropism [BIOLOGY] Orientation movement of a sessile organism in response to a stimulus. Also known as topotaxis. { 'trōˌpiz·əm }

tropocollagen [BIOCHEMISTRY] The fundamental units of collagen fibrils. { ¦träp·ō'kal·ə·jən }

tropomyosin [BIOCHEMISTRY] A muscle protein similar to myosin and implicated as being part of the structure of the Z bands separating sarcomeres from each other. { ˌträp·ə'mī·ə·sən }

troponin [BIOCHEMISTRY] A protein species located at specific stations every 36.5 nanometers on the actin helix in muscle sarcomere. { 'trō·pə·nən }

tropophytia [BOTANY] Plants that thrive in a climate that undergoes marked periodic changes. { ˌträp·ə'fī·shə }

trout [VERTEBRATE ZOOLOGY] Any of various edible fresh-water fishes in the order Salmoniformes that are generally much smaller than the salmon. { 'tràut }

Trucherognathidae [PALEONTOLOGY] A family of conodonts in the order Conodontophorida in which the attachment scar permits the conodont to rest on the jaw ramus. { ˌtrü·chə·räg'nath·əˌdē }

truffle [BOTANY] The edible underground fruiting body of various European fungi in the family Tuberaceae, especially the genus *Tuber*. { 'trəf·əl }

trumpeter [VERTEBRATE ZOOLOGY] A bird belonging to the Psophiidae, a family with three South American species; the common trumpeter (*Psophia crepitans*) is the size of a pheasant and resembles a long-legged guinea fowl. { 'trəm·pəd·ər }

truncate [BIOLOGY] Abbreviated at an end, as if cut off. { 'trəŋˌkāt }

truncus arteriosis [EMBRYOLOGY] The embryonic arterial trunk between the bulbous arteriosis and the ventral aorta in anamniotes and early stages of amniotes. { ¦trəŋ·kəs ärˌtir·ē'ō·səs }

trunk [ANATOMY] The main mass of the human body, exclusive of the head, neck, and extremities; it is divided into thorax, abdomen, and pelvis. [BOTANY] The main stem of a tree. { trəŋk }

Tryblidiidae [PALEONTOLOGY] An extinct family of Paleozoic mollusks. { ˌtrib·lə'dī·əˌdē }

Trypanorhyncha [INVERTEBRATE ZOOLOGY] An order of tapeworms of the subclass Cestoda; all are parasites in the intestine of elasmobranch fishes. { trəˌpan·ə'riŋ·kə }

Trypanosomatidae [INVERTEBRATE ZOOLOGY] A family of Protozoa, order Kinetoplastida, containing flagellated parasites which exhibit polymorphism during their life cycle. { trəˌpan·ə·sō'mad·əˌdē }

trypanosome [INVERTEBRATE ZOOLOGY] A flagellated protozoan of the genus *Trypanosoma*. { trə'pan·əˌsōm }

trypsin [BIOCHEMISTRY] A proteolytic enzyme which catalyzes the hydrolysis of peptide linkages in proteins and partially hydrolyzed proteins; derived from trypsinogen by the action of enterokinase in intestinal juice. { 'trip·sən }

trypsinogen [BIOCHEMISTRY] The zymogen of trypsin, secreted in the pancreatic juice. Also known as protrypsin. { trip'sin·ə·jən }

tryptophan [BIOCHEMISTRY] $C_{11}H_{12}O_2N_2$ An amino acid obtained from casein, fibrin, and certain other proteins; it is a precursor of indoleacetic acid, serotonin, and nicotinic acid. { 'trip·təˌfan }

tsetse fly [INVERTEBRATE ZOOLOGY] Any of various South African muscoid flies of the genus *Glossina*; medically important as vectors of sleeping sickness or trypanosomiasis. { 'setˌsē ˌflī }

TSH [BIOCHEMISTRY] *See* thyrotropic hormone.

tubal bladder [VERTEBRATE ZOOLOGY] A urine reservoir organ that is an enlargement of the mesonephric ducts in most fish; there are four types: duplex, bilobed, simplex with ureters tied, and simplex with separate ureters. { 'tü·bəl 'blad·ər }

tube [BIOLOGY] A narrow channel within the body of an animal or plant. { 'tüb }

tube cell [BOTANY] That nucleus of a pollen grain believed to influence the growth and development of the pollen tube. Also known as tube nucleus. { 'tüb ˌsel }

tube foot [INVERTEBRATE ZOOLOGY] One of the tentaclelike outpushings of the radial vessels of the water-vascular system in echinoderms; may be suctorial, or serve as stiltlike limbs or tentacles. { 'tüb ˌfùt }

tube nucleus [BOTANY] *See* tube cell. { 'tüb 'nü·klē·əs }

tuber [BOTANY] The enlarged end of a rhizome in which food accumulates, as in the potato. { 'tü·bər }

tuber cinereum [ANATOMY] An area of gray matter extending from the optic chiasma to the mammillary bodies and forming part of the floor of the third ventricle. { 'tü·bər si'ner·ē·əm }

tubercle [BIOLOGY] A small knoblike prominence. { 'tü·bər·kəl }

Tuberculariaceae [MYCOLOGY] A family of fungi of the order Moniliales having short conidia that form cushion-shaped, often waxy or gelatinous aggregates (sporodochia). { təˌbər·kyəˌla·rē'ās·ēˌē }

tuberculate [BIOLOGY] Having or characterized by knoblike processes. { tə'bər·kyə·lət }

tuberculin [IMMUNOLOGY] A preparation containing tuberculoproteins derived from *Mycobacterium tuberculosis* used in the tuberculin test to determine sensitization to tubercle bacilli. { tə'bər·kyə·lən }

tuberculin test [IMMUNOLOGY] A test for past or present infection with tubercle bacilli based on a delayed hypersensitivity reaction at the site where tuberculin or purified protein derivative was introduced. { tə'bər·kyə·lən ˌtest }

tuberosity [ANATOMY] A large or obtuse prominence, especially as on bone for muscle attachment. { ˌtü·bə'räs·əd·ē }

tuberous organ [PHYSIOLOGY] An electroreceptor most sensitive to high-frequency electric signals, and distributed over the body surface of electric fish. { 'tü·bə·rəs 'ȯr·gən }

Tubicola [INVERTEBRATE ZOOLOGY] An order of sedentary polychaete annelids that surround themselves with a calcareous tube or one which is composed of agglutinated foreign particles. { tü'bik·ə·lə }

Tubulanidae [INVERTEBRATE ZOOLOGY] A family of the order Palaeonemertini. { ˌtüb·yə'lan·əˌdē }

tubular gland [ANATOMY] A secreting structure

whose secretory endpieces are tubelike or cylindrical in shape. { 'tü·byə·lər 'gland }

tubule [ANATOMY] A slender, elongated microscopic tube in an anatomical structure. { 'tü ,byül }

Tubulidentata [VERTEBRATE ZOOLOGY] An order of mammals which contains a single living genus, the aardvark (*Orycteropus*) of Africa. { ,tü·byə·lə·den'täd·ə }

tubulin [BIOCHEMISTRY] A globular protein containing two subunits; 10-14 molecules are arranged to form a microtubule. { 'tü·byə·lən }

tubuloacinous gland [ANATOMY] *See* tubuloalveolar gland. { ¦tü·byə·lō'as·ə·nəs ,gland }

tubuloalveolar gland [ANATOMY] A secreting structure having both tubular and alveolar secretory endpieces. Also known as acinotubular gland; tubuloacinous gland. { ¦tü·byə·lō·al'vē·ə·lər ,gland }

tulip [BOTANY] Any of various plants with showy flowers constituting the genus *Tulipa* in the family Liliaceae; characterized by coated bulbs, lanceolate leaves, and a single flower with six equal perianth segments and six stamens. { 'tü·ləp }

tulip poplar [BOTANY] *See* tulip tree. { 'tü·ləp 'päp·lər }

tulip tree [BOTANY] *Liriodendron tulipifera*. A tree belonging to the magnolia family (Magnoliaceae) distinguished by leaves which are squarish at the tip, true terminal buds, cone-shaped fruit, and large greenish-yellow and orange-colored flowers. Also known as tulip poplar. { 'tü·ləp ,trē }

tumbleweed [BOTANY] Any of various plants that break loose from their roots in autumn and are driven by the wind in rolling masses over the ground. { 'təm·bəl,wēd }

tumid [BIOLOGY] Marked by swelling or inflation. { 'tü·məd }

tumor necrosis factor [IMMUNOLOGY] A monokine that induces leukocytosis, fever, weight loss, the acute-phase reaction, and necrosis of some tumors. { ¦tü·mər nə'krō·səs ,fak·tər }

tuna [VERTEBRATE ZOOLOGY] Any of the large, pelagic, cosmopolitan marine fishes which form the family Thunnidae including species that rank among the most valuable of food and game fish. { 'tü·nə }

tundra [ECOLOGY] An area supporting some vegetation between the northern upper limit of trees and the lower limit of perennial snow on mountains, and on the fringes of the Antarctic continent and its neighboring islands. Also known as cold desert. { 'tən·drə }

tung nut [BOTANY] The seed of the tung tree (*Aleurites fordii*), which is the source of tung oil. { 'təŋ ,nət }

tung tree [BOTANY] *Aleurites fordii*. A plant of the spurge family in the order Euphorbiales, native to central and western China and grown in the southern United States. { 'təŋ ,trē }

tunica [BIOLOGY] A membrane or layer of tissue that covers or envelops an organ or other anatomical structure. { 'tü·nə·kə }

tunica adventitia [ANATOMY] *See* adventitia. { 'tü·nə·kə ,ad·vən'tish·ə }

tunica intima [HISTOLOGY] *See* intima. { 'tü·nə·kə 'in·tə·mə }

tunica mucosa [HISTOLOGY] *See* mucous membrane. { 'tü·nə·kə myü'kō·zə }

Tunicata [INVERTEBRATE ZOOLOGY] A subphylum of the Chordata characterized by restriction of the notochord to the tail and posterior body of the larva, absence of mesodermal segmentation, and secretion of an outer covering or tunic about the body. { ,tü·nə'käd·ə }

Tupaiidae [VERTEBRATE ZOOLOGY] The tree shrews, a family of mammals in the order Insectivora. { tü'pī·ə,dē }

tupelo [BOTANY] Any of various trees belonging to the genus *Nyssa* of the sour gum family, Nyssaceae, distinguished by small, obovate, shiny leaves, a small blue-black drupaceous fruit, and branches growing at a wide angle from the axis. { 'tü·pə,lō }

Turbellaria [INVERTEBRATE ZOOLOGY] A class of the phylum Platyhelminthes having bodies that are elongate and flat to oval or circular in cross section. { ,tər·bə'lar·ē·ə }

turbidostat [MICROBIOLOGY] A device in which a bacterial culture is maintained at a constant volume and cell density (turbidity) by adjusting the flow rate of fresh medium into the growth tube by means of a photocell and appropriate electrical connections. { tər'bid·ə,stat }

turbinate [BOTANY] Shaped like an inverted cone. [INVERTEBRATE ZOOLOGY] Spiral with rapidly decreasing whorls from base to apex. { 'tər·bə·nət }

Turbinidae [INVERTEBRATE ZOOLOGY] A family of gastropod mollusks including species of top shells. { tər'bin·ə,dē }

Turdidae [VERTEBRATE ZOOLOGY] The thrushes, a family of passeriform birds in the suborder Oscines. { 'tər·də,dē }

turgor [BOTANY] Distension of a plant cell wall and membrane by the fluid contents. { 'tər·gər }

turgor movement [BOTANY] A reversible change in the position of plant parts due to a change in turgor pressure in certain specialized cells; movement of *Mimosa* leaves when touched is an example. { 'tər·gər ,müv·mənt }

turgor pressure [BOTANY] The actual pressure developed by the fluid content of a turgid plant cell. { 'tər·gər ,presh·ər }

turion [BOTANY] A scaly shoot, such as asparagus, developed from an underground bud. { 'tür·ē,än }

turkey [VERTEBRATE ZOOLOGY] Either of two species of wild birds, and any of various derived domestic breeds, in the family Meleagrididae characterized by a bare head and neck, and in the male a large pendant wattle which hangs on one side from the base of the bill. { 'tər·kē }

turmeric [BOTANY] *Curcuma longa*. An East Indian perennial of the ginger family (Zingiberaceae) with a short stem, tufted leaves, and short thick rhizomes; a spice with a pungent, bitter taste and a musky odor is derived from the rhizome. { 'tər·mər·ik }

Turnicidae [VERTEBRATE ZOOLOGY] The button quails, a family of Old World birds in the order Gruiformes. { tər'nis·ə,dē }

turnip [BOTANY] *Brassica rapa* or *B. campestris* var. *rapa*. An annual crucifer of Asiatic origin belonging to the family Brassiaceae in the order Capparales and grown for its foliage and edible root. { 'tər·nəp }

turnover [MOLECULAR BIOLOGY] The number of substrate molecules transformed by a single molecule of enzyme per minute, when the enzyme is operating at maximum rate. { 'tərn,ō·vər }

turnover number [BIOCHEMISTRY] The number of molecules of a substrate acted upon in a period of 1 minute by a single enzyme molecule, with the enzyme working at a maximum rate. { 'tərn·ō·vər ,nəm·bər }

turtle [VERTEBRATE ZOOLOGY] Any of about 240 species of reptiles which constitute the order Chelonia distinguished by the two bony shells enclosing the body. { 'tərd·əl }

tussock [ECOLOGY] A small hummock of generally solid ground in a bog or marsh, usually covered with and bound together by the roots of low vegetation such as grasses, sedges, or ericaceous shrubs. { 'təs·ək }

twin [BIOLOGY] One of two individuals born at the same time. { 'twin }

twiner [BOTANY] A climbing stem that winds about its support, as pole beans or many tropical lianas. { 'twī·nər }

Tylenchida [INVERTEBRATE ZOOLOGY] An order of soil-dwelling or phytoparasitic nematodes in the subclass Rhabdita. { tī'leŋ·kəd·ə }

Tylenchoidea [INVERTEBRATE ZOOLOGY] A superfamily of mainly soil and insect-associated nematodes in the order Tylenchida with a stylet for piercing live cells and sucking the juices. { tī·lən 'kȯid·ē·ə }

Tylopoda [VERTEBRATE ZOOLOGY] An infraorder of artiodactyls in the suborder Ruminantia that contains the camels and extinct related forms. { tī'läp·əd·ə }

tylose [BOTANY] A mass of parenchymal cells appearing somewhat frothlike in the pores of some hardwood trees. { 'tī,lōs }

tylostyle [INVERTEBRATE ZOOLOGY] A uniradiate spicule in Porifera with a point at one end and a knob at the other end. { 'tī·lə,stīl }

tylote [INVERTEBRATE ZOOLOGY] A slender sponge spicule with a knob at each end. { 'tī ,lōt }

tympanic cavity [ANATOMY] The irregular, air-containing, mucous-membrane-lined space of the middle ear; contains the three auditory ossicles and communicates with the nasopharynx through the auditory tube. { tim'pan·ik 'kav·əd·ē }

tympanic membrane [ANATOMY] The membrane separating the external from the middle ear. Also known as eardrum; tympanum. { tim'pan·ik 'mem,brān }

tympanum [ANATOMY] *See* tympanic membrane. [INVERTEBRATE ZOOLOGY] A thin membrane covering an organ of hearing in insects. { 'tim·pə·nəm }

type [SYSTEMATICS] A specimen on which a species or subspecies is based. { 'tīp }

Typhaceae [BOTANY] A family of monocotyledonous plants in the order Typhales characterized by an inflorescence of dense, cylindrical spikes and absence of a perianth. { tī'fās·ē,ē }

Typhales [BOTANY] An order of marsh or aquatic monocotyledons in the subclass Commelinidae with emergent or floating stems and leaves and reduced, unisexual flowers having a single ovule in an ovary composed of a single carpel. { tī'fā·lēz }

Typhlopidae [VERTEBRATE ZOOLOGY] A family of small, burrowing circumtropical snakes, suborder Serpentes, with vestigial eyes and toothless jaws. { tī'fläp·ə,dē }

Typhloscolecidae [INVERTEBRATE ZOOLOGY] A family of pelagic polychaete annelids belonging to the Errantia. { ,tif·lō·skō'les·ə,dē }

typhlosole [INVERTEBRATE ZOOLOGY] A dorsal longitudinal invagination of the intestinal wall in certain invertebrates serving to increase the absorptive surface. { 'tif·lə,sōl }

typhoid vaccine [IMMUNOLOGY] A type of killed vaccine used for active immunity production; made from killed typhoid bacillus (*Salmonella typhi*). { 'tī,fȯid vak'sēn }

Typotheria [PALEONTOLOGY] A suborder of extinct rodentlike herbivores in the order Notoungulata. { ,tī·pə'thir·ē·ə }

Tyranni [VERTEBRATE ZOOLOGY] A suborder of suboscine Passeriformes containing birds with limited song power and having the tendon of the hind toe separate and the intrinsic muscles of the syrinx reduced to one pair. { tə'ra,nī }

Tyrannidae [VERTEBRATE ZOOLOGY] The tyrant flycatchers, a family of passeriform birds in the suborder Tyranni confined to the Americas. { tə 'ran·ə,dē }

Tyrannoidea [VERTEBRATE ZOOLOGY] The flycatchers, a superfamily of suboscine birds in the suborder Tyranni. { tir·ə'nȯid·ē·ə }

tyrocidine [MICROBIOLOGY] A peptide antibiotic produced by *Bacillus brevis*; used to control fungi, bacteria, and protozoa. { ,tir·ə'sīd·ən }

tyrosinase [BIOCHEMISTRY] An enzyme found in plants, molds, crustaceans, mollusks, and some bacteria which, in the presence of oxygen, catalyzes the oxidation of monophenols and polyphenols with the introduction of —OH groups and the formation of quinones. { 'tir·ə·sə,nās }

tyrosine [BIOCHEMISTRY] $C_9H_{11}NO_3$ A phenolic alpha amino acid found in many proteins; a precursor of the hormones epinephrine, norepinephrine, thyroxine, and triiodothyronine, and of the black pigment melanin. { 'tir·ə,sēn }

tyrothricin [MICROBIOLOGY] A polypeptide mixture produced by *Bacillus brevis* and consisting of the antibiotic substances gramicidin and tyrocidine; effective as an antibacterial applied locally in infections due to germ-positive organisms. { ,tir·ə'thrīs·ən }

Tyson's gland [ANATOMY] A small scent gland in the human male which secretes the smegma. Also known as preputial gland. { 'tī·sənz ˌgland }

Tytonidae [VERTEBRATE ZOOLOGY] The barn owls, a family of birds in the order Strigiformes distinguished by an unnotched sternum which is fused to large clavicles. { tī'tän·əˌdē }

tyvelose [BIOCHEMISTRY] A dideoxy sugar found in bacterial lipopolysaccharides. { 'tī·vəˌlōs }

U

udder [VERTEBRATE ZOOLOGY] A pendulous organ consisting of several mammary glands enclosed in a single envelope; each gland has its own nipple; found in some mammals, such as the cow and goat. { 'əd·ər }

UDP [BIOCHEMISTRY] *See* uridine diphosphate.

UDPG [BIOCHEMISTRY] *See* uridine diphosphoglucose.

Uintatheriidae [PALEONTOLOGY] The single family of the extinct mammalian order Dinocerata. { yü‚win·tə·thə'rī·ə‚dē }

Uintatheriinae [PALEONTOLOGY] A subfamily of extinct herbivores in the family Uintatheriidae including all horned forms. { yü‚win·tə·thə'rī·ə‚nē }

Ulmaceae [BOTANY] A family of dicotyledonous trees in the order Urticales distinguished by alternate stipulate leaves, two styles, a pendulous ovule, and lack of a latex system. { əl'mās·ē‚ē }

ulna [ANATOMY] The larger of the two bones of the forearm or forelimb in vertebrates; articulates proximally with the humerus and radius and distally with the radius. { 'əl·nə }

ulnar deviation [BIOPHYSICS] A position of the human hand in which the wrist is bent toward the end finger. { ¦əl·nər ‚dē·vē'ā·shən }

Ulotrichaceae [BOTANY] A family of green algae in the suborder Ulotrichineae; contains both attached and floating filamentous species with cells having parietal, platelike or bandlike chloroplasts. { yü‚lä·trə'kās·ē‚ē }

Ulotrichales [BOTANY] A large, artificial order of the Chlorophyta composed mostly of freshwater, branched or unbranched filamentous species with mostly cylindrical, uninucleate cells having cellulose, but often mucilaginous walls. { yü‚lä·trə'kā·lēz }

Ulotrichineae [BOTANY] A suborder of the Ulotrichales characterized by short cylindrical cells. { yü‚lä·trə'kīn·ē‚ē }

ultimobranchial body [EMBRYOLOGY] One of the small, endocrine structures which originate as terminal outpocketings from each side of the embryonic vertebrate pharynx; can produce the hormone calcitonin. { ¦əl·tə·mō'braŋ·kē·əl 'bäd·ē }

ultrastructure [MOLECULAR BIOLOGY] The ultimate physiochemical organization of protoplasm. { ¦əl·trə'strək·chər }

Ulvaceae [BOTANY] A large family of green algae in the order Ulvales. { ‚əl'vās·ē‚ē }

Ulvales [BOTANY] An order of algae in the division Chlorophyta in which the thalli are macroscopic, attached tubes or sheets. { ‚əl'vā·lēz }

umbel [BOTANY] An indeterminate inflorescence with the pedicels all arising at the top of the peduncle and radiating like umbrella ribs; there are two types, simple and compound. { 'əm·bəl }

Umbellales [BOTANY] An order of dicotyledonous herbs or woody plants in the subclass Rosidae with mostly compound or conspicuously lobed or dissected leaves, well-developed schizogenous secretory canals, separate petals, and an inferior ovary. { ‚əm·bə'lā·lēz }

Umbelliferae [BOTANY] A large family of aromatic dicotyledonous herbs in the order Umbellales; flowers have an ovary of two carpels, ripening to form a dry fruit that splits into two halves, each containing a single seed. { ‚əm·bə'lif·ə‚rē }

umbilical artery [EMBRYOLOGY] Either of a pair of arteries passing through the umbilical cord to carry impure blood from the mammalian fetus to the placenta. { əm'bil·ə·kəl 'ärd·ə·rē }

umbilical cord [EMBRYOLOGY] The long, cylindrical structure containing the umbilical arteries and vein, and connecting the fetus with the placenta. { əm'bil·ə·kəl ‚kȯrd }

umbilical duct [EMBRYOLOGY] *See* vitelline duct. { əm'bil·ə·kəl ‚dəkt }

umbilical vein [EMBRYOLOGY] A vein passing through the umbilical cord and conveying purified, nutrient-rich blood from placenta to fetus. { əm'bil·ə·kəl 'vān }

Umbilicariaceae [BOTANY] The rock tripes, a family of Asco-lichenes in the order Lecanorales having a large, circular, umbilicate thallus. { ‚əm·bə·lə‚kar·ē'ās·ē‚ē }

umbilicus [ANATOMY] The navel; the round, depressed cicatrix in the median line of the abdomen, marking the site of the aperture through which passed the fetal umbilical vessels. { əm'bil·ə·kəs }

umbo [ANATOMY] A rounded elevation of the surface of the tympanic membrane. [INVERTEBRATE ZOOLOGY] A prominence above the hinge of a bivalve mollusk shell. { 'əm·bō }

umbonate [BIOLOGY] Having or forming an umbo. { 'əm·bə·nət }

UMP [BIOCHEMISTRY] *See* uridylic acid.

uncinate trophus [INVERTEBRATE ZOOLOGY] A trophus in rotifers characterized by a hooked or curved uncus. { 'ən·sə·nət 'träf·əs }

uncinus [INVERTEBRATE ZOOLOGY] In annelids, a minute, pectiniform neuroseta. { ən'sī·nəs }

uncompetitive enzyme inhibition [BIOCHEMISTRY] The prevention of an enzymic process as a result of the interaction of an inhibitor with the enzyme-substrate complex or a subsequent intermediate form of the enzyme, but not with the free enzyme. { ¦ən·kəm'ped·əd·iv 'en,zīm ,in·ə 'bish·ən }

underground stem [BOTANY] Any of the stems that grow underground and are often mistaken for roots; principal kinds are rhizomes, tubers, corms, bulbs, and rhizomorphic droppers. { ¦ən·dər¦graünd 'stem }

underwing [INVERTEBRATE ZOOLOGY] Either of a pair of posterior wings on certain insects, as the moth. { 'ən·dər,wiŋ }

ungula [VERTEBRATE ZOOLOGY] A nail, hoof, or claw. { 'əŋ·gyə·lə }

ungulate [VERTEBRATE ZOOLOGY] Referring to an animal that has hoofs. { 'əŋ·gyə·lət }

ungulicutate [VERTEBRATE ZOOLOGY] Having claws or nails. { ¦əŋ·gyə'lik·ə,tāt }

unguligrade [VERTEBRATE ZOOLOGY] Walking on hoofs. { 'əŋ·gyə·lə,grād }

unicellular [BIOLOGY] Composed of a single cell. { ¦yü·nə'sel·yə·lər }

unicellular gland [ANATOMY] A gland consisting of a single cell. { ¦yü·nə'sel·yə·lər ,gland }

unicuspid [ANATOMY] Having one cusp, as certain teeth. { ¦yü·nə'kəs·pəd }

unilateral hermaphroditism [ZOOLOGY] A type of hermaphroditism in which there is a combination of ovatestis on one side of the body with an ovary or testis on the other side. { ¦yü·nə'lad·ə·rəl hər'maf·rə·dī,tiz·əm }

unilocular [BIOLOGY] Having a single cavity. { ¦yü·nē'läk·yə·lər }

Unionidae [INVERTEBRATE ZOOLOGY] The freshwater mussels, a family of bivalve mollusks in the subclass Eulamellibranchia; the larvae, known as glochidia, are parasitic on fish. { ,yü·nē'än·ə ,dē }

Unipolarina [INVERTEBRATE ZOOLOGY] A suborder of the protozoan order Myxosporida characterized by spores with one to six (never five) polar capsules located at the anterior end. { ,yü·nə ,pō·lə'rīn·ə }

unitegmic [BOTANY] Referring to an ovule having a single integument. { ,yü·nə'teg·mik }

universal donor [IMMUNOLOGY] An individual of O blood group; can give blood to persons of all blood types. { ¦yü·nə¦vər·səl 'dō·nər }

universal recipient [IMMUNOLOGY] An individual of AB blood group; can receive a blood transfusion of all blood types, A, AB, B, or O. { ¦yü·nə ¦vər·səl ri'sip·ē·ənt }

unmyelinated [HISTOLOGY] Lacking myelin, either as a normal condition or as the result of a disease. { ¦ən'mī·ə·lə,nād·əd }

unstriated fiber [HISTOLOGY] *See* smooth muscle fiber. { ¦ən'strī,ād·əd 'fī·bərz }

Upupidae [VERTEBRATE ZOOLOGY] The Old World hoopoes, a family of birds in the order Coraciiformes whose young are hatched with sparse down. { yü'püp·ə,dē }

urachus [EMBRYOLOGY] A cord or tube of epithelium connecting the apex of the urinary bladder with the allantois; its connective tissue forms the median umbilical ligament. { 'yür·ə·kəs }

uracil [BIOCHEMISTRY] $C_4H_4N_2O_2$ A pyrimidine base important as a component of ribonucleic acid. { 'yür·ə,sil }

Uraniidae [INVERTEBRATE ZOOLOGY] A tropical family of moths in the superfamily Geometroidea including some slender-bodied, brilliantly colored diurnal insects which lack a frenulum and are often mistaken for butterflies. { ,yür·ə'nī·ə ,dē }

urceolate [BIOLOGY] Shaped like an urn. { ¦ər ¦sē·ə·lət }

urease [BIOCHEMISTRY] An enzyme that catalyzes the degradation of urea to ammonia and carbon dioxide; obtained from the seed of jack bean. { 'yür·ē,ās }

Urechinidae [INVERTEBRATE ZOOLOGY] A family of echinoderms in the order Holasteroida which have an ovoid test lacking a marginal fasciole. { ,yür·ə'kin·ə,dē }

Uredinales [MYCOLOGY] An order of parasitic fungi of the subclass Heterobasidiomycetidae characterized by the teleutospore, a spore with one or more cells, each of which is a modified hypobasidium; members cause plant diseases known as rusts. { yə,red·ən'ā·lēz }

uredinium [MYCOLOGY] The aggregation of sporebearing hyphae and urediospores of a rust fungus that forms beneath the cuticle or epidermis of a host plant. { ,yür·ə'din·ē·əm }

urediospore [MYCOLOGY] A thin-walled spore produced by rust fungi; gives rise to a vegetative mycelium which may produce more urediospores. { yə'rid·ē·ə,spör }

ureotelic [BIOLOGY] Referring to animals that produce urea as their main nitrogenous excretion. { yə¦rē·ə¦tel·ik }

ureter [ANATOMY] A long tube conveying urine from the renal pelvis to the urinary bladder or cloaca in vertebrates. { 'yür·əd·ər }

urethra [ANATOMY] The canal in most mammals through which urine is discharged from the urinary bladder to the outside. { yə'rē·thrə }

urethral gland [ANATOMY] One of the small, branched tubular mucous glands in the mucosa lining the urethra. { yə'rē·thrəl 'gland }

uric acid [BIOCHEMISTRY] $C_5H_4N_4O_3$ A white, crystalline compound, the excretory end product in amino acid metabolism by uricotelic species. { 'yür·ik 'as·əd }

uricase [BIOCHEMISTRY] An enzyme present in the liver, spleen, and kidney of most mammals except humans; converts uric acid to allantoin in the presence of gaseous oxygen. { 'yür·ə,kās }

uricotelism [PHYSIOLOGY] An adaptation of terrestrial reptiles and birds which effectively provides for detoxification of ammonia and also for efficient conservation of water due to a relatively low rate of glomerular filtration and active secretion of uric acid by the tubules to form a urine practically saturated with urate. { ,yür·ə'käd·əl ,iz·əm }

uridine [BIOCHEMISTRY] $C_9H_{12}N_2O_6$ A crystalline

nucleoside composed of one molecule of uracil and one molecule of D-ribose; a component of ribonucleic acid. { 'yu̇r·ə‚dīn }

uridine diphosphate [BIOCHEMISTRY] The chief transferring coenzyme in carbohydrate metabolism. Abbreviated UDP. { 'yu̇r·ə‚dīn dī¦fä‚sfāt }

uridine diphosphoglucose [BIOCHEMISTRY] A compound in which α-glucopyranose is esterified, at carbon atom 1, with the terminal phosphate group of uridine-5'-pyrophosphate (that is, uridine diphosphate); occurs in animal, plant, and microbial cells; functions as a key in the transformation of glucose to other sugars. Abbreviated UDPG. { 'yu̇r·ə‚dīn dī¦fä'sfō'glü‚kōs }

uridine monophosphate [BIOCHEMISTRY] *See* uridylic acid. { 'yu̇r·ə‚dīn ¦män·ə'fä‚sfāt }

uridine phosphoric acid [BIOCHEMISTRY] *See* uridylic acid. { 'yu̇r·ə‚dīn fä'sfȯr·ik 'as·əd }

uridylic acid [BIOCHEMISTRY] $C_9H_{13}N_2O_9P$ Water- and alcohol-soluble crystals, melting at 202°C; used in biochemical research. Also known as uridine monophosphate (UMP); uridine phosphoric acid. { ¦yu̇r·ə¦dil·ik 'as·əd }

urinary bladder [ANATOMY] A hollow organ which serves as a reservoir for urine. { 'yu̇r·ə‚ner·ē 'blad·ər }

urinary system [ANATOMY] The system which functions in the elaboration and excretion of urine in vertebrates; in humans and most mammals, consists of the kidneys, ureters, urinary bladder, and urethra. { 'yu̇r·ə‚ner·ē ‚sis·təm }

urination [PHYSIOLOGY] The discharge of urine from the bladder. Also known as micturition. { ‚yu̇r·ə'nā·shən }

urine [PHYSIOLOGY] The fluid excreted by the kidneys. { 'yu̇r·ən }

uriniferous tubule [ANATOMY] One of the numerous winding tubules of the kidney. Also known as nephric tubule. { ¦yu̇r·ə¦nif·ə·rəs 'tü ‚byül }

urn [BOTANY] The theca of a moss. { ərn }

urobilin [BIOCHEMISTRY] A bile pigment produced by reduction of bilirubin by intestinal bacteria and excreted by the kidneys or removed by the liver. { ‚yu̇r·ə'bī·lən }

urobilinogen [BIOCHEMISTRY] A chromogen, formed in feces and present in urine, from which urobilin is formed by oxidation. { ‚yu̇r·ə'bil·ə·jən }

urocanic acid [BIOCHEMISTRY] $C_6H_6N_2O_2$ A crystalline compound formed as an intermediate in the degradative pathway of histidine. { ¦yu̇r·ə ¦kan·ik 'as·əd }

Urochordata [INVERTEBRATE ZOOLOGY] The equivalent name for Tunicata. { ¦yu̇r·ə·kȯr 'däd·ə }

urochrome [BIOCHEMISTRY] $C_{43}H_{51}O_{26}N$ Yellow pigment found in normal urine. { 'yu̇r·ə‚krōm }

Urodela [VERTEBRATE ZOOLOGY] The tailed amphibians or salamanders, an order of the class Amphibia distinguished superficially from frogs and toads by the possession of a tail, and from caecilians by the possession of limbs. { ‚yu̇r·ə 'dē·lə }

urogenital diaphragm [ANATOMY] The sheet of tissue stretching across the pubic arch, formed by the deep transverse perineal and the sphincter urethrae muscles. Also known as triangular ligament. { ¦yu̇r·ə¦jen·əd·əl 'dī·ə‚fram }

urogenital system [ANATOMY] The combined urinary and genital system in vertebrates, which are intimately related embryologically and anatomically. Also known as genitourinary system. { ¦yu̇r·ə¦jen·əd·əl ‚sis·təm }

urogomphus [INVERTEBRATE ZOOLOGY] In certain insect larvae, a process on the terminal segment. Also known as pseudocercus. { ‚yu̇r·ə'gäm·fəs }

urokinase [BIOCHEMISTRY] An enzyme, present in human urine, that catalyzes the conversion of plasminogen to plasmin. { ‚yu̇r·ə'kī‚nās }

uropepsin [BIOCHEMISTRY] The end product of the secretion of pepsinogen into the blood by gastric cells; occurs in urine. { ¦yu̇r·ə'pep·sən }

uropod [INVERTEBRATE ZOOLOGY] One of the flattened abdominal appendages of various crustaceans that with the telson forms the tail fan. { 'yu̇r·ə‚päd }

uroporphyrin [BIOCHEMISTRY] Any of several isomeric, metal-free porphyrins, occurring in small quantities in normal urine and feces; molecule has four acetic acid (—CH_2COOH) and four propionic acid (—CH_2CH_2COOH) groups. { ¦yu̇r·ə 'pȯr·fə·rən }

Uropygi [INVERTEBRATE ZOOLOGY] The tailed whip scorpions, an order of arachnids characterized by an elongate, flattened body which bears in front a pair of thickened, raptorial pedipalps set with sharp spines and used to hold and crush insect prey. { ‚yu̇r·ə'pī‚jī }

uropygial gland [VERTEBRATE ZOOLOGY] A relatively large, compact, bilobed, secretory organ located at the base of the tail (uropygium) of most birds having a keeled sternum. Also known as oil gland; preen gland. { ¦yu̇r·ə¦pij·ē·əl ‚gland }

urostyle [VERTEBRATE ZOOLOGY] An unsegmented bone representing several fused vertebrae and forming the posterior part of the vertebral column in Anura. { 'yu̇r·ə‚stīl }

Urostylidae [INVERTEBRATE ZOOLOGY] A family of hemipteran insects in the superfamily Pentatomoidea. { ‚yu̇r·ə'stīl·ə‚dē }

Ursidae [VERTEBRATE ZOOLOGY] A family of mammals in the order Carnivora including the bears and their allies. { 'ər·sə‚dē }

ursolic acid [BIOCHEMISTRY] $C_{30}H_{48}O_3$ A pentacyclic terpene that crystallizes from absolute alcohol solution, found in leaves and berries of plants; used in pharmaceutical and food industries as an emulsifying agent. { ər'säl·ik 'as·əd }

Urticaceae [BOTANY] A family of dicotyledonous herbs in the order Urticales characterized by a single unbranched style, a straight embryo, and the lack of milky juice (latex). { ‚ərd·ə'kās·ē‚ē }

Urticales [BOTANY] An order of dicotyledons in the subclass Hamamelidae; woody plants or herbs with simple, usually stipulate leaves, and reduced clustered flowers that usually have a vestigial perianth. { ‚ərd·ə'kā·lēz }

use-dilution test [MICROBIOLOGY] A bioassay

method for testing disinfectants for use on surfaces where a substantial reduction of bacterial contamination is not achieved by prior cleaning; the test organisms *Salmonella choleraesuis* and *Staphylococcus aureus* are deposited in stainless steel cylinders which are then exposed to the action of the test disinfectant. { 'yüs də¦lü·shən ˌtest }

Usneaceae [BOTANY] The beard lichens, a family of Ascolichenes in the order Lecanorales distinguished by their conspicuous fruticose growth form. { ˌəs·nē'ās·ēˌē }

usnic acid [BIOCHEMISTRY] $C_{18}H_{16}O_7$ Yellow crystals, insoluble in water, slightly soluble in alcohol and ether, melts about 198°C; found in lichens; used as an antibiotic. Also known as usninic acid. { 'əs·nik 'as·əd }

Ustilaginaceae [MYCOLOGY] A family of fungi in the order Ustilaginales in which basidiospores bud from the sides of the septate epibasidium. { ˌəs·təˌlaj·ə'nās·ēˌē }

Ustilaginales [MYCOLOGY] An order of the subclass Heterobasidiomycetidae comprising the smut fungi which parasitize plants and cause diseases known as smut or bunt. { ˌəs·təˌlaj·ə'nā·lēz }

uterus [ANATOMY] The organ of gestation in mammals which receives and retains the fertilized ovum, holds the fetus during development, and becomes the principal agent of its expulsion at term. { 'yüd·ə·rəs }

uterus bicornis [ANATOMY] A uterus divided into two horns or compartments; an abnormal condition in humans but normal in many mammals, such as carnivores. { 'yüd·ə·rəs bī'kȯr·nəs }

utricle [ANATOMY] *See* utriculus. { 'yü·trə·kəl }

utriculus [ANATOMY] **1.** That part of the membranous labyrinth of the ear into which the semicircular canals open. **2.** A small, blind pouch extending from the urethra into the prostate. Also known as utricle. { yü'trik·yə·ləs }

uvea [ANATOMY] The pigmented, vascular layer of the eye: the iris, ciliary body, and choroid. { 'yü·vē·ə }

uvula [ANATOMY] **1.** A fingerlike projection in the midline of the posterior border of the soft palate. **2.** A lobe of the vermiform process of the lower surface of the cerebellum. { 'yü·vyə·lə }

V

vaccination [IMMUNOLOGY] Inoculation of viral or bacterial organisms or antigens to produce immunity in the recipient. { ˌvak·sə'nā·shən }

vaccine [IMMUNOLOGY] A suspension of killed or attenuated bacteria or viruses or fractions thereof, injected to produce active immunity. { vak'sēn }

vacuole [CELL BIOLOGY] A membrane-bound cavity within a cell; may function in digestion, storage, secretion, or excretion. { 'vak·yəˌwōl }

vagility [ECOLOGY] The ability of organisms to disseminate. { və'jil·əd·ē }

vagina [ANATOMY] The canal from the vulvar opening to the cervix uteri. { və'jī·nə }

vagina fibrosa tendinis [ANATOMY] A fibrous sheath surrounding the tendon of a muscle and usually confining the tendon in a bony groove. { və'jī·nə fī'brō·sə 'ten·də·nəs }

vaginate [BIOLOGY] Invested in a sheath. { 'vaj·əˌnāt }

vagotonine [BIOCHEMISTRY] An endocrine substance which is thought to be elaborated by cells of the pancreas and which regulates autonomic tonus. { və'gäd·əˌnēn }

vagus [ANATOMY] The tenth cranial nerve; either of a pair of sensory and motor nerves forming an important part of the parasympathetic system in vertebrates. { 'vā·gəs }

valence [BIOCHEMISTRY] The relative ability of a biological substance to react or combine. { 'vā·ləns }

valid [SYSTEMATICS] Describing a taxon classified on the basis of distinctive characters of accepted importance. { 'val·əd }

valine [BIOCHEMISTRY] $C_5H_{11}NO_2$ An amino acid considered essential for normal growth of animals, and biosynthesized from pyruvic acid. { 'vaˌlēn }

vallate papilla [ANATOMY] One of the large, flat papillae, each surrounded by a trench, in a group anterior to the sulcus terminalis of the tongue. Also known as circumvallate papilla. { 'vaˌlāt pə'pil·ə }

Valoniaceae [BOTANY] A family of green algae in the order Siphonocladales consisting of plants that are essentially unicellular, coenocytic vesicles, spherical or clavate, and up to 2.4 inches (6 centimeters) in diameter. { vəˌlō·nē'ās·ēˌē }

Valvatacea [PALEONTOLOGY] A superfamily of extinct gastropod mollusks in the order Prosobranchia. { ˌval·və'tā·shə }

valvate [BOTANY] Having valvelike parts, as those which meet edge to edge or which open as if by valves. { 'valˌvāt }

Valvatida [INVERTEBRATE ZOOLOGY] An order of echinoderms in the subclass Asteroidea. { ˌval·və'tīd·ə }

Valvatina [INVERTEBRATE ZOOLOGY] A suborder of echinoderms in the order Phanerozonida in which the upper marginals lie directly over, and not alternate with, the corresponding lower marginals. { ˌval·və'tīn·ə }

valve [ANATOMY] A flat of tissue, as in the veins or between the chambers in the heart, which permits movement of fluid in one direction only. [BOTANY] **1.** A segment of a dehiscing capsule or legume. **2.** The lidlike portion of certain anthers. [INVERTEBRATE ZOOLOGY] **1.** One of the distinct, articulated pieces composing the shell of certain animals, such as barnacles and brachiopods. **2.** One of two shells encasing the body of a bivalve mollusk or a diatom. { valv }

Valvifera [INVERTEBRATE ZOOLOGY] A suborder of isopod crustaceans distinguished by having a pair of flat, valvelike uropods which hinge laterally and fold inward beneath the rear part of the body. { val'vif·ə·rə }

valvula [BIOLOGY] A small valve. [INVERTEBRATE ZOOLOGY] One of the small processes forming a sheath for the ovipositor in certain insects. { 'val·vyə·lə }

valvulate [BIOLOGY] Having valvules. { 'val·vyə·lət }

vampire [VERTEBRATE ZOOLOGY] The common name for bats making up the family Desmodontidae which have teeth specialized for cutting and which subsist on a blood diet. { 'vamˌpīr }

Vampyrellidae [INVERTEBRATE ZOOLOGY] A family of protozoans in the order Proteomyxida including species which invade filamentous algae and sometimes higher plants. { ˌvam·pə'rel·əˌdē }

Vampyromorpha [INVERTEBRATE ZOOLOGY] An order of dibranchiate cephalopod mollusks represented by *Vampyroteuthis infernalis*, an inhabitant of the deeper waters of tropical and temperate seas. { ¦vam·pə·rō'mȯr·fə }

vancomycin [MICROBIOLOGY] A complex antibiotic substance produced by *Streptomyces orientalis*; useful for treatment of severe staphylococcic infections. { ˌvaŋ·kə'mīs·ən }

vane feather [VERTEBRATE ZOOLOGY] *See* contour feather. { 'vān ˌfeth·ər }

Vaneyellidae [INVERTEBRATE ZOOLOGY] A family of holothurian echinoderms in the order Dactylochirotida. { ˌvā·nē′el·əˌdē }

Vanhorniidae [INVERTEBRATE ZOOLOGY] A monospecific family of the Hymenoptera in the superfamily Proctotrupoidea. { ˌvanˌhȯr′nī·əˌdē }

Varanidae [VERTEBRATE ZOOLOGY] The monitors, a family of reptiles in the suborder Sauria found in the hot regions of Africa, Asia, Australia, and Malaya. { və′ran·əˌdē }

varicose vein [ANATOMY] An enlarged tortuous blood vessel that occurs chiefly in the superficial veins and their tributaries in the lower extremities. Also known as varicosity. { ′var·əˌkōs ′vān }

varicosity [ANATOMY] *See* varicose vein. { ˌvar·ə′käs·əd·ē }

variegate [BIOLOGY] Having irregular patches of diverse colors. { ′ver·ē·əˌgāt }

variegated position effect [GENETICS] A phenomenon observed in some cases when a chromosome aberration causes a wild-type gene from the euchromatin to be relocated adjacent to heterochromatin; the phenotypic expression of the wild-type allele will be unstable, producing patches of phenotypically mutant tissue that differ from the surrounding wild-type tissue. { ′ver·ē·əˌgād·əd pə¦zish·ən iˌfekt }

variety [SYSTEMATICS] A taxonomic group or category inferior in rank to a subspecies. { və′rī·əd·ē }

varix [INVERTEBRATE ZOOLOGY] A conspicuous ridge across each whorl of certain univalves marking the ancestral position of the outer lip of the aperture. { ′var·iks }

varnish tree [BOTANY] *Rhus vernicifera*. A member of the sumac family (Anacardiaceae) cultivated in Japan; the cut bark exudes a juicy milk which darkens and thickens on exposure and is applied as a thin film to become a varnish of extreme hardness. Also known as lacquer tree. { ′vär·nish ˌtrē }

vasa vasorum [ANATOMY] The blood vessels supplying the walls of arteries and veins. { ′vā·zə va ′sȯr·əm }

vascular [ANATOMY] Pertaining to blood vessels or other channels for the conveyance of a body fluid. { ′vas·kyə·lər }

vascular bundle [BOTANY] A strandlike part of the plant vascular system containing xylem and phloem. { ′vas·kyə·lər ′bənd·əl }

vascular cambium [BOTANY] The lateral meristem which produces secondary xylem and phloem. { ′vas·kyə·lər ′kam·bē·əm }

vascularization [PHYSIOLOGY] The formation of new blood vessels within tissue. { ˌvas·kyə·lə·rə′zā·shən }

vascular ray [BOTANY] A ray derived from cambium and found in the stele of some vascular plants, often separating vascular bundles. { ′vas·kyə·lər ′rā }

vascular tissue [BOTANY] The conducting tissue found in higher plants, consisting principally of xylem and phloem. { ′vas·kyə·lər ′tish·ü }

vas deferens [ANATOMY] The portion of the excretory duct system of the testis which runs from the epididymal duct to the ejaculatory duct. Also known as ductus deferens. { ′vas ′def·ə·rənz }

vasoconstrictor [PHYSIOLOGY] A nerve or an agent that causes blood vessel constriction. { ¦vā·zō·kən′strik·tər }

vasodilator [PHYSIOLOGY] A nerve or an agent that causes blood vessel dilation. { ¦vā·zō′dī ˌlād·ər }

vasomotion [PHYSIOLOGY] Change in the diameter of a blood vessel. Also known as angiokinesis. { ′vā·zəˌmō·shən }

vasomotor [PHYSIOLOGY] Pertaining to the regulation of the constriction or expansion of blood vessels. { ′vā·zəˌmōd·ər }

vasomotor center [PHYSIOLOGY] A large, diffuse area in the reticular formation of the lower brainstem; stimulation of different portions of this center causes either a rise in blood pressure and tachycardia (pressor area) or a fall in blood pressure and bradycardia (depressor area). { ¦vāz·ə′mōd·ər ˌsen·tər }

vasopressin [BIOCHEMISTRY] A peptide hormone which is elaborated by the posterior pituitary and which has a pressor effect; used medicinally as an antidiuretic. Also known as antidiuretic hormone (ADH). { ˌvā·zō′pres·ən }

vasotocin [BIOCHEMISTRY] A hormone from the neurosecretory cells of the posterior pituitary of lower vertebrates; increases permeability to water in amphibian skin and in bladder. { ˌvā·zə′tōs·ən }

vault [BIOLOGY] An anatomical structure that is arched or dome-shaped. { vȯlt }

v-body [CELL BIOLOGY] *See* nucleosome. { ′vē ˌbäd·ē }

vectorcardiogram [PHYSIOLOGY] The part of the pathway of instantaneous vectors during one cardiac cycle. Also known as monocardiogram. { ¦vek·tər′kärd·ē·əˌgram }

vectorcardiography [PHYSIOLOGY] A method of recording the magnitude and direction of the instantaneous cardiac vectors. { ¦vek·tərˌkärd·ē′äg·rə·fē }

vegetable [BOTANY] Resembling or relating to plants. { ′vej·tə·bəl }

vegetable diastase [BIOCHEMISTRY] *See* diastase. { ′vej·tə·bəl ′dī·əsˌtās }

vegetation [BOTANY] The total mass of plant life that occupies a given area. { ˌvej·ə′tā·shən }

vegetational plant geography [ECOLOGY] A field of study concerned with the mapping of vegetation regions and the interpretation of these in terms of environmental or ecological influences. { ˌvej·ə′tā·shən·əl ′plant jēˌäg·rə·fē }

vegetation and ecosystem mapping [BOTANY] An art and a science concerned with the drawing of maps which locate different kinds of plant cover in a geographic area. { ˌvej·ə¦tā·shən ən ′ek·ō¦sis·təm ′map·iŋ }

vegetation management [ECOLOGY] The art and practice of manipulating vegetation such as timber, forage, crops, or wild life, so as to produce a desired part or aspect of that material in higher quantity or quality. { ˌvej·ə′tā·shən ˌman·ij·mənt }

vegetation zone [ECOLOGY] **1.** An extensive, even

transcontinental, band of physiognomically similar vegetation on the earth's surface. **2.** Plant communities assembled into regional patterns by the area's physiography, geological parent material, and history. { ˌvej·ə'tā·shən ˌzōn }

vegetative [BIOLOGY] Having nutritive or growth functions, as opposed to reproductive. { 'vej·ə ˌtād·iv }

vegetative propagation [BOTANY] Production of a new plant from a portion of another plant, such as a stem or branch. { 'vej·əˌtād·iv ˌpräp·ə'gā·shən }

vegetative state [VIROLOGY] The noninfective state during which the genome of a phage multiplies and directs host synthesis of substances needed for production of infective particles. { 'vej·əˌtād·iv ˌstāt }

veil [BIOLOGY] *See* velum. { vāl }

Veillonella [MICROBIOLOGY] The type genus of the family Veillonellaceae; small cells occurring in pairs, chains, and clusters. { ˌvā·yō'nel·ə }

Veillonellaceae [MICROBIOLOGY] The single family of gram-negative, anaerobic cocci; characteristically occur in pairs with adjacent sides flattened; parasites of homotherms, including humans, rodents, and pigs. { ˌvā·yō·nə'lās·ēˌē }

vein [ANATOMY] A relatively thin-walled blood vessel that carries blood from capillaries to the heart in vertebrates. [BOTANY] One of the vascular bundles in a leaf. [INVERTEBRATE ZOOLOGY] **1.** One of the thick, stiff ribs providing support for the wing of an insect. **2.** A venous sinus in invertebrates. { vān }

velamen [BOTANY] The corky epidermis covering the aerial roots of an epiphytic orchid. { və'lā·mən }

velarium [INVERTEBRATE ZOOLOGY] The velum of certain scyphozoans and cubomedusans distinguished by the presence of canals lined with endoderm. { və'lar·ē·əm }

veld [ECOLOGY] *See* veldt. { velt }

veldt [ECOLOGY] Grasslands of eastern and southern Africa that are usually level and mixed with trees and shrubs. Also spelled veld. { velt }

veliger [INVERTEBRATE ZOOLOGY] A mollusk larval stage following the trochophore, distinguished by an enlarged girdle of ciliated cells (velum). { 'vē·lə·jər }

Veliidae [INVERTEBRATE ZOOLOGY] A family of the Hemiptera in the subdivision Amphibicorisae composed of small water striders which have short legs and a longitudinal groove between the eyes. { və'lī·əˌdē }

vellus [ANATOMY] Fine body hair that is present until puberty. { 'vel·əs }

Velocipedidae [INVERTEBRATE ZOOLOGY] A tropical family of hemipteran insects in the superfamily Cimicoidea. { vəˌläs·ə'ped·əˌdē }

velum [BIOLOGY] A veil- or curtainlike membrane. [INVERTEBRATE ZOOLOGY] A swimming organ on the larva of certain marine gastropod mollusks that develops as a contractile ciliated collar-shaped ridge. Also known as veil. { 'vē·ləm }

vena cava [ANATOMY] One of two large veins which in air-breathing vertebrates conveys blood from the systemic circulation to the right atrium. { ˌvē·nə 'kā·və }

venation [BOTANY] The system or pattern of veins in the tissues of a leaf. [INVERTEBRATE ZOOLOGY] The arrangement of veins in an insect wing. { ve'nā·shən }

venom [PHYSIOLOGY] Any of various poisonous materials secreted by certain animals, such as snakes or bees. { 'ven·əm }

venous pressure [PHYSIOLOGY] Tension of the blood within the veins. { 'vē·nəs ˌpresh·ər }

vent [ZOOLOGY] The external opening of the cloaca or rectum, especially in fish, birds, and amphibians. { vent }

venter [ANATOMY] The abdomen, or other body cavity containing organs. [BOTANY] The thickened basal portion of an archegonium. [INVERTEBRATE ZOOLOGY] **1.** The undersurface of an arthropod's abdomen. **2.** The outer, convex part of a curved or coiled gastropod or cephalopod shell. { 'ven·tər }

ventral [BOTANY] On the lower surface of a dorsiventral plant structure, such as a leaf. [ZOOLOGY] On or belonging to the lower or anterior surface of an animal, that is, on the side opposite the back. { 'ven·trəl }

ventral aorta [VERTEBRATE ZOOLOGY] The arterial trunk or trunks between the heart and the first aortic arch in embryos or lower vertebrates. { 'ven·trəl ā'ȯrd·ə }

ventralia [INVERTEBRATE ZOOLOGY] Paired sensory bristles on the ventral aspect of the head of gnathostomulids. { ven'tral·yə }

ventral light reflex [INVERTEBRATE ZOOLOGY] A basic means of orientation in aquatic invertebrates, such as shrimp, which swim belly up toward the light. { 'ven·trəl 'līt ˌrēˌfleks }

ventral rib [VERTEBRATE ZOOLOGY] Any of the ribs which lie in the septa dividing the trunk musculature into segments in fish. Also known as pleural rib. { 'ven·trəl 'rib }

ventricle [ANATOMY] **1.** A chamber, or one of two chambers, in the vertebrate heart which receives blood from the atrium and forces it into the arteries by contraction of the muscular wall. **2.** One of the interconnecting, fluid-filled chambers of the vertebrate brain that are continuous with the canal of the spinal cord. [ZOOLOGY] A cavity in a body part or organ. { 'ven·trə·kəl }

ventricose [BIOLOGY] Swollen or distended, especially on one side. { ˌven·trə·kōs }

ventricular septum [ANATOMY] *See* interventricular septum. { ven'trik·yə·lər 'sep·təm }

ventriculus [ZOOLOGY] A ventricle that performs digestive functions, such as a stomach or a gizzard. { ven'trik·yə·ləs }

ventromedial nucleus [ANATOMY] A central nervous system nucleus in the hypothalamus that appears to be the satiation center; bilateral surgical damage to this nucleus results in overeating. { ¦ven·trō'mēd·ē·əl 'nü·klē·əs }

venule [ANATOMY] A small vein. { 'ven·yül }

Venus' flytrap [BOTANY] *Dionaea muscipula.* An insectivorous plant (order Sarraceniales) of North and South Carolina; the two halves of a leaf blade can swing upward and inward as though hinged, thus trapping insects between the closing halves of the leaf blade. { 'vē·nəs 'flī,trap }

Verbeekinidae [PALEONTOLOGY] A family of extinct marine protozoans in the superfamily Fusulinacea. { ,ver,bā'kin·ə,dē }

Verbenaceae [BOTANY] A family of variously woody or herbaceous dicotyledons in the order Lamiales characterized by opposite or whorled leaves and regular or irregular flowers, usually with four or two functional stamens. { ,vər·bə 'nās·ē,ē }

Vermes [INVERTEBRATE ZOOLOGY] An artificial taxon considered to be a phylum in some systems of classification, but variously defined as including all invertebrates except arthropods, or including all vermiform invertebrates. { 'vər·mēz }

vermiform [BIOLOGY] Wormlike; resembling a worm. { 'vər·mə,fórm }

vermiform appendix [ANATOMY] A small, blind sac projecting from the cecum. Also known as appendix. { 'vər·mə,fórm ə'pen,diks }

Vermilingua [VERTEBRATE ZOOLOGY] An infraorder of the mammalian order Edentata distinguished by lack of teeth and in having a vermiform tongue; includes the South American true anteaters. { ,vər·mə'liŋ·gwə }

vermis [ANATOMY] The median lobe of the cerebellum. { 'vər·məs }

vernalization [BOTANY] The induction in plants of the competence or ripeness to flower by the influence of cold, that is, at temperatures below the optimal temperature for growth. { ,vərn·əl·ə 'zā·shən }

vernation [BOTANY] The characteristic arrangement of young leaves within the bud. Also known as prefoliation. { vər'nā·shən }

vernine [BIOCHEMISTRY] *See* guanosine. { 'vər ,nēn }

vernix caseosa [EMBRYOLOGY] A cheesy deposit on the surface of the fetus derived from the stratum corneum, sebaceous secretion, and remnants of the epitrichium. { 'vər·niks ,kā·sē'ō·sə }

verruca [BIOLOGY] A wartlike elevation on the surface of a plant or animal. { və'rü·kə }

Verrucariaceae [BOTANY] A family of crustose lichens in the order Pyrenulales typically found on rocks, especially in intertidal or salt-spray zones along rocky coastlines. { və,rü·kə·rē'ās·ē,ē }

Verrucomorpha [INVERTEBRATE ZOOLOGY] A suborder of the crustacean order Thoracica composed of sessile, asymmetrical barnacles. { və ,rü·kə'mór·fə }

verrucose [BIOLOGY] Having the surface covered with wartlike protuberances. { və'rü,kōs }

versatile anther [BOTANY] An anther whose attachment is near its middle, thus enabling it to swing freely. { 'vər·səd·əl 'an·thər }

vertebra [ANATOMY] One of the bones that make up the spine in vertebrates. { 'vərd·ə·brə }

vertebral arch [ANATOMY] An arch formed by the paired pedicles and laminas of a vertebra; the posterior part of a vertebra which together with the anterior part, the body, encloses the vertebral foramen in which the spinal cord is lodged in vertebrates. Also known as neural arch. { 'vərd·ə·brəl 'ärch }

vertebral column [ANATOMY] *See* spine. { 'vərd·ə·brəl 'käl·əm }

Vertebrata [VERTEBRATE ZOOLOGY] The major subphylum of the phylum Chordata including all animals with backbones, from fish to human. { ,vərd·ə'bräd·ə }

vertebrate zoology [ZOOLOGY] That branch of zoology concerned with the study of members of the Vertebrata. { 'vərd·ə·brət zō'äl·ə·jē }

vertical transmission [GENETICS] Passage of genetic information from one cell or individual organism to its progeny by conventional heredity mechanisms. { 'vərd·ə·kəl tranz'mish·ən }

verticillate [BOTANY] Whorled, in an arrangement resembling the spokes of a wheel. { ¦vərd·ə¦si ,lāt }

vesicle [BIOLOGY] A small, thin-walled bladderlike cavity, usually filled with fluid. { 'ves·ə·kəl }

Vespertilionidae [VERTEBRATE ZOOLOGY] The common bats, a large cosmopolitan family of the Chiroptera characterized by a long tail, extending to the edge of the uropatagium; almost all members are insect-eating. { ,ves·pər,til·ē 'än·ə,dē }

vespertine [VERTEBRATE ZOOLOGY] Active in the evening. { 'ves·pər,tīn }

Vespidae [INVERTEBRATE ZOOLOGY] A widely distributed family of Hymenoptera in the superfamily Vespoidea including hornets, yellow jackets, and potter wasps. { 'ves·pə,dē }

Vespoidea [INVERTEBRATE ZOOLOGY] A superfamily of wasps in the suborder Apocrita. { ve 'spóid·ē·ə }

vessel [BOTANY] A water-conducting tube or duct in the xylem. { 'ves·əl }

vessel segment [BOTANY] A single cell or unit of a plant vessel. { 'ves·əl ,seg·mənt }

vestibular apparatus [ANATOMY] The anatomical structures concerned with the vestibular portion of the eighth cranial nerve; includes the saccule, utricle, semicircular canals, vestibular nerve, and vestibular nuclei of the ear. { və'stib·yə·lər 'ap·ə ,rad·əs }

vestibular membrane of Reissner [ANATOMY] *See* Reissner's membrane. { və'stib·yə·lər 'mem ,brān əv 'rīs·nər }

vestibular nerve [ANATOMY] A somatic sensory branch of the auditory nerve, which is distributed about the ampullae of the semicircular canals, macula sacculi, and macula utriculi. { və'stib·yə·lər 'nərv }

vestibular reflexes [PHYSIOLOGY] The responses of the vestibular apparatus to strong stimulation; responses include pallor, nausea, vomiting, and postural changes. { və'stib·yə·lər 'rē,flek·səz }

vestibule [ANATOMY] **1.** The central cavity of the bony labyrinth of the ear. **2.** The parts of the membranous labyrinth within the cavity of the bony labyrinth. **3.** The space between the labia minora. **4.** *See* buccal cavity. { 'ves·tə,byül }

vestibulocerebellar [ANATOMY] Pertaining to the

vestibular fibers and the cerebellum. { və¦stib·yə·lō,ser·ə'bel·ər }

vestibulocochlear nerve [ANATOMY] *See* auditory nerve. { və¦stib·yə·lə'käk·lē·ər ,nərv }

vestibulospinal tract [ANATOMY] A tract of nerve fibers that originates principally from the lateral vestibular nucleus and descends in the anterior funiculus of the spinal cord. { və¦stib·yə·lə 'spīn·əl ,trakt }

vestige [BIOLOGY] A degenerate anatomical structure or organ that remains from one more fully developed and functional in an earlier phylogenetic form of the individual. { 'ves·tij }

vestigial [BIOLOGY] Of, being, or resembling a vestige. { və'stij·ē·əl }

vetch [BOTANY] Any of a group of mostly annual legumes, especially of the genus *Vicia*, with weak viny stems terminating in tendrils and having compound leaves; some varieties are grown for their edible seed. { vech }

V factor [MICROBIOLOGY] Phosphopyridine nucleotide, a growth factor required by the parasitic bacteria of the genus *Haemophilus*. { 'vē ,fak·tər }

viable [BIOLOGY] Able to live and develop normally. { 'vī·ə·bəl }

Vianaidae [INVERTEBRATE ZOOLOGY] A small family of South American Hemiptera in the superfamily Tingordea. { ,vē·ə'nā·ə,dē }

vibraculum [INVERTEBRATE ZOOLOGY] A specially modified bryozoan zooid with a bristlelike seta that sweeps debris from the surface of the colony. { və'brak·yə·ləm }

Vibrionaceae [MICROBIOLOGY] A family of gram-negative, facultatively anaerobic rods; cells are straight or curved and usually motile by polar flagella; generally found in water. { ,vib·rē·ō 'nās·ē,ē }

vibrissae [VERTEBRATE ZOOLOGY] Hairs with specialized erectile tissue; found on all mammals except humans. Also known as sinus hairs; tactile hairs; whiskers. { vī'bri,sē }

vicariants [ECOLOGY] Two or more closely related taxa, presumably derived from one another or from a common immediate ancestor, that inhabit geographically distinct areas. { vī'kar·ē·əns }

vicuna [VERTEBRATE ZOOLOGY] *Lama vicugna*. A rare, wild ruminant found in the Andes mountains; the fiber of the vicuna is strong, resilient, and elastic but is the softest and most delicate of animal fibers. { vī'kün·yə }

villous placenta [EMBRYOLOGY] *See* epitheliochorial placenta. { 'vil·əs plə'sen·tə }

villus [ANATOMY] A fingerlike projection from the surface of a membrane. { 'vil·əs }

vine [BOTANY] A plant having a stem that is too flexible or weak to support itself. { vīn }

vinegar eel [INVERTEBRATE ZOOLOGY] *Turbatrix aceti*. A very small nematode often found in large numbers in vinegar fermentation. Also known as vinegar worm. { 'vin·ə·gər ,ēl }

vinegar worm [INVERTEBRATE ZOOLOGY] *See* vinegar eel. { 'vin·ə·gər ,wərm }

Violaceae [BOTANY] A family of dicotyledonous plants in the order Violales characterized by polypetalous, mostly perfect, hypogynous flowers with a single style and five stamens. { ,vī·ə 'lās·ē,ē }

Violales [BOTANY] A heterogeneous order of dicotyledons in the subclass Dilleniidae distinguished by a unilocular, compound ovary and mostly parietal placentation. { ,vī·ə'lā·lēz }

viomycin [MICROBIOLOGY] A polypeptide antibiotic or mixture of antibiotic substances produced by strains of *Streptomyces griseus* var. *purpureus* (*Streptomyces puniceus*); the sulfate salt is administered intramuscularly for treatment of tuberculosis resistant to other therapy. { ,vī·ə 'mīs·ən }

viper [VERTEBRATE ZOOLOGY] The common name for reptiles of the family Viperidae; thick-bodied poisonous snakes having a pair of long fangs, present on the anterior part of the upper jaw, which fold against the roof of the mouth when the jaws are closed. { 'vī·pər }

Viperidae [VERTEBRATE ZOOLOGY] A family of reptiles in the suborder Serpentes found in Eurasia and Africa; all species are proglyphodont. { vī 'per·ə,dē }

viral shedding [VIROLOGY] Excretion of virus from a specific site in the body or from a lesion. { 'vī·rəl 'shed·iŋ }

Vireonidae [VERTEBRATE ZOOLOGY] The vireos, a family of New World passeriform birds in the suborder Oscines. { ,vir·ē'än·ə,dē }

virgate [BOTANY] Banded. { 'vər,gāt }

virgate trophus [INVERTEBRATE ZOOLOGY] A piercing type of trophus in rotifers that is thin and slightly toothed. { 'vər,gāt 'träf·əs }

virgo-forcipate trophus [INVERTEBRATE ZOOLOGY] A type of muscular chamber in rotifers containing jaws of a cuticular material intermediate between a piercing and grasping type of structure. { 'vər·gō 'fȯr·sə·pət ,träf·əs }

viridans streptococci [MICROBIOLOGY] A group of pathogenic and saprophytic streptococci including strains not causing beta hemolysis, although many cause alpha hemolysis, and none which elaborate a C substance. { 'vir·ə,danz ¦strep·tə'käk·ē }

virion [VIROLOGY] The complete, mature virus particle. { 'vir·ē,än }

viroid [MICROBIOLOGY] The smallest known agents of infectious disease, characterized by the absence of encapsidated proteins. { 'vī,rȯid }

virology [MICROBIOLOGY] The study of submicroscopic organisms known as viruses. { vī'räl·ə·jē }

virotoxin [BIOCHEMISTRY] One of a group of toxins present in the mushroom *Amanita virosa*. { ¦vī·rə¦täk·sən }

virulence [MICROBIOLOGY] The disease-producing power of a microorganism; infectiousness. { 'vir·ə·ləns }

virus [VIROLOGY] A large group of infectious agents ranging from 10 to 250 nanometers in diameter, composed of a protein sheath surrounding a nucleic acid core and capable of infecting all animals, plants, and bacteria; characterized by total dependence on living cells for reproduction and by lack of independent metabolism. { 'vī·rəs }

virus interference [MICROBIOLOGY] A phenomenon which may be defined as protection of host cells against one virus, conferred as a result of prior infection with a different virus. { 'vī·rəs ˌin·tər'fir·əns }

viscera [ANATOMY] The organs within the cavities of the body of an organism. { 'vis·ə·rə }

visceral arch [ANATOMY] One of the series of mesodermal ridges covered by epithelium bounding the lateral wall of the oral and pharyngeal regions of vertebrates; embryonic in higher forms, they contribute to formation of the face and neck. [VERTEBRATE ZOOLOGY] One of the first two arches of the series in gill-bearing forms. { 'vis·ə·rəl 'ärch }

visceral peritoneum [ANATOMY] That portion of the peritoneum covering the organs of the abdominal cavity. { 'vis·ə·rəl ˌper·ə·tə'nē·əm }

visceral pouch [EMBRYOLOGY] *See* pharyngeal pouch. { 'vis·ə·rəl 'pau̇ch }

viscid [BOTANY] Having a sticky surface, as certain leaves. { 'vis·əd }

viscus [ANATOMY] Singular of viscera. { 'vis·kəs }

vision [PHYSIOLOGY] The sense which perceives the form, color, size, movement, and distance of objects. Also known as sight. { 'vizh·ən }

visual acuity [PHYSIOLOGY] The ability to see fine details of an object; specifically, the ability to see an object whose angle subtended at the eye is 1 minute of arc. { 'vizh·ə·wəl ə'kyü·əd·ē }

visual pigment [BIOCHEMISTRY] Any of various photosensitive pigments of vertebrate and invertebrate photoreceptors. { 'vizh·ə·wəl 'pig·mənt }

visual projection area [PHYSIOLOGY] The receptive center for visual images in the cortex of the brain, located in the walls and margins of the calcarine sulcus of the occipital lobe. Also known as Brodmann's area 17. { 'vizh·ə·wəl prə'jek·shən ˌer·ē·ə }

visual purple [BIOCHEMISTRY] *See* rhodopsin. { 'vizh·ə·wəl 'pər·pəl }

visual-righting reflex [PHYSIOLOGY] A reflex mechanism whereby righting of the head and body is caused by visual stimuli. Also known as optical-righting reflex. { 'vizh·ə·wəl 'rīd·iŋ 'rē ˌfleks }

visual yellow [BIOCHEMISTRY] An intermediary substance formed from rhodopsin in the retina after exposure to light; it is ultimately broken down to retinene and vitamin A. { 'vizh·ə·wəl 'yel·ō }

Vitaceae [BOTANY] A family of dicotyledonous plants in the order Rhamnales; mostly tendril-bearing climbers with compound or lobed leaves, as in grapes (*Vitis*). { vī'tās·ēˌē }

vital capacity [PHYSIOLOGY] The volume of air that can be forcibly expelled from the lungs after the deepest inspiration. { 'vīd·əl kə'pas·əd·ē }

vitalism [BIOLOGY] The theory that the activities of a living organism are under the guidance of an agency which has none of the attributes of matter or energy. { 'vīd·əlˌiz·əm }

vitamer [BIOCHEMISTRY] One of several very similar chemical compounds that can perform a specific vitamin function. { 'vīd·ə·mər }

vitamin [BIOCHEMISTRY] An organic compound present in variable, minute quantities in natural foodstuffs and essential for the normal processes of growth and maintenance of the body; vitamins do not furnish energy, but are essential for energy transformation and regulation of metabolism. { 'vīd·ə·mən }

vitamin A [BIOCHEMISTRY] $C_{20}H_{29}OH$ A pale-yellow alcohol that is soluble in fat and insoluble in water; found in liver oils and carotenoids, and produced synthetically; it is a component of visual pigments and is essential for normal growth and maintenance of epithelial tissue. Also known as anti-infective vitamin; antixerophthalmic vitamin; retinol. { 'vīd·ə·mən ¦ā }

vitamin A aldehyde [BIOCHEMISTRY] *See* retinal. { 'vīd·ə·mən ¦ā 'al·dəˌhīd }

vitamin B_1 [BIOCHEMISTRY] *See* thiamine. { 'vīd·ə·mən ¦bē¦wən }

vitamin B_2 [BIOCHEMISTRY] *See* riboflavin. { 'vīd·ə·mən ¦bē¦tü }

vitamin B_2 phosphate [BIOCHEMISTRY] *See* riboflavin 5′-phosphate. { 'vīd·ə·mən ¦bē¦tü 'fä ˌsfāt }

vitamin B_3 [BIOCHEMISTRY] *See* pantothenic acid. { 'vīd·ə·mən ¦bē¦thrē }

vitamin B_6 [BIOCHEMISTRY] A vitamin which exists as three chemically related and water-soluble forms found in food: pyridoxine, pyridoxal, and pyridoxamine; dietary requirements and physiological activities are uncertain. { 'vīd·ə·mən ¦bē¦siks }

vitamin B_6 hydrochloride [BIOCHEMISTRY] *See* pyridoxine hydrochloride. { 'vīd·ə·mən ¦bē¦siks ¦hī·drə'klȯrˌīd }

vitamin B_{12} [BIOCHEMISTRY] A group of closely related polypyrrole compounds containing trivalent cobalt; the antipernicious anemia factor, essential for normal hemopoiesis. Also known as cobalamin; cyanocobalamin; extrinsic factor. { 'vīd·ə·mən ¦bē¦twelv }

vitamin B complex [BIOCHEMISTRY] A group of water-soluble vitamins that include thiamine, riboflavin, nicotinic acid, pyridoxine, panthothenic acid, inositol, *p*-aminobenzoic acid, biotin, folic acid, and vitamin B_{12}. { 'vīd·ə·mən ¦bē 'käm ˌpleks }

vitamin C [BIOCHEMISTRY] *See* ascorbic acid. { 'vīd·ə·mən ¦sē }

vitamin D [BIOCHEMISTRY] Either of two fat-soluble, sterol-like compounds, calciferol or ergocalciferol (vitamin D_2) and cholecalciferol (vitamin D_3); occurs in fish liver oils and is essential for normal calcium and phosphorus deposition in bones and teeth. Also known as antirachitic vitamin. { 'vīd·ə·mən ¦dē }

vitamin E [BIOCHEMISTRY] Any of a series of eight related compounds called tocopherols, α-tocopherol having the highest biological activity; occurs in wheat germ and other naturally occurring oils and is believed to be needed in certain human physiological processes. { 'vīd·ə·mən ¦ē }

vitamin G [BIOCHEMISTRY] *See* riboflavin. { 'vīd·ə·mən ¦jē }

vitamin K [BIOCHEMISTRY] Any of three yellowish oils which are fat-soluble, nonsteroid, and nonsaponifiable; it is essential for formation of prothrombin. Also known as antihemorrhagic vitamin; prothrombin factor. { 'vīd·ə·mən ¦kā }

vitamin P [BIOCHEMISTRY] A substance, such as citrin or one or more of its components, believed to be concerned with maintenance of the normal state of the walls of small blood vessels. { 'vīd·ə·mən ¦pē }

vitamin P complex [BIOCHEMISTRY] *See* bioflavonoid. { 'vīd·ə·mən ¦pē ‚käm‚pleks }

vitellarium [INVERTEBRATE ZOOLOGY] The part of the ovary in certain rotifers and flatworms that produces nutritive cells filled with yolk. Also known as yolk larva. { ‚vid·əl'ar·ē·əm }

vitelline artery [EMBRYOLOGY] An artery passing from the yolk sac to the primitive aorta in young vertebrate embryos. Also known as omphalomesenteric artery. { vī'tel‚ēn 'ärd·ə·rē }

vitelline duct [EMBRYOLOGY] The constricted part of the yolk sac opening into the midgut region of the future ileum. Also known as omphalomesenteric duct; umbilical duct. { vī'tel‚ēn 'dəkt }

vitelline membrane [CELL BIOLOGY] The cytoplasmic membrane on the surface of the mammalian ovum. { vī'tel‚ēn 'mem‚brān }

vitelline vein [EMBRYOLOGY] Any of the embryonic veins in vertebrates uniting the yolk sac and the sinus venosus; their proximal fused ends form the portal vein. Also known as omphalomesenteric vein. { vī'tel‚ēn 'vān }

vitellogenesis [PHYSIOLOGY] The process by which yolk is formed in the ooplasm of an oocyte. { vī‚tel·ə'jen·ə·səs }

Vitreoscillaceae [MICROBIOLOGY] Formerly a family of bacteria in the order Beggiatoales; included organisms which have a filamentous habit and move by gliding, but never store sulfur, and rely on organic nutrients in their metabolism. { ‚vi·trē‚äs·ə'lās·ē‚ē }

vitreous body [PHYSIOLOGY] *See* vitreous humor. { 'vi·trē·əs 'bäd·ē }

vitreous chamber [ANATOMY] A cavity of the eye posterior to the crystalline lens and anterior to the retina, which is filled with vitreous humor. { 'vi·trē·əs 'chām·bər }

vitreous humor [PHYSIOLOGY] The transparent gel-like substance filling the greater part of the globe of the eye, the vitreous chamber. Also known as vitreous body. { 'vi·trē·əs 'hyü·mər }

vittate [BOTANY] **1.** Having longitudinal stripes. **2.** Bearing specialized oil tubes (vittae). { 'vi ‚tāt }

Viverridae [VERTEBRATE ZOOLOGY] A family of carnivorous mammals in the superfamily Feloidea composed of the civets, genets, and mongooses. { vī'ver·ə‚dē }

Viviparidae [INVERTEBRATE ZOOLOGY] A family of fresh-water gastropod mollusks in the order Pectinibranchia. { ‚vī·və'par·ə‚dē }

viviparous [PHYSIOLOGY] Bringing forth live young. { vī'vip·ə·rəs }

vocal cord [ANATOMY] *See* vocal fold. { 'vō·kəl ‚kȯrd }

vocal fold [ANATOMY] Either of a pair of folds of tissue covered by mucous membrane in the larynx. Also known as vocal cord. { 'vō·kəl ‚fōld }

vocal sac [VERTEBRATE ZOOLOGY] An expansible pocket of skin beneath the chin or behind the jaws of certain frogs; may be inflated to a great volume and serves as a resonator. { 'vō·kəl ‚sak }

Vochysiaceae [BOTANY] A family of dicotyledonous plants in the order Polygalales characterized by mostly three carpels, usually stipulate leaves, one fertile stamen, and capsular fruit. { vō‚kizh·ē'ās·ē‚ē }

Voges-Proskauer test [MICROBIOLOGY] One of the four tests of the IMVIC test; a qualitative test for the formation of acetyl methylcarbinol from glucose, in which solutions of α-naphthol, potassium hydroxide, and creatinine are added to an incubated culture of test organisms in a glucose broth, and a pink to rose color indicates a positive reaction. { 'fō·gə 'präs‚kau̇·ər ‚test }

volar [ANATOMY] Pertaining to, or on the same side as, the palm of the hand or the sole of the foot. { 'vō·lər }

vole [VERTEBRATE ZOOLOGY] Any of about 79 species of rodent in the tribe Microtini of the family Cricetidae; individuals have a stout body with short legs, small ears, and a blunt nose. { vōl }

voluntary muscle [PHYSIOLOGY] A muscle directly under the control of the will of the organism. { 'väl·ən‚ter·ē 'məs·əl }

Volutidae [INVERTEBRATE ZOOLOGY] A family of gastropod mollusks in the order Neogastropoda. { və'lüd·ə‚dē }

volutin [BIOCHEMISTRY] A basophilic substance, thought to be a nucleic acid, occurring as granules in the cytoplasm and vacuoles of algae and other microorganisms. { 'väl·yəd·ən }

volva [MYCOLOGY] A cuplike membrane surrounding the base of the stipe in certain gill fungi. { 'väl·və }

volvent nematocyst [INVERTEBRATE ZOOLOGY] A nematocyst in the form of an unarmed, coiled tube that is closed at the end. { 'väl·vənt nə 'mad·ə‚sist }

Volvocales [BOTANY] An order of one-celled or colonial green algae in the division Chlorophyta; individuals are motile with two, four, or rarely eight whiplike flagella. { ‚väl·və'kā·lēz }

Volvocida [INVERTEBRATE ZOOLOGY] An order of the protozoan class Phytamastigophorea; individuals are grass-green or colorless, have one, two, four, or eight flagella, and thick cell walls of cellulose. { väl'väs·əd·ə }

Vombatidae [VERTEBRATE ZOOLOGY] A family of marsupial mammals in the order Diprotodonta in some classification systems. { väm'bad·ə ‚dē }

vomer [ANATOMY] A skull bone below the ethmoid region constituting part of the nasal septum in most vertebrates. { 'vō·mər }

vomeronasal cartilage [ANATOMY] A strip of hyaline cartilage extending from the anterior nasal spine upward and backward on either side of the septal cartilage of the nose and attached to the anterior margin of the vomer. Also known as Jacobson's cartilage. { ¦väm·ə·rō'nāz·əl 'kärt·lij }

vulture [VERTEBRATE ZOOLOGY] The common name for any of various birds of prey in the families Cathartidae and Accipitridae of the order Falconiformes; the head of these birds is usually naked. { 'vəl·chər }

vulva [ANATOMY] The external genital organs of women. { 'vəl·və }

vulval gland [ANATOMY] A scent gland in the vulval tissues of the human female. { 'vəl·vəl ˌgland }

W

wader [VERTEBRATE ZOOLOGY] *See* shore bird. { 'wād·ər }

wading bird [VERTEBRATE ZOOLOGY] Any of the long-legged, long-necked birds composing the order Ciconiiformes, including storks, herons, egrets, and ibises. { 'wād·iŋ ˌbərd }

Waldeyer's ring [ANATOMY] A circular arrangement of the lymphatic tissues formed by the palatine and pharyngeal tonsils and the lymphatic follicles at the base of the tongue and behind the posterior pillars of the fauces. { 'välˌdī·ərz ˌriŋ }

walking bird [VERTEBRATE ZOOLOGY] Any bird of the order Columbiformes, including the pigeons, doves, and sandgrouse. { 'wȯk·iŋ ˌbərd }

walnut [BOTANY] The common name for about a dozen species of deciduous trees in the genus *Juglans* characterized by pinnately compound, aromatic leaves and chambered or laminate pith; the edible nut of the tree is distinguished by a deeply furrowed or sculptured shell. { 'wȯl·nət }

walrus [VERTEBRATE ZOOLOGY] *Odobenus rosmarus.* The single species of the pinniped family Odobenidae distinguished by the upper canines in both sexes being prolonged as tusks. { 'wȯl·rəs }

wasp [INVERTEBRATE ZOOLOGY] The common name for members of 67 families of the order Hymenoptera; all are important as parasites or predators of injurious pests. { wäsp }

Wasserman test [IMMUNOLOGY] A complement-fixation test for syphilis using sensitized lipid extracts of beef heart as antigen. { 'was·ər·mən ˌtest }

water bug [INVERTEBRATE ZOOLOGY] Any insect which lives in an aquatic habitat during all phases of its life history. { 'wȯd·ər ˌbəg }

water conservation [ECOLOGY] The protection, development, and efficient management of water resources for beneficial purposes. { 'wȯd·ər ˌkän·sər'vā·shən }

watercress [BOTANY] *Nasturtium officinale.* A perennial cress generally grown in flooded soil beds and used for salads and food garnishing. { 'wȯd·ərˌkres }

waterfowl [VERTEBRATE ZOOLOGY] Aquatic birds which constitute the order Anseriformes, including the swans, ducks, geese, and screamers. { 'wȯd·ərˌfau̇l }

watermelon [BOTANY] *Citrullus vulgaris.* An annual trailing vine with light-yellow flowers and leaves having five to seven deep lobes; the edible, oblong or roundish fruit has a smooth, hard, green rind filled with sweet, tender, juicy, pink to red tissue containing many seeds. { 'wȯd·ərˌmel·ən }

water microbiology [MICROBIOLOGY] An aspect of microbiology that deals with the normal and adventitious microflora of natural and artificial water bodies. { 'wȯd·ər ¦mī·krō·bī'äl·ə·jē }

water moccasin [VERTEBRATE ZOOLOGY] *Agkistrodon piscivorus.* A semiaquatic venomous pit viper; skin is brownish or olive on the dorsal aspect, paler on the sides, and has indistinct black bars. Also known as cottonmouth. { 'wȯd·ər ˌmäk·ə·sən }

water pollution [ECOLOGY] Contamination of water by materials such as sewage effluent, chemicals, detergents, and fertilizer runoff. { 'wȯd·ər pəˌlü·shən }

water vascular system [INVERTEBRATE ZOOLOGY] An internal closed system of reservoirs and ducts containing a watery fluid, found only in echinoderms. { 'wȯd·ər 'vas·kyə·lər ˌsis·təm }

weasel [VERTEBRATE ZOOLOGY] The common name for at least 12 species of small, slim carnivores which belong to the family Mustelidae and which have a reddish-brown coat with whitish underparts; species in the northern regions have white fur during the winter and are called ermine. { 'wē·zəl }

web [VERTEBRATE ZOOLOGY] The membrane between digits in many birds and amphibians. { web }

Weberian apparatus [VERTEBRATE ZOOLOGY] A series of bony ossicles which form a chain connecting the swim bladder with the inner ear in fishes of the superorder Ostariophysi. { vā'bir·ē·ən ˌap·əˌrad·əs }

Weberian ossicle [VERTEBRATE ZOOLOGY] One of a chain of three or four small bones that make up the Weberian apparatus. { vā'bir·ē·ən 'äs·ə·kəl }

Weber's law [PHYSIOLOGY] The law that the stimulus increment which can barely be detected (the just noticeable difference) is a constant fraction of the initial magnitude of the stimulus; this is only an approximate rule of thumb. { 'vā·bərz ˌlȯ }

weed [BOTANY] A plant that is useless or of low

economic value, especially one growing on cultivated land to the detriment of the crop. { wēd }

weevil [INVERTEBRATE ZOOLOGY] Any of various snout beetles whose larvae destroy crops by eating the interior of the fruit or grain, or bore through the bark into the pith of many trees. { 'wē·vəl }

Weil-Felix test [IMMUNOLOGY] An agglutination test for various rickettsial infections based on production of nonspecific agglutinins in the blood of infected patients, and using various strains of *Proteus vulgaris* as antigen. { 'vīl 'fā·liks ,test }

Welwitschiales [BOTANY] An order of gymnosperms in the subdivision Geneticae represented by the single species *Welwitschia mirabilis* of southwestern Africa; distinguished by only two leaves and short, unbranched, cushion- or saucer-shaped woody main stem which tapers to a very long taproot. { wel,wich·ē'ā·lēz }

Westphal-Pilcz reflex [PHYSIOLOGY] *See* pupillary reflex. { west,fòl 'pils ,rē,fleks }

Westphal's pupillary reflex [PHYSIOLOGY] *See* pupillary reflex. { 'west,fòlz ¦pyü·pə,ler·ē 'rē ,fleks }

wetlands [ECOLOGY] An area characterized by a high content of soil moisture, such as a swamp or bog. { 'wet ,lanz }

whale [VERTEBRATE ZOOLOGY] A large marine mammal of the order Cetacea; the body is streamlined, the broad flat tail is used for propulsion, and the limbs are balancing structures. { wāl }

whalebone [VERTEBRATE ZOOLOGY] *See* baleen. { 'wāl,bōn }

wheat [BOTANY] A food grain crop of the genus *Triticum*; plants are self-pollinating; the inflorescence is a spike bearing sessile spikelets arranged alternately on a zigzag rachis. { wēt }

wheat germ [BOTANY] The embryo of a wheat grain. { 'wēt ,jərm }

whelk [INVERTEBRATE ZOOLOGY] A gastropod mollusk belonging to the order Neogastropoda; species are carnivorous but also scavenge. { welk }

whip grafting [BOTANY] A method of grafting by fitting a small tongue and notch cut in the base of the scion into corresponding cuts in the stock. { 'wip ,graft·iŋ }

whiskers [VERTEBRATE ZOOLOGY] *See* vibrissae. { 'wis·kərz }

white ant [INVERTEBRATE ZOOLOGY] *See* termite. { 'wīt 'ant }

white blood cell [HISTOLOGY] *See* leukocyte. { 'wīt 'bləd ,sel }

white corpuscle [HISTOLOGY] *See* leukocyte. { 'wīt 'kòr·pə·səl }

whitefish [VERTEBRATE ZOOLOGY] Any of various food fishes in the family Salmonidae, especially of the genus *Coregonus*, characterized by an adipose dorsal fin and nearly toothless mouth. { 'wīt,fish }

white potato [BOTANY] *See* potato. { 'wīt pə 'tād·ō }

whooping crane [VERTEBRATE ZOOLOGY] *Grus americana*. A member of a rare North American migratory species of wading birds; the entire species forms a single population. { 'hüp·iŋ ,krān }

whorl [ANATOMY] A fingerprint pattern in which at least two deltas are present with a recurve in front of each. [BOTANY] An arrangement of several identical anatomical parts, such as petals, in a circle around the same point. { wərl }

Whytt's reflex [PHYSIOLOGY] *See* pupillary reflex. { 'wī·əts ,rē,fleks }

Widal test [IMMUNOLOGY] A macroscopic or microscopic agglutination test for the diagnosis of typhoid fever and other *Salmonella* infections by using killed or preserved bacteria as the antigen. { we'däl ,test }

wide cross [GENETICS] A mating between individuals of different genera. { 'wīd 'kròs }

Wiesen [ECOLOGY] *See* meadow. { 'vēz·ən }

wild boar [VERTEBRATE ZOOLOGY] *Sus scrofa*. A wild hog with coarse, grizzled hair and enlarged tusks or canines on both jaws. Also known as boar. { 'wīld 'bòr }

wild cinnamon [BOTANY] *See* bayberry. { 'wīld 'sin·ə·mən }

wild type [GENETICS] The natural or unmutated organism or character. { 'wīld ,tīp }

Williamsoniaceae [PALEOBOTANY] A family of extinct plants in the order Cycadeoidales distinguished by profuse branching. { ,wil·yəm,sō·nē 'ās·ē,ē }

willow [BOTANY] A deciduous tree and shrub of the genus *Salix*, order Salicales; twigs are often yellow-green and bear alternate leaves which are characteristically long, narrow, and pointed, usually with fine teeth along the margins. { 'wil·ō }

wilting point [BOTANY] A condition in which a plant begins to use water from its own tissues for transpiration because soil water has been exhausted. { 'wilt·iŋ ,pòint }

windburn [BOTANY] Injury to plant foliage, caused by strong, hot, dry winds. { 'win,bərn }

wing [ZOOLOGY] Any of the paired appendages serving as organs of flight on many animals. { wiŋ }

Winteraceae [BOTANY] A family of dicotyledonous plants in the order Magnoliales distinguished by hypogynous flowers, exstipulate leaves, air vessels absent, and stamens usually laminar. { ,win·tə'rās·ē,ē }

Wirsung's duct [ANATOMY] The adult pancreatic duct in man, sheep, ganoid fish, teleost fish, and frog. { 'vir,zùŋz ,dəkt }

wobble pairing [MOLECULAR BIOLOGY] The ability of a transfer ribonucleic acid molecule to recognize more than one codon. { 'wäb·əl ,per·iŋ }

Wolbachieae [MICROBIOLOGY] A tribe of the family Rickettsiaceae; rickettsialike organisms found principally in arthropods. { wōl'bak·ē·ē }

wolf [VERTEBRATE ZOOLOGY] Any of several wild species of the genus *Canis* in the family Canidae which are fierce and rapacious, sometimes attacking humans; includes the red wolf, gray wolf, and coyote. { wùlf }

Wolffian duct [EMBRYOLOGY] *See* mesonephric duct. { 'wu̇l·fē·ən 'dəkt }

wolfsbane [BOTANY] *See* aconite. { 'wu̇lfs‚bān }

wolverine [VERTEBRATE ZOOLOGY] *Gulo gulo.* A carnivorous mammal which is the largest and most vicious member of the family Mustelidae. { ¦wu̇l·və¦rēn }

wood [BOTANY] The hard fibrous substance that makes up the trunks and large branches of trees beneath the bark. [ECOLOGY] A dense growth of trees, more extensive than a grove and smaller than a forest. { wu̇d }

woodland [ECOLOGY] *See* forest; temperate woodland. { 'wu̇d·lənd }

woodpecker [VERTEBRATE ZOOLOGY] A bird of the family Picidae characterized by stiff tail feathers and zygodactyl feet which enable them to cling to a tree trunk while drilling into the bark for insects. { 'wu̇d‚pek·ər }

wood ray [BOTANY] A vascular ray consisting of a radial row of parenchyma cells in secondary xylem. Also known as xylem ray. { 'wu̇d ‚rā }

wood sugar [BIOCHEMISTRY] *See* xylose. { 'wu̇d ‚shu̇g·ər }

wool [VERTEBRATE ZOOLOGY] The soft undercoat of various animals such as sheep, angora, goat, camel, alpaca, llama, and vicuna. { wu̇l }

worker [INVERTEBRATE ZOOLOGY] One of the neuter, usually sterile individuals making up a caste of social insects, such as ants, termites, or bees, which labor for the colony. { 'wər·kər }

work metabolism [PHYSIOLOGY] Metabolism in excess of resting metabolism that can be related to the performance of a specific task. { 'wərk mə‚tab·ə‚liz·əm }

work strain [PHYSIOLOGY] The response of the human body to work stress experienced in the performance of a task. { 'wərk ‚strān }

worm [INVERTEBRATE ZOOLOGY] **1.** The common name for members of the Annelida. **2.** Any of various elongated, naked, soft-bodied animals resembling an earthworm. { wərm }

wren [VERTEBRATE ZOOLOGY] Any of the various small brown singing birds in the family Troglodytidae; they are insectivorous and tend to inhabit dense, low vegetation. { ren }

Wright's inbreeding coefficient [GENETICS] *See* inbreeding coefficient. { ¦rīts 'in‚brēd·iŋ ‚kō·ə ‚fish·ənt }

wrist [ANATOMY] The part joining forearm and hand. { rist }

Wynyardiidae [PALEONTOLOGY] An extinct family of herbivorous marsupial mammals in the order Diprotodonta. { ‚win·yər'dī·ə‚dē }

X

Xanthidae [INVERTEBRATE ZOOLOGY] The mud crabs, a family of decapod crustaceans in the section Brachyura. { 'zan·thə,dē }

xanthine oxidase [BIOCHEMISTRY] A flavoprotein enzyme catalyzing the oxidation of certain purines. { 'zan,thēn 'äk·sə,dās }

xanthism [BIOLOGY] A color variation in which an animal's normal coloring is largely replaced by yellow pigments. Also known as xanthochroism. { 'zan,thiz·əm }

xanthochroism [BIOLOGY] *See* xanthism. { ¦zan·thrə'krō,iz·əm }

xanthomycin [MICROBIOLOGY] An antibiotic produced by a strain of *Streptomyces* and composed of two varieties, A and B; active in low concentrations against a number of gram-positive microorganisms. { ,zan·thə'mīs·ən }

xanthophore [CELL BIOLOGY] A yellow chromatophore. { 'zan·thə,fȯr }

Xanthophyceae [BOTANY] A class of yellow-green to green flagellate organisms of the division Chrysophyta; zoologists classify these organisms in the order Heterochlorida. { ,zan·thə 'fīs·ē,ē }

xanthophyll [BIOCHEMISTRY] $C_{40}H_{56}O_2$ Any of a group of yellow, alcohol-soluble carotenoid pigments that are oxygen derivatives of the carotenes, and are found in certain flowers, fruits, and leaves. Also known as carotenol; lutein. { 'zan·thə,fil }

xanthurenic acid [BIOCHEMISTRY] $C_{10}H_7NO_4$ Sulfur yellow crystals with a melting point of 286°C; soluble in aqueous alkali hydroxides and carbonates; excreted by pyridoxine-deficient animals after ingestion of tryptophan. { ¦zan·thyə¦ren·ik 'as·əd }

Xantusiidae [VERTEBRATE ZOOLOGY] The night lizards, a family of reptiles in the suborder Sauria. { ,zan·tə'sī·ə,dē }

X chromosome [GENETICS] The sex chromosome occurring in double dose in the homogametic sex and in single dose in the heterogametic sex. { 'eks 'krō·mə,sōm }

Xenarthra [VERTEBRATE ZOOLOGY] A suborder of mammals in the order Edentata including sloths, anteaters, and related forms; posterior vertebrae have extra articular facets and vertebrae in the hip, and shoulder regions tend to be fused. { zə 'när·thrə }

xenogamy [BOTANY] Cross-fertilization between flowers on different plants. { zə'näg·ə·mē }

xenogeneic [IMMUNOLOGY] Referring to cells, tissues, or organs used in transplantation that originate in a different species. { ¦zēn·ə·jə'nē·ik }

xenograft [IMMUNOLOGY] A graft performed between members of different species. { 'zēn·ə ,graft }

xenophyae [INVERTEBRATE ZOOLOGY] In xenophyophores, the inorganic portion of the test consisting of foreign elements, such as sponge spicules, foraminiferan or radiolarian tests, and mineral particles. { zə'näf·ē,ē }

Xenophyophorea [INVERTEBRATE ZOOLOGY] A class of giant marine benthic Rhizopoda. { zə ¦näf·ē·ə'fȯr·ē·ə }

Xenophyophorida [INVERTEBRATE ZOOLOGY] An order of Protozoa in the subclass Granuloreticulosia; includes deep-sea forms that develop as discoid to fan-shaped branching forms which are multinucleate at maturity. { ¦zē·nō,fī·ə'fȯr·əd·ə }

Xenopneusta [INVERTEBRATE ZOOLOGY] A small order of wormlike animals belonging to the Echiurida. { ,zēn·əp'nüs·tə }

Xenopterygii [VERTEBRATE ZOOLOGY] The equivalent name for Gobiesociformes. { zə,näp·tə'rij·ē,ī }

Xenosauridae [VERTEBRATE ZOOLOGY] A family of four rare species of lizards in the suborder Sauria; composed of the Chinese lizard (*Shinisaurus crocodilurus*) and three Central American species of the genus *Xenosaurus*. { ,zēn·ə'sȯr·ə,dē }

Xenungulata [PALEONTOLOGY] An order of large, digitigrade, extinct, tapirlike mammals with relatively short, slender limbs and five-toed feet with broad, flat phalanges; restricted to the Paleocene deposits of Brazil and Argentina. { zə ,nüŋ·gyə 'läd·ə }

xerarch succession [ECOLOGY] A type of succession that originates in a dry habitat. { 'zer,ärk sək ,sesh·ən }

xeric [ECOLOGY] **1.** Of or pertaining to a habitat having a low or inadequate supply of moisture. **2.** Of or pertaining to an organism living in such an environment. { 'zer·ik }

xerophyte [ECOLOGY] A plant adapted to life in areas where the water supply is limited. { 'zir·ə ,fīt }

xerosere [ECOLOGY] A temporary community in an ecological succession on dry, sterile ground such as rock, sand, or clay. { 'zir·ə,sir }

xiphidio cercaria [INVERTEBRATE ZOOLOGY] A digenetic trematode larva having a stylet in the oral sucker. { zə'fid·ē·ō sər'kar·ē·ə }

Xiphiidae [VERTEBRATE ZOOLOGY] The swordfishes, a family of perciform fishes in the suborder Scombroidei characterized by a tremendously pronounced bill. { zə'fī·ə,dē }

xiphisternum [ANATOMY] The elongated posterior portion of the sternum. { ¦zif·ə'stər·nəm }

Xiphodontidae [PALEONTOLOGY] A family of primitive tylopod ruminants in the superfamily Anaplotherioidea from the late Eocene to the middle Oligocene of Europe. { ,zif·ə'dänt·ə ,dē }

Xiphosura [INVERTEBRATE ZOOLOGY] The equivalent name for Xiphosurida. { ,zif·ə'su̇r·ə }

Xiphosurida [INVERTEBRATE ZOOLOGY] A subclass of primitive arthropods in the class Merostomata characterized by cephalothoracic appendages, ocelli, book lungs, a somewhat trilobed body, and freely articulating styliform telson. { ,zif·ə'su̇r·ə·də }

Xiphydriidae [INVERTEBRATE ZOOLOGY] A family of the Hymenoptera in the superfamily Siricoidea. { ,zif·ə'drī·ə,dē }

X organ [INVERTEBRATE ZOOLOGY] A cluster of neurosecretory cells of the medulla terminales, a portion of the brain lying in the eyestalk in stalk-eyed crustaceans. { 'eks ,ȯr·gən }

Xyelidae [INVERTEBRATE ZOOLOGY] A family of hymenopteran insects in the superfamily Megalodontoidea. { zī'el·ə,dē }

xylem [BOTANY] The principal water-conducting tissue and the chief supporting tissue of higher plants; composed of tracheids, vessel members, fibers, and parenchyma. { 'zī·ləm }

Xylocopidae [INVERTEBRATE ZOOLOGY] A family of hairy tropical bees in the superfamily Apoidea. { ,zīl·ə'käp·ə,dē }

Xylomyiidae [INVERTEBRATE ZOOLOGY] A family of orthorrhaphous dipteran insects in the series Brachycera. { ,zīl·ə'mī·ə,dē }

xylophagous [BIOLOGY] Referring to an organism which feeds on wood. { zī'läf·ə·gəs }

xylose [BIOCHEMISTRY] $C_5H_{10}O_5$ A pentose sugar found in many woody materials; combustible, white crystals with a sweet taste; soluble in water and alcohol; melts about 148°C; used as a nonnutritive sweetener and in dyeing and tanning. Also known as wood sugar. { 'zī,lōs }

Xyridaceae [BOTANY] A family of terrestrial monocotyledonous plants in the order Commelinales characterized by an open leaf sheath, three stamens, and a simple racemose head for the inflorescence. { ,zir·ə'dās·ē,ē }

Y

yak [VERTEBRATE ZOOLOGY] *Poephagus grunniens*. A heavily built, long-haired mammal of the order Artiodactyla, with a shoulder hump; related to the bison, and resembles it in having 14 pairs of ribs. { yak }

yam [BOTANY] **1.** A plant of the genus *Dioscorea* grown for its edible fleshy root. **2.** An erroneous name for the Puerto Rico variety of sweet potato; the edible, starchy tuberous root of the plant. { yam }

Y chromosome [GENETICS] The sex chromosome found only in the heterogametic sex. { 'wī 'krō·mə‚sōm }

yeast [MYCOLOGY] A collective name for those fungi which possess, under normal conditions of growth, a vegetative body (thallus) consisting, at least in part, of simple, individual cells. { yēst }

yellow fat cell [HISTOLOGY] A large, generally spherical fat cell with a thin shell of protoplasm and a single enlarged fat droplet which appears yellowish. { 'yel·ō 'fat ‚sel }

yellow-green algae [BOTANY] The common name for members of the class Xanthophyceae. { 'yel·ō ¦grēn 'al·jē }

yew [BOTANY] A genus of evergreen trees and shrubs, *Taxus*, with the fruit, an aril, containing a single seed surrounded by a scarlet, fleshy, cup-like envelope; the leaves are flat and acicular. { yü }

Y ligament [ANATOMY] *See* iliofemoral ligament. { 'wī ‚lig·ə·mənt }

yolk [BIOCHEMISTRY] **1.** Nutritive material stored in an ovum. **2.** The yellow spherical mass of food material that makes up the central portion of the egg of a bird or reptile. { yōk }

yolk larva [INVERTEBRATE ZOOLOGY] *See* vitellarium. { 'yōk ‚lär·və }

yolk sac [EMBRYOLOGY] A distended extraembryonic extension, heavy-laden with yolk, through the umbilicus of the midgut of the vertebrate embryo. { 'yōk ‚sak }

Y organ [INVERTEBRATE ZOOLOGY] Either of a pair of nonneural structures that are found in the anterior portion of the crustacean body; source of the molting hormone, ecdysone. { 'wī ‚ȯr·gən }

Young-Helmholtz theory [PHYSIOLOGY] A theory of color vision according to which there are three types of color receptors that respond to short, medium, and long waves respectively; primary colors are those that stimulate most successfully the three types of receptors. Also known as Helmholtz theory. { 'yəŋ 'helm‚hōlts ‚thē·ə·rē }

Younginiformes [PALEONTOLOGY] A suborder of extinct small lizardlike reptiles in the order Eosuchia, ranging from the Middle Permian to the Lower Triassic in South Africa. { ‚yəŋ·gə·nə'fȯr‚mēz }

Yponomeutidae [INVERTEBRATE ZOOLOGY] A heterogeneous family of small, often brightly colored moths in the superfamily Tineoidea; the head is usually smooth with reduced or absent ocelli. { ē‚pän·ə'myüd·ə‚dē }

Ypsilothuriidae [INVERTEBRATE ZOOLOGY] A family of echinoderms in the order Dactylochirotida having 8-10 tentacles, a permanent spire on the plates of the test, and the body fusiform or U-shaped. { ‚ip·sə·lō'thu̇r·ə‚dē }

Z

Zalambdalestidae [PALEONTOLOGY] A family of extinct insectivorous mammals belonging to the group Proteutherea; they occur in the Late Cretaceous of Mongolia. { zə‚lam·də'les·tə‚dē }

Zanclidae [VERTEBRATE ZOOLOGY] The Moorish idols, a family of Indo-Pacific perciform fishes in the suborder Acanthuroidei. { 'zaŋ·klə‚dē }

Zapodidae [VERTEBRATE ZOOLOGY] The Northern Hemisphere jumping mice, a family of the order Rodentia with long legs and large feet adapted for jumping. { zə'pȯid·ə‚dē }

zebra [VERTEBRATE ZOOLOGY] Any of three species of African mammals belonging to the family Equidae distinguished by a coat of black and white stripes. { 'zē·brə }

zebu [VERTEBRATE ZOOLOGY] A domestic breed of cattle, indigenous to India, belonging to the family Bovidae, distinguished by long drooping ears, a dorsal hump between the shoulders, and a dewlap under the neck; known as the Brahman in the United States. { 'zē·bü }

Zeiformes [VERTEBRATE ZOOLOGY] The dories, a small order of teleost fishes, distinguished by the absence of an orbitosphenoid bone, a spinous dorsal fin, and a pelvic fin with a spine and five to nine soft rays. { ‚zē·ə'fȯr‚mēz }

Zeoidea [VERTEBRATE ZOOLOGY] An equivalent name for Zeiformes. { zē'ȯid·ē·ə }

Zeomorphi [VERTEBRATE ZOOLOGY] An equivalent name for Zeiformes. { ‚zē·ə'mȯr·fī }

zero population growth [ECOLOGY] A theory which advocates that there be no increase in population, that each person replace only oneself, and that birth control be practiced in all nations. { 'zir·ō 'päp·yə‚lā·shən ‚grōth }

Ziehl-Neelsen stain [MICROBIOLOGY] A procedure for acid-fast staining of tubercle bacilli with carbol fuchsin. { 'zēl 'nēl·sən ‚stān }

zinc finger [BIOCHEMISTRY] A small structural domain that is organized around a zinc ion and is found in many gene-regulatory proteins. { ¦ziŋk ¦fiŋ·gər }

Zingiberaceae [BOTANY] A family of aromatic monocotyledonous plants in the order Zingiberales characterized by one functional stamen with two pollen sacs, distichously arranged leaves and bracts, and abundant oil cells. { ‚zin·jə·bə'rās·ē‚ē }

Zingiberales [BOTANY] An order of monocotyledonous herbs or scarcely branched shrubs in the subclass Commelinidae characterized by pinnately veined leaves and irregular flowers that have well-differentiated sepals and petals, an inferior ovary, and either one or five functional stamens. { ‚zin·jə·bə'rā·lēz }

Z line [HISTOLOGY] The line formed by attachment of the actin filaments between two sarcomeres. { 'zē ‚līn }

Zoantharia [INVERTEBRATE ZOOLOGY] A subclass of the class Anthozoa; individuals are monomorphic and most have retractile, simple, tubular tentacles. { ‚zō·ən'thar·ē·ə }

Zoanthidea [INVERTEBRATE ZOOLOGY] An order of anthozoans in the subclass Zoantharia; these are mostly colonial, sedentary, skeletonless, anemonelike animals that live in warm, shallow waters and coral reefs. { ‚zō·ən'thid·ē·ə }

Zoarcidae [VERTEBRATE ZOOLOGY] The eelpouts, a family of actinopterygian fishes in the order Gadiformes which inhabit cold northern and far southern seas. { zō'är·sə‚dē }

zoea [INVERTEBRATE ZOOLOGY] An early larval stage of decapod crustaceans distinguished by a relatively large cephalothorax, conspicuous eyes, and large, fringed antennae. { zō'ē·ə }

zonal centrifuge [BIOLOGY] A centrifuge that uses a rotating chamber of large capacity in which to separate cell organelles by density-gradient centrifugation. { 'zōn·əl 'sen·trə‚fyüj }

zona pellucida [HISTOLOGY] The thick, solid, elastic envelope of the ovum. Also known as oolemma. { 'zō·nə pə'lüs· əd·ə }

zonation [ECOLOGY] Arrangement of organisms in biogeographic zones. { zō'nā·shən }

zone of optimal proportion [IMMUNOLOGY] One of three zones considered to appear when antigen and antibody are mixed; it is that zone in which there is no uncombined antigen or antibody. Also known as equivalence zone. { 'zōn əv 'äp·tə·məl prə'pȯr·shən }

zone theory [PHYSIOLOGY] *See* stage theory. { 'zōn ‚thē·ə·rē }

zonite [INVERTEBRATE ZOOLOGY] A body segment in Diplopoda. { 'zō‚nīt }

zonula occludens [CELL BIOLOGY] *See* tight junction. { ¦zōn·yə·lə ə'klüd·ənz }

zoochlorellae [BIOLOGY] Unicellular green algae which live as symbionts in the cytoplasm of certain protozoans, sponges, and other invertebrates. { ¦zō·ə·klə'rel‚ē }

zoochory [BOTANY] Dispersal of plant disseminules by animals. { 'zō·ə‚klȯr·ē }

zooecium [INVERTEBRATE ZOOLOGY] The exoskeleton of a feeding zooid in bryozoans. { zō·'ē·shē·əm }

zoogeographic region [ECOLOGY] A major unit of the earth's surface characterized by faunal homogeneity. { ¦zō·ə¦jē·ə¦graf·ik 'rē·jən }

zoogeography [BIOLOGY] The science that attempts to describe and explain the distribution of animals in space and time. { ¦zō·ə,jē'äg·rə·fē }

zoogloea [MICROBIOLOGY] A gelatinous or mucilaginous mass characteristic of certain bacteria grown in organic-rich fluid media. { ,zō·ə 'glē·ə }

zooid [INVERTEBRATE ZOOLOGY] A more or less independent individual of colonial animals such as bryozoans and coral. { 'zō,òid }

zoology [BIOLOGY] The science that deals with knowledge of animal life. { zō'äl·ə·jē }

Zoomastigina [INVERTEBRATE ZOOLOGY] The equivalent name for Zoomastigophorea. { ,zō·ə ,mas·tə'jīn·ə }

Zoomastigophorea [INVERTEBRATE ZOOLOGY] A class of flagellate protozoans in the subphylum Sarcomastigophora; some are simple, some are specialized, and all are colorless. { ,zō·ə,mas·tə·gə'fór·ē·ə }

zoonoses [BIOLOGY] Diseases which are biologically adapted to and normally found in lower animals but which under some conditions also infect humans. { ,zō·ə'nō·sēz }

zooplankton [ECOLOGY] Microscopic animals which move passively in aquatic ecosystems. { ¦zō·ə'plaŋk·tən }

zoosphere [ECOLOGY] The world community of animals. { 'zō·ə,sfir }

zoosporangium [BOTANY] A spore case bearing zoospores. { ¦zō·ə·spə'ran·jē·əm }

zoospore [BIOLOGY] An independently motile spore. { 'zō·ə,spór }

zooxanthellae [BIOLOGY] Microscopic yellow-green algae which live symbiotically in certain radiolarians and marine invertebrates. { ,zō·ə·zan'thē,lē }

Zoraptera [INVERTEBRATE ZOOLOGY] An order of insects, related to termites and psocids, which live in decaying wood, sheltered from light; most individuals are wingless, pale in color, and blind. { zə'rap·tə·rə }

Zoroasteridae [INVERTEBRATE ZOOLOGY] A family of deep-water asteroid echinoderms in the order Forcipulatida. { ,zór·ō·a'ster·ə,dē }

Zorotypidae [INVERTEBRATE ZOOLOGY] The single family, containing one genus, *Zorotypus*, in the order Zoraptera. { ,zór·ə'tīp·ə,dē }

Zosteraceae [BOTANY] A family of monocotyledonous plants in the order Najadales; the group is unique among flowering plants in that they grow submerged in shallow ocean waters near the shore. { ,zäs·tə'rās·ē,ē }

Zosterophyllatae [PALEOBOTANY] *See* Zosterophyllopsida. { ¦zäs·tə·rō'fil·ə,tē }

Zosterophyllopsida [PALEOBOTANY] A group of early land vascular plants ranging from the Lower to the Upper Devonian; individuals were leafless and rootless. { ¦zäs·tə·rō·fə'läp·səd·ə }

Zygaenidae [INVERTEBRATE ZOOLOGY] A diverse family of small, often brightly colored African moths in the superfamily Zygaenoidea. { zī'jēn·ə,dē }

Zygaenoidea [INVERTEBRATE ZOOLOGY] A superfamily of moths in the suborder Heteroneura characterized by complete venation, rudimentary palpi, and usually a rudimentary proboscis. { ,zī·jə'nóid·ē·ə }

Zygnemataceae [BOTANY] A family of filamentous plants in the order Conjugales; they are differentiated into genera by chloroplast morphology, which may be spiral, bandlike, or cushionlike. { zig,nēm·ə'tās·ē,ē }

zygodactyl [VERTEBRATE ZOOLOGY] Of birds, having a toe arrangement of two in front and two behind. { ¦zī·gō¦dak·təl }

zygomatic bone [ANATOMY] A bone of the side of the face below the eye; forms part of the zygomatic arch and part of the orbit in mammals. Also known as malar bone. { ¦zī·gə¦mad·ik 'bōn }

zygomorphic [BIOLOGY] Bilaterally symmetrical. { ¦zī·gə¦mór·fik }

Zygomycetes [MYCOLOGY] A class of fungi in the division Eumycetes. { ,zī·gō,mī'sēd·ēz }

zygopophysis [ANATOMY] One of the articular processes of the neural arch of a vertebra. { ,zī·gə'päf·ə·səs }

Zygoptera [INVERTEBRATE ZOOLOGY] The damsel flies, a suborder of insects in the order Odonata; individuals are slender, dainty creatures, often with bright-blue or orange coloring and usually with clear or transparent wings. { zī'gäp·tə·rə }

zygospore [BOTANY] A thick-walled cell or resting spore that results from the fusion of similar reproductive cells, especially in organisms that reproduce by conjugation. { 'zī·gə,spór }

zygote [EMBRYOLOGY] **1.** An organism produced by the union of two gametes. **2.** The fertilized ovum before cleavage. { 'zī,gōt }

zygotene [CELL BIOLOGY] The stage of meiotic prophase during which homologous chromosomes synapse; visible bodies in the nucleus are now bivalents. Also known as amphitene. { 'zī·gə,tēn }

zygotic induction [VIROLOGY] Phage induction following conjugation of a lysogenic bacterium with a nonlysogenic one. { zī'gäd·ik in'dək·shən }

zymase [BIOCHEMISTRY] A complex of enzymes that catalyze glycosis. { 'zī,mās }

zymogen [BIOCHEMISTRY] The inactive precursor of an enzyme; liberates an active enzyme on reaction with an appropriate kinose. Also known as proenzyme. { 'zī·mə·jən }

zymogen granules [BIOCHEMISTRY] Granules of zymogen in gland cells, particularly those of the pancreatic acini and of the gastric chief cells. { 'zī·mə·jən 'gran,yülz }

zymogenic [MICROBIOLOGY] Obtaining energy by amylolitic processes. { ¦zī·mə¦jen·ik }

zymophore [BIOCHEMISTRY] The active portion of an enzyme. { 'zī·mə,fór }

zymosis [MICROBIOLOGY] *See* fermentation. { zī'mō·səs }

zymosterol [BIOCHEMISTRY] $C_{27}H_{43}OH$ An unsaturated sterol obtained from yeast fat; yields cholesterol on hydrogenation. { zī'mäs·tə,ról }

Zythiaceae [MYCOLOGY] A family of fungi of the order Sphaeropsidales which contains many plant and insect pathogens. { ,zith·ē·'ās·ē,ē }

Appendix

Appendix

Equivalents of commonly used units for the U.S. Customary System and the metric system

1 inch = 2.5 centimeters (25 millimeters)	1 centimeter = 0.4 inch	1 inch = 0.08 foot
1 foot = 0.3 meter (30 centimeters)	1 meter = 3.3 feet	1 foot = 0.3 yard (12 inches)
1 yard = 0.9 meter	1 meter = 1.1 yards	1 yard = 3 feet (36 inches)
1 mile = 1.6 kilometers	1 kilometer = 0.6 mile	1 mile = 5280 feet (1760 yards)
1 acre = 0.4 hectare	1 hectare = 2.47 acres	
1 acre = 4047 square meters	1 square meter = 0.0002 acre	
1 gallon = 3.8 liters	1 liter = 0.26 gallon	1 quart = 0.25 gallon (32 ounces; 2 pints)
1 fluid ounce = 29.6 milliliters	1 milliliter = 0.03 fluid ounce	1 pint = 0.125 gallon (16 ounces)
32 fluid ounces = 946.4 milliliters	1 liter = 1.1 quarts (0.3 gallon)	1 gallon = 4 quarts (8 pints)
1 quart = 0.9 liter	750 milliliters = 25.36 fluid ounces	
1 ounce = 28.4 grams	1 gram = 0.04 ounce	1 ounce = 0.6 pound
1 pound = 0.5 kilogram	1 kilogram = 2.2 pounds	1 pound = 16 ounces
1 ton = 907.18 kilograms	1 kilogram = 1.1 × 10^{-3} ton	1 ton = 2000 pounds
°F = (1.8 × °C) + 32	°C = (°F − 32) ÷ 1.8	

Appendix

Conversion factors for the U.S. Customary System, metric system, and International System

A. UNITS OF LENGTH

Units	cm	m	in	ft	yd	mi
1 cm	= 1	0.01*	0.39	0.033	0.01	6.21×10^{-6}
1 m	= 100.	1	39.37	3.28	1.09	6.21×10^{-4}
1 in	= 2.54	0.03	1	0.08...	0.03...	1.58×10^{-5}
1 ft	= 30.48	0.30	12.	1	0.33...	$1.89... \times 10^{-4}$
1 yd	= 91.44	0.91	36.	3.	1	$5.68... \times 10^{-4}$
1 mile	= 1.61×10^{5}	1.61×10^{3}	6.34×10^{4}	5280.	1760.	1

B. UNITS OF AREA

Units	cm^2	m^2	in^2	ft^2	yd^2	mi^2
1 cm^2	= 1	10^{-4}	0.16	1.08×10^{-3}	1.20×10^{-4}	3.86×10^{-11}
1 m^2	= 10^{4}	1	1550.00	10.76	1.30	3.86×10^{-7}
1 in^2	= 6.45	6.45×10^{-4}	1	6.94×10^{-3}...	7.72×10^{-4}	2.49×10^{-10}
1 ft^2	= 929.03	0.09	1.44.	1	0.11...	3.59×10^{-8}
1 yd^2	= 8361.27	0.84	1296.	9.	1	3.23×10^{-7}
1 mi^2	= 2.59×10^{10}	2.59×10^{6}	4.01×10^{9}	2.79×10^{7}	3.10×10^{6}	1

C. UNITS OF VOLUME

Units	m^3	cm^3	liter	in^3	ft^3	qt	gal
1 m^3	= 1	10^6	10^3	6.10×10^4	35.31	1.057×10^3	264.17
1 cm^3	= 10^{-6}	1	10^{-3}	0.061	3.53×10^{-5}	1.057×10^{-3}	2.64×10^{-4}
1 liter	= 10^{-3}	1000.	1	61.02374	0.03531467	1.056688	0.26
1 in^3	= 1.64×10^{-5}	16.39	0.02	1	5.79×10^{-4}	0.02	4.33×10^{-3}
1 ft^3	= 2.83×10^{-2}	28316.85	28.32	1728.	1	2.99	7.48
1 qt	= 9.46×10^{-4}	946.35	0.95	57.75	0.03	1	0.25
1 gal (U.S.)	= 3.79×10^{-3}	3785.41	3.79	231.	0.13	4.	1

D. UNITS OF MASS

Units	g	kg	oz	lb	metric ton	ton
1 g	= 1	10^{-3}	0.04	2.20×10^{-3}	10^{-6}	1.10×10^{-6}
1 kg	= 1000.	1	35.27	2.20	10^{-3}	1.10×10^{-3}
1 oz (avdp)	= 28.35	0.028	1	0.06	2.83×10^{-5}	$5. \times 10^{-4}$
1 lb (avdp)	= 453.59	0.45	16.	1	4.54×10^{-4}	0.0005
1 metric ton	= 10^6	1000.	35273.96	2204.62	1	1.10
1 ton	= 907184.7	907.18	32000.	2000.	0.91	1

Conversion factors for the U.S. Customary System, metric system, and International System (cont.)

E. UNITS OF DENSITY

Units	$g \cdot cm^{-3}$	$g \cdot L^{-1}$, $kg \cdot m^{-3}$	$oz \cdot in^{-3}$	$lb \cdot in^{-3}$	$lb \cdot ft^{-3}$	$lb \cdot gal^{-1}$
1 $g \cdot cm^{-3}$	= 1	1000.	0.58	0.036	62.43	8.35
1 $g \cdot L^{-1}$, $kg \cdot m^{-3}$	= 10^{-3}	1	5.78×10^{-4}	3.61×10^{-5}	0.06	8.35×10^{-3}
1 $oz \cdot in^{-3}$	= 1.729994	1730	1	0.06	108.	14.44
1 $lb \cdot in^{-3}$	= 27.68	27679.91	16.	1	1728.	231.
1 $lb \cdot ft^{-3}$	= 0.02	16.02	9.26×10^{-3}	5.79×10^{-4}	1	0.13
1 $lb \cdot gal^{-1}$	= 0.12	119.83	4.75×10^{-3}	4.33×10^{-3}	7.48	1

F. UNITS OF PRESSURE

Units	Pa, $N \cdot m^{-2}$	$dyn \cdot cm^{-2}$	bar	atm	kg (wt) $\cdot cm^{-2}$	mmHg (torr)	in Hg	lb (wt) $\cdot in^{-2}$
1 Pa, 1 $N \cdot m^{-2}$	= 1	10	10^{-5}	9.87×10^{-6}	1.02×10^{-5}	7.50×10^{-3}	2.95×10^{-4}	1.45×10^{-4}
1 $dyn \cdot cm^{-2}$	= 0.1	1	10^{-6}	9.87×10^{-7}	1.02×10^{-6}	7.50×10^{-4}	2.95×10^{-5}	1.45×10^{-5}
1 bar	= 10^{5}	10^{6}	1	0.99	1.02	750.06	29.53	14.50
1 atm	= 101325.0	1013250.	1.01	1	1.03	760.	29.92	14.70
1 kg (wt) $\cdot cm^{-2}$	= 98066.5	980665.	0.98	0.97	1	735.56	28.96	14.22
1 mmHg (torr)	= 133.32	1333.22	1.33×10^{-3}	1.32×10^{-3}	1.36×10^{-3}	1	0.04	0.02
1 in Hg	= 3386.39	33863.88	0.03	0.03	0.03	25.4	1	0.49
1 lb (wt) $\cdot in^{-2}$	= 6894.76	68947.57	0.07	0.07	0.07	51.71	2.04	1

G. UNITS OF ENERGY

Units	g mass	J	int J	cal	cal_{IT}	Btu_{IT}	kWh	hp h	ft-lb (wt)	cu ft-lb (wt) in^2	liter-atm
1 g mass	= 1	8.99×10^{13}	8.99×10^{13}	2.15×10^{13}	2.15×10^{13}	8.52×10^{10}	2.50×10^{7}	3.35×10^{7}	6.63×10^{13}	4.60×10^{11}	8.87×10^{11}
1 J	$= 1.11 \times 10^{-14}$	1	1.00	0.24	0.24	9.48×10^{-4}	$2.78... \times 10^{-7}$	3.73	0.74	5.12×10^{-3}	9.87×10^{-3}
1 int J	$= 1.11 \times 10^{-14}$	1.00	1	0.24	0.24	9.48×10^{-4}	2.78×10^{-7}	3.73×10^{-7}	0.74	5.12×10^{-3}	9.87×10^{-3}
1 cal	$= 4.66 \times 10^{-14}$	4.18	4.18	1	1.00	3.97×10^{-3}	$1.16... \times 10^{-6}$	1.56×10^{-6}	3.09	2.14×10^{-2}	0.04
1 cal_{IT}	$= 4.66 \times 10^{-14}$	4.19	4.19	1.00	1	3.97×10^{-3}	1.16×10^{-6}	1.56×10^{-6}	3.09	2.14×10^{-2}	0.04
1 Btu_{IT}	$= 1.17 \times 10^{-11}$	1055.06	1054.88	252.16	252	1	2.93×10^{-4}	3.93×10^{-4}	778.17	5.40	10.41
1 kWh	$= 4.01 \times 10^{-8}$	3600000.	3599406.	860420.7	859845.2	3412.14	1	1.34	2655224.	18439.06	35529.24
1 hp h	$= 2.99 \times 10^{-8}$	2684519.	2684077.	641615.6	641186.5	2544.33	0.75	1	1980000.	13750.	26494.15
1 ft-lb (wt)	$= 1.51 \times 10^{-14}$	1.36	1.36	0.32	0.32	1.29×10^{-3}	3.77×10^{-7}	$5.05... \times 10^{-7}$	1	$6.94... \times 10^{-3}$	0.01
1 cu ft-lb (wt) in^2	$= 2.17 \times 10^{-12}$	195.24	195.21	46.66	46.63	0.19	5.42×10^{-5}	$7.27... \times 10^{-5}$	144.	1	1.93
1 liter-atm	$= 1.13 \times 10^{-12}$	101.33	101.31	24.22	24.20	0.10	2.81×10^{-5}	3.77×10^{-5}	74.73	0.52	1

Appendix

Some representative enzymes, their sources, and reaction specificities

Enzyme	*Some sources*	*Reaction catalyzed*
Pepsin	Gastric juice	Hydrolysis of proteins to peptides and amino acids
Urease	Jack bean, bacteria	Hydrolysis of urea to ammonia and carbon dioxide
Amylase	Saliva, pancreatic juice	Hydrolysis of starch to maltose
Phosphorylase	Muscle, liver, plants	Reversible phosphorolysis of starch or glycogen to glucose-1-phosphate
Transaminases	Many animal and plant tissues	Transfer of an amino group from an amino acid to a keto acid
Phosphohexose isomerase	Muscle, yeast	Interconversion of glucose-6-phosphate and fructose-6-phosphate
Pyruvic carboxylase	Yeast, bacteria, plants	Decarboxylation of pyruvate to acetaldehyde and carbon dioxide
Catalase	Erythrocytes, liver	Decomposition of hydrogen peroxide oxygen and water
Alcohol dehydrogenase	Liver	Oxidation of ethanol to acetaldehyde
Xanthine oxidase	Milk, liver	Oxidation of xanthine and hypoxanthine to uric acid

ABO blood group system

Blood group	RBC *antigens*	*Possible genotypes*	*Plasma antibody*
A	A	AA or A/O	Anti-B
B	B	B/B or B/O	Anti-A
O	—	O/O	Anti-A and anti-B
AB	A and B	A/B	—

Values for a normal urine sample

Test or constituent	*Presence or concentration*
Volume	1000–1500 ml/day
Appearance	Clear; straw-colored
pH	Range 4.5–8; usually acid; diet-dependent
Protein	2–8 mg/dl per urine sample [less than 150 mg/(dl/day)]
Hemoglobin (blood)	Not normally detectable
Glucose (sugar)	0–20 mg/dl
Ketone bodies	0–2 mg/dl [less than 20 mg/(dl/day)]
Bilirubin	0.02 mg/dl
Urobilinogen	0.5–4 mg/day
Nitrite (bacterial metabolic product)	Not normally detectable
Leukocyte esterase (product of neutrophils)	Not normally detectable
Specific gravity	1.003–1.032

Appendix

Some functions of essential vitamins

Vitamins	Functions	Best sources	Deficiency	Daily recommended dietary allowance (RDA) for adults
FAT-SOLUBLE				
Vitamin A (retinoids, carotenes)	Maintenance of vision in dim light, growth, reproduction	Fish liver oils; liver; dairy products; yellow, orange, and green plants; carrots; sweet potatoes	Poor growth and night vision; blindness	1 mg (5000 IU) [as retinol]
Vitamin D (cholecalciferol)	Rickets-preventive factor, calcification of bones, calcium and phosphorus metabolism	Fish oils, fortified dairy products	Rickets, osteomalacia	5 μg (200 IU)
Vitamin E (tocopherols)	Antioxidant, membrane integrity and metabolism, heme synthesis	Grains and vegetable oils	Neuropathy	10 mg (15 IU)
Vitamin K	Blood-clotting factor	Green vegetables	Bleeding	80 μg
WATER-SOLUBLE				
Ascorbic acid (vitamin C)	Antiscorbutic (scurvy-preventive) factor, collagen formation, neurotransmitter synthesis	Citrus fruits, fresh vegetables, potatoes	Scurvy	60 mg
Thiamine (vitamin B-1)	Antiberiberi factor, energy utilization, particularly from carbohydrates	Pork, liver, whole grains	Beriberi	1.5 mg
Riboflavin (vitamin B-2)	Energy utilization, protein metabolism	Milk, egg white, liver, leafy vegetables	Cheilosis, glossitis	1.7 mg
Niacin	Antipellagra factor, energy release from carbohydrate, fat, and protein	Yeast, wheat germ, meats	Pellagra	19 mg
Vitamin B-6 (pyridoxine, pyridoxal, pyridoxamine)	Coenzyme for protein metabolism	Whole grains, yeast, egg yolk, liver	Skin disorders; convulsion in infants	2.0 mg
Pantothenic acid	Metabolism of protein, carbohydrate, fat	Liver, kidney, green vegetables, egg yolk		4–7 mg
Folate	Transfer of one-carbon units in metabolism	Liver, deep-green leafy vegetables	Macrocytic anemias	0.2 mg
Vitamin B-12 (cobalamin)	Blood formation, nervous tissue metabolism	Liver, kidney, yeast	Skin disorders	30–100 μg
Biotin	Synthesis and oxidation of fatty acids and carbohydrates	Liver, meats	Pernicious anemia	2.0 μg

Appendix

Human metabolic rate at typical activities

Activity			Metabolic rate, met units
Resting			
Sleeping			0.7
Reclining			0.8
Seated, quiet			1.0
Standing, relaxed			1.2
Walking (on the level)	mi/h	m/s	
	2	0.89	2.0
	3	1.34	2.6
	4	1.79	3.8
Miscellaneous occupations			
Bakery (such as cleaning tins, packing boxes)			1.4–2.0
Brewery (such as filling bottles, loading beer boxes onto belt)			1.2–2.4
Carpentry			
Machine sawing, table			1.8–2.2
Sawing by hand			4.0–4.8
Planing by hand			5.6–6.4
Foundry work			
Using a pneumatic hammer			3.0–3.4
Tending furnaces			5.0–7.0
Garage work (such as replacing tires, raising cars by jack)			2.2–3.0
General laboratory work			1.4–1.8
Machine work			
Light (such as, electrical industry)			2.0–2.4
Heavy (such as steel work)			3.5–4.5
Shop assistant			2.0
Teacher			1.6
Watch repairer, seated			1.1
Vehicle driving			
Car			1.5
Motorcycle			2.0
Heavy vehicle			3.2
Aircraft flying routine			1.4
Instrument landing			1.8
Combat flying			2.4
Domestic work, women			
Housecleaning			2.0–3.4
Cooking			1.6–2.0
Washing by hand and ironing			2.0–3.6
Shopping			1.4–1.8
Office work			
Typing			1.2–1.4
Miscellaneous office work			1.1–1.3
Drafting			1.1–1.3
Leisure activities			
Stream fishing			1.2–2.0
Calisthenics exercise			3.0–4.0
Dancing, social			2.4–4.4
Tennis, singles			3.6–4.6
Squash, singles			5.0–7.2
Basketball, half court, intramural			5.0–7.6
Wrestling, competitive or intensive			7.0–8.7
Golf, swinging and walking			1.4–2.6
Golf, swinging and golf cart			1.4–1.8

Appendix

Estimated sizes of human chromosomes*

Chromosome	*Length, 10^6 base pairs*	*Chromosome*	*Length, 10^6 base pairs*
1	263	14	109
2	255	15	106
3	214	16	98
4	203	17	92
5	194	18	85
6	183	19	67
7	171	20	72
8	155	21	50
9	145	22	56
10	144	X	164
11	144	Y	59
12	143		
13	114	TOTAL	3286

* From the Virtual Genome Center.

Classification of animal viruses

Family	*Prototype*	*Nucleic acid*	*Structure*
Poxviridae	Vaccinia virus	Double-stranded DNA	Complex, enveloped
Parvoviridae	Adeno-associated	Single-stranded DNA	Icosahedron, naked
Reoviridae	Reovirus	Double-stranded RNA, fragmented	Icosahedron, naked
Rhabdoviridae	Vesicular stomatitis virus rabies	Single-stranded RNA negative strand	Bullet-shaped, helical nucleocapsid, enveloped
Herpetoviridae	Herpes simplex	Double-stranded DNA	Icosahedron, enveloped
Adenoviridae	Adenovirus	Double-stranded DNA	Icosahedron, naked
Papovaviridae	Simian virus 40	Double-stranded DNA, circular	Icosahedron, naked
Retroviridae	Rous sarcoma	Single-stranded RNA	Complex, enveloped
Paramyxoviridae	Newcastle disease	Single-stranded RNA negative strand	Helical nucleocapsid, enveloped
Orthomyxoviridae	Influenza	Single-stranded RNA negative strand, fragmented	Helical nucleocapsid, enveloped
Togaviridae	Alpha: Sindbis Flavi: Yellow fever	Single-stranded RNA	Icosahedron, enveloped
Coronaviridae	Avian infectious bronchitis	Single-stranded RNA	Complex, enveloped
Arenaviridae	Lymphocytic choriomeningitis	Single-stranded RNA negative strand	Complex, enveloped
Picornaviridae	Polio	Single-stranded RNA	Icosahedron, naked
Bunyaviridae	Bunyamwera	Single-stranded RNA negative strand, fragmented	Complex, enveloped

Appendix

Some vaccines used in prevention of viral diseases of humans

Disease	*Condition of virus*	*Route of administration*
Poliomyelitis (Sabin)	Live attenuated	Oral
Poliomyelitis (Salk)	Inactivated	Subcutaneous
Measles	Live attenuated	Subcutaneous
Rubella	Live attenuated	Subcutaneous
Mumps	Live attenuated	Subcutaneous
Mumps (Finland)	Inactivated	Subcutaneous
Smallpox	Vaccinia virus, an apothogenic virus related to variola	Intradermal
Yellow fever	Live attenuated	Subcutaneous
Influenza	Live attenuated	Oropharyngeal
Influenza	Inactivated	Subcutaneous
Rabies	Inactivated	Subcutaneous
Adenovirus infection	Live attenuated	Oral by enteric coated capsules
Japanese B encephalitis	Inactivated	Subcutaneous
Eastern and western equine encephalitis	Inactivated	Subcutaneous
Chickenpox	Live attenuated	Subcutaneous
Hepatitis B (Heptavax-B)	Subunit (HBsAg)	Subcutaneous
Hepatitis B (Institut Pasteur)	Subunit (HBsAg) including Pre-S peptide	Subcutaneous
Hepatitis B	Subunit (recombinant DNA from yeast)	Subcutaneous
Cytomegalovirus disease	Live attenuated	Subcutaneous

Appendix

Common exogenous causes of disease

Causative agent	Disease
PHYSICAL	
Mechanical injury	Abrasion, laceration, fracture
Nonionizing energy	Thermal burns, electric shock, frostbite, sunburn
Ionizing radiation	Radiation syndrome
CHEMICAL	
Metallic poisons	Intoxication from methanol, ethanol, glycol
Nonmetallic inorganic poisons	Intoxication from phosphorus, borate, nitrogen dioxide
Alcohols	Intoxication from methanol, ethanol, glycol
Asphyxiants	Intoxication from carbon monoxide, cyanide
Corrosives	Burns from acids, alkalies, phenols
Pesticides	Poisoning
Medicinals	Barbiturism, salicylism
Warfare agents	Burns from phosgene, mustard gas
Hydrocarbons (some)	Cancer
NUTRITIONAL DEFICIENCY	
Metals (iron, copper, zinc)	Some anemias
Nonmetals (iodine, fluorine)	Goiter, dental caries
Protein	Kwashiorkor
Vitamins:	
A	Epithelial metaplasia
D	Rickets, osteomalacia
K	Hemorrhage
Thiamine	Beriberi
Vitamins (cont.):	
Niacin	Pellagra
Folic acid	Macrocytic anemia
B_{12}	Pernicious anemia
Ascorbic acid	Scurvy
BIOLOGICAL	
Plants (mushroom, fava beans, marijuana, poison ivy, tobacco, opium)	Contact dermatitis, systemic toxins, cancer, hemorrhage
Bacteria	Abscess, scarlet fever, pneumonia, meningitis, typhoid, gonorrhea, food poisoning, cholera, whooping cough, undulant fever, plague, tuberculosis, leprosy, diphtheria, gas gangrene, botulism, anthrax
Spirochetes	Syphilis, yaws, relapsing fever, rat bite fever
Virus	Warts, measles, German measles, smallpox, chickenpox, herpes, roseola, influenza, psittacosis, mumps, viral hepatitis, poliomyelitis, rabies, encephalitis, trachoma
Rickettsia	Spotted fever, typhus
Fungus	Ringworm, thrush, actinomycosis, histoplasmosis, coccidiomycosis
Parasites (animal)	
Protozoa	Amebic dysentery, malaria, toxoplasmosis, trichomonas vaginitis
Helminths (worms)	Hookworm, trichinosis, tapeworm, filariasis, ascariasis

Appendix

Selected environmental causes of human cancer

Agent	*Site or type of cancer*
Life-style risk factors	
Tobacco, smoking	Lung, larynx, mouth, pharynx, esophagus, bladder, pancreas, kidney, cervix, breast
Tobacco, smokeless	Mouth, oral cavity
Alcohol	Mouth, pharynx, esophagus, larynx, liver, breast, colon (?)
Sunlight	Skin, lip
Anabolic steroids	Liver
Phenacetin	Renal pelvis
Occupational risk factors	
Asbestos	Lung, pleura, peritoneum
Benzene	Leukemia
Benzo(a)pyrene	Lung, skin
Arsenic	Lung, skin, liver angiosarcoma
Aromatic amines	Bladder
Chromium	Lung
Nickel dust	Lung, nasal sinuses
Vinyl chloride	Liver angiosarcoma
Wood dust	Nasal sinuses, Hodgkin's disease
Herbicides	Lymphoma, soft tissue sarcoma
Leather	Nasal sinuses
Infectious risk factors	
Schistosoma haematobium	Squamous carcinoma of the bladder
Clonorchis sinensis	Cholangiocarcinoma of the liver
Epstein-Barr virus	Lymphomas, nasopharyngeal carcinoma
Hepatitis-B virus	Hepatocellular carcinoma
Human T-cell lymphotrophic virus	Adult T-cell leukemia
Papilloma virus	Cervix, skin, vulva
Iatrogenic risk factors	
Chronic alkylating agents	Leukemia, bladder
Radiation	Nearly any site
Estrogens, conjugated	Endometrium
Estrogens, synthetic	Transplacental cervix, vagina
Immunosuppressive agents	Lymphoma, skin, Kaposi's sarcoma
Thorium oxide	Liver angiosarcoma
Chlornaphazine	Bladder

Representative intracutaneous diagnostic skin tests

Name of test and use	*Readings and interpretation*
Schick test (for determining susceptibility to diphtheria); positive reaction presumptive of lack of immunity	Score as positive or negative after 48 h; positive reaction shows edema and usually scaling for 7 days; control permits evaluation of sensitivity to bacterial protein
Dick test (for determining susceptibility to scarlet fever); positive reaction presumptive of lack of immunity to the erythrogenic toxin	Read in 18–24 h; positive test requires an erythema over 0.4 in (10 mm) in diameter
Frei test (for lymphogranuloma venereum infection	Read at 48 and 96 h; positive reaction requires papule greater than 0.24 in (6 mm), with lesser reaction for control; reaction may be negative during first 1–6 weeks of infection; hypersensitivity may persist for life
Tuberculin test (for tuberculosis infection)	Read in 36–48 h, noting diameter of redness and swelling; hypersensitivity may be persistent
Ducrey test (for venereal disease due to *Haemophilus ducreyi*)	Read in 47 h; positive test requires an area of induration in excess of 0.28 in (7 mm)
Brucellergen test (for *Brucella* infection)	Read in 48–72 h; positive test requires erythema with an induration of at least 0.4 in (10 mm); should be evaluated in connection with other clinical data
Trichinella test (for trichinosis)	A positive reaction usually appears as a wheal with pseudopods within 20 min; may be a delayed reaction, evident only after 24 h

Appendix

Slow virus infections

Classification	*Virus*	*Disease [host]*
HUMANS		
RNA viruses		
Paramyxovirus	Measles variant	Subacute sclerosing panencephalitis (SSPE)
Rhabdovirus	Rabies	Rabies
Retrovirus	HTLV	T-cell leukemia, AIDS?
Picornavirus	Hepatitis A	Hepatitis
DNA viruses		
Papovavirus	JC, SV40-like	Progressive multifocal leukoencephalopathy (PML)
Unclassified	Hepatitis B	Hepatitis
ANIMALS		
RNA viruses		
Retrovirus	Visna	Meningoencephalitis [sheep]
	Maedi	Pneumonitis [sheep]
	Progressive pneumonia (PPV)	Pneumonitis [sheep]
	Equine infectious anemia	Hemolytic anemia, arteritis (equine infectious arteritis, or EIA) [horse]
	Gardner agent	Lower motor neuron [feral mouse]
	Xenotrophic C type	Hemolytic anemia, systemic lupus erythematosuslike [NZB mouse]
Arenavirus	Lymphocytic choriomeningitis (LCM)	Meningitis glomerulonephritis [mouse]
Paramyxovirus	Canine distemper	Encephalitis [dog]
Togavirus	Lactate dehydrogenase (LDV)	Elevated lactate dehydrogenase; mild nepthritis [mouse]
Picomavirus	Theiler's agent	Demyelination [mouse]
DNA viruses		
Parvovirus	Aleutian disease (ADV)	Arteritis, anemia, nephritis [mink]
Papovavirus	SV40	Progressive multifocal leukoencephalopathy [monkey]
UNCONVENTIONAL AGENTS		
		Creutzfeldt-Jakob disease (CJD) [human]
		Kuru [human]
		Scrapie [sheep, goat]
		Transmissible mink encephalopathy

Appendix

Sources of exposure and toxicological properties of selected metals

Substance	*Source of exposure*	*Acute effects in humans*	*Chronic effects in humans*	*Treatment*
Arsenic	Occupational, pesticidal, food, drinking water, air	Fever, anorexia, skin lesions, multiorgan effects	Liver injury, peripheral vascular disease, cancer	Chelating agents for pulmonary and skin effects
Beryllium	Occupational, air (coal combustion)	Chemical pneumonia	Chronic pulmonary disease, cancer	Cease exposure
Cadmium	Occupational, food, air	Nausea and vomiting (ingestion), chemical pneumonia (inhalation)	Pulmonary disease, kidney damage, hypertension, bone fragility, cancer	Cease exposure
Chromium	Occupational, food	Respiratory irritation (inhalation)	Irritation, skin reactions, cancer	Cease exposure
Cobalt	Occupational, food	Vomiting, diarrhea, cardiomyopathy	Goiter, dermatitis, lung irritation	Cease exposure
Copper	Occupational, water	Nausea, vomiting, liver damage	Wilson's disease	Chelating agents
Gold	Medicinal	None reported	Dermatitis, kidney disease	Cease exposure
Lead	Food, water, air, lead paint, occupational	Vomiting, spasms	Neurological effects (especially in children), kidney damage, anemia	Chelating agents, removal of sources
Manganese	Occupational, food, air	Inflammation of the lung	Respiratory disease, neurological effects	L-Dopa
Mercury	Occupational, food, air	Bronchitis (inhalation); abdominal cramps, bloody diarrhea, and kidney damage (ingestion)	Neurological effects, kidney damage	Hemodialysis, chelating agents
Molybdenum	Occupational	Irritation	—	Cease exposure
Nickel	Occupational	Headache, nausea, vomiting, fever	Dermatitis, cancer	Cease exposure
Silver	Occupational	Gastrointestinal irritation	Argyria, kidney and lung damage	Cease exposure
Thallium	Occupational, pesticidal	Gastrointestinal irritation, neurological effects, cardiovascular effects	Hair loss, cataracts, neurological effects, kidney damage, loss of vision	Cease exposure
Zinc	Occupational, food, water, air	Abdominal distress, diarrhea (ingestion); fever and chills (inhalation)	Pulmonary effects	Cease exposure

Appendix

Gestation periods of some common mammals

Animal	*Gestation, days*	*Animal*	*Gestation, days*
Armadillo	150	Guinea pig	68–71
Bear		Horse	330–380
Black	210	Human	270–295
Polar	240	Kangaroo	40–45
Cat	60	Lion	106
Chimpanzee	250	Mole	30
Cow	282	Mouse	20–21
Dog	58–65	Opossum	13
Donkey	365–380	Rabbit	30–43
Elephant		Whale	334–365
African	641	Wolf	63
Indian	607–641	Zebra	300–345
Giraffe	450		

Dental formulas of some mammals

	Teeth				
Animal	I	C	*Pm*	M	*Total*
Human	2/2	1/1	2/2	3/3	32
Cony	3/3	1/1	4/4	4/4	48
Beaver	1/1	0/0	1/1	3/3	20
Cat	3/3	1/1	3/2	1/1	30
Dog	3/3	1/1	4/4	2/3	42
Sheep	0/3	0/1	3/3	3/3	32
Lynx	3/3	1/1	2/2	1/1	28
Rat	1/1	0/0	0/0	3/3	16
Horse	3/3	1/1	4/4	3/3	44
Mole	3/3	1/1	4/4	3/3	44
Squirrel	1/1	0/0	2/1	3/3	22
Reindeer	0/3	0/1	3/3	3/3	32
Pig	3/3	1/1	4/4	3/3	44
Common seal	3/2	1/1	4/4	1/1	34
Skunk	3/3	1/1	3/3	1/2	34
Raccoon	3/3	1/1	4/4	2/2	40
Bear	3/3	1/1	4/4	2/3	42

Appendix

Cranial nerves of vertebrates

Number	*Name*	*Fiber types*	*Peripheral origin or destination*	*Vertebrates possessing this nerve*
—	Terminal	Somatic sensory	Anterior nasal epithelium	Almost all
I	Olfactory	Special sensory	Olfactory mucosa	All
—	Vomeronasal	Special sensory	Vomeronasal mucosa	Almost all
II	Optic	Special sensory	Retina of eye	All
III	Oculomotor	Somatic motor	Four extrinsic eye muscles	All
IV	Trochlear	Somatic motor	One extrinsic eye muscle	All
V	Trigeminal	Special visceral motor	Muscles of mandibular arch derivative	All
		Somatic sensory	Most of head	All
VI	Abducens	Somatic motor	One extrinsic eye muscle	All
—	Anterior lateral line	Special sensory	Lateral line organs of head	Fish and larval amphibians
VII	Facial	Special visceral motor	Muscles of hyoid arch derivative	All
		General visceral motor	Salivary glands	All
		Somatic sensory	Small part of head	All
		Visceral sensory	Anterior pharynx	All
		Special sensory	Taste, anterior tongue	All
VIII	Vestibulocochlear	Special sensory	Inner ear	All
	Posterior lateral line	Special sensory	Lateral line organs of trunk	Fish and larval amphibians
IX	Glossopharyngeal	Special visceral motor	Muscles of third branchial arch	All
		General visceral motor	Salivary gland	All
		Somatic sensory	Skin near ear	All
		Visceral sensory	Part of pharynx	All
		Special sensory	Taste, posterior tongue	All
X	Vagus	Special visceral motor	Muscles of arches 4–6	All
		General visceral motor	Most viscera of entire trunk	All
		Visceral sensory	Larynx and part of pharynx	All
		Special sensory	Taste, pharynx	All
XI	Spinal accessory	Special visceral motor	Some muscles of arches 4–6	Reptiles, birds, mammals
XII	Hypoglossal	Somatic motor	Muscles of tongue and anterior throat	Reptiles, birds, mammals